Mathematical Methods in Engineering and Physics

Publisher:	Petra Recter
Executive Editor:	Jessica Fiorillo
Market Solutions Assistant:	Kathryn Hancox
Executive Marketing Manager:	Christine Kushner
Associate Production Manager:	Joyce Poh
Production Management Services:	Laserwords
Cover Designer:	Kenji Ngieng
Cover Photo Credit:	diversepixel/Shutterstock

This book was set by Laserwords. Printed and bound by Courier Kendallville. This book is printed on acid free paper.

Founded in 1807, John Wiley & Sons, Inc. has been a valued source of knowledge and understanding for more than 200 years, helping people around the world meet their needs and fulfill their aspirations. Our company is built on a foundation of principles that include responsibility to the communities we serve and where we live and work. In 2008, we launched a Corporate Citizenship Initiative, a global effort to address the environmental, social, economic, and ethical challenges we face in our business. Among the issues we are addressing are carbon impact, paper specifications and procurement, ethical conduct within our business and among our vendors, and community and charitable support. For more information, please visit our website: www.wiley.com/go/citizenship.

Copyright © 2016 John Wiley & Sons, Inc. All rights reserved. No part of this publication may be reproduced, stored in a retrieval system or transmitted in any form or by any means, electronic, mechanical, photocopying, recording, scanning or otherwise, except as permitted under Sections 107 or 108 of the 1976 United States Copyright Act, without either the prior written permission of the Publisher, or authorization through payment of the appropriate per-copy fee to the Copyright Clearance Center, Inc. 222 Rosewood Drive, Danvers, MA 01923, website www.copyright.com. Requests to the Publisher for permission should be addressed to the Permissions Department, John Wiley & Sons, Inc., 111 River Street, Hoboken, NJ 07030-5774, (201)748-6011, fax (201)748-6008, website http://www.wiley.com/go/permissions.

Evaluation copies are provided to qualified academics and professionals for review purposes only, for use in their courses during the next academic year. These copies are licensed and may not be sold or transferred to a third party. Upon completion of the review period, please return the evaluation copy to Wiley. Return instructions and a free of charge return mailing label are available at www.wiley.com/go/returnlabel. If you have chosen to adopt this textbook for use in your course, please accept this book as your complimentary desk copy. Outside of the United States, please contact your local sales representative.

Library of Congress Cataloging-in-Publication Data

Felder, Gary, 1970- author.
   Mathematical methods in engineering and physics: introductory topics/Gary Felder, Smith College, Kenny Felder, Raleigh Charter High School.
      pages cm
   Includes index.
   ISBN 978-1-118-44960-8 (pbk.)
   1. Engineering mathematics—Textbooks. 2. Mathematical physics—Textbooks.
I. Felder, Kenny, 1966- author. II. Title.
   TA330.F45 2016
   510—dc23

                       2014032269

Printed in the United States of America

10 9 8 7 6 5 4 3 2 1

Mathematical Methods in Engineering and Physics

Gary N. Felder
Associate Professor, Smith College

Kenny M. Felder
Math and Physics Teacher, Raleigh Charter High School

WILEY

Contents

Preface xi

1 Introduction to Ordinary Differential Equations 1

- 1.1 Motivating Exercise: The Simple Harmonic Oscillator 2
- 1.2 Overview of Differential Equations 3
- 1.3 Arbitrary Constants 15
- 1.4 Slope Fields and Equilibrium 25
- 1.5 Separation of Variables 34
- 1.6 Guess and Check, and Linear Superposition 39
- 1.7 Coupled Equations (see felderbooks.com)
- 1.8 Differential Equations on a Computer (see felderbooks.com)
- 1.9 Additional Problems (see felderbooks.com)

2 Taylor Series and Series Convergence 50

- 2.1 Motivating Exercise: Vibrations in a Crystal 51
- 2.2 Linear Approximations 52
- 2.3 Maclaurin Series 60
- 2.4 Taylor Series 70
- 2.5 Finding One Taylor Series from Another 76
- 2.6 Sequences and Series 80
- 2.7 Tests for Series Convergence 92
- 2.8 Asymptotic Expansions (see felderbooks.com)
- 2.9 Additional Problems (see felderbooks.com)

3 Complex Numbers 104

- 3.1 Motivating Exercise: The Underdamped Harmonic Oscillator 104
- 3.2 Complex Numbers 105
- 3.3 The Complex Plane 113
- 3.4 Euler's Formula I—The Complex Exponential Function 117
- 3.5 Euler's Formula II—Modeling Oscillations 126
- 3.6 Special Application: Electric Circuits (see felderbooks.com)
- 3.7 Additional Problems (see felderbooks.com)

4 Partial Derivatives 136

- 4.1 Motivating Exercise: The Wave Equation 136
- 4.2 Partial Derivatives 137
- 4.3 The Chain Rule 145
- 4.4 Implicit Differentiation 153
- 4.5 Directional Derivatives 158
- 4.6 The Gradient 163
- 4.7 Tangent Plane Approximations and Power Series (see felderbooks.com)
- 4.8 Optimization and the Gradient 172
- 4.9 Lagrange Multipliers 181
- 4.10 Special Application: Thermodynamics (see felderbooks.com)
- 4.11 Additional Problems (see felderbooks.com)

5 Integrals in Two or More Dimensions 188

- 5.1 Motivating Exercise: Newton's Problem (or) The Gravitational Field of a Sphere 188
- 5.2 Setting Up Integrals 189
- 5.3 Cartesian Double Integrals over a Rectangular Region 204
- 5.4 Cartesian Double Integrals over a Non-Rectangular Region 211
- 5.5 Triple Integrals in Cartesian Coordinates 216
- 5.6 Double Integrals in Polar Coordinates 221
- 5.7 Cylindrical and Spherical Coordinates 229
- 5.8 Line Integrals 240
- 5.9 Parametrically Expressed Surfaces 249
- 5.10 Surface Integrals 253
- 5.11 Special Application: Gravitational Forces (see felderbooks.com)
- 5.12 Additional Problems (see felderbooks.com)

6 Linear Algebra I 266

- 6.1 The Motivating Example on which We're Going to Base the Whole Chapter: The Three-Spring Problem 266
- 6.2 Matrices: The Easy Stuff 276
- 6.3 Matrix Times Column 280
- 6.4 Basis Vectors 286
- 6.5 Matrix Times Matrix 294
- 6.6 The Identity and Inverse Matrices 303
- 6.7 Linear Dependence and the Determinant 312
- 6.8 Eigenvectors and Eigenvalues 325
- 6.9 Putting It Together: Revisiting the Three-Spring Problem 336
- 6.10 Additional Problems (see felderbooks.com)

7 Linear Algebra II 346

- 7.1 Geometric Transformations 347
- 7.2 Tensors 358
- 7.3 Vector Spaces and Complex Vectors 369
- 7.4 Row Reduction (see felderbooks.com)
- 7.5 Linear Programming and the Simplex Method (see felderbooks.com)
- 7.6 Additional Problems (see felderbooks.com)

8 Vector Calculus 378

- 8.1 Motivating Exercise: Flowing Fluids 378
- 8.2 Scalar and Vector Fields 379
- 8.3 Potential in One Dimension 387
- 8.4 From Potential to Gradient 396
- 8.5 From Gradient to Potential: The Gradient Theorem 402
- 8.6 Divergence, Curl, and Laplacian 407
- 8.7 Divergence and Curl II—The Math Behind the Pictures 416
- 8.8 Vectors in Curvilinear Coordinates 419
- 8.9 The Divergence Theorem 426
- 8.10 Stokes' Theorem 432
- 8.11 Conservative Vector Fields 437
- 8.12 Additional Problems (see felderbooks.com)

9 Fourier Series and Transforms 445

- 9.1 Motivating Exercise: Discovering Extrasolar Planets 445
- 9.2 Introduction to Fourier Series 447
- 9.3 Deriving the Formula for a Fourier Series 457
- 9.4 Different Periods and Finite Domains 459
- 9.5 Fourier Series with Complex Exponentials 467
- 9.6 Fourier Transforms 472
- 9.7 Discrete Fourier Transforms (see felderbooks.com)
- 9.8 Multivariate Fourier Series (see felderbooks.com)
- 9.9 Additional Problems (see felderbooks.com)

10 Methods of Solving Ordinary Differential Equations 484

- 10.1 Motivating Exercise: A Damped, Driven Oscillator 485
- 10.2 Guess and Check 485
- 10.3 Phase Portraits (see felderbooks.com)
- 10.4 Linear First-Order Differential Equations (see felderbooks.com)
- 10.5 Exact Differential Equations (see felderbooks.com)

viii   Contents

     10.6   Linearly Independent Solutions and the Wronskian (see felderbooks.com)
     10.7   Variable Substitution   494
     10.8   Three Special Cases of Variable Substitution   505
     10.9   Reduction of Order and Variation of Parameters (see felderbooks.com)
     10.10   Heaviside, Dirac, and Laplace   512
     10.11   Using Laplace Transforms to Solve Differential Equations   522
     10.12   Green's Functions   531
     10.13   Additional Problems (see felderbooks.com)

11   **Partial Differential Equations**   **541**

     11.1   Motivating Exercise: The Heat Equation   542
     11.2   Overview of Partial Differential Equations   544
     11.3   Normal Modes   555
     11.4   Separation of Variables—The Basic Method   567
     11.5   Separation of Variables—More than Two Variables   580
     11.6   Separation of Variables—Polar Coordinates and Bessel Functions   589
     11.7   Separation of Variables—Spherical Coordinates and Legendre Polynomials   607
     11.8   Inhomogeneous Boundary Conditions   616
     11.9   The Method of Eigenfunction Expansion   623
     11.10   The Method of Fourier Transforms   636
     11.11   The Method of Laplace Transforms   646
     11.12   Additional Problems (see felderbooks.com)

12   **Special Functions and ODE Series Solutions**   **652**

     12.1   Motivating Exercise: The Circular Drum   652
     12.2   Some Handy Summation Tricks   654
     12.3   A Few Special Functions   658
     12.4   Solving Differential Equations with Power Series   666
     12.5   Legendre Polynomials   673
     12.6   The Method of Frobenius   682
     12.7   Bessel Functions   688
     12.8   Sturm-Liouville Theory and Series Expansions   697
     12.9   Proof of the Orthgonality of Sturm-Liouville Eigenfunctions (see felderbooks.com)

- 12.10 Special Application: The Quantum Harmonic Oscillator and Ladder Operators (see felderbooks.com)
- 12.11 Additional Problems (see felderbooks.com)

13 Calculus with Complex Numbers 708

- 13.1 Motivating Exercise: Laplace's Equation 709
- 13.2 Functions of Complex Numbers 710
- 13.3 Derivatives, Analytic Functions, and Laplace's Equation 716
- 13.4 Contour Integration 726
- 13.5 Some Uses of Contour Integration 733
- 13.6 Integrating Along Branch Cuts and Through Poles (see felderbooks.com)
- 13.7 Complex Power Series 742
- 13.8 Mapping Curves and Regions 747
- 13.9 Conformal Mapping and Laplace's Equation 754
- 13.10 Special Application: Fluid Flow (see felderbooks.com)
- 13.11 Additional Problems (see felderbooks.com)

Appendix A Different Types of Differential Equations 765

Appendix B Taylor Series 768

Appendix C Summary of Tests for Series Convergence 770

Appendix D Curvilinear Coordinates 772

Appendix E Matrices 774

Appendix F Vector Calculus 777

Appendix G Fourier Series and Transforms 779

Appendix H Laplace Transforms 782

Appendix I Summary: Which PDE Technique Do I Use? 787

Appendix J Some Common Differential Equations and Their Solutions 790

Appendix K Special Functions 798

Appendix L Answers to "Check Yourself" in Exercises 801

Appendix M Answers to Odd-Numbered Problems (see felderbooks.com)

Index 805

PREFACE

The Four Students: A Math Methods Parable

The professor has just finished a first-semester lecture on definite integrals, possibly unaware that four different types of students are processing the information quite differently. The class comprises *aspiring physicists, aspiring engineers, aspiring mathematicians,* and *bad students.*

- The aspiring physicists are thinking, "Why does subtracting an antiderivative over here from an antiderivative over there give you the area under this curve?"
- The aspiring engineers are thinking, "Of what possible practical value is finding the area under a curve?"
- The aspiring mathematicians are thinking, "How can you claim to have found anything when you haven't rigorously defined 'area under a curve' in the first place?"
- The bad students are thinking, "Just tell me how to do the problems and get the answer you want on the test."

This book is written for aspiring physicists and engineers. For each topic, we hope to clearly answer the questions "Why does this mathematical technique work?" and "How is this used to solve practical problems?" We will fall short of the expectations of true mathematicians, and we hope to continually frustrate the bad students.

Exercises

In almost every section of our book you will find an "Exercise" and a set of "Problems." What's the difference?

The simple answer is that the Problems are, for the most part, independent of each other. You create an assignment that says "Do Section 18.3 Problems 2, 5, 9, 12, and 20." By contrast, an Exercise is an atomic block. You can assign a particular exercise, or you can skip that exercise, but you can't assign "Question 5" from the exercise because it only makes sense in context.

More importantly, the two serve different purposes. Problems are meant to follow a lecture, building and testing the students' skills and understanding of the topics you have discussed in class. Exercises are designed to facilitate active learning.

There are two types of exercises.

Motivating Exercises come at the beginning of each chapter. Their purpose is not to teach math, but to give a practical example of why the student needs the techniques in this chapter.

Discovery Exercises come at the beginning of (almost) every section. Their purpose is to step the student through a mathematical process, such as solving a differential equation or finding a Taylor series. Instead of just being told how to do it, the students do it for themselves.

Some frequently asked questions:

- *Do I need to assign all the exercises?* No. If you are uncomfortable with the process, you may want to try only one or two. We hope you will find them easy to use and valuable, and over time you will use them more, but you will probably never use them all.

- *At home or in class? Alone or in groups?* Mix it up. See what works for you. We sometimes assign them as homework due on the day we are going to cover the material, and sometimes as an in-class exercise to begin the lecture. You can have students do them individually or in groups, or a mix of the two. One professor we spoke to starts them in class, and then has her students finish them at home—an approach we never even thought of. You will probably keep your students' interest better if you vary your approach.
- *How long do they take?* Some are five minutes or less; some are twenty minutes or even more. Very few of them should take the students more than half an hour.
- *That was all pretty noncommittal. Do you have any solid advice at all?* Actually, we do. First, we hope you will use at least some of the exercises, because we believe they contribute a valuable part of the learning process. Second, exercises should almost always be used *before* you introduce a particular topic—not as a follow-up. You can start your lecture by taking questions and finding out where the students got stuck.

Problems

There are problems at the end of every section, and there are also "additional problems" at the end of every chapter. The "additional problems" give us an opportunity to ask questions where the students don't know exactly what's being covered. (When you look at a partial differential equation in "additional problems" you have to ask yourself: separation of variables? Method of transforms? What's the right approach here?)

We have resisted classifying the problems further, but you will find a general pattern something like this.

- A "walk-through" steps the students carefully through the process we want them to learn. We advise you to generally assign these problems.
- Next often comes a batch of straightforward, unmotivated calculation problems. "Evaluate the following triple integrals in spherical coordinates." They will generally move from easier to harder.
- Then come word problems. Some of these are practical applications; some fill in details that were left out of the explanation; some are just cool ideas that occurred to us while we were brainstorming over Bailey's Irish Cream.
- Finally, in some cases, there are "Explorations." These are harder, more involved, and often longer problems that may stretch beyond the presented material.

Problems without a computer icon (which are most of them) can be done entirely by hand, and should generally require no integration technique beyond u-substitution or integration by parts. A computer icon can mean anything from "This requires an integral that you can do on your calculator" to "This involves heavy use of a computer algebra program such as Mathematica, MATLAB, or Maple." The problems are written in a platform-independent way, and we provide no instruction on any of these computer tools in particular.

Chapter Order and Dependencies

One of the unusual things about Math Methods, as a course, is that it covers a broad variety of loosely (if at all) related topics. It can be taught as a sophomore level course with only two semesters of calculus as a prerequisite (so half the course becomes an introduction to multivariate calculus), or it can be taught as a first-year graduate-level course, or anything in between. It is taught in physics departments, engineering departments, economics departments, and occasionally math departments. All those different courses cover different topics.

So the textbook becomes what Stuart Johnson, our first editor at Wiley, called a "Chinese menu." You look it over and you decide you want this chapter, that chapter, skip three and pick up this one, and so on.

For this reason, we have endeavored as much as possible to make the chapters independent of each other. You don't need our vector calculus chapter to cover our linear algebra chapter, or vice versa. References from one section to another within a chapter are common; references from one chapter to another are rare.

That being said, there are exceptions. Most importantly, pretty much every chapter in the book relies on the information presented in Chapter 1, Introduction to Ordinary Differential Equations. The information is minimal: most of the techniques for *solving* ODEs are deferred to later chapters. But the students have to know what a differential equation is, whether they get it from our chapter or somewhere else. If you are not sure your students come into the course highly comfortable with this material, please start with this chapter before doing anything else. If you want to give it the most minimal treatment possibly, you can get away with doing the three sections "Overview of Differential Equations," "Arbitrary Constants," and "Guess and Check, and Linear Superposition" only. We believe strongly, however, that it is worth your class time to do more.

Beyond that, the first page of every chapter lists prerequisites. You should be able to use that to make sure your students are ready for any given chapter.

Last Word: Communicating Priorities to Students

Here is an experience that took me (Kenny) quite by surprise. I assigned the following problem. "Write the Maclaurin series expansion of e^{-x^2}; then use the first five terms of that series to approximate $\int_0^1 e^{-x^2} \, dx$." Many students came back saying "I had no trouble finding the Maclaurin series, but I didn't understand what you were asking me to do with it."

Just in case you're staring at that sentence with the same dumbstruck look I probably had, I want to stress that these were *not* weak students, and they *did* know how to integrate a polynomial.

And here's my point. We don't just want our students to learn methods; we want them to understand why those methods work, to view those methods in a larger mathematical context, and to be able to apply those methods to physical problems. But students don't develop those skills by being told "You should be able to think for yourself." They develop those skills, just like any others, with practice and feedback. This is particularly relevant for Math Methods, where the skill and the application may be separated by semesters or years. ("What do you mean, quantum mechanics students, you've never heard of a Fourier series? Didn't you take Math Methods?")

Everything in our book is structured to give your students that practice. A discovery exercise says "Don't just listen to me lecture about this; figure it out for yourself." A walk-through says "Let me help you with that important process." The problem after the walk-through, or the later problems in the section, often say "Let's think more deeply about that result" or "Let's see where that came from" or "Let's apply that technique to a circuit." If the explanation stepped through a particularly important derivation that you want your students to understand, there are almost certain to be some problems designed to make sure students followed the derivation. Give your students enough problems to master specific skills—evaluating a line integral, separating variables in a PDE, finding the coefficients in a Fourier series—but assign deeper problems in areas where you want deeper understanding.

Here is a specific example. In the section on Legendre polynomials in the special functions chapter, one problem steps the students through the solution of Hermite's differential equation. If you just want your students to be able to work with Legendre polynomials, you can certainly skip that problem. But if you want your students to follow the derivation of the Legendre polynomials, that problem will force them through every step of the process.

That brings up the more general topic of proofs. It's very rare for our book to prove a theorem before we use it. Much more commonly we present a theorem, show students how to use it, and then step them through the proof in a problem. The explanation will usually point to that problem with language such as "You'll show that this always works in Problem 14." In some cases the problem doesn't prove the theorem, but has the students show that it works for some important cases.

There are two reasons for this unusual structure. First, we believe students follow a proof better after they understand the result that is being proven. Second, we believe very strongly that students follow a proof better if they work through it themselves instead of just reading it.

One of the judgment calls you will have to make, therefore, is which of these proof problems to assign. Our own opinion on this matter, for whatever that's worth, is that the importance of a proof is *not* based on the importance of the result it proves, but on the technique that the proof demonstrates. It's tremendously important for all students to know that the derivative of $\sin x$ is $\cos x$, but very few students can prove it—and that's OK. On the other hand, the proof that $a_n = \frac{1}{L} \int_{-L}^{L} f(x) \cos\left(\frac{n\pi}{L} x\right) dx$ in a Fourier series involves an important trick that teaches students what orthogonal functions are and how to find the coefficients of many other such series, so it's worth some investment of class and homework time.

The Most Important Thing We Want to Tell You

Please see http://www.felderbooks.com for all the information you will find in this introduction, exercises formatted for printing, additional problems for every chapter, additional sections including "special applications," answers to odd-numbered problems, and a lot more.

Acknowledgements

Of all the tasks involved in writing a textbook, the one that we most drastically underestimated is writing solutions. We wanted to write the kind of detailed, carefully worked out solutions that we like to give to our students when they need help. Multiply that by 3,000 problems and you have a task that could have held up publication by a year.

So we turned to some of the most talented people we knew—our best current and former students. They stepped up and put in countless hours writing the solutions. (We read and edited all their solutions, so if you see a mistake, it's still our fault.) Along the way they became experts in both LaTeX and the mathematical techniques they were focused on. More importantly for us and for you, they provided the front line of feedback on which problems worked, which problems needed to be revised, and which problems needed to be scrapped. Even if you never look at the solutions, the book you're holding is far better for their dedicated work.

- Introduction to Ordinary Differential Equations: Olga Navros
- Taylor Series and Series Convergence: Szilvia Kiss
- Complex Numbers: Olga Navros, Amanda (Stevie) Bergman

- Partial Derivatives: Jonah Weigand-Whittier
- Integrals in Two or More Dimensions: Mariel Meier
- Linear Algebra I: Szilvia Kiss
- Linear Algebra II: Szilvia Kiss
- Vector Calculus: Meg Crenshaw
- Methods of Solving Ordinary Differential Equations: Alison Grady
- Calculus with Complex Numbers: Alison Grady

In addition, Kate Brenneman wrote computer solutions for the series chapter.

We can't possibly list all the other wonderful people who made valuable contributions to this book, but here are a few of the most important.

- Szilvia Kiss (yes, the same student who wrote solutions for *three chapters*), Isabel Lipartito, Emma Gould, Alison Grady, and Danika Luntz-Martin, all superstar students at Smith College, worked through half the book as an independent study. (Alison spent a full year at it and did the entire book.) They read the explanations, worked the exercises, chose which problems to do, and met with us weekly to go over questions—and to provide us feedback. "This was helpful, but that was confusing." "I learned in chemistry class about a great application of this topic." Every chapter benefited from their time and initiative.
- Professors Doreen Weinberger and Courtney Lannert at Smith College, Alexi Arango at Mt. Holyoke, Christine Aidala at the University of Michigan, Sean Bentley at Adelphi University, and Henry Rich at Raleigh Charter High School all chose our book as their primary Math Methods text before the book was published. We sent them PDFs, they printed out course-packs, and then they and their students provided us feedback. Because of their willingness to take that risk, the first users of our printed book will not be experimenting with untested material.
- Doreen Weinberger, in addition to teaching from the book, spent a great deal of time with Gary providing detailed feedback, suggesting additions or changes in focus, and describing applications that made great problems.
- Henry Rich of Raleigh Charter High School also read through and provided feedback on a number of the chapters and sections he *didn't* teach from. He is also our guiding light of linear algebra.
- Barbara Soloman read through and provided feedback on the first draft of our first chapter, "Introduction to Ordinary Differential Equations."
- Brian Leaf sent Gary an email saying "I know an editor at Wiley you should talk to," and thus got this entire project started. To this day we're not sure if that wound up being a favor.
- Dr. Steven Strogatz of Cornell gave us permission to use his wonderful "Romeo and Juliet" coupled differential equations example.
- Dr. Robert Scott at the University of Brest spent countless hours proofreading. He found mistakes and suggested improvements that made many of our chapters better and more accurate.
- Dr. Richard Felder—a chemical engineering professor, our father, and another proud Wiley author—has been with us through this process in more ways than we can possibly name. He was the one who said "The first chapter you write should be Partial Differential Equations." Then he read through that chapter and offered suggestions such as "Give students an opportunity to check their answers during these Exercises" and "Start each problem section with a problem that walks the students through the process you're modeling." This would be a very different book without his guidance.

The dedicated team at Wiley worked tirelessly to bring this project out of Gary and Kenny's imaginations and into a physical actual you-can-hold-it-in-your-hands printed book.

We would particularly like to thank Amanda Rillo and Kathyrn Hancox, Editorial Assistants; Jolene Ling and Joyce Poh, Senior Production Editors; Kristy Ruff and Christine Kushner, Marketing Managers; Krupa Muthu, Project Manager; Stuart Johnson, our first Executive Editor and the first person at Wiley to believe in this project; and Jessica Fiorillo, who jumped in with both feet to replace Stuart after his retirement. Jess always had a smile—you could hear it over the phone—and her "we'll make it work" attitude, even when we threw weird requests at her, made the whole process work.

Last, and always most, our wives. Rosemary McNaughton shared her expertise on typography and LaTeX; Joyce Felder worked through problems on Fourier series. But far more important is the support they provided during the three years that their husbands disappeared into our computers. Everything we do bears the mark of their time, support, talents, patience and understanding.

CHAPTER 1

Introduction to Ordinary Differential Equations

> *Before you read this chapter, you should be able to …*
> - evaluate a derivative.
> - evaluate an indefinite integral using u-substitution or integration by parts and use an "initial condition" to calculate the arbitrary constant (usually called C).
>
> *After you read this chapter, you should be able to …*
> - model real-world situations with ordinary differential equations.
> - given a differential equation and a proposed solution, determine if that proposed solution works.
> - solve differential equations by "separation of variables" or "guess and check."
> - identify a differential equation as "linear" or "non-linear," and identify a linear differential equation as "homogeneous" or "inhomogeneous."
> - find a "general solution" as a linear combination of individual solutions.
> - use initial conditions to determine the arbitrary constants in a general solution.
> - graphically represent the solutions to first-order differential equations using "slope fields."
> - graphically identify "equilibrium solutions" and classify them as "stable" or "unstable."
> - write, and identify solutions to, "coupled" differential equations.
> - use computers to analyze differential equations and their solutions.
> - most importantly, *interpret* the solution to a differential equation to model a physical problem.

Science is about predicting the behavior of systems. In many cases, we begin by knowing how the system will *change*. For instance, classical physics is built around the equation $F = ma$: the forces determine how the position changes, and from this information, physicists make predictions about the actual position. In a similar way, an economist might try to predict the price of gas by taking into account all the influences that will tend to drive the price up or drive it down.

Mathematically, knowledge about how a variable will change means knowledge of its derivatives. Therefore, many natural laws are expressible as equations about derivatives: "differential equations." Representing known laws as differential equations, and solving those equations, is one of the most important tools in modern science and engineering.

A function of one independent variable, such as the position of a particle as a function of time, may be modeled by an "ordinary differential equation" (or "ODE"). A function of multiple independent variables, such as temperature in a room as a function of x, y, z, and t, can be modeled by a "partial differential equation" (or "PDE"). This chapter, Chapter 10, and Chapter 12 discuss ODEs; Chapter 11 discusses PDEs.

1.1 Motivating Exercise: The Simple Harmonic Oscillator

Each chapter begins with a "motivating exercise" for the students to work through in class or as homework before studying the chapter. They are optional, but if you use them it can help the students understand why they are learning the material in the chapter.

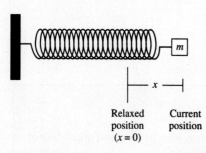

FIGURE 1.1
Relaxed position ($x = 0$)
Current position

Figure 1.1 shows an object of mass m attached to a spring. The spring obeys "Hooke's Law" which states that $F = -kx$, where…

- F is the force the spring exerts on the object.
- x is the object's displacement from its relaxed position. Note that x can be positive (if the spring is stretched) or negative (if the spring is compressed).
- k is a positive constant, the "spring constant."

In SI units F would be measured in Newtons and x in meters, and k would therefore be in N/m.

1. What does Hooke's Law predict about the *direction* of the force the spring exerts? Based on this, what kind of motion would you expect?

   See Check Yourself #1 in Appendix L

   Newton's Second Law tells us that $F = ma$; the force exerted by the spring on the object determines the acceleration of the object, where acceleration is the second derivative of position with respect to time.

2. Using Hooke's Law and Newton's Second Law, write an equation that relates the position of the object x to its acceleration a. Your equation will have two constants in it: m (the mass of the object) and k (the spring constant).

3. Replace a with its definition d^2x/dt^2 and rewrite your equation so that the second derivative is on the left side of the equation by itself.

   See Check Yourself #2 in Appendix L

4. One trivial solution to that equation is the simple function $x(t) = 0$. (If you take the second derivative, you do indeed get the same function back, times anything you like.) What does this solution tell us about one possible motion of the mass-on-spring system?

5. In order to find other possibilities, let's begin by supposing $k/m = 1$, so $d^2x/dt^2 = -x$. Which of the following functions is *the negative of its own second derivative*? One and only one of these functions works. (These solutions all appear to have incorrect units, but by setting $k/m = 1$ we've hidden the units. They will reappear in Part 6.)
   (a) $x = -1$
   (b) $x = -(1/6)t^3$
   (c) $x = e^t$
   (d) $x = e^{-t}$
   (e) $x = \sin t$

6. Now let's return to the original differential equation: $d^2x/dt^2 = -(k/m)x$. Which of the following functions is a solution to this equation? For each one, you need to test it by taking its second derivative and checking whether the answer you get is $-k/m$ times the original function. Once again, one and only one of these works.
   (a) $x = \sin t$
   (b) $x = \sin(kt/m)$
   (c) $x = \sin\left(\sqrt{k/m}\, t\right)$

(d) $x = \sin(mt/k)$
(e) $x = \sin\left(\sqrt{m/k}\ t\right)$

7. The function that you just found as a solution to $d^2x/dt^2 = -(k/m)x$ offers a prediction for the motion of the mass. Does this prediction correspond to the prediction you made on physical grounds in Part 1?

1.2 Overview of Differential Equations

In this section we introduce the *idea* of a differential equation. Our emphasis is on what a differential equation means, what a solution to a differential equation means, and how to use differential equations to model physical situations. Later, we will begin to look at the question of how to *find* solutions to differential equations.

1.2.1 Discovery Exercise: Overview of Differential Equations

Most sections begin with a "discovery exercise" that students can do in class or as homework before covering the section. They are optional, but if the students do them they will derive some of the key math ideas themselves.

1. A certain function $y(x)$ has the following property: its *derivative* is $6x$. We can express this mathematically by writing $dy/dx = 6x$.
   (a) Solve this equation; that is, find a function $y(x)$ that works.
   (b) Find a *different* function that also works. (*Hint*: it will only be slightly different from the first function.)

   See Check Yourself #3 in Appendix L

   (c) Test your solutions by taking their derivatives and making sure you get $6x$ both times.

2. A certain function has the following property: its *second derivative* is $6x$.
   (a) Write an equation that expresses the sentence "The second derivative of the function y is $6x$."
   (b) Solve the equation; that is, find a function y that works.
   (c) Find a *different* function that also works.
   (d) Test your solutions.

3. A certain function has the following property: *the function is its own derivative*.
   (a) Write an equation that expresses the sentence "Function y is its own derivative."

   See Check Yourself #4 in Appendix L

   (b) Solve the equation; that is, find a function y that works.
   (c) Test your solution by taking its derivative and making sure you get the same function you started with.
   (d) Find a *different* function that also works and test it. (*Hint*: adding a constant to your first solution will not work!)

4. A certain function has the following property: *when you take the derivative of the function, you get the same function back, times 2*.
   (a) Write an equation that expresses the sentence "When you take the derivative of function y, you get the original function y times 2."
   (b) Solve the equation; that is, find a function y that works.
   (c) Find a *different* function that also works.
   (d) Test your solutions.

5. A certain function $y(x)$ is the solution to the equation $dy/dx = xy^2$.
   (a) Express this equation in words. (This is the reverse of what you were doing before, where we gave you the words and you gave us the equation.)

(b) Which of the following functions could be $y(x)$? *Hint*: you should approach this by trying each function in that equation. One function will work, and the others will not. For the left side of the equation, find dy/dx. For the right side, square y and then multiply the answer by x. If the two sides come out the same, you have found a solution!

  i. $y = \sqrt{x^2 + 9}$
  ii. $y = e^{\sqrt{x+3}}$
  iii. $y = -2/(x^2 + 3)$
  iv. $y = x^3/3 + 6$

1.2.2 Explanation: Overview of Differential Equations

When you were first introduced to algebra, you learned that an equation such as $5x + 14 = x^2$ is not so much a statement as a question. It challenges you to find a number that has the following property: when you multiply this number by five and add fourteen, you get the original number squared. This particular equation happens to work if $x = -2$: you can confirm this by substituting -2 in for x, creating the true equation $-10 + 14 = 4$. When you learned a bit more algebra, you learned techniques for *solving* such equations; you could then figure out for yourself that $x = 7$ also works for this particular example. Most importantly, you learned how to *use* such equations to model problems and make predictions. ("What time would that train reach Chicago?")

Along the way, you also learned that there is not just one technique for solving all algebraic equations. For a linear equation such as $3 - 5x = 2x + 7a$ you would probably begin by getting all the terms with x on one side, and the remaining terms on the other (and you would expect to find exactly one answer). For a quadratic equation such as the one in the previous paragraph, it is more helpful to collect all the terms on the left (and you may find two answers, one answer, or none at all). So before you solve an equation, the first step is to figure out what *type* of equation it is: that is, to classify it.

Finally, you learned that some equations are difficult or impossible to solve analytically. The equation $10 \cos x = x + 1$ has many real solutions, but there's no easy way to find them without a computer.

A "differential equation" is similar to an algebraic equation in all of these respects, although its solution is a *function* instead of a *number*. Once again, however, you will learn to classify different equations so you can apply different approaches. You will learn to recognize when you have found all possible solutions. You will learn what kinds of equations you can solve analytically, and which are better approached with a computer. Most importantly, you will learn to use differential equations to model problems and make predictions.

Consider the motivating example at the beginning of this chapter, the simple harmonic oscillator. When we combine Newton's Second Law $F = ma$ with Hooke's Law for a spring $F = -kx$, recalling that $a = d^2x/dt^2$, we end up with the following equation:

$$\frac{d^2x}{dt^2} = -\frac{k}{m}x \qquad (1.2.1)$$

Mathematically, we are looking for some function $x(t)$ that has the following property: when you take the second derivative, you get the same function you started with, multiplied by $-k/m$. Physically, we understand that any function with this property may represent the motion of a mass on a spring. With a bit of experimentation, you may discover that all of the following functions work.

$$x(t) = 0, \qquad x(t) = 3\cos\left(\sqrt{\frac{k}{m}}\, t\right), \qquad x(t) = \sin\left(\sqrt{\frac{k}{m}}\, t + \frac{\pi}{3}\right)$$

There are three key questions you may be asking yourself about that equation and those solutions.

- **What do you mean, those functions all "work"?** You validate solutions the same way you would for an algebraic equation: by plugging them in. Let's try the third one.

$$x(t) = \sin\left(\sqrt{\frac{k}{m}}\, t + \frac{\pi}{3}\right) \to x'(t) = \sqrt{\frac{k}{m}} \cos\left(\sqrt{\frac{k}{m}}\, t + \frac{\pi}{3}\right)$$

$$\to x''(t) = -\frac{k}{m} \sin\left(\sqrt{\frac{k}{m}}\, t + \frac{\pi}{3}\right)$$

Thus we see that $x''(t)$ equals $x(t)$ multiplied by $-k/m$, so this function satisfies Equation 1.2.1.
- **What do these answers tell me about the mass on the spring?** Each of these solutions represents a possible motion for such a system. The actual motion depends on "initial conditions." For instance, if the spring begins at its relaxed position at rest, it will never move: $x(t) = 0$. If the spring is pulled out to position $x = 3$ and released from rest, it will begin to oscillate: $x(t) = 3 \cos(\sqrt{k/m}\, t)$.
- **How do I find such answers on my own?** We will begin to address this question in this chapter, and we will address it further in our chapter on methods of solving ordinary differential equations. For the moment, our emphasis is on understanding what it *means* to write and solve a differential equation.

A Close Look at a Very Simple Differential Equation

Shannon makes $8 an hour. If M is the total amount of money she has, and t is the number of hours she has been working, then the function $M(t)$ will obey the following equation.

$$\frac{dM}{dt} = 8 \tag{1.2.2}$$

If the units in that equation worry you—which they should!—hold onto that reservation; we will have more to say about that issue below. Right now we want to focus on the fact that Equation 1.2.2 can be understood in three different ways.

- *Analytically*, it says "Find a function $M(t)$ whose derivative is the constant function 8." If you find a function, take its derivative, and get 8, you have found a solution.
- *Graphically*, it says "The slope of this graph is 8 everywhere." (You may find this easier to think about as $dy/dx = 8$ but we urge you to become comfortable with other variable names.)
- *Logically*, it says "The rate of growth of Shannon's money is exactly 8 money-units (dollars) every time-unit (hour)." More concisely, Shannon makes $8.00/hour. It is this step—converting between equations and descriptions of reality—that requires human intervention. Computers generally do a better job than we do of the calculations and graphing.

These are not three different types of differential equations: they are three different perspectives on the *same* differential equation. You can measure your understanding by how well you can easily convert between the three.

It may occur to you, from any of these three points of view, that there is not just one solution to this equation. Analytically, the function $M = 8t$ has the correct derivative, but so do $M = 8t - 5$ and $M = 8t + \pi$. The "general solution" of this equation is $M = 8t + C$ where C is any constant. Graphically, you can visualize an infinite number of parallel lines, all with constant slopes of 8.

What does all that tell us logically? We know how fast Shannon is making money, but we don't know how much she started with! If she started with $100, then $M = 8t + 100$ is the correct function to describe her bank account. This idea—that a differential equation has an infinite number of solutions, and you choose one based on "initial conditions"—will be a recurring theme in this chapter.

An Equally Close Look at a More Complicated Differential Equation

Shannon retires and puts $1000 into a bank, where it earns 5% interest. This means that every year, the bank gives her 1/20 of her current balance. Now her $M(t)$ function obeys this equation.

$$\frac{dM}{dt} = \frac{1}{20} M \qquad (1.2.3)$$

Don't confuse this with the similar-looking equation $dM/dt = 1/20$. That would say "Every year, M goes up by $1/20$." This says something closer to "Every year, M goes up by $1/20$ of M." With this small step—multiplying the right side of the equation by M, the dependent variable—we have moved out of the world of introductory calculus. If you want to solve Equation 1.2.2, you simply integrate both sides. That won't work for Equation 1.2.3! Nonetheless, we can think about our new equation in the same three ways that we did the old.

- *Analytically,* Equation 1.2.3 says "Find a function $M(t)$ whose derivative is the original function divided by 20." You may want to think about this for a moment, and see if you can find a function that works.
- *Graphically,* it says "The slope of this graph is $1/20$ of the M-value." This will not be a line: as it goes higher, it will also get steeper. Once again, you may want to try your hand at sketching such a curve.
- *Logically,* it says "Each time-unit (year), Shannon's money grows by $1/20$ of Shannon's money." Your most important challenge is to convince yourself that this is exactly what is meant by the phrase "5% interest."

Let's take up the challenge analytically. We need a function that satisfies two criteria. Of course $M(t)$ must satisfy the differential equation, Equation 1.2.3. But it *also* has to satisfy the "initial condition" $M(0) = 1000$; in other words, it must correctly predict that Shannon starts with $1000. We approach such problems by first solving the differential equation, and then tweaking the answer to meet the initial condition. So let's start with this: how can you take the derivative of a function and get the same function back, multiplied by $1/20$?

The most obvious solution, perhaps, is $M(t) = 0$. Sure enough, the derivative is the original function, multiplied by $1/20$. Graphically, we get a horizontal line at zero; the slope (0) is always perfectly proportional to the height (0). Logically, we learn that if Shannon starts with no money, she will never get any money. It all works—but it doesn't tell us anything about Shannon's actual bank account. We have found a solution to the differential equation, but we cannot make it meet the initial condition.

After some trial and error you might hit on the next solution, $M = e^{t/20}$. If you take the derivative, you get $M'(t) = (1/20)e^{t/20}$, which is indeed $1/20$ of the original function. Graphically, an exponential function does exactly what we want: as it gets higher, it also gets steeper. This solves Equation 1.2.3, but unfortunately starts at $M(0) = 1$. So our job now is to modify this solution to meet the initial condition.

Many students next try $M = e^{t/20} + 999$. That gives us $M(0) = 1000$, which seems like a good start. But be careful: this doesn't solve the differential equation! $M'(t) = (1/20)e^{t/20}$ as before, but $M/20 = (1/20)(e^{t/20} + 999)$; Equation 1.2.3 is not satisfied unless $M'(t)$ and $M/20$ come out the same.

What *does* work is $M = 1000e^{t/20}$. $M'(t)$ and $M/20$ both give the same answer, $50e^{t/20}$. Furthermore, $M(0) = 1000$ so it correctly predicts that Shannon starts with $1000. This is the *only* function that satisfies both Equation 1.2.3 and the initial condition, so this is the function that correctly models Shannon's bank account.

If you were unable to find that solution on your own, don't worry about it. We are going to spend some of this chapter, plus two later chapters, learning how to find solutions. Focus now on what it *means* to say that a given function solves a differential equation and an initial condition.

Part of understanding "what it means" is testing your solutions to see if they make sense. Since Shannon is earning 5% each year, we might expect that in her first year, she will earn $\$1000 \times 5\% = \$50$. But in fact, our solution predicts that she ends up with $\$1000 e^{1/20} = \$1051.27$. Where did this small discrepancy come from? Our bank is not literally giving her 5% every year ("compounded annually"); it is giving her interest *all the time*, and then interest on her interest, and so on ("compounded continuously"), so after a year she has slightly more than 5% of what she started with. This distinction is explored in Problems 1.35 and 1.36. Apart from those two problems, we will ignore this distinction in this chapter; if we say something grows by 10% each year, you should translate that to $df/dt = .1f$ (with time measured in years of course).

EXAMPLE **Checking a solution to a differential equation**

Problem:
Consider the differential equation:

$$(x^2 + 1)\frac{dy}{dx} = 1 - 2xy \qquad (1.2.4)$$

with initial condition $y(0) = 6$. Which of these three equations is a valid solution?

1. $y = \dfrac{x+6}{x+1}$
2. $y = \dfrac{x}{x^2+1}$
3. $y = \dfrac{x+6}{x^2+1}$

Solution:
1. We substitute $y = (x+6)/(x+1)$ into both sides of the differential equation to see if they come out the same.

$$(x^2+1)\frac{dy}{dx} = (x^2+1)\frac{(x+1)-(x+6)}{(x+1)^2} = -\frac{5(x^2+1)}{(x+1)^2}$$

Meanwhile, on the other side of Equation 1.2.4,

$$1 - 2xy = \frac{x+1}{x+1} - \frac{2x(x+6)}{x+1} = \frac{-2x^2 - 11x + 1}{x+1}$$

These two functions are not the same: you can confirm this by plugging in numbers for x, or by plotting the two functions to confirm that they are genuinely different. Therefore, $y = (x+6)/(x+1)$ is *not* a solution to the differential equation.

Important note:
These two functions might be the same for *some* values of x, but that doesn't make $y = (x+6)/(x+1)$ a solution. The initial condition is about one particular x-value, but the differential equation itself must be satisfied for *all* x-values. (Don't set them equal and solve for x: it won't help.)

2. We can readily determine that $y = x/(x^2 + 1)$ does not meet the initial condition, since $y(0) = 0$. So this cannot be the solution we are looking for.
3. Once again we calculate both sides of the equation and compare. If $y = (x + 6)/(x^2 + 1)$ then we get this on the left.

$$(x^2 + 1)\frac{dy}{dx} = (x^2 + 1)\frac{x^2 + 1 - 2x(x + 6)}{(x^2 + 1)^2} = \frac{x^2 + 1 - 2x(x + 6)}{x^2 + 1}$$

Meanwhile, on the other side of the equation, we get this.

$$1 - 2xy = \frac{x^2 + 1}{x^2 + 1} - \frac{2x(x + 6)}{x^2 + 1} = \frac{x^2 + 1 - 2x(x + 6)}{x^2 + 1}$$

For the function $y = (x + 6)/(x^2 + 1)$, the quantities $(x^2 + 1)(dy/dx)$ and $1 - 2xy$ do come out the same; this function satisfies the differential equation. And what about the initial condition? We find $y(0) = 6$ as it should, so we have found a solution. Later in this chapter we will discuss guidelines for determining if this is the *only* solution. In this case, it is.

A Note About Units

You may quite correctly recoil in horror from $M = 8t + 100$, $M = 1000e^{t/20}$, and many of the other equations we have used in this section. Quantities with units such as M and t should never be added to unitless quantities such as 100, and the argument of an exponential should always be unitless. We can use these equations because 8, 100, 1000, and 20 in these equations only look unitless; for example, the 8 really means $8/h.

For this reason among others it is generally best to use letters instead of numbers. For example, suppose we had written Equation 1.2.3 (originally $dM/dt = M/20$) as $dM/dt = M/k$. To make the units work in that equation k must have units of time. Our solution would then have looked like $M = M_0 e^{t/k}$, and we could have checked that the argument of the exponential was unitless. If we had gotten $M = M_0 e^{kt}$ we would have caught our mistake since the argument of the exponential would have units of time squared.

Using letters in this way is a valuable habit, but we clearly are not doing so consistently. When you're learning unfamiliar math, sometimes specific numbers can be easier to work with than letters. When we feel this is the case, we dispense with units and just use numbers.

So what should you do with all this? When you first look at any problem, check to see if units are being used in the equations. If everything is in letters except for truly unitless quantities then you should make sure your answers have correct units; this painless step will save you from many common errors. If the problem has numbers where they don't belong in a physical problem then you must ignore units for that problem.

Stepping Back

We began this section with the differential equation that is most important in classical mechanics and mechanical engineering: Newton's Second Law, $F = ma$. That doesn't look like a differential equation, but it is because a simply means dv/dt, or equivalently d^2x/dt^2. So when we know the forces that act on an object, this equation tells us how fast the object's velocity will change, and that in turn tells us how its position will change. We then moved to economic examples, where terms such as "profit" and "interest" and "expenses" describe how a bank account changes (in both positive and negative ways); from that we can compute total amounts of money. In chemistry, "kinetics" describes the rate of a reaction, from which you can compute the amounts of the reactants. Calculations on electric circuits begin with "potential" that causes "current," the time derivative of charge.

All these examples, and many others in every field of science and engineering, ask the same fundamental question. "I know how fast this quantity is changing (because I have a model of the forces causing it to change). Now, what is the function?" That is the question that every differential equation asks. As you learn how to model physical situations with differential equations, solve those equations, and interpret the answers, you are learning one of the most useful tools for predicting the behavior of important systems.

1.2.3 Problems: Overview of Differential Equations

1.1 In this problem you are going to confirm that the solution to the differential equation $dy/dx = y \cos x$ with condition $y(0) = 1$ is $y = e^{\sin x}$.

(a) If $y = e^{\sin x}$, what is dy/dx?

(b) If $y = e^{\sin x}$, what is $y \cos x$?

If your answers to parts (a) and (b) are not the same, that would indicate that this function does *not* solve this differential equation. In this case, we are assuring you that it does—so if they did not come out the same, go back and find your mistake.

(c) Now plug $x = 0$ into the function to make sure it gives the answer specified in the initial condition.

1.2 One valid solution to the differential equation $f''(x) + 2f'(x) = 2e^x$ is $f(x) = (2/3)(e^{-2x} + e^x)$.

(a) Demonstrate that $f(x)$ does, in fact, solve the differential equation.

(b) If you add three to $f(x)$, is the resulting function also a solution?

(c) If you multiply $f(x)$ by 2, is the resulting function also a solution?

(d) You now have two valid solutions of this differential equation. One of those functions also satisfies the initial condition $f(0) = 13/3$. Which one?

For Problems 1.3–1.12, find *all* solutions that match the given differential equation. Remember that you test a solution by plugging it into the differential equation and seeing if both sides come out the same. Each problem will have at least one valid solution, but there may be several.

1.3 $dy/dx = 3$

(a) $y = 3$
(b) $y = 3x$
(c) $y = 3x + 9$
(d) $y = (3/2)x^2$
(e) $y = e^{3x}$

1.4 $dy/dx = 3x$

(a) $y = 3$
(b) $y = 3x$
(c) $y = 3x + 9$
(d) $y = (3/2)x^2$
(e) $y = e^{3x}$

1.5 $dy/dx = 3y$

(a) $y = (3/2)x^2$
(b) $y = e^{3x}$
(c) $y = 3e^x$
(d) $y = 7e^{3x}$
(e) $y = e^{3x+7}$
(f) $y = e^{3x} + 7$

1.6 $dz/dt = z^2$

(a) $z = (1/3)t^3$
(b) $z = e^{t^2}$
(c) $z = 1/t$
(d) $z = -1/t$
(e) $z = -1/t + 6$
(f) $z = -1/(t+6)$

1.7 $dP/dr = P - r$

(a) $P = r + 1$
(b) $P = e^r$
(c) $P = e^r + r + 1$
(d) $P = e^r + r + 2$
(e) $P = 2e^r + r + 1$

1.8 $dy/dx = (y^2 - xy)/(x^2 \ln x + 2xy)$

(a) $y = \ln x + x^2$
(b) $y = (\ln x)/x$
(c) $y = -x \ln x$

1.9 $d^2x/dt^2 = x$

(a) $x = e^t$
(b) $x = e^{-t}$
(c) $x = -e^t$
(d) $x = 3e^t - 5e^{-t}$
(e) $x = 0$

1.10 $d^2x/dt^2 = -x$
  (a) $x = e^t$
  (b) $x = e^{-t}$
  (c) $x = -e^t$
  (d) $x = \sin t$
  (e) $x = \cos t$
  (f) $x = 10 \sin t + 15 \cos t$

1.11 $d^2y/dx^2 - 6(dy/dx) + 9y = 0$
  (a) $y = e^{3x}$
  (b) $y = e^{-3x}$
  (c) $y = 3e^x$
  (d) $y = 2xe^{3x}$
  (e) $y = 7e^{3x} - 5xe^{3x}$

1.12 $d^2y/dx^2 - 6(dy/dx) + 9y = 18$
  (a) $y = 2$
  (b) $y = e^{3x}$
  (c) $y = e^{3x} + xe^{3x} + 2$

For Problems 1.13–1.16, find the only solution that matches the given differential equation and the given initial condition. (There will always be one and only one solution.)

1.13 $dh/dz = \ln z$, $h(1) = 5$
  (a) $h = 1/z + 4$
  (b) $h = \ln z + 5$
  (c) $h = z \ln z - z + 5$
  (d) $h = z \ln z - z + 6$

1.14 $dy/dx = y/x$, $y(3) = 12$
  (a) $y = x$
  (b) $y = 4x$
  (c) $y = x + 9$
  (d) $y = 12$

1.15 $dx/dt = t\sqrt{x}$, $x(0) = 9$
  (a) $x = t^2\sqrt{t} + 9$
  (b) $x = \sqrt{t^2 - 3\sqrt{t}} + 9$
  (c) $x = (t^2/4 + 3)^2$
  (d) $x = 9/(t^2 + 1)$

1.16 $dy/dt = 2y(1 - y)$, $y(0) = 1/10$
  (a) $y = e^{2t}/(9 + e^{2t})$
  (b) $y = e^{2t}/10$
  (c) $y = (1/10)(e^{2t} - 1)$
  (d) $y = 2t(1 - t) + 10$

1.17 You are running an online tournament with $P(t)$ players participating. Match each description below with the appropriate differential equation. (Nothing is changing other than what is described. For example, in part (a) no players are being eliminated, while in part (b) no new ones are joining.) Each equation should be matched to only one scenario.

  (a) Each day 50 new players join.   I. $dP/dt = 50$
  (b) Each day half the players in the tournament are eliminated.   II. $dP/dt = -50$
  (c) Each day 50 players are eliminated.   III. $dP/dt = 3P$
  (d) Each day everyone in the tournament convinces three friends to join.   IV. $dP/dt = 3t$
  (e) Three players join the first day, six the second day, nine the third day, and so on.   V. $dP/dt = -P/2$

1.18 Parts (a)–(e) below give five different scenarios for the evolution of a rabbit population $R(t)$. Match each one with the appropriate differential equation. Each equation should be matched to only one scenario.

  (a) The rabbit population is constant.   I. $R = 0$
  (b) 10 new rabbits are born every year.   II. $dR/dt = 10R$
  (c) On average, each rabbit produces 10 new rabbits per year.   III. $dR/dt = 10$
  (d) Each year each rabbit produces 10 new rabbits, but 100 rabbits are killed.   IV. $dR/dt = 0$
  (e) There are no rabbits.   V. $dR/dt = 10R - 100$

1.19 An object of mass m is moving along the x-axis (one dimension only). In each part of this problem you will be told the force acting on this object, and you will write a differential equation for the position $x(t)$ of the object. You do *not* need to solve the equation.

  (a) $F = -mg$ where g is a constant.
  (b) A charged ball experiences a force towards the origin proportional to one over its distance from the origin squared.
  (c) A boat experiences a constant forward force due to its motor and a backward drag force proportional to its speed.
  (d) A cyclist experiences a constant backward force due to friction and a forward force from pedaling. As the cyclist gets tired the pedaling force decreases linearly with time. (If the cyclist stops moving the frictional force will stop too, but ignore that here.)

1.20 The function $f(x)$ has the peculiar property that its *derivative* is the same function as its *square root*. (Assume $f(x) \geq 0$ for all x.)

(a) Express this sentence—"The derivative of $f(x)$ and the square root of $f(x)$ are the same function"—as a differential equation.

(b) Show that $f(x) = 0$ is a solution but $f(x) = 1$ is not.

(c) Now assume that $f(0) = 1$. Is $f(x)$ an increasing function, a decreasing function, or flat? How can you tell?

(d) Is $f(x)$ concave up, concave down, or neither? How can you tell?

1.21 The Explanation (Section 1.2.2) discussed two financial situations: one in which Shannon earns money at a consistent (linear) rate, and one in which she continuously earns interest on her money. Now suppose both are going on at the same time. Shannon makes $30,000/year salary, *and* her money always goes immediately into the bank, where it earns interest at an annual rate of 10%.

(a) Write the differential equation that says "Shannon's money increases every year by $30,000 plus 10% of whatever she has." Use M for the total of her money and t for time in years. (To make the answer look nicer, measure M in "thousands of dollars" instead of "dollars.")

(b) Assuming Shannon starts with no money, show that the function $M(t) = 300(e^{t/10} - 1)$ satisfies both the differential equation and the initial condition.

(c) We were able to write our solution in that form because we had already specified what units we were using for time and money. To write this solution with correct units, so that it would be valid for any units of time and money, we can write $M(t) = D(e^{t/k} - 1)$, where $D = 300$ and $k = 10$. What are the units of D and k?

1.22 Little Benjamin sneaks onto his father's computer every night. His father thinks Ben is playing video games. Actually, Ben is day trading, and he is very good at it. However much money he has at the beginning of the week, he always earns 10% of that in a week's time. Unfortunately, Ben also has a bubble-gum habit that costs him $15 a week.

(a) Write a differential equation that represents Ben's net worth. Use M for his money and t for time (measured in weeks).

(b) How much money does Ben need in order to break even every week? ("Break even" means his net worth is neither increasing nor decreasing; in other words, dM/dt is zero.)

(c) If Ben starts off with *more* money than you calculated in Part (b), what will happen to his money over time? Explain how your answer is based on your differential equation in Part (a).

(d) If Ben starts off with *less* money than you calculated in Part (b), what will happen to his money over time? Explain how your answer is based on your differential equation in Part (a).

1.23 Mary starts her career with an initial salary S_0. Each year she gets a 5% raise, so her salary one year later is $1.05 \times S_0$, her salary the following year is $1.05 \times 1.05 \times S_0 = 1.05^2 \times S_0$, and so on.

(a) Write an expression for Mary's salary $S(t)$. This will *not* be a differential equation, but a simple function of time.

(b) Assuming Mary never gains or loses money in any way other than earning her salary, write a differential equation for her amount of money $M(t)$. (You will need to use your answer to Part (a) here.)

(c) Now assume that in addition to earning her salary, Mary also spends an amount E each year, which starts at E_0 and increases by 2% each year (because of inflation). Write a new differential equation for $M(t)$.

(d) Finally, assume that Mary earns a salary $S(t)$, spends an amount $E(t)$, *and* is putting her money in the bank where it is earning 4% interest each year. Write a differential equation for $M(t)$.

1.24 The chemical tetrahydrofrefurol decomposes in the presence of a biological agent. Research shows that the rate of decomposition is proportional to the remaining mass. When the biological agent is first introduced and the decomposition begins ($t = 0$), there are 100 kg of tetrahydrofrefurol.

Use M for the mass of tetrahydrofrefurol and t for the amount of time since it began decomposing.

(a) Sketch a possible graph for the function $M(t)$ on purely physical grounds. You don't need to write any equations yet; just make sure your graph represents the idea that the more you have, the more you lose.

(b) Write a differential equation that represents the sentence "The amount of mass that disappears in any given hour is proportional to the mass at that hour." Your equation will introduce a constant of proportionality k.

(c) What are the units of k? Is it positive or negative? How do you know?

(d) The function $M(t) = 0$ should be a valid solution of your differential equation. Describe the scenario described by this solution.

(e) Find a function *other than* $M(t) = 0$ that solves your differential equation.

(f) Find a function that solves your differential equation *and* has the property that $M(0) = 100$. (If the graph of this function doesn't look more or less like the graph you drew in Part (a), one of them is wrong; figure out which one, and fix it!)

(g) A different chemical follows the same differential equation, but with a much higher value of k. Compare its decay process to the decay of tetrahydrofrefurol.

1.25 As a snowball melts, its mass m decreases at a rate proportional to its surface area. The surface area of a snowball is directly proportional to $m^{2/3}$.

(a) The problem tells us that dm/dt is proportional to $m^{2/3}$. Is the constant of proportionality positive or negative? How do you know?

(b) Write the differential equation. If the constant of proportionality is positive, call it k^2. If it is negative, call it $-k^2$.

(c) Show that $m = (5 - k^2 t/3)^3$ is a valid solution to your differential equation.

(d) What initial condition is implied by that solution?

1.26 A mass on a spring experiences a force $F_{\text{spring}} = -kx$ where x is the displacement from the relaxed position and k is a positive constant. It also experiences the force of friction from the table the mass is sliding along: the force is $F_{\text{friction}} = -3v/|v|$.

(a) $-3v/|v|$? That's a strange looking force. What is its magnitude? Which way does it push?

(b) Using Newton's Second Law $F_{\text{total}} = ma$ and the definitions of velocity (dx/dt) and acceleration (d^2x/dt^2), write a second-order differential equation for the mass's position as a function of time $x(t)$.

(c) In general, the best way to work with absolute values mathematically is to split into two cases. When $v \geq 0$ you can replace $|v|$ with v in your differential equation. Show that in this case one valid solution to the differential equation is:

$$x = 2 \sin\left(\sqrt{\frac{k}{m}}\, t + \frac{\pi}{4}\right) - \frac{3}{k}$$

(d) When $v < 0$ you can replace $|v|$ in your differential equation with $-v$. Find a solution to the differential equation that works in this case. (The technique of choice here is trial-and-error: play around until you find a function that works!)

1.27 Let $y = f(x)$ represent one particular graph (not the only one!) that follows the differential equation $dy/dx = y - x^2$.

(a) $f(x)$ goes through the point $(-1, 1)$. What is the slope of the curve at that point?

(b) $f(x)$ also goes through the point $(-3, 5)$. What is the slope of the curve at that point?

(c) $f(x)$ also goes through the point $(0, 2)$. What is the slope of the curve at that point?

(d) Based on all the information you have accumulated, draw a possible sketch of $f(x)$.

1.28 Let $y = f(x)$ represent one particular graph (not the only one!) that follows the differential equation $dy/dx = (x + 2y)/(x + 2)$.

(a) $f(x)$ goes through the point $(-3/2, 3/4)$. What is the slope of the curve at that point?

(b) Find d^2y/dx^2 as a function of x and y. In other words, find the derivative with respect to x of dy/dx. When you first apply the quotient rule, your answer will have dy/dx in it. You can then substitute $dy/dx = (x + 2y)/(x + 2)$ to obtain a function of just x and y. Simplify as much as possible.

(c) At the point $(-3/2, 3/4)$, is the function concave up or concave down?

(d) Based on your answers to (a) and (c), what can you conclude about this function at the point $(-3/2, 3/4)$?

1.29 A quantity Q evolves over time according to the differential equation $dQ/dt = Q(Q - 3)(Q - 5)$.

(a) If $Q(0) = 3$ then, at that point, $dQ/dt = 0$. So Q will stay 3, which means dQ/dt will stay 0 forever. Therefore $Q = 3$ is referred to as an "equilibrium solution" of the differential equation. What are the two other equilibrium solutions?

(b) If Q is between 0 and 3, is dQ/dt positive or negative? Based on this, how do you expect Q to evolve over time?

(c) If Q is between 3 and 5, is dQ/dt positive or negative? Based on this, how do you expect Q to evolve over time?

(d) If $Q > 5$, is dQ/dt positive or negative? Based on this, how do you expect Q to evolve over time?

1.30 Two different objects are both moving along the x-axis. Both experience forces that are proportional to their distance from $x = 0$, so we can write $F = kx$. Since $F = ma$ and $a = d^2x/dt^2$, we can write $d^2x/dt^2 = (k/m)x$. However, there is one big difference. For Object A, $k/m = 1$; for Object B, $k/m = -1$.

(a) For Object A, $d^2x/dt^2 = x$. Based directly on this differential equation, if the object starts at rest at $x = 0$, what will its acceleration be? What will happen over time?

(b) If Object A starts at rest at $x = 1$, will it accelerate to the right or left? Where will it go next, and what will happen to its acceleration? What will happen over time?

(c) If Object A starts at rest at $x = -1$, what will its acceleration be? Where will it go next, and what will happen to its acceleration? What will happen over time?

(d) For Object B, $d^2x/dt^2 = -x$. If the object starts at rest at $x = 0$, what will its acceleration be? What will happen over time?

(e) If Object B starts at rest at $x = 1$, will it accelerate to the right or left? Where will it go next, and what will happen to its acceleration? What will happen over time?

(f) If Object B starts at rest at $x = -1$, what will its acceleration be? Where will it go next, and what will happen to its acceleration? What will happen over time?

1.31 [*This problem depends on Problem 1.30.*] In Problem 1.30 you looked at two objects: Object A follows the differential equation $d^2x/dt^2 = x$, and Object B follows $d^2x/dt^2 = -x$.

(a) Solve the differential equation for Object A "by inspection": that is, think of a function $x(t)$ that is its own second derivative. (Do not use $x(t) = 0$.) Does this function match the behavior you predicted? (It may or may not, depending on the function you choose.)

(b) Solve the differential equation for Object B "by inspection": that is, think of a non-zero function $x(t)$ that is *negative* its own second derivative. (*Hint*: just throwing a negative into your answer for Object A doesn't work. Try it!) Does this function match the behavior you predicted? (It should.)

(c) Object C has $k/m = 9$. Solve its differential equation by inspection. When k/m is positive, what effect does its value have on the behavior of the system?

(d) Finally, Object D has $k/m = -9$. Solve its differential equation by inspection. When k/m is negative, what effect does its value have on the behavior of the system?

1.32 A quantity changes according to the differential equation $dy/dx = y^2$. You are going to draw three possible graphs that could represent such a quantity. All three graphs will be drawn on the domain $0 \leq x \leq 1$.

(a) The first graph starts at the point $(0, 0)$. Our differential equation promises us that the slope at that point is zero—the graph is horizontal—so move horizontally to the right (just a small distance) from that point.

(b) At the new point you reach, what is the slope, according to the differential equation? Based on that new slope, draw a little more of the graph.

(c) Keep going in this way to draw one possible graph.

(d) Now start over at the point $(0, 1)$. At this point, our differential equation promises us a slope of 1, so move up-and-right a bit at a 45° angle.

(e) At the new point you reach, is the slope the same, higher, or lower than 1? Based on that new slope, draw a bit more of the graph.

(f) Keep going in this way to draw a second possible graph.

(g) Now draw a third graph in the same way, starting at the point $(0, -1)$.

(h) For each of the three graphs you drew, is y increasing, decreasing, or staying constant?

You will explore this equation further in Section 1.9 Problem 1.198 (see felderbooks.com).

1.33 A "simple harmonic oscillator" (SHO) is any system that obeys a differential equation of the form $d^2x/dt^2 = -\omega^2 x$ where ω is a constant. The simplest possible SHO is one for which $d^2x/dt^2 = -x$.

(a) Show that $x = \sin t$ and $x = \cos t$ are both solutions to $d^2x/dt^2 = -x$.

(b) Each of these solutions represents an oscillation. The "period" of an oscillation is defined as the time it takes to complete one full cycle from its maximum value to its minimum value and back to the maximum again. What is the period of the oscillations described by $d^2x/dt^2 = -x$?

(c) If one object has position $x = \sin t$ and a different object $x = \cos t$, we have just seen that they obey the same differential equation and oscillate with the same period. If you observed the motion of both objects, how would they differ?

(d) Now suppose an SHO is described by $d^2x/dt^2 = -4x$. Write a solution to this equation and find its period.

(e) What is the period of an SHO that obeys $d^2x/dt^2 = -\omega^2 x$?

(f) What is the period of a 2 kg mass oscillating on a spring with spring constant $k = 200$ N/m? (Recall that the spring exerts a force $F = -kx$ on the mass.)

1.34 Constrained Population Growth

The *logistic equation* is used to model population growth in a constrained environment. As an example, the differential equation $dN/dt = (0.15)N(1 - N/7500)$ has been used to model the elephant population in South Africa's Kruger National Park.[1] $N(t)$ is the number of elephants and t is measured in years.

(a) According to this model, what is dN/dt when $N = 0$? Explain what this result means in terms of the elephant population.

(b) According to this model, what is dN/dt when $N = 7500$? Explain what this result means in terms of the elephant population.

(c) According to this model, what happens to dN/dt when $N > 7500$? Explain what this result means in terms of the elephant population.

(d) Make a table of N and dN/dt. Include $N = 10$, $N = 1000$, $N = 2000$, $N = 3000$, $N = 4000$, $N = 5000$, $N = 6000$, and $N = 7000$, plus the values you already calculated at $N = 0$ and $N = 7500$.

(e) Calculate the exact N-value at which the elephant population would grow the fastest. In other words, calculate the N-value that maximizes dN/dt.

(f) The initial condition in this case is that, in 1905 (which we might as well call $t = 0$), the population was $N = 10$. Based on your answers above and this initial condition, describe qualitatively the kind of growth you would expect to see over time. Where would the population head over time?

(g) The solution to this differential equation and initial condition is $N(t) = 7500e^{0.15t}/(e^{0.15t} + 749)$. Graph this function for $0 \le t \le 100$ and make sure it matches your qualitative prediction from Part (f). (If it doesn't, figure out what went wrong!)

1.35 Exploration: Compound Interest (by hand)

In the Explanation (Section 1.2.2), Shannon put $1000 into a bank account with 5% interest. Modeling her account with the equation $dM/dt = M/20$, we found that she earned slightly more than $1000 \times 5\% = \$50$ in her first year. Her balance is not incremented once a year, but incremented *continuously*, so she is always getting interest on the interest she just got. This distinction can be approached rigorously through a limit.

(a) Begin with an equation that literally models a 5% increase once a year. If Shannon begins the year with $\$M_0$, then after one year, she has $\$1.05M_0$. The following year she has $1.05 \times \$1.05M_0 = \$1.05^2 M_0$. How much money does she have after t years?

(b) Now, suppose Shannon's interest is "compounded" ten times per year. This means that after each 1/10 of the year, she receives 1/10 of 5% interest (her money multiplies by 1.005). If she receives such interest ten times per year, how much does she have after one year? After t years?

(c) The above thinking leads to the formula for compound interest. If Shannon starts with M_0 dollars and receives 5% interest compounded n times per year, then after t years she ends up with $M_0(1 + .05/n)^{nt}$ dollars. To compound *continuously*, take the limit of this formula as $n \to \infty$. (The most common trick for taking this limit starts by taking the natural log of the formula, using the laws of logs and l'Hôpital's rule to find the limit, and then raising e to your answer at the end.)

(d) Show that the answer to Part (c) is a valid solution to the differential equation $dM/dt = M/20$ with the initial condition $M(0) = M_0$.

(e) Now apply the same logic to a population growth scenario. If *every rabbit produces on average ten new rabbits per year*, then

---

[1] Gallego, Samy (2003). *Modelling Population Dynamics of Elephants*. PhD thesis, Life and Environmental Sciences, University of Natal, Durban, South Africa.

a starting population of 15 rabbits will produce 150 rabbits in their first year. What is $R(t)$ in this case? (Make sure your function predicts $R(1) = 15 + 150 = 165$.) If every rabbit produces on average 1 new rabbit every tenth of a year, what is $R(t)$? If every rabbit produces on average $10/n$ rabbits n times per year, what is $R(t)$? Take the limit of your answer as $n \to \infty$ and show that it satisfies the differential equation $dR/dt = 10R$ and the initial condition $R(0) = 15$.

(f) Based on your answer to Part (e), how long will it be until the Earth is covered with rabbits? Include whatever estimates you use or quantities you look up as part of your solution.

1.36 **Exploration: Compound Interest (by computer)**
In the Explanation (Section 1.2.2), Shannon put $1000 into a bank account with 5% interest. We solved the equation $dM/dt = M/20$ and got $M(t) = (\$1000)e^{t/20}$, and then noted that this causes her money to increase by more than 5% after a year because her balance keeps growing, and she immediately starts earning interest on that new balance. In this problem we take a numerical approach to the question: how much money does Shannon have after 20 years? In all cases, round off your answer to the nearest penny.

(a) If Shannon's money is "compounded annually"—that is, if she receives exactly 5% of her balance once a year—then the formula for her balance is $M(t) = 1000 \times 1.05^t$. Compute her balance after 20 years.

(b) If Shannon's money is "compounded monthly"—that is, if she receives $1/12$ of 5% of her balance every month—then the formula for her balance is $M(t) = 1000 \times (1 + .05/12)^{12t}$. Compute her balance after 20 years.

(c) If Shannon's money is "compounded daily"—that is, if she receives $1/365$ of 5% of her balance every day—then the formula for her balance is $M(t) = 1000 \times (1 + .05/365)^{365t}$. (Ignore leap years.) Compute her balance after 20 years.

(d) Many banks advertise that your money is "compounded continuously." Use a computer to take the limit of the above process as $n \to \infty$.

(e) Plug $t = 20$ into the solution $M(t) = (\$1000)e^{t/20}$. Comparing the result to your answers above, what kind of compounding does this solution represent?

1.3 Arbitrary Constants

In general one differential equation will have infinitely many solutions. To express these all in a compact form we write the equation with one or more "arbitrary constants," just like the $+C$ you write in the solution to an indefinite integral.

1.3.1 Discovery Exercise: Arbitrary Constants

1. For the differential equation $dy/dx = -3$ the solution can be written as $y = -3x + C$.
   (a) Plug in $C = 3$ and show that the resulting function is a valid solution of the differential equation.
   (b) Plug in $C = 0$ and show that the resulting function is a valid solution of the differential equation.
   (c) Plug in $C = x$ and show that the resulting function is *not* a valid solution of the differential equation.
   (d) What sorts of C-values are guaranteed to result in valid solutions?
   (e) What is the *only* C-value that satisfies the condition $y(-4) = 15$? (To find it, let $x = -4$ and $y = 15$ and solve for C.)
2. Consider the differential equation $dy/dx = e^y$.
   (a) Which of the following functions are valid solutions? (List all that apply.)
      i. $y = e^x$
      ii. $y = \ln x$
      iii. $y = -\ln(-x)$

iv. $y = -\ln(-x) + 4$
v. $y = -4\ln(-x)$
vi. $y = -\ln(-x + 4)$
vii. $y = -\ln(-x + 7)$

(b) Based on your answers, write a function that has a C in it, about which you can say, "This function is a valid solution to $dy/dx = e^y$ for any value of the constant C."
(c) Confirm that your solution works for $C = -3$.
(d) Find the value of C for which $y(0) = 0$.

1.3.2 Explanation: Arbitrary Constants

Consider the following equation.

$$\frac{dy}{dt} = t(y + 3) \quad \text{a differential equation} \tag{1.3.1}$$

You might stare at that for a long time without finding a working solution, but we'll save you the trouble and give you one.

$$y = e^{t^2/2} - 3 \quad \text{one particular solution} \tag{1.3.2}$$

We found a function and it works so we're done, right? Not so fast: here is another one.

$$y = 5e^{t^2/2} - 3 \quad \text{another solution} \tag{1.3.3}$$

Equations 1.3.2 and 1.3.3 are not the same function. They are not even simple permutations of each other, such as "add five" or "multiply by five." But they are both valid solutions to Equation 1.3.1. (Try them!) More generally we will show below that the function $Ce^{t^2/2} - 3$ (where C is a constant) is a valid solution. We call C in that formula an "arbitrary constant" because the solution works no matter what value of C you choose. We say that $Ce^{t^2/2} - 3$ is the "general solution" because it comprises all possible solutions. (We will say more about general solutions later.)

It's easy to verify mathematically that this general solution works, but what does it mean physically? A system governed by Equation 1.3.1 must have a particular y at any given t, right? Yes it must, but the differential equation alone is not enough to determine those values. We also need an "initial condition"—the value of y at some particular time, usually at $t = 0$. The differential equation *and* the initial condition together determine the state of the system.

EXAMPLE   A Solution with an Arbitrary Constant

Problem:
Show that $y = Ce^{t^2/2} - 3$ solves Equation 1.3.1 for any value of the constant C.

Solution:
For $y = Ce^{t^2/2} - 3$, $dy/dt = Ce^{t^2/2} \times t$. On the other side of Equation 1.3.1, $t(y + 3) = t(Ce^{t^2/2})$. They come out the same so this function satisfies the differential equation—no matter what constant C is.

Problem:
Find the particular solution that satisfies the initial condition $y(2) = 5$.

Solution:
Plugging $t = 2$ and $y = 5$ into our general solution, $5 = Ce^{2^2/2} - 3$ solves to give $C = 8/e^2$. So $y = (8/e^2)e^{t^2/2} - 3$ is the one and only function that satisfies both the differential equation and the given condition.

Problem:
Graph five different solutions to this differential equation. Include (and clearly mark) the solution that passes through the point $(2, 5)$.

Solution:

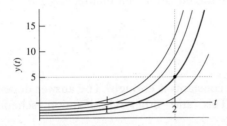

The plot shows the function $y = Ce^{t^2/2} - 3$ for values of C from 0 to $16/e^2$. The plot in blue is for $C = 8/e^2$, which passes through the point $(2, 5)$.

Different "Orders" of Equations, Different Numbers of Constants

Why does the solution to a differential equation have an arbitrary constant? Does every ODE solution have an arbitrary constant? Do you ever need more than one? These questions are all related; the answer to the first question will illuminate the others.

Consider a differential equation of the form $dx/dt = f(x, t)$. This is called a "first-order" differential equation. If you think of x as position and t as time, it says "here is the velocity for an object as a function of its position and time." That clearly is not enough to know where the object is, because it tells you nothing about where the object started. But if you know where the object started, then the differential equation tells you how it will move from there. So one arbitrary constant, filled in by one initial condition, is generally enough.

You can reach the same insight graphically. $dy/dx = f(x, y)$ says "here is the slope of a curve as a function of its location on the xy-plane." That is not enough to draw a specific curve, but if you know one point on the curve then you can fill in the rest.

Now consider a differential equation of the form $d^2x/dt^2 = f(x, v, t)$. Such a "second-order" differential equation says "here is the acceleration of an object as a function of its position, velocity, and time." (Newton's Second Law $F = ma$ often turns into such an equation.) Now there are two independent facts you don't know: the initial position tells you nothing about the initial velocity, and vice versa. But if you know both the initial position *and* the initial velocity, then you can predict the future of the system. Mathematically, the general solution to such an equation will have two arbitrary constants to be filled in by two initial conditions.

EXAMPLE   A Solution with Two Arbitrary Constants

Problem:
Find the particular solution to $x''(t) = x$ with initial position $x(0) = 1$ and initial velocity $v(0) = 2$.

Solution:
The functions $x = e^t$, $x = 0$, and $x = 3e^{-t}$ are all valid solutions to this differential equation. More generally, the following function is a valid solution for any values of the arbitrary constants C and D (as you should verify for yourself).

$$x(t) = Ce^t + De^{-t}$$

Plugging in $x(0) = 1$ gives $C + D = 1$. To plug in the second condition we first have to take the derivative, $x'(t) = Ce^t - De^{-t}$, and then plug in the initial velocity to give $C - D = 2$. Solving these two equations simultaneously gives $C = 3/2$, $D = -1/2$.

$$x(t) = \frac{3}{2}e^t - \frac{1}{2}e^{-t}$$

So how many arbitrary constants is enough? The answer depends on the "order" of the differential equation. (This and other classification schemes are summarized in Appendix A.)

Definition: Order of a Differential Equation
The "order" of a differential equation is the highest order derivative in the equation. For instance,

$$\frac{d^3z}{dt^3} + e^t \left(\frac{dz}{dt}\right)^{10} - 3z^7 = 4t^5$$

is a third order differential equation because it has a third derivative, and it has no fourth derivative, fifth derivative, etc.

The ideal solution to a differential equation is one that can match any possible initial conditions. Such a function is called the "general solution."

Definition: General Solution
The "general solution" to a differential equation is a formula that has one or more arbitrary constants in it. When you plug in specific values for these arbitrary constants you get specific solutions. To properly be called the general solution this formula must meet the following criteria.
1. For any value(s) of the arbitrary constant(s), the resulting functions are valid solutions of the differential equation.
2. By choosing the correct values of the arbitrary constants, you can generate *all possible* solutions to the differential equation. (In some cases this includes taking limits as the arbitrary constants approach ±∞.) This point will be clarified with several examples below.

We argued above, physically and graphically, that an nth-order differential equation requires n initial conditions. We can now state the same rule a different way: the general solution to an nth- order differential equation will typically have n arbitrary constants.

Rewriting Solutions with Arbitrary Constants
Three friends are solving the equation $dy/dx = 3$. Curly confidently answers $y = 3x + C$; his friend Larry slaps him on the head and argues that the answer is actually $y = 3x + 2C$; Moe pokes Larry in the eye and insists on $y = 3x + C - 5$. If you test these solutions, they all work. Which one is right?

They're all right! Remember that C means "Any number here will make a valid solution." $2C$ and $C-5$ also mean "any number" so they serve the same purpose.

Now our heroes are told to find the one-and-only solution that goes through the point $(1, 8)$. Plugging in $x = 1$ and $y = 8$, Curly concludes that $C = 5$. Larry argues that $C = 5/2$ while Moe gets $C = 10$. They can avoid further lamentable violence, however, if they notice that they all converged on the same solution, $y = 3x + 5$.

Please make sure you followed that last paragraph. In the solution to a differential equation, C is perfectly interchangeable with $C/4$ or C^3 or $\ln(C+17)$: you are free to choose whichever form is most convenient. When you match an initial condition you will find *different C-values*, but end up with the *same specific function*.

Here is a subtler example. Consider the following true fact.

$$\text{The general solution of } \frac{dy}{dx} = y \text{ is } y = Ce^x \qquad (1.3.4)$$

You can (and should) confirm for yourself that this function meets the first criterion of a general solution: it works for any value of C. We are also telling you that it meets the second criterion. *Any possible solution* of this differential equation can be written in the form Ce^x.

An apparent exception is the function $y = e^{x+7}$. It does work. (Try it!) And it doesn't look like Ce^x. (Does it?) It seems our general solution isn't very general after all.

But it is, and the key is a bit of algebra. $e^{x+7} = e^7 e^x$ so this does fit the template in Equation 1.3.4. The moral of this story is that any solution can be put into the form of the general solution, but it isn't always easy or obvious.

The Simple Harmonic Oscillator Equation

If Nature abhors a vacuum, she seems enamored of oscillation. Strings on an instrument wiggle up and down, a dropped weight vibrates, some stars grow and shrink in regular patterns, earthquakes are oscillations of the ground, and the list goes on. These systems all obey complicated differential equations. However, we'll see in Chapter 2 that for sufficiently small oscillations virtually any oscillator can be well modeled by the simple harmonic oscillator (SHO) equation below. This equation, together with its general solution, may be the most important-to-memorize formula in this book.

The Simple Harmonic Oscillator Equation

The following equation is known as the "simple harmonic oscillator equation," and any system that follows this equation is a "simple harmonic oscillator."

$$x''(t) = -\omega^2 x(t) \quad \text{where } \omega \text{ is a constant} \qquad (1.3.5)$$

The general solution to this equation can be written in either of the following equivalent forms:.

$$x(t) = A\cos(\omega t) + B\sin(\omega t) \qquad (1.3.6)$$
$$x(t) = C\cos(\omega t + \phi) \qquad (1.3.7)$$

An SHO equation could look like $x''(t) = -9x$ or $x''(t) = -\sqrt{2q\epsilon/5e^w}\, x$, but the system is an SHO as long as the equation sets the second derivative of a function equal to a negative constant times that function.

Equations 1.3.6 and 1.3.7 look very different but each one is the general solution, and remember what that means.

- Any solution to Equation 1.3.5 can be expressed in the form of Equation 1.3.6 with the right constants A and B.
- Any solution can also be expressed in the form of Equation 1.3.7 with the right constants C and ϕ.

This implies that the two functions can be rearranged to look like each other; you will do so in Problem 1.58.

In many cases Equation 1.3.6 is the simplest for matching initial conditions. For example, A is just $x(0)$. But Equation 1.3.7 more clearly shows the physical properties of the oscillator because the arbitrary constants C and ϕ represent the amplitude and phase, respectively.

In either representation ω is the angular frequency, so the period is $2\pi/\omega$. Notice that ω is not an arbitrary constant; it comes from the differential equation. For instance the equation $x'' = -9x$ always represents an oscillator with period $2\pi/3$, but that oscillator can have any amplitude and phase, which depend on the initial conditions.

1.3.3 Problems: Arbitrary Constants

1.37 Consider the linear differential equation $dy/dx = 3 - y$.
  (a) For the function $y = e^{-x} + 3$, calculate dy/dx.
  (b) For the function $y = e^{-x} + 3$, calculate and simplify $3 - y$.
  (c) Your answers to Parts (a) and (b) should have come out the same, demonstrating that $y = e^{-x} + 3$ is a solution to this differential equation. Using the same method, demonstrate that $y = 7e^{-x} + 3$ is *also* a solution to this differential equation.
  (d) Show that the function $y = Ce^{-x} + 3$ is a solution to this differential equation for any value of the constant C. (You do this the same way you did the previous problems: just let C work through the equations as a constant.)
  (e) What specific function does this general solution give you for the value $C = 0$? Show that this function is a solution to the differential equation.
  (f) Find the one and only function that satisfies both the differential equation $dy/dx = 3 - y$ and the initial condition $y(0) = 1$.
  (g) Show that the function $y = 3 - e^{10-x}$ is a solution to this differential equation. Then explain why this does not violate the rule that the solution in Part (d) is the general solution representing all possible solutions.

1.38 The general solution to the differential equation $dy/dx = 3 - y$ is $y = Ce^{-x} + 3$.
  (a) Graph the solutions to this equation for $C = -3$, $C = -2$, $C = -1$, $C = 0$, $C = 1$, $C = 2$, and $C = 3$. Limit your graphs to $x \geq 0$.
  (b) Discuss in words what all these functions have in common. How do they evolve as x increases? What is $\lim\limits_{x \to \infty} y(x)$?

Problems 1.39–1.45 give a differential equation, a solution with one or more arbitrary constants, and one or more initial conditions. In each case,
  (a) Show that the given solution solves the given differential equation for *any* values of the arbitrary constant(s).
  (b) Find the particular solution that matches the given condition(s).

1.39 $P'(t) = -10P(t)$
  Solution: $P(t) = Ce^{-10t}$
  Condition: $P(0) = 7$

1.40 $u'(t) = \sqrt{u}$
  Solution: $u(t) = (t/2 + C)^2$
  Condition: $u(0) = 5$

1.41 $\sqrt{x}\dfrac{dy}{dx} = y^2 e^{\sqrt{x}}$
  Solution: $y = \dfrac{-1}{2e^{\sqrt{x}} + C}$
  Condition: $y(0) = 5$

1.42 $dy/dx + y/x = 4\ln x$ $(x > 0)$
  Solution: $y = 2x\ln x - x + C/x$
  Condition: $y(1) = 10$

1.43 $dy/dx = (1 - 2xy)(x^2 + 1)$
  Solution: $y = (C + x)(x^2 + 1)$
  Condition: $y(1) = -1$

1.44 $d^2x/dt^2 = -9.8$
  Solution: $x = C_1 + C_2 t - 4.9t^2$
  Conditions: $x(0) = 6$, $x'(0) = -3$

1.45 $u''(t) + 4t^3 = 4u(t) + 6t$
  Solution: $u(t) = t^3 + C_1 e^{2t} + C_2 e^{-2t}$
  Conditions: $u(0) = 5$, $u'(0) = 6$

1.46 For each scenario write a differential equation to describe the situation, specify how many arbitrary constants would be in the general solution, and give one possible set of physical conditions that could be used to find the values of those arbitrary constants. For example, for a mass on a spring that exerts a force $F = -kx$ you might write the following:
- $m(d^2x/dt^2) = -kx$
- Since this is a second-order equation the general solution will have two arbitrary constants.
- To specify their values you could use the position and velocity of the mass at $t = 0$.

(a) A population of mice $M(t)$. Each year every mouse has on average 4 babies, and predators kill 100 of the mice.

(b) A sample of radioactive material $R(t)$. In some fixed interval of time half of the sample decays. (Since the half-life isn't specified here your differential equation will have an unknown constant in it, separate from the arbitrary constants that would appear in the solution if you were to solve it.)

(c) The temperature in a long hallway $T(x)$ obeys the differential equation $d^2T/dx^2 = 0$. (You don't have to write the differential equation for this one since we gave it to you.)

(d) A falling object experiences a constant gravitational force $-mg$ and an upward drag force proportional to the object's speed.

1.47 Consider the differential equation $x(dy/dx) = y \ln y$.

(a) Many differential equations can be solved for any given initial conditions. This equation, however, is restricted. For instance, we can see immediately that no solution is possible with the initial condition $y(1) = -1$. Why? Another impossible condition is $y(0) = 2$. Why? (Assume all variables are real.)

(b) One solution to this problem is the constant function $y = 1$. Show that this function solves the equation by finding $x(dy/dx)$ and $y \ln y$ and showing they are the same.

(c) The general solution to this equation is $y = e^{Cx}$. Show that this function solves the equation by finding $x(dy/dx)$ and $y \ln y$ and showing they are the same.

(d) If the function given in Part (c) is *the general* solution, it should include all solutions—including the one in Part (b). Show that it does by finding the value of C that leads to that solution.

(e) Of all the functions that solve this differential equation, only one of them has the property that $y(2) = 5$. By plugging $x = 2$ and $y = 5$ into the general solution, find the value of C that matches this condition.

1.48 For the differential equation $dy/dx = 1 - y^2$, the general solution is $y = (Ce^{2x} - 1)/(Ce^{2x} + 1)$.

(a) List all of the following functions that are valid solutions to this differential equation. (You do *not* have to test each one out in the differential equation, which would be tedious. Instead, see which functions fit the "template" provided by the general solution.)
  i. $y = 0$
  ii. $y = -1$
  iii. $y = e^{2x}$
  iv. $y = (3e^{2x} - 1)/(3e^{2x} + 1)$
  v. $y = (-5e^{2x} - 1)/(-5e^{2x} + 1)$
  vi. $y = (3e^{2x} - 1)/(-5e^{2x} + 1)$

(b) Choose *one* of the functions that you said was a valid solution, and demonstrate that it does indeed satisfy the differential equation. (Go ahead, choose the easy one!)

(c) Find the function that satisfies this differential equation with the initial condition $y(0) = 5$. (It is not one of the functions listed above.)

(d) Find $\lim_{C \to \infty}$ of the general solution given in the problem and show that the resulting function is also a valid solution.

1.49 Two particular solutions of the equation $4f''(z) + f(z) = 4f'(z)$ are $f_1(z) = e^{z/2}$ and $f_2(z) = ze^{z/2}$.

(a) Show that any linear combination $Af_1(z) + Bf_2(z)$ is also a valid solution to the differential equation.

(b) Find the particular solution that satisfies the conditions $f(2) = 14e$ and $f'(2) = 12e$.

1.50 A simple harmonic oscillator follows the differential equation $d^2x/dt^2 = -9x$.

(a) Show that the function $x = C\sin(3t)$ is a valid solution of this equation for any value of the constant C.

(b) Show that it is impossible, using the solution from Part (a), to meet the initial conditions $x(0) = 4$ and $x'(0) = 12$.

(c) Show that $x = C\sin(3t + \phi)$ is a valid solution to the differential equation for any values of the constants C and ϕ.

(d) Find the function that solves this differential equation and meets the initial conditions given in Part (b).

1.51 One of the simplest differential equations is the equation that says "The quantity x grows at a constant rate." Suppose that every one unit of time t, the quantity x grows by exactly k.

(a) Write a differential equation that expresses this rule.

(b) Write the general solution $x(t)$ to your differential equation. Note that your solution will have two unrelated constants in it: the k from the equation, and the C from the solution.

(c) Choosing $k = 3$, graph the different solutions for three different C-values.

(d) Choosing $k = -3$, graph the solutions for the same three C-values.

(e) Explain in words: what kind of growth results from this differential equation? How does k affect the behavior of the system and how does C affect it?

1.52 A simple model for population growth is "the more you have, the more you get": put more mathematically, "the rate of growth of the population is proportional to the population itself."

(a) Write a differential equation that expresses this model. Use P for the population, t for the time, and k for the constant of proportionality.

(b) Solve your differential equation "by inspection": that is, look at it until you can think of any function that works. (For the moment, avoid the "trivial" solution $P(t) = 0$. We'll come back to that one.)

(c) Now, start to modify your solution. If you add 5, does it still work? If you multiply by 3, does it still work? Your goal here is to arrive at the general solution: that is, a function that satisfies the original differential equation for any value of an arbitrary constant C.

(d) The solution $P(t) = 0$ should be a special case of your general solution. Find a value of C for which your general solution becomes $P(t) = 0$.

(e) Choosing $k = 2$, graph three different solutions of this differential equation, using different starting values $P(0)$. Make sure your graphs fit the claim that our differential equation began with: the higher the population, the faster the growth (i.e., the higher the slope)!

(f) Choosing $k = 1/2$, graph solutions for the same three starting values $P(0)$.

(g) Explain in words: what kind of growth results from this differential equation? How does k affect the behavior of the system and how does C affect it?

1.53 A tank contains 3 l of water. 30 g of salt are mixed evenly throughout the tank. Salt water is draining from the tank at a constant rate of 1 l/h. Let S be the amount of salt (measured in grams) remaining in the tank as a function of t (measured in hours).

(a) If nothing else is happening, write a differential equation for the function $S(t)$. Begin by asking yourself: how much salt disappears from the tank each hour?

(b) Find the solution to your differential equation in Part (a) with the initial condition $S(0) = 30$.

(c) Now assume that from the beginning fresh water is being dumped into the tank at a constant rate of 1 l/h, so the total volume of the tank is kept constant at 3 l, and the salt is always mixed evenly throughout the water. Write a different differential equation for $S(t)$ in this scenario, once again beginning with the question: how much salt disappears from the tank each hour?

(d) Find the solution to your differential equation in Part (c) with the initial condition $S(0) = 30$.

(e) Draw a quick sketch of your solution.

1.54 Water is leaking from a hole at the bottom of a tank. The height of water in the tank decreases at a rate proportional to the square root of the height.

(a) Why, physically, does the rate of leaking change over time, instead of being constant?

(b) Using h for the height of water in the tank and t for time, write the differential equation for $h(t)$. Your equation will contain a constant of proportionality k.

(c) Is your constant k positive or negative? How can you tell? What are the units of k?

(d) Based on your differential equation, describe in words how the height of the water will change over time.

(e) Now suppose it is raining into the top of the tank at a rate of 4 in/h. Write a new differential equation for $h(t)$.

(f) Your new differential equation has one special solution, an "equilibrium solution" where the amount of water flowing in and the amount of water flowing out are perfectly balanced. What is the height of the equilibrium solution? *Hint*: you can tell by looking at your differential equation and thinking about dh/dt at equilibrium!

1.55 You are running a scientific experiment on the bacteria Veribadocillin. The amount of bacteria in your container is constant, and every second it produces 2 g of the chemical Situnsene. You have filters in your container that remove 2% of the Situnsene each second.

(a) Write a differential equation for the amount of Situnsene $S(t)$ in the container.

(b) Verify that $S(t) = e^{-.02t} + 100$ is a solution to the equation you wrote.

(c) Use trial and error to find how you can modify this solution to write the general solution. You could try adding an arbitrary constant to S, multiplying S by an arbitrary constant, or adding or multiplying arbitrary constants to different parts of S. For each thing you try you should calculate both sides of the differential equation and check whether they match for *any value* of the arbitrary constant. Once you find a solution that works, you have the general solution.

(d) Find the solution $S(t)$ if the experiment started with 200 g of Situnsene.

(e) You should find from your equation that at late times the amount of Situnsene approaches a constant. Find the value of this constant and explain why the amount of Situnsene can stay constant at this value.

(f) Draw a quick sketch of your solution.

Problems 1.56–1.58 consider how *one general solution* can often be written in *many different forms*.

1.56 $dy/dx = y$ is a first-order linear differential equation.

(a) Show that $y = (C_1 + C_2)e^x$ is a solution for any constants C_1 and C_2.

(b) The general solution to a first-order linear differential equation should have only one independent arbitrary constant. Explain why the solution in Part (a) does not violate this rule.

(c) Show that $y = C_1 e^{x+C_2}$ is a solution for any constants C_1 and C_2.

(d) Explain how Part (c) *also* does not violate the rule that the general solution can have only one independent arbitrary constant.

(e) Write the general solution to this equation in a form containing only one arbitrary constant.

1.57 Consider the linear differential equation $d^2x/dt^2 = x$.

(a) The general solution to this equation is $x = Ae^t + Be^{-t}$, where A and B are arbitrary constants. Find the constants A and B given the initial conditions $x(0) = 0$, $x'(0) = 1$.

(b) An alternative way of expressing the general solution is $x = C\sinh(t) + D\cosh(t)$, where $\sinh(t) = (1/2)(e^t - e^{-t})$ and $\cosh(t) = (1/2)(e^t + e^{-t})$ and C and D are arbitrary constants. Find the constants C and D given the initial conditions $x(0) = 0$, $x'(0) = 1$.

(c) Show that your constants in Parts (a) and part (b) lead to the same solution (the only correct solution to this differential equation with these conditions).

1.58 An object moves according to the simple harmonic oscillator equation $d^2x/dt^2 = -x$.

(a) Show that $x = A\sin t + B\cos t$ is a valid solution.

(b) Show that $x = C\sin(t + \phi)$ is a valid solution.

(c) Show that both of these general solutions are in fact the same, by expressing A and B in terms of C and ϕ or vice-versa. You may need to look up some trig identities.

For Problems 1.59–1.61, you need to know the following definitions.

- Position is x (we're working strictly in one dimension here), and time is t.
- Velocity is $v = dx/dt$, and acceleration is $a = dv/dt = d^2x/dt^2$.
- Force is F.
- These quantities are related by Newton's Second Law, $F = ma$, where m, the mass, is generally a constant.

You may already know the solutions to some of these problems, but you're going to represent and solve them through differential equations.

1.59 A car is zooming down the road at a constant speed of 55 mph.

(a) Given that $v = 55$, and given the definition of v, write a differential equation for $x(t)$.

(b) Solve that differential equation: that is, write some function $x(t)$ that makes that differential equation true. You may be able to write many different $x(t)$ functions that work for this. Ultimately, you want to wind up with an answer that has an arbitrary constant C in it, to represent all possible solutions.

You now have an equation for $x(t)$. But it has an arbitrary constant C in it. What does that mean? Given the information we have—"a car moving at 55 mph"—we don't know where the car is, until we get more information.

(c) Here comes more information: at time $t = 0$, the car was at position $x = -5$. Use that fact to find C and write the real equation for $x(t)$: the one that will actually tell us where the car is.

(d) Describe in words the motion described by your $x(t)$ function. Does it make sense with the physical situation? Why or why not?

1.60 A piano is dropped from the top of a building. Neglecting friction for the moment, the force is given by $F = -mg$, where g is a positive constant.

(a) Write and solve an algebraic equation for $a(t)$.

(b) Given that, write a differential equation for $v(t)$. (The position x should not be in this equation.)

(c) Solve that differential equation. Once again, you may be able to write many different $v(t)$ functions that work, so use an arbitrary constant to represent all possible solutions. Call your constant C_1 instead of just C.

(d) Now, using your answer to Part (c) and the definition of velocity, write a differential equation for $x(t)$.

(e) Solve that. Of course, the constant C_1 will be in there, since it was part of v. But still, there will be many possible solutions. To represent them all, you will need a new arbitrary constant C_2.

You now have an equation for $x(t)$ that has two arbitrary constants, C_1 and C_2. We can find the first one based on the given information.

(f) Since the piano was "dropped" (not "thrown"), we can assume that $v(0) = 0$. Use this to find the constant C_1. Then rewrite your $x(t)$ formula with only one constant, C_2.

(g) Let's look at this graphically. Suppose $g = 10$. Pick a value for C_2 and draw a graph of your $x(t)$ function. (You can do this using a calculator or computer, but it should also be easy by hand.) Then pick a different value for C_2 and draw another graph. Continue until you have at least four graphs. What do they all have in common? How do they differ?

(h) To choose one specific function, we need one more piece of information that was not specified in the problem. What is it?

(i) Make up a reasonable answer and write the function $x(t)$ that matches it, with no arbitrary constants.

(j) Describe the motion of the piano in words. Does it make sense with the physical situation? Does it correspond to your answer?

1.61 A car is rolling along a flat highway. Neither the gas pedal nor the brakes are being pushed; the only force on the car is air resistance. A simple model is to say that air resistance is proportional to velocity: that is, the faster you go, the more resistance you feel. We can write that as $F = -bv$. *Hint:* if you have trouble with this problem you may want to work through Problem 1.60 as a guide.

(a) What is the sign of b? Explain why this equation would not make sense physically if b had the other sign.

(b) What are the SI units of b?

(c) Write a differential equation for v.

(d) Solve it in the most general terms possible. (This will introduce an arbitrary constant.)

(e) Now, use that answer to write a differential equation for x.

(f) Solve it in the most general terms possible. (This will introduce another constant.)

(g) Suppose the car started at position $x = x_0$ with velocity $v = v_0$. Find the values of the arbitrary constants and write the function $x(t)$ that describes the motion of the car.

(h) Check that each term in your equation has correct units.

(i) Take $m = 2000$, $b = 1000$, $x_0 = 0$, and $v_0 = 20$ in SI units. Draw quick sketches of $x(t)$ and $v(t)$.

(j) Describe in words the motion your sketch shows. Does it make sense with the physical situation?

(k) Sketch the solution again with $b = 2000$ (and all other numbers the same). How did that affect the motion?

1.62 For each SHO below find the period, amplitude, and phase, or say which ones cannot be determined from the given information.

(a) $x''(t) = -9x$

(b) $x''(t) = -25x$, $x(0) = 3$, $x'(0) = 0$

(c) $x''(t) = qx$ where q is a negative constant

(d) $x''(t) = -e^{2z}x$, $x(0) = 0$, $x'(0) = 1$ where z is a constant

1.63 The simple harmonic oscillator equation $x''(t) = -\omega^2 x(t)$ (where ω is a constant) has solution $x(t) = C\cos(\omega t + \phi)$. Solve for C (the amplitude of oscillation) and ϕ (the phase) in terms of the initial conditions $x(0) = x_0$ and $v(0) = v_0$ and the angular frequency ω.

1.64 For each SHO below find the period, amplitude, and phase, or say which ones cannot be determined from the given information.

(a) A 5 kg mass is on a spring with spring constant 10 N/m. The force exerted by a spring is $-kx$, where k is the spring constant and x is the displacement from equilibrium.

(b) The system in Part (a) is pulled out by a distance of 0.2 m and released from rest.

(c) A 2 m pendulum. A simple pendulum approximately obeys the equation $\theta''(t) = -(g/L)\theta$ where $g = 9.8$ m/s^2 and L is the length of the pendulum.

(d) A 2 m pendulum starts at $\theta = 0$ with initial angular velocity $\theta'(t) = 3$ s^{-1}.

1.4 Slope Fields and Equilibrium

We introduced differential equations by analogy to algebraic equations. In this section we take that analogy one step further by noting that a *graphical* approach provides a valuable complement to doing calculations.

For instance, you learned at some point to define $\cos \theta$ as adjacent/hypotenuse on a right triangle, or as an x-value on the unit circle. But when you graph an infinite periodic wave you gain a new understanding of the cosine function. It's difficult to "solve" an equation such as $10 \cos x = x + 1$, or even to figure out how many solutions there are—but with a few quick graphs you can find approximate values for all the solutions.

This section introduces a graphical approach to differential equations. Just like graphical approaches to algebraic equations, we will see that this visualization method offers insights that are difficult to achieve in any other way.

1.4.1 Discovery Exercise: Slope Fields

In this exercise, you will examine the differential equation:

$$\frac{dy}{dx} = |y| \tag{1.4.1}$$

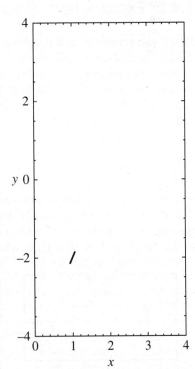

As you know, there is not only one function that solves such a differential equation; there are many different solutions, corresponding to different initial conditions. So when we set out to view this differential equation graphically we will not end up with "the right curve" but with a set of curves, sometimes called a "family of solutions," all represented on one graph.

1. Equation 1.4.1 tells us "For any given point that the graph of a function goes through, here is its slope at that point." Based on that formula, what is the slope of a function that follows this differential equation as it passes through the point $(1, -2)$?

You should have found that the slope at that point would be 2. We can represent this visually by drawing a small line through the point $(1, -2)$ with a slope of 2. The length of this line segment is irrelevant; all we are trying to show is the slope.

This drawing does *not* mean that the solution must go through the point $(1, -2)$. (There is in fact one solution that does go through that point, and infinitely many that don't.) The drawing also is *not* meant to suggest that the solution is linear, even for a short while. It simply says one solution rises with a slope of two through that point, and therefore momentarily looks something like the drawing.

2. On graph paper, draw lines similar to the one shown above for all integer points $0 \leq x \leq 4$ and $-4 \leq y \leq 0$ (a total of 25 points in the fourth quadrant).
3. Consider now the solution that begins at $(0, -4)$. The slope at that point is 4, so start moving to the right with a slope of four. As the curve rises, what happens to the slope? What happens to the function as a result? Draw the function that results.

You should have wound up with a drawing something like the one on the left.

Keep in mind that this is not ultimately about drawing curves, but about predicting behavior. If y represents some meaningful quantity and x represents time we can say the following: if y starts at -4 it will increase rapidly at first and then gradually slow down, approaching but never reaching zero. We figured all this out visually, without ever *solving* the differential equation.

4. Consider now the solution that begins at $(0, 0)$. Draw this function as you drew the previous one. Then write a brief paragraph describing this solution, just as we described the $(0, -4)$ solution.
5. Draw in slope lines for all the points on $0 \leq x \leq 4$ and $1 \leq y \leq 4$ (first quadrant).
6. Consider now the solution that begins at $(0, 1)$. Draw this function as you drew the previous one. Then write a brief paragraph describing this solution, just as we described the $(0, -4)$ solution.

1.4.2 Explanation: Slope Fields and Equilibrium

When you render x^2 as a parabola, or see $\ln x$ approaching its vertical asymptote, you learn things about those functions that are difficult to learn from calculations. Visualization gives us powerful insights into algebra, and there is tremendous benefit in applying the same idea to differential equations. "Slope fields" give us a graphical approach to first-order differential equations, allowing us to see the space of all possible solutions and to select one particular solution based on an initial condition (one point).

This process is best illustrated by an example.

The Problem

The picture below shows what electrical engineers call an "RC circuit": a simple circuit with a battery, a resistor, and a capacitor. (If you have never studied circuits then the physics in this problem won't make much sense to you, but you should still be able to follow the math, starting with Equation 1.4.2.) Over time, a charge Q will build up on the capacitor according to the equation $V = IR + Q/C$, where the current $I = dQ/dt$.

If the voltage of the battery is $V = 9$ V, the resistance of the resistor is $R = (9/2)\Omega$, and the capacitance of the capacitor is $C = (1/3)$F, then the equation for this accumulated charge (in Coulombs) becomes:

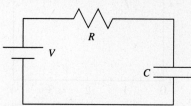

$$\frac{dQ}{dt} = 2 - \frac{2}{3}Q \qquad (1.4.2)$$

What behavior can we expect from this system over time?

1.4 | Slope Fields and Equilibrium

Drawing the Slope Field

Suppose that at time $t=0$ the capacitor has built up no charge at all. Equation 1.4.2 tells us that if $Q(0) = 0$ then $dQ/dt = 2$. We can represent this by drawing a small line with a slope of two at that point.

On the other hand, what if $Q = 0$ when $t = 1$? In this particular differential equation, dQ/dt depends only on Q, not on t. (The circuit reacts based on on how much charge has built up, regardless of how much time has elapsed.) So our differential equation also predicts a slope of 2 at the point $(1, 0)$, and at $(2, 0)$, and so on; we can go ahead and draw them all, as shown below.

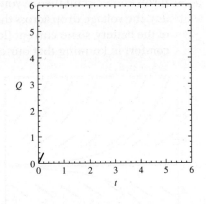

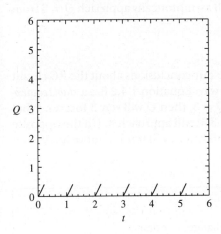

It's easy to get confused about what all that means. Certainly we aren't drawing a function! The key is to remember that a differential equation does not, by itself, specify a function: an infinite number of functions satisfy Equation 1.4.2. One of those functions happens to go through the point $(0, 0)$ and, at that point, it has a slope of 2. Another happens to go through $(1, 0)$ and it has a slope of 2 at that point…, and so on.

Meanwhile, what if $Q = 1$? Equation 1.4.2 now predicts a slope of $4/3$—once again, for all t-values. At $Q = 2$ the slope is $2/3$, at $Q = 3$ the slope is 0, at $Q = 4$ the slope is $-2/3$, and so on. When we draw lines at all integer points, we reach a drawing like Figure 1.2.

What Can We Learn from a Slope Field?

Figure 1.2 does not represent a function: it represents a "family of functions," the set of all solutions to Equation 1.4.2. As we have seen before, we can find a specific function to match any given initial condition.

The easiest case is $Q(0) = 3$. The drawing clearly shows that, from this point, the equation will progress straight to the right. As it does so, the slope will stay zero, so the function will continue straight to the right, and so on. In other words, one solution to Equation 1.4.2 is the simple function $Q = 3$. You can confirm this, as always, by plugging it into the equation. But we could have predicted this result without using the slope field, based on either the math or the physics.

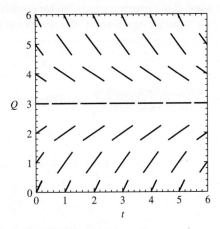

FIGURE 1.2 Slope field for $dQ/dt = 2 - (2/3)Q$.

- If you look at Equation 1.4.2 and ask the question "What makes $dQ/dt = 0$?" you quickly arrive at the answer $Q = 3$. More generally, whenever you find a Q-value for which dQ/dt is always zero, you have found a "constant solution" or "equilibrium solution." If Q reaches this value, it will stay there forever.

- If you know circuit theory, you know that when a charge $Q = 3$ builds up on the capacitor, the voltage drop across the capacitor is $Q/C = 9$ V, perfectly opposing the voltage of the battery, so no current flows. (If you have never studied circuits, you can still take comfort in knowing that our conclusion makes perfect physical sense.)

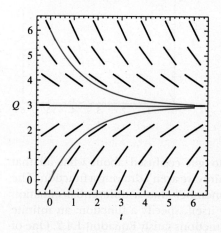

Now, suppose $Q(0) = 0$. The slope field shows clearly what will happen. Q will rise quickly at first, but as it rises, the slope will flatten out, so Q will asymptotically approach 3 from the bottom. Similarly, if Q starts out *higher* than 3, it will move downward more and more slowly, approaching 3 from the top.

We can now succinctly describe all the possible behaviors of the circuit: if the charge starts at $Q = 3$ it will stay that way forever, and if it starts at any other value it will asymptotically approach $Q = 3$ from above or below.

Equilibrium

We can rephrase our conclusions about the RC circuit in the following way: Equation 1.4.2 has a *stable equilibrium* at $Q = 3$. What makes it "equilibrium" is that, if $Q = 3$, then Q will stay 3 forever. What makes it "stable" is that, if Q is close to but not equal to 3, it will approach 3. (In the opposite situation, an "unstable equilibrium," a system near equilibrium will tend to move away from it: picture a pencil balanced perfectly on its point.)

Definitions: Equilibrium, Stable, Unstable

A system is in "equilibrium" if it is in a state that does not change over time.

Equilibrium states are classified by what happens when the system is near, but not in, the equilibrium state. If the system is near a "stable equilibrium" it will tend to move toward equilibrium. If it's near an "unstable equilibrium" the system will tend to move away from equilibrium.

EXAMPLE **Equilibrium**

Problem:
Draw a slope field for the differential equation

$$\frac{dy}{dt} = y(y-1)^2(y+1)$$

Identify the equilibrium states and classify each one as stable or unstable.

Solution:
The easiest way to start drawing a slope field is often to find the points where $dy/dt = 0$. In this case that's at $y = 0, \pm 1$, so we can start by filling in those points on a slope field.

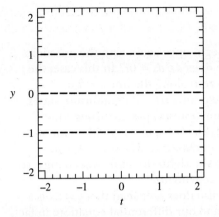

If y is any of the three values shown above, it will stay at that value forever; these are equilibrium states. To classify them and analyze the behavior, we consider other y-values.

- For $y > 1$ the derivative is positive. For values just above 1, the slope is positive but small, indicating a shallow incline; as y grows, the slope gets steeper.
- For $0 < y < 1$ the derivative is also positive. It is shallow near $y = 0$ and $y = 1$ and steeper toward the middle.
- The derivative is negative for $-1 < y < 0$—once again, most steeply somewhere in the middle.
- Finally, for $y < -1$ the derivative is positive, getting more so as you go lower.

Putting all of that together we get the following slope field.

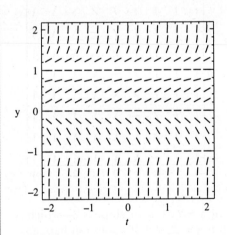

Look at the slope field and see if you can figure out how this system will behave if it *doesn't* start in one of its three equilibrium states.

- If the system starts anywhere below zero, it will approach $x = -1$ (from above or below).
- If the system starts between 0 and 1, it will approach 1 from below.
- If the system starts above 1, it will blow up.

$y = -1$ is a stable equilibrium, and $y = 0$ is unstable. We could say that $y = 1$ is stable from below but unstable from above; such a state is sometimes called "semi stable."

You may now think that "equilibrium means the slope is zero," but it isn't that simple: equilibrium occurs at a y-value for which the slope is *always* zero. The example below illustrates this distinction, and shows how you can still use a slope field to analyze a more complicated system.

EXAMPLE A Slope Field

Problem:
Draw a slope field for the differential equation

$$\frac{dy}{dx} = y - 2x$$

and describe the possible behaviors of the solutions.

Solution:

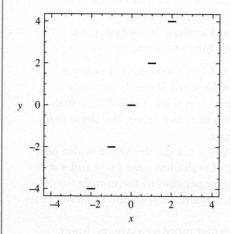

We begin once again with the question: "Where does $dy/dx = 0$?" In this case, the slope is zero along the line $y = 2x$:

These points are *not* equilibrium states. If the function $y(x)$ passes through the point $(1, 2)$, for example, y will momentarily be unchanging. However, as you move to the right, the slope will change, so y will not remain at 2.

This also does *not* mean that $y = 2x$ is a solution to our differential equation: in fact, we can clearly see that it isn't. If a solution crosses that line, it will do so horizontally, so it won't stay on the diagonal line.

But we *can* build our slope field by working out from that line. Consider a point one unit above the line—same x-value, one higher y-value. Since the function $y - 2x$ equals 0 on the line, adding 1 to the y-value will make the function $y - 2x$ equal 1 at such a point. This means the slope $dy/dx = y - 2x$ will be 1 at all such points. Two units above the line, the slope is 2; one unit below the line, the slope is -1; and so on. With this insight, it doesn't take long to fill in the graph.

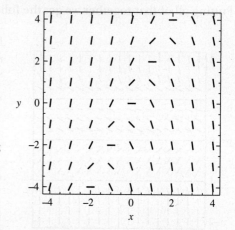

Looking at the slope field in this way can lead you to an important insight. All along the line $y = 2x$ the slope is zero. Along the line $y = 2x + 1$ the slope is 1. And all along the line $y = 2x + 2$ the slope is 2—which is precisely the slope required to keep you on the line $y = 2x + 2$. If a solution happens to go through that line, it will stay on that line. More succinctly, $y = 2x + 2$ is one solution of the differential equation! It isn't an equilibrium solution (this equation has none), but it is nice to find a simple line that works. You can probably use the slope field to convince yourself that $y = 2x + 2$ is the *only* straight line that can possibly solve this differential equation. (You will reach the same conclusion analytically in Problem 1.81.)

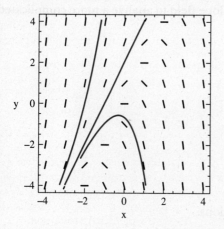

What happens to a solution that starts above that line? It will have a slope greater than 2, so it will rise faster than the line. This in turn will cause its slope to get even higher, so it will gradually pull away from the line. Conversely, a function slightly below $y = 2x + 2$ will have a slope lower than 2, so it will gradually fall below the line, eventually reaching a slope of zero and then turning downward.

1.4 | Slope Fields and Equilibrium

Stepping Back

A first-order differential equation gives you the slope y' as a function of the point (x, y) where you are; that is why slope fields represent such equations so well. A second-order differential equation gives you the concavity y'' as a function of both the point (x, y) and the slope y', so such equations cannot be represented by slope fields. In Chapter 10 we will introduce "phase portraits" to visually represent the solutions to higher order equations.

In the problems for this section you will draw many slope fields by hand. That's a useful skill because it forces you to think about the behavior of the system and about how the slope field is related to the equation. Once you have practiced that skill enough to be comfortable with the technique, you can learn a tremendous amount about a system by glancing over a computer-drawn slope field.

1.4.3 Problems: Slope Fields and Equilibrium

You may find graph paper helpful for the problems in this section.

1.65 Walk-Through: Slope Fields. The function $y(x)$ evolves according to the differential equation $dy/dx = xy^2/2$.

(a) At the point $(-2, -2)$, the differential equation predicts a slope of -4. Draw a small line with a slope of -4 at this point. Your slope will not be exact, and that's OK! You can't visually distinguish a slope of -4 from -5, but the slope should clearly be much steeper than -1.

(b) Draw similar lines with the appropriate slopes at all integer points on the domain $x \in [-2, 2]$, $y \in [-2, 2]$: a total of 25 points in all.

(c) Starting at the origin, draw a curve that extends both to the right and the left, using your slope lines as a guide.

(d) Draw three more curves, going through the points $(0, -2)$, $(0, -1)$, and $(0, 1)$. In each case, the detailed points you hit are not important, but the overall shape is: you should be able to see at a glance where the curve is increasing and decreasing, roughly where it reaches a local minimum, and so on.

(e) Based on your slope field, what is $\lim_{x \to \infty} y(x)$ if $y(0) = -1$?

In Problems 1.66–1.71, draw a slope field on all integer points $-3 \leq x \leq 3$, $-3 \leq y \leq 3$ (a total of 49 points in all). Then draw three sample curves through the slope field. It may help to first work through Problem 1.65 as a model.

1.66 $dy/dx = -1/2$

1.67 $dy/dx = y$

1.68 $dy/dx = -y$

1.69 $dy/dx = -x$

1.70 $dy/dx = 2^y$

1.71 $dy/dx = x/y$ (For this problem use the range $1 \leq y \leq 5$.)

In Problems 1.72–1.75 you will need to figure out the correct ranges of x and y-values you need to see the possible behaviors of the system. This may take some trial and error. For each problem:

(a) Begin by drawing small horizontal lines at all points on your plot where the slope is zero.

(b) Work out from there—left, right, up, and down—drawing slopes until you have filled in enough to understand the behavior of the system.

(c) Identify the "equilibrium solutions" if any, and classify them as "stable" or "unstable."

(d) Draw three sample curves through the slope field.

(e) Based on your drawing describe the possible behaviors of the system based on different initial conditions, assuming the dependent variable represents some meaningful quantity and the independent variable represents time.

1.72 $dy/dx = y + 2$

1.73 $dy/dx = -(y + 2)$

1.74 $dy/dx = 2y(3 - y)$

1.75 $dy/dx = x^2 + y^2 - 4$

1.76 Have a computer find an analytical solution to the equation $dy/dx = x^2 + y^2 - 4$. Explain why it is more useful to analyze this equation with slope fields (Problem 1.75) than with an analytical solution.

1.77 Draw a slope field for the differential equation $dy/dx = k(y - a)(y + b)$ where k, a, and b are all positive constants. Since you don't have numerical values for these constants you can't know exactly what the actual slopes are, but you can still make a slope field that shows where the slopes are zero, positive or negative, and where they tend to be large or small. Label the equilibrium points on the y-axis.

1.78 If you're not familiar with the wonderful function e^{-x^2}, there are two things you need to know for this problem. First, the equation $f'(x) = e^{-x^2}$ has no simple analytical solution. Second, the graph of $y = e^{-x^2}$ is shown below; it is always positive, reaches an absolute maximum at $(0,1)$, and approaches 0 as $x \to \pm\infty$.

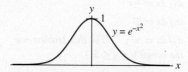

(a) Based on this graph, draw a slope field for the equation $f'(x) = e^{-x^2}$.

(b) Let $f(x)$ be the function that follows the differential equation $f'(x) = e^{-x^2}$ and contains the point $(-3, 0)$. Use your slope field to sketch a graph of $y = f(x)$.

(c) Use your sketch to answer the question: What is the x-value of the point of inflection of $f(x)$?

(d) Repeat Parts (a)–(b) for the differential equation $f'(x) = e^{-f^2}$. Then answer the question: what is the y-value of the point of inflection?

1.79 The equations $dy/dt = 2y - 1$ and $dy/dt = 1 - 2y$ look a great deal alike, but lead to very different types of behavior. In this problem, you will draw slope fields and analyze the behavior of both equations; this will require two separate graphs.

Note: because dy/dt has no t-dependence, you can ignore negative t-values without any loss of generality. So feel free to draw your slope fields only for positive t.

(a) For both of these graphs, $dy/dt = 0$ when $y = 1/2$. So begin both slope fields with horizontal lines at all points where $y = 1/2$.

(b) For $dy/dt = 2y - 1$, compute the slope when $y = 1$, and then draw in slope lines at all points where $y = 1$. Then repeat for $y = 2$, $y = 3$, $y = 0$, $y = -1$, and $y = -2$.

(c) Repeat Part (b) for $dy/dt = 1 - 2y$.

(d) Draw curves representing functions that begin at $(0, 0)$, $(0, 1/2)$, and $(0, 1)$ for both equations (six curves in all).

(e) For the equation $dy/dt = 2y - 1$, the function $y = 1/2$ represents an equilibrium solution: if y is ever $1/2$, it will stay $1/2$ forever. If y is "bumped" slightly above $1/2$, how will it evolve over time? If y is "bumped" slightly below $1/2$, how will it evolve over time? Does $y = 1/2$ represent a stable or unstable equilibrium?

(f) Repeat Part (e) for the equation $dy/dt = 1 - 2y$.

1.80 The equations $dy/dx = y - x$ and $dy/dx = x - y + 2$ both have the same linear solution $y = x + 1$, but behave quite differently otherwise. In this problem, you will draw slope fields and analyze the behavior of both equations; this will require two separate graphs.

(a) Show that $y = x + 1$ is a valid solution to both differential equations. (You will do this, not with a graph, but by plugging in.)

(b) If $y = x + 1$ is a solution to a differential equation, then dy/dx must equal 1 everywhere along that line. So begin both slope fields by drawing lines with a slope of 1 at the points $(-3, -2)$, $(-2, -1)$, and so on through $(3, 4)$.

(c) In the equation $dy/dx = y - x$, what happens to dy/dx if you increase y by one unit while leaving x unchanged? Based on your answer, draw in the slope lines at all points one unit higher than the points you drew in Part (b).

(d) Based on the same logic you used in Part (c), draw in the slope lines one unit higher. Then move down to one unit, and then two units, below the line $y = x + 1$.

(e) Repeat parts (c) and (d) on your other graph for $dy/dx = x - y + 2$.

(f) For the equation $dy/dx = y - x$, if the function starts on the line $y = x + 1$, it will stay on that line forever. How will a function evolve if it starts just above that line? Just below that line?

(g) For the equation $dy/dx = x - y + 2$, if the function starts on the line $y = x + 1$, it will stay on that line forever. How will a function evolve if it starts just above that line? Just below that line?

1.81 In the Explanation (Section 1.4.2), we used a slope field to determine that the equation $dy/dx = y - 2x$ has only one linear solution, which is $y = 2x + 2$. In this problem, you will prove the same result analytically.

(a) If there is a linear solution, it can be written in the form $y = mx + b$. So plug $y = mx + b$ into the differential equation $dy/dx = y - 2x$.

(b) The resulting equation sets two linear functions of x equal to each other. Put each function—the one on the left side of the equal sign, and the one on the right side—into the standard form of a line.

(c) If the left side of the equation is the same function as the right side, then their slopes (coefficients of x) must be equal, and their y-intercepts (constant parts) must also be equal. Use these two facts to solve for m and b.

(Incidentally, an alternative approach is to remember that a line is the only function with zero concavity. So starting with $dy/dx = y - 2x$ you can set $d^2y/dx^2 = 0$ and reach the same conclusion.)

1.82 The population of the planet Foom is described by a function $F(t)$. Every Foomian has, on average, 2 babies per year.

(a) Assuming no Foomians ever die, write a differential equation for $F(t)$.

(b) Sketch a slope field for the equation you wrote down in Part (a). Since the number of Foomians can never be negative you only need to include values $F \geq 0$. Identify any equilibrium points in this range.

(c) Using this slope field, describe all the possible long-term behaviors of $F(t)$.

1.83 A vicious rumor is spreading through a campus with a total population P. The number of students who have heard the rumor obeys the "logistic" differential equation $dS/dt = kS(P - S)$ where k is a positive constant.

(a) Draw a slope field for this equation. Because S cannot be smaller than 0 or larger than P you only need to include values of S in that range. Since you don't have numerical values for P and k you can't know exactly what the actual slopes are, but you can still make a slope field that shows where the slopes are zero, positive or negative, and where they tend to be large or small.

(b) For what values of S will $S(t)$ remain constant? (You can figure this out either from your slope field or directly from the differential equation.)

(c) Choose a value of $S(0)$ on your plot. (You don't need to pick a number; just mark a point on your plot.) Sketch the solution $S(t)$ with that initial condition on your slope field. Do *not* use a solution where $S(t)$ is constant.

(d) Based on your slope field, what is $\lim_{t \to \infty} S(t)$?

1.84 An object at $-200°C$ is placed in a $-1°C$ room. The object begins to warm, its temperature u following the differential equation $du/dt = u^4 - 1$.

(a) Explain why this equation, in this form, doesn't make sense for $u > 1$.

(b) Using a slope field, sketch the temperature $u(t)$. Be sure to show the correct concavity at all times and the correct limit $\lim_{t \to \infty} u(t)$.

1.85 Venusians, the plucky inhabitants of the planet Venus, reproduce by a process called "quatosis": every year, every Venusian splits into four little Venusians. Left unchecked, the population of Venusians would soon overwhelm their planet. However, Martians kidnap 12,000 Venusians from the planet every year.

(a) Write the differential equation that governs the number of Venusians $V(t)$. Your differential equation should take into account both the gain due to quatosis, and the loss due to kidnapping.

(b) Draw a slope field for that differential equation. You can safely ignore negative t and V-values, which confines your slope field to one quadrant. Beyond that, part of your job is to choose enough V-values to see the behavior of this differential equation.

(c) The equation has one equilibrium state; what is it?

(d) Is the equilibrium state stable or unstable? How can you tell?

(e) Explain in words to a non-mathematician (from Earth) what will happen if the Venusian population starts higher than the equilibrium value, and why it will happen.

1.86 The ancient Martian race is dying. The only way they keep their ancient culture going is by kidnapping; every year, they steal 12,000 Venusians. Through a high-tech process known as "Marsification" they transform their victims into Martians. Unfortunately, the process is unstable; the Martians cannot reproduce, and every year, one out of every five living Martians dies.

(a) Write the differential equation that governs the number of Martians $M(t)$. Your differential equation should take into account both the loss due to death, and the gain due to Marsification.

(b) Draw a slope field for that differential equation. You can safely ignore negative t and M-values, which confines your slope field to one quadrant. Beyond that, part of your job is to choose enough M-values to see the behavior of this differential equation.

(c) The equation has one equilibrium state; what is it?

(d) Is the equilibrium state stable, or unstable? How can you tell?

(e) Explain in words to a non-mathematician (from Earth) why your answer to Part (d) makes sense. In other words, how could you have looked at our description of Mars and predicted without doing any math whether the equilibrium would be stable or unstable?

1.87 Newton's law of heating and cooling says that the rate of change of the temperature Q of an object is proportional to the difference between the temperature of the object and the temperature Q_R of the room it's in.

(a) Write this law as a differential equation for Q. Your equation will have a constant of proportionality in it, which you can call k. Assume k is positive and make sure the signs in your equation make sense physically. (Does it behave correctly when $Q > Q_R$ and when $Q < Q_R$?)

(b) Let $k = 1$ s^{-1} and $Q_R = 20°$C. Draw a slope field for Q. Choose a range of values for Q and t that clearly shows the possible behaviors of the system.

Now suppose the same object is in the same room, but now the room is steadily heating up: $Q_R = 20 + t$ with temperature in degrees Celsius and time in seconds.

(c) Rewrite the differential equation. (Your answer should have t in it, not Q_R.)

(d) Draw the slope field for this new situation.

(e) Your slope field should show that there is an "attractor solution," a linear $Q = at + b$ that the temperature approaches no matter how it starts out. Using your slope field as a guide and some algebraic trial and error, find that linear solution and verify that it works in the differential equation.

(f) Describe what the temperature of the object does over time if $Q(0) = 20°$C.

1.5 Separation of Variables

We have discussed what differential equations mean, how to recognize and work with solutions, and how to visualize solutions. In this section we encounter our first formal technique for *solving* differential equations. No technique works on all equations, but we will see that "separation of variables" can find general solutions in many cases when these solutions are not immediately apparent. This technique only applies to first-order equations, however.

1.5.1 Discovery Exercise: Separation of Variables

Consider the equation:

$$\frac{dy}{dx} = x^2 y \qquad (1.5.1)$$

Before you begin the process, you may want to see if you can figure out the general solution to this equation by thinking about it. The answer is not immediately obvious, but it does make sense if you think about it in the right way! Below we show you one way to find the answer if you can't puzzle it out.

1. Begin by dividing both sides of Equation 1.5.1 by y.
2. Then multiply both sides of the equation by dx. The variables are now *separated*: all the x-dependence is on one side, and all the y-dependence on the other.
3. Write an integral sign in front of both sides.
4. Integrate both sides. Because this is an indefinite integral, you need a $+C$ on one side or the other. Traditionally, it is put on the right (the x side).
5. Solve the resulting equation for y.

See Check Yourself #5 in Appendix L
6. Check to make sure your solution solves Equation 1.5.1.
7. The solution to this equation is more commonly written as $y = Ce^{x^3/3}$. Explain why this is the same as the previous solution, even though they look quite different.

In this section you'll apply this technique, called "separation of variables," to a variety of differential equations.

1.5.2 Explanation: Separation of Variables

One of the most important techniques for solving differential equations is called "separation of variables." Below we demonstrate this technique on a population problem.

Assume that rabbits produce, on average, 5 babies per adult per year. If 100 rabbits are killed by hunters each year, the rabbit population $R(t)$ can be modeled by the equation:

$$\frac{dR}{dt} = 5R - 100 \qquad (1.5.2)$$

The technique comprises three steps: separate the variables, integrate both sides, and then solve for R.

1. **Separate the Variables.** Your goal here is to algebraically manipulate the equation into the form *(function of R) dR=(function of t) dt*. Equation 1.5.2 readily becomes:

$$\frac{1}{5R - 100} dR = dt$$

2. **Integrate both sides.** Make sure to put a $+C$ on the right side!

$$\int \frac{1}{5R - 100} dR = \int dt$$
$$\frac{1}{5} \ln|5R - 100| = t + C$$

Why only the right side? You could certainly write $+C_1$ on the left and $+C_2$ on the right. But then you could combine the two constants into $C_2 - C_1$ on the right. As we have discussed before, $C_2 - C_1$ simply means "any constant" so we lose nothing by calling it C.

On the other hand, many students make the mistake of leaving out the $+C$ entirely at this step, planning to toss it in later. This is an understandable habit left over from introductory calculus, where adding $+C$ to every answer is often the last step before turning in the "integrals" test. That doesn't work here. Add the C when you integrate, and then it will appear in the correct spot in the final function.

3. **Solve for R.**

$$\ln|5R - 100| = 5t + 5C$$
$$|5R - 100| = e^{5t+5C} = e^{5C}e^{5t}$$
$$5R - 100 = \pm e^{5C}e^{5t}$$
$$R = \pm \frac{1}{5} e^{5C} e^{5t} + 20$$

The expression $\pm(1/5)e^{5C}$ is just a complicated way of writing "any constant," so we replace it with C.

$$R = Ce^{5t} + 20$$

Another Look at C

The proper care and handling of your arbitrary constants is the subtlest part of this technique. In the example above we turned $C_1 - C_2$ into C and later turned $\pm(1/5)e^{5C}$ into C. As we discussed in Section 1.3, when you are finding a general solution, you can keep your constants as simple as possible.

What you *cannot* do, as we stressed above, is simply add $+C$ at the end of the problem. Test $R = (1/5)e^{5t} + C$ in Equation 1.5.2; it doesn't work. Adding the C when you integrate, and carrying it around the algebra from there, ensures that it ends up in the correct place in the final solution.

In addition, you cannot turn $5C$ or anything else into C when you are *checking* a solution. Remember that C in the equation means "any number works here"; for instance, $R = 7e^{5t} + 20$ is a valid solution to Equation 1.5.2. Checking should follow all the normal rules of algebra.

Quick Notes on Integration

You may have been confused by $\int dt$ in the example above, because there doesn't appear to be anything to integrate. You can think of this as $\int 1\, dt$: "The derivative of what function, with respect to t, gives the answer 1?" The answer is, of course, $t + C$.

The integral $\int dR/(5R - 100)$ can also cause confusion. The function $\ln|5R - 100|$ does *not* work; taking its derivative requires the chain rule, which kicks out an unwanted 5. We solved this problem by multiplying our answer by $1/5$. It may be easier to rewrite the question as $(1/5)\int dR/(R - 20)$. Then the integration step is less error-prone.

EXAMPLE **Separation of Variables**

Problem:
Find the general solution to the equation:

$$\frac{dy}{dx} = \frac{y}{e^x \ln y} \quad (y > 0) \tag{1.5.3}$$

Solution:

$$\frac{\ln y}{y} dy = \frac{1}{e^x} dx$$

$$\int \frac{\ln y}{y} dy = \int e^{-x} dx$$

$$\frac{1}{2}(\ln y)^2 = -e^{-x} + C$$

$$y = e^{\pm\sqrt{-2e^{-x}+C}}$$

You can readily verify that this solves Equation 1.5.3 for any value of C.

Limitations of the Technique

Separation of variables, as presented here, works only on *first-order* differential equations—and even there you may run into problems. Consider two ordinary, linear, first-order differential equations.

$$\frac{dy}{dx} = x + y \quad \text{and} \quad \frac{dy}{dx} = e^{\cos x}$$

Separation of variables fails in both cases. In the first example, there is no algebraic way to get all the *x*-dependence on one side and the *y*-dependence on the other. In the second, the variables are easy enough to separate, but you can't integrate the function!

Such problems may simply require a different technique. There are many to choose from, and you will see more as the book progresses. But many important differential equations turn out to be literally unsolvable. In that case, approximation techniques are used—either to simplify the differential equation itself, or to approach the answer numerically with a computer.

1.5.3 Problems: Separation of Variables

1.88 Walk-Through: Separation of Variables. In this problem, you are going to solve the equation $dx/dt = te^x$.

(a) Multiply both sides of the equation by $e^{-x}dt$. This "separates the variables": the left side has only *x*-dependence, and the right side *t*.

(b) Write an integral sign on both sides of the equation.

(c) Integrate both sides. Include a $+C$ in your answer on the right, but not on the left.

(d) Solve for *x*.

(e) Demonstrate that you have found a valid solution to the differential equation.

(f) Find the specific function that satisfies the differential equation $dx/dt = te^x$ and the initial condition $x(0) = 1$.

For Problems 1.89–1.101,

(a) Solve using separation of variables. (Even if you already know the solution, show how to use separation of variables to find it.) Your solution should have one arbitrary constant. It may help to first work through Problem 1.88 as a model.

(b) Demonstrate that your solution satisfies the differential equation.

You can do all relevant integrals in Problems 1.89–1.97 by inspection, by algebraic simplification, or with a *u*-substitution.

Problems 1.98–1.101 may require more advanced integration techniques such as integration by parts, partial fractions, and trig substitution.

1.89 $dy/dx = 2x$

1.90 $dy/dx = 2y$

1.91 $dy/dx = xy$

1.92 $dz/dt = 3z + 6$

1.93 $dz/dt = 8 - 5z$

1.94 $dx/dt = (t^2 + 1)/(2t)$

1.95 $dx/dt = (x^2 + 1)/(2x)$

1.96 $\dfrac{dA}{d\theta} = \dfrac{\tan\theta}{A^2\sqrt{A^3 + \pi}}$

1.97 $\dfrac{dr}{d\theta} = \dfrac{\sqrt{r}\,\cos(2\theta)}{e^{\sqrt{r}}}$

1.98 $dy/dx = x\sqrt{9 - y^2}$

1.99 $ds/dt = (s^3 + 3)/s^2 \times \ln t$

1.100 $dx/dt = \cos(\omega t)(x^2 + 25)$ (ω is a constant)

1.101 $dy/d\theta = y^2 \sin^3(\theta)$

1.102 $dy/dx = 1 - y^2$

(a) Create a slope field for all integer points $0 \leq x \leq 3$, $-3 \leq y \leq 3$.

(b) Based on your slope field, describe the behavior of $y(x)$ if…

  i. $y(0) > 1$
  ii. $y(0) = 1$
  iii. $-1 < y(0) < 1$
  iv. $y(0) = -1$
  v. $y(0) < -1$

(c) Solve the original equation to find a function $y(x)$. (The easiest way to do this by hand requires the technique of partial fractions for your integral. Alternatively you could set up the integrals by hand and then evaluate them with a computer.)

(d) Show that your resulting $y(x)$ function is a valid solution to the original differential equation.

(e) Find the particular solutions corresponding to the initial conditions $y(0) = 0$, $y(0) = 1$, and $y(0) = 2$. This will involve either finding the right value of the arbitrary constant or allowing it to approach ∞.

(f) 🖥 Plot the three particular solutions you just found. You should be able to see that they display the behaviors you described in Part (b).

1.103 In the Explanation (Section 1.5.2) we solved the equation $dR/dt = 5R - 100$ that describes a population of rabbits that is reproducing and being hunted. We found the solution $R = Ce^{5t} + 20$.

(a) Find the specific solutions for $R(0) = 10$, $R(0) = 20$, and $R(0) = 30$.
(b) Sketch the three solutions you found together on one plot.
(c) For each of these solutions, explain why the behavior you found (increasing, decreasing, or staying constant) makes sense. Don't answer in terms of equations and variables; explain it in terms of rabbits and hunters!
(d) Identify the equilibrium solution for this equation. Is it stable or unstable?

1.104 According to NASA, the Arctic sea ice extent is melting at a rate of 10% per decade. Suppose that at just the moment when the ice extent reaches 25 Million square kilometers, a crack team of engineers begins replenishing the ice at a constant rate of 1 Million km² per decade.

Let P equal the size of the Arctic sea ice extent, measured in millions of square kilometers. Let t equal the time, measured in decades.

(a) Write a differential equation for $P(t)$.
(b) Find the equilibrium solution to your equation. Does it represent a stable or unstable equilibrium? How can you tell?
(c) Solve the differential equation, with initial condition, to find the function $P(t)$.
(d) Verify that your solution solves the differential equation.

1.105 James has just jumped out of an airplane. After he opens his parachute he experiences two forces: the constant force of gravity, and a wind drag that is proportional to his velocity. His height may therefore follow the equation:

$$\frac{d^2h}{dt^2} = -9.8 - 2\frac{dh}{dt} \quad (1.5.4)$$

As a second-order differential equation, this is not technically solvable by separation of variables. However, because the variable h appears *only* in its derivatives, we can turn this into a first-order equation, and solve that by separation.

(a) Letting $v = dh/dt$, rewrite Equation 1.5.4 as a first-order differential equation in v.
(b) Your equation in Part (a) suggests that there is one velocity for which $dv/dt = 0$. What is this velocity?
(c) Solve your equation using separation of variables. Your solution should contain an arbitrary constant: call it C_1.
(d) Calculate $\lim_{t \to \infty} v(t)$ and use it to describe what is physically happening to James after he's been in the air for a long time.
(e) Now that you have a velocity function $v(t)$, integrate it with respect to t to find a position function $h(t)$. This will introduce a second constant C_2.
(f) Supposing James begins at a height of 3000 m with no initial velocity, what is his height 3 s later?

1.106 A frozen turkey placed in a 35°F refrigerator begins to thaw according to the equation $dQ/dt = (1/20)(35 - Q)$ where Q is the temperature of the turkey. Assume temperature is measured in degrees Fahrenheit and time is measured in hours.

(a) Draw a slope field for $0 \le t \le 5$ and $Q = -5, 15, 25, 35, 45, 55$.
(b) Based on your slope field, what is the equilibrium solution to this equation? Is it a stable or unstable equilibrium?
(c) Solve the differential equation by separation of variables, assuming the turkey started ($t = 0$) in a 0°F freezer.
(d) Based on your solution, what is $\lim_{t \to \infty} Q(t)$?
(e) How long will it take the turkey to reach 32°F?

1.107 The Expanding Universe
The universe is expanding according to "Hubble's Law":

$$\frac{da}{dt} = \sqrt{\frac{8\pi G}{3c^2}\rho}\, a \quad (1.5.5)$$

where a is the distance between two reference objects and ρ is the energy density of the universe. (Note: it will be easier to work with this equation if you collect the constants into one letter: say, $\beta = \sqrt{8\pi G/3c^2}$.)

(a) If we consider the universe to be primarily composed of *matter*, then $\rho = k/a^3$. (This makes sense if you think about it.) Solve the resulting differential equation.
(b) If we consider the universe to be primarily composed of *radiation*, then $\rho = k/a^4$. Solve the resulting differential equation.
(c) If we consider the universe to be primarily composed of *dark energy* (Einstein's "cosmological constant") then ρ is a constant. Solve the resulting differential equation.
(d) Describe in words the difference between how a universe expands when it is filled with matter or radiation vs. how it expands when it is filled with dark energy.

1.6 Guess and Check, and Linear Superposition

This section covers two separate but related topics. "Guess and check" is a way of finding one or more solutions to a differential equation; "linear superposition" is a set of rules for combining existing solutions to find the general solution. Together, they provide an analytical approach that works on many—though certainly not all!—differential equations.

1.6.1 Discovery Exercise: Guess and Check, and Linear Superposition

If we ask you to find a solution to the equation

$$\frac{d^2 u}{dx^2} - \left(\frac{1}{x} + 2x\right)\frac{du}{dx} - 8x^2 u = 0 \tag{1.6.1}$$

you might quite reasonably have no idea how to begin. So we're going to tell you something else: there is a solution of the form $u(x) = e^{(something)x^2}$. Of course there is no obvious way you could have figured that out from scratch, but now that you know it, your job is to find the *something*, which is much more doable.

1. Test $u(x) = e^{x^2}$ in Equation 1.6.1 and show that it *doesn't* work.

One down, right? But rather than checking every possible value of that *something*, you can check all of them at once.

2. Try the guess $u(x) = e^{px^2}$ where p is a constant. Plug this guess into Equation 1.6.1 and simplify the result until you get a quadratic equation for p.
3. Solve that quadratic equation to find the two values of p for which your guess works. Write down the two solutions $u_1(x)$ and $u_2(x)$ that you have found.

    See Check Yourself #6 in Appendix L

Note that p was not an *arbitrary constant* because the solution you tried only worked for certain values of p.

4. Verify that u_1 and u_2 are solutions to Equation 1.6.1.
5. Show that $4u_1(x)$ is also a solution.
6. Show that $Au_1(x) + Bu_2(x)$ is a solution for *any* constants A and B.

1.6.2 Explanation: Guess and Check, and Linear Superposition

Suppose you need to find a number x such that $100x = x^5 + 57$. After a bit of trial and error you stumble upon $x = 3$ and you have your solution. If someone challenges you to prove your "method" is valid you simply ask "Does $300 = 3^5 + 57$?" It does, and that's all the proof you need.

This suggests one of the most important methods for solving differential equations: guessing! With a bit of patience, and no loss of mathematical rigor, you can often try possibilities until you find one that works. We demonstrate this technique below on a typical dynamics problem.

Solving a Problem by Guessing

A mass on a spring experiences a force $F_{spring} = -kx$ where x is the displacement of the mass from equilibrium. If the mass is underwater it also experiences a damping force that opposes its motion: $F_{\text{damping}} = -cv$. Putting these together with Newton's Second Law gives the equation of motion for this mass, $ma = -cv - kx$. For our example we take $m = 1$, $c = 8$, and $k = 15$.

$$\frac{d^2x}{dt^2} + 8\frac{dx}{dt} + 15x = 0 \tag{1.6.2}$$

To solve this equation and determine the behavior of the mass, our strategy will be to guess an exponential function $x(t) = e^{pt}$. It's important to note that p is not an arbitrary constant! An arbitrary constant says: "This solution will work for *all* values of C." We are asking the question: "Is there *any* constant p for which $x = e^{pt}$ is a valid solution to Equation 1.6.2?" If there is, its derivatives look like this.

$$x = e^{pt} \quad \to \quad \frac{dx}{dt} = pe^{pt} \quad \to \quad \frac{d^2x}{dt^2} = p^2 e^{pt}$$

Now we plug all that into Equation 1.6.2 and see what we get.

$$p^2 e^{pt} + 8pe^{pt} + 15e^{pt} = 0 \quad \to \quad (p^2 + 8p + 15)e^{pt} = 0 \quad \to \quad p = -3, p = -5$$

Our guess has now justified itself—it has given us two specific solutions, $x = e^{-3t}$ and $x = e^{-5t}$.

Of course, you're never done with a real-world problem until you interpret the answer. If we had ended up with $x = e^{2t}$ we would know we had made a mistake; there's no way our underwater mass is going to go shooting off like a rocket. But negative exponential solutions make perfect sense. Our spring will ease down toward its relaxed position at $x = 0$.

Even assuming you followed all that, you may not feel that you could do the same thing on your own. The guess $x = e^{pt}$ came out of nowhere; how do you choose a good guess? The short answer is "try stuff, but be smart." There are only a few basic functions to choose from: constant, exponential, power, trig, logarithm. For Equation 1.6.2 most of these forms can be eliminated at a glance. If you try a power function such as $x = t^3$, the three different terms will all be different powers of t; they will add up to a polynomial. If you try $x = \sin t$, you will find yourself adding two sines and a cosine; again, they cannot possibly cancel. An exponential function seems at least a plausible guess, since you can see *before you try it* that it will create three different exponential terms that may, possibly, add up to zero.[2]

EXAMPLE Guess and Check

Problem:
Use guess and check to find a solution to $xy'(x) + 3y(x) = 0$.

Solution:
We'll try the guess $y = e^{kx}$. Plug that and its derivative $y'(x) = ke^{kx}$ into the differential equation.

$$xke^{kx} + 3e^{kx} = 0 \quad \to \quad e^{kx}(xk + 3) = 0 \tag{1.6.3}$$

A common student mistake is to solve that equation, but $k = -3/x$ will not provide a valid solution because our guess was $y = e^{kx}$ *for a constant k*. The assumption of a constant k was built into the process when we wrote $y'(x) = ke^{kx}$. No constant k solves Equation 1.6.3, so we conclude that e^{kx} was not a good guess.

---

[2] Section 10.2 will give more specific guidelines for a good guess.

We will therefore try a different guess: $y = x^k$ (again with a constant k). Plug that and its derivative $y'(x) = kx^{k-1}$ into the differential equation.

$$xkx^{k-1} + 3x^k = 0 \quad \rightarrow \quad x^k(k+3) = 0 \quad \rightarrow \quad k = -3$$

We now have a working solution, $y = x^{-3}$. You can verify this by testing it in the original equation. (In this case you can also arrive at the same solution by separating variables.)

Linear Differential Equations

Above we solved Equation 1.6.2 for a mass on an underwater spring and found two solutions, $x = e^{-3t}$ and $x = e^{-5t}$. Both functions work (you should check this) and make physical sense (as we discussed). But our goal is the *general* solution, with arbitrary constants to be filled in based on initial conditions such as $x(0)$ and $x'(0)$.

The strategy for finding the general solution depends on what type of differential equation you are solving. So before we return to our damp mass, we have to define a few terms.

Definitions: Linear and Homogeneous Differential Equations

In a "linear homogeneous" differential equation for a function $y(x)$ every non-zero term is a function of x or a constant, multiplied by y or one of its derivatives.

If in addition there is a term in which y doesn't appear at all, the equation is still linear but "inhomogeneous." (Some authors use "non-homogeneous." We shall non-use that term.)

The standard form for such equations puts all terms involving y on the left side.

Left side: Each term is a function of x or a constant, multiplied by y or one of its derivatives. That makes the equation "linear."

Right side: If this is zero the equation is "linear and homogeneous." If this is a constant or a function of x the equation is "linear and inhomogeneous."

$$3\frac{d^2y}{dx^2} + \frac{1}{x}\frac{dy}{dx} + (\cos x)y = \ln x \tag{1.6.4}$$

A linear, inhomogeneous equation

The point is that y and its derivatives do not appear in any form *other than* simply being multiplied by functions of x. For instance $\cos(y'')$, y^2, or $y(dy/dx)$ would render the equation non-linear. The distinction between homogeneous and inhomogeneous only applies to linear equations.

These definitions, along with several other classifications of differential equations, are in Appendix A.

These definitions classify every ODE into one of three categories.

- *Linear homogeneous equations*. These equations are generally the most straightforward, and we will show you how to solve them below.
- *Linear inhomogeneous equations*. Linear inhomogeneous equations require a bit more work but we can still usually handle them. We will discuss them after the homogeneous case below.

- *Non-linear equations.* Many important equations in the real world fall into this category, but these are generally difficult to solve. In many cases the best approach is to solve them numerically (Section 1.8) or approximate them with linear equations (Chapter 2). We will say nothing more about non-linear equations in this section.

Before you read further, see if you can properly categorize Equation 1.6.2. As we will see below, that is a key step in deciding how to solve it.

Linear, Homogeneous Differential Equations

You should have found that Equation 1.6.2 meets our definition of a linear, homogeneous equation. But it's not clear at this point what that classification has to do with solving the equation. Recall that we previoiusly found two independent solutions, $x = e^{-3t}$ and $x = e^{-5t}$. The following two results apply to those solutions, and more generally to any solutions of linear homogeneous equations.

- If we multiply either solution by a constant, the result is still a solution. (Try $7e^{-3t}$ and πe^{-5t} in Equation 1.6.2.)
- If we add our two solutions, the result is still a solution. (Try $e^{-3t} + e^{-5t}$.)

Together these observations provide the general solution we are looking for.

$$x(t) = Ae^{-3t} + Be^{-5t} \text{ where } A \text{ and } B \text{ are arbitrary constants} \tag{1.6.5}$$

Once you verify these results for this specific case, you may be able to see how they generalize—and in particular how they rely on beginning with an equation that is both linear and homogeneous. You have then arrived at the "principle of linear superposition."

The Principle of Linear Superposition

If $y_1(x)$ and $y_2(x)$ are both solutions to a linear, homogeneous differential equation, then any linear combination of the form $Ay_1(x) + By_2(x)$ is also a solution.

In Problem 1.130 you'll show why all linear homogeneous differential equations obey linear superposition, but here we illustrate the point with an example.

Given the equation... $\quad y'' - \dfrac{2}{x^2}y = 0$

A: Plug in $y = x^2 \quad \to \quad 2 - 2 = 0$

B: Plug in $y = \dfrac{1}{x} \quad \to \quad \dfrac{2}{x^3} - \dfrac{2}{x^3} = 0$

C: Plug in $y = x^2 + \dfrac{1}{x} \quad \to \quad \left(2 + \dfrac{2}{x^3}\right) - \left(2 + \dfrac{2}{x^3}\right) = 0$

The sum of the two individual solutions worked because each term in line C just gave the sum of the values in lines A and B. Similarly, multiplying a solution by a constant would just multiply each term by that constant, and the whole thing would still equal zero.

Now consider a non-linear example: $y' - y^2 = 0$.

Given the equation... $\quad y' - y^2 = 0$

A: Plug in $y = -\dfrac{1}{x} \quad \to \quad \dfrac{1}{x^2} - \dfrac{1}{x^2} = 0$

B: Plug in $y = \dfrac{1}{1-x} \quad \to \quad \dfrac{1}{(1-x)^2} - \dfrac{1}{(1-x)^2} = 0$

C: Plug in $y = -\dfrac{1}{x} + \dfrac{1}{1-x} \to \left(\dfrac{1}{x^2} + \dfrac{1}{(1-x)^2}\right) - \left(\dfrac{1}{x^2} + \dfrac{1}{(1-x)^2} - \dfrac{2}{x(1-x)}\right) = \dfrac{2}{x(1-x)}$

This time the sum of the two individual solutions didn't work because the y^2 in the differential equation introduced a cross-term in C that wasn't just the sum of what had been in A and B. Similarly, if you plugged in 2 times one of the solutions, the first term would get multiplied by 2 and the second term by 4, and they would no longer cancel.

We now have a two-step method for solving linear homogeneous differential equations.

1. Use guess and check to find n individual solutions for an nth order equation.
2. Use linear superposition to combine these solutions into a solution with n arbitrary constants, and you have the general solution.

EXAMPLE **Guess and Check on a Linear Homogeneous Differential Equation**

Problem:
Find the general solution to the equation:

$$2\frac{d^2y}{dx^2} - \frac{dy}{dx} - 6y = 0 \tag{1.6.6}$$

Solution:
When all the coefficients in your equation are constants, an exponential function is usually a good guess.

$$y = e^{kx} \quad \rightarrow \quad \frac{dy}{dx} = ke^{kx} \quad \rightarrow \quad \frac{d^2y}{dx^2} = k^2 e^{kx}$$

Plug all that into the differential equation.

$$2k^2 e^{kx} - ke^{kx} - 6e^{kx} = 0 \quad \rightarrow \quad e^{kx}(2k+3)(k-2) = 0 \quad \rightarrow \quad k = -3/2, \ k = 2$$

We now have two individual solutions to our second-order equation, $e^{-3x/2}$ and e^{2x}. A linear combination of solutions to a linear homogeneous equation is itself a solution.

$$y(x) = Ae^{-(3/2)x} + Be^{2x} \tag{1.6.7}$$

You can confirm that Equation 1.6.7 is a valid solution to Equation 1.6.6 for any values of the constants A and B, and it has the right number of arbitrary constants to be the general solution.

As a final note on that example, we never proved that we had *the general* solution to Equation 1.6.6. It's easy to show that our solution works for any values of A and B. It's much harder to prove that all possible solutions can be written in the form of Equation 1.6.7—and we're not going to. Here and elsewhere we will assume that when we have n independent arbitrary constants in our solution to an nth order linear ODE, we have the general solution. Exceptions (such as Problem 1.132) don't come up often.

Linear, Inhomogeneous Differential Equations

Let's return to our underwater mass-spring system and rotate the entire problem 90°. The forces of the spring and the water are unchanged but gravity now pulls with a constant force, leading to a linear-but-*inhomogeneous* differential equation.

$$\frac{d^2x}{dt^2} + 8\frac{dx}{dt} + 15x = -9.8 \tag{1.6.8}$$

Our first step is to start over with "guessing." Can you think of an function $x(t)$—any function at all—that solves Equation 1.6.8? *Hint*: there is a simple one!

Answer: $x(t) = -9.8/15$. Almost looks like cheating, doesn't it? Since this function is a constant, dx/dt and d^2x/dt^2 are both zero, so it clearly works. At the same time, this one-off solution doesn't seem to bring us any closer to finding the general solution that we need to solve problems.

Inhomogeneous equations do not obey superposition as we described it earlier. If we multiply our solution by three the equation will give us $3 \times (-9.8)$; if we add our solution to another solution the equation will give us $(-9.8) + (-9.8)$. In neither case will we have another working solution.

But because Equation 1.6.8 is still linear we can use superposition in a slightly different form. We will add our one-off solution to the general solution we found to the *homogeneous* Equation 1.6.2.

$$x(t) = Ae^{-3t} + Be^{-5t} - \frac{9.8}{15} \qquad \text{the solution to Equation 1.6.8}$$

Watch what happens when we plug that into the differential equation.

Plug $x = Ae^{-3t} + Be^{-5t}$ into $x'' + 8x' + 15x$ and you get 0
Plug $x = -9.8/15$ into $x'' + 8x' + 15x$ and you get -9.8
Plug $x = Ae^{-3t} + Be^{-5t} - 9.8/15$ into $x'' + 8x' + 15x$ and you get $0 + (-9.8) = -9.8$

The linear homogeneous equation $x'' + 8x' + 15x = 0$ is called the "complementary equation" to Equation 1.6.8. When we add the general solution of the complementary equation to a particular solution to Equation 1.6.8 we find a solution to Equation 1.6.8 that has the arbitrary constants it needs to meet any initial conditions. In other words, we find the general solution.

As before we can make immediate physical sense of this solution. Like the horizontal spring, this mass eases quickly toward equilibrium as the exponential terms decay. However, this does not occur at the relaxed position of the spring ($x = 0$) but at a lower point, where the upward pull of the spring perfectly balances the downward pull of gravity.

We now have a three-step method for solving linear inhomogeneous differential equations.

1. Find a specific solution to your equation: any function that works, no matter how simple. In the example above (and the example below) we find a constant solution by just looking, but sometimes this step is more complicated and involves a guess-and-check form of its own.
2. Write the complementary homogeneous equation by replacing the inhomogeneous term with zero. Find the general solution to this equation, as we discussed above.
3. The sum of these two solutions is the general solution to your inhomogeneous equation.

EXAMPLE **Guess and Check and Linear Superposition**

Problem:
Find the general solution to the equation:

$$2\frac{d^2y}{dx^2} - \frac{dy}{dx} - 6y = 30 \qquad (1.6.9)$$

Solution:
We begin with the complementary homogeneous equation.

$$2\frac{d^2y}{dx^2} - \frac{dy}{dx} - 6y = 0 \qquad (1.6.10)$$

We will approach Equation 1.6.10 by plugging in the guess $y = e^{kx}$.

$$2k^2 e^{kx} - k e^{kx} - 6 e^{kx} = 0$$

This factors as $e^{kx}(2k+3)(k-2) = 0$ so $k = -3/2$ or $k = 2$, corresponding to the individual solutions $e^{-3x/2}$ and e^{2x}. A linear combination of solutions to a linear homogeneous equation is itself a solution, so the general solution to Equation 1.6.10 is:

$$y_c(x) = A e^{-(3/2)x} + B e^{2x}$$

The solution works, and it has the right number of arbitrary constants, but it is the solution to the wrong equation. We want to solve Equation 1.6.9! Fortunately, it isn't hard to find a solution to that one just by looking:

$$y_p(x) = -5$$

Because y_p is a constant, its first and second derivatives are zero, so you can readily confirm that it does solve Equation 1.6.9. It is *one particular solution*. If we add it to our complementary solution the resulting function will still be a solution.

$$y = -5 + A e^{-(3/2)x} + B e^{2x}$$

That is the function we have been looking for. It solves Equation 1.6.9, and it has the two arbitrary constants it needs to meet any initial conditions.

Stepping Back

You now have a set of tools that can be used to find general solutions to many—although certainly not all—linear differential equations. But because the entire process is based on inspired guesses, it may not seem rigorous. Shouldn't we have to prove something?

In response, remember what we said about the algebra problem at the beginning of this section: *a working solution is its own proof.* If you plug it in and it works, it is a solution. And if it has the requisite number of independent arbitrary constants, it is the *general* solution. All the guessing and adding of solutions are just tricks to lead us to a solution that, when found, justifies itself.

1.6.3 Problems: Guess and Check, and Linear Superposition

1.108 Walk-Through: Guess and Check.
Consider the equation $(d^2y/dx^2) - 10(dy/dx) + 9y = 27$.

(a) Write the complementary homogeneous equation.

(b) Plug the "guess" $y = e^{kx}$ into the equation that you wrote in Part (a). The result should be an algebraic equation.

(c) Solve the resulting algebraic equation for k, and use the result to write two

different solutions to the complementary equation.
  (d) By multiplying your answers to Part (c) by arbitrary constants, and adding the two resulting solutions together, write the general solution to the complementary equation.
  (e) Find a simple solution—a constant—to the original inhomogeneous equation.
  (f) By adding your general solution from Part (d) to your particular solution from Part (e), write the general solution to the problem we started with.
  (g) Demonstrate that your function from Part (f) solves the differential equation.

1.109 Consider the problem $dy/dx = xy + y$. Suspecting an exponential solution, you decide to try $y = e^{kx}$.
  (a) Plug the trial solution $y = e^{kx}$ into the differential equation $dy/dx = xy + y$.
  (b) The resulting equation can easily be solved to yield $k = x + 1$. Have you found a solution? Test $y = e^{x(x+1)}$ in the differential equation $dy/dx = xy + y$.
  (c) It didn't work! Where did we go wrong?
  (d) Now solve this equation correctly using separation of variables.

1.110 $t^2\left(d^2r/dt^2\right) - 9t\left(dr/dt\right) + 16r = 4$ is an example of a "Cauchy–Euler equation." Such equations appear in a number of physics and engineering applications.
  (a) Write the complementary homogeneous equation.
  (b) Plug $r = e^{kt}$ into the equation you wrote in Part (a). Show that this solution will *not* work for any constant k: this equation has no exponential solution.
  (c) Plug the guess $r = t^n$ (where n is a constant) into the equation you wrote in Part (a). Solve the resulting algebraic equation for n: you should find two solutions.
  (d) Write the general solution to the equation you wrote in Part (a).
  (e) Find a specific solution to the original (inhomogeneous) equation.
  (f) Write the general solution to the inhomogeneous equation.
  (g) What is it about this particular equation that made $r = t^n$ work?

Problems 1.111–1.114 are all linear homogeneous differential equations with constant coefficients. In each case, try an exponential solution and find as many real solutions as you can. Then give the general solution, or explain why you cannot find the general solution by this method.

1.111 $d^2y/dx^2 - 7\left(dy/dx\right) + 12y = 0$
1.112 $d^2s/dt^2 + 3\left(ds/dt\right) + 7s = 0$
1.113 $d^2s/dt^2 + 5\left(ds/dt\right) - 4s = 0$
1.114 $d^2\rho/dz^2 + 10\left(d\rho/dz\right) + 25\rho = 0$

For Problems 1.115–1.119 find the general solution using guess-and-check. You may need to use trial and error to find the right kind of function to guess for the complementary and/or particular solutions.

1.115 $d^2y/dx^2 + (k^2 + p^2)y = 0$
1.116 $t^2(d^2s/dt^2) + 10t(ds/dt) + 20s = 0$
1.117 $d^4y/dx^4 - y = 0$
1.118 $d^2y/dx^2 + \omega^2 y = e^{kx}$ *Hint*: To find a particular solution, guess $y_p = ae^{kx}$ and solve for the constant a.
1.119 $x^2(d^2u/dx^2) + 4x(du/dx) - 4u = 4$

1.120 The equation $d^2x/dt^2 + 4(dx/dt) + 3x = \sin t$ can represent a harmonic oscillator that is damped (by friction or air resistance) but also *driven* by an oscillatory force.
  (a) Write the complementary homogeneous equation. Find the general solution by assuming an exponential form.
  (b) To find a specific solution to the original (inhomogeneous) equation, begin by guessing a solution of the form $x = A\sin t + B\cos t$. The numbers A and B do not represent arbitrary constants here; we are looking for one value that will work. Plug your guess into the differential equation.
  (c) Rearrange the resulting algebraic equation into the form:
  (bunch of constants) $\sin t$ + *(different bunch of constants)* $\cos t = \sin t$
  (d) To solve this equation, set the first "bunch of constants" to 1 and the second to 0. You can now solve for A and B.
  (e) Write the general solution to this differential equation.

1.121 Consider the differential equation $d^2y/dx^2 - 6\left(dy/dx\right) + 9y = 3$.
  (a) Find all exponential solutions to the complementary homogeneous equation.
  (b) Find a specific solution to the original equation.

(c) Write a solution to the original differential equation that has an arbitrary constant in it.

(d) Show that the function $y = xe^{3x}$ is a solution to the complementary homogeneous equation.

(e) Write the general solution (complete with two independent arbitrary constants) to the original differential equation.

1.122 The equation $x''(t) - 7x'(t) + 10x = 50 + e^{3t}$ has two inhomogeneous terms. You can solve it by approaching them separately.

(a) Find the general solution $x_c(t)$ to the complementary homogeneous equation.

(b) Find a particular solution $x_{p1}(t)$ to the equation $x''(t) - 7x'(t) + 10x = 50$. You can do this quickly, by just looking.

(c) Find a particular solution $x_{p2}(t)$ to the equation $x''(t) - 7x'(t) + 10x = e^{3t}$. To do this, plug in a solution of the form $x = Me^{3t}$ and find the value of M that works.

(d) Write the function $x_c(t) + x_{p1}(t) + x_{p2}(t)$ and show that it solves the original differential equation.

1.123 In the Explanation for slope fields (Section 1.4), we found that one specific solution to the equation $dy/dx = y - 2x$ is $y = 2x + 2$. We discussed the behavior of other solutions, but did not find them analytically.

(a) Begin by rewriting the differential equation in a more standard form, as $a_1(x)(dy/dx) + a_0(x)y = f(x)$.

(b) Write and solve the complementary homogeneous equation.

(c) Find a particular solution to the original equation by guessing a solution of the form $y_p = Ax + B$.

(d) Find the general solution to the original equation.

(e) Test your general solution.

(f) In the slope field explanation we concluded that if a function lies perfectly on the line $y = 2x + 2$ then it will stay there forever, but if it does not lie on that line it will move farther and farther away from that line. Explain how your general solution correctly predicts both of these behaviors.

1.124 Consider the equation $y^2 + (dy/dx)^2 = 1$. Important: don't confuse $(dy/dx)^2$ with d^2y/dx^2, which is something completely different!

(a) Find the two constant solutions to this equation.

(b) By experimenting with basic functions, find another solution.

(c) With a little more trial and error, rewrite your answer to Part (b) with an arbitrary constant in it. Make sure it still solves the differential equation.

(d) Is the solution you found in Part (c) the general solution? Why or why not?

1.125 Consider the equation $(dy/dx)^2 = y$. Important: don't confuse $(dy/dx)^2$ with d^2y/dx^2, which is something completely different!

(a) Plug in a solution of the form $y = kx^p$ where k and p are constants.

(b) You should now have an equation of the form $a_1 x^{b_1} = a_2 x^{b_2}$. The only way these can be the same function is if $a_1 = a_2$ and $b_1 = b_2$. Use these two facts to solve for k and p and write a solution to the differential equation. (You may find the solution $y = 0$; that function does work, but don't use it.)

(c) Show that your solution works. Then multiply your solution by an arbitrary constant and show that it *doesn't* work. Explain why this doesn't violate the principle of linear superposition.

1.126 The picture shows a standard diagram for a circuit with four elements: a battery that maintains a constant voltage V, a resistor with resistance R, a capacitor with capacitance C, and an inductor with inductance L.

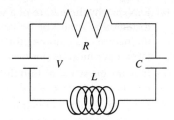

A charge Q will build up on the capacitor according to the equation $V = R(dQ/dt) + Q/C + L(d^2Q/dt^2)$.

(a) Find the function $Q(t)$ for a circuit with $V = 9$, $R = 3$, $C = 1/5$, and $L = 1/4$.

(b) Your solution should consist of a "transient" part that dies out quickly and a "steady-state" part that represents the long-term behavior of the circuit. Discuss the steady-state solution: what is the circuit doing? How could you have predicted this behavior—quantitatively, not just qualitatively—without solving the differential equation?

1.127 Newton's law of gravity says that an object in space falling directly towards Earth will experience an acceleration $r''(t) = -GM_E/r^2$, where G and M_E are constants representing the strength of gravity and the mass of the Earth, and r is the object's distance from the center of the Earth. (We assume here that the object's mass is much smaller than the Earth's.)

(a) Try a power law solution of the form $r(t) = a(t-b)^p$ and plug it in. As always, the result should be an algebraic equation.

(b) The equation you found in Part (a) should have been in the form *something* $(t-b)^{something} = something(t-b)^{something}$. The only way the two sides of that equation can represent the same function is if they both have the same exponent *and* the same coefficient in front, so you can rewrite this as two algebraic equations and find the values of two of the constants in your original guess. Write down the solution $r(t)$ that you find and verify that it does solve the original differential equation.

(c) Does your solution have any arbitrary constants in it? Is it the general solution?

1.128 In the Explanation (Section 1.6.2) we showed that a damped harmonic oscillator is modeled by the equation $ma = -cv - kx$. We then solved this equation for particular values of the constants—values that were chosen to work out nicely. In this problem you will take up the challenge more generally.

(a) Explain why the nature of a spring dictates that k must be positive, and the nature of a damping force dictates that c must be positive.

(b) Plug the guess $x = e^{pt}$ into the differential equation $m(d^2x/dt^2) + c(dx/dt) + kx = 0$.

(c) Solve the resulting equation for p.

(d) If the resulting equation has only one solution for p, the system is said to be "critically damped." What relationship between the constants c, k, and m leads to a critically damped oscillator? Express this relationship by writing a formula for c as a function of k and m.

(e) If c is greater than the formula you found in Part (d), the system is said to be "overdamped." How many real p-values do you find in this case, and are they positive or negative? What sort of behavior results from an overdamped system?

(f) If c is less than the formula you found in Part (d), how many real p-values do you find? What does that suggest about our guess? About the behavior of the system?

The rest of the problems in this section concern themselves with why and when the principle of linear superposition works.

1.129 Use Equation 1.6.4 to determine if each of the differential equations below is linear, and if so whether it is homogeneous. Then test the proposed solutions y_1 and y_2, and also their sum $y_1 + y_2$.

(a) $y'' - y = 0$, $y_1 = e^x$, $y_2 = e^{-x}$.

(b) $y'' - y = 2 \sin x$, $y_1 = e^x$, $y_2 = -\sin x$.

(c) $y'' - yy' = 0$, $y_1 = -2/x$, $y_2 = 2$.

1.130 In this problem you will prove the principle of linear superposition. Because we want a general proof, we begin with the general form for every linear homogeneous differential equation:

$$a_n(x)\frac{d^n y}{dx^n} + \ldots + a_3(x)\frac{d^3 y}{dx^3} + a_2(x)\frac{d^2 y}{dx^2} + a_1(x)\frac{dy}{dx} + a_0(x)y = 0$$
(1.6.11)

(a) Suppose $y = u_1(x)$ is a solution to Equation 1.6.11. Show that the function $y = Au_1(x)$ also works as a solution. (Remember that if A is a constant, $(d/dx)Au_1(x) = A(du_1/dx)$.)

(b) Suppose $u_1(x)$ and $u_2(x)$ are both solutions to Equation 1.6.11. Show that the function $u_1(x) + u_2(x)$ also works as a solution.

(c) Now consider a linear *inhomogeneous* differential equation:

$$a_n(x)\frac{d^n y}{dx^n} + \ldots + a_3(x)\frac{d^3 y}{dx^3} + a_2(x)\frac{d^2 y}{dx^2} + a_1(x)\frac{dy}{dx} + a_0(x)y = f(x) \quad (f(x) \neq 0)$$
(1.6.12)

Suppose $u_h(x)$ is a solution to Equation 1.6.11 and $u_i(x)$ is a solution to Equation 1.6.12. Show that the function $u_h(x) + u_i(x)$ solves Equation 1.6.12 but not 1.6.11.

1.131 To see why the principle of linear superposition only applies to linear equations, consider a non-linear equation of the form

$$a_2(x)\frac{d^2 y}{dx^2} + a_1(x)\frac{dy}{dx} + a_0(x)y^2 = 0 \quad (1.6.13)$$

(a) Suppose $u_1(x)$ and $u_2(x)$ are both solutions to Equation 1.6.13. Show that the function $u_1(x) + u_2(x)$ *does not work* as a solution.

(b) Repeat Part (a) assuming that dy/dx is squared but y is not. Once again assume that $u_1(x)$ and $u_2(x)$ are solutions to this new equation and show that $u_1(x) + u_2(x)$ is not.

1.132 Consider the linear differential equation $x'(t) = x(t)$.

(a) Show that $x(t) = |C|e^t$ is a solution for any value of the constant C.

(b) Since this is a first-order linear differential equation and we have a solution with an arbitrary constant you would normally expect it to be the general solution. Prove that it isn't by finding a solution that doesn't match this for any value of C.

The moral of the story is that it isn't always easy to tell if you have the "general solution" or not. But the truth is, we really had to contrive that example. In practice if you have n arbitrary constants for an nth order linear differential equation, you almost certainly have the general solution. (Be much more careful making generalizations about non-linear equations.)

1.133 In this problem you will show that the principle of linear superposition does not apply to the non-linear differential equation $y'' - y'y = 0$.

(a) What makes this equation non-linear?

(b) Assume you have found two solutions y_1 and y_2. Plug the guess $y = y_1 + y_2$ into the left side of this equation and simplify as much as possible using the fact that y_1 and y_2 are both solutions. What term are you left with that doesn't go away?

(c) Show that $y_1 = 2 \tan x$ is a solution.

(d) Show that $y_2 = 2$ is a solution.

(e) Plug in $y = 2 \tan x + 2$. Show that it doesn't work, and show that the term you are left with is the same term you found in Part (b).

1.134 Consider the equation $dy/dx = 1/y$.

(a) Explain how you can tell this equation is non-linear.

(b) Show that $y = \sqrt{2x + C}$ is a solution.

We have a solution with an arbitrary constant for a first-order equation, so it must be the general solution...right? That's a good rule of thumb but it does not always work, especially for non-linear equations.

(c) Show that $y = -\sqrt{2x + 1}$ is a solution, and that it cannot be written in the form given by Part (b). (This shows that Part (b) was not in fact the general solution.)

1.7 Coupled Equations (see felderbooks.com)

1.8 Differential Equations on a Computer (see felderbooks.com)

1.9 Additional Problems (see felderbooks.com)

CHAPTER 2

Taylor Series and Series Convergence

> *Before you read this chapter, you should be able to...*
>
> - take derivatives using the product rule and chain rule.
> - solve basic differential equations, including the simple harmonic oscillator equation (see Chapter 1).
> - use summation notation (the $\sum$ symbol) to represent series.
>
> *After you read this chapter, you should be able to...*
>
> - find a linear approximation to a given function at a given point, and use that line to find approximate solutions to problems involving that function near that point.
> - find a "Maclaurin series"—a polynomial approximation to a given function near $x = 0$—and use that polynomial to solve problems involving that function for small values of x.
> - find a "Taylor series"—a polynomial approximation to a given function near any given x-value—and use that polynomial to solve problems involving that function for nearby values of x.
> - from one given Taylor series, find many other Taylor series.
> - analyze a sum of infinitely many numbers, known as an "infinite series," to see under what conditions it "converges" (adds up to a finite number). In particular, determine what x-values allow a particular Taylor series to converge.
> - use "asymptotic expansions" to approximate functions with finite sums of divergent series.

Almost any function—there are exceptions, but they are rare—can be rewritten as a polynomial called a "Taylor series." In most cases this series has infinitely many terms. But a few of these terms often make a reasonable approximation to the function you started with, allowing you to find approximate solutions to problems that may be difficult or impossible to solve exactly.

Unfortunately, this technique does not work in every case: sometimes the Taylor series for a function does not approach any finite value as you keep adding more terms. Many Taylor series work for some values of x and not for others. In Sections 2.6–2.7 you will learn how to check whether an infinite series has a finite sum, thus determining when you can use a Taylor series.

2.1 Motivating Exercise: Vibrations in a Crystal

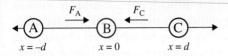

FIGURE 2.1 Nucleus B at $x = 0$ is pushed to the right by nucleus A at $x = -d$, and pushed to the left with equal force by nucleus C at $x = d$.

Consider a nucleus in a one dimensional crystal, held in place by the repulsive forces from its neighbors. This force obeys Coulomb's Law, $|\vec{F}| = (1/4\pi\varepsilon_0)(q_1 q_2/r^2)$, where r is the distance between two nuclei. We'll assume for simplicity that all the nuclei have the same charge so we can combine $q_1 q_2/(4\pi\varepsilon_0)$ into one constant κ and write $|\vec{F}| = \kappa/r^2$ (Figure 2.1).

1. Explain what will happen to nucleus B, assuming the surrounding nuclei are perfectly fixed in position.

Now suppose nucleus B is displaced slightly to a new position x (Figure 2.2).

FIGURE 2.2 When Nucleus B is displaced to the right a distance x, the two forces on it no longer balance.

2. Write the formula for F_A, the force that nucleus A exerts on nucleus B.
3. Write the formula for F_C, the force that nucleus C exerts on nucleus B.
4. Write the formula for the net force on nucleus B. (Be careful about signs!)
5. Recalling that $F_{net} = ma$ where $a = d^2x/dt^2$, write a second-order differential equation to model the motion of nucleus B.
6. Based on your understanding of the physical situation, describe in words the motion that this differential equation should describe for nucleus B.

If you rearrange your answer to Part 5, you should be able to write it as:

$$\frac{d^2 x}{dt^2} = \frac{\kappa}{m}\left(\frac{1}{(d+x)^2} - \frac{1}{(d-x)^2}\right)$$

The good news is, you now have the correct equation to describe the motion of a vibrating nucleus in a crystal. The bad news is, the equation is difficult to solve. (Try it on a computer if you don't believe us!)

Now comes the key step. If we assume that the displacement x is small compared to the distance between nuclei d, we can make the following approximation:

$$\frac{\kappa}{m}\left(\frac{1}{(d+x)^2} - \frac{1}{(d-x)^2}\right) \approx -\frac{4\kappa}{md^3}x \qquad (2.1.1)$$

We are not asking you to prove this approximation—we are waving our hands magically and pulling it out of a hat—although we will ask you, in Part 7, to confirm that it works pretty well.

7. Using the values $\kappa/m = 0.1$ N·m²/kg, $d = 3 \times 10^{-10}$ m, and $x = d/20$, compute both the original function (on the left side of Equation 2.1.1) and the approximation (right side).

8. The new differential equation $d^2x/dt^2 = -(4\kappa/md^3)x$ is much easier. Solve it by inspection. How does the resulting motion compare to your answer to Part 6?

The moral of the story should be clear: some equations are difficult or impossible to solve as given, but can be made reasonable by approximating difficult functions with simpler ones. But where did that particular approximation, $-(4\kappa/md^3)x$, come from? In this chapter you will use "linear approximations" to generate simple approximations to complicated functions, and then "Taylor series," which can give more accuracy when required.

2.2 Linear Approximations

The exercise at the beginning of this chapter illustrates a common problem: we can accurately model the physical situation with an equation, but the equation is difficult to solve analytically. Physicists and engineers often approach such problems with *approximations*, finding simpler problems that are more tractable. One straightforward approach that is surprisingly useful—if you know its limitations!—is a "linear approximation."

2.2.1 Discovery Exercise: Linear Approximations

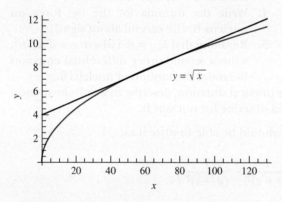

FIGURE 2.3 The function $y = \sqrt{x}$ (blue) and the tangent line to that curve at $x = 64$ (black).

Figure 2.3 shows a plot of a function and a line tangent to it at $x = 64$.

1. Suppose the curve $y = \sqrt{x}$ models some real-world function (Figure 2.3), and we use the tangent line to approximate this function. In general, for what kinds of x-values will this approximation work best? For what kinds of x-values will it work very poorly?

A tangent line, by definition, matches a curve at one point in two ways: it has the same y-value, and it has the same derivative.

2. At the point where $x = 64$, calculate the y-value of the curve, $f(64)$. (The line will have the same y-value at $x = 64$.)
3. At the point where $x = 64$, use a derivative to calculate the slope of the curve, $f'(64)$. (The line will have the same slope.)
4. Based on the point from Part 2 and the slope from Part 3, find the equation of the tangent line.
5. Plug $x = 69$ into both functions: the original curve $y = \sqrt{x}$, and the tangent line. If the tangent line value is used to approximate the real value, what is the percent error? Recall that the formula for percent error is $\left|\frac{\text{(real value)}-\text{(approximation)}}{\text{(real value)}}\right| \times 100$.
6. Use the tangent line to approximate $\sqrt{100}$. What is the percent error this time?
7. Use the tangent line to approximate $\sqrt{64.5}$. What is the percent error this time?

Now we're going to repeat the same exercise, but for a generalized function $y = f(x)$ at a point $x = x_0$.

8. At the point where $x = x_0$, the y-value of the curve—and therefore the line—is $f(x_0)$. The slope of the curve—and therefore the line—is $f'(x_0)$. Find the equation of the line. That is, find a general formula for the only function $y = mx + b$ that has a slope of $f'(x_0)$ and goes through the point $(x_0, f(x_0))$.
9. Plug the point $x = (x_0 + \Delta x)$ into the tangent line equation to calculate its y-value on the line, and therefore approximate its y-value on the curve.

See Check Yourself #8 in Appendix L

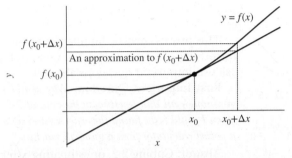

10. When we approximated $\sqrt{69}$, the linear approximation was higher than the actual value. For the curve in Figure 2.4, will the linear approximation be high or low?
11. *In general*, for what kinds of curves will the linear approximation come out high?

FIGURE 2.4 A function $f(x)$ and the tangent line at $x = x_0$.

2.2.2 Explanation: Linear Approximations

In introductory math courses, all the problems have exact answers. If the question is $2x^2 + 5x + 1 = 0$ then the answer is $x = \left(-5 \pm \sqrt{25-8}\right)/4$, exactly.

However, many problems—such as the crystal vibration equation at the beginning of this chapter—do not lend themselves to exact solutions. The bad news is, unsolvable equations come up all the time. The good news is, *approximate* solutions to such equations are often close enough to build bridges, design power plants, or send rockets to the moon.

One approximation technique is a *linear approximation:* we replace a complicated function with the tangent line to that function at a given point. One of the virtues of this technique is that *finding* a tangent line is fast and easy.

> **EXAMPLE** **Finding the Tangent Line to $f(x) = x^3$ at $x = 2$**
>
> 1. $f(2) = 8$ on the curve, and therefore the line must also go through the point $(2, 8)$.
> 2. $f'(2) = 12$ on the curve, and therefore the line must rise with a slope of 12. This means that the line goes *up* 12 units every time it moves 1 unit to the *right* of $x = 2$.
> 3. Therefore, at every point on the line, $y = 8 + 12(x - 2)$. If you want, you can rewrite this in the more familiar $y = mx + b$ form: in this case, $y = 12x - 16$.

Finding a tangent line is not a pointless geometrical exercise. The function $y = 12x - 16$ can serve as a stand-in for the function $y = x^3$, as long as you are working with x-values close to 2.

EXAMPLE Using the Tangent Line to Estimate the Value $(2.2)^3$

For x-values close to 2, the function $y = x^3$ can be approximated by the function $y = 12x - 16$. Plugging $x = 2.2$ into this line yields $y = 10.4$. We therefore conclude that $(2.2)^3$ is close to—but not exactly—10.4 (Figure 2.5).

The actual value is 10.648. If you use the linear approximation to find $(2.1)^3$ it gives you 9.2, compared to the actual value of 9.261. The closer you are to $x = 2$, the better the approximation works.

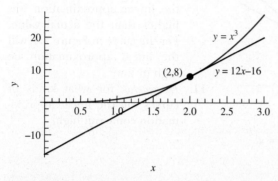

FIGURE 2.5 The function $y = x^3$ and the tangent line at $x = 2$.

This process may be familiar from your introductory calculus class, but we want to ask two important questions about it.

First Important Question: *Why would I go through all that to estimate the cube of 2.2, when I could have found the exact answer with a pocket calculator from a gumball machine?*

Answer: Cubing 2.2, or estimating $\sqrt{69}$ as you did in Exercise 2.2.1, are not real-world problems: they are trivial examples meant to introduce you to this important technique. The real importance of this technique is to replace complicated functions in the middle of difficult problems, such as the one in the Motivating Exercise (Section 2.1). In the problems you will use linear approximations to approximately solve some real, useful problems, as well as doing some "meaningless" examples like the one we did above for practice.

Second Important Question: *Why does a tangent line serve as a useful approximation to a curve, near the point of tangency?*

You've probably never wondered about that question, because it looks so obvious on a picture. The drawing above clearly shows that, for x-values close to 2, the function $y = 12x - 16$ is pretty darn close to $y = x^3$. But we urge you *not* to be satisfied with that explanation. There is a solid mathematical reason why tangent lines work well for this purpose. Understanding that reason may help deepen your understanding of the derivative itself, and can also help you to make more accurate, less obvious approximations.

Why the Tangent Line Approximates a Curve

Consider the function $f(x) = x^3$ at $x = 2$. Obviously, $f(2) = 8$. In the example above, we wanted to calculate $(2.2)^3$; that is, we wanted to know how much the function goes up as x goes from 2 to 2.2. Since we wanted to know how fast the function is changing, we looked at the derivative.

Since $f'(x) = 3x^2$, we know that $f'(2) = 12$. That's an easy calculation to do, but what does it really mean? It means that, when $x = 2$, the function is rising at a *rate* of 12 y-units per x-unit. If x advances by 3, y goes up by 36. Each time x advances by 1/10, y goes up by 1.2. *In general, if x advances by Δx, then y goes up by $12\Delta x$.*

That last sentence is so important to this whole process that we italicized it. Unfortunately, it's also a lie—because the slope of $y = x^3$ isn't a constant. The farther you get from $x = 2$, the farther the slope drifts away from 12. If the slope did stay 12 forever, our function would be a line. (Think about it.)

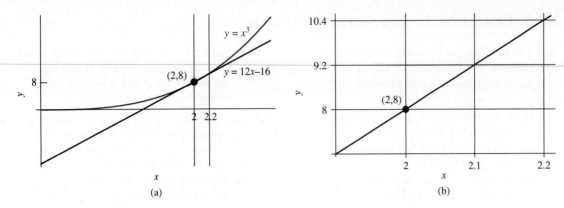

FIGURE 2.6 A tangent line approximation to $y = x^3$ (a) and a close-up view of the tangent line (b).

The left side of Figure 2.6 shows two different curves. The function $y = x^3$ goes through the point $(2, 8)$ with a slope of 12, but gets steeper as you move to the right. Its tangent line meets the curve at $(2, 8)$, and rises with a constant slope of 12. As the tangent line moves 0.2 to the right, it moves up $0.2 \times 12 = 2.4$, ending at the point $(2.2, 10.4)$.

Rising in this way, the black line goes through the point $(2.2, 10.4)$, as shown in the zoomed view on the right side of Figure 2.6. Rising in a similar way, the blue curve should come close to this point—which is to say, $(2.2)^3$ should be roughly 10.4, so that was our approximation. The same reasoning leads us to the equation for the line: instead of $x = 2.2$ we ask the same question about an arbitrary x-value. Rising from $y = 8$, the line rises with a slope of 12 as it moves to the right by a distance of $(x - 2)$, so it reaches the value $y = 8 + 12(x - 2)$. Remember that this formula—not the specific number 10.4, but the linear function itself—is the useful result we are looking for.

We are now ready to answer the question we started with: why is a tangent line a good approximation to a curve? It's because the tangent line starts at the same point as the curve, and rises at the same rate. In this example, both the curve and the tangent line move away from the point $(2, 8)$ with a slope of 12, so as x moves from 2 to 2.2, they each go up by roughly 2.4.

Why the Tangent Line Doesn't Get It Exactly Right

Above we used our linear approximation to estimate $(2.2)^3 \approx 10.4$. But we know this is not exactly correct, and we know why: it's because the slope of $y = x^3$ changes as we move to the right. If we want to know how far off our estimate is, we need to know *how much* the slope changes—and that question points us toward the *second* derivative.

$f''(x) = 6x$, so $f''(2) = 12$. A positive second derivative means the slope of the curve is increasing, which is why the actual cube of 2.2 is a little higher than our estimate, as we saw. You can reach the same conclusion visually: Figure 2.6 shows that the linear approximation undershoots because the curve is concave up.

The key point here is that we could make our answer more accurate if we factored in the way the derivative changes—that is, if we also considered the second derivative. In Problem 2.29 you'll use the second derivative of a function to make a *parabolic* approximation that is more accurate than the linear approximation. In Section 2.3 you'll continue the process, using higher derivatives to make ever-more-accurate polynomial approximations to functions.

2.2.3 Problems: Linear Approximations

2.1 The function $y = L(x)$ goes through the point $(0, 20)$ with a slope of $1/2$.
  (a) If the slope stayed $1/2$ forever, what would $L(1)$ be?
  (b) If the slope stayed $1/2$ forever, what would $L(-1)$ be?
  (c) If the slope stayed $1/2$ forever, what would $L(6)$ be?
  (d) If $L''(0) = 4$, so the slope does *not* in fact stay $1/2$ forever, which if any of your estimates in parts (a)–(c) would you expect to come out higher than the actual values? Explain.

2.2 The function $y = f(x)$ goes through the point $(10, 100)$ with a slope of -3.
  (a) Assuming the slope stays -3, what will the y-value be when $x = 11$?
  (b) Assuming the slope stays -3, what will the y-value be when $x = 12$?
  (c) Assuming the slope stays -3, what will the y-value be when $x = 20$?
  (d) Now, if $f''(10) < 0$, so the slope does not stay -3 forever, would you expect your estimates in parts (a)–(c) to be high or low?
  (e) Under what circumstances would your expectation in Part (d) be incorrect?

2.3 The functions $f(x)$ and $g(x)$ have the same tangent line at point P, where $x = a$ (Figure 2.7).

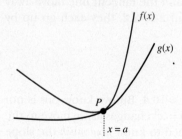

FIGURE 2.7

  (a) Copy this simple drawing into your homework, and draw in the tangent line at point P.
  (b) Is the information given in the problem enough for us to positively determine that $f'(a) = g'(a)$?
  (c) If you used the linear approximation to estimate $f(b)$ where b is an x-value near a, would your estimate come out high, or low?
  (d) For which function (f or g) does the linear approximation work better?
  (e) Which function (f or g) has a higher *second* derivative at point P? How does your answer to this question relate to your answers to Parts (c) and/or (d)?

2.4 $y = f(x)$ and $y = g(x)$ are two different non-linear functions, but they have something in common: their linear approximations at $x = 1$ are both the line $y = 2x$.
  (a) Draw two possible curves that could be $y = f(x)$ and $y = g(x)$.
  (b) Write one possible (nonlinear!) function that could be $y = f(x)$.

2.5 At time $t = 0$ the temperature is $60°C$ and at time $t = 2$ the temperature is $60.06°C$. Estimate the temperature at $t = 10$.

2.6 Suppose you are going to replace the function e^x with its linear approximation at $x = 0$.
  (a) Draw the function $y = e^x$ and the line tangent to it at $x = 0$.
  (b) What is the slope of the curve at that point?
  (c) What is the equation of the line?
  (d) Use the line to approximate $1/e$, $\sqrt{e}$, e, e^3, and e^6.
  (e) Use a calculator or graphing software to answer the question: how high does x have to become before the percent error in the approximation is greater than 20%?

2.7 In this problem we are going to estimate $\sqrt{23}$ by a linear approximation, but we are going to derive the linear approximation in two slightly different ways. In both cases, let $f(x) = \sqrt{x}$.
  (a) Calculate $f(25)$.
  (b) Calculate $f'(25)$.
  (c) Now: assume the slope *stays* at the value you calculated in Part (b).
    i. If x increases by 1, then how much does y increase by?
    ii. If x decreases by 1, then how much does y decrease by?
    iii. If x increases by Δx, then how much does y increase by?
    iv. Based on your answer to Parts (a) and (c)iii, estimate $f(23)$.
  (d) Let $g(x) = mx + b$ be the linear function such that $g(25)$ matches your answer to Part (a), and $g'(25)$ matches your answer to Part (b). Find the function $g(x)$.

(e) Use $g(23)$ to estimate $f(23)$.

(f) Your answers to Parts (c)iv and (e) should be exactly the same as each other. However, they are not exactly $\sqrt{23}$. Are they higher or lower than the actual value? Why?

In Problems 2.8–2.15, use a linear approximation at the *first* given point to estimate the value of the function at the *second* given point. (In each case, you may want to check the answer on your calculator. Was your estimate close? Was your estimate a bit high for concave down functions and a bit low for concave up functions?)

2.8 $y = x^2$ at $x = 10$ to estimate $(12)^2$

2.9 $y = 100/x$ at $x = 10$ to estimate $100/12$

2.10 $f(x) = (x^2 + 2x + 4)^5$ at $x = -1$ to estimate $f(-2)$

2.11 $y = \sin x$ at $x = 0$ to estimate $\sin(0.2)$

2.12 $y = \sin(2x)$ at $x = 0$ to estimate $\sin(0.2)$

2.13 $y = \ln x$ at $x = 1$ to estimate $\ln(1.1)$

2.14 $y = \ln x$ at $x = e$ to estimate $\ln 3$

2.15 $y = (\ln x)/x$ at $x = e$ to estimate $\ln 3$

2.16 Let $f(x) = \sin x$.
  (a) Use a linear approximation about $x = 0$ to estimate $\sin 3$.
  (b) Use a linear approximation about $x = \pi$ to estimate $\sin 3$.

2.17 (a) Use a linear approximation to the function $f(x) = xe^{x^2}$ at $x = 2$ to estimate $f(2.5)$.
  (b) Find the actual $f(2.5)$ on a calculator. The estimate is, in this case, nowhere near the actual answer. What happened? *Hint*: Find $f''(2)$.

2.18 You cannot use a linear approximation for the function $y = \sqrt[3]{x}$ about the point $x = 0$. Explain.

2.19 Find the linear approximation to the circle $x^2 + y^2 = 25$ at the point $(3, 4)$.

2.20 The "Fresnel integral," used in diffraction theory, is defined by the following two properties: $dC/dx = \cos(\pi x^2/2)$, and $C(0) = 0$. It is impossible to write a simple function in closed form that has these two properties, so approximation techniques are required.
  (a) Based on the two given properties, use a linear approximation to estimate $C(1/8)$.
  (b) Does your estimate fall above, or below, the actual $C(1/8)$? (Don't just guess: find the answer mathematically.)

2.21 Figure 2.8 depicts a familiar situation: a mass m swings at the end of a massless rope of length L, subject to the force of gravity and the tension force from the string. It can be shown that this pendulum will obey the differential equation $d^2\theta/dt^2 + (g/L)\sin\theta = 0$. (You can work that out yourself if you've taken an introductory mechanics class.) Unfortunately, this equation is not easy to solve!

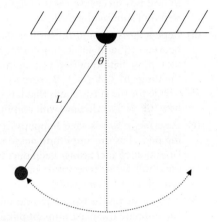

FIGURE 2.8 Simple pendulum.

  (a) Find the function that serves as a linear approximation to $\sin\theta$ around $\theta = 0$.
  (b) Replace the function $\sin\theta$ with its linear approximation.
  (c) Solve the resulting differential equation.
  (d) Describe in words the motion resulting from your solution.
  (e) For what kinds of θ-values is your solution valid?

2.22 [*This problem depends on Problem 2.21.*] A grandfather clock uses a pendulum similar to the one described in the previous problem to keep time. The pendulum swings in a narrow box so that it always has a small angle θ. The box is designed this way so that the linear approximation will be accurate. In this problem you will estimate how small those oscillations need to be. The parameter g is 9.8 m/s^2 and you can take $L = 1$ m.
  (a) Using your solution from the previous problem, find the period of the pendulum for small oscillations. You should find that your answer is independent of the amplitude, as long as it is small enough that your linear approximation can be considered valid.

(b) Use a computer to solve the original (not approximate) equation $d^2\theta/dt^2 + (g/L)\sin\theta = 0$ with initial conditions $d\theta/dt = 0$, $\theta = 5°$ and plot the result. (*Caution: Many computer programs expect angles to be in radians, so you may have to convert before entering θ.*) Use trial and error to find a final time that allows you to see at least two full oscillations.

(c) Have the computer find the period of the solution you found in Part (b). You should be able to do this by asking it to find two successive roots of the solution and subtracting them.

(d) Repeat Parts (b) and (c) for initial angles between 10° and 45° in steps of 5°. Make a plot of oscillation period vs. amplitude in the range $5° \leq A \leq 45°$. Be sure to choose a scale for the vertical axis that lets you see how the period changes with amplitude.

(e) Describe in words what happens to the period as the amplitude decreases. Does it keep increasing, keep decreasing, level off at some value, or do something else entirely?

(f) Using your results from Parts (a) and (d) estimate the maximum amplitude that a grandfather clock could swing at if you wanted its period of oscillation to be within 1% of the one given by the linear approximation.

2.23 The "Coleman–Weinberg equation" $d^2\phi/dt^2 = -\lambda\phi^3 \ln(\phi/v)$ plays an important role in some theories of particle physics.[1] The quantity $\phi(t)$ describes a "homogeneous scalar field." (Don't worry; you don't need to have any idea what that means to solve this problem.) The parameters λ and v are both positive constants.

(a) If you start with $\phi = v$, $d\phi/dt = 0$, will $d^2\phi/dt^2$ be positive, negative, or zero? Based on that answer, how will $\phi(t)$ behave with these initial conditions?

(b) If you start with $\phi > v$, $d\phi/dt = 0$, will $\phi(t)$ start increasing, start decreasing, or stay constant?

(c) If you start with $\phi < v$, $d\phi/dt = 0$, will $\phi(t)$ start increasing, start decreasing, or stay constant?

(d) Based on your answers so far, explain why, for initial conditions near $\phi = v$, you would expect $\phi(t)$ to oscillate about the point $\phi = v$.

Although $\phi(t)$ does oscillate, it is not a simple harmonic oscillator. To solve for it you need to use an approximation. Since $\phi(t)$ oscillates about $\phi = v$ you should expand about that point.

(e) Find a linear approximation to the right-hand side of the Coleman–Weinberg equation valid near $\phi = v$.

(f) Solve the resulting, simplified differential equation. You can do this with a computer or by hand. If you do it by hand you may find it useful to start by defining a new variable $u = \phi - v$ and plugging this into the differential equation.

(g) Find the period for small oscillations of a Coleman–Weinberg field.

(h) Explain why this period is only correct for small oscillations.

2.24 Consider the equation $dy/dt = (1 + \sin y)^{-2}$.

(a) Replace the right-hand side with its linear approximation around $y = 0$ and have a computer solve the resulting differential equation *analytically* with initial condition $y(0) = 0$.

(b) Numerically solve the original equation with the same initial condition and plot this together with your approximate solution out to time $t = 5$. You should find that the two solutions do *not* match well at late times.

(c) Explain why the linear approximation that you used was not a good approximation for this problem. (*Hint: Look at how y behaves at late times.*) Don't just say that the two solutions in Part (b) looked different; explain why the linear approximation, which works so well for many problems, was not very accurate for this one.

2.25 Consider the equation $dy/dt = 2 - e^y$.

(a) Replace the right-hand side with its linear approximation around $y = 0$ and have a computer solve the resulting differential equation *analytically* with initial condition $y(0) = 0$.

(b) Numerically solve the original equation with the same initial condition and plot this together with your approximate solution out to time $t = 5$. You

---

[1] S. R. Coleman and E. Weinberg, "Radiative Corrections As the Origin of Spontaneous Symmetry Breaking," Phys. Rev. D 7, 1888 (1973).

should find that the two solutions do *not* match well at late times.

The asymptotic value of y at late times (also known as the "steady-state" value) is the value of y such that $dy/dt = 0$.

(c) Find the steady-state value of y.

(d) Find a linear approximation to the right-hand side of the differential equation about the steady-state y-value.

(e) Now have the computer analytically solve the differential equation using this new linear approximation (still with initial condition $y(0) = 0$).

(f) Plot the three solutions together: the numerical solution to the original equation and the analytical solutions using the two different linear approximations.

(g) In which region of the plot does the approximation from Part (a) work better? In which region of the plot does the approximation from Part (d) work better? Why? Under what circumstances would each of these be a more useful approximation to use for a physical problem?

2.26 Consider the differential equation $d^2f/dt^2 = 2\left(e^{-3f} - 1\right)$.

(a) Replace the right-hand side with its linear approximation around $f = 0$ and solve the resulting differential equation. (You can do this by hand or with a computer.) Solve for the arbitrary constants with initial conditions $f(0) = 0$, $f'(0) = 0.2$.

(b) Numerically solve the original differential equation (without the approximation) using those same initial conditions. Plot the approximate and exact (numerical) solutions on the same plot, clearly marking which is which. Use trial and error to find a final time that allows you to see at least two full oscillations.

(c) Repeat Parts (a) and (b) for initial conditions $f(0) = 0$, $f'(0) = 2$.

(d) Describe in words the differences between your plots in Parts (b) and (c). In which case was the linear approximation more accurate? Why?

2.27 The linear approximation to the function $f(x)$ at the point $x = a$ is used to approximate the value of $f(b)$ where $b \neq a$. Write the general formula for the approximate value of $f(b)$.

2.28 *[This problem depends on Problem 2.27.]* Figure 2.9 shows two different functions $f(x)$ and $g(x)$, with $f(a) = g(a) = 0$. By replacing both functions with their linear approximations, derive l'Hôpital's rule, which states that $\lim\limits_{x \to a} \dfrac{f(x)}{g(x)} = \lim\limits_{x \to a} \dfrac{f'(x)}{g'(x)}$.

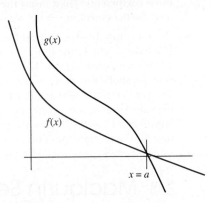

FIGURE 2.9

2.29 Exploration: Beyond Linear Approximations
A linear approximation of a function $y(x)$ at $x = a$ has the form $y = mx + b$. It matches the original function in two ways at $x = a$: it has the same y-value, and the same derivative. Generally, the farther x is from a, the worse the approximation.

In this problem you are going to find an approximation that for most functions is accurate over a wider range of values of x. The new approximate function is a parabola ($y = ax^2 + bx + c$) that matches the original function in three ways at a given point: same y-value, same first derivative, and same *second* derivative. Our example will return to our first linear approximation: approximating $y = \sqrt{x}$ around $x = 64$.

(a) Our "mystery parabola" must, of course, go through the same point as the original function. Write an equation that says "the function $y = ax^2 + bx + c$ goes through the point $(64, 8)$." Your equation will express a relationship between the coefficients a, b, and c.

(b) Our parabola must also have the same *slope* (first derivative) as the original function. Calculate the slope of $y = \sqrt{x}$ when $x = 64$, and calculate the slope of $y = ax^2 + bx + c$ when $x = 64$. When you set these two equal to each other, you get another relationship between the coefficients.

(c) Our parabola must also have the same *concavity* (second derivative) as the original function. Calculate the second derivative of $y = \sqrt{x}$ when $x = 64$, and calculate the second derivative of $y = ax^2 + bx + c$

when $x = 64$. When you set these two equal to each other, you get a third relationship between the coefficients.

(d) Solve the three equations to find the three coefficients. Then insert their values into the equation $y = ax^2 + bx + c$.

(e) Draw a graph showing three different functions: the original function $y = \sqrt{x}$, the linear approximation $y = (1/16)x + 4$, and your parabolic function $y = ax^2 + bx + c$. You can sketch these by hand, use a graphing calculator, or use a computer. Which approximation—the line, or the parabola—best fits the original function near $x = 64$?

(f) Plug $x = 69$ into the original curve $y = \sqrt{x}$ and the line $y = (1/16)x + 4$. If the linear approximation is used to approximate the real value, what is the percent error? Recall that the formula for percent error is $\left| \frac{\text{(real value)}-\text{(approximation)}}{\text{(real value)}} \right| \times 100$.

(g) Now plug $x = 69$ into both $y = \sqrt{x}$ and your parabolic approximation, and calculate the percent error. Is it higher or lower than the error for the line?

(h) Use your parabola to approximate $\sqrt{100}$. What is the percent error this time?

2.3 Maclaurin Series

If you have a function $f(x)$ and you are only interested in its behavior around $x = 0$, you can often replace $f(x)$ with its linear approximation.

$$f(x) \approx f(0) + f'(0)x$$

In Problem 2.29 you showed that you can get a more accurate approximation by using a quadratic function. We could use the same technique to arrive at the formula:

$$f(x) \approx f(0) + f'(0)x + \frac{1}{2}f''(0)x^2$$

In general, the more powers of x you add, the more accurate your approximation will be around $x = 0$. As you keep adding terms you will, in many circumstances, approach the exact value of $f(x)$. In the exercise below, you will find a polynomial approximation of arbitrary order for a function near $x = 0$. In Section 2.4 we'll generalize this procedure to find a polynomial approximation about values of x other than $x = 0$.

2.3.1 Discovery Exercise: A Polynomial Equivalent of the Sine

An infinitely long polynomial is referred to as a "power series":

$$c_0 + c_1 x + c_2 x^2 + c_3 x^3 + \ldots$$

Our goal in this exercise is to write a power series that equals $\sin x$ for all values of x: that is,

$$\sin x = c_0 + c_1 x + c_2 x^2 + c_3 x^3 + \ldots \tag{2.3.1}$$

In writing Equation 2.3.1, we have not actually solved anything. Equation 2.3.1 is merely the assertion that there exists a power series expansion of the function $\sin x$, if we can only find the right coefficients. In finding these coefficients for the sine function, we will develop a general process for finding the power series coefficients for any given function for which such a power series exists.

We start with a simple observation: if the equation is true for all x-values, then it must certainly be true for $x = 0$.

2.3 | Maclaurin Series

1. Plug the value $x = 0$ into both sides of Equation 2.3.1. Note that all the terms on the right except the first one vanish, enabling you to solve for the first coefficient, c_0. (As one student wryly observed, "One down, infinity to go.")
2. Now, take the derivative of both sides of Equation 2.3.1. (If two functions are equal everywhere, their derivatives must also be equal.)
3. Plug the value $x = 0$ into both sides of the equation you wrote down in the previous step. Once again, almost all the terms on the right disappear, enabling you in this case to solve for the second coefficient, c_1.
4. Repeat steps 2 and 3: take the derivative of both sides, and then plug $x = 0$ into both sides, until you have found all the coefficients through c_8.

If you've done everything correctly so far your first few terms should be:

$$\sin x = x - \frac{x^3}{3!} + \frac{x^5}{5!} + \dots$$

and you should have found at least one more term after that. The first term $f(x) = x$ is the linear approximation to $\sin(x)$ at $x = 0$. When you include the first two terms, $f(x) = x - x^3/(3!)$ should be a *better* approximation of $\sin(x)$ near $x = 0$. And $f(x) = x - x^3/(3!) + x^5/(5!)$ should be better still, and so on. For the function $\sin x$, and for many (though not all!) functions, we can continue adding terms to estimate the function as accurately as we need.

We refer to these successive approximations as "partial sums" of the series, and write:

first partial sum: $S_1 = x$
second partial sum: $S_2 = x - x^3/(3!)$
third partial sum: $S_3 = x - x^3/(3!) + x^5/(5!)$

Let's see how these partial sums work as approximations for the sine function.

5. Fill in all the numbers in the following table to at least five decimal places accuracy. These values of x are in radians.

x	$\sin x$	$S_1(x)$	$S_2(x)$	$S_3(x)$	$S_4(x)$
1					
2					
3					

6. Answer in words: looking at the *first row* in your table (for $x = 1$), what do we learn about successive sums of this series?
7. Answer in words: looking at *all three rows* of your table, what do we learn about this series as x gets farther from zero?
8. 💻 Now, use graphing software, a spreadsheet, or a graphing calculator to make *graphs* of all five functions: $\sin x$, $S_1(x)$, $S_2(x)$, $S_3(x)$, and $S_4(x)$, on the interval $-\pi \leq x \leq \pi$.
9. 💻 Answer in words: what do the five graphs tell us about successive sums of this series?

2.3.2 Explanation: Maclaurin Series

We have seen how to write a linear function $y = c_0 + c_1 x$ that will approximate a given function $f(x)$ near $x = 0$. As we noted previously, you can approximate $f(x)$ more accurately by writing a quadratic function.

$$f(x) \approx c_0 + c_1 x + c_2 x^2$$

Chapter 2 Taylor Series and Series Convergence

You can keep going, writing a third-order polynomial, then fourth order, and so on. If you want the polynomial to be within 0.01 of the correct value at $x = 1$, you should be able to get there by adding enough terms. If you want the polynomial to be within 0.001 of the correct value, you need to add more terms. In the *limit* as you add infinitely many terms, the "power series"

$$f(x) = c_0 + c_1 x + c_2 x^2 + c_3 x^3 + c_4 x^4 + \ldots$$

with appropriately chosen coefficients c_n will usually approach the original function exactly.

Can you find such a polynomial series, and use it to approximate a function to any desired degree of accuracy? The answer is…for some functions you can, and for some functions you can't. In this section, we discuss a method for answering the question: "If this function has a polynomial representation, what is it?" We discuss a bit later the question of: "For what functions does this work?"

Maclaurin Series

When we replace a function with a power series as described above, that infinite polynomial is called a "Maclaurin series." Every Maclaurin series has the form:

$$c_0 + c_1 x + c_2 x^2 + c_3 x^3 + c_4 x^4 + \ldots$$

What changes from one function to another are the coefficients c_0, c_1, and so on. In the following example, we are looking for the Maclaurin series for the function $y = 1/(1-x)$. The strategy goes as follows.

1. In the first step, we write $1/(1-x) = c_0 + c_1 x + c_2 x^2 + c_3 x^3 + c_4 x^4 + \ldots$. This step reflects nothing more than the unproven assumption that there exists a power series that equals $1/(1-x)$. As we go through the process, we will determine what the coefficients in this series must be, *if* that assumption is correct.
2. At this point, we plug $x = 0$ into both sides of the equation. On the right side, all the terms vanish except for the first (constant) term. We can therefore solve for that term.
3. To move to the next step, we take the derivative of both sides of the equation. (If two functions are the same, their derivatives must be the same.) Then, plugging in zero to both sides again, we find the next coefficient, and so on.

EXAMPLE **Building a Maclaurin Series**

Each equation is found by taking the derivative of the one above it.

Equation	Plug in $x = 0$
$\dfrac{1}{1-x} = c_0 + c_1 x + c_2 x^2 + c_3 x^3 + c_4 x^4 + \ldots$	$c_0 = 1$
$\dfrac{1}{(1-x)^2} = c_1 + 2c_2 x + 3c_3 x^2 + 4c_4 x^3 + \ldots$	$c_1 = 1$
$\dfrac{2}{(1-x)^3} = 2c_2 + (2 \times 3)c_3 x + (3 \times 4)c_4 x^2 + \ldots$	$c_2 = 1$
$\dfrac{2 \times 3}{(1-x)^4} = (2 \times 3)c_3 + (2 \times 3 \times 4)c_4 x + \ldots$	$c_3 = 1$
$\dfrac{2 \times 3 \times 4}{(1-x)^5} = (2 \times 3 \times 4)c_4 + \ldots$	$c_4 = 1$

By this point, the pattern is clear: all the coefficients come out 1. Plugging these coefficients into the original equation, we conclude that:

$$\frac{1}{1-x} = 1 + x + x^2 + x^3 + x^4 + \ldots$$

By determining the coefficients, we have found the Maclaurin series for the function $1/(1-x)$.

So Now That We've Done It, What Did We Do?
Even if you followed all the steps in that example, you may be mystified by the final result:

$$\frac{1}{1-x} = 1 + x + x^2 + x^3 + x^4 + \ldots \tag{2.3.2}$$

We can understand the meaning of such an equation in several different ways.

We can interpret Equation 2.3.2 as a *generalization about numbers* because it represents in one equation many different numerical statements. For instance, you can plug in $x = 1/3$ to get:

$$1.5 = 1 + (1/3) + (1/3)^2 + (1/3)^3 + \ldots \tag{2.3.3}$$

We will discuss the idea of infinite series more formally in Section 2.6 but one key point must be made from the outset. Equation 2.3.3 does *not* mean "when we finally add the infiniti-eth term, it will all add up to 1.5"; we never will, and it never will. Rather, this equation—and any infinite series—must be viewed as a limit, a limit of "partial sums." In this case:

first partial sum:	$S_1 = 1$	$= 1$
second partial sum:	$S_2 = 1 + (1/3)$	$= 1.33$
third partial sum:	$S_3 = 1 + (1/3) + (1/3)^2$	$= 1.44$
fourth partial sum:	$S_4 = 1 + (1/3) + (1/3)^2 + (1/3)^3$	$= 1.48$

and so on. The partial sums collect more and more terms as you go. Equation 2.3.3 is telling us that as we keep adding terms, the partial sums approach 1.5.

Similarly, plugging $x = -0.23$ into Equation 2.3.2 gives us:

$$\frac{1}{1+0.23} = 1 - 0.23 + (0.23)^2 - (0.23)^3 + (0.23)^4 + \ldots$$

Once again, we have a limit; if you add up the first hundred terms you will get S_{100}, which will be very close to $1/1.23$. The original power series uses an x to indicate that many different numbers can be substituted into the equation.

But we can also look at Equation 2.3.2 as describing a *function*. Remember that our interest in series began by looking for a way to replace one function (such as the complicated function in the vibrating crystal example) with another function that is easier to work with. We now know that the function $1/(1-x)$ can be reasonably replaced with S_1, the linear function $1 + x$. But S_2, the quadratic function $1 + x + x^2$, approximates the function more closely, and S_3 $(1 + x + x^2 + x^3)$ works better still, and so on. This insight suggests a general strategy: we can replace a difficult function with a polynomial, using as many terms as needed to obtain the desired accuracy (Figure 2.10).

You may have already noticed one obvious limitation: we could not have used this process at all if our original function, $1/(1-x)$ in this example, was not differentiable at $x = 0$.

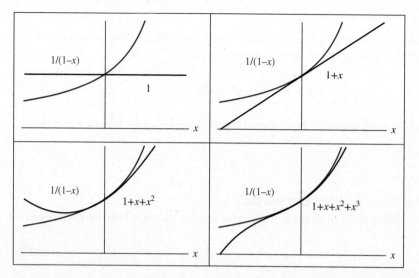

FIGURE 2.10 The plot of $1/(1-x)$ (blue) along with the first four partial sums of its Maclaurin series (black).

(You can't make a Maclaurin series for $|x|$, for example.) Below we address other limitations of this process. It turns out, however, that most functions that arise in engineering and physics problems can successfully be replaced with polynomials, making this one of the most common and powerful approximation techniques.

We should note one peculiar feature of this particular example. If you closely followed our algebra in "Example: Building a Maclaurin Series" above, you may have noticed factorials such as $2 \times 3 \times 4$ appearing on both sides of the equation. When you go through this process you will always find factorials on the right side; the 4th derivative of x^4 is 4! for instance. But in this unusual case the factorials also appeared on the left side, because the 4th derivative of $1/(1-x)$ is $4!/(1-x)^5$. The factorials therefore canceled on both sides, leaving very simple coefficients. For most functions this will not occur, so most Maclaurin series will have factorials in their coefficients.

When Does That Polynomial Work?

The polynomial $1 + x + x^2 + x^3 + x^4 + \ldots$ does approach the value of the function $1/(1-x)$ for $x = (1/3)$, and for many other x-values. But what happens if we plug in $x = 2$? Then the equation becomes:

$$-1 = 1 + 2 + 4 + 8 \ldots$$

We don't need any fancy limit-manipulation to tell us that this equation is absurd. The series $1 + 2 + 4 + 8 + \ldots$ clearly does not approach -1 (or any other finite number). So we see that our original equation works for some x-values, but not all of them.

More generally, a power series of the form $c_0 + c_1 x + c_2 x^2 + c_3 x^3 + \ldots$ will work best for "small" x-values, $x \approx 0$, because those values make the later terms diminish in significance quickly. (0.1^5 is a lot less than 0.1.) In the next section, we will address this problem by generalizing from Maclaurin series, which work best near zero, to "Taylor series," which can generally be created to work best near any chosen number.

Still later, we will take up the question of determining for which x-values a particular series works at all. A series can "not work" in the sense of blowing up (as in the example above), or—in rare cases—a power series can even add up to the wrong function. In Problem 2.52 you will see an example of the latter situation.

We will not take up the subject of proving that the power series you found is equivalent to the function you started with. We will confine our discussion to the following simple rule.

2.3 | Maclaurin Series

In the domain you are interested in, if the original function is everywhere differentiable (which is usually easy to check), and the power series you generate doesn't blow up (we'll see how to check that in Section 2.7), then the series *almost always* adds up properly to the original function.

How Small is Small Enough?

We said above that Maclaurin series work best for "small" x-values. It's tempting to suggest a rule of thumb such as "numbers less than 1 are small," but that immediately begs a question of units: we can't say that 8 hours is "big" and 1/3 of a day is "small," since they both represent the same amount of time. A time period is only "small" or "big" compared to another time period.

To take a more concrete example, let's return to our Motivating Exercise (Section 2.1): a nucleus in a crystal, displaced by a distance x from the midpoint of its two neighbors (Figure 2.11).

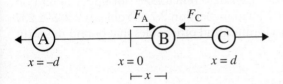

FIGURE 2.11 Nucleus B feels unequal forces from A and C.

Nucleus B moves according to the differential equation $d^2x/dt^2 = (\kappa/m)\left(1/(d+x)^2 - 1/(d-x)^2\right)$, which has no simple solution, but we can write a much simpler differential equation—one that we can actually solve—if we replace that function with an approximation for small values of x. We judge x as big or small *compared to d, the distance between nuclei*. If $x \ll d$, nucleus B feels almost perfectly balanced forces from A and C. As $x \to d$, nucleus B gets very close to nucleus C, and will feel a tremendous push to the left. (To an engineer or physicist, the symbols $\gg$ and $\ll$ mean "much greater than" and "much less than." To a mathematician, vague definitions like that are what's wrong with engineers and physicists.)

Because we're interested in how big x is compared to d, we work with the unitless quantity x/d. Now it doesn't matter if we're working in angstroms or light-years: if x/d is a lot closer to 0 than 1, the displacement is "small" in a physically meaningful sense. We can count on $(x/d)^n$ shrinking rapidly for large values of n.

In order to get there, note that we can rewrite $(d+x)$ as $d(1+x/d)$, and therefore $(d+x)^2$ as $d^2(1+x/d)^2$. So we can factor $1/d^2$ out of both parenthesized terms, rewriting the differential equation as:

$$\frac{d^2x}{dt^2} = \frac{\kappa}{md^2}\left(\frac{1}{(1+x/d)^2} - \frac{1}{(1-x/d)^2}\right)$$

It looks ugly, but it's not particularly difficult to work with. Since we're going to now treat x/d as our variable, let's call that u, and use it to build a Maclaurin series of the function in parentheses above. We're only going as far as the first derivative—the linear approximation—which turns out to be enough for most problems.

Equation	Plug in $u = 0$
$\dfrac{1}{(1+u)^2} - \dfrac{1}{(1-u)^2} = c_0 + c_1 u + \ldots$	$c_0 = 0$
$\dfrac{-2}{(1+u)^3} - \dfrac{2}{(1-u)^3} = c_1 + \ldots$	$c_1 = -4$

We conclude that for small values of u we can replace $1/(1+u)^2 - 1/(1-u)^2$ with $-4u$ which is $-4x/d$, making our differential equation:

$$\frac{d^2x}{dt^2} = -\frac{4\kappa}{md^3}x$$

You may recognize this as the equation that we presented without justification in the motivating exercise.

The point here is that you never declare a physical quantity with units to be "small" in some absolute sense: it is only meaningful to say that "quantity a is small *compared to* quantity b," where b has the same units as a. You can then describe the *ratio* a/b as small, which is often the first step in finding a useful approximation to a physical equation.

In many instances, though, we need something more specific than "x is much smaller than d." Suppose you need a given approximation to within 1%. Once you've estimated a function with a partial sum of a Maclaurin series, how can you know how far off your answer is? Appendix B gives a formula for the "Lagrange remainder," an upper bound on the difference between $f(x)$ and a series approximation. Section 2.7.4 Problems 2.233–2.237 (see felderbooks.com) explore the use of this formula and other rules that bound the error in series approximations.

The General Formula for Maclaurin Series

In Problem 2.48 you will apply the process described above to find the Maclaurin series for a generic function $f(x)$. With that result we can summarize everything we've said about Maclaurin series in one concise formula:[2]

$$f(x) = \sum_{n=0}^{\infty} \frac{f^n(0)}{n!} x^n = f(0) + f'(0)x + \frac{f''(0)}{2!}x^2 + \frac{f'''(0)}{3!}x^3 + \ldots \quad (2.3.4)$$

For most functions, the first few terms of this series will provide a good approximation to $f(x)$ for values of x sufficiently near zero, and in the limit where you keep adding more terms the sum will approach the exact value $f(x)$. We discuss later in the chapter how to tell for what values of x this statement will hold true.

2.3.3 Problems: Maclaurin Series

2.30 Walk-Through: Maclaurin Series. In this problem you'll find the first several terms of the Maclaurin series for $\cos(2x)$. The first step is to write a power series with generic coefficients.

$$\cos(2x) = c_0 + c_1 x + c_2 x^2 + c_3 x^3 + c_4 x^4 + \ldots \quad (2.3.5)$$

(a) Plug $x = 0$ into both sides of Equation 2.3.5 to find c_0.

(b) Take the derivative of both sides of Equation 2.3.5 to get a new equation that doesn't have c_0 in it.

(c) Plug $x = 0$ into both sides of your answer to Part (b) to find c_1.

(d) Repeat Parts (b)–(c) until you have all the coefficients through c_4.

(e) Write the fourth-order polynomial you just found.

(f) Plug $x = 0.3$ into $\cos(2x)$ and into your fourth-order polynomial approximation. To how many decimal places is the approximation accurate at this x-value?

2.31 Fill in the following table to find the Maclaurin series for e^x. Remember that you get from each step to the next by taking the derivative of both sides of the equation, and you get each c-value by plugging in $x = 0$ to both sides of the equation.

---

[2]Some people write the first term separately and start the series at $n = 1$, which avoids 0^0 appearing in the first term for $x = 0$.

2.3 | Maclaurin Series

Equation	Plug in $x = 0$
$e^x = c_0 + c_1 x + c_2 x^2 + c_3 x^3 + c_4 x^4 + \ldots$	$c_0 = 1$
$e^x = c_1 + 2c_2 x + 3c_3 x^2 + 4c_4 x^3 + \ldots$	$c_1 =$
	$c_2 =$
	$c_3 =$
	$c_4 =$

2.32 *[This problem depends on Problem 2.31.]*
In Problem 2.31 you found the Maclaurin series for the function e^x.

(a) When you add the first five terms of an infinite series, the result is called the "fifth partial sum" of the series. Write the fifth partial sum of this particular series.

(b) Take the derivative of the fifth partial sum of the series. What do you notice?

(c) Use the first five terms of the series to estimate e^5. What is the percent error in your calculation? Recall that the formula for percent error is $\left|\frac{\text{actual value} - \text{estimate}}{\text{actual value}}\right| \times 100$.

(d) Use the first five terms of the series to estimate $\sqrt{e}$. What is the percent error in your calculation?

(e) Draw graphs of e^x, and of the first three partial sums of this series, for $-1 \leq x \leq 1$. Does each partial sum approximate the original function better than the previous one?

In Problems 2.33–2.42 find the first three non-vanishing terms of the Maclaurin series for the given function $f(x)$. You may find it helpful to work through Problem 2.30 as a model.

2.33 e^{-x}

2.34 e^{-x^2}

2.35 $\sin x + \cos x$

2.36 $1/(1 + x)^2$

2.37 $\cos(x + x^2)$

2.38 $\ln(1 + x)$

2.39 e^{e^x}

2.40 $\sec x$ *Hint*: If you're not sure how to take the derivative of $\sec x$, rewrite it in terms of $\cos$ first.

2.41 $\tan x$ *Hint*: If you're not sure how to take the derivative of $\tan x$, rewrite it in terms of $\sin$ and $\cos$ first.

2.42 $\sinh x$. (Never heard of it? Look it up!)

In Problems 2.43–2.45 make a plot showing the given function and the first three partial sums of its Maclaurin series in the range $-5 \leq x \leq 5$. In calculating the partial sums you should skip terms that equal zero, so for example the second partial sum of $1 - x^2/2 + x^4/24$ would be $1 - x^2/2$. Make sure it's clear which curve is the function and which ones are the partial sums. Estimate visually the value of x where each partial sum stops being a good approximation. (There is no precise correct answer for this question; just make a reasonable estimate based on looking at your plot.)

2.43 $f(x) = \cos x$

2.44 $f(x) = 1/(2 + \cos x)$

2.45 $f(x) = \sin\left(e^{-x^2}\right)$

2.46 In this problem you'll use the function $f(x) = 0.5 \sin(x^2) + 1.5 \sin(x) + 0.2$ to visually demonstrate the convergence of a Maclaurin series to the function it is approximating.

(a) Plot $f(x)$ in the range $-5 \leq x \leq 5$. Describe its shape in 20 words or less.

(b) Display a graph of $f(x)$ showing $-5 \leq x \leq 5$, $-3 \leq y \leq 3$. On the same plot show its 10th order Maclaurin series—that is, all terms up through x^{10}. On the same plot show the 20th order series, the 30th, and so on up to the 100th. Plot $f(x)$ as a thick black curve and the partial sums as thin lines, each of a different color.

(c) Describe how the plot changes as you go from one partial sum to the next.

(d) Visually estimate the values of x (one low and one high) where the 100th-order partial sum stops being a good approximation to $f(x)$. How many local maxima and minima does this partial sum trace out nearly identically with $f(x)$?

2.47 $f(x) = (1 + x)^k$

(a) Find the Maclaurin series for $f(x)$ up to and including the x^4 term.

(b) Write a formula for the nth-power term of the series.

(c) Rewrite your answer to Part (b) in "closed form"—that is, with no "…" or series notation—under the assumption that k is a positive integer. This is a derivation of the famous "binomial series." (Recall that you are writing the nth term in closed form, not the entire nth partial sum.)

2.48 (a) Fill in the following table to find a general formula for the Maclaurin series for any function $f(x)$.

Equation	Plug in $x = 0$
$f(x) = c_0 + c_1 x + c_2 x^2 + c_3 x^3 + c_4 x^4 + \ldots$	$c_0 = f(0)$
$f'(x) = c_1 + 2c_2 x + 3c_3 x^2 + 4c_4 x^3 + \ldots$	$c_1 =$
	$c_2 =$
	$c_3 =$
	$c_4 =$

(b) Express the formula from Part (a) in summation notation.

2.49 In Einstein's theory of special relativity the total energy of a particle is given by $E = mc^2/\sqrt{1 - (v/c)^2}$ where m is the rest mass of the particle, v is the speed, and c, a constant, is the speed of light. (This problem does not require any knowledge of special relativity beyond that equation.)

For most everyday uses—such as racecars, supersonic jets, and rockets to the moon—the quantity v/c is very small. We can therefore make this formula easier to work with by creating a Maclaurin series for this function, using v/c as the variable and treating m and c as constants.

(a) Create a Maclaurin series for the function $1/\sqrt{1 - x^2}$ up to the second power of x.

(b) Use your answer to Part (a) to create a second-order approximation for the function $mc^2/\sqrt{1 - (v/c)^2}$ that is useful when v/c is close to zero (in other words when $v \ll c$, or the speed is much less than the speed of light).

(c) Your expression should consist of two familiar terms. What does each one represent physically?

2.50 A particle sits at a distance z away from the center of a bar of length L with a uniformly distributed charge Q. The electric field the particle experiences is given by $E = (2kQ/z)\left(1/\sqrt{4z^2 + L^2}\right)$ where k is a constant. (For those of you familiar with the laws of electricity, $k = 1/(4\pi\epsilon_0)$.)

We are going to simplify this electric field expression for two special cases. In the first case the particle is very close to the bar. This is the case where $z \ll L$, so the unitless quantity z/L is very small.

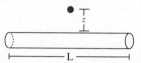

(a) We can rewrite the electric field as $E = (2kQ/zL)\left(1/\sqrt{4x^2 + 1}\right)$, where $x = z/L$. Expand this function in a Maclaurin series up to the second power of x. Plug $x = z/L$ back in to get an approximate expression for E valid when the particle is near the bar.

(b) The electric field at a distance z from an infinite bar of charge is $E = 2k\lambda/z$, where $\lambda = Q/L$ is the charge per unit length on the rod. The first term that you found should be equal to this expression, while the second term makes the approximation more accurate since the bar is not infinite. Explain why the second term has the sign that it does. In other words, would you expect the field from the finite bar to be larger or smaller than the field from an infinite bar with the same charge density, and does your result confirm your expectation?

Our second special case is a particle very *far* from the bar. For this case L/z will be very small.

(c) In Part (a) we rewrote E by factoring L out of the square root to make it a function of z/L. In a similar way, factor out $2z$ to make the square root a function of L/z.

(d) Create a Maclaurin series in L/z up to the second power for the resulting function.

(e) For this case your first term should reproduce the field from a point charge Q, namely $E = kQ/z^2$. The second term makes the expression more accurate since the bar is not an infinitely small point. Explain why the field from a bar of charge Q should be larger or smaller than the field from a point of charge Q and check that the sign of your second term matches this expectation.

2.51 In introductory physics you learn that a projectile moves in a parabola, but that is only true if you assume that the acceleration due to gravity is constant. When you throw a projectile it actually briefly enters a very narrow elliptical orbit about the center of the Earth, which is then interrupted when the projectile hits the Earth's surface. If the projectile is thrown horizontally and its distance from the center of the Earth is R_E (the radius of the Earth), the equation for this ellipse is:

$$y(x) = \frac{1}{\rho(2-\rho)}\left(V(1-\rho) + \sqrt{V^2 - \rho(2-\rho)x^2}\right)$$

The parameters are $V = v_0^2/g$ and $\rho = V/R_E$. These parameters are useful to keep the equations clean, but your answers should be expressed in terms of v_0, g, and R_E.

(a) Find the Maclaurin series for $y(x)$ up to the zeroth order (the constant term). Simplify your result as much as possible and explain what it means.

(b) Find the Maclaurin series for $y(x)$ up to the second order (the x^2 term).

(c) The equation for the ellipse is exact, but usually we use the simple approximation $F = -mg$ for motion near Earth's surface. Using this approximation find $x(t)$ and $y(t)$ and eliminate t to find $y(x)$. (Remember that the projectile is thrown horizontally at an initial height of R_E.) Your result should match your answer to Part (b).

(d) Make a rough sketch (not to scale!) of the exact elliptical trajectory and the parabolic approximation you found to it. Include a circle for the Earth in your sketch.

2.52 Do Maclaurin series always work? Consider the function:

$$f(x) = \begin{cases} e^{-1/x^2} & x \neq 0 \\ 0 & x = 0 \end{cases}$$

(a) Is this function continuous at $x = 0$? Why or why not?

(b) Find a Maclaurin series for this function. (After a few terms, you should be able to see the pattern for all the terms.)

(c) As you add more and more terms, does your series approximate the original function better and better?

2.53 (a) Find the Maclaurin series for $y = \cos^2 x$ up to the eighth power of x. (This gets algebraically messy.)

(b) Identify the pattern and write the general formula for all terms in this series. (The first term will not follow the pattern, so write it separately.)

(c) Find the Maclaurin series for $y = \sin^2 x$. (This should be a trivial one-step derivation based on your answer to Part (b).)

2.54 Calculating π. If you want to know the value of π to several decimal places you can of course just plug it into your calculator. But how were those decimals found, and how do computers nowadays calculate π to *billions* of decimal places? Many of the methods used involve Taylor series.

(a) Find the Maclaurin series for $\tan^{-1} x$ out to seventh order. (You can get the derivatives you need from a computer program, a table of derivatives, or just by searching the Web.) From there you should be able to spot the pattern and be able to write it down to any order you want. Use this to write an equation of the form $\tan^{-1} x = \sum_{n=0}^{\infty}$ <something>.

(b) Plug $x = 1$ into the equation you just wrote. This should give you an expression involving π on the left and a series involving simple fractions on the right. You can thus use this series to calculate π to any desired accuracy. Use your series up to the seventh-order term to find an estimate of π.

(c) 🖥 Now have a computer plug $x = 1$ into the 8th order polynomial, then the 9th, and so on up to the 1000th order polynomial. Generate a plot of the partial sum vs. n. The plot should converge toward the value of π.

(In practice this series is not a great way to calculate digits of π since it converges so slowly, as you saw in your plot. There are other, more complicated series that converge more quickly.)

2.4 Taylor Series

2.4.1 Explanation: Taylor Series

In the previous section, we found that many functions can be replaced by a *Maclaurin series*. Doing so allows us to replace difficult-to-use functions with simpler polynomial approximations that work best for *x*-values close to zero.

A "Taylor series" is the same idea, made more general: instead of creating a series that works best near zero, we can create a Taylor series to work well near any desired value. (Remember that when we wanted to estimate $\sqrt{69}$, we wrote a linear approximation for $\sqrt{x}$ that worked best near $x = 64$.) The goal is to build a power series of the form

$$f(x) = c_0 + c_1(x - a) + c_2(x - a)^2 + \ldots = \sum_{n=0}^{\infty} c_n(x - a)^n$$

In the example below, we once again approximate the function $1/(1 - x)$. In this case, however, instead of powers of x, we use powers of $(x - 3)$. Successive terms of the resulting series will decrease rapidly when x is close to 3. If the function is well behaved around $x = 3$ a few terms of this series should therefore approximate the function well in that neighborhood.

The strategy is the same as for Maclaurin series, except that we plug in $x = 3$ at each step to make all the terms of the series go to zero.

EXAMPLE **Building a Taylor Series**

Each equation is found by taking the derivative of the one above it.

Equation	Plug in $x = 3$
$\dfrac{1}{1-x} = c_0 + c_1(x-3) + c_2(x-3)^2 + c_3(x-3)^3 + c_4(x-3)^4 + \ldots$	$c_0 = 1/(-2)$
$\dfrac{1}{(1-x)^2} = c_1 + 2c_2(x-3) + 3c_3(x-3)^2 + 4c_4(x-3)^3 + \ldots$	$c_1 = 1/(-2)^2$
$\dfrac{2}{(1-x)^3} = 2c_2 + (2 \times 3)c_3(x-3) + (3 \times 4)c_4(x-3)^2 + \ldots$	$c_2 = 1/(-2)^3$
$\dfrac{2 \times 3}{(1-x)^4} = (2 \times 3)c_3 + (2 \times 3 \times 4)c_4(x-3) + \ldots$	$c_3 = 1/(-2)^4$
$\dfrac{2 \times 3 \times 4}{(1-x)^5} = (2 \times 3 \times 4)c_4 + \ldots$	$c_4 = 1/(-2)^5$

Plugging these coefficients into the original equation, we conclude that:

$$\frac{1}{1-x} = -\frac{1}{2} + \frac{1}{4}(x-3) - \frac{1}{8}(x-3)^2 + \frac{1}{16}(x-3)^3 - \frac{1}{32}(x-3)^4 + \ldots$$

This can be written more compactly in summation notation as

$$\frac{1}{1-x} = \sum_{n=0}^{\infty} \left(-\frac{1}{2}\right)^{n+1} (x-3)^n$$

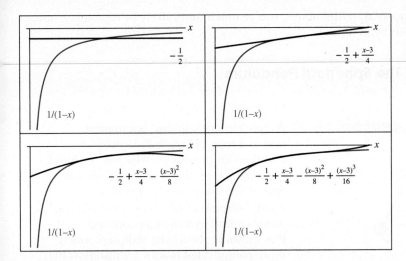

FIGURE 2.12 The plot of $1/(1 - x)$ (blue) along with the first four partial sums of its Taylor series about $x = 3$ (black).

Once again the more terms of this series we use, the closer it approximates the original function, as we see in Figure 2.12. In this case however the series approximates $1/(1 - x)$ best for x-values close to 3, since higher order terms vanish quickly for those values.

Taylor series are the culmination of everything we have done so far in this chapter.

- A linear approximation is a special case of a Taylor series. If we take the linear approximation to the function $1/(1 - x)$ at $x = 3$, using the techniques developed earlier in the chapter, we get $y = -1/2 + (1/4)(x - 3)$. With the Taylor series, however, we can add more terms to approximate the original function as accurately as we need.
- A Maclaurin series is also a special case of a Taylor series. A Taylor series can be built to approximate a function around any specified x-value. If that value happens to be $x = 0$, then the Taylor series is a Maclaurin series.

Polynomials are among the easiest functions to work with. We can add and subtract them, compare them, take their derivatives, and integrate them in straightforward ways. The ability to replace any function with a polynomial is therefore a tremendously powerful mathematical tool.

Once again, however, a cautionary note must be sounded. Just because you can build a Taylor series does not mean that it works! The Taylor series we just built for $1/(1 - x)$ works wonderfully at $x = 3.01$; it approximates the function very well with only a few terms. At $x = 4$ the series still works, but requires far more terms to create a reasonable approximation. And at $x = 6$ the series doesn't work at all. (Try it and see what the terms are doing.) So there are important questions of *domain* that are taken up in the final sections of this chapter.

A General Formula For Taylor Series

In Problem 2.68, you will apply the Taylor series process described above to the generic function $f(x)$. We can therefore write one concise formula[3] for the Taylor series of a function about $x = a$.

$$f(x) = \sum_{n=0}^{\infty} \frac{f^n(a)}{n!}(x - a)^n = f(a) + f'(a)(x - a) + \frac{f''(a)}{2}(x - a)^2 + \frac{f'''(a)}{6}(x - a)^3 \ldots \quad (2.4.1)$$

Applying this formula directly takes you through the same steps we used above—the same derivatives and substitutions—in a way that you may find a bit quicker. On the other hand,

---

[3] Some people write the first term separately and start the series at $n = 1$, which avoids 0^0 appearing in the first term for $x = a$.

if you write out the process as we did above, you're much more likely to remember how to find a Taylor series a year from now.

EXAMPLE: The Spherical Pendulum

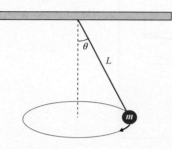

A "spherical pendulum" is a mass m tied to a rope of length L. Unlike a traditional pendulum, which is confined to swing in one direction, the spherical pendulum swings freely in both the x and y-directions (so it can, for instance, move around in a circle at constant θ as shown in the figure, although much more complicated modes are also possible).

It can be shown that a spherical pendulum will obey the differential equation:

$$\frac{d^2\theta}{dt^2} = h^2 \frac{\cos\theta}{\sin^3\theta} - \frac{g}{L}\sin\theta \qquad (2.4.2)$$

where g is the acceleration due to gravity and h is a constant, proportional to the angular momentum. We cannot explicitly solve this equation, so if we had no other information the best we could do would be to solve this numerically for lots of values of h and different initial conditions. Suppose, though, that we observed that the pendulum had nearly circular motion at $\theta = \pi/6$, but with slight deviations above and below that. Then we could expand the right side in a Taylor series around $\theta = \pi/6$. After two terms we would end up here:

$$h^2 \frac{\cos\theta}{\sin^3\theta} - \frac{g}{L}\sin\theta = 4h^2\sqrt{3} - \frac{g}{2L} - \left(40h^2 + \frac{\sqrt{3}g}{2L}\right)\left(\theta - \frac{\pi}{6}\right) + \ldots \qquad (2.4.3)$$

Looks awful, doesn't it? But most of the ugliness is just constants. If we define $a = 4h^2\sqrt{3} - g/(2L)$ and $b = 40h^2 + (\sqrt{3}g)/(2L)$ then the differential equation is just the following:

$$\frac{d^2\theta}{dt^2} = a - b\left(\theta - \frac{\pi}{6}\right) \qquad (2.4.4)$$

with solution:

$$\theta = \frac{a}{b} + \frac{\pi}{6} + C_1 \sin\left(\sqrt{b}\, t\right) + C_2 \cos\left(\sqrt{b}\, t\right) \qquad (2.4.5)$$

We have shunted all the algebra in this problem to you in Problem 2.70, which may leave you wondering why we bothered showcasing this example at all. But our point here is not to demonstrate how to find a Taylor series, or how to solve a differential equation, but how to *use* a Taylor series to turn an otherwise intractable problem into something we can handle.

Students often get confused in such problems about the role the variables play. In the differential equation and the solution, θ is the dependent variable and t the independent. But our Taylor series takes a piece of the differential equation—a function of θ—and expands it around a given θ-value. In that expression, θ momentarily acts as an independent variable.

And of course, we must not lose sight of the fact that our approximation is only useful if θ begins, and stays, close to $\pi/6$.

2.4 | Taylor Series

Stepping Back

We are now in a position to summarize everything we've said so far about Taylor series. (You'll find summaries of the basic formulas, as well as some commonly used Taylor series, in Appendix B.)

One of the most common techniques used to simplify intractable problems is to replace a complicated function with a polynomial that approximates the function within some range of values. The higher the order of the polynomial, the better the approximation will generally be, but scientists and engineers often use a first-order approximation because lines are so easy to work with.

The process begins by writing this equation, which simply asserts that there exists a power series replacement for $f(x)$ around $x = a$:

$$f(x) = c_0 + c_1(x - a) + c_2(x - a)^2 + c_3(x - a)^3 + \ldots$$

By setting $x = a$ on both sides you can find c_0. By taking the derivative of both sides and then setting $x = a$ again, you can find c_1, and so on. Equivalently, you can find the terms of the series from Equation 2.4.1.

This process does not work in all cases, however. Specifically:

- A function $f(x)$ can only have a Taylor series about $x = a$ if all of its derivatives exist at $x = a$.
- It is possible for the Taylor series for a function to diverge for some values of x, in which case it is simply not usable for those values. In Section 2.7 we discuss how to determine the values of x for which a given Taylor series converges.
- It is possible for a Taylor series to converge to the wrong function, but this is rare.

Strictly speaking, the only guarantee you get is that *if* a function $f(x)$ has a power series representation about $x = a$, the coefficients will be given by Equation 2.4.1. In practice, however, most Taylor series converge for values of x close to a, and the only cases where they converge to the wrong value are contrived cases dreamt up by clever mathematicians. (See for example Section 2.3, Problem 2.52.) Most scientists and engineers use Taylor series all the time without worrying about them converging to the wrong value, and we have never heard of a bridge collapsing as a result.

A more common problem is that, for many functions, the derivatives become complicated and it becomes difficult to calculate new terms. In the next section we will show you a set of methods that can often calculate Taylor series that would be tedious to find with the method described above.

2.4.2 Problems: Taylor Series

2.55 Walk-Through: Taylor Series. In this problem you'll find the first several terms of the Taylor series for $\cos(2x)$ about the point $x = \pi/4$. The first step is to write a power series with generic coefficients.

$$\cos(2x) = c_0 + c_1\left(x - \frac{\pi}{4}\right) + c_2\left(x - \frac{\pi}{4}\right)^2$$
$$+ c_3\left(x - \frac{\pi}{4}\right)^3 + c_4\left(x - \frac{\pi}{4}\right)^4 + \ldots \quad (2.4.6)$$

(a) Plug $x = \pi/4$ into both sides of Equation 2.4.6 to find c_0.

(b) Take the derivative of both sides of Equation 2.4.6 to get a new equation that doesn't have c_0 in it.

(c) Plug $x = \pi/4$ into your answer to Part (b) to find c_1.

(d) Repeat Parts (b)–(c) until you have all the coefficients through c_4.

(e) Write the fourth-order polynomial you just found.

(f) Plug $x = 1$ into $\cos(2x)$ and into your fourth-order polynomial approximation. To how many decimal places is the approximation accurate at this x-value?

2.56 (a) Find the Taylor series for $\sin x$ around $x = \pi$. Express your answer in summation notation.

(b) Use the first three non-zero terms of your series to estimate $\sin(3)$.

(c) Calculate the percent error in your estimate from Part (b).

(d) 🖥 Draw graphs of $\sin x$, and of the first three partial sums of your series, for $\pi/2 \leq x \leq 5\pi/2$. Does each partial sum approximate the original function better than the previous?

2.57 (a) Find the Taylor series for $\cos x$ around $x = \pi/3$. (*Hint*: $\cos(\pi/3) = 1/2$ and $\sin(\pi/3) = \sqrt{3}/2$.)

(b) Use the first three non-zero terms of your series to estimate $\cos(1)$.

(c) Calculate the percent error in your estimate from Part (b).

(d) 🖥 Draw graphs of $\cos x$, and of the first three partial sums of your series, for $-\pi/2 \leq x \leq \pi$. Does each partial sum approximate the original function better than the previous?

2.58 (a) Find the Taylor series for $\ln x$ around $x = 1$. Express your answer in summation notation.

(b) Use the first three non-zero terms of your series to estimate $\ln(1/2)$.

In Problems 2.59–2.67 find the specified Taylor series of the given function up to the third order. You may find it helpful to work through Problem 2.55 as a model.

2.59 $\sin(x^2)$ about $x = \sqrt{\pi}$

2.60 $1/\sqrt{x}$ about $x = 1$

2.61 $e^x + e^{-x}$ about $x = \ln 2$

2.62 $\sin^2 x$ about $x = \pi/2$

2.63 $\cos^2 x$ about $x = \pi/2$. (*Hint*: If you've solved the previous problem there's a quick way to do this one without taking any derivatives.)

2.64 e^{-x^2} about $x = 1$

2.65 $\sec x$ about $x = \pi$. *Hint*: If you're not sure how to take the derivative of $\sec x$, rewrite it in terms of $\cos$ first.

2.66 $\sqrt{\tan x}$ about $x = \pi/4$. *Hint*: If you're not sure how to take the derivative of $\tan x$, rewrite it in terms of $\sin$ and $\cos$ first.

2.67 $\sin x/x$ about $x = 2\pi$

2.68 (a) Fill in the following table to find the first five terms of the Taylor series for any function $f(x)$ around any given value $x = a$.

Equation	Plug in $x = a$
$f(x) = c_0 + c_1(x-a) + c_2(x-a)^2 + c_3(x-a)^3 + c_4(x-a)^4 + \ldots$	$c_0 = f(a)$
$f'(x) = c_1 + 2c_2(x-a) + 3c_3(x-a)^2 + 4c_4(x-a)^3 + \ldots$	$c_1 =$
	$c_2 =$
	$c_3 =$
	$c_4 =$

(b) Find the pattern in your answer to Part (a) and write a formula for the Taylor series of $f(x)$ around the value $x = a$. Your answer should be an infinite series.

2.69 The function $y = f(x)$ has a Taylor series around $x = -2$ given by:

$$f(x) = \frac{(x+2)^2}{4} + \frac{(x+2)^3}{9} + \frac{(x+2)^4}{16} + \frac{(x+2)^5}{25} + \ldots = \sum_{n=2}^{\infty} \frac{(x+2)^n}{n^2} \quad -3 < x < -1$$

(a) Show that this function has a critical point at $x = -2$.

(b) Does this critical point represent a relative minimum, a relative maximum, or neither? How can you tell?

(c) The function $g(x)$ is defined by two properties: $g'(x) = f(x)$, and $g(-2) = 5$. Find the Taylor series for $g(x)$ around $x = -2$ up to the fourth power of x.

2.70 In the Explanation (Section 2.4.1) we work through the example of a spherical pendulum modeled by Equation 2.4.2. For a given pendulum g and L will be fixed, but different values of h will correspond to different possible motions with different angular momenta. For each value of h there will still be different motions possible with different initial conditions θ and $d\theta/dt$.

(a) We began our solution by expanding the right side of the differential equation in a Taylor series around $\theta = \pi/6$. Show the steps required to find the first two terms of the relevant Taylor series. In other words, derive Equation 2.4.3.

(b) To solve Equation 2.4.4 we begin by rewriting it as:

$$\frac{d^2\theta}{dt^2} + b\theta = a + \frac{\pi}{6}b$$

i. Write down and solve the complementary homogeneous equation.
ii. Find a particular solution.
iii. Put your solutions together to find the general solution.

(c) For any given value of h Equation 2.4.2 has one "constant solution": one value where θ has no time-dependence. What must h be in order for $\theta = \pi/6$ to be a solution to Equation 2.4.2? What kind of motion does this solution describe?

(d) Plug in the value of h you found in Part (c) to express a and b in terms of g/L. Plug these in to write your solution $\theta(t)$ from Part (b) in terms of g/L.

(e) Our final solution represents an oscillatory θ. Express the period of this oscillation in terms of g and L.

2.71 A hydrogen molecule consists of two hydrogen atoms bound to each other by the interactions of their nuclei and electrons. While the details of the interactions are very complicated, it's been found empirically that their behavior can be modeled by saying their potential energy V as a function of their separation r is given by the "Morse potential": $V(r) = D\left(1 - e^{-b(r-r_0)}\right)^2$, where D, b and r_0 are positive constants.

(a) Sketch the Morse potential.
(b) Use a Taylor series to approximate $V(r)$ near the minimum at $r = r_0$. Keep only the first non-zero term.
(c) You should have found a potential of the form $V(r) = (1/2)k\left(r - r_0\right)^2$, where k is a constant that depends on D and b. It can be shown that this potential function will lead to oscillation with frequency $\sqrt{k/2m_p}/\pi$ (which of course depends on the k you found). Given that $D = 7.24 \times 10^{-19}$ J, $b = 2 \times 10^{10}$ m^{-1}, and the mass of a hydrogen nucleus is $m_p = 1.67 \times 10^{-27}$ kg, find the oscillation frequency of a hydrogen molecule.

2.72 Exploration: The Gravitational Attraction of the Earth
Based on Newton's Law of Universal Gravitation we can conclude that a body of mass m acting under the influence of the Earth's gravitational attraction (such as a satellite) experiences a force given by $F = -GM_E m/r^2$ where r is the object's distance from the center of the Earth. Since $F = ma$, the m terms on both sides cancel, leaving a differential equation:

$$\frac{d^2r}{dt^2} = \frac{-GM_E}{r^2}$$

where G is a universal constant, and M_E is also a constant, representing the mass of the Earth.

Unfortunately, this simple-looking differential equation cannot be solved in simple form. One approach is to replace the function $-GM_E/r^2$ with a Taylor series around $r = R_E$, the radius of the Earth.

(a) A "zeroth-order Taylor series" is the simplest approximation to any function: a horizontal line, corresponding to a constant function. Find the zero-order Taylor series approximation of the function $-GM_E/r^2$ at the point $r = R_E$.

(b) Where would you expect your answer in Part (a) to be most useful and accurate? Where would you expect it to be inaccurate and useless? (Your answer to this question should be expressed in non-technical and non-mathematical language.)

(c) Substituting that constant function for the original function, find a solution to Newton's Law $d^2r/dt^2 = -GM_E/r^2$.

(d) Now, using the values $G = 6.67 \times 10^{-11}$ m^3/(kg · s^2), $M_E = 5.97 \times 10^{24}$ kg, and $R_E = 6.37 \times 10^6$ m, estimate the acceleration due to gravity near the surface of the Earth. This should give you the familiar constant g from introductory mechanics.

(e) To increase our accuracy, create a *first*-order Taylor series (a linear approximation) for the function $-GM_E/r^2$ at the point $r = R_E$.

(f) The solution to Part (e) leads to a differential equation of the form $d^2r/dt^2 = a + br$. What are the constants a and b in terms of our more fundamental constants G, M_E, and R_E?

(g) The general solution to the equation in Part (f) is:

$$r(t) = C_1 e^{\sqrt{b}t} + C_2 e^{-\sqrt{b}t} - a/b$$

Confirm that this is a valid solution of the differential equation. (You could find it yourself using guess-and-check.)

(h) Solve for the arbitrary constants in this solution using the initial conditions $r(0) = R_E$, $r'(0) = 0$: in other words, an object falling from rest near Earth's surface. Use this to write an expression for the function $r(t)$ that depends only on G, M_E, R_E, and t.

(i) Expand the solution in Part (g) in a Maclaurin series around $t = 0$, going out to 4th order in t. Write the resulting solution $r(t)$ in terms of R_E and g (which you found in Part (a)). The second-order version of this series is the familiar equation from introductory mechanics, based on the approximation that the force of Earth's gravity is a constant. The fourth-order version of this series gives an equation that can be more accurate for long-distance falls.

2.5 Finding One Taylor Series from Another

In general, finding a Taylor series involves a lot of derivatives. Taking these derivatives is no great problem for a function like e^x or $\sin x$, but if you find yourself taking the fourth or fifth derivative of $e^{(x^2)}$, you might start to wonder if there is an easier way. In some cases, there is: you can find the Taylor series for one function from the Taylor series for a different function.

2.5.1 Explanation: Finding One Taylor Series from Another

In this section, we are going to find a number of Maclaurin series, all based on the four basic series expansions given in Appendix B. In the problems you'll show that these same techniques can be applied to find Taylor series that aren't Maclaurin series as well.

Here is our first example: suppose you want to find the Maclaurin series for the function $\sin(x^2)$. The standard Taylor series process is tedious for this series: the first derivative is $2x \cos(x^2)$, the second derivative is $2\cos(x^2) - 4x^2 \sin(x^2)$, and things only get worse from there.

We can avoid this process if we start with the Maclaurin series for $\sin x$. According to Appendix B, the series

$$\sin x = x - \frac{x^3}{3!} + \frac{x^5}{5!} - \frac{x^7}{7!} + \ldots \tag{2.5.1}$$

is *true for all x-values*. For instance, if we replace every occurrence of x with $1/2$, we get:

$$\sin(1/2) = (1/2) - \frac{(1/2)^3}{3!} + \frac{(1/2)^5}{5!} - \frac{(1/2)^7}{7!} + \ldots$$

Is that true? Does the series on the right converge to the number on the left? We don't have to work hard to answer that question: Equation 2.5.1 *means* that this is true. We can also replace each x with π, or with 0, or with q, or with any other number or variable we choose, and the resulting equations are guaranteed true.

So, we can replace every occurrence of x with x^2 and immediately get the result we are looking for:

$$\sin(x^2) = x^2 - \frac{x^6}{3!} + \frac{x^{10}}{5!} - \frac{x^{14}}{7!} \ldots = \sum_{n=0}^{\infty} \frac{(-1)^n x^{4n+2}}{(2n+1)!} \tag{2.5.2}$$

Stop at this point and convince yourself of two things.

1. The equation given above is still a Maclaurin series: that is, it is still a function of the form $c_0 + c_1 x + c_2 x^2 + \ldots$.
2. The Maclaurin series for $\sin x$ *implies* the equation given above. For instance, if you plug $x = 1/4$ into Equation 2.5.1, you get *exactly the same equation* that you get if you plug $x = 1/2$ into Equation 2.5.2. (Try it!)

Therefore, we have found the Maclaurin series for the function $\sin(x^2)$, with considerably less trouble than it would have taken to find the same series term by term.

As another example, suppose we want the Maclaurin series for the function $x \sin x$. We can return to Equation 2.5.1 and multiply both sides by x, yielding:

$$x \sin x = x^2 - \frac{x^4}{3!} + \frac{x^6}{5!} - \frac{x^8}{7!} \ldots = \sum_{n=0}^{\infty} \frac{(-1)^n x^{2n+2}}{(2n+1)!}$$

2.5 | Finding One Taylor Series from Another

It is also valid to *divide* both sides of the original equation by x, which tells us that:

$$\frac{\sin x}{x} = 1 - \frac{x^2}{3!} + \frac{x^4}{5!} - \frac{x^6}{7!} \cdots = \sum_{n=0}^{\infty} \frac{(-1)^n x^{2n}}{(2n+1)!} \qquad (2.5.3)$$

We have to be careful here: dividing both sides by x is not valid when $x = 0$, and the left side of Equation 2.5.3 is not defined in that case. However, it does work perfectly in the *limit*; you can show using l'Hôpital's rule that $\lim_{x \to 0} \sin x / x = 1$.

Now suppose we want the Maclaurin expansion of $\cos x$ (and suppose we didn't already have it). We can come back to Equation 2.5.1 and take the *derivative* of both sides, yielding:

$$\cos x = 1 - \frac{3x^2}{3!} + \frac{5x^4}{5!} - \frac{7x^6}{7!} \cdots$$

A nice bit of simplification occurs when we note that $5/(5!)$ is the same as $1/(4!)$: the numerators in each term reduce the denominators, bringing us to the series for $\cos x$ given in Appendix B.

$$\cos x = 1 - \frac{x^2}{2!} + \frac{x^4}{4!} - \frac{x^6}{6!} \cdots$$

It may occur to you that there is another way to get from sine to cosine: by integrating! This is often useful, and it is perfectly valid as long as you are careful about one thing: the first (constant) term will be missing, since integrating adds one to each power of x. That first term appears again as part of the $+C$ that follows any indefinite integral; you solve for the C by plugging in $x = 0$.

$$\sin x = x - \frac{x^3}{3!} + \frac{x^5}{5!} - \frac{x^7}{7!} \cdots$$

Integrating both sides gives:

$$\int \sin x \, dx = \int \left(x - \frac{x^3}{3!} + \frac{x^5}{5!} - \frac{x^7}{7!} \cdots \right) dx$$

$$-\cos x = \frac{x^2}{2} - \frac{x^4}{4!} + \frac{x^6}{6!} - \frac{x^8}{8!} \cdots + C$$

$$\cos x = -\frac{x^2}{2} + \frac{x^4}{4!} - \frac{x^6}{6!} + \frac{x^8}{8!} \cdots + C$$

Plugging in $x = 0$ to both sides, we conclude that $C = 1$.

While it is valid to replace x on both sides of an equation with *any* function, the result is not always a power series. For instance, suppose you wanted to find the Maclaurin series of $\sin(\sin x)$. You might be tempted to return to Equation 2.5.1 and replace every x with $\sin x$, yielding:

$$\sin(\sin x) = \sin x - \frac{\sin^3 x}{3!} + \frac{\sin^5 x}{5!} - \frac{\sin^7 x}{7!} \cdots$$

This equation is valid for all x-values, but it is *not* the Maclaurin series for $\sin(\sin x)$ or anything else, because it is not a polynomial. It does not have the properties that make Taylor series so useful to us: for instance, the first two terms do not define a line.

To get a Taylor series, we can plug the Maclaurin series for the sine directly into *itself*.

$$\sin x = x - \frac{x^3}{3!} + \frac{x^5}{5!} - \cdots$$

$$\sin(\sin x) = \left(x - \frac{x^3}{3!} + \frac{x^5}{5!} \cdots \right) - \frac{1}{3!} \left(x - \frac{x^3}{3!} + \frac{x^5}{5!} \cdots \right)^3 + \frac{1}{5!} \left(x - \frac{x^3}{3!} + \frac{x^5}{5!} \cdots \right)^5 \cdots$$

We can determine from this expression the coefficients of the actual Maclaurin series for sin(sin x). For example, the first expression in parentheses starts with x. The lowest power of x in the next part, $(x - x^3/(3!) + x^5/(5!)\ldots)^3$, is x^3. The lowest power in the next part is x^5, and so on, so the only term with x raised to the power 1 is that first x, and the Maclaurin series thus begins

$$\sin(\sin x) = x + \ldots$$

In Problem 2.85 you'll expand out different parts of this expression to derive the next several terms of the Maclaurin series.

This being a chapter on series, it's appropriate to sum up. It is often easiest to generate a Taylor series by starting with one that you already know. You can replace each occurrence of x with another function, multiply both sides by a power of x, take the derivative, or integrate (remembering the $+C$). These tricks can turn basic Taylor series into a wide variety of series and avoid a lot of messy derivatives.

But there is one other rule to bear in mind: algebra is always allowed! For instance, we have seen that:

$$\frac{1}{1-x} = 1 + x + x^2 + x^3 \ldots$$

Suppose we want the Maclaurin series for $1/(2-x)$? We start by dividing the top and bottom by 2.

$$\frac{1}{2-x} = \frac{(1/2)}{1-(1/2)x} = \frac{1}{2}\left(\frac{1}{1-(1/2)x}\right)$$

$$\frac{1}{2-x} = \frac{1}{2}\left(1 + \frac{1}{2}x + \left(\frac{1}{2}x\right)^2 + \left(\frac{1}{2}x\right)^3 \ldots\right) = \frac{1}{2} + \frac{1}{4}x + \frac{1}{8}x^2 + \ldots$$

2.5.2 Problems: Finding One Taylor Series from Another

For most of the problems in this section, you should *not* have to start from first principles to find the given series. Instead, find these series as variations of the ones given in Appendix B.

2.73 $e^{\sqrt{x}} = 1 + \sqrt{x} + x/(2!) + x^{3/2}/(3!) + x^2/(4!) + \ldots$

(a) Explain in your own words why this equation follows directly from one of the common Maclaurin series listed in Appendix B.

(b) Plug $x = 9$ into both sides of this equation. If you use the fifth partial sum of the series to estimate the actual e^3, how close do you come?

(c) If you find the Maclaurin series for $e^{\sqrt{x}}$, you will *not* get this series. Why not?

In Problems 2.74–2.82, find the first five non-zero terms of the Maclaurin series. Then express the entire (infinite) series in summation notation.

2.74 $2\cos(3x)$

2.75 $\sin x + \cos x$

2.76 e^{3x}

2.77 xe^{x^2}

2.78 $1/(1+x)$

2.79 $1/(2+x)$

2.80 $x^2\sqrt{1+x}$ (For this problem it is not necessary to write the entire series in summation notation.)

2.81 $\sqrt{4+x^2}$ (For this problem it is not necessary to write the entire series in summation notation.)

2.82 $\ln(1-x)$

2.83 Use one of the common Maclaurin series listed in Appendix B to find the Maclaurin series for $(1 + x)^2$. What happens to all the terms that are higher order than x^2?

2.84 Use one of the common Maclaurin series listed in Appendix B to find the Maclaurin series for $(1 + x)^3$.

2.85 Find the Maclaurin series for $\sin(\sin x)$ through the x^5 term.

2.86 Find the Maclaurin series for $\cos(\sin x)$ through the x^5 term.

2.87 Find the Maclaurin series for $e^{\sin x}$ through the x^5 term.

2.88 Find the Maclaurin series for $\cos(\cos x)$ through the x^3 term. Why is this harder than Problems 2.85–2.87?

2.89 In this problem, we're going to find the Taylor series for $\ln(x^2)$ around $x = 1$.

  (a) Begin by finding the Taylor series for $\ln x$ around $x = 1$. (This will not follow from our basic Maclaurin series: you will generate this Taylor series "from scratch.") You should be able to take the first few derivatives and then spot the pattern so you can write the series in summation notation.

  (b) Explain why it is *not* valid to plug x^2 into your answer to Part (a) to find the Taylor series for $\ln(x^2)$.

  (c) Find and apply an alternative method to quickly and easily find the Taylor series for $\ln(x^2)$ from the (already found) Taylor series for $\ln x$.

2.90 In this problem, we're going to find the Taylor series for $\sqrt[3]{x^2}$ around $x = 1$.

  (a) Begin by finding the 3rd-order Taylor series for $\sqrt[3]{x}$ around $x = 1$.

  (b) Explain why it is *not* valid to plug x^2 into your answer to Part (a) to find the Taylor series for $\sqrt[3]{x^2}$.

  (c) Find and apply an alternative method to quickly and easily find the 3rd-order Taylor series for $\sqrt[3]{x^2}$ from the (already found) Taylor series for $\sqrt[3]{x}$.

2.91 In the Motivating Exercise (Section 2.1), we looked at an atomic nucleus in a crystal. Its motion is subject to the differential equation:

$$\frac{d^2x}{dt^2} = \frac{\kappa}{m}\left(\frac{1}{(d+x)^2} - \frac{1}{(d-x)^2}\right)$$

In the section on Maclaurin series, we replaced that function with a simpler one. In this problem we are going to repeat that exercise—and get the same answer!—with considerably less work.

  (a) Begin by factoring out $1/d^2$ from this force law. This re-expresses the force as a function of x/d, a unitless quantity that we are assuming is small.

  (b) Now, replace each of the two terms with a first-order Maclaurin series, with x/d as the variable. You can do this quickly and easily by using the $(1 + x)^p$ expansion.

  (c) Solve the resulting differential equation to show that the nucleus will undergo simple harmonic oscillation. Find the period of this oscillation as a function of k, d, and the mass m of the nucleus.

2.92 The "Gaussian" function e^{-x^2} is defined for all x-values. Therefore, for any two given x-values, we can ask "What is the area under this curve between these two x-values?" and the answer will be finite and well-defined. Unfortunately, we cannot *find* the answer in the traditional way, because the function e^{-x^2} has no antiderivative that can be written in terms of elementary functions.

  (a) Write the Maclaurin series for e^{-x^2} through the x^8 term.

  (b) Use your result from Part (a) to approximate $\int_0^1 e^{-x^2}\,dx$.

  (c) Check your answer by integrating on a calculator or computer.

2.93 (a) Find the Maclaurin series for $1/(1 + x^2)$.

  (b) Find the Maclaurin series for $\tan^{-1} x$. (*Hint*: the derivative of $\tan^{-1} x$ is $1/(1 + x^2)$.)

  (c) By plugging $x = 1$ into your answer to Part (b), find a series that converges to $\pi/4$.

  (d) Add the first five terms of your answer to Part (c) and multiply the answer by 4 to approximate π.

2.94 The term $\mathcal{O}(x^n)$ means "terms including x raised to the n and higher powers." For example, the Maclaurin series for $\sin x$ can be written as $\sin x = x - x^3/3! + \mathcal{O}(x^5)$, meaning the terms that weren't written out explicitly include terms of order 5 and above. You can use this to help you keep track of what you're neglecting in complicated calculations. As an example, consider finding the Maclaurin series for $f(x) = (\sin x - x\cos x)/(3\sin x - x\cos x)$.

  (a) Write out $f(x)$ as a fraction, using the Maclaurin series for $\sin x$ and $\cos x$.

Keep all terms through 6th order on top and through 4th order on bottom. Your numerator and denominator should end with $\mathcal{O}(x^7)$ and $\mathcal{O}(x^5)$ respectively.

(b) Divide top and bottom of the fraction by $2x$. Dividing $\mathcal{O}(x^5)$ by x turns it into $\mathcal{O}(x^4)$. Explain why dividing $\mathcal{O}(x^4)$ by 2 doesn't do anything to it.

(c) Use the binomial expansion to turn the fraction into a product of the following form.

$$f(x) = \left(\text{some stuff} + \mathcal{O}(x^6)\right)\left(\text{some other stuff} + \mathcal{O}(x^{\text{for you to figure out}})\right)$$

(d) Multiply out the two series to get the Maclaurin series for $f(x)$. Based on the $\mathcal{O}$ symbols you've been carrying through the calculations, to what order is your answer accurate?

2.6 Sequences and Series

Our work with Taylor series begs the more fundamental questions of infinite series in general. What does it mean to say that an infinitely long series "adds up to" a finite number? How can we determine when or if it does so?

The remainder of this chapter will be devoted to these questions. We begin with one of the most important examples, the "geometric" series. Along the way, we will develop some key concepts that can be applied to finite and infinite series. All this work will prepare us to rigorously answer the question, "for which x-values does this Taylor series actually add up to anything at all?"

2.6.1 Discovery Exercise: Geometric Series

Consider the sum: $5 + 15 + 45 + 135 + 405$. This is called a "geometric series," meaning each term is obtained by multiplying the previous term by a constant. The constant is known as the "common ratio," or r. In this example, r is 3.

You could add up those five terms in a minute, by hand or with a calculator. You could also presumably add up the first six terms of that series, since you can readily figure out that the next term is $405 \times 3 = 1215$. But what if we asked you to add up the first twenty terms, or the first hundred?

Amazingly, there is a trick—we call it the "geometric series trick"—that quickly adds up all the terms of any geometric series. You're going to use it to add up just the five-term series we gave above, but you'll see that it would be equally quick for a series of any length. Later we'll show you how to use it to add up an infinite series.

You begin by assigning the letter S to the sum you are looking for:

$$S = 5 + 15 + 45 + 135 + 405$$

1. What is $3S$? (Write it out term by term; don't add them up.)
2. Now, write the equation $3S =$ (what you just said), and below that, write our original equation for S. Then subtract the two equations. You will find that, on the right side of the equal sign, all but two of the terms cancel out.

$$3S =$$
$$S =$$

3. Solve the resulting equation for S.

See Check Yourself #9 in Appendix L

4. Repeat this process with the geometric series $1000 + 100 + 10 + 1 + 1/10 + 1/100$. Note that in this case $r = 1/10$, so you will have to begin by writing a formula for S and one for $(1/10)S$. Once again, make sure you get the correct answer (which is 1111.11 in this case).

5. Now let's generalize this process. The general geometric series is…

$$S = a_1 + a_1 r + a_1 r^2 + a_1 r^3 + a_1 r^4 + \ldots a_1 r^{n-1}$$

Multiply this whole equation (both sides) by r. Then, subtract that new equation from the original equation. You should once again find that all the terms on the right but two cancel, leaving a formula you can solve for S. Do that. When you are done, you should have a general formula for the sum of any geometric series that starts at a_1 and goes on for n terms with common ratio r.

See Check Yourself #10 in Appendix L

2.6.2 Discovery Exercise: Infinite Series

In this exercise you'll apply the geometric series trick from the previous exercise to find the values of x that a particular Taylor series works for.

Our example is a Maclaurin series we have worked with many times:

$$\frac{1}{1-x} = 1 + x + x^2 + x^3 + x^4 + \ldots$$

We have discussed the fact that this works very well for $x = 1/3$, and does not work at all for $x = 2$. This brings up the practical question: how can we determine when a Taylor series works? But underlying that question is a more theoretical concern: what does it mean for an infinite series to add up to anything at all?

As an example consider the series $\sum_{n=1}^{\infty} (1/2)^n$, or $(1/2) + (1/4) + (1/8) + \ldots$.

We begin our examination of this series visually, on the number line below.

1. The series begins at $1/2$. Mark this point on the number line, and label it S_1 for the first partial sum.
2. After the second term, the series reaches $(1/2) + (1/4)$. This is the second partial sum: label it S_2 on the number line. (You are not labeling the number $1/4$; you are labeling $(1/2) + (1/4)$, the series after two terms.)
3. Now label S_3 which is $(1/2) + (1/4) + (1/8)$. (You can of course punch this into a calculator, but you will find the pattern more easily if you work with fractions instead of decimals.)
4. Find S_4, S_5, and S_6. Write down their values and label them on the number line.
5. As we add more and more terms, where are we heading?

This brings us to the mathematical definition of an infinite series: the sum of an infinite series is defined as the limit as $n \to \infty$ of the partial sums S_n. Let's see how that definition applies to a few different series.

6. Consider the series $\sum_{n=1}^{\infty} (1/3)^n$.

(a) Write out the first five terms of the series.
(b) Write out the first five partial sums of the series.
(c) Calculate the 20th partial sum of the series. (*Hint*: don't do it by hand!)
(d) Where does it look like this is headed?
(e) Calculate the nth partial sum of the series.
(f) Calculate $\lim_{n\to\infty} S_n$.

See Check Yourself #11 in Appendix L

7. For a geometric series whose first term is a_1 and whose common ratio is r, the sum of the first n terms is given by the formula $S_n = a_1(1 - r^n)/(1 - r)$.
   (a) If $r = 2$, what is the sum of the *infinite* geometric series? (That is, what is $\lim_{n\to\infty} S_n$?)
   (b) If $r = 1/2$, what is $\lim_{n\to\infty} S_n$?
   (c) If $r = 1$, what is $\lim_{n\to\infty} S_n$? (*Hint*: Rather than applying l'Hôpital's rule, write down a geometric series with $r = 1$ and it's easy to see where the partial sums are headed.)
   (d) Now, generalize: for what values of r ($r > 0$) does an infinite geometric series approach a finite sum?
   (e) Finally, do the same for $r < 0$.

8. The Maclaurin series for $1/(1 - x)$ is, in fact, a geometric series. Based on your answer to Part 7, for which values of x does that series add up to a finite number? (This range of values is called the "interval of convergence" of the series.)

2.6.3 Explanation: Sequences and Series

A Taylor series is a particularly important case of the more general concept of a *series*. Now that we have discussed Taylor series, we are going to step back and explore series in general. We do this for two reasons.

First, an understanding of series provides mathematical rigor to Taylor series (and other useful series such as Fourier series, which we will encounter later). When we say that an *infinitely long* Taylor series adds up to a *finite* function, what exactly do we mean? A rigorous definition exists: without it, our hand-waving may get us into trouble.

Second, we have seen that Taylor series do not always work! The Maclaurin series for $1/(1 - x)$ only works if x is between -1 and 1. This is an important fact in itself, but even more important as a reminder that any Taylor series we generate may not always work. By examining the basic theory of infinite series, we can approach the question of whether a particular Taylor series adds up to some finite sum ("converges").

Mathematicians will tell you that it is always dangerous to use a series without first proving that it converges. Engineers and physicists, on the other hand, tend to ignore this step, and simply assume that their series will converge. Because this book is intended primarily for engineers and physicists, we have taken the unusual step of placing this section at the *end* of the chapter. Some readers may choose to ignore it entirely. They can do so because, as it turns out, many of the most commonly used Taylor series (most notably the Maclaurin expansions for sine, cosine, and exponentials) do in fact converge for all x-values. But the expansion of $(1 + x)^p$ does *not* converge for all x; engineers know they should only use it for small values of x. This underscores the point that if you ignore this topic, you sacrifice mathematical rigor—and run the risk of using a series that doesn't work.

Sequences and Geometric Sequences

A "sequence" is any list of numbers. For instance 8, 6, 7, 5, 3, 0, 9 is a sequence made famous by Tommy Tutone in 1982.

A "geometric sequence" is a list where each number is generated by multiplying the previous number by a constant. An example is 2, 6, 18, 54, 162, for which the "common ratio"

(r)—that is, the ratio between any two adjacent numbers—is 3. Another example is 162, 54, 18, 6: in this case $r = 1/3$.

A "recursive definition" of a sequence defines each term as a function of the previous one. So the recursive definition of a geometric sequence is $a_n = ra_{n-1}$. (Read, "to generate any given term, multiply the previous term by the common ratio.")

An "explicit definition" of a sequence defines the nth term without reference to the previous term. This type of definition is more useful, because it enables you to find (for instance) the 20th term without finding 19 other terms first.

To find the explicit definition of a geometric sequence, we start listing the terms. Say the first term is a_1. The second term is, of course, the first term multiplied by the common ratio, or ra_1. The third term multiplies *that* by the common ratio, yielding $r^2 a_1$. So we get a list like this:

a_1	a_2	a_3	a_4	a_5
a_1	ra_1	$r^2 a_1$	$r^3 a_1$	$r^4 a_1$

and so on. From this you can see the generalization that $a_n = r^{n-1} a_1$, which is the explicit definition we were looking for. Read this as: "The nth term is obtained by starting with the first term and multiplying it by the common ratio $(n - 1)$ times."

Series and Geometric Series

A "series" is the sum of all the numbers in a sequence: for instance, $8 + 6 + 7 + 5 + 3 + 0 + 9$. If you add up all the terms of a geometric sequence, you have a "geometric series."

As you saw if you worked the exercise at the beginning of this section, there is a beautiful trick that can be used to add up all the terms of any geometric series, no matter how long. For instance, $2 + 6 + 18 + 54 + 162$ is a geometric series with common ratio $r = 3$. Let's call the sum of this series S. So what is $3S$? Multiply each term by 3, and you get

$$3S = 6 + 18 + 54 + 162 + 486 \quad \text{(confirm this for yourself)}$$
$$S = 2 + 6 + 18 + 54 + 162$$

Subtracting the first equation minus the second, we see that $2S = 486 - 2$, so $S = 242$. We were thus able to add up the series without actually adding up all the terms. That may seem like a lot of trouble to add up five numbers, but remember that we could have added up a hundred numbers just as easily! (See, for instance, Problem 2.99 in this section.)

We can generalize this to an arbitrary geometric series. Recall that a generalized geometric series looks like $a_1 + a_1 r + a_1 r^2 + a_1 r^3 \ldots a_1 r^{n-1}$. So we can write

$$S = a_1 + a_1 r + a_1 r^2 + a_1 r^3 + \ldots a_1 r^{n-1}$$
$$rS = a_1 r + a_1 r^2 + a_1 r^3 + a_1 r^4 \ldots a_1 r^{n-1} + a_1 r^n \text{ (confirm this!)}$$

Again, subtracting and solving, we get

$$S - rS = a_1 - a_1 r^n$$
$$S(1 - r) = a_1 (1 - r^n)$$
$$S = a_1 \frac{1 - r^n}{1 - r} \quad (2.6.1)$$

This is a general formula for the sum of any finite geometric series with the first term a_1, common ratio r, and a total of n terms. This would be a good place to stop and check yourself: can you reproduce these quick derivations without looking? Can you apply them to a new sequence or series? Most importantly, do all the steps make sense to you?

Infinite Series, and "Convergence"

Consider the series $\sum_{n=1}^{\infty} (1/3)^n$, or $1/3 + 1/9 + 1/27 + 1/81 + \ldots$

We can easily ask "What do the first three terms of this series add up to?" Or the first five terms, or the first ten terms. Because this is a Geometric series we can even find a quick answer to the question "What do the first *hundred* terms of this series add up to?" as discussed above.

But before we ask the question, "What do all *infinity* terms of this series add up to?" we should pause. If we say that the series adds up to 1/2, we cannot possibly mean that "Once I finally add up all infinity of these terms, the sum will have reached exactly 0.5": such a claim would be, not only wrong, but meaningless. Before we talk about adding up an infinite number of terms, we have to clearly define what we mean!

Nineteenth-century mathematicians struggled with this question, grappling with infinite series such as:

$$1 - 1 + 1 - 1 + 1 - 1 \ldots$$

Some argued that the series must add up to 0, by grouping it like this:

$$\underbrace{1-1}_{0} + \underbrace{1-1}_{0} + \underbrace{1-1}_{0} \ldots$$

Others argued that the series adds up to 1, by a slightly different grouping:

$$1 \underbrace{-1+1}_{0} \underbrace{-1+1}_{0} \underbrace{-1+1}_{0} \ldots$$

Still others with different definitions argued that the series added up to 1/2! These mathematicians were not just disagreeing about the "right answer" to this sum: they were disagreeing about what an infinite sum *means*. Once you settle on a definition, you can begin to look for the answers.

The most commonly accepted definition today is based on partial sums.

Definition: Series Convergence

The "nth partial sum" of a series, often designated as S_n, is defined as the sum of the first n terms of the series.

The sum of an infinite series is defined as $\lim_{n \to \infty} S_n$. If that limit exists the series is said to "converge." If the limit does not exist the series is said to "diverge."

Once this definition is accepted, the answer to the riddle of $1 - 1 + 1 - 1 + 1 - 1 \ldots$ becomes obvious. The partial sums are 1, 0, 1, 0, 1, 0, so they are not approaching anything: the series diverges.

More importantly, it clarifies what we mean when we say that $\sum_{n=1}^{\infty} (1/3)^n$ adds up to 1/2. If you add up the first three terms, you get roughly 0.48. If you add up the first five terms, you get roughly 0.498. The first ten terms add up to slightly more than 0.49999. The sum will get *arbitrarily close* to 0.5 if we add up enough terms. We say that the series "converges to 1/2."

We can prove this by remembering that a *finite* geometric series adds up to $a_1(1-r^n)/(1-r)$. So for the series $\sum (1/3)^n$ we can find the sum of the first n terms.

$$S_n = \left(\frac{1}{3}\right) \frac{1-(1/3)^n}{1-1/3} = \frac{1}{2}\left[1 - \left(\frac{1}{3}\right)^n\right]$$

In the limit as $n \to \infty$ this sum approaches 1/2.

The definition $\sum a_n = \lim_{n\to\infty} S_n$ gives us—in some cases—a useful tool for determining whether a series converges or diverges. Below we give two examples, each important in its own right, but also useful to show how we use partial sums to evaluate an infinite series.

> **EXAMPLE** **Convergence and Divergence of Geometric Series**
>
> A geometric series is any series of the form $\sum a_1 r^{n-1}$ where a_1 is the first term and r is the common ratio. By using the "geometric series trick" it can be shown that the nth partial sum of such a series is given by $S_n = a_1 (1 - r^n)/(1 - r)$.
>
> The sum of the entire (infinite) series is therefore, by definition, $\lim_{n\to\infty} a_1 (1 - r^n)/(1 - r)$, Since the only n-dependence in this formula appears in the r^n in the numerator, we can evaluate that limit by inspection. The limit of this function as $n \to \infty$ depends on the value of r, and can be summarized as follows:
>
> - If $|r| < 1$ (in other words, if r is *between* -1 and 1), the series converges to $a_1/(1 - r)$.
> - If $|r| \geq 1$, the series diverges.

(Equation 2.6.1 is not valid when $r = 1$, but it is easy to check that the infinite geometric series diverges in this case.)

> **EXAMPLE** **"Telescoping" Series**
>
> Consider the series $\sum_{n=2}^{\infty} (1/(n-1) - 1/n)$. Based on the discussion above of partial sums, can you figure out what this series does?
>
> The strategy is to start writing out terms. For instance, the fifth partial sum is:
>
> $$S_5 = \left(\frac{1}{1} - \frac{1}{2}\right) + \left(\frac{1}{2} - \frac{1}{3}\right) + \left(\frac{1}{3} - \frac{1}{4}\right) + \left(\frac{1}{4} - \frac{1}{5}\right) + \left(\frac{1}{5} - \frac{1}{6}\right)$$
>
> Note that the "middle" elements, such as the $-1/2$ in the first term and the $1/2$ in the second term, cancel out. So $S_5 = 1 - 1/6$, or $5/6$. More generally,
>
> $$S_n = \left(\frac{1}{1} - \frac{1}{2}\right) + \left(\frac{1}{2} - \frac{1}{3}\right) + \ldots + \left(\frac{1}{n} - \frac{1}{n+1}\right) = 1 - \frac{1}{n+1} = \frac{n}{n+1}$$
>
> Such a series is referred to as a "telescoping" series. (The metaphor refers to the way some telescopes collapse in on themselves.) When a series telescopes, you can evaluate it directly by taking the limit of S_n: in this case, the infinite series converges to $\lim_{n\to\infty} [n/(n+1)] = 1$.

When You Can't Find the nth Partial Sum

Telescoping series and geometric series have the wonderful property that you can write a closed-form expression for the nth partial sum. You can therefore directly use the definition of series convergence to find what the infinite series do.

In most cases, however, it is impossible to write such an expression, so more indirect means must be used.

EXAMPLE: The Harmonic Series

The series $\sum_{n=2}^{\infty}(1/n)$ is given a special name, the "harmonic series." On casual inspection, it is not obvious whether this series converges or diverges: for instance, there is no obvious way to find the nth partial sum. However, the series can be shown to diverge by a clever trick. Shown below are the harmonic series and a completely different series obtained by replacing the denominator of each term with the next power of two.

Harmonic: $\frac{1}{2} + \frac{1}{3} + \frac{1}{4} + \frac{1}{5} + \frac{1}{6} + \frac{1}{7} + \frac{1}{8} + \frac{1}{9} + \frac{1}{10} + \frac{1}{11} + \frac{1}{12} + \frac{1}{13} + \frac{1}{14} + \frac{1}{15} + \frac{1}{16} + ...$

Alternate: $\frac{1}{2} + \frac{1}{4} + \frac{1}{4} + \frac{1}{8} + \frac{1}{8} + \frac{1}{8} + \frac{1}{8} + \frac{1}{16} + \frac{1}{16} + \frac{1}{16} + \frac{1}{16} + \frac{1}{16} + \frac{1}{16} + \frac{1}{16} + \frac{1}{16} + ...$

The second series can easily be shown to diverge. It has two 1/4 terms, which add up to 1/2. It has four 1/8 terms, which also add up to 1/2, and so on. The second series is an infinite string of halves, which will therefore grow without bound.

But term for term, the elements in the harmonic series are *at least as big* as the elements in this alternate series! So when the alternate series reaches 2, the harmonic series will be higher than 2. When the alternate series reaches 100, the harmonic series will be higher than 100, and so on. Since the alternate series grows without bound, the harmonic series must also diverge.

The nth-Term Test

At the end of this chapter, we will present many general rules for determining whether a series converges or diverges. But one rule is simple enough, and important enough, to present from the start.

The nth-Term Test

Given a series $\sum a_n$: if $\lim_{n \to \infty} a_n \neq 0$ then the series diverges.

Important notes:

1. This test refers to a_n, the nth term of the series, *not* to S_n, the nth partial sum.
2. This test can prove that a series diverges, but it can never prove that a series converges.

EXAMPLE: The nth-Term Test

Consider the series $\sum_{n=1}^{\infty} n/(n+1) = (1/2) + (2/3) + (3/4) + ...$ The individual terms are approaching 1, not zero: therefore, we can conclude immediately that the series diverges.

The nth-term test gives us a tool for examining a series without calculating the nth partial sum. As with all such tools, however, the validity of the test itself must ultimately be proven *based on* the S_n definition of convergence. (By analogy, remember that we rarely use the $\lim_{h \to 0}$ definition to take a derivative—but the rules we *do* use, such as "derivative of sine is cosine" or the product rule, must be proven from that definition.)

So in order to understand why the nth-term test works, consider the fact that a series only converges if the partial sums approach some finite value. That can't possibly happen for the series in our example above: no matter what value it seems to be approaching by the 100th term, it will jump *past* that value in the 101st term, adding a number that is almost 1. The series will continue to grow without bound.

If you follow that logic for this one example, you can probably see why it generalizes to the statement that the partial sums can *never* approach a finite value unless the individual terms are approaching zero.

The converse of the nth-term test would state that "For any series in which $\lim_{n\to\infty} a_n = 0$, the series converges." If that were true, proving that any series converges or diverges would be as easy as taking a limit! Unfortunately, no such guarantee can be given. For instance, in the "harmonic series" presented above, $\lim_{n\to\infty} a_n = 0$ but the series diverges anyway. In Section 2.7 we will present other methods that can be used when the nth-term test fails.

2.6.4 Problems: Sequences and Series

2.95 Consider $\sum_{n=1}^{\infty} \sin\left(\frac{\pi n}{2}\right)$

   (a) Compute the 22nd partial sum of this series.
   (b) Does this infinite series converge or diverge?

2.96 *True or False?* ("True" means it will be true for *any series*.) In each case, offer a brief explanation of your answer.

   (a) $\sum (a_n + 2) = \left(\sum a_n\right) + 2$
   (b) $\sum (2a_n) = 2\sum a_n$
   (c) $\sum (a_n)^2 = \left(\sum a_n\right)^2$

2.97 Consider the geometric series $1 + 10 + 100 + 1000 + \ldots$

   (a) What is the 20th term of this series?
   (b) What is the sum of the first 20 terms in this series?
   (c) Express Part (b) in summation notation.

2.98 Consider the geometric series $2 + 8 + 32 + 128 + \ldots$

   (a) What is the 32nd term of this series?
   (b) What is the sum of the first 32 terms in this series?
   (c) Express Part (b) in summation notation.

2.99 Consider the geometric series $100 + 20 + 4 + \ldots$

   (a) What is r, the common ratio?
   (b) What is the 100th term of this series?
   (c) What is the sum of the first 100 terms in this series?
   (d) Express Part (c) in summation notation.

2.100 Consider the geometric series with $a_1 = 2$ and $r = -3$.

   (a) List the first six terms of the series.
   (b) What is the 15th term of this series?
   (c) What is the sum of the first 15 terms in this series?
   (d) Express Part (c) in summation notation.

2.101 Compute $\sum_{n=1}^{10} (10 \times 1.1^n) \ldots$

   (a) by listing all ten terms and adding them
   (b) by using the geometric series "trick"
   (c) by using the formula $a_1(1 - r^n)/(1 - r)$

2.102 Consider the geometric series $\sum_{n=1}^{\infty} 9 \times (2/3)^n$.

   (a) Write out the first five terms.
   (b) Compute the sum of the first five terms.
   (c) Compute the sum of the first twenty terms.
   (d) Compute the sum of the first n terms.
   (e) Compute the infinite sum, by taking the limit of your answer to Part (d) as $n \to \infty$.
   (f) Life would have been simpler if you could have just "pulled the 9 out" like this: $9\sum_{n=1}^{\infty} (2/3)^n$. Is that valid, or not? Why?

2.103 Consider the geometric series $2 + 2(1.1) + 2(1.1)^2 + 2(1.1)^3 + 2(1.1)^4 + \ldots$

   (a) Write this in summation notation
   (b) Compute the sum of the first five terms.
   (c) Compute the sum of the first twenty terms.
   (d) Compute the sum of the first n terms.
   (e) Compute the infinite sum by taking the limit of your answer to Part (d) as $n \to \infty$.

2.104 Consider the series $\sum_{n=1}^{\infty} \frac{1}{n(n+1)}$.

   (a) Write out the first five terms of the series.
   (b) Write out the first five *partial sums* of the series. Compute them as fractions, not as decimals.

(c) Based on spotting a pattern in your answers to Part (b), write a closed-form expression for the nth partial sum of this series.

(d) Compute the infinite sum by taking the limit of your answer to Part (c) as $n \to \infty$.

2.105 *[This problem depends on Problem 2.104.]* The steps you have just taken do not constitute a proof because you have no evidence that the pattern you spotted in Part (b) will continue forever. Prove this pattern in either of the following two ways. One way is to rewrite the expression $1/[n(n+1)]$ using *partial fractions*: in this form the series can be shown to telescope. The other way is to use a *proof by induction*. Neither of these techniques is discussed in this book.

2.106 The function $6/(n^2 + 4n + 3)$ can be rewritten using partial fractions as $3/(n+1) - 3/(n+3)$. Using the function *in this rewritten form*,

(a) Write out the first five terms of the series $\sum_{n=1}^{\infty} 6/(n^2 + 4n + 3)$.

(b) Compute the fifth partial sum of the series $\sum_{n=1}^{\infty} 6/(n^2 + 4n + 3)$.

(c) Compute the nth partial sum of the series $\sum_{n=1}^{\infty} 6/(n^2 + 4n + 3)$.

(d) Does $\sum_{n=1}^{\infty} 6/(n^2 + 4n + 3)$ converge or diverge? If it converges, what does it converge to?

2.107 Consider the series $\sum_{n=1}^{\infty} \ln\left(\frac{n}{n+1}\right)$.

(a) Let's start where we should always start with any unfamiliar series: the nth-term test. Does this test (*choose one of the following*):
 i. guarantee that the series converges,
 ii. guarantee that the series diverges, or
 iii. Give us no guarantee of any kind of what this series does?

(b) What is the third partial sum S_3?

(c) What is the nth partial sum S_n? (Your answer should be a formula in closed form, not a series.)

(d) Based on your answer to Part (c), say whether the series converges or diverges. If it converges, say what it converges to.

2.108 Show that $\sum_{n=3}^{\infty} \frac{n^2 - 4}{n^2 - 1}$ diverges.

2.109 Show that $\sum_{n=1}^{\infty} \frac{1}{2n}$ diverges.

2.110 Does the geometric series $\sum_{n=1}^{\infty} \frac{5^{n+1}}{3^{2n}}$ converge or diverge? If it converges, what does it converge to?

Problems 2.111–2.113 are about "repeating decimals." For instance the decimal $0.23232323\ldots$, often written $0.\overline{23}$, means $23/100 + 23/10,000 + \ldots$

2.111 Consider the repeating decimal $0.33333\ldots$ or $0.\overline{3}$.

(a) Write the first three terms of this series.

(b) What is r, the common ratio?

(c) Find the sum of the infinite series.

2.112 Convert the decimal given above, $0.\overline{23}$, into a fraction by computing the sum of an infinite series.

2.113 Students often ask whether $0.\overline{9}$ actually *equals* 1 or only *approximates* 1. Answer this question with a convincing argument.

2.114 A ball is thrown up 90 feet in the air. it comes down, bounces, and then rises up to a height of 81 feet. Then it falls to its second bounce, and so on. Each time it falls, it bounces back to 90% of its previous height.

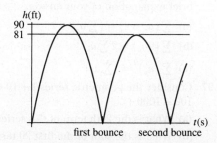

(a) How high does the ball rise before the 30th bounce? (Please read the question carefully. If the question read "How high does the ball rise before the second bounce?" the answer would be 81 feet.)

(b) How far has the ball traveled when it hits the ground at the 30th bounce? (Again, read carefully. If the question read "How far has the ball traveled when it hits the ground at the second bounce?" the answer would be 90 feet up plus 90 feet down plus 81 feet up plus 81 feet down, for a total of 342 feet.)

(c) How far does the ball travel in the limit as the number of bounces approaches infinity?

2.115 According to an ancient story—very famous and almost certainly not true—a man did a great service for the king, and the king asked the man to name his reward. The man pointed to a chess board. (Chess boards, then as now, had 64 squares.) "Your majesty, place one grain of wheat in the first square of this board. On the next square place two grains. On the next square four grains, and so on, each square getting twice as many grains as the square before it. When you have placed the last grains on the 64th square, I will take my reward and leave." The king thought he was getting a great deal.

(a) How many grains of wheat would be placed on the final square?
(b) How many grains, total, would be placed on the chess board?

2.116 You buy a used car for $5000. You pay for it by borrowing $5000 on a credit card. The credit card company charges 15% interest: that is, every year, your debt multiplies by 1.15. (This assumes that you are not paying any of the money back.)

(a) How much do you owe after four years?
(b) How much do you owe after ten years?
(c) How much do you owe after n years?
(d) How long will it be before you owe $100,000?

2.117 In 1975 Intel founder Gordon Moore predicted that *every two years, the number of transistors that can be inexpensively placed on an integrated circuit would double.* This famous prediction is now known as "Moore's Law." At the time the Intel 8080 chip held 4500 transistors and Moore's law has so far proven quite accurate.

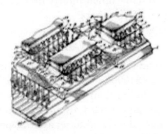

In the fictitious Museum of Random Computer Stuff a special exhibit contains an integrated circuit from 1975, one from 1977, one from 1979, and so on. Each circuit follows Moore's law precisely. There are 4500 transistors on the 1975 circuit, 9000 on the 1977 circuit, and so on, ending with a circuit from 2011.

(a) How many circuits are in the exhibit?
(b) How many transistors are on the last circuit (the one from 2011)?
(c) How many transistors are there in the entire exhibit?
(d) Real life inventor and futurist Raymond Kurzweil has very publicly predicted that by the year 2020 a personal computer will have roughly the same computational capacity as a human brain. How many transistors will be on one circuit in that year?

2.118 On graduating from college, Johnny gets himself a good job and starts saving for retirement (good). His savings plan is that, each year, he stuffs some money under his mattress (bad). The first year, he puts aside $200. The second year, as his income goes up, he is able to put aside $220. The third year he puts aside $242. Each year, the amount he saves goes up by 10%.

(a) How much does Johnny put aside in the 30th year?
(b) In what year does Johnny first put aside over $10,000?
(c) After 40 years, as he is nearing retirement, Johnny needs to know how much money has piled up under his mattress over the years. Express this question in summation notation.
(d) Now answer the question.

2.119 One important use of series is in *Riemann sums*, which approximate an integral as a series of rectangles. In a left-hand Riemann sum the height of each rectangle is determined by the height of the function at the left-most point in the rectangle.

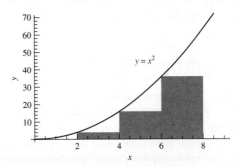

(a) The picture shows the function $y = x^2$. The area under this function between $x = 0$ and $x = 8$ is being approximated by a left-hand Riemann sum using four rectangles of equal width. Calculate the sum of the areas of these four rectangles.

(b) Use summation notation to represent the question in Part (a).

(c) Calculate the percent error if your answer in Part (a) is used to approximate $\int_0^8 x^2\, dx$.

(d) Suppose the area under the function $y = \ln x$ between $x = 1$ and $x = 4$ is calculated by using a left-hand Riemann sum with six rectangles of equal width. Use summation notation to write the resulting sum, and then calculate the answer.

(e) Suppose the area under the function $y = (1.2)^x$ between $x = 0$ and $x = 50$ is calculated by using a left-hand Riemann sum with fifty rectangles of equal width. Use summation notation to write the resulting sum, and then calculate the answer.

2.120 One definition of "area under a curve" is the limit of a Riemann sum as the number of rectangles approaches infinity. In this problem you will approach the integral $\int_0^{50} 1.2^x\, dx$ using this definition. (If you're not familiar with Riemann sums they are defined in Problem 2.119.)

(a) Use summation notation to represent the area under this curve if you use a Riemann sum with n rectangles of equal width.

(b) Rewrite your answer in "closed form": not a series, just a function of n, the number of rectangles.

(c) Use a calculator or computer to estimate the limit of your area as $n \to \infty$.

(d) Evaluate $\int_0^{50} 1.2^x\, dx$ and see how close your estimate was. *Hint*: To find the antiderivative of $(1.2)^x$ it is helpful to rewrite the integrand as $(1.2)^x = e^{x \ln(1.2)}$.

2.121 The series $\sum (3^{n-1}/4^n)$ does not, on the face of it, appear to be a geometric series—but it is.

(a) Rewrite it in such a way that it is obviously geometric.

(b) Find the first term a_1 and the common ratio r.

(c) Does the series converge or diverge? If it converges, what does it converge to?

2.122 An "arithmetic sequence" *adds* a constant to each term to generate the next term. An arithmetic sequence has a "common difference" d instead of a "common ratio" as a geometric sequence has. As an example, consider the arithmetic sequence 10, 13, 16, 19, 22.

(a) Find a recursive formula for generating this sequence: that is, a formula for the nth term a_n as a function of a_{n-1}.

(b) What would be the 100th term in this sequence?

(c) Find an explicit formula for generating this sequence: that is, a formula for a_n as a function of n. (Make sure your formula correctly predicts that $a_5 = 22$.)

2.123 Just as the Greek capital letter Sigma, Σ, is used to indicate repeated *addition*, the Greek capital letter Pi, Π, is used to indicate repeated *multiplication*.

(a) Compute $\prod_{n=1}^{5}(n+1)$.

(b) Write the product $2 \times 6 \times 18 \times 54$ in product notation.

(c) Rewrite the sum $\sum_{n=1}^{50} \ln n$ using product notation.

2.124 You are saving money in an Individual Retirement Account (IRA). At the beginning of each year you deposit $1000 into your account. At the end of each year your IRA gains 6% interest; that is, the total amount of money in the IRA is multiplied by 1.06.

At the beginning of the first year you deposit $1000. At the end of the first year interest brings your total to $1060. At the beginning of the second year you deposit another $1000, bringing the total to $2060. At the end of the second year interest brings that total up to $2183.60.

(a) How much money do you have after ten years? (*Hint*: Work backward. Your last deposit was made after the ninth year. How much was it worth after the tenth year? Your next-to-last deposit was made after the eighth year. How much was it worth after the tenth year?)

(b) How much money do you have after y years?

(c) How much money do you have after thirty years?

2.125 If an insurance company is going to pay you $1000 in one year they need to set aside less than $1000 right now. Assuming they make 1.4% interest on their savings, they need to set aside approximately $1000 × 0.986 today to make a payment of $1000 next year. If they need to pay you $1000 two years from today they need to set aside roughly $1000 × $(0.986)^2$ today.

(a) How much money do they need to set aside today if they need to pay you $1000/year for the next thirty years? (This is called the "net present value" of the payout.)

(b) How much do they need to set aside if they need to pay you $1000/year *forever*?

2.126 The ancient Greek philosopher Zeno attempted to prove that all motion is impossible by means of the following paradox. Suppose you are walking to someone's house. Before you can get there you must of course travel halfway there. Once that's accomplished you just have to cover that same distance again, right? But not so fast—before you can do that you have to cover half the *remaining* distance. And after that step you have a fourth of the original distance left. Before you can cover that you have to travel half *that* distance, and so on. Because you have to cross an infinite number of distances, Zeno concluded that you can never get anywhere.

Translate Zeno's paradox into an infinite series. Then, by showing that this series converges, show that it is indeed possible to take Zeno's infinite number of steps in a finite time.

2.127 The series $\sum_{n=1}^{\infty} 1/n^4$ converges, although we won't be able to prove it until Section 2.7. *Estimate* the sum of this series by computing successive partial sums on your calculator or a computer and seeing what they tend to converge to.

2.128 Determine $\sum_{n=0}^{99} \int_{n}^{n+1} x^3 \, dx$.

2.129 The series $1 + 2x + 4x^2 + 8x^3 + 16x^4 + \ldots$ is geometric. However, we cannot ask the question "does it converge or diverge?" because the answer depends on the value of x. Rather, we ask the question: "For what values of x does this series converge?" Make sure your answer takes into account both positive and negative values of x.

2.130 You flip a coin over and over. The probability that the *first time you gets heads* is on the third flip is 1/8, because it requires starting with the sequence tails, tails, heads. More generally the odds that the first heads will occur on the nth flip is $P(n) = 1/2^n$.

(a) Add up all the probabilities $P(n)$ from $n = 1$ to ∞. Explain why your answer makes sense for this scenario.

Your goal for the rest of this problem is going to be to find the "expectation value" of n. If you did this experiment many times and wrote down the value of n each time (i.e., how many flips it took you to get heads), the expectation value is what all of those results would tend to average to over time. The formula for it is $\sum_{n=1}^{\infty} n/2^n$. This is not an easy sum to evaluate, but we're going to show you a trick for making it easier.

(b) We define $f(t) = \sum_{n=1}^{\infty} e^{nt}/2^n$. (The function $f(t)$ is called a "moment generating function.") Rewrite $f(t)$ as a geometric series and find its value. Write your answer in the form of an equation: $\sum_{n=1}^{\infty} e^{nt}/2^n = \text{<something>}$.

(c) Take the derivative with respect to t of both sides of the equation you just wrote. The result should be a new equation that will still have an infinite sum on the left and a normal expression on the right, both functions of t.

(d) By evaluating both sides of this equation at $t = 0$, find the expectation value of n as defined above.

2.131 **Exploration: Chaos Theory**

Biologists model constrained population growth by the recursive sequence $p_{n+1} = kp_n(1 - p_n)$, where p_n is the population after n generations and k is a positive constant called the "growth rate." p_n is always between 0 (no population) and 1 (full capacity): $p = 1/2$ does not mean the population is half an animal, but that the population is at half its capacity.

(a) Letting $k = 1.2$ and $p_1 = 0.1$, use a computer program to rapidly generate at least thirty generations. You will see the population approaching an equilibrium point. What is it?

(b) With k still set to 1.2, experiment with different values of p_1 ranging from 0.1 to 0.9. Note that the resulting numbers differ each time but they approach the same limiting value.

(c) Repeat the experiment with k set to 1.5, 2.0, 2.3, and 2.7. Record in a table the k-values and the resulting equilibrium points. For each value of k be sure to try at least two values of p_1.

(d) Repeat the experiment with $k = 3.1$. What happened?

(e) Repeat the experiment with $k = 3.2$. The results are similar, but not identical, to the results in Part (d). Describe them.

(f) Repeat the experiment with $k = 3.5$. What happened now?

(g) Finally, repeat the experiment with $k = 3.7$.

With $k = 3.7$ the sequence represents what mathematicians call "chaos." There is no discernable pattern to the numbers. If you repeat the experiment with the exact same starting value, of course you will get the same results. However, if you change the starting value, even by a very small amount, the resulting numbers will be completely different. (This property goes by the imposing name "sensitive dependence on initial conditions.")

(h) Chaotic behavior—including sensitive dependence on initial conditions—has been shown to apply to real-world systems including the weather and the stock market. What implications does this have for scientists attempting to predict the behavior of such systems?

2.7 Tests for Series Convergence

We have discussed the creation of Taylor series, and the question of series convergence or divergence in general. We are now ready to establish a battery of tests that can be applied to determine whether a series converges.

2.7.1 Explanation: The Ratio Test

This section discusses the "Ratio Test," the most important test for determining the convergence of a Taylor series. Other tests are discussed in Section 2.7.3.

What Do We Know So Far?
We have already encountered three rules of series convergence:

1. *The definition of convergence:* We define S_n (the "nth partial sum") of a series as the sum of the first n terms, and we define $\sum_{n=1}^{\infty} a_n$ as $\lim_{n \to \infty} S_n$. For series where we can find a closed-form expression for the nth partial sum, such as "telescoping" series, this definition enables us to determine if the series converges, and if so, what it converges to.
2. *The "nth-term test":* For any series $\sum_{n=1}^{\infty} a_n$, if $\lim_{n \to \infty} a_n \neq 0$, then the series diverges. (This is the limit of the *individual terms*, not of the partial sums!) This test follows directly from the definition of series convergence: if the individual terms are not approaching zero, then the partial sums cannot be "closing in" on any finite value. Remember, however, that this test can only show *divergence*: if $\lim_{n \to \infty} a_n = 0$, the series may converge or diverge.
3. *Geometric series:* For a geometric series with common ratio r, if $|r| \geq 1$, then the series diverges; if $|r| < 1$, then the series converges to $a_1/(1-r)$. We proved this directly from the definition of convergence, using the "geometric series trick" to show that $S_n = a_1(1-r^n)/(1-r)$, and then taking the limit. Note, incidentally, that for any power series $\sum c_n(x-a)^n$, if the coefficients c_n form a geometric sequence (including the case where they are all equal), then the power series itself is a geometric series.

These three facts form a good start, but they are not sufficient for many of the series that we most want to address. For instance, we know that the Maclaurin series for e^x is $\sum (x^n/n!)$, but how do we determine which x-values this series converges for? $\lim_{n \to \infty} (x^n/n!) = 0$ for any x-value, so the nth-term test is no help here. The series is not geometric, and there is no easy way to write a closed-form expression for the nth partial sum, so our other two techniques are equally useless. Other approaches will be required.

The Ratio Test

The ratio test is a formula that looks at the ratio of each term in the series to the previous term.

> **The Ratio Test**
>
> If $\lim_{n\to\infty} \left|\dfrac{a_{n+1}}{a_n}\right|$ is ... $\begin{cases} < 1 & \text{The series } \sum a_n \text{ converges} \\ > 1 & \text{The series } \sum a_n \text{ diverges} \\ = 1 & \text{The ratio test is inconclusive for this series} \end{cases}$

EXAMPLE Applying the Ratio Test to a Numerical Series

Does the series $\dfrac{2}{1} + \dfrac{4}{1\times 2} + \dfrac{8}{1\times 2\times 3} + \dfrac{16}{1\times 2\times 3\times 4} + \ldots$ converge or diverge?

This series can also be written as $\sum_{n=1}^{\infty} 2^n/(n!)$. Applying the ratio test...

$$\left|\frac{a_{n+1}}{a_n}\right| = \left|\frac{2^{n+1}/(n+1)!}{2^n/(n!)}\right| = \left|\left(\frac{2^{n+1}}{(n+1)!}\right)\left(\frac{n!}{2^n}\right)\right| = \left|\left(\frac{2^{n+1}}{2^n}\right)\left(\frac{n!}{(n+1)!}\right)\right|$$

Now, $2^{n+1}/2^n$ reduces to 2 by the rules of exponents. $n!/(n+1)!$ may be less familiar, but if you write out what each part means, it looks like $\dfrac{1\times 2\times 3 \ldots \times n}{1\times 2\times 3 \ldots \times n\times (n+1)}$ and everything cancels but the last term, leaving $1/(n+1)$. So, returning to our ratio test, we have... $\lim_{n\to\infty} |2/(n+1)| = 0$. Since the limit is less than 1, the series converges.

Pay close attention to those steps, because they are the same almost every time you apply the ratio test. First, you write out the ratio. Then you rearrange the fractions so that similar terms are grouped together: in this case, the two exponential terms, and the two factorial terms. Then you simplify and take the limit.

EXAMPLE Applying the Ratio Test to the Maclaurin Series for e^x

The function e^x can be represented by the power series $\sum x^n/(n!)$. Recall that for a power series, our question is not *does it converge*, but *for which x-values does it converge*?

$$\left|\frac{a_{n+1}}{a_n}\right| = \left|\frac{x^{n+1}/(n+1)!}{x^n/n!}\right| = \left|\left(\frac{x^{n+1}}{x^n}\right)\left(\frac{n!}{(n+1)!}\right)\right| = \left|\frac{x}{n+1}\right|$$

Now we take the limit: $\lim_{n\to\infty} |x/(n+1)| = 0$ for all x-values, so this series converges for all x-values.

EXAMPLE Applying the Ratio Test to Another Power Series

Consider the power series $\sum n! x^n$. Applying the ratio test as before:

$$\left|\frac{a_{n+1}}{a_n}\right| = \left|\frac{(n+1)! x^{n+1}}{n! x^n}\right| = |(n+1)x|$$

Now we take the limit $\lim_{n\to\infty} |(n+1)x|$. If $x = 0$ then this limit approaches zero, which is less than 1, so the series converges. For *any x-value other than zero*, the limit approaches infinity, which is considerably more than 1, so the series diverges.

EXAMPLE Applying the Ratio Test to Yet Another Power Series

(We promise it's the last one for the moment.)

Consider the power series $\frac{2(x-5)}{1} + \frac{4(x-5)^2}{2} + \frac{8(x-5)^3}{3} + \frac{16(x-5)^4}{4} + \ldots = \sum_{n=1}^{\infty} \frac{2^n(x-5)^n}{n}$.

Once again, applying the ratio test:

$$\left|\frac{a_{n+1}}{a_n}\right| = \left|\frac{2^{n+1}(x-5)^{n+1}/(n+1)}{2^n(x-5)^n/n}\right| = \left|\left(\frac{2^{n+1}(x-5)^{n+1}}{n+1}\right)\left(\frac{n}{2^n(x-5)^n}\right)\right|$$

$$= \left|\left(\frac{2^{n+1}(x-5)^{n+1}}{2^n(x-5)^n}\right)\left(\frac{n}{n+1}\right)\right|$$

Since $\lim_{n\to\infty} [n/(n+1)] = 1$, this reduces to $|2(x-5)|$. So after all that algebra, the ratio test promises that this series will converge whenever $2|x-5| < 1$, or $|x-5| < 1/2$. Now what?

One approach to such absolute value inequalities is to say that if the absolute value of *some number* is less than $1/2$, then *some number* must be between $-1/2$ and $1/2$. So $-1/2 < (x-5) < 1/2$, so $4.5 < x < 5.5$.

For this power series we conclude that:

1. If $4.5 < x < 5.5$, then the ratio test promises *convergence*.
2. If $x < 4.5$ or $x > 5.5$, then the ratio test promises *divergence*.

But what of $x = 4.5$ and $x = 5.5$? When the limit is exactly 1 the ratio test is inconclusive. At these two points we need other tests, which will be discussed in the next section.

The three examples above illustrate the *process* for applying the ratio test to power series. They also illustrate the types of series where the ratio test is most powerful—those involving exponential functions and factorials—since the ratio a_{n+1}/a_n simplifies naturally for these functions.

Less obviously, these three examples illustrate the only three possible results of applying the ratio test to a power series.

1. Some power series *converge everywhere*.
2. Some power series *diverge everywhere* except at one point: the *x*-value around which the power series was built. (At this value, a power series obviously converges to C_0, since all the other terms are zero!)
3. Some power series converge in a given interval, and diverge outside that interval. The center of the interval is always the *x*-value that the series was built around: for instance, a power series built around $x = 5$ (i.e., $\sum c_n (x-5)^n$) may converge for all *x*-values between 3 and 7, or between -1 and 11. At the end points of this interval, the ratio test is inconclusive, and other techniques (discussed in the following section) must be used.

One final question must be raised. We proved earlier that the series $\sum x^n/(n!)$ converges for all *x*-values, but not that it converges to e^x. Is it possible that it converges to the wrong thing?

It turns out that $\sum x^n/(n!)$ does in fact converge to e^x for all *x*-values. However, there is no mathematical guarantee for all convergent Taylor series: for instance, we saw in Section 2.3 Problem 2.52 that the function e^{-1/x^2} can be expanded into a Maclaurin series that does not converge back to e^{-1/x^2}. For full mathematical rigor, *after* you prove that a Taylor series converges, you must then prove that it converges to the function you want. You can find a more thorough treatment of this issue in many calculus texts. We ignore it here for a purely pragmatic reason: it is possible in principle, but extremely rare in practice, for a Taylor series to converge to the wrong function.

Why Does the Ratio Test Work?

The ratio test says roughly "If this *were* a geometric series, what geometric series would it be?" If the ratio a_{n+1}/a_n is a constant, then the series is actually geometric. If this ratio is not constant, but is *approaching* a given value as $n \to \infty$, then the series approaches a geometric series. Since we know that geometric series converge if $|r| < 1$, the result of the ratio test is at least plausible.

For example, consider a series where $\lim_{n \to \infty} |a_{n+1}/a_n| = 0.9$. This means that—at some point, as you advance through the series—the ratio will be below 0.91, and will stay below 0.91 forever. For all *n*-values beyond that point, we can write:

$$\frac{|a_{n+1}|}{|a_n|} < 0.91 \quad \to \quad |a_{n+1}| < 0.91 \, |a_n|$$

Now consider a series with $a_{n+1} = 0.91 a_n$. Such a series would be geometric with $r = 0.91$, so it would converge. And the individual terms of this series would always be *higher than* the terms of our actual series. We therefore conclude that our original series cannot possibly blow up, and must converge.

(This intuitive notion can be expressed more rigorously in terms of the "basic comparison test" discussed in the next section. If you're concerned about the absolute values in that equation, that will be addressed in "absolute convergence," also in the next section. But even without all that analysis you should be able to convince yourself that since the geometric series converges, our always-closer-to-zero series converges too.)

This example proves half of the ratio test: if $\lim_{n \to \infty} |a_{n+1}/a_n| < 1$ then the series must converge. Proving the other half is left for you to tackle in Problem 2.146.

2.7.2 Problems: The Ratio Test

In Problems 2.132–2.140, determine whether the ratio test proves the series convergent or divergent, or if the ratio test is inconclusive.

2.132 $\sum_{n=1}^{\infty}(1/n!)$

2.133 $\sum_{n=1}^{\infty}(1/2^n)$

2.134 $\sum_{n=1}^{\infty}2^n/(n!)$

2.135 $\sum_{n=1}^{\infty}(n!/2^n)$

2.136 $\sum_{n=1}^{\infty}(1/n^5)$

2.137 $\sum_{n=1}^{\infty}(1/\sqrt{n})$

2.138 $\sum_{n=1}^{\infty}(\ln n/n^3)$

2.139 $\sum_{n=2}^{\infty}1/(n\ln n)$

2.140 $\sum_{n=1}^{\infty}n^5/(2n)!$

In Problems 2.141–2.145, use the ratio test to determine whether the given power series converges for all x-values, converges for only one x-value, or converges in an interval. In the latter case, you will determine the interval, but you will not be able (yet!) to determine what happens at the end points of the interval.

2.141 $\sum_{n=0}^{\infty}\dfrac{(-1)^n x^{2n+1}}{(2n+1)!}$ (This is the Maclaurin series for the sine function.)

2.142 $\sum_{n=1}^{\infty}(-1)^{n-1}(x-1)^n/n$

2.143 $\sum_{n=2}^{\infty}(2x-6)^n/(n^2-1)$

2.144 $\sum_{n=1}^{\infty}\dfrac{(-1)^n 2^{2n-1}}{(2n)!}x^{2n}$

2.145 $\sum_{n=1}^{\infty}(x+1)^n/n^n$

2.146 In the Explanation (Section 2.7.1) we proved half of the ratio test: if $\lim\limits_{n\to\infty}|a_{n+1}/a_n| < 1$, then the series must converge. Construct a similar argument to show the other half: if $\lim\limits_{n\to\infty}|a_{n+1}/a_n| > 1$, then the series must diverge.

2.7.3 Explanation: Other Tests for Series Convergence

As we have seen, the ratio test has an important limitation: if the limit comes out 1, then the test is inconclusive. Many power series converge on a finite interval, and the ratio test provides no hint of whether these series converge at the *end points* of this interval. We therefore need to apply other tests.

Below we list the most common tests used to find whether a given series converges or diverges and give pointers on how to choose an appropriate test for a given series. The reasons *why* these tests work are explored in the problems.

A few brief notes are in order before the tests are listed.

1. Convergence and divergence happen at the *tail* of a series, not at the beginning. For instance, if you ignore the first five-hundred terms of a series, and prove that the *remaining* (still infinitely long) series converges, then the original series must still converge. (Of course, those first terms that you ignored, although they are not relevant to *whether* the series converges, are often the most important in estimating *what* the series converges to!)
2. Remember that we have already seen a few key ways of determining if a series converges: the *n*th-term test, the rule for geometric series, the trick for "telescoping" series, and of course the ratio test.

The tests for series convergence that we describe below, along with the ones we have already covered, are summarized in Appendix C.

The Integral Test

To apply the integral test, you replace a series with an *improper integral* and determine if the integral converges (has a finite value). For instance, given the series $\sum_{n=1}^{\infty}(1/n^2)$, we would evaluate the integral $\int_{1}^{\infty}(1/x^2)dx$.

The Integral Test

Consider a continuous, positive-valued, decreasing function $f(x)$ such that $f(n) = a_n$ for all integer n. If the integral $\int_1^\infty f(x)dx$ converges, then the series $\sum_{n=1}^\infty a_n$ converges. If the integral diverges, then the series diverges.

You can see why the integral test works in general by viewing the series as a Riemann Sum for the integral: see Problem 2.181.

Note, however, that the improper integral and the series are not the same. For instance, you can easily determine that $\int_1^\infty (1/x^2)dx$ converges to 1. This does *not* mean that $\sum_{n=1}^\infty (1/n^2)$ converges to 1; it only proves that the series converges. (In fact it converges to $\pi/6$.)

This test works for almost any function you can integrate. The primary limitation is that some functions—for instance, anything involving factorials—cannot be integrated.

EXAMPLES The Integral Test

- $\sum_{n=1}^\infty (\ln n / n)$ Be careful! The integral test requires a function that is everywhere decreasing. The function $\ln x / x$ increases until $x = e$ and decreases thereafter. But we can prove convergence or divergence after a certain point, so we can start at $n = 3$ and prove convergence or divergence from there. The first two terms will not invalidate our result.

$$\int_3^\infty \frac{\ln x}{x} dx = \frac{(\ln x)^2}{2} \Big]_3^\infty$$

which diverges. The integral test therefore promises us that the original series diverges.

- $\sum_{n=1}^\infty (n/e^n)$

$$\int_1^\infty \frac{x}{e^x} dx = \frac{-x-1}{e^x} \Big]_1^\infty = \frac{2}{e}$$

The integral test therefore shows that the original series converges. It does not, however, tell us what the original series converges to.

As a final note, the lower limit of your integral need not always be 1. Remember that (assuming no terms are undefined) you can prove convergence or divergence of a series starting with any convenient term.

p-Series

A *p*-series is any series of the form $\sum (1/n^p)$. These series follow a simple convergence rule.

> **p-Series**
>
> If $p \leq 1$, the series $\sum (1/n^p)$ diverges; if $p > 1$, then the series converges.

You can prove this rule with the integral test: see Problem 2.182.

> **EXAMPLES** p-series
>
> - $\sum_{n=1}^{\infty} (1/n^2)$ is a p-series with $p > 1$. It therefore converges.
> - $\sum_{n=1}^{\infty} (1/\sqrt{n})$ is a p-series with $p < 1$. It therefore diverges.

This is one of the easiest tests to apply: if you recognize the form of a series as a p-series, then you know immediately what the series does. In that sense, working with p-series is very similar to working with geometric series (which, as you may recall, converge if and only if $|r| < 1$). In fact, the two types are so similar that the biggest danger is confusing them! Be careful about this: $\sum (1/2^n)$ is a geometric series; $\sum (1/n^2)$ is a p-series.

The Basic Comparison Test

Consider how we showed that the harmonic series diverges (Section 2.6). We found a completely different series that diverges. We showed that each term of the harmonic series is *greater than or equal to* the corresponding term of the original series. Since the partial sums of the "other" series grow without bound, the partial sums of the harmonic series must do the same.

This is an example of the "basic comparison test." If you find a divergent series with terms that are always *smaller* than the terms of your series, then your series must also diverge. Conversely, if you find a convergent series whose terms are always *larger* than the terms in your series, then your series must also converge. Note, however, that the comparison test applies only to series in which all the terms are positive.

> **The Basic Comparison Test**
>
> Consider two positive-term series $\sum a_n$ and $\sum b_n$.
>
> - If $a_n \leq b_n$ for all n, and $\sum b_n$ converges, then $\sum a_n$ converges.
> - If $a_n \geq b_n$ for all n, and $\sum b_n$ diverges, then $\sum a_n$ diverges.

Be careful to avoid using the basic comparison test backward. If your series is *smaller* than some other series that *diverges*, that doesn't prove anything about your series one way or the other. If your series is *larger* than a *convergent* series, that doesn't help either. In order to use this test, you must be able to say "My series is smaller than a convergent series" or "My series is larger than a divergent series."

The basic comparison test is usually used to compare your series with a geometric series or a *p*-series, since those have simple convergence rules.

> **EXAMPLES** **The Basic Comparison Test**
>
> - $\sum_{n=2}^{\infty} 1/(\sqrt{n} - 1)$. We can readily see that $1/(\sqrt{n}-1) \geq 1/\sqrt{n}$ for all n values. Since $\sum 1/\sqrt{n}$ is a *p*-series with $p = 1/2$, it diverges. The basic comparison test therefore proves that original series diverges as well.
> - $\sum_{n=1}^{\infty} 1/(\sqrt{n} + 1)$. Now $1/(\sqrt{n}+1) \leq 1/\sqrt{n}$ for all n-values. However, being *smaller* than a known *divergent series* does not prove anything; therefore the basic comparison test does not help with this example. (Below we will see a different test that can be used on this series.)
> - $\sum_{n=1}^{\infty} \left(\ln n / \sqrt{n} \right)$. Here we are going to use the inequality $\ln n > 1$. Of course that is not true for the first few n-values, but we can prove convergence or divergence after a certain point and those first couple of terms will not invalidate our result. With than in mind we write $\left(\ln n / \sqrt{n} \right) > \left(1/\sqrt{n} \right)$. The latter is a divergent p series ($p = 1/2$) which assures us that the original series diverges as well.

When using the basic comparison test, you may want to recall these inequalities:

$$-1 \leq \sin n \leq 1$$

$$-1 \leq \cos n \leq 1$$

$$-\pi/2 < \tan^{-1} n \leq \pi/2$$

$$-\pi/2 < \sin^{-1} n \leq \pi/2$$

$$0 \leq \cos^{-1} n \leq \pi$$

$$1 \leq \ln n \leq n \text{ (for sufficiently large } n\text{)}$$

The Limit Comparison Test

Like the basic comparison test, this test requires finding some other series to compare with your original series. Unlike the basic comparison test, this test does not say "I am greater than that divergent series" or "I am less than that convergent series"; instead, the limit comparison test says "I am *like* that other series."

> **The Limit Comparison Test**
>
> If $\lim_{n \to \infty}(a_n/b_n)$ exists and is not zero, then $\sum a_n$ and $\sum b_n$ either *both converge* or *both diverge*.

In practice, you use the limit comparison test when you begin with an intuition that "for very high values of n, my (complicated) series is really a lot like this other (simpler) series."

EXAMPLES **The Limit Comparison Test**

- $\sum_{n=1}^{\infty} 1/(\sqrt{n}+1)$. We saw above that the basic comparison test doesn't help us with this one, because being *less than* a *divergent series* does not prove anything. We intuitively note, however, that when n is very large, $\sqrt{n}$ is much bigger than 1, so this series should behave a lot like $\sum(1/\sqrt{n})$. We can make this intuition rigorous with the limit comparison test.

$$\lim_{n\to\infty} \frac{1/\sqrt{n}}{1/(\sqrt{n}+1)} = 1$$

Since the result exists and is not zero, the two series must either *both converge* or *both diverge*. Since $\sum_{n=1}^{\infty}(1/\sqrt{n})$ is known to diverge, $\sum_{n=1}^{\infty} 1/(\sqrt{n}+1)$ must do the same.

- $\sum_{n=2}^{\infty}(n+1)/(2n^3 + 3n^2 - 5n)$. Once again the problem begins with intuition: for large values of n this function will approximate $n/2n^3 = 1/2n^2$. So we note that

$$\lim_{n\to\infty} \frac{(n+1)/(2n^3 + 3n^2 - 5n)}{1/n^2} = \frac{1}{2}$$

Since $1/n^2$ is a convergent *p*-series, the original series must also converge.

Alternating Series

An "alternating series" alternates between positive and negative values. Such series can be written by multiplying the terms of a positive-term series by $(-1)^n$. For instance,

$$\sum_{n=1}^{\infty} \frac{(-1)^n}{n^2} = -1 + \frac{1}{4} - \frac{1}{9} + \frac{1}{16}\cdots$$

Such a series has three possible fates.

- The series may diverge. While this is certainly possible, it is much less common than it is for positive-term series: alternating series tend to converge, since the partial sums are going back and forth instead of always moving up.
- The series may *conditionally converge*. This means that the alternating series converges *because* it is alternating. More formally, we can say that a series "conditionally converges" if $\sum a_n$ converges, but $\sum |a_n|$ diverges.
- The series may *absolutely converge*. This means that the series would converge even if it did not alternate; in other words, $\sum |a_n|$ converges. If the series $\sum |a_n|$ converges then the series $\sum a_n$ converges as well.

There are two common approaches to determining the convergence of an alternating series. The first is to test for absolute convergence. The other approach is the alternating series test:

> **The Alternating Series Test**
> If the alternating series $\sum_{n=1}^{\infty}(-1)^{n-1}b_n = b_1 - b_2 + b_3 - b_4 + b_5 - b_6 \ldots (b_n > 0)$ satisfies …
> - $b_{n+1} \leq b_n$ for all n, and
> - $\lim_{n\to\infty} b_n = 0$
>
> then the series is convergent.

Note what the b_n terms represents in this formulation: not the terms in the original (alternating) series, but their absolute values. If these terms are everywhere *decreasing and approaching zero*, then the original (alternating) series converges.

EXAMPLE — Absolute Convergence

Consider the alternating series presented above, $\sum_{n=1}^{\infty}(-1)^n/n^2$. If we remove the alternator we get $\sum_{n=1}^{\infty}(1/n^2)$ which converges because it is a p-series with $p > 1$. Therefore, we can conclude that $\sum_{n=1}^{\infty}(-1)^n/n^2$ absolutely converges.

EXAMPLES — Conditional Convergence

- $\sum_{n=1}^{\infty}(-1)^n/n$ is called the "alternating harmonic series." Since $1/(n+1) \leq (1/n)$ for all n-values, and $\lim_{n\to\infty}(1/n) = 0$, we can conclude that this series converges by the alternating series test. Since the positive-term harmonic series has been shown divergent, we conclude that $\sum_{n=1}^{\infty}(-1)^n/n$ is conditionally convergent.
- $\sum_{n=1}^{\infty}(-1)^{n+1}n/(n^2+1)$ is more difficult to analyze. If we remove the alternator we find the series $\sum_{n=1}^{\infty} n/(n^2+1)$ which diverges. (Prove this using the limit comparison test with $1/n$.) So we know that our series does not absolutely converge; we will apply the alternating series test to determine if it converges at all.
  It is not difficult to prove (using l'Hôpital's Rule, for instance) that $\lim_{n\to\infty} n/(n^2+1) = 0$. But what of the other part of the alternating series test: how do we show that the series is everywhere decreasing? By using the derivative! The derivative of $n/(n^2+1)$ is $(1-n^2)/(n^2+1)^2$. Since the denominator is always positive and the numerator is negative for all $n > 1$, the derivative is negative, which means the original function is decreasing. We can therefore pronounce $\sum_{n=1}^{\infty}(-1)^{n+1}n/(n^2+1)$ "conditionally convergent" by the alternating series test.

Chapter 2 Taylor Series and Series Convergence

Putting it Together: Intervals of Convergence

The question "for which x-values does this power series converge?" generally begins with the ratio test. In many cases, the ratio test tells you that the series converges inside a certain interval, and diverges outside it—but is inconclusive at the end points of the interval.

Each of these end points gives you a different numerical series. By using other tests for convergence, we can evaluate the behavior of those two series, and thus determine the complete interval of convergence of the power series.

> **EXAMPLE** **Power Series**
>
> In the ratio test section we considered the power series $\sum_{n=1}^{\infty} 2^n(x-5)^n/n$. Using the ratio test, we were able to conclude that:
>
> - If $4.5 < x < 5.5$, then the ratio test indicates *convergence*.
> - If $x < 4.5$ or $x > 5.5$, then the ratio test indicates *divergence*.
>
> At $x = 4.5$ and $x = 5.5$ the ratio test yielded 1, which is inconclusive. We can now return to these series to find out what the power series does at the end points of its interval of convergence.
>
> For $x = 4.5$ the power series becomes $\sum_{n=1}^{\infty} 2^n(-1/2)^n/n = \sum_{n=1}^{\infty}(-1)^n/n$. This series was shown above to converge by the alternating series test.
>
> For $x = 5.5$ the power series becomes $\sum_{n=1}^{\infty} 2^n(1/2)^n/n = \sum_{n=1}^{\infty}(1/n)$. The harmonic series diverges.
>
> We conclude that this series converges for $4.5 \leq x < 5.5$.

The two end points of an interval of convergence are, in general, unrelated to each other: they may both converge, they may both diverge, or (as in the example above) one may converge and the other diverge.

2.7.4 Problems: Other Tests for Series Convergence

2.147 (a) Use the basic comparison test to show that $\sum_{n=1}^{\infty} \frac{1}{n^2+1}$ is convergent by comparing it with the series $\sum_{n=1}^{\infty} \frac{1}{n^2}$.

(b) Explain why the same strategy cannot be used as readily to evaluate $\sum_{n=2}^{\infty} \frac{1}{n^2-1}$.

(c) Now, use the limit comparison test to show that $\sum_{n=2}^{\infty} \frac{1}{n^2-1}$ converges.

2.148 Identify each of the following series as a *convergent geometric series*, a *divergent geometric series*, a *convergent p-series*, or a *divergent p-series*.

(a) $\sum 1/n^3$
(b) $\sum 1/3^n$
(c) $\sum 1/\sqrt[3]{n}$
(d) $\sum 3^{2n}/5^{n-1}$
(e) $\sum 1/n$

In Problems 2.149–2.153, use the integral test to establish whether the given series converges or diverges.

2.149 $\sum_{n=1}^{\infty} 1/n^2$

2.150 $\sum_{n=1}^{\infty} ne^{-n^2}$

2.151 $\sum_{n=1}^{\infty} ne^{-n}$

2.152 $\sum_{n=2}^{\infty} 1/(n \ln n)$

2.153 $\sum_{n=2}^{\infty} n/\sqrt{n^2-1}$

In Problems 2.154–2.157, use the basic comparison test to establish whether the given series converges or diverges.

2.154 $\sum_{n=1}^{\infty} 1/(2^n+3)$

2.155 $\sum_{n=1}^{\infty} |\sin n|/n^{1.2}$

2.156 $\sum_{n=3}^{\infty} 3/\sqrt{n-2}$

2.157 $\sum_{n=1}^{\infty} (\ln n)/n^3$

2.7 | Tests for Series Convergence

In Problems 2.158–2.162, use the limit comparison test to establish whether the given series converges or diverges.

2.158 $\sum_{n=2}^{\infty} 1/(n^2 - 1)$
2.159 $\sum_{n=1}^{\infty} (n^2 - 2n + 5)/(2n^3 - 5n^2 + 4n - 10)$
2.160 $\sum_{n=1}^{\infty} \sqrt{n^2 + 3}/(n^2 + 3n + 3)$
2.161 $\sum_{n=2}^{\infty} 3/(3^{n-1} - 1)$
2.162 $\sum_{n=1}^{\infty} \sin(1/n)$

Problems 2.163–2.166 are alternating series. For each series ...

(a) If $\lim_{n\to\infty} a_n \neq 0$, indicate that the series diverges by the nth-term test.

(b) If the positive-term series $\sum |a_n|$ can be shown to converge, indicate that the series is absolutely convergent.

(c) If the positive-term series $\sum |a_n|$ diverges, $\lim_{n\to\infty} a_n = 0$, and $|a_{n+1}| \leq |a_n|$ for all n, indicate that the series "conditionally converges" by the alternating series test.

2.163 $\sum_{n=1}^{\infty} (-1)^n / 3^n$
2.164 $\sum_{n=1}^{\infty} (-1)^n / \sqrt{n}$
2.165 $\sum_{n=1}^{\infty} (-1)^n / \sin n$
2.166 $\sum_{n=1}^{\infty} (-1)^n (\ln n)/n$

In Problems 2.167–2.175, determine whether each series is convergent or divergent by using the nth-term test, the ratio test, the integral test, the basic comparison test, or the limit comparison test.

2.167 $\sum_{n=1}^{\infty} \ln n / \sqrt{n}$
2.168 $\sum_{n=7}^{\infty} \sqrt{(n^2 + 2n + 2)/(n^2 - 5n - 10)}$
2.169 $\sum_{n=1}^{\infty} 2^n/(n!)$
2.170 $\sum_{n=2}^{\infty} (n-1)/(n^2 - 1)$
2.171 $\sum_{n=1}^{\infty} \left(e^{-\sqrt{n}}\right)/\sqrt{n}$
2.172 $\sum_{n=1}^{\infty} \dfrac{1}{\ln(10^n)}$
2.173 $\sum_{n=1}^{\infty} |\cos n|/(e^n + 2)$
2.174 $\sum_{n=2}^{\infty} 1/\left[n(\ln n)^2\right]$
2.175 $\sum_{n=1}^{\infty} (1 + 1/n)^{-n}$

In Problems 2.176–2.179, determine the *interval of convergence* for the given power series. This process always begins with the ratio test. However, if the ratio test determines that the series converges within a given *interval*, then other tests will be needed to determine the convergence or divergence of the two series at the end points of this interval.

2.176 $\sum_{n=1}^{\infty} (x-2)^n/3^n$
2.177 $\sum_{n=1}^{\infty} (x+1)^n/n^2$
2.178 $\sum_{n=1}^{\infty} (3^n x^n/n!)$
2.179 $\sum_{n=1}^{\infty} (n^n x^n/n!)$

In Problems 2.180–2.182 you will look at why the various tests in this section work.

2.180 Suppose the positive-term series $\sum_{n=1}^{\infty} a_n$ is known to diverge. And suppose that, for all n-values, the terms b_n are *greater than* the terms a_n. Explain why $\sum_{n=1}^{\infty} b_n$ must diverge. (Don't say "because of the basic comparison test": the point is to explain why the basic comparison test works. Instead, consider the *definition of series convergence*, which involves partial sums.)

2.181 The following drawing shows a function $y = f(x)$. The gray region is a *circumscribed Riemann Sum* for the area under the curve between $x = 1$ and $x = 6$.

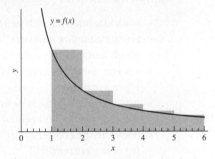

(a) Explain why the area inside the first shaded rectangle is $f(1)$.

(b) Give a formula, in summation notation, for the area under the four shaded rectangles.

(c) The drawing demonstrates half of the integral test. Specifically, it demonstrates that *if the integral $\int_1^{\infty} f(x)\,dx$ diverges, then the series $\sum_{n=1}^{\infty} f(n)$ diverges*. Why? (*Hint*: basic comparison test.)

(d) Create a different drawing that demonstrates the other half of the integral test: *if the integral $\int_1^{\infty} f(x)\,dx$ converges, then the series $\sum_{n=1}^{\infty} f(n)$ converges.*

(e) Create a drawing that illustrates the following restriction on the integral test: *if the function $f(x)$ is not everywhere decreasing, then the integral test may not hold.*

2.182 Use the integral test to prove that the p-series $\sum_{n=1}^{\infty} (1/n^p)$ converges if and only if $p > 1$. *Hint*: $1/n^p = n^{-p}$, and p is a constant.

2.8 Asymptotic Expansions (see felderbooks.com)

2.9 Additional Problems (see felderbooks.com)

CHAPTER 3

Complex Numbers

> *Before you read this chapter, you should be able to...*
> - use Ordinary Differential Equations (ODEs) to model physical situations (see Chapter 1).
> - solve ODEs by using separation of variables or "guess and check" and linear superposition (see Chapter 1).
> - graph functions in polar coordinates, and convert between polar and Cartesian representations.
> - in addition, you should have some basic familiarity with imaginary numbers.
>
> *After you read this chapter, you should be able to...*
> - work algebraically with imaginary and complex numbers.
> - rewrite complex numbers, including complex exponentials, in $a + bi$ format.
> - graph and multiply complex numbers on the complex plane.
> - convert complex numbers between "real and imaginary" and "modulus and phase" representations.
> - use complex numbers to find real-valued answers to real-valued problems, especially (but not exclusively) differential equations.
> - compute the complex impedance of a circuit and use it to predict the amplitude and phase lag of the current.

When René Descartes coined the term "imaginary" for the square root of a negative number, he intended the word as an insult. Like most mathematicians of the time he viewed any attempt to take the square root of a negative number as a beginner's error, like dividing by zero or starting a land war in Asia.

Today these "imaginary numbers" are indispensable in electrical engineering, quantum mechanics, and many other fields. In this chapter we will discuss the nature of such numbers, how to work with them, and how to visualize them on a graph. But we will also see how these numbers help solve important problems.

If you have never worked with complex numbers before, the introduction in Sections 3.2 and 3.3 will be too brief; we recommend you begin by consulting an algebra text to gain some experience with the basics. On the other hand, if it's been a few years since you encountered i, those sections may be useful to polish some slightly rusted skills and get you ready for the rest of the chapter.

3.1 Motivating Exercise: The Underdamped Harmonic Oscillator

A "damped harmonic oscillator," such as an underwater mass on a spring, follows the force law $ma = -cv - kx$.

1. Remembering that $a = d^2x/dt^2$ and $v = dx/dt$, rewrite this equation as a second-order differential equation for $x(t)$.
2. Consider the constants $m = 1/2$ kg, $c = 7$ kg/s, $k = 20$ N/m. Plug in a guess of the form $x(t) = e^{pt}$ and solve for p. You should find two possible values for p.
3. Using the values of p you just found, write the general solution to this differential equation for the given values of the constants. Remember that the general solution to a second-order differential equation must have two arbitrary constants.

   See Check Yourself #12 in Appendix L

The damping force from the water prevents the mass from oscillating at all; it decays exponentially toward equilibrium. Such an oscillator is "overdamped."

4. Now consider the same mass on the same spring, but with a weaker damping force: $c = 2$. Rewrite the differential equation with this constant changed, plug in the guess $x(t) = e^{pt}$, and solve for p. You will find, in your solutions, the square root of a negative number. Just leave it there as part of your solutions, but simplify everything as much as possible.
5. Write the general solution to this differential equation, with the square root of a negative number as a part of your answer.

   See Check Yourself #13 in Appendix L

6. Choose one of the two independent solutions you found and confirm that it solves the differential equation.

A solution of this form, even if it works mathematically, does not seem to tell us anything about the motion of our mass-on-spring system. But in fact it tells us everything—once we know how to interpret it! We will return to this example in Section 3.4 to see how such a solution describes sinusoidal oscillation with decreasing amplitude, exactly as you would expect from such a system.

3.2 Complex Numbers

This section and Section 3.3 are intended as a review of a topic that you may have not seen for a few years.

3.2.1 Discovery Exercise: Imaginary Numbers

1. Explain, using words and equations, why the equation $x^2 = -1$ has no real answer, but $x^3 = -1$ does.

But suppose there *were* an answer to $x^2 = -1$? It couldn't be a real number such as 5, $-3/4$, or π. So we give it a new name: i for the *imaginary* number.
If $i^2 = -1$ then what is

2. $i + 5i$?
3. $(i + 1)^2$? (*Hint:* it isn't zero!)
4. $(5 + 3i)(5 - 3i)$?

   See Check Yourself #14 in Appendix L

5. $(3i)^2$?
6. $\left(\sqrt{2}\,i\right)^2$?

7. $\left(\sqrt{2i}\right)^2$?
8. $\sqrt{-25}$?
9. $\sqrt{-3}$?

If you rewrite $-i$ as $-1 \times i$, then what is

10. $i(-i)$?
11. $(-i)^2$?
12. $(-3i)^2$?

3.2.2 Explanation: Complex Numbers

This explanation and Section 3.3 cover a week or two of math. If you have seen most of this material, and only need a refresher, these sections will remind you of the key terms and concepts that we need to begin our chapter. But if this is your first encounter with i, we suggest you go through these two sections over a period of several days, reading slowly and working problems as you go. Or better yet, find a high school algebra source and go through that material before coming back to this chapter.

We begin with some definitions.

Definitions of Key Terms: Complex Numbers

IMPORTANT: For the purpose of these definitions, the variables a and b designate real numbers.

In Words	Symbolically
Since no real number can square to give a negative answer, we define a new number i that does.	$i \equiv \sqrt{-1}$ or $i^2 = -1$
A "real number" has no i in it. If you square a real number, you get a non-negative real number.	$a^2 \geq 0$
An "imaginary number" is a real number times i (or) an "imaginary number" is the square root of any negative real number.	$(bi)^2 \leq 0$
A "complex number" is the sum of a real number and an imaginary number.	$z = a + bi$
"Re" designates the "real part" of a complex number, the part that is not multiplied by i.	$Re(a + bi) = a$
"Im" designates the "imaginary part" of a complex number, the *real number* that is multiplied by i.	$Im(a + bi) = b$

Algebra involving i is generally no different from algebra involving any other number. If you multiply i by 5, you get $5i$. If you add i to $5i$, you get $6i$. But you have to keep in mind that $i^2 = -1$, and that your answer can always be written in the form $a + bi$.

> **EXAMPLES** **Algebra with *i***
>
> **Question:** Simplify the expression $(3+2i)^2$.
>
> **Solution:**
>
> $(3+2i)^2 = 9 + 12i + 4i^2 = 9 + 12i - 4 = 5 + 12i.$
>
> **Question:** Simplify $\sqrt{-20}$.
>
> **Solution:**
>
> $\sqrt{-20} = \sqrt{4 \times 5 \times -1} = \sqrt{4} \times \sqrt{5} \times \sqrt{-1} = 2\sqrt{5}\,i.$
>
> **Question:** Check the previous answer.
>
> **Solution:**
> The way to verify a square root is to square it back. $\left(2\sqrt{5}\,i\right)^2 = 4 \times 5 \times -1 = -20.$ It works.

Who Wants Complex Numbers?

Science is about solving real problems. "If *this* much voltage is applied to a wire *that* thick, *how long* will it be before the wire breaks?" We measure certain quantities and try to predict other quantities, and *all those quantities are real numbers*. You cannot own i sheep, exert i Newtons of force, earn $\$i.00$, expend i Joules of energy, or travel i miles in i hours. The mathematical consistency of i is beautiful, but beyond that, why bother with something so … imaginary?

The answer is that the road from real-valued questions to real-valued answers often goes through complex numbers. This remarkable fact was discovered by Rafael Bombelli in 1572. Bombelli was using a well-known algorithm to solve the equation $x^3 = 15x + 4$, and he found that the square root of a negative number appeared in the middle of his calculations. Mathematicians of the time would typically have given up at this point, but Bombelli continued and found that the impossible quantity canceled out later in the process, leaving the very real answer $x = 4$.

In the Motivating Exercise (Section 3.1) you used the differential equation $(1/2)(d^2x/dt^2) + 2(dx/dt) + 20x = 0$ to model a damped harmonic oscillator. This real-valued question has a real-valued solution: $x = e^{-2t}[A\sin(6t) + B\cos(6t)]$. Like Bombelli's famous "4," this answer is easy to verify by simply plugging it into the original equation. But also like Bombelli's 4, this answer is difficult to *arrive* at without using a few complex numbers along the way. You began that process in the motivating exercise; we will finish it in Section 3.4. But first, in this section, we lay down some basics.

Powers of *i*

The first two powers of i are easy: $i^1 = i$ (of course) and $i^2 = -1$ (by definition). To find i^3 we multiply i^2 by i. We get $-1 \times i = -i$.

To find i^4, we multiply by i again: $-i \times i = -1 \times i \times i = -1 \times -1 = 1$.

Multiplying by i again, we find that $i^5 = i$; we're back where we started! You don't need to do any more hard thinking to figure out the next powers: i^6 will be -1 and so on.

i^1	i^2	i^3	i^4	i^5	i^6	i^7	i^8	i^9	…
i	-1	$-i$	1	i	-1	$-i$	1	i	…

i^4 and i^8 are both 1, and you can probably see that i^{12} and i^{16} will come out the same way. Raising i to a multiple of 4 is easy: i^{40} and i^{68} and i^{4352} are all 1.

And what about, say, i^{31}? The key is figuring out where you are in the cycle of four. We know that $i^{32} = 1$, so we are right before that in the cycle, which puts us at $-i$.

i^{30}	i^{31}	i^{32}	i^{33}
-1	$-i$	1	i

Whenever you are asked to find i to a power, find a nearby multiple of four, and use that to locate yourself in the cycle.

Equality and Inequality in Complex Numbers

What does it mean for two complex numbers to be equal? $7 + 3i$ does *not* equal 7, or $3 + 7i$; it does not equal anything except $7 + 3i$.

> **Definition: Complex Equality**
>
> Two complex numbers are equal to each other if and only if their real parts are equal, *and* their imaginary parts are equal.

So whenever we assert that two complex numbers equal each other we are actually making two separate, independent statements.

EXAMPLE **Complex Equality**

Question: If $3x + 4yi + 7 = 4x + 8i$, where x and y are real numbers, what are x and y?

Solution:
Normally it is impossible to solve one equation for two unknowns, but this is really two separate equations.
  Real part on the left = real part on the right: $\quad 3x + 7 = 4x$
  Imaginary part on the left = imaginary part on the right: $\quad 4y = 8$
  We can solve these equations to find $x = 7$, $y = 2$.

And what about inequalities? The answer may surprise you: *there are no inequalities with complex numbers.*

The real numbers have a property that for any two real numbers a and b, exactly one of the following three statements must be true: $a = b$, $a > b$, or $a < b$. This may seem too obvious to glorify with the name "property," but it becomes more interesting when you realize that it is *not* true for complex numbers. i is not equal to 1, greater than 1, or less than 1; it is just different from 1. (This corresponds to the fact that real numbers can be represented on a number line. Choose any two distinct points, and one of them is to the left of the other. We shall see in Section 3.3 that complex numbers can be represented on a plane, which offers no such comparison.)

The Cartesian Form $a + bi$ and the Complex Conjugate

Throughout this section we have emphasized complex numbers in the form $a + bi$, which is sometimes called the "Cartesian" or "rectangular" form. It is an important fact that all complex numbers can be written in that form. However, complex numbers are often presented in other forms, and it's up to you to convert them.

For instance, consider $z = (6 - i)^2$. The real part of that number is not 6, nor is it 6^2; in fact there is no way to tell, in this form, what the real part is. But if we square it out we get $36 - 12i + i^2 = 35 - 12i$. Now we can readily see that $Re(z) = 35$ and $Im(z) = -12$. (As always, be aware that $Im(z)$ is a *real* number.)

For the fraction $1/i$ we accomplish the same thing by multiplying the top and bottom of the fraction by $-i$ to find that the real part is 0 and the imaginary part is -1.

How about $10/(3-4i)$? Your first impulse might be to multiply the numerator and denominator by i, or by $3-4i$. Give these a try: they don't help! What does help is multiplying by $3+4i$.

EXAMPLE Putting a fraction into $a + bi$ form

Problem:
Find the real and imaginary parts of the number $\frac{10}{3-4i}$.

Solution:

$z = \dfrac{10}{3-4i}$ A complex number not currently in the form $\sqrt{a+bi}$

$= \dfrac{10(3+4i)}{(3-4i)(3+4i)}$ Multiply the top and bottom by the same (carefully chosen) number

$= \dfrac{30+40i}{3^2 - (4i)^2}$ Remember that $(x+a)(x-a) = x^2 - a^2$

$= \dfrac{30+40i}{9+16}$ $(4i)^2 = 4^2 i^2 = 16(-1) = -16$, which we are *subtracting* from 9

$= \dfrac{30+40i}{25}$ The bottom no longer has i. The top does, but this is easy to deal with.

$= \dfrac{30}{25} + \dfrac{40i}{25}$ Break up the fraction

$= \dfrac{6}{5} + \dfrac{8}{5}i$ So we're there! $Re(z) = \dfrac{6}{5}$ and $Im(z) = \dfrac{8}{5}$

Why did we choose $3+4i$, and why did it work? This trick is a specific example of using a *complex conjugate*.

Definition: Complex Conjugate

The complex conjugate of any number $a + bi$ is $a - bi$. (The real part stays the same, and the imaginary part changes sign.)

The complex conjugate of z is sometimes written as $\bar{z}$ and sometimes as z^*. In this text we use z^*.

When you multiply any complex number by its complex conjugate, the result is a non-negative real number. The square root of that number is called the "modulus" of z, and is written $|z|$. We'll have much more to say about moduli in Section 3.3, but for now the result we need is simply that zz^* is real.

In the example above we multiplied the top and bottom of a fraction by the complex conjugate of the denominator. Because this gave us a real number in the denominator, it was easy to separate the resulting fraction into real and imaginary parts.

Our definition tells you how to find the complex conjugate of a number that is in $a + bi$ form. If a number is expressed in a different form, you can of course *put* it in $a + bi$ form and then switch the sign of b, but there is generally an easier way: replace i with $-i$ everywhere in the expression. For example, the complex conjugate of $\sin^2(3/i)$ is $\sin^2(3/-i)$. This is not guaranteed to work in all cases, but as a practical matter it works for virtually any expression you are likely to work with.

Other Complex Forms

Converting other forms to $a + bi$ can be trickier. Below we demonstrate an algebraic approach to $\sqrt{i}$.

EXAMPLE **Putting a square root into $a + bi$ form**

Problem:
Rewrite $\sqrt{i}$ in Cartesian form.

Solution:
Since we know that $\sqrt{i}$ can be written in the form $a + bi$, we start by writing this fact as an equation.

$a + bi = \sqrt{i}$	This is the assertion that the number $\sqrt{i}$ can somehow be written in $a + bi$ form. Our challenge is to find the real numbers a and b.
$(a + bi)^2 = i$	You can think of this as "squaring both sides" but we prefer to think of it as the *definition* of a square root; we are looking for the number (in $a + bi$ format) that squares to give i.
$a^2 + 2abi - b^2 = i$	Expand.
$a^2 - b^2 = 0$ and $2ab = 1$	As we discussed above, if two complex numbers equal each other, the real parts must be equal and the imaginary parts must be equal.
$a = \pm b$	We get this from $a^2 = b^2$.
$a = b$	We eliminate the possibility that $a = -b$ because the *other* equation, $2ab = 1$, can only be satisfied if a and b have the same sign.
$2a^2 = 1$	Substitute $a = b$ into $2ab = 1$.
$a = \pm\sqrt{1/2}$	… and b is the same.
$\sqrt{i} = \sqrt{1/2} + \sqrt{1/2}\, i$ or $-\sqrt{1/2} - \sqrt{1/2}\, i$	Test both answers by squaring them; make sure you get i both times!

You know that $x^2 = 9$ has two answers (3 and -3), and $x^2 = -25$ does too ($5i$ and $-5i$). So it isn't terribly surprising that we just found two answers for $x^2 = i$. These are specific examples of a very general result: the function $\sqrt{z}$ has exactly two answers[1] except at $z = 0$. Furthermore, $\sqrt[3]{z}$ gives three answers for all $z \neq 0$, and so on. You'll explore this idea further in the problems.

You might think that we will next tackle e^i and $\sin i$ and every other function you can think of. These functions are all meaningful, and can all be converted to $a + bi$ form. But before we can find answers, we have to define the questions! There was no struggle to define $\sqrt{i}$; we were looking for a complex number $a + bi$ that multiplied by itself to get the answer i. But e^i is less clear—it certainly isn't "multiply e by itself i times"—and we cannot set out to find its real and imaginary parts until we define what it means.

One very general definition comes from Taylor series. From the Maclaurin series for exponentials we know that for any real number x,

$$e^x = 1 + x + \frac{1}{2}x^2 + \frac{1}{6}x^3 + \ldots = \sum_{n=0}^{\infty} \frac{1}{n!}x^n$$

If we replace x with a complex number z, every term in this series can be expressed in the form $a + bi$, so we *define* the complex function e^z by that series. Similar definitions can be applied to the sine, cosine, and many other functions. You will experiment with such definitions in Problems 3.44 and 3.45.

Using Taylor series to define complex functions is powerful, general, and consistent. It allows us to write virtually any expression involving i in the form $a + bi$. It preserves the algebraic properties of these functions such as trig identities, rules for multiplying exponentials, and even rules for differentiating and integrating these functions.

Working with Complex Functions

Complex functions such as $\sin z$ and $\cos z$ can be defined by plugging a complex z into the Maclaurin series for the real functions. But even without doing that we can say confidently that these functions follow the same rules as their real counterparts, such as

- $\frac{d}{dz}(\sin z) = \cos z$
- $\sin^2 z + \cos^2 z = 1$
- $\sin z = \cos(\pi/2 - z)$

We will explore the behavior of these complex functions and the meaning of their derivatives in more depth in Chapter 13 but here it is important to know that all the usual rules of algebra and calculus apply.

This definition still leaves open the all-important question of interpretation; we still don't know what it means to find $x = e^{\left(-2+\sqrt{-36}\right)t}$ as an equation of motion. Section 3.4 focuses on the questions of how to interpret complex functions and understand their behavior, focusing in particular on the exponential function e^{ix}.

---

[1] When we work with real numbers we make the function $\sqrt{x}$ single-valued by fiat: we simply declare that "only the positive answer counts" so $\sqrt{9}$ is 3, not -3. With complex numbers there is no easy way to pick out one answer to "count" so the function is inherently multivalued. We will discuss this issue further in Chapter 13.

3.2.3 Problems: Complex Numbers

In Problems 3.1–3.19, find the real and imaginary parts of the given numbers.

- **3.1** 3
- **3.2** i
- **3.3** 0
- **3.4** $3 - 4i$
- **3.5** $7 + 2i + 8 - 5i$
- **3.6** $(7 + 2i) - (8 - 5i)$
- **3.7** $(2 + 3i)/4$
- **3.8** $(7i)^2$
- **3.9** $-(7i)^2$
- **3.10** $(-7i)^2$
- **3.11** $\sqrt{-9}$
- **3.12** $\sqrt{-8}$
- **3.13** $(8 - 5i)^2$
- **3.14** $(8 + 6i)(8 - 6i)$
- **3.15** $(3 - 7i)(4 - i)$
- **3.16** $2/i$
- **3.17** $3/(5 + 12i)$
- **3.18** $(7 - i)/(3 + 4i)$
- **3.19** $(a + bi)(a - bi)$

3.20 The number $7/(2 - i) - 14/(3 + i)$ can be simplified in two different ways.

(a) One approach is to rewrite both fractions with a common denominator of $(2 - i)(3 + i)$. After you subtract the numerators, simplify both the numerator and the denominator into $a + bi$ form. Then multiply the top and bottom by the complex conjugate of the denominator to put the entire fraction into $a + bi$ form.

(b) The alternative approach is to put the individual fractions $7/(2 - i)$ and $14/(3 + i)$ into $a + bi$ form first. Then subtract the real parts and subtract the imaginary parts to find the answer.

If the two methods did not arrive at the same answer, something went wrong. Find it!

3.21 Find the two solutions to the equation $x^2 = -100$.

3.22 Use the quadratic formula to find the two solutions to $2x^2 + 6x + 9 = 0$. Put your answers in $a + bi$ form.

3.23 Use the quadratic formula to find the two solutions to $x^2 - 4x + 5 = 0$. Put your answers in $a + bi$ form.

3.24 Show that if a quadratic equation $ax^2 + bx + c = 0$ (with a, b, and c all real) has non-real solutions, these solutions are complex conjugates of each other.

3.25 Fill in the following table.

i^1	i^2	i^3	i^4	i^5	i^6	i^7	i^8	i^9	i^{10}	i^{11}	i^{12}

3.26 Fill in the following table.

i^{51}	i^{52}	i^{53}	i^{54}	i^{55}	i^{56}	i^{57}	i^{58}	i^{59}	i^{60}	i^{61}	i^{62}

3.27 $i^{400} =$

3.28 $i^{241} =$

3.29 $i^{94} =$

3.30 $i^{77} =$

3.31 Find the second derivative of the function $f(z) = z^i$.

3.32 Find the second derivative of the function $f(z) = \sin(iz)$.

3.33 Find the second derivative of the function $f(z) = \ln(iz)$.

3.34 Find the general solution to the differential equation $d^2x/dt^2 = -4x$ by plugging in a guess of the form $x = e^{kt}$ and solving for the constant k.

3.35 Find the general solution to the differential equation $5(d^2y/dx^2) + 2(dy/dx) + y = 0$ by plugging in a guess of the form $y = e^{kx}$ and solving for the constant k.

In Problems 3.36–3.38 solve for r and s assuming they are both real numbers.

3.36 $r + ri - 6s + 2si = 1 - 3i$

3.37 $s + 2si + 3ri = 6r + 9 + 8i$

3.38 $(1/2)r^2 + 2r^2i + (1/2)s + 3si + 7i = 7$

3.39 Find two numbers in $z = a + bi$ format that satisfy $z^2 = 8i$.

3.40 Find two numbers in $z = a + bi$ format that satisfy $z^2 = -4i$.

3.41 Find two numbers in $z = a + bi$ format that satisfy $z^2 = 3 + 4i$.

3.42 Find three numbers in $z = a + bi$ format that satisfy $z^3 = -1$.

3.43 Consider two complex numbers c and d.

(a) Prove that $(c + d)^* = c^* + d^*$.

(b) Prove that $(cd)^* = c^* d^*$.

3.44 Here are the Maclaurin series for the sine and cosine.

$$\sin x = x - \frac{x^3}{3!} + \frac{x^5}{5!} - \frac{x^7}{7!} + \ldots$$
$$\cos x = 1 - \frac{x^2}{2!} + \frac{x^4}{4!} - \frac{x^6}{6!} + \ldots$$

(a) Write the first four non-zero terms of the Maclaurin series for $\sin(ix)$. Assuming x is a real number, write your final answer in $a + bi$ form.

(b) Write the first four non-zero terms of the Maclaurin series for $\cos(ix)$. Assuming x is a real number, write your final answer in $a + bi$ form.

(c) Use your answers to Parts (a) and (b) to write $\sin^2(ix) + \cos^2(ix)$. Expand it out and simplify keeping all terms up to fourth order in x, and show that (up to fourth order at least) $\sin^2(ix) + \cos^2(ix) = 1$.

(d) Use your answers to Parts (a) and (b) to show that $\frac{d}{dx}\sin(ix)$ is $i\cos(ix)$.

3.45 The Maclaurin series for the exponential function is:

$$e^x = 1 + x + \frac{x^2}{2!} + \frac{x^3}{3!} + \frac{x^4}{4!} + \frac{x^5}{5!} + \ldots$$

(a) Write the first six non-zero terms of the Maclaurin series for e^{ix}. Assuming x is a real number, write your final answer in $a + bi$ form.

(b) Use your answer to show that $\frac{d}{dx}e^{ix}$ is ie^{ix}.

(c) Plug $x + iy$ into the Maclaurin series above to find the first four non-zero terms of the Maclaurin series for e^{x+iy}. Assuming x and y are real numbers, write your final answer in $a + bi$ form.

(d) Multiply the Maclaurin series for e^x by the Maclaurin series for e^{iy}, keeping all terms up to the combined third order. (For example, keep a term with xy^2 but not one with x^2y^2.) Write your answer in Cartesian form and show, using this and your previous answer, that $e^{x+iy} = e^x e^{iy}$.

3.3 The Complex Plane

The real numbers can be represented graphically on a number line. Every point on this line corresponds to exactly one real number, and every real number has exactly one point. Complex numbers require a two-dimensional representation, the "complex plane." Every point on the plane corresponds to exactly one complex number, and every complex number has exactly one point.

3.3.1 Discovery Exercise: The Complex Plane

Every point on the "complex plane" represents one complex number. The location on the horizontal axis represents the real part of the number; the vertical axis is the imaginary part.

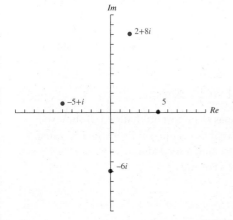

1. Draw the complex plane and label the three points corresponding to the numbers -3, $9i$, and $-3 + 9i$.
2. A "pure imaginary" number is a real number times i. (In other words it has an imaginary part but no real part.) Where are the pure imaginary numbers located on the complex plane?
3. Draw an arbitrary point somewhere in the first quadrant, and label it $a + bi$. Then draw and label its complex conjugate.
4. If you graph all the points that can be written $x - 3i$ where x is a real number, what is the resulting shape?

3.3.2 Explanation: The Complex Plane

A number line gives us a visual "map" of the set of all real numbers. If you need to explain why −5 is less than −3, or why there must be a number between any two distinct numbers, you may well find yourself drawing a picture. The line represents negative and positive numbers, integers and fractions, rational and irrational numbers. But you will not find i anywhere on that line. Applying our visual minds to complex numbers requires a different representation.

That representation is called the "complex plane" (also known as the "Argand plane"). Remember that every complex number can be represented in the form $z = x + iy$ where x and y are real numbers. So we put the x (or $Re(z)$) on the horizontal axis and y (or $Im(z)$) on the vertical axis to find the unique location for the number z.

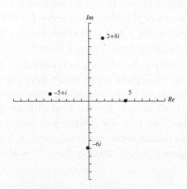

You are used to two-dimensional graphs, but this is quite different. When you graph real numbers on a plane, every point represents two different numbers. For instance if you graph the function $y = x^2$ the point $(3, 9)$ represents the 3 that went into the function, and the 9 that came out. It is on your graph because these two numbers have the relationship $9 = 3^2$.

On the other hand, with complex numbers that same point represents only one number, the number $3 + 9i$. It does not imply any relationship between the 3 and the 9 except that, together, they make up one complex number.

Modulus and Phase

Writing $z = x + iy$ specifies a point on the complex plane in Cartesian coordinates. You can (as always) identify the same point in polar coordinates. The distance to the origin is called the "modulus" and the angle off the positive real axis the "phase" (although other names are often used).

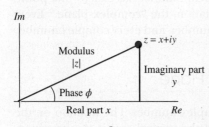

The modulus is written $|z|$ because it fulfills the same function for a complex number that absolute value plays for a real number or magnitude for a vector: it's a non-negative real number that measures how far z is from zero. (The two most commonly used other names for modulus are "absolute value" and "magnitude.") You will prove in Problem 3.56 that defining $|z|$ as the distance from the origin to z on the complex plane is equivalent to saying $|z|^2 = zz^*$, as we asserted in Section 3.2. There is no generally accepted symbol for the phase; we will use the same symbol ϕ that we use for the angle in polar coordinates.

Modulus and phase greatly simplify the process of multiplying and dividing complex numbers.

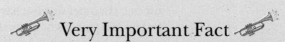

Very Important Fact

Consider two complex numbers. z_1 has modulus $|z_1|$ and phase ϕ_1; z_2 has modulus $|z_2|$ and phase ϕ_2. If you multiply these two numbers, you get a number with modulus $|z_1| \times |z_2|$ and phase $\phi_1 + \phi_2$: the moduli multiply and the phases add.

(This fact is so important that we will show you how to prove it twice: once in Problem 3.58 using trigonometry, and a second time in Section 3.4 using Euler's Formula.)

3.3 | The Complex Plane

Depending on your background this may be the first unfamiliar fact you've encountered in this chapter, so it's worth slowing down a bit to take it in. Let's multiply the numbers $z_1 = \sqrt{3} + i$ and $z_2 = 2 + 2i$ twice: once using a Cartesian representation, and once polar.

Cartesian	Polar
$z_1 z_2 = \left(\sqrt{3} + i\right)(2 + 2i)$	$\|z_1\| = \sqrt{3+1} = 2$
$\quad = 2\sqrt{3} + 2i\sqrt{3} + 2i + 2i^2$	$\phi_1 = \tan^{-1}(1/\sqrt{3}) = \pi/6$
$\quad = 2\sqrt{3} + 2i\sqrt{3} + 2i - 2$	$\|z_2\| = \sqrt{4+4} = 2\sqrt{2}$
$\quad = \left(2\sqrt{3} - 2\right) + \left(2\sqrt{3} + 2\right)i$	$\phi_2 = \tan^{-1}(1) = \pi/4$
	To find the product, multiply the moduli and add the phases.
	Modulus: $\|z_1 z_2\| = 2(2\sqrt{2}) = 4\sqrt{2}$
	Phase: $\phi = \pi/6 + \pi/4 = 5\pi/12$

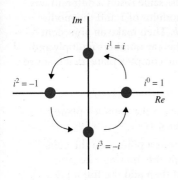

FIGURE 3.1 Multiplying a number by i rotates it 90° counterclockwise in the complex plane.

We leave it to you to confirm that we just got the same answer in two different forms. You may well object that the polar approach is just as hard, if not harder, than the Cartesian. The example below shows a problem that is much easier to solve in the polar representation, but our main purpose here is to introduce you to a rule and a technique. Your work with complex numbers will depend tremendously on your ability to move back and forth between algebraic and graphical representations.

Here's another example. In Section 3.2 we examined the powers of i; let's return to that question graphically. The number i has a modulus of 1 and a phase of $\pi/2$, so multiplying any number by i rotates it 90° around the complex plane. Starting with $i^0 = 1$, Figure 3.1 shows where repeated multiplications by i take us. You can readily see the cycle of 4 that will repeat itself forever. In Section 3.2 this may have looked like a curious coincidence; on the graph it is obvious.

We close with one more example, a problem that would be difficult (or at least tedious) with algebra but is a snap on the complex plane.

EXAMPLE   A Complex Number to a Large Power

Problem:
Simplify $(1 + i)^{10}$.

Solution:
If you plot the point $1 + i$ on the complex plane you can quickly see that its phase is $\pi/4$ and its magnitude is $\sqrt{2}$.

If we multiply this number by itself ten times, the modulus will multiply ten times, so the new modulus will be $\left(\sqrt{2}\right)^{10} = 32$. The phase will *add* to itself ten times, so the new phase will be $10 \times \pi/4 = 5\pi/2$, which places us on the positive imaginary axis. We conclude that $(1 + i)^{10}$ is the imaginary number $32i$.

3.3.3 Problems: The Complex Plane

3.46 Graph the following points on the complex plane, labeling each one.
  (a) $4i$
  (b) 3
  (c) $2 + 6i$
  (d) The complex conjugate of $2 + 6i$
  (e) $3/i$

3.47 Consider the product $(3 - 3i)(-1 - 5i)$.
  (a) Find the modulus and phase of the number $3 - 3i$.
  (b) Find the modulus and phase of the number $-1 - 5i$.
  (c) To find the product, multiply the magnitudes from Parts (a) and (b) and add the phases.
  (d) Convert your result from Part (c) to Cartesian coordinates to find the real and imaginary parts of the product.
  (e) Multiply $(3 - 3i)(-1 - 5i)$ algebraically to confirm your result.

3.48 The drawing shows two complex numbers z_1 and z_2 on the complex plane; $|z_2| = 1$ and $|z_1|$ is unspecified but greater than 1.

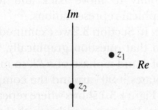

  (a) Is z_1 a real number (such as 5), an imaginary number (such as $3i$), or neither (a number $a + bi$ where $a \neq 0$ and $b \neq 0$)?
  (b) Is z_2 a real number, an imaginary number, or neither?
  (c) Answer in words: if I take any complex number in the world (not necessarily z_1) and multiply it by z_2, what happens to its magnitude? What happens to its angle? Based on your answers, what happens to the point?
  (d) Copy the drawing into your homework. Then add to the drawing the labeled point $z_1 + z_2$.
  (e) Add to your drawing the labeled point $z_1 z_2$.

3.49 Consider the number $12 + 5i$.
  (a) If you locate this number on the complex plane, what is its *distance* from the origin?
  (b) If you multiply this number by its complex conjugate, what do you get?
  (c) What is the relationship between your answers to Parts (a) and (b)?

3.50 Graph the following points on the complex plane, labeling each one.
  (a) $(2 + i)/(3 - i)$
  (b) $\left(\dfrac{-2 + 2i}{\sqrt{8}}\right)^{20}$

3.51 In this problem you will find the modulus of the number $1/(a + ib)$ where a and b are real numbers.
  (a) Break the number $1/(a + ib)$ into its real and imaginary parts and then find the modulus.
  (b) Now find the same result a different way. Find the modulus of 1 and the modulus of $a + ib$. Then make an argument based on the way numbers multiply and divide on the complex plane that leads to $|1/(a + bi)|$.
  (c) If $|z| = k$ then what is $|1/z|$?

3.52 Draw the region on the complex plane for which $|z - 2| \leq 1$.

3.53 The drawing shows a point z on the complex plane. Copy the drawing into your homework, and then add the following labeled points to the drawing.

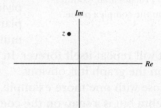

  (a) $z/2$
  (b) $-z$
  (c) z^*
  (d) iz

3.54 Write each of the following numbers in Cartesian form. These will all be easiest to do geometrically.
  (a) $(-2 + 2i)^8$
  (b) $-5i/(-1 + i)$
  (c) $(-16)^{1/4}$ (Only give the answer that's in the first quadrant.)

3.55 Let $z = p/(p + i)$ where p is a real number. For what values of p would you find this number in the first quadrant (upper-right-hand corner) of the complex plane? In the second quadrant? In the third? In the fourth?

3.56 The product of any number and its own complex conjugate is a non-negative real number that represents the square of the modulus: $zz^* = |z|^2$.

(a) To prove this result algebraically, let $z = a + bi$. Algebraically write the expression for z^* and then multiply it by z. When you simplify your answer, it should have no imaginary part. This proves that it is real.

(b) We defined $|z|$ as the distance from the origin to the point $z = a + ib$ on the complex plane. What is that distance in terms of a and b? How does this answer compare to what you found for zz^* in Part (a)?

(c) To prove the same result graphically on the complex plane, begin by drawing an arbitrary point z and its complex conjugate z^*.

(d) If z has a phase of ϕ, what is the phase of z^*? Since phases add, what is the phase of zz^*? What does that tell us about the number?

(e) Note on your drawing that z and z^* have the same modulus $|z|$. What does that tell us about the modulus of zz^*?

3.57 For each number z below, is there a number c that, when multiplied by its own complex conjugate, gives z? If so, find such a number c. If not, explain why not.

(a) $z = 25$
(b) $z = -13$
(c) $z = 3 + 4i$

3.58 Consider two complex numbers $r = a + bi$ and $s = c + di$, where a, b, c, and d are all real numbers.

(a) Find the phase and modulus of both r and s.

(b) Multiply $a + bi$ times $c + di$ to get the product $p = rs$.

(c) Use the expression you found for p to find its phase and modulus.

(d) Show that the phase you found for p is the sum of the phases that you found for r and s and that the modulus you found for p is the product of the moduli you found for r and s. This will require the use of some trig identities.

3.4 Euler's Formula I—The Complex Exponential Function

One of the most important uses of complex numbers is to describe oscillatory functions. In this section we will relate complex exponentials to real sines and cosines, and throughout the rest of the chapter we'll use that to solve physical problems.

3.4.1 Discovery Exercise: Euler's Formula I—The Complex Exponential Function

Some of the simplest oscillatory functions result from the following differential equation:

$$\frac{d^2y}{dx^2} = -y \tag{3.4.1}$$

1. Show that $y = A \cos x + B \sin x$ is a valid solution to Equation 3.4.1 for any values of the constants A and B.
2. Explain how we know that $y = A \cos x + B \sin x$ is the *general* solution to Equation 3.4.1.
3. Show that $y = e^{ix}$ is a valid solution to Equation 3.4.1.

Since $y = A \cos x + B \sin x$ is the general solution to Equation 3.4.1, any other solution can be rewritten in that form. So we must be able to write:

$$e^{ix} = A \cos x + B \sin x \tag{3.4.2}$$

It is vital to note that Equation 3.4.2 is not true for *all* values of A and B; e^{ix} represents a particular solution to Equation 3.4.1, which means that Equation 3.4.2 must be true for *some particular* values of A and B. Our job is now to find them.

4. Plug $x = 0$ into both sides of Equation 3.4.2 to find one of the two constants.
5. Take the derivative with respect to x of both sides of Equation 3.4.2 and then plug in $x = 0$ to find the other constant.

   See Check Yourself #15 in Appendix L

6. The function e^{kt} is used to describe a growing quantity when k is a positive real number, and a decaying quantity when k is a negative real number. Based on the results you've found in this exercise, what kind of quantity does e^{kt} describe when k is imaginary?

3.4.2 Explanation: Euler's Formula I—The Complex Exponential Function

In the previous section we rewrote many complex functions in $a + bi$ form. We also briefly discussed the fact that other functions such as exponents and sines can be written in this form once they are properly defined (for instance by a Taylor series). The real and imaginary parts of complex exponentials are given by "Euler's Formula."

Euler's Formula

$$e^{ix} = \cos x + i \sin x \tag{3.4.3}$$

It is difficult to overstate the importance of this simple-looking equation. Richard Feynman called it "one of the most remarkable, almost astounding, formulas in all of mathematics." In Problem 3.74 you will prove this formula using Maclaurin series, and you proved it a different way in the discovery exercise (Section 3.4), but here we focus on what we can *do* with this formula.

To start with, we can use it to find the real and imaginary parts of numbers.

- **Question:** What is $e^{i\pi}$?

  Answer: $\cos(\pi) + i\sin(\pi) = -1$.

- **Question:** Express 2^i in $a + bi$ form.

  Answer: $2^i = \left(e^{\ln 2}\right)^i = e^{i \ln 2} = \cos(\ln 2) + i \sin(\ln 2)$.

- **Question:** Express $\ln(i)$ in $a + bi$ form.

  Answer: We can see from Euler's Formula that $e^{(\pi/2)i} = i$. Since $\ln(i)$ asks the question "e to what power gives i?" we now have an answer: $\ln(i) = (\pi/2)i$.

- **Question:** Simplify e^{-ix}.

  Answer: From Euler's formula $e^{-ix} = \cos(-x) + i\sin(-x)$. Recalling that $\cos(-x) = \cos x$ and $\sin(-x) = -\sin x$, we get $e^{-ix} = \cos x - i\sin x$. In other words e^{-ix} is the complex conjugate of e^{ix}, as it should be since we just replaced i with $-i$.

Sines and cosines are related by dozens of trig identities. Exponential functions, with three simple rules of exponents, are generally much easier to work with. So one of the great benefits of Euler's formula is allowing us to substitute exponential functions for sines and cosines. As our first example, we show below how easy Euler's formula makes deriving the double-angle formulas.

3.4 | Euler's Formula I—The Complex Exponential Function

EXAMPLE Deriving the Double-Angle Formulas

From the laws of exponents $e^{2ix} = \left(e^{ix}\right)^2$, so:

$$\cos(2x) + i\sin(2x) = (\cos x + i\sin x)^2 = \cos^2 x + 2i\sin x\cos x - \sin^2 x$$

Equating real and imaginary parts on both sides:

$$\cos(2x) = \cos^2 x - \sin^2 x$$
$$\sin(2x) = 2\sin x\cos x$$

An Alternative Form for Writing Complex Numbers

One way to identify a particular complex number is to say "Here is its real part a and here is its imaginary part b." You specify both those parts by writing the number in the form $z = a + bi$.

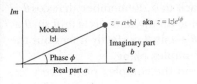

$a^2 + b^2 = |z|^2 \qquad a = |z|\cos\phi$
$\tan\phi = b/a \qquad b = |z|\sin\phi$

As we have seen, another way to identify a particular complex number is to say "Here is its modulus $|z|$ and here is its phase ϕ." Thanks to Euler's formula you can specify both *those* parts by writing the number in the form $z = |z|e^{i\phi}$. You will show this result generally in Problems 3.72 and 3.73.

Section 3.2 stated when you multiply two complex numbers, their moduli multiply and their phases add. We emphasized the importance of this rule, but did not prove it, because the easiest proof is one quick line of algebra using the exponential form we have now introduced.

$$|z_1|e^{i\phi_1}|z_2|e^{i\phi_2} = |z_1||z_2|e^{i(\phi_1 + \phi_2)}$$

EXAMPLE Alternative Representations

Problem:
Express the number $z = 3 + 4i$ in complex exponential form.

Solution:
The modulus is 5 and the phase is $\tan^{-1}(4/3) \approx 0.927$. We conclude that $3 + 4i = 5e^{0.927i}$.

Remember that these are not two different numbers, but two different ways of expressing the same number. The first representation makes it clear that $Re(z) = 3$ and $Im(z) = 4$, while the second shows us that $|z| = 5$ and $\phi_z \approx 0.927$.

Problem:
Express the number $10e^{(2\pi/3)i}$ in $a + bi$ form.

Solution:
Euler's formula does the conversion for us.

$$10e^{(2\pi/3)i} = 10\cos\frac{2\pi}{3} + 10i\sin\frac{2\pi}{3} = -5 + 5\sqrt{3}\,i$$

If you follow the number $3 + ti$ as t goes from 0 to 1 you trace out a line segment on the complex plane. What does the number $5e^{it}$ trace out? Consider that question for a moment before proceeding to the next section.

The Complex Exponential Function Represents Oscillations

In the Motivating Exercise (Section 3.1), a function of the form e^{ikt} came up in solving the equation for a damped oscillator. How does such a function behave? You can trace out the function e^{it} by looking at the "easy" points.

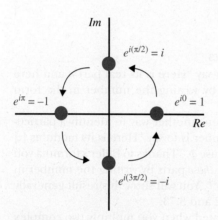

FIGURE 3.2 Adding $\pi/2$ to the argument of a complex exponential rotates the number by 90° counterclockwise in the complex plane.

At $t = 0$ we have $e^0 = 1$. (You can see this by plugging $t = 0$ into Euler's formula, or just remember that *anything* raised to the zero is 1.) As we discussed above, $e^{(\pi/2)i} = i$ and $e^{\pi i} = -1$. Finally, $e^{(3\pi/2)i} = -i$. If you're not seeing a pattern to these numbers, look at the complex plane. We see on Figure 3.2 that the function e^{it} neither grows nor decays, but cycles. You can confirm that by calculating $e^{2\pi i}$ and verifying that you get back to 1. Remember that $\cos\theta$ is defined as the x-value on the unit circle and $\sin\theta$ as the y-value, so $\cos\theta + i\sin\theta$ perfectly traces out the unit circle on the complex plane.

If e^{it} cycles around the complex plane every 2π, what does e^{2it} do? It makes the exact same cycle, still with a radius of 1, but it makes it twice as quickly; at $t = \pi$ it arrives back at 1 where it started. The number multiplied by it is the angular frequency, as you would expect from $e^{ikt} = \cos(kt) + i\sin(kt)$.

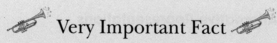 **Very Important Fact**

Whenever you see a function of the form e^{ikt} where k is a real constant and t is the real independent variable, you are looking at a periodic function with period $2\pi/k$.

When we started writing this chapter, we asked each other, "If the students only remember one thing from this chapter, what do we want it to be?" Our answer: "We want them to know how complex exponential functions represent oscillatory behavior." We can summarize this behavior with three rules, using a and b to represent real numbers:

- A function of the form e^{at} represents a growing ($a > 0$) or decaying ($a < 0$) function.
- A function of the form e^{ibt} represents an oscillatory function with period $2\pi/b$.
- A function of the form $e^{(a+ib)t}$ represents an oscillatory function with period $2\pi/b$, but with a growing ($a > 0$) or decaying ($a < 0$) amplitude, as demonstrated in the example below.

3.4 | Euler's Formula I—The Complex Exponential Function

> **EXAMPLE** **The Complex Exponential Function**
>
> **Problem:**
> Sketch the real and imaginary parts of the function $f(t) = e^{(2+5i)t}$.
>
> **Solution:**
> We begin by using the rules of exponentials to rewrite this as $f(t) = e^{2t}e^{5it}$. Next we can use Euler's formula to break $f(t)$ into real and imaginary parts: $f(t) = e^{2t}\cos(5t) + ie^{2t}\sin(5t)$.
>
>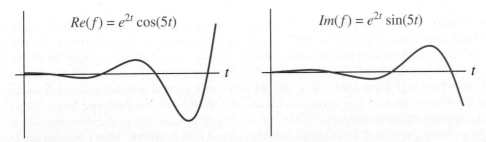
>
> Stop and ask yourself what those graphs tell us, about this function and about complex exponentials in general. We started with $f(t) = e^{(2+5i)t}$. The imaginary part of the exponent (5) gave us the frequency of oscillation, or equivalently gave us a period of $2\pi/5$. The real part of the exponent (2) determined the rate of exponential increase of the amplitude.

A Damped Oscillator

In the Motivating Exercise (Section 3.1) you wrote the equation for a damped harmonic oscillator such as an underwater mass on a spring. Newton's Second Law for such a system becomes $ma = -cv - kx$, and with the constants $m = 1/2$ kg, $c = 2$ kg/s and $k = 20$ N/m the differential equation becomes:

$$\frac{1}{2}\frac{d^2x}{dt^2} + 2\frac{dx}{dt} + 20x = 0 \tag{3.4.4}$$

Guessing $x = e^{pt}$ leads to $\left((1/2)p^2 + 2p + 20\right)e^{pt} = 0$, and the quadratic formula yields two solutions. Because this is a linear homogeneous differential equation, we can combine them to find the general solution:

$$x = Ae^{(-2+6i)t} + Be^{(-2-6i)t} \tag{3.4.5}$$

This solution satisfies Equation 3.4.4 (try it!), but where's that soggy mass (at any given time t)? We can approach that question two ways.

We have stressed that complex numbers act as a bridge from real-valued questions to real-valued answers, so one approach is to turn this complex answer into a real one. We begin with the laws of exponents and, of course, Euler's Formula.

$$x = Ae^{-2t}e^{6it} + Be^{-2t}e^{-6it} = e^{-2t}\left[Ae^{6it} + Be^{-6it}\right]$$
$$= e^{-2t}[A\cos(6t) + Ai\sin(6t) + B\cos(-6t) + Bi\sin(-6t)]$$

Cosine is an "even function" ($\cos(-x) = \cos x$) and sine is an "odd function" ($\sin(-x) = -\sin x$), so:

$$x = e^{-2t}[A\cos(6t) + Ai\sin(6t) + B\cos(6t) - Bi\sin(6t)]$$

(We could have gotten there a bit more quickly by using $e^{-ix} = \cos x - i \sin x$ in the previous step.) We can now group sine and cosine terms:

$$x = e^{-2t}[(A+B)\cos(6t) + (Ai - Bi)\sin(6t)]$$

Finally, recalling that $A + B$ just means "any constant" and $Ai - Bi$ also means "any constant" we can call them a new A and B, and write a general—and real—answer:

$$x = e^{-2t}[A\cos(6t) + B\sin(6t)]$$

Are we really allowed to replace $Ai - Bi$ with a new arbitrary constant with no i in it? As always, the final test of any approach to differential equations is to check the answer. Take a few moments to confirm that this solution satisfies Equation 3.4.4. No imaginary numbers will appear in your calculation; they helped us find the answer, but the answer stands on its own.

Done checking? Then notice that this solution, with all real numbers, does tell us clearly how the system will act. The spring oscillates, as it would with no damping force (although with a slightly different frequency), but the amplitude decays due to the damping force.

So we have a real and meaningful solution to our real problem. That's wonderful, but it took a fair bit of algebra and it isn't always necessary: you can directly interpret the complex exponential solution, Equation 3.4.5, based on the rules we developed above. $Ae^{(-2+6i)t}$ represents an oscillatory function. The imaginary part of the exponent tells us that the angular frequency of that oscillation is 6 (which is equivalent to saying the period is $2\pi/6$). The real part of the exponent represents a decay in the amplitude of that oscillation: e^{-2t} of course decays faster than e^{-t} would, but not as fast as e^{-3t}. The arbitrary A tells us that the *initial* amplitude can be anything at all, depending on how far back we pulled the spring.

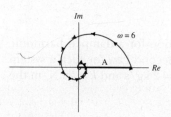

The function $Ae^{(-2+6i)t}$ on the complex plane

And what about the other solution, $Be^{(-2-6i)t}$? The frequency and decay rate are the same; only the phase, and possibly the initial amplitude, are different. (You will see this in Problem 3.68, but you may want to also think about a more familiar ODE solution, $x = A\cos t + B\sin t$. How do the two parts of that solution compare to each other?)

Our point—and it's a big one—is that once you are comfortable with complex exponential functions, you can interpret them without needing to isolate their real and imaginary parts. We will return to this point in the problems, and again in the next section.

3.4.3 Problems: Euler's Formula I—The Complex Exponential Function

In Problems 3.59–3.64 find the real and imaginary parts of the given complex number.

3.59 $e^{(\pi/6)i}$

3.60 $e^{(-\pi/6)i}$

3.61 $e^{2\pi i}$

3.62 $3e^{(\pi/3)i}$

3.63 $e^{(\pi/3)i-2}$

3.64 $e^{(\pi/3)i} - 2$

3.65 Convert each of the following complex numbers to the form $z = |z|e^{i\phi}$.

(a) 7

(b) -7

(c) i

(d) $1 + i$

(e) $5 + 12i$

(f) $5 - 12i$

3.66 The function $x(t) = 3e^{(-2+4i)t}$ represents a decaying oscillatory function. In this problem you will examine the effect of the three constants (3, −2, and 4). You can graph by hand or with a computer, but you should clearly show the period, amplitude, and phase of each function.

(a) Use Euler's formula to rewrite this function in the form $x(t) = a(t) + b(t)i$, where $a(t)$ and $b(t)$ are real-valued functions.

(b) Plot the real and imaginary parts of this function on two separate graphs. You will need to consider how to choose your limited domain to convey the behavior of these functions.

(c) Plot the real and imaginary parts of the function $x(t) = e^{(-2+4i)t}$. How do they differ from the original function?

(d) Plot the real and imaginary parts of the function $x(t) = 3e^{(-1+4i)t}$. How do they differ from the original function?

(e) Plot the real and imaginary parts of the function $x(t) = 3e^{(-2+i)t}$. How do they differ from the original function?

(f) Plot the real and imaginary parts of the function $x(t) = 3e^{(-2-4i)t}$. How do they differ from the original function?

3.67 In the Explanation (Section 3.4.2) we visualize a complex function by plotting its real and imaginary parts separately. Another approach we demonstrated is to plot the trajectory the function follows in the complex plane. For example, you can plot the function e^{it} by drawing a point for e^0, then a point for e^i, and so on, connecting the points with a smooth curve, and then drawing arrows on that curve to show the direction the trajectory is moving in. The result is shown below.

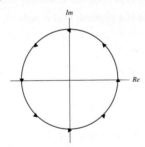

Draw trajectories in the complex plane for the following functions:

(a) e^{-it}
(b) $e^{(-1+i)t}$
(c) $e^{(-2+i)t}$
(d) $e^{(-2+4i)t}$
(e) $e^{(-2-4i)t}$
(f) $3e^{(-2-4i)t}$
(g) e^t

3.68 The solution to the simple harmonic equation $d^2x/dt^2 = -4x$ can be written $x(t) = Ae^{2it} + Be^{-2it}$. For this problem assume A and B are real.

(a) Plug this solution into the ODE and show that it works.

(b) Find the amplitude, period, and phase of $Re(x(t))$. Note that some of them will be functions of A and B and some will not. The phase of $\cos(\omega t + \phi)$ is defined to be ϕ.

(c) Find the amplitude, period, and phase of $Im(x(t))$. (To find the phase of a sine function you can rewrite it as a cosine.)

(d) Choose constants A and B such that the real part oscillates with an amplitude of 22 and the imaginary part with an amplitude of 8.

(e) The solution to the damped oscillator equation $x''(t) + 4x'(t) + 8x(t) = 0$ is $x(t) = Ae^{(-2+2i)t} + Be^{(-2-2i)t}$. Describe qualitatively how the behavior of this system is like, and unlike, the behavior described by the previous function.

3.69 The picture on Page 115 shows what happens when you repeatedly multiply a complex number by i; it rotates 90° counterclockwise each time. The picture on Page 120 shows what happens when you repeatedly add $\pi/2$ to the argument of the function e^{ix}. Explain, using Euler's formula, why these two pictures look the same.

3.70 Prove that the point $e^{i\theta}$ on the complex plane is exactly 1 unit away from the origin for any real θ.

3.71 Find the distance from the point $Ae^{i\theta}$ to the origin (A and θ real and $A > 0$).

3.72 Complex numbers are often written in the form $z = \rho e^{i\phi}$ where ρ and ϕ are real numbers and $\rho \geq 0$. In this problem you will prove that, for a number written in this way, ρ represents the modulus and ϕ the phase.

(a) Use Euler's formula to rewrite this number in the form $z = a + bi$. (Of course, a and b will be functions of ρ and ϕ.)

(b) Using the form you found in Part (a), find the modulus and phase of this number.

3.73 In this problem you will prove that *any* complex number can be written as $\rho e^{i\phi}$ where ρ and ϕ are the modulus and phase of z, respectively.

(a) We begin with $z = a + bi$ which represents all complex numbers by definition. Find the modulus ρ and phase ϕ of this number in terms of a and b.

(b) Of course you can write a number $\rho e^{i\phi}$ with the two values you found in Part (a). But does it really represent the number we started with? To find out, use Euler's formula to break this $\rho e^{i\phi}$ into its real and complex parts, and simplify as much as possible. You may need to look up some trig identities to do this.

3.74 In this problem you will prove Euler's formula using Maclaurin series.

(a) Write the first three non-zero terms of the Maclaurin series for sin x and for cos x.

(b) Write the first six non-zero terms of the Maclaurin series for e^z.

(c) Plug $z = ix$ into the Maclaurin series you just wrote for e^z. Show that the first six terms of the Maclaurin series for e^{ix} equals the first three terms of the series for cos x plus i times the first three terms of the series for sin x.

(d) Now write the entire infinite Maclaurin series for sin x, cos x, and e^z in summation notation. Plug $z = ix$ into the infinite series for e^z and show that the result is the series for cos x plus i times the series for sin x.

3.75 Triple-Angle Formulas In the Explanation (Section 3.4.2) we used Euler's formula to derive the "double-angle formulas" for sin(2x) and cos(2x). Using a similar process, derive expressions for the sine and cosine of 3x.

3.76 Trig Identities for Sums In the Explanation (Section 3.4.2) we used Euler's formula to derive the "double-angle formulas" for sin(2x) and cos(2x). Using a similar process, derive expressions for sin($a + b$) and cos($a + b$).

3.77 Walk-Through: Using Complex Exponentials to Solve a Homogeneous ODE. In this problem, you will solve the differential equation $5(d^2y/dx^2) - 6(dy/dx) + 9y = 0$. You will first find a complex solution, and then a real-valued solution.

(a) Substitute a guess of the form $y = e^{px}$ into the differential equation.

(b) Solve the resulting algebra equation for p. You should find two solutions.

(c) Write down two functions $y(x)$ that solve the differential equation.

(d) Combine your solutions to write the *general* solution to this differential equation.

(e) Rewrite your solutions using the rule $e^{a+b} = e^a e^b$, and then factor out the common factor from both terms.

(f) Rewrite the complex exponential terms using Euler's Formula.

(g) Using the fact that the cosine is an even function and the sine is an odd function, combine terms so you have one cosine and one sine.

(h) Finally, combine arbitrary constants. (Remember that i is a constant, not a variable!) You should now have the general solution to the differential equation in a form that has no imaginary numbers.

(i) Verify that your solution works.

(j) Describe in words the behavior of this function as x grows larger.

In Problems 3.78–3.82 find the general, real-valued solution to the given differential equation. (It may help to first work through Problem 3.77 as a model.)

3.78 $d^2x/dt^2 - 2(dx/dt) + 2x = 0$

3.79 $d^2x/dt^2 + 4(dx/dt) + 8x = 0$

3.80 $4(d^2x/dt^2) + 4(dx/dt) + 5x = 0$

3.81 $9(d^2x/dt^2) - 6(dx/dt) + 2x = 0$

3.82 $3(d^2x/dt^2) + 6(dx/dt) + 7x = 0$

3.83 For a system described by the given complex function, answer the following two questions.
- Does the system oscillate? If so, with what period?
- Does the system gradually increase (toward $\pm\infty$), decay (toward zero), or neither?

(a) e^{7t}
(b) e^{-7t}
(c) e^{5it}
(d) $e^{(7+5i)t}$
(e) $e^{(7-5i)t}$
(f) $e^{(-7+5i)t}$

3.84 For a real number x, what is $\lim_{x \to \infty} e^{kx}$ if

(a) k is a positive real number?
(b) k is a negative real number?
(c) $k = 0$?
(d) $k = i$?
(e) $k = -1 + i$?

3.85 An underwater mass on a spring experiences two forces: $F_{spring} = -kx$ and $F_{water} = -cv$. Newton's Second Law for this system is $ma = -cv - kx$. For this problem, assume the mass $m = 2$ kg and the spring constant $k = 5$ N/m. Higher values of the constant c indicate a stronger damping force.

(a) If $c = 6$ kg/s the system is "underdamped." Solve the differential equation in this case and show that the resulting motion is oscillatory.

(b) If $c = 7$ kg/s the system is "overdamped." Solve the differential equation in this case and show that the resulting motion is *not* oscillatory.

(c) Somewhere between $c = 6$ kg/s and $c = 7$ kg/s the system is "critically damped." For any c-value below this critical value, the system will oscillate. Find the c-value that leads this system to be critically damped.

3.86 [This problem depends on Problem 3.85.] Have a computer generate three plots on the same set of axes, one for the underdamped case, one for the overdamped case, and one for the critically damped case. You may choose any initial conditions you wish other than starting at rest at the origin, but you should use the same initial conditions for all three cases. Clearly indicate which plot is which.

Problems 3.87–3.88 concern a circuit with four elements: a battery that maintains a constant voltage V, a resistor with resistance R, a capacitor with capacitance C, and an inductor with inductance L.

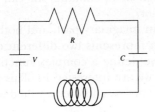

A charge Q will build up on the capacitor according to the equation $V = R(dQ/dt) + Q/C + L(d^2Q/dt^2)$ where V, R, C, and L are positive constants.[2]

3.87 Consider a circuit with no battery ($V = 0$), an inductor with $L = 8$ H, a capacitor with $C = 1$ F, and a resistor with $R\,\Omega$.
  (a) Write the differential equation for this circuit.
  (b) Write the general solution to your differential equation.
  (c) Write the general solution for a resistor $R = 6\,\Omega$.
  (d) Write the general solution for a resistor $R = 4\,\Omega$. Just plugging in $R = 4$ should give you a solution with imaginary numbers in it. Rewrite that solution so it has no imaginary numbers.
  (e) What is $\lim_{t \to \infty} Q(t)$ in both cases? Explain this result physically.
  (f) Qualitatively contrast your answers to Parts (c) and (d). What effect does increasing the resistance have on the circuit?

3.88 [This problem depends on Problem 3.87.]
  (a) Write the general solution for a circuit with a 9-Volt battery, an inductor

with $L = 8$ H, a capacitor with $C = 1$ F, and a resistor with $R = 6\,\Omega$.
  (b) Write the general solution for a circuit with a 9-Volt battery, an inductor with $L = 8$ H, a capacitor with $C = 1$ F, and a resistor with $R = 4\,\Omega$.
  (c) What is $\lim_{t \to \infty} Q(t)$ in both cases? Explain this result physically.

In Problems 3.89–3.90 you will break complex functions into their real and imaginary parts based on the general form of a differential equation, just as you did in the Discovery Exercise (Section 3.4).

3.89 In this problem you will examine the function e^{-ix} by considering the differential equation:
$$\frac{d^2y}{dx^2} = -y \qquad (3.4.6)$$
  (a) Show that $y = A\cos x + B\sin x$ is a valid solution to Equation 3.4.6 for any values of the constants A and B.
  (b) Explain how we know that $y = A\cos x + B\sin x$ is the *general* solution to Equation 3.4.6; that is, that all possible solutions can be written in that form.
  (c) Show that $y = e^{-ix}$ is a valid solution to Equation 3.4.6.

  We now know that e^{-ix} can be written as $A\cos x + B\sin x$ for some constants A and B: that is,
  $$e^{-ix} = A\cos x + B\sin x \qquad (3.4.7)$$
  (d) Plug $x = 0$ into both sides of Equation 3.4.7 to find one of the two constants.
  (e) Take the derivative with respect to x of both sides of Equation 3.4.7 and then plug in $x = 0$ to find the other constant.
  (f) In conclusion, break the function e^{-ix} into its real and imaginary parts.

3.90 The general solution to the equation $d^2y/dx^2 = y$ is $y = Ae^x + Be^{-x}$.
  (a) Show that the function $y = \sin(ix)$ is a solution to the same differential equation.
  (b) Find the constants A and B such that $\sin(ix) = Ae^x + Be^{-x}$.
  (c) The function $y = \cos(ix)$ solves the same differential equation. Find the constants C and D such that $\cos(ix) = Ce^x + De^{-x}$.
  (d) Find the real and imaginary parts of $\tan(ix)$.

---

[2] Electrical engineers tend to write their equations in terms of the current I, which is easier to directly measure than the charge Q, but the charge works better for our example here.

3.91 *[This problem depends on Problem 3.90.]*
A system is modeled by the differential equation $d^2x/dt^2 = 9x$.

(a) Show that $x = A\sin(3it) + B\cos(3it)$ is a valid solution of this differential equation for any constants A and B.

(b) Break $x(t)$ from Part (a) into its real and imaginary parts, using the formulas for $\sin(ix)$ and $\cos(ix)$ that you found in Problem 3.90.

(c) Rearrange your answer to collect the e^{3t} terms and e^{-3t} terms together.

(d) By creating two new arbitrary constants A_2 and B_2 that are functions of the old A and B, write a completely real solution to this differential equation.

(e) Explain why this is not the best way to find the solution to this equation.

3.5 Euler's Formula II—Modeling Oscillations

In the previous section we saw that complex exponential functions oscillate. But physical quantities—the values that we measure, predict, and care about—are always real. The function e^{ikt} oscillates, and pendulums oscillate, but the angle of a pendulum still can't be e^i. So how do we use complex numbers to model real behavior?

Well, every complex number has a real part and an imaginary part, and both of them are real numbers. So an oscillating complex function represents two different oscillating real functions. This suggests an important strategy: we define a complex function whose real or imaginary part is the actual physical quantity we are interested in. This process is demonstrated mathematically in the discovery exercise below, and then applied to physical examples in the ensuing explanation and problems.

3.5.1 Discovery Exercise: Euler's Formula II—Modeling Oscillations

The integral

$$\int \sin(2x)e^x\, dx \qquad (3.5.1)$$

conventionally requires integration by parts (twice) and an extra trick. In this exercise you will solve it in an easier way. Begin by considering a different integral.

$$\int e^{2ix} e^x\, dx \qquad (3.5.2)$$

1. Use the properties of exponentials to combine the two exponentials in Equation 3.5.2 into one exponential. Do *not* expand the complex exponential with Euler's formula.
2. Evaluate the integral to find $\int e^{2ix} e^x\, dx$. Don't forget to include an arbitrary constant C.

   See Check Yourself #16 in Appendix L

3. Rewrite your solution in the standard form $a + bi$ with real a and b. This will require using Euler's formula as well as the trick from Section 3.2 for getting i out of the denominator of a fraction. To split your arbitrary constant you can just write it as $C = A + iB$. Your answer should be in the form

$$\int e^{2ix} e^x\, dx = (\text{Some real function of } x) + i(\text{Some other real function of } x) \qquad (3.5.3)$$

4. Use Euler's formula to rewrite $\int e^{2ix}e^x dx$ as the integral of a real function plus i times the integral of another real function $\left(\int a(x)dx + i\int b(x)dx\right)$. For this part you are *not* using the results you've derived so far; just split the integrand into real and imaginary parts.
5. Recall that every complex equation is two real equations in disguise: one that says the real part of the left-hand side equals the real part of the right-hand side and one that says the same for the imaginary parts. Using your results from Parts 3 and 4 rewrite Equation 3.5.3 as two real equations.

It should now be clear why we set out to solve a complex integral when what we really wanted was the solution to a real integral. One of the two real equations you just wrote down should be in the form $\int \sin(2x)e^x dx = (\text{some real function of } x)$. In other words, by solving the relatively simple integral $\int e^{2ix}e^x dx$ you were able to find the solution to the more difficult integral $\int \sin(2x)e^x dx$.

6. Take the derivative of the solution you found for $\int \sin(2x)e^x dx$ and verify that it does give you $\sin(2x)e^x$.

In this exercise you solved a purely mathematical problem with no physical context, but in doing so you've learned a technique for solving many physical problems: define a sinusoidal quantity as the real or imaginary part of a complex quantity, and you replace a math problem involving trigonometry with an easier math problem involving exponentials.

3.5.2 Explanation: Euler's Formula II—Modeling Oscillations

Sinusoidal oscillations arise in applications ranging from circuit theory to string theory. In all but the simplest cases working with sine and cosine functions leads to a confusing mass of algebra with trig identities. The great power of Euler's formula is that it allows us to do that algebra using the much simpler rules for exponential functions. We illustrate this point below with several physical examples.

A Driven Oscillator

Consider a damped oscillator that is being pushed (or "driven" in the lingo) by a sinusoidally oscillating force:

$$x''(t) + x'(t) + 5x = 10\cos(2t) \tag{3.5.4}$$

This is an inhomogeneous differential equation so the general solution will be the sum of a particular and a complementary solution. In the previous section we solved the complementary equation (an undriven damped oscillator) and found solutions that decay exponentially. So if we are interested in late-time behavior, the complementary solution will become negligible and we can focus on the particular solution. An oscillating function seems likely on both mathematical and physical grounds, so we could try guessing sine and/or cosine solutions. You'll solve the problem this way in Problem 3.102, but it's a pain in the neck.

To make things easier, consider a different equation:

$$\mathbf{X}''(t) + \mathbf{X}'(t) + 5\mathbf{X} = 10e^{2it} \tag{3.5.5}$$

Here $\mathbf{X}(t)$ is a complex function. (We will generally use **boldface letters** for complex-valued functions and non-boldface letters for real-valued functions.) Convince yourself that Equation 3.5.4 is the real part of Equation 3.5.5, provided we identify $x(t)$ as the real part of $\mathbf{X}(t)$.

Plugging in the guess $\mathbf{X}(t) = X_0 e^{2it}$ leads to $X_0 = 10/(-4 + 2i + 5)$, which simplifies to:

$$\mathbf{X}(t) = (2 - 4i)e^{2it} \tag{3.5.6}$$

This looks a lot like the work we did in Section 3.4, but with some important differences. There we solved a real-valued differential equation by guessing an exponential solution $x = e^{pt}$, and when we did the math, p wound up complex. We interpreted the answer by converting it to a real solution with Euler's formula, or by thinking about the properties of complex exponentials.

By contrast, a guess-and-check approach to Equation 3.5.4 requires a messy combination of sines and cosines. So we took a less direct approach, creating a complex-valued differential equation whose *real part* was Equation 3.5.4. That changes the last step: the function we're looking for is the real part of the solution. Expanding the complex exponential and multiplying everything out, we find that $Re(\mathbf{X}(t))$ is:

$$x(t) = 2\cos(2t) + 4\sin(2t) \qquad (3.5.7)$$

You can verify that Equation 3.5.7 is a valid solution to Equation 3.5.4, and imaginary numbers will never come up in the process. As always, we treat i as an indirect path from a real-valued question to a real-valued answer.

But just as in Section 3.4, we need to discuss an alternative ending to this story: one in which we skip the process of isolating the real part, and directly interpret the complex function.

Interpreting Complex Exponential Solutions to Differential Equations

In Section 3.4 we saw that the function $e^{(a+bi)t}$ oscillates with period $2\pi/b$ and amplitude e^{at}. Of course if we multiply the entire function by 3, that just scales up the amplitude. But Equation 3.5.6 doesn't quite look like that; the multiplier out front, $(2 - 4i)$, is itself complex, and it's not obvious what effect that will have on the behavior.

Let's look at $(2 - 4i)$ and e^{2it} separately, and then see what happens when we multiply them.

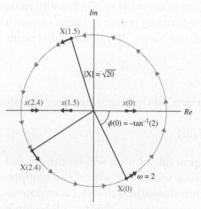

FIGURE 3.4 A rotating function $\mathbf{X}(t)$ in the complex plane and its real part $x(t)$ on the real axis. The functions are shown at $t = 0$ and two representative later times.

- The number $(2 - 4i)$ is a constant with respect to time. It has modulus $\sqrt{20}$ and phase $(-\tan^{-1} 2)$, and they never change.
- The function e^{2it} traces out a circle on the complex plane. Its modulus is always 1. Its phase starts at $\phi = 0$ and rotates at 2 radians/sec (assuming time is measured in seconds).
- When we multiply two complex numbers, moduli multiply and phases add. So $(2 - 4i)e^{2it}$ starts at an angle of $(-\tan^{-1} 2)$ and rotates at 2 rad/sec around a circle with a radius (modulus) of $\sqrt{20}$.

This oscillation is shown in Figure 3.4. Your first impulse may be to recoil from the picture, but we urge you to stay with it until you can see how it represents the conclusion we drew above: $\mathbf{X}(t)$ traces out a circle on the complex plane, starting down in the fourth quadrant, while its real part $x(t)$ oscillates sinusoidally between $-\sqrt{20}$ and $\sqrt{20}$. The more comfortable you are with these kinds of pictures on the complex plane, the more comfortable you will be with complex functions.

Generalizing, the complex function $\mathbf{X}(t) = X_0 e^{i\omega t}$ traces out a circle in the complex plane where the phase and modulus of X_0 determine the radius of the circle and where we start on it, and ω is the angular frequency of the oscillation.[3]

---

[3] When we write $x = A\cos(\omega t)$ we use ω for the "angular frequency" of the motion. When an object is moving around a circle, we use ω for the "rotational velocity" $d\phi/dt$. This may seem like two different uses for the same letter—which does happen sometimes of course—but if you look at the sinusoidal motion as the real part of a complex oscillation, the two uses are the same.

Of course, we don't want to lose sight of our original objective. The real part of this function is the x-value on that circle; it's a sinusoidal wave whose amplitude, frequency, and phase are determined by X_0 and ω. But you will save yourself a lot of algebra, and build your intuition, if you work directly with the complex form as long as possible.

EXAMPLE: Using Euler's Formula to Solve an Inhomogeneous Differential Equation

Question: Find a particular solution to the following equation, where y and t both represent real numbers.

$$y'''(t) + 3y''(t) + 4y = 5\cos(2t) \qquad (3.5.8)$$

Solution:
The problem asks only for a particular solution, so the answer will have no arbitrary constants. To find the general solution, we would begin by solving the complementary homogeneous equation and add the two solutions.

To find the particular solution we begin instead by writing a new differential equation involving complex numbers.

$$\mathbf{Y}'''(t) + 3\mathbf{Y}''(t) + 4\mathbf{Y} = 5e^{2it} \qquad (3.5.9)$$

Equation 3.5.8 is the real part of Equation 3.5.9 (confirm this for yourself), so the $y(t)$ we want will be the real part of the $\mathbf{Y}(t)$ we find.

We now guess a solution of the form $\mathbf{Y} = Y_0 e^{2it}$. Plugging this into Equation 3.5.9 gives

$$-8iY_0 - 12Y_0 + 4Y_0 = 5$$

$$Y_0 = \frac{5}{-8-8i} = -\frac{5}{8}\frac{1}{1+i} = -\frac{5}{8}\frac{1-i}{1+1} = -\frac{5}{16}(1-i)$$

$$\mathbf{Y} = -\frac{5}{16}(1-i)e^{2it}$$

You can check that this is a solution to Equation 3.5.9. The figure below shows the initial value of $\mathbf{Y}$ on the complex plane. From there it will rotate with angular frequency $\omega = 2$, which means its real part $y(t)$ will oscillate on the real axis with that same angular frequency.

The function $Y_0 e^{2it}$ describes a sinusoidal oscillation with amplitude $|Y_0| = 5\sqrt{2}/16$, angular frequency 2, and initial phase $\phi(0) = 3\pi/4$. For most purposes we would stop there.

If necessary you could take the real part of this function to get the real solution $y(t) = -(5/16)(\cos(2t) + \sin(2t)) = 5\sqrt{2}/16\cos(2t + 3\pi/4)$, which matches what we learned from the complex solution. You can confirm that this solution satisfies Equation 3.5.8.

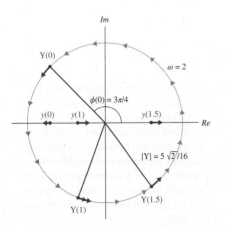

Adding Waves

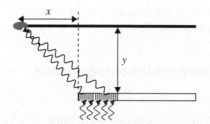

As a final example for this section, consider different waves with equal frequencies and amplitudes but different phases. This could describe a diffraction grating, in which a wave passes through many small slits and then travels to a receiver. The phase of the wave from each slit is proportional to the distance from the slit to the receiver.

In Problem 3.118 you'll show that the signal at the receiver is a sum of sinusoidal waves with equal frequency and evenly spaced phases. In that problem you'll use cosines, so just for variety we'll do it here with sines.

$$S = \sum_{n=1}^{N} \sin(\omega t + n\delta)$$

This sum can be calculated using trig identities, but the math is tedious. Instead we define the following complex quantity. (Because we used a sine instead of a cosine, S is the imaginary part of $\mathbf{S}$.)

$$\mathbf{S} = \sum_{n=1}^{N} e^{i(\omega t + n\delta)} = e^{i\omega t} \sum_{n=1}^{N} \left(e^{i\delta}\right)^n$$

The sum of a geometric series is $\sum_{n=1}^{N} r^n = r\left(1 - r^N\right)/(1 - r)$, so this sum becomes

$$\mathbf{S} = e^{i(\omega t + \delta)} \frac{1 - e^{iN\delta}}{1 - e^{i\delta}} \qquad (3.5.10)$$

We conclude that the signal S is an oscillation with angular frequency ω, as we would expect, and the amplitude of that oscillation is given by

$$|\mathbf{S}| = \sqrt{\mathbf{S}\mathbf{S}^*} = \sqrt{e^{i(\omega t + \delta)} \frac{1 - e^{iN\delta}}{1 - e^{i\delta}} e^{-i(\omega t + \delta)} \frac{1 - e^{-iN\delta}}{1 - e^{-i\delta}}} = \sqrt{\frac{1 - \cos(N\delta)}{1 - \cos\delta}}$$

That information is sufficient for most purposes. If we need to know the exact form of the signal S and not just its magnitude, we can rewrite $\mathbf{S}$ in the form $a + bi$ and take its imaginary part. You'll do that in Problem 3.114.

You can also use the complex form of series like this to add up sine waves graphically in the complex plane like vectors. See Problem 3.117.

3.5.3 Problems: Euler's Formula II—Modeling Oscillations

3.92 Walk-Through: Using Complex Exponentials to Find a Particular ODE Solution. In this problem you will find a particular solution to this differential equation.

$$x''(t) + 5x'(t) + 6x(t) = 4\sin(2t) \qquad (3.5.11)$$

But you will begin with a very different equation.

$$\mathbf{X}''(t) + 5\mathbf{X}'(t) + 6\mathbf{X}(t) = 4e^{2it} \qquad (3.5.12)$$

where $\mathbf{X}(t)$ is a complex function.

(a) Like any complex function, $\mathbf{X}(t)$ can be broken into real and imaginary parts. Plug $\mathbf{X}(t) = y(t) + ix(t)$ into Equation 3.5.12 and take all indicated derivatives. Then write the resulting complex equation as two different equations: real part on the left equals real part on the right, and imaginary part on the left equals imaginary part on the right.

(b) Now go back to Equation 3.5.12 in its original form. Guess a solution of the form

$\mathbf{X}(t) = pe^{2it}$ and find the value of p. Express your answer in the form $p = a + bi$.

(c) What is the modulus of p? What does that tell you about the behavior of $x(t)$?

(d) What is the period of oscillation of $x(t)$?

(e) Plot p on the complex plane. Using that plot, and recalling that $x = Im(\mathbf{X})$, what is the initial value of the particular solution $x(t)$ that you are finding?

(f) You have a particular solution to Equation 3.5.12 in the form pe^{2it}. Use Euler's formula and a little bit of algebra to rewrite this solution in the form $\mathbf{X}(t) = y(t) + ix(t)$.

(g) Based on your answers to Parts (a) and (f), write the real-valued answer to Equation 3.5.11.

(h) Plug your answer for $x(t)$ into Equation 3.5.11 and check that it works.

In Problems 3.93–3.96 find a particular solution to the given differential equation. You will use complex exponentials to find your solutions, but your final answer should have no complex numbers in it. It may help to first work through Problem 3.92 as a model.

3.93 $y''(x) + 2y'(x) + y(x) = \sin x$

3.94 $y''(t) + 3y'(t) + 2y(t) = 3\cos(2t)$

3.95 $f''''(x) + 6f''(x) + f(x) = 2\cos x$

3.96 $f''(x) + 4f'(x) + 2f(x) = \cos(x) + \sin(2x)$
Hint: One way to approach this is to write two separate functions whose *sum* solves this differential equation.

3.97 Figure 3.5 shows a complex oscillation $\mathbf{X}(t)$ and its real part $x(t)$ at a few different times. The radius of the circle is 3.

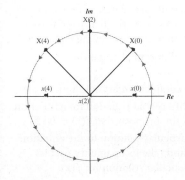

FIGURE 3.5

(a) From the picture, estimate $\mathbf{X}(0)$ and $x(0)$.
(b) Copy the picture and add $\mathbf{X}(9)$ and $x(9)$.

(c) Write a possible function $\mathbf{X}(t)$ that matches Figure 3.5.

3.98 Figure 3.5 shows a complex oscillation $\mathbf{X}(t)$ and its real part $x(t)$ at a few different times.

(a) Copy the picture and on the same plot draw a function $\mathbf{Y}(t)$ with the same angular frequency and initial phase as $\mathbf{X}$ but twice the modulus. Include $\mathbf{Y}(0)$ and $\mathbf{Y}(4)$ and their real parts $y(0)$ and $y(4)$.

(b) Copy Figure 3.5 again and on the same plot draw a function $\mathbf{Z}(t)$ with the same modulus and initial phase as $\mathbf{X}$ but twice the angular frequency. Include $\mathbf{Z}(0)$ and $\mathbf{Z}(4)$ and their real parts $z(0)$ and $z(4)$.

3.99 For each scenario draw a picture like Figure 3.5 where the real function $x(t)$ represents the position of the given oscillator. Your picture should show $\mathbf{X}(t)$ and $x(t)$ at time $t = 0$ and at least one other time. Unlike Figure 3.5 your pictures should indicate the radius of the circle traced out by $\mathbf{X}$.

(a) A mass on a spring is pulled out to a distance .2 m to the right and released from rest, after which it oscillates with period 3 s.

(b) A mass on a spring starts at the equilibrium position with an initial velocity to the right and then oscillates with amplitude .3 m and period 2 s.

(c) A 2 kg mass on a spring with spring constant 3 N/m is pulled out to a distance of .3 m to the left and released from rest.

3.100 Each plot below shows a pair of complex oscillations: $\mathbf{X}(t) = |X|e^{i\phi_{0x}}e^{i\omega_x t}$ and $\mathbf{Y}(t) = |Y|e^{i\phi_{0y}}e^{i\omega_y t}$. For each pair, answer the following questions. Is $|X|$ greater than, less than, or equal to $|Y|$? Is ϕ_{0x} greater than, less than, or equal to ϕ_{0y}? Is ω_x greater than, less than, or equal to ω_y? Assume in all cases that ϕ and ω are between 0 and 2π.

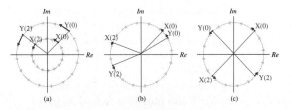

3.101 Suppose you solved for the motion of a damped, driven oscillator and got the solution $\mathbf{X} = (3 - 4i)e^{3it}$.

(a) The function $\mathbf{X}$ represents an oscillation in the complex plane. Find the amplitude and frequency of that oscillation.

(b) Find the phase of the initial value $\mathbf{X}(0)$.

(c) Suppose the driving force had been $F = 2\cos(3t)$. In the complex equation for **X** you would have replaced this with $\mathbf{F} = 2e^{3it}$. Plot the initial values of the complex driving force $\mathbf{F}(t)$ and the complex position $\mathbf{X}(t)$ on the complex plane.

(d) The system winds up oscillating at the same frequency as the driving force, but out of phase with it. Is the position oscillating behind the driving force, or ahead of it? By how much?

3.102 In the Explanation (Section 3.5.2) we solved Equation 3.5.4 by using complex exponentials. You can also solve it directly with trig functions. Plug in a guess of the form $x(t) = p\cos(2t) + q\sin(2t)$ and find the values of p and q. Write your answer in the form of a particular solution to Equation 3.5.4.

3.103 An oscillator obeys the equation $x''(t) + 2x'(t) + 2x(t) = 10\cos t$.

(a) Write a differential equation for a complex function $\mathbf{X}(t)$ whose real part is $x(t)$.

(b) Guess a solution of the form $\mathbf{X}(t) = X_0 e^{it}$ and plug it in to find X_0. Write X_0 in the form $a + bi$ where a and b are real.

(c) What is the amplitude of oscillations of $x(t)$?

(d) What is the period of oscillations of $x(t)$?

(e) Plot X_0 on the complex plane. Recalling that $x = Re(\mathbf{X})$, use your plot to find the initial value $x(0)$.

3.104 Consider a more general damped, driven oscillator than we considered in the Explanation (Section 3.5.2): $Ax''(t) + Bx'(t) + Cx(t) = D\cos(\omega t)$. All constants and variables in this equation represent real quantities.

(a) Write a differential equation for a complex function $\mathbf{X}(t)$ whose real part is $x(t)$. The right-hand side of the complex differential equation must be $De^{i\omega t}$, as the real part matches the right-hand side of the given differential equation. The left-hand side will be of the same form as the given equation, with $x(t)$ replaced by $\mathbf{X}(t)$.

(b) Guess a solution of the form $\mathbf{X}(t) = X_0 e^{i\omega t}$ and plug it in to find X_0 in terms of A, B, C, and D.

(c) What is the amplitude of oscillations of $x(t)$?

(d) What is the period of oscillations of $x(t)$?

(e) Find the real solution $x(t)$.

3.105 Calculate the integral $\int e^x \cos x \, dx$ using the method outlined in the Discovery Exercise (Section 3.5.2).

3.106 A metal ball of mass 2 kg is attached to a spring with spring constant 2 N/m. As it oscillates it experiences a drag force $F_{drag} = -cv$, where v is the box's velocity and $c = 5$ kg/s. It is also being pushed back and forth by an oscillating magnetic field that exerts a force $F_m = F_0 \cos(\omega t)$, where $F_0 = 3$N and $\omega = 2\text{s}^{-1}$.

(a) Write the equation of motion for this ball.

(b) Solve the complementary homogeneous equation. This will involve guessing an exponential solution, finding that solution, and then writing its real and imaginary parts. Your final answer to this part should be a completely real-valued function with two arbitrary constants.

(c) Find a particular solution. This will involve writing a complex differential equation of which the original differential equation is either the real or the imaginary part, solving, and isolating the appropriate part. Your final answer to this part should be a real-valued function.

(d) Write the general solution to the equation of motion, and verify it. No imaginary numbers should be involved in this part of the problem.

(e) What happens to the ball in the long run?

3.107 A driven, *undamped* oscillator obeys the differential equation $x''(t) + 4x(t) = \cos(3t)$.

(a) Find the complementary solution to this differential equation.

(b) Find the particular solution.

(c) Write the general solution.

(d) Find the solution if the object starts at rest at position $x = 4$. Verify that your answer is a solution to the differential equation.

3.108 Consider the differential equation $f'(x) + f(x) = \sin x$.

(a) Find the complementary solution.

(b) Write a differential equation for a function $\mathbf{F}(x)$ with the property that $f(x) = Im(\mathbf{F})$.

(c) Find a particular solution to the equation for $\mathbf{F}(x)$ and take its imaginary part to find a particular solution for $f(x)$.

(d) Find the general solution for $f(x)$ and plug it in to verify that it works.

(e) Find $f(x)$ if $f(0) = 0$.

3.109 The technique you learned in this section—replacing an equation for a real function $f(x)$ with an easier-to-solve equation for a complex function $\mathbf{F}(x)$—only works if the equation is linear. To see why that is, you will attempt to solve the real-valued equation $(f')^2 + 13f^2 = \cos(6x)$ by first solving the complex equation $(\mathbf{F}')^2 + 13\mathbf{F}^2 = e^{6ix}$.

(a) Find the values of k for which $\mathbf{F} = ke^{3ix}$ is a valid solution to the differential equation for $\mathbf{F}$. You should find two answers.

(b) Take the real part of the resulting $\mathbf{F}(x)$ functions to find two guesses for $f(x)$.

(c) Plug in those guesses into the differential equation for $f(x)$ and show that they *do not* work.

(d) Explain why this method doesn't work for a non-linear equation. In other words, look at the equations for f and $\mathbf{F}$ and explain why you would not necessarily expect that a solution to the f equation would be the real part of a solution to the $\mathbf{F}$ equation.

3.110 Consider the complex function $\mathbf{X_1}(t) = 5e^{\pi it}$.

(a) Draw a curve indicating the path this function traces out on the complex plane. Use arrows to indicate the direction of travel, and label the time values of a few key points.

(b) If $Im(\mathbf{X_1}(t))$ represents the behavior of a real-world quantity, describe the behavior of that quantity. (You should be able to provide a full description based on your plot, without ever writing the real-valued function explicitly.)

(c) Repeat Parts (a) and (b) for the function $\mathbf{X_2}(t) = (3+4i)e^{\pi it}$. Your description should focus on how the behavior is similar to, and different from, $\mathbf{X_1}(t)$.

(d) Repeat Parts (a) and (b) for the function $\mathbf{X_3}(t) = (3+4i)e^{2\pi it}$. Your description should focus on how the behavior is similar to, and different from, $\mathbf{X_2}(t)$.

3.111 (a) Write a complex function $\mathbf{X_1}(t)$ that follows a circle of radius 13 in the complex plane, moving counterclockwise, starting when $t=0$ at $13i$, and covering a full circle every 3 s.

(b) Write a complex function $\mathbf{X_2}(t)$ that follows a circle of radius 13 in the complex plane, moving counterclockwise, starting at $5+12i$, and covering a full circle 30 times per second.

(c) Write a complex function $\mathbf{X_3}(t)$ that follows a circle of radius 13 in the complex plane, moving *clockwise*, starting at $5-12i$ and moving ever faster.

3.112 Consider the function $\mathbf{X}(t) = \mathbf{X_0}e^{(a+bi)t}$, where $\mathbf{X_0}$ is a complex number and a and b are real numbers.

(a) If you change $\mathbf{X_0}$, how does the oscillation around the complex plane change?

(b) If you change a, how does the oscillation around the complex plane change?

(c) If you change b, how does the oscillation around the complex plane change?

(d) One system is modeled by $Re(\mathbf{X}(t))$ and a different system is modeled by $Im(\mathbf{X}(t))$. How are these two systems alike, and how do they differ?

3.113 In the Explanation (Section 3.5.2) we analyzed the function $\mathbf{X}(t) = X_0 e^{i\omega t}$ by looking at its modulus and phase. In this problem you will look at the real and imaginary parts of the same function.

(a) Writing X_0 as $a+bi$, find the real and imaginary parts of $X_0 e^{i\omega t}$. Your answer will of course be a function of a, b, and ω as well as of t.

(b) Find the modulus of your answer to Part (a), and show that $|\mathbf{X}(t)| = |X_0|$.

(c) All dependence on t dropped out of your calculations: although $\mathbf{X}(t)$ depends on t, its modulus $|\mathbf{X}(t)|$ does not. What does that fact alone tell you about $\mathbf{X}(t)$ on the complex plane?

3.114 In the Explanation (Section 3.5.2) we calculated the sum $S = \sum_{n=1}^{N} \sin(\omega t + n\delta)$ by defining it to be the imaginary part of the sum $\mathbf{S} = e^{i\omega t} \sum_{n=1}^{N} e^{in\delta}$, which we found was given by Equation 3.5.10. Write this solution in the form $a+bi$ and take its imaginary part to find the real signal S.

3.115 In this problem you will examine the sum $\sum_{n=1}^{N} \cos(\omega t + n\delta + \phi_0)$.

(a) Write this sum as the real part of a complex exponential sum.

(b) Using the laws of exponents, rewrite that sum in the form $a \sum (r^n)$.

(c) Using the fact that $\sum_{n=1}^{N} (r^n) = r(1-r^N)/(1-r)$, write a closed-form expression for the sum.

(d) How did ϕ_0 affect the final answer? (Consider possible changes in amplitude, frequency, and phase. You do *not* have to find the real and imaginary parts to answer this question.)

3.116 In this problem you will examine the sum $\sum_{n=1}^{N} 2^n \cos(\omega t + n\delta)$.

(a) Write this sum as the real part of a complex exponential sum.

(b) Using the laws of exponents, rewrite that sum in the form $a \sum (r^n)$.

(c) Using the fact that $\sum_{n=1}^{N} (r^n) = r(1 - r^N)/(1 - r)$, write a closed-form expression for the sum.

(d) In the Explanation (Section 3.5.2) we described $\sum_{n=1}^{N} \sin(\omega t + n\delta)$ as a receiver at a particular location receiving signals from several identical, evenly spaced sources of electromagnetic radiation. Changing from sine to cosine doesn't make a substantial difference, but inserting 2^n does. How would you change that scenario to match the sum in this problem?

(e) Did the new factor of 2^n in this problem change the frequency of the resulting oscillations? If so, how?

3.117 Consider the sum $\sum_{n=1}^{5} \sin(n\pi/8)$, which we can consider as the imaginary part of $\sum_{n=1}^{5} e^{in\pi/8}$. In the Explanation (Section 3.5.2) we calculated a similar sum using a geometric series, but you can also do it graphically by thinking of the terms in the complex sum as vectors in the complex plane. This problem will be easiest with a protractor and a piece of graph paper.

(a) What are the modulus and phase of the first term in the complex series?

(b) Draw the complex plane and draw an arrow from the origin to this first term. The modulus will tell you the length of this arrow and the phase will give the direction.

(c) Draw another arrow representing the second term, once again using its proper modulus and phase. To add the two numbers, place the beginning of this second arrow at the end of the first one.

(d) Continue in this way until you have included all five arrows. Then draw an arrow from the origin to the end of the arrow for the last term. This long arrow represents the total sum.

(e) Recalling that the real sum we want is the imaginary part of this complex sum, use your drawing to estimate $\sum_{n=1}^{5} \sin(n\pi/8)$.

(f) Add up those five numbers on your calculator and compare it to what you got graphically.

(g) Find N such that $\sum_{n=1}^{5} \cos(n\pi/8) = 0$. Notice that we're using cosines here, so you want the *real* part of the sum to be zero. Explain how you got your answer based on the graphical method, with no calculation of trig functions or exponentials needed.

3.118 Exploration: Diffraction Grating
In a diffraction experiment light passes through an array of N thin slits, lined up in a row with spacing δ between adjacent slits. The light has wavelength λ. You may assume the light beams all have zero phase and equal amplitude A as they pass through the slits. The light is detected on a wall parallel to the wall with the slits, a distance y behind it. The detector is placed on the back wall at a distance x beyond the first slit.

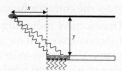

(a) How far does the light travel as it moves from the nth slit to the detector?

(b) Use a Maclaurin series to approximate your answer to Part (a) as a linear function in δ. Under what circumstances is this a reasonable approximation to use?

(c) Using the approximate distance you found, what phase change does the light experience as it goes from the nth slit to the detector? (The phase difference can be defined as 2π times the number of wavelengths between the slit and the detector.)

(d) The signal received from a single slit is $A\cos(\omega t + \phi)$, where ω is the angular frequency of the light beam and ϕ is the phase change from the slit to the detector. Write a sum representing the total signal received at the detector.

(e) The total signal can be rewritten in the form $\sum A\cos(\omega t + c + dn)$. Find c and d in terms of the constants given in the problem.

(f) Rewrite the signal as the real or imaginary part of a complex exponential function.

Find the sum of this complex series. Give your answer in terms of c and d, which will be much shorter and easier than writing it in terms of the original constants in the problem.

Your sum should have been in the form $Ze^{i\omega t}$, where Z is a complex constant. The total signal received at the detector is the real part of your answer, but without calculating that you can see that the signal will oscillate with an amplitude $|Z|$. Even finding that modulus is a fairly tedious calculation, however, so let's turn to a computer. Let $A = 1$, $y = 9$ m, $\delta = 7 \times 10^{-4}$ m, and $\lambda = 10^{-6}$ m. Our goal is to see the pattern of light and dark that will emerge on the wall, so we need to leave x as a variable.

(g) 💻 Plug your solution from Part (f) into a computer using $N = 2$ for a double-slit experiment. Ask the computer to plot the amplitude of the signal as a function of x. What pattern emerges on the wall?

(h) 💻 Repeat Part (g) for 5 slits, 10 slits, and 100 slits. How does the pattern change as the number of slits increases?

3.6 Special Application: Electric Circuits (see felderbooks.com)

3.7 Additional Problems (see felderbooks.com)

CHAPTER 4

Partial Derivatives

> *Before you read this chapter, you should be able to...*
> - take derivatives using the product rule, quotient rule, and chain rule.
> - graph single-variable functions in Cartesian (rectangular) and polar coordinates.
> - work with vectors, including computing the dot product.
> - graph and interpret "multivariate" functions (one dependent variable, multiple independent variables).
>
> *After you read this chapter, you should be able to...*
> - evaluate and interpret "partial derivatives" of multivariate functions.
> - use the chain rule to find derivatives of multivariate composite functions.
> - use "implicit differentiation" to relate derivatives of functions with respect to a common independent variable.
> - evaluate and interpret the "directional derivatives" and the "gradient" of a multivariate function.
> - create power series approximations of multivariate functions.
> - find minima and maxima of multivariate functions using gradients and/or Lagrange multipliers.
> - write and interpret partial derivatives written in the format $(\partial U/\partial T)_P$ and apply them to thermodynamics problems.

"How far will the meteoroid descend into the atmosphere before burning up entirely?" An introductory calculus problem might claim that the distance d is a function of the meteoroid's speed v, or of its mass m, or of the atmospheric temperature T. Any of these functions might be valid under the right circumstances, but all are oversimplifications because the meteoroid's penetration actually depends on *all* these variables—among others.

Functions of one variable are the exception; most important quantities depend on many different variables. In this chapter we assume you have some basic familiarity with multivariate functions, but have not necessarily done much *calculus* with them. We look at derivatives of multivariate functions, known as "partial derivatives," and discuss some of their many applications.

4.1 Motivating Exercise: The Wave Equation

A string that can vibrate up and down is described by a height function $y(x, t)$. The drawing below shows a string at some instant that we'll call $t = 10$ and labels a point indicating that $y(4, 10) = 2$. Assume throughout this exercise (except Part 4) that the ends of the string at $x = 0$ and $x = 10$ are tied down, meaning $y = 0$ at those points.

1. The slope of the string at the point marked in Figure 4.1 is "the derivative of y with respect to x," sometimes written $\partial y/\partial x$. Estimate $\partial y/\partial x$ at the point shown by looking at the drawing.
2. You cannot estimate "the derivative of y with respect to t" by looking at the drawing, but suppose we told you that at this particular point on the string at this particular moment, $\partial y/\partial t = -3$. Estimate where that piece of string would be at $t = 10.5$.

See Check Yourself #17 in Appendix L

An "equation of motion" relates the acceleration of an object to its position and/or velocity. For example, a mass attached to a rusted spring might obey the equation $d^2x/dt^2 = -(b/m)(dx/dt) - (k/m)x$. If you know the position and velocity at any given moment, you can use this equation to calculate the acceleration and thereby predict the motion over time.

The equation of motion for our string is the "wave equation":

$$\frac{\partial^2 y}{\partial t^2} = c^2 \frac{\partial^2 y}{\partial x^2} \qquad (4.1.1)$$

On the left is the second derivative of y with respect to t: the vertical acceleration. On the right is the second derivative of y with respect to x: the concavity. This equation asserts that if you know the concavity at any given point—any particular x and t—you can determine the acceleration at that point. If you know the shape of the entire string at a particular moment, as well as the velocity at each point, you can use this equation to predict the motion of the string over time.

3. Consider a string that begins at rest stretched into a horizontal line. (We can express this mathematically by writing $y(x, 0) = 0$.) Based on Equation 4.1.1, what will happen to this string over time? Your answer will not require calculations, but you shouldn't just say what you would expect physically. Explain what behavior Equation 4.1.1 predicts by thinking about concavity and acceleration.
4. Now consider a string that begins at rest stretched into a diagonal line from the point $(0, 0)$ to the point $(10, 20)$, so $y(x, 0) = 2x$. Based on Equation 4.1.1, what will happen to this string over time?

See Check Yourself #18 in Appendix L

5. Now consider the string whose initial position $y(x, 0)$ is represented in Figure 4.1. Based on Equation 4.1.1, which parts of the string will accelerate downward? Which parts will accelerate upward?

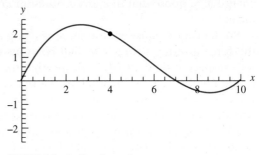

FIGURE 4.1 A vibrating string.

4.2 Partial Derivatives

For a single-variable function, we write dy/dx: "the derivative of y with respect to x." A multivariate function has different derivatives "with respect to" its different independent variables. Calculating these derivatives is not difficult, but some thought is needed to properly interpret their meanings.

138 Chapter 4 Partial Derivatives

4.2.1 Discovery Exercise: Partial Derivatives

The drawing below shows a function $z(x, y)$, with one point on the plot marked.

1. If you start at the marked point and move in the positive x-direction, holding y constant, is z increasing, decreasing, or staying constant?
2. If you start at the marked point and move in the positive y-direction, holding x constant, is z increasing, decreasing, or staying constant?

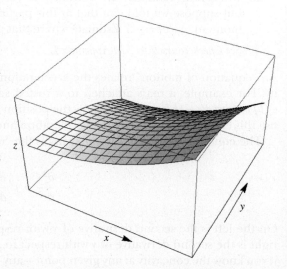

The rate of change of z in the x-direction, holding y constant, is "the derivative of z with respect to x," usually written $\partial z/\partial x$.

3. Based on your answers above, is $\partial z/\partial x$ positive, negative, or zero at the marked point?

   See Check Yourself #19 in Appendix L

4. What about $\partial z/\partial y$?
5. Looking at the plot, is $\partial^2 z/\partial x^2$ at the marked point positive, negative, or zero? Explain what about the surface lets you know.
6. Suppose that z represents the concentration of salt in a lake, y represents depth in that lake, and x represents time. Explain what each of your answers to Parts 3–5 tells you physically about the lake.

4.2.2 Explanation: Partial Derivatives

The attractive force F that an atomic nucleus exerts on an electron is a function of the nuclear charge q. Suppose that in a given situation $dF/dq = 500$ in SI units. What exactly does that mean?

Students often answer "the force is rising at a rate of 500," but what does *that* mean? Is the force growing stronger by 500 Newtons every second? No, that would be a derivative with respect to time; dF/dq indicates that the force will rise by 500 Newtons if you add one Coulomb of charge to the nucleus. More generally, a derivative always says "the dependent variable will rise by this much *per unit increase in the independent variable.*" You can never have "the derivative of this" without "with respect to that."

A multivariate function, then, has different derivatives with respect to its different independent variables. These are called *partial* derivatives.

Definition: Partial Derivatives

Consider a variable y that depends on multiple independent variables x_1, x_2, *etc.* The "partial derivative of y with respect to x_n"—that is, with respect to one of the independent variables—represents how much y changes per unit change in x_n, with all other independent variables held constant. Partial derivatives can be written like ordinary derivatives, but replacing d with ∂.

For instance, a function $y(x, t)$ has two partial derivatives.

$$\frac{\partial y}{\partial x} = \lim_{\Delta x \to 0} \frac{y(x + \Delta x, t) - y(x, t)}{\Delta x} \quad \text{and} \quad \frac{\partial y}{\partial t} = \lim_{\Delta t \to 0} \frac{y(x, t + \Delta t) - y(x, t)}{\Delta t}$$

As usual, the best explanation is an example.

> **EXAMPLE** **Traffic Density**
>
> Interstate 40 stretches 2500 miles from Barstow, California to Wilmington, North Carolina. Let c represent the traffic density, measured in cars per mile. The density c is a function of x, your position on the road, measured in driving miles from the Western tip. The density is also a function of time of day, t, measured in hours since midnight.
>
> Question: What do the partial derivatives $\partial c/\partial x$ and $\partial c/\partial t$ represent, and what are their units?
>
> Answer:
> $\partial c/\partial x$ measures the change with respect to position, holding time constant. Imagine taking a picture from a traffic helicopter over Memphis. You see a relatively low traffic density to the West of Memphis, gradually increasing as you move into the city, and then decreasing as you move Eastward out of the city again. Hence, $\partial c/\partial x$, measured in cars/mile per mile (or cars/mile^2), is positive on the West side of the city and negative on the East.
> $\partial c/\partial t$ measures the change with respect to time, holding position constant. Imagine a traffic camera stationed under a bridge. As rush hour begins, it sees a steady increase in traffic density, so $\partial c/\partial t$ is positive. As evening wears on into night, the density decreases so $\partial c/\partial t$ is negative. The units are cars/mile per hour.
> The photograph captures a single moment in time, so it shows only a change due to position. The traffic camera stays at one position, and sees only changes with respect to time. The partial derivatives isolate the change due to each variable, holding the other variable constant.

Evaluating Partial Derivatives

Evaluating partial derivatives is easy if you know how to evaluate regular derivatives: you take the derivative with respect to one variable, treating the other variables as constants.

> **EXAMPLE** **Evaluating Partial Derivatives**
>
> Question: Find the partial derivatives of $z(x, y) = \sin(2x + y^2) + \ln x$.
>
> Answer:
> $\partial z/\partial x = 2\cos(2x + y^2) + 1/x$, and $\partial z/\partial y = 2y\cos(2x + y^2)$.

Visualizing Partial Derivatives and Second Derivatives

Any function of two variables can be visualized as a surface, where z is the dependent variable and x and y are the independent variables. Viewed this way, $\partial z/\partial x$ represents the slope of the surface if you move in the positive x-direction, leaving your y-coordinate unchanged. Likewise, $\partial z/\partial y$ represents the slope of the surface if you move in the positive y-direction.

140 Chapter 4 Partial Derivatives

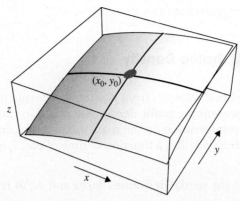

At the point (x_0, y_0), the slope of the black curve is $\partial z/\partial x$ and the slope of the blue curve is $\partial z/\partial y$.

But another visualization, more useful in many circumstances, comes from the "height of a string" function $y(x, t)$ from Section 4.1.

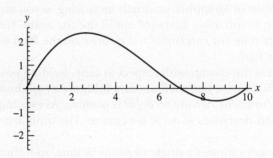

This function has two different first derivatives with different meanings.

- $\partial y/\partial x$ asks the question "how much does the height increase per unit distance as we move to the right?" In a word, it gives the *slope*.
- $\partial y/\partial t$ asks the question "how much does the height increase per unit time if we watch one point on the string?" In a word, it gives the *velocity*.

Roughly speaking, $\partial y/\partial x$ is how much higher the string is one unit further to the right, and $\partial y/\partial t$ is how much higher the string will be at the point you're looking at one unit of time later.

Since we can take the derivative of either of these derivatives with respect to either variable, there are four second derivatives.

- $(\partial/\partial x)(\partial y/\partial x)$ asks the question "How much does the slope change per unit distance if we move to the right?"
- $(\partial/\partial t)(\partial y/\partial x)$ asks the question "How much does the slope change per unit time if we watch one point and wait?"
- $(\partial/\partial x)(\partial y/\partial t)$ asks the question "How much does the velocity change per unit distance if we move to the right?"
- $(\partial/\partial t)(\partial y/\partial t)$ asks the question "How much does the velocity change per unit time if we watch one point and wait?"

The first item in this list, generally written as $\partial^2 y/\partial x^2$, represents the *concavity* at a particular point on the rope and moment in time. The fourth item in the list, $\partial^2 y/\partial t^2$, represents the *acceleration* of one point on the rope at one instant.

The middle two, $\partial^2 y/\partial t \partial x$ and $\partial^2 y/\partial x \partial t$, are called "mixed partial derivatives," or "mixed partials" for short. Surprisingly, partial derivatives generally commute: that is, if you take the same derivatives in a different order, you get the same result.

> **Schwarz' Theorem (also known as "Clairaut's Theorem"): Equality of Mixed Partials**
> If $(\partial/\partial x)(\partial z/\partial y)$ and $(\partial/\partial y)(\partial z/\partial x)$ are both continuous at a given point (x, y), then they are equal to each other at that point.
> $$\frac{\partial^2 z}{\partial x \, \partial y} = \frac{\partial^2 z}{\partial y \, \partial x}$$

As an example of Schwarz' theorem, consider the function $z(x, y) = xe^{xy^2}$.

$$\partial z/\partial x = e^{xy^2} + xy^2 e^{xy^2}$$
$$\partial^2 z/\partial y \partial x = 2xye^{xy^2} + 2xye^{xy^2} + 2x^2 y^3 e^{xy^2} = 4xye^{xy^2} + 2x^2 y^3 e^{xy^2}$$
$$\partial z/\partial y = 2x^2 y e^{xy^2}$$
$$\partial^2 z/\partial x \partial y = 4xye^{xy^2} + 2x^2 y^3 e^{xy^2}$$

The "continuous second derivatives" requirement is important for mathematical accuracy, but for most functions that engineers and physicists use you can assume the mixed partials are equal. Problems 4.20–4.22 offer one look at why mixed partials come out the same.

Notation

The most common way to write a partial derivative is the one we've been using: $\partial z/\partial x$. An alternative is a subscript, with or without a comma: $z_{,x} = z_x = \partial z/\partial x$. It's also common in physics to use a dot over a variable to mean a derivative with respect to time: $\dot{z} = \partial z/\partial t$.

In cases where there is potential ambiguity about which variable is being held constant, you can indicate that with a subscript on the entire partial derivative:

$$\left. \frac{\partial z}{\partial x} \right|_y \text{ or } \left(\frac{\partial z}{\partial x} \right)_y \text{ means partial derivative of } z \text{ with respect to } x, \text{ holding } y \text{ constant}$$

You'll work through an example with ambiguity like this in Problem 4.27, and we'll show some important applications of this idea in Section 4.10 (see felderbooks.com).

4.2.3 Problems: Partial Derivatives

A number of the problems in this section ask you to explain the meaning of different partial derivatives and/or their signs. In each case you should give answers that an average 12 year old could understand. A poor answer would be "Traffic density decreases with respect to position, holding time constant." A good answer would be "There's more traffic on the highway inside Memphis than there is outside the city."

4.1 A meteoroid descending into the atmosphere travels a distance x before burning up entirely.[1] This distance is a function of (among other things) the meteoroid's initial speed v and its mass m.

In each of the parts below, explain the meaning of the given partial derivative in terms a 12 year old could understand. Also give the units of the derivative and whether you would expect it to be positive or negative.

(a) $\partial x/\partial v$
(b) $\partial x/\partial m$
(c) List one other factor that the meteoroid's descent distance depends on. Then answer

---

[1] It's a common mistake to refer to the rock as a "meteor," which actually means the visible trail left by a meteoroid as it burns up. If a meteoroid reaches the ground it becomes a "meteorite." Now don't you feel smart?

the same questions for the partial derivative with respect to that variable.

4.2 The amount you pay in taxes depends on your income I, your total number of dependents d, and the amount you give to charity c.

In Parts (a)–(c), explain the meaning of the given partial derivative in terms a 12 year old could understand. Also give the units of the derivative and whether you would expect it to be positive or negative.

(a) $\partial T/\partial I$
(b) $\partial T/\partial d$
(c) $\partial T/\partial c$
(d) List one other factor that your taxes depend on. Then answer the same questions for the partial derivative with respect to that variable.

4.3 Your puppy's weight W depends on its caloric intake c and how many hours per week you walk it, h. In Parts (a)–(b), explain the meaning of the given partial derivative in terms a 12 year old could understand. Also give the units of the derivative and whether you would expect it to be positive or negative.

(a) $\partial W/\partial c$
(b) $\partial W/\partial h$
(c) List one other factor that your puppy's weight depends on. Then answer the same questions for the partial derivative with respect to that variable.

4.4 Consider a function $u(x, y, z, t)$ that gives the temperature of the air in a room as a function of time.

(a) What would it mean physically if $\partial u/\partial z > 0$ at some point (x_0, y_0, z_0, t_0)?
(b) What would it mean physically if $\partial u/\partial t = 0$ at some point (x_0, y_0, z_0, t_0)?
(c) Is it possible for both of the above statements to be true at the same place and time?

4.5 The temperature u on Skullcrusher Mountain depends on height h and time t.

(a) Everywhere on the mountain $\partial u/\partial h$ is negative. What does that tell you about the mountain?
(b) Throughout the morning $\partial u/\partial t$ is positive. What does that tell you about the mountain?
(c) Suppose $\partial^2 u/\partial h^2$ is negative. What would that mean? (Explain in the clearest and least technical language you can, for the benefit of a mountain-climber who knows no calculus.)

(d) Suppose $\partial^2 u/\partial t^2$ is negative. What would that mean? (Same comment.)
(e) Suppose $\partial^2 u/(\partial t \partial h)$ is zero. What would that mean? (Same comment.) *There are two equally valid answers you could give to this question.*

For Problems 4.6–4.9 find the partial derivatives $\partial f/\partial x$, $\partial f/\partial y$, and $\partial^2 f/\partial x \partial y$.

4.6 $f = xy^2$
4.7 $f = x/y$
4.8 $f = a\cos(bxy) + ce^{d/y}$
4.9 $f = \sin x / \cos y$

4.10 If $f(x, y, t) = ae^{bx^2 y}\sin(\omega t)$ find $\partial f/\partial t$ and $\partial^2 f/\partial x \partial y$.

4.11 For the function $f(x, y) = x/\cos y$ calculate both mixed partial derivatives and verify that they are equal.

4.12 For the function $f(x, y) = xe^{xy}$ calculate both mixed partial derivatives and verify that they are equal.

4.13 You are standing on a surface whose height is given by the function $z(x, y)$. At the spot where you are standing, $\partial z/\partial x = 5$ and $\partial z/\partial y = -5$. Are you facing uphill or downhill if you face…

(a) East (positive x-direction)?
(b) North (positive y-direction)?
(c) West (negative x-direction)?
(d) South (negative y-direction)?

4.14 You are standing on a surface whose height is given by the function $z = e^{x+y}/y^2$. Your position is $(1, 1, e^2)$, and you are facing in the positive x-direction.

(a) Do a calculation that proves that you are looking uphill.
(b) If you move in the positive x-direction, will the uphill slope in the x-direction increase or decrease?
(c) If instead you move in the positive y-direction—still facing in the positive x-direction—will the uphill slope in the x-direction increase or decrease?

4.15 If a battery with potential V is hooked across a resistor of resistance R, a current flows whose magnitude is $I = V/R$. (V, I, and R are all positive quantities.) In Parts (a) and (b) your final answers should be in language that would make sense to someone who has never taken calculus.

(a) Calculate $\partial I/\partial V$. Is it positive or negative? What does this tell us about the current through such a circuit?

(b) Calculate $\partial I/\partial R$. Is it positive or negative? What does this tell us about the current through such a circuit?

(c) Demonstrate that $\partial^2 I/\partial V \partial R = \partial^2 I/\partial R \partial V$.

4.16 A light source of strength S shining on an object d meters away provides an illumination given by $I = kS/d^2$ where k is a positive constant and S and d are positive variables.

(a) Without doing any calculations, would you expect $\partial I/\partial S$ to be positive, negative, or zero? Explain your answer. Then make a similar prediction and explanation for $\partial I/\partial d$.

(b) Calculate $\partial I/\partial S$. Does its sign match your prediction?

(c) Calculate $\partial I/\partial d$. Does its sign match your prediction?

(d) Suppose you were looking at a 100 watt light bulb and someone suddenly replaced it with a 150 watt bulb. Would that have a bigger effect on how much light you see if the bulb were 1 m away from you or if it were 100 m away? Based on your answer, would you expect $\partial^2 I/\partial S \partial d$ to be positive or negative?

(e) Calculate $\partial^2 I/\partial S \partial d$. Does its sign match your prediction?

(f) Calculate $\partial^2 I/\partial d \partial S$ and verify that it equals $\partial^2 I/\partial S \partial d$.

4.17 The angle ϕ in polar coordinates equals $\tan^{-1}(y/x)$.

(a) Draw a picture showing ϕ at the point $(1, 1)$ and a nearby point $(1 + dx, 1)$. Using that picture predict whether you expect $\partial \phi/\partial x$ to be positive or negative at that point.

(b) Add to your picture a point at $(1, 1 + dy)$, and use it to predict the sign of $\partial \phi/\partial y\, (1, 1)$.

(c) Using similar pictures, predict the signs of the each of the derivatives $\partial \phi/\partial x$ and $\partial \phi/\partial y$ at the following points: $(0, 1)$, $(-1, 1)$, $(-1, -1)$.

(d) Calculate the derivatives $\partial \phi/\partial x$ and $\partial \phi/\partial y$ at the point $(1, 1)$ and check that the signs match your predictions.

(e) Based on the signs you predicted for $\partial \phi/\partial y$ at the points $(-1, 1)$, $(0, 1)$, and $(1, 1)$, what would you expect for the sign of $\partial^2 \phi/\partial x \partial y$ at the point $(0, 1)$? Explain your prediction, *then* calculate $\partial^2 \phi/\partial x \partial y$ and check if it matches your prediction.

4.18 In special relativity, the length of an object is given by the formula $L = L_0 \sqrt{1 - v^2/c^2}$ where L_0 is the "rest length" of the object, v is its speed, and c, the speed of light, is a constant. (Note that $v \geq 0$.)

(a) Calculate $\partial L/\partial L_0$. Is it positive or negative? What does this tell you about the length of high-speed objects in special relativity?

(b) Calculate $\partial L/\partial v$. Is it positive or negative? What does this tell you about the length of high-speed objects in special relativity?

(c) Calculate $\lim\limits_{v \to c^-} (\partial L/\partial v)$. Based on the result, explain how you can guess the sign of $\partial^2 L/\partial v^2$ without taking another derivative.

4.19 Planck's Law of blackbody radiation states that a completely black object will emit radiation according to the formula:

$$I(\nu, T) = \frac{2h\nu^3}{c^2 \left(e^{\frac{h\nu}{kT}} - 1 \right)}$$

where I is radiated power, ν is the frequency of emitted radiation, T is temperature in Kelvin, and h, c, and k are positive constants.

(a) Calculate $\partial I/\partial T$. Is it positive or negative? What does this tell you about the blackbody radiation emitted by different objects?

(b) Calculate $\partial I/\partial \nu$.

(c) Pick any three values for the ratio $h/(kT)$ and plot $\partial I/\partial \nu$ as a function of ν for each of those values. (Assume you're working in units where $h = c = 1$.)

(d) The three plots should all have the same basic shape. Describe this shape and explain what it tells you about the blackbody radiation emitted by an object.

(e) For each plot there should be a positive value $\nu = \nu^*$ for which $\partial I/\partial \nu = 0$. Estimate that value for each temperature. How does ν^* depends on temperature? What does that tell you physically about the blackbody radiation emitted by an object?

4.20 One way to represent a function of two variables is with a table of values. For instance, the following table shows that $z(3, 1) = 70$.

$z(x, y)$

$y \backslash x$	1	2	3	4	5
1	10	50	70	80	85
2	9	49	69	79	84
3	7	47	67	77	82
4	4	44	64	74	79
5	0	40	60	70	75

To estimate $\partial z/\partial x(3, 1)$, you can look at the average rate of change as x goes from 2 to 3 (holding y constant): $(70 - 50)/(3 - 2) = 20$. Then repeat the calculation as x goes from 3 to 4: $(80 - 70)/(4 - 3) = 10$. Averaging the two, we approximate $\partial z/\partial x(3, 1) = 15$.

(a) Estimate $\partial z/\partial y(2, 3)$.
(b) Estimate $\partial z/\partial y(3, 3)$.
(c) Estimate $\partial z/\partial y(4, 3)$.
(d) Based on your three answers, estimate $\partial^2 z/\partial x \partial y(3, 3)$.
(e) Use a similar method to estimate $\partial^2 z/\partial y \partial x(3, 3)$. The equality of mixed partials tells us that your final answer should come out the same as Part (d), although the intermediate calculations will be completely different.

4.21 *[This problem depends on Problem 4.20.]* Another way to estimate a derivative from a table is to *skip* the value you are interested in, and find the average around it. For instance, to estimate $\partial z/\partial x(3, 1)$ in Problem 4.20, you look at the two values around the 70: $(80 - 50)/(4 - 2) = 15$.

(a) Estimate $\partial z/\partial y(2, 3)$ using this quicker method.

To show that this method will always work, consider the more generic table of values below.

$z(x, y)$

$y \backslash x$	1	2	3	4	5
1	z_{11}	z_{21}	z_{31}	z_{41}	z_{51}
2	z_{12}	z_{22}	z_{32}	z_{42}	z_{52}
3	z_{13}	z_{23}	z_{33}	z_{43}	z_{53}
4	z_{14}	z_{24}	z_{34}	z_{44}	z_{54}
5	z_{15}	z_{25}	z_{35}	z_{45}	z_{55}

(b) Compute $\partial z/\partial x(3, 2)$ in this table by using the "find two different rates of change and average them" method from Problem 4.20.
(c) Compute $\partial z/\partial x(3, 2)$ in this table by using the quicker method demonstrated in this problem.

4.22 *[This problem depends on Problems 4.20 and 4.21.]* Calculate $\partial^2 z/\partial x \partial y(3, 3)$ and $\partial^2 z/\partial y \partial x(3, 3)$ for the table of values given in Problem 4.21, to demonstrate the equality of mixed partials more generally.

A differential equation is an equation involving derivatives. If it involves partial derivatives, it's called a "partial differential equation." In Problems 4.23–4.26 you should plug each of the given solutions into the partial differential equation given in the problem to see whether it's a valid solution. In some cases more than one of the solutions may work. (Any letters that don't appear in the derivatives are constants.)

4.23 $\partial f/\partial t = c\, \partial f/\partial x$
(a) $f(x, t) = cxt$
(b) $f(x, t) = c$
(c) $f(x, t) = (x + ct)^2$
(d) $f(x, t) = (x - ct)^2$
(e) $f(x, t) = x^2 + ct^2$

4.24 "The Wave Equation": $\partial^2 y/\partial t^2 = c^2\, \partial^2 y/\partial x^2$. This describes (among other things) waves on a stretched string.
(a) $y(x, t) = cxt$
(b) $y(x, t) = c$
(c) $y(x, t) = (x + ct)^2$
(d) $y(x, t) = (x - ct)^2$
(e) $y(x, t) = x^2 + ct^2$

4.25 "The Heat Equation": $\partial u/\partial t = c^2\, \partial^2 u/\partial x^2$. This describes (among other things) temperature along a thin rod.
(a) $u(x, t) = x^2 + 2c^2 t$
(b) $u(x, t) = cx$
(c) $u(x, t) = (x + ct)^2$
(d) $u(x, t) = e^{x+c^2 t}$

4.26 💻 $\partial f/\partial t = (1 - x^2)(\partial^2 f/\partial x^2) - 2x(\partial f/\partial x)$. Note that several of the proposed solutions below involve functions that you may not have heard of, but you can still use a computer algebra program to take their derivatives and check them in the given equation.
(a) $f(x, t) = \sin x e^{-2t}$
(b) $f(x, t) = J_1(x) e^{-2t}$ ("Bessel function")
(c) $f(x, t) = P_1(x) e^{-2t}$ ("Legendre polynomial")
(d) $f(x, t) = Ai(x) e^{-2t}$ ("Airy function")
(e) $f(x, t) = H_1(x) e^{-2t}$ ("Hermite polynomial")

4.27 In cases where a function depends on several variables that are related to each other, the notation $\partial f/\partial x$ can be ambiguous without a specification of what other variable is being held constant. (We will take up this issue in more detail in Section 4.10 (see felderbooks.com).) To illustrate how that can happen, suppose you are designing a box with no top for your company to sell. The volume is given by $V = wlh$ where w, l, and h are the width, length, and height respectively. You want to use a fixed amount of material, given by the surface area $S = wl + 2wh + 2lh$.
(a) Using the constraint that the surface area S must remain constant, express V as a function of w and l.

(b) Find $(\partial V/\partial w)_l$, meaning the rate of change of the volume with respect to w, holding the length l constant.

(c) Now express V as a function of w and h and find $(\partial V/\partial w)_h$.

(d) Compute $(\partial V/\partial w)_l$ and $(\partial V/\partial w)_h$ for the case $l = 1$ m, $h = 1$ m, and $S = 5$ m².

(e) You should have found in this case that one of the two derivatives is zero and the other is positive. Explain this result in terms that could easily be understood by a student who has never taken calculus.

4.3 The Chain Rule

In this section we will see how the single-variable "chain rule" extends to multiple variables, and use partial derivatives to estimate the total change of a function.

4.3.1 Discovery Exercise: The Chain Rule

We begin by considering a single-variable function $y(x)$, with $dy/dx = 3$.

1. Suppose $y(40) = 100$.
   (a) What is $y(41)$?
   (b) What is $y(50)$?

   See Check Yourself #20 in Appendix L

2. In general, if you add an arbitrary amount Δx to x, how much does y go up?

We now consider a function of two variables $z(x, y)$, with $\partial z/\partial x = 3$ and $\partial z/\partial y = -5$.

3. If x increases by 10 while y remains constant, what happens to z?
4. Suppose $z(40, 40) = 100$.
   (a) What is $z(50, 40)$?
   (b) What is $z(40, 50)$?
   (c) What is $z(50, 50)$? *Hint: you can get there from your answers to Part 4a or Part 4b. You might want to try both in order to check your answer.*
5. If $z(3, 4) = 0$, what is $z(5, 7)$?

4.3.2 Explanation: The Chain Rule

A pipe carries a stream of chocolate from one part of the Wonka Chocolate Factory to the next. If you dip in a spoon, you will taste a sweetness S determined by the concentration of cocoa c and the concentration of sugar g. So we will model the sweetness by a function $S(c, g)$. (Cocoa is bitter, so $\partial S/\partial c$ is negative, while $\partial S/\partial g$ is obviously positive.)

The concentration of cocoa varies at different places in the pipe, and also changes over time, so c is a function of x and t. The same is true of g.

Now imagine a chocolate-tasting fish swimming through the pipe, so its position x is a function of t. The changes in x and t both lead to changes in c and g, which in turn lead to changes in the sweetness S that the fish tastes. How can we calculate the rate at which the fish will experience sweetness increasing or decreasing?

This is a complicated problem, but we can approach it systematically with the multivariate chain rule. We will begin by reminding you of the single-variable chain rule, and then show the multivariate chain rule in some simpler scenarios, before plunging into the river of chocolate.

The Single-Variable Chain Rule

In single-variable calculus, you learned to use the chain rule to find the derivative of a composite function. For instance, if $z = \sin(x^2)$ then $dz/dx = \cos(x^2) 2x$.

To understand the chain rule more generally, consider the breakdown of functions in that example. We can write the problem like this:

$$z(y) = \sin y$$
$$y(x) = x^2$$

The chain rule then tells us that:

$$\frac{dz}{dx} = \left(\frac{dz}{dy}\right)\left(\frac{dy}{dx}\right) \tag{4.3.1}$$

Equation 4.3.1 *looks* good—the dy factors in the numerator and denominator cancel, leaving dz/dx—but many students have trouble connecting it to the chain rule. Our focus here is not on "how to take the derivative of $\sin\left(x^2\right)$" but on how Equation 4.3.1 expresses the steps in that process, as surely as $f(x)g'(x) + f'(x)g(x)$ expresses the steps in the product rule.

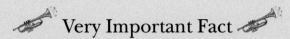

Very Important Fact

When you write the chain rule, you write a series of fractions that multiply so their numerators and denominators cancel. This is not a new chain rule; it is just a (possibly) new way of writing the chain rule you learned in your first calculus class.

Take a moment to calculate dz/dy and dy/dx and convince yourself that Equation 4.3.1 does represent the steps you take when you use the chain rule to find the derivative of $z = \sin\left(x^2\right)$. (We're serious; you'll get a lot more out of this section if you stop now and work through this. We're just waiting...done? Okay, let's move on.)

We use the chain rule whenever we have a chain of dependence among several variables. The above example demonstrates the simplest possible case:

The chain rule tells us to move down the picture from z at the top to x at the bottom, multiplying as we go.

A slightly more complicated case that might appear in an introductory calculus class is the chain of functions: $a(b) = e^b$, $b(c) = \cos(c)$, $c(d) = \ln(d)$. The chain rule tells us that:

$$\frac{da}{dd} = \left(\frac{da}{db}\right)\left(\frac{db}{dc}\right)\left(\frac{dc}{dd}\right) \tag{4.3.2}$$

Once again, we urge you to convince yourself of two things.

1. Equation 4.3.2 makes perfect sense if you think of each derivative as a fraction, with numerators and denominators canceling.

2. This is really nothing new. If someone had asked you yesterday to take the derivative of the function $e^{\cos(\ln d)}$, you would have gone through precisely the steps represented by Equation 4.3.2—even though you might never have written the equation like that—in order to end up with $e^{\cos(\ln d)}(-\sin(\ln d))(1/d)$.

Adding Contributions from Different Variables

Consider a simplified version of the chocolate problem, in which the fish doesn't swim; it just rests on the bottom, experiencing a dS/dt but not a dS/dx. The dependency tree in this case looks like this:

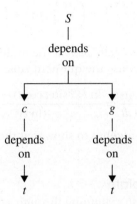

You can calculate dS/dt by following the chain rule down the tree from S to t, but in which direction? The answer—and this is the key to applying the chain rule in any multivariate situation—is that you go in all possible directions, and you add their contributions.

$$\frac{dS}{dt} = \left(\frac{\partial S}{\partial c}\right)\left(\frac{dc}{dt}\right) + \left(\frac{\partial S}{\partial g}\right)\left(\frac{dg}{dt}\right) \qquad (4.3.3)$$

You can convince yourself of Equation 4.3.3 with common sense. If the concentration of cocoa is increasing by 3 units every minute, and every additional unit of cocoa decreases the sweetness by 5, then the cocoa is decreasing the sweetness by 15 every minute. The effect of $(\partial S/\partial g)(dg/dt)$ similarly measures the increase (or decrease) based on the changing sugar concentration. The new element, the "plus" between the two terms, indicates that you experience both effects at once.

A different example makes this result more visual: consider graphing a function $z(x, y)$, so z is the height above the xy-plane. As you move along the surface, you change your x- and y-coordinates simultaneously, and your z-coordinate changes in response (since you are staying on the surface). How fast does z change? We answer this question by replacing one diagonal step across the xy-plane with two perpendicular steps. If you take a step in the positive x-direction, the height increases by $(\partial z/\partial x)dx$. A step in the positive y-direction increases the height by $(\partial z/\partial y)dy$. The two effects combine when you take a diagonal step, so $dz = (\partial z/\partial x)dx + (\partial z/\partial y)dy$. (The Discovery Exercise, Section 4.3.1, was designed to bring you to this conclusion on your own.)

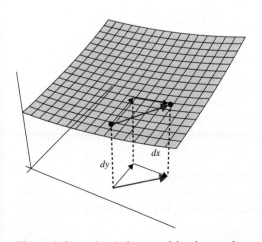

The total change in z is the sum of the changes due to changing x and y.

EXAMPLE The Chain Rule

Question: Consider the surface represented by the function $z = x^2/y$. A point moves along that surface with $x = \sin t$ and $y = \ln t$. How fast is the point's height changing as a function of time?

Answer:
The chain rule tells us that:

$$\frac{dz}{dt} = \left(\frac{\partial z}{\partial x}\right)\left(\frac{dx}{dt}\right) + \left(\frac{\partial z}{\partial y}\right)\left(\frac{dy}{dt}\right) = \left(\frac{2x}{y}\right)\cos t + \left(-\frac{x^2}{y^2}\right)\left(\frac{1}{t}\right) = \frac{2\sin t \cos t}{\ln t} - \frac{\sin^2 t}{t(\ln t)^2}$$

If we approach the same problem *without* the multivariate chain rule, we begin by writing $z(t) = \sin^2 t/\ln t$ and use the quotient rule:

$$\frac{dz}{dt} = \frac{\ln t (2 \sin t \cos t) - \sin^2 t/t}{(\ln t)^2}$$

It doesn't take much more algebra to show that the two results are the same.

Back to the River of Chocolate

We began this section with a fish swimming through a chocolate river. The scenario describes the chain of dependence shown in Figure 4.2.

No matter how complicated such a chain is, once you have drawn it, it isn't hard to use. If you want the change in sweetness with respect to time, trace down all the paths from S to t, adding all their effects:

$$\frac{dS}{dt} = \frac{\partial S}{\partial c}\frac{\partial c}{\partial x}\frac{dx}{dt} + \frac{\partial S}{\partial c}\frac{\partial c}{\partial t} + \frac{\partial S}{\partial g}\frac{\partial g}{\partial x}\frac{dx}{dt} + \frac{\partial S}{\partial g}\frac{\partial g}{\partial t} \qquad (4.3.4)$$

If there were any branches of the dependency tree that didn't end in t, then dS/dt wouldn't be a meaningful quantity. For example, we could define a quantity $R(x, t)$ as the sweetness of the river. (At any given place and time, the river has a certain sweetness, regardless of the fish.) We could speak of $\partial R/\partial t$: at a given place and time, the river's sweetness is increasing by this much. But we could not speak of a dR/dt because the river doesn't have a single sweetness at each time. Recalling that S is the sweetness experienced by the fish, dS/dt exists because the fish *does* experience a particular sweetness at each time. In terms of dependency trees, R depends on x, y, z, and t, but none of those are functions of the other ones.

With a bit of practice, you can easily write a chain rule for any given dependency diagram. If we now gave you specific functions that describe the fish's position as a function of time, and the concentration of cocoa as a function of time and position, and so on, you could calculate change in sweetness as seen by the fish. Of course this problem is deliberately silly, but this technique is important in many areas of physics and engineering; you will see some serious examples in the problems.

Total Derivatives and Partial Derivatives

Focus for a moment on the left-hand branch of Figure 4.2. The tree allows us to find the rate of change of c, the concentration of cocoa:

$$\frac{dc}{dt} = \frac{\partial c}{\partial x}\frac{dx}{dt} + \frac{\partial c}{\partial t} \qquad (4.3.5)$$

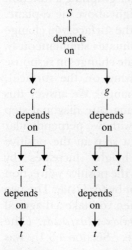

FIGURE 4.2

It isn't hard to follow the diagram and write the correct equation. Here's the hard part: how do we make sense of an equation that contains two different "derivative of c with respect to t" quantities? Discussing what these two quantities mean, and how they differ from each other, offers us a window into the more general issue of total derivatives (dc/dt) and partial derivatives ($\partial c/\partial t$).

Remember that the cocoa concentration is changing over time for two reasons. First, c is different in different places, which matters because the fish is moving. Second, c is different at different times. Both of these factors contribute to the overall change the fish experiences. If both factors cause increases, the fish will experience an extra-large cocoa increase; if one causes a decrease, they may cancel each other out.

The quantity $\partial c/\partial t$ isolates one of these changes, the part that is *directly* due to the fact that concentration depends on time; it ignores the *indirect* effect due to the fact that the concentration depends on position, which in turn depends on time. To put it another way, $\partial c/\partial t$ represents the rate of concentration change the fish would see if it stopped swimming. (Note from Equation 4.3.5 that dc/dt and $\partial c/\partial t$ are the same if $dx/dt = 0$.) On the other hand, dc/dt takes into account both the direct and indirect effects.

So whenever you see a total derivative, you know that "If this variable changes by this much, that variable will change by that much." If dc/dt is a constant 5, wait 3 s and c will go up by 15 for the fish. A partial derivative tells you much less: "If this variable changes while all other variables remain constant" is, in many cases, an impossible scenario. (If the fish is swimming, you cannot literally change t without also changing x.) The partial derivative is still useful, but often only as a step toward finding the total derivative.

Differentials (or "Our Friends, dx and ∂x")

When you write dx/dt, what do the dx and dt parts by themselves mean?

Many mathematicians will quickly tell you that they don't mean anything at all. The expression dx/dt is not actually a fraction; it is an atomic piece of notation that means "the derivative of x with respect to t." But engineers and scientists don't treat it that way at all. We call dt a "differential" and view it as a small change in time, a little cousin of Δt. The statement $dx/dt = 5$ means that if t changes by a small dt, that will cause a small change dx, and the ratio "dx divided by dt" will be 5.

With this perspective the following statements are easily explainable, instead of looking like a collection of coincidences.

- The chain rule $da/dc = (da/db)(db/dc)$ looks like fractions canceling.
- In general, dx/dy is the reciprocal of dy/dx. For instance, if $dy/dx = 2x$ then $dx/dy = 1/(2x)$. (Go on, try it. You know you want to.)
- You can solve some differential equations by multiplying both sides by dx and then integrating.
- The units of dy/dx are the units of y divided by the units of x. For instance, if x is in meters and t in seconds, then dx/dt is meters/second.

This is not just a philosophical issue. Understanding dx in this way is vital to setting up integrals, as we will discuss extensively in Chapter 5. Differentials are used extensively in thermodynamics, where you will be expected to work with equations such as the "thermodynamic identity" $dU = T\, dS - P\, dV$. ("If S changes by this much while V changes by that much, then how much will U change?")

Differentials can also be viewed as margins of error. For instance, the kinetic energy of a Newtonian object is $E = (1/2)mv^2$. Based on this equation, we can write:

$$dE = mv\, dv + (1/2)v^2\, dm \qquad (4.3.6)$$

Divide both sides by dt and you have an equation relating dE/dt to dv/dt and dm/dt. But you can also take Equation 4.3.6 on its own, to say that if the velocity is $10 \pm .1$ m/s and the mass is $30 \pm .01$ kg, then the energy is (1500 ± 30.5) J.

But if it is vital to understand that dx can be treated as a variable, it is also important to know that ∂x cannot! If you try to multiply both sides of an equation by ∂t or cancel ∂t out of the top and bottom of a fraction you will at best get something meaningless and at worst get an equation that is simply wrong. You will explore this issue in Problem 4.50.

4.3.3 Problems: The Chain Rule

4.28 Consider a function $z(x, y)$ where x depends on a and b, y depends on b, and a depends on b.
  (a) Draw the dependency tree described in the problem.
  (b) Using the dependency tree you just drew, write the chain rule that tells you dz/db.
  (c) Take $z = x^2 - y^2$, $x = a^3 e^b$, $da/db = 2e^{2b}$ and $dy/db = 14e^{7b}$. Use the chain rule you just wrote to calculate dz/db. Use the properties of exponents to simplify your answer as much as possible.

4.29 *[This problem depends on Problem 4.28.]* Make up two functions $a(b)$ and $y(b)$ that have the derivatives given in Problem 4.28. Plug these and the functions given in that problem to write a function $z(b)$. Take the derivative of this function and verify that you get the same answer you got in Problem 4.28.

4.30 Consider a function $f(x, y)$ where x depends on p and q; y depends on b; q depends on p, a, and b; p depends on a; and a depends on b.
  (a) Draw the dependency tree that the problem just described.
  (b) Using the dependency tree you just drew, write the chain rule that tells you df/db.

4.31 $z = x^e e^y$, $x = 4t + 5$, and $y = 2 - t$.
  (a) Draw the dependency diagram for this situation.
  (b) Calculate dz/dt using your dependency diagram.
  (c) Find the function $z(t)$ and take its derivative directly. Show that your two results are equal.

4.32 $z = x(y^2 + 1)$, $x = uv$, and $y = (u + v)^3$.
  (a) Draw the dependency diagram for this situation.
  (b) Calculate $\partial z/\partial u$ and $\partial z/\partial v$ using your dependency diagram.
  (c) Find the function $z(u, v)$ and take its derivatives directly.

4.33 The buoyancy B of a hot air balloon depends on the balloon's volume V, the density of the air in the balloon ρ_b, and the density of the surrounding atmosphere ρ_{atm}. The density of the atmosphere depends on the altitude h. The density of the air in the balloon depends on the altitude h and the balloon's temperature T. Both h and T depend on the time t.
  (a) Draw the dependency tree for B.
  (b) Write the chain rule for dB/dt, assuming the volume of the balloon stays constant.
  (c) In the Explanation (Section 4.3.2), we said that dB/dt can only exist if the dependency tree for B ends at t on every branch. This problem seems to violate that rule; explain why it doesn't.

4.34 If a potential V is applied across a resistor R, the power lost across the resistor is given by the formula $P = V^2/R$.
  (a) Find the derivatives $\partial P/\partial V$ and $\partial P/\partial R$.
  (b) As the potential and resistance both change, how fast does the power change? (In other words, find dP/dt as a function of V, R, dV/dt, and dR/dt.)
  (c) If $V = V_0 e^{-t}$ and $R = R_0 \sin t$, find dP/dt.

4.35 If a basketball team makes p three-point shots, g field goals, and f free throws, its score is $s = 3p + 2g + f$. Of course, as the game proceeds, all of these variables are functions of time.
  (a) Draw the dependency diagram for this situation.
  (b) Calculate ds/dt as a function of dp/dt, dg/dt, and df/dt.
  (c) If the team scores 1 3-point shot every 5 min, 1 field goal every 2 min, and 1 free point shot every 3 min, how fast does its score change per minute?

4.36 A snowball rolls down a hill, its mass m and velocity v both changing with time.
  (a) Momentum is given by the formula $p = mv$. Write an equation for $\dot{p}$ as a function of m,

v, $\dot{m}$, and $\dot{v}$. (Remember that a dot indicates a derivative with respect to time.)

(b) Kinetic energy is given by the formula $E = (1/2)mv^2$. If the snowball is gaining 0.2 kg of mass each second, it's speeding up by 0.1 m/s each second, and its current mass and speed are 12.5 kg and 3.2 m/s, how fast is its kinetic energy increasing?

4.37 According to the Stefan-Boltzmann law, the power emitted by a star is given by $P = \sigma A T^4$, where A is the star's surface area ($4\pi R^2$), T is its surface temperature, and $\sigma = 5.67 \times 10^{-8}$ J s^{-1} m^{-2} K^{-4}. Near the end of its life our sun will expand to become a "red giant." As it expands its radius R will increase at approximately 10^{-5} m/s while its surface temperature will decrease at roughly 10^{-13} K/s. Assuming it starts at its current state with $R \approx 7 \times 10^8$ m and $T \approx 6000$ K, how fast will its total power output change as it expands toward the red giant phase?

4.38 Consider a function $F(x, y)$ with known derivatives $\partial F/\partial x$ and $\partial F/\partial y$. Find formulas for the derivatives $\partial F/\partial \rho$ and $\partial F/\partial \phi$, where ρ and ϕ are the polar coordinates.

4.39 Object A is moving with velocity v_{AB} relative to object B, and B is moving with velocity v_{BC} (in the same direction) relative to object C. According to special relativity, the velocity of A with respect to C is:

$$v_{AC} = \frac{v_{BC} + v_{AB}}{1 + v_{BC} v_{AB}/c^2}$$

where c, the speed of light, is a constant. If $v_{AB} = 0.7c$, $dv_{AB}/dt = .1c/s$, $v_{BC} = 0.8c$, and $dv_{BC}/dt = -.2c/s$, how fast is the velocity of A with respect to C changing? (If you've studied relativity you might wonder what reference frame t is measured in. Physically interpreting these derivatives can be tricky, but this has no effect on the calculations we're asking for.)

4.40 The temperature at a point (x, y) is $T(x, y)$, measured in degrees Celsius. The temperature function is independent of time and satisfies $\partial T/\partial x(2, 3) = 4$ and $\partial T/\partial y(2, 3) = 3$, where x and y are measured in centimeters. A bug crawls so that its position after t seconds is given by $x = \sqrt{1+t}$, $y = 2 + t/3$. How fast is the temperature rising on the bug's path after 3 s?

4.41 A small metal ball with charge q at a distance r from a large charge Q experiences a force $F = -kqQ/r^2$, where k is a constant.

The ball is moving away from the large charge at a rate dr/dt while charge is being added to it. If you measure that the force on it is changing at a rate dF/dt, how fast must the charge be getting added to the ball? (Assume the large charge Q stays constant.)

4.42 A can is manufactured as a right circular cylinder. The radius is $r = 3''$ with a margin of error of $dr = .01''$. The height is $h = 6''$ with a margin of error of $dh = .001''$.

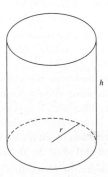

(a) Calculate dV, the margin of error of the volume.

(b) Calculate dA, the margin of error of the surface area.

4.43 In order to measure the total force F on a particle you measure its mass and acceleration to be $m \pm dm$ and $a \pm da$. What is the uncertainty in your measurement of F?

4.44 When an object is at a distance o from a thin lens with focal length f, it projects an image at a distance i from the lens according to the lens equation: $(1/o) + (1/i) = (1/f)$. You place an object at (3.2 ± 0.3) cm from a lens with focal length (1.7 ± 0.2) cm. Predict the image distance i, including the uncertainty in the prediction.

4.45 If you stand outside, the perceived temperature T you feel depends on the actual outside temperature u and the amount of clothing c you are wearing. However, the amount of clothing you wear depends on the outside temperature.

(a) Draw the dependency diagram for this situation.

(b) Your dependency diagram should allow you to conclude that $dT/du = (\partial T/\partial c)(dc/du) + (\partial T/\partial u)$. Explain the meanings of the quantities dT/du and $\partial T/\partial u$. Your explanation should explain how these two quantities are different from each other, and how the former depends on the latter.

4.46 Explain why we wrote dx/dt and not $\partial x/\partial t$ on the right-hand side of Equation 4.3.4.

4.47 Consider a fly buzzing around a room. The temperature T_{fly} that the fly feels at any given moment depends on the fly's position (because temperature varies throughout the room) and the time (because the temperature throughout the room is changing). The fly's position is, of course, a function of time.

(a) Describe in terms accessible to an average 12 year old what $\partial T_{fly}/\partial t$ and dT_{fly}/dt represent physically.

(b) Describe a situation in which dT_{fly}/dt is positive and $\partial T_{fly}/\partial t$ is negative. Give enough detail to explain why the two derivatives have the signs they do. Your explanation should not contain any math terms like "with respect to" or "holding … constant," but should just be a description of what's happening in the room and what the fly is doing.

(c) Consider the function $T_{room}(x, y, z, t)$. Explain why dT_{room}/dt doesn't mean anything.

(d) We use T_{room} and T_{fly} to represent two different variables. However, $\partial T_{room}/\partial t$ and $\partial T_{fly}/\partial t$ are equal. Why?

(e) We said in the Explanation (Section 4.3.2) that you can only calculate a total derivative df/dx if all the branches of the dependency tree for f end with x. Draw the dependency trees for T_{fly} and T_{room} and use them to verify that dT_{fly}/dt is a meaningful quantity and dT_{room}/dt isn't.

4.48 In older physics textbooks it was common to talk about "relativistic mass," which increases as an object goes faster. In such a formulation, the energy of a moving object depends on its velocity v and its relativistic mass m, while m in turn depends on v.

(a) Draw a dependency tree for the energy E.

(b) Write the chain rule for dE/dv.

(c) For each of the derivatives in your answer to Part (b) (including dE/dv itself) explain whether you would expect it to be positive or negative, and why.

4.49 A mountain climber feels a temperature u that depends on the climber's height h and the time t. Of course the climber's height also depends on the time.

(a) Draw a dependency tree for the temperature u.

(b) Write the chain rule for du/dt.

(c) Assume the climber is ascending the mountain sometime around sunrise. For each of the derivatives in your answer to Part (b) (including du/dt itself) explain whether you would expect it to be positive or negative, and why.

4.50 Exploration: The Meaning of Δx, dx and ∂x
The local cheese factory is growing. They are adding two new cheese presses every day, and each cheese press can make 300 lb of cheese per day. In this part of the problem you will be working with only three variables: Δt (time interval), Δp (change in number of cheese presses), and Δc (change in daily cheese output).

(a) How quickly is the daily cheese output increasing? (Your answer will be a number, with units.)

(b) Express the statement "they are adding two new cheese presses every day" in terms of our three variables. *Hint*: It isn't just $\Delta p = 2$.

(c) Express the statement "each cheese press can make 300 lb of cheese per day" in terms of our three variables.

(d) Write an equation that shows how you got the answer to Part (a). Your equation should be based on our three variables, and should not contain any specific numbers.

Unfortunately, mice have gotten into the factory. Every day brings five more mice, and every mouse eats 1/10 lb. of cheese every day. In addition to the three variables above, you will now also work with Δm.

(e) Now how quickly is the daily cheese output increasing?

(f) Write an equation that shows how you got the answer to Part (e). Your equation should be based on our four variables with no numbers. Your equation should contain Δc in three places, but it won't mean the same thing in each place.

(g) In each place Δc appears in your equation in Part (f), what does it mean? (Answer separately for each of the three places.)

(h) In the limit as $\Delta t \to 0$, all the Δ variables become d or ∂. Rewrite your answer to Part (f) in this case.

(i) Explain, based on this problem, why dc is a meaningful variable but ∂c is not.

4.4 Implicit Differentiation

When an equation expresses a relationship between two or more variables, "implicit differentiation" allows you to find the rate of change of one of these variables with respect to another, *without* having to solve for one of the variables first.

4.4.1 Discovery Exercise: Implicit Differentiation

The equation $x^2 + y^2 = 25$ describes a circle with radius 5 centered on the origin.
Answer questions 1–6 by looking at the picture.
(No algebra should be required.)

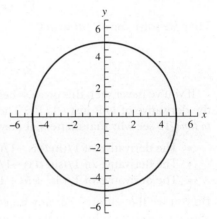

1. What is the slope of this curve at the point $(0, 5)$?
2. What is the slope of this curve at the point $(5, 0)$?
3. What is the slope of this curve approaching as you approach the point $(5, 0)$ from above?

    See Check Yourself #21 in Appendix L

4. What is the slope of this curve approaching as you approach the point $(5, 0)$ from below?
5. What is the slope of this curve at the point $(5/\sqrt{2}, 5/\sqrt{2})$?
6. What is the slope of this curve at the point $(5/\sqrt{2}, -5/\sqrt{2})$?

For questions 7–10, *approximate* the answers by looking at the picture. No two of your answers should be the same.

7. What is the slope of this curve at the point $(4, 3)$?
8. What is the slope of this curve at the point $(3, 4)$?
9. What is the slope of this curve at the point $(4, -3)$?
10. What is the slope of this curve at the point $(3, -4)$?
11. You can write the slope of this curve as a function of x and y. Guess at such a function for this curve. Your function $m(x, y)$ should exactly match your answers to questions 1–6, and should approximately match your answers to questions 7–10.

4.4.2 Explanation: Implicit Differentiation

The pressure, volume, and temperature of gas inside an expandable container can in some circumstances be modeled by the equation[2]

$$\frac{an^2}{V} + PV = nRT \qquad (4.4.1)$$

Here n is the number of moles of gas and a and R are constants. Suppose we change the temperature and pressure at controlled rates dT/dt and dP/dt and we want to know how quickly the volume will change in response.

---

[2] You may recognize this as a variation of the more familiar "ideal gas law" $PV = nRT$. This equation takes into account some surface effects that are neglected by the ideal gas law. This is a simplified version of the "van der Waals" equation, which you will use in Problem 4.72.

One approach to this problem would be to first solve Equation 4.4.1 for the volume, and then find dV/dt with the multivariate chain rule. However, that first step can be messy (as it is in this case) or impossible. We therefore find dV/dt without first solving for V, a process called "implicit differentiation": we take the derivative of both sides of the equation with respect to time.

The right-hand side of Equation 4.4.1 is simply a constant times the function $T(t)$, so its derivative is $nR(dT/dt)$. To take the derivative of the left-hand side we have to use the chain rule and the product rule.

$$-\frac{an^2}{V^2}\frac{dV}{dt} + P\frac{dV}{dt} + V\frac{dP}{dt} = nR\frac{dT}{dt}$$

Then we solve for dV/dt to get

$$\frac{dV}{dt} = \frac{nR(dT/dt) - V(dP/dt)}{P - (an^2/V^2)}$$

If you've never seen this process before, those derivatives are likely to throw you. Why isn't the derivative of $1/V$ just $-1/V^2$? The answer is that we are taking the derivative *with respect to time*. To see why that matters, it may help to consider a few specific functions.

- The derivative of $1/(\ln t)$ is $-1/(\ln t)^2 \times (1/t)$.
- The derivative of $1/(\sin t)$ is $-1/(\sin^2 t) \times (\cos t)$.
- The derivative of $1/(3t^2 + 5t + 12)$ is $-1/(3t^2 + 5t + 12)^2 \times (6t + 5)$.

Do you see the pattern? For any function $f(t)$, the derivative of $1/f$ is $-(1/f^2) \times (df/dt)$. We applied the chain rule as always, but in this case we applied it to a function $V(t)$ that we don't know, so when we did the step that says "multiply by the derivative of what's inside" we wrote dV/dt instead of some specific function of t.

EXAMPLE Implicit Differentiation

Problem:
The quantities $x(t)$, $y(t)$, and $z(t)$ are related by $e^{xy/z^2} = 2$. Find $\dot{z}$ in terms of $\dot{x}$ and $\dot{y}$. (Remember that a dot means "derivative with respect to time.")

Solution:
Take the derivative of both sides with respect to t. First use the chain rule to take the derivative of the exponential times the derivative of what's inside. Then use the quotient rule on the fraction. Finally, use the product rule for the derivative of the numerator. That sounds like a lot of steps, but each one is straightforward.

$$e^{xy/z^2}\left(\frac{z^2(x\dot{y} + y\dot{x}) - 2xyz\dot{z}}{z^4}\right) = 0$$

This can only be satisfied if the numerator of the fraction is zero. That gives us a simple equation we can solve for $\dot{z}$.

$$\dot{z} = \frac{z}{2xy}(x\dot{y} + y\dot{x})$$

4.4.3 Problems: Implicit Differentiation

4.51 Walk-Through: Implicit Differentiation. The functions $x(t)$, $y(t)$, and $z(t)$, are related to each other and to t by the following equation.

$$x^2y - e^x + z^2t = 0$$

You are going to find the derivative dx/dt.

(a) Take the derivative of both sides of this equation with respect to t. You'll need to use the product rule where two functions are multiplied by each other, and you'll need to use the chain rule for composite functions such as x^2.

(b) Solve the resulting equation for dx/dt. Your answer should depend on the values of all three functions, both constants, the derivatives dy/dt and dz/dt, and t itself.

(c) At $t = 1$ s, $x = 0$, $y = 0$, and $z = 1$. If y is increasing at a rate of 5 per second and z is increasing at a rate of 3 per second, estimate $x(1.1)$.

(d) Why is your answer to Part (c) only approximate?

In Problems 4.52–4.56 you will be given an equation relating the functions $f(t)$, $g(t)$, and $h(t)$. Find the derivative df/dt in terms of f, g, h, dg/dt and dh/dt. Assume any letters other than f, g, h, and t represent constants.

4.52 $f^2g^3 - h/f = 0$
4.53 $f\cos(gh) + e^{f/h} = h$
4.54 $fgh = te^{af}$
4.55 $ag + bh + cf = t/f$
4.56 $t^2f + ag/t = he^{bf}$

4.57 Consider the curve $x^4 + x^2 + y^2 + y^4 = 1$.

(a) Take the derivative of both sides of this equation with respect to x. *Hints:* The derivative of y^2 is not just $2y$, and the derivative of 1 is not 1.

(b) Solve the resulting equation to find a formula for dy/dx as a function of x and y.

(c) Find the slope of this curve at the point $(0.2, 0.77)$.

(d) Find the slope of this curve at the point $(0.2, -0.77)$.

4.58 In the Discovery Exercise (Section 4.4.1), you found the slope of the curve $x^2 + y^2 = 25$ at various points. Use implicit differentiation to find the slope of this curve as a function of x and y. Verify that your formula matches the shape of the circle at the points $(0, 5)$, $(5, 0)$, and $(5/\sqrt{2}, 5/\sqrt{2})$.

4.59 *[This problem depends on Problem 4.58.]* In Problem 4.58 you found a formula for the slope of the curve $x^2 + y^2 = 25$ as a function of x and y.

(a) To find d^2y/dx^2, take the derivative of your answer with respect to x. Note that this will require implicit differentiation once again. (Use the quotient rule, remembering that the derivative of y is dy/dx.)

(b) Where you wrote dy/dx in your answer to Part (a), substitute the formula that you found in Problem 4.58. You now have a formula for the concavity of the circle as a function of x and y.

(c) Simplify your answer as much as possible. The fact that $x^2 + y^2 = 25$ on this curve will be useful here.

(d) Where does your formula predict that this curve is concave up, and where does it predict concave down? Does this result match the shape of the circle?

4.60 The function $x^3 + y^3 = 1$ describes a curve.

(a) Take the derivative of both sides with respect to x, using y' to represent the derivative of y with respect to x.

(b) Solve the resulting equation for $y'(x, y)$. Simplify as much as possible.

(c) Based on your answer, where is this curve increasing? Where is it decreasing?

(d) Take the derivative of your answer from Part (b) with respect to x. Your answer will be a function $y''(x, y, y')$.

(e) Substitute your answer from Part (b) for y' into your answer to Part (d). Simplify the resulting expression as much as possible. For the final simplification, you will replace $x^3 + y^3$ with 1.

(f) Based on your answer to Part (e), where is this curve concave up? Where is it concave down?

(g) Solve the equation $x^3 + y^3 = 1$ for y, and take the first and second derivatives of the resulting function. Confirm that your answers match your answers to Parts (b) and (e).

(h) 🖥 Graph the function. Confirm that the resulting function matches your predictions.

156 Chapter 4 Partial Derivatives

4.61 A curve is defined by the relation
$y^3 + y = e^x + x$.

(a) Use implicit differentiation to find the slope of this curve. Call the resulting function $m(x, y)$.

(b) Is $m(x, y)$ positive or negative? Explain in words what this tells us about the shape of the curve.

(c) 🖥 Plot $y^3 + y = e^x + x$ on a computer and confirm that it matches you said in Part (b).

In Problems 4.62–4.64, find the slope and concavity of the given curve as functions of x and y. It may help to first work through Problem 4.60 as a model. Simplify your answers as much as possible.

4.62 $x^{3/2} + y^{3/2} = 1$

4.63 $x + y = xy$

4.64 $\ln y / y = 3x$

In Problems 4.65–4.69 you will use implicit differentiation to find the derivatives of "inverse functions."

4.65 In this problem you will show that the derivative of $\ln x$ is $1/x$.

(a) Starting with the equation $y = \ln x$, solve for x.

(b) Take the derivative of both sides of the resulting equation with respect to x.

(c) Solve the resulting equation for dy/dx. The answer will be a function of y.

(d) Finally, use substitution to express your answer in terms of x. (Remember the relationship between x and y in this problem!)

4.66 [*This problem depends on Problem 4.65.*] Find the derivative of $y = \sqrt{x}$ by following the same steps you took in Problem 4.65.

4.67 In this problem you will find the derivative of $\sin^{-1} x$.

(a) Starting with the equation $y = \sin^{-1} x$, solve for x.

(b) Take the derivative of both sides of the resulting equation with respect to x.

(c) Solve the resulting equation for dy/dx. The answer will be a function of $\cos y$.

(d) Convert your answer to a function of $\sin y$, using the identity $\sin^2 y + \cos^2 y = 1$.

(e) Finally, use substitution to express your answer in terms of x. (Remember the relationship between x and y in this problem!)

4.68 [*This problem depends on Problem 4.67.*] Find the derivative of $y = \cos^{-1} x$ by following the same steps you took in Problem 4.67.

4.69 [*This problem depends on Problem 4.67.*] Find the derivative of $y = \tan^{-1} x$ by following the same steps you took in Problem 4.67. You will need to know that the derivative of $\tan x$ is $\sec^2 x$, and the identity $\tan^2 x + 1 = \sec^2 x$.

4.70 The amount of *something* (S) in a vat is related to the amount of *goop* (G) and *yucky glop* (Y) by the equation

$$S^2 = aGY + be^{c(G/Y)}$$

where a and b are constants. If you've measured the rates of change dS/dt and dG/dt, find the rate dY/dt.

4.71 An airplane passes a distance h over a control tower heading upward at a 40° angle, as shown below. D is the distance from the control tower to the airplane and s is the distance the airplane has traveled since passing over the control tower.

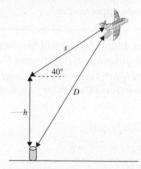

(a) Write an equation relating D to s and h. (*Hint*: The quickest way involves the law of cosines.)

(b) Differentiate the equation you just wrote and solve for dD/dt as a function of D, s, h, and ds/dt. (Remember that h is the height of the airplane at the moment it passes over the control tower, so it's a constant.)

(c) If the airplane passed 3 km over the control tower and has traveled 5 km since then at a constant speed of 600 km/h, how fast is its distance from the control tower increasing?

4.72 The van der Waals equation of state relates the pressure, volume, and temperature of a gas by the following equation where a, b, n, and R are all constants.

$$PV + \frac{an^2}{V} - nbP - \frac{abn^3}{V^2} = nRT$$

(a) Assuming dP/dt and dT/dt are known, find a formula for dV/dt.

(b) The Explanation (Section 4.4.2) solved this problem for the case $b = 0$. Check your answer by verifying that it reduces to our answer in this case.

4.73 When an object is placed near a thin magnifying glass it can project an image onto a piece of paper on the other side of the glass. The distance from the glass to the object o and the distance from the glass to the image i are related by the "thin lens formula."

$$\frac{1}{f} = \frac{1}{o} + \frac{1}{i}$$

Here f is the "focal length" of the magnifying glass. Assume a magnifying glass has focal length 3 cm and you are holding a piece of paper at a distance of 8 cm from it. If the object is moving away from the glass at 2 cm/s how fast do you need to move the piece of paper away to keep the image in focus on the paper?

(a) First answer this question by using implicit differentiation to take the derivative of the thin lens formula with respect to time.

(b) Next answer by solving the thin lens formula for i and then taking the derivative directly. Make sure you get the same answer both ways.

4.74 An economic rule of thumb states that to maximize profit you should set the price P for a product so that it is related to the "marginal cost" M (cost to produce one more unit) and the "price elasticity of demand" E (a measure of how much demand drops as the price increases) according to the following equation.

$$\frac{P - M}{P} = -\frac{1}{E}$$

(Note that E is a negative number; otherwise this formula would make no sense.) Assume you are selling widgets that have a marginal cost of $2 and you have found the optimal price to be $4. If your marginal cost is going up 50 cents every month and your market research shows that the price elasticity of demand is decreasing (becoming more negative) at a rate of 0.2 per month, how fast should you change your price to keep it at the optimal level?

4.75 The Cobb–Douglas production function models the production of a commodity with the equation $Y = AL^\alpha K^\beta$ where Y is the total value of the commodity produced, L and K are the input of labor and capital respectively, and A, α, and β are constants related to overall productivity (e.g., the technology level).

(a) If you want to increase production at a specified rate dY/dt, but the available labor is decreasing at a rate dL/dt, how fast must you increase the capital input?

(b) In the previous part you took A, α, and β to be constants. Now suppose that increased technology is improving the productivity of labor at a rate $d\alpha/dt$. Still treating A and β as constants, how fast must you now increase capital input to keep productivity rising at the desired rate dY/dt?

4.76 Exploration: A Formula for Implicit Differentiation. Any implicit differentiation problem with two variables begins with an equation relating those two variables. This equation can always be expressed in the form $F(x, y) = k$ where k is a constant. By solving that generic equation for dy/dx, you derive a formula for the derivative of any such function.

The dependency tree in this situation looks like this:

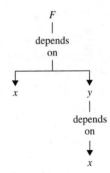

(a) Starting with the equation $F(x, y) = k$, take the derivative of both sides with respect to x. Use the dependency tree drawn above to expand the left side of your result.

(b) Solve the resulting equation for dy/dx.

(c) Use the formula you found in Part (b) to find the slope of the circle $x^2 + y^2 = 25$.

(d) If you haven't already solved Problem 4.58 use implicit differentiation directly on the formula $x^2 + y^2 = 25$ to find dy/dx and check that it matches your answer to Part (c).

Chapter 4 Partial Derivatives

4.5 Directional Derivatives

We have seen that a multivariate function has two or more "partial derivatives," and we have discussed how to use them to answer certain types of questions about the rate of change of such a function. In this section we begin to look more generally at the question of how fast a multivariate function is changing.

4.5.1 Discovery Exercise: Directional Derivatives

The following picture shows the function $z = y^2 - x$ on the domain $x \in [0, 3]$, $y \in [0, 3]$. The point $(1, 1, 0)$ is labeled.

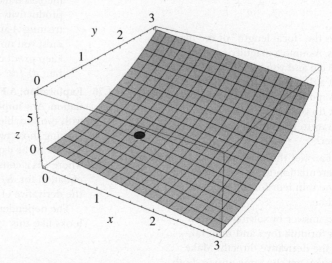

In all the questions that follow, assume that you begin at the point $(1, 1, 0)$. As you change your x- and y-coordinates, your z-coordinate changes to keep you on the surface.

Parts 1–4 can be answered exactly using partial derivatives. Make sure, however, that your answers make sense with the picture.

1. If you move in the positive *x*-direction, are you moving up or down? With what slope?
2. If you move in the negative *x*-direction, are you moving up or down? With what slope?
3. If you move in the positive *y*-direction, are you moving up or down? With what slope?
4. If you move in the negative *y*-direction, are you moving up or down? With what slope?

See Check Yourself #22 in Appendix L

Parts 5–8 should be answered approximately, based on the picture.

5. Now suppose you move diagonally, following the vector $\hat{i}+\hat{j}$ along the *xy*-plane (but allowing *z* to change as always so you stay on the surface). Are you moving up or down? With approximately what slope?
6. Suppose you follow the vector $\hat{i}-\hat{j}$ along the *xy*-plane. Are you moving up or down? With approximately what slope?
7. What direction would you move along the *xy*-plane if you wanted to go *upward* as steeply as possible? (In Section 4.6 you'll learn how to use something called a "gradient" to easily calculate this.)
8. If you placed a ball on this surface at the point $(1, 1, 0)$, which way would it roll?

4.5.2 Explanation: Directional Derivatives

The Discovery Exercise, Section 4.5.1, presented you with a series of questions about how fast a function of x and y changes as you move in different directions. The tool for mathematically answering such questions is called a "directional derivative." This section will first look at what a directional derivative *means*, and then present the formula for *finding* one.

What Do Directional Derivatives Mean?

If you look around the room you are sitting in right now, every point in the air has a certain temperature. We can represent this mathematically with a "scalar field" $f(x, y, z)$. The phrase "scalar field" simply indicates that there is a certain number, or "scalar," at every point in space. For instance, you might find that the temperature at the point $(3, 2, 7)$ is $4°$ below zero, so you would write $f(3, 2, 7) = -4$.

If we ask the question "What is the derivative of f?" or "How fast is this function changing?" you should object immediately that the question is ambiguous: changing with respect to position, or time? In this section we will not be interested in changes with respect to time; we want to know how the temperature distribution varies at different points throughout the room at one frozen (no pun intended) moment in time.

So you make another measurement and determine that $f(3.1, 1.9, 7.02) = 0$. You might reasonably conclude that the derivative of f at the point $(3, 2, 7)$ is very high, since f changes so much over a short distance. But another measurement reveals that $f(3.1, 1.9, 6.9)$ is exactly -4, suggesting a derivative of 0. Which one is right? Our point is that the question "how fast is this function changing as you move from $(3, 2, 7)$ to another point?" is still ambiguous: you need to specify what direction you are moving in! For any given direction, the function is changing at a particular rate.

> **The Meaning of the Directional Derivative**
>
> Given a scalar field $f(x, y, z)$, a point (x_0, y_0, z_0), and a direction specified by the vector $\vec{u}$, the "directional derivative" $D_{\vec{u}}f(x_0, y_0, z_0)$ gives the rate of change of f as you start at the point (x_0, y_0, z_0) and move in the direction of $\vec{u}$.

As always, don't let the smooth-sounding phrase "rate of change" pass you by too quickly. We might say, somewhat loosely, that $D_{\vec{u}}f$ gives the amount that f will change if you move by precisely one unit in the $\vec{u}$-direction. It would be more accurate to say that if you take a small step (magnitude ds) in the $\vec{u}$-direction, $D_{\vec{u}}f$ represents the change in f per unit change in position, df/ds. In the limit as $ds \to 0$, this ratio becomes the actual directional derivative. This can be expressed in the following equation, where $\hat{u}$ represents a unit vector in the direction of $\vec{u}$.

$$D_{\vec{u}}f(\vec{r}) = \lim_{h \to 0} \frac{f(\vec{r} + h\hat{u}) - f(\vec{r})}{h}$$

The partial derivatives we have seen are special cases of directional derivatives. For instance, for the direction $\vec{u} = \hat{i}$, $D_{\vec{u}}f = \partial f/\partial x$. But a scalar field has an infinite number of directional derivatives—not just two or three pointing along the coordinate axes. If you look in a particular direction $\vec{u}$ and see a constant derivative $D_{\vec{u}}f = 3$, that means the temperature is rising at a rate of 3 degrees for each unit step you take in that direction.

EXAMPLE: Using a Directional Derivative

Question: Consider the scalar field f with $f(3, 2, 7) = -4$. For the direction vector $\vec{u} = 3\hat{i} + 4\hat{k}$, $D_{\vec{u}}f = 2$. Estimate $f(6, 2, 11)$ and explain why your answer is only approximate.

Answer:
The journey from $(3, 2, 7)$ to $(6, 2, 11)$ is taken in the direction of $\vec{u}$, so we have the derivative we need. A derivative of 2 means that f is increasing by 2 units *per unit traveled*; since we travel 5 units, f will increase by a total of 10. We estimate that $f(6, 2, 11) = 6$.

This answer is approximate because it assumes that $D_{\vec{u}}f = 2$ throughout the journey. For most functions, the directional derivative changes as you move from one point to another. Approximations of this sort work well over short distances; for longer distances we need a "line integral" (Chapter 5) to add up all the small displacements.

How Do We Find Directional Derivatives?

Suppose we measure the function f at the point (x_0, y_0, z_0). Then we move away from that point by vector $\vec{u}$, and measure the function again. How much does the function change?

We learned in Section 4.2 how to answer such a question for the easiest special case, which is the case of $\vec{u}$ pointing directly along one of the coordinate axes. For instance, if we take a step of distance Δx along the x-axis, leaving y and z unchanged, then the change df is given by $(\partial f / \partial x)\Delta x$, where $\partial f / \partial x$ is generally easy to compute.

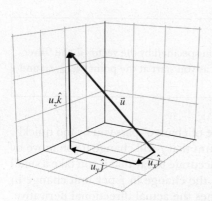

So instead of moving diagonally along our vector $\vec{u}$, we take three on-axis steps: a distance u_x along the x-axis, a distance u_y along the y-axis, and a distance u_z along the z-axis. Each step gives us a change in f that is easy to compute; together, they give us the total df.

That answers the question "how much does the function change as we move by $\vec{u}$?" but that isn't quite the question we want. A directional derivative is a *rate* of change; roughly speaking, we want to find how much f changes per unit distance. So we begin by finding the "unit vector" $\hat{u}$, a vector in the direction of $\vec{u}$ with magnitude 1. (You obtain such a vector by dividing $\vec{u}$ by its magnitude, a process called "normalizing" the vector.) The amount f changes as we take *that* step is the directional derivative we are looking for.

The Formula for a Directional Derivative

Given a scalar field $f(x, y, z)$ and a direction specified by the *unit* vector $\hat{u} = u_x\hat{i} + u_y\hat{j} + u_z\hat{k}$, the directional derivative is given by:

$$D_{\vec{u}}f = \frac{\partial f}{\partial x}u_x + \frac{\partial f}{\partial y}u_y + \frac{\partial f}{\partial z}u_z \qquad (4.5.1)$$

You'll show in Problem 4.80 why this formula only works when $\hat{u}$ is a unit vector. You'll show in Problem 4.81 that this formula is really another version of the chain rule.

EXAMPLE Finding a Directional Derivative

Question: Find the derivative of the function $f(x, y, z) = x^4 y/z$ at the point $(3, 2, 1)$ in the direction $\vec{u} = 5\hat{i} - 12\hat{k}$.

Answer:
We begin by "normalizing" $\vec{u}$: $|\vec{u}| = 13$, so $\hat{u} = 5/13\hat{i} - 12/13\hat{k}$.
$\partial f/\partial x = 4x^3 y/z$, so $\partial f/\partial x(3, 2, 1) = 216$. Similarly, $\partial f/\partial y(3, 2, 1) = 81$ and $\partial f/\partial z(3, 2, 1) = -162$.
So $D_{\vec{u}}f(3, 2, 1) = (216)(5/13) + (81)(0) + (-162)(-12/13) \approx 232.6154$.

Directional Derivatives in Two Dimensions

Our discussion above focused on a function of three variables, such as the temperature as a function of position, but directional derivatives apply to any number of variables. For a function of two variables, you can think about everything exactly as we did for three—for instance, $f(x, y)$ might represent the temperature at every point on a plane. But there is another, more visual interpretation, and that is to consider a surface where $z(x, y)$ gives the height.

As you stand at a given point on the surface, facing the positive x-direction, you see the surface ahead of you rising with a slope of $\partial f/\partial x$. As you rotate your body around (without changing position!) you see infinitely many different slopes, corresponding to infinitely many viewing angles. If you look in a particular direction $\vec{u}$ and see a derivative of 3, that means you are looking up a steep hill.

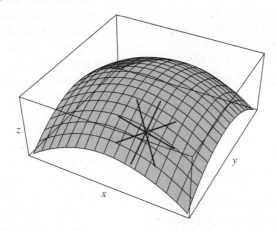

The point of this exercise is not simply that "you were taught to think of the derivative as a slope and you still can," and it is certainly not that "finding the slopes of surfaces in various directions has tremendous real-world importance." The point is to engage your visual mind in understanding the math. It's difficult to imagine looking in a certain direction and "seeing the temperature go up"; it's comparably easy to imagine a mountain path rising up in front of you.

If you visualize surfaces in this matter, you can solve some problems without doing any calculations at all. For instance, suppose that for a particular surface S at a particular point

(x_0, y_0), the derivative $D_{\vec{u}}z$ in a particular direction $\vec{u}$ is 3. What can you say about $D_{-\vec{u}}z$? A picture here is worth a thousand calculations: if you are looking up a slope of 3 in one direction, then turning around 180° you will see a slope of −3.

4.5.3 Problems: Directional Derivatives

4.77 The temperature distribution in a room is given by $f(x, y, z) = \sin x + 3\ln(z + 2)$. At the point $(\pi, 1, 1)$ in this room...

(a) find the derivative $D_{\vec{u}}f$ in the direction $\vec{u} = \hat{i}$. Explain what your result tells you about the function.

(b) find the derivative $D_{\vec{u}}f$ in the direction $\vec{u} = \hat{j}$. Explain what your result tells you about the function.

(c) find the derivative $D_{\vec{u}}f$ in the direction $\vec{u} = \hat{k}$. Explain what your result tells you about the function.

(d) find the derivative $D_{\vec{u}}f$ in the direction $\vec{u} = \hat{j} + \hat{k}$. Explain what your result tells you about the function.

4.78 For the function $f(x, y, z) = (x + 2y)/z$ at the point $(4, 1, 3)$...

(a) find the derivative in the direction $\hat{i} + 2\hat{j} - 2\hat{k}$.

(b) find the derivative in the direction $3\hat{i} + 6\hat{j} - 6\hat{k}$.

(c) find the derivative in the direction $-\hat{i} - 2\hat{j} + 2\hat{k}$.

(d) Explain how your answers to Parts (b) and (c) could have been predicted *without* doing any calculations, based solely on knowing your answer to Part (a).

4.79 For the function $f(x, y, z) = xe^y + x^2z$ at the point $(3, 0, -1)$...

(a) find the derivative in the direction $-5\hat{i} + 3\hat{j} + 9\hat{k}$.

(b) find the derivative in the direction $-5\hat{i} + 2\hat{j} + 8\hat{k}$.

(c) find the derivative in the direction $-5\hat{i} + 4\hat{j} + 10\hat{k}$.

(d) Your answer to Part (a) should have come out bigger than the other two. In fact, as you will be able to prove after the next section, that particular direction gives the highest possible derivative for this particular function at this particular point. Assuming the function f represents temperature, express the fact we just stated as a statement about the temperature in the room, using no mathematical terminology.

4.80 Explain why Equation 4.5.1 only represents a directional derivative if the direction is specified as a unit vector.

4.81 In this problem you will derive a way to physically interpret Equation 4.5.1 for a directional derivative. Assume a function $f(x, y, z)$ is defined in the region around you, and you are moving through that region. (You can think of f as representing temperature if you'd like something more concrete to imagine, but we're leaving it open-ended to emphasize that the result works for any function.)

(a) We've already said that f depends on x, y, and z, and as you move your position coordinates depend on time. Draw those facts as a dependency diagram and use it to write the chain rule for df/dt.

(b) Assume you are moving with a velocity $\hat{v} = v_x\hat{i} + v_y\hat{j} + v_z\hat{k}$, where your total speed is given by $|\hat{v}| = 1$. Explain why, in this particular case, the rate of change df/dt that you experience must be equal to the directional derivative $D_{\vec{u}}f$.

(c) Using your results above and the definition of velocity as the rate of change of position, derive Equation 4.5.1.

4.82 $g(4, 5, 6) = 10$ and for $\vec{u} = 3\hat{i} + 4\hat{j}$, $D_{\vec{u}}g = 3$.

(a) Estimate $g(7, 9, 6)$. (*Hint*: the answer is not 13.)

(b) Estimate $g(10, 13, 6)$.

(c) Which of your two estimates would you expect to be the more accurate? Why?

(d) Estimate $g(1, 1, 6)$.

4.83 $h(10, 10, 10) = 3$ and $h(9.7, 10.1, 10.2) = 2$.

(a) Use this information to estimate $D_{\hat{u}}h$ for a particular $\hat{u}$. Your answer should indicate what your $\hat{u}$ is, as well as its $D_{\hat{u}}h$.

(b) Use your answer to Part (a) to estimate h at another point. Your answer should indicate both the point and your estimated h for this point.

4.84 You are standing on Skullcrusher Mountain, which is shaped like the graph of $z = x^4 e^{-y^2}$, at the point $(1,1)$.

(a) If you set out in the positive *x*-direction, will you be hiking up or sliding down?

(b) If you set out in the positive *y*-direction, will you be hiking up or sliding down?

(c) If you set out at a 45° angle between the positive *x*- and *y*-directions, will you be hiking up or sliding down?

(d) Find an angle between the positive *x*- and *y*-directions such that, if you walk in that direction, you will be following a "level curve"—momentarily moving neither up nor down.

4.85 Standing at the origin on the plane $4x + 6y - 2z = 0$, you begin to climb. The positive *z*-axis points straight up but you cannot move directly that way: you can travel in a direction along the *x*-axis, along the *y*-axis, or along the direction $y = x$, always staying on the plane $4x + 6y - 2z = 0$. Order these three climbs from most to least steep.

4.86 You are in a space station that is being flooded with poison gas. The concentration of the gas is given by $e^{-x^2 y^2}$. You are currently at the point $(1, 1)$ and there are hallways leading off in the directions of the lines $y = x$, $y = -x$, and the *x*- and *y*-axes. Of the eight possible ways you could go (along any of the four lines in either direction), which way will reduce the concentration of poison gas you are breathing the fastest?

4.87 **Exploration: The Direction of Fastest Increase** Survey teams have determined that the density of gold deposits in your area roughly follows the function $\rho(x, y) = \sin(\pi x/2) \sin(\pi y^2/2) e^{-x^2 + y^2}$. Your team is currently digging at the position $(1, 1)$ and tomorrow you want to start hiking in the direction in which the density will increase the fastest.

(a) For a given direction ϕ (measured counterclockwise from the positive *x*-axis), find the unit vector $\hat{u}$ pointing in the direction of ϕ. *Hint*: The answer is not related to your current position.

(b) Find the directional derivative $D_{\hat{u}}\rho$ in the direction of the angle ϕ. Your answer should be a function of *x*, *y*, and ϕ.

(c) Evaluate $D_{\hat{u}}\rho$ at the point $(1, 1)$. The result should be a function of ϕ.

(d) Using the result you found for $D_{\hat{u}}\rho$, find the direction your team should hike in so as to increase the density of gold as fast as possible, and the directional derivative in that direction.

In the next section you will learn an easier way to solve problems like this.

4.6 The Gradient

The gradient is a higher-dimensional generalization of the derivative. Like a derivative, its simple formula underlies a powerful new way of looking at and understanding functions—in this case, multivariate functions. Also like the derivative, it is an idea that takes a considerable amount of time and thought to get used to, but it is well worth the effort.

4.6.1 Discovery Exercise: The Gradient

Consider a function $f(x, y)$ with constant partial derivatives $\partial f/\partial x = 1$ and $\partial f/\partial y = 2$.

1. Find the directional derivative of f in the direction $\hat{i} + c\hat{j}$. Your answer will depend on the unknown constant *c*.

   See Check Yourself #23 in Appendix L

2. Find the value of *c* that maximizes the directional derivative of f.
3. In what direction does f increase the fastest? Give your answer in the form of a vector in the *xy*-plane that points in the direction of fastest increase of f.
4. Normalize your answer to give a unit vector $\hat{u}$ that points in the direction of fastest increase of f.

Now we'll generalize the problem to a function $f(x, y)$ for which $\partial f/\partial x = a$ and $\partial f/\partial y = b$.

Chapter 4 Partial Derivatives

5. Repeat the steps above. Find the directional derivative of f in the direction $\hat{i} + c\hat{j}$ and use that to find the direction in which f increases the fastest. Once again, express your final answer as a unit vector $\hat{u}$ that points in the direction of fastest increase of f.
6. What is the rate of change of f in the direction you just found? In other words, what is the fastest rate of increase that f can have in any direction?
7. Multiply the vector you found in Part 5 by the answer you found in Part 6.

Your final result was a vector that points in the direction in which f increases the fastest and whose magnitude is the rate of change of f in that direction. That vector is called the "gradient" of f. We explore its properties in this section.

4.6.2 Explanation: The Gradient

Recall Equation 4.5.1 for a directional derivative:

$$D_{\hat{u}}f = \frac{\partial f}{\partial x}u_x + \frac{\partial f}{\partial y}u_y + \frac{\partial f}{\partial z}u_z$$

This equation can be written as a dot product.

$$D_{\hat{u}}f = \left(\frac{\partial f}{\partial x}\hat{i} + \frac{\partial f}{\partial y}\hat{j} + \frac{\partial f}{\partial z}\hat{k}\right) \cdot \hat{u}$$

The left-hand vector in this dot product, made up of all the partial derivatives of f, is called the *gradient* and is generally written $\vec{\nabla}f$. (You'll sometimes see grad(f), and you'll also sometimes see ∇f without the vector sign.)

The Definition of the Gradient

The *gradient* of a scalar field is defined as:

$$\vec{\nabla}f = \frac{\partial f}{\partial x}\hat{i} + \frac{\partial f}{\partial y}\hat{j} + \frac{\partial f}{\partial z}\hat{k} \qquad (4.6.1)$$

Using the gradient, we can define the directional derivative of f in the direction of a unit vector $\hat{u}$ more concisely.

$$D_{\hat{u}}f = \vec{\nabla}f \cdot \hat{u} \qquad (4.6.2)$$

Equation 4.6.2 can be expressed in words as "The directional derivative of f in the direction $\hat{u}$ is the *component of the gradient vector in that direction*." If you think about this interpretation, you may begin to get a feeling for what the gradient represents.

The example below computes a directional derivative that we already found in the previous section. Of course we get the same answer, since we do all the same calculations in the same order! Note, however, that the gradient separates the calculations that involve the function in general from the calculations that involve this direction in particular. If we wanted to find derivatives of the same function in many different directions, we would not keep repeating the same work.

4.6 | The Gradient

EXAMPLE **Using the Gradient to Find a Directional Derivative**

Question: Find the gradient of the function $f(x, y, z) = x^4 y/z$ at the point $(3, 2, 1)$, and use it to find the derivative in the direction $\vec{u} = 5\hat{i} - 12\hat{k}$.

Answer:
$$\vec{\nabla} f = (\partial f/\partial x)\,\hat{i} + (\partial f/\partial y)\,\hat{j} + (\partial f/\partial z)\,\hat{k} = (4x^3 y/z)\hat{i} + (x^4/z)\hat{j} - (x^4 y/z^2)\hat{k}$$

At the point $(3,2,1)$, the gradient is the vector $216\hat{i} + 81\hat{j} - 162\hat{k}$. Note that this calculation has nothing to do with any particular direction $\vec{u}$; it is a property of f at that point that can be used to find the derivative in any direction.

For this particular direction, $\hat{u} = (5/13)\hat{i} - (12/13)\hat{k}$, so
$$D_{\vec{u}} f(3, 2, 1) = (216)(5/13) + (81)(0) + (-162)(-12/13) \approx 232.6154.$$

What does the Gradient Mean?

We began our study of directional derivatives by considering the temperature function $f(x, y, z)$: a "scalar field," meaning a specific number assigned to every point in the room.

When we take the gradient of that function, we get $\vec{\nabla} f(x, y, z)$: a "vector field." You can imagine a little arrow at every point in the room, each with its own magnitude and direction. Whenever you find a gradient, you are finding a vector field that somehow contains information about a given scalar field.

But what information? Suppose I tell you that "at this particular point in space, $\vec{\nabla} f$ has a magnitude of 7 and points in precisely that direction." What have you learned about the scalar field $f(x, y, z)$—about the temperature?

To approach this question, consider the following thought experiment. You are planted firmly in the middle of a scalar field. You can face in any direction $\vec{u}$ you like, but your position (x_0, y_0, z_0) never changes.

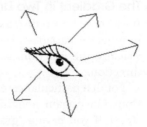

Question: What direction would you face if your goal was to *maximize* the directional derivative? (Try to answer this question yourself, based on Equation 4.6.2.)

Answer: We noted above that the component of the gradient pointing in any direction tells us the directional derivative in that direction. So the directional derivative is zero in directions perpendicular to the gradient, it's negative in the direction opposite the gradient, and it takes its largest positive value in the direction of the gradient.

We conclude that the function increases most rapidly in the direction the gradient points in. Turning that around, we can say that the gradient points in the direction in which the function increases most rapidly. In our temperature-filled room, the gradient is like a heat-seeking bug, always pointing in the direction of the most warmth.

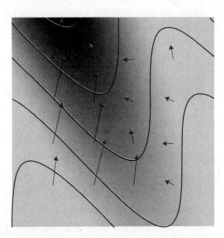

FIGURE 4.3 The arrows show the gradient of the function *density of ink*. Each arrow points in the direction where the page is getting dark most rapidly from that point. The curves are contours of constant ink density.

And what about the magnitude of the gradient?

Once again we turn to Equation 4.6.2. If we face in the direction of the gradient, then $D_{\hat{u}}f$ is the magnitude of the gradient vector, times the magnitude of $\hat{u}$ (which is 1), times $\cos(0) = 1$. So the magnitude of the gradient vector gives the directional derivative in that particular direction.

> **The Meaning of the Gradient**
>
> The gradient of a scalar field is a vector field. At any given point in the scalar field, the gradient points in the direction of steepest increase. The magnitude of the gradient gives the rate of change of the field in that direction.

This is a good place to stop and make sure you've followed everything. We began with a mathematical definition of the gradient—a minor change in notation around our already-existing idea of a directional derivative. From there, we have arrived at an interpretation of what the gradient means. If you calculate that for a particular field at a particular point the gradient has a magnitude of 7 and points in the direction $\hat{i} + 5\hat{j} - 3\hat{k}$, can you explain what that means about the field at that point? Can you explain how that physical interpretation comes from the math?

The Gradient in Two Dimensions

In Section 4.5 we discussed the fact that directional derivatives apply to functions of any number of variables, but for the particular case of functions of *two* variables, we can visualize the function as a surface and the derivative represents the slope in a particular direction.

For that particular case, then, we can "see" the gradient more clearly than in higher dimensions. Once again, imagine standing on a mountainside whose height is given by the function $z(x, y)$. If you have a "gradient compass" it will point in the direction of steepest possible incline. The magnitude of the gradient will give you the slope in that direction.

How do you build a gradient compass? It's easier than you think: just put a ball down at your feet. The ball will roll in the direction of steepest *decline*, which is $-\vec{\nabla}z$.

EXAMPLE **Gradient and Directional Derivative as Slope**

Question: For the surface described by $z(x, y) = e^{x-y}$ at the point $(1, 1)$, what is the gradient and what does it mean?

Answer:

$\vec{\nabla}z = e^{x-y}\hat{i} - e^{x-y}\hat{j}$, so $\vec{\nabla}z(1, 1) = \hat{i} - \hat{j}$. This indicates that if you want to move up as quickly as possible you should move at a 45° angle South of East, and that in that direction the surface will rise with a slope of $\sqrt{2}$.

The closer your angle is to that particular direction, the more steeply uphill you are looking. If you let go of a ball at this point on the surface, it would roll along the steepest decline, which is always $-\vec{\nabla}z$, or $-\hat{i} + \hat{j}$ in this case.

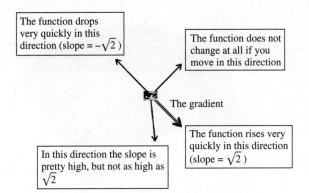

Question: On that same surface, what is $D_{\vec{u}}z(1,1)$ for $\vec{u}$ pointing 60° North of East, and what does it mean?

Answer:
We begin with a quick drawing and a bit of trig to find that $\hat{u} = (1/2)\hat{i} + (\sqrt{3}/2)\hat{j}$.

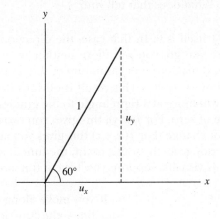

$$D_{\hat{u}}z(x,y) = (\hat{i}-\hat{j}) \cdot \left(\frac{1}{2}\hat{i} + \frac{\sqrt{3}}{2}\hat{j}\right) = \frac{1-\sqrt{3}}{2}$$

If you take a small step of horizontal distance ds along that 60° angle, your height (z-coordinate) will go down by roughly $0.366\,ds$. This makes sense if you look at the relative contributions of dx and dy. As x increases, z increases. As y increases, z decreases. The total dz is negative because, at a 60° angle, y is changing faster than x.

A final word of warning is in order. When we talk about the gradient as the slope of a surface, we are talking about a function with two independent variables, x and y. The gradient points in the direction of steepest increase, but the gradient does not point *up* the hill; the gradient has no z-component at all! The gradient points in the direction *in the xy-plane* that you would travel to obtain the steepest incline.

168 Chapter 4 Partial Derivatives

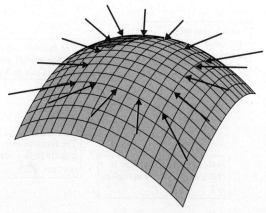

For a function $f(x,y)$ the gradient points horizontally, not "up the hill."

A Directional Derivative of Zero

Suppose you find that, for a particular function at a particular point, the derivative in a particular direction is zero. What does that tell you?

$$|\vec{\nabla}f||\hat{u}|\cos\theta = 0$$

One possibility is that $\vec{\nabla}f$ itself is $\vec{0}$. In that case, the directional derivative is zero in all directions: no matter where you go, you will move neither up nor down. This often (not always) corresponds to a local maximum or minimum of the function.

If $\vec{\nabla}f$ is not zero, then θ must be 90°. This result is obvious from the math, but it is not obvious visually: whenever you move at a right angle to the gradient, you move along a path with a directional derivative of zero. For $f(x,y)$, this gives you two different directions along which the function does not change. For $f(x,y,z)$ this gives you an infinite number of such directions: an entire plane along which, at this point, the function does not change.

To better understand this visually, suppose you have a function $f(x,y)$ represented as a contour map.

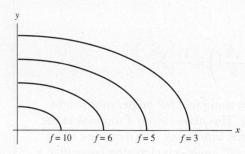

If you move along a contour line—any contour line—the function does not change. (For instance, on the third contour line in the drawing above, the function $f(x,y)$ stays 5.) Therefore, you must be moving in a direction in which $D_{\hat{u}}f = 0$. This gives us another visual interpretation of the gradient: *at any given point, the gradient points perpendicular to the contour lines of a function.* The same concept applies in higher dimensions: for a function $f(x,y,z)$ the gradient points perpendicular to the "level surfaces," and so on.

4.6.3 Problems: The Gradient

4.88 Consider the scalar field $f(x,y) = \sqrt{x^2 + y}$.

(a) Find the gradient $\vec{\nabla}f$ as a function of x and y.

(b) Find the gradient $\vec{\nabla}f(3,7)$. Express your answer as the magnitude and direction of a vector.

(c) If $f(x,y)$ represents the temperature at every point on a plane, give a physical description of the meaning of your answer to Part (b). Your explanation should include both the magnitude and the direction of the vector.

(d) If $f(x,y)$ represents the height (z-coordinate) at every point on a surface, give a physical description of the meaning of your answer to Part (b). Your explanation should include both the magnitude and the direction of the vector.

(e) Find the directional derivative $D_{\vec{u}}f$ at the point $(3, 7)$ in the direction of the vector $\hat{i} + \hat{j}$. (If you get the answer 7/8, try again.)

(f) Based on your answer to Part (e), estimate the value of $f(3.02, 7.02)$. Then calculate the actual $f(3.02, 7.02)$ for comparison.

4.89 Walking into a study session, you hear your friend Shannon saying "So for this particular field at this point, the gradient is −7." Explain how you can tell Shannon, with no further information, that she is wrong.

4.90 Find the gradient of the function $f(x) = 3x^2$. Which way does it point, and why?

4.91 The plot below shows the contour lines of a function $z(x, y)$. Copy this plot and add to it vectors at a variety of points (at least five, in different parts of the plot) showing the gradient at those points. Your vectors won't be precise, but they should all point in the correct direction and they should be larger in places with big gradients than they are in places with small gradients.

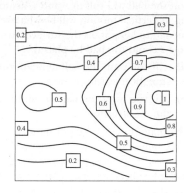

(If you want to print the picture from a computer instead of copying it, make a contour plot of $z(x, y) = e^{-(x-1)^2-y^2} + 0.5e^{-(x+1)^2-y^2}$ in the range $-1.2 \le x \le 1.2$, $-1.2 \le y \le 1.2$.)

4.92 A plot of the function $z(x, y)$ is shown at the top of the next column. (As usual, z is the vertical axis.) Draw a set of x- and y-axes and sketch the gradient vectors $\vec{\nabla}z$ at a variety of points (x, y) (at least five points). Note that your image will be 2D, unlike the 3D plot of z shown here. Your vectors won't be precise, but they should all point in the correct direction and should be larger in places with big gradients than they are in places with small gradients.

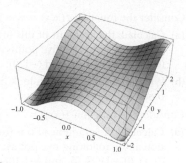

4.93 A scalar field is given by $f(x, y) = e^{-(x+1)^2-y^2} + e^{-(x-1)^2-y^2}$.

(a) Calculate the gradient $\vec{\nabla}f$.

(b) Have a computer generate a 3D plot of the field with x and y on the horizontal axes and f on the vertical axis.

(c) Have the computer generate either a contour plot of f (showing lines of constant f) or a density plot (shading regions of the plot according to the value of f). This plot will be 2D, with only x- and y-axes.

(d) Have the computer plot the vectors $\vec{\nabla}f$ on the same plot as your contour or density plot.

(e) Explain why the image you generated looks the way it does. In other words, looking at the contour lines or shading, how could you have predicted what the gradient vectors would be doing?

4.94 The function $z = 2x + 5y - 10$ represents a plane.

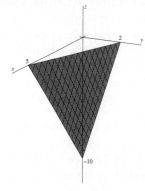

(a) Find the gradient $\vec{\nabla}z(x, y)$.

(b) The gradient is a constant—that is, it is not a function of x or y. What does that tell you about the surface?

(c) A "level curve" is a curve (in this case a line) along the surface that maintains a constant z-value. What direction would you walk to follow a level curve—in other words, to have a directional derivative of zero?

4.95 Consider the function $f(x, y) = x^2 + y^2$.

(a) Calculate the gradient of this function at the following five points: $(0, 0)$, $(0, 1)$, $(0, -1)$, $(1, 0)$, and $(-1, 0)$.

(b) What might this function look like based on your five gradients?

4.96 Consider the function $f(x, y) = x^2 - y^2$.

(a) Calculate the gradient of this function at the following five points: $(0, 0)$, $(0, 1)$, $(0, -1)$, $(1, 0)$, and $(-1, 0)$.

(b) What might this function look like based on your five gradients?

4.97 The function $z(x, y)$ at the point $(0, 0)$ has a gradient $4\hat{i} + 6\hat{j}$.

(a) What is the derivative in the direction pointing toward $(2, 3)$?

(b) What is the derivative in the direction pointing toward $(-2, -3)$?

(c) If you put a ball down on this surface at the point $(0, 0)$, which way would it roll?

(d) Find a unit vector (magnitude of 1) for which $D_{\hat{u}} = 0$.

(e) For how many *other* unit vectors does $D_{\hat{u}} = 0$?

4.98 Each part below gives information about a function of x and y. Describe each function. Your answer may include a description in words, a 3-D drawing, or anything else that will help understand the function. Suggesting a possible function $f(x, y)$ may be part of your answer, but it is not a complete answer.

(a) For function $f(x, y)$, the gradient $\vec{\nabla} f = 2\hat{i}$ everywhere.

(b) For function $g(x, y)$, the gradient $\vec{\nabla} g = x\hat{i}$ everywhere. (So at all points where $x = 2$, the gradient is $2\hat{i}$ and so on.)

(c) For function $h(x, y)$, the gradient $\vec{\nabla} h = -\rho \hat{\rho}$ everywhere, where ρ is the distance from the z-axis and $\hat{\rho}$ points directly away from the z-axis at each point.

4.99 The figure below shows a hemisphere of radius 5. The point $(0, 4, 3)$ on that hemisphere is indicated.

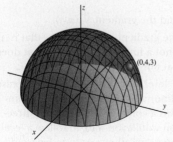

(a) One way to mathematically represent that hemisphere is $z = f(x, y)$ where $f(x, y) = \sqrt{25 - x^2 - y^2}$. Calculate $\vec{\nabla} f$ at the indicated point and describe in words which way it points.

(b) Another way to mathematically represent that hemisphere is with the equation $g(x, y, z) = 25$ where $g(x, y, z) = x^2 + y^2 + z^2$. Calculate $\vec{\nabla} g$ at the indicated point and describe in words which way it points.

(c) Neither of your above answers seems to point "up the hill" in the drawing. Explain how both of them follow the rule that the gradient always points in the direction in which the function increases most steeply.

4.100 You stand in a valley described by the equation $z = x^2 - xy - 2y$ at the point $(2, 1, 0)$.

(a) If you put a ball down at your feet, in what direction will the ball roll? (Your answer will be a vector in three-space.)

(b) If you want to walk a level path, in what direction should you walk? (Your answer will be a vector with only x- and y-components.)

4.101 A surface fluctuates according to the equation $z = \cos(xt) + \sin(yt)$. A ball is to be placed at the point $(1, 1, \cos t + \sin t)$ at some time t. The descriptions below are about the direction the ball will roll in x and y, assuming its vertical motion will be whatever is needed to keep it on the surface. At what time t could the ball be placed…

(a) …if you want the ball to begin rolling straight toward the z-axis?

(b) …if you want the ball to roll in the +x-direction?

4.102 The magnitude of the gravitational force between two objects is given by $F = Gm_1 m_2/r^2$ where G is a constant and the other three letters are variables.

(a) Find the gradient $\vec{\nabla} F(m_1, m_2, r)$.

(b) Based on your answer to Part (a), which of the three variables should increase, and which should decrease, if you set out to increase the gravitational force?

4.103 One cause of air flow is the so-called "pressure gradient force" given by the formula $\vec{F} = -k\vec{\nabla} P$ where P is the air pressure and k is a positive constant. Explain why it makes sense to expect air to move in the direction of this vector.

4.104 Cecelia the depth-seeking crab is on a lake bed shaped like $z = x^2/4 - y/10 + 10$. Assuming

Cecelia starts at the point $(4, 10, 13)$ and always crawls to lower depths as quickly as possible, describe with words and pictures the path she will follow. (You do not need to give a mathematical equation that expresses her path.)

Electrostatic theory is based on a scalar field called "electric potential" $V(x, y, z)$. Problems 4.105–4.108 require you to know the following facts about electric potential.

- The proximity of positive charges increases the electric potential; the proximity of negative charges decreases the potential.
- The electric field is computed as $\vec{E} = -\vec{\nabla} V$.
- Positively charged particles tend to flow in the direction given by the electric field.

4.105 The electric potential in a region of space is given by $V = x^2 + y^2 - z^2$. If a positive charge were placed at the point $(1, 2, 3)$ what direction would it move in? (Your answer should be a vector, but you don't have to write it as a unit vector.)

4.106 In the presence of a single positively charged particle located at the origin, the potential field is given by $V = k/\sqrt{x^2 + y^2 + z^2}$ where k is a positive constant.
(a) Calculate the electric field $\vec{E} = -\vec{\nabla} V$.
(b) Use your answer to Part (a) to predict the motion of a positively charged particle that starts at rest in this field.
(c) In the presence of a single *negatively* charged particle at the origin, the potential field is given by $V = -k/\sqrt{x^2 + y^2 + z^2}$. How does this change your answers?

4.107 An "electric dipole" consists of two equal and opposite charges. If a dipole in two dimensions consists of a positive charge at position $(x_0, 0)$ and a negative charge at position $(-x_0, 0)$ then the potential field produced by the dipole is

$$V = \frac{k}{\sqrt{(x - x_0)^2 + y^2}} - \frac{k}{\sqrt{(x + x_0)^2 + y^2}}$$

(a) Calculate the electric field $\vec{E} = -\vec{\nabla} V$.
(b) A positive charge feels a force pushing it in the direction of the electric field. If you place a positive charge on the positive x-axis at $x > x_0$, in what direction will it be pushed?
(c) If you place a positive charge at position $(5x_0, 2x_0, 0)$, in what direction will it be pushed? (Give your answer as an angle.)

4.108 A "conductor" is a material with free electrons that can easily move in response to electric fields. If you generate an electric field near a conductor the electrons will move in response, thus changing their own electric fields, until the electric field within the conductor is zero. This happens so quickly that for most purposes you can assume that the electric field is always zero inside a conductor.
(a) Explain why every conductor is an "equipotential," meaning a region of constant potential.
(b) A thin metal sheet is in the shape of a spherical shell: $x^2 + y^2 + z^2 = R^2$. Knowing that the entire surface of the sphere has a constant potential, in what direction must the electric field point at $(R, 0, 0)$? (There are actually two possible directions.)

4.109 Exploration: An Alternative Basis for Directional Derivatives For the function $f(x, y)$ at the point $(2, 3)$, the derivative in the direction of the vector $3\hat{i} + 4\hat{j}$ is 2, and the derivative in the direction of the vector $5\hat{i} + 12\hat{j}$ is -2.
(a) Find $\partial f / \partial x$ and $\partial f / \partial y$ at this point.
(b) Give one possibility for what the function $f(x, y)$ might be.
(c) If this function were a hill, and you released a ball at the point $(2, 3)$, which way would the ball initially roll?

We are used to starting from the derivatives in the x- and y-directions, and using those to figure out the derivatives in other directions. In this case we started with two other directional derivatives but got the same information. In general, the derivatives of $f(x, y)$ in almost any two directions are enough to find them in all other directions.

(d) Let $\hat{u} = a\hat{i} + b\hat{j}$ and $\hat{v} = c\hat{i} + d\hat{j}$ be two unit vectors and let the directional derivatives of f at a point be $D_{\hat{u}} f = \alpha$ and $D_{\hat{v}} f = \beta$. Find $\partial f / \partial x$ and $\partial f / \partial y$ at that point in terms of a, b, c, d, α, and β.
(e) We said that the derivatives in the directions of *almost* any two vectors are good enough to find all others. What would have to be true of the vectors $\hat{u}$ and $\hat{v}$ for you to be unable to find $\partial f / \partial x$ and $\partial f / \partial y$ from their directional derivatives? Answer this twice. First, give an algebraic answer based on your calculations, and then say what that would imply geometrically about the two vectors.

4.110 *[This problem depends on Problem 4.109, and on some knowledge of linear algebra.]* Generalizing your answer to Problem 4.109, what would have to be true of n vectors in an n-dimensional space if the derivatives of f along those directions weren't sufficient for you to know the derivatives in all directions?

4.7 Tangent Plane Approximations and Power Series (see felderbooks.com)

4.8 Optimization and the Gradient

The term "optimization" refers to a broad spectrum of techniques that are used to maximize or minimize important variables. Entire courses are given on linear and non-linear optimization algorithms that are beyond the scope of this book. But in this section and Section 4.9 we will present two core techniques—the first based on the gradient, the second called "Lagrange multipliers"—that can be used to find minima and maxima of multivariate functions.

A third technique, the "simplex method," will be presented in Chapter 7.

4.8.1 Discovery Exercise: Optimization and the Gradient

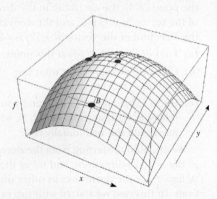

FIGURE 4.4

Figure 4.4 shows three points on a plot of a function $f(x, y)$. At point A, $\partial f / \partial x$ is positive and $\partial f / \partial y = 0$.

1. At point B, is $\partial f / \partial x$ positive, negative, or zero? What about $\partial f / \partial y$?
2. Point C is a local maximum of f. At that point are $\partial f / \partial x$ and $\partial f / \partial y$ positive, negative, or zero?
3. For a smooth one variable function $f(x)$ a local maximum or minimum always occurs where $df/dx = 0$. Based on your answers above, how would you generalize that rule to a two-variable function $f(x, y)$?

4.8.2 Explanation: Optimization and the Gradient

You learned in introductory calculus how to optimize a function of one variable. It's a little more complicated than saying "a function reaches a maximum or minimum where the derivative is zero." In fact, it is possible to have a derivative of zero without reaching an extremum (think of $y = x^3$ at the origin), and it is possible to reach an extremum without a derivative of zero ($y = |x|$ at the origin). With a bit more care, however, we can make a few confident statements.

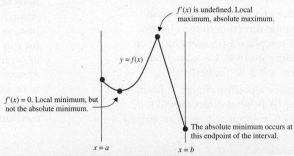

- A "critical point" is defined as a point where the derivative is either zero or undefined.
- All local minima and maxima will occur at critical points.
- A continuous function on a closed interval will achieve an absolute maximum and an absolute minimum. These will occur at critical points *or* at the endpoints of the interval.

As an example, consider the picture above. On the closed interval $[a, b]$ this function attains an absolute maximum and an absolute minimum. On the open interval (a, b) it attains an absolute maximum but does *not* attain an absolute minimum.

Before we extend these rules to higher dimensions, let's define our terms. We present these definitions in terms of a function of two variables $f(x, y)$ but they extend naturally to more (or fewer) variables.

Definitions: Local and Absolute Maximum and Minimum

A function $f(x, y)$ attains a "local maximum" (or "relative maximum") at the point (a, b) if $f(a, b) \geq f(x, y)$ for all points (x, y) in a disk with radius $r > 0$ centered on (a, b).

A function $f(x, y)$ attains an "absolute maximum" within region R at the point (a, b) if $f(a, b) \geq f(x, y)$ for all points (x, y) in R. (When we make a statement such as "$3 - x^2 - y^2$ attains an absolute maximum at $(0, 0)$" the implied region is "all points." In such a case the term "global maximum" is sometimes used.)

The definitions for local and absolute minimum are analogous. For functions of n variables the "disk" around the local maximum or minimum is an n-dimensional "neighborhood" around that point.

Take a moment to convince yourself that these technical definitions reflect a simpler intuitive understanding of these concepts. One subtlety is the use of $\geq$ instead of $>$ in the definitions. You might think that the plane $z(x, y) = 7$ never attains a maximum, but in fact its global maximum is 7 and it attains it everywhere!

With those definitions in place, we can articulate three rules parallel to the single-variable rules we stated above.

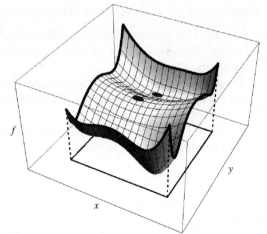

FIGURE 4.5 The function shown above has three critical points in the indicated domain: two shown by dots and another at the minimum in the lower right. Two of the critical points are local minima, including the one in the lower right that is also an absolute minimum. The absolute maximum occurs on the boundary at the upper left corner. The critical point near the center is a "saddle point," which we will discuss shortly.

- A critical point occurs where all the partial derivatives are zero, or where any of the partial derivatives is undefined. Put more succinctly, a critical point occurs where the gradient is zero or undefined.
- All local minima and maxima occur at critical points.
- A continuous function in a closed, bounded region will achieve an absolute maximum and an absolute minimum, and these will occur at critical points or on the boundary.

One approach to optimization, therefore, is to start with the gradient. The first example below demonstrates most of the elements of the basic process. The second example adds the issue of a closed boundary.

The Biggest Storage Chest

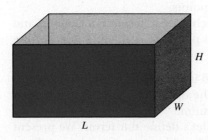

A storage chest is to be built to the following specifications. The left side, right side, back, and bottom are made of a material that costs $1/\text{ft}^2$. The front is made of a more decorated material that costs $2/\text{ft}^2$. There is no top. The total cost of the chest is therefore $C(L, W, H) = LW + 2HW + 3LH$. The volume, of course, is $V = LWH$.

Question: Find the largest-volume chest that can be made for $20.

Thinking: As always, we urge you to think about the question before you begin to solve. Will there be a minimum volume? Will there be a maximum volume?

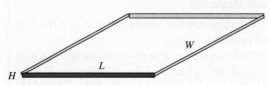

FIGURE 4.6 You can make the volume as small as you like while maintaining a cost of $20.

First, there is an implicit constraint on the domain of L, W, and H; they all have to be positive. It simply isn't a chest otherwise. Given that, with a bit of thought, you can convince yourself that there is no minimum volume. Consider, for instance, allowing H to approach zero while the area of the bottom, LW, approaches 20. (Figure 4.6.)

On the other hand, it is not possible to make the volume arbitrarily large. If you make any one of the dimensions extremely large you will need to make *both* of the other ones correspondingly small to keep the surface area reasonable, so in the extreme limit as any one dimension gets too large the total volume will have to shrink. That suggests that there is some optimal set of dimensions that maximizes the volume.

Solving: Our two equations play different roles in the problem. $V = LWH$ is the "objective function," the quantity that we want to optimize. Our strategy will involve setting $\vec{\nabla} V$ to zero to find critical points.

The cost equation defines a "constraint." We are not trying to optimize cost, so we will not be taking its gradient. Instead, we will use the fixed cost to eliminate one of the variables in our objective function.

So, we solve the equation $LW + 2HW + 3LH = 20$ for one of the three variables. We choose (quite arbitrarily) to write $W = (20 - 3LH)/(L + 2H)$. Plugging this into the volume, we get:

$$V = LH(20 - 3LH)/(L + 2H)$$

This "constrained" volume function $V(L, H)$ will attain a maximum. We begin by taking its gradient.

$$\vec{\nabla} V = \frac{H^2(40 - 3L^2 - 12LH)}{(L + 2H)^2}\hat{L} + \frac{2L^2(10 - 3H^2 - 3LH)}{(L + 2H)^2}\hat{H}$$

Recall that a critical point is one where the gradient is zero *or undefined*. We would therefore have to consider all the points with $L + 2H = 0$ as critical points, but we can ignore them because the lengths must be positive quantities. We ignore $H^2 = 0$ and $L^2 = 0$ for the same reason. We are therefore left finding critical points where $40 - L^2 - 12LH = 0$ *and* $10 - 3H^2 - 3LH = 0$. Solving these two equations simultaneous leads to $H = \sqrt{10}/3$ and $L = 2\sqrt{10}/3$. We can use these values to find $W = \sqrt{10}$, which gives a volume of $V = 20\sqrt{10}/9$.

So we have found a critical point: what does it tell us about our box? If we were just using mathematical brute force, we could use the "second derivatives test" described below to check if this is a minimum or a maximum. Then we would also have to check the boundaries $L = 0$, $W = 0$, and $H = 0$. In this case those steps are unnecessary because we've already argued that the volume reaches a maximum and no minimum, and clearly the volume would be zero if any of the dimensions were zero, so we conclude that $V = 20\sqrt{10}/9$ ft^3 is the largest possible box we can build for $20.

EXAMPLE: Optimization on a Bounded Region

Question: Find the absolute minimum and maximum of the function $f(x, y) = x^2 + y^2 - 2x - 4y + 10$ on the closed region R shown below:

Solution:
$\vec{\nabla} f(x, y) = \langle 2x - 2, 2y - 4 \rangle$ is never undefined, and reaches zero at the point $(1, 2)$. So that is the only critical point. But just as with functions of one variable, the absolute minimum and maximum can occur at the critical points *or on the boundary*, so our next task is to look on the boundary. In this case, the boundary is made up of three lines.

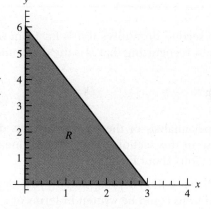

The first boundary line is $y = 0$ and $f(x, 0) = x^2 - 2x + 10$ reaches a critical point when $2x - 2 = 0$ so $x = 1$. So we will have to consider the point $(1, 0)$.

The next boundary line is $x = 0$ and $f(0, y) = y^2 - 4y + 10$ reaches a critical point when $2y - 4 = 0$, so we consider the point $(0, 2)$.

The third boundary line is $y = -2x + 6$. Along this line, $f = x^2 + (-2x + 6)^2 - 2x - 4(-2x + 6) + 10$, which reaches a critical point at $(9/5, 12/5)$.

Finally, there are the boundary points of the boundary lines to consider: $(0, 0)$, $(0, 6)$, and $(3, 0)$.

We are guaranteed to find an absolute maximum and minimum among the points we have listed. To find them, find the value of the objective function at each point.

$$f(1, 2) = 5, \quad f(1, 0) = 9, \quad f(0, 2) = 6, \quad f(9/5, 12/5) = 5.8,$$
$$f(0, 0) = 10, \quad f(0, 6) = 22, \quad f(3, 0) = 13.$$

So within this region the function attains an absolute minimum of 5 at $(1, 2)$ and an absolute maximum of 22 at $(0, 6)$.

Different Types of Critical Points and the Second Derivatives Test

A given critical point can represent a local maximum, a local minimum, or neither. For functions of one variable, the "second derivative test" is one way of distinguishing these three cases. Once you have established that $f'(c) = 0$ at some point $x = c$, you check the second derivative at that value. If $f''(c) < 0$ you have found a local maximum; $f''(c) > 0$ indicates a local minimum.

For a function of two variables the "second derivatives test" provides a similar way of classifying critical points.

The Second Derivatives Test for Two Independent Variables

Suppose the function $f(x, y)$ has been found to have a critical point at $(x, y) = (a, b)$. We define a new function D as follows.

$$D(x, y) = \left(\frac{\partial^2 f}{\partial x^2}\right)\left(\frac{\partial^2 f}{\partial y^2}\right) - \left(\frac{\partial^2 f}{\partial x \, \partial y}\right)^2 \qquad (4.8.1)$$

Evaluate the function D at the critical point (a, b).

- If $D(a, b) > 0$ and $\partial^2 f / \partial x^2 (a, b) > 0$ then $f(x, y)$ attains a local minimum at (a, b).
- If $D(a, b) > 0$ and $\partial^2 f / \partial x^2 (a, b) < 0$ then $f(x, y)$ attains a local maximum at (a, b).
- If $D(a, b) < 0$ then $f(x, y)$ attains neither a minimum nor a maximum at (a, b).

If $D(a, b) = 0$ the test is inconclusive.

The second derivatives test is far from obvious. A deep understanding of why it works begins by recognizing that D is the determinant of a matrix called the "Hessian."

$$D(x, y) = \begin{vmatrix} \partial^2 f / \partial x^2 & \partial^2 f / (\partial x \, \partial y) \\ \partial^2 f / (\partial y \, \partial x) & \partial^2 f / \partial y^2 \end{vmatrix}$$

A proper analysis of the Hessian and its determinant would take us far afield from our purpose in this section. Even without linear algebra, however, we can make a few salient observations about Equation 4.8.1.

- If $D > 0$ then $\partial^2 f / \partial x^2$ and $\partial^2 f / \partial y^2$ must have the same sign. Hence, the first two conditions could be written in terms of $\partial^2 f / \partial y^2$ instead of $\partial^2 f / \partial x^2$ without changing the results.
- If $\partial^2 f / \partial x^2$ and $\partial^2 f / \partial y^2$ are both positive then the function may or may not attain a local minimum, but it cannot possibly attain a local maximum. This makes sense if you think about it visually. Similarly, if both are negative then you cannot attain a minimum.

Perhaps most interestingly, what happens if $\partial^2 f / \partial x^2$ and $\partial^2 f / \partial y^2$ are different signs? In that case D *must* be negative, signifying neither a maximum or a minimum. Such a point is called a "saddle point" and it has no real equivalent among functions of one variable. The center point of a saddle represents a local minimum as you move in one direction and a local maximum as you move in a different direction, so it is neither a local minimum nor maximum of the surface. See Figure 4.5 for an example.

For a function of more than two variables $f(x_1, x_2, \ldots)$ the Hessian matrix is defined as $H_{ij} = \partial^2 f / \partial x_i \partial x_j$, and the second derivatives test involves the eigenvalues of the Hessian. If all the eigenvalues are positive the critical point is a minimum, if they are all negative it's a maximum, and if some are positive and some negative it's a saddle point. In all other cases the test is inconclusive. See Problem 4.131.

EXAMPLE **The Second Derivatives Test**

Problem:
Find and classify the critical points of $f(x, y) = x^3 + 2y^3 - 6x^2 y - 60x$.

Solution:
We begin by finding the critical points.

$$\vec{\nabla} f = (3x^2 - 12xy - 60)\hat{i} + (6y^2 - 6x^2)\hat{j} = 3(x^2 - 4xy - 20)\hat{i} + 6(y^2 - x^2)\hat{j}$$

This gradient is never undefined, so all critical points will occur at points where $\vec{\nabla} f = \vec{0}$. Setting $y^2 - x^2 = 0$ yields $y = \pm x$. Plugging $y = x$ into $3x^2 - 4xy - 20 = 0$ gives no real answers, but plugging $y = -x$ into the same equation gives $x = \pm 2$.

Critical points: $(2, -2)$ and $(-2, 2)$
To classify these points we begin by finding D.

$$D(x, y) = \begin{vmatrix} \partial^2 f/\partial x^2 & \partial^2 f/(\partial x\, \partial y) \\ \partial^2 f/(\partial y\, \partial x) & \partial^2 f/\partial y^2 \end{vmatrix} = \begin{vmatrix} 6x - 12y & -12x \\ -12x & 12y \end{vmatrix}$$
$$= 72xy - 144y^2 - 144x^2 = 72(xy - 2x^2 - 2y^2)$$

Plugging in our two critical points we find that $D(2, -2) < 0$ and $D(-2, 2) < 0$. We conclude that this function has no local maxima or minima anywhere!

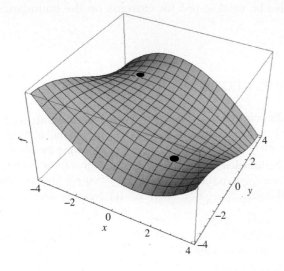

The plot confirms that the two critical points are both saddle points.

Stepping Back

A typical optimization problem involves one objective function that you want to optimize and any number of constraints on the independent variables. The constraints come in two broad categories: inequalities and equations.

An inequality defines a boundary to the domain on which you are trying to optimize the objective function. In the "optimization on a bounded region" problem above, the variables have to satisfy $x \geq 0$, $y \geq 0$, and $y \leq 6 - 2x$. These three inequalities restrict $f(x, y)$ to the triangular domain shown at the beginning of that problem. In the storage chest example, the inequalities $L > 0$, $W > 0$, and $H > 0$ restrict the domain to the first octant of the 3D region (L, W, H).

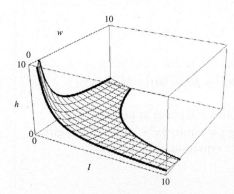

FIGURE 4.7 The constraint equation for the storage chest problem confines the domain to the 2D surface shown here. The requirement that the dimensions can't be negative confines the domain to be within the boundaries shown in bold.

The storage chest problem also has an equation constraint, $LW + 2HW + 3LH = 20$. That constraint restricts the domain to lie on the surface defined by that equation. A constraint equation, unlike an inequality, reduces the dimension of the domain.

Inequalities make an optimization problem easier in one way and harder in another. When the independent variables are constrained to a certain domain you may discard any critical points you find outside that domain. If you are looking for absolute maxima and minima, however, you must also consider all the boundaries in addition to the critical points.

To deal with a constraint *equation*, one approach is to solve the equation for one of the variables and plug that into the objective function to reduce the number of independent variables. In the next section we will show you another approach using "Lagrange multipliers." Either of these methods can also be used to test for extrema on the boundaries of a problem with inequalities.

4.8.3 Problems: Optimization and the Gradient

4.129 Find all critical points of the function $4x^3 + y^3 + 3x^2 - 90x - 48y + 15$. Use the second derivatives test to classify each critical point.

4.130 Find all critical points of the function $2x^3 - 4x^2y + 8xy - 56x$. Use the second derivatives test to classify each critical point.

4.131 💻 Find all critical points of the function $x^4 + 2x^3 - x + y^2 + z^2 + xyz$. Use the second derivatives test to classify each critical point.

4.132 Walk-Through: Optimization on a Closed Bounded Region. In this problem you will find the absolute maximum and minimum of the function $f(x, y) = x^2 + 3y^2 + 4xy + 2x$ on the region bounded by the line $y = x - 13$ and the coordinate axes.

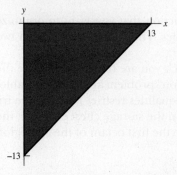

(a) Find the gradient $\vec{\nabla}f$.

(b) Critical points occur where $\vec{\nabla}f = \vec{0}$. To find them, set $\partial f/\partial x = 0$ and $\partial f/\partial y = 0$ and solve the resulting equations simultaneously for x and y. (In this problem you will find only one critical point.)

(c) Next consider the first bounding curve.
  i. Substitute $y = x - 13$ into the function $f(x, y)$ to find a formula for the function along this line that only depends on x. Simplify your answer as much as possible.
  ii. Set the derivative equal to zero and solve to find points along this line that represent potential minima and maxima. (Once again, in this case you will find only one.)

(d) Repeat Part (c) along the bounding curve $x = 0$. (You will find one point.)

(e) Repeat Part (c) along the bounding curve $y = 0$. You will find one point, but then you will ignore that point: explain why.

(f) List all the boundary points of the bounding curves.

(g) Plug all the critical points you have found and boundary points of the bounding curves into the original function.

(h) What are the absolute maximum and minimum of this function on this region?

4.133 *[This problem depends on Problem 4.132.]*
(a) Use the second derivatives test on the critical point you found in Problem 4.132. What does it tell you about that critical point? Does its result agree with what you found in that problem?
(b) Plot $f(x,y)$ on the domain given in the problem. Mark the critical point and check that it is of the type you found in Part (a). Check that the absolute minimum and maximum occur at the points you found them at.

In Problems 4.134–4.137 find the absolute maximum and minimum values of the given function in the given closed, bounded region. You may find it helpful to work through Problem 4.132 as a model.

4.134 $f(x,y) = x^2 - 6x - y^2$ on the region bounded by the curves $x = \sqrt{36 - y^2}$ and $x = 0$.

4.135 $f(x,y) = (2x - 2)\cos y - x$ on $0 \leq x \leq 2\pi$, $0 \leq y \leq 2\pi$.

4.136 $f(x,y) = xe^{-y}$ on the region bounded by the curves $y = x^2 - 3$ and $y = 2x$.

4.137 $f(x_1, x_2, x_3, x_4) = 4x_1 x_4 + 3x_1 x_2 - 3x_2 x_3$ subject to the constraints $x_1 - 4x_2 - x_3 = 0$, $x_1^2 + x_3 - x_4 = 0$, $0 \leq x_1 \leq 2$, $0 \leq x_2 \leq 2$. *Hint:* You can make this problem somewhat easier by choosing the right variables to eliminate.

4.138 In the Explanation (Section 4.8.2) we considered a storage chest with three independent dimensions. The volume of the storage chest is given by the formula $V = LWH$ and the cost to build the chest is $C = LW + 2HW + 3LH$. In this problem we will ask the question: if your goal is to build a 24 ft³ chest, what dimensions minimize the cost? Note that the roles of the two equations in this problem are reversed from the Explanation: the cost is now the function we want to minimize, and the volume provides the constraint.
(a) Use the constraint to write an equation solving for one of the three variables as a function of the other two.
(b) Substitute your formula from Part (a) into $C(L, W, H)$ to find the cost as a function of two variables.
(c) Find the gradient of your function from Part (b).
(d) A critical point can occur if the gradient is undefined. Explain why that does not apply in this case.
(e) A critical point can also occur where the gradient is zero, which means both partial derivatives must be zero. Set them equal to zero and solve to find the dimensions of the box.
(f) Make a simple argument—no calculus required—that your dimensions cannot possibly *maximize* the cost, because you can make the cost arbitrarily high within the given constraints.

4.139 A box with no top is to be made all of the same material, so its cost is $C = C_0(LW + 2HW + 2LH)$. The post office puts a limit of 130" on the linear dimension $D = H + L + W$ that can be sent by first class mail. What is the most expensive such box that can be sent?

4.140 Prove that if you construct a rectangular solid of surface area A the largest volume it can have occurs when it's a cube.

In Problems 4.141–4.144 your job is to find the point on the given surface S that is closest to the given point P. To put it another way, your job is to find the point with the minimum distance to P, subject to the constraint that your answer must lie on S. *Hint:* Rather than minimizing "distance to P" it may be easier to minimize "distance to P squared," which will give you the same point.

4.141 Point P is the origin and surface S is the plane $z = 2x + 5y - 10$.

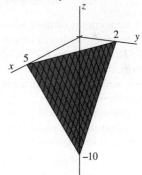

4.142 Point P is $(1, 0, 0)$ and surface S is the paraboloid $z = x^2 + y^2$.

4.143 Point P is $(1, 2, 0)$ and surface S is the paraboloid $z = x^2 + y^2$.

4.144 Point P is the origin and surface S is the surface $x^2y^2z = 1$.

4.145 A box is inscribed inside the paraboloid $z = x^2 + 2y^2$ as follows. Start with a point (x, y, z) in the first octant with $z < 1$. Extend left to $-y$, back to $-x$, and up to $z = 1$. Find the starting point (x, y, z) of the largest possible box.

4.146 The entropy of a gas is proportional to $f \ln U$ where f is the number of degrees of freedom of the gas (which depends on how many molecules there are and on the properties of those molecules), and U is its energy. So if you have three types of gas in a closed container the total entropy will be:

$$S = k \left[f_1 \ln U_1 + f_2 \ln U_2 + f_3 \ln U_3 \right]$$

The total energy of the system is fixed, which corresponds to a constraint $U_1 + U_2 + U_3 = U_T$ where U_T is a constant. This energy can move between the different gases, and it will do so in a way that maximizes the total entropy of the system. Show that this maximum will be attained when the energy of each gas is proportional to its number of degrees of freedom.

4.147 The "moment of inertia" I of a particle about a point is a measure of how hard it is to rotate that particle about that pivot point. It's given by the formula $I = mr^2$ where m is the particle's mass and r is the distance from the particle to the pivot. The moment of inertia of a collection of particles is simply the sum of their individual moments of inertia. Consider a system made of three particles, with the masses and locations shown below.

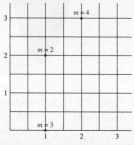

About what pivot point does this system have the smallest possible moment of inertia?

4.148 *[This problem depends on Problem 4.147.]* The "center of mass" of a collection of particles with masses m_1, m_2, ...located at positions (x_1, y_1), (x_2, y_2), ...is a point with coordinates

$$x_{COM} = \frac{m_1 x_1 + m_2 x_2 + \ldots}{m_1 + m_2 + \ldots},$$

$$y_{COM} = \frac{m_1 y_1 + m_2 y_2 + \ldots}{m_1 + m_2 + \ldots}$$

Prove that the pivot point that minimizes the moment of inertia for a collection of particles is the center of mass of those particles. (For this problem you are only working in two dimensions, but it's easy to generalize the proof to three or more dimensions.)

4.149 Coulomb's law says that a point charge q produces an electric field $\vec{E}$ whose magnitude at each point equals $|q|/(4\pi\varepsilon_0 r^2)$, where ε_0 is a constant and r is the distance from the charge. The direction of $\vec{E}$ is away from the charge if $q > 0$ and toward it if $q < 0$. When you have more than one charge the electric field at any given point is the *vector* sum of the fields produced by each charge, so you have to break the field from each charge into components and then add them. Suppose a charge $q = 5$ is placed at the point $(0, 1)$. Your goal is to place two negative charges $q = -3$ and $q = -4$ somewhere on the upper half $(y > 0)$ of the unit circle so as to minimize the magnitude of $\vec{E}$ at the origin.

(a) Solve for the optimal location of the two negative charges using Cartesian coordinates. *Hint 1*: it's easier to minimize $|\vec{E}|^2$ rather than $|\vec{E}|$, but in the end it amounts to the same thing. *Hint 2*: start with four variables for the positions of the two negative charges and use the constraints from the problem to eliminate the y-coordinates for the charges and get two simultaneous equations for x_3 and x_4 at the critical points. Once you use these equations to find a relation between x_3 and x_4 and get a single equation for x_3, it may be easier to solve that equation by expressing it in terms of y_3.

(b) Rewrite the electric field so that instead of $E_x(x_3, y_3, x_4, y_4)$ and $E_y(x_3, y_3, x_4, y_4)$ you have $E_x(\phi_3, \phi_4)$ and $E_y(\phi_3, \phi_4)$. Then set *those* partial derivatives to zero to find the same locations you found in Part (a).

Problems 4.150–4.154 are about the "least squares method" for finding a function $y_{fit}(x)$ that approximates ("fits") a set of data points. A perfect fit would be one that exactly went through every data point, but that usually makes the fitting functions too complicated. To measure the accuracy of the fit you measure the vertical distance from each point to the curve, $y_{data}(x) - y_{fit}(x)$, square them, and add them together to get the "squared error." For example, if you fit the points (1, 1), (2, 7), (3, 30) with the curve $y_{fit} = x^3$ then the squared error would be $(1 - 1)^2 + (7 - 8)^2 + (30 - 27)^2 = 10$. The smaller the squared error is, the better the fit. (Squaring is important because it means you are counting every error as positive whether the data point is above or below the line.)

4.150 Assume you want to fit the data shown below to a linear function $y_{fit} = mx + b$.

x	0	1	2	3
y	0.2	2	4.1	5.6

(a) If your goal is to find the best-fit line to these data, what two variables are you trying to find?

(b) Write the function that you are trying to minimize. It should only include the two variables you just identified, plus a lot of numbers.

(c) Minimize that function and find the best-fit line. Your final answer should be in the form of an equation $y = \ldots$ for the best-fit line for these data.

4.151 (a) Find the best-fit line $y_{fit} = mx + b$ for the pair of points $(0, y_0)$ and $(1, y_1)$. Explain why your answer makes sense.

(b) Find a formula for the best-fit line for three points: $(0, y_0)$, $(1, y_1)$, and $(2, y_2)$.

4.152 Find the best-fit quadratic function $y_{fit} = ax^2 + bx + c$ for the following data points:

x	0	1	2	3	4
y	2	1	3	6	10

4.153 Consider the data shown below.

x	1	2	3	4	5
y	1.4	2.7	7.2	5.6	8.1

(a) Find the best-fit cubic function $y_{fit} = ax^3 + bx^2 + cx + d$ to these data.

(b) Find the best-fit cubic function $y_{fit} = ax^3 + bx^2 + cx + d$ to these data subject to the constraint $a + b + c + d = 1$.

(c) Plot the original data points and the two fitting curves you found together on the same plot.

4.154 In this problem you will see how a moderately large data set can be fit by different types of curves. (In practice engineers optimize functions with as many as several thousand variables, but this one is at least large enough that you wouldn't want to do it by hand.) Your data set will consist of 21 evenly spaced points from $x = 0$ to $x = \pi$ on the plot of $y = \cos x$: $(0, 1)$, $(\pi/20, \cos(\pi/20))$, $(2\pi/20, \cos(2\pi/20))$, $(3\pi/20, \cos(3\pi/20))$, $\ldots(\pi, -1)$.

(a) Have the computer calculate the squared error for a linear fit $y_{fit} = mx + b$ to these data points. Your answer should be a function of m and b.

(b) Have the computer minimize that function of m and b (either by setting the partial derivatives equal to zero or by using the program's built-in minimization function). The result should be a best-fit line y_{fit}. What is the squared error for this fit?

(c) Plot the data and the line together on one plot.

(d) Find the best-fit quadratic function for these data. Give the squared error and plot the data and the fitting function together on one plot.

(e) Find the best-fit cubic function for these data. Give the squared error and plot the data and the fitting function together on one plot.

4.9 Lagrange Multipliers

4.9.1 Explanation: Lagrange Multipliers

In Section 4.8 we solved the following problem:

> A storage chest is to be built to the following specifications. The left side, right side, back, and bottom are made of a material that costs $1/ft^2$. The front is made of a more decorated material that costs $2/ft^2$. There is no top. Find the largest-volume chest that can be made for $20.

The problem contains an objective function $V = LWH$ and a constraint $C(L, W, H) = LW + 2HW + 3LH = 20$. In the previous section, we approached such a problem in two steps: use the constraint to eliminate one variable in the objective function, and then find the points where the gradient is zero or undefined. Lagrange multipliers provide an alternative approach, based on the following formula.

> **Lagrange Multipliers**
> Given a multivariate function f to be optimized, and a constraint $g = k$, where g is another function and k is a constant, the extrema will be found at points where $\vec{\nabla} f = \lambda \vec{\nabla} g$ where λ is a scalar constant. In the case of multiple constraints $g_i = k_i$, the extrema will occur where $\vec{\nabla} f = \sum_i \lambda_i \vec{\nabla} g_i$.

Below we demonstrate how to use Lagrange multipliers to solve the storage chest problem. Following that we offer a justification for the Lagrange multiplier formula.

Using Lagrange Multipliers to Solve the Storage Chest Problem

The objective function in this case—the function we want to optimize—is $V(L, W, H) = LWH$. Its gradient is $\vec{\nabla} V = <WH, LH, LW>$. The constraint is $LW + 2HW + 3LH = 20$; the gradient of the function is $\vec{\nabla} C = <W + 3H, L + 2H, 2W + 3L>$. Setting $\vec{\nabla} V = \lambda \vec{\nabla} C$ yields three equations:

$$WH = \lambda(W + 3H) \tag{4.9.1}$$

$$LH = \lambda(L + 2H) \tag{4.9.2}$$

$$LW = \lambda(2W + 3L) \tag{4.9.3}$$

We have only three equations to solve for four unknowns. The final equation is always the constraint itself:

$$LW + 2HW + 3LH = 20 \tag{4.9.4}$$

If these were all linear equations we could approach them systematically with matrices. However, there is no general, systematic approach for solving n non-linear equations for n unknowns. Our best advice, as is so often the case, is that the more you do it the better you get at it.

One approach that is often useful is to divide two equations, so λ cancels. Dividing Equation 4.9.1 by Equation 4.9.2 yields $W/L = (W + 3H)/(L + 2H)$. Cross-multiplying and simplifying turns this into $2HW = 3HL$ which means that either $H = 0$ or $2W = 3L$. We discard $H = 0$ on physical grounds.

Similarly, dividing Equation 4.9.2 by Equation 4.9.3 becomes $3H = W$. We can then substitute both of these results into Equation 4.9.4 to replace both L and H with W, producing: $(2/3)W^2 + (2/3)W^2 + (2/3)W^2 = 20$

Solving, $W = \sqrt{10}$. It is now a trivial matter to find L and H and then the maximum possible volume. Of course we get the same answer we got in Section 4.8.

Lagrange multipliers can be used for many different types of constrained optimization problems, but one common use of them is for optimization in a bounded region. Recall from Section 4.8 that if you want to optimize a function in a bounded region you first find the critical points inside the region, and then you optimize the function along each of the boundaries. The example below shows how you can do this last step with Lagrange multipliers.

EXAMPLE Lagrange Multipliers

Problem:
In Section 4.8 we found the absolute maximum and minimum of the function $f(x, y) = x^2 + y^2 - 2x - 4y + 10$ on a triangular region. As part of the problem, we had to find its extreme along the line $y = -2x + 6$. Find the same point using Lagrange multipliers.

Solution:
$\vec{\nabla} f = \langle 2x - 2, 2y - 4 \rangle$. The constraint can be written as $g(x, y) = 6$ where $g(x, y) = y + 2x$, so $\vec{\nabla} g = \langle 2, 1 \rangle$. So the Lagrange multiplier formula yields the following two equations:

$$2x - 2 = 2\lambda \quad (4.9.5)$$

$$2y - 4 = \lambda \quad (4.9.6)$$

The constraint provides the final equation as always:

$$y = -2x + 6 \quad (4.9.7)$$

In this trivial case, Equations 4.9.5 and 4.9.6 immediately yield $2x - 2 = 2(2y - 4)$ or $x = 2y - 3$. Substituting into Equation 4.9.7 we find that $y = 12/5$, just as we found before. From there we can quickly get $x = 9/5$ and $f(9/5, 24/5) = 29/5$. We can also get $\lambda = 4/5$; we don't need that number to optimize the function, but it can be useful, as we will discuss below.

Note that Lagrange multipliers in this problem are *not* a substitute for the entire process of finding critical points and checking boundaries. They help with one part of the process, finding critical points along a given boundary.

The "Lagrange Multiplier" in Lagrange Multipliers

When you are optimizing a function of n variables with one constraint, this method gives you $n + 1$ equations for $n + 1$ unknowns: the original independent variables, and the "Lagrange multiplier" λ. In many cases you solve for all these variables and then ignore the λ-value you found. Your goal was to find the variables that optimize the objective function, and you've found them.

But λ does have a meaning that can be useful. If we increase the value of the constraint, λ tells us how fast the objective function will change.

As an example, λ in our storage chest example above works out to be $\sqrt{10}/6$. If we increase the maximum cost of our box by some small amount dC, the maximum volume will increase by $dV = \left(\sqrt{10}/6\right) dC$.

Our second example above involved optimizing a function along the line $y + 2x = 6$. We found that $\lambda = 0.8$. We conclude that along the line $y + 2x = 6.1$ the function's optimum value would be higher by about 0.08.

Why Do Lagrange Multipliers Work?

This is far from a proof, but we can offer an intuitive justification for Lagrange multipliers in two dimensions. We begin by representing the objective function $f(x, y)$ by a contour map, which might look something like this.

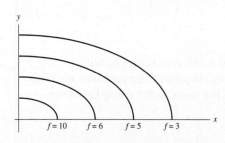

Remember how this two-dimensional map displays a function of two variables. For all points on the lower-left-hand curve, $f(x, y) = 10$. All points on the next curve over have $f(x, y) = 6$, and so on. You might reasonably conclude that if our goal is to maximize this function, we should move down and left; if our goal is to minimize the function, we should move up and right.

Remember also that the gradient $\vec{\nabla} f$ points perpendicular to these contour lines at all points. When you move perpendicular to the gradient the directional derivative is zero, which means the function doesn't change, which defines a contour line.

Now we add the constraint $g(x, y) = k$ to our drawing. This constraint is a single curve that represents one contour line of the function $g(x, y)$.

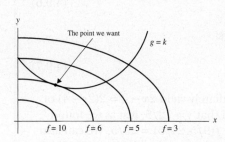

Our goal is to follow along the curve $g = k$ until we reach the highest possible value of f. The key insight is that, at that point, the two contour lines will be parallel. If they are not parallel, you are still cutting through contour lines of f, moving up to higher values. Because the two curves are parallel at this point, their gradient vectors are also parallel. This is captured in the Lagrange multiplier formula: $\vec{\nabla} f = \lambda \vec{\nabla} g$, where λ is a scalar, mathematically indicates that $\vec{\nabla} f$ is parallel to $\vec{\nabla} g$.

Linear Programming and the Simplex Method

Both optimization methods we have discussed, the gradient and Lagrange multipliers, solve similar types of problems. In these problems the objective function and the constraints may be complicated, but there are relatively few of them.

Engineers often confront problems in which all the relevant functions are linear, but the variables and constraints number in the hundreds or thousands. Such problems require sophisticated algorithms for moving through the space of possibilities in a reasonable amount of time. This is referred to as "linear programming" or "linear optimization." In Section 4.11 Problems 4.220–4.222 (see felderbooks.com) you will solve a few linear optimization problems small enough to be done without a computer. In Chapter 7 we will introduce the most important linear programming algorithm, the "simplex method."

4.9.2 Problems: Lagrange Multipliers

4.155 Walk-Through: Lagrange Multipliers. One part of Section 4.8 Problem 4.132 required you to find the critical point(s) of $f(x, y) = x^2 + 3y^2 + 4xy + 2x$ along the line $y = x - 13$. In this problem you will repeat that exercise using Lagrange multipliers.

(a) Find the gradient of the objective function.

(b) Write the constraint in the form $g(x, y) = k$ where k is a constant. Then find the gradient of the function $g(x, y)$.

(c) The equation $\vec{\nabla} f = \lambda \vec{\nabla} g$ gives you two equations to solve for x, y, and λ. Write these two equations.

(d) The third equation relating the three unknown variables is the constraint itself. Solve these three equations to find the (x, y) coordinates of all critical points.

In Problems 4.156–4.165 you will redo the given problems from Section 4.8. If you already did a given problem in that section, much of the work—using the second derivatives test to classify critical points, or plugging in values to find maxima and minima—will not have to be redone here. The only step you need to redo is finding the critical points of the objective function subject to the constraint in the problem, which you will now do using Lagrange multipliers. You may find it helpful to work through Problem 4.155 as a model.

4.156 Problem 4.138.

4.157 Problem 4.139.

4.158 Problem 4.140.

4.159 Problem 4.141.

4.160 Problem 4.142.

4.161 Problem 4.143.

4.162 Problem 4.144.

4.163 Problem 4.145.

4.164 Problem 4.146.

4.165 Problem 4.153 Part (b).

In Problems 4.166–4.168 you will redo the given bounded-region problems from Section 4.8. If you already did a given problem in that section you can use your results from that for the critical points in the interior of the region. However, you will now find the critical points on the boundaries by using Lagrange multipliers with the boundary equations as constraints.

4.166 Problem 4.132.

4.167 Problem 4.134.

4.168 Problem 4.136.

4.169 Section 4.8 Problem 4.137 asked for the critical points of $f(x_1, x_2, x_3, x_4) = 4x_1 x_4 + 3x_1 x_2 - 3x_2 x_3$ subject to the constraints $x_1 - 4x_2 - x_3 = 0$ and $x_1^2 + x_3 - x_4 = 0$. (That problem also involved two other constraints that we will ignore here.) Find those critical points using Lagrange multipliers. *Hint*: The box on Page 182 gives the formula for Lagrange multipliers with multiple constraints.

4.170 This problem is based on Section 4.8 Problem 4.149. If you haven't done that problem you should read it, but you do not need to have done that problem to solve this one.

(a) In Section 4.8 Problem 4.149 Part (a) the two negative charges were confined to the upper half of the unit circle. Now suppose those charges could be anywhere on the unit circle. Explain why loosening this restriction would make the problem considerably more difficult using the method of Section 4.8.

(b) Now allowing the two negative charges to be anywhere on the unit circle, you will solve this problem using Lagrange multipliers. Write, but do not yet solve, the six equations for the six unknowns (the x and y positions of the two particles, plus two Lagrange multipliers for the two constraints).

(c) Solve your equations to find the positions of the two negative charges.

4.171 The Explanation (Section 4.9.1) justified the formula $\vec{\nabla} f = \lambda \vec{\nabla} g$ by arguing that at such a point the curve $g(x, y) = k$ is parallel to the level curves of f. But this logic does not work if $\lambda = 0$. Does that special case also lead to a critical point? Why or why not?

4.172 There is a classic optimization problem known as the "milkmaid problem." A milkmaid is at the location (x_M, y_M) and she needs to milk a cow at location (x_C, y_C). Before milking the cow, however, she needs to stop by the river to wash her pail.

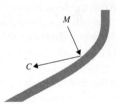

The path of the river is described by the equation $g(x, y) = 0$ for some function g. Her goal is to find the point (x, y) on the river that will allow her to wash her pail and then get to the cow with as little travel distance as possible.

(a) Using Lagrange multipliers, write the equations that need to be solved to find the optimal point (x, y) where the milkmaid should wash her pail.

(b) Write and solve those equations assuming the river flows long the x-axis, the milkmaid's initial position is $(3, 2)$, and the cow's position is $(1, 2)$. Explain why your answer makes sense.

(c) Write and solve those equations assuming the river flows long the line $y = x$, the milkmaid's initial position is (1, 3), and the cow's position is (3, 5). Explain why your answer makes sense.

(d) 🖥 Assume the river is described by the curve $y = x^3$, the milkmaid's initial position is (0, 2), and the cow's position is (1, 4). Have a computer numerically calculate where the milkmaid should wash her pail. Make a plot showing the river, the two points given here, and the point you just found.

4.173 Your spaceship needs to gather rock samples from a planet's surface and deliver them to a nearby space station. You are currently at location (50, 10, 20) and the space station is at (30, 20, 30). (All coordinates are measured in thousands of miles in a coordinate system with the origin at the center of the planet.) You need to choose where to gather the rocks so that you can reach the space station with as little distance traveled as possible.

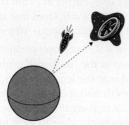

(a) Write the function you are trying to minimize in terms of the coordinates (x, y, z) of the point where you will gather the rocks.
(b) The planet's surface is a sphere of radius 10 centered on the origin. Write the constraint equation that x, y, and z must satisfy.
(c) Set up, but do not yet solve, the equations to find the coordinates of the point where you should gather the rocks.
(d) 🖥 Numerically solve the equations you wrote to find the coordinates where you should land on the planet.

4.174 The squared distance from some point (a, b) to some other point (x, y) is $(x - a)^2 + (y - b)^2$. For simplicity assume throughout this problem that both points are in the first quadrant.

(a) Use Lagrange multipliers to find the point (x, y) that minimizes this squared distance subject to the constraint that (x, y) is on the line $y - x = 0$.

(b) Find the value of λ in the equations you used. Explain what this value represents.
(c) Repeat Parts (a)–(b) with the constraint $y^2 - x^2 = 0$.
(d) You should have found that the point (x, y) (and thus the distance) was the same for the two constraints, but λ was different. Explain why you should expect both of those to be true.

4.175 A projectile launched from the ground with initial velocity components v_x and v_y will travel a distance equal to $2v_x v_y / g$. The energy that you initially give to the projectile is $(1/2)m(v_x^2 + v_y^2)$.

(a) Use the method of Lagrange multipliers to find the maximum possible distance a projectile can go subject to the constraint that its initial energy is E.
(b) Find the value of λ from the equations you wrote.
(c) Using that value of λ, how much farther can the projectile go if you increase the energy you launch it with by dE?

4.176 The Cobb–Douglas production function models the production of a commodity with the equation $Y = AL^\alpha K^\beta$ where Y is the total value of the commodity produced, L and K are the input of labor and capital respectively, and A, α, and β are constants related to overall productivity (e.g., the technology level). For this problem take $A = 20$, $\alpha = 0.6$, and $\beta = 0.4$, where all monetary values are measured in hundreds of thousands of dollars and times in months. The costs of labor and materials are $C = 30L + 50K$, and your total budget is 200.

(a) Use Lagrange multipliers to find the values of L and K that maximize your production subject to the constraint $C = 200$.
(b) Find the value of λ from your equations.
(c) Explain what λ tells you about the production of your firm. Use only words, no formulas.
(d) Based on the value of λ you found, how much could you increase your production if your budget increased by $10.00?

4.177 The number of thneeds your company can produce is $L^\alpha K^\beta$ where L is the number of workers and K is the number of thneed-making machines they operate. Your total profit is $P = TL^\alpha K^\beta - sL - cK$ where T, s, and c are the price of a thneed, the wage of a worker, and the operating costs of a thneed-making machine, respectively.[4] However, labor laws require that each worker may only oversee one machine: $L = K$.

(a) Use Lagrange multipliers to find the value of L that maximizes your profit subject to the constraint $L = K$.

(b) Solve for λ in the equations you wrote.

(c) A government worker offers to bend the regulations, allowing you to use fewer workers than the law requires, in exchange for a modest "compensation" (bribe). Based on the value of λ you found, what is the maximum bribe you should be willing to offer in order to have w fewer workers than machines?

(d) Just to get specific, suppose $T = 1000$, $\alpha = 0.8$, $\beta = 0.2$, $s = 10$, and $c = 1$. How much should your company be willing to pay in order to be allowed to hire 1000 fewer workers?

4.10 Special Application: Thermodynamics (see felderbooks.com)

4.11 Additional Problems (see felderbooks.com)

---

[4]Thanks to Martin Osborne, from whose idea this problem was adapted with permission.

CHAPTER 5

Integrals in Two or More Dimensions

> *Before you read this chapter, you should be able to...*
> - evaluate definite and indefinite integrals, including those that require u-substitution or integration by parts.
> - graph points and functions in polar coordinates.
> - find the sum, dot product, and cross product of two vectors.
>
> *After you read this chapter, you should be able to...*
> - set up integrals based on word problems.
> - set up and evaluate "double integrals" and "triple integrals," and use them to solve problems involving volume, surface area, center of mass, moment of inertia, and others.
> - set up and evaluate integrals in polar, cylindrical, and spherical coordinates.
> - interpret parametric representations of surfaces.
> - set up and evaluate "line integrals" and "surface integrals."

This chapter is *not* about how to evaluate an integral. We assume you can find basic antiderivatives, including those that require u-substitution or integration by parts. You may also know more advanced techniques such as partial fractions and trig substitution—but if you don't, you won't find them here.

Instead, we're going to begin with a couple of sections that explain the connection between real-world problems and integrals. ("Find the area under this curve" hardly counts as an important engineering application.) We encourage you to pay special attention to those first few sections. If you can adopt our perspective on *what an integral is*, then the rest of the chapter, in which we extend the idea into multivariate functions, should follow naturally.

5.1 Motivating Exercise: Newton's Problem (or) The Gravitational Field of a Sphere

Isaac Newton had a problem. His new law of universal gravitation did a great job of explaining apples falling from trees. It also accounted perfectly for Kepler's observationally based laws of planetary motion, as long as he assumed that the planets, the sun, and the moon were all point particles. The problem was, they weren't. Newton needed to show that a full sphere would *act like* a point particle located at its center.

In this exercise you will take the first steps toward the proof that Newton needed. (Spheres will not come into this exercise at all, but we will come back to them later.) The only physics you need to know is that a particle with mass M creates a "gravitational potential" at every point around it, given by the formula:

$$V(x, y, z) = -GM/r \qquad (5.1.1)$$

where G is a constant and r is the distance from the particle to the point. If there is more than one particle around, the gravitational potential at each point is the sum of the potentials created by all the nearby particles. (Gravitational potential is used to figure out gravitational force, but we won't be doing that here.)

In all the problems below, the question is the same: *Use Equation 5.1.1 to calculate the gravitational potential at the point $(0, d)$ on the xy-plane.*

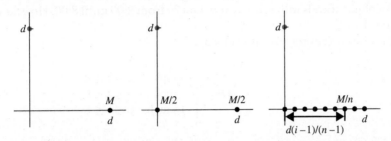

1. Suppose the only object around is a particle of mass M at the point $(d, 0)$ (the first image above). Find the gravitational potential.
2. Now suppose there are two objects, each of mass $M/2$, located at $(0, 0)$ and $(d, 0)$. (See the second image.) Calculate the gravitational potential.
   See Check Yourself #25 in Appendix L
3. Now suppose there are n objects, each of mass M/n, spread evenly between $(0, 0)$ and $(d, 0)$ (third image). Object number i is located at position $x_i = d(i-1)/(n-1)$.
   (a) Check that this gives the correct positions for x_1 and x_n.
   (b) Calculate the gravitational potential created by object i.
   (c) Write a sum representing the total potential at $(0, d)$ created by this collection of objects. (You do not need to evaluate the sum; just write it.)

In the limit as $n \to \infty$ your final answer becomes the total gravitational potential generated by a mass M spread evenly between $x = 0$ and $x = d$. But calculating such a quantity by adding up a series and then taking a limit is generally not practical. Our biggest goal in this chapter is to get you to think of integrals as the solution to *that problem:* a fast method for calculating the limit of the sum.

In Problem 5.13 you will properly finish this problem using an integral. In the later sections, you will see how to extend such an integral into two or three dimensions, and in Section 5.7 Problem 5.188 you will sum up the gravitational potential due to a sphere. Finally in Section 5.11 we will return to, and solve, Newton's original problem.

5.2 Setting Up Integrals

One of the main themes of this book is that "solving equations" has never been the heart of mathematics, and it is even less so with modern software. Today most high school students have pocket calculators that can calculate integrals faster and more accurately than a college professor can. But figuring out what to integrate is a human skill, and learning that skill requires learning what an integral represents.

So this chapter is focused on setting up integrals, rather than evaluating them. Any integrals we expect you to take by hand should be possible using no technique more advanced than u-substitution or integration by parts. Any problem with a more difficult integral will be marked with a computer icon ( ). You can try to evaluate such an integral by using any advanced techniques you know, or you can ask a computer to evaluate it analytically. If the computer can't do that—and there are many functions that have no antiderivatives—then you can ask a computer to approximate the integral numerically, which is essentially a Riemann sum with many slices. Of course a numerical integral is only possible if all the constants in the integral have been expressed as numbers.

5.2.1 Discovery Exercise: Adding Up the Pieces

1. An object travels with velocity $v = 4$ mi/h from 2:00 until 8:00. How far does the object travel?
2. Another object travels with velocity:

$$v(t) = \begin{cases} 4 \text{ mph} & 2 \leq t < 4 \\ 8 \text{ mph} & 4 \leq t < 6 \\ 16 \text{ mph} & 6 \leq t \leq 8 \end{cases}$$

(All times are measured in hours.) How far does the object travel between $t = 2$ and $t = 8$?

See Check Yourself #26 in Appendix L

3. A thin metal bar has a linear density given by $\lambda = 4$ kg/m. How much mass lies within a 6-meter length of this bar?
4. Another bar has a linear density given by:

$$\lambda(x) = \begin{cases} 4 \text{ kg/m} & 2 \leq x < 4 \\ 8 \text{ kg/m} & 4 \leq x < 6 \\ 16 \text{ kg/m} & 6 \leq x \leq 8 \end{cases}$$

(All distances are measured in meters.) What is the mass of the part of the object that lies between $x = 2$ and $x = 8$?

5.2.2 Explanation: Setting Up Integrals in One Dimension

What is an integral?

If you answered "the area under a curve" then you know how to find the area under a curve (just in case anyone asks), but perhaps not much else. If you answered "an antiderivative" then you know how to *evaluate* integrals, but you don't necessarily know what you're finding.

This section presents a different way of thinking about integrals. The test of your understanding will be your ability to set up integrals that represent solutions to real-world problems.

The Discovery Exercise (Section 5.2.1) presents two common applications for the integral: finding distance traveled from velocity, and finding mass from density. In both cases, the problem is a simple multiplication if the variable (velocity or density) is constant. If the variable is a step function (sequential horizontal lines), you do the multiplication on separate pieces, and then add up the results (distances or masses) to find the total.

If the function is not a step function, you can approximate it with a step function. As the number of steps approaches infinity, the approximation becomes exact. That is what an integral—every integral!—represents.

We illustrate this process below. The first two examples show the basic process of setting up an integral as a sum of infinitely many infinitesimal pieces. The remaining examples apply this process to a variety of physical problems. Finally the section labeled "Stepping Back" reviews the steps involved. We strongly encourage you to read the "Basic Examples" and "Stepping Back" sections carefully. On the other hand, you might find it more helpful to skim the advanced examples for the moment, and then come back to them as you encounter those types of physical examples later in the chapter. For instance, it may be easier to follow a center of mass calculation in 2D or 3D if you first review our 1D center of mass example.

Basic Examples

Mass and Density If you have a 2 meter bar with linear density $\lambda = 3$ kg/m then the bar has a total mass of 6 kg. That's just the definition of density in 1D:

$$m = \lambda L \quad \text{if } \lambda \text{ is constant} \tag{5.2.1}$$

Now consider a different bar, of length L, whose density is proportional to the distance from its left edge. We set our origin at the left edge, so $\lambda = cx$ where c is a positive constant. How do we find the mass of this bar? We cannot use Equation 5.2.1 directly and write $m = cxL$; the mass of the entire bar cannot depend on x, the position along the bar! Do we need two different equations, one for constant densities and an unrelated one for variable densities?

No, there is a way to apply Equation 5.2.1 to our new bar. Imagine breaking the bar into tiny little pieces, like the shaded one shown below.

The variable x designates the location of one particular piece. dx is the change in x as you go from the left to the right side of that piece: like Δx, but smaller. How small? So small that we can reasonably treat λ as constant within that one little piece, and therefore we can use Equation 5.2.1: $dm = \lambda \, dx$. (Just as dx is a tiny change in position, dm is a tiny change in mass.) The next piece has a different λ, but also constant, leading to a different dm. To approximate the mass of the bar, we add up the masses of all the pieces.

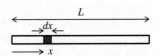

FIGURE 5.1 The smaller dx gets, the closer the small blue slice comes to constant density.

How accurate is that approximation? The smaller each dx is, the more reasonable our constant-λ assumption becomes, and therefore the more accurate our final answer will be. In the limit as $dx \to 0$ (or as the number of pieces approaches ∞) the mass becomes exact.

And that—this is the moment we have been building to—that is what an integral does. When we write the integral below, we mean "multiply cx (the density of one piece) by dx (the length of one piece), and then add all of them up as x goes from 0 to L, in the limit as $dx \to 0$."

$$M = \int_0^L cx \, dx = c \left[\frac{x^2}{2} \right]_0^L = \frac{1}{2} cL^2 \tag{5.2.2}$$

As a final check, we can consider the units of this result. We first introduced c when we said that the density was cx. Since density in 1D has units of mass per length, c must have units of mass per length squared to make cx be a density. Looking at our final answer, the expression we got for M thus has units of mass, which is reassuring.

Distance Traveled You already know that velocity is the derivative of position, so position is the integral of velocity. But let us briefly walk-through the argument in terms of what an integral is.

First we ask, what if velocity is constant? If you travel 60 mi/h for 3 h, then your total distance traveled is 180 miles: $\Delta x = v \, \Delta t$.

We approach the more general case of a function $v(t)$ by breaking the journey into small "slices," a series of trips of duration dt. For each such trip, we assume that the velocity is relatively constant; we can therefore use the same product we wrote above, and write $dx = v(t) dt$. The total distance traveled is the sum of all such trips, in the limit as $dt \to 0$, which we write as:

$$\Delta x = \int v(t) dt$$

Notice once again how the dt appeared in this calculation. We did not put it there because "you can't have an integral without it"; we multiplied it by velocity to find distance.

Advanced Examples

Center of Mass You may recall from introductory physics that every object has a special point called its "center of mass." For instance, consider an object dangling partly over a ledge: will it stay put or will it fall? The answer depends on which side of the ledge contains the object's center of mass.

The center of mass of a collection of n point masses in 1D is given by:

$$x_{COM} = \frac{x_1 m_1 + x_2 m_2 + x_3 m_3 + \ldots + x_n m_n}{M} = \frac{1}{M} \sum_{i=1}^{n} x_i m_i \quad (5.2.3)$$

Here $M = m_1 + m_2 + \ldots + m_n$ is the total mass of the collection. Note that this formula is a weighted average of the positions, influenced more by heavier objects than lighter. For example, if you have a mass $m_1 = 2$ at $x = 1$ and a mass $m_2 = 3$ at $x = 4$ then the center of mass is $(2 \cdot 1 + 3 \cdot 4)/5 = 14/5$, which is somewhere between the two but closer to the heavier one.

Now suppose we want the center of mass of the rod of length L and density $\lambda = cx$ from the example above. Our strategy is to break the rod into pieces so small that we can treat each piece as a point mass (Figure 5.1). The position of our slice is just x, and its mass is $dm = cx\,dx$. So what we are adding up for each slice, based on the $x_i m_i$ in Equation 5.2.3, is $x(cx\,dx)$. Adding them all up with an integral, the center of mass formula becomes

$$x_{COM} = \frac{1}{M} \int x\,dm = \frac{1}{M} \int_0^L cx^2\,dx$$

Evaluating this integral and dividing by $M = (1/2)cL^2$ (from the mass and density example above) gives $x_{COM} = 2L/3$. Clearly this has correct units. It also makes sense that the answer is somewhere between $L/2$ and L, since the right side of the bar is heavier than the left.

Note that if we hadn't already found the mass (Equation 5.2.2) we would have to do a separate integral to calculate that. Center of mass problems often involve an integral for the numerator (sum of $x_i m_i$) and another one for the denominator (total mass).

Moment of Inertia The "moment of inertia" measures how hard it is to start an object rotating about a given axis. The moment of inertia of a point particle is mr^2 where m is the mass of the particle and r is the distance of the particle from the axis. If you have more than one particle their combined moment of inertia is simply the sum of their individual ones. For example, if you have a dumbbell with masses $m = 2$ at positions $x = 3$ and $x = -3$ then the moment of inertia of that dumbbell about a perpendicular axis through its center is $2(3^2) + 2(3^2) = 36$. As you've surely figured out by now, to find the moment of inertia of an extended object you break it into slices so small you can consider each one to be a point particle, find the moment of inertia of a single slice, and integrate to add them all up.

$$I = \sum mr^2 \quad \text{or} \quad I = \int r^2\,dm$$

We want to calculate the moment of inertia of a uniform rod of length L and mass M about a perpendicular axis through one of its ends, as shown below. We've also included, as usual, a thin slice with thickness dx and position x.

Since the bar is uniform you might think we can just multiply instead of integrating, but even though each slice has the same mass, they do not all have the same moment of inertia. To find the moment of inertia of a slice we need to know its mass dm and its distance r from the axis. (Stop to make sure you understand why we wrote the mass as a differential and the distance as a normal variable.) Normally we would have labeled r on our picture, but in this case it's just the same thing as x.

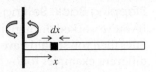

The mass of our slice is the total mass of the bar multiplied by the fraction of the bar represented by our slice: $dm = M(dx/L)$. Pause to make sure you follow that piece of the calculation, and that you see that it only works because the bar is uniform![1] For a variable bar we would find dm from a density function that would have to be specified in the problem.

So the total moment of inertia is:

$$I = \int_0^L \frac{M}{L} x^2 \, dx = \frac{1}{3} ML^2$$

The units are correct: moment of inertia is mass times distance squared. We can also easily check in a physics book or online that this is the correct moment of inertia for a thin rod about a perpendicular axis through one end.

Electric Potential A set of charges creates at each point in space an "electric potential," which measures how much energy another charge would have if you put it at that point. If you haven't studied electricity you're welcome to think deeply about that definition or ignore it. But for the purposes of this chapter there are two things you need to know about electric potential.

1. A charge q creates a potential at every point in space equal to kq/r where k is a constant of nature and r is the distance from the charge to the point where you're calculating the potential. (This is one form of a physics principle known as "Coulomb's Law.")
2. If there is more than one charge in some region the potential at each point is the sum of the potentials created by each charge individually.

For example, if you have a charge $q_1 = 2$ at $x = 0$ and a charge $q_2 = 3$ at $x = 2$, then the potential at $x = 7$ is $(2k/7) + (3k/5)$. Simple enough, right?

Now suppose you have a bar of length L with charge Q uniformly distributed across the length of the bar. We want to find the electric potential at a distance w from the left end of the bar. We've shown the bar below, and as always we've also drawn a small slice and labeled its position x and its thickness dx.

Because the charge density is uniform, the charge of our tiny slice is the total charge of the bar multiplied by the fraction of the bar represented by our slice: $dq = Q(dx/L)$. The distance from the slice to the point where we are calculating the potential is $r = x + w$. So the potential at that point is:

$$V = \int k \frac{dq}{r} = \int_0^L k \frac{Q}{L} \frac{1}{x+w} dx = \frac{kQ}{L} [\ln(x+w)]_0^L = \frac{kQ}{L} (\ln(L+w) - \ln(w)) = \frac{kQ}{L} \ln\left(\frac{L+w}{w}\right)$$

Checking units is especially easy for this example. The argument of the logarithm is unitless and the coefficient in front is in the form k times a charge divided by a distance, which we know is correct for potential. We can also check this answer by thinking about a limit where we know what the answer must be. See Problem 5.12.

---

[1] You can also think of similar ratios: $dm/M = dx/L$. Or you can say that M/L is the density, and multiply that by the length dx. All of these of course get you to the same place, but remember to use them only for uniform density!

Stepping Back: Setting Up 1D Integrals

If you've worked through the basic examples above and at least looked through the advanced ones then you should have a very good idea of what's involved in setting up integrals. All these different examples follow the same basic steps.

- Break the problem into small intervals, or slices.
- For each interval or slice, you should be able to calculate the quantity you are looking for: the mass, the electric potential, *etc*. This calculation will involve an approximation, of the form "this slice is so small that I can assume such-and-such a quantity will not change within this slice, although it may change between this slice and the next." Given that assumption, you should be able to calculate the quantity you are looking for by multiplying something by the length of the slice, which will introduce a dx into the formula.
- Summing over all the slices gives you an approximation for the total quantity you are looking for. In the limit as the size of each slice approaches zero—or, as the number of slices approaches infinity—the approximation becomes exact. This limit-of-a-sum is what the integral computes for you.
- Check your answer for units, for behavior in appropriate limits, and for any other tests you can find to see if it makes sense.

If you make a mistake in setting up your problem, you may come to the part where you are supposed to integrate and realize you don't have a differential there. The typical novice reaction is to simply stick dx at the end of your formula so you can integrate. This never works. The dx has to come out of your calculations, e.g. when you find that the mass of your slice is its density times its thickness dx. If you get to an integral without a differential, go back and find what you did wrong.

The same logic applies to the canonical integration problem, the area under a curve. (Think about Riemann sums!) You will explore that connection in Problem 5.14; watch, as always, how the dx comes from a multiplication.

As a final note, it's often helpful to stop and check units before completely finishing a problem to avoid wasting time working with a wrong formula. What are the units of $\int f(x)dx$, though? Because dx represents a small change in x, a small version of Δx, it has the same units as x. And because $f(x)dx$ is a multiplication, it has the units of $f(x)$ times the units of x. The integral sign just means "add these up," so it doesn't affect the units. We urge you to develop the habit of checking units on all answers unless the problem was given in terms of pure numbers.

5.2.3 Problems: Setting Up Integrals in One Dimension

5.1 Walk-Through: Setting Up an Integral in One Dimension. A long thin bar has a *linear density* λ. If linear density is a constant, you can multiply it by length to get mass. In this problem, we will use the variable x to represent the position on the bar, measured from the left.

(a) If a bar has a linear density $\lambda = 10$, what is the mass of the part of the bar that extends from $x = 1$ to $x = 3$?

A different bar has a linear density given by $\lambda = 3x$ (so the density starts at 0 on the left and increases steadily as you move to the right). To find the mass, you divide the bar into small segments. Each segment does have a constant density, more or less.

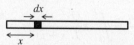

(b) Briefly explain the meaning of the variables x and dx as they apply to the shaded segment in the drawing above.

(c) Find the mass dm of the shaded segment. (This step will *not* involve integrating anything.)

(d) Write $\int_1^3$ in front of your answer to Part (c) to indicate that you are adding up all the segments from $x = 1$ to $x = 3$ in the limit as $dx \to 0$.

(e) Evaluate the integral to find the total mass of the bar between $x = 1$ and $x = 3$.

5.2 *[This problem depends on Problem 5.1.]* In Problem 5.1 you found the exact mass of the region of a bar between $x = 1$ and $x = 3$ with a linear density given by $\lambda = 3x$. In this problem you are going to approximate the same mass using Riemann sums.

(a) We begin with a Riemann sum with two intervals, so $\Delta x = 1$. We will use a left-hand Riemann sum, so λ for the first integral is $\lambda(1) = 3$. What is the mass of the first interval?

(b) The second interval, stretching from $x = 2$ to $x = 3$, has the same width ($\Delta x = 1$) but a different density ($\lambda(2) = 6$). What is the mass of this interval?

(c) Add your two answers to estimate the mass of the bar.

(d) Now repeat this exercise, dividing the region from $x = 1$ to $x = 3$ into four equal intervals with $\Delta x = 1/2$. Add the masses of the four individual segments to estimate the mass of the bar.

(e) In Problem 5.1 you found the exact mass of this region of the bar. Why are your estimates in this problem lower than the actual mass? Why is your second estimate more accurate than your first?

5.3 *[This problem depends on Problems 5.1 and 5.2.]* In Problem 5.2 you estimated the mass of a bar using a Riemann sum with two intervals, and then a more accurate sum with four intervals. This time you will use n intervals and take a limit, which is exactly the calculation that an integral performs for you. Your answers to all of the parts in this problem should depend on nothing but n (the number of intervals) and i (which goes from 1 to n over the different intervals).

(a) If the bar has n intervals, what is the width Δx for each interval?

(b) If the bar has n intervals, what is the x-value at the left of the ith interval? (To test your answer, make sure that your formula gives $x = 1$ for the first interval and $x = 3 - \Delta x$ for the last.)

(c) Calculate the density of each interval (based on a left-hand Riemann sum).

(d) Calculate the mass of each interval.

(e) When you add up all the small masses to find the mass of the bar, the resulting sum is an arithmetic sequence. The sum of an arithmetic sequence is $(n/2)(m_{first} + m_{last})$. Find the sum of all the small masses as a function of n, the number of intervals.

(f) Find $\lim_{n \to \infty} m(n)$. The result should match your answer to Problem 5.1.

5.4 A thin metal bar has linear density $\lambda = 1/(x + 1)^2$ where x is the distance from the left side of the bar.

(a) What is the mass of the segment of the bar that stretches from $x = 0$ to $x = 3$?

(b) What is the mass of the segment of the bar that stretches from $x = 3$ to $x = 5$?

(c) What is the mass of the segment of the bar that stretches from $x = 0$ to $x = 5$?

(d) If the bar begins at $x = 0$ on the left and stretches infinitely far to the right, what is its total mass?

5.5 A car starts at the origin when $t = 0$ and moves thereafter with velocity $v(t) = 1 - (t - 1)^2$ (with time measured in seconds and distance in meters).

(a) How far has the car moved after one second?

(b) What is the largest x-value the car will reach? (*Hint*: The answer is not 5/3.)

For Problems 5.6–5.9 you will be given information about a rod that stretches from $x = 0$ to $x = L$. For each problem calculate the total mass (unless it's already given), the center of mass, and the moment of inertia about a perpendicular axis through $x = 0$. As always, you should use the information you were given to find the units of all constants in the problem and use that to check the units on your answers. All letters other than λ and x designate positive constants.

5.6 The rod is uniform with total mass M.

5.7 The mass density is $\lambda = c \sin(\pi x/L)$. Explain why your answer for center of mass makes sense physically.

5.8 The mass density is $\lambda = c/(x + w)$. Explain why your answer for center of mass makes sense physically in the limits $w \to 0$ and $w \to \infty$. (You can take these limits by hand, or you may plug them into a computer.) *Hint*: all of the integrals you'll need can be evaluated with the substitution $u = x + w$.

5.9 The mass density is $\lambda = ae^{-bx}$.

5.10 Find the moment of inertia of a uniform rod of length L and mass M about a perpendicular axis through its center.

5.11 A rod goes from $x = 0$ to $x = L$ with charge density $\lambda = cx^2$. Find the electric potential at $x = 0$.

5.12 In the Explanation (Section 5.2.2) we derived the electric potential at a distance w from a uniform rod of charge Q and length L. Take the limit of that answer as $L \to 0$ and explain why it's the answer you would expect physically.

5.13 A uniform rod of charge Q goes from the origin to the point $(d, 0)$. Find the electric potential produced by this rod at the point $(0, h)$. (This completes the exercise you began in Section 5.1; although that was gravity and this is electricity, the math is the same. Note that you will need to use a picture to express the distance from a slice to the point in terms of x.)

5.14 In this problem you're going to calculate the area under a curve using the same kind of logic we've been using to set up other integrals in this section.

  (a) First consider the function $h(x) = 4$. Sketch the function and find the area under the curve between $x = 2$ and $x = 5$. *This should be quick and easy and should not involve an integral.*

  (b) Part (a) was easy because h was constant. Now sketch a function $f(x)$ that is not constant. (Just to keep things simple, make the function continuous and make f positive everywhere in your sketch.)

  (c) Label a thin slice on the x-axis at position x with thickness dx.

  (d) Because we are assuming dx is extremely small, we can treat f as constant across the entire slice. Write the area under the curve just in the thin region above your slice. You should do this exactly the same way you did Part (a).

  (e) Take an integral of your result to get an expression for the total area under the curve of $f(x)$ between $x = 2$ and $x = 5$.

5.15 A curve rises with a slope of m. (Remember, as always, that slope is defined as rise/run!)

  (a) We begin with a curve of constant m, *i.e.* a line. If $m = 10$, how much does the y-value rise between $x = 13$ and $x = 15$? (No calculus required!)

  (b) Now consider a curve with a variable function $m(x)$. But we will assume that the slope does not change too much as you move a short distance, so $m(10 + dx) \approx m(10)$. Under that assumption, how much does the y-value rise between $x = 10$ and $x = 10 + dx$? (Your answer will not involve an integral.)

  (c) How much does the y-value rise between $x = 10$ and $x = 13$?

5.16 The "profit" of a business is defined as "revenue" (money received from customers) minus "cost" (money spent). (Throughout this problem time will be measured in days.)

  (a) If a business makes a steady profit of \$20,000/day, how much money does it make in a (7-day) week?

  (b) Negative profit is "loss." If a business has a steady loss of \$5000/day, how much money does it make in a week?

  (c) If a business makes \$1000/day for three days, then \$3000/day for three days, and then −\$500/day for three days, what is the total profit for this nine-day period?

  (d) If a business makes $p(t)$ dollars/day during a small time period dt, what is its total profit?

  (e) If a business makes $p(t)$ dollars/day from $t = 32$ to $t = 59$ (the entire month of February), what is its total profit?

  (f) If the net worth of the business is the function $m(t)$, explain why $\int_{32}^{59} p(t)\,dt = m(59) - m(32)$.

  (g) If the profit function is given by $p(t) = t^3 + t^2 - 8t - 12$, find the value of t such that $\int_0^t p(t)\,dt = 0$.

  (h) Explain to the president of the company—who understands business, but doesn't know much math—the significance of the t-value you found in Part (g).

5.17 A plant is growing. The variable g represents its rate of growth in inches per week.

  (a) If $g = 2$ inches/week, how much does the plant grow during any given 3-week period?

  (b) If $g = g(t)$ inches/week, how much does the plant grow during a small time period of duration dt weeks? Assume dt is small enough that the growth can be considered constant for that interval.

  (c) If $g = g(t)$ inches/week, how much does the plant grow from Week 7 to Week 10?

  (d) If the height of the plant is the function $h(t)$, explain why $\int_7^{10} g(t)\,dt = h(10) - h(7)$.

5.18 Water is flowing into a bucket. Let r represent the flow rate of the water, measured in gallons/minute.

  (a) If $r = 1/2$, how much water fills the bucket between $t = 4$ and $t = 10$?

(b) If $r = t^2$, how much water fills the bucket between t and $t + dt$? (Your answer will not involve an integral.)

(c) If $r = t^2$, how much water fills the bucket between $t = 4$ and $t = 10$?

5.19 In a circuit, the current I measures how many Coulombs of charge pass through the wire every second.

(a) If $I = 6$ Coulombs/sec, how many Coulombs of charge pass through the circuit between $t = 9$ and $t = 13$?

(b) If $I(t)$ represents the current as a function of time, how many Coulombs of charge pass through the wire in dt seconds, assuming dt is small enough to treat the current as constant?

(c) If $I(t) = 5\sin(100\pi t)$, how many Coulombs of charge pass through the wire between $t = 0$ and $t = 1/100$?

(d) If $I(t) = 5\sin(100\pi t)$, how many Coulombs of charge pass through the wire between $t = 0$ and $t = 1/50$?

(e) Your answer to Part (d) suggests that no current was flowing at all, but of course this is not true. So what does your answer mean, physically, about the circuit?

5.20 When a constant force acts on an object the "impulse" produced by that force is defined as the amount of force times the amount of time over which it acts. So if I push on a desk with a 12 N force for 10 s I impart $120\,\text{N} \cdot \text{s}$ of impulse to the desk. Suppose an electromagnet is slowly turned on so that it exerts a force on an iron filing that starts at zero and grows linearly with time until it is turned off at $t = t_f$. Find the total impulse exerted by the electromagnet on the iron filing. Your answer should contain an unknown constant of proportionality.

5.21 Figure 5.2 shows a dam of height H, acting as a wall on the right side of a body of water. The width W of the dam goes into the page. The pressure of the water at a depth D below the surface is $\rho g D$, where ρ is the density of water. The force exerted by the water pressure on a surface of the dam of area A is the water pressure multiplied by A over any region for which the water pressure is roughly constant. This force exerts a torque equal to the force times the height above the bottom of the dam. Find the total torque exerted by the water on the dam.

FIGURE 5.2

5.22 The "work" done by a constant force on an object equals the amount of force times the distance the object travels (assuming the movement is in the same direction as the force). An asteroid falling toward Earth experiences a gravitational force $-Gm_A m_E/r^2$ where G is a constant, m_A and m_E are the masses of the asteroid and the Earth (also constant), and r is the distance of the asteroid from the center of the Earth. The minus sign indicates that the force is toward the Earth.

(a) Find the work done by gravity on the asteroid as it falls from a height H above the Earth's surface to the surface of the Earth, at a distance R_E from the Earth's center.

(b) The work done by a force tells you how much kinetic energy that force gives to an object. A 1 km wide asteroid would have a mass of roughly 10^{12} kg. If such an asteroid fell from a height of 10,000 km above the Earth's surface, starting at rest, how much energy would it have when it reached the Earth's surface? (*Hint*: You will need to look up some information to answer this question.)

(c) The atomic bomb that was dropped on Hiroshima released about $6 \times 10^{13} J$ of energy. How many Hiroshima equivalents would such an asteroid be? (An actual asteroid impact would be much smaller than what you calculated because of air resistance, but the difference in energy would still be released into the atmosphere, causing some pretty catastrophic effects there as well.)

5.23 The "work" done by a constant force on an object equals the amount of force times the distance the object travels (assuming the movement is in the same direction as the force). An object's potential energy is the work that a force would do if that object moved from the origin (an arbitrary reference point) to its current location. For each part below, you'll be given a force acting on an object, the position of the object, and the position of the reference point. Find the potential energy of the object.

(a) A rock at height h above the ground experiences a constant gravitational force $-mg$. The reference point is the ground.

(b) A block on a spring experiences a force $-kx$ where k is a constant and x is the amount the spring is stretched. The reference point is the position where the spring isn't stretched at all, and the block is currently at a distance L from that position.

(c) An asteroid is currently at a distance R from the center of the Earth. The reference point is a point infinitely far away; in other words it's at the limit $r \to \infty$.

5.24 A short, straight wire of length ds carrying a current I produces the following magnetic field at a nearby point P.

$$d\vec{B} = \frac{\mu_0 I}{4\pi} \frac{d\vec{s} \times \vec{r}}{r^3}$$

Here $d\vec{s}$ is a vector of length ds that points in the direction of the current flow and $\vec{r}$ goes from the wire to P. We use r for the magnitude of $\vec{r}$. A straight wire goes from $(0, -H)$ to $(0, H)$ with a current in the positive y-direction.

(a) What is the direction of $d\vec{B}$ at the point $(L, 0)$? *Hint*: it's the same no matter where $d\vec{s}$ is on the wire.

(b) Set up an integral to find the magnetic field at $(L, 0)$.

(c) Evaluate your integral. You can do this with a computer or using the trig substitution $y = L \tan \theta$.

(d) The magnetic field produced by an *infinite* wire with a current I is $\mu_0 I/(2\pi L)$. Check that your answer reduces to this in the limit $H \to \infty$.

5.2.4 Explanation: Single Integrals in Multiple Dimensions

The process we outlined above can also be used to solve some problems in more than one dimension. The hardest part is choosing the correct shape for your "slices." For most problems, there is only one correct slicing; if you try it a different way, you will not be able to set up the integral. So as you work through the examples below, pay particular attention to why we chose the slice we did in each case.

A Disk with Density as a Function of ρ

A disk of radius R has density (mass per unit area) given by $\sigma = c\rho^2$, where ρ is the distance from the center of the disk. Find the mass of this disk.

For a two-dimensional region with constant density, the mass is $m = \sigma A$. Our strategy is therefore to break the disk into "slices" such that each slice has a constant density (more or less), write $dm = \sigma \, dA$ for each slice, and then sum the masses of all the slices. We choose as our slice a "ring," the region between two concentric circles. (More on this choice below.)

FIGURE 5.3 The large disk can be built from small rings like this.

The independent variable in all such problems is the variable that distinguishes one slice from another. Each ring in this disk is identified, not by an x- or y-value, but by its distance from the center of the disk: the polar coordinate ρ. The width of the ring is $d\rho$, a small change in ρ as we go from the inside of the ring to the outside.

To see that this is the correct slicing, you have to convince yourself of two things. First, the density of the ring is more or less constant, provided $d\rho$ is sufficiently small. (This is true, in this case, because the density is a function of ρ.) Second, if we add all the concentric rings from $\rho = 0$ at the center to $\rho = R$ at the edge, we can build up the entire disk. (We don't "miss" anything.)

Once you are convinced, the next step is to find the mass of our small ring by multiplying its density $(c\rho^2)$ times its area. What is the area of this ring? You might answer $\pi(\rho + d\rho)^2 - \pi\rho^2$, and that is technically correct, but we can find a useful approximation to this if we rely on the fact that $d\rho$ is small. Imagine that the ring is made of rubber; we cut the rubber at the bottom, and bend the remaining shape up into a rectangle. The length of this rectangle is

$2\pi\rho$, and its width is $d\rho$. We therefore write that the area of our ring is $dA = 2\pi\rho\,d\rho$ and its mass is $dm = (2\pi\rho\,d\rho)c\rho^2$.

To find the mass of the entire disk, we sum up all the pieces by integrating:

$$M = \int_0^R 2\pi c\rho^3\,d\rho = \frac{1}{2}\pi cR^4$$

We would like to draw your attention to three key features of this process.

- We could have used the same strategy for any density $\sigma(\rho)$, provided that the density was solely a function of ρ.
- As we stressed in the previous section, $d\rho$ entered the calculation as a part of a multiplication: it was not "tacked on" at the end.
- Finally, the ring that we used as a slice cannot actually be bent up into a rectangle, since the outer edge is longer than the inner edge. This is another approximation—like the constant density of the ring—that gets better the smaller we make $d\rho$. If we computed the mass of this disk by using five rings, we would get a reasonable approximation; if we used ten rings, our approximation would be better. As the number of rings approaches infinity, the approximation approaches the mass of the disk. And that is exactly what the integral does for us: it adds up all the slices in the limit as $d\rho \to 0$. If you're not comfortable with this argument, you will derive this approximation in a different way in Problem 5.26.

Finally, we check our answer. To check units notice that the density $c\rho^2$ has units of mass per distance squared, so c has units of mass per distance to the fourth, and thus the final answer has correct units for mass. You can also notice that the largest density is on the outer rim with $\sigma = cR^2$. If the entire disk had that density the total mass would be $(cR^2)(\pi R^2) = \pi cR^4$. The actual answer should be smaller than that, which it is.

A Disk With Density as a Function of ϕ

Next we consider a disk with density given by $\sigma = c\sin^2\phi$ where ϕ is the angle off the positive x-axis. What is the mass of this disk?

The slice we used in the previous example does not work in this case. No matter how small we make $d\rho$, the density of the ring does not approach a constant: it goes from 0 all the way to c. The fundamental strategy of "slices of constant density" is still the right approach, but this time we need a slice that looks more like a slice of pie.

Once again, we define the variable that distinguishes one slice from another. In this case it is ϕ, the angle off the positive x-axis. Therefore, $d\phi$ is the angle subtended by the slice: the change in angle from its bottom to its top.

Because σ is a function of ϕ, the density of such a slice is more or less constant, provided $d\phi$ is sufficiently small. If we add up the masses of slices going all the way around the circle, we can build up the mass of the entire disk.

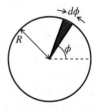

The area of such a slice is the area of the entire circle, multiplied by the *fraction* of the circle represented by our slice. (For instance, a 7° angle represents 7/360 of a full circle.) So we write $dA = (\pi R^2) \times (d\phi/2\pi) = (R^2/2)d\phi$ and $dm = (R^2/2)c\sin^2\phi\,d\phi$. To find the mass of the entire disk, we sum the masses of the individual slices by integrating:

$$M = \int_0^{2\pi} c\frac{R^2}{2}\sin^2\phi\,d\phi = \frac{c\pi R^2}{2}$$

(You can easily look up the integral of $\sin^2\phi$ or you can find it using trig identities.) Since angles measured in radians are unitless, the formula for σ implies that c has units of density, so c times R^2 has correct units for mass. Once again you can check that this is a fraction of what the mass would be if the highest density on the disk, c, were the density throughout.

It may occur to you to wonder: how would we find the mass of such a disk if its density were a function of both ρ and ϕ? Although the basic strategy is the same, the steps are a bit more involved and lead to a "double integral," a topic we will take up in the next section.

Moment of Inertia of a Disk

Find the moment of inertia of a uniform disk of mass M and radius R about an axis through its center.

Recall from the previous section that the moment of inertia of a point-sized object is mr^2. We need a slice that has a well-defined r, which leads us once again to rings where r, the distance from the axis, is the polar ρ.

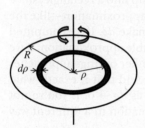

The mass of such a ring is the total mass of the disk (M) multiplied by the fraction of the disk that the ring represents. Since we calculated above that the area of a ring is $2\pi\rho\, d\rho$, our ring has mass $dm = M(2\pi\rho\, d\rho)/(\pi R^2)$. The total moment of inertia is therefore:

$$I = \int_0^R \frac{2M}{R^2} \rho^3 d\rho = \frac{1}{2} MR^2$$

This has correct units for a moment of inertia, and once again it's easy to look up that this is the correct formula.

Center of Mass of a Cone

We'll end this section with a 3D example. A uniform right circular cone of height H makes a 45° angle from the horizontal. Find its center of mass.

In two dimensions, finding the center of mass of an object is generally two separate problems: finding its x-component, and finding its y-component. In three dimensions, the center of mass is three separate problems. In this case, however, the x- and y-components are obvious from symmetry, so we need worry only about the z-component. As we saw in the previous section, the formula is:

$$z_{COM} = \frac{z_1 m_1 + z_2 m_2 + z_3 m_3 + \ldots}{M}$$

For our cone, the slice with a roughly constant z-coordinate is a very short cylinder.

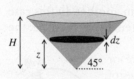

The variable that distinguishes one slice from the next is the height z. So the volume of our cylindrical slice is $\pi r^2\, dz$ where r is the radius of the cylinder, but we need to rewrite this with z as the only variable. The 45° angle means that $r = z$. (Make sure you can see that. For a different angle, we would use trig to get the relationship between z and r.) Therefore, the volume of the slice is $\pi z^2 dz$. The mass of the slice is the total mass of the cone divided by the fraction of the cone's volume taken up by the slice. You'll calculate in Problem 5.39 that the volume of the entire cone is $(1/3)\pi H^3$, so the mass of the slice is $dm = M(\pi z^2\, dz)/\left[(1/3)\pi H^3\right]$. To plug this into the formula for center of mass we multiply the mass dm by the position z, integrate over all the slices, and then divide by M.

$$z_{COM} = \frac{1}{M} \int_0^H \frac{3M}{H^3} z^3 dz = \frac{3}{4} H$$

Notice that this has units of height. Also, just looking at the cone the center of mass had to be somewhere between $H/2$ and H.

5.2.5 Problems: Single Integrals in Multiple Dimensions

5.25 Walk-Through: Single Integral in Multiple Dimensions. A disk with radius R has a density given by $\sigma = a/(\rho^2 + b^2)$ where ρ is the distance from the center.

(a) Draw the disk. Within your drawing, draw a small ring at radius ρ with width $d\rho$. Explain why such a slice can be assumed to have a roughly constant density for this particular density function.

(b) Explain why the area of your small ring can be approximated as $2\pi\rho\, d\rho$ for small $d\rho$.

(c) Calculate the mass dm of the small ring. Your answer will not involve an integral.

(d) Now use an integral to add up the masses of all the concentric rings to find the exact mass of the ring.

(e) Use the equation $\sigma = a/(\rho^2 + b^2)$ to determine the units of the constants a and b. Then use these units to confirm that your final answer has the units of mass.

5.26 In the Explanation (Section 5.2.4) we needed the area of a thin ring of inner radius ρ and thickness $d\rho$ (Figure 5.3). We estimated the area as $2\pi\rho\, d\rho$ by imagining that the ring could be stretched up into a rectangle.

(a) Explain why such a ring, no matter how thin, cannot actually be stretched up into a rectangle.

(b) Calculate the exact area of such a ring as a function of ρ (the distance of its inner circle from the center) and $d\rho$ (the width of the ring).

(c) Simplify your formula as much as possible. Then use your formula to explain why $2\pi\rho\, d\rho$ is a good approximation for small values of $d\rho$.

5.27 Given that the circumference of a circle is $2\pi R$, prove that the area of a circle is πR^2. (In doing so, you show that the values of π in these two formulas are exactly the same number—a result that the ancient Greeks suspected, but could never quite prove.)

5.28 A disk with radius R has a density given by $\sigma = c\sin(\phi/2)$ where c is a constant and ϕ is the angle from the positive x-axis. Find the mass of this disk.

5.29 A disk with radius 3, centered on the origin, has a density given by $\sigma = |y|$.

(a) Find the mass of the top half of the disk.
(b) Find the mass of the entire disk.

5.30 The drawing shows a washer with outer radius r_2 and inner radius r_1.

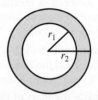

(a) Calculate its mass if its density is given by $\sigma = a\sqrt{\rho^2 + b^2}$ where ρ is the distance to the center.

(b) Calculate its mass if its density is given by $\sigma = c\phi$ where ϕ is the angle off the positive x-axis.

5.31 A triangle is bounded by the x-axis, the vertical line $x = 2$, and the line $y = 2x$. The density (mass per unit area) is ky^2, where k is a constant. You're going to find the height of the center of mass.

(a) Draw the triangle.

(b) To set up an integral, you're going to need to make a thin slice through your triangle, but first you need to figure out which way to slice it. In order to calculate y_{COM} what will you need to calculate about your slice once you draw it?

(c) Using your answer to Part (b), explain whether your slice should be horizontal or vertical and why.

(d) Draw the slice and set up and evaluate the integral to find y_{COM}.

5.32 A flat piece of wood of uniform density is cut into the shape of the trapezoid shown below.

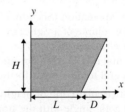

(a) Without doing any calculations, predict whether x_{COM} will be greater than or less than $(L + D)/2$ and explain how you know.

(b) Make a similar prediction for y_{COM}.

(c) Find the center of mass of the trapezoid (both x_{COM} and y_{COM}). Make sure your answers match your predictions.

5.33 A uniform flat piece of metal of mass density σ in the xy-plane has boundaries $y = 0$, $x = L$, and $y = cx^2$ (where L and c are positive constants). Find its center of mass (both components).

5.34 Find the moment of inertia of a uniform square of mass M and edge length L about one of its edges.

5.35 A square with side length a has density $\sigma = kx$ where x is the distance from the left side and k is a positive constant.

(a) Calculate the mass of the square.

(b) Calculate the center of mass of the square.

(c) Calculate the moment of inertia of the square about its left edge.

(d) Calculate the moment of inertia of the square about a vertical line drawn a distance b from the left edge.

(e) Show that the moment of inertia is smallest around an axis drawn at the center of mass.

5.36 Given that the surface area of a sphere is $4\pi R^2$, show that the volume of a sphere is $(4/3)\pi R^3$. *Hint*: build the sphere from concentric spherical "shells."

5.37 Given that the surface area of a sphere is $4\pi R^2$, find the mass of a sphere of radius R whose density is given by $\rho = \rho_0 e^{-(r/R)^3}$ where r is the distance from the center of the sphere. *Hint*: It may be helpful to work Problem 5.36 first.

5.38 (a) Find the electric potential at the origin produced by a uniform *hollow* sphere of radius R and charge Q centered on the origin.

(b) Find the electric potential at the origin produced by a uniform *solid* sphere of radius R and charge Q centered on the origin. *Hint*: It may be helpful to work Problem 5.36 first.

(c) Which answer came out bigger? Why?

5.39 In the Explanation (Section 5.2.4) we asserted that the volume of a right circular cone of height H that makes a 45° angle from the horizontal is $(1/3)\pi H^3$. Set up and evaluate an integral to derive this formula.

5.40 A right circular cone of height H that makes a 45° angle with the horizontal has a density $a/(H^3 + z^3)$ where z is the distance from the ground.

(a) Find the mass of the cone.

(b) 🖥 Find the height of the center of mass of the cone.

5.41 Calculate the volume of a right circular cone of height H that makes a 30° angle with the horizontal.

5.42 Calculate the volume of a right circular cone of height H and upper radius R.

5.43 A very simple model of the atmosphere (ignoring differences in temperature and humidity) says that the density of air falls off exponentially with altitude h. The density at sea level ($h = 0$) is approximately 1.2 kg/m³ and it falls to half that value at an altitude of roughly $h = 6000$ m.

(a) Write a function for the density of air as a function of altitude, using the values given in the problem.

(b) Using this simplified model, find the total amount of air in a one square meter column extending from sea level up infinitely high.

(c) The temperature of the air, like the density, falls off roughly exponentially with altitude. Explain why it does *not* make sense to integrate the temperature with respect to altitude.

5.44 *[This problem depends on Problem 5.43.]* Consider the Earth as a sphere, and the atmosphere to be a spherical shell around the Earth. In this problem you will assume that the atmospheric density follows the same model as in Problem 5.43, but you will have to rewrite that model as a function of r (distance from the center of the earth) instead of h (height above sea level). Use R_0 to represent the radius of the Earth (so $r = R_0$ when $h = 0$).

(a) Write h as a function of r, and then use this to write the density of the atmosphere as a function of r instead of h.

(b) Calculate the total amount of air, where the atmosphere starts at $r = R_0$ and continues up to infinity.

5.45 A four-sided pyramid has height H and four bottom edges of length L.

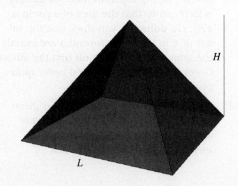

(a) Draw a pyramid with a horizontal slice through it at a distance y from the base. What shape is your slice?

(b) What is the volume of your slice in terms of its width w and its thickness dy?

(c) If you look at the pyramid directly facing one of the sides, it looks like an isosceles triangle with base L and height H. Draw that triangle and show your slice. Use that drawing to find the width w in terms of y, L, and H.

(d) Use an integral to add up the volumes of all the slices and find the total volume of the pyramid.

5.46 The Great Pyramid of Cheops is a four-sided pyramid (see the picture in Problem 5.45) with bottom edges $L = 230$ m long and height $H = 147$ m. Its volume is $(1/3)L^2 H$. Find its center of mass, assuming uniform density.

5.47 Find the moment of inertia of a uniform cylinder of mass M, height H and radius R about an axis drawn parallel to the height, through the center.

For Problems 5.48–5.50 find the center of mass of the given object, and find its moment of inertia about the z-axis. For the center of mass you need to specify all of its coordinates, but you can often identify some (or all) of them by symmetry without any calculations. For example, a uniform cone centered on the z-axis has $x_{COM} = y_{COM} = 0$, but z_{COM} requires an integral.

5.48 A uniform half-disk of mass M in the xy-plane defined by $x^2 + y^2 \leq R^2$, $y \geq 0$.

5.49 A disk in the xz-plane defined by $x^2 + z^2 \leq R^2$, with mass density $\sigma = kx^2$

5.50 A uniform solid sphere of mass M and radius R centered on the origin.

5.51 A uniform disk of radius R centered on the origin in the xy-plane has total charge Q. In this problem you will find the electric potential produced by this disk at the point $(0, 0, h)$.

(a) Draw the disk with a ring-shaped slice at a distance ρ from the center.

(b) Find the distance of this ring from the point $(0, 0, h)$. (The answer will be essentially the same for all points on the ring.)

(c) Find the potential produced by this ring at the point $(0, 0, h)$.

(d) Integrate to find the total potential at $(0, 0, h)$ produced by the disk.

(e) As a reality check, find $\lim_{R \to 0} V$ by applying l'Hôpital's rule to your answer. Explain why the result you got makes sense.

5.52 In this problem you are going to calculate the volume of a sphere, starting only with the knowledge that the area of a circle is πr^2. The drawing shows a sphere of radius R and a small "slice"—a disk at height y from the center of the sphere, with width dy. The radius r of the disk is shown: don't confuse it with the radius R of the sphere!

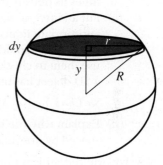

(a) A cylinder of radius r and height h has volume $\pi r^2 h$. Use this fact to find the volume dV of the disk shown, as a function of r and dy.

(b) Based on the drawing, rewrite your function dV so that it depends on the variables y and dy and the constant R, but not the variable r.

(c) Integrate to add up the volumes of all such disks. (Remember that y is the variable and R is a constant!) You should conclude at the end that the volume of a sphere is $V = (4/3)\pi R^3$.

5.53 Find the height of the center of mass of a hemisphere of radius R and constant density σ. (If you haven't done Problem 5.52, it may help to work through that one first.)

5.3 Cartesian Double Integrals over a Rectangular Region

This section begins the main focus of the chapter, which is the application of integrals to multivariate functions.

5.3.1 Discovery Exercise: Cartesian Double Integrals over a Rectangular Region

1. A horizontal plank with height H and width W has a density given by $\sigma = kx$, where x is the distance from the left side of the strip and k is a constant. You want to calculate the mass of the plank.
   (a) You begin by drawing a thin vertical strip on your box, as shown below. What is the area dA of this thin strip?

   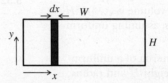

   (b) This strip is at position x in the horizontal direction. Multiply the area of the strip times its density to find the mass dm of this thin strip.
   (c) Put an integral sign in front of the expression you just wrote for dm and fill in the appropriate limits of integration. Evaluate this integral to find the mass M of the plank. You should be able to find the units of k from the expression $\sigma = kx$ and use them to check that your answer has units of mass. (Remember that density for a 2D object has units of mass per area.)

   See Check Yourself #27 in Appendix L

   (d) Explain why the procedure you just followed would not have worked if you had started by drawing a thin horizontal strip instead of a vertical one.

Now you will redo this problem assuming the density is given by $\sigma = qxy$. (You should figure out the units of the constant q so you will be able to check the units on your final answer.) Once again you begin by drawing a thin strip as shown in the picture above. Its area dA is the same as what you calculated above.

2. Explain why you cannot find the mass of the strip dm just by multiplying density times area as you did above.
3. Supposing our thin strip were drawn at $x = 1$, set up and evaluate an integral with respect to y for the mass of the strip. Your integral will involve dx and dy and q, but after you integrate with respect to y your final answer will be a function of dx and q.
4. Now supposing our thin strip were drawn at $x = 2$, set up and evaluate an integral with respect to y for the mass of the strip.
5. Now generalize: our thin strip is drawn at some fixed x-value. Set up and evaluate an integral with respect to y for the mass of the strip. Your final answer will be a function of x, dx, and q.
6. Put an integral sign in front of the expression you just wrote for dm and fill in the appropriate limits of integration. Evaluate this integral to find the mass M of the plank (and check that your answer has correct units).

See Check Yourself #28 in Appendix L

7. In Part 1, just as in many problems in the previous sections, you calculated the mass of a two-dimensional object using only one integral. What was it about the second problem that made it require two integrals?

5.3.2 Explanation: Cartesian Double Integrals over a Rectangular Region

In the previous section we showed you how to set up an integral for a quantity Q in some region by breaking up the region into thin strips, calculating the amount dQ in that thin strip, and then integrating to add up the contributions of all the strips. For example, suppose a team of exterminators has carefully measured the rat population on Farmer Brown's land and found that the density of rats (rats per unit area) falls off linearly as you move north, away from the barn where the food is stored: $\sigma = k(H - y)$. To find the total rat population on the farm you would draw a thin, horizontal strip at position y, as shown below.

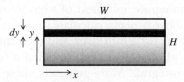

The area of this strip is $dA = W\,dy$. The number of rats in this strip is the density of rats times the area, or $dR = [k(H - y)][W\,dy]$. To find the total number of rats we put an integral sign in front of dR. Since the integration variable is y, the limits are 0 and H, so

$$R = \int dR = kW \int_0^H (H - y)\,dy = \frac{1}{2}kWH^2$$

As a quick check we can note that $\sigma = k(H - y)$ has units of rats per distance squared, so k must have units of rats per distance cubed, which means the final answer has correct units for rats.

We've revisited this type of problem to point out that this process only worked because of a special property of the density function: σ did not depend on x. To show why this matters, let's redo the problem assuming that there are food sources to the south and west, so the rat population has density $\sigma = q(H - y)(W - x)$. We begin the problem by again drawing the thin strip in the picture above with area $dA = W\,dy$. The next step is the one where we multiplied the area by the density to find the number of rats in this strip. Don't read on until you clearly see why we can't do that with our new density function.

Did you see it? No matter how thin you make dy, the density of our slice isn't constant; it's high on the left and low on the right. Rats!

So if we can't multiply "the density of the strip" (which doesn't exist) by its area (which does), how do we count rats in the strip? Well, that's exactly the sort of problem we solved in the last section: we break this horizontal strip up into tiny little boxes, find the number of rats in each box, and then integrate that to find the number of rats in the entire horizontal strip. Then we'll integrate *that* to find the whole rat population of the farm.

The area of the little box is $dA_{box} = dx\,dy$. Because the box is small in *both* directions we consider it to have just one value of x and one value of y, so the number of rats in the box is density times area: $dR_{box} = q(H - y)(W - x)\,dx\,dy$. The number of rats in the strip is the integral of dR_{box} as x goes from 0 to W.

$$dR_{strip} = \int_{x=0}^{x=W} q(H - y)(W - x)\,dx\,dy$$

Finally, the rat population on the farm is the integral of dR_{strip} as y goes from 0 to H.

$$R = \int_{y=0}^{y=H} \left(\int_{x=0}^{x=W} q(H-y)(W-x)dx \right) dy$$

We evaluate this from the inside out.

$$R = \int_{y=0}^{y=H} \left(q(H-y) \left[Wx - \frac{1}{2}x^2 \right]_{x=0}^{x=W} \right) dy = \int_{y=0}^{y=H} \left(\frac{1}{2}qW^2(H-y) \right) dy$$

$$= \frac{1}{2}qW^2 \left[Hy - \frac{1}{2}y^2 \right]_{y=0}^{y=H} = \frac{1}{4}qW^2H^2$$

(We'll leave it to you to check the units on this answer. Remember to start by figuring out the units of q!)

We are going to proceed to more complicated situations where the function to integrate or the limits of integration are not obvious, and we're going to use different coordinate systems for some of our integrals, but most of what we do in the rest of this chapter will be some variation on the above process: calculate some quantity dQ for a little box, integrate it in one direction to find dQ for a thin strip, and then integrate *that integral* in some other direction to find Q for a whole shape. This is called a "double integral." For a 3D shape you end up with a "triple integral." More generally, these are called "multiple integrals" or "iterated integrals."

You should be aware, though, that some of the ways we wrote things in this example were for pedagogical purposes and are not standard. Usually people write dR where we wrote dR_{box}, and don't write dR_{strip} at all. They *do* draw a picture with the box and strip and with x, y, dx, and dy clearly labeled, and you should make that a habit.

Finally, people don't generally write $x =$ or $y =$ in the limits of integration, and they don't put big parentheses around the inner integral. So the integral we wrote above would commonly be written as:

$$R = \int_0^H \int_0^W q(H-y)(W-x)\, dx\, dy \qquad (5.3.1)$$

Equation 5.3.1 is a set of instructions that we read from the inside out. First integrate the function as x goes from 0 to W, treating x as the only variable and y as a constant. This will give you a function of y only, which you will then integrate from 0 to H. (Physicists often avoid the need for this rule by writing each differential next to its integral, as in $R = \int_0^H dy \int_0^W dx\, q(H-y)(W-x)$. This notation tends to annoy mathematicians.)

You may be wondering whether we could have drawn a vertical strip and integrated dR_{box} with respect to y first and then integrated the result of that with respect to x. (If you weren't before, we hope you're wondering now.) Yes, we could have, and we would have gotten the same answer in the end. "Fubini's Theorem" tells us that the order of integration can be reversed without changing the final result, provided the function being integrated is "absolutely convergent" in the region of integration. (That means that the integral converges even if you replace the integrand with its absolute value.) For the vast majority of cases choosing the order of integration is a matter of convenience. As the following example shows, sometimes one order is noticeably more convenient than the other.

EXAMPLE **Double Integral**

Question: A plastic rectangle with corners at $(0,0)$ and $(3,5)$ has an electric charge with the charge density (charge per unit area) $\sigma = x^3/(1+x^2y)^4$. Calculate the total charge on the rectangle.

Answer: We can begin this problem by slicing the rectangle into horizontal strips as shown.

Beginning students often think such a diagram means "integrate y first" (after all, the strip has a height dy), but it means the opposite: first we integrate dx to add up the mass *along* the slice, and then dy to add all the slices.

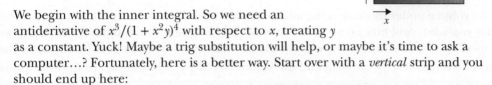

$$Q = \int_0^5 \int_0^3 \frac{x^3}{(1+x^2 y)^4} \, dx \, dy \quad \text{(correct, but hard)}$$

We begin with the inner integral. So we need an antiderivative of $x^3/(1+x^2 y)^4$ with respect to x, treating y as a constant. Yuck! Maybe a trig substitution will help, or maybe it's time to ask a computer...? Fortunately, here is a better way. Start over with a *vertical* strip and you should end up here:

$$Q = \int_0^3 \int_0^5 \frac{x^3}{(1+x^2 y)^4} \, dy \, dx \quad \text{(also correct, and easier)}$$

Start again with the inner integral, now treating x as a constant. This time, a simple u-substitution does the trick:

$$\int_0^5 \frac{x^3}{(1+x^2 y)^4} \, dy = \left. \frac{-x}{3(1+x^2 y)^3} \right|_0^5 = \frac{-x}{3(1+5x^2)^3} - \frac{-x}{3}$$

The outer integral is also not too hard.

$$\int_0^3 \left(\frac{-x}{3(1+5x^2)^3} + \frac{x}{3} \right) dx = \left[\frac{1}{60(1+5x^2)^2} + \frac{x^2}{6} \right]_0^3 = 1.483$$

Notice that the units seem to have disappeared entirely, because many of the units were implicit in the original problem. The upper limit for x was given as 3, which had to be shorthand for 3 meters or inches or some other unit of distance, but that unit was hidden in the calculations. The constant σ also appeared to have wrong units, which means it must really be ae^{bxy} where $a = b = 1$ in whatever units the problem was using. If you start with unitless numbers, you can't check units at the end.

Volume Under a Surface

When you first learned about integrals you learned that the integral of a function is equal to the area under the curve. Hopefully you now think of an integral as much more than that, but area can still serve as a useful visualization of how an integral adds infinitely many small contributions to give one finite result. So we want to conclude this section by showing you how, in a similar way, a double integral can be interpreted as the volume under a surface.

Consider, then, a surface defined by the function $z(x, y)$. We want to find the volume below this surface above a finite rectangle $x \in [a, b]$, $y \in [c, d]$.

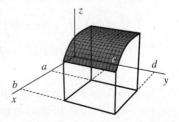

Our strategy, as always, is to divide the problem into infinitesimal "slices." The picture below shows a strip at a fixed y-value, extended along the x-direction, with a small thickness dy. (We could just as easily have chosen the opposite.)

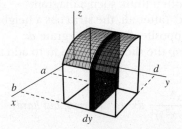

The volume under this slice is the thickness (dy) times the two-dimensional area (parallel to the xz-plane). And how do we find that area? We integrate! This is the sort of integral we are used to: y is constant, so we just want the area under a curve as x goes from a to b. The area is therefore $A = \int_a^b z(x,y)dx$, and the volume $dV = \left(\int_a^b z(x,y)dx\right)dy$. To find the entire volume, we add up the differential volumes of all such slices; that is, we integrate as y goes from c to d.

$$V = \int_c^d \int_a^b z(x,y)dx\, dy$$

You should be able to see how this process maps onto the double integral examples we worked above. For example, in the rat problem x and y represented distances and the function $z(x,y)$ represented rats per unit area, so the "volume" under each slice was the number of rats in that region. Adding all of those up gave us the total "volume," which was the number of rats on the farm.

Separable Integrals

A "separable" function of x and y is one that can be written as the product of two functions: one function of x only, and one of y only. If you are taking a double integral of a separable function and the limits of integration are all constants, you can evaluate each integral independently, and then multiply the answers. (The phrase "the limits of integration are constants" may not seem restrictive right now, but in later sections you'll see problems where they aren't. In such cases you cannot separate the integral even if the integrand is separable.)

EXAMPLE **Separable Double Integral**

Question: Find the volume under the function $z = e^x \cos y$ over the region $0 \leq x \leq 2$, $\pi/2 \leq y \leq \pi$.

Answer: Following the previous example, we could write $\int_0^2 \int_{\pi/2}^{\pi} e^x \cos y\, dy\, dx$ and evaluate the y integral first; or, we could write $\int_{\pi/2}^{\pi} \int_0^2 e^x \cos y\, dx\, dy$ and evaluate the x integral first.

Instead, we choose to separate the function:

$$\left(\int_0^2 e^x dx\right)\left(\int_{\pi/2}^{\pi} \cos y\, dy\right)$$

Evaluating separately and multiplying leads to the answer $1 - e^2$.

In Problem 5.79 you will prove that separating an integral in this way leads to the same answer as the "inside-out" approach.

5.3.3 Problems: Cartesian Double Integrals over a Rectangular Region

5.54 Walk-Through: Cartesian Double Integral over a Rectangular Region. Find the mass of a rectangular object that extends from $(0, 0)$ to (W, H) with density $\sigma = cxe^{-y/H}$.

(a) From what you are given in this problem, what are the units of c? (The answer should refer to types of units, not specific ones. For example, the units of W are "distance," not "meters" or "inches." It would take on specific distance units if you gave it a numerical value.)

(b) Draw this rectangle. Inside it draw a tiny box at some typical position (which in this case just means anywhere that isn't right on an edge). Label the distance from the tiny box to the y-axis as x, and similarly label y. Also label the thickness of the box in both directions as dx and dy.

(c) What is the area dA of the tiny box?

(d) What is the mass dm of the tiny box?

(e) Draw a horizontal strip that extends your tiny box across the entire rectangle. Write (but don't yet evaluate) a single integral with respect to x to calculate the mass of that strip. Include the limits of integration. The quantity you're integrating should have both dx and dy in it, but so far you're only integrating it with respect to x.

(f) Now write an integral with respect to y *of the integral you just wrote* to calculate the mass of the entire rectangle. This second integral adds up the masses of all the horizontal strips to give you the total mass.

(g) Evaluate the inner integral (the one with respect to x), treating y as a constant. This should leave you with a formula for the total mass that is a single integral with respect to y.

(h) Evaluate that integral to find the total mass.

(i) Using the units you found for c, check the units on your answer.

For Problems 5.55–5.59 indicate whether the integral as written represents summing up over vertical or horizontal slices, and evaluate the double integral.

5.55 $\int_0^1 \int_0^1 x^2 y^3 \, dx \, dy$

5.56 $\int_0^1 \int_0^1 x^2 y^2 \, dy \, dx$

5.57 $\int_0^W \int_0^H \alpha e^{k(x+y)} \, dx \, dy$

5.58 $\int_1^e \int_0^1 x/\left(x^2 + y\right)^2 \, dx \, dy$

5.59 $\int_{-L}^L \int_{-H}^H ay^2 \sin(bx) \cos\left(cy^3\right) \, dy \, dx$

5.60 The region S is symmetric about the y-axis.

We're not giving you an equation for the shape, so you have to approach this question visually. Indicate whether each integral will come out positive, negative, or zero.

(a) $\iint_S x \, dA$

(b) $\iint_S x^2 \, dA$

(c) $\iint_S y \, dA$

(d) $\iint_S y^2 \, dA$

(e) $\iint_S (x - y) \, dA$

5.61 If the region $a \leq x \leq b$, $c \leq y \leq d$ is filled with a material of density $\sigma(x, y)$, you can use a double integral to find the total amount of the material.

(a) One way to evaluate this is as $\int_a^b \left(\int_c^d \sigma(x, y) dy \right) dx$. Draw a picture showing what the inner integral (in parentheses) represents and explain in words what the outer integral represents.

(b) Another way to evaluate this is as $\int_c^d \left(\int_a^b \sigma(x, y) dx \right) dy$. Draw a picture showing what the inner integral (in parentheses) represents and explain in words what the outer integral represents.

5.62 In this problem, you will take three different approaches to find the volume under the surface $z = e^{3x+y}$ over the rectangle $-2 \leq x \leq 2$, $-1 \leq y \leq 1$.

(a) Begin by writing $\int_{-2}^2 \int_{-1}^1 e^{3x+y} \, dy \, dx$. Evaluate the inside integral, $\int_{-1}^1 e^{3x+y} dy$. Find an antiderivative with respect to y, and then plug the limits of integration in for y, treating x as a constant throughout. At the end you should have a simple function of x.

(b) Finish the solution you began in Part (a) by taking the outside integral with respect to x.

(c) Now approach the same problem as $\int_{-1}^{1} \int_{-2}^{2} e^{3x+y} \, dx \, dy$, evaluating the inside (x) integral first, and then the outside. The individual steps will look different, but the final result should be the same that you found in Part (b).

(d) Finally, approach the same problem by separating it into $\left(\int_{-2}^{2} e^{3x} dx\right) \left(\int_{-1}^{1} e^{y} dy\right)$. Evaluate both integrals and multiply the answers, and confirm once again that you get the same answer you got in Part (b).

5.63 A square with corners at $(0, 0)$ and $(2, 2)$ has a charge density given by $\sigma = \sin(\pi x/2 + \pi y/4)$.

(a) Draw the square, and draw a thin vertical strip at $x = 1$ with thickness dx. Find the total charge on this strip. Note that your answer will be a function of dx, but not of x or y.

(b) Draw the square again, and draw a thin horizontal strip at $y = 1$ with thickness dy. Find the total charge on this strip.

(c) Find the total charge on the square.

5.64 Find the volume under the surface $z = e^{y/x}$ on $x \in [1, 3]$, $y \in [0, 4]$.

(a) Write (but do not evaluate) the double integral such that the integral with respect to x should be taken first.

(b) Write (but do not evaluate) the double integral such that the integral with respect to y should be taken first.

(c) One of the two integrals can easily be evaluated by hand; the other cannot. Based on this, choose whether to use the form from Part (a) or Part (b).

(d) Evaluate the first (inner) integral. The result should be a single integral that is impossible to evaluate by hand.

(e) ![computer] Evaluate the second (outer) integral using a computer or calculator.

5.65 A rectangular region extending from $(0, 0)$ to $(7, 10)$ is filled with a charge density (charge per unit area) $\sigma = kye^{xy}$ where k is a positive constant. Your job is to find the total charge.

(a) Write the double integral with the x integral first and evaluate it.

(b) ![computer] Write the double integral with the y integral first. Evaluating this is considerably harder; you can do the inner integral using integration by parts, but will still need a computer for the outer integral. Use one and verify that you get the same answer you got in Part (a).

5.66 Evaluate $\int_{1}^{\infty} \int_{1}^{\infty} e^{-x-2y} \, dx \, dy$

5.67 The Discovery Exercise (Section 5.3.1) had you find the mass of a plank of height H and width W with density $\sigma = kx$ by drawing a thin vertical strip and taking a single integral over those strips. Find the mass of that strip starting with a thin *horizontal* strip. Doing it this way will require a double integral.

5.68 A rectangular square of plastic that goes from $(0, 0)$ to (L, L) has an electric charge on it with the charge density (charge per unit area) $\sigma = kye^{qx}$. Find the total charge on the plastic.

5.69 Consider $\int_{0}^{2\pi} \int_{0}^{5} y^2 \sin x \, dy \, dx$.

(a) Evaluate the double integral.

(b) What does the answer tell you about the surface $z = y^2 \sin x$ on the specified domain?

5.70 The integral $\int_{c}^{d} \int_{a}^{b} k \, dx \, dy$ (where k is a constant) gives the volume of a rectangular solid.

(a) What is the dimension of this solid in the x-, y-, and z-directions?

(b) Evaluate the integral to show that it gives the correct volume for this rectangular solid.

5.71 Consider $\int_{-3}^{3} \int_{-5}^{5} x^2 e^y \, dx \, dy$.

(a) Evaluate the integral as written.

(b) Reevaluate the integral as x goes from 0 to 5 (instead of -5 to 5), and then double the result. Do you get the same answer you got in Part (a)?

(c) Reevaluate the original integral as y goes from 0 to 3 (instead of -3 to 3), and then double the result. Do you get the same answer you got in Part (a)?

(d) In general, under what circumstances is it valid to consider half the region and then double the result?

5.72 If $\int_{a}^{b} \int_{0}^{1} (3x + 2y) \, dx \, dy = 0$, find a function $a(b)$.

5.73 ![computer] Find the electric potential at the origin produced by a uniform square of charge Q with corners at $(0, 0)$ and (L, L).

5.74 ![computer] A square with corners at $(0, 0)$ and $(2, 2)$ has a charge density given by $\sigma = \sin(\pi x + 2\pi y)$, where all quantities are measured in SI units (in which $k = 9 \times 10^9$). Set up and numerically evaluate an integral for the electric potential at the origin.

5.75 A survey team has measured the density of gold (mass of gold per unit area of desert) in the Cartesian Desert. Defining an origin at one corner of the desert they found the gold density in the region $0 \leq x \leq L$,

$0 \leq y \leq H$ to be $k\sin(\alpha x)e^{-\beta y}$, where k and β are positive constants and $L \leq \pi/\alpha$.

(a) Find the total amount of gold in the Cartesian desert.

(b) Take $L = 100$ mi, $H = 150$ mi, $k = .7$ g/mi^2, $\alpha = .02$ mi^{-1}, and $\beta = .003$ mi^{-1}. How much gold is in the desert?

5.76 A two meter by two meter square of paper is colored such that the amount of ink per unit area is proportional to the distance from one of the corners. Write the density function and find the total amount of ink on the paper. (The answer will contain a constant of proportionality.)

5.77 A rectangle goes from $(0,0)$ to $(2,3)$ with density xy^2. Find the moment of inertia of that rectangle around the x-axis.

5.78 A rectangular holding tank is designed to store pollutants from a factory, but the pollutants don't mix evenly through the stagnant water. A team of workers has measured what fraction f of the liquid in the tank is made of pollutants in different spots, and their results can be modeled by the function $f = 1/(x + y + 2)$ where x and y are measured in meters. The tank covers the region $0 \leq x \leq 10$, $0 \leq y \leq 15$. Your boss has asked you to find what fraction of the entire tank is made of pollutant.

(a) As a warm-up, consider the easier problem of a tank divided into three parts: one of area A_1 made of 1/3 pollutant, one of area A_2 made of 1/2 pollutant, and one of area A_3 made of 3/4 pollutant. What fraction of *that* tank is made of pollutant? (If you're stuck, try making up some numbers for the total amount of liquid in the tank and the areas of each of the three parts. Once you can answer it with specific numbers, go back and do it with letters for those quantities instead.)

(b) Explain why you couldn't answer that question by adding the three fractions.

(c) Now, for the actual tank in this problem, break it into tiny boxes of size $dx\,dy$ and apply the same logic you applied to the simpler problem above.

(d) Pop your final answer into a computer to numerically approximate the fraction of the tank that is polluted. (As a sanity check, your final answer for the fraction should be between 0 and 1.)

5.79 In this problem you will show that it is valid to "separate" a separable integral. To prove this result in full generality, you will evaluate $\int_c^d \int_a^b f(x)g(y)\,dx\,dy$ but you will *not* separate the variables. Assume the existence of a function $F(x)$ such that $F'(x) = f(x)$, and a function $G(y)$ such that $G'(y) = g(y)$; these functions will figure in your answers.

(a) Evaluate the inside integral $\int_a^b f(x)g(y)dx$, remembering to treat $g(y)$ as a constant.

(b) Now integrate your answer from Part (a) as y goes from c to d.

(c) Confirm that your answer from Part (b) matches the answer you would get by separating variables.

5.4 Cartesian Double Integrals over a Non-Rectangular Region

When you integrate over a non-rectangular region, the overall technique and principles from the previous section still apply, but more care must be taken in finding the proper limits of integration.

5.4.1 Discovery Exercise: Cartesian Double Integrals over a Non-Rectangular Region

A half-disk of radius R above the x-axis has a density given by $\sigma = kyx^2$ where k is a constant. Your goal is to find the total mass.

1. The drawing below shows a small rectangle at a specific x- and y-value with width dx and height dy. Write the area dA and the mass dm of this small region as a function of x, y, dx, and dy.

2. In the next drawing this small rectangle has been extended into a vertical strip. Write a function for the *y*-value at the *top* of this strip as a function of *x*.

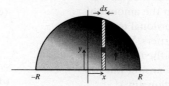

3. To compute the mass of the entire vertical strip, integrate your dm from Part 1 as y goes from the bottom ($y = 0$) to the top (the function you found in Part 2). The result will be a function of x and dx.
4. Integrate your answer from Part 3 as x goes from $-R$ to R to compute the mass of the entire half-disk. Check that your answer has correct units.

   See Check Yourself #29 in Appendix L

5.4.2 Explanation: Cartesian Double Integrals over a Non-Rectangular Region

The shaded triangle below has a density given by $\sigma = kxy$. What is the total mass?

Students often write $\int_0^{10} \int_0^{20} (kxy) \, dy \, dx$: an understandable mistake, since y clearly goes from 0 to 20 and x from 0 to 10. But those limits of integration would give the mass of an entire rectangle, not of our triangle. You can see this clearly if you draw several different "slices."

The two slices both have the same width dx, but different heights. The slice on the left has a height of 4, and therefore a mass $dm = \int_0^4 (2ky) \, dy \, dx$; for the slice on the right, $dm = \int_0^{12} (6ky) \, dy \, dx$.

Generalizing, a slice at an arbitrary x-position in the middle has a mass given by $dm = \int_0^{2x} (kxy) \, dy \, dx$. In this generic form we can add up the masses of all the slices, from the slice on the far left ($x = 0$) to the slice on the far right ($x = 10$).

$$m = \int_0^{10} \int_0^{2x} (kxy) \, dy \, dx \quad (5.4.1)$$

In general, if all your limits of integration are constants, you are describing a rectangular region. For other shapes, the inner limits of integration are functions. The outer (last) limits of integration are still constants, so the answer comes out as a constant.

Just as in the previous section, you can line up your strips along either axis. If we break this triangle into horizontal strips, each is bounded on the left by the line $y = 2x$, so the x-value there is $y/2$. Each strip is bounded on the right by $x = 10$, so the mass is:

$$m = \int_0^{20} \int_{y/2}^{10} (kxy) \, dx \, dy \tag{5.4.2}$$

For a finite problem, the two approaches will always produce the same answer as each other. (You may want to try your hand at the two integrals above; both should give you $5000k$.) However, there are now two different considerations when choosing which variable to integrate first. The integrand still matters, as it may be difficult or impossible to find an antiderivative with respect to one of the variables. But the shape of the region also plays into your decision, as the example below demonstrates.

EXAMPLE Double Integral over a Non-Rectangular Region

Question: Find the volume under the function $z = x + y$ over the region bounded by the curves $x = 4y - y^2$ and $x = 8y - 2y^2$.

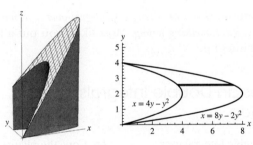

The image on the left shows the volume in 3D. The image on the right shows the shadow of this region in the xy-plane.

Solution:
The first step in any such problem is deciding which way to integrate first. We choose horizontal strips, as drawn above. (More on this decision below.) The next step is to integrate *along* each such strip, summing up the function as x goes from $4y - y^2$ on the left to $8y - 2y^2$ on the right. Then we add up all the strips to get the correct double integral.

$$\int_0^4 \int_{4y-y^2}^{8y-2y^2} (x + y) \, dx \, dy$$

Note the order of those limits; x is $4y - y^2$ on the left, $8y - 2y^2$ on the right. In general, the lower value should always be the first (lower) limit of integration, or the sign of your answer will be wrong.

To evaluate, we begin as always with the inner integral:

$$\int_{4y-y^2}^{8y-2y^2} (x+y) dx = \left. \frac{x^2}{2} + xy \right|_{4y-y^2}^{8y-2y^2} = \frac{(8y - 2y^2)^2}{2} + (8y - 2y^2)y - \left[\frac{(4y - y^2)^2}{2} + (4y - y^2)y \right]$$

The rest of the problem—simplifying all that, and integrating it as y goes from 0 to 4—we leave as an exercise for the reader. Of course you're not going to actually do it (who would?), but it is worth taking a moment to convince yourself that you could if you had to.

In the mean time, we turn our attention to one final question: what would have happened if we had approached this particular problem with vertical strips instead of horizontal ones?

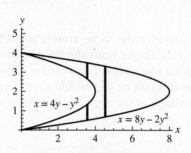

To find the limits of integration, you have to find the y-values at the top and bottom of each strip. This presents two problems. First, you have to solve the equations $x = 4y - y^2$ and $x = 8y - 2y^2$ for y, which is doable but unpleasant. Second, the three strips shown are all bounded by *different curves,* so you have to break the problem up into three different regions and evaluate the integrals separately. See Problems 5.98–5.99.

This example makes a strong case for beginning each problem with the question: "Which will be easier, horizontal slices (dx first) or vertical (dy first)?" A few seconds of thought up front can save you a lot of agony on the back end.

As a final word of caution: *if you are doing an integral over a region in the xy-plane and your answer has x or y in it you did something wrong.* Most likely you put a function as the limit of integration for your last integral.

5.4.3 Problems: Cartesian Double Integrals over a Non-Rectangular Region

5.80 Walk-Through: Cartesian Double Integral over a Non-Rectangular Region. Integrate the function $x + y$ over the region bounded by $y = x^2$, the x-axis, and the line $x = 3$. You will solve this problem two different ways. (In this particular problem, the two are of comparable difficulty.)

(a) Copy the picture above, and draw a vertical strip in the middle. The x-value where your strip is located should simply be labeled x (not a specific number); the width of the strip is of course dx.

(b) What is the y-value at the bottom of your strip? (The answer will be a number.) What is the y-value at the top of your strip? (The answer will be a function of x.)

(c) Set up an integral with respect to y to represent the integral along the strip.

(d) Now set up the outer integral to add up all these strips as x goes from a number at the left to a number at the right.

(e) Evaluate the double integral.

(f) Copy the picture again, and this time draw a horizontal strip in the middle, at an arbitrary y-value with height dy.

(g) What is the x-value at the left side of your strip? (The answer will be a function of y.) What is the x-value at the right side? You can check your answers by making sure that x goes from 0 to 3 at the bottom, and from 3 to 3 at the top.

(h) Set up an integral with respect to x to represent the integral along the strip.

(i) Now set up the outer integral to add up all these strips as y goes from a number at the bottom to a number at the top.

(j) Evaluate the double integral. You should get the same answer that you got in Part (e)!

In Problems 5.81–5.86, find the volume under the surface $z(x, y)$ in the specified region. In some cases, horizontal and vertical slices will be of comparable difficulty; in other cases, one approach will be much easier than the other.

5.81 $z = xy$ above the part of the disk $x^2 + y^2 \leq R^2$ that lies in the first quadrant.

5.82 $z = x^2 + y^2$ above the region between the x-axis, the y-axis, and the line $y = 3 - x$.

5.83 $z = 2x^5 y \sqrt{xy^2 + 1}$ above the region between the x-axis, the line $x = 2$, and the parabola $y = x^2$.

5.84 $z = x + y$ above the region between the lines $y = -1$, $y = 1$, and $x = 3$ and the right half of the circle $x^2 + y^2 = 1$.

5.85 $z = e^{x-y}$ above the region in the second quadrant (upper left) that's to the right of the line $y = -x$.

5.86 $z = x^2 - y^2 + 3$ above the region between the x-axis, the lines $x = 2$ and $x = -2$, and the parabola $y = x^2 + 1$.

In Problems 5.87–5.96, the given integral describes a particular region in the xy-plane. Draw that region, and then write an integral that expresses the same region with the x- and y-order reversed. (For instance, the region in Equation 5.4.1 can be reexpressed as Equation 5.4.2.)

5.87 $\int_3^7 \int_4^6 f(x, y) \, dx \, dy$

5.88 $\int_0^8 \int_0^{2y} f(x, y) \, dx \, dy$

5.89 $\int_0^8 \int_{2y}^{16} f(x, y) \, dx \, dy$

5.90 $\int_{-8}^8 \int_{2y}^{16} f(x, y) \, dx \, dy$

5.91 $\int_0^{\pi/2} \int_0^{\cos x} f(x, y) \, dy \, dx$

5.92 $\int_0^1 \int_{x^2}^{\sqrt{x}} f(x, y) \, dy \, dx$

5.93 $\int_0^5 \int_{\sqrt{25-x^2}}^{25-x^2} f(x, y) \, dy \, dx$

5.94 $\int_{-\sqrt{24}}^{\sqrt{24}} \int_{\sqrt{25-x^2}}^{25-x^2} f(x, y) \, dy \, dx$

5.95 $\int_0^3 \int_0^x f(x, y) \, dy \, dx + \int_3^{9/2} \int_0^{9-2x} f(x, y) \, dy \, dx$

5.96 $\int_0^3 \int_x^{9-2x} f(x, y) \, dy \, dx$

5.97 In this problem you'll find the area under the curve $y = x^3$ from $x = 0$ to $x = 2$ in two ways.

(a) Draw a small box in the middle of the area. Find the area of that small box and take a double integral of that area to find the area under the curve.

(b) Find the area in the usual way as a single integral.

5.98 The region bounded by the curves $x = -y^2$ and $y = x/2 + 12$ is filled with a uniform density σ.

(a) Sketch the two curves and the region between.

(b) Find the x- and y-coordinates of the points of intersection of the two curves. (These will be essential for finding the limits of integration.) Label them on your graph.

Your first job is to find the total mass of this region.

(c) The easiest (and therefore best) way to set up this problem is with horizontal slices. The lower limit on x is the same function of y for all slices, and the upper limit on x is the same function of y for all slices. Write the double integral this way and evaluate it to find the total mass.

(d) Vertical slices in this case require breaking the region up into two subregions. The left-hand region is bounded below by $y = -\sqrt{-x}$ and above by $y = x/2 + 12$; the right-hand region has different bounds. Write both integrals and evaluate them separately; then add them to find the total mass. You should of course get the same answer you got in Part (c)!

Your next job is to find the y-coordinate of the center of mass.

(e) First, without any calculations, make a guess based on your drawing. Do you expect y_{COM} to be positive or negative? What's another constraint you can make (either an upper or a lower bound)?

(f) Now set up and evaluate the proper double integral, using either horizontal or vertical slices. If your answer differs wildly from your guess, double-check.

5.99 The region bounded by the curves $y = (x-2)^2$ and $y = \sqrt{x}$ is filled with a charge density $\sigma = (\sin y)/x$. Find the total charge in this region.

(a) Sketch the two curves and the region between.

(b) Find the x- and y-coordinates of the points of intersection of the two curves. (These will be essential for finding the limits of integration.) One of these points can be found just by thinking about it, but you will need a computer or calculator to find the other.

(c) The easiest (and therefore best) way to set up this problem is with vertical slices. The lower limit on y is the same for all slices, and the upper limit on y is the same for all slices. Write the double integral this way and evaluate it numerically to find the total charge.

(d) Horizontal slices in this case require breaking the region up into two subregions. In the lower region, each slice is bounded on the left by $x = 2 - \sqrt{y}$ and on the right by $x = 2 + \sqrt{y}$. In the upper region, each slice is bounded on the left by $x = y^2$ and on the right by $x = 2 + \sqrt{y}$. Find the total charge of these two regions separately, and add them to find the total charge. You should of course get the same answer you found with vertical slices!

5.100 A plastic triangle is bounded by the x-axis, the line $x = 2$, and the line $y = 2x$. It has mass density xy^2 and charge density xy.

(a) Find the total mass of the triangle.

(b) Find the total charge of the triangle.

(c) Find the center of mass of the triangle.

(d) Find the moment of inertia of the triangle about the y-axis.

(e) Find the electric potential produced by the triangle at the origin.

5.101 A quarter-disk, radius R, sits in the first quadrant of the xy-plane with its corner at the origin. Its mass M and charge Q are uniformly distributed.

(a) Set up, but do not yet evaluate, an integral for the electric potential produced by this quarter-disk at the origin.

(b) Set up, but do not yet evaluate, an integral for the moment of inertia of this disk around the z-axis.

(c) Evaluate the integrals you wrote in Parts (a)–(b).

In Section 5.6 you'll learn a much easier way to solve integrals like this without fancy integration techniques or computers.

5.102 A uniform circle of plastic of radius R has total charge Q. Find the electric potential produced by the circle at a point on its edge.

5.103 Find the moment of inertia of a uniform half-disk of mass M and radius R about its flat edge.

5.5 Triple Integrals in Cartesian Coordinates

The ideas of the previous sections on double integrals can easily be extended to triple integrals. You write an expression for something you want to calculate in a tiny box (the mass, the charge, or whatever else), integrate in one direction to find the value of that quantity for a tiny strip, integrate in another direction to find that quantity for a thin slice, and finally integrate in the third direction to get your final answer for the whole 3D shape. If the region you're interested in is a rectangular solid then the limits of integration are constants. Other shapes can be more challenging, because you may have to picture the three-dimensional shape to correctly find the limits of integration.

5.5.1 Explanation: Triple Integrals in Cartesian Coordinates

We're going to find the volume of the shape shown below. The bottom and sides are the xy-, xz-, and yz-planes, and the top is the plane $x + y + z = L$. You could approach this with the techniques of Section 5.2: break the volume into thin slices, use geometry to find the

5.5 | Triple Integrals in Cartesian Coordinates

volume of each slice, and take a single integral to find the volume. Or, using the methods of Section 5.4, you could find the "volume under a surface" using a double integral over the triangular region at the bottom. We will approach it, however, by integrating the volume of a little box across all three dimensions. We've drawn a little blue box $dV = dx\, dy\, dz$ below. We've also drawn a strip for the first integral, which we've arbitrarily chosen to take in the z-direction.

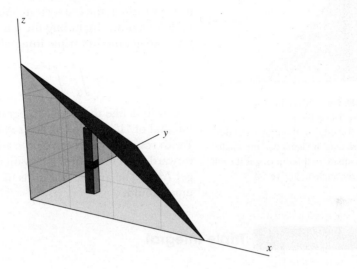

The bottom of the strip is at $z = 0$ and the top of the strip is on the plane $x + y + z = L$, so the upper limit is at $z = L - x - y$.

$$dV_{strip} = \int_0^{L-x-y} dz\, dx\, dy$$

(Note that we've put dz on the left since this is the first integral we're taking.)

Next we choose, again arbitrarily, to extend this strip in the y-direction. Figure 5.4 shows the original strip and the extended slice.

It isn't easy to look at that picture and see the limits of integration for y, and it becomes more difficult as the regions become more complicated. So here comes a valuable trick that we will return to many times. Since only the x and y integrals remain, let's see what that last picture looks like when viewed directly from overhead: its "shadow" on the xy-plane. (See Figure 5.5.)

Make sure you see the relationship between these two representations of the same region. The little box in Figure 5.5 represents the vertical strip in Figure 5.4. If you integrate over x and y to add up all the little boxes in the 2D triangle, you're adding up all the strips in the 3D region—which will perfectly fill in the volume we're

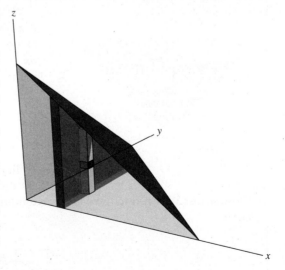

FIGURE 5.4 The small rectangular block in the middle is $dx\, dy\, dz$. It is in a vertical column extending in the z-direction, which is in a larger triangular region extending in the y-direction.

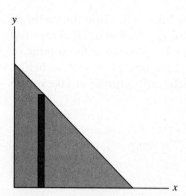

FIGURE 5.5 This is the domain of Figure 5.4 on the *xy*-plane. The tall thin rectangle in this drawing is the big triangle in Figure 5.4; the small blue square in this drawing is the tall thin rectangle in Figure 5.4.

looking for. The diagonal line at the top of the 2D picture is found in the 3D picture as the intersection of $x + y + z = L$ with $z = 0$, so its equation is $x + y = L$.

Once you see what Figure 5.5 has to tell us about Figure 5.4, you can find the limits of integration of x and y—the limits of integration over a 2-dimensional region—just as we did in the last section. (You should take a moment to do that now.) Including the limits we have already found for z, we get the full triple integral:

$$V = \int_0^L \int_0^{L-x} \int_0^{L-x-y} dz\, dy\, dx$$

It may look like there's no integrand, but think of it as the integral of "1" multiplied by $dz\, dy\, dx$. The antiderivative of 1 with respect to z is of course z, and the process is straightforward from there. You may want to give it a try; you should get $L^3/6$. The answer has units of distance cubed, as a volume should.

EXAMPLE   Triple Integral

Question: A right circular cone has height H and edges defined by the equation $z^2 = x^2 + y^2$. Its charge density (charge per unit volume) is cx^2yz. Find the total charge in the cone.

Solution:
On Page 200 we broke a similar cone into small circular "slices" with height dz, and it's worth taking a moment to note why we *can't* use that strategy here. The charge density in this cone is a function of x and y as well as z, so such a slice—no matter how thin—would not have a uniform density. We need to integrate over every variable.

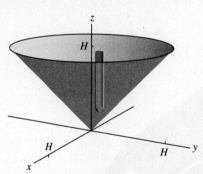

We choose to begin with z, which means drawing a tall thin strip. This strip exists at a particular spot (x, y), covering a small area $dx\, dy$, and extending from $z = \sqrt{x^2 + y^2}$ at the bottom to $z = H$ at the top.

Our first integral is therefore:

$$\int_{\sqrt{x^2+y^2}}^{H} cx^2 yz\, dz$$

To find the limits on the other two variables, ask the question: where do we need to draw all those strips? We need to draw them extending down from every point along the top circle.

Using this drawing, we can fill in the limits on x and y.

$$\int_{-H}^{H} \int_{-\sqrt{H^2-x^2}}^{\sqrt{H^2-x^2}} \int_{\sqrt{x^2+y^2}}^{H} cx^2 yz \, dz \, dy \, dx$$

You can evaluate that by hand or with a computer. You can also take a sneaky shortcut by evaluating the z integral by hand and then converting the remaining integral to polar coordinates (Section 5.6). However you do it, you end up with the answer 0. Well, of course you do! The charge density is odd in y, and the region of integration is symmetric in y. Every positive charge at $y > 0$ is perfectly balanced by a negative charge at $y < 0$, so the total charge of zero was predictable at the outset.

5.5.2 Problems: Triple Integrals in Cartesian Coordinates

5.104 The paraboloid $x = (y^2 + z^2)/L$, from the origin to $x = L$, is filled with a gas with density $k\sqrt{L^2 - y^2}$.

(a) Draw the paraboloid. You may want to use a computer to help you visualize it.

(b) Add to your drawing a small box somewhere in the middle. Don't put it anywhere special like on the axis or at the edge. What is the mass dm of the box?

(c) If you extend your little box into a strip parallel to the x-axis, what is the highest x-value your strip reaches (a constant)? What is the lowest value of x your strip reaches (a function of y and z)?

(d) Draw the 2D shape in the yz-plane that's left after you do the x integral. Include a little box in the shape indicating the strip you used above in the 3D shape.

(e) Write the formula for the boundary of the 2D shape you drew. The formula should include y, z, and constants.

(f) Extend your box into a strip in the z-direction. What are the upper and lower limits of that strip (both functions of y)?

(g) Write and evaluate a triple integral for the total mass of the paraboloid.

5.105 On pg. 218 we determined that the total charge on the given cone was zero, due to symmetry in the y-direction. What is the total charge on the $y > 0$ half of the cone? Make sure to show that your answer has correct units.

In Problems 5.106–5.110 integrate the function $f(x, y, z)$ in the given region.

5.106 A rectangular solid with corners at the origin and the point (W, L, H); $f = \sin(\pi x/W) \cos(\pi y/(2L)) e^{-z}$.

5.107 The $x > 0$ half of a cylinder of radius R, centered on the z-axis and extending from the xy-plane on the bottom to $z = H$ at the top; $f = x \sin(\pi z/H)$.

5.108 A right circular cone centered on the z-axis. The vertex is the origin and the top is a circle of radius R at $z = H$; $f = xyz$.

5.109 A sphere of radius R centered on the origin; $f = x^2 + y^2 + z^2$.

5.110 The region above the surface $z = x^2 + y^2$ and below the surface $x^2 + y^2 + z^2 = 2$; $f = z$.

5.111 A cube has corners at the origin and (L, L, L) and density $c \sin(\pi x/L) \cos(\pi y/(2L)) e^{-kz}$.

(a) Find the mass of the cube.

(b) Find the height of the cube's center of mass.

5.112 Find the moment of inertia of a uniform cube of mass M and length L about an axis through one of its edges.

5.113 Region R_1 is bounded by the surface $z = y^2$ and by the planes $z = 25$, $x = 0$ and $x = 3$. The mass density is z. Find the mass, and the center of mass, of this region.

5.114 Region R_2 is bounded by the surface $z = y^2$ and by the planes $z = 25$, $y = x$, and $x = 10$. The mass density is z. Find the mass, and the center of mass, of this region.

5.115 Find the volume of the region bounded by $z = x^2 + y^2$ and $z = 2x + 8$.

5.116 The region directly above the sphere $x^2 + y^2 + z^2 = 1$ going infinitely high up is filled with a uniform charge density ze^{-z^2}.
  (a) Set up a triple integral to find the total charge in this region. The first (innermost) integral should be with respect to z.
  (b) Evaluate the z integral. Look at the x and y integrals long enough to convince you that you wouldn't want to deal with them.
  You'll return to this problem in Section 5.7, where you'll learn a much easier way to approach it.

5.117 The curve $y = \sin x$ for $0 \leq x \leq \pi$ is rotated around the x-axis, producing the figure $y^2 + z^2 = \sin^2 x$. What limits of integration describe this figure?

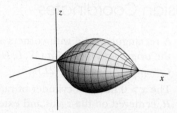

  (a) One approach is to begin by letting x go from $\sin^{-1}\sqrt{y^2 + z^2}$ to $\pi - \sin^{-1}\sqrt{y^2 + z^2}$. If that is your first integral, what are the limits of integration for y and z?
  (b) Another approach is to begin by letting z go from $-\sqrt{\sin^2 x - y^2}$ to $\sqrt{\sin^2 x - y^2}$. If that is your first integral, what are the limits of integration for x and y?

5.118 A solid hemisphere of radius R sits with its flat end on the xy-plane, centered at the origin. Its mass M and charge Q are uniformly distributed.
  (a) Find the moment of inertia of this hemisphere about the z-axis.
  (b) Find the electric potential at the origin.
  (c) If we add another identical hemisphere underneath, we have a full sphere with mass $2M$ and charge $2Q$. What are its moment of inertia about the z-axis and potential at the origin?

5.119 Find the moment of inertia of a uniform cube of mass M and length L about an axis that goes from one corner to the opposite corner. (*Hint:* Align the axes with the edges of the cube, not with the axis of rotation.) A possibly useful formula: Given a point (x_0, y_0, z_0) and a line $z = ax = by$, the position on that line closest to the point is at $z = (ax_0 + by_0 + z_0)/(1 + a^2 + b^2)$.

5.120 Find the electric potential created by a uniform cube of charge Q and length L at one of its corners.

5.121 A pyramid has density $\rho = kz$. The sides are given by $z + x = L$, $z - x = L$, $z + y = L$, and $z - y = L$. Find the total mass.

5.122 A uniform three-sided pyramid has length L, height h, and mass M.
  (a) If you place the z-axis through the center of the pyramid then the top of the pyramid is at $(0, 0, h)$ and the base forms an equilateral triangle with its center at the origin. Put one of the vertices of the triangle on the positive x-axis and find the coordinates of all three vertices.
  (b) The next step is to find the equations of the three planes that bound the pyramid (not counting the bottom, which is the xy-plane). For each plane, we know three points: two that you found in Part (a), and the top of the pyramid at $(0, 0, h)$. One of those planes is defined by a simple linear equation relating x and z. Find the equation for this plane.
  (c) The other two planes are $z = (-h\sqrt{3}/L)x - (3h/L)y + h$ and $z = (-h\sqrt{3}/L)x + (3h/L)y + h$. Confirm that each of these planes contains the three points that it should.
  (d) You're going to take a triple integral to find the total volume of the pyramid. Explain which integral it will be easiest to take first and why. Find the limits for that integral.
  (e) Draw the 2D shape you are left with after taking that first integral. Include formulas for all the lines on your drawing.
  (f) Finish setting up the triple integral and evaluate it to find the volume of the pyramid.
  (g) Check that your answer has correct units. Find at least one other way you can check that your answer makes sense.

5.123 Find the moment of inertia of the four-sided pyramid shown below about a vertical axis through its center. The sides are given by $z + x = L$, $z - x = L$, $z + y = L$, and $z - y = L$, and the mass M is distributed uniformly throughout the pyramid. The volume can be looked up, derived geometrically, or integrated, but instead we'll go ahead and tell you that $V = (4/3)L^3$.

5.6 | Double Integrals in Polar Coordinates

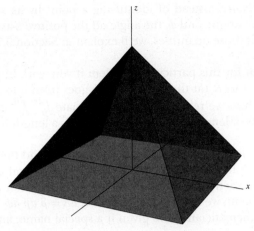

5.124 *[This problem depends on Problem 5.122.]* Find the center of mass of the pyramid in Problem 5.122. (You should have already done nearly all of the work for this problem when you did that one.)

5.125 🖳 Find the integral of the function $f(x, y, z) = x^2 + z^2$ in the region between the surfaces $x^2 + z^2 = (y-1)^2$ and $x^2 + z^2 = (y+1)^2$.

5.126 🖳 A uniform object of charge 3×10^{-9} fills an ellipsoid with semi principal axis lengths 1, 2, and 3. Find the electric potential at the center of the ellipsoid. All numbers are given in SI units, in which $k = 9 \times 10^9$.

5.127 🖳 Find the electric potential at a point $(0, 0, 5)$ due to a uniform sphere of radius 2 with charge 10^{-7} centered on the origin. All numbers are given in SI units, in which $k = 9 \times 10^9$.

5.6 Double Integrals in Polar Coordinates

The chapter so far has been about integrals in Cartesian coordinates. For many problems, other coordinate systems are easier to work with. We introduce that in this section with the familiar example of polar coordinates. In the next section we'll introduce two 3D coordinate systems that are useful for many problems.

5.6.1 Discovery Exercise: Double Integrals in Polar Coordinates

The drawing below shows a small region within some small $d\rho$ at a distance ρ from the origin, and some small $d\phi$ at an angle ϕ off the positive x-axis. Your job is to find the area dA of this region.

1. We begin by ignoring ϕ and looking at the entire ring between the two dotted lines. If this ring were cut at the bottom and stretched up into a rectangle, what would be the width and length of this rectangle? What would be its area?
2. What fraction of the ring is represented by our region?
3. Based on those two answers, what is the area of the region dA?

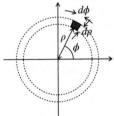

5.6.2 Explanation: Double Integrals in Polar Coordinates

Suppose you set out to use a double integral to find the area of a circle of radius R. The area of a tiny box is $dx \, dy$, and the lower and upper limits for y are given by the formula for a circle, $x^2 + y^2 = R^2$.

$$A = \int_{-R}^{R} \int_{-\sqrt{R^2-x^2}}^{\sqrt{R^2-x^2}} dy \, dx = 2 \int_{-R}^{R} \sqrt{R^2 - x^2} \, dx$$

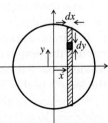

You may remember how to use a trig substitution to turn this into an integral of $\cos^2 \theta$ which you can then simplify with a double-angle formula, but you're probably not looking forward to it. When you're working with a circle, it's often easier to use *polar coordinates:* instead of identifying a point by its x- and y-coordinates you use ρ, the distance to the origin, and ϕ, the angle off the positive x-axis. (If you prefer the letters r and θ to represent those quantities, we'll explain in Section 5.7 why we don't.)

It's easy to see the limits of integration for this particular problem if you work in polar coordinates: ρ goes from 0 (at the center) to R (at the edge), and ϕ goes from 0 to 2π to trace all around the circle. So you might think we're simply going to evaluate $\int_0^{2\pi} \int_0^R d\rho\, d\phi$. But that integral can't be right; it has units of length, not area, because $d\rho$ is a length but $d\phi$ is an angle (and thus unitless).

So where did we go wrong? We got the limits of integration correct, but to find the total area we have to add up the areas of all the little boxes that make up the circle, and $d\rho\, d\phi$ is not that area. Below we will show you exactly how to define a differential box in polar coordinates and find its area. For the moment, we'll jump to the answer: $dA = \rho\, d\rho\, d\phi$. That extra factor of ρ is so important that mathematicians have given it a special name, and we are giving it its own special box.

Jacobian in Polar Coordinates

To convert a double integral from Cartesian coordinates to polar coordinates you have to throw in an extra factor of ρ, called the "Jacobian in polar coordinates."

$$\iint f(x, y)\, dx\, dy = \iint f(\rho, \phi) \rho\, d\rho\, d\phi$$

For this equation to work you also have to find the correct limits of integration for ρ and ϕ so you are covering the same 2D region you were in the Cartesian integral.

This equation assumes ρ is positive.

With that little trivia fact in hand, we're ready to return to the area of our circle. In Cartesian coordinates the integrand when you're performing a double integral to find area is just 1, so in polar coordinates we get the following.

$$\int_0^{2\pi} \int_0^R \rho\, d\rho\, d\phi = \int_0^{2\pi} \frac{R^2}{2} d\phi = \pi R^2$$

That was easy. But where did that extra ρ come from?
We were hoping you'd ask.

We begin as always with a little box. Our traditional dA has been defined by a small dx and a small dy; this one, of course, is defined by a small $d\rho$ and a small $d\phi$. Unlike our usual box, this one is not literally a rectangle—the lines on the left and right are not quite parallel to each other, and the top and bottom are not strictly lines at all—but it does become more rectangular as $d\phi \to 0$, so we will make that approximation. The "height" of the rectangle is $d\rho$.

But in no approximation is $d\phi$ the width; $d\phi$ is the *angle subtended* by the box. What we need is the length of an arc: a part of a circle of radius ρ, subtending an angle $d\phi$. Since the whole circle has circumference $2\pi\rho$ and this arc takes up a fraction $d\phi/(2\pi)$ of that circle, the length of the arc is $\rho\, d\phi$.

So the area of the box is $\rho\, d\rho\, d\phi$. (The Discovery Exercise, Section 5.6.1, was designed to guide you to this conclusion.) This box does not only apply to the "area of a circle" problem we worked above; it is the fundamental area unit in polar coordinates. That is the logic underlying the rule in the box above.

5.6 | Double Integrals in Polar Coordinates

> **EXAMPLE** **Potential of a Disk**

Question: A disk of radius R centered on the origin in the xy-plane has charge density $c\cos^2\phi$ where c is a constant. Find the electric potential at the point $(0,0,h)$.

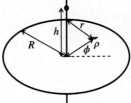

Solution:
The two things we need to know about our differential box in order to find the potential it produces are its charge and its distance from the point $(0,0,h)$. The charge is charge density times area: $dq = (c\cos^2\phi)(\rho\,d\rho\,d\phi) = c\rho\cos^2\phi\,d\rho\,d\phi$. From the picture we can see that the distance r is given by the Pythagorean theorem as $r = \sqrt{\rho^2 + h^2}$. Recalling that potential due to a point charge is kq/r, the total potential is thus

$$V = \int_0^R \int_0^{2\pi} kc\frac{\rho\cos^2\phi}{\sqrt{\rho^2+h^2}}\,d\phi\,d\rho = \pi kc \int_0^R \frac{\rho}{\sqrt{\rho^2+h^2}}\,d\rho = \pi kc\left(\sqrt{R^2+h^2} - \sqrt{h^2}\right)$$

(In this example, integrating with respect to ρ before ϕ would have turned out just as easy. Remember, however, that sometimes one order will be easier than the other.) From the formula for charge density c must have units of charge per distance squared, so this has units of k times charge per distance, as it should.

In Problem 5.139 you'll set this same problem up in Cartesian coordinates, which should dispel any doubts about how useful other coordinate systems can be.

Non-constant Limits of Integration

The computation $\int_{x_1}^{x_2} \int_{y_1}^{y_2} f(x,y)\,dy\,dx$ applies only to a rectangular region; for more complicated regions, the limits of integration on the inside integral become functions. (The final limits are always constants.)

Similarly, $\int_{\phi_1}^{\phi_2} \int_{\rho_1}^{\rho_2} f(\rho,\phi)\rho\,d\rho\,d\phi$ defines a region called a "polar rectangle." (You can see why it's called that, but keep in mind that it is not a rectangle at all.) For any other region, once again, the inner limits of integration (usually ρ) become functions. The key is to express the bounding region in the form $\rho(\phi)$, as shown in the example below.

> **EXAMPLE** **Polar Integral with Non-constant Limits**

Question: The triangle between the lines $y = x$, $y = 2x$, and $y = c$ is filled with a charge density $\sigma = k\rho\phi$, where $c = 2$ m and $k = 0.5$ C/m^3. Find the total charge in this region.

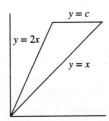

Solution:
Your first impulse might be to attack the problem in Cartesian coordinates. Horizontal slices would make this region easier, but the integrand, $k\rho\phi$, would be a nightmare. So there are good reasons for wanting to represent this region in polar coordinates.

Imagine shooting rays out from the origin through this region. Each ray would continue until it reached the line $y = c$; it would never bump into the other

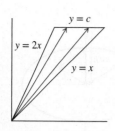

two lines. This tells us that the line $y = c$ always sets the boundary for ρ in this region.

But in polar coordinates we need to express that boundary in the form $\rho(\phi)$, so instead of $y = c$ we will write $\rho \sin \phi = c$. We see then that if ρ goes from 0 to $c/(\sin \phi)$, it will cover the entire region. Meanwhile, ϕ goes from $\pi/4$ on the right to $\tan^{-1} 2$ on the left. So the charge of our region is given by:

$$\int_{\pi/4}^{\tan^{-1} 2} \int_0^{c/(\sin \phi)} (k\rho\phi)\rho \, d\rho \, d\phi = k \int_{\pi/4}^{\tan^{-1} 2} \int_0^{c/(\sin \phi)} \rho^2 \phi \, d\rho \, d\phi$$

It's easy enough to evaluate the first integral by hand, and the second with a computer or calculator (or just do both on a computer). Putting in $c = 2$ and $k = 0.5$ it comes out to 0.78 C. The "needs-a-person" math is all in the setup.

Suppose you had forgotten the Jacobian and written this.

$$\int_{\pi/4}^{\tan^{-1} 2} \int_0^{c/(\sin \phi)} (k\rho\phi) \, d\rho \, d\phi \quad \text{mistake!}$$

You know from the problem that $k\rho\phi$ is a charge per unit area (that's what charge density is in 2D), so to get the total charge it must be multiplied by an area. Since $d\rho$ has units of length and $d\phi$ is unitless, you can see that this must be wrong. You couldn't do this units checking if you just replaced c and k with numbers at the beginning.

Jacobians for Other Coordinate Systems

In Figure 5.6 we started at a polar point (ρ, ϕ) and increased the coordinates by $d\rho$ and $d\phi$. The resulting shape was an odd-looking curved region sometimes called a "polar rectangle." The area of that polar rectangle defined the Jacobian in polar coordinates.

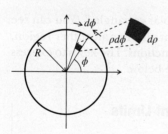

FIGURE 5.6 Differential polar area.

We can repeat that process for any new coordinate system we define, but amazingly, we don't have to. We can instead write a general formula that will find the Jacobian of any coordinate system.

Consider an arbitrary system with coordinates u and v. We define this coordinate system by the functions $x(u, v)$ and $y(u, v)$ that convert it to Cartesian coordinates. (So instead of saying "ρ is the distance to the origin and ϕ is the angle" we would define polar coordinates by the equations $x = \rho \cos \phi$ and $y = \rho \sin \phi$.) You can think of these functions as a mapping that turns each point (u, v) into one corresponding point (x, y).

Figure 5.7 shows a simple region in uv space being mapped to a more complicated region in xy space. The dotted lines show gridlines of constant v in both pictures, the solid lines gridlines of constant u. The blue boxes in both pictures stretch from some specific (u, v) by a small du and dv. To integrate a function across this entire region we need to multiply the value of the function in each such box by the area of the box. Therefore, the area of one box is the Jacobian.

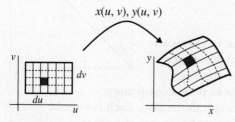

FIGURE 5.7 Differential area for a generic 2D system.

We begin with two observations about this small box.

- A line in *uv* space will generally correspond to a curve in *xy* space. However, for sufficiently small *du* and *dv*, each small motion can be considered a line in *xy* space. (Any sufficiently small curve looks like a line.)
- However, no matter how small *du* and *dv* are, the resulting motions are *not* guaranteed to be perpendicular.

In Figure 5.8 vector $\vec{a}$ represents the change in position caused by adding *du* to *u*, and vector $\vec{b}$ represents the *dv* change. The resulting shape is a parallelogram.

If that is the generic shape for all coordinate systems then why did Figure 5.6 (polar coordinates) look so different? First, we showed the $d\phi$ change as a curve—which it is, but we can treat it as a line for a small enough $d\phi$. Second, the $d\rho$ and $d\phi$ changes were perpendicular to each other, creating in effect a rectangle. A parallelogram is the more general case.

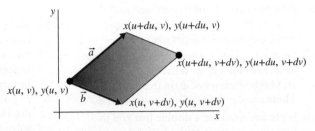

FIGURE 5.8 Differential area for a generic 2D system.

In Problem 5.152 you will use Figure 5.8 to derive the following formula.

$$\left| \frac{\partial x}{\partial u}\frac{\partial y}{\partial v} - \frac{\partial x}{\partial v}\frac{\partial y}{\partial u} \right| \qquad \text{The Jacobian for an arbitrary 2D coordinate system} \qquad (5.6.1)$$

In Problems 5.153–5.156 you will apply this formula to find Jacobians for a number of 2D coordinate systems, and in Problems 5.157–5.162 you will use these Jacobians to evaluate integrals.

If you have never studied any linear algebra, skip this paragraph and go right to the "Very Important Fact" below. You can come back to this after Chapter 6. You can write the formula for the Jacobian more concisely in a way that generalizes to coordinate systems in more than two dimensions. The Jacobian is the absolute value of the determinant of the matrix formed from the partial derivatives of the Cartesian coordinates with respect to the new coordinates. You'll apply this equation to calculate some 3D Jacobians in Section 5.7.

$$\left\| \begin{array}{cc} \partial x/\partial u & \partial x/\partial v \\ \partial y/\partial u & \partial y/\partial v \end{array} \right\| \qquad \left\| \begin{array}{ccc} \partial x/\partial u & \partial x/\partial v & \partial x/\partial w \\ \partial y/\partial u & \partial y/\partial v & \partial y/\partial w \\ \partial z/\partial u & \partial z/\partial v & \partial z/\partial w \end{array} \right\| \qquad (5.6.2)$$

Jacobian in 2D Jacobian in 3D

The double-bars in Equation 5.6.2 are not a new notation but a confusing one: they indicate the absolute value (outside bars) of the determinant (inside bars) of a matrix. The determinant of a matrix can be negative but a Jacobian never can.

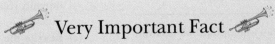

Very Important Fact

If you write an integral in polar or any other non-Cartesian system it's easy to forget the Jacobian, but you'll spot the mistake right away if you *check units*. For this reason among others we strongly encourage you to check your units before calculating complicated integrals.

5.6.3 Problems: Double Integrals in Polar Coordinates

5.128 Walk-Through: Double Integral in Polar Coordinates. A half-disk of radius R in the upper plane $y > 0$, centered on the origin, has density $\sigma = c\rho \sin\phi$.

(a) Draw the semicircle with a differential box at a random point inside it (not on one of the axes or at the edge). The differential box should extend over a small $d\rho$ and $d\phi$.

(b) Find the mass dm of that differential box as a function of ρ and ϕ (its location) and $d\rho$ and $d\phi$ (its size).

(c) Draw a half-ring extending that box along different values of ϕ. What are the lowest and highest values of ϕ on that half-ring? These are the limits of integration for ϕ.

(d) Write and evaluate a double integral to find the total mass of the half-disk.

(e) The SI units of density in 2D are kg/m². Use the equation $\sigma = c\rho \sin\phi$ to find the units on the constant c. Then use these units to make sure your answer to Part (d) has correct units.

(f) What is the maximum density on the semicircle? Find what the mass would be if the entire semicircle had that density and make sure your answer in Part (d) is smaller than this.

For Problems 5.129–5.134 integrate the function $f(\rho, \phi)$ in the indicated region.

5.129 A circle of radius R centered on the origin; $f = \rho^3 \sin\phi$.

5.130 A semicircle of radius R in the lower half-plane $y < 0$; $f = \rho/\phi$.

5.131 A ring centered on the origin with inner radius R_1 and outer radius R_2; $f = e^{\rho^2} \cos\phi$.

5.132 The triangle with edges $x = 0$, $y = 2$, and $y = x$; $f = (\cos\phi)/\rho$.

5.133 🖥 The triangle with edges $x = 0$, $y = 0$, and $x + y = 1$; $f = 1/\rho$.

5.134 The region above the x-axis and below the curve $y = 1 - x^2/4$; $f = (\cos^2 \phi)/\rho$.

In Problems 5.135–5.138 you will redo problems from Section 5.4. You did them the first time in Cartesian coordinates; now you will use polar.

5.135 Problem 5.81.

5.136 🖥 Problem 5.82.

5.137 Problem 5.101. (In polar coordinates you can not only set up these integrals, you can evaluate them without a computer.)

5.138 Problem 5.103. (You should not need a computer to evaluate this integral in polar coordinates, but you will need to look up the integral of $\sin^2 \phi$ or derive it using trig identities.)

5.139 Redo the example "Potential of a Disk" from the Explanation (Section 5.6.2), only in Cartesian coordinates. You'll need to start by expressing the charge density in terms of x and y, simplifying as much as possible. You only need to set up the double integral; life is too short to try to solve it.

5.140 A 2D object has density σ. You can calculate the mass of the object as $\iint \sigma \, dx \, dy$, but in polar coordinates it is $\iint \sigma\rho \, d\rho \, d\phi$. Explain in your own words why you need that extra factor of ρ in the polar integral.

5.141 A piece of wood occupies the upper half of a disk of radius R centered on the origin, with density (mass per unit area) $k\rho^2 \sin\phi$. Find its mass.

5.142 Set up and evaluate a double integral in polar coordinates to find the moment of inertia of a uniform disk of radius R and mass M about a perpendicular axis through its center. (You may have already solved this problem using a 1D integral.)

5.143 A flat (effectively 2D) object is in the shape of a semicircle of radius R. The density of the object is proportional to distance from the flat edge of the semicircle.

(a) Draw the object with x- and y-axes and express the density function in Cartesian coordinates. Your answer should have an unspecified constant of proportionality.

(b) Set up and evaluate a double integral for the mass of the object in Cartesian coordinates. Your answer should only include R and the constant of proportionality you introduced in the density function.

(c) Rewrite the density function in polar coordinates.

(d) Set up and evaluate a double integral for the mass in polar coordinates. Your answer should be the same as the one you got in Cartesian coordinates.

5.6 | Double Integrals in Polar Coordinates

5.144 A uniform triangle with mass M lies in the xy-plane with edges $y = 0$, $x = 3$, and $x = y$. Find the moment of inertia of this triangle about the z-axis.

5.145 A triangle lies in the xy-plane with edges $x = 0$, $y = 0$, and $y = y_0 - 2x$ ($y_0 > 0$). The density of the triangle is proportional to distance from the z-axis. Find the moment of inertia of this triangle about the z-axis. Your answer should contain a constant of proportionality from the density function. Be sure to check the units of that constant and use them to check the units of your answer.

5.146 A rectangle with corners at $(0, 0)$ and $(2\sqrt{3}, 2)$ is filled with a material of density $\sigma = 1/(x^2 + y^2 + 1)$.

(a) Write a double integral to represent the mass of this region in Cartesian coordinates.

(b) Draw the rectangle, and draw in a diagonal line from $(0, 0)$ to $(2\sqrt{3}, 2)$. This diagonal line divides the rectangle into two triangles.

(c) Write a double integral to represent the mass of the upper triangle in polar coordinates.

(d) Write a double integral to represent the mass of the lower triangle in polar coordinates.

(e) Compute the mass of this rectangle using both your Cartesian and your polar integrals. Make sure you get the same answer!

5.147 The polar equation $\rho = \sin(3\phi)$ is called a "three-petaled rose." You're going to find its area.

(a) Draw a quick sketch of the rose. You can do this by hand, or get some help from a computer or calculator.

(b) The three petals are at different places, but it is still valid to find the area of just one petal and multiply it by three. Why?

(c) Because every petal begins and ends at the origin, you can solve the equation $\rho = 0$ to find the ϕ-values that begin and end a petal. Using that equation and your drawing, find the limits of integration that define one of the petals.

(d) Find the area of the entire rose.

5.148 The polar equation $\rho = 1 - \sin\phi$ is called a "limaçon" or a "cardioid." Suppose it is filled with a charge density $\sigma = \cos\phi$.

(a) Draw a quick sketch of the shape. You can do this by hand, or get some help from a computer or calculator.

(b) Find the total charge of the region.

(c) Find the total charge of the right half of the region only.

5.149 Region R is bounded by two circles of radius 3, one centered at the origin and the other at the point $(3, 0)$. An object fills region R with a constant density σ.

(a) Draw region R.

(b) Write polar equations for both circles. (One way to do this is the write them in Cartesian and then convert to polar. Both equations should be pretty simple when you're done.)

(c) Find the ϕ-values where the two circles intersect.

(d) Add to your drawing lines from the origin at the two ϕ-values you just calculated. These lines divide region R into three subregions. Two of those subregions have the same area, and therefore the same mass: which two?

(e) Calculate the mass of one of those two subregions. Pay careful attention to the limits of integration: which of the two circles forms a boundary for ρ in this subregion? You can do this integral with a computer, look it up online or in a table, or do it by hand using the trig identity $\cos^2 x = (1/2)(1 + \cos(2x))$.

(f) Calculate the mass of the third subregion. Once again, start by asking the question: which of the two circles forms a boundary for ρ in this subregion?

(g) Find the mass of region R.

5.150 Region R is bounded by two parabolas, $y = 1/4 - x^2$ and $y = -1/4 + x^2$. An object fills region R with a constant density σ. Your job is to find the moment of inertia of this object around a perpendicular axis through the origin (the z-axis).

(a) Draw region R.

(b) Set up the appropriate integral in Cartesian coordinates.

(c) Set up the appropriate integral in polar coordinates.

(d) Evaluate whichever integral looks easier to find the answer.

5.151 A flat object in the xy-plane fills the region between $y = x^2$ and $y = x$. The charge density on the object is $\sigma = \rho/\sin\phi$. Find the electric potential at the origin.

5.152 Exploration: Jacobians in Alternative Coordinate Systems. If you move from (ρ, ϕ) to $(\rho + d\rho, \phi + d\phi)$ you sweep out an area of $\rho\, d\rho\, d\phi$, which is why ρ is the polar Jacobian. To generalize this result to arbitrary coordinate systems we need to find the area of the parallelogram in Figure 5.8 on Page 225.

(a) Let dy equal the change in the y-coordinate between the top and bottom of the vector labeled $\vec{a}$ in Figure 5.8. Explain why $dy = (\partial y/\partial u)du$.

(b) Write the $\hat{i}$- and $\hat{j}$-components of $\vec{a}$.

(c) Write the $\hat{i}$- and $\hat{j}$-components of $\vec{b}$.

(d) The area of a parallelogram with sides $\vec{a}$ and $\vec{b}$ is the magnitude of the cross product, $|\vec{a} \times \vec{b}|$. (See Section 5.10 Problem 5.271.) Find the area of the $du\, dv$ parallelogram. (In order to take this cross product you will have to treat both sides as three-dimensional vectors with $\hat{k}$-components of zero.)

(e) The Jacobian is a function $f(u, v)$ such that the area of that parallelogram is $f(u, v)\, du\, dv$. Write the Jacobian for the uv-coordinate system.

Equation 5.6.1 gives a general formula for the Jacobian in two dimensions. (If you worked Problem 5.152 you derived that formula, but these problems do not depend on the derivation.) In Problems 5.153–5.156 use that formula to evaluate the Jacobian for the given coordinate system. You will need to start by writing x and y in terms of the new coordinates if those formulas aren't given in the problem.

5.153 Use the formula to verify the Jacobian for polar coordinates.

5.154 A coordinate system rotated 45° relative to the x- and y-axes can be defined by $x = (u - v)/\sqrt{2}$, $y = (u + v)/\sqrt{2}$. In addition to finding the Jacobian, explain why the result makes sense. In other words, how could you have predicted this Jacobian without calculating it?

5.155 $x = \sqrt{2}\, u$, $y = (u + v)/\sqrt{2}$

5.156 It's possible to define a set of coordinates R and ϕ that show the infinite 2D plane in a finite circle. The angle ϕ is the usual polar angle and the distance R is related to the polar coordinate ρ by $\rho = R/(1 - R)$. You can see from this definition that the domain $0 \le R < 1$ is equivalent to the domain $0 \le \rho < \infty$.

5.157 Region R is bounded by the curves $y^2 = 4 + 4x$ and $y^2 = 4 - 4x$ and the constraint $y \ge 0$.

(a) Draw the region.

(b) Integrate the function $f(x, y) = 3y$ in region R.

Now you will reframe the region by using the coordinate system $x = u^2 - v^2$, $y = 2uv$, subject to the restriction $u \ge 0$.

(c) Explain why we can assume $v \ge 0$ even though this restriction was not specified.

(d) One of the bounding curves of R is $y^2 = 4 + 4x$. Rewrite this curve as an equation in terms of u and v. (It may help to remember that u and v must be real numbers.)

(e) Similarly rewrite $y^2 = 4 - 4x$ in terms of u and v.

(f) What is the final bounding curve? Write the equation for this curve in xy space. Then write the corresponding equations (two of them!) in uv space.

(g) Draw the resulting region in uv space.

(h) Calculate the Jacobian of the uv coordinate system.

(i) Convert the function $f(x, y) = 3y$ into the uv system and then integrate it through the region in uv space. You should get the same answer you got in Part (b).

(j) Integrate the function $f(x, y) = 1/\sqrt{x^2 + y^2}$ on region R. (Unlike our first example, this is really tough to do in Cartesian coordinates; use the uv transformation instead!)

5.158 In this problem you will integrate the function $f(x, y) = x^2 + y^2$ in the diamond-shaped region with corners at $(0, 1)$, $(1, 0)$, $(0, -1)$, and $(-1, 0)$.

(a) Draw the region.

(b) Evaluate the integral in Cartesian coordinates. You will have to be careful about the limits of integration!

(c) Evaluate the integral with the substitution $x = u + v$, $y = u - v$. Remember that any coordinate transformation requires three changes: the integrand, the region (or the limits of integration), and the Jacobian.

5.159 Integrate $(y - x)e^{(2x+y)(y-x)}$ within the region bounded by the lines $y = -2x$, $y = x$, $y = 1 - 2x$, and $y = x + 1$. *Hint*: The integrand suggests a coordinate system that might make this problem easier.

5.160 Integrate the function $f(x, y) = \sqrt{x^2 - y^2}$ in the triangular region bordered between the lines $y = x$, $y = 1 - x$, and $y = 0$ by using the transformation $x = u^2 + v^2$, $y = u^2 - v^2$. Assume that $u \geq 0$ and $v \geq 0$.

5.161 Integrate $\sqrt{x^2 - y^2}$ in the region with $x > 0$ bounded by the lines $x = \pm 2y$ and the hyperbola $x^2 - y^2 = 1$. Evaluate the integral in the uv coordinates defined by $x = u \cosh v$ and $y = u \sinh v$. You can find definitions of cosh and sinh in Appendix J. You will need to know that $\cosh^2 x - \sinh^2 x = 1$, $d/dx(\cosh x) = \sinh x$, and $d/dx(\sinh) = \cosh x$.

5.162 In this problem you will be evaluating integrals within the semi ellipse $25x^2 + 4y^2 = 900$, $y > 0$.

(a) Draw the region. *Hint*: Find the two x-values where $y = 0$ and the y-value where $x = 0$ and sketch the semi ellipse from there.

(b) Integrate the function x in this region. *Hint*: You can answer this question without doing any calculations.

(c) Integrate the function y in this region. *Hint*: Start with the substitution $x = 2u$, $y = 5v$.

5.7 Cylindrical and Spherical Coordinates

You saw in the last section that polar coordinates can be easier than Cartesian coordinates for some 2D problems. Many 3D problems also depend on angles and distances that aren't easy to express in Cartesian coordinates. Coordinate systems with angular variables are generically called "curvilinear coordinate systems," and there are two that are commonly used in 3D problems.

5.7.1 Discovery Exercise: Cylindrical and Spherical Coordinates

In cylindrical coordinates, the position of an object is specified by the following three numbers:

- ρ gives the distance from the z-axis
- ϕ gives the angle around the z-axis, starting at the positive x-axis
- z gives the distance from the xy-plane

Note that ρ and ϕ are the two-dimensional *polar* representation of the xy position of an object. The third variable, z, comes from Cartesian coordinates.

1. What are the Cartesian (x, y, z) coordinates of the point whose cylindrical coordinates are $\rho = 2$, $\phi = \pi/3$, $z = -2$?
2. What are the cylindrical (ρ, ϕ, z) coordinates of the point whose Cartesian coordinates are $(3, 4, 10)$?

   See Check Yourself #30 in Appendix L

3. What shape is described by the equation $\rho = 2$? (It isn't a circle.)
4. What shape is described by the equation $\phi = \pi/3$? (It isn't a line.)

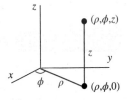

FIGURE 5.9

In spherical coordinates, the position of an object is specified by the following three numbers:

- r gives the distance from the origin
- θ gives the angle down from the z-axis
- ϕ gives the angle around the z-axis, starting at the positive x-axis

Note that ϕ is the same variable in cylindrical and spherical coordinates. Figure 5.10 shows both the cylindrical and the spherical coordinates for a point.

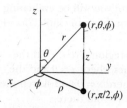

FIGURE 5.10

5. In Figure 5.10, the angle between the blue line labeled r and the z-axis is labeled θ. The angle between the blue line labeled r and the black line labeled z is not labeled. What is that angle in terms of spherical coordinates?

6. What are the Cartesian (x, y, z) coordinates of the point whose spherical coordinates are $r = 2$, $\theta = \pi/6$, $\phi = \pi/3$? *Hint*: begin by calculating the cylindrical ρ even though it is not your final goal.

7. What are the spherical (r, θ, ϕ) coordinates of the point whose Cartesian coordinates are $(3, 4, 10)$?

   See Check Yourself #31 in Appendix L

8. What shape is described by the equation $r = 2$?
9. What shape is described by the equation $\theta = \pi/3$?

5.7.2 Explanation: Cylindrical and Spherical Coordinates

Cylindrical Coordinates

You are producing a gas that is mixing with the air in a cylindrical tank of height H and radius R. The concentration of gas decreases exponentially with height because it is heavier than air. The gas is being produced from a vertical pipe in the center and removed by filters on the outside edge, so its concentration is inversely proportional to distance from the central vertical axis. We want to find the amount of gas in the tank.

What is the natural coordinate system for this problem? The concentration varies with the height in the pipe, which is the Cartesian z. But the concentration also depends on the distance from the central vertical axis; $\sqrt{x^2 + y^2}$ will make for painful integration, but the polar ρ is exactly what we're looking for.

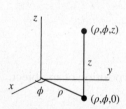

So this scenario leads us to *cylindrical coordinates* (sometimes called "circular cylinder coordinates"). Horizontal position is given by the polar coordinates ρ and ϕ, and height by the good old Cartesian z. Note that ρ no longer represents "distance from the origin" as it did in polar coordinates; it is "distance from the z-axis." The three variables are usually written in the order (ρ, ϕ, z).

EXAMPLE **Positions in Cylindrical Coordinates**

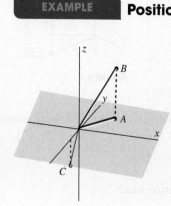

Question:
The picture to the left shows three points A, B, and C. Their Cartesian coordinates are $(1, 1, 0)$, $(1, 1, 2)$, and $(0, -1, -1)$ respectively. What are their cylindrical coordinates?

Solution:
Points A and B are at $\phi = \pi/4$ and $\rho = \sqrt{2}$. The fact that they are at different heights has no effect on their ρ and ϕ coordinates. Point C is at $\phi = 3\pi/2$ and $\rho = 1$.

In all cases the cylindrical z coordinate is the same as the Cartesian one, so you have

$$A : (\sqrt{2}, \pi/4, 0)$$
$$B : (\sqrt{2}, \pi/4, 2)$$
$$C : (1, 3\pi/2, -1)$$

We deliberately presented that example without conversion equations because it's important to think about coordinate systems visually. When you do need the equations, they're not hard: the conversions between (x, y) and (ρ, ϕ) are the same as they are for polar, and z remains unchanged. See Appendix D.

The final piece for taking integrals with cylindrical coordinates is the Jacobian. We must find the volume of the differential box shown in the picture below.

The vertical side has height dz. The other two sides can be seen from the shadow of the box on the xy-plane, which is the same as the differential box in polar coordinates. The volume of the 3D box is thus:

$$dV_{cylindrical} = \rho \, d\rho \, d\phi \, dz$$

Equivalently we can say the Jacobian in cylindrical coordinates is ρ, just as it was in polar coordinates.

We are now ready to return to our cylinder of gas. The verbal description of the concentration leads readily to a cylindrical function: $c = ke^{-\omega z}/\rho$. The limits of integration are also easy to write in this system. (You may want to figure them out before reading our answer below.) So the total amount of gas in the tank is:

$$n = \int_0^H \int_0^R \int_0^{2\pi} k(e^{-\omega z}/\rho) \rho \, d\phi \, d\rho \, dz = \frac{2\pi k R}{\omega}\left(1 - e^{-\omega H}\right)$$

Those limits of integration were constants in cylindrical coordinates because the region was, conveniently, a cylinder. For more difficult regions we use the same trick we emphasized in Cartesian 3-dimensional integrals: find the limits for one variable, thus reducing the problem to a two-dimensional region to find the remaining limits.

Which variable first? It depends on the region, but most often integrals that you do in cylindrical or spherical coordinates are over regions that you can make by taking a shape and rotating it around the z-axis, either partway or entirely. In that case the ϕ integral will have constant limits of integration and the 2D shape you are left with in ρ and z is simply the shape that was rotated around the z-axis, as shown to the right.

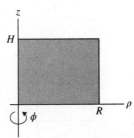

Each tiny box in this 2D shape represents a ring around the z-axis in the 3D region. Adding up all of those boxes in 2D means you are adding up all of those rings in 3D, which exactly fills the cylinder. With this 2D image it's easy to see that the limits are $0 \leq \rho \leq R, 0 \leq z \leq H$.

EXAMPLE   Moment of Inertia of a Cone

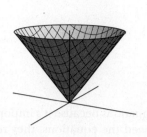

Question: A right circular cone of height H has its vertex at the origin and its side defined by the equation $z^2 = c\left(x^2 + y^2\right)$. The cone has density $k\rho$ where ρ is distance from the z-axis. Find its moment of inertia about the z-axis.

Solution:
Even though the formula for the cone is in Cartesian coordinates, the shape of the region, the density function, and the formula for moment of inertia all point toward cylindrical coordinates as the easiest choice for the problem. A differential box in cylindrical coordinates has volume $dV = \rho \, d\rho \, d\phi \, dz$, so its mass in this case is $dm = k\rho^2 \, d\rho \, d\phi \, dz$. The distance from the axis of rotation is ρ, so the moment of inertia of the differential box is $dI = k\rho^4 \, d\rho \, d\phi \, dz$.

In this region, ϕ goes all the way around from 0 to 2π. As we discussed above, that leaves us with a 2D region that we can use to find the limits for ρ and z. The region that you have to rotate about the z-axis to make a cone is the triangle shown below.

To find the equation for the right edge of the triangle we convert the equation for the cone into cylindrical coordinates: $z^2 = c\rho^2$. Since ρ and z are both positive in this region this is just $z = \sqrt{c}\rho$. You can integrate ρ and z in that triangle in either order; we arbitrarily chose ρ first.

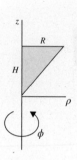

$$M = \int_0^H \int_0^{z/\sqrt{c}} \int_0^{2\pi} k\rho^4 \, d\phi \, d\rho \, dz = \frac{\pi k H^6}{15 c^{5/2}}$$

From the density expression k has units of mass per distance to the fourth, and from the formula for the cone c is unitless, so the units are mass times distance squared, as they should be.

Spherical Coordinates

Our last system is "spherical coordinates" (sometimes called "spherical polar coordinates"). The first coordinate r is distance from the origin. The other two coordinates are angles. The "azimuthal angle" ϕ is the same variable we used in cylindrical coordinates, the angle around the z-axis going counterclockwise from the positive x-axis. The "polar angle" θ is the angle down from the positive z-axis. Spherical coordinates are written in the order (r, θ, ϕ).

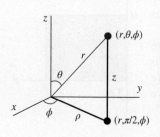

The drawing to the left shows both the cylindrical and the spherical coordinates for a point. Note that the domains $r \geq 0$, $\phi \in [0, 2\pi]$, and $\theta \in [0, \pi]$ are enough to cover all possible points.

On the Earth's surface θ and ϕ represent latitude and longitude respectively, with one difference. Latitude is zero at the equator and $\pm\pi/2$ at the poles. (It's usually given in degrees, but spherical coordinates are always given in radians so we've converted latitude to radians as well.) The spherical coordinate θ, however, is 0 at the North pole, $\pi/2$ at the equator, and π at the South pole.

5.7 | Cylindrical and Spherical Coordinates

EXAMPLE | Positions in Spherical Coordinates

Question:
The picture to the right shows three points A, B, and C. Their Cartesian coordinates are $(1, 1, 0)$, $(1, 1, 1)$, and $(-1, -1, -1)$ respectively. What are their spherical coordinates?

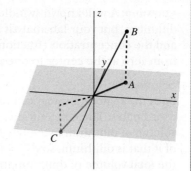

Solution:
The angle ϕ only depends on horizontal position. Ignoring height (z), points A and B are in the middle of the first quadrant and point C is in the middle of the third quadrant, so $\phi_A = \phi_B = \pi/4$ and $\phi_C = 5\pi/4$. Starting from the positive z-axis you have to go down $\pi/4$ to reach point B, $\pi/2$ to reach point A, and $3\pi/4$ to reach point C. Finally, we can get r using the distance formula.

$$A : (\sqrt{2}, \pi/4, \pi/2)$$
$$B : (\sqrt{3}, \pi/4, \pi/4)$$
$$C : (\sqrt{3}, 5\pi/4, 3\pi/4)$$

See Appendix D for the equations for converting between Cartesian and spherical coordinates. (They are not hard to derive once you realize that the cylindrical ρ equals $r \sin \theta$.)

Figure 5.11 shows a differential box in spherical coordinates. Its volume is $r^2 \sin \theta \, dr \, d\theta \, d\phi$, so the Jacobian in spherical coordinates is $r^2 \sin \theta$. You'll derive this formula in Problem 5.181.

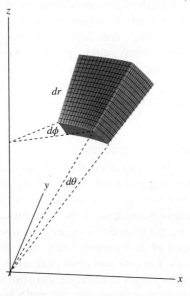

FIGURE 5.11

EXAMPLE — Never Trust a Ferengi

Question: A well-known swindler named Quark attempts to sell you a sphere of pure dilithium, but your lab analysis indicates that only the outer edge is pure dilithium, and the concentration (fraction of dilithium by volume) rises linearly with radius from zero at the center to one at the edge. What fraction of this sphere is dilithium?

Solution:
Both the region of integration and the integrand are perfect candidates for spherical coordinates. The concentration is $c = r/R$, where R is the radius of the sphere. The volume of dilithium in a differential box is the volume of the box times the fraction of it that is dilithium, so $dV_d = (r/R) r^2 \sin\theta \, dr \, d\theta \, d\phi$. You can integrate dV_d to find the total volume of dilithium in the sphere, and divide that by the volume of the sphere to find the fraction of dilithium. To cover all directions, ϕ should go from 0 to 2π and θ should go from 0 to π. To cover the entire sphere r should go from 0 to R.

$$f = \frac{1}{V_{sphere}} \int_0^{2\pi} \int_0^{\pi} \int_0^R \frac{1}{R} r^3 \sin\theta \, dr \, d\theta \, d\phi = \frac{3}{4}$$

The answer comes out unitless, which makes sense for volume of dilithium per volume of the sphere.

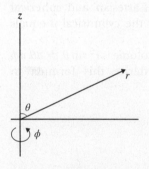

Our example above featured the simplest possible shape in spherical coordinates, a sphere, leading to constant limits of integration. For more challenging regions, the trick we used in cylindrical coordinates applies here as well: you can often start with simple limits on ϕ. This once again leaves you with whatever 2D region you rotated to get your 3D shape, but this time the remaining coordinates cover that region in something much like *polar* coordinates: r in this space is like the polar ρ, and θ is like (but not identical to) the polar ϕ.

Finding the limits of integration to express a two-dimensional region in polar coordinates is a skill you have practiced, and it will serve you here. You'll see examples of this in the problems, but the good news is that most real-world integrals where you need to use spherical coordinates are either over spheres or simple portions of spheres (e.g. hemispheres) where the limits of integration are easy to find.

The Awkward Variable-Naming Discussion

Mathematicians generally use r for a point's distance from the z-axis, and ρ for its distance to the origin. They use θ for the angle around the z-axis, and ϕ for the angle down from the z-axis. You probably first encountered these variables in a math class, meeting r and θ as the polar coordinates, and you may well feel that all our variables are backward.

However, the conventions we presented in the sections above—r from the origin and ρ from the z-axis, ϕ going around while θ comes down—are the most commonly used ones among physicists and engineers. You can find texts that use ρ and r the other way around, ones that use r to mean different things in different coordinate systems, ones that use θ to mean different things in different coordinate systems, and a host of other possible combinations. Almost no physics or engineering texts use all the same conventions as math books, however.

To make matters worse, both mathematicians and physicists write spherical coordinates with θ before ϕ. Since the meanings of θ and ϕ are reversed, the coordinates $(1, \pi/2, \pi/6)$ represent two completely different points in the two different systems. And just to add one more bit of confusion, the angle around the z-axis is called the "azimuthal" angle and the angle down from the z-axis the "polar" angle, even though the azimuthal angle is the one that shows up in polar coordinates.

We are writing a book for physics and engineering students, and we will consistently follow the conventions presented in this section. If you are taking another math class while going through this book, you may have to change letters on a daily basis. In response to the confusion this may engender, we offer the following sentiment from the famous philosopher Han Solo:

It's not our fault.

Stepping Back: Setting up 3D Integrals

The first question in any multidimensional integration is what coordinate system will make the problem easiest. That depends on two things: the function you are integrating (for instance, it's easier to integrate r than $\sqrt{x^2 + y^2 + z^2}$), and the region of integration (for instance, shapes that are symmetric around the z-axis lend themselves to cylindrical or spherical coordinates).

The function and the shape will often push toward the same coordinate system. (Things in cylinders don't tend to depend on x and y.) But when these two criteria conflict, pick one coordinate system, and if it gets too messy consider going back and trying another.

Once you've chosen your coordinate system, don't forget to include the volume of a differential box in that system when writing the integrand. These areas and volumes are listed for the common coordinate systems in Appendix D. You use that area or volume when calculating a property of a differential box such as its volume, mass, or moment of inertia.

Having found your integrand, choose which integral to take first, usually choosing whichever looks easiest. The limits on ϕ are often constant, making it a good first choice in many cases. Whatever integral you do first, you can then use the 2D shape that covers the remaining two variables to find your limits of integration.

What about other coordinate systems? Should we invent "conical coordinates" and "dodecahedral coordinates" for problems with those shapes? Generally not: for most purposes Cartesian, polar, cylindrical, and spherical will suffice. But some circumstances, including many relativistic calculations, do call for new coordinate systems. Equation 5.6.2 gives the formula for finding the Jacobian in any coordinate system.

5.7.3 Problems: Cylindrical and Spherical Coordinates

5.163 Each of the following points is specified in Cartesian, cylindrical, or spherical coordinates. Convert each point to both of the other two systems.
(a) $(x, y, z) = (1, -1, 4)$
(b) $(x, y, z) = (-1, 1, 4)$
(c) $(\rho, \phi, z) = (5, 2\pi/3, -1)$
(d) $(r, \theta, \phi) = (1, 0, 0)$

5.164 Describe and/or draw each of the following surfaces or regions.
(a) $z = \rho^2$
(b) $\rho^2 + z^2 = 25$
(c) $\rho^2 \leq z \leq 4$
(d) $\theta = \pi/6$
(e) $0 \leq \theta \leq \pi/6, 0 \leq r \leq 2$

5.165 Each of the formulas below specifies a surface or region in Cartesian, cylindrical, or spherical coordinates. Rewrite the formula for each surface or region in both of the other two systems.
(a) $z = 5$ (This is the same in Cartesian and cylindrical, so you only need to rewrite it in spherical coordinates.)
(b) $\theta = \pi/4$
(c) $x^2 + y^2 = 4$

236 Chapter 5 Integrals in Two or More Dimensions

5.166 *This problem requires that you know how to calculate the determinant of a matrix. See Chapter 6. Use Equation 5.6.2 to find the Jacobians for cylindrical and spherical coordinates. Make sure your answers match the ones we gave.*

5.167 The Earth has a radius of 4000 miles, and rotates about its axis once every 24 h. We take the axis of rotation as the z-axis. A dead girl named Lucy starts out at time $t = 0$ at the (Cartesian) position $(0, 3464, 2000)$ and lies in the ground, rotating with the Earth.

(a) Write spherical functions $r(t)$, $\theta(t)$, $\phi(t)$ for her position as a function of time.

(b) Convert your functions to $x(t)$, $y(t)$, $z(t)$.

5.168 The Earth has a radius of approximately 6000 km and rotates around its axis every 24 h. James begins at the South Pole and flies, drives, and boats so that he is always going due North at 50 km/h. (Note that he maintains a distance of 6000 km from the center of the Earth—the origin—and that while he travels North, he is also spinning with the Earth.) We take the axis of rotation as our z-axis.

(a) How long does it take James to travel from the South Pole to the North Pole?

(b) Give the spherical functions $r(t)$, $\theta(t)$, $\phi(t)$ that represent James' motion.

(c) Give the cylindrical functions $\rho(t)$, $\phi(t)$, $z(t)$ that represent his motion. (Your answers will be easier to interpret if you simplify them with a few trig identities.)

(d) At what time does James reach the equator? Explain how you can tell that your ρ and z functions make sense at this time.

(e) At what time does James reach the North Pole? Explain how you can tell that your ρ and z functions make sense at this time.

(f) Give the Cartesian functions $x(t)$, $y(t)$, $z(t)$ that represent his motion.

5.169 Walk-Through: Integral in Cylindrical or Spherical Coordinates. An object enclosed by the paraboloid $z = c\left(x^2 + y^2\right)$ from $z = 0$ to $z = H$ has density $D = k\sqrt{x^2 + y^2}/z$.

(a) What are the units of c and k?

(b) Sketch the region. Your sketch doesn't have to be a work of art, but it should be clear enough to give you a clear sense of what region you're integrating over.

(c) Explain why the limits of integration for this region will be simpler in cylindrical coordinates than in Cartesian or spherical, having nothing to do with what function you are integrating. Since the function f is also simplest in cylindrical coordinates, that's the system you'll use.

(d) Rewrite the formulas for the paraboloid and the density in terms of cylindrical coordinates.

(e) Write the mass dm of a differential volume. Be sure to include the Jacobian in cylindrical coordinates.

(f) What are the limits of integration for ϕ?

(g) Draw the 2D shape that remains after you've written the ϕ integral. The drawing should have a ρ-axis and a z-axis. Include a differential box in your drawing. (This small box in ρz space represents a ring in xyz space; take a moment to visualize that.)

(h) Using your 2D drawing to guide you, find the limits of integration for ρ and z. You can take these two integrals in either order, but for this problem the $d\rho \, dz$ integration turns out easier than $dz \, d\rho$. (We know because we tried both ways; sometimes that's what you have to do.)

(i) Write and evaluate the triple integral to find the mass of the object.

(j) Check the units of your answer.

For Problems 5.170–5.175 integrate the given function over the indicated region.

5.170 $f(\rho, \phi, z) = \rho^3 \cos\phi/z$. The region is the cylinder with boundaries $z = H_1$, $z = H_2$, and $\rho = R$.

5.171 $f(\rho, \phi, z) = e^{\rho^2} z^2$. The region is defined by taking the cylinder bounded by $z = 0$, $z = H$, and $\rho = R$ and only considering the half of it with $x \geq 0$.

5.172 $f(\rho, \phi, z) = z\cos(\phi/4 + \rho/R)$. The region is a right circular cone of height H and radius R pointing upward with its vertex at the origin.

5.173 $f(r, \theta, \phi) = r^2 \cos^2\theta$. The region is a ball of radius R centered on the origin.

5.174 $f(r, \theta, \phi) = e^{r/R} \cos\phi$. The region is the $x \geq 0$ half of a ball of radius R centered on the origin.

5.175 $f(r, \theta, \phi) = r^2$. The region is a right circular cone of height H and radius R pointing upward with its vertex at the origin.

In Problems 5.176–5.179 you will redo problems from Section 5.5. You did them the first time in Cartesian coordinates; now you will use cylindrical or spherical. You should be able to set up and evaluate all necessary integrals without a computer.

5.176 Problem 5.107.

5.177 Problem 5.109.

5.178 Problem 5.116.

5.179 Problem 5.118.

5.180 In Cartesian coordinates all three coordinates must go from $-\infty$ to ∞ in order to cover all space. In cylindrical coordinates, you can cover all possible points with $0 \leq \rho < \infty$, $0 \leq \phi < 2\pi$, and $-\infty < z < \infty$. In spherical coordinates, we traditionally cover all possible points with $r \geq 0$, $0 \leq \theta \leq \pi$, and $0 \leq \phi < 2\pi$, but there are other possibilities.

  (a) Describe the region covered with the more restrictive $0 \leq \phi \leq \pi$ (using the same range of values given above for the other coordinates).

  (b) Describe the region covered in spherical coordinates if $-\infty < r < \infty$, $0 \leq \theta \leq \pi$, and $0 \leq \phi \leq \pi$.

  (c) Describe the region covered in spherical coordinates if $0 \leq r < \infty$, $0 \leq \theta \leq 2\pi$, and $0 \leq \phi \leq \pi$.

5.181 Figure 5.11 shows a differential box in spherical coordinates. We assume that in the limit where the box is small enough we can treat this box as a rectangular solid, so finding its volume is reduced to finding the lengths of its three sides. We'll even give you one for free; the one labeled dr in the picture has length dr.

  (a) One of the sides of the box is an arc (a segment of a circle) that subtends an angle $d\theta$. What is the radius of the circle that this arc is part of? Give your answer in terms of r, θ, and/or ϕ.

  (b) What is the length of the arc in Part (a)?

  (c) Repeat Parts (a) and (b) for the side that subtends an angle $d\phi$. Give the radius of the circle it is part of (this is not entirely obvious), and then the length of the arc.

  (d) Multiply the lengths of the three sides to find the volume of the differential box.

5.182 In the Explanation (Section 5.7.2) we took an integral of a cylinder of radius R and height H in cylindrical coordinates. After integrating with respect to ϕ we claimed that we had reduced the problem to a 2D rectangle of height H and width R. Why did that rectangle only have width R even though the cylinder has a width $2R$?

5.183 A right circular cone of height H and upper radius R with its vertex at the origin has density $k\rho$ where ρ is distance from the z-axis. Find its total mass.

5.184 An Hourglass Figure The figure below shows an object shaped like an hourglass, with the z-axis going through its center. The top and bottom are at $z = \pm H$ and the sides are described by the function $x^2 + y^2 = z^2 + a^2$. The density of the object is $c\rho^2$, where ρ is distance from the z-axis.

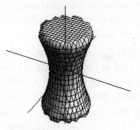

  (a) Find the object's mass.

  (b) Write, but do not evaluate, a triple integral that represents the moment of inertia of this shape about the z-axis.

  (c) Write, but do not evaluate, a triple integral that represents the moment of inertia of this shape about the x-axis.

5.185 You're making a giant sculpture of an ice cream cone. The side of the sculpture is the right circular cone $z^2 = 2\left(x^2 + y^2\right)$ from $z = 0$ to $z = H$. The top of the cone is a portion of the sphere $x^2 + y^2 + z^2 = R^2$. The density of the sculpture is kr, where r is distance from the vertex of the cone. Find the total mass of the sculpture.

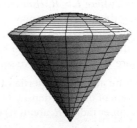

  (a) The constants R (radius of the sphere) and H (height of the cone) are not independent, because the two shapes have to meet at the boundary. Find the relationship between these two constants.

(b) Find the total mass of the sculpture. Your answer should contain either H or R but not both.

5.186 Find the moment of inertia of a uniform sphere of radius R and mass M about an axis through its center.
(a) Write, but do not evaluate, the triple integral that represents this problem in Cartesian coordinates.
(b) Write, but do not evaluate, the triple integral that represents this problem in cylindrical coordinates.
(c) Write, but do not evaluate, the triple integral that represents this problem in spherical coordinates.
(d) Evaluate one of those triple integrals—whichever one looks easiest!—to find the answer.

5.187 A spherical shell of mass M has inner radius R_1 and outer radius R_2. (In other words it's a sphere of radius R_2 with a spherical hole in the middle of radius R_1.)
(a) Find the moment of inertia of the shell about an axis through its center.
(b) Evaluate your answer for the case $R_1 = 0$. What does that answer represent?
(c) Evaluate your answer in the limit $R_1 \to R_2$. What does that answer represent?

5.188 In Section 5.1 we presented Isaac Newton's problem: show that a planet can be mathematically treated as if it were a point particle at its center. We will approach this problem for gravitational *force* in Section 5.11 (see felderbooks.com), but here you will tackle it with gravitational *potential*. Every particle of mass m creates a potential $V = -Gm/r$ at a distance r away from itself. Find the total potential created by a uniform sphere of mass M and radius R at a point L away from the center, where $L > R$. *Hint*: At one point in taking the integral you may come across the expression $\sqrt{(a-b)^2}$. Figure out whether a or b in that expression is larger so that you can be sure to take the positive square root.

5.189 Find the moment of inertia of a right circular cone of uniform density D, with height H and upper radius R, about its central axis.

5.190 A sphere of radius R has density proportional to distance from its center. Find the moment of inertia of the sphere about an axis through its center. (Your answer will contain a constant of proportionality. Make sure you evaluate the units of that constant and use them to check the units of your answer.)

5.191 A cylindrical container with radius R and height H is rotating about its central axis. The air inside the cylinder is effectively pushed to the outside, so the density grows exponentially with distance from the axis of rotation. Meanwhile the cylinder is tall enough that gravity causes the density to fall off exponentially with height. (In Problem 5.196 you will analyze a more realistic model of density in a spinning cylinder.)
(a) Write the density function. The function should depend on the coordinates in whatever coordinate system you are using and should also include some unknown constants.
(b) Identify the units of all of the constants you introduced.
(c) Set up a triple integral to find the moment of inertia of the cylinder about its central axis.
(d) Evaluate your integral.
(e) Check that your answer has correct units.

5.192 A cylinder of radius 1 m and height 2 m has a uniformly distributed charge 10^{-8} C. Find the electric potential produced by the cylinder at a point on the axis of the cylinder a distance 4 m from the center of the cylinder.

5.193 Imagine two right circular cylinders of radius 1, one centered on the x-axis and the other on the z-axis. A uniform object of mass 1 fills the space where the two cylinders intersect. You're going to find the moment of inertia of that object about the z-axis.

(a) Explain why this problem is easiest to solve in cylindrical coordinates.
(b) Explain why it would be hard to set up the limits of integration for either the ϕ or ρ integrals first. You will therefore start with the z integral.
(c) We won't ask you to draw this shape with a box and a strip, but even without that you can see from the picture that a vertical strip will go from the bottom of the x-axis cylinder to the top. To

turn that fact into limits of integration for z, write the formula for the x-axis cylinder *in Cartesian coordinates.*

(d) Now rewrite the formula for the x-axis cylinder in cylindrical coordinates, and use that to find the limits of integration for the z integral.

(e) With the z integral specified, the remaining 2D shape you need to integrate over in ρ and ϕ is the shadow of the 3D object on the xy-plane. Draw that shadow. (*Hint*: If you're stuck, think about the z-axis cylinder that goes up through this shape.)

(f) Using your 2D drawing, find the limits of integration for ϕ and ρ and write an expression for the moment of inertia of the object. Your expression should be a fraction with a triple integral in the numerator and in the denominator.

(g) 💻 Evaluate your integral numerically to find I.

5.194 An object fills the ellipsoid $x^2 + y^2 + 2z^2 = R^2$ with a density of kz^2. Find the moment of inertia of the object about the z-axis.

(a) What are the limits of integration for ϕ?

(b) Draw the 2D shape that would rotate about the z-axis to make the ellipsoid. Explain why the horizontal axis in this drawing only needs a positive side.

(c) Rewrite the equation for the ellipsoid in cylindrical coordinates.

(d) Find the limits of integration for ρ and z and write the triple integral for the moment of inertia of the object.

(e) Evaluate your integral to find I.

5.195 An object fills the ellipsoid $x^2 + y^2 + 2z^2 = 1$. Its density is given by the spherical $r^2 \sin\theta$. Find the mass of the object.

(a) What are the limits of integration for ϕ?

(b) Draw the 2D shape that would rotate about the z-axis to make the ellipsoid. Explain why the horizontal axis in this drawing only needs a positive side.

(c) Rewrite the equation for the ellipsoid in spherical coordinates. After simplifying, your answer should be a function $r(\theta)$ with no ϕ dependence.

(d) Find the limits of integration for r and θ and write the triple integral for the mass of the object.

(e) 💻 Evaluate your integral numerically to find M.

5.196 **Exploration: Babylon 5**
You are the chief design engineer for the space station "Babylon 5." The station is a cylinder with radius 500 m and length 9000 m. It rotates about its central axis once every 45 s, which creates Earth-like artificial gravity on the edge of the station. One of your most important jobs is to design the engines that start the station spinning and correct its rotation if it gets perturbed (e.g. during an alien attack). To know how strong you need to make the engines, you need to know the station's moment of inertia.

The rotation has the effect of pushing the air toward the outer edges of the station. Specifically, the density of air D as a function of distance ρ from the central axis obeys the differential equation

$$\frac{dD}{d\rho} = \frac{4\pi^2 m N_A}{RTt^2}\rho D$$

where m is the mass of a nitrogen molecule, N_A and R are Avogadro's number and the gas constant, T is the temperature, and t is the period of rotation (given above as 45 s).

(a) Solve this differential equation to find $D(\rho)$, using the boundary condition that the density of air at the edge of the station, $\rho = \rho_{max}$ (i.e. 500 m), is equal to D_E, the density of air at sea level on Earth. *Use letters at this point. You'll plug in numbers later.*

(b) Using the density function you just found, write and evaluate a triple integral for the moment of inertia of Babylon 5. Your calculation will only include the air, not the metal that the space station is made of. Once again your answer should only include letters: use L for the station's 9000m length. Check that the units of your answer are correct.

(c) Now plug in numbers. Use a comfortable room temperature for T (and be sure to put it in Kelvins) and look up any other information you need. Your final answer should be a numerical value for the moment of inertia of the air in Babylon 5.

(d) Now consider the metal. Babylon 5 contains 2.5 million tons of metal, or roughly 2.3 billion kilograms. Assuming that the metal is uniformly distributed around the outer edge of the station, find the moment of inertia of the metal frame of Babylon 5.

(e) Combining your answers, find the total moment of inertia of the station.

(f) A collision with a heavy freighter has slowed the rotation of Babylon 5 to a period of 47 s, so the engines must get it rotating at its usual speed again. Using the approximation that the moment of inertia remains at the value you found in Part (e), calculate how much torque the engines must exert to get Babylon 5 back to its proper rotation speed in one hour.

5.8 Line Integrals

A "line integral" adds up the contributions of a field along a curve.

5.8.1 Discovery Exercise: Line Integrals

We begin with a fact from physics: if a force $\vec{F}$ acts upon an object that moves through a linear displacement $\vec{ds}$, the work done by the force on the object is given by the dot product $W = \vec{F} \cdot \vec{ds}$.

1. Calculate the work done by a force $\vec{F} = 2\hat{i} + 3\hat{j}$ on an object as it moves along a line from the point $(1, 1)$ to $(5, 25)$.

   See Check Yourself #32 in Appendix L

Part 1 does not require an integral, because a constant force is acting along a straight line. If the force or the path varies, you employ the usual strategy: break the journey into small parts, compute each part with a simple dot product like Part 1, and use an integral to add up all the contributions.

Consider, for instance, a bead moving along a wire in the shape of the curve $y = x^2$. A force $\vec{F} = (x + y)\hat{i} + (xy)\hat{j}$ acts on the bead as it moves from $(1, 1)$ to $(5, 25)$. Your job in this exercise is to calculate the work done by the force on the bead.

We begin with a small part of the journey—small enough to treat the force as a constant and the journey as a line.

2. The force is expressed as a function of x and y. Re-express the force—both its x- and y-components—as a function of x only. Remember what curve we are moving along!

3. The figure to the left represents a small displacement $\vec{ds} = dx\hat{i} + dy\hat{j}$ at some point on the curve $y = x^2$. Express the relationship between dy and dx, based once again on the curve. (*Hint*: It's not $dy = dx^2$.)
4. Calculate the work done along this small displacement. Your answer should be expressed as a function of x and dx only.
5. Integrate your answer as x goes from 1 to 5.

See Check Yourself #33 in Appendix L

5.8.2 Explanation: Line Integrals

Line Integrals of Vectors

You may recall from introductory physics that when you push or pull an object the "work" you do on it is defined as the force you exert times the distance the object moves in the direction of that force. Mathematically, if a constant force $\vec{F}$ pushes an object through a linear displacement $\vec{s}$, the work is $W = \vec{F} \cdot \vec{s}$. Work is important because it tells you how much kinetic energy you impart to an object.

If the force is not constant or the path is not straight, you break the displacement into small trips for which you can treat the force as constant and displacement as straight, and add up the work done on all of those tiny trips. Hopefully by this point you recognize that as a description of integration: $W = \int \vec{F} \cdot \vec{ds}$.

5.8 | Line Integrals

Mechanical work is a specific example of the mathematical idea of a "line integral." All line integrals perform the following operation.

- You have a curve C. In the example of work, that curve represents the path along which an object travels.
- You also have a vector field $\vec{v}$. In the example of work, that field represents a force acting on the object as it travels along C.
- You break the curve into small displacements $d\vec{s}$. For each such displacement, you calculate *length of displacement* multiplied by *component of the vector field in the direction of the displacement*. This is the dot product $\vec{v} \cdot d\vec{s}$. In the example of work, that dot product represents the amount of mechanical work done by that (relatively constant) force over that (relatively linear) displacement.
- Then you add up those products all along the curve. This is of course the job of an integral, $\int_C \vec{v} \cdot d\vec{s}$. In the example of work, this finds the total work done by that particular force as the object travels along that particular path.

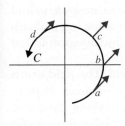

FIGURE 5.12 A vector $\vec{v}$ shown at several points along a curve C. At point a, $\vec{v}$ is parallel to the curve, so $\vec{v} \cdot d\vec{s} = v\, ds$, the largest it can be. At point b the curve is vertical, so only the vertical component of $\vec{v}$ contributes to $\vec{v} \cdot d\vec{s}$. At point c, $\vec{v}$ is perpendicular to C, so $\vec{v} \cdot d\vec{s} = 0$. Finally, at point d the vector is antiparallel to the curve, so $\vec{v} \cdot d\vec{s} = -v\, ds$.

More concisely, the line integral asks the question: "How much does vector field $\vec{v}$ point along curve C?" For instance, Figure 5.12 shows a curve C and a constant vector field $\vec{v} = 2\hat{i} + \hat{j}$. Because the line integral adds up the component of the vector that points *in the direction of the curve*, this particular vector contributes differently at different points along C.

As a way of testing your understanding, try to answer the questions in the example below before you read our answers.

> **EXAMPLE** **Line Integrals That Don't Require Integration**
>
> For these problems, curve C is a straight line from $(1, 3)$ to $(4, 7)$. Note that the length of curve C is 5 and that, in each example, the magnitude of the force is 10.
>
> 1. Evaluate $\int_C \vec{f} \cdot d\vec{s}$ where $\vec{f} = 6\hat{i} + 8\hat{j}$.
>
>   The key to this problem is recognizing that $\vec{f}$ always points in the same direction as the curve. If we ask "how much of this function points along that curve?" the answer is "all of it." When two vectors are parallel, a dot product becomes simple multiplication; for instance, if this force pushes an object through this displacement, the work done is force × displacement. This force adds 50 to the kinetic energy of the object (in whatever units are being used).
>
> 2. Evaluate $\int_C \vec{g} \cdot d\vec{s}$ where $\vec{g} = -6\hat{i} - 8\hat{j}$.
>
>   In this case, the force always points directly *opposite* the displacement. When two vectors are antiparallel, a dot product is negative. The line integral is −50.
>
> 3. Evaluate $\int_C \vec{h} \cdot d\vec{s}$ where $\vec{h} = 8\hat{i} - 6\hat{j}$.
>
>   This force always points *perpendicular* to the displacement, so the dot product, and thus the line integral, is zero. If $\vec{h}$ is a force and C a displacement, the force neither speeds up nor slows down the motion.
>
> 4. Evaluate $\int_C \vec{m} \cdot d\vec{s}$ where $\vec{m} = 8\hat{i} + 6\hat{j}$.
>
>   This force is neither parallel nor perpendicular to the motion. This is where the dot product really comes into play, to isolate the component of the force that points along the direction of motion. $(8\hat{i} + 6\hat{j}) \cdot (3\hat{i} + 4\hat{j}) = 48$: slightly less than in Part 1, because the perpendicular component of the force is wasted.

For a constant vector field along a straight line, you don't actually have to integrate to find a line integral. So the examples above should help you understand what a line integral tells you about a vector field and a curve, but they won't help you evaluate line integrals in less trivial situations.

Consider, therefore, a charged particle in a non-uniform electric field. The field exerts a force $\vec{F} = kx\,\hat{i} + (qy/x)\hat{j}$ while the particle moves along the path $y = cx^2$ from $(0,0)$ to (x_f, cx_f^2). (The particle must be experiencing other forces to make it move along this curve, but you don't need to know them to solve the problem.) What work would that force do on the particle?

To answer that question, we have to clearly define the differential displacement. In two dimensions we can always write $d\vec{s} = dx\,\hat{i} + dy\,\hat{j}$. But dx and dy are not independent in that expression; they are related by the equation for the curve, so in this case $dy = 2cx\,dx$. The equation for the curve also allows us to replace every occurrence of y in the formula for $\vec{F}$ with cx^2, so the work done during one differential displacement can be written:

$$dW = \vec{F} \cdot d\vec{s} = kx\,dx + \frac{q(cx^2)}{x}(2cx\,dx) = \left(kx + 2qc^2x^2\right)dx$$

The total work is thus

$$W = \int_0^{x_f} \left(kx + 2qc^2x^2\right) dx = \frac{k}{2}x_f^2 + \frac{2qc^2}{3}x_f^3$$

From the equations for $\vec{F}$ and the equation of the curve $y = cx^2$ we see that k has units of force over distance, q has units of force, and c has units of one over distance, so both terms have units of force times distance. Notice that the work came out finite even though the force was divergent at the starting point of the path. (Think about the dot product at that point.)

We can summarize what we've said so far in the following definition.

Definition: Line Integral

A "line integral" of a vector $\vec{V}(x, y, z)$ along a curve C is the integral:

$$\int_C \vec{V} \cdot d\vec{s}$$

where $d\vec{s} = dx\,\hat{i} + dy\,\hat{j} + dz\,\hat{k}$ is a differential vector pointing tangent to the curve at each point.

In the special case where C is a closed loop (a curve that ends where it began) we use the symbol $\oint$ as in $\oint_C \vec{V} \cdot d\vec{s}$.

In the example we worked above the curve was expressed as $y(x)$, allowing us to rewrite the line integral (both $\vec{F}$ and $d\vec{s}$) as functions of x. It is often more convenient to express a curve parametrically. For example, the two functions $x = 5\cos\theta$, $y = 5\sin\theta$ for $0 \leq \theta \leq 2\pi$ describe a circle, which cannot be written as a function $y(x)$. Equivalently, those two functions can be combined into a position vector $\vec{r} = (5\cos\theta)\hat{i} + (5\sin\theta)\hat{j}$. You evaluate line integrals along such a curve by rewriting everything in terms of the parameter, as illustrated in the example below.

5.8 | Line Integrals

> **EXAMPLE** **Line Integral That Does Require Integration**
>
>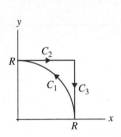
>
> **Question:** Evaluate the line integral of $\vec{V} = (x/y)\hat{i} + k\hat{j}$ along the path shown below: starting at the point $(R, 0)$, tracing out quarter-circle C_1 up to $(0, R)$, then moving straight across C_2 to (R, R), and then back down C_3 to the start.
>
> **Solution:**
> The quarter-circle C_1 can be parameterized as $x = R\cos\phi$, $y = R\sin\phi$ with ϕ going from 0 to $\pi/2$. From these equations $dx = -R\sin\phi\, d\phi$ and $dy = R\cos\phi\, d\phi$, so
>
> $$d\vec{s} = -R\sin\phi\, d\phi\, \hat{i} + R\cos\phi\, d\phi\, \hat{j}$$
> $$\vec{V} = \frac{\cos\phi}{\sin\phi}\hat{i} + k\hat{j}$$
> $$\int_{C_1} \vec{V}\cdot d\vec{s} = \int_0^{\pi/2}\left(-\frac{\cos\phi}{\sin\phi}R\sin\phi + kR\cos\phi\right)d\phi = \int_0^{\pi/2} R(-1+k)\cos\phi\, d\phi = R(-1+k)$$
>
> The horizontal line C_2 is defined by the equation $y = R$. Along such a line only the x-component of the vector matters. (We hope you can see that fact visually, but for a more formal approach, write $d\vec{s} = dx\,\hat{i}$ and see what happens with the dot product.)
>
> $$\int_{C_2} V\cdot d\vec{s} = \int_0^R (x/R)dx = R/2$$
>
> The final stretch C_3 is defined by $x = R$. Here we only care about the y-component of the vector, but be careful: if k is positive then the line integral along this path is negative because we are going down. (Formally, $d\vec{s}$ here is $-dy\,\hat{j}$.) But no integral is required to find $\int_{C_3} \vec{V}\cdot d\vec{s} = -kR$. So the total line integral along the path is:
>
> $$\oint_{C_{total}} \vec{V}\cdot d\vec{s} = R(-1+k) + R/2 - kR = -R/2$$
>
> Since $\vec{V}$ is unitless, $\vec{V}\cdot d\vec{s}$ should have units of distance, which our answer does. It's also not surprising to see k cancel out at the end; consider that the work done by the force $k\hat{j}$ in moving up from 0 to R perfectly cancels the *negative* work done in moving back down again.

Line Integrals of Scalars

Most line integral problems involve vector functions, but in some cases you want to integrate a scalar along a curve. For example, suppose we know the linear density λ of a curved piece of wire and we want to know its total mass. As you know, we approach this by breaking the curve into pieces so small that we can consider the density to be constant in each one. By the definition of linear density the mass of a tiny segment of length ds is λds where ds is a scalar representing the length of a short segment. We then integrate to sum up the entire mass.

As in the vector case, the math plays out a bit differently depending on whether the curve is specified as $y(x)$ or parametrically. In Problems 5.232 and 5.233 you will derive the following formulas.

Line Integral of a Scalar Function

The line integral $\int_C f(x, y)\, ds$ where C is given as a function $y(x)$ is:

$$\int_{x_1}^{x_2} f(x, y(x))\sqrt{1 + (y'(x))^2}\, dx \tag{5.8.1}$$

The line integral $\int_C f(x, y, z)\, ds$ where C is given as $x(t), y(t), z(t)$ is:

$$\int_{t_1}^{t_2} f(x(t), y(t), z(t))\sqrt{(x'(t))^2 + (y'(t))^2 + (z'(t))^2}\, dt \tag{5.8.2}$$

EXAMPLE **Line Integral of a Scalar Along a Curve**

Question: Find the line integral of the function $f = xy$ along the curve $y = x^4$ from $x = 0$ to $x = 1$.

Solution:

$$\int f(x, y)\, ds = \int_0^1 x(x^4)\sqrt{1 + (4x^3)^2}\, dx = \int_0^1 x^5 \sqrt{1 + 16x^6}\, dx$$

This can easily be evaluated to give $(17^{3/2} - 1)/144 \approx 0.48$.

Stepping Back

There are some topics in math (partial derivatives are a vivid example) where the "mechanics" are easy to learn, but the concept takes a lot of hard thought. For other topics, the difficulty may lie more with the process. But line integrals, of vector fields in particular, have the unfortunate property that the concept and the mechanics both take hard work to master. Worse still, they don't seem to have much to do with each other.

For understanding the concept, begin with the question "How much does this vector field point *along* this curve?" Imagine sailing a boat and asking, not "how hard is the wind blowing?", but "how much is the wind helping me go?" That question doesn't seem to have a multiplication in it, but it does. (If you double the strength of the wind, or if you double the length of the journey, the answer will generally double.) So for each differential $d\vec{s}$ along the curve, you are multiplying the length of $d\vec{s}$—not by the magnitude of the vector, but by the component of the vector in the direction of $d\vec{s}$.

That should make you think of a dot product, which brings us to the mechanics.

When you first meet a line integral, it is in the intimidating-looking form $\int_C \vec{v} \cdot d\vec{s}$ with two or three independent variables. But those variables are related by the fact that they are confined to lie on the curve C. So the curve—whether it is expressed as $y(x)$ or as $x(t), y(t), z(t)$—acts as a conversion formula, allowing you to rewrite both $\vec{v}$ and $d\vec{s}$ in terms of one independent variable. At that point, you have a single integral $\int f(t)\, dt$ to evaluate. A line integral is always a single integral in the end, and you know how to handle that.

5.8.3 Problems: Line Integrals

Problems 5.197–5.199 can be calculated without actually taking any integrals. Think about what a line integral means!

5.197 Line L goes from the point $(2,3)$ straight up to $(2,10)$.
  (a) Vector field $\vec{f}$ is $10\hat{j}$. Calculate $\int_L \vec{f} \cdot d\vec{s}$.
  (b) Vector field $\vec{g}$ is $-10\hat{j}$. Calculate $\int_L \vec{g} \cdot d\vec{s}$.
  (c) Vector field $\vec{h}$ is $10\hat{i}$. Calculate $\int_L \vec{h} \cdot d\vec{s}$.
  (d) Vector field $\vec{v}$ is $3\hat{i} + 4\hat{j}$. Calculate $\int_L \vec{v} \cdot d\vec{s}$.

5.198 Curve P is the polar curve $\rho = \cos(3\phi)$ in the xy-plane, and vector $\vec{F} = x^2 y \hat{k}$. Calculate $\oint_P \vec{F} \cdot d\vec{s}$.

5.199 Curve C is a circle of radius 3 in the xy-plane, traversed counterclockwise.
  (a) Vector field $\vec{f}$ has magnitude 10 everywhere, and always points around the circle, counterclockwise. Calculate $\oint_C \vec{f} \cdot d\vec{s}$.
  (b) Vector field $\vec{g}$ has magnitude 10 everywhere, and always points around the circle, clockwise. Calculate $\oint_C \vec{g} \cdot d\vec{s}$.
  (c) Vector field $\vec{h}$ has magnitude 10 everywhere, and always points radially away from the origin. Calculate $\oint_C \vec{h} \cdot d\vec{s}$.
  (d) Vector field $\vec{v} = 10\hat{j}$. Calculate $\oint_C \vec{v} \cdot d\vec{s}$.

5.200 Walk-Through: Line Integrals. A force $\vec{F} = e^x \hat{i} + y^2 \hat{j}$ acts upon an object as it moves along the curve $y = \sin x$ from $x = 0$ to $x = 2\pi$. Your job is to find the work done by this force on this object.
  (a) Using the equation of the curve, rewrite the force as a function of x only (instead of x and y).
  (b) A differential displacement is $d\vec{s} = dx\,\hat{i} + dy\,\hat{j}$ as always. Using the equation of the curve, rewrite this displacement as a function of x and dx only (no y or dy).
  (c) Write the dot product $dW = \vec{F} \cdot d\vec{s}$ and simplify as much as possible.
  (d) Integrate as x goes from 0 to 2π.

For Problems 5.201–5.206 evaluate the line integral of the indicated vector field over the given curve. You may find it helpful to first work through Problem 5.200 as a model.

5.201 $\vec{f}(x,y) = y\hat{i} + x\hat{j}$. The curve is the parabola $y = x^2$ from $x = 0$ to $x = 2$.

5.202 $\vec{f}(x,y,z) = x\hat{i} + yz^2\hat{k}$. The curve is $z = y^2 = x^4$ from the origin to the point $(1,1,1)$.

5.203 $\vec{f}(x,y,z) = (x+y)\hat{i} + z\hat{j}$. The curve is $x(\alpha) = \alpha^2$, $y(\alpha) = \sqrt{\alpha}$, $z(\alpha) = 1/\alpha$ from $\alpha = 1$ to $\alpha = 2$.

5.204 $\vec{f}(x,y) = x^2\hat{i} + y^3\hat{j}$. The curve is a circle of radius 1 centered on the origin, traversed counterclockwise.

5.205 $\vec{f}(x,y) = x\hat{i} + z\hat{j} + y\hat{k}$. The curve is a circle of radius 1 centered on the z-axis at $z = 2$, traversed counterclockwise.

5.206 $\vec{f}(x,y) = x\hat{i} - y^2\hat{k}$. The curve is the intersection of the plane $z = x$ with the paraboloid $z = x^2 + y^2$, going from the origin to the point $(1,0,1)$ on the $y > 0$ side of the curve.

For Problems 5.207–5.213 evaluate the line integral of the indicated scalar field over the given curve.

5.207 $f(x,y) = x + y$. The curve is $y = x^3$ from $x = 0$ to $x = 2$.

5.208 $f(x,y) = xy^2$. The curve is the semicircle $x^2 + y^2 = 4$, $x < 0$.

5.209 $f(x,y) = \sin x$. The curve is $y = \ln(\cos x)$ as x goes from 0 to $\pi/3$.

5.210 $f(x,y) = 3y/x$. The curve is the parabola $y = x^2$ as x goes from 1 to 2.

5.211 $f(x,y,z) = 1/(x+z)^2$. The curve is $x = e^{2t}$, $y = 2e^{2t}$, $z = 3e^{2t}$ as t goes from -1 to 1.

5.212 $f(x,y,z) = xy + z$. The curve is $x = y = z$ from the origin to the point $(2,2,2)$.

5.213 $f(x,y,z) = (x+y)/z$. The curve is $x(t) = t^2$, $y(t) = t^3$, $z(t) = 1/t$ from $(1,1,1)$ to $(9,27,1/3)$.

5.214 A thin tube of gas is bent into the shape $y = cx^3$ from $y = 0$ to $y = H$. The density (mass per unit length) of gas in the tube is ky, where k is a constant. Find the total mass of the gas in the tube.

5.215 A ball is released from rest at the origin and slides down a frictionless track. From conservation of energy the ball's speed at any point on the track is given by $v = \sqrt{-2gy}$, where the height y is always negative since the ball started at $y = 0$. Recall that in any short interval the time it takes to travel a distance ds is $dt = ds/v$. If the track's shape is $y = -kx^{6/5}$ (with k a constant), find the time it takes to slide from the origin to x_f.

5.216 A particle moves from the origin to the point $(1,1)$ while being acted on by the force $\vec{F} = x\hat{i} + (x+y)\hat{j}$.

(a) Find the work done by this force if the particle moves along the x-axis from the origin to $(1,0)$ and then in the y-direction to $(1,1)$.

(b) Find the work if instead the particle moves in a straight line from the origin to $(1,1)$.

(c) Find the work if the particle moves along the curve $y = x^2$ from the origin to $(1,1)$.

5.217 A force $\vec{F} = 5\hat{i} - 12\hat{j}$ acts upon an object that moves in a straight line from the point $(3,9)$ to $(6,13)$.

(a) To calculate the work done by this force on this object, $W = \int_C \vec{F} \cdot d\vec{s}$, it is not necessary to perform an integral. Why not?

(b) Calculate the work done by this force on this object.

(c) Of all the constant forces with magnitude 13, which one would have the *greatest possible* dot product along this line? (*Hint*: this is not a calculus-based optimization problem. Just think about how line integrals work.) Calculate the work done in this case.

(d) Of all the constant forces with magnitude 13, which one would have the *lowest possible* dot product along this line? Calculate the work done in this case.

(e) Find a constant force with magnitude 13 that would do no work at all ($W = 0$) along this line.

5.218 A charged particle experiences an electric force $\vec{F} = 2x\hat{i} - \hat{j}$ as it moves along the path $y = x^2$ from $x = 0$ to $x = 1$. Find the work done by the electric force on the particle.

The gravitational force of the Earth on an object has magnitude k/r^2 where r is the object's distance from the Earth's center and k is a constant that depends on the object's mass. The force points directly toward the center of the Earth. In Problems 5.219–5.224 find the work done on an object as it moves along the indicated path. The center of the Earth is at the origin. Assume all letters other than x, y, z, and t are constants.

5.219 An asteroid falls in a straight line from a distance R_1 to a distance R_2. (Begin by defining your axes.)

5.220 A satellite goes a fourth of the way around Earth in a circle of radius R.

5.221 A particle moves along the path $x = R\cos(\omega t)$, $y = R\sin(\omega t)$, $z = H - pt$ from $t = 0$ to $t = 4\pi/\omega$.

5.222 A rocket goes a fourth of the way around Earth in an elliptical path $x^2/a^2 + z^2/b^2 = 1$, $y = 0$, starting at the point $(0, 0, b)$. (This is an ellipse centered on the Earth, which is not what an orbiting object would do, but a rocket could be made to for reasons that we can't begin to imagine.) *Hint*: If you end up with an integral that looks impossible you should be able to evaluate it with some simple trig identities and a u-substitution.

5.223 A rocket goes all the way around Earth in the same ellipse as Problem 5.222.

5.224 A rocket can go wherever its engines take it, but for an orbiting satellite the center of the Earth is not at the center of a satellite's orbit, but at one focus. Do Problem 5.222 with the x coordinate of the orbit shifted by $\sqrt{a^2 - b^2}$ ($a > b$). For this problem you just need to set up the integral, not evaluate it.

A wire carrying a current I causes a magnetic field $\vec{B}$ to point along a circle around the wire. Problems 5.225–5.230 are about "Ampère's Law," which states that around any closed curve C, $\oint_C \vec{B} \cdot d\vec{s} = \mu_0 I_{enclosed}$ where μ_0 is a positive constant.

5.225 Consider a single wire with current I.

(a) For a loop of radius ρ around the wire, the magnetic field $\vec{B}$ has a constant magnitude. Calculate $\oint_C \vec{B} \cdot d\vec{s}$ around such a loop as function of $|\vec{B}|$ and ρ.

(b) Use your answer and Ampère's Law to find $|\vec{B}|$ as a function of I.

5.226 *[This problem depends on Problem 5.225.]* In Problem 5.225 you found the field from a single wire carrying current I.

(a) Express your answer as a vector $\vec{B}$ in unit vector notation.

(b) Consider a circle of radius ρ surrounding the wire. Write functions $x(\phi)$ and $y(\phi)$ that parametrize that circle in terms of an angle and then use those functions to evaluate $\oint_C \vec{B} \cdot d\vec{s}$ around the circle. Your answer should confirm Ampère's law.

(c) Still using the magnetic field around a single wire, set up the integrals you need to evaluate $\oint_C \vec{B} \cdot d\vec{s}$ around a rectangle that surrounds the wire. You'll have to choose where to put your axes and choose a specific rectangle to integrate around.

(d) Evaluate the integral you found in Part (c) and show that it obeys Ampère's law.

5.227 *[This problem depends on Problem 5.225.]* A long, thin sheet of metal covers the region $-L \leq x \leq L$, $-\infty < y < \infty$. It is carrying a total current I in the positive y-direction, with the current uniformly spread out across the width of the sheet.

(a) Using the result you found in Problem 5.225 for the field from a single wire, set up and evaluate a single integral (not a line integral) for the magnetic field produced at a point (x, y, z) by this sheet of metal.

(b) Take the line integral of the magnetic field you found in Part (a) around the rectangle with corners at $(-2L, 0, -L)$ and $(2L, 0, L)$ and verify that it obeys Ampère's law.

(c) Take the line integral of the magnetic field you found in Part (a) around a circle of radius $2L$ surrounding the sheet and verify that it also obeys Ampère's law.

5.228 A thick wire of radius R carries a current I uniformly spread out throughout the cross-sectional area of the wire. The magnetic field at any point (x, y) inside the wire is $\vec{B} = \mu_0 I / (2\pi R^2) (-y\hat{i} + x\hat{j})$.

(a) How much current is enclosed by a circle of radius ρ centered on the central axis of the wire (with $\rho < R$)?

(b) Evaluate $\oint_C \vec{B} \cdot d\vec{s}$ around the circle from Part (a) and verify that it obeys Ampère's law.

(c) Repeat Parts (a)–(b) for a square of side length $R/2$ centered on the central axis of the wire.

5.229 The xy-plane is covered by an infinite sheet of metal that carries a current in the y-direction with current per unit length in the x-direction equal to J. The magnetic field is in the positive x-direction for points above the plane and the negative x-direction for points below it. You are going to calculate the magnetic field from the sheet of metal by considering a rectangle in the xz-plane that extends through the sheet.

(a) By symmetry the field must be independent of x and y. By considering more than one such rectangle, use Ampère's law to explain how you can know that the magnitude of the field is independent of z (aside from the direction of the field being different for $z > 0$ and $z < 0$).

(b) Evaluate $\oint_C \vec{B} \cdot d\vec{s}$ around one of these rectangles. Your answer should include the as-yet-unknown magnitude $|\vec{B}|$.

(c) Use your answer and Ampère's law to find $|\vec{B}|$.

5.230 A thick wire is carrying a current that is *not* uniformly distributed throughout the wire. Instead the current density is a function of ρ, distance from the central axis of the wire. You measure the magnetic field at various points throughout the wire and find it to be $\vec{B} = k(x^2 + y^2)(-y\hat{i} + x\hat{j})$. Evaluate $\int_C \vec{B} \cdot d\vec{s}$ around a circle of radius ρ around the central axis of the wire and use your result and Ampère's law to find $J(\rho)$, the current per unit area as a function of radius.

5.231 A wire bent into the shape $z = 2x = y^2$ from $(2, 2, 4)$ to $(8, 4, 16)$ has linear density $\lambda = y/z$. Find the total mass of the wire.

5.232 In this problem you will derive Equation 5.8.2, the line integral of a scalar function along a parametrically expressed curve. Our curve is expressed as two functions, $x = x(t)$ and $y = y(t)$. (We will work in two dimensions for the sake of the drawing, but the idea doesn't change in three.) We begin as always with a differential piece of the curve small enough to be considered a line.

(a) Express ds in terms of dx and dy based on the picture.
(b) Express dx in terms of dt based on the formula for the curve, $x = x(t)$ and $y = y(t)$.
(c) Express dy in terms of dt.
(d) Express $\int f(x, y)\,ds$ as a single integral with respect to t.

5.233 Derive Equation 5.8.1, the line integral of a scalar function along a curve expressed as $y(x)$. You can follow the derivation from Problem 5.232, putting everything in terms of dx instead of dt. Alternatively, you can start with Equation 5.8.2, and think of $y(x)$ as a parametric representation with x as the independent variable.

5.234 **Exploration: Vector Line Integrals in Polar Coordinates**
In many physical situations, vector fields are simplest in curvilinear coordinate systems. As an example, consider an object moving around a quarter-circle from $(R, 0)$ to $(0, R)$ under the influence of a force:

$$\vec{F} = \frac{\tan^{-1}(y/x)}{\sqrt{x^2 + y^2}} \left(-y\hat{i} + x\hat{j} \right)$$

(a) Express the curve as a function $y(x)$, and write an integral with respect to x to calculate the work done by the force $\vec{F}$ on the particle. You do not have to calculate this integral; we just want to scare you a bit.
(b) Draw the quarter-circle, and at several points along the path, draw the direction of the force.
(c) The force was given in x- and y-components, but we could express it instead as a ρ-component (pointing directly away from the origin) and a ϕ-component (pointing along the circle). Based on your drawing, would the ρ-component be positive, negative, or zero? Would the ϕ-component be positive, negative, or zero?
(d) Find the magnitude of $\vec{F}$, simplifying as much as possible.
(e) Having determined the magnitude and direction of $\vec{F}$, express $\vec{F}$ in polar coordinates.
(f) Express the path in polar coordinates.
(g) Using the fact that $d\vec{s} = d\rho\,\hat{\rho} + \rho\,d\phi\,\hat{\phi}$, write and evaluate an integral with respect to ϕ for the work done by the force on the particle.

5.235 **Exploration: Scalar Line Integrals in Polar Coordinates**

In Cartesian coordinates, we have seen that the differential ds can be written as $\sqrt{1 + (dy/dx)^2}\,dx$. This comes from a Pythagorean triangle where one leg is dx and the other is $(dy/dx)dx$. In polar coordinates we represent ds with a different Pythagorean triangle. We've drawn this "triangle" above, with the hypotenuse labeled ds. One leg is made by holding ρ constant while varying ϕ, and vice versa for the other.

(a) The "triangle" in the picture is not a real triangle because two of the sides are curved. Why is it still okay to treat it mathematically as if it were a triangle?
(b) Label one leg in the drawing $d\rho$.
(c) Label the other leg in the drawing. Hint: $d\phi$ is not the length of this leg, it is the angle it subtends. You want its length!
(d) Use the Pythagorean theorem to find ds. Using the fact that the length $d\rho$ can be rewritten as $(d\rho/d\phi)d\phi$, express your answer in terms of ρ, ϕ, $d\rho/d\phi$, and $d\phi$. (These might not all show up in the answer.)
(e) Using the ds you just found, write and numerically evaluate a line integral in polar coordinates to find the arc length of the curve $\rho = \sin(3\phi)$ as ϕ goes from 0 to $\pi/3$.

5.9 Parametrically Expressed Surfaces

If x, y, and z are all expressed as functions of one variable (such as time t), the result is a curve. In this section we will express x, y, and z as functions of *two* independent variables, which results in a surface. As we will see, this allows us to mathematically describe surfaces that may not be expressible as $z(x, y)$ functions.

5.9.1 Discovery Exercise: Parametrically Expressed Surfaces

A surface is described by the equations $x = u + v^2$, $y = u^2$, $z = v^2$ on the domain $u \geq 0$, $v \geq 0$. For instance, $u = 7$ and $v = 5$ correspond to the point $(32, 49, 25)$. Your job is to find *all* points that can result from plugging (u, v) points into these functions.

1. For $(u, v) = (0, 0)$, find the values of x, y, and z. Plot the corresponding point (x, y, z). (Note that your drawing is based only on x, y, and z, *not* on u and v.)
2. For $(u, v) = (1, 0)$, plot the appropriate point (x, y, z).
3. For $(u, v) = (1, 1)$, plot the appropriate point (x, y, z).

We can build up a lot of points this way, but let's try a faster method. We'll start by seeing what this graph looks like when $v = 0$. Because $z = v^2$, this means that $z = 0$; that is, we are seeing what this surface looks like on the xy-plane.

4. For $v = 0$ find the functions $x(u)$ and $y(u)$.
5. Use those two functions to find a function $y(x)$ that has no u or v in it.
6. Graph the resulting function. The result is a picture of what this surface looks like on the xy-plane.

   See Check Yourself #34 in Appendix L

So instead of building up our graph one point at a time, we can go one curve at a time.

7. Plug $v = 1$ into the function for z to find what plane we are working on.
8. Plug $v = 1$ into the equations for x and y to find functions $x(u)$ and $y(u)$. Then use those functions to find a function $y(x)$.
9. Draw the curve that you found in Part 8 that shows what this surface looks like on the plane you found in Part 7.
10. Repeat Questions 7–9 for $v = 2$.

    See Check Yourself #35 in Appendix L

11. Repeat Questions 7–9 for $v = 3$.
12. Draw and/or describe the surface you get when you include all real values of u and v.

5.9.2 Explanation: Parametrically Expressed Surfaces

A curve can generally be described by parametric functions $x(t)$, $y(t)$, and (in three dimensions) $z(t)$. We tend to think of the independent variable in such a formulation as representing time, so we are saying "At 2:00 the object was here and by 2:01 it had moved there."

But let's step away from "time" as an interpretation: t is one variable, a point on a number line. The functions $x(t)$ and $y(t)$ provide a mapping from the one-dimensional universe of t to the two-dimensional universe of x and y. For a particular range of t-values, the result is a curve in the xy-plane.

Chapter 5 Integrals in Two or More Dimensions

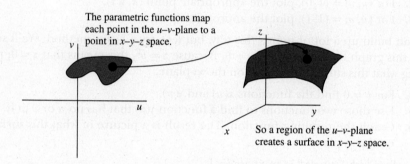

The parametric functions provide a mapping from each "t"-value (point on a number line) to an "x–y" point in a two-dimensional space.

So a range of "t"-values create a curve on the x–y-plane.

Once you think about parametrically expressed curves in this more abstract "mathematical" way, you can step up to parametrically expressed surfaces. You express x, y, and z as functions of *two* independent variables u and v. This maps a region on the uv-plane to a surface in three dimensions.

The parametric functions map each point in the u–v-plane to a point in x–y–z space.

So a region of the u–v-plane creates a surface in x–y–z space.

Expressing surfaces in this way takes some getting used to, but it's important. The simplest alternative, $z(x, y)$, is constrained to obey a vertical line test—any given (x, y) pair can lead to only one z-value—so even simple shapes such as a sphere cannot be expressed that way.

EXAMPLE Parametrically Expressed Surface

Question: Describe the surface $x = v \cos u$, $y = v \sin u$, $z = v^2$ for all values in $0 \leq u \leq 2\pi$, $0 \leq v \leq 3$.

Solution:
Let's begin by making sure you understand the mapping. You can choose any point (u, v) in the domain, and it corresponds to a particular point (x, y, z). The collection of *all* points so generated is the surface we are looking for. For instance, $(u, v) = (\pi/6, 2)$ becomes the point $(x, y, z) = \left(\sqrt{3}, 1, 4\right)$.

If we generate enough points that way, we will see what the surface looks like, but it will take a long time. So we look for shortcuts and patterns. For this particular set of functions, we can begin by examining specific values of v.

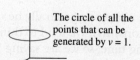

The circle of all the points that can be generated by $v = 1$.

- At $v = 0$ we have $x = 0$, $y = 0$, $z = 0$: the origin.
- At $v = 1$ we have $x = \cos u$, $y = \cos u$, $z = 1$. As u goes from 0 to 2π this traces out a circle of radius 1 at $z = 1$.

- At $v = 2$ we have $x = 2\cos u$, $y = 2\cos u$, $z = 4$. As u goes from 0 to 2π this traces out a circle of radius 2 at $z = 4$.
- At $v = 3$ we have a circle of radius 3 at $z = 9$.

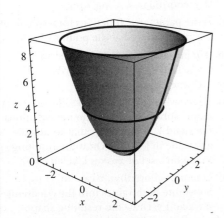

As we move up from the xy-plane, we have circles of increasing radius. The circles we used are called "gridlines of constant v."

We can get more insight into the shape by looking at the gridlines of constant u. For instance, when $u = 0$ we have $x = v$, $y = 0$, and $z = v^2$, so it traces out the parabola $z = x^2$ on the xz-plane. As u goes around, the parabola spins around, so the whole surface is shaped like a bowl.

As a quick check to make sure you're following us, how would you change the function $z(v)$ to make the surface a cone? See Problem 5.244. In this case it would be clearer to call the independent variables ϕ and ρ; there's no law saying they have to be called u and v!

The three functions $x(u, v)$, $y(u, v)$, and $z(u, v)$ can be combined together into the position vector $\vec{r} = x\,\hat{i} + y\,\hat{j} + z\,\hat{k}$. For instance, the example above could be written $\vec{r}(u, v) = v\cos u\,\hat{i} + v\sin u\,\hat{j} + v^2\,\hat{k}$. This is of course the same thing written in a slightly different way. We will see in Section 5.10 that the $\vec{r}$ form is sometimes more useful for plugging into formulas.

5.9.3 Problems: Parametrically Expressed Surfaces

5.236 Walk-Through: Parametrically Expressed Surface. A surface is represented by the equations $x = 2\cos u + v^2$, $y = v$, $z = \sin u$ for $0 \le u \le 2\pi$, $-3 \le v \le 3$.

(a) For each (u, v) pair given below, give the corresponding (x, y, z) point.
  i. $(0, 1)$
  ii. $(\pi/6, -2)$
  iii. $(\pi/2, 3)$

(b) Plug $v = 0$ into the equation for y. What plane contains all the points for which $v = 0$?

(c) Plug $v = 0$ into the equations for x and z. Draw a quick sketch of the ellipse that the resulting $x(u)$ and $z(u)$ functions describe. (If you're not used to drawing ellipses, start with the unit circle $x = \cos\phi$, $y = \sin\phi$ and then double the x-value at each point.)

(d) Repeat Parts (b) and (c) for $v = -2$.

(e) Repeat Parts (b) and (c) for $v = 3$.

(f) Draw and/or describe the entire surface. Include gridlines of constant u and constant v on your drawing, or describe what they look like if you are describing the surface.

(g) How would the surface change if we restricted u to the domain $[0, \pi]$?

(h) How would the surface change if we extended the domain of v to $(-\infty, \infty)$?

5.237 [*This problem depends on Problem 5.236.*]

(a) Write parametric equations for a surface that looks exactly like the surface in Problem 5.236, but moved three units in the $+y$-direction.

(b) Write parametric equations for a surface that looks exactly like the surface in Problem 5.236, but reflected to the other side of the yz-plane.

(c) Write parametric equations for a surface that looks exactly like the surface in Problem 5.236, but with smaller ellipses.

5.238 [This problem depends on Problems 5.236 and 5.237.] Use a computer to graph the surfaces in Problems 5.236 and 5.237. If the drawings do not match your own work, figure out what went wrong!

In Problems 5.239–5.243, draw and/or describe the given surface. Include gridlines of constant u and v in your drawing or describe what they look like. Unless otherwise specified, assume that both u and v span $(-\infty, \infty)$.

5.239 $x = u + v$, $y = 2u + v$, $z = 3u - v$

5.240 $x = 2v \cos u$, $y = 6 - 2v$, $z = 3v \sin u$, $0 \leq u \leq 2\pi$

5.241 $x = u$, $y = v^2$, $z = u^2 - v^4$

5.242 $\vec{r} = u\,\hat{i} + v\,\hat{j} + \sqrt{u^2 + v^2}\,\hat{k}$

5.243 $x = u$, $y = e^{-u} \cos v$, $z = e^{-u} \sin v$, $u \geq 0$, $0 \leq v \leq 2\pi$

5.244 In the example "Parametrically Expressed Surface" on Page 250 we gave parametric equations for a paraboloid.

(a) How could you change the function $z(v)$, leaving everything else the same, to make that surface into a cone instead?

(b) What would happen to the resulting cone if you changed the domain of v to $(-\infty, \infty)$?

5.245 Surface S is a disk of radius 3 centered on the z-axis at $z = 5$.

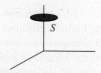

(a) Represent surface S in Cartesian coordinates. (This will require one equation and one inequality.)

(b) Represent surface S in cylindrical coordinates. (Ditto.)

(c) Represent surface S parametrically.

5.246 Surface S is a right circular cylinder of radius 4 centered on the x-axis. One end of the cylinder is on the yz-plane, and the length of the cylinder is 10 units.

(a) Represent surface S parametrically. Make sure to specify the domains of your u and v variables.

(b) How would you change your answer if you only wanted to represent the quarter-cylinder where y and z are both negative?

5.247 (a) Write parametric functions for a sphere of radius 7 centered on the origin. *Hint*: Call your independent variables θ and ϕ instead of u and v, and write out the conversions from spherical to Cartesian coordinates, using 7 for r.

(b) Write parametric functions for a sphere of radius 7 centered at the point $(3, 7, -4)$.

5.248 The equations $x = uv$, $y = u$, $z = v^2$ for $-1 \leq u \leq 1$ and $-2 \leq v \leq 2$ describe the "Whitney Umbrella."

(a) To see what this shape looks like on the xy-plane, substitute $v = 0$ into the equations for x and y. (Remember that u can take any value in $[-1, 1]$.) Draw the resulting shape on a set of xy-axes labeled $z = 0$.

(b) One unit above that, v can be -1 or 1. Substitute both values into the equations for x and y. Draw the resulting shape on a set of xy-axes labeled $z = 1$.

(c) Do a similar drawing for $z = 4$.

(d) Describe and/or draw the Whitney Umbrella.

5.249 The equations $x = (a + b \cos v) \cos u$, $y = (a + b \cos v) \sin u$, $z = b \sin v$ where u and v both range from 0 to 2π define a torus. Graph this shape on a computer for different values of a and b. Describe in words what the shape looks like, and what a and b represent about the shape.

5.250 The function $y = \sqrt{x}$ in the xy-plane is rotated around the x-axis. In this problem you will write parametric functions for the resulting surface.

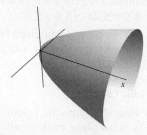

(a) Explain why this surface cannot possibly be expressed by a function $z(x, y)$.

(b) Rather than u and v, the two parameters will be x (position on the x-axis) and θ (an angle rotating around the x-axis). Describe and/or draw the gridlines of constant x and constant θ.

(c) At $x = 9$ we want a circle around the x-axis of radius 3. Write parametric functions $y(\theta)$ and $z(\theta)$ for this circle.

(d) Write the parametric equations $x(x, \theta)$, $y(x, \theta)$ and $z(x, \theta)$ for this surface.

5.251 *[This problem depends on Problem 5.250.]*
(a) Write parametric equations for the surface that results when the curve $z = \sin x$ on the xz-plane is rotated about the x-axis.
(b) 🖥 Use a computer to graph your equations from Part (a) and confirm that the resulting surface looks right.

5.252 *[This problem depends on Problem 5.250.]* Write parametric equations for the surface that results when the function $y = f(x)$ on the xy-plane is rotated about the x-axis.

5.253 If $x(u, v)$, $y(u, v)$, and $z(u, v)$ are all linear functions, the resulting surface is generally a plane.
(a) To demonstrate this result, let $x = au + bv + c$ and similarly for y and z (but with different constants). Solve the first two equations for u and v and plug them in to find an equation $z(x, y)$, which should be a plane.
(b) Give an example of three linear parametric functions that do not describe a plane. What do they describe?

5.254 A parametric surface can be expressed as the vector-valued function $\vec{r}(u, v)$. So the partial derivative $\partial \vec{r}/\partial u$ is a vector that points in a certain direction for any given values of u and v.

(a) In general, which way would you expect this vector to point? (Think about what partial derivatives mean!)
  i. Along a gridline of constant u.
  ii. Along a gridline of constant v.
  iii. Neither of the above.
  Briefly explain your answer.

Now you're going to test it on the surface we examined in the Explanation (Section 5.9.2): $\vec{r}(u, v) = v\cos u\,\hat{i} + v\sin u\,\hat{j} + v^2\hat{k}$.

(b) Draw the surface. Find the point represented by $(u, v) = (0, 1)$ and mark it on your drawing.
(c) From that point, which way do the gridlines of constant u and v go out? Explain your answer visually, in terms of the way this surface draws itself.
(d) Calculate $\partial \vec{r}/\partial u$ at that point. Does it point along the gridline of constant u, of constant v, or neither?
(e) Calculate $\partial \vec{r}/\partial v$ at that point. Does it point along the gridline of constant u, of constant v, or neither?

Make sure all your answers agree. By the end you should have a simple explanation of which way $\partial \vec{r}/\partial u$ points—not just for this surface, but in general—and why. This result will be important to our work in Section 5.10.

5.10 Surface Integrals

A line integral adds up a field along a curve, asking at each point "how much does this field point along the curve?" A surface integral adds up a field along a surface, asking at each point "how much does this field point *through* this surface?"

5.10.1 Discovery Exercise: Surface Integrals

Rainfall is often measured in inches per hour. If that seems strange, imagine placing these three buckets side by side in a rainstorm. The rising waterline in each bucket, measured in inches per hour, will be exactly the same.

1 in.²

2 in.²

3 in.²

254 Chapter 5 Integrals in Two or More Dimensions

1. Although (as stated above) the three buckets will see the same rising waterline, their total accumulation of water (measured in gallons/hour for instance, or in^3/h) will be very different. Which of the following is primarily responsible for this difference? (Assume none of the buckets fills up.)
   (a) The tops, through which the rain enters, have different surface areas.
   (b) The bottoms, on which the rain lands, have different surface areas.
   (c) The buckets are different heights.
   (d) The buckets have different volumes.

See Check Yourself #36 in Appendix L

In the three buckets above, the accumulation will again be different. We are interested in measuring the total accumulation of water in in^3/h.

2. A rainfall of 0.2 inches per hour is considered "moderate." After one hour of such a rain, the middle bucket shown above will have a waterline at 0.2 inches. How much water (measured in in^3) will it have accumulated?
3. After 1 h of the same rain, how much water (measured in in^3) will each of the *other* two buckets accumulated? *Hint*: "accumulated water" is the same as "total amount of water that has passed through the top."
4. A rainfall of 0.4 in/h is considered "heavy." After 1 h of such a rain, how much water has accumulated in each bucket?
5. Write a formula for the total amount of water that accumulates in a bucket after 1 h. Your formula should have two independent variables.

See Check Yourself #37 in Appendix L

Now we're going to change the rate of accumulation *without* changing either the rate of the rainfall, or the area of the top of the bucket. How? Tilt the bucket!

6. How fast does the water accumulate if we tilt the bucket 90° from its starting position?
7. Let r equal the rate of rainfall, A the area of the top of the bucket, and θ the angle of tilt, going from 0 in the original vertical position to $\pi/2$ in a completely horizontal position. Which of the following is a reasonable formula for the rate of water accumulation W?

   (a) $W = rA$
   (b) $W = rA\theta$
   (c) $W = rA\cos\theta$
   (d) $W = rA\sin\theta$
   (e) $W = rA\tan\theta$
   (f) none of the above is reasonable

5.10.2 Explanation: Surface Integrals

If you understood Section 5.8 on line integrals you have a head start on this section because the concepts are analogous.

A line integral says "Add up this function all along this **curve**."	A surface integral says "Add up this function all along this **surface**."
The line integral of a vector function is a scalar that measures how much this function points **along the curve**.	The surface integral of a vector function is a scalar that measures how much this function points **through the surface**.
You can often compute a line integral without actually integrating anything; you find the right component (dot product) and multiply it by the **length of the curve**.	You can often compute a surface integral without actually integrating anything; you find the right component (dot product) and multiply it by the **area of the surface**.
If you do have to integrate, use the formula for the curve to translate the entire problem to a **single integral**.	If you do have to integrate, use the formula for the surface to translate the entire problem to a **double integral**.
You can also take the line integral of a scalar function—for instance, to add up density along a **curve**.	You can also take the surface integral of a scalar function—for instance, to add up density along a **surface**.

The Discovery Exercise (Section 5.10.1) is entirely about a surface integral, although it never uses the word. The vector field is the rainfall, presumed in this case to point straight down, with a magnitude measured in in/h. The surface is the top of the bucket. The goal is to measure the total extent to which the rain vector flows through the surface.

And what did we find? The accumulation is proportional to both the magnitude of the rainfall and the area of the surface, but it also depends on the angle between them. The final answer to the Discovery Exercise is $W = rA\cos\theta$. This can be expressed more concisely by using a convention, common in such cases, of expressing area as a vector: the magnitude of this vector is the area itself, and its direction is perpendicular to the surface. If we let $\vec{A}$ represent the top of the bucket in this way, we can write $W = \vec{r} \cdot \vec{A}$. As with any dot product, it depends on the angle θ between the rain vector and the area vector: an angle of 0 maximizes the result, and an angle of $\pi/2$ makes the result zero.

By this point in the chapter we hope you can recite the next paragraph in your sleep. If the vector field (rainfall) varies, or if the surface twists, we write $dW = \vec{r} \cdot d\vec{A}$ to find the rainfall through a small area over which we consider everything constant. Then we use an integral—a double integral in this case—to add up the rainfall passing through the entire surface.

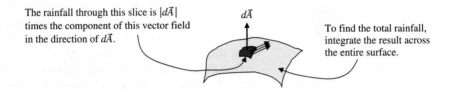

The rainfall through this slice is $|d\vec{A}|$ times the component of this vector field in the direction of $d\vec{A}$.

$d\vec{A}$

To find the total rainfall, integrate the result across the entire surface.

Definition: Surface Integral

Equation 5.10.1 defines the "surface integral" of vector $\vec{V}(x, y, z)$ along surface S.

$$\iint_S \vec{V} \cdot d\vec{A} \qquad (5.10.1)$$

The vector $d\vec{A}$ is a differential area vector. Its magnitude is the differential area dA and its direction is normal to the surface at this point. The integrand can therefore be written in the following forms.

$$\vec{V} \cdot d\vec{A} = \vec{V} \cdot (dA\,\hat{n}) \quad \text{or} \quad \vec{V} \cdot d\vec{A} = \left|\vec{V}\right| dA \cos\theta$$

Here $\hat{n}$ represents a unit vector normal to the surface at a given point, and θ represents the angle between $\vec{V}$ and $\hat{n}$.

In the special case where S is a closed surface (a surface that completely surrounds a 3D region) we write $\oiint_S \vec{V} \cdot d\vec{A}$.

To understand $\iint \vec{V} \cdot d\vec{A}$ you need to understand three important physical quantities, all with different meanings and different units.

- $\vec{V}(x, y, z)$ could be the momentum density of a fluid, meaning its density times velocity, with units of $(kg/m^3)(m/s)$. We prefer to think of that as $kg/m^2/s$ because it represents how much fluid is flowing through the surface around (x, y, z) per unit area per second.
- $d\vec{A}$ is the area vector of a surface. Its units are of course m^2.
- If $\vec{V}$ is a momentum density then $\vec{V} \cdot d\vec{A}$ is a "mass flow rate." That quantity has units of kg/s because it measures how much fluid is flowing through this surface every second. When you write $\vec{V} \cdot d\vec{A}$ you are calculating the flow rate through one tiny differential surface. When you integrate it you are calculating the flow rate through a larger surface, but this does not change the meaning or units of the quantity.

The word we are dancing around in this discussion is "flux." In fluid flow "flux" does not exactly refer to any of the three quantities we have just mentioned: the flux is the component of the momentum density in the direction of the area. Like $\vec{V}$ it has units of $kg/m^2/s$, and also like $\vec{V}$, it is defined at a point rather than over a surface.

What makes this particularly confusing is that the word is used quite differently in electromagnetism. The local quantity (analogous to $\vec{V}$ in fluid flow) is an electric or magnetic field. The word flux is generally used for the integrated quantity: for example "magnetic flux" means "the surface integral of the magnetic field."

When talking about electromagnetism we will use "flux" to mean the surface integral of $\vec{E}$ or $\vec{B}$. When talking about fluid flow we will use "momentum density" or "velocity" (as appropriate) for the field and "flow rate" for the integrated quantity, and we'll only use "flux" when the distinction between the local and integrated quantities doesn't matter, e.g. "The direction of the flux through this hole is upward."

And speaking of direction, there's an ambiguity in the sign of flux. Is the wind blowing through your door a positive flux or a negative one? The decision is arbitrary; you can flow through a surface in two directions, and neither one is inherently positive or negative, so just be clear how you're defining it. In the particular case of a closed surface like a sphere or a box, however, there is a convention that is widely accepted enough to be useful, and that is to take the positive direction pointing *out* of the surface. With that convention we can interpret the integral along a closed surface as the net flow out of the region that it surrounds (Figure 5.13).

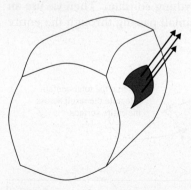

FIGURE 5.13 If this vector field represents the momentum density of the air, then its surface integral tells us how rapidly the container is losing air.

5.10 | Surface Integrals

As we did with line integrals, we will begin with a few examples where you do not have to integrate before diving into the algebra. In each case think mathematically about $\iint_S \vec{f} \cdot d\vec{A}$, but also think about the physical scenario: the vector field represents the momentum density of a flowing fluid, and our goal is to measure the rate at which this fluid passes *through* the surface.

EXAMPLES — Surface Integrals That Don't Require Integration

The first three problems take place on the disk $x^2 + y^2 \le 9$, $z = 5$. We choose to define flux through this surface as being positive in the upward direction, although we could just as easily define it the other way.

1. Evaluate $\iint_S \vec{f}_1 \cdot d\vec{A}$ where $\vec{f}_1 = 2z\hat{k}$.

    The vector field is perpendicular to the surface at every point, so it all "counts." We multiply the magnitude of the vector field (which is 10 everywhere on this disk) times the area of the surface (which is 9π) to find $\iint_S \vec{f}_1 \cdot d\vec{A} = 90\pi$. If you think of this vector field as representing the momentum density of a fluid, 90π measures the total rate at which fluid passes through the disk.

2. Evaluate $\iint_S \vec{f}_2 \cdot d\vec{A}$ where $\vec{f}_2 = (x+y)\hat{i} + (e^x/z^2)\hat{j}$.

    This looks complicated, but it's easy when you realize that $\vec{f}_2$ has no $\hat{k}$-component. In formal mathematical terms, we note that $d\vec{A}$ always points in the $\hat{k}$-direction for S, so $\vec{f}_2 \cdot d\vec{A}$ is always zero. Visually, you can picture a fluid whose flow is always horizontal; it does not move *through* the disk at all. The answer is 0.

3. Evaluate $\iint_S \vec{f}_3 \cdot d\vec{A}$ where $\vec{f}_3 = 4\hat{i} + 7\hat{k}$.

    The $\hat{i}$-component does not contribute (like Example 2) and the $\hat{k}$-component fully contributes (like Example 1), so the answer is $7 \times 9\pi = 63\pi$. Pause on this result and think about why we take a dot product: the more directly the fluid flows in the direction of the surface, the more fluid flows through the surface.

The remaining problems take place on a different surface: S_2 is a sphere of radius 3 centered on the origin. Because this is a closed surface, convention dictates that we define positive flux as outward. (Convention also dictates that a surface integral on a closed surface is written with a loop cutting through the integral signs: $\oiint$.)

4. Evaluate $\oiint_{S_2} \vec{f}_4 \cdot d\vec{A}$ where $\vec{f}_4 = -3\hat{k}$.

    The integral of this field through the top half of our sphere might represent the total amount of fluid entering the spherical region; the integral through the bottom half would be the total amount leaving that same region. We can see without any calculation that the two results cancel each other out, so the total amount of fluid in the region is not changing. The surface integral is zero. Generalizing, you can see that the integral of a *constant vector field* through a *closed surface* is always zero.

5. Evaluate $\oiint_{S_2} \vec{f}_5 \cdot d\vec{A}$ where $\vec{f}_5 = \vec{r}$ or, equivalently, $\vec{f}_5 = x\hat{i} + y\hat{j} + z\hat{k}$.

    This vector field always points directly away from the origin, so it always points directly through the sphere. Its magnitude is $|\vec{f}_5| = \sqrt{x^2 + y^2 + z^2} = r$, which is 3 everywhere on this sphere. So the surface integral is 3 times the surface area of the sphere, which is $3 \times 4\pi r^2 = 108\pi$.

Evaluating Surface Integrals

What happens when you have to do the math? Remember how we evaluate line integrals: use the equation for the curve to express x, y, dx and dy in terms of a single variable such as t, and then just evaluate a single integral. The strategy is similar here, using the equation of the surface to find a double integral in u and v. The $d\vec{A}$ part of the surface integral depends on how the surface is expressed.

The Differential Area Element

For a surface expressed parametrically ($\vec{r} = x(u,v)\hat{i} + y(u,v)\hat{j} + z(u,v)\hat{k}$) a surface integral is evaluated as a double integral over u and v.

$$d\vec{A} = \left(\frac{\partial \vec{r}}{\partial u} \times \frac{\partial \vec{r}}{\partial v}\right) du\, dv \tag{5.10.2}$$

For a surface expressed as $z(x, y)$ the surface is evaluated as a double integral over x and y.

$$d\vec{A} = \left[-\left(\frac{\partial z}{\partial x}\right)\hat{i} - \left(\frac{\partial z}{\partial y}\right)\hat{j} + \hat{k}\right] dx\, dy \tag{5.10.3}$$

These formulas aren't obvious. We'll work an example to show you how to use the parametric one, and then outline an argument as to why it works. You will derive the non-parametric formula in Problem 5.268.

EXAMPLE **Surface Integral that Does Require Integration**

Question: Section 5.9 discussed the surface $x(u,v) = v\cos u$, $y(u,v) = v\sin u$, $z(u,v) = v^2$ on $0 \leq u \leq 2\pi$, $0 \leq v \leq 3$. Find the integral of $\vec{f} = (z+3)\hat{k}$ through this surface.

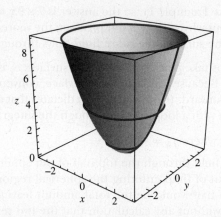

Answer: We begin by representing the vector field as a function of u and v: in this case, $\vec{f} = (v^2 + 3)\hat{k}$.

Then we use Equation 5.10.2. Writing the surface as $\vec{r} = v\cos u\,\hat{i} + v\sin u\,\hat{j} + v^2\hat{k}$ we get $\partial\vec{r}/\partial u = -v\sin u\,\hat{i} + v\cos u\,\hat{j}$ and $\partial\vec{r}/\partial v = \cos u\,\hat{i} + \sin u\,\hat{j} + 2v\hat{k}$, so:

$$d\vec{A} = \left(2v^2\cos u\,\hat{i} + 2v^2\sin u\,\hat{j} - v\,\hat{k}\right) du\, dv$$

Taking the dot product:

$$\iint_S \vec{f} \cdot d\vec{A} = \int_0^3 \int_0^{2\pi} (-v^3 - 3v)\, du\, dv$$

You can easily evaluate that integral to find the answer $-135\pi/2$.

Why did it come out negative? If you look at our $d\vec{A}$ you can see that it pointed out-and-down through the bowl. Since $\vec{f}$ points upward, the surface integral came out negative. This does not tell us anything about the problem; remember that direction is arbitrary, and $d\vec{A}$ could just as easily have pointed in-and-up if we had parametrized the surface differently.

Now where did that cross product come from?

The first step toward deriving Equation 5.10.2 is remembering what we're trying to accomplish. (Understanding the question in this case is at least as difficult, and more important, than understanding the answer.) We have parametrically expressed a surface as $\vec{r}(u, v)$ and now we're looking for the vector $d\vec{A}$. How would we know it if we saw it?

The direction is easy: $d\vec{A}$ should point perpendicular to the surface at all points. It may not be obvious how to accomplish that, but at least we know what we want.

The magnitude is the area of a differential element. Imagine starting at a particular point (u_0, v_0) that corresponds to a particular (x_0, y_0, z_0). As you let u range between u_0 and $u_0 + du$, and v range between v_0 and $v_0 + dv$, you're going to trace out a small region on the $\vec{r}(u, v)$ surface. That's what we're looking for: the small area you cover as you range through du and dv. (By analogy, you may want to look back at Figure 5.6; the area swept out by the polar $d\rho$ and $d\phi$ is $\rho\, d\rho\, d\phi$.)

To see how Equation 5.10.2 accomplishes all that, consider the vector $(\partial \vec{r}/\partial u)du$. It represents the small change in x, y, and z when you increment u a small amount while leaving v unchanged. That means it points tangent to the surface in the direction of a gridline of constant v. Similarly the vector $(\partial \vec{r}/\partial v)dv$ points tangent to the surface along a gridline of constant u. The fact that they are both tangent to the surface means their cross product must be normal to it, which is the direction we want.

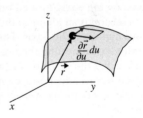

For a sufficiently small du, we can as always treat $(\partial \vec{r}/\partial u)du$ as a straight line. But it is not necessarily perpendicular to $(\partial \vec{r}/\partial v)dv$ no matter how small they get, so the area they trace out is a parallelogram. The magnitude of the cross product of two vectors gives you the area of the parallelogram formed by the vectors (as you will show in Problem 5.271), so the vector $d\vec{A}$ given by Equation 5.10.2 has both the magnitude and the direction we want. (This derivation should remind you of our derivation of the generic Jacobian in Section 5.6.)

Scalar Surface Integrals

Just as with line integrals, we have emphasized vector functions, but it is also possible to take the surface integral of a *scalar* function. For instance, if the function represents the density of a surface, the surface integral gives the total mass. If the surface is expressed parametrically as $\vec{r}(u, v)$, then the surface integral of the scalar function $f(x, y, z)$ is:

$$\iint_S f(x, y, z)\, dA = \iint f\left(x(u, v), y(u, v), z(u, v)\right) \left| \frac{\partial \vec{r}}{\partial u} \times \frac{\partial \vec{r}}{\partial v} \right| du\, dv \qquad (5.10.4)$$

(That formula should make sense if you followed our explanation of the vector formula above.) If the surface is expressed as $z(x, y)$, then the surface integral of the scalar function $f(x, y, z)$ is:

$$\iint_S f(x, y, z) dA = \iint f\left(x, y, z(x, y)\right) \sqrt{1 + \left(\frac{\partial z}{\partial x}\right)^2 + \left(\frac{\partial z}{\partial y}\right)^2}\, dx\, dy \qquad (5.10.5)$$

You will derive this formula in Problem 5.268.

5.10.3 Problems: Surface Integrals

Problems 5.255–5.257 can be done without integrating anything; just think about what a surface integral means.

5.255 Surface S in this problem is the rectangle defined by $y \in [0, 2]$, $z \in [0, 3]$ on the yz-plane. Treat the positive x-direction as the positive direction through this surface.
(a) Evaluate $\iint_S \vec{f}_1 \cdot d\vec{A}$ where $\vec{f}_1 = 10\hat{i}$.
(b) Evaluate $\iint_S \vec{f}_2 \cdot d\vec{A}$ where $\vec{f}_2 = -10\hat{i}$.
(c) Evaluate $\iint_S \vec{f}_3 \cdot d\vec{A}$ where $\vec{f}_3 = 10\hat{i} + 6\hat{j} - 18\hat{k}$.
(d) Write a non-zero vector $\vec{f}_4$ such that $\iint_S \vec{f}_4 \cdot d\vec{A} = 0$.

5.256 Surface S in this problem comprises three subsurfaces. One is a disk of radius 2, centered on the z-axis, located on the xy-plane. The second is an identical disk located 10 units higher on the z-axis. The final piece is a cylinder connecting the two. So the entire surface has the shape of a can. As with any closed surface, the positive direction will be taken as pointing everywhere *out* of the surface.
(a) Evaluate $\oiint_S \vec{f}_1 \cdot d\vec{A}$ where $\vec{f}_1 = 10\hat{k}$.
Hint: You can consider each of the three surfaces separately.
(b) Evaluate $\oiint_S \vec{f}_2 \cdot d\vec{A}$ where $\vec{f}_2 = 10\hat{i}$.
(c) Evaluate $\oiint_S \vec{f}_3 \cdot d\vec{A}$ where $\vec{f}_3 = z\hat{k}$.
(d) Evaluate $\oiint_S \vec{f}_4 \cdot d\vec{A}$ where $\vec{f}_4 = 10\hat{\rho}$. The unit vector $\hat{\rho}$ points directly away from the z-axis.

5.257 Surface S in this problem is a sphere of radius 5 centered on the origin. As with any closed surface, the positive direction will be taken as pointing everywhere *out* of the surface.

(a) Evaluate $\oiint_S \vec{f}_1 \cdot d\vec{A}$ where $\vec{f}_1 = 10\hat{k}$.
(b) Evaluate $\oiint_S \vec{f}_2 \cdot d\vec{A}$ where $\vec{f}_2 = x\hat{k}$.
(c) Evaluate $\oiint_S \vec{f}_3 \cdot d\vec{A}$ where $\vec{f}_3 = 10\hat{r}$. The vector $\hat{r}$ points directly away from the origin.
(d) Evaluate $\oiint_S \vec{f}_4 \cdot d\vec{A}$ where $\vec{f}_4 = x\hat{i} + y\hat{j} + z\hat{k}$.
(e) Evaluate $\oiint_S \vec{f}_5 \cdot d\vec{A}$ where $\vec{f}_5 = x\hat{i}$.
Hint: Think about how this is related to your answer to Part (d).

5.258 Walk-Through: Surface Integral. Surface S is defined by the parametric equations $x = u + v$, $y = u^2$, $z = u(v + 1)$ for $0 \le u \le 2$, $0 \le v \le 5$. Find $\iint_S \vec{f} \cdot d\vec{A}$ for $\vec{f} = z^2\hat{i} + z^2\hat{j} + (xy)\hat{k}$.
(a) Begin by using the equation for surface S to rewrite $\vec{f}$ as a function of u and v.
(b) Rewrite surface S as a vector-valued function $\vec{r}(u, v)$.
(c) Use Equation 5.10.2 to find $d\vec{A}$ for surface S.
(d) Write a double integral for $\iint_S \vec{f} \cdot d\vec{A}$.
(e) Evaluate the double integral.

For Problems 5.259–5.261 find the surface integral of the given vector field through the surface S. You may find it helpful to first work through Problem 5.258 as a model.

5.259 The equations $x = uv$, $y = u$, $z = v^2$ where $-1 \le u, v \le 1$ describe the "Whitney Umbrella." The vector field is $\vec{f} = (x + z)\hat{i} + (y + z)\hat{j} + (x + y)\hat{k}$.

5.260 The equations $x = \sin u$, $y = \sin v$, $z = \sin(u + v)$ where $0 \le u, v \le \pi/2$ describe a "Lissajous Surface." The vector field is $\vec{f} = \sqrt{1 - x^2}\,\hat{i} - \sqrt{1 - y^2}\,\hat{j} + xy\,\hat{k}$.

5.261 The equations $x = \sinh u \cos v$, $y = \sinh u \sin v$, $z = \cosh u$ where $0 \leq u \leq 2$ and $\pi/6 \leq v \leq \pi/2$ describe part of a hyperboloid. The vector field is $\vec{f} = (1/y)\hat{i} + z\hat{k}$.

5.262 Rain is falling straight down. Will you get wetter (measured in number of raindrops per minute that hit your body) if you are standing still, or zooming forward on a bicycle? Why?

5.263 In this problem, S represents a closed, permeable surface, and $\vec{V}$ represents the momentum density of air. At time $t = 0$ the surface surrounds 1 kg of air. An hour later, you come back and measure the amount of air surrounded by the surface. What might you expect to find if...

(a) $\oiint_S \vec{V} \cdot d\vec{A} = 0$?

(b) $\oiint_S \vec{V} \cdot d\vec{A} > 0$?

(c) $\oiint_S \vec{V} \cdot d\vec{A} < 0$?

5.264 The equation $z = \sqrt{1 - x^2 - y^2}$ represents the top half of a sphere. Your job is to find the integral of $y\hat{i} + x\hat{j} + z\hat{k}$ upward through this surface.

(a) Find the answer by parameterizing the hemisphere and using Equation 5.10.2. *Hint*: Think about spherical coordinates.

(b) 🖥 Find the answer directly by using Equation 5.10.3.

5.265 The equation $z = \sqrt{1 - x^2 - y^2}$ represents the top half of a sphere. The density of this hemisphere is given by $\sigma = (x^2 + y^2)$ kg/m². Your job is to find the total mass of this hemisphere.

(a) Find the answer by parameterizing the hemisphere and using Equation 5.10.4.

(b) 🖥 Find the answer directly by using Equation 5.10.5.

5.266 In the example "Surface Integral that Does Require Integration" on Page 258 we calculated the surface integral of a vector field through the bowl-shaped surface given by $\vec{r} = v \cos u\,\hat{i} + v \sin u\,\hat{j} + v^2 \hat{k}$ on $0 \leq u \leq 2\pi$, $0 \leq v \leq 3$. In this problem you're going to consider the same surface with a simpler vector field, the constant $\vec{V} = 3\hat{k}$.

(a) Calculate the integral of $\vec{V}$ through the surface S.

(b) Write a parametric expression for the disk that would cover the top of that bowl.

(c) Calculate the integral of $\vec{V}$ through the top of the bowl. Use normal vectors pointing upward from the disk. This part should not require any calculus.

(d) Using your answers to Parts (a) and (c), find the integral through the closed surface consisting of the bowl and disk together. Once again, this should not require actually integrating anything.

(e) In Parts (a), (c), and (d) you calculated the integral of $\vec{V}$ through three different surfaces. If you interpret $\vec{V}$ as the momentum density of a fluid, what does each of those three integrals tell you about the flow of the fluid? Explain why your answer to Part (d) makes sense physically.

5.267 In the example "Surface Integral that Does Require Integration" on Page 258 we calculated the integral of a vector field $\vec{f} = (z + 3)\hat{k}$ through the bowl-shaped surface given by $\vec{r} = v \cos u\,\hat{i} + v \sin u\,\hat{j} + v^2 \hat{k}$ on $0 \leq u \leq 2\pi$, $0 \leq v \leq 3$.

(a) Calculate the integral of $\vec{f}$ through a disk of radius 3 in the xy-plane, centered on the z-axis.

(b) Calculate the integral of $\vec{f}$ through a disk of radius 3 in the $z = 9$ plane, centered on the z-axis.

(c) Rank the three surface integrals by magnitude: through the bowl, through the lower disk, and through the upper disk. Looking at the field $\vec{f}$ and the shape and location of the surfaces, explain why it makes sense which is largest and which is smallest.

5.268 How do we find the formulas for $d\vec{A}$ or dA if the surface is expressed as $z(x, y)$? We could derive them from scratch, essentially replicating the process we used for a parametric representation, but it's easier to *use* the formulas we already derived. Any $z(x, y)$ function is a special case of a parametric representation, which we can write as $\vec{r} = x\hat{i} + y\hat{j} + z(x, y)\hat{k}$.

(a) Use this form and Equation 5.10.2 to derive Equation 5.10.3.

(b) Use this form and Equation 5.10.4 to derive Equation 5.10.5.

5.269 In Equation 5.10.2 we take the cross product of two partial derivatives. Putting $\partial \vec{r}/\partial u$ first is arbitrary, and we could just have easily put $\partial \vec{r}/\partial v$ first. But the cross product is not commutative; we would get a different answer. How do the two answers differ, and why do we get two different answers for the same question?

5.270 On page 250 we considered the surface $\vec{r} = v\cos u\,\hat{i} + v\sin u\,\hat{j} + v^2\,\hat{k}$ with $0 \leq u \leq 2\pi,\ 0 \leq v \leq 3$.

(a) $\partial \vec{r}/\partial u$ points along the gridline of constant v. Which way does it point? (i.e. which way do you move along the surface as u increases?)

(b) $\partial \vec{r}/\partial v$ points along the gridline of constant u. Which way does it point?

(c) If we use $\left((\partial \vec{r}/\partial u) \times (\partial \vec{r}/\partial v)\right) du\,dv$ to define $d\vec{A}$, which way does $d\vec{A}$ point? (*Hint*: don't do any math. Use your previous answers and the right-hand rule.)

(d) Based on your previous answer—and again, without doing any math—if we integrate the function $\hat{k}$ through this surface, will the answer come out positive or negative?

(e) How do your answers change if we instead define $d\vec{A}$ as $\left((\partial \vec{r}/\partial v) \times (\partial \vec{r}/\partial u)\right) du\,dv$?

5.271 The picture shows the parallelogram defined by vectors $\vec{a}$ and $\vec{b}$ with angle θ between them. We want the area of this parallelogram.

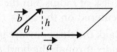

(a) If the triangle on the left were cut out and put on the right, the resulting figure would be a rectangle. What are the width and height of this rectangle, in terms of letters given in the drawing?

(b) If your answer included the letter h, rewrite h so that you can express the area of the rectangle, and therefore the parallelogram, in terms of only three quantities: $|\vec{a}|, |\vec{b}|,$ and θ.

(c) Recalling that the magnitude of the cross product of two vectors is $|\vec{a}|\,|\vec{b}|\sin\theta$, you should have shown that the area of the parallelogram is $|\vec{a} \times \vec{b}|$. When we used that fact in this section, what were $\vec{a}$ and $\vec{b}$ and what did the parallelogram represent?

5.272 A positively charged particle produces an electric field $\vec{E}$ around itself. The magnitude of the electric field is kq/r^2 where k is a constant, q is the particle's charge (also constant), and r is distance from the particle. The direction of the field is directly away from the particle. (This can all be expressed by writing $\vec{E} = kq\vec{r}/|\vec{r}|^3$ where $\vec{r} = x\hat{i} + y\hat{j} + z\hat{k}$.) For this problem we will consider a particle of charge q at the origin.

(a) Write parametric equations for a disk of radius w centered on the z-axis at height H. For the rest of this problem we'll call this disk D.

(b) Find the electric flux: that is, the surface integral of $\vec{E}$ through D.

(c) The surface $x = R\sin\theta\cos\phi$, $y = R\sin\theta\sin\phi$, $z = R\cos\theta$, $0 \leq \phi \leq 2\pi$, $0 \leq \theta \leq \theta_{max}$ describes a portion of a sphere. We will call this surface P. Find R and θ_{max} so that the boundary of P is the same circle that forms the boundary of D.

(d) Take a scalar surface integral to find the area of P. Express your answer in terms of w and H.

(e) Using your answer to Part (d) find the flux of $\vec{E}$ through P. This part should not require an integral.

(f) What relationship did you find between the flux through D and the flux through P? If you pretend for a moment that $\vec{E}$ is the mass flow rate of a fluid, what would this result tell you about that fluid?

In electrostatics, "Gauss's Law" promises that if you draw a closed surface around any collection of charges, the electric flux—that is, the integral of the electric field through that surface—is proportional to the charge enclosed.

$$\oiint_S \vec{E} \cdot d\vec{A} = \frac{Q_{enclosed}}{\varepsilon_0} \qquad (\varepsilon_0 \text{ is a constant})$$

In Problems 5.273–5.277 you will use Gauss's Law to find the magnitude of the electric field in various charge configurations. You should be able to do these problems without integrating anything.

5.273 A sphere of radius R has a uniformly distributed charge Q.

(a) Draw a Gaussian sphere at a radius $r > R$ around this sphere. Based on Gauss's Law, what is the total flux through this sphere?

(b) We know from symmetry that the electric field through your sphere always points directly outward, perpendicular to the sphere, and that its magnitude $|\vec{E}|$ is the same everywhere on the sphere. Based on those facts, find the flux of $\vec{E}$ through the sphere as a function of $|\vec{E}|$. (In this part you will ignore Part (a).)

(c) Putting your two answers together, calculate $|\vec{E}|$ outside the sphere of charge as a function of r.

5.274 *[This problem depends on Problem 5.273.]* Repeat Problem 5.273 for a Gaussian sphere of radius $r < R$ to find the electric field inside the sphere of charge. *Hint*: the difference comes in the very first step, when you have to calculate $Q_{enclosed}$.

5.275 *[This problem depends on Problem 5.273.]* The novel "Flatland" by Edwin Abbott describes a 2 dimensional world inhabited by creatures shaped as squares, triangles, and other polygons for whom "up" and "down" are synonymous with "North" and "South." Suppose a Flatland physicist figured out that the flux of the electric field through any closed loop in Flatland was proportional to the charge enclosed. He writes flux as $\oint \vec{E} \cdot d\vec{\lambda}$ where $d\vec{\lambda}$ is an outward-pointing vector normal to the curve with magnitude equal to the arc length $d\lambda$. (Note that $d\vec{\lambda}$ is perpendicular to our usual $d\vec{s}$.)

(a) Using a similar argument to the one you used in Problem 5.273, find the flux through a circle of radius r around a point charge Q and use the 2D version of Gauss's law to find the electric field in Flatland as a function of r.

(b) Some versions of string theory predict that space is 10 dimensional. The theory also predicts that gravity should obey Gauss's law: the flux of the gravitational field through any closed surface is proportional to the mass/energy enclosed. Using a similar argument to the one in Part (a), find how the gravitational field from a point mass depends on r in a 10 dimensional space. You don't need to find the constant coefficient, just the r dependence. *Hint*: You can figure out how the surface area of an n dimensional sphere depends on its radius just by thinking about units.

You should have concluded that gravity in a 10 dimensional universe does not obey an inverse square law. There are explanations in string theory for why the world appears 3D to us and why gravity appears to obey an inverse square law, but they are too complicated for us to get into here. Go forth and study.

5.276 An infinitely long straight wire carries a uniform charge density λ.

(a) Begin by drawing the wire. Around it you will draw a Gaussian cylinder, consisting of three parts: a tube at a distance ρ from the wire with length L, and circular caps on each end of the tube. (Remember that Gauss's Law applies only to a closed surface, so we need those caps!)

(b) Find the charge enclosed by the surface. Based on this and Gauss's Law, calculate the flux of the electric field through the surface.

(c) Based on symmetry considerations, which way does the electric field point?

(d) Calculate the flux of the electric field through the caps at the top and bottom of the Gaussian surface.

(e) Calculate the flux of the electric field through the cylindrical part of the surface. Your answer will be a function of ρ, L, and $|\vec{E}|$.

(f) Putting your answers together, calculate $|\vec{E}|$ as a function of distance to the wire.

5.277 An infinite plane carries a uniform charge density σ.

(a) Begin by drawing the plane. Around it you will draw a Gaussian "pillbox": a circle of radius R at a distance $L/2$ above the plane, an identical circle at distance $L/2$ below the plane, and a cylinder connecting them.

(b) Find the charge enclosed by the surface. Based on this and Gauss's Law, calculate the flux of the electric field through the surface.

(c) Based on symmetry considerations, which way does the electric field point?

(d) Calculate the flux of the electric field through the circles at the top and bottom of the Gaussian surface.

(e) Calculate the flux of the electric field through the cylindrical part of the surface.

(f) Putting your answers together, calculate $|\vec{E}|$ as a function of distance to the plane.

Problems 5.278–5.281 are based on Faraday's law of magnetic induction. In words, Faraday's law states that a changing magnetic flux—that is, a change in the surface integral of the magnetic field—causes (or "induces") an electric field. Mathematically, the law states that if you draw any closed loop,

$$\oint_{\text{loop}} \vec{E} \cdot d\vec{s} = \frac{d}{dt} \iint_{\text{area}} \vec{B} \cdot d\vec{A}$$

We define $\varepsilon = -\oint \vec{E} \cdot d\vec{s}$ as "electromotive force," or "emf" to its friends. (The name is deceptive: ε has units of electric potential, not of force, although it can *cause* a force on charged particles.) We also often use Φ_B for magnetic flux. With those changes, Faraday's law can be written more concisely:

$$\varepsilon = -\frac{d\Phi_B}{dt}$$

In these questions we will refer to a "flow" of magnetic field. There is nothing physically flowing anywhere, but a magnetic field is mathematically analogous to a flowing fluid, and we take advantage of that analogy when we find its flux.

5.278 A wire is bent into a circle of radius R, and a constant magnetic field B_0 flows directly (perpendicularly) through the enclosed disk.
 (a) Calculate the flux Φ_B through the wire.
 (b) Now the magnetic field starts increasing, so $B = B_0 + at$ where a is a positive constant. Calculate the induced emf around the wire.
 (c) Now the magnetic field is constant again, but the loop is shrinking, so its radius $r = R - bt$ where b is a positive constant. Calculate the induced emf around the wire.
 (d) Now the magnetic field is increasing and the loop is shrinking at the same time. Find the value of b required to make the emf equal zero at a specific time $t = \tau$ (when the radius is still positive).
 (e) Find the units of R, B_0, a, and b from the equations given in the problem and use them to check that your answer to Part (d) has correct units.

5.279 *[This problem depends on Problem 5.278.]* The B-field in Problem 5.278 is again constant, and the loop is not shrinking, but the loop is rotating about its diameter with a constant angular velocity $\omega = d\theta/dt$, where θ is the angle between the B-field and a vector normal to the loop.

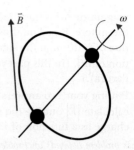

$\vec{B}$ goes perpendicularly through the circle when $\theta = 0$ and doesn't go through the circle at all when $\theta = 90°$.
 (a) Calculate the induced emf as a function of θ.
 (b) At what value of θ does the flux reach a maximum? At what value of θ does the emf reach a maximum?

5.280 An "inductor" is a common element in electric circuits, created by twisting a wire into a helical shape usually called a solenoid. As a current I runs through the wire it creates a magnetic field running directly through the center of the solenoid with magnitude $|\vec{B}| = \mu_0 n I$ where n is the number of loops per unit length in the solenoid. (So a solenoid of length L contains nL loops.) Let A be the area of each loop.

 (a) Calculate Φ_B through one loop of the solenoid.
 (b) The total flux through the solenoid is the sum of the fluxes through all the loops. (This is sometimes called the "magnetic flux linkage.") Calculate Φ_B through the entire solenoid.
 (c) If the current I is not changing, how much emf is induced in the solenoid?
 (d) If the current I is changing, calculate the emf through the solenoid as a function of all the given variables and dI/dt.
 (e) In circuit analysis, we say that the voltage drop (the emf) across an inductor is $L(dI/dt)$ where L is the "inductance" of the inductor. Give a formula for L in terms of the physical properties of the solenoid.

5.281 An infinite straight wire carrying a current I will produce a magnetic field $\vec{B}$ with magnitude $\mu_0 I/(2\pi\rho)$ where ρ is distance from the wire and μ_0 is a constant. The direction of the field points tangentially along a circle centered on the wire. For this problem we will consider a wire along the z-axis with a current I moving upward.

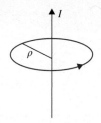

(a) Find the flux of $\vec{B}$ through a square with corners at $(L, 0, 0)$ and $(2L, 0, L)$.
(b) If you increase the current I in the straight wire at a constant rate, rising from 0 to I_f in time τ, find the emf around the square loop in Part (a).
(c) Take $L = .04$ m. In what time τ would you have to raise the current in the wire from 0 to $I_f = (1/10)$ A in order to produce an emf equivalent to 9 V in the square loop?

5.11 Special Application: Gravitational Forces (see felderbooks.com)

5.12 Additional Problems (see felderbooks.com)

CHAPTER 6

Linear Algebra I

> *Before you read this chapter, you should be able to ...*
> - find the components of a vector and use them to represent the vector in terms of the basis vectors $\hat{i}$ and $\hat{j}$.
> - solve two linear equations for two unknowns.
> - find the general solution of a linear differential equation by the method of "guess and check" (see Chapter 1).
>
> *After you read this chapter, you should be able to ...*
> - multiply a matrix by a column matrix to transform information from one form to another, or to visually transform a vector.
> - express any vector as a linear combination of any set of basis vectors.
> - multiply matrices to create a single matrix that performs multiple transformations.
> - find and use an inverse matrix to reverse a transformation.
> - find the eigenvectors and eigenvalues of a matrix, and use them to define a natural basis for a transformation.
> - use matrices to rewrite and solve coupled differential equations.

Not knowing your personal history with matrices, we're going to guess that it looks something like this. Your high school teacher told you that you multiply matrices by taking the rows of this matrix, turning them sideways, and multiplying them by the columns of that matrix, "because I said so." It didn't make any sense and it certainly didn't seem to have any point, but you dutifully mastered the algorithm, passed the test, and promptly forgot all about it.

We request that you set aside what you learned previously. (If you never have learned about matrices, you're that much ahead of the game.) This small step back to the very beginning of matrices will pay big dividends as we make our way through linear algebra.

6.1 The Motivating Example on which We're Going to Base the Whole Chapter: The Three-Spring Problem

Each chapter of this book starts with a "motivating exercise": an important problem that you can solve with the techniques introduced in the chapter. This section serves that purpose, but it's also a roadmap. We're going to set up and solve a fairly complicated mechanical problem. At several points throughout the solution, we will pull a mathematical result out of a hat. Each time we do so, we will tell you what section of the chapter will teach you the linear algebra you need for that step.

6.1 | The Motivating Example on which We're Going to Base the Whole Chapter

If you skip this section, the rest of the chapter should still make sense, and you can skip the parts that say "Application to the Three-Spring Problem." But if you do take the time up front to work through this solution then the individual pieces will be more coherent, because you will see how they fit together as you're moving through them.

At the end of the chapter we'll return to this same problem and walk you back through it with all the steps filled in.

6.1.1 Explanation: All I Really Need to Know about Matrices I Learned from the Three-Spring Problem

Figure 6.1 shows two balls connected to each other and to the walls by three springs. We will assume throughout this section that the balls begin at rest, so our initial conditions are only their starting positions $x_1(0)$ and $x_2(0)$. As an example the drawing shows Ball 1 at its equilibrium position and Ball 2 displaced to the right.

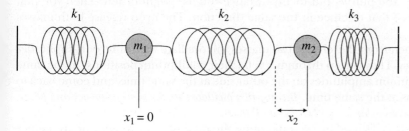

FIGURE 6.1 The three-spring problem with Ball 1 at equilibrium and Ball 2 displaced to the right.

With a bit of introductory physics (Problem 6.10) you can derive the appropriate differential equations. For $m_1 = 2$ kg, $m_2 = 9$ kg, $k_1 = 2$ N/m, $k_2 = 6$ N/m, and $k_3 = 21$ N/m, the equations become:

$$\frac{d^2 x_1}{dt^2} = -4x_1 + 3x_2 \tag{6.1.1}$$

$$\frac{d^2 x_2}{dt^2} = \frac{2}{3}x_1 - 3x_2 \tag{6.1.2}$$

We begin by noticing that Equation 6.1.1 writes $\ddot{x}_1$ as a *linear combination* of x_1 and x_2, and Equation 6.1.2 does the same for $\ddot{x}_2$. (If you're not familiar with that notation, $\dot{x}$ means dx/dt and $\ddot{x}$ means d^2x/dt^2 and so on.) In general, a linear combination of x and y is any function $ax + by$ where a and b are constants. Linear algebra is all about working with that particular kind of function, using a mathematical tool called a "matrix."

> Section 6.3 will discuss rewriting linear equations with matrices. In Section 6.3 Problem 6.40 you will rewrite Equations 6.1.1 and 6.1.2 as a single matrix equation. This simple change in notation will allow us to bring all the tools of matrix mechanics to bear on this problem.

Throughout this section, you will see other arrows like the one above. Each one will point to a section of this chapter where you will learn how to use matrices to solve part of this problem. In some cases your first reaction may be "I don't need matrices to solve this." But this example has only two equations with two unknowns: engineers frequently work with hundreds of equations and unknowns, and matrices give us the approach that scales up to that size.

Two Simple Solutions for the Just-Right Initial Conditions

For any value of A the following functions provide a valid solution to Equations 6.1.1 and 6.1.2.

$$x_1(t) = A\cos\left(\sqrt{2}\,t\right)$$
$$x_2(t) = \frac{2}{3}A\cos\left(\sqrt{2}\,t\right) \tag{6.1.3}$$

You can easily verify that Equation 6.1.3 works by plugging it into Equations 6.1.1 and 6.1.2, but this isn't just any solution: it's a particularly important type of solution called a "normal mode." The whole strategy of solving problems like this is based on knowing how to find the normal modes, and how to use them once you find them. We will provide sketchy outlines of both those processes in this section, and flesh them out throughout the chapter. But first it is essential for you to understand the kind of behavior that Equations 6.1.3 describe.

Let's start by considering the initial conditions. Equation 6.1.3 implies four of them: $x_1(0) = A$, $x_2(0) = (2/3)A$, and both $\dot{x}_1(0)$ and $\dot{x}_2(0)$ are zero.[1] This tells us how to get the system into this mode. You pull m_1 out to any distance you like, right or left. Then you pull m_2 out to exactly $2/3$ of that distance in the same direction. Then you release both masses from rest.

And what will happen from there? The two masses will oscillate with different amplitudes but the same frequency. They will go through $x = 0$ (their equilibrium positions) at the same time, reach their maximum amplitudes on the other side at the same time, and come back to their starting positions at the same time. *Because they oscillate with the same frequency and phase, they will maintain the relationship $x_2 = (2/3)x_1$ at all times.*

So this normal mode is defined by two quantities: the ratio of x_2 to x_1, and the frequency with which the system oscillates. If the two positions are in that ratio, they will stay in that ratio and oscillate with that frequency forever. All this behavior is based on the fact that the two masses are oscillating with the same frequency as each other.

Definition: Normal Modes

A normal mode of this system is a solution in which the ratio x_2/x_1 stays constant over time. This implies that the two balls are oscillating with the same frequency, so the motion of the entire system is periodic.

As we will see below, most possible behaviors of this system do not qualify as normal modes. For most initial conditions the ratio x_2/x_1 will not remain constant and the balls will not oscillate with the same frequency (or with any constant frequency). But there is one other normal mode, described by the following solution.

$$x_1(t) = B\cos\left(\sqrt{5}\,t\right)$$
$$x_2(t) = -\frac{1}{3}B\cos\left(\sqrt{5}\,t\right) \tag{6.1.4}$$

Once again this solution represents a very specific set of initial conditions: you pull Ball 1 a certain distance from equilibrium, pull Ball 2 a third of that distance in the *opposite* direction, and release them from rest. The balls move inward at the same time, and then outward. But once again the frequencies are the same, the motion is simple and periodic, and the ratio $x_2 = -(1/3)x_1$ persists.

---

[1] If the initial velocities were non-zero we would need to add sine solutions as well as cosines: see Problem 6.12.

6.1 | The Motivating Example on which We're Going to Base the Whole Chapter

You might now expect that for every initial x_2/x_1 there is a certain normal mode with a certain frequency associated with it, but in fact Equations 6.1.3 and 6.1.4 represent the only two normal modes of this particular system. There are no more.

So we've learned a lot about our system, but you're probably not ready to sound the victory bells. First of all, what if x_2 doesn't happen to start at $(2/3)x_1$ or $-(1/3)x_1$? And second of all, how did we find those solutions in the first place? Below we will outline the answers to both of those questions, but we will have to leave a number of holes that will be filled in with matrices.

More Complicated Solutions for Other Initial Conditions

Let's see how our system responds to a variety of initial conditions.

FIRST EXAMPLE $x_1(0) = 12$ and $x_2(0) = -4$

This fits our second normal mode, where $B = 12$.

$$x_1(t) = 12\cos\left(\sqrt{5}\,t\right)$$
$$x_2(t) = -4\cos\left(\sqrt{5}\,t\right)$$

The balls will oscillate with a frequency of $\sqrt{5}$ and with amplitudes 12 and 4. Their displacements will always be in opposite directions (out of phase by π).

SECOND EXAMPLE $x_1(0) = 4$ and $x_2(0) = 5/3$

There is no way to get *either* of our two solutions to fit this initial condition, and there is no normal mode with $x_2/x_1 = 5/12$. It seems that our solutions so far are only useful in a few very carefully selected cases.

But Equations 6.1.1 and 6.1.2 are linear, homogeneous differential equations, so any combination of solutions is a solution. If we choose $A = 3$ in the first normal mode and $B = 1$ in the second normal mode, and then *add the solutions*, we get a new solution.

$$x_1(t) = 3\cos\left(\sqrt{2}\,t\right) + \cos\left(\sqrt{5}\,t\right)$$
$$x_2(t) = 2\cos\left(\sqrt{2}\,t\right) - \frac{1}{3}\cos\left(\sqrt{5}\,t\right) \tag{6.1.5}$$

In Problem 6.13 you will plug these in to verify that they solve the differential equations and our initial condition. But it's more important to see that they work *without* plugging them in, by following this logic.

- The solution $x_1 = 3\cos(\sqrt{2}\,t)$, $x_2 = 2\cos(\sqrt{2}\,t)$ solves the differential equations; we know this because it fits our first normal mode. It meets the initial conditions $x_1(0) = 3$, $x_2(0) = 2$.

- The solution $x_1 = \cos(\sqrt{5}\,t)$, $x_2 = -(1/3)\cos(\sqrt{5}\,t)$ solves the differential equations because it is our second normal mode. It meets the initial conditions $x_1(0) = 1$, $x_2(0) = -1/3$.
- When we add these two solutions the result *must* still satisfy the differential equations, and must meet initial conditions $x_1(0) = 4$, $x_2(0) = 5/3$.

The "Second Example" above is the most important part of this section, capturing an entire solution in the two numbers $A = 3$ and $B = 1$. In other words the system starts out in a state that can be represented as 3 of the first normal mode plus 1 of the second normal mode. We know how each of these normal modes evolves in time, so we know how the entire system will evolve. Hence, the question "how do I find the behavior of this system?" is replaced with "how do I express the initial state as a linear combination of the two normal modes?"

Section 6.3 will use matrix multiplication to convert from normal modes to positions. Section 6.4 will reframe that process as a "change of basis" and give the conditions under which such a conversion can reliably be done.

We need to make one more point about the above solution: the resulting motion is not simple. If you were to watch the two balls the motion would look almost random, and it would not be periodic no matter how long you looked. But hidden under the randomness would be two simple, periodic systems being added together.

The moral of Figure 6.2 is that if you look at the system and ask "Where is Ball 1 and where is Ball 2 over time?" the answer looks chaotic. You see the underlying order if you ask instead "How much of the first normal mode and how much of the second normal mode do I have?"

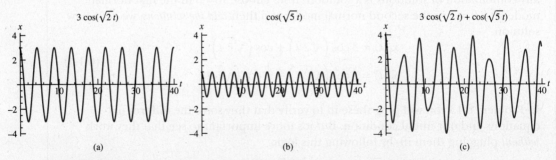

FIGURE 6.2 Plots (a) and (b) show normal modes, which are periodic. Plot (c) shows the motion of $x_1(t)$ in the second example above. It's a sum of periodic motions, but it is not periodic.

6.1 | The Motivating Example on which We're Going to Base the Whole Chapter

THIRD EXAMPLE $x_1(0) = 5$ and $x_2(0) = 7$

We chose those numbers precisely because there doesn't seem to be any obvious way to get there from our two normal modes. But once again, you can do so if you choose the right values of A and B and then add solutions. In this case, the right values are $A = 26/3$ and $B = -11/3$.

$$x_1 = \frac{26}{3}\cos\left(\sqrt{2}\,t\right) - \frac{11}{3}\cos\left(\sqrt{5}\,t\right)$$
$$x_2 = \left(\frac{2}{3}\right)\frac{26}{3}\cos\left(\sqrt{2}\,t\right) - \left(-\frac{1}{3}\right)\frac{11}{3}\cos\left(\sqrt{5}\,t\right)$$

Finding the numbers 26/3 and −11/3 in the last example required us to find just the right linear combination of solutions to match the desired initial conditions. In other words, it requires us to translate from "positions of the balls" to "amplitudes of the normal modes." Section 6.6 will find the inverse of the normal-modes-to-positions matrix to accomplish this.

Now, How Did We Come Up with Those Normal Modes?

Remember that this is going to be an outline, not a full derivation. Many of the key steps will have to be filled in later in the chapter when we have learned more about matrices.

As we usually do with differential equations, we begin by thinking about what kind of solution we can guess. You can imagine trying any pair of functions $x_1(t)$ and $x_2(t)$, but we're going to restrict our guess in two important ways. First, given that we're describing balls on springs, we're going to guess oscillatory solutions. Second, less obviously, we're going to guess solutions for which both balls will oscillate with the same frequency: that is, normal modes. As we have seen, they are simple to understand physically and to work with mathematically, and they can combine to give us the more complicated behavior that may also arise.

Let's start by guessing a very simple normal mode: $x_1 = C_1 \cos t$ and $x_2 = C_2 \cos t$. Plugging these into Equations 6.1.1 and 6.1.2 and simplifying a little gets us here:

$$3C_1 - 3C_2 = 0$$
$$\frac{2}{3}C_1 - 2C_2 = 0 \tag{6.1.6}$$

Section 7.4 (see felderbooks.com) will discuss how to solve simultaneous linear equations. (Remember again that we need to be able to do this with 100 equations, not just with two as in this example.)

You can probably solve Equations 6.1.6 in a minute or two, but we'll save you the trouble and tell you that $C_1 = C_2 = 0$. (Looks obvious now, doesn't it?) That leads us to $x_1(t) = x_2(t) = 0$, sometimes called the "trivial solution": it does satisfy Equations 6.1.1 and 6.1.2, but it isn't very helpful. We want to know how the balls move, not how they stand still.

So "cos t" didn't work out too well. What if we change the frequency? In Problem 6.16 you will guess oscillatory functions with many different frequencies, but they will lead back to the same place.

So here comes a particularly lucky guess: what if we just happen to try $x_1 = C_1 \cos(\sqrt{2}\,t)$ and $x_2 = C_2 \cos(\sqrt{2}\,t)$? Then we end up here.

$$2C_1 - 3C_2 = 0 \qquad (6.1.7)$$
$$-\frac{2}{3}C_1 + C_2 = 0 \qquad (6.1.8)$$

Once again $C_1 = C_2 = 0$ fits, but this time it isn't the only solution. Equation 6.1.7 can be rewritten as $C_2 = (2/3)C_1$. Equation 6.1.8 can be written the same way. That means any numbers C_1 and C_2 with that particular ratio will solve both equations. So this time we have a useful solution—an infinite number of them, in fact—which are the first normal mode we saw above.

$$x_1(t) = A\cos\left(\sqrt{2}\,t\right)$$
$$x_2(t) = (2/3)A\cos\left(\sqrt{2}\,t\right)$$

Why was it that Equations 6.1.6 got us nowhere, but Equations 6.1.7 and 6.1.8 worked so well? Because the first set of equations had only one solution; the second set of equations was "linearly dependent" (or "redundant"), and had infinitely many solutions.

> Section 6.7 will explain how to determine if a set of simultaneous equations has only one solution, as opposed to being "linearly dependent" (infinitely many solutions) or "inconsistent" (no solutions). It will show how you can find the frequencies ($\sqrt{2}$ and $\sqrt{5}$ in this example) if you assume cosine solutions.

Finally, what led us to expect that we could find a normal mode solution in the first place? It turns out that in almost every case linear differential equations for coupled oscillators have normal mode solutions, and the general solutions can be built up from linear combinations of these solutions.

> If you've seen linear algebra before, you may remember finding "eigenvectors" and "eigenvalues." In Section 6.8 you'll see that the eigenvectors of a matrix tell you the normal modes of a system, and the eigenvalues tell you the frequencies of those normal modes.

Stepping Back
One of our goals in this section is to have you understand normal modes as a way of thinking about the behavior of a coupled oscillator. Equations 6.1.3 represent a simple, periodic, and *self-perpetuating* solution to the differential equation: that is, if the system is ever in that state, it will stay in that state forever. Equations 6.1.4 represent another self-perpetuating solution. Any combination of these solutions is also a solution (because the differential equations are linear and homogeneous). Most remarkably, any state of the system can be represented as

6.1 | The Motivating Example on which We're Going to Base the Whole Chapter

a linear combination of these solutions—so once you know the combination, you know just how it will evolve.

But we also want to set up the two key places where matrices play into the process.

- We can use a matrix to convert from A and B to $x_1(t)$ and $x_2(t)$. ("If the first coefficient is 7 and the second is -4, what are the balls actually doing?") As you will see this is a job ideally suited for matrices, because it is a linear transformation from one multivariable representation to a different multivariable representation of the same state. An inverse matrix performs the same conversion in the other direction, finding the right combination of normal modes to match any given initial conditions.
- We can also use a matrix to find the normal modes in the first place. The "eigenvalues" of a matrix give the frequency of each normal modes, and the "eigenvectors" give the ratio of x_1 to x_2.

These two uses of matrices are not unrelated, but that won't be clear until you have learned about eigenvalues. Focus now on how two numbers such as "$A = 26/3$ and $B = -11/3$" can describe the entire behavior of the system—the initial conditions and how they will evolve through time—and you will begin the chapter on a solid foundation.

6.1.2 Problems: The Three-Spring Problem

6.1
(a) If the three-spring system in the Explanation (Section 6.1.1) begins at rest with $x_1(0) = 3$ and $x_2(0) = 2$ then it is in the first normal mode. Describe in words how this system will evolve over time.

(b) If the three-spring system begins at rest with $x_1(0) = 3$ and $x_2(0) = -1$ then it is in the second normal mode. How will the resulting behavior be like Part (a), and how will it be different?

(c) Now—suppose you solved the differential equations with the initial conditions in Part (a), and then you solved them again with the initial conditions from Part (b), and then you added those two solutions. Would the resulting functions necessarily solve the differential equations? Why or why not? What initial conditions would they represent?

(d) Describe in words how the system in Part (c) will evolve over time.

6.2 The general solution to Equations 6.1.1 and 6.1.2 with $\dot{x}_1(0) = \dot{x}_2(0) = 0$ is any combination of Equations 6.1.3 and Equations 6.1.4.

(a) Write the solution represented by the constants $A = -10$, $B = 0$. What initial conditions $x_1(0)$ and $x_2(0)$ does this solution represent?

(b) Write the solution represented by the constants $A = 10$, $B = -9$. What initial conditions $x_1(0)$ and $x_2(0)$ does this solution represent?

(c) What constants A and B lead to the initial conditions $x_1(0) = 13.5$ and $x_2(0) = -3$? What is the solution to the differential equations with these initial conditions?

6.3 Which of the following represents a normal mode? We are not asking if they are solutions to the differential equations in this section, or to any other particular equations. We are simply saying: if some system moved as described by these functions, would that be considered a normal mode of the system? For each one explain in one brief sentence why it is or isn't a normal mode.

(a) $x_1 = A \sin t$, $x_2 = 2A \sin t$
(b) $x_1 = A \sin t$, $x_2 = 2A \sin(2t)$
(c) $x_1 = Ae^{2it}$, $x_2 = 3Ae^{2it}$
(d) $x_1 = Ae^{2it}$, $x_2 = Ae^{4it}$

6.4 Consider a system defined by the positions of two objects, $a(t)$ and $b(t)$. We're not going to tell you anything about what those two objects are, or the differential equations that represent them. We will tell you that the positions oscillate sinusoidally, and they both start at rest at $t = 0$.

(a) Suppose we tell you that "If $b = 6a$ then both positions will oscillate with a frequency of 10 rad/s." If you start with $a = 2$ and $b = 12$ and watch the system go for a few seconds, will b still equal $6a$?

(b) Suppose we tell you that "If $b = 7a$ then a will oscillate at 2 cycles per second and b at 3 cycles per second." If you start with

$a = 2$ and $b = 14$ and watch the system go for a few seconds, will b still equal $7a$?

6.5 Consider a system of balls on springs like the ones we've been discussing in this section and assume the system is in one of its normal modes.

(a) Is it possible for one of the balls to be to the right of its equilibrium point while another one is to the left of its equilibrium point? Why or why not?

(b) Is it possible for one of the balls to pass through its equilibrium point at a time when another one is not at its equilibrium point? Why or why not?

(c) How would your answers change if the system were not in a normal mode?

6.6 In this problem you will consider the three-spring problem with both masses equal and all three spring constants equal. (This is *not* the situation described by Equations 6.1.1 and 6.1.2.) You should be able to answer all of the following questions by thinking about the physical situation without doing any calculations.

(a) First consider initial conditions in which you pulled both balls to the right by the same amount and let go.
  i. Spring 1 (the leftmost spring) would pull Ball 1 to the left. Spring 3 would push Ball 2 to the left. Which of these forces would be larger, or would they be equal? Why?
  ii. Would the force of Spring 2 on Ball 1 be to the left, to the right, or zero? Why?
  iii. Describe the long term behavior of the system given these initial conditions. Explain why this would be a normal mode solution.

(b) Describe the other normal mode of the system. Specify both the initial conditions and the long term behavior associated with this second normal mode.

6.7 [*This problem depends on Problem 6.6.*] Problem 6.6 described the initial conditions that would lead to the two normal modes of a coupled system. For example, if you started with the initial conditions $x_1(0) = x_2(0) = 1$ that would put the system in the first normal mode, oscillating with amplitude 1.

(a) In a similar way, give a set of initial conditions $x_1(0)$ and $x_2(0)$ that would put the system in the second normal mode, with the balls again oscillating with amplitude 1.

(b) Now suppose the system starts with initial conditions $x_1 = 0$, $x_2 = 1$. Write this initial condition as a combination of the two normal modes you found in Problem 6.6. Use A and B for the normal modes. Your answer should be in the form $aA + bB$, but with specific numbers for a and b.

(c) Is the solution in Part (b) a normal mode? Why or why not?

6.8 [*This problem depends on Problem 6.6.*] With $k = 8$ N/m and $m = 4$ kg the problem described in Problem 6.6 is described by the equations $\ddot{x}_1 = -4x_1 + 2x_2$, $\ddot{x}_2 = 2x_1 - 4x_2$.

(a) Solve these equations numerically with initial conditions corresponding to each of the normal modes you found. Plot the solutions and describe the behavior of x_1 and x_2 in each case.

(b) Repeat Part (a) with initial conditions corresponding to 2 times one normal mode plus 3 times the other.

(c) Is the solution you plotted in Part (b) a normal mode solution? Why or why not?

6.9 The picture shows a "double pendulum": a pendulum hanging from a pendulum. Consider the following possibilities.

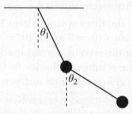

(a) The top pendulum swings with a period of 2 s, and the bottom pendulum swings from the top pendulum with a period of 2 s. Is this a normal mode? Is the resulting motion periodic?

(b) The top pendulum swings with a period of 2 s, and the bottom pendulum swings from the top pendulum with a period of 3 s. Is this a normal mode? Is the resulting motion periodic?

(c) The top pendulum swings with a period of 2 s, and the bottom pendulum swings from the top pendulum with a period of π s. Is this a normal mode? Is the resulting motion periodic?

6.10 In this problem you will derive the equations of motion for the three-spring system. The only physics you need is $F = ma$ and the fact that a spring exerts a force $F = -kx$ where k is the spring constant and x is the displacement.

6.1 | The Motivating Example on which We're Going to Base the Whole Chapter

(a) The position $x_1 = x_2 = 0$ represents the equilibrium position. Now imagine that Ball 1 is at this position precisely, but Ball 2 is slightly to the right of this position, as shown in Figure 6.1. Which springs now push on Ball 1, and in which directions? Which springs now push on Ball 2, and in which directions?

(b) Now imagine that Ball 1 is displaced to the right by a distance x_1, and Ball 2 is displaced to the right by x_2. This time the answers will be quantitative, and they will include the three spring constants k_1, k_2, and k_3. We begin with Ball 1.
  i. The force of Spring 1 on Ball 1 depends only on the position x_1 (the other position is irrelevant). Write a formula for this force.
  ii. How much is Spring 2 stretched? Your answer should be a function of x_1 and x_2. For example, if both balls are displaced the same amount to the right then Spring 2 isn't stretched at all. As a check on your answer, make sure it gives a positive answer when Spring 2 is longer than its equilibrium length, a negative answer when it is shorter, and 0 when Spring 2 is at its equilibrium length (neither stretched nor compressed.)
  iii. Using your answer to Part (ii), write a formula for the force of Spring 2 on Ball 1.
  iv. Putting the two forces together and using $F = ma$, write a differential equation for $x_1(t)$. Be sure you have the sign of each force correct. *Hint*: One way to check yourself is to plug in numbers and make sure you get Equation 6.1.1!

(c) Repeat Part (b) for x_2.

6.11 Plug Equations 6.1.3 into Equations 6.1.1 and 6.1.2 to confirm that they are valid solutions for any constant A.

6.12 Our treatment of the three-spring problem was incomplete because we looked only at the cosine parts of the solutions, ignoring the sines.

(a) Show that the following equations are valid solutions to Equations 6.1.1 and 6.1.2 for any constants A_1 and A_2.

$$x_1(t) = A_1 \cos\left(\sqrt{2}\,t\right) + A_2 \sin\left(\sqrt{2}\,t\right)$$
$$x_2(t) = \frac{2}{3}A_1 \cos\left(\sqrt{2}\,t\right) + \frac{2}{3}A_2 \sin\left(\sqrt{2}\,t\right)$$
(6.1.9)

(b) Show that the initial conditions $\dot{x}_1(0) = \dot{x}_2(0) = 0$ lead to $A_2 = 0$ and therefore to the solution we used in the Explanation (Section 6.1.1).

(c) Do Equations 6.1.9 represent a normal mode of the system?

(d) The three-spring problem has four initial conditions: the initial position and velocity of Ball 1 and the initial position and velocity of Ball 2. What *must* be true of all the initial conditions for the balls to follow Equations 6.1.9?

6.13 In our second example we created a new solution by combining the two normal modes, along with carefully choosing values of the arbitrary constants.

(a) Show that Equations 6.1.5 are a valid solution to Equations 6.1.1 and 6.1.2.

(b) Use Equations 6.1.5 to find $x_1(0)$ and $x_2(0)$ and confirm that they meet the intended initial conditions.

6.14 Arguably the most important point in our treatment of the three-spring problem—both for understanding normal modes, and for setting up linear algebra—is creating new solutions as linear combinations of the two solutions we found. (This works for any linear homogeneous differential equation.)

(a) Show that if you add Equations 6.1.3 to Equations 6.1.4 the resulting functions are valid solutions to Equations 6.1.1 and 6.1.2 for any values of A and B.

(b) What are the initial positions $x_1(0)$ and $x_2(0)$ for these generalized solutions? Your answers will be functions of A and B.

(c) Now turn it around; starting with your answers to Part (b), solve for A and B as functions of $x_1(0)$ and $x_2(0)$. In doing so, you find a direct way to meet any given initial positions, as we did in our third example. (Note that we still do not have a fully general solution to the three-spring problem. We left off all the sine terms based on our initial assumption that both balls start at rest, so we don't have to worry about $v_1(0)$ and $v_2(0)$.)

(d) Write the solution if you pull Ball 1 to the left by a distance of 1 and Ball 2 to the right by the same amount and release them.

6.15 🖥 A key point in our explanation is that a normal mode represents simple oscillation with one frequency, but that *combinations* of normal modes are generally not simple or periodic.

Often they do not appear to be built from sines and cosines (even though they are). Graph each function below from $t = 0$ to $t = 30$.
   (a) $\sin\left(\sqrt{2}\, t\right)$.
   (b) $\sin\left(\sqrt{5}\, t\right)$.
   (c) $\sin\left(\sqrt{5}\, t\right) + \sin\left(\sqrt{2}\, t\right)$.
   (d) $3\sin\left(\sqrt{5}\, t\right) + 2\sin\left(\sqrt{2}\, t\right)$.

6.16 In this problem you will solve Equations 6.1.1 and 6.1.2 by guess and check.
   (a) Plug in the solution $x_1 = C_1 \cos(2t)$, $x_2 = C_2 \cos(2t)$. Write down and solve the resulting two linear equations for C_1 and C_2. What motion does the resulting solution describe?
   (b) Plug in the solution $x_1 = C_1 \cos\left(\sqrt{3}\, t\right)$, $x_2 = C_2 \cos\left(\sqrt{3}\, t\right)$. Write down and solve the resulting two linear equations for C_1 and C_2. What motion does the resulting solution describe?
   (c) Plug in the solution $x_1 = C_1 \cos(\omega t)$, $x_2 = C_2 \cos(\omega t)$ where ω is a constant. Show that for any value of ω, you can solve the resulting linear equations with $C_1 = C_2 = 0$.
   (d) Now plug in the solution $x_1 = C_1 \cos\left(\sqrt{5}\, t\right)$, $x_2 = C_2 \cos\left(\sqrt{5}\, t\right)$. Show that the resulting linear equations both reduce to the *same* equation, and are therefore solved as long as C_1 and C_2 are in the correct ratio. This ratio should lead you to our second normal mode.

6.17 In this problem you will walk through the entire three-spring problem, from the beginning, with different masses and spring constants. The differential equations you will solve are:

$$\frac{d^2 x_1}{dt^2} = -5x_1 + 8x_2 \quad (6.1.10)$$

$$\frac{d^2 x_2}{dt^2} = x_1 - 3x_2 \quad (6.1.11)$$

   (a) The functions $x_1 = A\cos t$, $x_2 = kA \cos t$ provide a valid solution to Equations 6.1.10 and 6.1.11 for *any* value of A, provided you choose the right value of k. Show that this works by finding the right value of k.
   (b) Show that the functions $x_1 = C\cos(2t)$, $x_2 = kC\cos(2t)$ do not provide a valid solution unless $C = 0$.
   (c) Show that the functions $x_1 = D\cos\left(\sqrt{2}\, t\right)$, $x_2 = kD\cos\left(\sqrt{2}\, t\right)$ do not provide a valid solution unless $D = 0$.
   (d) The functions $x_1 = B\cos\left(\sqrt{7}\, t\right)$, $x_2 = kB\cos\left(\sqrt{7}\, t\right)$ provide a valid solution to Equations 6.1.10 and 6.1.11 for *any* value of B, provided you choose the right value of k. Show that this works by finding the right value of k.
   (e) Parts (a) and (d) represent solutions, but are they normal modes? How can you tell?
   (f) What initial conditions are represented by the solutions to Parts (a) and (d)?
   (g) Find a solution to match the initial conditions $x_1 = -2$, $x_2 = 2$.
   (h) Find a solution to match the initial conditions $x_1 = 10$, $x_2 = 2$.

6.18 Each of the scenarios below refers to a solution to Equations 6.1.1–6.1.2. For each one have a computer create an animation of two moving balls. Draw them as balls oscillating about their equilibrium positions, which you should separate by a distance of 3.
   (a) Show the two balls oscillating in the first normal mode, with $A = 2$.
   (b) Show the two balls oscillating in the second normal mode, with $B = 2$.
   (c) Show the two balls oscillating in a sum of the two normal modes with $A = B = 1$.
   (d) Describe the motion in each of the three cases.

6.2 Matrices: The Easy Stuff

Throughout this chapter we are going to apply matrices to the problem in Section 6.1: first to represent the problem, and then to simplify and solve different parts of it. But before we get there we have to introduce some basic terminology and operations.

6.2 | Matrices: The Easy Stuff

6.2.1 Discovery Exercise: The Easy Matrix Stuff

The following matrix, stolen from a rusted lockbox in the back of a large, dark lecture hall, is the gradebook for Professor Snape's class in potions.

$$G = \begin{array}{c} \\ \text{Granger, H} \\ \text{Longbottom, N} \\ \text{Malfoy, D} \\ \text{Potter, H} \\ \text{Weasley, R} \end{array} \begin{pmatrix} \text{Poison} & \text{Cure} & \text{Love Philter} & \text{Invulnerability} \\ 100 & 105 & 99 & 100 \\ 80 & 90 & 85 & 85 \\ 95 & 90 & 10 & 85 \\ 70 & 75 & 70 & 75 \\ 85 & 90 & 95 & 90 \end{pmatrix}$$

When we say this is a "matrix" we're referring to the grid of numbers. The labels (such as "Granger, H" or "Poison") explain what this particular matrix represents, but they are not part of the matrix itself.

The matrix can be viewed as a list of horizontal *rows* or as a list of vertical *columns*.

1. Is "all Longbottom's grades" a row or a column?
2. Is "all grades on Invulnerability" a row or a column?

The "dimensions" of a matrix are the number of rows and columns...in that order. So a 10×20 matrix means 10 rows and 20 columns.

3. What are the dimensions of the gradebook matrix?

For two matrices to be "equal" they must be exactly the same in every way: same dimensions, and every element the same. If everything is not precisely the same, the two matrices are not equal.

4. What must x and y be, in order to make the following matrix equal to Professor Snape's gradebook matrix?

$$\begin{pmatrix} 100 & 105 & 99 & 100 \\ 80 & x+y & 85 & 85 \\ 95 & 90 & 10 & 85 \\ 70 & 75 & x-y & 75 \\ 85 & 90 & 95 & 90 \end{pmatrix}$$

If two matrices do not have exactly the same dimensions, you cannot add or subtract them. If they do have the same dimensions, you add and subtract them just by adding or subtracting each individual element.

As an example: Professor Snape has decided that his grades are too high, so he decides to subtract the following grade-curving matrix from his original grade matrix.

$$\begin{pmatrix} 5 & 0 & 10 & 0 \\ 5 & 0 & 10 & 0 \\ 5 & 0 & 10 & 0 \\ 5 & 0 & 10 & 0 \\ 5 & 0 & 10 & 0 \end{pmatrix}$$

5. Write the new gradebook after the curve.
6. Now, suppose the professor wants to curve everyone's grades down (as before), but curve the "Potter, H" grades down twice as much. What grade-curving matrix would he subtract from his original gradebook?

Multiplying something by 3 always means adding it to itself three times, so once you know how to add matrices, you can figure out how to multiply a matrix by a number.

7. If we call the gradebook matrix **G**, what is matrix 3**G**? In other words, what is **G** + **G** + **G**?

6.2.2 Explanation: The Easy Matrices Stuff

The following is an example of a matrix.

$$\mathbf{M} = \begin{pmatrix} 13 & -7 & 4 \\ 10 & \pi & 0 \end{pmatrix}$$

The plural of matrix is "matrices" (not "matrixes.") The singular of matrices is "matrix" (not "matrice").[2]

Basic Properties of Matrices
- A row is a horizontal list of numbers; a column is a vertical list of numbers. (Picture columns in Greek architecture. They don't have much to do with linear algebra, but they are columns, and they are vertical.) So the above example has two rows and three columns.
- The dimensions of a matrix are the number of rows, and the number of columns, in that order. So above is a 2×3 matrix.
- Each element is referred to by its row followed by its column, written as a subscript. So if the above matrix is called **M**, then $M_{23} = 0$.

Basic Operations on Matrices
- To add or subtract matrices you add or subtract each individual element. This operation is only allowed when the two matrices have the same dimensions; adding a 2×3 matrix to a 3×2 matrix is illegal.
- For two matrices to be equal they must have the same dimensions, and all their corresponding elements must be equal.
- When you multiply a matrix by a number you multiply each element in the matrix by that number.

A Few More Random Definitions
- A "column matrix" is a matrix with just one column. We will use column matrices to represent many different kinds of information, such as spatial vectors.
- A "row matrix" is a matrix with just one row.
- A "square matrix" is a matrix with the same number of rows as columns. These turn out to be particularly important as transformers of information.
- The "main diagonal" of a square matrix[3] is the list of items going from the upper left to the lower right. So the main diagonal of $\begin{pmatrix} 1 & 2 \\ 3 & 4 \end{pmatrix}$ consists of the numbers 1 and 4.
- A "diagonal matrix" is one where every element is zero except the ones on the main diagonal. So $\begin{pmatrix} 1 & 0 \\ 0 & 4 \end{pmatrix}$ is a diagonal matrix.

---

[2] And don't even get us started on the word "series."
[3] Also called "principal diagonal," "primary diagonal," or "leading diagonal."

You can see that the matrix at the beginning of this section fits none of these categories, and that a 1×1 matrix fits all of them; it's a column matrix, row matrix, square matrix, and diagonal matrix.

All of these properties and definitions, as well as others we will introduce throughout the chapter, are collected in Appendix E. If we refer to the "main diagonal" (introduced here) or to an "orthogonal matrix" (introduced later) and you don't remember what that is, be sure to have that appendix conveniently bookmarked.

Typography for Variables

A variable in this chapter can stand for many different types of objects, so we'll use the following conventions to keep them straight.

- A boldfaced capital letter is a matrix, like matrix **M** shown above.
- A boldfaced lowercase letter like **c** is a column matrix. (In Section 6.4 we will more formally discuss these objects as "generalized" or "non-spatial" vectors.)
- A letter with an arrow over it like $\vec{v}$ represents a spatial vector; a hat like $\hat{\imath}$ represents a unit vector. Spatial vectors may be either lowercase or uppercase.
- A non-boldfaced no-arrow no-hat letter like x represents a number.

6.2.3 Problems: The Easy Matrix Stuff

6.19 Matrix **M** is shown below.

$$\mathbf{M} = \begin{pmatrix} 3 & 2 & -4 & 9 & 0 \\ 6 & 10 & -3 & 12 & x \\ 7 & p & 3 & 8 & 15 \\ 2 & 23 & 3y & 20 & f \end{pmatrix}$$

(a) What are the dimensions of matrix **M**?
(b) Copy the third row of matrix **M**.
(c) Write the matrix $(1/2)\mathbf{M}$.
(d) What is $M_{23} + M_{14}$?

6.20 For each of the operations indicated below give the answer or explain why the operation cannot be carried out.

$$\mathbf{A} = \begin{pmatrix} 3 & -3 & 0 \\ 2 & 4 & -1 \end{pmatrix} \quad \mathbf{B} = \begin{pmatrix} 2 & 6 & 4 \\ 9 & n & 8 \end{pmatrix}$$

$$\mathbf{C} = \begin{pmatrix} 5 & 7 \\ 9 & -n \end{pmatrix}$$

(a) $\mathbf{A} + \mathbf{B}$
(b) $2\mathbf{C}$
(c) $\mathbf{B} + \mathbf{C}$
(d) $2\mathbf{A} - \mathbf{B}$
(e) $A_{12} + C_{21}$

6.21 For each of the operations indicated below give the answer or explain why the operation cannot be carried out.

$$\mathbf{A} = \begin{pmatrix} 2 & 7 \\ 1 & -2 \end{pmatrix} \quad \mathbf{B} = \begin{pmatrix} 4 & -3 \\ 9 & -8 \end{pmatrix}$$

$$\mathbf{c} = \begin{pmatrix} 1 \\ 5 \end{pmatrix}$$

(a) $\mathbf{A} + \mathbf{B}$
(b) $\mathbf{A} - \mathbf{c}$
(c) $A_{21} + c_{12}$
(d) $\sum_{i=1}^{2} A_{1i} c_{i1}$

6.22 For each equation below, solve for the values of all the letters or explain why it's not possible.

(a) $\begin{pmatrix} 2x \\ 5y \end{pmatrix} + \begin{pmatrix} x+y \\ -6x \end{pmatrix} = \begin{pmatrix} 6 \\ 2 \end{pmatrix}$

(b) $\begin{pmatrix} x+y \\ 3x-2y \end{pmatrix} + \begin{pmatrix} 4x+y \\ x+5y \end{pmatrix} = \begin{pmatrix} 10 & 6 \\ 8 & 9 \end{pmatrix}$

(c) $x \begin{pmatrix} 2 \\ 3 \end{pmatrix} = \begin{pmatrix} 4 \\ 6 \end{pmatrix}$

6.23 For each equation below, solve for the values of all the letters or explain why it's not possible.

(a) $\begin{pmatrix} 1 & 2 \\ 3 & 4 \end{pmatrix} + \frac{1}{2} \begin{pmatrix} w & x \\ y & z \end{pmatrix} = \begin{pmatrix} 1 & 1 \\ 0 & 0 \end{pmatrix}$

(b) $\begin{pmatrix} x+y \\ 2 \end{pmatrix} - \begin{pmatrix} 3 \\ x-y \end{pmatrix} = \begin{pmatrix} 1 & 2 \\ 3 & 4 \end{pmatrix}$

(c) $x \begin{pmatrix} 2 \\ 3 \end{pmatrix} = \begin{pmatrix} 4 \\ 8 \end{pmatrix}$

6.24 Write the 3×3 matrix defined by $M_{ij} = i^2 - j^2$.

6.25 Write a formula M_{ij} for the elements of the matrix $\begin{pmatrix} 3 & 5 \\ 4 & 6 \end{pmatrix}$. A formula like that is how you write a matrix in "index notation." (See e.g. Problem 6.24.)

6.26 In order to stock a small hotel bathroom, you need one bottle of shampoo, two bars of soap, and four towels. A large hotel bathroom requires two bottles of shampoo, three bars of soap, and five towels. We represent this information with column matrices as follows. (Remember that the labels are not part of the matrices *per se*, but they tell us what the elements represent.)

$$\mathbf{S} = \begin{matrix} shampoo\ bottles \\ soap\ bars \\ towels \end{matrix} \begin{pmatrix} 1 \\ 2 \\ 4 \end{pmatrix}$$

$$\mathbf{L} = \begin{matrix} shampoo\ bottles \\ soap\ bars \\ towels \end{matrix} \begin{pmatrix} 2 \\ 3 \\ 5 \end{pmatrix}$$

Answer the following questions in matrix form. For instance, for the question "what supplies do you need to stock three small bathrooms?" you would not answer "three shampoo bottles, six…" but rather:

$$3\mathbf{S} = \begin{matrix} shampoo\ bottles \\ soap\ bars \\ towels \end{matrix} \begin{pmatrix} 3 \\ 6 \\ 12 \end{pmatrix}$$

(a) What supplies would you need to stock four large bathrooms?

(b) What supplies would you need to stock three small bathrooms and four large bathrooms?

(c) What supplies would you need to stock n small bathrooms and m large bathrooms?

6.3 Matrix Times Column

This section and Section 6.5 build up matrix multiplication as a compact notation for representing linear combinations. Along the way we will present many different situations that call for such a process.

In this section we consider the special case of matrix multiplication where the second matrix has only one column. The more general algorithm presented in Section 6.5 will be seen as a simple extension of this specific case.

6.3.1 Discovery Exercise: Multiple Value Problems

This exercise has no matrices in it, and contains no trick questions: it really is as straightforward as it seems. But if you take a few minutes to answer these questions, it will help set up our explanation of matrix multiplication in the rest of the section.

1. Every book in my collection has 200 pages.
   (a) If I have 7 books, how many pages do I have?
   (b) If I have b books, how many pages do I have?
   (c) If I have p pages, how many books do I have?
2. Every book in my collection has 10 pictures and 20,000 words.
   (a) If I have 7 books, how many pictures do I have, and how many words?
   (b) If I have b books, how many pictures do I have, and how many words?
3. Now I have two kinds of books in my collection. Each "kiddy book" has 30 pictures and 500 words. Each "chapter book" has 10 pictures and 20,000 words.
   (a) If I have 4 kiddy books and 6 chapter books, how many pictures do I have? How many words?

   See Check Yourself #38 in Appendix L

   (b) If I have k kiddy books and c chapter books, how many pictures do I have? How many words?

6.3.2 Explanation: Matrix Times Column

Books and Pages

If you haven't looked at the Discovery Exercise (Section 6.3.1), please do so quickly now. The arithmetic is straightforward, but the succession of problems points to a particular way of thinking.

In the first problem you start with one fact (number of books) and you want a different fact (number of pages). In order to convert you need one piece of information: the number of pages per book. We write:

$$\underset{p}{\text{number of pages}} = \underset{200}{\text{contents of a book}} \; \underset{b}{\text{number of books}}$$

Think of the 200 as a *conversion* from "number of books" (on the right) to "number of pages" (on the left). The last part of this problem indicates that the conversion is reversible, a very useful property of some—but not all—conversions.

We can express problem 2 the same way, but we want to convert a single-fact (number of books) to a double-fact (number of words and number of pictures). The conversion therefore requires two pieces of information: words per book, and pictures per book. We use a 2×1 column matrix to represent each double-fact and write, analogously to our equation above:

$$\underset{\text{total collection}}{\begin{matrix}\text{pictures}\\\text{words}\end{matrix}\begin{pmatrix} 70 \\ 140,000 \end{pmatrix}} = \underset{\text{contents of one book}}{\begin{matrix}\text{pictures}\\\text{words}\end{matrix}\begin{pmatrix} 10 \\ 20,000 \end{pmatrix}} \; \underset{\text{number of books}}{7}$$

There's nothing new here mathematically—just a reminder that multiplying a matrix by a number multiplies each element of the matrix by that number. The point is that the matrix $\begin{pmatrix} 10 \\ 20,000 \end{pmatrix}$ is being used to convert "number of books" to "number of pictures and number of words."

Problem 3 begins with a double-fact (number of kiddy books and number of chapter books) and ends with a different double-fact (number of words and number of pictures). The conversion therefore requires four pieces of information.

$$\begin{matrix}\text{pictures}\\\text{words}\end{matrix}\begin{pmatrix} ? \\ ? \end{pmatrix} = \begin{matrix}\text{pictures}\\\text{words}\end{matrix}\overset{\text{kiddy books} \;\; \text{chapter books}}{\begin{pmatrix} 30 & 10 \\ 500 & 20,000 \end{pmatrix}} \; \begin{matrix}\text{kiddy books}\\\text{chapter books}\end{matrix}\begin{pmatrix} 4 \\ 6 \end{pmatrix}$$

We have left question marks where the answers go, although they aren't hard to figure out. But focus on the transformation process this equation represents.

- On the far right is a column matrix representing the information we start with. Its labels go on its left.
- The matrix in the middle acts as a conversion *from* its labels on top, *to* its labels on the left—in this case, from kiddy and chapter books to pictures and words.
- On the far left, on the other side of the equal sign, is a column matrix representing the information we end up with, also labeled on the left.

Once you understand the setup of the problem, it becomes obvious how to solve it. We can represent the contents of 4 kiddy books and 6 chapter books as follows.

$$4\begin{pmatrix} 30 \\ 500 \end{pmatrix} + 6\begin{pmatrix} 10 \\ 20,000 \end{pmatrix} \qquad \text{(read "4 kiddy books plus six chapter books")}$$

This gives us the numbers we need to fill in the question marks in our previous equation. There are 180 pictures and 122,000 words in total. More importantly, this example gives us the general algorithm for multiplying a matrix (on the left) by a column matrix (on the right).

Multiplying a Matrix Times a Column Matrix

Consider a matrix as a sequence of columns **cn** multiplied by a column matrix with elements E_n. So we write:

$$(\mathbf{c1}\ \mathbf{c2}\ \mathbf{c3}\ \ldots\ \mathbf{cn}) \begin{pmatrix} E_1 \\ E_2 \\ E_3 \\ \ldots \\ E_n \end{pmatrix} \text{ where each } \mathbf{c} \text{ is a column; each } E \text{ is a number}$$

If the number of columns in the left-hand matrix does not equal the number of elements in the right-hand matrix, then the operation is illegal. Otherwise, the product is given by the matrix:

$$E_1 \mathbf{c1} + E_2 \mathbf{c2} + E_3 \mathbf{c3} + \ldots + E_n \mathbf{cn}$$

In other words, the column matrix on the right supplies the coefficients for a linear combination of the vectors represented by the left-hand matrix. As usual, an example is worth a thousand explanations.

EXAMPLE **Matrix Times Column**

Question: Multiply $\begin{pmatrix} 1 & 2 & 3 \\ 4 & 5 & 6 \end{pmatrix} \times \begin{pmatrix} 7 \\ 8 \\ 9 \end{pmatrix}$

Answer: $7 \begin{pmatrix} 1 \\ 4 \end{pmatrix} + 8 \begin{pmatrix} 2 \\ 5 \end{pmatrix} + 9 \begin{pmatrix} 3 \\ 6 \end{pmatrix} = \begin{pmatrix} 50 \\ 122 \end{pmatrix}$

A minor variation of this problem, $\begin{pmatrix} 1 & 2 & 3 \\ 4 & 5 & 6 \end{pmatrix} \begin{pmatrix} 7 \\ 8 \end{pmatrix}$, is an illegal matrix multiplication. Since the left-hand matrix has three columns, the right-hand (column) matrix must specify exactly three coefficients: no more and no less.

Whenever you multiply matrices you are creating a linear combination of *somethings*. Each column in the left-hand matrix represents one "something"; the right-hand matrix represents the coefficients. In our example above we multiplied matrices to combine 4 kiddy books with 6 chapter books. Section 6.4 will discuss another important example, spatial vectors. You will work through many other examples in the problems.

"But I was taught to think of the left-hand matrix as rows, not columns," we hear some of you cry. If you know that method you should be able to convince yourself that it is equivalent to the one we're presenting (see Section 6.5 Problem 6.91), and that either one takes about as much work as the other. But we hope to convince you that thinking of the column on the right as a series of coefficients for the columns of the matrix on the left gives you insight into the meaning of matrix multiplication that the other method doesn't.

Application to the Three-Spring Problem

Section 6.1 presented two coupled differential equations. At any given moment we can express the position of the system using a column matrix: $\mathbf{x} = \begin{pmatrix} x_1 \\ x_2 \end{pmatrix}$. In this expression x_1 and x_2 are numbers (the positions of the two balls), and $\mathbf{x}$ is the state of the system (which requires both numbers).

We found two normal mode solutions, each corresponding to a particular initial condition. If $x_2(0) = (2/3)x_1(0)$ the system oscillates according to its first normal mode, whereas $x_2(0) = -(1/3)x_1(0)$ leads to the second normal mode. For all other cases we rewrite the initial conditions as a linear combination of these two special states.

$$\begin{pmatrix} x_1(0) \\ x_2(0) \end{pmatrix} = A \begin{pmatrix} 1 \\ 2/3 \end{pmatrix} + B \begin{pmatrix} 1 \\ -1/3 \end{pmatrix}$$

For instance, in our third example (p. 271) we saw that $A = (26/3)$ and $B = -(11/3)$ allowed us to match the initial conditions $x_1(0) = 5$, $x_2(0) = 7$. We can now write that relationship as a matrix equation.

$$\begin{matrix} \text{position of} \\ \text{Ball 1} \\ \text{Ball 2} \end{matrix} \begin{pmatrix} \begin{matrix} \text{first normal} \\ \text{mode} \\ 1 \\ 2/3 \end{matrix} & \begin{matrix} \text{second normal} \\ \text{mode} \\ 1 \\ -1/3 \end{matrix} \end{pmatrix} \begin{matrix} \text{1st normal mode} \\ \text{2nd normal mode} \end{matrix} \begin{pmatrix} 26/3 \\ -11/3 \end{pmatrix} = \begin{matrix} \text{position of} \\ \text{Ball 1} \\ \text{Ball 2} \end{matrix} \begin{pmatrix} 5 \\ 7 \end{pmatrix}$$

Stop and make sure that you can clearly see all the following information in that one equation.

- The first normal mode represents initial conditions $x_1 = 1$ and $x_2 = 2/3$; the second normal mode, $x_1 = 1$ and $x_2 = -1/3$.
- The matrix multiplication says "I am multiplying the entire first normal mode (the first column) by $26/3$, the second normal mode by $-11/3$, and adding the results."
- The end result converts the given information from "this much of the first normal mode and that much of the second" to "this initial position for the first ball and that for the second."

So the matrix multiplication above allows us to answer the question: given any values of A and B (the arbitrary constants in the solution), what are the corresponding initial conditions? In Section 6.6 we will use the "inverse matrix" to answer the opposite question: given a set of initial conditions to meet, what should we choose for A and B? (That's how we got the numbers $26/3$ and $-11/3$ in the first place.)

Stepping Back

We began this section with $p(b) = 200b$ because you can think of a matrix as a kind of function, transforming each input into an output. For a regular function the input and output are numbers. For a matrix the input and output are lists of numbers, not necessarily the same length, which we write in vertical columns. In that sense matrices are more general than functions, but they can only represent transformations where each output is a linear combination of the inputs. If $x = a^2 - e^b$ and $y = \ln(ab)$, you can't use matrices to convert from a and b to x and y. Hence the name "linear algebra."

The way you apply a matrix to a list of numbers is through matrix multiplication:

$$\begin{pmatrix} 1 & 3 \\ 2 & 4 \end{pmatrix} \begin{pmatrix} 5 \\ 6 \end{pmatrix} = 5 \begin{pmatrix} 1 \\ 2 \end{pmatrix} + 6 \begin{pmatrix} 3 \\ 4 \end{pmatrix} = \begin{pmatrix} 23 \\ 34 \end{pmatrix}$$

The 2×2 matrix in the above example might represent the following information: "Each small structure contains 1 ton of steel and 2 tons of concrete, while each large structure contains 3 tons of steel and 4 tons of concrete." We could then interpret the above matrix multiplication as a conversion from "5 small and 6 large structures" to "23 tons of steel and 34 tons of concrete."

$$\begin{matrix} & \text{small} & \text{large} \\ \text{steel} \\ \text{concrete} \end{matrix} \begin{pmatrix} 1 & 3 \\ 2 & 4 \end{pmatrix} \begin{matrix} \text{small} \\ \text{large} \end{matrix} \begin{pmatrix} 5 \\ 6 \end{pmatrix} = \begin{matrix} \text{steel} \\ \text{concrete} \end{matrix} \begin{pmatrix} 23 \\ 34 \end{pmatrix}$$

Watch how the matrix multiplication, when you view the process as we presented it, explicitly says "5 of these and 6 of those." Watch how this problem, the "books" example above, the normal modes of the three-spring problem, and the many different examples you will encounter in the problems all use linear combinations to convert information from one form to another. This single idea is powerful enough to make its presence felt in all areas of physics and engineering.

A transformation matrix always converts from the labels on its top to the labels on its left. That rule will help you keep your variables straight even when the variables are not explicitly written, which they usually aren't. You know that the statement "every book has 200 pages" would generally be represented by an equation like $p = 200b$. Of course that equation doesn't mean anything unless you know that p represents the number of pages and b the number of books, but we generally write the equation without the labels and let the interpretation come later. Similarly with matrices: we are labeling them all explicitly right now to emphasize how the information in the top labels is transformed to the information in the left-hand labels. (Small and large structures to steel and concrete.) But it is more common to do the math with unlabeled numbers and let the interpretation come at the end.

6.3.3 Problems: Matrix Times Column

6.27 At the Smelli Deli, a large sandwich has 1/2 lb of meat and 1/3 lb of vegetables. A regular sandwich has 1/4 lb of meat and 1/3 lb of vegetables. A child's sandwich has 1/5 lb of meat and (of course) no vegetables. You are gathering the ingredients for 3 large sandwiches, 1 regular, and 2 children's. We represent this situation as follows:

$$\begin{matrix} & \text{large} & \text{regular} & \text{child} \\ \text{meat} \\ \text{vegetables} \end{matrix} \begin{pmatrix} 1/2 & 1/4 & 1/5 \\ 1/3 & 1/3 & 0 \end{pmatrix} \begin{matrix} \text{large} \\ \text{regular} \\ \text{child} \end{matrix} \begin{pmatrix} 3 \\ 1 \\ 2 \end{pmatrix}$$

$$= \begin{matrix} \text{meat} \\ \text{vegetables} \end{matrix} \begin{pmatrix} ? \\ ? \end{pmatrix}$$

(a) When you perform the matrix multiplication, the first step is to multiply 3 $\begin{pmatrix} 1/2 \\ 1/3 \end{pmatrix}$. Explain what the result of this step represents—not in terms of matrices and columns, but of ingredients.

(b) Perform the complete matrix multiplication to determine how much of each ingredient you need.

In Problems 6.28–6.33, perform the given matrix multiplication, or identify it as illegal.

6.28 $\begin{pmatrix} 10 & 3 \\ -2 & 0 \end{pmatrix} \begin{pmatrix} 6 \\ 2/3 \end{pmatrix}$

6.29 $\begin{pmatrix} 10 & x & 1 \\ -2 & 0 & -4 \end{pmatrix} \begin{pmatrix} 6 \\ 2/3 \\ 8 \end{pmatrix}$

6.30 $\begin{pmatrix} 10 & 3 \\ -2 & 0 \end{pmatrix} \begin{pmatrix} 6 \\ 2/3 \\ 8 \end{pmatrix}$

6.31 $\begin{pmatrix} -1 & 2 & 1 \end{pmatrix} \begin{pmatrix} 3 \\ 0 \\ 4 \end{pmatrix}$

6.32 $\begin{pmatrix} 5 & 0 & 2 \\ -1 & 4i & 0 \\ 3 & -2 & 1 \end{pmatrix} \begin{pmatrix} 4 \\ -3 \\ 0 \end{pmatrix}$

6.33 $\begin{pmatrix} 1 & 7 & 0 \\ 6 & -3 & w \\ 9 & 0 & 1/2 \end{pmatrix} \begin{pmatrix} 2 \\ 8 \end{pmatrix}$

6.34 $\begin{pmatrix} 2 & -3 \\ 3 & 1 \end{pmatrix} \begin{pmatrix} x^2 \\ y \end{pmatrix} = 1/2 \begin{pmatrix} -2 \\ 30 \end{pmatrix}$.
Solve for x and y.

6.35 To trace the background of the Icelandic population blood types in Iceland have been compared to those in Sweden, with the following results.
- Iceland: 56% type O, 31% type A, 11% type B, 2% type AB
- Sweden: 38% type O, 44% type A, 12% type B, 6% type AB

Suppose you have a random sample of I Icelanders and S Swedes.
  (a) Create a column matrix to represent the number of Icelanders and Swedes in your sample. Your matrix should be labeled on the left.
  (b) Create a matrix to convert from that matrix to a column matrix of different blood types. Your matrix should have labels on the top (countries) and on the left (blood types).
  (c) Perform the matrix multiplication.
  (d) Explain what your result means in terms of blood types in the sample.

6.36 In order to stock a small hotel bathroom, you need one bottle of shampoo, two bars of soap, and four towels. A large hotel bathroom requires two bottles of shampoo, three bars of soap, and five towels. What supplies would you need to stock three small bathrooms and four large bathrooms? Express this question and the answer as a multiplication of labeled matrices.

6.37 James is studying the diversity of oribatid and gamasid soil mites in two different samples. From the first sample, taken from forest floor leaf litter, he estimates 240,000 oribatid mites/m² and 86,000 gamasid mites/m². His estimation from the second sample, collected from a pile of grass clippings, is 127,000 oribatid mites/m² and 102,000 gamasid mites/m². James gives one of his graduate students a sample made up of 3 m² of the forest floor sample and 8 m² of the grass clipping sample. Use a multiplication of labeled matrices to determine the total number of oribatid and gamasid mites in the sample.

6.38 In the famous Henry Rich Armed Forces, a cavalry division contains 1 infantry battalion and 4 tank battalions, while an armored division contains 2 infantry battalions and 2 tank battalions.
  (a) If there are 3 cavalry divisions and 5 armored divisions, how many infantry battalions and tank battalions are there? (In this part it is not necessary to use matrices explicitly; just think it through and get the right answer.)
  (b) Express your answer to Part (a) as the product of two labeled matrices resulting in a third labeled matrix.
  (c) Each infantry battalion contains 100 officers, 1000 men, and 10 tanks. Each tank battalion contains 200 officers, 800 men, and 300 tanks. Express all this information in a labeled matrix that will convert numbers of battalions to officers, men, and tanks.
  (d) If there are 3 cavalry divisions and 5 armored divisions, how many officers, men, and tanks does that represent?

6.39 Create a word problem that requires a matrix multiplication. Your word problem should not involve any of the scenarios we have used so far: books, hotel bathrooms, sandwiches, military units, *etc.*

6.40 Rewrite Equations 6.1.1–6.1.2 (on Page 267) as a single matrix equation. One of the matrices in your equation should be the column $\begin{pmatrix} x_1 \\ x_2 \end{pmatrix}$, which represents the state of the system.

6.41 One solution to the differential equations in Section 6.1 is obtained by choosing $A = -10$ and $B = 3$; that is, -10 of the first normal mode and $+3$ of the second. What initial condition does this correspond to?
  (a) Express this question as a matrix multiplication.
  (b) Answer the question.

6.42 A solution to the differential equations in Section 6.1 is specified as "A times the first normal mode plus B times the second." For example, if $A = 3$ and $B = 2$ then the solution, expressed in terms of x_1 and x_2, is:

$$\begin{pmatrix} x_1(t) \\ x_2(t) \end{pmatrix} = \begin{pmatrix} 3\cos(\sqrt{2}\,t) + 2\cos(\sqrt{5}\,t) \\ 2\cos(\sqrt{2}\,t) - (2/3)\cos(\sqrt{5}\,t) \end{pmatrix}$$

Write a matrix to convert from the vector $\begin{pmatrix} A \\ B \end{pmatrix}$ to the vector $\begin{pmatrix} x_1(t) \\ x_2(t) \end{pmatrix}$. As a check on your answer, if you multiply your matrix by $\begin{pmatrix} 3 \\ 2 \end{pmatrix}$ it should give you the vector shown on the right above.

6.4 Basis Vectors

This section discusses an important application of matrix multiplication, which is converting between different representations of a vector. We focus here on vectors as you have generally used them: quantities with magnitude and direction, or equivalently arrows. Toward the end we will begin using the mathematics of vectors for non-spatial quantities, a topic to which we will return in Section 7.3.

6.4.1 Discovery Exercise: Basis Vectors

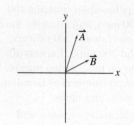

The drawing to the left shows two vectors $\vec{A}$ and $\vec{B}$.

You can express any vector in the plane as $a\vec{A} + b\vec{B}$ where a and b are carefully chosen numbers. For instance, if you pick a vector somewhere between the two, it might be $(1/3)\vec{A} + (1/2)\vec{B}$. A vector pointing up-and-left might be $\vec{A} + (-\vec{B})$.

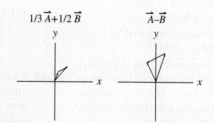

For this exercise your job is entirely visual: no calculations are involved, so your answers will be approximate.

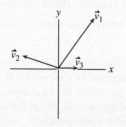

1. Draw the vector $-\vec{A} + \vec{B}$.
2. Draw the vector $(1/2)\vec{A} + 2\vec{B}$.
3. Draw the vector $0\vec{A} + 0\vec{B}$.
4. The graph on the left shows three vectors. For each vector, estimate how many $\vec{A}$s and $\vec{B}$s would add up to that vector. In other words, find a and b to express each vector as $a\vec{A} + b\vec{B}$.

See Check Yourself #39 in Appendix L

6.4.2 Explanation: Basis Vectors

No introductory physics class would be complete without a rousing discussion of a block sliding down a ramp.

To approach such a system you break the force of gravity F_G into two components. The "parallel" (to the ramp) component $F_\parallel$ causes acceleration; the "perpendicular" component $F_\perp$ causes a normal force that in turn causes friction. In effect you are creating a new set of axes that is more natural to this problem than the traditional horizontal and vertical axes. In the drawing below we label these new axes x' and y'.

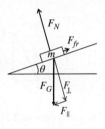

In the xy-coordinate system F_G is the simplest kind of vector: it points directly along one axis so it has only one component, $-mg\hat{j}$. But the friction and normal forces are more complicated, with both $\hat{i}$ and $\hat{j}$ components.

In the $x'y'$ system the situation is reversed. Now F_{fr} and F_N are simple: one points directly in the $\hat{i}'$-direction and the other in the $\hat{j}'$-direction. Of course F_G is no longer as simple, having both $\hat{i}'$- and $\hat{j}'$-components. (The real reason the $x'y'$ system is better for this problem is because the acceleration vector $\vec{a}$ is simple in that system.)

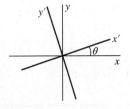

Section 7.1 will return to the idea of rotated axes, and use matrices to effect such a transformation. Right now our point is more general. You are used to expressing every vector as a linear combination of $\hat{i}$ and $\hat{j}$, but you can equivalently express every vector in other ways such as a combination of $\hat{i}'$ and $\hat{j}'$. This shift in perspective does not change the physics of the situation, but the numbers that *describe* the situation are all different.

The Words "Span" and "Basis"

You know that instead of saying "the vector with magnitude 10 at 30°" you can say "$5\sqrt{3}\,\hat{i} + 5\hat{j}$." This familiar notation expresses a linear combination: take this much of the vector $\hat{i}$ and that much of the vector $\hat{j}$ and add them.

Because we can create *any* vector in the xy-plane from these two vectors, we say that $\hat{i}$ and $\hat{j}$ "span" the plane. Of course, we could add a few more to that list: we can build any vector in the xy-plane by combining the three vectors $\hat{i}$, $\hat{j}$, and $\hat{i} + 2\hat{j}$ if we want to, so those three vectors also span the plane. But three vectors is more than we need. When we pare down to a minimal set, that set defines a "basis."

Definitions: Span and Basis

If a set of vectors $\vec{v}_1, \vec{v}_2 \ldots \vec{v}_n$ has the property that any vector in a space can be built as a linear combination of these vectors, then these vectors "span" the space.

If all n vectors are *required* to span the space—in other words, if taking any one of them away means the remaining set does not span the space—then the vectors are said to be a "basis" for that space.

In the Discovery Exercise (Section 6.4.1) you combined two vectors $\vec{A}$ and $\vec{B}$ to build a wide variety of vectors. But can they create any vector, or are some vectors beyond their reach? We know that we can span the xy-plane with the two vectors $\hat{i}$ and $\hat{j}$, or we can use $\hat{i}'$ and $\hat{j}'$. Is it important that our two basis vectors are unit vectors? Is it crucial that they are perpendicular?

You may be surprised at how generally we can answer that question.

Any two non-parallel, non-zero vectors in a plane form a basis for that plane.

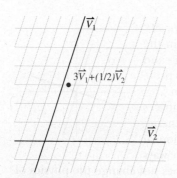

FIGURE 6.3 The point $3\vec{V}_1 + (1/2)\vec{V}_2$ represented on $\vec{V}_1\vec{V}_2$ graph paper.

We can represent $\vec{A}$ as $4\hat{i} + 9\hat{j}$, or as $3\vec{V}_1 + (1/2)\vec{V}_2$.

You will prove that in Problem 6.55 by finding a general formula for combining any two vectors to create any third vector. But right now, let's play with an example. We're going to span the xy-plane with these two vectors.

$$\vec{V}_1 = \hat{i} + 3\hat{j}, \quad \vec{V}_2 = 2\hat{i}$$

So when we describe a vector we won't give its $\hat{i}$- and $\hat{j}$-components. Instead we will describe one vector as $12\vec{V}_1 + \vec{V}_2$, another as $\pi\vec{V}_1 - 3\vec{V}_2$, and so on. For any vector you can draw in the xy-plane, there is a way—and there is only one way—to write it as a combination of $\vec{V}_1$ and $\vec{V}_2$.

Don't think of the $\hat{i}\hat{j}$ representation as the "real vector" and the $\vec{V}_1 \vec{V}_2$ one as simply a shorthand for expressing so many $\hat{i}$s and $\hat{j}$s: these are two perfectly equivalent representations of the same space. We can do all our graphing on $\vec{V}$ paper.

Since we have two different ways of expressing the same vector, there must be a way to convert between them. Converting from the $\vec{V}_1 \vec{V}_2$ to the $\hat{i}\hat{j}$ representation is easy; the other way takes a bit of algebra.

EXAMPLE   Converting Between Bases

Let $\vec{V}_1 = \hat{i} + 3\hat{j}$ and $\vec{V}_2 = 2\hat{i}$.

Question: Convert the vector $10\vec{V}_1 - 7\vec{V}_2$ to the $\hat{i}\hat{j}$ representation.

Answer: $10(\hat{i} + 3\hat{j}) - 7(2\hat{i}) = -4\hat{i} + 30\hat{j}$.

Question: Convert the vector $4\hat{i} + 5\hat{j}$ to the $\vec{V}_1\vec{V}_2$ representation.

Answer: If $4\hat{i} + 5\hat{j} = a(\hat{i} + 3\hat{j}) + b(2\hat{i})$ then:
$4 = a + 2b$
$5 = 3a + 0b$
Solving, we find our components are $a = 5/3$ and $b = 7/6$.

Can you see how this topic lends itself to matrices? Any vector can be expressed as a linear combination of basis vectors, and "linear combination" is the phrase that always points us to matrices. Using a column matrix to represent each vector, the first calculation in the example above can be written:

$$\begin{array}{c} \phantom{\hat{i}} \vec{V}_1 \ \vec{V}_2 \\ \begin{array}{c}\hat{i}\\\hat{j}\end{array}\!\begin{pmatrix} 1 & 2 \\ 3 & 0 \end{pmatrix} \end{array} \begin{array}{c} \vec{V}_1 \\ \vec{V}_2 \end{array}\!\begin{pmatrix} 10 \\ -7 \end{pmatrix} = \begin{array}{c}\hat{i}\\\hat{j}\end{array}\!\begin{pmatrix} -4 \\ 30 \end{pmatrix}$$

The first column of the transformation matrix is $\vec{V}_1$ and the second is $\vec{V}_2$. So multiplying that matrix by the vector in the middle says "I want 10 of $\vec{V}_1$ and -7 of $\vec{V}_2$."

And what about the second calculation in the example? We can of course express the question with matrices:

$$\hat{i} \begin{pmatrix} \vec{V_1} & \vec{V_2} \\ 1 & 2 \\ 3 & 0 \end{pmatrix} \vec{V_1} \begin{pmatrix} a \\ b \end{pmatrix} = \hat{i} \begin{pmatrix} 4 \\ 5 \end{pmatrix}$$

But that doesn't seem to get us any closer to a solution. We will readily answer such questions when we introduce the "inverse matrix"—which is exactly what it sounds like, a matrix designed to go the other way—in Section 6.6.

Different Representations of a Transformation

The picture below shows the effect of a transformation that we shall call **T** on a few vectors.

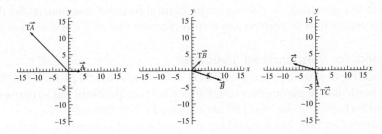

You might now expect us to give you "the matrix" that performs transformation **T**, but be careful. The matrix that performs this transformation depends on the coordinate system—the basis—you use to span the space.

- If all points are represented by giving their x- and y-components (in other words in the $\hat{i}\hat{j}$ basis), then the matrix $\begin{pmatrix} -3 & -10 \\ 3 & 8 \end{pmatrix}$ represents the transformation **T**, as shown in all three pictures above.
- Above we looked at the basis vectors $\vec{V_1} = \hat{i} + 3\hat{j}$ and $\vec{V_2} = 2\hat{i}$. If you describe any point in this basis, then transformation **T** is performed by the matrix $\begin{pmatrix} 9 & 2 \\ -21 & -4 \end{pmatrix}$. The pictures don't change, although the algebra does.
- If you express all points as combinations of $\vec{B_1} = -5\hat{i} + 3\hat{j}$ and $\vec{B_2} = 2\hat{i} - \hat{j}$, then the matrix that performs transformation **T** is $\begin{pmatrix} 3 & 0 \\ 0 & 2 \end{pmatrix}$.

These are obviously three different matrices, but they represent only one transformation, meaning that the "before-and-after" picture above looks exactly the same in all three cases. It is the numbers used to describe the drawings, and therefore the numbers that effect the transformation, that differ. We describe these as three "similar matrices."

> **Definition: Similar Matrices**
> Two matrices are "similar" if they represent the same transformation in two different coordinate systems.

If this definition seems confusing, think back to the inclined plane example at the beginning of the section. The force of gravity on a particular block is a vector that has a well defined magnitude (mg) and direction (down). The *representation* of that vector, however, is $\begin{pmatrix} 0 \\ -mg \end{pmatrix}$ in the $\hat{\tilde{i}}\hat{\tilde{j}}$ basis and $\begin{pmatrix} -mg\sin\theta \\ -mg\cos\theta \end{pmatrix}$ in the $\hat{i}'\hat{j}'$ basis.[4] Likewise, when we talk about similar matrices we are talking about two different sets of numbers that represent the same physical thing.

The third matrix above is "diagonal," meaning that all the elements off the main diagonal are zero. This makes its effect particularly easy to describe: "multiply the $\vec{B}_1$-component by 3 and the $\vec{B}_2$-component by 2." If we had to work with transformation **T** a lot, it would help to do all our work in the $\vec{B}_1 \vec{B}_2$ basis, which is referred to as the "natural basis" for this particular transformation. A different transformation would have a different natural basis.

Our main point here is the difference between a transformation and a matrix: *one transformation* is represented by *different matrices* in different bases. Section 6.8 will discuss how to find the natural basis where a given transformation is represented by a diagonal matrix. Section 6.9 will represent the equations for the three-spring problem using a matrix and solve them by expressing everything in the natural basis for that matrix. Section 7.2 will take up very generally how to convert any transformation matrix from one basis to another.

Non-spatial (or "Generalized") Vectors

Section 6.3 began with the example: "Each 'kiddy book' has 30 pictures and 500 words. Each 'chapter book' has 10 pictures and 20,000 words." Our problem was to convert from "4 kiddy books and 6 chapter books" to "180 pictures and 122,000 words."

The current section has focused on the problem of converting a vector from one basis representation to another. But we have seen that this very different looking problem can be accomplished with all the same operations, encapsulated in the "matrix times column" algorithm. We can bring the two problems closer together by considering the book scenario as another use of basis vectors.

Suppose we use a 2×1 matrix to represent the number of kiddy books and the number of chapter books in a library. So **k** below represents "1 kiddy book and no chapter books" and **c** represents the opposite.

$$\mathbf{k} = \begin{matrix} kiddy\ books \\ chapter\ books \end{matrix} \begin{pmatrix} 1 \\ 0 \end{pmatrix} \qquad \mathbf{c} = \begin{matrix} kiddy\ books \\ chapter\ books \end{matrix} \begin{pmatrix} 0 \\ 1 \end{pmatrix}$$

This **k** is of course not a vector in the sense of "a quantity with magnitude and direction" but it behaves in a similar way mathematically, so we look on it as a generalized version of a vector.[5] The contents of any library can be expressed as $a\mathbf{k} + b\mathbf{c}$ for the correct choices of the numbers a and b. So instead of saying "I have four kiddy books and six chapter books" I could describe my library as $4\mathbf{k} + 6\mathbf{c}$.

Alternatively, we can use a vector (a different 2×1 matrix) to represent "number of pictures" and "number of words." The vector **p** represents 1 picture and no words, and **w** the opposite. Any library can be expressed as a linear combination $p\mathbf{p} + w\mathbf{w}$. This is a different way of expressing the contents of the same library. With that mental shift, we can look on the following matrix multiplication as a change from the **kc** basis to the **pw** basis.

$$\begin{matrix} pictures \\ words \end{matrix} \begin{pmatrix} \overset{kiddy\ book}{30} & \overset{chapter\ book}{10} \\ 500 & 20{,}000 \end{pmatrix} \begin{matrix} kiddy\ books \\ chapter\ books \end{matrix} \begin{pmatrix} 4 \\ 6 \end{pmatrix} = \begin{matrix} pictures \\ words \end{matrix} \begin{pmatrix} 180 \\ 122{,}000 \end{pmatrix}$$

---

[4]To paraphrase the Buddha, do not confuse a pair of numbers representing a force with the actual force.
[5]We refer to "magnitude-and-direction" quantities as "spatial vectors" and designate them with arrow variables $\left(\vec{v}\right)$; we designate generalized vectors with boldfaced lowercase variables (**k**).

This multiplication shows that describing my library as $4\mathbf{k} + 6\mathbf{c}$ is equivalent to describing it as $180\mathbf{p} + 122,000\mathbf{w}$.

We began this section by showing how you can create any vector as a linear combination of basis vectors, and how matrix multiplication allows you to express this operation. We have now arrived at the same conclusion from the other end: when you use matrix multiplication to represent a linear combination, you can think of the operation as converting a vector from one basis to another. We will return to this theme throughout the chapter.

A final note: our discussion of spatial vectors was mostly based on spanning a plane *i.e.* putting together vectors to create any vector in two dimensions. We applied this above to the *xy*-plane, but we could similarly express any vectors in any other plane as linear combinations of two basis vectors in that plane. In three dimensions, of course, we require three vectors to span the space. In a more abstract sense, then, we refer to the "dimensionality" of a space as the number of vectors required to span that space[6]. The space of books described above is a two-dimensional space and "kiddy books" and "chapter books" form one basis for it, while "pictures" and "words" form another one.

Application to the Three-Spring Problem

In the three-spring problem you can specify the initial conditions in two ways. The obvious way is to give the initial positions of Ball 1 and Ball 2. Alternatively, you can specify how much of each normal mode you started with. (This second approach is less intuitive, but it connects more easily with writing a solution.) Either of these is a valid basis for the space of initial conditions. In Section 6.3 we saw how to use a 2×2 matrix to convert between the two representations.

Since each basis consists of two "vectors," the space of possible initial conditions for the three-spring problem has "two dimensions" (or two "degrees of freedom"), even though in physical space it all takes place along the *x*-axis.

The three-spring problem actually has four degrees of freedom, and the space of initial conditions is four dimensional, because you have to specify initial positions and velocities for both balls. We noted when we introduced the problem that we were only considering cases with zero initial velocity, and we will generally continue to do so to keep the algebra simpler. We treat non-zero initial velocities in some of the problems, and it's not especially different conceptually.

6.4.3 Problems: Basis Vectors

6.43 Consider the vectors $\vec{A}$ and $\vec{B}$ defined by the drawing. The magnitude of $\vec{A}$ is 1; $\vec{B}$ is obviously somewhat larger. You can express any vector as $a\vec{A} + b\vec{B}$ where a and b are carefully chosen numbers.

(a) Draw pictures of the vectors $2\vec{A} + \vec{B}$, $\vec{A} - \vec{B}$, and $-\vec{A}$.
(b) Draw the vectors $3\hat{i}$, $-3\hat{j}$, and $-2\hat{i} + \hat{j}$, and $\vec{0}$. For each one, estimate the coefficients that would be used to express this vector in the form $a\vec{A} + b\vec{B}$.

In Problems 6.44–6.48, $\vec{A} = \hat{i} + 2\hat{j}$ and $\vec{B} = -4\hat{i} + \hat{j}$.

6.44 Vector $\vec{V}_1$ can be represented as a combination of these two vectors as follows: $\vec{V}_1 = 5\vec{A} + 10\vec{B}$. Use a matrix multiplication to convert from this form to a representation of $\vec{V}_1$ as a combination of $\hat{i}$ and $\hat{j}$.

6.45 Vector $\vec{V}_2$ can be represented as a combination of these two vectors as follows: $\vec{V}_2 = \vec{A} - \vec{B}$. Use a matrix multiplication to

---

[6]In 2D the requirement that the two vectors be non-parallel amounts to saying they can't be on the same line. In 3D the requirement is that they can't be on the same plane. See Problem 6.57.

convert from this form to a representation of $\vec{V}_2$ as a combination of $\hat{i}$ and $\hat{j}$.

6.46 Explain how to interpret the matrix multiplication $\begin{pmatrix} 1 & -4 \\ 2 & 1 \end{pmatrix} \begin{pmatrix} 3 \\ 2 \end{pmatrix}$ as a linear combination of $\vec{A}$ and $\vec{B}$, and draw the result.

6.47 (a) Write a matrix multiplication that makes the statement "The vector $-24\hat{i} - 3\hat{j}$ can be expressed as $a\vec{A} + b\vec{B}$."

(b) Solve for a and b to convert this vector from $\hat{i}\hat{j}$ representation to $\vec{A}\vec{B}$ representation.

6.48 Draw a sheet of $\vec{A}\vec{B}$ graph paper (See Figure 6.3 for an example.). Use it to locate the points $(\vec{A}, \vec{B}) = (2, 1), (1, 3),$ and $(-1, 3)$.

6.49 The matrix $\mathbf{D} = \begin{pmatrix} 4 & 0 \\ 0 & -1 \end{pmatrix}$ performs a very simple transformation.

(a) Multiply $\mathbf{D}$ by the generic vector $\begin{pmatrix} x \\ y \end{pmatrix}$.

(b) Describe in words the effect of matrix $\mathbf{D}$ on any vector.

6.50 One possible basis for the *xy*-plane consists of the vectors $\vec{H}_1 = \hat{i} + \hat{j}$ and $\vec{H}_2 = \hat{i} - \hat{j}$. The image below shows H graph paper, with the *x*- and *y*-axes also shown for reference.

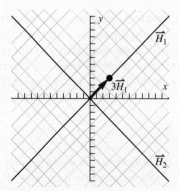

The transformation $\mathbf{E}$ is represented in the $\hat{i}\hat{j}$ basis by the matrix $\mathbf{E}_\mathbf{i} = \begin{pmatrix} 3/2 & 1/2 \\ 1/2 & 3/2 \end{pmatrix}$, and in the $\vec{H}_1\vec{H}_2$ basis by the matrix $\mathbf{E}_\mathbf{h} = \begin{pmatrix} 2 & 0 \\ 0 & 1 \end{pmatrix}$. (We chose the matrix $\mathbf{E}_\mathbf{h}$ and figured out that $\mathbf{E}_\mathbf{i}$ is what it looks like in the $\hat{i}\hat{j}$ basis. Don't worry if you have no idea how we figured out what this matrix would look like in a different basis; we'll teach you that in Section 7.2.)

(a) The vector $\vec{P} = 3\vec{H}_1$, shown in our drawing above, can be represented in the $\vec{H}_1\vec{H}_2$ basis as $\begin{pmatrix} 3 \\ 0 \end{pmatrix}$.

i. Calculate $\mathbf{E}_\mathbf{h}\vec{P}$ (working entirely in the $\vec{H}_1\vec{H}_2$ basis).

ii. Copy our drawing of the $\mathbf{H}$ graph paper and the vector $\vec{P}$, and on the same drawing show $\mathbf{E}_\mathbf{h}\vec{P}$.

iii. In words, what did the transformation $\mathbf{E}$ do to the vector $\vec{P}$?

iv. Now, write the vector $\vec{P}$ in the $\hat{i}\hat{j}$ basis.

v. Act on your new vector with $\mathbf{E}_\mathbf{i}$. Draw the $\hat{i}\hat{j}$ vectors before and after. Your picture should look exactly like your previous picture, showing that the same vector has been acted on by the same transformation.

(b) Repeat Part (a) on a different vector of your own choosing. Remember to start by choosing a vector in the $\vec{H}_1\vec{H}_2$ basis.

6.51 One possible basis for the *xy*-plane consists of the vectors $\vec{S}_1 = \hat{i} + \hat{j}$ and $\vec{S}_2 = \hat{j}$. The transformation $\mathbf{F}$ is represented in the $\hat{i}\hat{j}$ basis by the matrix $\mathbf{F}_\mathbf{i} = \begin{pmatrix} 0 & 2 \\ 1 & 0 \end{pmatrix}$, and in the $\vec{S}_1\vec{S}_2$ basis by the matrix $\mathbf{F}_\mathbf{s} = \begin{pmatrix} 2 & 2 \\ -1 & -2 \end{pmatrix}$.

(a) Begin with the vector $\vec{U} = \vec{S}_1 + 2\vec{S}_2$.

i. Calculate $\mathbf{F}_\mathbf{s}\vec{U}$ (working entirely in the $\vec{S}_1\vec{S}_2$ basis).

ii. Draw $\vec{S}_1\vec{S}_2$ graph paper, as we did for a different basis in Figure 6.3. On that paper show $\vec{U}$ and $\mathbf{F}_\mathbf{s}\vec{U}$.

iii. Now, use the matrix $\begin{pmatrix} 1 & 0 \\ 1 & 1 \end{pmatrix}$ to convert vector $\vec{U}$ to the $\hat{i}\hat{j}$ basis.

iv. Act on your new vector with $\mathbf{F}_\mathbf{i}$. Draw the $\hat{i}\hat{j}$ vectors before and after. (Your picture should look exactly like your previous picture, showing that the same vector has been acted on by the same transformation.)

(b) Repeat Part (a) on a different vector of your own choosing. Remember to start by choosing a vector in the $\vec{S}_1\vec{S}_2$ basis.

6.52 One possible basis for the *xy*-plane consists of the vectors $\vec{S}_1 = \hat{i} + 2\hat{j}$ and $\vec{S}_2 = 2\hat{i} + 3\hat{j}$. The transformation $\mathbf{F}$ is represented in the $\hat{i}\hat{j}$ basis by the matrix $\mathbf{F}_\mathbf{i} = \begin{pmatrix} 1 & 1 \\ 1 & 0 \end{pmatrix}$, and in the $\vec{S}_1\vec{S}_2$ basis by the matrix $\mathbf{F}_\mathbf{s} = \begin{pmatrix} -7 & -11 \\ 5 & 8 \end{pmatrix}$.

(a) Begin with the vector $\vec{U} = \vec{S}_1 + 2\vec{S}_2$.

i. Calculate $\mathbf{F}_\mathbf{s}\vec{U}$ (working entirely in the $\vec{S}_1\vec{S}_2$ basis).

ii. Draw $\vec{S}_1\vec{S}_2$ graph paper, as we did for a different basis in Figure 6.3. On that paper show $\vec{U}$ and $\mathbf{F}_s\vec{U}$.

iii. Now, use the matrix $\begin{pmatrix} 1 & 2 \\ 2 & 3 \end{pmatrix}$ to convert vector $\vec{U}$ to the $\hat{ij}$ basis.

iv. Act on your new vector with $\mathbf{F}_i$. Draw the $\hat{ij}$ vectors before and after. (Your picture should look exactly like your previous picture, showing that the same vector has been acted on by the same transformation.)

(b) Repeat Part (a) on a different vector of your own choosing. Remember to start by choosing a vector in the $\vec{S}_1\vec{S}_2$ basis.

6.53 A transformation matrix is represented in the $\hat{ij}$ basis by the matrix $\mathbf{T} = \begin{pmatrix} 2 & 1 \\ 1 & 2 \end{pmatrix}$.

(a) Draw $\hat{i}$ and $\mathbf{T}\hat{i}$.
(b) Draw $\hat{j}$ and $\mathbf{T}\hat{j}$.
(c) Draw $(\hat{i}+\hat{j})$ and $\mathbf{T}(\hat{i}+\hat{j})$.
(d) Draw $(\hat{i}-\hat{j})$ and $\mathbf{T}(\hat{i}-\hat{j})$.
(e) Your first answers may look confusing, but Parts (c) and (d) should make it clear what the natural basis for this transformation is. (The matrix performs a stretch along one of them, and leaves the other alone.) What is that basis?
(f) Write the matrix that represents this transformation in the basis you described in Part (e). This should not require any calculations.

6.54 A vector $\vec{V}$ is expressed in the form $p\vec{P} + q\vec{Q}$ where $\vec{P}$ is the vector $2\hat{i} - 4\hat{j}$, and $\vec{Q}$ is the vector $8\hat{i} + 4\hat{j}$. In other words p and q are the components of $\vec{V}$ in the $\vec{P}\vec{Q}$ basis.

(a) Write a transformation matrix that will stretch your vector by a factor of 6 in the $\vec{P}$-direction and a factor of 7 in the $\vec{Q}$-direction. (This is difficult if the vector is specified in the $\hat{ij}$ basis, but easy in the $\vec{P}\vec{Q}$ basis, so you should write it in the $\vec{P}\vec{Q}$ basis.)
(b) Write a transformation matrix that will reverse the $\vec{P}$-component of your vector, leaving the $\vec{Q}$-component unchanged. (Same note.)
(c) Write a matrix to convert the vector $\vec{V}$ to the $\hat{ij}$ basis; in other words the matrix should convert $\begin{pmatrix} p \\ q \end{pmatrix}$ to $\begin{pmatrix} x \\ y \end{pmatrix}$.

6.55 The Explanation (Section 6.4.2) claimed that "Any two non-parallel, non-zero vectors form a basis for a plane." In this problem you will prove it. Let $\vec{A} = A_x\hat{i} + A_y\hat{j}$ and $\vec{B} = B_x\hat{i} + B_y\hat{j}$ represent two vectors that we want to use as a basis; your goal is to express vector $\vec{V} = V_x\hat{i} + V_y\hat{j}$ in the form $\vec{V} = a\vec{A} + b\vec{B}$.

(a) Solve for the variables a and b in terms of all the constants given in the problem.
(b) Under what circumstances are your answers invalid? Answer by looking at your formulas, and then show that the answers match the criterion we described.

6.56 The three vectors $\vec{P} = 3\hat{i} - 2\hat{j} + \hat{k}$, $\vec{Q} = 7\hat{i} + 4\hat{j} - \hat{k}$, and $\vec{R} = (1/2)\hat{k}$ form a basis for three-space.

(a) Use a matrix multiplication to translate the vector $2\vec{P} + 10\vec{Q} - 6\vec{R}$ to the $\hat{ij}\hat{k}$ basis.
(b) Translate the vector $2\hat{i} - 10\hat{j} + 6\hat{k}$ into the $\vec{P}\vec{Q}\vec{R}$ basis.

6.57 The three vectors $\vec{P} = 3\hat{i} - 2\hat{j} + \hat{k}$, $\vec{Q} = 7\hat{i} + 4\hat{j} - \hat{k}$, and $\vec{R} = \hat{i} + 8\hat{j} - 3\hat{k}$ do *not* form a basis for three-space. Show that you cannot use these three vectors in any combination to create the vector $\vec{i} + \vec{j}$. (These three vectors do not form a basis—even though they are not zero and not parallel—because they are "linearly dependent," a topic we will discuss further in Section 6.7.)

6.58 The two vectors $\vec{A} = \hat{i} + \hat{j} + 2\hat{k}$ and $\vec{B} = \hat{i} + \hat{k}$ form a basis for the plane $z = x + y$.

(a) Express the vector $3\vec{A} + 2\vec{B}$ in the $\hat{ij}\hat{k}$ representation.
(b) Express the vector $3\hat{i} + \hat{j} + 4\hat{k}$ in the $\vec{A}\vec{B}$ representation. (You'll just have to play around with this one.)
(c) Prove that the vector $\hat{i}$ cannot be expressed in the $\vec{A}\vec{B}$ representation.
(d) Prove that any vector $a\vec{A} + b\vec{B}$ lies on the plane $z = x + y$.

6.59 The vectors $\hat{i}$ and $\hat{i} + \hat{k}$ span a space.

(a) What space do they span, and what is the dimensionality of that space?
(b) Give one example of a third vector you could add to make a basis for a higher dimensional space.
(c) Give one example of a third vector you could add that would not make a basis for a higher dimensional space.

6.60 For each set of vectors below say whether or not it is a basis. If so, what is the dimensionality of the space that it's a basis for?

(a) $\vec{A} = \hat{i}$, $\vec{B} = \hat{j} + \hat{k}$
(b) $\vec{A} = \hat{i}$, $\vec{B} = \hat{j}$, $\vec{C} = 3\hat{j}$

(c) $\vec{A} = \hat{i}, \vec{B} = \hat{j}, \vec{C} = 3\hat{k}$

(d) **a** represents 1 house and 2 apartments, **b** represents 3 houses and 5 apartments.

6.61 An object is free to move about in the xy-plane with no forces on it.

(a) What physical information would you have to give about the object to fully specify its initial conditions?

(b) What is the dimensionality of the space of initial conditions for this object?

(c) Write one possible set of basis vectors for the initial conditions for the object. *Hint*: Every basis vector must have the same dimensionality as the space itself. That is, your answer to Part (b) tells you how many components each basis vector must have.

6.62 An object is confined to move along the curve $y = x^2$ in the xy-plane. The space S is defined as the set of all possible positions for the object along that curve. What is the dimensionality of S?

Section 6.3 showed how a matrix multiplication can be used to find the initial positions associated with a combination of normal modes in the three-spring problem.

$$\begin{array}{c} \text{position} \\ \text{of} \\ \text{Ball 1} \\ \text{Ball 2} \end{array} \begin{pmatrix} \text{1st} & \text{2nd} \\ \text{normal} & \text{normal} \\ \text{mode} & \text{mode} \\ 1 & 1 \\ 2/3 & -1/3 \end{pmatrix} \begin{array}{c} \text{1st normal mode} \\ \text{2nd normal mode} \end{array} \begin{pmatrix} 26/3 \\ -11/3 \end{pmatrix}$$

$$= \begin{array}{c} \text{position of Ball 1} \\ \text{position of Ball 2} \end{array} \begin{pmatrix} 5 \\ 7 \end{pmatrix}$$

Problems 6.63–6.66 refer to this conversion.

6.63 (a) What basis vectors do we use to express the state of the system as the positions of the two balls?

(b) What basis vectors do we use to express the state of the system as a combination of normal modes?

6.64 Represent the initial condition "3 of the first normal mode and 6 of the second one" in the basis of initial ball positions.

6.65 Represent the initial condition "a of the first normal mode and b of the second one" in the basis of initial ball positions.

6.66 Represent the initial condition "Ball 1 at position 2 and Ball 2 at position -1" in the basis of normal modes.

6.5 Matrix Times Matrix

Section 6.3 built up an algorithm for multiplying an $(\text{anything} \times n)$ matrix times an $(n \times 1)$ column matrix. In this section we build up a more general algorithm for multiplying two matrices. It is an easy extension of the previous algorithm, but it serves a powerful purpose: creating one matrix that represents two or more consecutive transformations.

6.5.1 Discovery Exercise: Matrix Times Matrix

1. An LCD screen contains 10,000,000 transistors and 3000 feet of wire; a circuit board contains 100,000,000 transistors and 100 feet of wire; a knob contains 1 transistor and no wire. Write a labeled matrix designed to convert from "number of screens, circuit boards, and knobs" to "number of transistors and wires."

2. A computer monitor contains 1 LCD screen, 3 circuit boards, and 1 knob. A TV contains 1 LCD screen, 5 circuit boards, and 7 knobs. Write a labeled matrix designed to convert from "number of monitors and TVs" to "number of screens, circuit boards, and knobs."

3. If you have 1 monitor and 1 TV, how many transistors and wires do you have?

   See Check Yourself #40 in Appendix L

4. If you have M monitors and T TVs, how many transistors and wires do you have?

5. Write a labeled matrix designed to convert from "number of monitors and TVs" to "number of transistors and wires."

6.5.2 Explanation: Matrix Times Matrix

A matrix is a kind of function. A "normal" function turns one single-valued variable (a number) into another single-valued variable; a matrix turns one multiple-valued variable (a column matrix or vector) into another multiple-valued variable.

Two functions can be chained together into a composite function. The notation $f(g(x))$ represents one function that means "plug x into the g function, and then plug the answer into the f function." Note that the notation is read right-to-left, so $f(g(x))$ does the g function first; if you want to do f first, you write $g(f(x))$.

A similar principle holds with matrices. If $\mathbf{x}$ is a column matrix and $\mathbf{A}$ and $\mathbf{B}$ are transformation matrices, then $\mathbf{ABx}$ means "multiply matrix $\mathbf{B}$ times $\mathbf{x}$, and then multiply matrix $\mathbf{A}$ by the result." The matrix $\mathbf{AB}$ is a single function that does both of these steps in right-to-left order. Matrix $\mathbf{BA}$ applies both transformations with $\mathbf{A}$ coming first, which will generally lead to a different result.

Molecules, Atoms, and Particles (Oh My!)

In this example we're going to write a matrix to convert from molecules to atoms, and another matrix to convert from atoms to particles. Each of these is an example of the "Matrix Times Column" information conversion we discussed in Section 6.3. Our new goal is to write a single matrix that accomplishes both steps at once, converting molecules directly to subatomic particles. Once you follow the initial steps where we lay out the goal, you may be able to figure out the new matrix for yourself. You will then have figured out how to multiply two conversion matrices to create a new "composite" conversion matrix. If you worked the Discovery Exercise (Section 6.5.1) then you've already calculated such a composite matrix once yourself.

A water molecule, H_2O, consists of two hydrogen atoms and one oxygen atom. A hydrogen peroxide molecule, H_2O_2, consists of two hydrogen atoms and two oxygen atoms. A hydrogen molecule is just two hydrogen atoms, H_2. We can represent this information in a matrix that converts from numbers of molecules to numbers of atoms.

$$\mathbf{X} = \begin{array}{c} \\ hydrogen \\ oxygen \end{array} \begin{array}{c} H_2O \ \ H_2O_2 \ \ H_2 \\ \left(\begin{array}{ccc} 2 & 2 & 2 \\ 1 & 2 & 0 \end{array}\right) \end{array}$$

Remember as always that the purpose of such a matrix is to convert from the top labels to the side labels: in the above case, from molecules to atoms.

A hydrogen atom has 1 proton and no neutrons, while an oxygen atom has 8 protons and 8 neutrons.

$$\mathbf{Y} = \begin{array}{c} \\ protons \\ neutrons \end{array} \begin{array}{c} hydrogen \ \ oxygen \\ \left(\begin{array}{cc} 1 & 8 \\ 0 & 8 \end{array}\right) \end{array}$$

If you have 10 water molecules, 20 hydrogen peroxide molecules, and 30 hydrogen molecules, how many protons and neutrons do you have?

We can solve this problem with two matrix multiplications. Let $\mathbf{m}$ be the vector for the number of molecules, $\mathbf{a}$ be a vector for the number of atoms, and $\mathbf{p}$ be a vector for the number of particles.

$$\begin{array}{c} hydrogen \\ oxygen \end{array} \left(\begin{array}{c} 120 \\ 50 \end{array}\right) = \begin{array}{c} \\ hydrogen \\ oxygen \end{array} \begin{array}{c} H_2O \ \ H_2O_2 \ \ H_2 \\ \left(\begin{array}{ccc} 2 & 2 & 2 \\ 1 & 2 & 0 \end{array}\right) \end{array} \begin{array}{c} H_2O \\ H_2O_2 \\ H_2 \end{array}\left(\begin{array}{c} 10 \\ 20 \\ 30 \end{array}\right)$$

$$\mathbf{a} \quad = \quad\quad\quad \mathbf{X} \quad\quad\quad\quad\quad \mathbf{m}$$

$$\begin{array}{c} \textit{protons} \\ \textit{neutrons} \end{array}\begin{pmatrix} 520 \\ 400 \end{pmatrix} = \begin{array}{c} \textit{protons} \\ \textit{neutrons} \end{array}\begin{pmatrix} \overset{\textit{hydrogen}}{1} & \overset{\textit{oxygen}}{8} \\ 0 & 8 \end{pmatrix}\begin{array}{c} \textit{hydrogen} \\ \textit{oxygen} \end{array}\begin{pmatrix} 120 \\ 50 \end{pmatrix}$$

$$\mathbf{p} \qquad = \qquad\qquad\qquad \mathbf{Y} \qquad\qquad\qquad\qquad \mathbf{a}$$

More concisely, we can combine the equations $\mathbf{p} = \mathbf{Ya}$ and $\mathbf{a} = \mathbf{Xm}$ into the equation $\mathbf{p} = \mathbf{Y(Xm)}$.

$$\begin{array}{c} \textit{protons} \\ \textit{neutrons} \end{array}\begin{pmatrix} 520 \\ 400 \end{pmatrix} = \begin{array}{c} \textit{protons} \\ \textit{neutrons} \end{array}\begin{pmatrix} \overset{\textit{hydrogen}}{1} & \overset{\textit{oxygen}}{8} \\ 0 & 8 \end{pmatrix}\left[\begin{array}{c} \text{H} \\ \text{O} \end{array}\begin{pmatrix} \overset{H_2O}{2} & \overset{H_2O_2}{2} & \overset{H_2}{2} \\ 1 & 2 & 0 \end{pmatrix}\begin{array}{c} H_2O \\ H_2O_2 \\ H_2 \end{array}\begin{pmatrix} 10 \\ 20 \\ 30 \end{pmatrix}\right]$$

$$\mathbf{p} \qquad = \qquad\qquad \mathbf{Y} \qquad\qquad [\qquad\qquad \mathbf{X} \qquad\qquad\qquad \mathbf{m}\ ]$$

You must read this equation from right to left: $\mathbf{m}$ gets acted on by $\mathbf{X}$, and then the result gets acted on by $\mathbf{Y}$, to produce $\mathbf{p}$.

Now we want to define one matrix that performs the molecules-to-particles transformation. We will call this new matrix the product $\mathbf{YX}$. Reading from right to left again the matrix $\mathbf{YX}$ will represent "Do transformation $\mathbf{X}$ and then transformation $\mathbf{Y}$."

$$\begin{array}{c} \textit{protons} \\ \textit{neutrons} \end{array}\begin{pmatrix} 520 \\ 400 \end{pmatrix} = \begin{array}{c} \textit{protons} \\ \textit{neutrons} \end{array}\begin{pmatrix} \overset{H_2O}{?} & \overset{H_2O_2}{?} & \overset{H_2}{?} \\ ? & ? & ? \end{pmatrix}\begin{array}{c} H_2O \\ H_2O_2 \\ H_2 \end{array}\begin{pmatrix} 10 \\ 20 \\ 30 \end{pmatrix}$$

$$\mathbf{p} \qquad = \qquad\qquad \mathbf{YX} \qquad\qquad\qquad \mathbf{m}$$

If you've followed our train of thought to this point, you can reason out what goes into those question marks. For instance, the first column represents the number of protons and neutrons in a water molecule. A water molecule contains 2 hydrogen atoms (each with 1 proton and no neutrons), and 1 oxygen atom (with 8 protons and 8 neutrons), so it contains a total of 10 protons and 8 neutrons.

But you should also see that this is exactly the operation represented by multiplying matrix $\mathbf{Y}$ by the first column—the "water molecule" column—of matrix $\mathbf{X}$. You compute $2\begin{pmatrix}1\\0\end{pmatrix} + 1\begin{pmatrix}8\\8\end{pmatrix}$ to find that the first column of matrix $\mathbf{YX}$ is $\begin{pmatrix}10\\8\end{pmatrix}$. You then repeat the process for hydrogen peroxide and hydrogen molecules, putting each in its own column of matrix $\mathbf{YX}$.

Matrix Times Matrix

Summarizing what we did above, we now have the general algorithm for multiplying two matrices:

> **Multiplying a Matrix Times a Matrix**
>
> Consider the matrix product $\mathbf{AB}$ where matrix $\mathbf{B}$ is considered as a sequence of columns ($\mathbf{b1\ b2\ b3\ \ldots\ bn}$). Then the product $\mathbf{AB}$ is given by the matrix:
>
> $$\mathbf{AB} = (\mathbf{A\,b1}\quad \mathbf{A\,b2}\quad \mathbf{A\,b3}\quad \ldots \quad \mathbf{A\,bn})$$

Each individual term in this sequence is the product of matrix **A** times a column of matrix **B**, as defined in Section 6.3. These individual products are *not* added together, but instead form different columns of the result.

In words: "You already know how to multiply a matrix times a single column. When multiplying matrix **A** by multicolumn matrix **B**, multiply **A** by each column of **B** to produce the corresponding column of the product **AB**." As usual, your best strategy is probably to follow the example below, and then go back and try to make sense of the definition above.

EXAMPLE Matrix Times Matrix

Question: Multiply $\begin{pmatrix} 1 & 2 & 3 \\ 4 & 5 & 6 \end{pmatrix} \begin{pmatrix} 7 & 10 \\ 8 & 11 \\ 9 & 12 \end{pmatrix}$

Answer: We find the answer by moving column-by-column through the right-hand matrix.

$\begin{pmatrix} 1 & 2 & 3 \\ 4 & 5 & 6 \end{pmatrix} \begin{pmatrix} 7 \\ 8 \\ 9 \end{pmatrix}$ gives us the first column of the answer, $\begin{pmatrix} 50 \\ 122 \end{pmatrix}$.

$\begin{pmatrix} 1 & 2 & 3 \\ 4 & 5 & 6 \end{pmatrix} \begin{pmatrix} 10 \\ 11 \\ 12 \end{pmatrix}$ gives us the second column of the answer, $\begin{pmatrix} 68 \\ 167 \end{pmatrix}$.

We conclude that $\begin{pmatrix} 1 & 2 & 3 \\ 4 & 5 & 6 \end{pmatrix} \begin{pmatrix} 7 & 10 \\ 8 & 11 \\ 9 & 12 \end{pmatrix} = \begin{pmatrix} 50 & 68 \\ 122 & 167 \end{pmatrix}$.

Notes on this process:

- The right-hand matrix can have as few or as many columns as you like; each one becomes a column in the answer.
- The left-hand matrix can have as few or as many rows as you like; that determines the size of the columns in the answer.
- However, the number of *rows* in the right-hand matrix must perfectly match the number of *columns* in the left-hand matrix, or the multiplication is illegal.
- Reversing the order will in many cases turn a legal multiplication into an illegal one. In this case the multiplication remains legal, but the answer is completely different: $\begin{pmatrix} 7 & 10 \\ 8 & 11 \\ 9 & 12 \end{pmatrix} \begin{pmatrix} 1 & 2 & 3 \\ 4 & 5 & 6 \end{pmatrix} = \begin{pmatrix} 47 & 64 & 81 \\ 52 & 71 & 90 \\ 57 & 78 & 99 \end{pmatrix}$.

This process is designed to create one matrix that does the work of two. The example above shows that multiplying by the matrix $\begin{pmatrix} 50 & 68 \\ 122 & 167 \end{pmatrix}$ is equivalent to multiplying by $\begin{pmatrix} 7 & 10 \\ 8 & 11 \\ 9 & 12 \end{pmatrix}$ and then $\begin{pmatrix} 1 & 2 & 3 \\ 4 & 5 & 6 \end{pmatrix}$. (Right to left!)

Given that goal, how do you know that this is the right algorithm? One way is to go back to the molecules-to-particles example above and consider again what the elements of the new matrix *must* be, and why. A more formal proof is presented in Problem 6.90.

If your matrices are labeled, the new matrix picks up the top labels of the right-hand matrix, and the left labels of the left-hand matrix. The example below illustrates this, as well as giving another example of a composite transformation.

EXAMPLE Matrix Times Matrix with Labels

We have used this example before: "Each 'kiddy book' has 30 pictures and 500 words. Each 'chapter book' has 10 pictures and 20,000 words." Now we add to that scenario: every college apartment has 5 kiddy books and 20 chapter books. Every family household has 40 kiddy books and 35 chapter books. Multiplying these two matrices gives us a single matrix that converts from "apartments and houses" to "pictures and words."

$$\begin{matrix} & \text{kiddy} & \text{chapter} \\ \text{pictures} & \begin{pmatrix} 30 & 10 \\ 500 & 20{,}000 \end{pmatrix} \\ \text{words} & \end{matrix} \begin{matrix} & \text{apartment} & \text{house} \\ \text{kiddy} & \begin{pmatrix} 5 & 40 \\ 20 & 35 \end{pmatrix} \\ \text{chapter} & \end{matrix}$$

$$= \begin{matrix} & \text{apartment} & \text{house} \\ \text{pictures} & \begin{pmatrix} 350 & 1550 \\ 402{,}500 & 720{,}000 \end{pmatrix} \\ \text{words} & \end{matrix}$$

Make sure you understand why we multiplied those two matrices *in that order*, why the labels came out the way they did, and what the final matrix accomplishes. This example summarizes everything we have said in this section, if you follow it carefully.

Point Matrices

Computer graphics software provides the underlying engine for modern computer games and animated movies. A drawing is represented in such systems by a "point matrix."[7] Other matrices effect transformations such as rotating, scaling, or moving such a drawing.

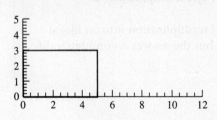

For instance, the matrix $\begin{pmatrix} 0 & 5 & 5 & 0 & 0 \\ 0 & 0 & 3 & 3 & 0 \end{pmatrix}$ represents a line from $(0, 0)$ to $(5, 0)$, and then a line from there to $(5, 3)$, and so on. It therefore creates the drawing on the left.

If we multiply matrix $\begin{pmatrix} 2 & 0 \\ 0 & 2/3 \end{pmatrix}$ by this matrix (try it!), the resulting matrix renders as Figure 6.4.

---

[7] The term "point matrix" is not in common use, but as far as we can tell no other term for such a matrix is in common use either. This term was coined by Henry Rich, to whom we are in debt not only for "point matrix" but for a great deal of the way we present linear algebra.

So we can describe matrix $\begin{pmatrix} 2 & 0 \\ 0 & 2/3 \end{pmatrix}$ by describing its effect on any point matrix: it stretches it by a factor of 2 in the *x*-direction and a factor of 2/3 in the *y*-direction.

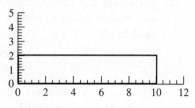

FIGURE 6.4

Point matrices will become a recurring theme in this chapter, not just because they are important in computer animation, but because they give us a visual representation of transformations. For instance, suppose matrix **P** is a point matrix representing a rectangle, **S** is a transformation that stretches objects in the *x*-direction, and **R** is a transformation that rotates objects 30° clockwise. The picture below shows **RSP** (stretch **P** and then rotate) and **SRP** (rotate **P** and then stretch). We end up with very different figures at the end of these two processes. This makes the claim "matrix multiplication is not commutative" visually obvious: **RS** is one transformation, and **SR** is a different one.

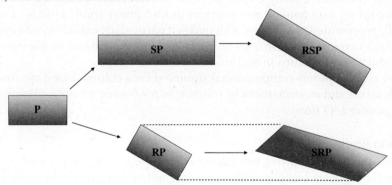

As we discuss matrix transformations throughout this chapter, we will sometimes show their effects on individual vectors and sometimes on shapes represented by point matrices. The two are equivalent because we define a shape by the position vectors of each of its vertices.[8] When we multiply a transformation matrix by a point matrix, we are using the "matrix times matrix" algorithm to act on all these different position vectors at once.

Alternative Interpretations

Our motto for matrix multiplication is "it's all about the columns." We think of both the left-hand and the right-hand matrices as a series of columns that might represent spatial vectors or other collections of related information.

$$\begin{pmatrix} a & b \\ c & d \end{pmatrix} \begin{pmatrix} e & f \\ g & h \end{pmatrix} \qquad \text{First step: } e \begin{pmatrix} a \\ c \end{pmatrix} + f \begin{pmatrix} b \\ d \end{pmatrix} \qquad (6.5.1)$$

That step becomes the first column of the solution, and you proceed similarly along all the columns of the right-hand matrix. But two other interpretations are available.

1. "It's all about the rows." Think of both matrices as collections of rows. For Equation 6.5.1 you would start here.

$$\text{First Step: } a \begin{pmatrix} e & f \end{pmatrix} + b \begin{pmatrix} g & h \end{pmatrix}$$

That step becomes the first row of the solution, and you proceed similarly down all the rows of the left-hand matrix.

---

[8]When you use matrices to transform shapes that don't have vertices, such as circles, you approximate the shapes with polygons so you can represent them with point matrices.

2. "Rows on the left, columns on the right." This method is based on the fact that a row matrix times a column matrix gives a 1×1 answer. For Equation 6.5.1 you would start here.

$$\text{First Step: } \begin{pmatrix} a & b \end{pmatrix} \begin{pmatrix} e \\ g \end{pmatrix} = (ae + bg)$$

This number becomes one element of the solution. You will experiment with this method in Problem 6.91.

We call these different "interpretations" (you could also use the word "methods") because they are all perfectly equivalent. The rule of when you are allowed to multiply matrices, and the matrix you get after you perform the multiplication, are the same in all three cases. The last of these methods is perhaps the most convenient way to multiply matrices in your head.

But we will continue to emphasize the column interpretation. From toy examples such as chapter books and kiddy books, to point matrices and geometric transformations, to the three-spring problem, we use matrices to find linear combinations of vectors that are generally represented as columns. This makes it particularly natural to view every matrix multiplication as a transformation from one basis to another, a subject we discussed in Section 6.4 and to which we will return in Section 7.3.

There is also a purely computational argument for a column-based approach. Computers typically access and store matrices by column, so processing a matrix column-by-column can save expensive I/O time.

Stepping Back

The algorithm for multiplying two matrices may appear both complicated and arbitrary. The solution for "appears complicated" is practice, and in this case we do recommend what may feel like mindless drill; keep multiplying matrices until the process is comfortable. It comes up often enough to be worth your time.

The solution for "appears arbitrary" is to understand that while matrix multiplication is not generally commutative ($\mathbf{AB} \neq \mathbf{BA}$), it is always associative:

$$\mathbf{A(BC)} = \mathbf{(AB)C}$$

The left side of that equation says "act on **C** with **B**, and then act on the result with **A**." The right side describes a new matrix **AB** that, when acting on **C**, produces the same result in one step. Our goal in this section has been to convince you that the product **AB** as we defined it here is the correct and only matrix that accomplishes this.

6.5.3 Problems: Matrix Times Matrix

6.67 A Regular Valentine Basket contains 10 chocolates, 3 flowers, and a card. A Deluxe Valentine Basket contains 20 chocolates, 5 flowers, and a card.

(a) Write a *labeled matrix* to convert from "number of regular baskets and deluxe baskets" to "number of chocolates, flowers, and cards." Call this matrix **B**.

(b) Now, every drug store carries 20 regular baskets and 10 deluxe baskets. Every grocery store carries 50 regular baskets and 30 deluxe baskets. Write a labeled matrix to convert from "number of drug stores and grocery stores" to "number of regular baskets and deluxe baskets." Call this matrix **S**.

(c) Multiply to find matrix **BS**.

(d) Use your answer to Part (c) to answer the question: If you buy out all the gift baskets from 4 drug stores and 6 grocery stores, how many chocolates, flowers, and cards do you have?

6.68 Every math class in Sunnydale High School has 1 teacher and 19 students.

Every science class has 1 teacher, 1 teaching assistant, and 15 students.

(a) Write a labeled matrix to convert from "number of math and science classes" to "number of teachers, students, and assistants." Call this matrix **P**.

(b) Every teacher has 3 textbooks, 15 pens, and 5 calculators. Every student has 1 textbook, 2 pens, and 1 calculator. Every assistant has just a pen. Write a labeled matrix to convert from "number of teachers, students, and assistants" to "number of textbooks, pens, and calculators." Call this matrix **T**.

(c) Which matrix product would be useful in this scenario, **PT** or **TP**? Why?

(d) Multiply to find the product you specified in Part (c).

(e) Create a simple question that could be answered by using the matrix you found in Part (d). (See Problem 6.67(d) for the type of question we're looking for.)

6.69 A cool young family has two sports cars. A boring older middle class family has a minivan, a sports car, and a smart car.

(a) Write labeled matrix **F** to transform number of young and old families to number of sports cars, minivans, and smart cars.

(b) A sports car has 6 cylinders and 4 seats; a minivan has 4 cylinders and 7 seats; a smart car has 3 cylinders and 2 seats. Write labeled matrix **C** to transform sports cars/minivans/smart cars to cylinders and seats.

(c) Both **FC** and **CF** are legal matrix multiplications, but only one of them is useful in this scenario. Which one, and why?

(d) Multiply to find the product you specified in Part (c).

(e) Create a simple question that could be answered by using the matrix you found in Part (d). (See Problem 6.67(d) for the type of question we're looking for.)

6.70 A swank, sophisticated graveyard has 50 vampires and only 3 zombies. But your trashy, low-class graveyard, it has only 3 deadbeat vampires and 200 zombies. Now, as everyone knows, every vampire has two eyes, two teeth (sharp and pointy), and one brain. Every zombie has one eyeball, a full set of 32 teeth, and 3 (partially digested) brains. Write and evaluate a matrix multiplication to answer the question: if there are s sophisticated graveyards and t trashy graveyards in town, how many eyes, teeth, and brains does that represent?

In Problems 6.71–6.80, perform the given matrix multiplication, or identify it as illegal.

6.71 $\begin{pmatrix} 1 & 7 \\ 3 & 2 \end{pmatrix} \begin{pmatrix} 5 & 9 \\ 4 & 8 \end{pmatrix}$

6.72 $\begin{pmatrix} 5 & 9 \\ 4 & 8 \end{pmatrix} \begin{pmatrix} 1 & 7 \\ 3 & 2 \end{pmatrix}$

6.73 $\begin{pmatrix} 1 & 2 \\ -3 & 1 \end{pmatrix} \begin{pmatrix} 0 & 4 & 1 & 0 \\ 0 & 1 & 3 & 0 \end{pmatrix}$

6.74 $\begin{pmatrix} 0 & 4 & 1 & 0 \\ 0 & 1 & 3 & 0 \end{pmatrix} \begin{pmatrix} 1 & 2 \\ -3 & 1 \end{pmatrix}$

6.75 $\begin{pmatrix} 2 & 4 & 6 \end{pmatrix} \begin{pmatrix} 1 \\ 3 \\ 5 \end{pmatrix}$

6.76 $\begin{pmatrix} 1 \\ 3 \\ 5 \end{pmatrix} \begin{pmatrix} 2 & 4 & 6 \end{pmatrix}$

6.77 $\begin{pmatrix} 3 & 1 & 0 \\ 5 & 2 & 4 \\ 10 & 9 & 8 \end{pmatrix} \begin{pmatrix} 1 & 0 & 0 \\ 0 & 1 & 0 \\ 0 & 0 & 1 \end{pmatrix}$

6.78 $\begin{pmatrix} 1 & 0 & 0 \\ 0 & 1 & 0 \\ 0 & 0 & 1 \end{pmatrix} \begin{pmatrix} 3 & 1 & 0 \\ 5 & 2 & 4 \\ 10 & 9 & 8 \end{pmatrix}$

6.79 $\begin{pmatrix} 3 & 1 & 0 \\ 5 & 2 & 4 \\ 10 & 9 & 8 \end{pmatrix} \begin{pmatrix} 0 & 0 & 1 \\ 0 & 1 & 0 \\ 1 & 0 & 0 \end{pmatrix}$

6.80 $\begin{pmatrix} 3 & 1 & 0 \\ 5 & 2 & 4 \\ 10 & 9 & 8 \end{pmatrix} \begin{pmatrix} 1 & 1 & 1 \\ 1 & 1 & 1 \\ 1 & 1 & 1 \end{pmatrix}$

6.81 (a) Write two matrices that can be added to each other, and can also be multiplied by each other.

(b) Write two matrices that can be added, but cannot be multiplied.

(c) Write two matrices that can be multiplied, but cannot be added.

(d) Write two matrices that cannot be added or multiplied.

6.82 (a) Calculate
$\begin{pmatrix} 1 & 3 \\ 2 & 4 \end{pmatrix} \begin{pmatrix} 5 & 7 \\ 6 & 8 \end{pmatrix}$.

(b) Calculate $\begin{pmatrix} 5 & 7 \\ 6 & 8 \end{pmatrix} \begin{pmatrix} 1 & 3 \\ 2 & 4 \end{pmatrix}$.

(c) Based on your answers, does multiplication of 2×2 matrices follow the commutative property, or not?

6.83 (a) Calculate
$\begin{pmatrix} 4 & 1 \\ -2 & 3 \\ 0 & 4 \end{pmatrix} \left[\begin{pmatrix} 1 & -2 \\ 2 & 1 \end{pmatrix} + \begin{pmatrix} 1 & 0 \\ 0 & 3 \end{pmatrix} \right]$.

(b) Calculate $\begin{pmatrix} 4 & 1 \\ -2 & 3 \\ 0 & 4 \end{pmatrix} \begin{pmatrix} 1 & -2 \\ 2 & 1 \end{pmatrix} +$
$\begin{pmatrix} 4 & 1 \\ -2 & 3 \\ 0 & 4 \end{pmatrix} \begin{pmatrix} 1 & 0 \\ 0 & 3 \end{pmatrix}.$

(c) Based on your answers, does matrix multiplication follow the distributive property, or not?

6.84 Let $\mathbf{A} = \begin{pmatrix} 10 & x \\ -3 & i \end{pmatrix}$, $\mathbf{B} = \begin{pmatrix} 4 & 2 \\ i & 0 \end{pmatrix}$ and $\mathbf{C} = \begin{pmatrix} 8 & -5 \\ y & 7 \end{pmatrix}$.

(a) Calculate $(\mathbf{AB})\mathbf{C}$.
(b) Calculate $\mathbf{A}(\mathbf{BC})$.
(c) Based on your answers, does matrix multiplication follow the associative property, or not?

6.85 Polygon **P** is a square with vertices at $(0, 0)$, $(1, 0)$, $(1, 1)$, and $(0, 1)$.

(a) Draw polygon **P**.
(b) Write a point matrix to represent polygon **P**. *Hint*: it will be a 2×5 matrix.
(c) Apply the transformation $\begin{pmatrix} 3 & 0 \\ 0 & 1 \end{pmatrix}$ to polygon **P**. Draw the resulting figure, and then describe in words the transformation effected by this matrix.
(d) Apply the transformation $\begin{pmatrix} 1 & 0 \\ 0 & 3 \end{pmatrix}$ to polygon **P**. Draw the resulting figure, and then describe in words the transformation effected by this matrix.
(e) Apply the transformation $\begin{pmatrix} 3 & 0 \\ 0 & 3 \end{pmatrix}$ to polygon **P**. Draw the resulting figure, and then describe in words the transformation effected by this matrix.
(f) Apply the transformation $\begin{pmatrix} 2 & 1 \\ 1 & 2 \end{pmatrix}$ to polygon **P**. Draw the resulting figure, and then describe in words the transformation effected by this matrix. *Hint*: you might find it helpful to figure out what this does to a generic vector $x\hat{i} + y\hat{j}$.

6.86 The point matrix $\mathbf{P} = \begin{pmatrix} 0 & 2 & 8 & 5 & 0 \\ 0 & 5 & 8 & 2 & 0 \end{pmatrix}$ represents a shape, by giving a computer instructions to draw a line from the point $(0, 0)$ to $(2, 5)$, and then a line from $(2, 5)$ to $(8, 8)$, and so on.

(a) Draw the resulting shape.

(b) Matrix $\mathbf{A} = \begin{pmatrix} 0 & 1 \\ -1 & 0 \end{pmatrix}$ represents a transformation. Find matrix **AP** and draw the resulting shape.

(c) Looking at your drawing, the shape **AP** could be described as "the original shape **P** reflected across the x-axis" or as "the original shape **P** rotated clockwise." These are not the same transformation—**A** does either one or the other—but you can't yet tell which one. So pick a different matrix (either a vector or another point matrix) and see what **A** does to it to figure out which of these two transformations matrix **A** performs.

6.87 The matrix
$\mathbf{N} = \begin{pmatrix} 0 & 0 & 1 & 2 & 3 & 2 & 1 & 0 \\ 0 & 1 & 0 & 0 & 1/2 & 1 & 1 & 0 \end{pmatrix}$
represents the shape formed by a line from $(0, 0)$ to $(0, 1)$, a line from $(0, 1)$ to $(1, 0)$, and so on. We will call this shape Nemo.

(a) Draw Nemo.
(b) Find the matrix $(2/3)\mathbf{N}$ and draw the resulting shape. Look! Nemo swam further away!
(c) Matrix $\mathbf{R} = \begin{pmatrix} \sqrt{3}/2 & -1/2 \\ 1/2 & \sqrt{3}/2 \end{pmatrix}$ represents a transformation. Find matrix **RN** and draw the resulting shape. What did Nemo do this time?
(d) Matrix $\mathbf{T} = (2/3)\mathbf{R}$ is a single matrix that performs both of these transformations. Compute this matrix and apply it to matrix **N**.
(e) Now apply matrix T again to your *answer* to Part (d). Then apply matrix **T** again to that answer. Repeat the process 10 times and draw the matrices **N**, **TN**, **TTN**,…**T**^{10}**N** together on one plot (or better yet in an animation) to see Nemo slowly spiral down into a whirlpool.

6.88 (a) A 2×2 diagonal matrix can always be represented as $\begin{pmatrix} a & 0 \\ 0 & b \end{pmatrix}$. If this matrix acts on a point matrix, what effect does it have on the resulting shape?

(b) A three-dimensional shape can be described by a 3×n point matrix where the first row is the x-coordinate of each point, the second y, and the third z. If a 3×3 diagonal matrix acts upon such a point matrix, what effect does it have on the resulting shape?

6.89 Matrix $\mathbf{M} = \begin{pmatrix} M_{11} & M_{12} \\ M_{21} & M_{22} \end{pmatrix}$ transforms the point $(1, 0)$ into $(2, 1)$ and transforms the point $(0, 1)$ into $(-2, 2)$.

(a) Write the equation that says that $\mathbf{M}$ transforms the point $(1, 0)$ into $(2, 1)$. Solve to find two of the elements of $\mathbf{M}$.

(b) Write the equation that says that $\mathbf{M}$ transforms the point $(0, 1)$ into $(-2, 2)$. Solve this equation to find the remaining two elements.

(c) Let $\mathbf{D}$ equal the point matrix $\begin{pmatrix} 0 & 1 & 1 & 0 & 0 \\ 0 & 0 & 1 & 1 & 0 \end{pmatrix}$. Draw the shapes represented by $\mathbf{D}$ and $\mathbf{MD}$.

6.90 Matrix multiplication is designed to be *associative*; $(\mathbf{AB})\mathbf{x}$ should give the same answer is $\mathbf{A}(\mathbf{Bx})$, which is another way of saying "matrix $\mathbf{AB}$ does the work of matrix $\mathbf{B}$ followed by matrix $\mathbf{A}$." In this problem you will demonstrate that the only definition of matrix multiplication that meets this goal is the method described in this section. Let matrix $\mathbf{A} = \begin{pmatrix} A_{11} & A_{12} \\ A_{21} & A_{22} \end{pmatrix}$, $\mathbf{B} = \begin{pmatrix} B_{11} & B_{12} \\ B_{21} & B_{22} \end{pmatrix}$, and $\mathbf{x} = \begin{pmatrix} 1 \\ 0 \end{pmatrix}$.

(a) Multiply $\mathbf{B}$ times $\mathbf{x}$ to find $\mathbf{Bx}$.

(b) Multiply $\mathbf{A}$ times your answer from Part (a) to find $\mathbf{A}(\mathbf{Bx})$.

(c) The goal is to find a single matrix $\mathbf{C}$ that performs the work of matrices $\mathbf{A}$ and $\mathbf{B}$ together. So let matrix $\mathbf{C} = \begin{pmatrix} C_{11} & C_{12} \\ C_{21} & C_{22} \end{pmatrix}$ and write the equation $\langle answer\ to\ Part\ (b) \rangle = \mathbf{Cx}$. Solve this equation to find two of the elements of matrix $\mathbf{C}$.

(d) Repeat the process, replacing $\mathbf{x}$ with the matrix $\mathbf{y} = \begin{pmatrix} 0 \\ 1 \end{pmatrix}$ to find the other two elements of $\mathbf{C}$.

(e) Confirm that matrix $\mathbf{C} = \mathbf{AB}$ based on the definition of matrix multiplication given in the Explanation (Section 6.5.2) matches the result you found in this problem.

6.91 If you have learned about matrices before, you probably learned to multiply them as follows. To multiply a row (on the left) by a column (on the right), multiply all the corresponding elements and add the result. To multiply matrices more generally, multiply the nth row on the left by the mth column on the right to produce the nth-row mth-column element of the answer.

(a) Multiply $\begin{pmatrix} -3 & 7 \\ 6 & 8 \\ 2 & 5 \end{pmatrix} \begin{pmatrix} -10 & 4 \\ -3 & 9 \end{pmatrix}$ by this rows×columns method.

(b) Perform the same matrix multiplication by the column-based method we discuss.

(c) The two methods require the same number of calculations and always produce the same answer. Why do you suppose we stress the column-based method?

6.92 The rule for multiplying two matrices $\mathbf{A}$ and $\mathbf{B}$ with result $\mathbf{C}$ can be written compactly using index notation: $C_{ij} = \sum_k A_{ik} B_{kj}$, where the sum over k runs over all of its possible values in the matrices $\mathbf{A}$ and $\mathbf{B}$. Consider the matrices $\mathbf{A} = \begin{pmatrix} A_{11} & A_{12} & A_{13} \\ A_{21} & A_{22} & A_{23} \end{pmatrix}$, $\mathbf{B} = \begin{pmatrix} B_{11} & B_{12} \\ B_{21} & B_{22} \\ B_{31} & B_{32} \end{pmatrix}$.

(a) Find the product $\mathbf{AB}$ using the method described in this section.

(b) Calculate C_{12} using the method given in this problem and show that it matches your answer from Part (a).

6.6 The Identity and Inverse Matrices

In this section we meet two very important matrices, the "identity matrix" and the "inverse matrix."

6.6.1 Discovery Exercise: The Identity and Inverse Matrices

For the purposes of this exercise, let $\mathbf{A} = \begin{pmatrix} 2 & 1 \\ 5 & 3 \end{pmatrix}$.

1. Find a matrix $\mathbf{I}$ such that $\mathbf{AI} = \mathbf{A}$; that is, when you multiply our $\mathbf{A}$ by your $\mathbf{I}$, the result is our matrix $\mathbf{A}$ again. (We haven't given you any method for doing this, but you can probably get there pretty quickly by trial and error. Don't go farther until you confirm that your answer works!)

2. We know that matrix multiplication is not generally "commutative"; that is, $\mathbf{AB} \neq \mathbf{BA}$. Show that the multiplication in Part 1 *is* commutative, meaning $\mathbf{AI} = \mathbf{IA}$.
3. Your goal is now to find a matrix $\mathbf{B}$ such that $\mathbf{AB} = \mathbf{I}$ (where $\mathbf{I}$ is the matrix you found in Part 1).
   (a) Explain why matrix $\mathbf{B}$ must have dimensions 2×2.
   Now that we know the dimensions, we will let $\mathbf{B} = \begin{pmatrix} w & y \\ x & z \end{pmatrix}$; in other words, use letters to represent all the unknown numbers. The defining equation $\mathbf{AB} = \mathbf{I}$ therefore becomes:

$$\begin{pmatrix} 2 & 1 \\ 5 & 3 \end{pmatrix} \begin{pmatrix} w & y \\ x & z \end{pmatrix} = \mathbf{I} \qquad (6.6.1)$$

   (b) Perform the matrix multiplication on the left side of Equation 6.6.1. The result should be another 2×2 matrix that you are setting equal to $\mathbf{I}$.
   (c) Recall that if two matrices equal each other, all the corresponding elements must be equal. Use this fact to write four equations to solve for the unknown w, x, y, and z.
   (d) Solve your equations to find matrix $\mathbf{B}$.

   See Check Yourself #41 in Appendix L

4. Show that the multiplication in Part 3 is commutative, meaning $\mathbf{AB} = \mathbf{BA}$.

6.6.2 Explanation: The Identity and Inverse Matrices

We have said that a matrix is like a function, because it transforms information from one form to another. We have said that multiplying two matrices is like creating a composite function, one operation that effects two or more transformations in a row. The inverse matrix is analogous to an inverse function, reversing a transformation.

Equations 6.6.2 demonstrate why we call 2^x and $\log_2 x$ "inverse functions."

$$\begin{array}{lllll} \text{When} & 2^x & \text{acts on} & 3 & \text{it gives you} & 8 \\ \text{When} & \log_2 x & \text{acts on} & 8 & \text{it gives you} & 3 \end{array} \qquad (6.6.2)$$

The numbers 3 and 8 make a simple demonstration but the point is that 2^x and $\log_2 x$ will always reverse each other in this way. One turns -5 into $1/32$, the other turns $1/32$ into -5. One turns 0 into 1, the other turns 1 into 0.

Equations 6.6.3 offer an analogous matrix equation.

$$\begin{array}{llllll} \text{When} & \begin{pmatrix} 2 & 1 \\ 5 & 3 \end{pmatrix} & \text{multiplies (or transforms)} & \begin{pmatrix} 4 \\ -1 \end{pmatrix} & \text{it gives you} & \begin{pmatrix} 7 \\ 17 \end{pmatrix} \\ \text{When} & \begin{pmatrix} 3 & -1 \\ -5 & 2 \end{pmatrix} & \text{multiplies (or transforms)} & \begin{pmatrix} 7 \\ 17 \end{pmatrix} & \text{it gives you} & \begin{pmatrix} 4 \\ -1 \end{pmatrix} \end{array} \qquad (6.6.3)$$

Once again, $\begin{pmatrix} 4 \\ -1 \end{pmatrix}$ and $\begin{pmatrix} 7 \\ 17 \end{pmatrix}$ are just one simple example. The more general point is that $\begin{pmatrix} 2 & 1 \\ 5 & 3 \end{pmatrix}$ and $\begin{pmatrix} 3 & -1 \\ -5 & 2 \end{pmatrix}$ always reverse each other's effects in this way: they are "inverse matrices."

Below we present two formal definitions that may not seem to have anything to do with the reversing effect described above. Then we will wind our way back to the idea of inverting a transformation.

6.6 | The Identity and Inverse Matrices

Definitions: Identity and Inverse Matrices

An "identity matrix" $\mathbf{I}$ obeys the equation $\mathbf{IA} = \mathbf{A}$ for any matrix $\mathbf{A}$ for which that matrix multiplication is legal.

The inverse matrix of a matrix $\mathbf{A}$ is defined by the following property.

$$\mathbf{AA}^{-1} = \mathbf{I} \text{ and } \mathbf{A}^{-1}\mathbf{A} = \mathbf{I}$$

These definitions may be easiest to understand by analogy to numbers. $\mathbf{I}$ is the matrix equivalent of the number 1, the multiplicative identity; when you multiply it by another number, you get that other number back. $\mathbf{A}^{-1}$ is analogous to the reciprocal of a number; 2/3 and 3/2 are reciprocals because they multiply to 1.

Based on these definitions, see if you can convince yourself of the following facts.

- If $\mathbf{A}$ is not a square matrix it has no inverse matrix.
- An identity matrix is always a square matrix with 1s along the main diagonal and 0s everywhere else. For instance, the identity for any 3×3 matrix is $\begin{pmatrix} 1 & 0 & 0 \\ 0 & 1 & 0 \\ 0 & 0 & 1 \end{pmatrix}$.
- If $\mathbf{A}$ is a square matrix, it commutes with the identity matrix: $\mathbf{AI} = \mathbf{IA} = \mathbf{A}$.

The Discovery Exercise (Section 6.6.1) presents a straightforward, but time-consuming, process for finding the inverse of a matrix: you write the equation $\mathbf{AA}^{-1} = \mathbf{I}$, substitute letters for the unknown elements of $\mathbf{A}^{-1}$, and solve. In Problem 6.94 you will practice this method on a specific matrix. In Problem 6.123 you will apply this method to the generic 2×2 matrix, and show that:

$$\text{The inverse of } \begin{pmatrix} a & b \\ c & d \end{pmatrix} \text{ is } \frac{1}{ad-bc} \begin{pmatrix} d & -b \\ -c & a \end{pmatrix} \tag{6.6.4}$$

For instance, in the Discovery Exercise, you found the inverse of the matrix $\mathbf{A} = \begin{pmatrix} 2 & 1 \\ 5 & 3 \end{pmatrix}$. Using Equation 6.6.4 we can see readily that $\mathbf{A}^{-1} = \frac{1}{6-5} \begin{pmatrix} 3 & -1 \\ -5 & 2 \end{pmatrix} = \begin{pmatrix} 3 & -1 \\ -5 & 2 \end{pmatrix}$. You may want to confirm for yourself that this matrix multiplies by $\mathbf{A}$ (in either order!) to produce the identity matrix $\mathbf{I}$.

In practice, you never find an inverse matrix by using the definition directly; you use a formula or, better yet, a computer. The important thing is to understand what an inverse matrix is, and when to use it.

Based on that Definition, What Does an Inverse Matrix Do?

Question: If $\mathbf{X}$ is a matrix, and $\mathbf{T}$ is a transformation, then what is $\mathbf{T}^{-1}\mathbf{TX}$? (Try to answer this question for yourself before you read on.)

Answer: $\mathbf{T}^{-1}\mathbf{T} = \mathbf{I}$ (by the definition of $\mathbf{T}^{-1}$), and $\mathbf{IX} = \mathbf{X}$ (by the definition of $\mathbf{I}$), so $\mathbf{T}^{-1}\mathbf{TX} = \mathbf{X}$.

Question: So what?

Answer: The equation $\mathbf{T}^{-1}\mathbf{TX} = \mathbf{X}$ tells us that $\mathbf{T}^{-1}$, acting on the matrix $\mathbf{TX}$, produces the matrix $\mathbf{X}$. In other words, $\mathbf{T}^{-1}$ must undo whatever $\mathbf{T}$ did.[9]

That little bit of algebra connects the definition $\mathbf{T}^{-1}\mathbf{T} = \mathbf{I}$ with our original goal of creating an "inverse function," reversing the inputs and the outputs. Consider the following examples.

- $\mathbf{X}$ is a point matrix that describes a shape, and $\mathbf{T}$ is a transformation that rotates the shape 30° clockwise and then stretches it by a factor of 2 in the x-direction. So what does $\mathbf{T}^{-1}$ do? Since it must turn $\mathbf{TX}$ back into $\mathbf{X}$, it must shrink it by a factor of 2 in the x-direction and then rotate it 30° counterclockwise.

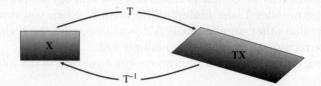

- $\mathbf{x}$ contains the number of offices and conference rooms, and $\mathbf{T}$ converts this into the number of chairs and tables contained therein. Since $\mathbf{T}^{-1}$ must turn $\mathbf{Tx}$ back into $\mathbf{x}$, it can be used to answer the question: "If you have c chairs and t tables, then how many offices and conference rooms must you have?" (This calculation will sometimes lead to fractional offices or conference rooms, which you should interpret as "no possible combination would lead to exactly that many chairs and tables.")

> **EXAMPLE** **Particles to Molecules**
>
> **Question:** A molecule of methane (CH_4) has 1 carbon atom and 4 hydrogen atoms. A molecule of butane (C_4H_{10}) has 4 carbon atoms and 10 hydrogen atoms. Each carbon atom has 6 protons and 6 neutrons, while each hydrogen atom has 1 proton and no neutrons. If a sample contains 3192 protons and 2214 neutrons, how many molecules each of methane and butane does it contain?
>
> **Answer:** We start by writing a matrix $\mathbf{X}$ to answer the easier question: given a certain number of molecules of methane and butane, how many protons and neutrons are there? We can get that by multiplying the molecules-to-atoms matrix by the atoms-to-particles matrix.
>
> $$\mathbf{X} = \begin{array}{c} \\ protons \\ neutrons \end{array}\begin{array}{c} carbon \quad hydrogen \\ \begin{pmatrix} 6 & 1 \\ 6 & 0 \end{pmatrix} \end{array} \begin{array}{c} \\ carbon \\ hydrogen \end{array}\begin{array}{c} CH_4 \quad C_4H_{10} \\ \begin{pmatrix} 1 & 4 \\ 4 & 10 \end{pmatrix} \end{array}$$
>
> $$= \begin{array}{c} \\ protons \\ neutrons \end{array}\begin{array}{c} CH_4 \quad C_4H_{10} \\ \begin{pmatrix} 10 & 34 \\ 6 & 24 \end{pmatrix} \end{array}$$

---

[9] We are implicitly using a fact that we emphasized in Section 6.5, the associativity of matrix multiplication: that is, we are assuming $\left(\mathbf{T}^{-1}\mathbf{T}\right)\mathbf{X}$ is the same as $\mathbf{T}^{-1}\left(\mathbf{TX}\right)$.

Next we can use Equation 6.6.4 to find $\mathbf{X}^{-1}$, which will convert from protons and neutrons to molecules of methane and butane.

$$\mathbf{X}^{-1} = \frac{1}{36} \begin{pmatrix} 24 & -34 \\ -6 & 10 \end{pmatrix} = \begin{matrix} CH_4 \\ C_4H_{10} \end{matrix} \begin{pmatrix} \overset{protons}{2/3} & \overset{neutrons}{17/18} \\ -1/6 & 5/18 \end{pmatrix}$$

Finally, applying this matrix to the numbers of particles given:

$$\begin{matrix} CH_4 \\ C_4H_{10} \end{matrix} \begin{pmatrix} \overset{protons}{2/3} & \overset{neutrons}{17/18} \\ -1/6 & 5/18 \end{pmatrix} \begin{matrix} protons \\ neutrons \end{matrix} \begin{pmatrix} 3192 \\ 2214 \end{pmatrix} = \begin{matrix} CH_4 \\ C_4H_{10} \end{matrix} \begin{pmatrix} 37 \\ 83 \end{pmatrix}$$

The sample contains 37 molecules of methane and 83 of butane.

Section 6.4 gave the example of using $\vec{V}_1 = \hat{i} + 3\hat{j}$ and $\vec{V}_2 = 2\hat{i}$ as a basis. If you understand matrix multiplication, it's easy to see how the following matrix multiplication converts from a $\vec{V}_1 \vec{V}_2$ representation to an $\hat{i}\hat{j}$ representation of a vector.

$$\begin{matrix} \hat{i} \\ \hat{j} \end{matrix} \begin{pmatrix} \overset{\vec{V}_1}{1} & \overset{\vec{V}_2}{2} \\ 3 & 0 \end{pmatrix} \begin{matrix} \vec{V}_1 \\ \vec{V}_2 \end{matrix} \begin{pmatrix} 10 \\ -7 \end{pmatrix} = \begin{matrix} \hat{i} \\ \hat{j} \end{matrix} \begin{pmatrix} -4 \\ 30 \end{pmatrix}$$

The first column of the transformation matrix is $\vec{V}_1$ and the second is $\vec{V}_2$. So multiplying that matrix by the vector in the middle says "I want 10 of $\vec{V}_1$ and -7 of $\vec{V}_2$." The answer tells you that you end up with $-4\hat{i} + 30\hat{j}$.

Inverse matrices reverse this operation: given any vector represented in the $\hat{i}\hat{j}$ basis, you can find its $\vec{V}_1$- and $\vec{V}_2$-components.

$$\begin{matrix} \vec{V}_1 \\ \vec{V}_2 \end{matrix} \begin{pmatrix} \overset{\hat{i}}{0} & \overset{\hat{j}}{1/3} \\ 1/2 & -1/6 \end{pmatrix} \begin{matrix} \hat{i} \\ \hat{j} \end{matrix} \begin{pmatrix} -4 \\ 30 \end{pmatrix} = \begin{matrix} \vec{V}_1 \\ \vec{V}_2 \end{matrix} \begin{pmatrix} 10 \\ -7 \end{pmatrix}$$

You will work with this use of inverse matrices in Problems 6.116 and 6.117.

Linear Equations and Matrix Algebra

Consider the following matrix equation.

$$\underset{\mathbf{A}}{\begin{pmatrix} 3 & 2 \\ 7 & -10 \end{pmatrix}} \underset{\mathbf{x}}{\begin{pmatrix} x \\ y \end{pmatrix}} = \underset{\mathbf{b}}{\begin{pmatrix} -8 \\ -26 \end{pmatrix}} \tag{6.6.5}$$

You could write this out as two regular algebra equations and then solve for x and y using substitution or elimination, but we want to use this example to show you how you can do algebra directly with matrices. In this case, we're going to solve this equation for $\mathbf{x}$.

If matrices were numbers we would solve $\mathbf{Ax} = \mathbf{b}$ by dividing both sides by $\mathbf{A}$, but there is no such thing as matrix division. How would you solve the numerical equation $5x = 12$ if we told you that you were allowed to multiply but not divide? Of course you would multiply both sides by $1/5$. In an analogous way—and for analogous reasons—you can solve matrix equations by multiplying by the inverse matrix.

But there is one limitation to this analogy, because matrix multiplication is not commutative. To do the same thing to both sides of the equation we must "left-multiply" both sides by $\mathbf{A}^{-1}$, or "right-multiply" both sides. If we left-multiply one side while right-multiplying the other side, we have not done the same thing to both sides.

In the case of Equation 6.6.5, left-multiplying is the operation we want.

$$\mathbf{A}^{-1}\mathbf{A}\mathbf{x} = \mathbf{A}^{-1}\mathbf{b}$$

Note that we have legally done the same thing to both sides. But now look what we get! $\mathbf{A}^{-1}\mathbf{A}$ becomes $\mathbf{I}$ and then $\mathbf{I}\mathbf{x}$ becomes $\mathbf{x}$.

$$\mathbf{x} = \mathbf{A}^{-1}\mathbf{b}$$

Find the inverse of $\mathbf{A}$ and multiply it by $\mathbf{b}$ and you find the matrix we were looking for.

$$\mathbf{x} = \begin{pmatrix} -3 \\ 1/2 \end{pmatrix}$$

As a quick check, our original matrix equation was equivalent to the two equations $3x + 2y = -8$ and $7x - 10y = -26$, and you can easily verify that $x = -3, y = 1/2$ is the correct solution. The value of this technique is not for solving systems of linear equations; row reduction (Section 7.4 at felderbooks.com) is more efficient for that. However, there are situations where it is important to do algebra directly with matrices, often without knowing numerical values for them, and this provides a simple illustration of the technique.

Non-invertible Transformations

We noted before that only square matrices can be inverted. You'll show in Problem 6.235 why this makes sense in terms of what matrix transformations do; roughly speaking the output of a 2×3 matrix doesn't have enough information to be inverted and the output of a 3×2 matrix has too much. Square matrices have the same amount of information in their input and output, so they generally can be inverted.

Generally, but not always. If we tell you a terrarium contains 300 ants (6 legs each and no hands) and 20 spiders (8 legs each and no hands), you can easily tell us that the terrarium contains 1960 legs and zero hands. The matrix for doing that conversion is $\begin{pmatrix} 6 & 8 \\ 0 & 0 \end{pmatrix}$. If we just told you that the terrarium contained 1960 legs and no hands, though, you wouldn't know if it had 300 ants and 20 spiders, or 220 ants and 80 spiders, or any of countless other possible combinations. This matrix simply has no inverse. In this case it's true for a trivial reason (neither animal has hands, so saying "zero hands" doesn't give you any useful information). There are much less obvious non-invertible transformations, however. We will return to this issue in Section 6.7.

Application to the Three-Spring Problem

In Section 6.3 we discussed how a matrix can convert from "A of the first normal mode and B of the second" to $x_1(0)$ and $x_2(0)$. To solve the three-spring problem for a given set of initial conditions, we need to go the other way. For example, in Section 6.1 we gave you the initial conditions $x_1(0) = 5$ and $x_2(0) = 7$ and asserted without explanation that you could make these by adding $26/3$ times the first normal mode to $-11/3$ times the second normal mode. You now know enough to understand how we got those numbers. We inverted normal modes-to-positions matrix that we derived in Section 6.3 to get a positions-to-normal modes matrix, which we then multiplied by the column matrix $\begin{pmatrix} 5 \\ 7 \end{pmatrix}$ to get $\begin{pmatrix} 26/3 \\ -11/3 \end{pmatrix}$. You'll carry out the details in Problem 6.115.

6.6.3 Problems: The Identity and Inverse Matrices

6.93 The Pernicious Puzzle company makes two kinds of gift baskets. Basic baskets come with 5 easy puzzles and 2 hard puzzles. Deluxe baskets come with 7 easy puzzles and 6 hard ones. Each easy puzzle has 20 regular pieces and 10 edge pieces, while each hard puzzle has 500 regular pieces and 100 edge pieces.

(a) Find a matrix X that converts from number of basic and deluxe baskets to number of regular and edge pieces.

(b) Use the matrix X to answer the question: if you need to ship 50 basic baskets and 20 deluxe baskets to a supplier, how many regular pieces and how many edge pieces do you need to make?

The new intern James was getting ready to do inventory when he knocked over a shelf full of gift baskets, spilling everything onto the floor. He diligently picked everything up and before trying to reassemble it all he counted that there were 286,420 regular pieces and 61,910 edge pieces. Now he needs to know how many of each type of basket there were.

(c) Use Equation 6.6.4 to find the inverse matrix X^{-1}.

(d) We said above that X converts from numbers of basic and deluxe baskets to numbers of regular and edge pieces. In a similar way, describe what X^{-1} does.

(e) Use the matrix X^{-1} to find how many basic baskets and how many deluxe baskets James knocked over.

6.94 Let $A = \begin{pmatrix} 3 & 5 \\ 2 & 4 \end{pmatrix}$.

(a) Show that $I = \begin{pmatrix} 1 & 0 \\ 0 & 1 \end{pmatrix}$ fulfills the definition of the identity matrix for matrix A.

(b) Explain why, in order to fulfill the definition of the inverse, A^{-1} must be a 2×2 matrix.

(c) Using $\begin{pmatrix} a & c \\ b & d \end{pmatrix}$ to represent the unknown A^{-1}, write the equation $AA^{-1} = I$, filling in the 2×2 elements of each of the three matrices.

(d) Perform the multiplication on the left side of your equation in Part (c).

(e) You are now setting two 2×2 matrices equal to each other, resulting in four separate equations. Solve these four equations to find the unknowns a, b, c, and d.

(f) You have now found matrix A^{-1} to satisfy the equation $AA^{-1} = I$. To show that it fills the definition of an inverse matrix, show that it also satisfies the equation $A^{-1}A = I$.

6.95 *[This problem depends on Problem 6.94.]* For matrix A as defined in Problem 6.94, find matrix A^{-1} by using Equation 6.6.4. (If you get a different answer, something has gone wrong somewhere!)

6.96 *[This problem depends on Problem 6.94.]* For matrix A as defined in Problem 6.94, find matrix A^{-1} on a calculator or computer. (If you get a different answer, something has gone wrong somewhere!)

6.97 M is a 2×2 matrix that multiplies the x-component of vectors by 2.

(a) Describe in words what transformation M^{-1} performs.

(b) Write down matrix M.

(c) To verify your answer to Part (b) multiply your matrix M by the generic vector $\begin{pmatrix} x \\ y \end{pmatrix}$ and make sure you get the answer you should.

(d) Find M^{-1} using Equation 6.6.4 and verify that it performs the transformation you described.

6.98 M is a 2×2 matrix that multiplies all vectors by -1. Equivalently you could say it reflects them across the origin.

(a) Describe in words what transformation M^{-1} performs.

(b) Write down matrix M.

(c) To verify your answer to Part (b) multiply your matrix M by the generic vector $\begin{pmatrix} x \\ y \end{pmatrix}$ and make sure you get the answer you should.

(d) Find M^{-1} using Equation 6.6.4. Explain how you could have predicted the answer without doing any calculations.

6.99 Based on the definition of the identity matrix $IA = A$, many people's first guess for a 2×2 identity matrix is $\begin{pmatrix} 1 & 1 \\ 1 & 1 \end{pmatrix}$. Show that this *only* fulfills the definition of the identity matrix for $A = \begin{pmatrix} 0 & 0 \\ 0 & 0 \end{pmatrix}$. *Hint*: start by assuming that it works for some unknown matrix A, and then prove that all four elements of A must be 0.

6.100 Matrix $\mathbf{M} = \begin{pmatrix} 2 & 9 & -24 \\ -1 & -4 & 11 \\ -3 & -15 & 40 \end{pmatrix}$.

Show that $\mathbf{M}^{-1} = \begin{pmatrix} 5 & 0 & 3 \\ 7 & 8 & 2 \\ 3 & 3 & 1 \end{pmatrix}$.

6.101 Matrix $\mathbf{N} = \begin{pmatrix} 3 & -6 \\ -2 & x \end{pmatrix}$. For what value(s) of the variable x does this matrix have *no inverse matrix*?

6.102 Let $\mathbf{v} = \begin{pmatrix} 1 \\ 1 \end{pmatrix}$ and $\mathbf{F} = \begin{pmatrix} 1/2 & 1/2 \\ 1/2 & 1/2 \end{pmatrix}$.

(a) Show that $\mathbf{Fv} = \mathbf{v}$.

(b) Part (a) seems to show that $\mathbf{F}$ is an identity matrix. Explain why it isn't.

6.103 Matrix $\mathbf{D}$ acts as a transformation. If you multiply it by a point matrix $\mathbf{X}$, the result is a new matrix $\mathbf{DX}$ whose shape looks like matrix $\mathbf{X}$ rotated 20° clockwise and shrunk by a factor of 2. *Note:* Nothing in this problem requires you to find the elements of matrix $\mathbf{D}$.

(a) What is the matrix $\mathbf{D}^{-1}\mathbf{D}$?

(b) What is the matrix $(\mathbf{D}^{-1}\mathbf{D})\mathbf{X}$?

(c) What is the matrix $\mathbf{D}^{-1}(\mathbf{DX})$?

(d) In words, what effect does multiplying by $\mathbf{D}^{-1}$ have on a point matrix $\mathbf{Y}$?

6.104 Section 6.3 presented the scenario, "Each kiddy book has 30 pictures and 500 words; each chapter book has 10 pictures and 20,000 words." In this question we ask: "If your collection contains a total of 2360 pictures and 376,500 words, how many of each type of book do you have?"

(a) Answer this question with no reference to matrices by setting up and solving two simultaneous equations.

(b) Now, recall that we used matrix $\mathbf{P}$ to transform from kiddy-and-picture books to pictures-and-words:

$$\mathbf{P} = \begin{matrix} \\ pictures \\ words \end{matrix} \begin{pmatrix} kiddy\ books & chapter\ books \\ 30 & 10 \\ 500 & 20{,}000 \end{pmatrix}$$

Find matrix $\mathbf{P}^{-1}$ and show how you can use it to answer the same question you answered in Part (a) (and hopefully get the same answer!).

6.105 Section 6.5 Problem 6.70 began: "A swank, sophisticated graveyard has 50 vampires and only 3 zombies. But your trashy, low-class graveyard, it has only 3 dead-beat vampires and 200 zombies."

(a) Write a matrix $\mathbf{G}$ that can be used to answer the question: "If our town has 39 swank graveyards and 97 trashy graveyards, how many vampires and zombies do we have?"

(b) Write a question that could be answered by matrix $\mathbf{G}^{-1}$.

6.106 (a) Write a matrix $\mathbf{H}$ that stretches any shape by a factor of 3 in the x-direction, leaving the y-direction unchanged.

(b) Write a matrix $\mathbf{V}$ that shrinks any shape by a factor of 3 in the x-direction, leaving the y-direction unchanged.

(c) Explain how you can tell—just from the verbal descriptions, with no math—that matrices $\mathbf{H}$ and $\mathbf{V}$ are inverses of each other.

(d) Now demonstrate mathematically that they are.

6.107 Walk-Through: Solving Simultaneous Equations. In this problem you will use inverse matrices to solve the following two equations.

$$3x + 2y = 6$$
$$-12x + 5y = 28$$

(a) These two equations can be written as a single matrix equation: $\mathbf{Ax} = \mathbf{b}$, where $\mathbf{x}$ contains nothing but the variables we are looking for. Write out the matrices $\mathbf{A}$, $\mathbf{x}$, and $\mathbf{b}$.

(b) If you right-multiply both sides of the equation by $\mathbf{A}^{-1}$ you get $\mathbf{AxA}^{-1} = \mathbf{bA}^{-1}$. If you left-multiply both sides of the equation by $\mathbf{A}^{-1}$ you get $\mathbf{A}^{-1}\mathbf{Ax} = \mathbf{A}^{-1}\mathbf{b}$. Which of these equations represents legal matrix multiplication?

(c) Use the definition of the inverse matrix to simplify the equation you said was legal.

(d) Find $\mathbf{A}^{-1}$ and then use the equation you wrote to find $\mathbf{x}$.

(e) Based on your answer to Part (d), what values of x and y solve our original equations? Verify that your answer works.

In Problems 6.108–6.111 use inverse matrices to solve the given sets of equations. It may help to first work through Problem 6.107 as a model.

6.108 $5x - y = 13$, $4x + 2y = 2$

6.109 $2m + 3n + 17 = 5$, $(1/3)(m + 6n) = -3$

6.110 $9p^2 - r = 6$, $6p^2 + r/3 + 1 = 0$

6.111 🖥️ $9a + 3b - 5c + 7d = 8$, $3a - 4b + c = 10$, $10a + 3b + c + 6d = 3$, $a - b + 2c - d = 69$. (You can have a computer solve these equations to check your answer, but to solve the problem you need to show the steps of using an inverse matrix to solve them, using the computer to do the matrix algebra for you.)

6.112 Attempt to solve the equations $4x + 6y = 12$, $6x + 9y = 14$ using the method described in the Explanation (Section 6.6.2) and Problem 6.107. Explain what went wrong in terms of the matrix operations; then explain what went wrong in terms of the original equations, with no reference to matrices.

6.113 Begin with the matrix equation **ABC = D**. Assume **A**, **B**, **C**, and **D** are all square matrices of equal size. In each part below solve for the indicated matrix in terms of the other matrices and their inverses.

(a) **C**

(b) **A**

(c) **B**

6.114 In each part below solve for the matrix **M** in terms of the other matrices and their inverses. In each case assume that all the matrices are square matrices of equal size.

(a) $\mathbf{AM} + \mathbf{B} = \mathbf{C}$

(b) $\mathbf{MA} + \mathbf{B} = \mathbf{CD}$

(c) $\mathbf{A}^{-1}\mathbf{MA} = \mathbf{B}$

6.115 Section 6.1 presented the three-spring scenario; Section 6.3 re-expressed the scenario as a matrix problem, and presented the following multiplication.

$$\begin{array}{c} \textit{position of Ball 1} \\ \textit{position of Ball 2} \end{array} \begin{pmatrix} \textit{first normal mode} & \textit{second normal mode} \\ 1 & 1 \\ 2/3 & -1/3 \end{pmatrix}$$

$$\begin{array}{c} \textit{first normal mode} \\ \textit{second normal mode} \end{array} \begin{pmatrix} 26/3 \\ -11/3 \end{pmatrix} = \begin{array}{c} \textit{position of Ball 1} \\ \textit{position of Ball 2} \end{array} \begin{pmatrix} 5 \\ 7 \end{pmatrix}$$

This shows how you can use a 2×2 matrix to convert from "A of the first normal mode and B of the second" to the initial positions $x_1(0)$ and $x_2(0)$. But when we solved the three-spring problem, we had to go the other way; we wanted $x_1(0) = 5$ and $x_2(0) = 7$, and we had to find A and B accordingly.

(a) Find the inverse of the above 2×2 matrix.

(b) Use your inverse matrix to show that $x_1(0) = 5$ and $x_2(0) = 7$ lead to $A = 26/3$ and $B = -11/3$.

(c) Find the solution to the three-spring problem that meets the initial conditions $x_1(0) = 8$ and $x_2(0) = -3$.

6.116 Section 6.4 used a matrix multiplication to convert a vector from $\vec{V}_1 \vec{V}_2$ representation to $\hat{i}\hat{j}$ representation, where $\vec{V}_1 = \hat{i} + 3\hat{j}$ and $\vec{V}_2 = 2\hat{i}$.

(a) Write a matrix to convert from $\hat{i}\hat{j}$ representation to $\vec{V}_1 \vec{V}_2$.

(b) Use your matrix to convert the vector $4\hat{i} + 5\hat{j}$ to $\vec{V}_1 \vec{V}_2$ representation.

6.117 Basis P represents all vectors as a linear combination of the two vectors $\vec{P}_1 = 3\hat{i} - 4\hat{j}$ and $\vec{P}_2 = 2\hat{i} + \hat{j}$. Basis Q represents all vectors as a linear combination of the two vectors $\vec{Q}_1 = \hat{i} + 5\hat{j}$ and $\vec{Q}_2 = 4\hat{j}$. In this problem you are going to create a matrix that converts vectors directly from basis P to basis Q.

(a) Write a matrix that converts from basis P to an $\hat{i}\hat{j}$ representation. (No calculations are needed here; you can just write down the matrix.)

(b) Write a matrix that converts from basis Q to an $\hat{i}\hat{j}$ representation.

(c) Write a matrix that converts from an $\hat{i}\hat{j}$ representation to basis Q.

(d) Write a single matrix that converts from basis P to basis Q.

(e) Use your matrix to convert $\vec{P}_1 - \vec{P}_2$ to basis Q. Draw the vector and show its $\vec{P}_1$- and $\vec{P}_2$-components and its $\vec{Q}_1$- and $\vec{Q}_2$-components.

6.118 Matrix $\mathbf{S} = \begin{pmatrix} 1 & 1 \\ 0 & 1 \end{pmatrix}$ is called a "shear matrix."

(a) Apply matrix **S** to the vertices of polygon **P**, whose vertices are $(0,0)$, $(1,0)$, $(1,1)$, and $(0,1)$ in that order. Draw **P** on one set of coordinate axes, and the transformed polygon **P'** on a different graph.

(b) Describe in words what **S** did to **P**.

(c) The inverse transformation matrix $\mathbf{S}^{-1}$ should undo the effect of **S**; in other words, it should transform **P'** back to **P**. Describe in words what it would need to do to make that happen.

(d) Think about what $\mathbf{S}^{-1}$ would do if it were applied to **P** (not **P'**—we know what would happen then!). Draw a picture of how you expect **P** will be transformed by $\mathbf{S}^{-1}$.

(e) Calculate S^{-1} and apply it to P. Draw the resulting figure.

6.119 Matrix X interchanges the axes; in other words, it turns every point (x, y) into (y, x).

(a) An easy way to find matrix X is to note the effect it will have on the point matrix $P = \begin{pmatrix} 1 & 0 \\ 0 & 1 \end{pmatrix}$. Determine the point matrix that should result when X acts on P. Then write the resulting equation, and you can write down matrix X immediately.

(b) Explain, based on our *verbal* description of the effect of matrix X, how we know that matrix X should be its own inverse.

(c) Now prove mathematically that it is, by showing that $XX = I$.

6.120 *[This problem depends on Problem 6.119.]* In Problem 6.119 you found a matrix X that is its own inverse. You found it by considering a verbal description of a matrix that will undo itself; you proved it by showing that it satisfied the equation $XX = I$.

(a) Show mathematically that the 2×2 identity matrix I is its own inverse.

(b) Find 4 other 2×2 matrices (for a total of 6) that are their own inverses. For each one say in words what the matrix does and demonstrate algebraically that it is its own inverse.

6.121 Matrix AB means "First do matrix B, and then do matrix A." Is matrix $(AB)^{-1}$ the same as $A^{-1}B^{-1}$, or $B^{-1}A^{-1}$, or neither? Explain the reasoning that leads you to your conclusion.

6.122 Let matrix $C = \begin{pmatrix} 1 & i \\ i & 1 \end{pmatrix}$ where $i \equiv \sqrt{-1}$. Find C^{-1} and demonstrate that $CC^{-1} = I$.

6.123 In this problem you will derive the formula for the inverse of the generic 2×2 matrix $A = \begin{pmatrix} a & c \\ b & d \end{pmatrix}$.

(a) Using $A^{-1} = \begin{pmatrix} w & y \\ x & z \end{pmatrix}$ to represent the inverse matrix, write out the equation $AA^{-1} = I$.

(b) Perform the requisite multiplication and set the two resulting matrices equal to each other. You should now have four separate linear equations.

(c) Solve for the unknown variables w, x, y, and z. Remember that each one will end up being a function of a, b, c, and d.

(d) In the end, with possibly a bit of rewriting, you should be able to factor $1/(ad - bc)$ out of all four answers, resulting in Equation 6.6.4.

(e) A "singular matrix" has no inverse. Based on your result, what must be true for a 2×2 matrix to be "singular?"

6.124 The inverse of a diagonal matrix is a diagonal matrix with each element inverted. In other words, if $A = \begin{pmatrix} c_1 & 0 & & \\ 0 & c_2 & & \\ & & \ldots & \\ & & & c_n \end{pmatrix}$ then $A^{-1} = \begin{pmatrix} 1/c_1 & 0 & & \\ 0 & 1/c_2 & & \\ & & \ldots & \\ & & & 1/c_n \end{pmatrix}$.

(a) Prove this fact mathematically.

(b) Now explain it verbally by saying what effect A and A^{-1} have as transformations of an n-dimensional vector.

6.7 Linear Dependence and the Determinant

Linear algebra gives us a systematic way of working with sets of linear equations. With three or more equations it can be difficult to figure out if a unique solution even exists—and in some cases answering that question is enough. The existence of unique solutions hinges on whether your equations are "linearly dependent."

When the equations are represented as a matrix, we can use the "determinant" of that matrix to determine if they are linearly dependent. This section will begin by looking at the concept of linear dependence, and will then move on to other uses of the determinant.

6.7 | Linear Dependence and the Determinant

6.7.1 Discovery Exercise: Linear Dependence and the Determinant

1. Solve the following equations for x and y.

$$2x + 3y = 7$$
$$4x + 2y = 21$$

   See Check Yourself #42 in Appendix L

2. Try to solve the following equations for x and y. Explain what went wrong.

$$2x + 3y = 7$$
$$4x + 6y = 21$$

3. Find at least two pairs of numbers (x, y) that solve the following equations.

$$2x + 3y = 7$$
$$4x + 6y = 14$$

You may have been taught that n linear equations with n unknowns have exactly one solution, but this isn't always the case. In some cases they have one solution, in some cases they have none, and in some cases they have infinitely many solutions. With two equations and two unknowns you can figure out which case you're dealing with by playing with the equations for a bit, but larger sets of equations demand a systematic approach.

6.7.2 Explanation: Linear Dependence and the Determinant

Associated with any square matrix is a number called its "determinant." This section will tell you how to compute the determinant of a matrix and show you some of the ways to interpret it. In particular, if the determinant of a matrix is zero, that implies the following things.

- If the matrix represents the coefficients of a system of n linear equations with n unknowns, a zero determinant means those equations don't have a unique solution.
- If the matrix represents n vectors with n components each, a zero determinant means those vectors cannot provide a basis for an n-dimensional space.
- If the matrix represents a transformation, a zero determinant means it has no inverse.

These three items may appear unrelated, but we hope to convince you that they are different ways of saying the same thing.

Linearly Independent Equations

When you solve two linear equations with two unknowns, three different types of outcome are possible.

The left plot in Figure 6.5 represents the "typical" case: the two equations have one unique solution. The only other possibilities are "inconsistent" (no solutions) or "linearly dependent" (infinitely many solutions because the two equations represent the same line). Linear equations can never have 2 solutions, or 3, or 17: they always have 1 solution, none, or infinitely many.

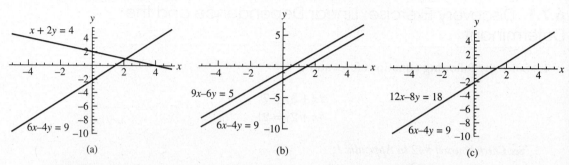

FIGURE 6.5 The three possibilities for two linear equations: intersect at a point, never intersect, or intersect everywhere on both lines.

But if it is typical to have one unique solution, what makes the lines in the middle and right plots atypical? In both cases, the left side of one equation is a multiple of the left side of the other: $9x - 6y = (3/2)(6x - 4y)$, and $12x - 8y = 2(6x - 4y)$. With that insight we can look at such equations in their generic form.

$$ax + by = r_1$$
$$cx + dy = r_2 \qquad (6.7.1)$$

These equations have a unique solution *unless* the left side of the second equation is a multiple of the left side of the first equation. That's equivalent to saying $c/a = d/b$, which we can rearrange as follows.

> **For two simultaneous linear equations in the form given by Equations 6.7.1:**
>
> - If $ad - bc = 0$ then the equations are inconsistent or linearly dependent.
> - If $ad - bc \neq 0$ then the equations have one unique solution.

As you can see, the quantity $ad - bc$ determines if a set of equations has a unique solution, but it does not distinguish "inconsistent" from "linearly independent" sets of equations. That difference depends on the constant terms r_1 and r_2, a subject to which we will return more generally in Section 7.4 (see felderbooks.com). Here we need address only one special case: what if r_1 and r_2 are both zero?

An equation of the form $ax + by = 0$ is called "homogeneous." A set of homogeneous equations *always* has the solution $x = y = 0$, known as the "trivial solution." (Graphically, all homogeneous lines go through the origin.) Therefore, n homogeneous equations with n unknowns can never be inconsistent. Either they have only one unique solution (the origin) or they are linearly dependent and have infinitely many solutions.

EXAMPLE   **Linear Dependence**

Question: What does λ have to equal in order for the equations

$$x + \lambda y = 0$$
$$2x + 3y = 0$$

to have any non-trivial solutions?

Answer: You can easily verify that $x = y = 0$ works no matter what λ is. If $ad - bc$ is anything other than zero, then that trivial solution will be *the unique* solution. For there to be any other solutions, the equations must be linearly dependent, so $3 - 2\lambda = 0$, meaning $\lambda = 3/2$.

Conclusion: For $\lambda = 3/2$ the two equations both reduce to the same equation; any pair of numbers (x, y) that satisfies $x = -3y/2$ will solve both equations. For any other λ, the only solution is the trivial one: $x = y = 0$.

We have restricted our discussion so far to 2 equations with 2 unknowns. The first and most important result—that every set will have 0 solutions, 1 solution, or infinitely many solutions—applies to any set of linear equations. (You will prove this for 3 equations with 3 unknowns in Problem 6.148.) But spotting linear dependence becomes more complicated in a case like this.

$$A : 2x - 3y + 4z = 3$$
$$B : x + 4y - z = 1$$
$$C : 4x - 17y + 14z = 7$$

You can easily check that none of these equations is a multiple of any of the others. What you can't tell so easily is that equation C is 3 times equation A minus 2 times equation B. That means equation C gives us no information that was not already implicit in the first two equations, so these equations are linearly dependent (infinitely many solutions). The following definition applies to any set of 2 or more equations.

Linearly Dependent Equations

A set of equations is linearly dependent if any one of them can be formed from a linear combination of any of the others.

Matrices and Determinants

We're going to reframe our results above in matrix language.

The process starts when we take a set of equations and write down the coefficients of all the variables—in other words, all the numbers on the left side of the equation, ignoring the constant terms on the right side as well as the variables themselves. For n equations with n unknowns, the result is a square matrix. For instance:

For the equations
$$\begin{aligned} 3x + 7y + 2z &= 10 \\ 4x - 8y + 6z &= 11 \\ 5x \quad\quad - z &= 13 \end{aligned}$$
the matrix of coefficients is $\begin{pmatrix} 3 & 7 & 2 \\ 4 & -8 & 6 \\ 5 & 0 & -1 \end{pmatrix}$

We saw above that these coefficients determine whether the equations have a unique solution. So we define a new function of these numbers, called the "determinant"; it determines the types of results we expect to find. We have seen what that function looks like for a 2×2 matrix.

Chapter 6 Linear Algebra I

Definition: The Determinant of a 2×2 Matrix

The determinant of matrix $\begin{pmatrix} a & c \\ b & d \end{pmatrix}$ is $ad - bc$.

Determinants are often indicated with absolute value signs, so we write $\begin{vmatrix} a & c \\ b & d \end{vmatrix} = ad - bc$.

Above we stated a rule for determining if two simultaneous equations have a unique solution. We can now restate that rule in terms of matrices. This new rule will generalize to more than two equations, provided we define the determinants of higher-level matrices correctly.

What the Determinant tells us about a Set of Equations

Given a set of n linear equations with n unknowns, let **C** be the matrix of coefficients for those equations, and let |**C**| be the determinant of that matrix.

- If |**C**| = 0 then the equations are either inconsistent (no solution) or linearly dependent (infinitely many solutions).
- If |**C**| is anything other than zero then the equations have one unique solution.

If the equations are all homogeneous then they cannot be inconsistent, so in that case |**C**| = 0 implies they are linearly dependent.

We've seen that these rules works for a 2×2 matrix if we define the determinant as $ad - bc$, and we have also seen that finding the right relationship of coefficients is harder for more equations. Fortunately, that work has been done once and for all, and the formula for the determinant of larger matrices has been defined to make those rules work. We will present that definition below, but in practice you will rarely if ever need to use it. You need to know how to find the determinant of a 2×2 matrix (which we told you above), and sometimes even a 3×3 matrix (which we tell you below), by hand. For any higher-level matrix, just get the determinant from a computer. In all cases, though, you need to know what to do with the determinant once you have it!

EXAMPLE: Linear Dependence With More Than Two Equations

Question: What does λ have to equal in order for the equations

$$(1 - \lambda)a + b - 3c + 2d = 0$$
$$2a + (3 - \lambda)b + c - 4d = 0$$
$$5a + 2b + (-1 - \lambda)c + 3d = 0$$
$$2a - 2b + 2c + (3 - \lambda)d = 0$$

to have any non-trivial solutions?

6.7 | Linear Dependence and the Determinant

Answer: The only way they could have any other solutions in addition to the trivial one is if they are linearly dependent. We asked a computer to find the determinant of the matrix

$$\begin{pmatrix} (1-\lambda) & 1 & -3 & 2 \\ 2 & (3-\lambda) & 1 & -4 \\ 5 & 2 & (-1-\lambda) & 3 \\ 2 & -2 & 2 & (3-\lambda) \end{pmatrix}$$

and got the result $\lambda^4 - 6\lambda^3 + \lambda^2 + 3\lambda - 39$. We then asked the computer to set that equal to zero and solve for λ, with the results $\lambda = -1.75$, 5.93, or $0.006 \pm 1.71i$. The equations will have non-trivial solutions if and only if λ has one of these four values.

If this example seems pointless, hold on. This is the type of calculation you need to do to find "eigenvalues" of a matrix. We'll define what that means in Section 6.8 and by the end of the chapter you'll have used eigenvalues in solving the three-spring problem among others.

Calculating Determinants

We present here the algorithm for finding determinants with no justification at all. We're not even going to say, as we usually do, "you will prove in thus-and-such a problem that this is the right process." We're just going to show you how to find the determinant of a 3×3 matrix, and how the process generalizes up from there.

Note that the determinant is only defined for a *square matrix:* one with the same number of rows as columns. Ask a computer for the determinant of a 2×3 matrix you and should get an error.

The process is called "expansion by minors." A "minor" of a matrix is a smaller matrix, obtained by crossing out the row and column that contain a particular element. The picture shows a 3×3 matrix with the row and column containing the "2" crossed out. What's left is the 2×2 matrix $\begin{pmatrix} 4 & 6 \\ 7 & 9 \end{pmatrix}$, the "minor" of that element. The determinant of this particular minor is $(4 \times 9) - (6 \times 7) = -6$.

The process works like this.

1. Move across the top row of the matrix. For each element, multiply that number times the determinant of its minor.
2. Take all the numbers you get by this process and *alternately add and subtract them.*

Got all that? Of course you don't, but it will be clearer with an example.

EXAMPLE **Finding a Determinant**

Question: If matrix $\mathbf{A} = \begin{pmatrix} 3 & 7 & 2 \\ 4 & -8 & 6 \\ 5 & 0 & -1 \end{pmatrix}$, find its determinant $|\mathbf{A}|$.

Answer: Begin in the upper-left-hand corner. The number there is a 3, and its minor is $\begin{pmatrix} -8 & 6 \\ 0 & -1 \end{pmatrix}$. You multiply the number 3 times the determinant of that 2×2 matrix, and you get a number.

Then you repeat this process all along the top row, multiplying the 7 by the determinant of its minor, and the 2 by the determinant of its minor. This gives you three numbers, which you alternately add and subtract.

$$3\begin{vmatrix} -8 & 6 \\ 0 & -1 \end{vmatrix} - 7\begin{vmatrix} 4 & 6 \\ 5 & -1 \end{vmatrix} + 2\begin{vmatrix} 4 & -8 \\ 5 & 0 \end{vmatrix}$$
$$= 3(-8 \times (-1) - 6 \times 0) - 7(4 \times (-1) - 6 \times 5) + 2(4 \times 0 - (-8) \times 5)$$

The determinant is 342. And so what? Because $|\mathbf{A}| \neq 0$ we can say that any set of equations with these coefficients would have one unique solution. You will see other valuable interpretations of that number as we go.

The determinant of any square matrix can be defined inductively in this way.
- The determinant of a 1×1 matrix is simply the number.
- The determinant of any higher square matrix is defined by "expansion by minors" as explained above.

This process gives you the correct determinant of a 2×2 matrix, $ad - bc$. It also tells you how to find the determinant of a 4×4 matrix, but the process involves four separate 3×3 minors, and each of those involves three 2×2 minors! For very large matrices—and yes, they do come up all the time—expansion by minors is too slow even for a computer, so many shortcuts have been developed. We will pass over that topic, content that our computers can find determinants for us.

It is sometimes convenient to be able to quickly calculate 3×3 determinants by hand, however. Problem 6.145 illustrates a shortcut for doing so.

Linearly Independent Vectors

We have discussed the determinant as a tool for determining if a set of equations is linearly dependent. We're now going to use the determinant in other situations, all based on the same idea of linear dependence.

Section 6.4 discussed the issue of basis vectors. Here is a more general version of the rule we presented there.

What the Determinant tells us about a Set of Vectors

Given a set of n vectors with n components each, let $\mathbf{C}$ be a matrix in which each column is one of those vectors, and let $|\mathbf{C}|$ be the determinant of that matrix.

- If $|\mathbf{C}| = 0$ then the vectors are linearly dependent. In that case, at least one of those vectors can be expressed as a linear combinations of some of the others.
- If $|\mathbf{C}|$ is anything other than zero then the vectors are linearly independent. In that case, the vectors form a basis for an n-dimensional space. You can create any vector in n-dimensional space by choosing the right combination of these vectors.

6.7 | Linear Dependence and the Determinant

In two dimensions the only way for two non-zero vectors $\vec{V}_1$ and $\vec{V}_2$ to be linearly dependent is if $\vec{V}_1 = k\vec{V}_2$, which is to say they are parallel.

But consider these three vectors:

$$\vec{V}_1 = 2\hat{i} - 3\hat{j} + 4\hat{k}$$
$$\vec{V}_2 = \hat{i} + 4\hat{j} - \hat{k}$$
$$\vec{V}_3 = 4\hat{i} - 17\hat{j} + 14\hat{k}$$

Any two of these vectors are non-parallel, which is to say, linearly independent. But the set of three is *not* linearly independent: you can build any one of these vectors from the other two. For instance, $\vec{V}_3 = 3\vec{V}_1 - 2\vec{V}_2$.

So imagine trying to use these three vectors as a basis, which is to say, trying to build all other vectors as $a\vec{V}_1 + b\vec{V}_2 + c\vec{V}_3$. If we replace $\vec{V}_3$ with $3\vec{V}_1 - 2\vec{V}_2$ then we have expressed all vectors using only $\vec{V}_1$ and $\vec{V}_2$! It is impossible for two vectors to span 3-space, so it must be impossible for three linearly dependent vectors to act as a basis.

We can express this same result visually by saying that $\vec{V}_1$, $\vec{V}_2$, and $\vec{V}_3$ all lie in a plane; they can be used only to build other vectors that lie in that same plane. But remember that the same result applies in higher dimensions that we cannot visualize.

The Determinant and the Inverse Matrix

In Section 6.6 we saw that the inverse of $\begin{pmatrix} a & c \\ b & d \end{pmatrix}$ is $\frac{1}{ad-bc}\begin{pmatrix} d & -c \\ -b & a \end{pmatrix}$. One immediate consequence of this formula is that if $ad - bc = 0$ then there is no inverse matrix. This principle generalizes to larger matrices.

> **What the Determinant tells us about the Inverse Matrix**
>
> If $|\mathbf{A}| = 0$ then $\mathbf{A}$ has no inverse matrix. If $|\mathbf{A}|$ is any non-zero value, then an inverse matrix $\mathbf{A}^{-1}$ exists.
>
> (Remember that this whole discussion applies to square matrices only. A non-square matrix has neither a determinant nor an inverse matrix.)

This result may seem unrelated to the sets of equations and vectors for which we've used the determinant before. Once again, though, the key is linear dependence. Recall how we used matrix multiplication to convert one vector basis to another:

$$\begin{array}{c} \\ \hat{i} \\ \hat{j} \end{array}\!\!\begin{array}{c} \vec{V}_1\ \vec{V}_2 \\ \begin{pmatrix} 1 & 2 \\ 3 & 0 \end{pmatrix} \end{array} \quad \begin{array}{c} \\ \vec{V}_1 \\ \vec{V}_2 \end{array}\!\!\begin{pmatrix} 10 \\ -7 \end{pmatrix} = \begin{array}{c} \\ \hat{i} \\ \hat{j} \end{array}\!\!\begin{pmatrix} -4 \\ 30 \end{pmatrix}$$

We interpret this equation as defining $\vec{V}_1 = \hat{i} + 3\hat{j}$ and $\vec{V}_2 = 2\hat{i}$, and then multiplying to find that $10\vec{V}_1 - 7\vec{V}_2$ is the same as $-4\hat{i} + 30\hat{j}$. The inverse matrix converts back from $\hat{i}\hat{j}$ representation to $\vec{V}_1\vec{V}_2$.

But what happens if $\vec{V}_1$ and $\vec{V}_2$ are linearly dependent, such as $\vec{V}_1 = \hat{i} + 3\hat{j}$ and $\vec{V}_2 = 2\hat{i} + 6\hat{j}$? Both lie on the line $y = 3x$, so any linear combination of them lies on that same line. Of course you can still use a matrix to convert any linear combination of those two vectors to an $\hat{i}\hat{j}$ representation, but going the other way is not possible. So the matrix $\begin{pmatrix} 1 & 2 \\ 3 & 6 \end{pmatrix}$ doesn't have an inverse. (Take a moment to check that its determinant is 0.)

The same conclusion holds in higher dimensions. If three vectors are linearly dependent they all lie on a plane, so you can't convert from a general 3D vector to a combination of those three vectors.

With that visual image in mind, you can generalize this to non-geometric transformations. If a small book has 100 pages and 50,000 words and a big book has 200 pages and 100,000 words, and we know that we have 1000 pages and 500,000 words, there's no way to tell how many of each kind of book we have. It could be 10 small books and no big books, 5 big books and no small books, or many other combinations. Because $<100, 50,0000>$ and $<200, 10,0000>$ are linearly dependent—that is, because one big book is exactly two small books—the transformation "small books and big books" to "pages and words" can't be inverted.

Application to the Three-Spring Problem

In Section 6.1 we set out to solve the differential equations:

$$\frac{d^2 x_1}{dt^2} = -4x_1 + 3x_2$$
$$\frac{d^2 x_2}{dt^2} = \frac{2}{3}x_1 - 3x_2 \tag{6.7.2}$$

As we explained in that section, we want to find "normal modes"—solutions in which the two balls oscillate at the same frequency, so $x_1 = C_1 \cos(\omega t)$ and $x_2 = C_2 \cos(\omega t)$. But these functions don't work unless you pick just the right frequency. For instance, we began with the simple case $\omega = 1$. Plugging $x_1 = C_1 \cos t$ and $x_2 = C_2 \cos t$ into Equations 6.7.2 gives

$$-C_1 = -4C_1 + 3C_2$$
$$-C_2 = \frac{2}{3}C_1 - 3C_2$$

You should be able to quickly convince yourself that these equations have only one solution, namely the "trivial" $C_1 = C_2 = 0$. Both balls sitting at rest forever is not the solution we want to find. Trying other frequencies leads to a similar dead end.

At this point in Section 6.1 we magically pulled the right frequencies out of a hat, but here's how you can find them for yourself. If you plug in the generic guess $x_1 = C_1 \cos(\omega t)$, $x_2 = C_2 \cos(\omega t)$ and gather like terms, you end up here.

$$\left(-4 + \omega^2\right) C_1 + 3C_2 = 0$$
$$\frac{2}{3}C_1 + \left(-3 + \omega^2\right) C_2 = 0 \tag{6.7.3}$$

These are homogeneous linear equations, so the trivial solution works for any ω value. For there to be any other solutions the equations must be linearly dependent, which means the determinant of the matrix of coefficients must be 0. That gives the equation $\left(-4 + \omega^2\right)\left(-3 + \omega^2\right) - 2 = 0$, which can be solved to give $\omega^2 = 2$ or 5.

Here's what we conclude from that little bit of math.

- For most values of ω Equations 6.7.3 have only one solution, which is $C_1 = C_2 = 0$, which corresponds to the springs sitting eternally at their relaxed positions. They cannot vibrate at those frequencies.
- For $\omega = \sqrt{2}$ and $\omega = \sqrt{5}$ Equations 6.7.3 are linearly dependent: they have *infinitely many* solutions. For instance, for $\omega = \sqrt{2}$ these equations both become $C_2 = (2/3)C_1$. If you pull back the first mass by any amount, and pull back the second mass 2/3 that far in the same direction, the system will oscillate with frequency $\sqrt{2}$.

Stepping Back

In this section we have talked about the idea of linear dependence in relation to equations, spatial vectors, and more general "vectors." In its most general form, any set of *things* is linearly dependent if one of those things can be written as a linear combination of others of those things. (You can even have a linearly dependent, or linearly independent, set of matrices!)

The determinant of a square matrix is a number you can calculate from the elements of that matrix. The determinant is zero if and only if the rows of the matrix are linearly dependent, or equivalently if the columns are linearly dependent. (It's possible to prove that either of those implies the other.) All of the applications of the determinant we've seen in this section are different uses for knowing whether a set of things is linearly dependent.

In this section the only thing we've done with the determinant is to ask whether or not it is zero. The determinant tells you more than that. For instance, when you consider a square matrix to be a transformation acting on a shape, the determinant of that square matrix is the factor by which the area or volume of that shape is multiplied. We will discuss this issue in Section 7.1.

If the determinant is zero, however, it doesn't tell you whether a set of equations is inconsistent or linearly dependent, how many of the variables are linearly independent if they all aren't, or what the solution to the equations is if there is a unique solution. In Section 7.4 (see felderbooks.com) we'll teach you a method called "row reduction" that can answer these questions and is more efficient than calculating large determinants.

6.7.3 Problems: Linear Dependence and the Determinant

6.125 Consider the following linear equations for x and y, with an unknown parameter λ.

$$2x + y = 0$$
$$3x + (5 - \lambda)y = 0$$

(a) Let $\lambda = 1$ and solve the resulting equations for x and y.

(b) Let $\lambda = 2$ and solve the resulting equations for x and y.

(c) Let $\lambda = 3$ and solve the resulting equations for x and y.
This is getting boring, isn't it?

(d) Show that for any value of λ, these equations will be solved by letting $x = y = 0$.

For these equations to have any *other* solution, they must be linearly dependent.

(e) Write the matrix of coefficients **C** for these two equations. Note that x and y are the variables, so they are not coefficients: your matrix will contain only numbers and the letter λ.

(f) Compute the determinant $|\mathbf{C}|$.

(g) Find the value of λ for which $|\mathbf{C}| = 0$.

(h) Substitute that value of λ into the original two equations, and show that the resulting equations are the same.

(i) Write three ordered pairs (x, y) that satisfy these equations using the value of λ that you found.

In Problems 6.126–6.130 find the given determinant by hand. (Of course there's nothing unethical about repeating the exercise on a computer to check your answer!)

6.126 $\begin{vmatrix} 3 & 8 \\ 4 & 2 \end{vmatrix}$

6.127 $\begin{vmatrix} 8 & 3 \\ 4 & 2 \end{vmatrix}$

6.128 $\begin{vmatrix} 1 & 2 & 3 \\ 4 & 5 & 6 \\ 7 & 8 & 9 \end{vmatrix}$

6.129 $\begin{vmatrix} -3 & 1 & 0 \\ 2 & 4 & 8 \\ 3 & 0 & 5 \end{vmatrix}$

6.130 $\begin{vmatrix} a & b & c \\ d & e & f \\ g & h & i \end{vmatrix}$

In Problems 6.131–6.138 you will be given a set of equations.

- Classify the set of equations as "all homogeneous" or "not all homogeneous."

- If the equations are all homogeneous, classify them as having a unique solution, or as being linearly dependent (infinitely many solutions).
- If the equations are not all homogeneous, classify them as having a unique solution, or as being "inconsistent or linearly dependent." (Section 7.4 at felderbooks.com discusses distinguishing these two cases.)

6.131 $3x - 4y = 7, 2x - 5y = 3$

6.132 $5x - 2y = 2, 15x - 6y = -3$

6.133 $x + 3y = 4, -2x - 6y = -8$

6.134 $x - y = 0, 3x + 3y = 0$

6.135 $x - 2y + 3z = 5, 2x - y + 3z = 3, -x + 4y - 2z = 1$

6.136 $x + 2y - z = 0, -2x + 3y + 2z = 0, x + y - z = 0$

6.137 💻 $-6x + 3y + 4z - 2q + 5r = 2, -x + 2y - z + q + 2r = -3, -4x + 2y + 3z - 2q - r = -1, 2x - 2y - z + 2q + r = 0, 5x - 5y - 3q = 2$

6.138 💻 $x + 2y + 3z + 4q + r = 0, 5x + 6y + 7z + 8r = 0, 9x + 10y + 11q + 12r = 0, 13x + 14z + 15q + 16r = 0, 17y + 18z + 19q + 20r = 0$

For Problems 6.139–6.143 find the value(s) of λ for which the given homogeneous equations have non-trivial solutions.

6.139 $\lambda x + y = 0, 3x - y = 0$

6.140 $(1 - \lambda)x - 2y = 0, 3x + (3 - \lambda)y = 0$

6.141 $\lambda x + y = 0, (-2 - \lambda^2)x - 3y = 0$

6.142 💻 $(2 - \lambda)x + 3y - 2z = 0, -x + (1 - \lambda)y + 2z = 0, 3x + 3y + (-1 - \lambda)z = 0$

6.143 💻 $(1 - \lambda)x + 2y + 3z + 4q + r = 0, 5x + (6 - \lambda)y + 7z + 8r = 0, 9x + 10y - \lambda z + 11q + 12r = 0, 13x + 14z + (15 - \lambda)q + 16r = 0, 17y + 18z + 19q + (20 - \lambda)r = 0$

6.144 The following is not technically a matrix because $\hat{i}, \hat{j}$ and $\hat{k}$ are not numbers, but let's pretend it is.

$$\begin{pmatrix} \hat{i} & \hat{j} & \hat{k} \\ A_x & A_y & A_z \\ B_x & B_y & B_z \end{pmatrix}$$

If A_x, A_y and A_z represent the three components of vector $\vec{A}$, and similarly for vector $\vec{B}$, show that the determinant of this "matrix" gives the cross product $\vec{A} \times \vec{B}$. (If you're comfortable with determinants, this winds up being a useful mnemonic for finding cross products.)

6.145 A shortcut for finding the determinant of a 3×3 matrix is illustrated below.

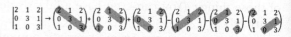

You consider six diagonals, including ones that wrap around from one side to the other. For each one you multiply the three numbers. Then you add the three "upper left to lower right" diagonals and subtract the "upper-right to lower left." So the determinant of the sample matrix shown above is

$$2 \cdot 3 \cdot 3 + 1 \cdot 1 \cdot 1 + 2 \cdot 0 \cdot 0 \\ - 2 \cdot 3 \cdot 1 - 1 \cdot 0 \cdot 3 - 2 \cdot 1 \cdot 0 = 13$$

(a) Use this trick to find the determinant of the matrix

$$\begin{pmatrix} 1 & 0 & -1 \\ 2 & 1 & 3 \\ 0 & -2 & 2 \end{pmatrix}$$

(b) Use this trick to find the determinant of the generic 3×3 matrix

$$\begin{pmatrix} a & b & c \\ d & e & f \\ g & h & i \end{pmatrix}$$

(c) Find the determinant of the generic 3×3 matrix given above using expansion by minors. Show that you get the same answer both ways.

Important note:

This trick only works for 3×3 matrices! See Problem 6.146.

6.146 *[This problem depends on Problem 6.145.]* Problem 6.145 showed a shortcut for finding determinants. In this problem you will show that the shortcut works *only* for 3×3 matrices.

(a) Show that if you apply this diagonal method to any 2×2 matrix, you get a determinant of zero.

(b) Show that if you apply this diagonal method to the following 4×4 matrix, you get the wrong answer.

$$\begin{pmatrix} 1 & 1 & 0 & 0 \\ 1 & 1 & 0 & 0 \\ 0 & 0 & 1 & 0 \\ 0 & 0 & 0 & 1 \end{pmatrix}$$

6.147 If two linear equations have different slopes, then they intersect in a unique point. If two linear equations have the same slope, then they represent either the same line (linearly dependent) or two parallel lines (inconsistent). With that in mind, consider the equations:

$$ax + by = r_1$$
$$cx + dy = r_2$$

Find the slopes of these two lines. Use the result to prove that if the determinant of the matrix of coefficients is zero, then the equations must be dependent or inconsistent.

6.148 In the Explanation (Section 6.7.2) we noted that two linear equations with two unknowns can be graphed as two lines on the xy-plane. The only three possibilities for the intersection of those two lines are that they are parallel (no intersection, inconsistent equations), they are the same line (infinitely many intersection points, linearly dependent equations), or they intersect at a single point. Put succinctly, their regions of intersection can be either nothing, a point, or a line. In this problem you will extend this logic to three equations with three unknowns, which can be graphed as three planes.

(a) First consider two planes. What are the three possibilities for their regions of intersection?

(b) Now consider the intersection of a third plane with each of the regions you described above. For example, one possibility for two planes is that they intersect on a line. So describe all the possible regions of intersection of the third plane with that line. Do the same with the other possible regions you found in Part (a). You should find four possibilities in all for the region of intersection of three planes.

(c) Using your answers to Part (b), explain why three linear equations with three unknowns must either have zero solutions (inconsistent), infinitely many solutions (linearly dependent), or one solution.

6.149 For each set of vectors shown below, say whether it can form a basis (either for a plane or a 3D space), and how you know. (The answer in each case only needs to be a few words long.)

(a) $\vec{V}_1 = 3\hat{i} + 4\hat{j}$, $\vec{V}_2 = 3\hat{i} - 4\hat{j}$
(b) $\vec{V}_1 = 3\hat{i} + 4\hat{j}$, $\vec{V}_2 = -6\hat{i} - 8\hat{j}$
(c) $\vec{V}_1 = 3\hat{i} + 4\hat{j} - \hat{k}$, $\vec{V}_2 = \hat{i} - \hat{j} + 7\hat{k}$, $\vec{V}_3 = 5\hat{i} + 9\hat{j} - 9\hat{k}$

6.150 For each set of vectors shown below, say whether it can form a basis, and how you know. (The answer in each case only needs to be a few words long.)

(a) $\vec{V}_1 = <1, 2, 3>$, $\vec{V}_2 = <4, 5, 6>$, $\vec{V}_3 = <1, 0, 1>$
(b) $\vec{V}_1 = <1, 2, 3, 4>$, $\vec{V}_2 = <-1, 3, -2, 0>$, $\vec{V}_3 = <2, 9, 7, 12>$, $\vec{V}_4 = <-2, 3, -1, 2>$
(c) $\vec{V}_1 = <1, 2, 0, -1, 3>$, $\vec{V}_2 = <1, 3, 0, 0, 2>$, $\vec{V}_3 = <2, -1, 3, 4, 1>$, $\vec{V}_4 = <6, -3, 0, 2, 1>$

6.151 Matrix $\mathbf{M} = \begin{pmatrix} 3 & -2 & 6 & 5 \\ 1 & 4 & 0 & 7 \\ 8 & 1 & 2 & -4 \\ -1 & -1 & 10 & 21 \end{pmatrix}$.

(a) Find $|\mathbf{M}|$.
(b) Find $\mathbf{M}^{-1}$.

6.152 $\mathbf{M}$ is a square matrix whose top row is all zeros. You should be able to answer the following questions without knowing anything else about $\mathbf{M}$.

(a) What does having all zeros in the top row tell you about the transformation that $\mathbf{M}$ performs?
(b) Explain, based on your answer to Part (a), why $\mathbf{M}$ cannot have an inverse.
(c) Using the algebraic rule for finding determinants, prove that $|\mathbf{M}| = 0$.

6.153 Matrix $\mathbf{Z} = \begin{pmatrix} 2 & 6 & 1 \\ 5 & 3 & -4 \\ a & b & c \end{pmatrix}$. Find non-zero numbers a, b, and c so that $\mathbf{Z}$ has no inverse.

6.154 A carbon nucleus has 6 protons and 6 neutrons. An oxygen nucleus has 8 protons and 8 neutrons.

(a) Write a matrix $\mathbf{X}$ that converts from "Number of carbon and oxygen nuclei" to "Number of protons and neutrons."
(b) A nuclear reaction involves 48 protons and 48 neutrons. Write at least two possible combinations of carbon and oxygen nuclei that could emerge from this reaction.
(c) Based on your answers to Parts (a) and (b), how can you predict without doing any calculations whether or not $\mathbf{X}$ is invertible?
(d) Calculate $|\mathbf{X}|$ and (hopefully) verify the prediction you made in Part (c).

In Problems 6.155–6.159 you will be shown a diagram illustrating a chemical reaction. For example, the reaction

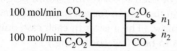

represents 100 mol/min each of CO_2 and C_2O_2 coming in and unknown rates of C_2O_6 and CO coming out. For each reaction…

- Write down a set of equations setting the total input rate of each type of atom equal to its output rate. In our example above, setting the input and output rates of carbon equal gives $100 + 2(100) = 2\dot{n}_1 + \dot{n}_2$ and setting the input and output rates of oxygen equal gives $2(100) + 2(100) = 6\dot{n}_1 + \dot{n}_2$.
- Rewrite the equations in matrix form and use a determinant to figure out if they have a unique solution.
- If they do have a unique solution find it. Otherwise simply state that the equations are either inconsistent (meaning the reaction is impossible) or linearly dependent (meaning the output rates can't be determined from the input rates). You don't need to say which of those two is the case—we'll teach you how to distinguish inconsistent from linearly dependent equations in Section 7.4 (see felderbooks.com)—but you should note that the equations are "either inconsistent or linearly dependent." Our example above produces 25 mol/min of C_2O_6 and 250 mol/min of CO.

These are not necessarily realistic reactions, which would require uglier math than we want to get into in these problems.

6.155

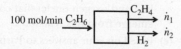

6.156

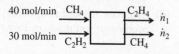

6.157

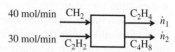

6.158

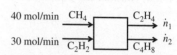

6.159

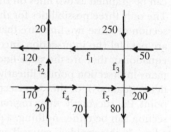

6.160 The diagram below shows a map of one way streets. The numbers and letters represent known and unknown rates of traffic flow, measured in cars per hour.

(a) For each intersection, write an equation that says "The total rate of cars coming in equals the total rate of cars going out."

(b) Is there a unique solution for all of the unknown flow rates? Explain how you got your answer, including what you asked a computer or calculator to calculate for you and what answer it gave.

6.161 Following the method outlined in the Explanation (Section 6.7.2), find the normal modes of the system:

$$\frac{d^2 x_1}{dt^2} = -4x_1 - 4x_2$$
$$\frac{d^2 x_2}{dt^2} = -x_1 - 4x_2$$

You may assume that the normal modes are sinusoidal. (You'll verify that assumption by finding them.) You may also assume the initial velocities are zero so you can use only cosines and no sines.

6.162 [*This problem depends on Problem 6.161.*] For the system in Problem 6.161 find the solution that matches the initial conditions $x_1(0) = 7.5$, $x_2(0) = 6.25$. *Hint*: you can do this with simple algebra, but it's faster with an inverse matrix.

6.8 Eigenvectors and Eigenvalues

In Section 6.4 we discussed using vectors other than $\hat{i}$ and $\hat{j}$ as a basis, and we said that a given transformation has a "natural basis" that makes it easy to work with. If the transformation is represented as a matrix, that natural basis is defined by the "eigenvectors" of the matrix. In this section we will see what eigenvectors are, how to find them, and (most importantly) how to use them to solve problems.

6.8.1 Discovery Exercise: Eigenvectors and Eigenvalues

Matrix **S** is a point matrix: $\mathbf{S} = \begin{pmatrix} 0 & 1 & 1 & 0 & 0 \\ 0 & 0 & 1 & 1 & 0 \end{pmatrix}$.

1. Draw the shape that matrix **S** describes.

Matrix **B** represents a transformation: $\mathbf{B} = \begin{pmatrix} 3 & 0 \\ 0 & 5 \end{pmatrix}$.

2. Find the matrix **BS** and draw the resulting shape.
3. Describe in words: what effect does matrix **B** have on this (or any other) shape?
   See Check Yourself #43 in Appendix L

Matrix **T** represents a different transformation: $\mathbf{T} = \begin{pmatrix} 3 & 1 \\ 3 & 5 \end{pmatrix}$.

4. Find the matrix **TS** and draw the resulting shape.

It's not at all clear what **T** does, is it? It seems to move everything around almost randomly! But let's try it on a new shape: $\mathbf{C} = \begin{pmatrix} 0 & -1 & 0 & 1 & 0 \\ 0 & 1 & 4 & 3 & 0 \end{pmatrix}$.

5. Find the matrix **TC**.
6. Draw the shapes described by matrix **C** and matrix **TC**.
7. For this particular shape, it should be clear what effect **T** had—and in fact, it has the same effect on any shape. (If you sketch this too casually it might not be clear, but if you graph it carefully on graph paper or a computer you should be able to see it.) Fill in the blanks in the following sentence:

   Matrix **T** stretches a shape by a factor of ____ in the direction _____, and it also stretches a shape by a factor of ____ in the direction _____.

6.8.2 Explanation: Eigenvectors and Eigenvalues

We are now ready to put in place the last piece we need for a proper analysis of the three-spring problem from Section 6.1. In our initial treatment of that problem we found that the right initial conditions led to simple behavior called normal modes. Other initial conditions led to more complicated behavior, but we could describe that behavior by combining the normal modes.

When we write the problem with matrices, we can describe those normal modes as the "eigenvectors"—the natural basis—of a transformation matrix. We will begin with a visual description of eigenvectors, and end by applying them to the three-spring problem.

Throughout this section we will focus on 2×2 matrices but the concepts scale up to higher dimensional square matrices.

A Visual Look at Eigenvectors

Figure 6.6 shows the effect of the transformation matrix $\mathbf{T} = \begin{pmatrix} 2 & 3 \\ 4 & 1 \end{pmatrix}$ on three different vectors. Look at them for a second and try to summarize visually what $\mathbf{T}$ does.

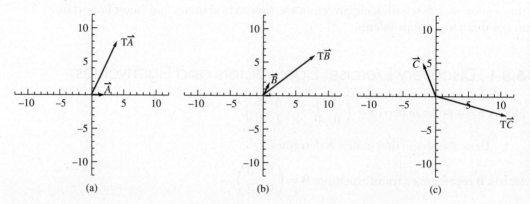

FIGURE 6.6 The effect of matrix $\mathbf{T}$ on the vectors $2\hat{i}$ (a), $\hat{i} + 2\hat{j}$ (b), and $-2\hat{i} + 5\hat{j}$ (c).

It apparently make things bigger, but beyond that it's hard to say much. $\mathbf{T}$ seems to stretch things out and turn them around randomly. But let's look at the effect that $\mathbf{T}$ has on two other (carefully chosen) vectors.

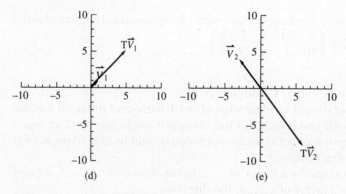

FIGURE 6.7 The effect of matrix $\mathbf{T}$ on the vectors $\vec{V}_1 = \hat{i} + \hat{j}$ (d) and $\vec{V}_2 = -3\hat{i} + 4\hat{j}$ (e).

Those two pictures are more important than you might suppose, so make sure you see what they tell us.

- When $\mathbf{T}$ acts on $\vec{V}_1 = \hat{i} + \hat{j}$ it stretches it by a factor of 5, and does not change its direction. (You don't have to take our word for that. We gave you $\mathbf{T}$ above, so you can multiply it by $\hat{i} + \hat{j}$ and see what you get.) Mathematically we write $\mathbf{T}\vec{V}_1 = 5\vec{V}_1$.
- When $\mathbf{T}$ acts on $\vec{V}_2 = -3\hat{i} + 4\hat{j}$ it stretches it by a factor of 2 and reverses its direction. Mathematically we write $\mathbf{T}\vec{V}_2 = -2\vec{V}_2$.

And so what? Well, remember how we approached the three-spring problem in Section 6.1. We found two special initial conditions for which we could easily predict the behavior of the springs. That gave us a complete understanding of the system because we could write any other initial condition as a combination of those two. The situation here is perfectly analogous to that one. We have two special vectors for which we can easily describe the effect of matrix **T**. That gives us a complete understanding of the transformation because we can describe any other vector as a combination of those two.

The two special initial conditions for the three-spring problem were called the "normal modes" of that system. The two special vectors for **T** are called the "eigenvectors" of that transformation.

Definition: Eigenvectors and Eigenvalues

Given a square matrix **M**, the "eigenvalue equation" for the matrix is:

$$\mathbf{M}\mathbf{v} = \lambda \mathbf{v} \tag{6.8.1}$$

where **v** is a vector and λ is a scalar. For any non-zero solution to this equation, **v** is an "eigenvector" of **M** and λ is an "eigenvalue."

(The phrase "non-zero" is important here. Equation 6.8.1 will *always* be satisfied if **v** is all zeros, but that doesn't count.)

Equation 6.8.1 does not imply that there is anything particularly special about the matrix **M**. It says that for almost any square matrix, there are a few vectors that the matrix will act on in this simple way. (The prefix "eigen" comes from the German for "one's own"; every matrix has a few vectors of its own.)

We said that the eigenvectors give you a complete understanding of a particular transformation. Let's see how that works for the transformation **T** we described above.

EXAMPLES   **Eigenvectors and Eigenvalues**

In these examples the matrix **T** and the vectors $\vec{V}_1$ and $\vec{V}_2$ are as defined above.

Problem:
Multiply **T** times the vector $\vec{V}_1$.

Solution:
$\mathbf{T}\vec{V}_1 = 5\vec{V}_1$. This mathematical equation expresses the same idea that we expressed in words above: **T** stretches $\vec{V}_1$ by 5 while leaving its direction unchanged. This defines $\vec{V}_1$ as an eigenvector of **T** with eigenvalue 5.

Problem:
Multiply **T** times the vector $\vec{V}_2$.

Solution:
$\mathbf{T}\vec{V}_2 = -2\vec{V}_2$. The negative sign indicates the reversal of direction. So we see that $\vec{V}_2$ is also an eigenvector of **T**, with an eigenvalue of -2.

Problem:
Multiply **T** times the vector $\vec{D} = 10\vec{V}_1$.

Solution:
$\mathbf{T}\left(10\vec{V}_1\right) = 10\left(\mathbf{T}\vec{V}_1\right) = 10\left(5\vec{V}_1\right) = 50\vec{V}_1$. Notice that the answer is $5\vec{D}$. This tells us that $\vec{D}$ is an eigenvector of $\mathbf{T}$ with an eigenvalue of 5.

Problem:
Multiply $\mathbf{T}$ times the vector $\vec{E} = 10\vec{V}_1 + 7\vec{V}_2$.

Solution:
$\mathbf{T}\left(10\vec{V}_1 + 7\vec{V}_2\right) = 10\left(\mathbf{T}\vec{V}_1\right) + 7\left(\mathbf{T}\vec{V}_2\right) = 10\left(5\vec{V}_1\right) + 7\left(-2\vec{V}_2\right) = 50\vec{V}_1 - 14\vec{V}_2$. The product is not a scalar multiple of $\vec{E}$, so $\vec{E}$ is not an eigenvector of $\mathbf{T}$—the transformation changed the direction as well as the magnitude of $\vec{E}$. But we were still able to perform the multiplication easily because $\vec{E}$ was expressed as a combination of our eigenvectors.

Problem:
Multiply $\mathbf{T}$ times the vector $\vec{F} = -11\hat{i} + 24\hat{j}$.

Solution:
This problem doesn't seem to have anything to do with our eigenvectors. But $\vec{V}_1$ and $\vec{V}_2$ form a basis; any vector can be expressed in terms of them, if we want to. In this case, a bit of work (using an inverse matrix for instance) tells us that $\vec{F} = 4\vec{V}_1 + 5\vec{V}_2$. In that form we can do the multiplication trivially and find that $\mathbf{T}\vec{F} = 20\vec{V}_1 - 10\vec{V}_2$.

Vector $\vec{D}$ above imparts a lesson that you don't want to miss. We saw that because $\vec{V}_1$ is an eigenvector with eigenvalue 5, $10\vec{V}_1$ is also an eigenvector with eigenvalue 5. More generally, any scalar multiple of an eigenvector is itself an eigenvector, with the same eigenvalue. You will often hear that a 2×2 matrix has two eigenvectors, but this is not strictly accurate: it generally has two *linearly independent* eigenvectors, or two special directions, each with its own eigenvalue.

The last example above (vector $\vec{F}$) shows why eigenvectors are so important. You can express any vector as a linear combination of the eigenvectors of a particular transformation. When a vector is so expressed, you can easily determine the effect of that transformation on that vector. This is called the "natural basis" for the transformation, and we shall return to it below.

Finding Eigenvectors and Eigenvalues

When you need the eigenvalues and eigenvectors of a matrix you will generally ask a computer to give them to you. But there are important insights to be gained in finding them by hand a few times, and you already know all the key steps in the process.

It begins, as you might expect, by writing the eigenvalue equation for the matrix. When you set out to solve that equation, you immediately find that one solution is a vector full of zeros, which doesn't count. You can only have *other* solutions if the equations are linearly dependent, so we set the determinant of the matrix of coefficients equal to zero and use that to solve for the eigenvalue λ. Finally, plug each eigenvalue back into the linear equations to find the eigenvectors.

As usual, an example will illustrate this more clearly than an explanation.

EXAMPLE   Finding Eigenvectors and Eigenvalues

Question: Find the eigenvectors and eigenvalues of the matrix $\begin{pmatrix} 9 & 4 \\ 3 & 8 \end{pmatrix}$.

Answer: By definition the eigenvectors and eigenvalues satisfy the eigenvalue equation

$$\begin{pmatrix} 9 & 4 \\ 3 & 8 \end{pmatrix} \begin{pmatrix} A \\ B \end{pmatrix} = \lambda \begin{pmatrix} A \\ B \end{pmatrix}$$

Expanding this out and gathering like terms

$$\begin{aligned} (9 - \lambda) A + 4B &= 0 \\ 3A + (8 - \lambda) B &= 0 \end{aligned} \qquad (6.8.2)$$

One solution is obviously $A = B = 0$. These linear equations can only have *more* solutions if they are linearly dependent, which is to say, the determinant of the matrix of coefficients must be zero.

$$\begin{vmatrix} 9 - \lambda & 4 \\ 3 & 8 - \lambda \end{vmatrix} = (9 - \lambda)(8 - \lambda) - (4)(3) = 0$$

This is a quadratic equation with solutions $\lambda = 12$ and $\lambda = 5$, our two eigenvalues.

Now, plug $\lambda = 12$ into our equations for A and B and both equations reduce to the same relationship: $4B = 3A$. That tells us that $\begin{pmatrix} 4 \\ 3 \end{pmatrix}$, *or any multiple thereof,* is the eigenvector that corresponds to this eigenvalue. (Remember that any multiple of an eigenvector is another eigenvector.)

We'll leave it to you to find the eigenvector that corresponds to $\lambda = 5$.

For almost any 2×2 matrix the process looks like that example. It leads to a quadratic equation, meaning that we are setting a second-order polynomial—the "characteristic polynomial" of the matrix—equal to zero. A 3×3 matrix has a third-order characteristic polynomial, and so on. Since we know a great deal about the roots of polynomials, we can say a great deal about the eigenvalues of a matrix.

- Most $n \times n$ matrices will have n eigenvalues, corresponding to n linearly independent eigenvectors. This is what we would hope for, since we want to use the resulting vectors as the basis for an n-dimensional space!
- It is possible in some cases to have fewer eigenvalues than n, but never more.
- It is also possible for some eigenvalues to be complex. When a real matrix has complex eigenvalues, they will come in complex conjugate pairs.

Diagonalizing a Matrix

Section 6.4 discussed using different basis vectors to represent a space. When you switch from the $\hat{i}\hat{j}$ basis to the $\vec{V}_1 \vec{V}_2$ basis, all your vectors look the same (the same little arrows on a graph), but the numbers used to represent those vectors are different. All your transformations are the same but the matrices you use to represent those transformations are different.

In the table below, $\vec{V}_1 = \hat{i} + \hat{j}$ and $\vec{V}_2 = -3\hat{i} + 4\hat{j}$.

This object in the $\hat{i}\hat{j}$ basis…	…is represented in the $\vec{V}_1 \vec{V}_2$ basis as
The vector $11\hat{i} - 3\hat{j}$ or $\begin{pmatrix} 11 \\ -3 \end{pmatrix}$	$5\vec{V}_1 - 2\vec{V}_2$ or $\begin{pmatrix} 5 \\ -2 \end{pmatrix}$
The eigenvector $-3\hat{i} + 4\hat{j}$ or $\begin{pmatrix} -3 \\ 4 \end{pmatrix}$, with eigenvalue -2	$\vec{V}_2$ or $\begin{pmatrix} 0 \\ 1 \end{pmatrix}$, still with eigenvalue -2
The transformation matrix $\mathbf{T} = \begin{pmatrix} 2 & 3 \\ 4 & 1 \end{pmatrix}$	$\mathbf{D} = \begin{pmatrix} 5 & 0 \\ 0 & -2 \end{pmatrix}$

We could make a table like that for any two bases, and include any vectors and transformations we like. The last row of the table shows that **T** and **D** are "similar matrices": they represent the same transformation in two different bases.

But we didn't choose that particular basis arbitrarily. We chose it because $\vec{V}_1$ and $\vec{V}_2$ are the eigenvectors of the transformation we called **T**. The eigenvectors of a matrix form the "natural basis" for the transformation effected by that matrix, because in that basis the transformation is easy to work with.

Definition: Natural Basis and Diagonalization

The eigenvectors of a transformation define the "natural basis" for that transformation. When you express everything in the natural basis, two notable simplifications happen.

1. The eigenvectors of the transformed matrix are $\begin{pmatrix} 1 \\ 0 \end{pmatrix}$ and $\begin{pmatrix} 0 \\ 1 \end{pmatrix}$. We saw this in the table above, and it makes perfect sense: in the basis defined by the eigenvectors, each eigenvector is "1 of this basis vector and none of the other basis vector(s)."
2. The transformation itself, expressed in its own natural basis, is a "diagonal matrix." (Didn't remember that term, did you? We defined it all the way back in Section 6.2.) For matrix **D** in the above table the numbers along the principal diagonal are the eigenvalues, and all the other numbers are zeros. For this reason, translating to the natural basis is often called "diagonalizing" the matrix.

Why does working in the natural basis give you a diagonal matrix? And why is it such a big deal to have a diagonal matrix anyway?

Well, remember what eigenvectors mean. When a transformation acts on an eigenvector, it multiplies it by a scalar (the eigenvalue). When a transformation acts on a different vector that is represented as a linear combination of the eigenvectors—that is, in the basis of the eigenvectors—it multiplies each component by the associated eigenvalue. And that's exactly what happens when you multiply a diagonal matrix (below we use λ_1 and λ_2 for the eigenvalues) by an arbitrary point (x, y).

$$\begin{pmatrix} \lambda_1 & 0 \\ 0 & \lambda_2 \end{pmatrix} \begin{pmatrix} x \\ y \end{pmatrix} = \begin{pmatrix} \lambda_1 x \\ \lambda_2 y \end{pmatrix}$$

In general, when you multiply a matrix by a vector, the new *x*-component depends on both the old *x*-component and the old *y*-component. But when the matrix is diagonal, the new x-component depends only on the old x-component and the new y depends only on the old y. That is why diagonal matrices are so important: they "decouple" the variable dependencies.

In Section 6.9 we will see that diagonalization is ultimately the right way to look at the three-spring problem. The original problem is two coupled differential equations, meaning that x_1'' depends on both x_1 and x_2, and so does x_2''. When you rewrite the entire problem in the natural basis of the transformation matrix the equations decouple. We'll define new variables f and g that are the amplitudes of the normal modes and write differential equations for them: f'' will depend only on f and g'' will depend only on g, so we will have two ordinary differential equations that we know how to solve.

Below we take a less ambitious approach. But as you will see, the eigenvectors are still the key to finding the normal modes.

Application to the Three-Spring Problem

The three-spring problem began with the following two differential equations.

$$\frac{d^2 x_1}{dt^2} = -4x_1 + 3x_2$$
$$\frac{d^2 x_2}{dt^2} = \frac{2}{3}x_1 - 3x_2$$

The positions of the balls form a vector: not a vector in the sense of "a quantity with magnitude and direction" but a vector in the more general sense of "a variable that must be represented with two numbers." And the equations are based on linear combinations of these positions. All this suggests rewriting these two equations as one matrix equation:

$$\begin{pmatrix} \ddot{x}_1 \\ \ddot{x}_2 \end{pmatrix} = \begin{pmatrix} -4 & 3 \\ 2/3 & -3 \end{pmatrix} \begin{pmatrix} x_1 \\ x_2 \end{pmatrix} \quad (6.8.3)$$

The right-hand matrix in that equation is a 2×1 vector that represents *the state of the system:* a description of what the system looks like at any given time and, ultimately, the function we hope to solve for. The middle matrix represents a transformation that acts upon this state.

Recall from Section 6.1 that our entire solution was based on normal modes, and that each normal mode was defined by two properties: a particular ratio of x_2 to x_1, and a frequency of oscillation. Now that you know how to work with matrices, we can tell you in four words how to find the normal modes. **Just find the eigenvectors!**

How to Find the Normal Modes of the Three-Spring Problem

Every eigenvector of the transformation matrix represents a normal mode of the system. The frequency of oscillation for that normal mode is the square root of the absolute value of the associated eigenvalue.

For instance, one eigenvector of the matrix $\begin{pmatrix} -4 & 3 \\ 2/3 & -3 \end{pmatrix}$ is $\begin{pmatrix} 1 \\ 2/3 \end{pmatrix}$ with eigenvalue -2. (You can find this and the other eigenvector by hand, as discussed above, or by computer.) This particular state means $x_1 = 1$ and $x_2 = 2/3$ but remember that any multiple of an eigenvector is also an eigenvector, so $\begin{pmatrix} A \\ (2/3)A \end{pmatrix}$ represents a normal mode of the system for any A. Physically that means that if $x_2 = (2/3)x_1$ then the system is in that normal mode and will maintain that ratio forever. Because the eigenvalue is -2 this normal mode oscillates with frequency $\sqrt{2}$. The arbitrary A means this normal mode can occur with any amplitude.

But why do the eigenvectors of the matrix give us the normal modes of the system? The answer follows directly from the definition of an eigenvector. Look what happens if we plug $x_2 = (2/3)x_1$ into Equation 6.8.3.

$$\begin{pmatrix} \ddot{x}_1 \\ (2/3)\ddot{x}_1 \end{pmatrix} = \begin{pmatrix} -4 & 3 \\ 2/3 & -3 \end{pmatrix} \begin{pmatrix} x_1 \\ (2/3)x_1 \end{pmatrix} = \begin{pmatrix} -2x_1 \\ -(4/3)x_1 \end{pmatrix} \tag{6.8.4}$$

Equation 6.8.4 provides the vital link between eigenvectors and normal modes, so be sure to follow it in three ways.

1. First follow the algebra. Look what substitutions we made in Equation 6.8.3, and why. Do the matrix multiplication to make sure you get what we got.
2. Second, notice that when we multiplied the transformation matrix by the vector, we got the same vector multiplied by -2. That's what eigenvectors do, and it's why they are the key to this whole process.
3. Finally, notice that we have *decoupled* the two variables. That is, we now have the ordinary differential equation $\ddot{x}_1 = -2x_1$. (In fact, we have it twice; as you will see in Problem 6.190, if you start with the wrong ratio, you end up with contradictory equations.) And of course we have the promise that $x_2 = (2/3)x_1$, so one solution gives us both variables.

We can write down the solutions by inspection. They are of course a mix of sines and cosines, but as we did in Section 6.1 we're going to leave out the sine terms so both balls start at zero velocity. With that restriction, the solution to $\ddot{x}_1 = -2x_1$ is $x_1 = A\cos(\sqrt{2}\, t)$. Remember this equation is only valid when $x_2 = (2/3)x_1$, so it can be written

$$\begin{pmatrix} x_1 \\ x_2 \end{pmatrix} = \begin{pmatrix} A\cos(\sqrt{2}\, t) \\ (2/3)A\cos(\sqrt{2}\, t) \end{pmatrix} \quad \text{The first normal mode solution}$$

In a similar way we can plug in the eigenvector $x_2 = -(1/3)x_1$ to get the equation $\ddot{x}_1 = -5x_1$, which leads to

$$\begin{pmatrix} x_1 \\ x_2 \end{pmatrix} = \begin{pmatrix} B\cos(\sqrt{5}\, t) \\ -(1/3)B\cos(\sqrt{5}\, t) \end{pmatrix} \quad \text{The second normal mode solution}$$

We discussed in Section 6.1 how to combine these normal modes to match initial conditions. There is no need to repeat that here.

Summing up, if we write a set of coupled, linear differential equations as a matrix equation, the eigenvectors of the matrix tell us the normal modes of the system, and the eigenvalues tell us the frequencies of those normal modes. Once we have those normal modes, we can write arbitrary initial conditions as a sum of normal modes and thus find their behavior. All of this is parallel to the use of eigenvectors in geometric transformations.

6.8 | Eigenvectors and Eigenvalues

Geometric Transformation	Three-Spring Problem
If the components of a vector happen to be in precisely the ratio of an eigenvector of the transformation matrix, then the transformation produces a very simple result: it multiplies both components by the same scalar.	If the initial positions of two objects happen to be in precisely the ratio of an eigenvector of the differential equation matrix, then the differential equation produces a very simple result: it multiplies both initial conditions by the same cosine function.
If the vector components are not in the ratio of an eigenvector, then you rewrite the vector as a combination of eigenvectors. You then act on each eigenvector independently (by just multiplying), and combine the results.	If the initial positions are not in the ratio of an eigenvector, then you rewrite them as a combination of eigenvectors. You then act on each eigenvector independently (by just multiplying), and combine the results.

6.8.3 Problems: Eigenvectors and Eigenvalues

6.163 The eigenvectors of the matrix $\mathbf{T} = \begin{pmatrix} -1 & 8 \\ 0 & 5 \end{pmatrix}$ include $\vec{V}_1 = \begin{pmatrix} 4 \\ 3 \end{pmatrix}$ with eigenvalue $\lambda_1 = 5$ and $\vec{V}_2 = \begin{pmatrix} -1 \\ 0 \end{pmatrix}$ with eigenvalue $\lambda_2 = -1$.

(a) Confirm that $\vec{V}_1$ works as an eigenvector, and that its eigenvalue is λ_1.

That question may have taken you a minute or two, but the questions below should take no more than (say) fifteen seconds each.

(b) $\mathbf{T}\left(10\vec{V}_1\right) =$

(c) $\mathbf{T}\left(10\vec{V}_2\right) =$

(d) $\mathbf{T}\left(6\vec{V}_1 - 5\vec{V}_2\right) =$

(e) $\mathbf{T}\left(a\vec{V}_1 - b\vec{V}_2\right) =$

6.164 *[This problem depends on Problem 6.163.]* In this problem you will use $\vec{V}_1$ and $\vec{V}_2$, the eigenvectors of $\mathbf{T}$ from Problem 6.163, as a basis for representing vectors and transformations.

(a) Express the vector $\vec{V}_1$ in the $\vec{V}_1\vec{V}_2$ basis.

(b) Express the vector $-4\vec{V}_2$ in the $\vec{V}_1\vec{V}_2$ basis.

(c) Express transformation $\mathbf{T}$ as a matrix in the $\vec{V}_1\vec{V}_2$ basis.

(d) Multiply the matrix you wrote in Part (c) by the vector $\begin{pmatrix} a \\ b \end{pmatrix}$. Explain what we learn from your results.

6.165 Consider the matrix $\mathbf{M} = \begin{pmatrix} 2 & 1 \\ 3 & 4 \end{pmatrix}$ as a transformation acting on vectors.

(a) Find $\mathbf{M}\vec{A}$ where $\vec{A} = \begin{pmatrix} 1 \\ 1 \end{pmatrix}$. Is the result a scalar multiple of $\vec{A}$? If it is not, indicate that $\vec{A}$ is not an eigenvector of $\mathbf{M}$. If it is, indicate that $\vec{A}$ is an eigenvector of $\mathbf{M}$, and find its associated eigenvalue. (This last step should require no work that you have not already done.)

(b) Repeat Part (a) for the vector $\vec{B} = \begin{pmatrix} 1 \\ -1 \end{pmatrix}$.

(c) Repeat Part (a) for the vector $\vec{C} = \begin{pmatrix} 1 \\ 3 \end{pmatrix}$.

(d) Repeat Part (a) for the vector $\vec{D} = \begin{pmatrix} 4 \\ 1 \end{pmatrix}$.

6.166 Determine whether each vector is an eigenvector of $\begin{pmatrix} 2 & 6 \\ 2 & 3 \end{pmatrix}$. If it is an eigenvector, indicate its eigenvalue.

(a) $2\hat{\imath} - \hat{\jmath}$

(b) $3\hat{\imath} + 2\hat{\jmath}$

(c) $5\hat{\imath} - \hat{\jmath}$

(d) $3\hat{\imath} + 4\hat{\jmath}$

(e) $6\hat{\imath} + 4\hat{\jmath}$

6.167 *[This problem depends on Problem 6.166.]* A 2×2 matrix will have two linearly independent eigenvectors: sometimes less, but never more. Why did you appear to find three in Problem 6.166?

6.168 The figure shows vector $\vec{V}$, and the results of four different transformations of vector $\vec{V}$.

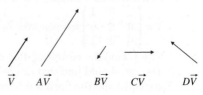

For each transformation, indicate whether $\vec{V}$ is an eigenvector of the transformation. If so, estimate its eigenvalue.

6.169 Matrix $S = \begin{pmatrix} 2 & 1 \\ 1 & 2 \end{pmatrix}$ represents a transformation.

  (a) Matrix $P_1 = \begin{pmatrix} 0 & 1 & 1 & 0 & 0 \\ 0 & 0 & 1 & 1 & 0 \end{pmatrix}$. Draw the figures represented by P_1 and SP_1.

  (b) Matrix $P_2 = \begin{pmatrix} 0 & 2 & 1 & -1 & 0 \\ 0 & 2 & 3 & 1 & 0 \end{pmatrix}$. Draw the figures represented by P_2 and SP_2.

  (c) Describe in words the effect of matrix S: it stretches any figure how far, in what direction?

  (d) Which shape made it easier to see the effect of matrix S? Why?

6.170 The matrix $\begin{pmatrix} 5 & 1 \\ 6 & 4 \end{pmatrix}$ has two eigenvalues.

  (a) One eigenvector is $\vec{v}_1 = \begin{pmatrix} 1 \\ 2 \end{pmatrix}$. Find the associated eigenvalue λ_1.

  (b) The other eigenvalue is $\lambda_2 = 2$. Find the associated eigenvector $\vec{v}_2$.

6.171 The matrix $\begin{pmatrix} 7 & -2 \\ 5 & 1 \end{pmatrix}$ consists entirely of real numbers, but its eigenvectors and eigenvalues are complex.

  (a) One eigenvector is $\vec{v}_1 = \begin{pmatrix} 3+i \\ 5 \end{pmatrix}$. Find the associated eigenvalue λ_1.

  (b) The other eigenvalue is $\lambda_2 = 4 - i$. Find the associated eigenvector $\vec{v}_2$.

6.172 What are the eigenvectors of the identity matrix I? Why?

6.173 We've said that if $\vec{v}$ is an eigenvector of M with eigenvalue λ then any multiple of $\vec{v}$ is also an eigenvector with the same eigenvalue.

  (a) Show that if two vectors $\vec{v}_1$ and $\vec{v}_2$ are both eigenvectors of M with the same eigenvalue λ, then any linear combination of $\vec{v}_1$ and $\vec{v}_2$ is as well.

  (b) As a trivial example, what are all of the eigenvectors of the 2×2 identity matrix and what are their eigenvalues?

  (c) As a less trivial example, the matrix
  $T = \begin{pmatrix} 2 & 0 & 1 \\ 0 & 2 & 0 \\ 0 & 0 & 1 \end{pmatrix}$ has eigenvectors $\hat{i}, \hat{j}$, and $-\hat{i} + \hat{k}$. Find the corresponding eigenvalues. You should find that two of them are equal and the third is different.

  (d) What are all the eigenvectors of T and what are their eigenvalues?

6.174 Walk-Through: Finding Eigenvectors and Eigenvalues. In this problem you're going to find the eigenvectors and eigenvalues of the matrix $M = \begin{pmatrix} 9 & 6 \\ 8 & 1 \end{pmatrix}$.

  (a) Write the eigenvalue equation $Mv = \lambda v$ for this matrix, using the unknowns x and y for the elements of the eigenvector, and multiply it out to two separate algebra equations.

  (b) Collect like terms until both equations have the form $ax + by = 0$.

  Note that $x = y = 0$ is a valid solution to these equations. This represents the "trivial" solution that doesn't count as an eigenvector.

  (c) For these two linear equations to have more solutions, they must be linearly dependent: that is, the determinant of their matrix of coefficients must be zero. Write the resulting equation for λ.

  (d) Solve for λ. The result should give you two eigenvalues.

  (e) For each eigenvalue, find the eigenvector:
   i. Plug the eigenvalue into one of the equations of the form $ax + by = 0$. Solve to find a relationship between x and y.
   ii. Check that the other equation gives you the same relationship. That tells you that you correctly found an eigenvalue, and that any vector (x, y) that satisfies that relationship is an eigenvector with this eigenvalue.
   iii. Write an eigenvector $\begin{pmatrix} x \\ y \end{pmatrix}$ that satisfies the relationship you found.
   iv. Confirm that your eigenvalue and eigenvector satisfy the eigenvalue equation for this matrix.

In Problems 6.175–6.178, find the eigenvectors and eigenvalues of the given matrix. It may help to first work through Problem 6.174 as a model.

6.175 $\begin{pmatrix} 2 & 6 \\ 4 & 7 \end{pmatrix}$

6.176 $\begin{pmatrix} 3 & 3 \\ 2 & 8 \end{pmatrix}$

6.177 $\begin{pmatrix} 4 & 2 \\ 6 & 5 \end{pmatrix}$

6.178 $\begin{pmatrix} 6 & 2 \\ 4 & 5 \end{pmatrix}$

6.8 | Eigenvectors and Eigenvalues

6.179 Explain why we cannot use the techniques in this section to find a natural basis for the matrix $\begin{pmatrix} 6 & -1 \\ 4 & 2 \end{pmatrix}$. *Hint*: try it!

6.180 The matrix $\mathbf{T} = \begin{pmatrix} 4 & 2 \\ 9 & 1 \end{pmatrix}$ has eigenvectors $\vec{V}_1 = \begin{pmatrix} 2 \\ 3 \end{pmatrix}$ with eigenvalue $\lambda_1 = 7$, and $\vec{V}_2 = \begin{pmatrix} 1 \\ -3 \end{pmatrix}$ with eigenvalue $\lambda_2 = -2$.

(a) Find the $\hat{i}$- and $\hat{j}$-components of the vector $5\vec{V}_1 + 4\vec{V}_2$.

(b) Write a matrix $\mathbf{B}$ that will convert any vector from a $\vec{V}_1\vec{V}_2$ representation to an $\hat{i}\hat{j}$ representation.

(c) Find the $\vec{V}_1$- and $\vec{V}_2$-components of the vector $5\hat{i} + 4\hat{j}$.

(d) Evaluate $\mathbf{T}\left(5\vec{V}_1 + 4\vec{V}_2\right)$. *Hint*: work entirely in the $\vec{V}_1\vec{V}_2$ basis without converting anything to the $\hat{i}\hat{j}$ basis.

(e) Matrix $\mathbf{T}$ represents a particular transformation on vectors in the $\hat{i}\hat{j}$ basis. Write a different matrix that performs the same transformation on vectors in the $\vec{V}_1\vec{V}_2$ basis. (*Hint*: think about what happened in Part (d)!)

6.181 The matrix $\mathbf{T} = \begin{pmatrix} 5 & 1 \\ 6 & 6 \end{pmatrix}$ has eigenvector $\vec{V}_1 = \begin{pmatrix} 1 \\ 3 \end{pmatrix}$ with eigenvalue $\lambda_1 = 8$ and eigenvector $\vec{V}_2 = \begin{pmatrix} -1 \\ 2 \end{pmatrix}$ with $\lambda_2 = 3$.

(a) Matrix $\mathbf{T}$ performs a certain transformation on any vector expressed in the form $x\hat{i} + y\hat{j}$. Write a different matrix that performs the same transformation on any vector expressed in the form $a\vec{V}_1 + b\vec{V}_2$.

(b) Draw the vectors $\hat{i}$ and $\mathbf{T}\hat{i}$.

(c) Represent the vector $\hat{i}$ in the $\vec{V}_1\vec{V}_2$ basis, and then act on the resulting vector with the matrix you wrote in Part (a).

(d) Draw both your beginning and ending vectors from Part (c). Your axes will still be x and y, so you will want to express vectors in the $\hat{i}\hat{j}$ basis to draw them.

6.182 The matrix $\mathbf{N} = \begin{pmatrix} 7 & 2 \\ 3 & 2 \end{pmatrix}$ has eigenvector $\vec{V}_1 = \begin{pmatrix} 2 \\ 1 \end{pmatrix}$ with eigenvalue $\lambda_1 = 8$ and eigenvector $\vec{V}_2 = \begin{pmatrix} -1 \\ 3 \end{pmatrix}$ with $\lambda_2 = 1$.

(a) Matrix $\mathbf{N}$ performs a certain transformation on any vector expressed in the form $x\hat{i} + y\hat{j}$. Write a different matrix that performs the same transformation on any vector expressed in the form $a\vec{V}_1 + b\vec{V}_2$.

(b) Vector $\vec{F} = 2\hat{i} + 3\hat{j}$. Draw the vectors $\vec{F}$ and $\mathbf{N}\vec{F}$.

(c) Represent the vector $\vec{F}$ in the $\vec{V}_1\vec{V}_2$ basis, and then act on the resulting vector with the matrix you wrote in Part (a).

(d) Draw both your beginning and ending vectors from Part (c). Your axes will still be x and y, so you will want to express vectors in the $\hat{i}\hat{j}$ basis to draw them.

6.183 Construct a matrix with the following two eigenvectors: $\vec{V}_1 = \begin{pmatrix} 2 \\ 3 \end{pmatrix}$ has $\lambda_1 = 7$, and $\vec{V}_2 = \begin{pmatrix} 5 \\ 4 \end{pmatrix}$ has $\lambda_2 = -6$. (A computer is not essential for this problem, but you may want to use one for some of the algebra.)

6.184 In this problem you will find the eigenvectors and eigenvalues of the matrix
$$\mathbf{M} = \begin{pmatrix} -10 & -10 & 1 \\ 2 & 1 & 3 \\ 4 & 10 & 2 \end{pmatrix}.$$

(a) Write the statement "$x\hat{i} + y\hat{j} + z\hat{k}$ is an eigenvector of $\mathbf{M}$ with eigenvalue λ" as a matrix equation.

(b) Multiply the matrices to turn this into three homogeneous linear equations. Put all of the non-zero terms on the left side.

(c) The only way 3 homogeneous linear equations for 3 variables can have non-trivial solutions is for the determinant of the matrix of coefficients to be zero. Use that condition to write the "characteristic polynomial" for this matrix: the polynomial that must be zero for λ to be an eigenvalue.

(d) 💻 Set the characteristic polynomial equal to zero and solve for λ.

(e) Plug each of the values of λ you found into the equations for x, y, and z. For each value of λ you should find, not a specific set of values, but a set of ratios between them that must be satisfied for the equations to work.

(f) What are the eigenvectors and eigenvalues of $\mathbf{M}$?

6.185 Walk-Through: Solving Coupled Differential Equations with Eigenvectors and Eigenvalues. In this problem you are going to solve the differential equations:

$$\frac{d^2 x_1}{dt^2} = -8x_1 - 8x_2$$

$$\frac{d^2 x_2}{dt^2} = -x_1 - 6x_2$$

(a) Begin by rewriting these two coupled differential equations as one differential equation involving matrices.
(b) Find the eigenvectors and eigenvalues of the transformation matrix in your equation.
(c) Plug in the first eigenvector and solve the resulting differential equations for x_1 and x_2 to find one normal mode of the system.
(d) Rewrite and solve the differential equation for the second eigenvector.
(e) Combine your two solutions. The result should be a solution with two arbitrary constants.
(f) Find the solution to this system if both objects start at rest with $x_1 = 1$ and $x_2 = 1$.

In Problems 6.186–6.189 use eigenvectors and eigenvalues to solve the given differential equations with the given initial and/or boundary conditions. You may find it helpful to first work through Problem 6.185 as a model.

6.186 $d^2 x_1/dt^2 = -11x_1 + 6x_2$, $d^2 x_2/dt^2 = 6x_1 - 6x_2$, $x_1(0) = 1$, $x_2(0) = 8$, $x_1'(0) = x_2'(0) = 0$

6.187 $dx_1/dt = 8x_1 + 2x_2$, $dx_2/dt = 9x_1 + 5x_2$, $x_1 = 1$, $x_2 = 2$. *Notice that these are first-order differential equations.*

6.188 $d^2 x_1/dt^2 = 7x_1 + 2x_2$, $d^2 x_2/dt^2 = 6x_1 + 8x_2$, $x_1(0) = 2$, $x_2(0) = 1$. Assume $\lim_{t \to \infty} x_1 = \lim_{t \to \infty} x_2 = 0$. (You'll apply that condition during the same part of the process where you apply the initial conditions.)

6.189 $d^2 x_1/dt^2 = 4x_1 + 9x_2$, $d^2 x_2/dt^2 = 2x_1 + x_2$, $x_2'(0) = \sqrt{2}$, $x_1(0) = x_2(0) = x_1'(0) = 0$.

6.190 In the Explanation (Section 6.8.2) we found that plugging $x_2 = (2/3)x_1$ into Equation 6.8.3 gave us a differential equation that we could solve for x_1. In fact, it gave us the same differential equation twice. What happens if we try an initial state that is not an eigenvector? Plug $x_2 = 2x_1$ into Equation 6.8.3 and show that you end up with two contradictory differential equations.

6.9 Putting It Together: Revisiting the Three-Spring Problem

Section 6.1 walked through the problem of two balls connected by three springs, but left a lot of gaps in the math. In this section we circle back to that problem with the tools of linear algebra now firmly in hand to fill in the gaps.

6.9.1 Explanation: Revisiting the Three-Spring Problem

In this explanation we are going to solve two "coupled differential equations" problems. The first solution will closely parallel the process we used in Section 6.1; it makes use of matrices and eigenvalues, but fundamentally works in the basis of the two positions x_1 and x_2. At the very end it uses the normal-mode basis to find the solution for a particular set of initial conditions. The second process translates the entire problem into the basis defined by the eigenvectors—the normal modes—right from the start. In this natural basis we find that the problem almost solves itself.

In both cases we have changed the original differential equations so the answers will come out different. Following both solutions carefully is a great way to make sure you understand the issues we have discussed throughout this chapter.

The first approach, illustrated in the example on Page 337, uses matrices in three key steps.

6.9 | Putting It Together: Revisiting the Three-Spring Problem

1. We begin by rewriting the problem as a matrix equation. Two characteristics of this problem suggest such an approach: the state of the system requires multiple variables, and the derivatives are based on linear combinations of those variables.
2. We then find the normal modes. Each normal mode is characterized by a particular ratio of x_2 to x_1 and a particular frequency of oscillation; the former is given by an eigenvector, and the latter by its associated eigenvalue (Section 6.8). We are not emphasizing here "how to find eigenvectors and eigenvalues" (although that's a good thing to know). Focus instead on how the definition of an eigenvector transforms the coupled equations into two decoupled differential equations that we can solve individually.
3. Finally we combine those normal modes, which amounts to a change of basis (Section 6.4). A matrix converts "this much of the first normal mode and that much of the second" to "this is x_1 and this is x_2" (Section 6.3), and its inverse matrix (Section 6.6) converts the other way to find the particular solution for given initial conditions.

EXAMPLE **Eigenvectors and Coupled Equations**

Question: Find the solution to the equations

$$\ddot{x}_1 = -3x_1 - 2x_2$$
$$\ddot{x}_2 = -x_1 - 2x_2$$

with initial conditions $x_1(0) = 3$, $x_2(0) = -9$, $\dot{x}_1(0) = \dot{x}_2(0) = 0$.

Answer: We begin by rewriting the problem as a matrix equation.

$$\begin{pmatrix} \ddot{x}_1 \\ \ddot{x}_2 \end{pmatrix} = \begin{pmatrix} -3 & -2 \\ -1 & -2 \end{pmatrix} \begin{pmatrix} x_1 \\ x_2 \end{pmatrix} \quad (6.9.1)$$

The next step is to find the two eigenvectors of the 2×2 matrix in Equation 6.9.1. The first, which you can find by hand (see Section 6.8) or on a computer, is $\begin{pmatrix} 1 \\ 1/2 \end{pmatrix}$ with eigenvalue -4. Any multiple of an eigenvector is also an eigenvector, so that first eigenvector tells us that if you plug in any state in which $x_2 = (1/2)x_1$, the matrix will simply multiply that state by -4. So if we replace x_2 with $(1/2)x_1$ the equation becomes:

$$\begin{pmatrix} \ddot{x}_1 \\ (1/2)\ddot{x}_1 \end{pmatrix} = -4 \begin{pmatrix} x_1 \\ (1/2)x_1 \end{pmatrix}$$

So we now have the equation $\ddot{x}_1 = -4x_1$. (In fact we have this same equation twice.) The solution to this equation becomes the first normal mode.

$$x_1(t) = A_1 \cos(2t) + A_2 \sin(2t), \quad x_2(t) = \frac{1}{2}x_1(t) \quad (6.9.2)$$

The other eigenvector is $\begin{pmatrix} 1 \\ -1 \end{pmatrix}$ with eigenvalue -1, which becomes the other normal mode.

$$x_1(t) = B_1 \cos t + B_2 \sin t, \quad x_2(t) = -x_1(t) \quad (6.9.3)$$

The general solution to our initial problem is a sum of these two normal modes. Next we apply our initial conditions. Knowing that the initial velocities are zero we discard the sine terms. We can easily write a matrix that converts normal mode amplitudes to initial positions; inverting that matrix allows us to find the normal mode amplitudes that match our given initial positions. As with the eigenvectors above, we're not showing you here the process of finding an inverse matrix, which is in Section 6.6; we're showing you what to do with it.

$$\text{The inverse of } \begin{matrix} x_1(0) \\ x_2(0) \end{matrix} \begin{pmatrix} A_1 & B_1 \\ 1 & 1 \\ 1/2 & -1 \end{pmatrix} \text{ is } \begin{matrix} A_1 \\ B_1 \end{matrix} \begin{pmatrix} x_1(0) & x_2(0) \\ 2/3 & 2/3 \\ 1/3 & -2/3 \end{pmatrix}$$

Applying this matrix to the initial conditions $x_1(0) = 3$, $x_2(0) = -9$ gives $A_1 = -4$, $B_1 = 7$, so the solution is

$$x_1(t) = -4\cos(2t) + 7\cos t$$
$$x_2(t) = -2\cos(2t) - 7\cos t$$

If there are any steps in that solution that you didn't follow, we hope you can find the appropriate part of the chapter to go back and review. Pay particular attention to the way each eigenvector of the matrix became a normal mode of the system; because the matrix multiplication became a scalar multiplication, the solution was two different oscillations with the same frequency.

The process we just modeled is all you need to solve this type of coupled oscillator problem. In the rest of this section we are going to show you a slightly different way of writing the steps in that solution. This second approach illustrates the meaning of basis states and diagonalization more directly than the method we just used.

Above we approached Equation 6.9.1 by plugging in the eigenvectors one at a time, and seeing what they did to the equation. Instead, let's rewrite the entire equation in the basis defined by the eigenvectors. So the state of the system—the positions of the balls, which we have previously expressed as x_1 and x_2—we will now express like this.

$$\begin{pmatrix} x_1(t) \\ x_2(t) \end{pmatrix} = f(t) \begin{pmatrix} 1 \\ 1/2 \end{pmatrix} + g(t) \begin{pmatrix} 1 \\ -1 \end{pmatrix} \quad (6.9.4)$$

You can interpret Equation 6.9.4 physically. As we have said all along, instead of asking "where is each ball?" we ask "how much of each normal mode do we have?" But right now let's focus on two mathematical questions.

Can we do that? Yes we can, for absolutely any functions x_1 and x_2, and here's why. At any given time t, the values x_1 and x_2 represent a vector. Equation 6.9.4 tells us that we can express that vector as a linear combination of the two eigenvectors. We know we can do that, *not* because they are eigenvectors, but just because they are two non-parallel vectors and therefore form a basis. The numbers f and g represent the coefficients in this new basis. At a later time t the values x_1 and x_2 are different; we are converting them to the *same* basis with *different* coefficients. That is why f and g are functions of t.

What happens when we do that? When we express all our states in a different basis, the entire equation—including the transformation matrix itself—changes. When you use the eigenvectors as a basis, the transformation is represented by a diagonal matrix of the eigenvalues.

6.9 | Putting It Together: Revisiting the Three-Spring Problem

(This fact comes very directly from the definition of an eigenvector.) And a diagonal matrix always *decouples the equations*. Equation 6.9.1 becomes:

$$\begin{pmatrix} \ddot{f} \\ \ddot{g} \end{pmatrix} = \begin{pmatrix} -4 & 0 \\ 0 & -1 \end{pmatrix} \begin{pmatrix} f \\ g \end{pmatrix} \qquad (6.9.5)$$

Our hope is that you understand why this equation must be true given that $\begin{pmatrix} f \\ g \end{pmatrix}$ is the vector $\begin{pmatrix} x_1 \\ x_2 \end{pmatrix}$ expressed in the natural basis of the transformation matrix, but you can derive it more directly by plugging Equation 6.9.4 into Equation 6.9.1. See Problem 6.206.

This gives us the two equations $\ddot{f} = -4f$ and $\ddot{g} = -g$. Solving those, and plugging the solutions into Equation 6.9.4, leads us back to the solutions we found before.

When we used this approach to solve the three-springs problem the normal modes ended up being sine waves, but once you reduce the problem to a set of decoupled ODEs you can solve them to find the normal modes no matter what kind of function they are. Also, instead of solving for a particular set of initial conditions as we have you can always choose to add the normal modes with their arbitrary constants to get the general solution. Both of these points are illustrated in the example below.

EXAMPLE **Diagonalization and Coupled Equations**

Question: Find the general solution to the equations

$$\begin{pmatrix} \dot{x} \\ \dot{y} \end{pmatrix} = \begin{pmatrix} -4 & -2 \\ -3 & 1 \end{pmatrix} \begin{pmatrix} x \\ y \end{pmatrix} \qquad (6.9.6)$$

(Notice that these are first derivatives, so this is different from our previous example, but the same techniques apply.)

Answer: The eigenvectors and eigenvalues of this transformation matrix are $\begin{pmatrix} 1 \\ 1/2 \end{pmatrix}$ and $\begin{pmatrix} 1 \\ -3 \end{pmatrix}$ with corresponding eigenvalues -5 and 2 respectively. We can write x and y in this basis as

$$\begin{pmatrix} x(t) \\ y(t) \end{pmatrix} = f(t) \begin{pmatrix} 1 \\ 1/2 \end{pmatrix} + g(t) \begin{pmatrix} 1 \\ -3 \end{pmatrix} \qquad (6.9.7)$$

Equation 6.9.7 says "you can express vector (x, y) at any given time as a linear combination of these two basis vectors, with f and g as the coefficients." You can see what this conversion does to Equation 6.9.6 by directly plugging in these expressions for $x(t)$ and $y(t)$ (Problem 6.207) or by "converting the transformation" (Section 7.2) but you don't have to: we already know that in the basis defined by the eigenvectors, a transformation becomes a diagonal matrix of the eigenvalues. (Such a matrix says "multiply the $\vec{V}_1$-component by λ_1" and so on.) So Equation 6.9.6 becomes:

$$\begin{pmatrix} \dot{f} \\ \dot{g} \end{pmatrix} = \begin{pmatrix} -5 & 0 \\ 0 & 2 \end{pmatrix} \begin{pmatrix} f \\ g \end{pmatrix}$$

The "natural basis" for this transformation has done its job, giving us the *decoupled* differential equations $\dot{f} = -5f$ and $\dot{g} = 2g$ with solutions $f = Ae^{-5t}$ and $g = Be^{2t}$. Finally, we can go back to the xy representation to write the general solution.

$$x = Ae^{-5t} + Be^{2t}, \quad y = \frac{1}{2}Ae^{-5t} - 3Be^{2t}$$

Stepping Back

As you may have guessed, we want you to be able to solve the three-spring problem. Coupled oscillators come up a lot. If you boil the whole thing down to "the resonant frequencies of coupled oscillators are the square roots of the absolute values of the eigenvalues," you will have a tidbit that may prove useful.

But beyond that, of course, our strategy in this chapter has been to use the three-spring problem as both introduction and capstone to your understanding of linear algebra. It is for that reason that we just solved the same problem twice. In both cases, the takeaway is "the eigenvectors of the matrix define the normal modes of the system."

In the first approach, this fact came very directly from the eigenvalue equation. When you replace the state of the system with an eigenvector, the two variables get separate differential equations but with the same frequency, which defines a normal mode. If you try replacing the state of the system with something that isn't a normal mode (say, $x_2 = 3x_1$) you will end up with two contradictory differential equations.

In the second approach, we rewrote the entire problem in the basis defined by the normal modes. Instead of solving directly for the positions of the balls, we solved for the amplitudes of the normal modes. This is less direct, but it gives you perhaps the clearest view of why these normal modes work and no others would; only in the basis of the normal modes does the transformation matrix become diagonal, thus decoupling the two equations.

If you find the second approach confusing start by getting comfortable with the first approach. You should be able to understand that and solve problems with it, but once you have worked with it a while come back to the second approach and see if you can see how it is doing the same thing as the first one, but in a different basis.

6.9.2 Problems: Revisiting the Three-Spring Problem

6.191 The figure below shows a set of coupled oscillators. For this problem take $k = 8$ N/m, $m = 2$ kg.

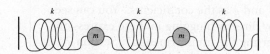

In later problems you'll solve for the normal modes of systems like this, but for this one we're just going to tell you that one of them is represented by the eigenvector $\begin{pmatrix} 1 \\ 1 \end{pmatrix}$ with eigenvalue 2, and the other is represented by the eigenvector $\begin{pmatrix} 1 \\ -1 \end{pmatrix}$ with eigenvalue $2\sqrt{3}$.

(a) In the first normal mode both balls move to the right and left in unison, always maintaining the same displacement from their equilibrium points. Explain why it makes sense physically that the two balls could move together in this way. *Hint*: think about the forces on the two balls, and in particular think about whether the middle spring is stretched, compressed, or neither in this normal mode.

(b) In Part (a) we described what one normal mode physically represents. Give a similar description of the other normal mode. (You don't have to do what you did in that part; just do the part we did.)

(c) Explain why the normal mode in Part (b) has a higher frequency than the one in Part (a). Your answer shouldn't be about eigenvalues or normal mode frequencies, but about springs and forces.

6.192 The image below shows three coupled oscillators.

You should be able to write down their equations of motion and solve them to find the normal modes, but in this problem we want to focus on the physical interpretation so we're going to give you the eigenvectors of the transformation matrix.

$$\begin{pmatrix} x_1 \\ x_2 \\ x_3 \end{pmatrix} = \begin{pmatrix} 1 \\ \sqrt{2} \\ 1 \end{pmatrix}, \begin{pmatrix} -1 \\ 0 \\ 1 \end{pmatrix}, \begin{pmatrix} 1 \\ -\sqrt{2} \\ 1 \end{pmatrix}$$

(These come out the same regardless of the values of k and m.) The first of these eigenvectors describes a normal mode in which you pull all three balls to the right (or left), pulling the middle one slightly farther than the outer two, and then let go. They will then oscillate that way forever, always moving to the right and left together, with the middle one oscillating with a larger amplitude than the other two. Using that description as a guide, write similar descriptions for what the other two eigenvectors physically represent.

6.193 **Walk-Through: Coupled Differential Equations, First Method.** In this problem you will find the solution to the equations

$$\ddot{x} = -6x + y$$
$$\ddot{y} = -4x - y$$

using the approach outlined in the example on Page 337.

(a) Rewrite these differential equations as a single matrix equation. One side will be the column matrix $\begin{pmatrix} \ddot{x} \\ \ddot{y} \end{pmatrix}$. The other side will be a square matrix of all numbers times a column matrix that represents the state of the system.

(b) Find the eigenvectors and eigenvalues of the square matrix in your equation. You may do this with a computer or by using the method described in Section 6.8.

(c) Choose one of the eigenvectors you found and use it to write y as a function of x. Plug this function into your matrix equation from Part (a) and multiply the matrices. The result should give you two equivalent differential equations for x.

(d) Solve your equation for x. The solution should contain two arbitrary constants. Then use the relationship between x and y in this eigenvector to write the solution for y, which will contain the same two constants.

(e) Repeat Parts (c)–(d) for the other eigenvector.

(f) Combine your answers to find the general solution for x and y.

(g) Plug your general solution into the original differential equations and verify that it works.

6.194 [*This problem depends on Problem 6.193.*]

(a) Assume that $\dot{x}(0) = \dot{y}(0) = 0$. Use this constraint to simplify your solution. (Several terms should drop out.) Assume these initial velocities are zero throughout the rest of this problem.

(b) Write a set of initial conditions for x and y for which the system will be in one of its normal modes. Write the solution $x(t)$, $y(t)$ corresponding to those initial conditions. This should require no calculations.

(c) Suppose your system is in a mix of the first normal mode with amplitude A plus the second normal mode with amplitude B. Write the matrix that converts the vector $\begin{pmatrix} A \\ B \end{pmatrix}$ to the initial conditions $x(0)$ and $y(0)$.

(d) Using the matrix you found in Part (c), write the initial conditions that would give you $A = B = 2$.

(e) Write the matrix that converts from the initial conditions $\begin{pmatrix} x(0) \\ y(0) \end{pmatrix}$ to the normal mode amplitudes A and B.

(f) The system starts with $x(0) = 3$, $y(0) = 4$. Use the matrix you wrote in Part (e) to find A and B and use that to write the solution $x(t)$, $y(t)$ for these initial conditions.

6.195 **Walk-Through: Coupled Differential Equations, Second Method.** In this problem you will find the solution to the equations

$$\ddot{x} = -3x - 4y$$
$$\ddot{y} = -3x + y$$

using the approach illustrated in the example on Page 339.

(a) Rewrite these differential equations as a single matrix equation. One side will be the column matrix $\begin{pmatrix} \ddot{x} \\ \ddot{y} \end{pmatrix}$. The other side will be a square matrix of all numbers times a column matrix that represents the state of the system.

(b) Find the eigenvectors and eigenvalues of the square matrix in your equation. You may do this with a computer or by using the method described in Section 6.8.

(c) Write the vector $\begin{pmatrix} x(t) \\ y(t) \end{pmatrix}$ as a sum of the eigenvectors you wrote in Part (b). The coefficients in this sum will be functions of time, which you can call $f(t)$ and $g(t)$.

(d) The equation you wrote in Part (a) should have had a column matrix on the left and a square matrix and a column matrix on the right. All of these matrices were written in the xy basis. Rewrite the equation in the fg basis. The result should be two decoupled differential equations.

(e) Solve the equations to find $f(t)$ and $g(t)$. Include all arbitrary constants.

(f) Use your solutions for f and g and your answer to Part (c) to write the general solution $x(t)$, $y(t)$.

6.196 Solve the differential equations in Problem 6.193 using the approach illustrated in the example on Page 339.

6.197 Solve the differential equations in Problem 6.195 using the approach outlined in the example on Page 337.

In Problems 6.198–6.203 find the normal mode solutions for the given sets of equations. Then find the amplitude of each normal mode if the system starts with the given initial conditions. When writing the solutions you should include all arbitrary constants, but when you plug in initial conditions for second-order equations assume the initial first derivatives are zero. In all cases you may use a computer to find eigenvectors and eigenvalues or do it by hand, but in the case of larger matrices we've marked the problems with a computer icon to indicate that doing it by hand would be tedious.

6.198 $\ddot{x} = -7x + 2y$, $\ddot{y} = -4x - y$, $x(0) = 1$, $y(0) = 3$

6.199 $\ddot{x} = -3x - y$, $\ddot{y} = -4x - 3y$, $x(0) = 1$, $y(0) = -5$

6.200 $\ddot{x} = -4x + y$, $\ddot{y} = x - 4y$, $x(0) = 1$, $y(0) = -1$

6.201 $\dot{x} = 2x + y$, $\dot{y} = 4x + 5y$, $x(0) = 2$, $y(0) = 1$. *Notice that these are first derivatives.*

6.202 $\ddot{x} = 3x + y + z$, $\ddot{y} = x + y + 3z$, $\ddot{z} = -x + 3y + z$, $x(0) = 1$, $y(0) = 2$, $z(0) = 3$ *Hint*: For positive eigenvalues you can solve the ODEs using exponentials, but it's easier to apply the condition that the initial derivatives are zero if you write the solution in terms of sinh and cosh. If you have no idea what those are just go ahead and use exponentials and it won't be too bad, and you can learn about sinh and cosh in Section 12.3.

6.203 $\dot{x} = -2x - y + z$, $\dot{y} = -2x + y - 3z$, $\dot{z} = x + y - 2z$, $x(0) = 1$, $y(0) = 1$, $z(0) = 1$. *Notice that these are first derivatives.*

6.204 In the example on Page 337 we solved a coupled oscillator problem on the assumption that the balls started with zero velocity. In this problem you will relax that assumption. Your starting point will be Equations 6.9.2–6.9.3.

(a) The four initial conditions are $x_1(0)$, $x_2(0)$, $\dot{x}_1(0)$, and $\dot{x}_2(0)$. If $A_1 = 1$ and the other three arbitrary constants are all zero, what are the values of these four initial conditions? Answer the same question for each of the other three arbitrary constants.

(b) Using your answers to Part (a), write a 4×4 matrix to convert from the four arbitrary constants to the four initial conditions.

(c) Invert that matrix to find a matrix that converts from initial conditions to arbitrary constants. (If you think about this there's a relatively simple way to do it by hand, but you're welcome to use a computer if you prefer.)

(d) Write down the solution for the initial conditions $x_1(0) = 3$, $x_2(0) = 1$, $\dot{x}_1(0) = 1$, $\dot{x}_2(0) = -2$.

6.205 In the example on Page 337 our solution required finding the eigenvectors and eigenvalues of $\begin{pmatrix} -3 & -2 \\ -1 & -2 \end{pmatrix}$. Find those eigenvectors and eigenvalues (without a computer) using the method described in Section 6.8.

6.206 In the Explanation (Section 6.9.1) we said that when you rewrote x_1 and x_2 as a sum of the eigenvectors of the transformation matrix the matrix would become diagonal. That must be true by the definition of eigenvectors, but Equation 6.9.5 may still look a bit mysterious. In this problem you'll derive that equation more rigorously.

(a) Rewrite the right-hand side of Equation 6.9.4 as a square matrix times a column matrix. This should only require looking at the right-hand side and remembering what matrix times column means.

(b) Plug Equation 6.9.4, with the right-hand side rewritten as a matrix times column, into Equation 6.9.1. The result should be a matrix equation with a square matrix times a column matrix on the left, and a square matrix times a square matrix times a column matrix on the right.

(c) Multiply the two square matrices on the right, so that your equation is now a square matrix times a column matrix on each side.

(d) Multiply both sides of the equation by the inverse of the square matrix on the left of the equation, and carry out the multiplication with the square matrix on both sides. The resulting equation should be Equation 6.9.5.

6.207 In the example on Page 339 we converted coupled differential equations for x and y into decoupled differential equations for f and g based on the properties of eigenvectors. For the moment, let's pretend you had never heard of an eigenvector. Starting with the equations $\dot{x} = -4x - 2y$ and $\dot{y} = -3x + y$, make the substitutions $x = f + g$ and $y = (1/2)f - 3g$ and see what you get. (We hope it will look like what we got!)

6.208 The techniques in this section can be applied to a variety of differential equations that don't look like the ones we've solved so far. As an example, you'll use them in this problem to solve the single damped oscillator equation $\ddot{x} + 4\dot{x} + 3x = 0$.

(a) Define a variable $v(t) = \dot{x}(t)$. Rewrite the damped oscillator equation as an equation for $\dot{v}$ in terms of x and v.

(b) The equation you just wrote, along with the equation $\dot{x} = v$, are a pair of linear, coupled differential equations. Use the methods of this section to find the general solution $x(t)$.

6.209 In this problem you will solve the equations

$$\dot{x} = -9x - 10y$$
$$\ddot{y} = -7x - 6y$$

(a) Define a variable $v(t) = \dot{y}(t)$. Rewrite the equation for $\ddot{y}$ as an equation for $\dot{v}$.

(b) Including the equation $\dot{y} = v$ you should now have three coupled first-order differential equations for x, y, and v. Use the methods of this section to find the general solution to these equations. (The only part you should need the computer for is finding the eigenvectors and eigenvalues of a 3×3 matrix.)

6.210 Explain why you cannot use the techniques of this section to solve the following differential equations:

$$\ddot{x} = -x^2 - 3y$$
$$\ddot{y} = x + 2y$$

In Problems 6.211–6.216, write the equations of motion for the system and find the normal modes. In each case you'll need to start by writing expressions for the force on each ball from each of the two springs touching it. That force will depend on how stretched or compressed the spring is, which in turn will depend on the positions of the two balls touching the spring (or just one if the spring is connected to the wall). As an example, consider the three-spring problem we solved in the Explanation (Section 6.9.1). The leftmost spring is stretched by x_1, so it exerts a force $-k_1 x_1$ on Ball 1. The middle spring is stretched by $x_2 - x_1$, but if it is stretched it pulls Ball 1 to the right, so its force on Ball 1 is $k_2(x_2 - x_1)$. Putting those together and dividing force by mass to get acceleration gives $\ddot{x}_1 = -(k_1/m_1)x_1 + (k_2/m_1)(x_2 - x_1)$. If you plug in the numbers given in the Explanation this will give you the differential equation we wrote for x_1. You can similarly derive the equation for x_2.

6.211 $k = 10$ N/m, $m = 3$ kg

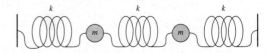

6.212 Leave k and m as letters in your solution.

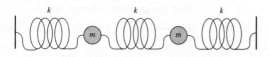

6.213 $k_1 = 16$ N/m, $k_2 = 4$ N/m, $k_3 = 10$ N/m, $m_1 = 2$ kg, $m_2 = 2$ kg

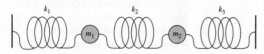

6.214 $k_1 = 3$ N/m, $k_2 = 2$ N/m, $m_1 = 1$ kg, $m_2 = 1$ kg

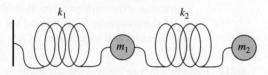

6.215 🖥 $k_1 = 24$ N/m, $k_2 = 6$ N/m, $m_1 = 3$ kg, $m_2 = 2$ kg. *The only thing you should need a computer for is the eigenvectors and eigenvalues of a 3×3 matrix.*

6.216 🖥 $k_1 = 24$ N/m, $k_2 = 6$ N/m, $m_1 = 3$ kg, $m_2 = 1$ kg, $m_3 = 6$ kg. *The only thing you should need a computer for is the eigenvectors and eigenvalues of a 3×3 matrix.*

6.217 A double pendulum is shown below.

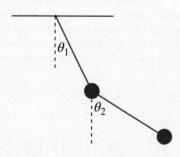

The equations describing this system depend on the angles θ_1 and θ_2, the gravitational acceleration g, and the length L of the strings.

$$2\ddot{\theta}_1 + \ddot{\theta}_2 \cos(\theta_1 - \theta_2) + \dot{\theta}_2^2 \sin(\theta_1 - \theta_2) + 2\frac{g}{L}\sin\theta_1 = 0$$

$$\ddot{\theta}_2 + \ddot{\theta}_1 \cos(\theta_1 - \theta_2) - \dot{\theta}_1^2 \sin(\theta_1 - \theta_2) + \frac{g}{L}\sin\theta_2 = 0$$

These equations are non-linear and cannot be solved using the methods in this section. For small oscillations, however, you can assume θ_1, θ_2, and all of their derivatives remain small, and thus approximate this with a set of linear equations.

(a) Replace all the trig functions with the linear terms of their Maclaurin series expansions. Then eliminate any remaining non-linear terms. The result should be two coupled, linear differential equations.

(b) Solve the equations algebraically for $\ddot{\theta}_1$ and $\ddot{\theta}_2$ so you can write them in the form we used in this section.

(c) Find the normal mode frequencies of this system for small oscillations.

6.218 🖥 *[This problem depends on Problem 6.217.]* For this problem take $g = 9.8$ m/s and $L = 1.0$ m and assume in all cases that the two pendulums start at rest.

(a) Choose one of the two normal modes of the system and numerically solve the original equations (not the linearized ones) for $\theta_1(0) = 0.1$, with $\theta_2(0)$ chosen to be whatever it needs to be for that normal mode. Plot $\theta_1(t)$ and $\theta_2(t)$ together on one plot. Include enough time on your plot to see at least 3 full oscillations of the pendulums.

(b) Repeat Part (a) for values of $\theta_1(0) = 0.2$, 0.3, and so on up to 1. In each case adjust $\theta_2(0)$ to whatever the normal mode solution says it should be.

(c) Describe what the system's behavior looks like for small and large amplitudes using these initial conditions. Explain why the different behavior in the different cases makes sense.

6.219 The figure below shows a circuit with capacitors and inductors.

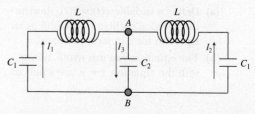

6.9 | Putting It Together: Revisiting the Three-Spring Problem

Conservation of charge requires that $I_3 = I_1 + I_2$, so there are only two variables needed to describe the current in the system. You can get equations for those currents by setting the voltage drop from point A to point B equal along all three paths and differentiating: $I_1/C_1 + L\ddot{I}_1 = I_2/C_1 + L\ddot{I}_2 = -(I_1 + I_2)/C_2$. (Don't worry if you didn't follow how we got that; your job is going to be to solve it.) Remember that $\ddot{I}$ means d^2I/dt^2.

(a) Write these as a pair of equations for $\ddot{I}_1$ and $\ddot{I}_2$.

(b) Find the normal modes of the system and use them to write the general solution.

(c) In one of the normal modes the current on both sides of the circuit is equal, so at some point in time a current I is flowing up on the left, an equal current I is flowing up on the right, and a current $2I$ is flowing down the middle. Then they reverse—down on both sides and up in the middle—and go back and forth like that forever. Using this description as a guide, describe the physical state represented by the other normal mode.

6.10 Additional Problems (see felderbooks.com)

CHAPTER 7

Linear Algebra II

> *Before you read this chapter, you should be able to…*
> - do basic algebra with matrices, including matrix multiplication.
> - use a "point matrix" to represent a shape, and a transformation matrix to change that shape.
> - find eigenvectors and eigenvalues of a matrix.
> - use matrices to convert vectors from one basis to another.
> - find the determinant of a matrix and use it to determine if a set of vectors is independent, or if a set of linear equations is independent.
>
> (See Chapter 6 for all of these topics.)
>
> *After you read this chapter, you should be able to…*
> - find the matrix that performs a given geometric transformation, or find the transformation performed by a given matrix.
> - identify scalars, vectors, and tensors as defined by their transformation properties.
> - convert tensors, including transformation matrices, from one basis to another.
> - identify vector spaces and find their dimensions.
> - express abstract vectors in terms of different bases.
> - use "row reduction" to find the solution to a set of linear equations, or to determine if they are inconsistent or linearly dependent.
> - use the "simplex method" to solve problems in linear programming.

Chapter 6 introduced the use of matrices to express problems in different bases. As an example, it showed how you can solve a set of coupled differential equations by finding the natural basis for the problem—the normal modes—and rewriting the problem in that basis. The skills you learned in that chapter are useful for almost any application of linear algebra, which is to say almost any problem involving sets of linear relationships between quantities.

This chapter introduces some more advanced linear algebra tools. Describing geometric transformations with matrices is crucial in graphical programming (e.g., video games). Converting transformation matrices to other bases is useful in mechanics and relativity. The notion of vector spaces generalizes the ideas of linear algebra to a broader range of abstract mathematical spaces. Row reduction is the method many computer programs use to solve linear equations. And finally, the simplex method is a computationally efficient approach to optimization problems involving large numbers of variables and linear constraints.

A word about dependencies. The material in this chapter depends on the material in Chapter 6. For the most part the different sections in this chapter do not depend on each other; you can go through vector spaces without tensors, or row reduction without geometric transformations, etc. But there are two exceptions: the tensor section (7.2) depends on some of the ideas in the geometric transformations section (7.1), and the simplex method (7.5) depends heavily on row reduction (7.4).

7.1 Geometric Transformations

Matrix multiplication is a way of transforming a vector, where "vector" is a very general concept that can be applied to non-spatial quantities such as the positions of two balls along a line or the numbers of different types of molecules. In this section, however, we will focus on spatial vectors and visual transformations. We do this not only because such transformations are important in everything from computer games to astronomy, but because you can understand transformations better when you think about them visually.

7.1.1 Discovery Exercise: Geometric Transformations

The picture below shows three ants on a turntable. A short time later the turntable has rotated counterclockwise by an angle $\Delta\phi$. You're going to find the new position of the ants.

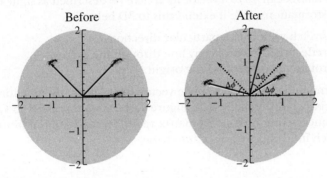

1. Start by considering an ant at some arbitrary position $x_0\hat{i} + y_0\hat{j}$. Express x_0 and y_0 in terms of the polar coordinates ρ_0 and ϕ_0.
2. Similarly express x_f and y_f, the coordinates after rotation, in terms of ρ_f and ϕ_f.
3. Express ρ_f and ϕ_f in terms of ρ_0, ϕ_0, and the rotation angle $\Delta\phi$. Plug this into your answer for Part 2 to get expressions for x_f and y_f in terms of the variables ρ_0, ϕ_0, and $\Delta\phi$.

   See Check Yourself #44 in Appendix L

4. Your formulas for x_f and y_f should have expressions of the form $\sin(a+b)$ and $\cos(a+b)$. Use the sine and cosine addition formulas to expand these out.
5. Now you should be able to recognize your expressions for x_0 and y_0 from Part 1 showing up inside your expressions for x_f and y_f. Substitute x_0 and y_0 in for them. The result will be an expression for x_f and y_f that only depends on x_0, y_0, and $\Delta\phi$.
6. Write your answer in the form of a matrix equation, filling in the question marks below with formulas that include $\Delta\phi$.

$$\begin{pmatrix} x_f \\ y_f \end{pmatrix} = \begin{pmatrix} ? & ? \\ ? & ? \end{pmatrix} \begin{pmatrix} x_0 \\ y_0 \end{pmatrix}$$

   You now have a matrix that can be multiplied by any vector $x_0\hat{i} + y_0\hat{j}$ to determine where that vector will be after the rotation.

7. In the picture above the initial positions are $\hat{i}$, $\hat{i}+\hat{j}$, and $-\hat{i}+\hat{j}$ and the rotation angle is 30°. Use your matrix to determine where the three ants end up after rotation. Make sure your answers seem to roughly match the final positions shown in the figure.
8. Without doing any calculations, predict where each of the three ants would have ended up if the turntable had been rotated counterclockwise by 90°. Check your answer by replacing $\Delta\phi$ with 90° in your matrix, and then acting with that matrix on the ants' initial positions.

What you've just derived is a "transformation matrix," a matrix that can be used to geometrically transform vectors. In this section you'll encounter transformation matrices that stretch and reflect vectors as well as rotating them.

7.1.2 Explanation: Geometric Transformations

The software that runs video games and computer animation is all based on matrices: point matrices to represent objects, and transformation matrices that act on those objects. A matrix transformation changes the coordinates of a vector (x, y) into a new set of coordinates $(\tilde{x}, \tilde{y})$ related to the originals by a set of linear equations:

$$\begin{pmatrix} \tilde{x} \\ \tilde{y} \end{pmatrix} = \begin{pmatrix} a & b \\ c & d \end{pmatrix} \begin{pmatrix} x \\ y \end{pmatrix} \quad \leftrightarrow \quad \begin{array}{l} \tilde{x} = ax + by \\ \tilde{y} = cx + dy \end{array} \quad (7.1.1)$$

Everything a matrix can do to a vector in 2D can be described as some combination of three types of transformations. (We'll extend this to 3D below.)

- It can stretch vectors in a particular direction.
- It can reflect vectors across any line through the origin.
- It can rotate vectors around the origin.

A transformation can act on a single vector, or it can act on an entire shape described by a point matrix (as in the pictures in Figure 7.1). As we discussed in Chapter 6 the two are equivalent, since acting on a point matrix really means acting individually on the position vectors of each individual vertex of the shape.

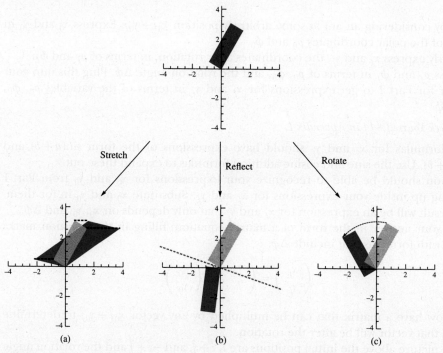

FIGURE 7.1 A rectangle being stretched by a factor of 2.5 in the x-direction (a), reflected across the line $y = -x/3$ (b), and rotated $60°$ counterclockwise about the origin (c).

But to describe a transformation we must describe its effects on *all* vectors and shapes, not just on one. For instance, Figure 7.2 shows one vector being transformed into another.

7.1 | Geometric Transformations

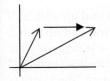

FIGURE 7.2 A vector transformation.

The simplest description of this particular transformation would be that the vector was stretched in the x-direction. (Yes, stretching a vector can change its direction!) But we might also say that the vector was rotated clockwise and then stretched up-and-right. Or perhaps it was rotated counterclockwise, stretched up-and-left, and then reflected. If you want to know which transformation this particular matrix actually causes, you would have to see its effects on other vectors.

There are also many transformations that a matrix *cannot* cause. Most obviously, Equation 7.1.1 is linear. You could define a new set of coordinates $\tilde{x} = x^3 + y^3$, $\tilde{y} = x - y$, but you could not use a matrix to find the $\tilde{x}\tilde{y}$-coordinates from the xy-coordinates because the transformations aren't linear. That's why this field is called "linear algebra."

Another limitation is that any matrix leaves the vector $(0, 0)$ unchanged, as you will show in Problem 7.25. You cannot write a matrix multiplication that performs a simple "translation" such as moving an object three units to the right, and you cannot reflect or rotate around an axis that doesn't go through the origin. (There is a work-around used in computer graphics to represent translations with matrix multiplication. See Section 7.6 Problem 7.141 at felderbooks.com.)

Finding the Matrix for a Particular Transformation

The Discovery Exercise (Section 7.1.1) was designed to bring you to the matrix that rotates a shape by an angle of $\Delta\phi$ counterclockwise.

$$\begin{pmatrix} \cos(\Delta\phi) & -\sin(\Delta\phi) \\ \sin(\Delta\phi) & \cos(\Delta\phi) \end{pmatrix} \quad (7.1.2)$$

You can see that this makes sense by trying some simple examples. For instance, the matrix that rotates all vectors 90° counterclockwise is $\begin{pmatrix} 0 & -1 \\ 1 & 0 \end{pmatrix}$. You should take a moment to check that this produces the results you would expect if you multiply it by the vectors $\begin{pmatrix} 1 \\ 0 \end{pmatrix}$, $\begin{pmatrix} 0 \\ 1 \end{pmatrix}$, and $\begin{pmatrix} 1 \\ 1 \end{pmatrix}$. Then write the rotation matrix for a rotation by 45° counterclockwise and plug in a few simple vectors to verify that it gives you what you would expect.

A rotation matrix in 2D always looks like Equation 7.1.2, but reflection matrices and stretching matrices do not all look so similar. Nonetheless, in simple cases it isn't hard to come up with a matrix that stretches or reflects a vector. For example, the matrix $\begin{pmatrix} 2 & 0 \\ 0 & 1 \end{pmatrix}$ stretches the x-component of all vectors by a factor of 2 and the matrix $\begin{pmatrix} -1 & 0 \\ 0 & 1 \end{pmatrix}$ reflects vectors across the y-axis. (Reversing the x-component reflects any point across the y-axis, not across the x-axis.)

What if you wanted to stretch vectors by a factor of 3 along the line $y = 2x$? You can find the appropriate matrix if you remember that the eigenvalues of a matrix describe how far it stretches objects in the specific directions given by the eigenvectors. Here you want a matrix that has an eigenvector pointing along $y = 2x$ (that is, in the direction $\hat{i} + 2\hat{j}$) with an eigenvalue of 3. You also want the second eigenvector to be perpendicular to the first, and with an eigenvalue of 1, so there is no other stretch. Problems 7.20–7.22 examine eigenvectors of stretching matrices. Similarly, if you wanted to reflect about a line you would need an eigenvector with eigenvalue 1 along that line, and one with eigenvalue -1 perpendicular to it.

Commutative Transformations

Recall that the matrix $\mathbf{C} = \mathbf{AB}$ performs transformation $\mathbf{B}$ and then transformation $\mathbf{A}$. (You read right-to-left because $\mathbf{AB}\vec{V}$ means "$\mathbf{A}$ acting on $\mathbf{B}\vec{V}$.") Order generally matters. For example, Figure 7.3 shows that if you start with the vector $\hat{j}$, stretch it by 2 in the y-direction,

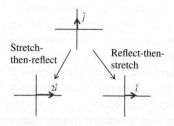

FIGURE 7.3 Order matters.

and then reflect it about the line $y = x$, you end up with $2\hat{i}$. If you do the reflection and then the stretch you end up with $\hat{i}$.

In the last paragraph we made a visual statement: "the order in which you apply these two transformations matters." This corresponds to a mathematical statement: "the matrices that represent these two transformations do not commute."

$$\begin{pmatrix} 1 & 0 \\ 0 & 2 \end{pmatrix} \begin{pmatrix} 0 & 1 \\ 1 & 0 \end{pmatrix} \neq \begin{pmatrix} 0 & 1 \\ 1 & 0 \end{pmatrix} \begin{pmatrix} 1 & 0 \\ 0 & 2 \end{pmatrix}$$

When two matrices *do* commute mathematically, that means their transformations can be taken in either order. For instance, any two 2D rotation matrices commute. This is not obvious looking at the matrices, but it is obvious if you think about it physically: rotating by 30° clockwise and 50° counterclockwise results in a 20° counterclockwise rotation, no matter which order you do them in.

Transformations in Three Dimensions

Our discussion above focused on 2D examples, but most of the same ideas apply in 3D. Transformations are still represented by square matrices, 3×3 in this case. Shapes can still be represented by point matrices, with the vectors defining the vertices of polygons (such as triangles and rectangles) that themselves define the surfaces of three-dimensional objects. (See Problem 7.27.) Reflections in 3D go across planes instead of lines, and rotations go around lines through the origin instead of just going around the origin as they do in 2D. Stretches still occur along lines.

Consider the relatively simple example of rotating about the z-axis. The z-components of vectors will remain unchanged, while the x- and y-components will mix in the same way they do in a 2D rotation.

$$\begin{pmatrix} \cos(\Delta\phi) & -\sin(\Delta\phi) & 0 \\ \sin(\Delta\phi) & \cos(\Delta\phi) & 0 \\ 0 & 0 & 1 \end{pmatrix} \quad \textit{Rotation about the z-axis}$$

In 2D a rotation matrix has no real eigenvectors. In 3D it has one, which is the axis of rotation. Of course a rotation about any line other than the coordinate axes will look more complicated, but the axis of rotation will still be the only real eigenvector, and it will have an eigenvalue of 1.

A more interesting case comes up when you consider a stretch. Take a moment before reading on to try to figure out the eigenvectors and eigenvalues of the transformation "stretch by a factor of 2 in the z-direction." Don't keep reading until you've come up with your answer.

The most obvious one is the z-axis, with eigenvalue 2, which is to say that if you stretch everything in the z-direction then vectors on the z-axis just get stretched. Give yourself full credit if you said the two other eigenvectors are $\hat{i}$ and $\hat{j}$, both with eigenvalue 1. That's correct—vectors along the x- and y-axes are unchanged by this transformation—but that's not the whole story. The vector $\hat{i} + 3\hat{j}$ is also unchanged, so it must also be an eigenvector. In fact any vector in the xy-plane is an eigenvector of this transformation with eigenvalue 1. This transformation is one example of a general rule about eigenvectors:

Degenerate Eigenvalues

If a matrix **M** has two or more linearly independent eigenvectors with the same eigenvalue, that eigenvalue is said to be "degenerate." In that case any linear combination of those eigenvectors is also an eigenvector with the same eigenvalue.

Geometrically, two linearly independent eigenvectors with a degenerate eigenvalue define a plane on which all of the vectors are eigenvectors.

See Problem 7.23 for another example of degenerate eigenvectors.

There is one important difference worth noting between 2D and 3D rotations; unlike in 2D, 3D rotations generally do not commute. Consider the following simple example. Start with the vector $\hat{j}$ and rotate it 90° counterclockwise about the z-axis and you'll get $-\hat{i}$. Rotate that by 90° counterclockwise about the y-axis and you'll get $\hat{k}$. Perform the same two operations in the other order and you'll end up at $-\hat{i}$.

The Determinant and Geometric Transformations

The simple transformation matrix $\mathbf{D} = \begin{pmatrix} 3 & 0 \\ 0 & -4 \end{pmatrix}$ stretches all shapes by a factor of 3 in the x-direction and −4 in the y-direction. If you use $\mathbf{D}$ to transform any point matrix, the *area* of the resulting shape will multiply by 3 and also by 4, so the new shape will have 12 times the area of the original. We can state this result generally: any diagonal transformation matrix multiplies areas (or volumes in 3D) by the absolute value of the product of its diagonal elements. (The absolute value tells us that whether the product is 12 or −12, the area will multiply by 12. We will discuss the effect of a sign change below.)

But most matrices aren't diagonal. How does the matrix $\mathbf{T} = \begin{pmatrix} 2 & 3 \\ 4 & 1 \end{pmatrix}$ affect areas? The two eigenvalues of this matrix are 5 and −2, so it stretches vectors by a factor of 5 in one direction, and in another direction it reverses them and stretches them by a factor of 2. That means this matrix increases areas by a factor of 10. In general, a transformation matrix multiplies areas (or volumes) by the absolute value of the product of its eigenvalues. (Our conclusion in the previous paragraph was a special case of this one, since the diagonal elements of a diagonal matrix *are* its eigenvalues.)

It turns out that you can find this quantity, the product of all the eigenvalues, without calculating the individual eigenvalues themselves. You will show in Problem 7.30 that the determinant of a diagonal matrix is the product of the diagonal entries. More generally, it can be shown that the determinant of any matrix equals the product of its eigenvalues. Chapter 6 discussed three interpretations of the determinant; we now add a fourth.

What the Determinant Tells Us about Transforming a Shape

When a shape represented by a point matrix $\mathbf{P}$ is transformed by a transformation matrix $\mathbf{T}$, the area (in 2D) or volume (in 3D) of the shape is multiplied by the absolute value of the determinant $|\mathbf{T}|$.

This is an important addition to our understanding of determinants, and can lead to many important insights. We offer three below.

- Given two matrices with $|\mathbf{A}| = 3$ and $|\mathbf{B}| = 5$, what is the determinant of matrix $\mathbf{AB}$? We can answer this question with no algebra if we remember that $\mathbf{AB}$ means "first multiply by $\mathbf{B}$ and then by $\mathbf{A}$." So the area will multiply first by 5 and then by 3, resulting in a total increase by a factor of 15. This train of logic leads us to the following generalization about determinants:

$$|\mathbf{AB}| = (|\mathbf{A}|)(|\mathbf{B}|)$$

- Our other uses for the determinant were based only on whether the determinant was zero. This is our first reason to care what non-zero number the determinant is! But the special case of zero is still worth noting, because a matrix with a determinant of zero will transform any shape into a shape with zero area. For instance, the matrix $\begin{pmatrix} 2 & 3 \\ 0 & 0 \end{pmatrix}$ drops all points onto the x-axis. The resulting figure (a line segment) does not contain

enough information to recreate the original figure (a 2D shape). This connects our new interpretation to one of our old ones: a matrix with a determinant of zero cannot be inverted.
- When you apply a transformation to a shape, it multiplies each eigenvector by the associated eigenvalue, thus multiplying the entire area by the determinant. Those eigenvectors, eigenvalues, and determinant are properties of the transformation itself, not just of the matrix that represents it. In a different coordinate system that same transformation will be represented by a different matrix ("similar" to the original). We conclude that any two similar matrices—that is, any two matrices that represent the same transformation in different bases—will have the same eigenvalues and the same determinant. The eigenvectors will also be the same, although their representations in the new basis will be different.

If we tell you that a given 2 × 2 matrix has a determinant of 10, you can conclude that it represents two linearly independent equations and/or two linearly independent vectors. You can conclude that it has an inverse matrix. You can now also conclude that it will multiply the area of any point matrix by 10. But if we tell you that the determinant is −10 you can come to all those same conclusions. So what does the negative determinant mean?

Remember that the determinant is the product of the eigenvalues. Consider a matrix with eigenvalues 5 and 2, and a different matrix with eigenvalues 5 and −2. The first will stretch any object by a factor of 5 in one direction and 2 in another direction. The second will do the same, but then it will reflect shapes in the second direction, so the shapes come out backward. The sign of the determinant therefore tells us about "orientation." A matrix with a positive determinant may stretch and rotate an object shaped like a lowercase 'b', but it will still be a lowercase 'b' in the end. A matrix with a negative determinant will turn it into a lowercase 'd'. See Problems 7.31–7.32.

EXAMPLES — Geometric Transformations

Question: Describe the effects of the following two transformation matrices. You may use a computer to help analyze **B**.

$$\mathbf{A} = \begin{pmatrix} 2 & 0 & 0 \\ 0 & \sqrt{3}/2 & -1/2 \\ 0 & 1/2 & \sqrt{3}/2 \end{pmatrix} \qquad \mathbf{B} = \begin{pmatrix} 3/5 & -4/5 \\ -4/5 & -3/5 \end{pmatrix}$$

Answer A:
You can figure out what this matrix does with no calculations. The new x-component only depends on the old x-component, and the new y and z only depend on the old y and z (not the old x). If there are two groups of variables that are not mixed like that then you say the matrix is "block diagonal."[a] What it does to x is trivial: it doubles it. What it does to y and z is clear if you write $\sqrt{3}/2$ as cos 30° and 1/2 as sin 30°. The lower right block is a 2D rotation matrix that rotates vectors 30° about the x-axis. Using that description, try to predict what this matrix will do to the unit vectors $\hat{i}, \hat{j}$, and $\hat{k}$, and then test them to see if your predictions bear out.

---

[a]Technically a matrix is only block diagonal if the variables in each group are next to each other, so a matrix that mixes x and z but doesn't mix either one with y isn't block diagonal, although in most ways it acts like one. See Appendix E for a definition of "block diagonal."

Answer B:
You can't tell what this one does by inspection. (At least we can't.) When in doubt, the best way to tell what a matrix does is often to find its eigenvectors and eigenvalues. You can do this by hand or by a computer, and either way you find that $\hat{i} + 2\hat{j}$ has eigenvalue -1 and $-2\hat{i} + \hat{j}$ has eigenvalue 1. Since all vectors along the line $y = -x/2$ stay the same and all vectors perpendicular to that line are reversed, this is a reflection about that line.

(Bonus questions: What are the eigenvectors and eigenvalues of matrix **A**? What are the determinants of both matrices? Hint: matrix **A** has only one real eigenvector.)

Orthogonal Matrices

Matrices that only rotate and reflect vectors, but leave all vectors with the same length, are called "orthogonal matrices." All the eigenvalues of an orthogonal matrix must have magnitude 1. Being orthogonal implies having a determinant of ± 1, but the converse is not true. A matrix that stretches by a factor of 2 in the x-direction and $1/2$ in the y-direction would still have determinant 1.

To check if a matrix is orthogonal you could find all of its eigenvalues, but there's a faster way. It requires us to introduce a new term: the "transpose" of a matrix is obtained by turning all its rows into columns. For instance, the transpose of $\mathbf{M} = \begin{pmatrix} 1 & 2 & 3 \\ 4 & 5 & 6 \end{pmatrix}$ is $\mathbf{M}^T = \begin{pmatrix} 1 & 4 \\ 2 & 5 \\ 3 & 6 \end{pmatrix}$. For a square matrix the transpose is a mirror image across the main diagonal: $\begin{pmatrix} a & b \\ c & d \end{pmatrix}^T = \begin{pmatrix} a & c \\ b & d \end{pmatrix}$. A matrix is orthogonal if $\mathbf{M}^T = \mathbf{M}^{-1}$. In other words, to see if a square matrix **M** is orthogonal multiply it by its transpose; it's orthogonal if and only if you get the identity matrix.

The name "orthogonal" comes from the fact that all of the columns (or equivalently all the rows) of an orthogonal matrix represent orthogonal (aka perpendicular) vectors of length 1, just in case you were wondering.

Stepping Back

When Charles Darwin studied the finches of the Galápagos and Cocos islands in 1835, he was impressed by the diverse shapes and sizes of their beaks, which were remarkably well adapted to their food gathering needs. 175 years later a group of evolutionary biologists and applied mathematicians used a set of transformation matrices to determine that nearly all the beaks can be classified into two groups.[1] Within each group, every beak can be transformed to every other beak by a stretch (diagonal matrix). Furthermore, each group can be transformed to the *other* group by a shear matrix.[2] This result suggests that the wide variation in beaks can be accounted for by only two or three genetic parameters, which one biologist observed "makes it much more feasible that you can get that much change in a relatively short time."

---

[1] Campas *et al.*, "Scaling and shear transformations capture beak shape variation in Darwins finches," Proceedings of the National Academy of Sciences 2010; published ahead of print February 16, 2010, doi:10.1073/pnas.0911575107.

[2] In 2D a shear matrix is either $\begin{pmatrix} 1 & k \\ 0 & 1 \end{pmatrix}$ or $\begin{pmatrix} 1 & 0 \\ k & 1 \end{pmatrix}$ where k is a non-zero constant. Visually it tilts shapes, e.g., turning a rectangle into a parallelogram.

So visual transformations—and the matrices that represent them—have important applications in any field that represents information with vectors. In many cases those vectors are used to build geometrical constructs such as computer graphics, simulations of rotating galaxies, force vectors in a tilted coordinate system, or bird beaks. In other cases the vectors represent abstract quantities such as commodity prices, population statistics, or kiddy books and chapter books. Even in these abstract cases, however, drawing the effects of a matrix as if it were a geometric transformation can help build your intuition for the system.

The way you interpret a transformation depends on whether it is "active" or "passive." An active transformation changes the objects represented by the vectors. If you use a rotation matrix to advance the particles in a tornado from their current positions to their positions one tenth of a second later then you are performing an active transformation. A passive transformation views the unchanged objects from a different coordinate system. If you find out how the constellations appear to move across the night sky as we watch them from a rotating Earth, you are performing a passive transformation.

Computer graphics requires frequent use of both active and passive transformations. The same transformation on the screen could represent the object you are looking at rotating 20° clockwise or your character's head turning 20° counterclockwise. The fact that these two transformations look identical held up our understanding of the night sky for well over 1000 years.

7.1.3 Problems: Geometric Transformations

7.1 (a) Write a matrix **A** that stretches any 2D vector by 8 in the *x*-direction and 3 in the *y*-direction.

(b) Write a matrix **B** that reflects any 2D vector across the *x*-axis.

(c) Write a matrix **C** that rotates any 2D vector 90° clockwise.

(d) Write a matrix **M** that performs the stretch from Part (a), then the reflection from Part (b), and finally the rotation from Part (c).

(e) Draw the vector $\hat{\imath}$ and show how it goes through those three steps in that order. Then show that $M\hat{\imath}$ gives you the answer that it should.

(f) Write a matrix **N** that does the rotation, then the reflection, and then the stretch.

(g) Draw the vector $\hat{\imath}$ and show how it goes through those three steps in *that* order. Then show that $N\hat{\imath}$ gives you the answer that it should.

(h) Calculate the determinants of matrices **M** and **N**. Explain in terms of areas why they came out the way they did.

7.2 *[This problem depends on Problem 7.1.]* Find the inverse of matrix **M**. Describe visually what it does. Multiply it by the vector you got in Part (e). Why did the answer come out the way it did?

7.3 The "point matrix" $\mathbf{P} = \begin{pmatrix} 0 & 5 & 8 & 2 & 0 \\ 0 & 0 & 2 & 2 & 0 \end{pmatrix}$ represents a shape by giving a computer instructions to draw a line from the point (0,0) to (5,0), and then a line from (5,0) to (8,2), and so on.

(a) Draw the resulting shape.

(b) Matrix $\mathbf{A} = \begin{pmatrix} 0 & 1 \\ -1 & 0 \end{pmatrix}$ represents a transformation. Find matrix **AP** and draw the resulting shape. Then answer in words: what effect does **A** have on a shape?

(c) Matrix $\mathbf{B} = \begin{pmatrix} 0 & 1 \\ 1 & 0 \end{pmatrix}$ represents a different transformation. Find matrix **BP** and draw the resulting shape. Then answer in words: what effect does **B** have on a shape?

(d) Multiply **A** times **B** to create the matrix **AB**, and then apply that matrix to the original matrix to create **ABP**. What effect does **AB** have?

(e) Multiply **B** times **A** to create the matrix **BA**, and then apply that matrix to the original matrix to create **BAP**. What effect does **BA** have?

(f) Matrix $\mathbf{C} = \begin{pmatrix} 1 & 0 \\ 0 & -2 \end{pmatrix}$ represents a transformation. Find matrix **CP** and draw the resulting shape. Then answer in words: what effect does **C** have on a shape?

(g) Calculate the determinants $|\mathbf{A}|$, $|\mathbf{B}|$, and $|\mathbf{C}|$ and explain how they are reflected in the transformations you drew.

7.4 Write a matrix that rotates 3D vectors 30° counterclockwise about the y-axis, as viewed from positive y. Check your answer by acting with it on the vector $\hat{\imath}$ and seeing that it ends up where you wanted.

Problems 7.5–7.10 make a statement in words, such as "The determinant of a product is the product of the determinants." In each case:

- Write the statement mathematically. In our example, you would write $|\mathbf{AB}| = (|\mathbf{A}|)(|\mathbf{B}|)$.
- Explain verbally why the statement must be true. We discussed our example in the Explanation (Section 7.1.2) based on the "area multiplier" interpretation of the determinant.
- Demonstrate the statement mathematically. In our example we would multiply two matrices with arbitrary components (letters instead of numbers), compute the determinants, and show that the answers come out the same.

Unless otherwise specified assume all vectors and matrices in these problems are 2D.

7.5 If $\mathbf{R}$ represents a rotation matrix of any amount, its determinant will always be 1.

7.6 If $\mathbf{R}$ is a rotation matrix, its inverse is a rotation matrix with minus the angle.

7.7 If you act on a vector with two different rotation matrices $\mathbf{R}_1$ and $\mathbf{R}_2$, you end up in the same place no matter which order they act in.

7.8 If you multiply a 2×2 matrix by a constant, the determinant multiplies by the square of that constant.

7.9 If you multiply a 3×3 matrix by a constant, the determinant multiplies by the cube of that constant.

7.10 For any square matrix $\mathbf{M}$ with a non-zero determinant, the determinant of its inverse is the inverse of its determinant. (You only need to demonstrate this for a 2×2 matrix.)

7.11 For each of the transformations below, say how many real, linearly independent eigenvectors there are, and what their eigenvalues are. Remember that when you have two degenerate eigenvalues you have a whole plane of eigenvectors, but that plane still only represents two linearly independent eigenvectors. No calculations should be required for this problem.

(a) A reflection in 2D
(b) A reflection in 3D
(c) Switching the x- and y-coordinates of a 2D vector
(d) The identity matrix in 3D
(e) The matrix $\begin{pmatrix} 2 & 0 \\ 0 & 2 \end{pmatrix}$
(f) A 30° rotation about the z-axis followed by a reflection across the xy-plane.

7.12 Write a transformation matrix that reflects 2D vectors across the line $y = -x$. *Hint*: Think about what happens to the x- and y-components of the vectors.

7.13 You can tell by inspection that the matrix $\begin{pmatrix} 1 & 0 \\ 0 & 7 \end{pmatrix}$ stretches vectors by a factor of 7 in the y-direction. With a matrix like $\mathbf{M} = \begin{pmatrix} 2 & 2 \\ 2 & 5 \end{pmatrix}$ you can't tell as easily, but you can figure it out from the eigenvectors and eigenvalues.

(a) Find the eigenvectors and eigenvalues of $\mathbf{M}$, either by hand or with a computer.
(b) Say in words what each of the eigenvectors does when you act on it with $\mathbf{M}$.
(c) What transformation does $\mathbf{M}$ perform?

For Problems 7.14–7.19 describe in words what transformation the matrix performs. For some you can tell by inspection. For others you will need to start by finding the eigenvectors and eigenvalues.

7.14 $\begin{pmatrix} 1 & 0 \\ 0 & -2 \end{pmatrix}$

7.15 $\begin{pmatrix} \sqrt{3}/2 & -1/2 \\ 1/2 & \sqrt{3}/2 \end{pmatrix}$

7.16 $\begin{pmatrix} -8 & -6 \\ 3 & 3 \end{pmatrix}$

7.17 $\begin{pmatrix} -3 & -6 \\ 5 & 8 \end{pmatrix}$

7.18 $\begin{pmatrix} 1 & 0 & 0 \\ 0 & \sqrt{2}/2 & \sqrt{2}/2 \\ 0 & -\sqrt{2}/2 & \sqrt{2}/2 \end{pmatrix}$

7.19 💻 $\begin{pmatrix} 4/3 & 1/3 & 1/3 \\ 1/3 & 4/3 & 1/3 \\ 1/3 & 1/3 & 4/3 \end{pmatrix}$

7.20 In this problem you'll find the matrix $\mathbf{M}$ that stretches vectors by a factor of 3 along the line $y = 2x$.

(a) Just from this description, what are the eigenvectors and eigenvalues of $\mathbf{M}$?

(b) Write two matrix equations that express mathematically what you concluded in Part (a).

(c) Write out those matrix equations as four algebraic equations, using a, b, c, and d for the unknown elements of **M**, and solve the equations to find **M**.

7.21 In this problem you'll find the matrix **M** that stretches vectors by a factor of 3 along the axis $x = y$ and reflects vectors across that same axis.

(a) Just from this description, what are the eigenvectors and eigenvalues of **M**?

(b) Write two matrix equations that express mathematically what you concluded in Part (a).

(c) Write out those matrix equations as four algebraic equations, using a, b, c, and d for the unknown elements of **M** and solve the equations to find **M**.

7.22 If you want to stretch a vector by a factor of 3 in direction $\vec{d}$ then your matrix's first eigenvector should be $\vec{d}$ with eigenvalue 3, and its second eigenvector should be perpendicular to $\vec{d}$ with eigenvalue 1. (Problems 7.20 and 7.21 illustrated this point.) The least obvious part of that rule is that the second eigenvector must be perpendicular to $\vec{d}$.

To investigate that claim, we begin by considering matrix $\mathbf{T}_1$ whose eigenvectors are $\hat{j}$ (with eigenvalue 3) and $\hat{i}$ (with eigenvalue 1). *Note: this entire problem can be solved with almost no calculations if you understand at each step what the question is asking.*

(a) Let $\vec{v} = \hat{i} + \hat{j}$. Calculate the vector $\mathbf{T}_1 \vec{v}$.

(b) Draw $\vec{v}$ and $\mathbf{T}_1 \vec{v}$ on the same graph. Now consider matrix $\mathbf{T}_2$ whose eigenvectors are $\vec{A} = \hat{j}$ (with eigenvalue 3) and $\vec{B} = \hat{i} + \hat{j}$ (with eigenvalue 1).

(c) What are the components of $\vec{v}$ in the $\vec{A}\vec{B}$ basis?

(d) What is the effect of matrix $\mathbf{T}_2$ on $\vec{v}$?

(e) Draw $\vec{v}$ and $\mathbf{T}_2 \vec{v}$ on the same graph.

(f) Did $\mathbf{T}_1$ stretch v by 3 in the $\hat{j}$-direction? Did $\mathbf{T}_2$?

7.23 Matrix **T** has eigenvectors $\hat{i}$ and $\hat{j}$, both with eigenvalue 3. It also has eigenvector $\hat{k}$ with eigenvalue 1.

(a) For $\vec{v}_1 = 2\hat{i} - 5\hat{j}$ find $\mathbf{T}\vec{v}_1$.

(b) Is $\vec{v}_1$ an eigenvector of **T**? If so, find its eigenvalue.

(c) For $\vec{v}_2 = 2\hat{i} - 5\hat{k}$ find $\mathbf{T}\vec{v}_2$.

(d) Is $\vec{v}_2$ an eigenvector of **T**? If so, find its eigenvalue.

(e) In general, what vectors are eigenvectors of **T**?

7.24 You are looking for a transformation **M**, and initially all you know about it is that it turns the vector $\hat{i} - \hat{j}$ into the vector $-\hat{i} - \hat{j}$.

(a) Write a rotation matrix that could be **M**, given this information.

(b) Write a reflection matrix that could be **M**, given this information.

(c) What would each of the matrices you just wrote do to the vector $\hat{i} + \hat{j}$?

7.25 Prove that any transformation matrix moves the origin to itself.

7.26 Can you write a matrix that converts the Cartesian coordinates of a point into the spherical coordinates of that point? If so, write the matrix. If not, explain why not.

7.27 Point matrices are a bit more complicated in 3D than in 2D, but the basic idea is the same. Let **P** be the following big-but-not-so-bad-when-you-look-at-it-carefully matrix.

$$\mathbf{P} = \begin{pmatrix} 0 & 1 & 1 & 0 & 0 & 0 & 1 & 1 & 0 & 0 & 0 & 0 & 1 & 1 & 1 & 0 \\ 0 & 0 & 1 & 1 & 0 & 0 & 0 & 1 & 1 & 0 & 1 & 1 & 1 & 0 & 0 & 0 \\ 0 & 0 & 0 & 0 & 0 & 1 & 1 & 1 & 1 & 1 & 1 & 0 & 0 & 1 & 1 & 0 & 0 \end{pmatrix}$$

(a) Describe the shape represented by **P**. What kind of shape is it, how big is it, and where is it?

(b) Write a matrix that reflects vectors across the yz-plane and stretches them by 2 in the z-direction.

(c) Act with the transformation matrix you wrote on **P**. Describe the resulting shape.

7.28 In this problem you will show both physically and mathematically that 3D rotation matrices do not commute. Stand facing your desk. Define your right to be the positive x-direction, in front of you to be the positive y-direction, and up to be the positive z-direction. Your upper body is the vector you're transforming, pointing through the top of your head, and initially it points in the positive z-direction.

(a) Bend forward 90°. What axis did you rotate about and in what direction did your head end up pointing (x, y, or z, and positive or negative)?

(b) Still bent over, turn left 90°. What axis did you rotate about and in what direction did your head end up pointing?

(c) Stand up straight facing your desk again to reset back to your original vector. Now do the same rotation you did in Part (b). In what direction did your head end up pointing?

(d) Now do the rotation from Part (a). Be sure to rotate about the same axis you did before. This time it will not be a forward bend. (If you're not flexible enough to bend sideways 90°, and who is really, just go as far as you can and pretend.) In what direction did your head end up pointing?

(e) Write the matrices for the rotations in Parts (a) and (b). Multiply them in both orders to see if they commute. How could you have predicted the answer without doing any calculations?

7.29 Matrix $\mathbf{S}$ stretches all vectors by a factor of 5 in direction $\vec{D}$ (and that's all it does). When this matrix acts on most vectors, it changes both their magnitudes and their directions.

(a) When $\mathbf{S}$ acts on the vector $\vec{A}$, it changes its magnitude but not its direction. What can you conclude about $\vec{A}$?

(b) When $\mathbf{S}$ acts on the vector $\vec{B}$, it does not change its magnitude or direction. What can you conclude about $\vec{B}$?

7.30 In this problem you will prove that the determinant of a diagonal matrix is the product of the numbers along the main diagonal. (For instance, $\begin{vmatrix} 3 & 0 & 0 \\ 0 & 5 & 0 \\ 0 & 0 & -2 \end{vmatrix} = 3 \times 5 \times -2 = -30$.)

(a) Prove that this is true for any diagonal 2×2 matrix.

(b) Prove that if this result holds for an $n \times n$ matrix, then it also holds for an $(n+1) \times (n+1)$ matrix.

7.31 The figure below shows a right triangle with vertices at $(0,0)$, $(2,0)$, and $(2,1)$. We've labeled the sides A, B, and C, and we'll refer to the triangle as R. You can define the "orientation" of R by the fact that if you start at the origin (the vertex AC) and move along A, you then have to turn left to turn onto B. In this problem you'll explore how different transformations affect this orientation.

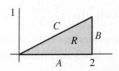

(a) Transform each of the vectors of R with the matrix $\mathbf{T}_1 = \begin{pmatrix} 0 & 1 \\ -1 & 0 \end{pmatrix}$ and draw the resulting shape. Side A of the old triangle becomes one of the sides of your new triangle; mark that side A_1, and similarly for the other two sides. Does the transformed triangle have the same orientation as R (turn left from A to B) or the opposite orientation (turn right from A to B)?

(b) Repeat Part (a) for each of the following matrices.

i. $\mathbf{T}_2 = \begin{pmatrix} -2 & 0 \\ 0 & -2 \end{pmatrix}$

ii. $\mathbf{T}_3 = \begin{pmatrix} 0 & 1 \\ 1 & 0 \end{pmatrix}$

iii. $\mathbf{T}_4 = \begin{pmatrix} 1 & 0 \\ 0 & -3 \end{pmatrix}$

(c) For each of the matrices $\mathbf{T}_1$ to $\mathbf{T}_4$, calculate the determinant.

(d) Based on your results, what does the determinant of a matrix tell you about the orientation of shapes?

7.32 You perform a transformation on the shape **a**. For each of the resulting shapes shown below, is the determinant positive or negative, and is its absolute value less than or greater than 1?

(a) ɑ

(b) 𝑎

(c) ꭤ

(d) ɐ

7.33 The matrices $\begin{pmatrix} 1 & 4 \\ 2 & 7 \end{pmatrix}$ and $\begin{pmatrix} 3 & 2 \\ 6 & 1 \end{pmatrix}$ are *not* similar; that is, they do not represent the same transformation as each other in any possible coordinate systems. How we can tell that for sure?

7.34 For each matrix below, determine if it is orthogonal. If it is not, find at least one vector whose magnitude is changed by the matrix.

(a) $\begin{pmatrix} 1 & 1 \\ 1 & -1 \end{pmatrix}$

(b) $\begin{pmatrix} 1/2 & \sqrt{3}/2 \\ -\sqrt{3}/2 & 1/2 \end{pmatrix}$

(c) $\begin{pmatrix} 1/\sqrt{3} & -1/\sqrt{2} & 1/\sqrt{6} \\ -1/\sqrt{3} & 1/\sqrt{3} & -1/\sqrt{3} \\ 1/\sqrt{3} & -1/\sqrt{6} & 1/\sqrt{2} \end{pmatrix}$

7.35 Prove that any 2D rotation matrix is orthogonal. Explain why this result makes sense. Explain without a mathematical proof why you would or wouldn't expect the same result to hold in 3D.

7.36 For each transformation matrix described below, say whether the determinant is positive, negative, or zero, or if there isn't enough information to tell.
  (a) The identity matrix
  (b) It has three eigenvalues: 5, −5, and −2.
  (c) It transforms the vector $\hat{i}$ into $-\hat{i}$.
  (d) It transforms the unit square with corners at $(0,0)$ and $(1,1)$ into a unit square with corners at $(0,0)$ and $(-1,1)$.
  (e) The vectors $\hat{i}+\hat{j}$ and $\hat{i}-\hat{j}$ get transformed into each other.

7.37 3D objects are often represented on computers by lists of polygons that can be drawn to approximate the surface of the shape. In this problem you're going to represent and manipulate a sphere.
  (a) Define a set of polygons representing a sphere with 8 lines of longitude and 8 lines of latitude. The first set should be a set of triangles going from the point $(0,0,1)$ to points with $\theta = \pi/8$ and ϕ going from 0 to 2π in 8 steps. Each triangle will use two different values of ϕ. The next polygons should be quadrilaterals connecting points with $\pi/8$ to points with $\theta = \pi/4$, again with two values of ϕ per polygon. Continue until you get a set of triangles that reach the south pole. Have the computer draw all these polygons.
  (b) Redefine the list and redraw the sphere with ϕ and θ taking on 64 values each instead of 8.
  (c) Write a matrix **S** that stretches vectors by a factor of 2 in the y-direction. Apply **S** to each polygon vertex and draw the result. What does it look like?
  (d) Define a matrix **R** that rotates vectors 45° around the z-axis. Apply R to the vertices of the original polygons and draw the result. What does it look like?
  (e) Apply the matrix **RS** to the original vertices and redraw the shape. Explain why it looks the way it does.

7.2 Tensors

You know that a physical quantity such as velocity or electric field can be represented by a vector. That vector in turn can be specified by giving a sequence of numbers (components), but those components change if you represent the same vector in a different basis. In this section we introduce the "tensor," a physical quantity that can be represented by a matrix. Like a vector, a tensor is specified using different numbers in different bases, where these representations all describe the same thing in different ways.

We first introduced this idea in Chapter 6, where we defined "similar matrices" as ones that represent the same transformation in different bases. In this section we convert a transformation matrix from one basis to another, and we also generalize the idea to tensors other than transformations.

7.2.1 Discovery Exercise: Tensors

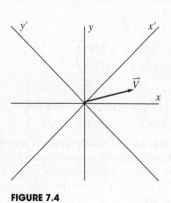

FIGURE 7.4

Figure 7.4 shows a vector $\vec{V}$ and two sets of axes. The exact components of the vector are not important for this problem, but it is important that the $x'y'$-axes are at a 45° angle relative to the xy-axes. Let V_x and V_y be the components of $\vec{V}$ in the xy basis, and let $V_{x'}$ and $V_{y'}$ be the components of $\vec{V}$ in the $x'y'$ basis.

1. Write a matrix **C** that converts vectors from the xy basis to the $x'y'$ basis. In other words, write the components of **C** such that $\begin{pmatrix} V_{x'} \\ V_{y'} \end{pmatrix} = \mathbf{C} \begin{pmatrix} V_x \\ V_y \end{pmatrix}$. *Hint*: Try to see what the vector $<1,1>$ in the xy basis should become in the $x'y'$ basis and make sure your matrix **C** gets that one right.

See Check Yourself #45 in Appendix L

2. Write a matrix that converts vectors from the $x'y'$ basis to the xy basis. How is this matrix related to the matrix **C** that you just wrote?
3. Write a matrix **M′** that takes a vector in the $x'y'$ basis and stretches it by a factor of 2 in the x'-direction. In other words $\mathbf{M'}\begin{pmatrix} V_{x'} \\ V_{y'} \end{pmatrix} = \begin{pmatrix} 2V_{x'} \\ V_{y'} \end{pmatrix}$.
4. Using your answers so far, write a matrix **M** that takes a vector in the xy basis, converts it to the $x'y'$ basis, stretches it by two in the x'-direction, and then converts back to the xy basis.

   See Check Yourself #46 in Appendix L

5. Describe in words what the matrix **M** does to a vector $\vec{V}$ in the xy basis. Your answer should make no reference to the $x'y'$ basis, but should just say what kind of stretching or reflection it performs. *Hint*: The answer does *not* involve a rotation.
6. Choose a vector on which the transformation you described should look very simple, and check that the matrix **M** you wrote down does what you would expect to that vector.

7.2.2 Explanation: Tensors

Consider the classic "block sliding down a ramp" problem. The key to such a problem is translating between two different coordinate systems, or bases. In one system the two axes are y (defined by the force of gravity) and x (perpendicular to y). In the other system the two axes are x' (the direction of the ramp) and y' (perpendicular to x'). The two systems are offset by an angle θ.

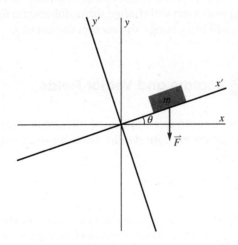

A typical ramp problem begins with two given quantities.

- The mass of the block in the xy system is the scalar m. In the rotated coordinate system, the mass of the block is still m.
- The force of gravity in the xy system is the vector $\vec{F} = -mg\hat{j}$. In the rotated coordinate system, $\vec{F} = -mg\sin\theta \hat{i}' - mg\cos\theta \hat{j}'$.

How did we determine those conversions? The mass was easy since a scalar is the same in any coordinate system. The force can be converted by drawing right triangles, but we tackled it by using a rotation matrix—the same matrix that converts any point from one system to another. (In Problem 7.51 you'll show that this gives the same results as drawing the triangles.)

If this all sounds too obvious (or too familiar) to discuss, here comes something new: those two conversions are the *definitions* of the terms "scalar" and "vector."

Chapter 7 Linear Algebra II

> **Definitions: Scalar and Vector**
>
> Consider two different coordinate systems and an orthogonal matrix **C** that converts any set of coordinates—any single point—from the first system to the second. (The restriction to orthogonal conversions means we are considering rotation and reflection but not stretching.)
>
> - In the first coordinate system, a certain value is specified by the number s. If this is a "scalar" quantity then its value in the second coordinate system is given by:
>
> $$s' = s \qquad (7.2.1)$$
>
> - In the first coordinate system, a different value is specified by the sequence of numbers $\vec{v}$. If this is a "vector" quantity then its value in the second coordinate system is given by:
>
> $$\vec{v}' = \mathbf{C}\vec{v} \qquad (7.2.2)$$
>
> Be cautious about the distinction between active and passive transformations when converting vectors. If your coordinate system rotates 30° counterclockwise, you must apply a 30° clockwise rotation to every object!

The definition tells us that a scalar is not just "a quantity that can be expressed by a number"—it is a specific type of quantity that retains its value across different coordinate systems. Less obviously, the definition tells us the same thing about a vector. You might stand on our ramp and declare that "the block experiences a gravitational force of 98 N in this direction (point your finger), and it accelerates at 4.9 m/s² in that direction (point a different finger)." Changing your basis will change the numbers you use to specify the components of those vectors, but it will not change the vectors themselves.

EXAMPLES **Scalars and Vector Fields**

Air is moving around a room in complicated ways, so at any given point there is an air velocity. In a particular coordinate system this is represented by the numbers $\vec{v} = v_1 \hat{x}_1 + v_2 \hat{x}_2 + v_3 \hat{x}_3$.

Question: Is the air velocity a vector?

Answer:
Yes. The vector itself is independent of the coordinate system used: at a particular point the air is moving *this* fast in *this* direction. But the coordinates v_1, v_2, and v_3 depend on the coordinate system. If you switch to a different coordinate system, the new components will be given by $\vec{v}' = \mathbf{C}\vec{v}$ where **C** is the matrix that converts positions from the first coordinate system to the second.

Question: Is the magnitude of the air velocity a scalar?

Answer:
Yes. Imagine that at one particular point the air is moving at 5 m/s. In any coordinate system, the air at that point will still be moving at 5 m/s.

Question: Is the x-component of the air velocity a scalar?

Answer:
No, and this is a good example of why "a quantity that can be represented by a number" is not an adequate definition of scalar. In different coordinate systems the x-component will be different. To put it another way, if I tell you that the x-component at a particular spot is 4 m/s that doesn't tell you much if you don't know what coordinate system I'm using.

Question: The color of an object can be specified as a list of how much red, green, and blue make it up. Is that list a vector?

Answer:
No. If you change coordinates each of those colors stays the same, so that is a list of three scalars rather than a vector.

Here is a subtler example: you probably think of the cross product of two vectors as a vector, but you'll show in Problem 7.50 that it doesn't obey Equation 7.2.2 under a reflection. Given two vectors $\vec{A}$ and $\vec{B}$, the quantity $\vec{A} \cdot \vec{B}$ is a scalar, but $\vec{A} \times \vec{B}$ is not actually a vector. (It's not a scalar either, of course. Technically it's a "pseudovector.")

Converting a Transformation

Jack is playing a video game and his character has just encountered "Rubber Woman," who has the power to stretch her body vertically. When Rubber Woman uses her power Jack sees the following transformation occur on the screen.

Suppose you are the programmer writing this video game. The position of each point on Rubber Woman's body is stored as a 2D vector. You have the computer multiply each of those vectors by the stretching matrix $\mathbf{S} = \begin{pmatrix} 1 & 0 \\ 0 & 2 \end{pmatrix}$ and then redraw the screen.[3]

Suppose, however, that before Rubber Woman performs this stretch Jack rotates his character's head 30° to the left. On his screen he sees everything in the scene rotate 30° to the right. Once again, you the programmer know what to do: the matrix $\mathbf{C} = \begin{pmatrix} \cos 30° & \sin 30° \\ -\sin 30° & \cos 30° \end{pmatrix}$ converts each position vector from the old coordinates to the new rotated ones. ("Conversion" in this context is synonymous with "passive transformation.")

Now comes the interesting part. When Rubber Woman uses her power you have to stretch her body exactly as before but from a different perspective. How do you implement that transformation?

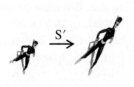

You can answer that question with a three-step process. First convert each point's position back to the original basis. (This is accomplished by the matrix $\mathbf{C}^{-1}$.) Then use matrix $\mathbf{S}$ to find the coordinates of the stretched Rubber Woman—still in the original basis. Finally convert those coordinates into the new basis using matrix $\mathbf{C}$. Remembering the right-to-left order of matrix multiplication, we write:

$$\mathbf{S}' = \mathbf{CSC}^{-1} \tag{7.2.3}$$

---

[3]Here and below we are assuming for simplicity that the origin of your coordinate system is at Rubber Woman's feet. We are also ignoring the fact that many video games define and transform objects in a three-dimensional space and then use complicated algorithms to render them properly on a two-dimensional screen.

We are *not* suggesting that you need to go through a computationally inefficient three-step process for each point on Rubber Woman's body. The entire process of converting to the original basis, performing the stretch in that basis, and then converting back to the new basis can be effected by one matrix, and Equation 7.2.3 gives the formula for finding that matrix. In short, **S'** performs the stretch in the head-rotated basis just as **S** does in the original basis.

> **EXAMPLE** **Converting a Transformation**
>
> Question: Transformation **T** stretches any object by a factor of 2 in the *x*-direction and 6 in the *y*-direction, and then rotates the object by 30° counterclockwise. What matrix performs transformation **T** in a coordinate system defined by the basis vectors $\vec{V}_1 = 3\hat{i} + \hat{j}$ and $\vec{V}_2 = 10\hat{i} + 4\hat{j}$?
>
> Answer:
> First let's figure out what matrix performs transformation **T** in $\hat{i}\hat{j}$ space. The stretch is $\begin{pmatrix} 2 & 0 \\ 0 & 6 \end{pmatrix}$ and the rotation is $\begin{pmatrix} \sqrt{3}/2 & -1/2 \\ 1/2 & \sqrt{3}/2 \end{pmatrix}$. So the matrix that performs both operations—remember right-to-left order!—is:
>
> $$\mathbf{T}_{\hat{i}\hat{j}} = \begin{pmatrix} \sqrt{3}/2 & -1/2 \\ 1/2 & \sqrt{3}/2 \end{pmatrix} \begin{pmatrix} 2 & 0 \\ 0 & 6 \end{pmatrix} = \begin{pmatrix} \sqrt{3} & -3 \\ 1 & 3\sqrt{3} \end{pmatrix}$$
>
> The matrix that converts from a $\vec{V}_1 \vec{V}_2$ representation to $\hat{i}\hat{j}$ is $\begin{pmatrix} 3 & 10 \\ 1 & 4 \end{pmatrix}$. (This is $\mathbf{C}^{-1}$ in the formula above.) The inverse of that matrix, which goes back to $\vec{V}_1 \vec{V}_2$, is $\frac{1}{2}\begin{pmatrix} 4 & -10 \\ -1 & 3 \end{pmatrix}$. So the matrix that performs transformation **T** in $\vec{V}_1 \vec{V}_2$ space is:
>
> $$\mathbf{T}_{\vec{V}_1 \vec{V}_2} = \frac{1}{2}\begin{pmatrix} 4 & -10 \\ -1 & 3 \end{pmatrix} \begin{pmatrix} \sqrt{3} & -3 \\ 1 & 3\sqrt{3} \end{pmatrix} \begin{pmatrix} 3 & 10 \\ 1 & 4 \end{pmatrix}$$
>
> $$= \begin{pmatrix} -21 - 9\sqrt{3} & -74 - 40\sqrt{3} \\ 6 + 3\sqrt{3} & 21 + 13\sqrt{3} \end{pmatrix}$$

You could of course have a computer do that last multiplication for you. But what a computer can't do—what you need to be able to do—is see that last step as saying "first convert from $\vec{V}_1 \vec{V}_2$ to $\hat{i}\hat{j}$ representation, then perform the transformation in that space, and then convert back." The final matrix therefore performs transformation **T** in $\vec{V}_1 \vec{V}_2$ space.

Tensors
The transformation in our computer game example has some important features in common with a vector. Like a vector, it has an essential nature that is independent of a coordinate system (it stretches objects by a certain amount in a certain direction). Also like a vector, it is specified by a set of numbers that must be converted as we switch from one coordinate system to another.

Remember that Equation 7.2.2 tells us how to convert a vector between bases, and that conversion is the *definition* of a vector. Because **S** follows a different conversion rule, it is a different type of quantity. We call such a quantity a "rank two tensor."

Definition: Tensor

Consider two different coordinate systems and an orthogonal conversion matrix **C** that converts any single point from the first system to the second. In the first coordinate system, a certain value is specified by the sequence of numbers **T**. If this is a "rank two tensor" (often simply called a "tensor") then its value in the second coordinate system is given by:

$$\mathbf{T}' = \mathbf{CTC}^{-1} \qquad (7.2.4)$$

A scalar is a "rank 0 tensor" and a vector is a "rank 1 tensor." Ranks higher than two are important in the theory of general relativity, but are less common in most other fields. Our explanation above is designed to convince you that a transformation is a rank 2 tensor. But the concept is broader than that—that is, there are quantities other than transformation matrices that convert according to Equation 7.2.4.

As an example, suppose you want to know how the air velocity changes from one place in the room to another. Since $\vec{v}$ is given by three components v_x, v_y, and v_z and we can measure the rates of change of each component independently in three different directions, there are *nine* independent spatial derivatives we can calculate. For instance, if you held a pinwheel[4] facing the *x*-direction, $\partial v_x / \partial z$ would tell you how much it would speed up or slow down as you moved it slowly upward. The nine components of this "velocity gradient" can be arranged in a matrix, and you'll show in Problem 7.56 that this matrix transforms like a tensor under coordinate transformations. That means the velocity gradient is a tensor.

The transformation rule for tensors tells you how its *components* change from one basis to another, but the physical thing that the tensor represents is the same regardless of basis. Imagine standing at a given point in the room, with your left hand pointing at the ceiling and your right hand at the door. You could ask "How much is *this* (left hand) component of the velocity changing as you move in *that* (right hand) direction?" The velocity gradient gives you a way of answering that question, and the answer is the same no matter what coordinate system you use.

Index Notation

The components of a matrix **M** can be expressed in "index notation" as M_{ij}, which refers to the element in the *i*th row and *j*th column. For example, a 2×2 matrix can be written as $\mathbf{M} = \begin{pmatrix} M_{11} & M_{12} \\ M_{21} & M_{22} \end{pmatrix}$. Expressions involving matrix multiplication can often be expressed compactly with this notation. For example, the equation $A_{ij} = \sum_k B_{ik} C_{kj}$ completely describes the matrix product $\mathbf{A} = \mathbf{BC}$. You'll show in Problem 7.52 that for bases related by orthogonal transformations the rules for transforming vectors and tensors can be expressed in index notation as:

$$v'_i = \sum_j C_{ij} v_j \qquad (7.2.5)$$

$$T'_{ij} = \sum_k \sum_l C_{ik} C_{jl} T_{kl} \qquad (7.2.6)$$

Index notation makes it easier to generalize the conversion rule to higher-rank tensors.

---

[4]It's a toy that spins when wind blows on it.

> **Transforming Tensors: Index Form**
>
> If a rank n tensor has components $T_{abc...}$ in frame F and $\mathbf{C}$ is an orthogonal matrix that converts coordinates from F into F':
>
> $$T'_{abc...} = \sum_\alpha \sum_\beta \sum_\gamma \cdots C_{a\alpha} C_{b\beta} C_{c\gamma} \cdots T_{\alpha\beta\gamma...} \qquad (7.2.7)$$

In fields such as relativity that use a lot of tensors, expressions such as these can be expressed more easily with "Einstein summation notation." Whenever a single index appears twice in the same term, it is automatically summed over all its possible values. For instance the rule for transforming a rank 2 tensor would be written $T'_{ij} = C_{ik} C_{jl} T_{kl}$, with an implied sum over k and l since they each appear twice in the term on the right-hand side. It should be clear from context what the limits of those sums are, e.g., from 1 to 3 if these are 3×3 matrices. We will not use Einstein notation in this text, but you should be aware of it if you read articles or books with tensor equations.

Stepping Back

The terms scalar, vector, and tensor are defined by their behavior under coordinate transformation. All three terms therefore have meaning only with respect to specific transformations. To illustrate that point, let's return to a couple of our examples from Page 360.

- The color of an object is the same in any coordinate system, so color is not a vector but three scalars—with respect to *spatial* transformations. On the other hand color transforms like a generalized vector when we convert from "red, green, and blue" to "hue, saturation and lightness."
- The speed of an object is the same in any coordinate system, so speed is a scalar—provided those coordinate systems are not moving with respect to each other. But an observer on the ground and an observer on a passing train will not agree on the speeds of various objects they can both see. Under that transformation, speed is not a scalar.

For orthogonal transformations of the spatial variables in Cartesian coordinates the definitions we've given here are adequate, but in more general cases it's necessary to define different types of vectors and tensors that obey different transformation laws. Nonetheless, the concepts you've learned here will provide a strong basis if you go on to study more advanced tensor analysis.

We should note that Equation 7.2.3 for converting a transformation is very general because it doesn't depend on what type of conversion the matrix $\mathbf{C}$ represents. The index notation form, Equation 7.2.7, does not generalize to all situations, however.

7.2.3 Problems: Tensors

7.38 Walk-Through: Converting a Transformation. In this problem you will find a matrix that stretches vectors by a factor of 4 along the axis $y = 2x$.

(a) The first step is to define a set of axes such that y' points along the line $y = 2x$ and x' points perpendicular to it. Sketch the x-, y-, x'-, and y'-axes on one plot.

(b) In the $x'y'$ basis, write the matrix that stretches vectors by a factor of 4 in the y'-direction.

(c) Write a matrix that converts vectors from the xy basis to the $x'y'$ basis.

(d) Write a matrix that converts vectors from the $x'y'$ basis to the xy basis.

(e) Write a matrix that takes a vector in the xy basis, converts it to the $x'y'$ basis, stretches it by 4 in the y'-direction, and then converts it back to the xy basis.

(f) Suppose the matrix you wrote in Part (e) acts upon the vector $\vec{V}_1 = \hat{i} + 2\hat{j}$. Predict the result of this operation without doing any calculations, and explain your reasoning.

(g) Act with your matrix on $\vec{V}_1$ and verify that it did what you predicted. If not, figure out whether your calculations or your matrix were wrong.

(h) Suppose the matrix you wrote in Part (e) acts upon the vector $\vec{V}_2 = 2\hat{i} - \hat{j}$. Predict the result of this operation without doing any calculations, and explain your reasoning.

(i) Act with your matrix on $\vec{V}_2$ and verify that it did what you predicted. If not, figure out whether your calculations or your matrix were wrong.

7.39 In an example in Section 6.9 we converted the transformation matrix $\begin{pmatrix} -4 & -2 \\ -3 & 1 \end{pmatrix}$ from an "x is this and y is that" representation to "this much of the first normal mode and that much of the second." Do this conversion using the method described in this section. *Hint*: Equation 6.9.7 gives you the conversion between the two coordinate systems.

7.40 Using the methods described in this section, write a matrix that stretches all vectors by a factor of 2 along the line $y = -x$. Verify that your matrix works by testing it on at least two vectors. It may help to first work through Problem 7.38 as a model.

7.41 The Explanation (Section 7.2.2) introduced Rubber Woman, the spectacular heroine who stretches her body in a video game. It also introduced Jack, the player whose head is rotated so that the entire game is seen at a 30° tilt.

(a) Find the matrix S' that stretches Rubber Woman in Jack's head-rotated basis.

(b) What are the eigenvalues and eigenvectors of the matrix you calculated in Part (a)? Explain how you know the answer, *not* based on the methods described in Chapter 6, but based on what the matrix does.

7.42 (a) Using the methods described in this section, write a matrix that reflects all vectors about the line $y = x/\sqrt{3}$. (*Hint*: In what coordinate system is this an easy transformation?) Verify that your matrix works by testing it on at least two vectors.

(b) What is the determinant of the matrix you calculated in Part (a)? Explain how you know the answer, *not* based on the formula in Chapter 6, but based on what the matrix does. (You can of course use the formula to make sure you're right!)

7.43 The matrix $\mathbf{T} = \begin{pmatrix} 1 & 2 \\ 3 & 4 \end{pmatrix}$ performs a transformation on any vector that is represented as a linear combination of $\hat{i}$ and $\hat{j}$. Write a matrix that performs the same transformation on a vector that is represented as a linear combination of $\vec{V}_1 = (1/2)\hat{i} + (\sqrt{3}/2)\hat{j}$ and $\vec{V}_2 = -(\sqrt{3}/2)\hat{i} + (1/2)\hat{j}$.

7.44 (a) The matrix $\mathbf{T} = \begin{pmatrix} -19 & 12 \\ 12 & -12 \end{pmatrix}$ performs a transformation on any vector that is represented as a linear combination of $\hat{i}$ and $\hat{j}$. Write a matrix that performs the same transformation on a vector that is represented as a linear combination of $\vec{V}_1 = (3/5)\hat{i} + (4/5)\hat{j}$ and $\vec{V}_2 = -(4/5)\hat{i} + (3/5)\hat{j}$.

(b) In this case you get a very interesting matrix. What does that tell you about this particular transformation in that particular basis?

7.45 Let T be a transformation that stretches any vector by a factor of 3 horizontally and by a factor of 4 vertically.

(a) The xy-coordinate system is the natural basis for this transformation because the horizontal stretch lies directly along the x-axis and the vertical stretch directly along the y-axis. Write a matrix $\mathbf{M}$ that represents the transformation T in the xy-coordinate system. (This should be very quick and easy.)

(b) Write a matrix $\mathbf{M}'$ that expresses the same transformation in a coordinate system rotated 45° with respect to the x- and y-axes.

(c) To test your matrix, convert the vector $\hat{i} + \hat{j}$ into the new space, and then use your matrix. The result should be translatable back to $3\hat{i} + 4\hat{j}$.

7.46 The matrix $\begin{pmatrix} 0 & 1 \\ 1 & 0 \end{pmatrix}$ reverses the x- and y-coordinates of a point. What does this transformation look like in a coordinate system that is rotated 45° from the regular axes? Why?

7.47 Prove using a tensor transformation that a 2D rotation matrix is unaffected by a rotation of the axes. Explain why this makes sense.

7.48 In this problem you'll find a matrix that rotates 3D vectors about the line L, defined as $z = x$ in the xz-plane.

(a) You can define a set of axes such that the z'-axis points along L, the x'-axis points along the line perpendicular to L in the xz-plane, and the y'-axis is the same as the y-axis. In the $x'y'z'$ basis, write a matrix that rotates vectors 30° around the z'-axis, counterclockwise as viewed from above.

(b) Write a matrix that converts vectors from the xyz basis to the $x'y'z'$ basis.

(c) Write a matrix that converts vectors from the $x'y'z'$ basis to the xyz basis.

(d) Using all of your previous results, write a matrix that rotates vectors in the xyz basis 30° about the line L, counterclockwise as viewed from above.

7.49 Write a matrix that stretches 3D vectors by a factor of 2 along the line $y = x$ in the xy-plane and then rotates them 20° about the line $y = -x$ in the xy-plane (counterclockwise as viewed from the first octant).

7.50 If we describe a matrix by saying (for instance) that it rotates vectors by a certain angle around some line or reflects them across a certain plane, this transformation should have precisely that effect on *all* vectors. However, some quantities behave like vectors under rotation but not under reflection. Such quantities are not technically vectors, so we call them "pseudovectors." In this problem you will show that the cross product of two vectors is a pseudovector. Let $\vec{A} = \hat{i} + \hat{j}$ and $\vec{B} = -\hat{i} + \hat{j}$.

(a) Draw the vectors $\vec{A}$ and $\vec{B}$. From the right-hand rule, does $\vec{A} \times \vec{B}$ point up or down?

(b) Now consider a simple transformation that reverses the sign of every vector: $\vec{V} \to -\vec{V}$. Draw the transformed vectors $\vec{A}'$ and $\vec{B}'$. (This is an active transformation, so the axes remain the same as they were before.) From the right-hand rule, does $\vec{A}' \times \vec{B}'$ point up or down?

(c) What happened to the cross product of the two vectors under this transformation? What would have happened to it if the cross product were an actual vector?

7.51 A block is at rest on a plane tilted at an angle θ above the horizontal. It experiences a gravitational force mg straight down, a friction force F_{fr} parallel to the plane, a normal force perpendicular to the plane, and a tension force from a string pulling it at a 45° angle between horizontal and vertical. (If θ happened to be 45° the tension force would be parallel to the normal force, but you should not assume that.)

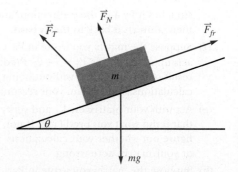

(a) Write all the forces in component form in a basis with horizontal and vertical axes. For all but one of the forces this will require drawing a triangle with the force vector as the hypotenuse and the components as the legs.

(b) In a similar way, draw triangles to find all the forces in component form in a basis with axes parallel and perpendicular to the plane.

(c) Write a rotation matrix that converts vectors from the horizontal-vertical basis to the parallel-perpendicular basis. Use this matrix to rotate all the vectors you found in Part (a) and verify that you get the same answers you got in Part (b).

7.52 The Explanation (Section 7.2.2) gave the rules for transforming vectors and matrices in two forms: first with abstract quantities, and then with index notation. In this problem you will show that they are equivalent.

(a) Show that Equation 7.2.2 for vector transformations is expressed in index notation as Equation 7.2.5.

(b) Show that Equation 7.2.4 for tensor transformations is expressed in index notation as Equation 7.2.6 for an orthogonal transformation.

7.53 The Kronecker Delta You may be familiar with the Kronecker delta: δ_{ij} equals 1 if $i = j$ and 0 otherwise.

(a) Remember that one way to define any matrix is by using index notation to define all its elements. Assuming i and j each run from 1 to 3, write the matrix defined by the Kronecker delta.

(b) You should have recognized your answer. In words, what matrix does the Kronecker delta represent? (You wrote a 3 × 3 matrix, but this answer would be the same in any number of dimensions.)

(c) The definition of a tensor requires it to be convertible between different coordinate systems by Equation 7.2.4. For instance, if we define a tensor by saying "it effects this transformation" then as we convert it from one coordinate system to another it should continue to effect the same transformation. In this problem we have defined a matrix by giving all its elements. For this matrix to be a tensor it must retain this definition in different coordinate systems. Show that the tensor transformation law leaves the components of the Kronecker delta unchanged. *Hint*: it will be easier to use the transformation law in matrix form rather than index form.

7.54 **The Levi-Civita Symbol** The "Levi-Civita symbol" ϵ_{ijk} is often used to simplify complicated expressions. The indices all run from 1 to 3, and ϵ_{ijk} is equal to 0 if any two of them are equal, 1 if they are in the order 123, 231, or 312, and −1 if they are in the order 132, 321, or 213.

(a) As an example of the use of the Levi-Civita symbol, the determinant of a 3×3 matrix can be written $|\mathbf{M}| = \sum_i \sum_j \sum_k \epsilon_{ijk} M_{1i} M_{2j} M_{3k}$. Write out the terms of this sum and show that it reproduces the usual formula for a determinant. (The definition of the Levi-Civita symbol and this formula for determinants can be generalized to more than three dimensions.)

(b) Assume that in some basis B, ϵ_{ijk} is defined as we have described. Let $\mathbf{C}$ be the matrix that transforms vectors from B to another basis B'. Assuming ϵ_{ijk} transforms according to Equation 7.2.7 write ϵ'_{abc} in terms of the components of $\mathbf{C}$ and ϵ_{ijk}. Your answer should contain three sums.

(c) Use the transformation you wrote in Part (b) to write all the non-zero terms in the sums for ϵ'_{112}. You should find that all the terms cancel.

(d) Similarly, write out the terms for ϵ'_{123}. Simplify your answer. You should assume that $\mathbf{C}$ is a rotation matrix, meaning $|\mathbf{C}| = 1$.

(e) Still assuming $|\mathbf{C}| = 1$, find ϵ'_{321}.

(f) Based on the three examples that you have gone through by hand, write a general rule for what happens to any element of the Levi-Civita matrix under a rotation.

(g) The Levi-Civita symbol is not defined by the tensor transformation laws, but rather by the rule that in any basis ϵ_{ijk} has the values described above. You should have found so far that this is consistent with the tensor transformation laws for a rotation. Show that it is *not* consistent with those laws for a reflection. (This should require no new calculations.) An object that obeys tensor transformation laws under rotation but not reflection is called a "pseudotensor." See Problem 7.50.

7.55 **Velocity Gradient** A gas is flowing throughout a region with velocity $\vec{v} = ay^2 \hat{i}$.

(a) Calculate the components of the velocity gradient tensor. Write your answer as a matrix.

(b) Define a new set of axes rotated 30° counterclockwise from the original set. Express the velocity gradient tensor in this new basis by using a tensor transformation. Your answer will be a function of y.

(c) Express x and y in terms of x' and y' and use that to write the transformed velocity gradient as a function of x' and y'.

(d) Express the velocity field in this new basis. Give your answer in terms of x' and y'.

(e) Recalculate the velocity gradient from the new velocity field and verify that it matches your answer to Part (b).

7.56 In this problem you will prove that the velocity gradient $V_{ij} = \partial v_i / \partial x_j$ is a tensor. Consider an orthogonal matrix $\mathbf{C}$ that converts coordinates from one basis to another. Since $\vec{v}$ and $\vec{x}$ are both vectors, $\vec{v}' = \mathbf{C}\vec{v}$ and $\vec{x}' = \mathbf{C}\vec{x}$. Your goal here is to find the components $V'_{ij} = \partial v'_i / \partial x'_j$ and show that they are related to the components of $\mathbf{V}$ by Equation 7.2.7.

(a) Write the equation $\vec{v}' = \mathbf{C}\vec{v}$ in index notation. Use that equation to rewrite V'_{ij} as a sum over the derivatives $\partial v_a / \partial x'_j$. (Notice that x is primed and v is not. We're getting the primes off one step at a time.)

(b) Use the chain rule to write $\partial v_a / \partial x'_j$ as a sum over the derivatives $\partial v_a / \partial x_b$. Your answer will involve derivatives of unprimed x-components with respect to primed x-components.

(c) Write a matrix equation giving $\vec{x}$ in terms of $\vec{x}'$. Using the fact that $\mathbf{C}$ is orthogonal write your matrix equation in terms of $\mathbf{C}^T$ instead of $\mathbf{C}^{-1}$. Then write this equation in index notation.

(d) Because $x'_i = \sum_a C_{ia} x_a$, the derivative $\partial x'_i / \partial x_j$ just equals C_{ij}. In a similar way, use your answer to Part (c) to find $\partial x_b / \partial x'_j$.

(e) Putting it all together, show that V'_{ij} is related to V_{ab} in the way you would expect for a tensor.

Problems 7.57–7.62 have to do with Einstein's special theory of relativity. In relativity we treat time as a fourth coordinate, so we label "events" in spacetime with four-dimensional vectors (t, x, y, z). (By convention these components are numbered 0 to 3, with time being the 0th component.) The "Lorentz transformations" relate the coordinates that different observers measure for events if the observers are in uniform motion relative to each other. In particular, if observer O measures the coordinates (t, x, y, z) for an event and observer O' is moving past O at velocity v in the x-direction, then O' measures the coordinates (t', x', y', z'), where:

$$t' = \gamma \left(t - \frac{vx}{c^2} \right)$$
$$x' = \gamma(x - vt)$$
$$y' = y$$
$$z' = z$$

The "Lorentz factor" is defined as $\gamma = 1/\sqrt{1 - v^2/c^2}$. The point of view of each observer is called a "reference frame" and the transformation from one reference frame to another moving at constant velocity relative to the first is called a "boost." The transformation shown above is for a boost in the x-direction.

7.57 (a) Write the 4×4 matrix $\mathbf{C}$ that converts vectors from the O reference frame to the O' reference frame.

(b) Write the 4×4 matrix $\mathbf{C}^{-1}$ that converts vectors from the O' reference frame to the O reference frame. *Hint*: O' is defined by the fact that it is moving with velocity $+v$ from the O reference frame. What does O look like from the O' reference frame?

(c) Not every element of $\mathbf{C}$ has the same units. Explain why it makes sense for each element to have the units that it does.

7.58 Galilean Relativity

(a) Once again assume observer O' is moving past observer O at speed v in the x-direction. Observer O measures that at time t a certain event occurs at position (x, y, z). Ignoring Einstein's theory and just using Galilean ideas of motion, what would be the coordinates (t', x', y', z') measured by observer O'?

(b) Write a matrix converting vectors in frame O to frame O' using Galilean relativity.

The rest of the problems in this block involve Einstein's relativity and the conversion you calculated in Problem 7.57. You may want to consider in each case how the answers would differ in a Galilean (classical) world.

7.59 *[This problem depends on Problem 7.57.]*

(a) Write a matrix $\mathbf{M}$ that rotates vectors in the O frame 45° around the z-axis, counterclockwise as viewed from above. (This matrix will be 4×4, but it should leave the time unchanged.)

(b) Convert that rotation matrix to a reference frame moving at $\vec{v} = c/\sqrt{2}\,\hat{i}$ relative to O.

(c) Describe a scenario in which the matrix you found in Part (b) might be useful. Be sure to include what object is being observed, what is happening to it, and who is observing it.

7.60 *[This problem depends on Problem 7.57.]*
Four-velocity
An object is at rest at the origin in frame O, so its spacetime coordinates are $t = t$, $x = y = z = 0$.

(a) The coordinates (t, x, y, z) are sometimes referred to as an object's "four-position" because they represent the four-dimensional extension of its coordinate position. For this object the four-position is $\vec{r} = (t, 0, 0, 0)$. Calculate the four components of the velocity $\vec{v} = d\vec{r}/dt$.

(b) Suppose the frame O' is moving at velocity w in the x-direction relative to frame O. (Note that v in this problem is the velocity of the object, so w is the letter that should appear in the transformation matrix.) Use the Lorentz transformations to find the four-position $\vec{r}\,'$ for this object. Your answer will be a function of t.

(c) The first component of $\vec{r}\,'$ is t'. Using that fact, rewrite $\vec{r}\,'$ as a function of t' instead of t.

(d) Take the derivative of $\vec{r}\,'$ with respect to t'.

(e) In Part (a) you calculated $\vec{v}$, the velocity of the object in frame O. Convert this to the O' frame, just as you did earlier for the position. You should find that the result does *not* match your answer from Part (d).

What you've just shown is that the derivative of four-position with respect to time is not a vector. The correct way to define four-velocity is the derivative of four-position with respect to "proper time," which is the time coordinate in the object's own reference frame (the one in which it's at rest). In this case that reference frame is O.

(f) Using your answer to Part (b), calculate $d\vec{r}\,'/dt$. Check that your answer matches your answer to Part (e).

7.61 *[This problem depends on Problem 7.57.]* **Length Contraction and Time Dilation**

(a) In frame O, object A is at rest at the origin and object B is at rest a distance L away in the x-direction. Write four-vectors for the time and position of A and B, leaving t as a variable in the vectors.

(b) Convert your position vectors to a frame O' moving at speed v with respect to O in the x-direction. What is the distance between objects A and B in frame O'?

(c) A clock is at rest at the origin in frame O. Event C occurs when the clock shows time 0 and event D occurs when it shows a later time $t = \tau$. Write four-vectors for the time and position of events C and D in frame O. Then convert the vectors to frame O'. What is the time difference between the two events in frame O'? What is the spatial distance?

7.62 *[This problem depends on Problem 7.57.]* **The Stress-Energy Tensor**

The "stress-energy tensor" **T** describes the density and pressure of a fluid. The upper-left-hand element (the "time-time" component T_{00}) represents the density of the fluid and the other diagonal elements equal the "normal stress." If the normal stress is the same in all three directions then it is simply the pressure. For a "perfect fluid," meaning an ideal gas in equilibrium, the stress-energy tensor is diagonal in its rest frame: $T_{00} = \rho$, $T_{11} = T_{22} = T_{33} = P$ and all other components are zero. Find the stress-energy tensor for a perfect fluid in a frame moving at speed v relative to the fluid's rest frame and use that to find the density of the fluid in that moving frame.

7.63 The trace of a matrix is the sum of the elements along the main diagonal: $Tr(\mathbf{M}) = \sum_i M_{ii}$. Prove that the trace is invariant to an orthogonal transformation. *Hint*: remember that if **C** effects an orthogonal transformation then $\mathbf{C}^T = \mathbf{C}^{-1}$.

7.3 Vector Spaces and Complex Vectors

We have seen that the concept of a "vector" can be used in a spatial sense (a quantity with magnitude and direction), but it can also be used in a more general sense—quantities such as "kiddy books and chapter books" or "position of each object in the three-spring system" can be mathematically manipulated as vectors. In this section we provide a more formal definition of a vector and a vector space. We discuss translating between different bases within the same vector space, and we consider complex-valued vector spaces and matrices.

7.3.1 Explanation: Vector Spaces and Complex Vectors

Suppose we are thinking of a third-order polynomial, and we want to tell you—as concisely as possible—what third-order polynomial we are thinking of. We can accomplish this by giving you four numbers, as long as we have all previously agreed on what those four numbers mean. Here is one way we could do it.

Representation 1

Rule: The numbers a, b, c, and d represent coefficients in the polynomial $ax^3 + bx^2 + cx + d$.

Example: $<5, -2, \pi, 3/4>$ represents $5x^3 - 2x^2 + \pi x + 3/4$.

Here is a much less obvious coding. We define four special polynomials as follows.

$$P_0(x) = 1, \quad P_1(x) = x, \quad P_2(x) = \frac{1}{2}(3x^2 - 1), \quad P_3(x) = \frac{1}{2}(5x^3 - 3x)$$

These are not the polynomial we want to tell you about; these are the code we will use to communicate our polynomial, as follows.

> **Representation 2**
> Rule: A_0, A_1, A_2 and A_3 represent coefficients in the polynomial $A_0 P_0 + A_1 P_1 + A_2 P_2 + A_3 P_3$.
> Example: $<5, -2, \pi, 3/4>$ represents $5 - 2x + (\pi/2)(3x^2 + 1) + (3/8)(5x^3 - 3x)$.

Representation 2 is based on the "Legendre polynomials." You will encounter these functions in another form in Problem 7.82, and we will discuss them in more detail in Chapter 12. But here we are less interested in the particular properties of Representation 2 than in pointing out that both representations accomplish the same goal: they give us a way of describing any third-order polynomial using four numbers. They also give us an excuse for introducing some important terms.

- The set of all third-order polynomials is an example of a "vector space." Any given third-order polynomial is a "vector" in this space.
- The functions x^3, x^2, x, and 1 are themselves third-order polynomials—that is, they are vectors in this space. Any other vector in this space can be expressed as a linear combination of those four vectors (that's what we did in Representation 1). Those vectors are therefore said to "span" this vector space.
- None of the functions x^3, x^2, x, or 1 can be expressed as a linear combination of the other three. Those four vectors are therefore said to be "linearly independent."
- If you remove any vector (x^3, x^2, x or 1) from that list, the resulting list of three vectors no longer spans the space; we need all four. These four vectors are therefore said to provide a "basis" for this vector space. Each of these four vectors is a "basis vector" in Representation 1.
- The four vectors P_0–P_3 provide an alternative basis for the space of third-order polynomials. You could come up with many such other bases, but each would require four basis vectors. The space of third-order polynomials is therefore said to have "dimension" four.

We must emphasize that "vector" is not a fancy word for "list of numbers." A list of numbers can *represent* a vector, but the representation depends on the basis. Above we saw one list $<5, -2, \pi, 3/4>$ represent two completely different vectors in different bases. Similarly, $<5, 3, -1, 1>$ in Representation 1 and $<2, 2, 2, 2>$ in Representation 2 are two different lists of numbers, but they represent the same vector.

Third-order polynomials are just one example of the very broad topic of vector spaces. Below we provide a general definition of that term. As you read each requirement in the definition we encourage you to see how it applies to our example of third-order polynomials, and also to the more familiar example of spatial vectors.

The reason for defining "vector space" in the abstract is that you can prove important results from those ten properties. Such results are then guaranteed to be true for all vector spaces.

For instance, suppose you find a set of n linearly independent vectors $\mathbf{v}_1, \mathbf{v}_2 \ldots \mathbf{v}_n$ that span a given vector space $\mathcal{V}$. This is enough to prove that $\mathcal{V}$ is an n-dimensional space, which in turn guarantees two facts.

- It is impossible to span $\mathcal{V}$ with fewer than n vectors.
- Any set of n linearly independent non-zero vectors in $\mathcal{V}$ define a basis for $\mathcal{V}$.

These results can be proven from the existence of any set of n linearly independent vectors that span $\mathcal{V}$. (See Problem 7.69.) And while these results are important in themselves, they

7.3 | Vector Spaces and Complex Vectors

> **Definition: Vector Space**
>
> Consider a set of objects called "vectors." An operation called "addition" can act on any two of these vectors to produce a third vector. (This is often the operation that we traditionally call addition, but it doesn't have to be.) The operation "multiplication of a vector by a scalar" is also defined. The objects in this set form a "vector space" $\mathcal{V}$ if all the following criteria are met.
>
> 1. *Closure under addition:* The sum of any two vectors $\mathbf{a} + \mathbf{b}$ in $\mathcal{V}$ gives a vector in $\mathcal{V}$.
> 2. *Commutativity of addition:* $\mathbf{a} + \mathbf{b} = \mathbf{b} + \mathbf{a}$.
> 3. *Associativity of addition:* $(\mathbf{a} + \mathbf{b}) + \mathbf{c} = \mathbf{a} + (\mathbf{b} + \mathbf{c})$.
> 4. *Additive identity:* There exists a vector $\mathbf{0}$ such that $\mathbf{0} + \mathbf{a} = \mathbf{a}$ for all $\mathbf{a}$ in $\mathcal{V}$.
> 5. *Additive inverse:* For each vector $\mathbf{a}$ in $\mathcal{V}$ there exists a vector $-\mathbf{a}$ such that $\mathbf{a} + (-\mathbf{a}) = \mathbf{0}$.
> 6. *Closure under scalar multiplication:* For any vector $\mathbf{a}$ in $\mathcal{V}$ and any scalar k, $k\mathbf{a}$ is a vector in $\mathcal{V}$.
> 7. *Distributivity of scalar multiplication with respect to vector addition:* $k(\mathbf{a} + \mathbf{b}) = k\mathbf{a} + k\mathbf{b}$.
> 8. *Distributivity of scalar multiplication with respect to scalar addition:* $(k + p)\mathbf{a} = k\mathbf{a} + p\mathbf{a}$.
> 9. *Associativity of scalar multiplication:* $k(p\mathbf{a}) = (kp)\mathbf{a}$.
> 10. *Scalar identity:* There exists a scalar 1 such that $1\mathbf{a} = \mathbf{a}$ for all $\mathbf{a}$ in $\mathcal{V}$.
>
> If the set of scalars is taken to be the real numbers then $\mathcal{V}$ is a "real vector space." If it's the complex numbers then $\mathcal{V}$ is a "complex vector space."

also provide a more general example of the use of a concept such a "vector space." We do not have to prove these results for one particular vector space or one particular kind of vector space; we prove it once for all of them.

Inner Product and Norm

An "inner product space" is a vector space for which you have defined an "inner product" that multiplies two vectors to produce a scalar. For spatial vectors the inner product is the dot product: $\vec{a} \cdot \vec{b} = a_x b_x + a_y b_y + a_z b_z$. This formula can be generalized to any number of dimensions. (The cross product, by contrast, is peculiar to three dimensions.) We will reserve the "dot" notation for dot products of spatial vectors, and denote general inner products by $(\mathbf{a}, \mathbf{b})$.

For spatial vectors, the dot product can also be used to define a vector's magnitude: $|\vec{a}| = \sqrt{\vec{a} \cdot \vec{a}}$. We generalize that rule by defining the "norm" of any vector as $||\mathbf{a}|| = \sqrt{(\mathbf{a}, \mathbf{a})}$. The inner product must satisfy three conditions.

1. *Linearity:* $(k\mathbf{a} + p\mathbf{b}, \mathbf{c}) = k(\mathbf{a}, \mathbf{c}) + p(\mathbf{b}, \mathbf{c})$.
2. *Symmetry:* $(\mathbf{a}, \mathbf{b}) = (\mathbf{b}, \mathbf{a})$ for real inner product spaces. (We'll discuss complex vector spaces below.)
3. *Positive-definiteness:* $(\mathbf{a}, \mathbf{a}) \geq 0$. The only vector with $(\mathbf{a}, \mathbf{a}) = 0$ is the "zero vector" (defined in Property 4 of our definition of a vector space).

Orthonormal Bases

Two vectors are "orthogonal" if their inner product is zero. (For spatial vectors the dot product is proportional to the cosine of the angle between the vectors, so "orthogonal" is essentially synonymous with "perpendicular.") A set of basis vectors that are all mutually orthogonal is called an "orthogonal basis." A vector with norm 1 is "normalized." An orthogonal basis with all vectors normalized is an "orthonormal basis." (Think of $\hat{i}, \hat{j}$, and $\hat{k}$.)

In an orthonormal basis, the inner product looks like the usual dot product. That is, consider two vectors **a** and **b** represented in the base of vectors $e_1, e_2 \dots e_n$.

$$\begin{aligned}\mathbf{a} =&< a_1, a_2, \dots >= a_1 \mathbf{e}_1 + a_2 \mathbf{e}_2 + \dots \\ \mathbf{b} =&< b_1, b_2, \dots >= b_1 \mathbf{e}_1 + b_2 \mathbf{e}_2 + \dots\end{aligned} \quad (7.3.1)$$

If that base is orthonormal, then $(\mathbf{a}, \mathbf{b}) = a_1 b_1 + a_2 b_2 + \dots + a_n b_n$. You will prove this result in Problem 7.72.

Problems 7.81–7.82 outline the "Gram-Schmidt" process for deriving an orthonormal basis for a vector space.

Complex Vector Spaces

Spatial vectors are real, but generalized vectors can be complex. Complex vectors play important roles in fields such as optics, circuit theory, and quantum mechanics.

The rules and definitions we've discussed above hold for complex vector spaces with one important exception. We said above that an inner product should have the properties of "symmetry" and "positive-definiteness." You will show in Problem 7.73 that a complex vector space cannot have both if we use the definitions we gave above. We therefore modify our symmetry requirement so that we can retain positive-definiteness.

2. *Symmetry:* $(\mathbf{a}, \mathbf{b}) = (\mathbf{b}, \mathbf{a})^*$

3. *Positive-definiteness:* $(\mathbf{a}, \mathbf{a})$ is a positive real number for any non-zero vector **a**.

Our new symmetry requirement stipulates that if you reverse the order of the two vectors, the new inner product is the complex conjugate of the old. For real vector spaces this simply reproduces the original symmetry condition.

For complex vectors $(\mathbf{a}, k\mathbf{b}) = k^*(\mathbf{a}, \mathbf{b})$. In other words, when you pull a scalar out of the second position in an inner product you have to take its complex conjugate. That in turn implies that for two complex vectors expressed in an orthonormal basis, the inner product does not exactly follow the usual dot product rule, but instead has to be modified by taking the complex conjugate of the components of the second vector: $(\mathbf{a}, \mathbf{b}) = a_1 b_1^* + a_2 b_2^* + \dots$. You will show all of this in Problem 7.74.

Complex vectors can be transformed by matrices just as real vectors can. Many of the ideas we've developed about real vector transformations can be generalized to complex vector spaces. As you've seen in some of the examples above, these generalizations involve inserting complex conjugates into some of the formulas, but in all cases those modifications stem from the original requirement of positive-definiteness.

Real matrices	**Complex matrices**
The "transpose" of matrix **M** (usually written $\mathbf{M}^T$) comes from exchanging its rows for columns.	The "adjoint" of matrix **M** (usually written $\mathbf{M}^\dagger$ which is pronounced "M dagger") comes from taking its transpose and taking the complex conjugate of every element.
If a matrix equals its own transpose it is "symmetric" which implies that its eigenvectors are orthogonal to each other.	If a matrix equals its own adjoint it is "Hermitian" which implies that its eigenvectors are orthogonal to each other.
A matrix is "orthogonal" if it preserves the norm of real vectors when it transforms them. Geometrically this corresponds to rotations and reflections but not stretching.	A matrix is "unitary" if it preserves the norm of complex vectors. As often happens with complex numbers, the simple geometrical interpretation we had for real matrices doesn't apply here.

7.3 | Vector Spaces and Complex Vectors

Real matrices	**Complex matrices**
A matrix is orthogonal if and only if its transpose is its inverse: $\mathbf{MM}^T = \mathbf{I}$.	A matrix is unitary if and only if its adjoint is its inverse: $\mathbf{MM}^\dagger = \mathbf{I}$.
A matrix is orthogonal if and only if all its eigenvectors have modulus 1. So an orthogonal matrix must have a determinant of ± 1 but the converse is not true: a non-orthogonal matrix may also have a determinant of ± 1.	A matrix is unitary if and only if all its eigenvectors have modulus 1. So a unitary matrix must have a determinant with modulus 1 but the converse is not true: a non-unitary matrix may also have a determinant of modulus 1.

Appendix E provides a glossary of these and other matrix-related terms.

EXAMPLE Hermitian and Unitary Matrices

Question: Which of the following matrices are Hermitian? Which are unitary?

$$\mathbf{A} = \begin{pmatrix} 3 & 2-i \\ 2+i & -2 \end{pmatrix} \quad \mathbf{B} = \begin{pmatrix} i & 0 \\ 0 & -1 \end{pmatrix} \quad \mathbf{C} = \begin{pmatrix} -(1/2)+(i/2) & (1/2)+(i/2) \\ (1/2)+(i/2) & -(1/2)+(i/2) \end{pmatrix}$$

Answer:
We can see if they are Hermitian by inspection. The elements on the main diagonal must all be real, and each other element must be the complex conjugate of the one across the main diagonal from it. Thus **A** is Hermitian and **B** and **C** are not.

To see if a matrix is unitary is usually harder, but in the case of **B** it can be done by inspection. Since the matrix is diagonal we can read off that the eigenvalues are i and -1, so **B** is unitary. For the other two it's easiest to directly check if their adjoints are their inverses. To take the adjoint switch each element with the one across the main diagonal and then take its complex conjugate. Since **A** is Hermitian (also known as "self-adjoint") $A^\dagger = A$.

$$\mathbf{AA}^\dagger = \begin{pmatrix} 3 & 2-i \\ 2+i & -2 \end{pmatrix} \begin{pmatrix} 3 & 2-i \\ 2+i & -2 \end{pmatrix} = \begin{pmatrix} 14 & 2-i \\ 2+i & 9 \end{pmatrix}$$

Since this isn't the identity matrix **A** is not unitary.

$$\mathbf{CC}^\dagger = \begin{pmatrix} -(1/2)+(i/2) & (1/2)+(i/2) \\ (1/2)+(i/2) & -(1/2)+(i/2) \end{pmatrix} \begin{pmatrix} -(1/2)-(i/2) & (1/2)-(i/2) \\ (1/2)-(i/2) & -(1/2)-(i/2) \end{pmatrix}$$
$$= \begin{pmatrix} 1 & 0 \\ 0 & 1 \end{pmatrix}$$

Since this is the identity matrix, $\mathbf{C}^\dagger$ is the inverse of **C**, so **C** is unitary.

7.3.2 Problems: Vector Spaces and Complex Vectors

7.64 Which of the following is a vector space? For each one that isn't, show at least one criterion that it fails to meet. Assume addition and multiplication by a scalar are defined in the obvious ways.

(a) The set of all position vectors (x, y, z) that lie on the plane $z = x + y$
(b) The set of all 3-vectors with $v_x > 0$
(c) The set of all real numbers

(d) The set of all possible box sizes (height, width, length)

7.65 Explain why each of the following sets does *not* qualify as a vector space.
  (a) The set of all integers from 0 through 10, where the vector space operation of "addition" is regular numerical addition.
  (b) The set of all real numbers, where the vector space operation of "addition" is actually multiplication.
  (c) The set of all resistors, where the vector space operation of "addition" creates a new resistor which sums the two resistances.
  (d) The set of all spatial vectors in a plane that have magnitudes greater than or equal to one, where the vector space operation of "addition" is regular vector addition.

7.66 Set E is the set of all even integers (positive, negative, and zero). Set D is the set of all odd integers (positive and negative). Set Z is set D plus the number 0. Prove that only one of these three sets is a vector space, using the usual definitions of addition and multiplication.

7.67 (a) The space of third-order polynomials, with addition defined in the usual way for polynomials, is a vector space. For example, it obeys closure under addition because

$$(a_1 x^3 + b_1 x^2 + c_1 x + d_1)$$
$$+ (a_2 x^3 + b_2 x^2 + c_2 x + d_2)$$
$$= (a_1 + a_2) x^3 + (b_1 + b_2) x^2$$
$$+ (c_1 + c_2) x + (d_1 + d_2)$$

is still a third-order polynomial. In a similar way, show that the set of third-order polynomials obeys all the other requirements for a vector space.
  (b) What is the dimension of this space?
  (c) Write two bases for this space.
  (d) Express the polynomial $1 - x + x^3$ in both of the bases you wrote.

7.68 The space of all real 2D matrices is a real vector space with addition and multiplication by a scalar defined in the usual way for matrices.
  (a) What is the dimension of this space?
  (b) Write two separate bases for this space.
  (c) Write the matrix that rotates all 2D vectors 90° clockwise and express it as an ordered list in each of the two bases you wrote.

7.69 In the Explanation (Section 7.3.1) we claimed that a vector space has a unique dimension n. For example, suppose the vectors **x** and **y** form a basis for the vector space V. You will prove in this problem that no single vector can be a basis for V, and any three vectors in V must be linearly dependent. Together, these two statements imply that any basis for V must consist of exactly two vectors. You'll start with the second one.
  (a) Consider three vectors in space V: **f**, **g**, and **h**. By the definition of a basis it must be possible to write **f**, **g**, and **h** as linear combinations of **x** and **y**. Write these out explicitly, using the letters α, β, γ, δ, ϵ, and ζ for the six unknown coefficients of these linear combinations. Prove that **h** can be written as a linear combination $r\mathbf{f} + s\mathbf{g}$ by solving for r and s in terms of α–ζ.
  (b) Prove that if a single vector **z** were a basis for V then **x** and **y** couldn't be linearly independent.

  You've just shown that if a vector space has a basis with two vectors, then any possible basis for the space must have exactly two vectors. The argument can be generalized in a similar way to higher dimensions.

7.70 Prove that a real vector space cannot have exactly two vectors. Can it have just one?

7.71 In relativity the position and time of each event are assigned a four-vector (t, x, y, z). The "spacetime interval" between two events is $-(t_2 - t_1)^2 + (x_2 - x_1)^2 + (y_2 - y_1)^2 + (z_2 - z_1)^2$. Is the spacetime interval an inner product? Explain.

7.72 The Explanation (Section 7.3.1) claims that the inner product for any real-valued orthonormal basis looks like a dot product. In this problem you will prove this result.
  (a) Equations 7.3.1 show two vectors **a** and **b** written in components in a particular basis. Use the property of linearity to express the inner product $(\mathbf{a}, \mathbf{b})$ in terms of the given components and inner products of the basis vectors.
  (b) Assuming that the given basis is orthogonal, simplify the resulting formula.
  (c) Now assuming that the given basis is orthonormal, further simplify the formula. The end result should look like the familiar formula for a dot product.

7.73 In this problem you will show that for complex inner product spaces, the requirement of positive-definiteness demands that we include a complex conjugate in the requirement of symmetry. Consider a complex inner product space with some vector **a**, with norm $||\mathbf{a}|| = k$. Assuming **a** isn't the zero vector k must be a positive real number.

(a) Take the symmetry condition in its real form: $(\mathbf{a}, \mathbf{b}) = (\mathbf{b}, \mathbf{a})$. Use linearity and symmetry to prove that $||i\mathbf{a}|| = -k$, which violates the condition of positive-definiteness.

(b) Now use the correct symmetry condition $(\mathbf{a}, \mathbf{b}) = (\mathbf{b}, \mathbf{a})^*$ (along with linearity) to show that $||i\mathbf{a}|| = k$.

7.74 (a) The linearity condition for inner products only specifies what happens when the first vector in the inner product is a linear combination: $(k\mathbf{a} + p\mathbf{b}, \mathbf{c}) = k(\mathbf{a}, \mathbf{c}) + p(\mathbf{b}, \mathbf{c})$. Use this condition and the symmetry condition to prove that for a real inner product space the same rule applies to linear combinations in the second position: $(\mathbf{a}, k\mathbf{b} + p\mathbf{c}) = k(\mathbf{a}, \mathbf{b}) + p(\mathbf{a}, \mathbf{c})$.

(b) Use linearity and symmetry to prove that for a *complex* vector space linear combinations in the second position follow a slightly modified rule that involves complex conjugates: $(\mathbf{a}, k\mathbf{b} + p\mathbf{c}) = k^*(\mathbf{a}, \mathbf{b}) + p^*(\mathbf{a}, \mathbf{c})$.

(c) Use your result from Part (b) to prove that if $\mathbf{a}$ and $\mathbf{b}$ are expressed in terms of an orthonormal basis $\mathbf{e}_1, \mathbf{e}_2, \ldots$, their inner product is given by $a_1 b_1^* + a_2 b_2^* + \ldots$.

7.75 Prove that the inner product of any vector with the zero vector must be zero.

7.76 Which of the following matrices are symmetric? Which ones are orthogonal?

(a) $\mathbf{A} = \begin{pmatrix} 1 & 1 \\ 1 & 1 \end{pmatrix}$

(b) $\mathbf{B} = \begin{pmatrix} 1/\sqrt{2} & 1/\sqrt{2} \\ 1/\sqrt{2} & -1/\sqrt{2} \end{pmatrix}$

(c) $\mathbf{C} = \begin{pmatrix} 1 & 0 & 0 \\ -1 & -2 & 1 \\ 2 & 0 & -1/2 \end{pmatrix}$

7.77 Which of the following matrices are Hermitian? Which ones are unitary?

(a) $\mathbf{D} = \begin{pmatrix} 1 & 2i \\ -2i & -3 \end{pmatrix}$

(b) $\mathbf{E} = \begin{pmatrix} i/\sqrt{2} & 1/\sqrt{2} \\ -i/\sqrt{2} & 1/\sqrt{2} \end{pmatrix}$

(c) $\mathbf{F} = \begin{pmatrix} i & 0 & 0 \\ -1 & -2 & 1 \\ 2i & 0 & i/2 \end{pmatrix}$

7.78 Let $\mathcal{F}$ be the space of all real, square-integrable functions $f(x)$. ("Square-integrable" means $\int_{-\infty}^{\infty} [f(x)]^2 \, dx$ exists.)

(a) Show that $\mathcal{F}$ meets all the criteria for a vector space.

(b) Define the inner product of two functions f and g to be $\int_{-\infty}^{\infty} f(x)g(x) \, dx$. Show that with this definition $\mathcal{F}$ is an inner product space. Remember that f and g are real-valued functions when you apply the test for symmetry.

7.79 **Jones Calculus** A light beam traveling in the z-direction generally contains an oscillating electric field in both the x- and y-directions. These oscillating waves are typically represented by complex exponentials, where the actual field is understood to be either the real or imaginary part of the complex function: $E_x = C_x e^{i(kz-\omega t)}$, $E_y = C_y e^{i(kz-\omega t)}$. Since the complex exponential piece at the end is the same for both x and y it is often not written, and the light beam is represented by a "Jones vector" $<C_x, C_y>$. The constants C_x and C_y are generally complex. The intensity of the light beam is proportional to $|C_x|^2 + |C_y|^2$. The phases of C_x and C_y contain the information about the phase lag between the x and y oscillations.

(a) When light passes through a polarizing filter oriented in the x-direction, it blocks any y-component of the electric field while allowing the x-component through unchanged. Write a matrix that represents the transformation a Jones vector undergoes when passing through such a filter.

(b) In general, a matrix that represents the transformation experienced by a Jones vector when it passes through a medium is called a "Jones matrix." If the Jones matrix for a particular medium is unitary, what does that tell you about the physical effect it has on the light beam?

(c) Which of the following transformations are unitary?

i. $\begin{pmatrix} 0 & i \\ i & 0 \end{pmatrix}$

ii. $\begin{pmatrix} i & 1 \\ i & -1 \end{pmatrix}$

iii. A brick wall that blocks all light.

iv. A polarizing filter oriented in the x-direction.

v. A "multiplier" that doubles the x- and y-components of the electric field.

vi. A crystal that rotates the electric field vector 45° in the xy-plane without changing its magnitude.

7.80 **Quantum States** In classical physics a particle has a definite energy at each moment. In quantum mechanics a particle can be in a "superposition" in which it has energy E_1 and energy E_2 simultaneously. That doesn't mean its energy is the sum of the two. In fact if you measure its energy you will find one of those answers, E_1

or E_2, but never both. Before the measurement its state can be written $c_1\mathbf{E}_1 + c_2\mathbf{E}_2$, where $\mathbf{E}_1$ represents the state of definitely having energy E_1. When you measure the energy $|c_n|^2$ is the probability that you will find it with energy E_n.

(a) A particle is in a superposition of two energy states, as described above. What must $|c_1|^2 + |c_2|^2$ equal and why?

(b) The particle passes through a device that changes its state to some other superposition of the two possible energies. The matrix M represents the transformation that the state undergoes in this device. In other words $\begin{pmatrix} c_{1,\text{final}} \\ c_{2,\text{final}} \end{pmatrix} = M \begin{pmatrix} c_{1,\text{initial}} \\ c_{2,\text{initial}} \end{pmatrix}$. Explain why M must be a unitary matrix.

7.81 Exploration: The Gram-Schmidt Process and Spatial Vectors

The "Gram-Schmidt" process starts from any set of basis vectors $\mathbf{u}_1, \mathbf{u}_2 \ldots \mathbf{u}_n$, and uses them to derive a an orthonormal basis $\mathbf{e}_1, \mathbf{e}_2 \ldots \mathbf{e}_n$ for that space. In this problem we'll illustrate how the process works for spatial vectors. In Problem 7.82 you'll generalize it to a non-spatial vector space. As our example, we'll consider the space of 3D vectors, with the initial basis $\vec{u}_1 = \hat{i} + \hat{j}$, $\vec{u}_2 = 2\hat{i} - 3\hat{k}$, $\vec{u}_3 = \hat{j} - 5\hat{k}$.

(a) Show that these three vectors are neither orthogonal nor normalized. Show that they are, however, linearly independent (and thus form a basis).

(b) To define the first vector of our orthonormal basis, $\vec{e}_1$, simply take $\vec{u}_1$ and divide it by its own magnitude to get a normalized vector.

(c) The second basis vector $\vec{e}_2$ has to meet a stricter criterion; in addition to being normalized it must be orthogonal to $\vec{e}_1$. The formula $\vec{u}_2 \cdot \vec{e}_1$ gives us the component of $\vec{u}_2$ in the direction of $\vec{e}_1$. You can define $\vec{v}_2$ as $\vec{u}_2 - (\vec{u}_2 \cdot \vec{e}_1)\vec{e}_1$ and it's guaranteed to be orthogonal to $\vec{e}_1$ because you've subtracted off the part of $\vec{u}_1$ pointing along $\vec{e}_1$. Then you can define $\vec{e}_2$ by dividing $\vec{v}_2$ by its own norm.

(d) Derive $\vec{e}_3$ in a similar way. Starting with $\vec{u}_3$, subtract its components along $\vec{e}_1$ and $\vec{e}_2$ and then normalize it.

(e) Verify that your resulting set of basis vectors is orthonormal.

(f) The vectors $\vec{q}_1 = \hat{i} + \hat{j} - 3\hat{k}$ and $\vec{q}_2 = \hat{j} + 5\hat{k}$ form a basis for a 2D vector space. Use the Gram-Schmidt process to derive an orthonormal basis for that space.

7.82 Exploration: The Gram-Schmidt Process and Legendre Polynomials

Problem 7.81 illustrated the "Gram-Schmidt Process" for deriving an orthonormal basis set for a vector space. In that problem you used the process on spatial vectors. In this one you will use it on a non-spatial vector space. You do not need to have done Problem 7.81 to do this one.

An "analytic" function is one that can be represented by a Maclaurin series. For instance, the series $1 + x + x^2 + x^3 + \ldots$ converges to $1/(1-x)$ on the domain $-1 < x < 1$ but not outside that interval. We therefore describe $1/(1-x)$ as being analytic on $-1 < x < 1$.

Let $\mathcal{F}$ be the space of all functions that are analytic on the domain $[-1, 1]$. (So our example $1/(1-x)$ doesn't quite make it, since it is not analytic on the boundaries of that interval.) Since these functions can all be written as Maclaurin series the set of functions $u_0(x) = 1$, $u_1(x) = x$, $u_2(x) = x^2 \ldots$ form a basis for $\mathcal{F}$. You can define an inner product for $\mathcal{F}$ as $(f, g) = \int_{-1}^{1} f(x)g(x)\,dx$. In this problem you will derive an orthonormal basis for $\mathcal{F}$. The resulting set of basis functions are called "Legendre polynomials."

Warning: Each Legendre polynomial is multiplied by a constant. In this problem you will be choosing the constants required to create an orthonormal basis; for other applications other constants are preferable. Therefore the polynomials you find here will not match the Legendre polynomials listed in Chapter 12 and Appendix J.

(a) Show that the basis $u_n(x)$ is not orthonormal.

(b) Define the first Legendre polynomial $P_0(x)$ by dividing $u_0(x)$ its own norm.

(c) For the second Legendre polynomial you need a function that is orthogonal to $P_0(x)$. To get that, define $v_1 = u_1 - (u_1, P_0)P_0$. Then divide v_1 by its own norm to get $P_1(x)$, a function that is normalized and orthogonal to $P_0(x)$.

(d) Define $v_2 = u_2 - (u_2, P_0)P_0 - (u_2, P_1)P_1$, and then find $P_2 = v_2/||v_2||$.

(e) Find P_3.

This problem brings us back to where we started the section: you can span the analytic functions with a basis of $1, x, x^2, x^3 \ldots$, or you can span the same space with a basis of the Legendre polynomials. In Problem 7.83 you will see a significant advantage of the latter basis—an advantage which is entirely due to the orthonormal basis vectors.

7.83 Exploration: Polynomial Approximation. In Problem 7.82 you derived an orthonormal form of the Legendre polynomials. You do not need to have done that problem in order to do this one. You just need to know that these polynomials are orthonormal: that is, $\int_{-1}^{1} P_n(x)P_m(x)\,dx$ is 0 if $n \neq m$ and 1 if $n = m$.

Consider a function $f(x)$. Your goal is to find the third-order polynomial $\hat{f}(x)$ that best approximates $f(x)$ between $x = -1$ and $x = 1$. Our definition of "best approximation" will be the function that minimizes $\int_{-1}^{1} [f(x) - \hat{f}(x)]^2 \, dx$. A Taylor series is the best polynomial approximation at and around a given point, but this technique can give you a better approximation on a given interval.

For reasons that will soon be clear, we will look for an answer in the form $\hat{f}(x) = c_0 P_0(x) + c_1 P_1(x) + c_2 P_2(x) + c_3 P_3(x)$. Your goal is to find the constants c_0–c_3.

(a) Write the integral that represents the quantity you are trying to minimize.

(b) Your integrand is the square of a 5-term series. We're not asking you to write all 25 resulting terms, but write the term that involves $[P_2(x)]^2$.

(c) Now write the term that involves $P_2(x)P_3(x)$.

(d) Evaluate those two integrals. Remember that the Legendre polynomials are orthonormal!

(e) Now that you see the pattern, expand the square. All the integrals that involve only Legendre polynomials can be replaced with 0 or 1, as appropriate, leaving only integrals that involve $f(x)$.

(f) Two of the terms in your answer involve c_0. Find the value of c_0 that minimizes those two terms. (Your answer will have the functions $f(x)$ and $P_0(x)$ in it, and you can just leave them like that.)

7.84 [*This problem depends on Problem 7.83.*] The first four normalized Legendre polynomials are $P_0 = 1/\sqrt{2}$, $P_1 = \sqrt{3/2}\, x$, $P_2 = (1/2)\sqrt{5/2}(3x^2 - 1)$, and $P_3 = (1/2)\sqrt{7/2}(5x^3 - 3x)$.

(a) Find the best third-order polynomial approximation to e^x between $x = 1$ and $x = -1$. (This will involve either computer integration or some tedious integration by parts. The exact coefficients will be complicated expressions, but you can just give numerical values for them.)

(b) Show on one plot the original function e^x between $x = -1$ and $x = 1$ and each of the first four partial sums of its polynomial expansion. (The first partial sum is the zero-order approximation and the fourth partial sum is the third-order approximation.)

(c) Find the fourth-order Maclaurin series for e^x.

(d) Show on one plot the original function e^x between $x = -1$ and $x = 1$, the second-order approximation you derived using Legendre polynomials, and the second-order Maclaurin series. Which approximation looks better?

(e) Show on one plot the original function e^x between $x = -0.1$ and $x = 0.1$, the second-order approximation you derived using Legendre polynomials, and the second-order Maclaurin series. Which approximation looks better?

(f) In what kinds of situations would each of these approximations be most useful?

7.4 Row Reduction (see felderbooks.com)

7.5 Linear Programming and the Simplex Method (see felderbooks.com)

7.6 Additional Problems (see felderbooks.com)

CHAPTER 8

Vector Calculus

> *Before you read this chapter, you should be able to...*
>
> - work with vectors, including conversion between magnitude-and-direction and component forms, addition and subtraction, and the dot and cross products.
> - evaluate and interpret derivatives and integrals in functions of one variable.
> - evaluate and interpret partial derivatives of multivariate functions, including use of the multivariate chain rule (see Chapter 4).
> - set up and evaluate multiple integrals in two and three dimensions, working in Cartesian, polar, cylindrical, and spherical coordinates (see Chapter 5).
> - evaluate and interpret line integrals of vector functions (see Chapter 5).
> - evaluate and interpret surface integrals of vector functions (see Chapter 5).
>
> *After you read this chapter, you should be able to...*
>
> - interpret various visual representations of scalar and vector fields.
> - predict the behavior of a system based on a graph of its potential function.
> - evaluate and interpret the gradient and Laplacian of a scalar field.
> - evaluate and interpret the divergence and curl of a vector field.
> - identify conservative vector fields.
> - use the gradient theorem, the divergence theorem, and Stokes' theorem to simplify calculations.

A single-variable function starts with one independent variable that might, for instance, represent position on a number line. It tells you how another variable, such as a height or density, depends on that one number.

A "field" is a function of position in any number of dimensions, so it can depend on 2 or 3 variables. The field tells you how another variable, such as a density or a force, depends on that position.

All the mathematics of functions, from basic algebra through derivatives and integrals, can be generalized to these higher dimensional spaces. This chapter will begin with a brief introduction to fields, and will then focus on derivatives of fields and fundamental theorems in more than one dimension.

8.1 Motivating Exercise: Flowing Fluids

An explosion at the Glogrene nuclear facility has released a cloud of radioactive waste into the surrounding area. The grounds of the facility cover a disk of radius 3 miles, and the waste is uniformly distributed throughout that area with density $\sigma = 1,000,000$ terabecquerels per square mile. (It's a unit used for measuring radioactive materials. Trust us.) Weather satellites have made detailed measurement of wind velocities in the area. They have determined that at any given point (x, y) the wind velocity is roughly $\vec{v} = x^3 \hat{i} + (x^2 + y^3)\hat{j}$, where x and y are in miles and $\vec{v}$ is in miles per hour.[1] For instance, at the point $(1, 1)$ the wind velocity is $\hat{i} + 2\hat{j}$. The Glogrene plant is at the origin.

1. What is the wind velocity at the origin?
2. What is the wind velocity at the point $(1, 2)$?
   See Check Yourself #48 in Appendix L
3. What is the wind velocity at the point $(3, 0)$?

The president has called you in to calculate how quickly radioactive waste is leaving the Glogrene grounds.

4. Draw a diagram of the Glogrene facility. (Just draw a circle with a radius of 3.) Label at the point $(3, 0)$ a small segment of the circumference with height dy.
5. Which component of the wind velocity is causing waste to flow out through this segment?
   See Check Yourself #49 in Appendix L
6. Multiply that component of the velocity times the density of waste, times the height dy, to get the rate at which waste is flowing out through that infinitesimal segment.

To find the total rate at which waste is leaving, you get ready to repeat the above calculation for a differential segment at an arbitrary point on the circle, which will involve finding the component of $\vec{v}$ perpendicular to the segment. Then you will express the amount of waste flowing through that segment per unit time in terms of just one variable (probably the polar angle ϕ) so that you can integrate it to find the total. This is not an approach you want to take if you can avoid it.

As you may have guessed, you can. We assert with no justification that the total rate at which waste is flowing out of the grounds is given by the density of waste times the double integral over the disk of $(\partial v_x/\partial x) + (\partial v_y/\partial y)$ where v_x is the x-component of $\vec{v}$.

7. Calculate $(\partial v_x/\partial x) + (\partial v_y/\partial y)$ for the wind velocity given above.
8. This integral will be easiest in polar coordinates. Rewrite your answer to Part 7 in polar coordinates.
9. Set up and evaluate the double integral of the function you found in Part 7 over the disk. This integral should be straightforward in polar coordinates. Multiply your answer by the density to find the rate at which waste is leaving the Glogrene grounds.

The good news is that you've calculated the rate at which waste is flowing into the community, and using your result the government has created a sensible evacuation plan. The bad news is that you calculated something that we threw at you with no justification. How would this trick apply to other situations? When are you allowed to use it? Why does it work? What is it called?

The last question (which you probably weren't wondering anyway) is easy. We used something called "The Divergence Theorem" to get that formula. For the other answers, you'll have to read this chapter.

8.2 Scalar and Vector Fields

The word "field" has different definitions in different areas of math. For our purposes, a "field" is a function of position. Examples include the temperature in a large room, the magnetic field around the Earth, and the water velocity in the ocean. In many cases the field is changing, so the dependent variable is a function of both position and time.

This section introduces some tools for visualizing and working with scalar and vector fields.

---

[1] For simplicity we will assume there is negligible vertical flow, so you can treat this as a two-dimensional problem.

8.2.1 Discovery Exercise: Scalar and Vector Fields

The xy-plane is covered with sand. Wandering the plane in your bare feet, you conclude that the depth of sand at any given point can be modeled by the equation $z = x^2 \sin^2 y$.

1. What is the minimum depth of the sand? List three points where you would find this depth.
2. Explain why no point can claim to have the maximum depth of sand.
3. Starting at the point $(1, \pi/2)$, you take a long walk in the positive x-direction. Describe what you see (or feel), over time, happening to the sand depth under your feet.

   See Check Yourself #50 in Appendix L

4. Returning to $(1, \pi/2)$ you take another walk, this time in the positive y-direction. Describe your experience during this different walk.

Six months later the sand has been swept away by seven maids with seven mops, and has been replaced with a uniform shallow layer of water. Unlike the sand, which was stationary, this water is moving around. The velocity of the water at any point is given by $\vec{v} = x^2 \hat{i} + 3\hat{j}$.

5. If you drop a leaf on the water at the point $(0, 2)$, which way will it head, and how fast? Describe its motion over time.
6. If you drop a leaf on the water at the point $(1, 2)$, which way will it head, and how fast? Describe its motion over time. Your description should include its approximate heading in the short term (soon after you drop it) and the long term (much later).

   See Check Yourself #51 in Appendix L

7. If you drop a leaf on the water at the point $(-1, 2)$, which way will it head, and how fast? Describe its motion over time. Your description should include its approximate heading in the short term (soon after you drop it) and the long term (much later).

8.2.2 Explanation: Scalar and Vector Fields

Scalar Fields

Any given point in the room around you is at a certain temperature. This is an example of a "scalar field": a scalar function of one or more spatial variables. Other examples of scalar fields in the room are the air's density, the air pressure, and the brightness.

You can picture a scalar field as a number at every point in space: a 17 here, a −5 there, and so on. This number is not the coordinates of the point, but it is a function of those coordinates. For instance, at the point $(3, 2, 8)$ the temperature in the room may be 30°.

A function of three variables is easy (and important) to talk about, but difficult to visualize. So let's talk about fewer dimensions. A function of one variable can be visualized in two dimensions: $y = f(x)$ uses one dimension for the independent variable, and the other dimension for the dependent. Similarly, a function of two variables can be visualized in three dimensions, traditionally using x and y for the independent variables and z for the dependent. As an example, Figure 8.1 represents the function $z = x^2 \sin^2 y$. Such a function may represent the height of a mountain, in which case the graph is a fairly literal picture. But it could also represent the density of a window, the temperature of a floor, or the thickness of the ozone layer as a function of latitude and longitude. The picture of a surface allows us to talk about a

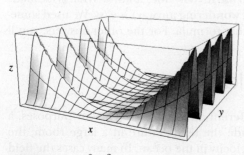

FIGURE 8.1 $z = x^2 \sin^2 y$.

"slope" or a "minimum" of the function in visual terms, but the meaning of such phenomena might be very abstract.

Contour Plots

Another way to visualize a scalar field is with a *contour plot*. For a function of two variables, a contour plot shows the "level curves" of the function. For instance, the Figure 8.2 shows the function $f(x, y) = x^2 \sin^2 y$. You can see the lines where $f = 0$, and the curves where $f = 1$, and the curves where $f = 2$, and so on.

Contour plots can represent any function of two variables, but one particularly common use—as well as a great metaphor—is a topographical (or "topo") map, where the contour lines represent curves of constant altitude. Following a curve along such a map keeps you at a constant height, while moving from the "1000" to the "1100" curve is climbing. A common convention in such maps is that the curves represent evenly spaced altitudes, such as one curve every 100 ft or every 250 ft. With that convention, the density of the curves corresponds to the slope of the incline. See Problem 8.7.

A function of three variables has "level surfaces" that are harder to draw, but still useful to picture. For the function $f(x, y, z) = 3x - 2y + 7z$, every level surface would be a plane ($3x - 2y + 7z = 0$, $3x - 2y + 7z = 1$, and so on). For a more interesting example, consider the function $f(x, y, z) = x^2 + y^2 + z^2$. Every level surface of this function is a sphere. What does that mean? If you move around one of those spheres, the value of the function does not change; whenever you move away from the origin, the value of the function increases.

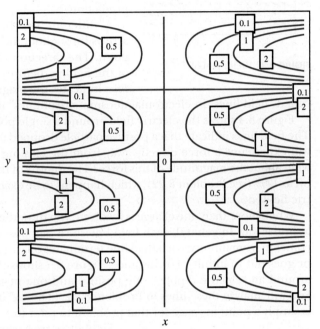

FIGURE 8.2 $z = x^2 \sin^2 y$.

Vector Fields

If a scalar field is a number at every point in space, a vector field is a vector at every point. Once again, the field is not the coordinates of the point, but it is a function of those coordinates; for instance, at the point $(3, 2, 8)$ the value of the electric field may be $(12 \text{V/m})\hat{i} - (7.9 \text{V/m})\hat{k}$. As you look at the room around you, every point has an air velocity; the collection of all these velocities is a vector field. Other examples include magnetic and gravitational fields.

Mathematically, we represent a vector field as one function for each component. For instance, we could write $\vec{v}(x, y, z) = (2x + y)\hat{i} + xyz^2\hat{j} + 4\hat{k}$. At the point $(1, 1, 1)$ this field would have the value $3\hat{i} + \hat{j} + 4\hat{k}$. But just as with scalar fields, we will focus on two-dimensional examples so we can draw them. The concepts, once mastered, will scale up.

Two techniques are commonly used to visualize vector fields. The first, sometimes called a "vector plot," places one arrow at every point on the field.[2] The direction of the arrow

---

[2] Of course you can't literally draw one arrow at every point, so you draw a whole lot of them and hope. This is part of what makes this method cumbersome.

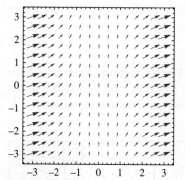

FIGURE 8.3 Vector plot of $\vec{v} = x^2 \hat{i} + 3\hat{j}$.

indicates the direction of the field and the length of the arrow indicates the magnitude of the field at each point. Figure 8.3 represents the field $\vec{v} = x^2\hat{i} + 3\hat{j}$ in this way. The Discovery Exercise (Section 8.2.1) asked you to describe the path of a leaf caught up in this particular velocity field. You can solve this problem visually by following the arrows in the diagram on the right, remembering that longer arrows mean faster travel. Do the results match your mathematical predictions?

A different visualization technique draws continuous arrows throughout the space instead of small arrows at each point. The direction of each arrow indicates the direction of the field, as before. This time, however, the *density of the lines* indicates the magnitude of the field. These arrows are called "field lines" in electromagnetism, "streamlines" in fluid flow.

Figure 8.4 shows the electric field around a single positive charge. The field lines indicate that a positive particle dropped into this space will experience a force directly away from the existing charge. And what about the magnitude of this force? The density of the lines drops off as $1/r^2$ (see Problem 8.22), which happens to be exactly what electric fields do.

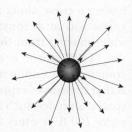

FIGURE 8.4 Field lines representing the electric field from a positive charge.

When more than one charge is involved, you draw the field lines in a similar way. In general, each field line starts at a positive charge or comes in from infinity; each field line terminates on a negative charge or goes out to infinity. For instance, Figure 8.5 shows the field lines around an "electric dipole," one positive and one negative charge. You will analyze these lines in Problem 8.24. (The "A" in the picture is simply a point in space for that problem to refer to.)

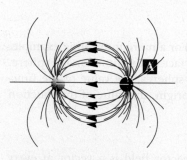

FIGURE 8.5 Field lines for the electric field around a dipole. The lines start on the positive charge on the right and end on the negative one on the left.

Field lines are a wonderful tool. They are easier to draw and interpret than a vector plot, and they sometimes provide quick answers with a minimum of calculation, as you will see in the Problems 8.22–8.23. But they don't always work. You draw a set of field lines in the *direction* of the vector field, and you want their density to tell you how the *magnitude* of the vector field is changing. If the vector field's magnitude doesn't happen to change in the way the field lines imply, then you simply can't draw field lines for that vector field. (See Problem 8.28.) Section 8.9 will discuss what fields can be represented with field lines. Such fields are a very important special case, including electric, magnetic, and gravitational fields, as well as many examples in fluid flow. But for other vector fields, the more cumbersome method must be used.

Field Derivatives

The rest of this chapter is primarily about derivatives of fields.

For a scalar field you can take partial derivatives along any of the coordinate axes. One particularly useful combination of those partial derivatives is called the "gradient," and we'll introduce it in Section 8.4. There is a useful combination of second derivatives called the Laplacian, which we will introduce in Section 8.6.

A vector field has nine different partial derivatives. For instance, there is the partial derivative of the *y*-component of the vector as you move in the *x*-direction. This is *not* a second derivative; it is the first derivative of one component of a vector. The partial derivative of the *x*-component with respect to *y* would generally be completely different. Two especially useful combinations of these nine partial derivatives, called the "divergence" and "curl," will be introduced in Section 8.6.

8.2.3 Problems: Scalar and Vector Fields

8.1 The *xy*-plane is covered with ink. The density of the ink at any given point is $\sigma(x, y) = x^2/e^y$.

(a) What is the density at the point $(1, 0)$?

(b) What is the rate of change of the density as you move through the point $(1, 0)$ in the positive *x*-direction?

(c) If you start at the point $(1, 0)$ and walk in the positive *x*-direction, what will you see over time? Where will the density head as $x \to \infty$?

(d) If you start at the point $(1, 0)$ and walk in the negative *x*-direction, what will you see over time? Where will the density head as $x \to -\infty$?

(e) What is the rate of change of the density as you move through the point $(1, 0)$ in the positive *y*-direction?

(f) If you start at the point $(1, 0)$ and walk in the positive *y*-direction, what will you see over time? Where will the density head as $y \to \infty$?

(g) If you start at the point $(1, 0)$ and walk in the negative *y*-direction, what will you see over time? Where will the density head as $y \to -\infty$?

(h) Draw all the level curves where $\sigma = 1$ — that is, the curves in the *xy*-plane that represents all points where the ink density is exactly one.

(i) Draw the level curves of this function for three other values, all on the same graph. Label each with its σ-value.

(j) How much total ink is contained in the rectangle $x \in [-2, 2]$, $y \in [0, 1]$?

8.2 [*This problem depends on Problem 8.1.*] Use a computer to draw the surface $z = x^2/e^y$. Use the drawing to confirm that the slopes in the *x*- and *y*-directions match your predictions from Problem 8.1. Based on your picture, if you were starting at the point $(-1, .5)$, estimate the direction in which the density of ink would increase the fastest.

8.3 List at least four examples of fields, not including any of the examples mentioned in this section. For each one say whether it is a scalar field or a vector field. Your list should include at least one of each.

8.4 The scalar field *T* represents the temperature, in a room with $0 \le x \le 10$, $0 \le y \le 10$, and $0 \le z \le 10$. For each of the following temperature distributions, describe in words the temperature a fly would experience as it buzzes around the room. Each description should be just a few sentences of non-technical English.

(a) $T(x, y, z) = 70$

(b) $T(x, y, z) = 70 - z$

(c) $T(x, y, z) = 70 + x + y - z$

(d) $T(x, y, z) = 70 + 5 \sin x$

(e) $T(r, \theta, \phi) = 70/(r + 1)$

8.5 One of the most important examples of a scalar field is electric potential. Every point in space has a particular electric potential, and positively charged particles feel a force pushing them from regions of high potential to low. As an improbable example, consider a region of space filled with potential $V = xe^{-y^2 - z^2}$.

(a) A positively charged particle placed into this potential field will feel a force. Is the *x*-component of that force positive, negative, or zero? Or can you not say because the answer depends on where the particle is placed? Explain your answer.

(b) For any given *x*-value, this potential decreases as *y* and *z* increase. What does that tell us about the force that positive charges will experience in this space?

(c) Find the sign of $\partial V/\partial x$, $\partial V/\partial y$, and $\partial V/\partial z$ at the point $(0, 1, 0)$. What does each one tell you about the force that a positively charged particle will experience at that point?

8.6 Contour lines are often used on geological maps. For instance, the drawing below indicates one line where the ground is 17' above sea level, two curves where the ground is 13' above sea level, and so on. Call this function $h(x, y)$ and assume that you are standing in the middle of the curve labeled 5 (right around the number "5" in the drawing).

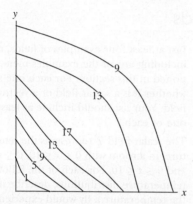

(a) At the point where you are standing, is $\partial h/\partial x$ positive or negative? Is $\partial h/\partial y$ positive or negative?

(b) If you drop a rock on the ground, in roughly what direction will it roll?

(c) Looking at the entire map, where does the ground seem to slope most steeply? How can you tell?

8.7 The figure below shows a "topo map" of a small area near Seattle. Each of the lines in the drawing represents a contour line of constant altitude, with an elevation gain of roughly 250 feet between neighboring lines.

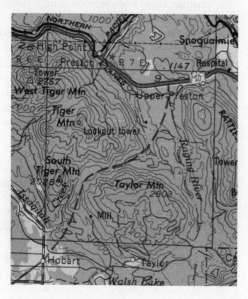

(a) If you were at the top of Taylor Mountain (near the bottom of the map) and wanted to descend as quickly as possible, would you go north or south? Explain how you can tell from the map.

(b) If you were driving east on highway 10, which is the road at the top of the drawing, briefly describe what the landscape would look like on your right and left.

8.8 The plot below shows the density of Starbucks coffeehouses in a part of downtown Seattle. (As a reasonable approximation we can treat this as a continuous function.) The arrow indicates north and the dot indicates your current position.

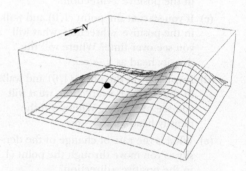

(a) Draw a contour plot of this same function. You don't need to label values on your contour lines, but you should show their shapes and you should include enough of them to clearly see how the function behaves in different places.

(b) What direction would you move if you wanted to immediately lower the number of coffeehouses around you as quickly as possible?

(c) What direction would you move if you wanted to immediately increase the number of coffeehouses around you as quickly as possible?

(d) What direction would you move if you weren't worried about immediate gratification, but wanted to go in a straight line toward the highest density of coffeehouses in the neighborhood?

8.9 Figure 8.6 shows cell phone reception in downtown Dinkerville. The arrow indicates north and the dot indicates your current position. You are currently getting poor reception and you want to move to a better spot.

8.2 | Scalar and Vector Fields

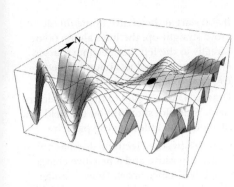

FIGURE 8.6

(a) In what direction should you walk to get the best *immediate* improvement to your reception?

(b) In what direction should you walk to get to the best reception possible in the downtown area?

8.10 The average rainfall in Flobbertown follows the strange pattern $R = (1 + \sin x)\, e^{-x^2 y^2}$, where x and y are distances north and east from one corner of the town.

(a) Pick at least 2 values of x and sketch the rainfall function $R(y)$ at that value. Label your plots with their x-values.

(b) Pick at least 2 values of y and sketch the rainfall function $R(x)$ at that value. Label your plots with their x-values.

(c) 🖥 Have a computer generate a 3D plot of $R(x, y)$. Make sure you plot over a region that clearly shows the general behavior of the function, and includes all the x and y-values you used for your sketches. Check if the computer plot matches your constant-latitude and constant-longitude sketches. (If it doesn't, figure out what you did wrong.)

(d) Based on your computer plot, if you were to start at the position $(\pi/4, 1)$ roughly what direction could you move in to keep R constant? *Hint*: You may find it easier to answer this if you make a second plot that zooms in on a small region around this point.

8.11 🖥 The town of Chewandswallow has been buried in piles of bread. The depth of bread is given by $B = \cos(x + y) + \sin(x^2 + y^2)$, where the town covers the region $0 \le x \le \pi/2,\, 0 \le y \le \pi$.

(a) Have a computer make a contour plot of $B(x, y)$.

(b) Describe the shape of the contours.

(c) Based on your plot, if you were in the center of town and wanted to slide down a mountain of bread, in approximately what direction would you get the steepest descent?

8.12 Describe the level curves of each of the following fields.

(a) $f(x, y) = 2x - 3y$
(b) $f(x, y) = \sin(2x - 3y)$
(c) $f(x, y) = 1/\left(y - 2x^2\right)$
(d) $f(x, y) = e^{-x^2 - y^2}$
(e) $f(x, y) = e^{-x^2 - 2y^2}$

8.13 Describe the level surfaces of each of the following fields.

(a) $f(x, y, z) = 2x + 3y - 4z$
(b) $f(x, y, z) = \cos(2x + 3y - 4z)$
(c) $f(x, y, z) = e^{x^2 + y^2}$
(d) $f(x, y, z) = e^{x^2 + y^2 + z^2}$
(e) $f(x, y, z) = e^{-(x^2 + y^2 + 2z^2)}$

8.14 A square region has temperature 0 on the bottom and right edges and temperature $T_0 > 0$ at the left edge. The top is insulated, which means $\partial T/\partial y$ (the derivative perpendicular to the top edge) is 0 at the top edge. Sketch a possible set of "isotherms," curves of constant temperature, inside this square region.

The xy-plane is covered with water. At any given point, the velocity of the water is given by the vector field $\vec{v}(x, y)$. For each of Problems 8.15–8.20, draw a vector plot of the velocity field. Then describe the motion of a leaf that is dropped into the water at the point $(1, 0)$. Your description should say what it does initially and what its long-term motion will look like. (Don't worry about the fact that some of these are not physically possible velocity fields.)

8.15 $\vec{v}(x, y) = \hat{i} - \hat{j}$
8.16 $\vec{v}(x, y) = x\hat{i} + y\hat{j}$
8.17 $\vec{v}(x, y) = y\hat{i} + x\hat{j}$
8.18 $\vec{v}(x, y) = y\hat{i} - x\hat{j}$
8.19 $\vec{v}(x, y) = \sin x\, \hat{i}$
8.20 $\vec{v}(x, y) = \sin x\, \hat{j}$

8.21 Figure 8.4 shows the electric field lines emanating from a positively charged particle.

(a) Draw a 2D plot showing the field lines in the xy-plane. (This plot should look like a slice through Figure 8.4.)

(b) Draw a 2D vector plot of the same electric field.

(c) Draw the field lines in the xy-plane around a single *negatively* charged particle.

8.22 Figure 8.4 shows the electric field lines emanating from a positively charged particle. The direction of the arrows indicates that the electric field, and therefore the force on a positive charge, points outward. The density of the lines indicates the magnitude of the field. You don't get to choose their density, but you can calculate it. Assume you draw n evenly spaced field lines emanating from the charge. So if you draw a sphere centered on the charge, at any radius, n evenly spaced field lines will penetrate that sphere.

(a) What is the surface area of a sphere at a distance r from the center?

(b) So what is the density, in lines/m², of the lines penetrating such a sphere?

(c) What does this tell us about the electric field around a point particle?

8.23 The figure shows the electric field lines emanating from an infinite line of charge with a uniform charge density λ Coulombs/meter. Assume you draw evenly spaced field lines emanating from the line of charge, as shown below. If you draw a cylinder of height H centered on the line, some number n of field lines will penetrate that cylinder, regardless of its radius.

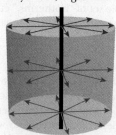

(a) What is the area of a cylinder of height H and radius r centered on the line?

(b) So what is the density, in lines/m², of the lines penetrating such a cylinder?

(c) What does this tell us about the electric field around a line of charge?

(d) The field lines from an infinite plane of positive charge point straight away from the plane. As these field lines penetrate other planes, parallel to the original but increasingly far off, what will happen to the density of the field lines? What does this tell us about the electric field around a plane of charge?

8.24 Figure 8.5 shows the electric field lines around a "dipole," a positive and a negative charge kept a fixed distance apart.

(a) What does the figure indicate is the direction of the electric field at the point labeled "A"? Why does this make sense, physically?

(b) If you start at the center of the dipole and move straight up, the lines always point directly to the left, but become sparser. What does this tell us about the electric field directly above a dipole?

(c) Draw a 2D vector plot of the same electric field. Your plot should show the field in a plane that includes the two particles.

8.25 Figure 8.5 shows the electric field lines around a "dipole," a positive and a negative charge kept a fixed distance apart. Draw a similar plot for the electric field lines around two positive charges a small distance apart. Your plot can be in 2D, on a plane that includes both particles. Remember that field lines generally begin at positively charged particles or at infinity and end at negatively charged particles or at infinity. (If you're careful you can find a few exceptions in this case.)

8.26 The figure below shows three positively charged particles set at the vertices of an equilateral triangle.

(a) Draw the field lines between and around this configuration. Remember that all field lines begin at positively charged particles and terminate at negatively charged particles, or infinity.

(b) What does your drawing predict about the electric field directly in the center of the triangle?

(c) What does your drawing predict about the electric field outside the triangle?

8.27 Is it possible, in a field line representation of a vector field, for two of the field lines to cross each other? If so, what does it tell us about the vector field at that point? If not, why not?

8.28 The field line image below shows a vector field pointing entirely in the x-direction.

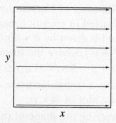

(a) Which of the following fields could be represented by those field lines? (Only one of these answers is correct.)
  i. $3\hat{i}$
  ii. $x\hat{i}$
  iii. $y\hat{i}$
(b) Of the two incorrect choices given above, one of them could not be represented by field lines. The other could, but the picture would not look like the one shown above. Say which one of those could have field lines and draw its field lines. Explain why the other one couldn't.

8.3 Potential in One Dimension

We have discussed scalar fields such as temperature and air pressure. This section introduces another scalar field, more abstract but no less important, called "potential." Potentials, like forces, provide a mathematical description of the effects of one particle on another. In Section 8.4 our discussion of potential will lead us to the first field derivative in the chapter, the gradient.

8.3.1 Discovery Exercise: Potential in One Dimension

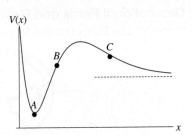

The drawing shows a "potential function" $V(x)$. An object that experiences this potential will tend to move from high potential to low potential. The function is therefore conventionally described as a hill, and the object as a ball that is pulled downhill.

1. Ball B begins at rest at position B.
   (a) Which direction will ball B roll at first?
   (b) What will happen to ball B over time? Assume it rolls without losing energy.

   See Check Yourself #52 in Appendix L

2. Ball C begins at rest at position C.
   (a) Which direction will ball C roll at first?
   (b) What will happen to ball C over time? Assume it rolls without losing energy.
3. Ball A begins at rest at position A, a local minimum of $V(x)$. What will happen to ball A over time?
4. Which of the following most accurately describes the relationship of $V(x)$ to the force on a ball? (choose one)
   (a) Where $V(x)$ is positive, the force is positive (and vice versa).
   (b) Where $V(x)$ is positive, the force is negative (and vice versa).
   (c) Where dV/dx is positive, the force is positive (and vice versa).
   (d) Where dV/dx is positive, the force is negative (and vice versa).
   (e) Where d^2V/dx^2 is positive, the force is positive (and vice versa).
   (f) Where d^2V/dx^2 is positive, the force is negative (and vice versa).

8.3.2 Explanation: Potential in One Dimension

Our goal in this section is to introduce the mathematical and physical idea of "potential." Our strategy is to show how "gravitational potential" relates to the familiar idea of a gravitational field: same physics, different math.

But if the gravitational field isn't all that familiar then we will just be relating one confusing idea to another. So we begin here by relating the gravitational field to yet another idea that we hope is familiar, the gravitational force.

We can express Newton's law of gravitation by saying that a mass M exerts a force $\vec{F}$ on a mass m. The direction of this force is toward M, and its magnitude is $|\vec{F}| = GMm/x^2$ where G is a constant and x is the distance between the two objects. In this formulation we imagine every object in the world exerting a direct influence on every other object.

But it is mathematically more convenient—and, as it turns out, physically more correct—to break the process into two steps. We say that mass M creates a "gravitational field" $\vec{g}$, an invisible vector field in the space around the object. The direction of this field is toward M and its magnitude is $|\vec{g}| = GM/x^2$. Then we say that mass m feels a force from that field given by $\vec{F} = m\vec{g}$. So the calculations lead back to $|\vec{F}| = GMm/x^2$, but with the gravitational field acting as a middleman between the object that creates the force and the object that experiences it. The gravitational field is "the force per unit mass that an object would feel if it were at that point."

The "gravitational field" formulation introduces concepts that are more abstract than the "direct force" approach, but it pays us back by making calculations easier. Gravitational field to gravitational potential is another step in the same direction. In this section we will show how potential gives us yet another mathematical formulation for the same physics; in Section 8.4 we will see how it simplifies calculations.

Gravitational Fields and Gravitational Potential

Consider the Earth (mass $6M$, position $x = 0$) and the moon (mass M, position $x = 400$).

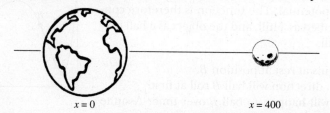

Our goal is to calculate the motion of a third object placed into this system, neglecting any motion of the Earth and moon themselves. Because we are working in this section in one dimension, we will assume that the third object is on the axis between the Earth and moon; we can therefore express both vectors and scalars as numbers. Newton's laws of gravitation can then be expressed in three rules.

1. Each body creates a "gravitational field" $g(x)$ that points toward itself. (That is, the field is positive-valued to the left of the body, and negative to the right.)
2. At any given point, the total gravitational field is the sum of the fields created by both bodies. (Sign is important here; positive and negative values can cancel.)
3. The mass at a point x will feel a force proportional to the magnitude of g and with a direction given by the sign of g (e.g. leftward if g is negative).

We can turn these general ideas into precise predictions with the right equations. The magnitude of the gravitational field from body i is $|g_i| = GM_i/x_i^2$ where G is a positive constant, M_i is the mass of the body, and x_i is the distance to the body. (You have to put in the appropriate sign for g by hand.) The total $g = \sum g_i$. A mass m inside a gravitational field g feels a force $F = mg$.

Take a moment to convince yourself that these rules express the familiar idea that any mass placed into the system feels pulled toward both the Earth and the moon—or, more generally, that all masses are attracted to each other. But this same idea can be expressed in a very different way.

1. Each body creates a "potential field" $V(x)$ that is low at points near itself, and high at points far away.
2. At any given point, the total potential is the sum of the potentials created by both bodies.
3. A mass is pulled from regions of high potential to low.

The first rule can be mathematically expressed in the equation $V_i = -GM_i/|x_i|$ (negative everywhere). The second rule is of course $V = \sum V_i$. We will return below to the mathematical expression of the third rule, and to the connection between the potential V and the gravitational field g. First, however, we encourage you to consider again how these rules express the idea that all objects are attracted to each other. In fact, the ultimate test of the "potential" formulation of gravity is that it yields the same predictions as the "gravitational field" formulation; it is not a different physical theory, just a different mathematical expression of the same theory.

"*Why go to all this trouble to create a new, harder-to-understand expression of a theory that I already understood the first way?*" The main advantage is that the gravitational field is a vector, potential a scalar. That makes potential much easier to compute, graph, and interpret than the gravitational field. Physicists trying to understand a complicated system such as an atomic nucleus or star cluster often begin by drawing a potential function for it. In this section we can treat the gravitational field as a signed scalar number because we are working in only one dimension; this helps you understand what potential means, but obscures some of its real power, which will be more apparent in the two- and three-dimensional examples in Section 8.4.

Using Potential to Predict Short-Term Behavior

We said above that the Earth (mass $6M$, position $x = 0$) and the moon (mass M, position $x = 400$) each creates a potential $V_i = -GM_i/|x_i|$. The resulting potential between the two bodies is thus:

$$V = -GM\left(\frac{6}{x} + \frac{1}{400-x}\right) \quad 0 < x < 400 \qquad (8.3.1)$$

Figure 8.7 shows a graph of this function. We must emphasize that this graph does not represent two or three dimensions of space; it describes only the line directly connecting the Earth to the moon. The horizontal axis on this graph represents position along that line, and the vertical axis represents gravitational potential. The graph shows that $V(x)$ is everywhere negative and that it reaches a maximum value of $V = -0.3$ at a point roughly 284 miles from the Earth. You will explore the region beyond the moon in Problem 8.38, and the vertical asymptotes that bound this graph in Problem 8.50. But focus right now on the rule "objects are pulled from high to low potential." How can we use Figure 8.7 to predict the behavior of an object—say, a rocket—that finds itself between the Earth and moon?

To answer this question, think of the potential graph as a hill, and the rocket as a ball rolling on the hill. For instance, a ball placed at $x = 200$ on the above hill will roll to the left. This tells us that a rocket located half-way between the Earth and moon will feel a net force toward the Earth. (Remember that the "hill" is only a metaphor; the rocket is pulled to the left, but not downward, and in fact the entire scenario takes place along one dimension.) A rocket at $x = 300$ feels a slight pull to the right. The local maximum at $x \approx 284$ represents an unstable equilibrium where the forces pulling toward Earth and moon are in balance.

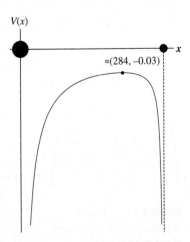

FIGURE 8.7 Gravitational potential between the Earth and moon.

Students unused to potential often try to predict behavior based on whether the potential is high or low, negative or positive. But as you can see from the discussion above, *what matters is whether the potential is increasing or decreasing*. In regions where the potential graph slopes up, the body is pulled to the left, and vice versa:

$$g(x) = -dV/dx \tag{8.3.2}$$

Equation 8.3.2 provides the link between the potential V and the gravitational field g. You can translate a potential field to a gravitational field, and then into a force, to create an equation of motion. But you can also predict motion directly from potential, without ever mentioning the gravitational field.

EXAMPLE Parallel Plate Capacitor

Question: The graphs show the electric potential V and the electric field E around an idealized parallel plate capacitor. How do they relate, and what are they telling us?

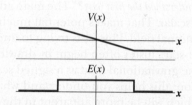

Answer:
The question "How do they relate?" is easy to answer mathematically: $E = -dV/dx$. So E is zero where V is flat, and positive where V slopes down.

They are telling us that a positively charged particle will feel no force if it is outside the capacitor (where V is flat and E is zero). Inside the capacitor, the particle will feel a force to the right. You can see this because E is positive, or you can see it because the potential field slopes downward.

Students often incorrectly conclude that there is no force where the potential is zero, but this is not correct: there is no force where the potential is *flat*. A potential of zero does not signify anything in particular, as we discuss below.

Using Potential to Predict Long-Term Behavior

Whenever you have a potential function, you have a force that conserves energy. This is a bit tricky, because potential is not the same thing as potential energy, although the two are closely related: we will explore the connection more closely in Section 8.11. But here we need only to know that as you move higher on a potential graph, you move to a region of higher potential energy. As you move lower, that energy is converted to kinetic energy. As always, the image of a ball rolling around a hill serves well.

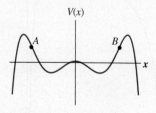

FIGURE 8.8 A potential function: An object that starts at rest at A will oscillate between A and B.

So consider an object at rest at position A on the potential graph in Figure 8.8. As it "rolls down the hill" to the right, it picks up speed; as it starts back up again, it slows down. It has enough energy to make it over the first hill, but not the second: it will reach position B momentarily at rest, and then will begin slowly back to the left. If the force associated with this potential is the only force acting on the object, it will move between A and B forever. If another force is acting, the behavior may be different. For instance, a "damping" force such as friction or air resistance may drain off some of the energy, so the object will eventually come to rest at one of the local minima.

Finding V from a Field

Equation 8.3.2 can be reversed to go from the gravitational field to the potential: $V = -\int g(x)\,dx$. For example, we started with Newton's law of gravitation, which says $g(x) = -GM/x^2$ for positive x. Integrating, we see that for positive x, $V(x) = -GM/x + C$. You'll treat negative values of x in Problem 8.36 to see why we used an absolute value when we wrote $V(x) = -GM/|x|$ above.

But let's not lose sight of that $+C$. If we wrote instead $V = -GM/x + 5$, moving the potential graph of Equation 8.3.1 uniformly up, what changes would that cause? The rule "objects roll down the potential hill" would predict the same behavior in all circumstances. And the calculation $g = -dV/dx$ would throw away the additional constant. So whether you approach the problem visually or computationally, you conclude that nothing in the system would change.

This is an important insight into the nature of potential fields. "The potential is 17 at Point A" tells you nothing, but "The potential is 17 at Point A and 19 at Point B" tells you that an object will tend to move from B to A. Potential is only meaningful as a relative value. You can express this mathematically by saying that every potential function includes an arbitrary $+C$, but the same idea is generally expressed a different way: we say that you can choose any position you like to be $V = 0$, and measure all other potentials relative to that reference point.

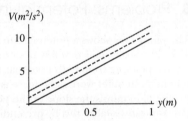

FIGURE 8.9 Three physically equivalent functions $V(y)$ for gravitational potential in my office

Consider the gravity near Earth's surface, where the potential V is a simple function of height y. If a book falls from my hand to the floor, we say it has moved from high potential to low. But does $V = 0$ represent the floor of my office, or the ground three stories below, or sea level several hundred feet below that? It doesn't matter. Starting from $g = -9.8$ we can find V with an indefinite integral, and write $V = -\int g(y)\,dy = 9.8y + C$ where C can be any constant you like. We can also use a definite integral $V = -\int_{y_0}^{y} g(\tilde{y})\,d\tilde{y}$ where y_0 is any constant you like: the arbitrary place that we choose to represent $V = 0$. (We needed a new variable $\tilde{y}$ inside the integral since y is now being used for the limit of integration.)

You may be more accustomed to the potential energy $PE = mgy$ (often written mgh). This is the same as $V(y)$ (including the arbitrary choice of a $y = 0$ location) except for that factor of m. An elephant and a mouse on the roof are both at the same gravitational *potential*, but they have very different *potential energy*. We will return to the relationship between potential and potential energy in Section 8.11.

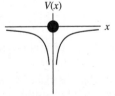

Coming back to the idea of a reference point where $V = 0$, there is one special case worth mentioning. In our discussion of the Earth and moon we wrote $V = -GM/|x|$ with no arbitrary constant; the reference point is implicit in the equation, but it's not obvious what the reference point is. If you look at the plot of gravitational potential around Earth, you can see the choice we have made: we defined the $V = 0$ point to be "at infinity" (technically a limit as $x \to \infty$). Having made that choice, we calculate all other potentials from there: $V(x) = -\int_{\infty}^{x} g(\tilde{x})\,d\tilde{x}$.

Stepping Back

Newtonian gravitation can be described two equivalent ways. We can say that every mass creates a "gravitational field" that points toward itself, and every other mass is pulled in the direction of the gravitational field. Or, we can say that every mass creates a "potential field" that is low around itself and high far away, and every other mass is pulled from regions of high potential to low. These two formulations lead to the same physical predictions, so choosing one over the other is a matter of mathematical convenience.

The rules of electrostatics can similarly be framed in terms of electric fields or of electric potential. In this case, however, the sign of a charge (positive or negative) determines the type of potential field it causes, and the sign of another charge determines how it responds to a potential field; positive charges seek low potential, and negative charges high. In Problems 8.46–8.48 you will develop the rules of electric potential that lead to the familiar behavior of charged particles.

Although potential does not represent a new physical theory, it is an indispensable mathematical tool. Circuit analysis is done entirely in terms of potential, with the electric field rarely mentioned at all. In quantum mechanics, the behavior of a system is described in terms of the potential field around it, rather than the forces acting upon it. Newtonian mechanics is approached the same way in the "Lagrangian" and "Hamiltonian" formulations, which are often preferred to $F = ma$.

Remember that an object placed into a gravitational field will be pulled toward the lowest potential it can find. This insight is the key to understanding potential in one dimension, and also forms the bridge to higher dimensions, as we will see in the next section.

8.3.3 Problems: Potential in One Dimension

8.29 The plot below shows a potential function $V(x)$. Physically, objects at any position feel a force pointing from high potential to low potential. In other words, the force is proportional to minus the slope of the potential. You can visualize this by pretending that the potential plot is a hill and objects experiencing this force are balls rolling on that hill.

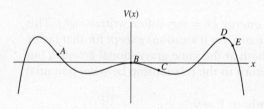

(a) Is the slope of V at position A positive or negative?
(b) If an object is placed at position A, will the force on it be positive or negative?
(c) If an object is placed at rest at position A, the force on it will cause it to move until it reaches a point with the same potential as where it started. Describe the long-term behavior of such an object.
(d) Describe the long- and short-term behaviors of objects placed at rest at positions B, C, D, and E. (Point D is at the maximum of the potential.)

8.30 [This problem depends on Problem 8.29.] Sketch the field whose potential is shown in Problem 8.29.

8.31 Consider a potential function $V(x) = kx$ where k is a positive constant. An object in this field will be pulled "downhill"—that is, pulled to the left. If it starts at rest, will it move leftward with a constant speed, or will it gradually slow down, or will it continually speed up? Why?

8.32 Consider a potential function $V(x) = kx^2$ where k is a positive constant. Describe the behavior of an object in such a potential field.

8.33 Figure 8.10 shows a potential function. Describe the short and long-term behavior of an object started at rest at each of the indicated points. Points C and D are at a local minimum and maximum of $V(x)$.

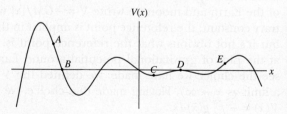

FIGURE 8.10

8.34 Figure 8.11 shows a potential function. Describe the short and long-term behavior of an object started at rest at each of the indicated points. Points C and D are at a local maximum and minimum of $V(x)$.

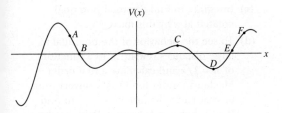

FIGURE 8.11

8.35 The $V(x)$ diagram below is sometimes referred to as a "potential well."

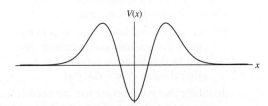

(a) An object starts at the local minimum, moving to the right slowly (low energy). Describe what will happen to that object over time, assuming it does not lose any energy.

(b) Describe what will happen to the object over time if it *does* lose energy in motion.

(c) A different object starts at the local minimum, moving to the right quickly (high energy). Describe what will happen to that object over time.

8.36 Consider a one-dimensional universe with an object, mass M, fixed at the position $x = 0$. The equation $|g| = GM/x^2$ gives the magnitude of the gravitational field, but you have to supply the sign to make the field pull leftward or rightward as appropriate.

(a) For what x-values does g have a positive sign, and for what x-values does it have a negative sign?

(b) For positive values of x, write $g(x)$ with the correct sign and integrate it to find $V(x)$. Your answer should have an arbitrary constant in it.

(c) For negative values of x, write $g(x)$ with the correct sign and integrate it to find $V(x)$.

(d) Write a formula for the potential $V(x)$ such that the gravitational field is always given by $g = -dV/dx$. Your answer should have one arbitrary constant.

(e) Choose the arbitrary constant so that $\lim_{x \to \infty} V(x) = 0$.

(f) Draw a quick sketch of the resulting potential field.

(g) Your potential should be negative for all x-values. Explain why this must be true, based on the rules "objects move from regions of high potential to low potential" and $\lim_{x \to \infty} V(x) = 0$.

8.37 Consider a potential field $V(x) = \sin x$.

(a) An object begins at rest in this potential field at $x = 3\pi/4$. Describe the subsequent short-term and long-term behavior of this object. (Hint: draw the potential field. Then imagine a ball that starts at this point on the curve and is free to roll with no loss of energy.)

(b) An "equilibrium point" means an x-value where an object feels no net force. An object at rest at an equilibrium point will stay there. Identify four equilibrium points in this potential field.

(c) A "stable equilibrium" means that if an object is slightly displaced from the equilibrium point, it will tend to return to the equilibrium point. Identify two points of stable equilibrium in this potential field. (You may use answers from Part (b) if they are stable equilibrium points.)

(d) An "unstable equilibrium" means that if an object is slightly displaced from the equilibrium point, it will tend to move away from the equilibrium point. Identify two points of unstable equilibrium in this potential field. (You may use answers from Part (b) if they are unstable equilibrium points.)

8.38 In the Explanation (Section 8.3.2) we considered the gravitational field and potential for an object between the Earth and the moon ($0 < x < 400$). In this problem you will consider an object beyond the moon ($x > 400$).

(a) The formula $|g_i| = GM_i/x_i^2$ gives the magnitude of g for body i; you have to add the sign (negative for a leftward pull, positive for rightward) yourself. Calculate the gravitational fields from the moon and Earth at a point $x > 400$, and sum them to find the total gravitational field in this region.

(b) The potential field from body i is given by $V_i = -GM_i/|x_i|$. Calculate the gravitational potentials from the moon and Earth, using the fact that $x > 400$ to write formulas that do not require absolute values, and sum them to find the total potential in this region.

(c) Confirm, based on your two answers, that $g = -dV/dx$ and that $\lim_{x \to \infty} V(x) = 0$.

(d) Draw a graph of the total potential field V on the domain $x > 400$. Use this graph to qualitatively describe the motion of a body that begins at rest at $x = 1000$.

For Problems 8.39–8.41 sketch a potential field that leads to the indicated behavior. You do not have to write any equations; the answer is a graph.

8.39 All objects are pulled to the left. The farther left you go, the stronger the pull.

8.40 All objects are pulled to the left. The farther left you go, the weaker the pull.

8.41 All objects are pushed away from $x = 2$. As you move away from $x = 2$ to the left, the force gets weaker. As you move away from $x = 2$ to the right, the force gets stronger.

8.42 Two objects of equal mass M are fixed in place at $x = 0$ and $x = 3$. This question is about the effect of those objects on a third, movable object.

  (a) In what positions x will the third object feel a net pull to the right? In what positions will it feel a net pull to the left? In what positions will it feel no net pull?

  (b) Sketch $V(x)$ in the region $-5 < x < 5$ and make sure it matches your predictions from Part (a).

8.43 Three objects of equal mass M are fixed in place at $x = 0$, $x = 3$, and $x = 6$. This question is about the effect of those three fixed objects on a fourth, movable object. In the region $0 < x < 3$ the gravitational field is $g(x) = GM[-1/x^2 + 1/(3-x)^2 + 1/(6-x)^2]$. At $x = 1.46$, $g(x) = 0$.

  (a) What would happen to the fourth object if it were placed initially at rest at $x = 1.46$?

  (b) What would happen to the fourth object if it were placed initially at rest at $x = 1.3$?

  (c) What would happen to the fourth object if it were placed initially at rest at $x = 1.6$?

  (d) Sketch the potential in the region $0 < x < 3$, making sure it matches all your predictions.

  (e) Sketch the potential in the region $-10 < x < 10$. This should require no new calculations.

  (f) Write the formula for the potential field $V(x)$ in each region, using the formula $V = \sum(-GM_i/|r_i|)$. Confirm that it also matches your sketch from Part (e).

8.44 An object of mass M is evenly distributed between $x = 0$ and $x = L$.

  (a) Consider a small "slice" of this mass, extending from some $x = a$ ($0 \le a \le L$) to $a + da$. What is the mass dm of that slice?

  (b) What is the gravitational potential dV caused by that slice at some x-value ($x > L$)?

  (c) Integrate to find the total potential created at x by this mass.

  (d) To see the behavior of the potential at large distances rewrite V as a function of $u = 1/x$ and calculate a first order Maclaurin series for $V(u)$. Convert your answer back to a function $V(x)$ to find how V behaves at large x. Explain why your result makes physical sense.

8.45 An object whose mass is extended from $x = 0$ to $x = L$ has density $\lambda = kx$ where k is a positive constant.

  (a) Set up and evaluate an integral to calculate the gravitational potential created by this object at a point $x > L$, using the rule $V = -GM/|x|$.

  (b) Show that your answer has the correct units for potential. (In order to do this, you will have to first figure out the units for k.)

  (c) Set up and evaluate an integral to calculate the gravitational field created by this object at a point $x > L$, using the rule $g = -GM/x^2$.

  (d) Confirm that your potential and gravitational fields have the correct relationship to each other.

The Explanation (Section 8.3.2) expresses Newton's law of gravitation in two ways: first in terms of gravitational fields, and then in terms of potential. In Problems 8.46–8.48 you will do the same for Coulomb's law of electrostatics. The basic idea is the same, but instead of "all masses attract each other" the rule is "like charges repel and opposite charges attract."

8.46 A positive charge Q creates an electric field with magnitude $E = kQ/x^2$ pointing everywhere *away* from itself.

  (a) Sketch a graph of the electric potential around a positive charge located at $x = 0$. Your graph should follow two rules. First, a positive charge placed into this potential field should be pulled from regions of high potential to low. Second, $\lim_{x \to \infty} V(x) = 0$. (You can base this on the gravitational potential around a single mass, but the two drawings are *not* identical.)

  (b) Write a formula $V(x)$ such that $E(x) = -dV/dx$. Your formula should, of course, match your drawing from Part (a).

  (c) Sketch a graph of the electric potential around a negative charge located at $x = 0$. Your graph should follow the same two rules as in Part (a).

(d) How can you use a potential plot to predict the behavior of a *negative* charge?

8.47 *[This problem depends on Problem 8.46.]* Consider an electric field that can be represented by the following potential graph.

(a) A positively charged particle begins at the point labeled P. Describe its motion over both the short and long term.
(b) A negatively charged particle begins at the point labeled P. Describe its motion over both the short and long term.

8.48 *[This problem depends on Problem 8.46.]* Consider an "electric dipole" comprising a positive charge $Q = 1$ fixed at $x = 0$ and a negative charge $Q = -1$ fixed at $x = 5$.

(a) Calculate and draw the field of electric potential around these charges. Remember that the potential at any given point is the sum of the potentials created by the two charges.
(b) Using your potential plot, predict the motion of a positive charge placed at $x = -1$, a second positive charge placed at $x = 2.5$, a third positive charge placed at $x = 3$, and a final positive charge placed at $x = 10$ in this field.
(c) Using your potential plot, predict the motion of a *negative* charge placed at each of the same four spots.

8.49 In the Explanation (Section 8.3.2) we showed that the gravitational potential near Earth's surface is gy, where y is height off the ground and g is the constant 9.8 m/s². The diagram below shows the profile of some hills.

(a) As a ball rolls up and down these hills it experiences a different potential at each point because its height is changing. Sketch a graph of the potential $V(x)$ that the ball experiences.
(b) Explain, using your sketch, why the "ball rolling on a hill" metaphor works for any potential curve, even if it isn't for a gravitational force.

8.50 **What about that vertical asymptote?** In this problem you will consider the potential due to the Earth, which we take to be at $x = 0$. There are no other masses in the problem (no moon, for example). You will only consider the right side of the axis ($x \geq 0$). The equations are a bit different on the left, but the graphs are perfectly symmetrical.

(a) Consider a simple system in which the Earth is a point particle with mass M located at $x = 0$. Calculate and sketch the potential field $V(x)$.

You should have found a vertical asymptote at $x = 0$. This is not a physical reality; it is due to the simplified assumption of the Earth as a point particle. To create a more realistic picture, consider the Earth's mass to be spread between $x = -R$ and $x = R$. For $x > R$, the gravitational field is still $g = -GM/x^2$. But inside the Earth, the gravitational field has magnitude $g = -Kx$ where K is a positive constant.

(b) Explain why it makes sense physically in this scenario that $g = 0$ at $x = 0$ (the center of the Earth).
(c) Calculate the potential field for $0 \leq x \leq R$ such that $-Kx = -dV/dx$. Your answer should have an arbitrary constant.
(d) Calculate $V(R)$ based on your formula from Part (a), and then choose the arbitrary constant in Part (c) to make V continuous at $x = R$. (Your answer for the arbitrary constant should depend on K, G, M, and R.)
(e) Write a piecewise function that represents $V(x)$ for all $x \geq 0$, and sketch it.

8.51 A solar cell works by creating a region of fixed negative charge next to a region of fixed positive charge. The resulting electric field is shown in the plot below. It is zero on either side of the cell (where wires lead to the other parts of the circuit) and negative in the middle. We are taking "right" to be the positive direction as usual.

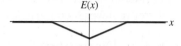

(a) In the region just to the left of the origin, is the field pointing to the left or right?
(b) In the region just to the right of the origin, is the field pointing to the left or right?
(c) If we choose to set $V = 0$ to the left of the solar cell, sketch the potential $V(x)$ throughout this region.
(d) Sunlight shining on the cell causes an atom to release a positive and a negative charge. Assume this happens at $x = 0$. The positive

charge is pushed toward low potential and the negative charge is pushed toward high potential. Which way do they each go?

Once you've gotten positive and negative charges to go in different directions you can use them to store energy in a battery, so the mechanism you've analyzed above allows solar cells to generate electricity from sunlight.

8.52 An electron in a crystal experiences a periodic potential that moves up and down depending on how close the electron is to one of the positively charged nuclei. In the simplest case the potential looks roughly sinusoidal. (In this problem you will treat the electron as a small negatively charged Newtonian object. Real electrons obey the laws of quantum mechanics, which leads to very different behavior.) Remember to take into account the fact that the electron's charge is negative.

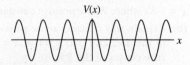

(a) If an electron starts at rest at a local minimum of the potential describe its long-term behavior.

(b) If an electron starts at rest somewhere between a minimum and a maximum of the potential, describe its long-term behavior.

(c) If an electron starts at a maximum of the potential moving rapidly to the right, describe its long-term behavior.

(d) Sum up your answers by describing all of the possible behaviors of an electron in this potential.

Many crystals have more than one kind of nucleus, which attract the electrons with different strengths. In a simple alternating lattice, the potential for the electron might look like the following:

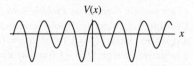

(e) For the alternating lattice shown above, describe all of the possible behaviors of the electron.

(f) Sketch the electric fields corresponding to the two potentials shown above.

8.4 From Potential to Gradient

In the previous section, we saw that the rules of gravitation and electrostatics can be framed in terms of a potential function that is related to the gravitational or electric field. When we take the same concept to higher dimensions, the advantages of this extra step become clear. However, we have to use a higher dimensional generalization of the derivative called the "gradient" to relate the potential function to the field.

If you've read our chapter on partial derivatives you have seen us introduce the gradient in another context. We are going to approach the topic very differently in this case, but everything you have already learned about the gradient will still be useful. If you have never seen a gradient before, rest assured we will not assume any prior knowledge of it in this chapter. If you've never seen partial derivatives, though, you're pretty much doomed in this chapter.

8.4.1 Discovery Exercise: From Potential to Gradient

You have seen in one dimension how a potential function predicts motion: every object is pushed from high potential toward low potential. Now consider the graph below in which the x and y axes represent position in a plane, and the vertical axis represents the potential at each point (x, y).

Just as in 1D, you can imagine the potential as a hill with a ball rolling on it. The ball will be pulled in the downhill direction, and the steeper the slope the more it will accelerate. For the problems below you don't need to calculate any numbers; you should be able to see the answers from the plot.

1. If a ball were at rest at location A, would it roll in the positive x-direction, the negative x-direction, or neither? Would it roll in the positive y-direction, the negative y-direction, or neither? (Remember that this entire plot takes place in a two-dimensional universe; there is no z-direction because the vertical axis represents the potential $V(x, y)$.)

   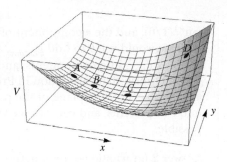

   See Check Yourself #53 in Appendix L

2. Repeat Part 1 for balls that begin at rest at locations B, C, and D.
3. Which ball will accelerate faster, ball A or ball B?
4. Now, answer with a brief equation (no words required): what must be true of the function $V(x, y)$ at a given point for a ball to feel a pull in the positive x-direction?

8.4.2 Explanation: From Potential to Gradient

In Section 8.3 we showed how the gravitational force between two objects can be expressed in two steps: an object creates a gravitational field g that points toward itself, and another object experiences a force $F = mg$ from that field. We then discussed a different mathematical way to express the same physical law: an object creates a potential field $V(x)$ that is lowest at points close to itself, and another object is pulled from regions of high potential to low potential.

The same two approaches can be used in higher dimensional spaces. The big difference is that the gravitational field is now a vector $\vec{g}(x, y, z)$. To sum the contributions of many bodies, you have to sum (or integrate) their x-, y-, and z-components separately. But the potential field is still a scalar $V(x, y, z)$, which makes it much easier to work with.

> **EXAMPLES** **Calculating the Potential in Two Dimensions**
>
> **Question:** In the previous section we considered the Earth (mass $6M$) at the point $(0, 0)$ and the moon (mass M) at $(400, 0)$. Calculate the gravitational field and potential created by these bodies at point P at $(300, 50)$. (Even in 3D space this is effectively a 2D problem since no relevant vectors point in or out of the page.)
>
>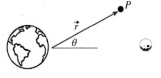
>
> **Answer 1 (gravitational field):**
> The calculations rely on a vector $\vec{r}$ that points from the body exerting the force to the point in question. For instance, from the Earth to point P, $\vec{r} = 300\hat{i} + 50\hat{j}$. The gravitational field $\vec{g}$ has magnitude $GM/|\vec{r}|^2$ and points back toward the body. (We can express this more concisely by writing $\vec{g} = \left(-GM/|\vec{r}|^2\right)\hat{r}$.) So the magnitude of $\vec{g}$ at point P from the Earth is $6GM/92,500$.
>
> It isn't enough to know the magnitude, though; we need to sum the $\vec{g}$ vectors from the Earth and moon, which requires breaking them into components. The gravitational field produced by the Earth points along the same line as $\vec{r}$ (in the opposite direction), so we use θ to find its components. The angle θ in the drawing is

$\tan^{-1}(1/6)$, and the x-component of the Earth's field is $g_{Earth,x} = -6GM/92{,}500 \cos\theta$. Next we would use θ to find $g_{Earth,y}$, and make another drawing with a different angle to break $\vec{g}_{moon}$ into components, and so on…

You'll finish that calculation in Problem 8.64. It's tedious. Now imagine if we added a third object off the plane of the page, thus requiring breaking all the 3D vectors into components, and you can see why we want to avoid these vector sums if at all possible.

Answer 2 (gravitational potential):
The potential $V = -GM/|\vec{r}|$. We take the potential at point P from the Earth, the potential at point P from the moon, and add them.

$$V = -6\frac{GM}{\sqrt{92{,}500}} - \frac{GM}{\sqrt{12{,}500}}$$

That's all there is to it!

The above example should decisively convince you that the gravitational potential is easier to find than the gravitational field. When we calculate gravitational effects from extended objects instead of point sources (so we're integrating instead of just adding) the math gets messier still, and the advantages of potential are even greater. But what do we do with the potential once we have found it?

Getting from V to $\vec{g}$

In one dimension, the rule $g = -dV/dx$ tells us two things about a potential field. First: an object feels a force toward a region of lower potential. Second: the faster the potential is changing, the greater the force.

In higher dimensions, $\vec{g}$ is a vector that follows those same two rules. It points in the direction in which the potential is decreasing the fastest, and its magnitude indicates how quickly the potential is decreasing in that direction. Consider the $V(x, y)$ function in Figure 8.12.

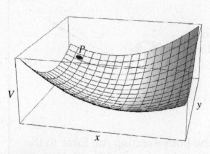

FIGURE 8.12

Remember that this surface represents a dependent variable (V) of two independent variables (x and y): a potential field in two dimensions. A ball placed at point P will roll to the right and toward us. The slope is shallow in this direction, indicating a relatively weak force.

In three dimensions we lose our ability to draw a graph, but the principle of moving in the direction of steepest descent is the same. (It is still possible to graph level surfaces of the potential in 3D. See Problem 8.13.)

To express all this mathematically we define an operator called the "gradient," a multidimensional generalization of the derivative. If you have worked through Section 4.6 you have already seen one introduction to the gradient, starting with the idea of a "directional derivative." Here we are approaching the same topic starting with potential functions. You can work through either section without the other, but if you do both, take the time to convince yourself that they are both talking about the same topic. This will deepen your understanding of this very important operator.

Definition of the Gradient, Part I: The Interpretation

The "gradient" of a scalar field is a vector field. At any given point, the gradient points in the direction of steepest increase of the scalar field. The magnitude of the gradient is the rate of increase of the scalar field in that direction. The gradient of scalar field f is written $\vec{\nabla} f$.[a]

Yes, we said steepest *increase* even though the gravitational field points in the direction of steepest *decrease*. So just as $g = -dV/dx$ in one dimension, $\vec{g} = -\vec{\nabla} V$ in higher dimensions.

---

[a]Sometimes $\vec{\nabla} f$ is written grad f, mostly in older texts

And how do you do the calculations? Given a scalar field, you have to find three values: the x-component of the gradient, the y-component, and the z-component. But if you focus for a moment on the first of these, you might realize that we solved precisely this problem in the previous section: we found the x-component of the gravitational field by writing $g = -dV/dx$. Similar considerations apply to the y- and z-directions, which leads to the formula.

Definition of the Gradient, Part II: The Formula

The "gradient" of a scalar field $f(x, y, z)$ is:

$$\vec{\nabla} f = \frac{\partial f}{\partial x}\hat{i} + \frac{\partial f}{\partial y}\hat{j} + \frac{\partial f}{\partial z}\hat{k} \tag{8.4.1}$$

We've said that the gradient tells you the direction of steepest increase and the rate of change in that direction, but more generally each component of the gradient in any direction tells you the rate of change of a function in that direction. The direction opposite the gradient represents the steepest possible *decrease* of the function. If you move in any direction perpendicular to the gradient the function doesn't change. Mathematically this can be expressed by saying the rate of change of a scalar field f in the direction of any unit vector $\hat{u}$ is $\left(\vec{\nabla} f\right) \cdot \hat{u}$. (The rate of change of a function in an arbitrary direction is called a "directional derivative." We discuss them in Chapter 4.)

We have introduced the gradient as a way of finding the gravitational or electric field based on potential, and this is indeed one of its main purposes. But given any scalar field, it is possible—and often useful—to evaluate its gradient.

EXAMPLE   The Gradient

Question: The temperature in a room follows the distribution $T(x, y, z) = x/y - 2\ln(z)$. Find the gradient of this field at the point $(3, 2, 5)$ and explain what this particular vector tells us about the temperature at that point. Assume all distances are in meters and temperatures in degrees Celsius.

Answer:
$\vec{\nabla} T = (1/y)\hat{i} - (x/y^2)\hat{j} - (2/z)\hat{k}$, so $\vec{\nabla} T(3, 2, 5) = (1/2)\hat{i} - (3/4)\hat{j} - (2/5)\hat{k}$

This tells us that if a fly at the point $(3, 2, 5)$ looks around in every possible direction, it will find the temperature increasing most rapidly in the direction $(1/2)\hat{i} - (3/4)\hat{j} - (2/5)\hat{k}$. If the fly sets out in that direction, it will see the temperature increasing at a rate of $\sqrt{1/4 + 9/16 + 4/25}$ degrees per meter traveled.

The opposite direction, $(-1/2)\hat{i} + (3/4)\hat{j} + (2/5)\hat{k}$, represents the direction of steepest decrease at that point. In any direction perpendicular to this the fly would be moving along an "isotherm," meaning a surface of constant temperature.

Equipotential Lines and Surfaces

Section 8.2 discussed two ways to visualize a scalar field. One is by graphing: you can represent a function of one variable with a two-dimensional curve $y(x)$, and a function of two variables with a three-dimensional surface $z(x, y)$. We have seen, in this section and the previous, that if such a graph represents gravitational potential then the gravitational field points in the direction of steepest decrease, or "down the hill."

But we also discussed "level curves" of a function of two variables, or "level surfaces" of a function of three. When the scalar field is a potential, these are called "equipotential" lines or surfaces, and they are often the most convenient representation. To start with a simple example, consider the gravitational potential around a point mass. The potential is lowest at points close to the mass and higher farther away, and it is only a function of your distance to the mass. Therefore, the equipotential surfaces are concentric spheres: a sphere for $V = -100$, a more distant sphere for $V = -50$, and so on.

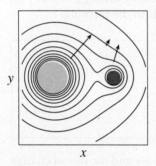

FIGURE 8.13 Equipotential lines and some gradient vectors for the Earth–moon system. The potential gets higher as you move away from the two objects. The gradient vectors are opposite the direction of the gravitational force.

Our Earth–moon system provides a more interesting example, shown in Figure 8.13. How do you go from such a potential graph to the gradient? At any given point, if you move along the curve, you move in a direction that keeps the potential constant. If your goal is to increase the potential as much as possible, you take the steepest path to the next potential curve up, which means you move perpendicular to the potential curve you're on.

So in general, given any graph of the contour lines of a scalar function, the gradient of that function points perpendicular to those curves at any given point, towards higher values. If the scalar function represents equipotential lines, then those perpendicular vectors represent *minus* the gravitational (or electric) field. For instance the arrows in Figure 8.13 show the gradient at a few places, but the gravitational field points opposite these arrows.

8.4.3 Problems: From Potential to Gradient

8.53 The gravitational potential from a single mass M is given by $V = -GM/r$ where r is the distance to the object. In Cartesian coordinates we can rewrite this as $V = -GM/\sqrt{x^2 + y^2 + z^2}$.

(a) Calculate the gradient $\vec{\nabla} V$ and use it to find the gravitational field $\vec{g}$.

(b) The gravitational field is more commonly expressed by giving its magnitude $\left(|\vec{g}| = GM/r^2\right)$ and its direction (directly toward the object creating the field). Show that your answer matches this expression.

In Problems 8.54–8.57, find the gradient of the given function at the given point.

8.54 $f(x, y) = x^2 - 5y$ at $(3, 2)$.

8.55 $f(x, y) = 17$ at $(6, 2)$

8.56 $f(x, y, z) = x \ln(y + 3z)$ at $(5, 7, -2)$

8.57 $f(x, y, z) = \sin x + \cos^2 y + z^2 - 6z + 9$ at $(\pi/2, 3\pi/2, 3)$

8.58 The temperature inside a room is modeled by $T(x, y, z) = e^{3z} \sin(2x + y)$.

(a) A fly at position $(0, 0, 0)$ is attracted to warmth, so it flies in the direction of steepest temperature increase. Give a vector that tells the direction of its flight.

(b) What is the magnitude of the gradient vector at position $(0, 0, 0)$?

(c) Explain in non-technical language what your answer to Part (b) means to the fly. Your answer should include the units of the gradient of T.

8.59 The height of the mountains on one side of the land of Mordor can be modeled by $h(x, y) = e^{-(3x-y+1)^2}$.

(a) A rock is dropped at the point $(0, 0, 1/e)$. Which way will it roll?

(b) Find the curve in the xy-plane that satisfies the equation $\vec{\nabla} h = \vec{0}$.

(c) An orc is walking along the path you found in Part (b), in the direction of increasing x, and a hobbit is sneaking along a path parallel to him, but lower on the hill. Are the orc and hobbit moving uphill, downhill, or neither? *Hint*: the answer is the same for both.

(d) The orc's path has $\vec{\nabla} h = 0$ and the hobbit's doesn't. What does that tell you about what each of their paths looks like? *Hint*: it can't be whether they are moving up, down, or neither since we already told you the answer is the same for both of them.

8.60 The electric field $\vec{E}(x, y, z)$ inside a conductor is generally zero. (If it isn't, the electrons redistribute until a field of zero is reestablished.) What does that tell us about the potential field $V(x, y, z)$ inside a conductor?

8.61 The deeper you descend into the ocean, the greater the water pressure. Pressure tells you the force per unit area that a fluid exerts on any object, so if there is greater pressure on one side of the object than the other that will create a net force on the object.

(a) If $P(x, y, z)$ represents the water pressure, give a possible function that could represent $\vec{\nabla} P(x, y, z)$. (Based on the information we have given you, there are many possible right answers to this question. But there are also plenty of wrong ones!)

(b) Suppose at one particular spot in the ocean, $\vec{\nabla} P$ has a positive component in the x-direction. What would happen to an object in the water? What would happen to the water itself?

8.62 A supermarket is shaped like a large rectangular box. Somewhere in the middle of that rectangular box is a smaller rectangular box with the frozen foods. The closer you get to the frozen foods, the colder the air gets.

(a) Draw a simple two-dimensional schematic diagram of the supermarket with the frozen food shelf somewhere in the middle. Draw a vector plot that could represent the gradient of the temperature field, $\vec{\nabla} T$, everywhere in the supermarket. Remember to follow both rules: the direction of each arrow should indicate the direction of the vector field (the direction of steepest increase of T), and the length of each arrow should represent the magnitude of the field (the rate of increase of T).

(b) We're not going to ask you to draw a three-dimensional representation of the supermarket, but say a few words about what your vector field would look like in the third dimension. Assume the freezer shelf extends all the way to the ceiling.

8.63 The brightness of the light on a table is given by a function $B(x, y)$. Consider a light source placed at $(0, 0)$ and a brighter light source placed at $(0, 5)$. Draw a 2D vector plot that could represent the gradient $\vec{\nabla} B$ throughout the room.

8.64 The Explanation (Section 8.4.2) presents the following scenario: the Earth (mass $6M$) is at position $(0, 0)$ and the moon (mass M) is at position $(400, 0)$. In this problem you will calculate the gravitational field at point P at $(300, 50)$, using the formula $\vec{g} = \left(-GM/|\vec{r}|^2\right) \hat{r}$.

(a) The vector $\vec{r}$ from the Earth to P is $300\hat{i} + 50\hat{j}$. Calculate the magnitude and direction of the gravitational field at P due to the Earth, and then calculate the x- and y-components of this field. (If you don't know where to start, we outline the first steps in the Explanation.)

(b) Write the vector $\vec{r}$ that points from the moon to point P.

(c) Calculate the magnitude and direction of the gravitational field at P due to the moon, and then calculate the x- and y-components of this field.

(d) Sum the contributions of the Earth and the moon to find $\vec{g}$ at this point.

Now, let's approach the problem the other way. The gravitational potential of a body is $V = -GM/|\vec{r}|$.

(e) The vector $\vec{r}$ from the Earth to an arbitrary point is $x\hat{i} + y\hat{j}$. Calculate the gravitational potential at an arbitrary point (x, y) due to the Earth.

(f) Write the vector $\vec{r}$ from the moon to an arbitrary point (x, y).

(g) Calculate the gravitational potential at an arbitrary point (x, y) due to the moon.

(h) Calculate the gravitational potential $V(x, y, z)$ caused by the Earth and moon combined.

(i) Now use the formula $\vec{g} = -\vec{\nabla} V$ to find the gravitational field due to the Earth and moon.

(j) Plug the point $(300, 50)$ into your formula for $\vec{g}(x, y)$. Make sure your answer matches your answer to Part (d)!

8.65 Consider an "electric dipole" comprising a positive charge $Q = 1$ fixed at $(0, 0, 0)$ and a negative charge $Q = -1$ fixed at $(5, 0, 0)$.

(a) This problem takes place in three dimensions. However, we can calculate, draw, and work with the potential and electric fields in two dimensions without any loss of generality. Explain why.

(b) Calculate the electric potential field $V(x, y)$.

(c) Calculate the electric field $\vec{E}(x, y)$.

(d) Sketch a contour plot of $V(x, y)$. On that same plot draw arrows in at least three places representing $\vec{E}$.

8.66 The following plot shows equipotential lines due to a configuration of positive charges.

(a) Copy the sketch and draw at least 3 electric field arrows on it.

(b) What would look the same and what would look different in the plot if the charges were negative?

8.67 For each of the functions below, make a 3D plot and a contour plot in the region $-1 \le x \le 1$, $-1 \le y \le 1$. On the contour plot show at least three arrows representing the gradient at different points.

(a) $f(x, y) = e^{-(x-1)^2 - y^2} + e^{-(x+1)^2 - y^2}$

(b) $f(x, y) = \sin(xy)$

(c) $f(x, y) = \sin(x^2 y)$

8.68 You are running an experiment in a large, flat Petri dish, which you can treat as effectively 2D. A heater is heating the center of the dish, so the temperature throughout the dish falls off exponentially with distance from the heater: $T = e^{-x^2 - y^2}$, as shown below.

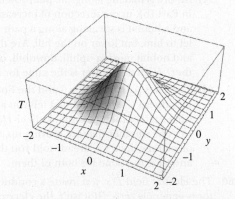

(a) At the point $(1, 1)$ on the plot, in what direction does the gradient point? Explain why the direction is not "up the hill."

(b) Draw a contour plot of $T(x, y)$. On that same plot draw arrows representing $\vec{\nabla} T$ at three different points, including $(1, 1)$.

(c) In general, what is the relationship between the gradient arrows and the contour lines?

8.5 From Gradient to Potential: The Gradient Theorem

In one dimension, $g = -dV/dx$. You can express the same relationship the opposite way as $V = -\int g \, dx$. So you know how to get from V to g, and vice versa.

In more than one dimension, $\vec{g} = -\vec{\nabla} V$. If you want to find the potential function for a given gravitational field, how do you run *that* equation backward?

You usually can't. $\vec{g} = y\hat{i} + 2x\hat{j}$ is a pretty simple vector field, but it has no potential function—which is to say, there is no scalar whose gradient happens to be that vector. If you start arbitrarily making up vector fields, most of them won't have potential functions. So the ones that do get a special name.

8.5 | From Gradient to Potential: The Gradient Theorem

First Definition: Conservative Field

The vector field $\vec{v}$ is "conservative" if it can be written as the gradient of some scalar field F.

$$\vec{v} = \vec{\nabla} F \qquad (8.5.1)$$

If no such scalar field exists, $\vec{v}$ is "non-conservative."

Conservative fields will be discussed in more detail in Section 8.11. There we will give some equivalent definitions of the term (which is why we called that our "first definition"). We will discuss important physical properties of conservative fields, and give shortcuts for determining if a given field is conservative.

In this section, however, we are introducing the term only so that we can present an important relationship between gradients and line integrals, known as the "gradient theorem."[3]

The Gradient Theorem

If $\vec{v}$ represents a conservative vector field—that is, if you can write $\vec{v} = \vec{\nabla} F$ for some scalar field F—then:

$$\int_C \vec{v} \cdot d\vec{s} = F(P_2) - F(P_1) \qquad (8.5.2)$$

where P_1 and P_2 are the endpoints of curve C.

The gradient theorem gives you a fast way to evaluate line integrals. You don't bother with dot products, integrate anything, or even worry about the path; just plug two points into one scalar function and you're done. But of course, it only works if that scalar function exists! For non-conservative fields, you still have to integrate along the curve.

The gradient theorem is also useful in the other direction: you can evaluate the line integral in order to find the potential function. But that begs two immediate questions.

- *Along what path do you evaluate the line integral?* The surprising answer, implied by the gradient theorem, is that it doesn't matter: you choose the path that will make your work easiest. Does that mean that the result of a line integral depends only on the endpoints, regardless of the path you take between them? Yes, it does—but only for conservative vector fields! For non-conservative fields, integrals along different paths will give different answers.
- *The gradient theorem doesn't tell me the potential anywhere, just the potential difference between two points.* That's true, but remember that potential is only defined up to an arbitrary constant. So we pick any reference point where we define $F = 0$ and then calculate $-\int_C \vec{v} \cdot d\vec{s}$ from the reference point to an arbitrary point (x, y) to get the potential $F(x, y)$.

In the problems you will use the gradient theorem in both directions: going from the potential function to the line integral, and going from the line integral to the potential function. Below we demonstrate the second use, and then discuss why the gradient theorem works.

---

[3] The gradient theorem goes by many other names, such as "The fundamental theorem for line integrals" and "The fundamental theorem for gradients." While we prefer more concise verbiage, there is something to be said for names that remind us that all such theorems are variations of the fundamental theorem of calculus.

EXAMPLE Finding the Potential of a Vector Field

Question: Find the potential $F(x, y)$ of the field $\vec{v} = e^y \hat{i} + xe^y \hat{j}$, setting $F = 0$ at the origin.

Answer:
To find the potential at (x, y) we evaluate a line integral from the origin to (x, y). Of course we want to choose a path that will make our calculations easy; we will go straight across to $(x, 0)$ and from there to (x, y).

Along the first segment, y is the constant zero. But x is a variable, starting at zero and ending at the x-value toward which we are integrating. We cannot use the letter x to mean both "the variable that keeps changing along this line" and "the constant value at the end of this line." There is no universal convention for how to handle this situation, but we have chosen to use $\tilde{x}$ for the variable and x for the constant.

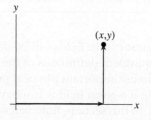

$$\int_0^x v_x d\tilde{x} = \int_0^x 1 \, d\tilde{x} = x$$

Along the second segment, $\tilde{x}$ is the constant x and $\tilde{y}$ varies from 0 to y. We are therefore integrating v_y instead of v_x.

$$\int_0^y xe^{\tilde{y}} d\tilde{y} = xe^y - x$$

Combining the two segments we find $\int_C \vec{v} \cdot d\vec{s} = xe^y$. Remember that the gradient theorem just relates line integrals to gradients, but potential includes an additional minus sign, so $F(x, y) = -xe^y$.

Having found this answer, it's easy to check it:

$$-\vec{\nabla} F = -\frac{\partial F}{\partial x}\hat{i} - \frac{\partial F}{\partial y}\hat{j} = e^y \hat{i} + xe^y \hat{j}$$

Since $-\vec{\nabla} F = \vec{v}$, our answer is correct.

This would be a good time to stop and confirm the gradient theorem (or your understanding of it). The example above gave us a vector field and its potential function. So pick any two points, and any path between them, and verify that the two sides of the gradient theorem give you the same answer. If something doesn't match, you've made a mistake somewhere!

Why does The Gradient Theorem Work?

The gradient theorem looks wildly improbable when you think about it. It evaluates a line integral along any path—straight or curved, long or short—simply by looking at the endpoints of that path, without regard to what happens in between. In that respect it should remind you of the fundamental theorem of calculus, which finds the area under an entire curve by simply looking at the antiderivative at the two endpoints of the interval.

To see why the gradient theorem works, we begin (as always) by carefully framing the problem. We have a scalar field F. Its gradient was called $\vec{v}$ in Equation 8.5.2, but here let's just call it $\vec{\nabla} F$. Our goal is to evaluate the line integral $\int_C \vec{\nabla} F \cdot d\vec{s}$ along some curve C.

8.5 | From Gradient to Potential: The Gradient Theorem

As with any integral, we begin our analysis with a small slice. $d\vec{s}$ is a step along the path, small enough to be considered a line. At that particular spot the vector $\vec{\nabla}F$ points at some angle (which could be anything) off the path.

What does the dot product of those two vectors represent?

$$\vec{\nabla}F \cdot d\vec{s} = \left(\frac{\partial F}{\partial x} \hat{i} + \frac{\partial F}{\partial y} \hat{j} \right) \cdot \left(dx\, \hat{i} + dy\, \hat{j} \right) = \frac{\partial F}{\partial x} dx + \frac{\partial F}{\partial y} dy$$

$(\partial F/\partial x)dx$ is the amount the function would change if you took a small step dx in the x-direction. Add this to the change along dy (and also dz in three dimensions) and you get the total change in F as you move through the vector $d\vec{s}$. (If you've studied directional derivatives you should recognize that what we are calculating here is the directional derivative in the direction of $d\vec{s}$, multiplied by the length ds, to give the amount that F changes along $d\vec{s}$. But directional derivatives aren't necessary to follow this logic if you understand partial derivatives.)

This is the key point in the derivation: $\vec{\nabla}F \cdot d\vec{s}$ tells you how much F increases (or decreases) as you move along a small displacement $d\vec{s}$. When you integrate, you add up all the incremental changes in F as you move along the entire curve. So you end up with the total change in F as you move from the beginning of the curve (P_1) to the end (P_2)—and there's the gradient theorem.

Stepping Back

The past two sections have been all about the relationship between a scalar field and its gradient, or equivalently between a vector field and its potential. You know that in one dimension the following two statements are mathematically equivalent.

- If $f(x) = x^2$ then its derivative is $f'(x) = 2x$.
- If $g(x) = 2x$ then its integral is $\int g(x)dx = x^2 + C$.

The gradient and potential give us a higher dimensional analog, with a negative sign thrown in.

- If $F(x, y) = x^2 + 3y$ then its gradient is $\vec{\nabla}F = 2x\hat{i} + 3\hat{j}$.
- If $\vec{v} = 2x\hat{i} + 3\hat{j}$ then its potential function is $-\int_C \vec{v} \cdot d\vec{s} = -x^2 - 3y + C$.

In many math texts, and in hydrodynamics, no negative sign is used so that "gradient" and "potential function" are perfect opposites. We will retain the negative sign that is used in electricity and gravity, which simplifies the relationship to potential energy.

8.5.1 Problems: From Gradient to Potential

8.69 In this problem you will find the potential function F for the vector field $\vec{v} = 3x^2y\hat{i} + (x^3 + y^2)\hat{j}$. You are always free to set $F = 0$ at any point you want, so for simplicity take $F(0,0) = 0$.

(a) Find the potential $F(0, 1)$ by evaluating $-\int_C \vec{v} \cdot d\vec{s}$ from $(0, 0)$ to $(0, 1)$.

(b) Find the potential $F(1, 1)$ by evaluating $-\int_C \vec{v} \cdot d\vec{s}$ from $(0, 0)$ to $(1, 1)$. You may choose any path you wish to integrate along.

(c) Find the potential at an arbitrary point (x, y) by evaluating $-\int_C \vec{v} \cdot d\vec{s}$, with the integration variables $(\tilde{x}, \tilde{y})$ going from $(0, 0)$ to (x, y). You may choose any path you wish to integrate along. The result should be a function $F(x, y)$.

(d) Calculate $-\vec{\nabla}F$ using the potential function you just found and verify that it equals $\vec{v}$.

(e) Suppose you had chosen a different point to set $F = 0$ and evaluated all of your integrals from that point. How would that have changed the potential function $F(x, y)$ that you found in Part (c)?

In Problems 8.70–8.76 you will find a function F such that $\vec{\nabla}F = -\vec{v}$. In some cases you will be given a point to use as $F = 0$; in other cases you will choose your own point. Integrate from that point to (x, y, z) along the easiest possible path to find the potential function F. Then evaluate $\vec{\nabla}F$ to confirm that it works.

8.70 $\vec{v} = 3\hat{i} + 4\hat{j}$. Let $F = 0$ at $(0, 0)$.

8.71 $\vec{v} = 2x\,\hat{i} + \cos y\,\hat{j}$. Let $F = 0$ at $(0, 0)$.

8.72 $\vec{v} = 2xy\,\hat{i} + (x^2 + 2y\ln z)\hat{j} + (y^2/z)\hat{k}$. Let $F = 0$ at $(0, 0, 0)$.

8.73 $\vec{v} = (1/x)\hat{i}$

8.74 $\vec{v} = (1/x)\hat{i} + (1/y)\hat{j}$

8.75 $\vec{v} = \dfrac{3}{3x+y^2}\hat{i} + \dfrac{2y}{3x+y^2}\hat{j}$

8.76 $\vec{v} = 1/(x+y)\hat{i} + 1/(x+y)\hat{j} + \left(1/z^2\right)\hat{k}$

In Problems 8.77–8.79…

(a) Find the potential function F for the vector field $\vec{v}$. As always, you should do that by choosing the easiest line integral you can from an $F = 0$ point to (x, y, z).

(b) Evaluate the given line integral by using the gradient theorem, based on the potential function you found.

(c) Evaluate the given line integral again, this time *not* using the gradient theorem, and confirm that you get the same answer.

8.77 $\vec{v} = 3\hat{i} + 2y\hat{j}$ along the curve $y = x^2$ from $(1, 1)$ to $(3, 9)$.

8.78 🖥 $\vec{v} = 2x\sin(3z)e^{x^2+y}\hat{i} + \sin(3z)e^{x^2+y}\hat{j} + 3\cos(3z)e^{x^2+y}\hat{k}$ along the curve $x = 2t$, $y = e^t$, $z = t^2$ as t goes from 0 to 1. *The integral to find F can be done easily by hand, but the line integral will require a computer or integral table.*

8.79 🖥 $\vec{v} = (x+1)e^{x+y}\hat{i} + xe^{x+y}\hat{j}$ along the curve $x = \cos t$, $y = \sin t$ from $(1, 0)$ to $(0, 1)$. *The integral to find F can be done with integration by parts or on a computer, but the line integral will require a computer or integral table.*

8.80 A sphere of radius R is filled with charge that drops linearly to zero toward the edge: that is, the charge density is $D = (k/R)(R-r)$ where r is distance from the center. This creates an electric field pointing radially outward, with magnitude given by:

$$|\vec{E}| = \begin{cases} \dfrac{kR^3}{12\varepsilon_0 r^2} & r \geq R \\[6pt] \dfrac{k}{R\varepsilon_0}\left(\dfrac{1}{3}Rr - \dfrac{1}{4}r^2\right) & r < R \end{cases}$$

where ε_0 is a positive constant. (You can verify this electric field using Gauss's law, or just take our word for it.) The force on a charge q in such a field is $\vec{F} = q\vec{E}$.

(a) Find the work required to bring a charge q from infinity to a distance r from the center of the sphere, where $r > R$ (outside the sphere).

(b) Find the work required to bring a charge q from infinity to a distance r from the center of the sphere, where $r < R$ (inside the sphere). Each term in this answer has the same units as the one we already checked above.

(c) Electric potential can be defined as the work per unit charge required to bring charges in from infinity, or $V = W/q$. Draw the potential field $V(r)$ that you have calculated, including both regions $r < R$ and $r \geq R$. Choose any convenient value you like for the constant k that appeared in the density function.

8.81 The gradient theorem, as expressed in Equations 8.5.1 and 8.5.2, applies in any number of dimensions.

(a) In two dimensions F is $F(x, y)$, v is $P(x, y)\hat{i} + Q(x, y)\hat{j}$, and the point P_1 is (x_1, y_1). Rewrite Equations 8.5.1 and 8.5.2 for this specific case without mentioning gradients, potentials, or conservative fields.

(b) Rewrite Equations 8.5.1 and 8.5.2 in *one* dimension. The result should look familiar.

8.82 **Exploration: Finding a Potential by Brute Force**
In the Explanation (Section 8.4.2) we showed you how to find the potential function for a vector field by using a line integral. There is another, more straightforward approach. As usual, which one is easier to use varies from one case to the next.

The gradient of a scalar function $F(x, y, z)$ is given by:

$$\vec{\nabla}F = 2x\cos\left(x^2+z\right)\hat{i} + 2yz\hat{j} + \left[\cos\left(x^2+z\right) + y^2 + 3z^2\right]\hat{k}$$

Your job is to find the scalar function.

(a) Write the simplest function $F(x, y, z)$ you can that satisfies $\partial F/\partial x = 2x\cos\left(x^2+z\right)$.

(b) You can of course add a constant to your answer, but you can do more than that; you can add any function of y and

z without changing $\partial F/\partial x$. So add an arbitrary function $G(y, z)$ to your function from Part (a). You now have the general solution to the partial differential equation $\partial F/\partial x = 2x\cos(x^2 + z)$.

(c) Take the partial derivative of your answer to Part (b) with respect to y, and set it equal to $2yz$. You can now solve for the function $G(y, z)$, but your answer will still have an arbitrary function in it.

(d) Finish finding the function F. In this part your answer should have an arbitrary *constant*, but not an arbitrary function.

(e) Take the gradient of your answer to Part (d) and verify that it works.

(f) Try to use the same technique to find the potential function for $\vec{v} = x\hat{i} + xy\hat{j}$. What went wrong?

8.6 Divergence, Curl, and Laplacian

Section 8.4 introduced the "gradient," a powerful tool for analyzing the behavior of any scalar field. In this section we encounter two different derivatives that are used to analyze vector fields, and one more for scalar fields. All of these field derivatives are summarized in Appendix F.

8.6.1 Discovery Exercise: Divergence, Curl, and Laplacian

The vector field $\vec{V}(x, y)$ represents how much mass passes by per unit cross-sectional area per unit time. The "divergence" of $\vec{V}$ represents how much the gas is flowing *away from* a given point. To approach the divergence, we consider a square region. As the gas flows through this region, we ask the question: Is gas flowing out of this region faster than it is flowing in? (Generally losing gas, positive divergence.) Or is gas flowing into this region faster than it is flowing out? (Generally gaining gas, negative divergence.) Or is gas flowing in and out at the same rate? (Neither gaining nor losing gas, zero divergence.)

In all our examples, the gas is flowing to the right. Assume the constant k in the equations below is positive. In our first example, the flow of the gas is constant. This field could be represented as $\vec{V}_1 = k\hat{i}$.

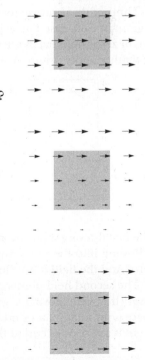

1. Compare the rates at which gas is entering our region (from the left) and exiting (to the right). Which rate is faster, or are they the same?
2. Is this region generally losing gas, gaining gas, or neither?
3. Is the divergence of $k\hat{i}$ positive, negative, or zero?

In our second example, the flow of the gas increases as you move up. This field could be represented as $\vec{V}_2 = ky\hat{i}$.

4. Compare the rates at which gas is entering our region (from the left) and exiting (to the right). Which rate is faster, or are they the same?
5. Is this region generally losing gas, gaining gas, or neither?

   See Check Yourself #54 in Appendix L

6. Is the divergence of $ky\hat{i}$ positive, negative, or zero?

In our third example, the flow of the gas increases as you move to the right. This field could be represented as $\vec{V}_3 = kx\hat{i}$.

7. Compare the rates at which gas is entering our region (from the left) and exiting (to the right). Which rate is faster, or are they the same?
8. Is this region generally losing gas, gaining gas, or neither?

9. Is the divergence of $kx\hat{i}$ positive, negative, or zero?
10. Create a vector flow where the divergence is negative. Represent your flow with both a drawing and a formula.
11. Now think about examples such as $e^x\hat{i}$, $(1/x)\hat{i}$, $(x-y)\hat{i}$, and, more generally, $f_x(x,y)\hat{i}$. For vector flows that are strictly in the x-direction (as all our examples were), what has to be true of f_x for the divergence to be positive?
12. Finally, think about two-dimensional vector fields. What has to be true of f_x and f_y for the divergence to be positive?

8.6.2 Explanation: Divergence, Curl, and Laplacian

Our customary way of describing a vector field asks three questions at each point: "How much does this field point in the x-direction," "How much does it point in the y-direction," and "How much in the z-direction?" The answers offer a complete description of the field, but not always in a form that is easy to work with.

Here are two very different questions: "How much does this vector field point *away from* this point," and "How much does this vector field swirl *around* this point?" These two, the "divergence" and "curl" respectively, offer a very different (and often more useful) breakdown of how a vector field is flowing and changing.

A First Intuitive Look

The momentum density of a fluid is its density times velocity. The surface integral of momentum density tells you the "flow rate": how much of a given fluid is passing through a given surface per unit time.[4]

Imagine a vector field representing the momentum density of a fluid in some region. The divergence at each point answers the question: "Is the fluid generally rushing away from this point?" The curl answers the question: "If I stick a paddle wheel in this fluid, which way (if any) will it spin?"

Think of the three vector plots below as representing momentum density of a fluid in a 2D region. In each case figure out if the divergence is positive (flowing out), negative (flowing in), or zero (neither) at the origin. Then figure out if the curl is clockwise, counterclockwise, or neither at the origin. After you think through your answers, read ours below the pictures.

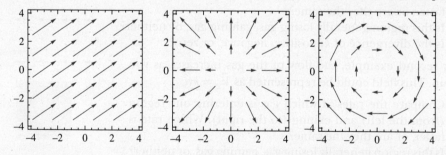

The first drawing shows a constant vector field. At the origin (or at any other point), the field is flowing into the point and out of the point with the same strength, so the divergence is zero. And the field is not flowing around the point, so the curl is zero.

The second field points ever away from the origin, so the divergence at the origin is positive. (If it were moving toward the origin, the divergence would be negative.) Once again, there is no rotation around the origin, so the curl is zero. The third drawing is the opposite: there is a clockwise curl at the origin but a zero divergence.

---

[4]Momentum density and flow rate were discussed more carefully in Chapter 5.

The drawings below are less obvious; you might think both represent zero-divergence zero-curl fields, but in fact neither one does. To help guide you we've drawn a small box and paddle wheel centered on the origin. Once again, think about your own answers before reading ours below.

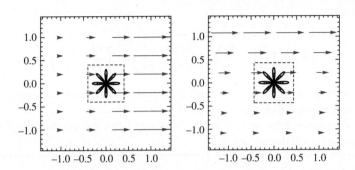

Consider the boxed region in the first field. You see the field moving into this region from the left, and back out again on the right. But the outward flow is *greater* than the inward flow, so the net divergence is positive; over time, the total amount of fluid in that region would decrease. You should be able to convince yourself that this would be true for *any* region you could draw in this flow, so the divergence is positive everywhere. The curl of this field is zero, by the way.

Now consider the boxed region in the second field. The inward and outward flows are equal, so there is no divergence. There does not seem to be any rotation either, but look what will happen to the paddle wheel: the push on top is stronger than the one on bottom, so the paddle wheel will rotate clockwise. Once again you should be able to convince yourself that the same would be true about a paddle wheel placed anywhere in this field, so the curl is clockwise everywhere. Since you can have a non-zero curl at a point even when the field vectors don't rotate around that point, we use the term "circulation" for the property being measured by curl. We'll define that term more rigorously later, but for now you can just think of it in terms of the paddle wheel.

In three dimensions we can't describe the curl by simply saying "clockwise" or "counter-clockwise"; the field might curl around the *x*-axis, or around the *y*-axis, or around any other line you can draw. So we specify the curl with a direction vector, following a right-hand rule: if you wrap the fingers of your right hand in the direction the paddle wheel would rotate, your thumb points in the direction of the curl. So the curl for the vector field above points into the page. Of course it wouldn't matter if we used a *left*-hand rule instead, but it does matter that we all use the same rule, and the right-hand rule is universally agreed upon. We can thus use a single vector to communicate an arbitrary direction of circulation in three dimensions.

The following facts are critical for understanding the divergence and curl. *The divergence is a scalar; the curl is a vector.* One number tells you everything you need to know about the divergence, noting that the number can be positive (net outward flow), negative (net inward flow), or zero (balance). But curl has a magnitude (how much circulation) and a direction (what line the field is circulating about, and which way it flows around that line).

The Formulas for Divergence and Curl

Now that you know what divergence and curl measure, we can give you the formulas for calculating them. The connection between the visual descriptions above and the formulas below are not obvious! The Discovery Exercise (Section 8.6.1) attempted to lead you to the formula for the divergence; we will build up both formulas more carefully in Section 8.7.

The Formulas for the Divergence and Curl

Given a vector field $\vec{f} = f_x\hat{i} + f_y\hat{j} + f_z\hat{k}$ the divergence (written $\vec{\nabla}\cdot\vec{f}$) and curl (written $\vec{\nabla}\times\vec{f}$) are given by:

$$\vec{\nabla}\cdot\vec{f} = \frac{\partial f_x}{\partial x} + \frac{\partial f_y}{\partial y} + \frac{\partial f_z}{\partial z} \tag{8.6.1}$$

$$\vec{\nabla}\times\vec{f} = \left(\frac{\partial f_z}{\partial y} - \frac{\partial f_y}{\partial z}\right)\hat{i} + \left(\frac{\partial f_x}{\partial z} - \frac{\partial f_z}{\partial x}\right)\hat{j} + \left(\frac{\partial f_y}{\partial x} - \frac{\partial f_x}{\partial y}\right)\hat{k} \tag{8.6.2}$$

Remember that divergence is a scalar. Those three partial derivatives are not components of a vector; they are simply being added. Curl is a vector.

You will sometimes see these written as div $\vec{f}$ and curl $\vec{f}$ but the notation $\vec{\nabla}\cdot\vec{f}$ and $\vec{\nabla}\times\vec{f}$ has mnemonic value, as we will discuss below.

The cross product and the curl are often written as the determinant of a matrix. It's not a real matrix, but if you've already memorized the determinant of a 3×3 matrix then you can leverage that brain space to remember the formula for the curl.

$$\vec{\nabla}\times\vec{f} = \begin{vmatrix} \hat{i} & \hat{j} & \hat{k} \\ \frac{\partial}{\partial x} & \frac{\partial}{\partial y} & \frac{\partial}{\partial z} \\ f_x & f_y & f_z \end{vmatrix}$$

If you're not comfortable calculating determinants you can ignore this matrix and use Equation 8.6.2 directly. Below we illustrate the use of this trick, but if you skip over the matrix you should still be able to follow the example.

EXAMPLE: Calculating Divergence and Curl

Question: Calculate the divergence and curl of the field $\vec{f} = e^{6x-4y}\hat{i} + (x^2/2 - 2xy)\hat{j} + (\sin(z)/x)\hat{k}$ at the point $(2, 3, \pi/2)$. What do they tell us about the field at that point?

Answer:

$$\vec{\nabla}\cdot\vec{f} = 6e^{6x-4y} - 2x + \cos(z)/x$$

$$\vec{\nabla}\cdot\vec{f}(2, 3, \pi/2) = 6 - 4 = 2$$

$$\vec{\nabla}\times\vec{f} = \begin{vmatrix} \hat{i} & \hat{j} & \hat{k} \\ \frac{\partial}{\partial x} & \frac{\partial}{\partial y} & \frac{\partial}{\partial z} \\ e^{6x-4y} & x^2/2 - 2xy & (\sin z)/x \end{vmatrix}$$

$$= (0 - 0)\hat{i} + \left(0 + \sin(z)/x^2\right)\hat{j} + \left(x - 2y + 4e^{6x-4y}\right)\hat{k}$$

$$\vec{\nabla}\times\vec{f}(2, 3, \pi/2) = (1/4)\hat{j}$$

The positive divergence tells us that the field is generally moving away from the point $(2, 3, \pi/2)$ near that point. The curl in the $\hat{j}$-direction indicates a circulation around this point. An observer looking at this point from the positive y-direction would measure that circulation as counterclockwise.

Less Obvious Examples of Divergence and Curl

At the beginning of this section we presented some simple pictures that showed what divergence and curl tell you about a field. The examples below are trickier: think hard about your answers before you read ours!

EXAMPLE **Less Obvious Examples of Divergence and Curl**

Question: The pictures below show the fields $\vec{f} = x/\left(x^2+y^2\right)^2 \hat{i} + y/\left(x^2+y^2\right)^2 \hat{j}$ and $\vec{g} = y/\left(x^2+y^2\right)^2 \hat{i} - x/\left(x^2+y^2\right)^2 \hat{j}$. First, look at the pictures and try to predict the sign of the divergence and the direction of the curl at the point $(0, 1)$. Then calculate the divergence and curl to check your answers.

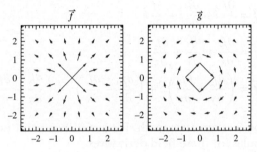

Answer:
Looking at the pictures it's clear that $\vec{f}$ (on the left) has zero curl and $\vec{g}$ on the right has zero divergence. Since all the vectors are pointing outward on the left and clockwise on the right, it seems like $\vec{f}$ has positive divergence and $\vec{g}$ has a curl pointing into the page. If you calculate them, though, you find just the opposite. At the point $(0, 1)$, $\vec{\nabla} \cdot \vec{f} = -2$ and $\vec{\nabla} \times \vec{g} = 2\hat{k}$, which points out of the page. What's going on?

As always, imagine a small box around the point $(0, 1)$ and a small paddle wheel at that point. The vectors of $\vec{f}$ around that point are pointing straight up, but they are rapidly getting smaller, so there's more flow into the box than out of it. Similarly, the vectors of $\vec{g}$ are pointing to the right there, and the ones above are much smaller than the ones below, so the curl is counterclockwise. What makes these examples unintuitive is that we naturally tend to focus on the overall flow, which looks outward for $\vec{f}$ and clockwise for $\vec{g}$. At any particular point *other than the origin*, the flow is not doing that, however.

The point of this example is not that you shouldn't trust your intuition about divergence and curl. On the contrary, you can build up a good intuition for them that works even in unusual circumstances. You need to get in the habit of thinking about what the vectors are doing in the vicinity of a particular point.

Are You Ever Going to Tell Us What That $\vec{\nabla}$ Thing Is?

You have now seen a $\vec{\nabla}$ in the notation used for the gradient, the divergence, and the curl. It's called "del"[5] (because it looks like an upside-down Delta), and it is defined as follows:

$$\vec{\nabla} \equiv \left(\hat{i}\frac{\partial}{\partial x} + \hat{j}\frac{\partial}{\partial y} + \hat{k}\frac{\partial}{\partial z}\right)$$

---

[5] "Del" is actually the name of the mathematical object we define here. The typographical symbol ∇ is called "nabla."

We write the $\hat{i}$ to the left of the $\partial/\partial x$ so it won't look like we're taking the derivative of $\hat{i}$ which would of course be zero. With that cleared up, you're probably still thinking "What is that?" It's kind of like a vector—it has $\hat{i}$-, $\hat{j}$-, and $\hat{k}$-components—but it isn't a real vector because those components aren't numbers. The symbol $\vec{\nabla}$ by itself doesn't mean anything. It's a set of instructions that takes on meaning when you use it to act on a function.

A set of instructions for turning one function into another one is called an operator. For example, we could define an operator $O = d/dx$. Then if $f(x) = x^3$ we could write $Of = 3x^2$. When it looks like O is multiplying f, it's really acting on it.

But $\vec{\nabla}$ is more complicated because it's a vector operator. We can multiply a vector three ways: multiply it by a scalar, find its dot product with another vector, or find its cross product with another vector. These three operations with $\vec{\nabla}$ represent the gradient, divergence, and curl, respectively.

$$\vec{\nabla} f = \frac{\partial f}{\partial x}\hat{i} + \frac{\partial f}{\partial y}\hat{j} + \frac{\partial f}{\partial z}\hat{k}$$

$$\vec{\nabla} \cdot \vec{f} = \frac{\partial f_x}{\partial x} + \frac{\partial f_y}{\partial y} + \frac{\partial f_z}{\partial z}$$

$$\vec{\nabla} \times \vec{f} = \left(\frac{\partial f_z}{\partial y} - \frac{\partial f_y}{\partial z}\right)\hat{i} + \left(\frac{\partial f_x}{\partial z} - \frac{\partial f_z}{\partial x}\right)\hat{j} + \left(\frac{\partial f_y}{\partial x} - \frac{\partial f_x}{\partial y}\right)\hat{k}$$

Nothing about $\vec{\nabla}$ introduces any new math. It's just a shorthand notation for writing and remembering the formulas for these field derivatives.

The Laplacian

If the gradient, divergence, and curl represent first derivatives of fields, what are the second derivatives? You can think of many possible answers to this question, but one of them turns out to be particularly useful: the divergence of the gradient of a scalar field S is called the "Laplacian" of S.

> **Definition: The Laplacian**
>
> The "Laplacian" of a scalar field S is the divergence of the gradient:
>
> $$\nabla^2 S = \vec{\nabla} \cdot \left(\vec{\nabla} S\right) = \frac{\partial^2 S}{\partial x^2} + \frac{\partial^2 S}{\partial y^2} + \frac{\partial^2 S}{\partial z^2}$$

The Laplacian serves roughly the same role in higher dimensions that the second derivative fills in one. For instance, if $f''(x) = 0$ everywhere, then $f(x)$ has no minima or maxima; if $\nabla^2 f(x, y, z) = 0$ everywhere, then $f(x, y, z)$ has no minima or maxima.

More importantly, the Laplacian comes up in a variety of important equations. For instance, in electrostatics, a charge distribution with density $D(x, y, z)$ creates an electric potential V that obeys Poisson's equation $\nabla^2 V = -D/\epsilon_0$.

The Laplacian is usually written $\nabla^2 S$ because it's the expression you would get if you used the rules above to find the "squared magnitude" ∇^2 and then acted with that on S. Some texts write ΔS instead.

A Word About Dimensions

You may recall that for basic vector multiplication, $s\vec{v}$ (scalar times vector) and $\vec{v}_1 \cdot \vec{v}_2$ (the dot product) are defined in any number of dimensions, but $\vec{v}_1 \times \vec{v}_2$ (the cross product) is only defined in exactly three dimensions.

8.6 | Divergence, Curl, and Laplacian

Here again the analogy of $\vec{\nabla}$ to a vector holds up. You can take the gradient of a scalar function of n variables, and end up with an n-dimensional vector. You can take the divergence of a vector field in n dimensions, and end up with a scalar function of n variables. The curl, however, is an inherently three-dimensional entity. You can take the curl of a two-dimensional vector field by treating the third component as zero, but the curl itself will point into the third dimension. And while there are ways to generalize the idea into four or more dimensions, you're not talking about the curl any more. Three-dimensional vector fields define a special case where the curl is meaningful, but they are after all a pretty important special case in our universe.

8.6.3 Problems: Divergence, Curl, and Laplacian

8.83 The vector field $\vec{f}(x, y, z) = \ln(2x + 3y)\hat{i} + (2xy/z)\hat{j} + (x+1)yz\,\hat{k}$.

(a) Calculate the divergence of this field as a function of x, y, and z.

(b) Calculate $\vec{\nabla} \cdot \vec{f}(0, 1, 2)$.

(c) Explain briefly what your answer to Part (b) tells you about the field at this point.

(d) Calculate the curl of this field as a function of x, y, and z.

(e) Calculate $\vec{\nabla} \times \vec{f}(0, 1, 2)$.

(f) Suppose $\vec{f}$ is the momentum density of a fluid and you place a paddle wheel at the point $(0, 1, 2)$ oriented parallel to the xy-plane. Which way will it turn, if either? Answer the same question for paddle wheels at that point parallel to the xz- and yz-planes.

In Problems 8.84–8.90, for the given field at the point $(2, 2)$, say if the divergence is positive, negative, or zero. Then say if the curl at $(2, 2)$ points into the page, out of the page, or is zero.

8.85

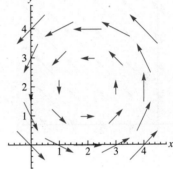

8.86

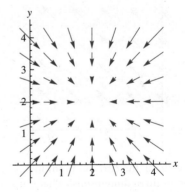

8.84

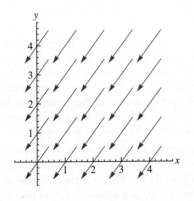

8.87

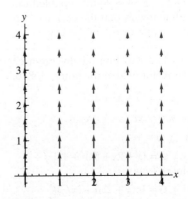

8.88

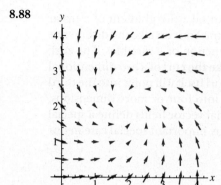

8.89

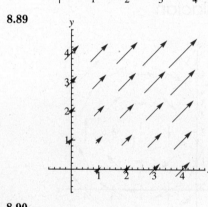

8.90

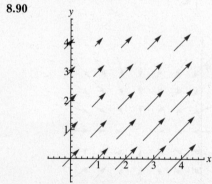

8.91 In general, the curl can point in any direction in three dimensions. What is it about Problems 8.84–8.90 that enabled us to confidently say that the curl would point directly into or out of the page?

In Problems 8.92–8.97 find the divergence and curl of the given vector field at the point $(3, 4, 0)$.

8.92 $\vec{k}(x, y, z) = 3\hat{i} - \pi\hat{j} + 16\hat{k}$

8.93 $\vec{L}(x, y, z) = x\hat{i}$

8.94 $\vec{U}(x, y, z) = y\hat{i}$

8.95 $\vec{f}(x, y, z) = (x + 2y + 3z)\hat{i} + 4xyz\hat{j} + x\sin(\pi y + z)\hat{k}$

8.96 $\vec{g}(x, y, z) = \ln(x + 2z)\hat{i} + (x/y)\hat{k}$

8.97 $\vec{h}(x, y, z) = \left[(x + z)/y\right]\hat{i} + \cos^2 z\hat{j} + x^5\hat{k}$

8.98 The function $\vec{v}(x, y, z)$ describes the velocity of the air in each case described below. For each one, say whether the divergence is positive, negative, or zero. Then say which way the curl points, or that it is zero.
(a) A steady wind is blowing at 5 mph Eastward across a field.
(b) The same wind blows Eastward across the same field, but weakens as it goes. (The farther East you go, the slower the wind.)
(c) The same wind blows Eastward across the same field, and does *not* weaken as it goes, but does move more slowly at higher altitudes.

8.99 $\vec{f} = x^2\hat{i} + y^2\hat{j}$.
(a) Draw a vector plot of this field in the domain $-3 \leq x \leq 3$, $-3 \leq y \leq 3$. Your drawing does not have to be perfectly to scale, but it should show direction accurately, and give some sense of where the field is larger and smaller.
(b) Calculate $\vec{\nabla} \cdot \vec{f}$ at the point $(2, 2)$. Explain based on your drawing why the answer has the sign it has.
(c) Calculate $\vec{\nabla} \cdot \vec{f}$ at the point $(-2, -2)$. Explain based on your drawing why the answer has the sign it has.
(d) Calculate $\vec{\nabla} \cdot \vec{f}$ at the point $(2, -2)$. Explain based on your drawing why the answer has the sign it has.

8.100 The figure shows a vector field $\vec{p}$, along with five marked points. The field $\vec{p}$ equals $\hat{j}$ in the region $0 \leq x \leq 1$, $0 \leq y \leq 1$, and drops to 0 very rapidly outside that region.

For each marked point, indicate whether the divergence is positive, negative or zero. Then indicate whether the curl points into the page, out of the page, or zero. You should be able to answer all these questions without doing any calculations. No two points will have both answers the same.

8.101 A positive charge Q at the origin creates an electric field whose magnitude is $|\vec{E}| = kQ/r^2$, where r is the distance to the origin, and whose direction points directly away from the origin.

(a) Write the field $\vec{E}$ in the form $E_x\hat{i} + E_y\hat{j} + E_z\hat{k}$. (One approach is to begin by writing $\vec{E}$ in terms of $\vec{r}$ and then converting to Cartesian.)

(b) Make a rough sketch of $\vec{E}$ in two dimensions. You may use a vector plot or field lines.

(c) Calculate $\vec{\nabla} \cdot \vec{E}$. Is the result positive, negative, or zero?

(d) Your answer to the last part may have surprised you. Draw a small box around some point other than the origin and explain why it makes sense from your plot that the divergence at that point behaves the way you found in Part (c).

(e) Calculate $\vec{\nabla} \times \vec{E}$. Explain why the result makes sense based on your drawing.

8.102 A wire carrying a current I in the positive z-direction creates a magnetic field whose magnitude is $|\vec{B}| = \mu_0 I/(2\pi\rho)$ where μ_0 is a constant and ρ is the distance from the wire. The direction of the field points around the wire, counterclockwise as viewed from above.

(a) Write the field $\vec{B}$ in the form $B_x\hat{i} + B_y\hat{j} + B_z\hat{k}$. (Just play around with this until you get something that works at key obvious points and then check that it works more generally.)

(b) Make a rough sketch of $\vec{B}$ in two dimensions for some constant z.

(c) Calculate $\vec{\nabla} \cdot \vec{B}$. Is the result positive, negative, or zero? Explain why this result makes sense based on your drawing.

(d) Calculate $\vec{\nabla} \times \vec{B}$. Your result may surprise you: you will examine why it makes sense in Section 8.7 Problem 8.116.

8.103 Harry and Ron are hunched over a long homework scroll. Harry says "The Laplacian here obviously points straight up!" Ron replies "You stupid git, it points down!" Without glancing at the problem, Hermione cheerily calls over her shoulder "You're both wrong." How does she know?

8.104 Prove that for any scalar field $S(x, y, z)$, $\vec{\nabla} \times \vec{\nabla} S$ ("the curl of the gradient") is $\vec{0}$.

8.105 Prove that for any vector field $\vec{v}(x, y, z)$, $\vec{\nabla} \cdot \left(\vec{\nabla} \times \vec{v}\right)$ ("the divergence of the curl") is 0.

8.106 The Laplacian is the divergence of the gradient. Problems 8.104 and 8.105 asked you about the curl of the gradient and the divergence of the curl. Why can't we ask you about the curl of the divergence? Why can't we ask you about the gradient of the curl?

8.107 (a) Prove that any linear function $f(x, y, z) = ax + by + cz + d$ is a solution to Laplace's equation $\nabla^2 f = 0$.

(b) Find another (non-constant, non-linear) solution to Laplace's equation.

8.108 $f(x, y) = e^{-(x-1)^2 - y^2} - e^{-(x+1)^2 - y^2}$

(a) Have a computer make a 3D plot of $f(x, y)$. Choose a domain for your plot that lets you clearly see how the function behaves.

(b) Based on your plot, predict the sign of the Laplacian at the points $(-1, 0)$, $(0, 0)$, and $(1, 0)$. For each one explain why you would expect the answer you gave based on the plot.

(c) Calculate $\nabla^2 f$ (by hand or on a computer) and check your predictions.

8.109 The electric potential V is related to the density of charge ρ in a region by Poisson's equation: $\nabla^2 V = \rho/\epsilon_0$. You measure the potential in a room to be $k\left(x^4 - y^4\right)$ where k is a positive constant. In what parts of the room is there positive charge? Negative charge? No net charge?

8.110 The steady-state temperature T in a region of space with no external sources of heating or cooling approximately obeys Laplace's equation: $\nabla^2 T = 0$.

(a) Assume a 1D rod has fixed temperatures at the ends: $T(0) = T_1$, $T(L) = T_2$. Solve Laplace's equation in 1D to find $T(x)$ along the rod.

(b) Which of the following steady-state temperature distributions is possible for a 2D region without sources of heating or cooling? (There may be more than one correct answer.)
  i. $T(x, y) = x - y$
  ii. $T(x, y) = \sin(x - y)$
  iii. $T(x, y) = e^{-x} \sin y$
  iv. $T(x, y) = e^{-x-y}$

8.111 What does the formula $\vec{\nabla} \cdot \vec{v}$ have to do with the question "how much does the vector field flow away from this point?" We will confine ourselves mostly to the simplest case: a vector that is defined only on the positive x-axis, and has only an x-component.

(a) Begin with a constant vector, $\vec{f} = 2\hat{i}$. For any small segment of the x-axis, the vector flow into that segment is the same as the vector flow out of that point. What does this imply for the divergence? What is df_x/dx?

(b) Now consider a growing vector, $\vec{g} = 2x\hat{\imath}$. For any given segment, which is greater, the vector flow into that segment or the flow out? What does this imply for the divergence? What is dg_x/dx?

(c) Now consider a shrinking vector, $\vec{h} = (1/x)\hat{\imath}$. For any given segment, which is greater, the vector flow into that segment or the flow out? What does this imply for the divergence? What is dh_x/dx?

(d) Repeat this exercise for vectors pointing to the left (still on the positive x-axis). Create simple functions that are growing and shrinking, predict their divergences, and then calculate their derivatives. Does the generalization "the formula for divergence tells how much the vector field flows away from a small region around this point" always hold up?

(e) How would we extend this argument to two or three dimensions?

8.7 Divergence and Curl II—The Math Behind the Pictures

We're going to do two things in this section. First, we're going to define more carefully what we mean when we say the divergence measures "how much the field points away from a point" and the curl measures "circulation around a point." Second, we will show how you can use those definitions to derive the formulas we gave for the divergence and curl.

8.7.1 Explanation: Divergence and Curl II—The Math Behind the Pictures

Divergence

Describing divergence as "how much a vector field flows away from a given point" sounds good until you think hard about it. But at any given point, the vector field just has a certain magnitude and a certain direction. It certainly doesn't point away from or toward itself; what matters is how much the *nearby* vectors point toward or away from our point. But how nearby is "near enough," and how do we quantify all that?

The answer begins with: draw a closed surface around the point, and measure the integral of the vector field through that surface. Do you remember the convention for surface integrals through a closed surface? Every vector (or component thereof) pointing out through the surface counts positive toward the total. Vectors pointing inward count negative, and vectors pointing along the surface don't count at all. All that is exactly what we want for the divergence. Take a moment to look at Figure 8.14 before reading further.

(Wait a minute: we said we would draw a surface *around* the point, and then we put our point in the corner of the box! That is a legitimate objection, but it winds up making no difference, as you will see in Problem 8.112. This way just makes the math a bit simpler.)

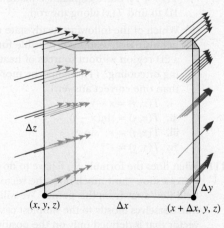

FIGURE 8.14 A vector field flows through a box around the point (x, y, z).

To better approach the divergence at the point (x, y, z), we shrink the surface around that point. In the limit as the volume ΔV enclosed by the surface approaches zero, the details of the surface disappear and only the vector field immediately around our point matters.

8.7 | Divergence and Curl II—The Math Behind the Pictures

Are we there yet? Not quite. A surface integral is field strength times area, so as our region shrinks to zero, the integral will do the same. So instead of looking at the integral itself, we look at the integral per unit volume. This gives us a useable definition:

The divergence of a vector field $\vec{f}$ at a point P is the surface integral per unit volume of $\vec{f}$ through a surface surrounding P in the limit where the volume of that surface shrinks to zero.

$$\vec{\nabla} \cdot \vec{f} = \lim_{\Delta V \to 0} \left(\frac{\oiint \vec{f}(x,y,z) \cdot d\vec{A}}{\Delta V} \right) \tag{8.7.1}$$

This somewhat hand-waving definition can be made mathematically rigorous, but like the definition of the derivative itself, Equation 8.7.1 is not something people actually use for calculations. Our goal is to translate it into Equation 8.6.1 by going back to Figure 8.14 and calculating the surface integral.

First we consider the left and right faces of the box. Since surface integrals depend only on the component of a vector perpendicular to a surface, we are only interested in the value of f_x on those faces.

Now, suppose f_x were equal on both faces, and suppose it pointed to the right. That would create a positive (outward) flow on the right face, and an equivalent negative (inward) flow on the left face, adding up to zero.[6] The only way these two faces can contribute to the divergence is if f_x changes as we move to the right. This should be sounding like a buildup to a derivative, but let's keep going and see how that happens.

At the right face of the box, the perpendicular component of the field is $f_x(x + \Delta x, y, z)$. The surface integral is that component multiplied by the area ($\Delta y \Delta z$). The left side is similar, but with two differences. First, the perpendicular component is $f_x(x, y, z)$. Second, if f_x is positive (rightward) then the flux through this face is negative (inward), and vice versa, so we need a negative sign. So the total integral between these two faces is $\left(f_x(x + \Delta x, y, z) - f_x(x, y, z) \right) \Delta y \Delta z$. Equation 8.7.1 then instructs us to divide by the volume ($\Delta x \Delta y \Delta z$) and take the limit:

$$\lim_{\Delta x \to 0} \frac{f_x(x + \Delta x, y, z) - f_x(x, y, z)}{\Delta x}$$

The contribution of the left and right faces of the cube to the divergence

Remember when it sounded like we were describing a derivative above? This is the definition of a partial derivative, so we can write more succinctly that the left and right faces of the cube make a net contribution to the divergence of $\partial f_x/\partial x$. Similar arguments show that the front and back faces contribute $\partial f_y/\partial y$ and the top and bottom ones contribute $\partial f_z/\partial z$. Put it all together and you get Equation 8.6.1.

Curl

The curl can be defined similarly. Instead of a surface we draw a closed loop around the point, and take the line integral of the vector field around that loop. (We've been bandying the term "circulation" about when talking about the curl; the circulation of a vector field around a closed loop means the line integral of that field around that loop.)

The line integral matches the curl just as well as the surface integral matches the divergence, because it measures how much the field is circling around the point, ignoring the components that radiate in or out. But remember that curl is a vector! If our loop lies flat in the xy

---

[6]We are following the standard convention in fluid flow where "flux" refers to the component of the vector field pointing through the surface and "flow" refers to that component multiplied by the area, i.e. the value of the surface integral.

(horizontal) plane, then it will measure the z-component of the curl. (Remember the right-hand rule.) So we need three different loops to get the three different components of the curl.

Other than that, the outline is similar to what we did for the divergence. Just as before, we shrink our loop down to focus on the region immediately around the point. Just as before, a smaller loop will lead to a smaller line integral, so we look at the circulation per unit area enclosed to reach a definition:

The z-component of the curl of a vector field $\vec{v}$ at a point P is the circulation per unit area of $\vec{v}$ around a loop surrounding P, in a plane perpendicular to $\hat{k}$, in the limit where the area of that loop shrinks to zero.

$$\lim_{\Delta A \to 0} \left(\frac{\oint \vec{f}(x,y,z) \cdot d\vec{s}}{\Delta A} \right) \quad \text{the component of the curl of } \vec{f} \text{ perpendicular to the plane of the loop}$$

(8.7.2)

The steps in turning this into Equation 8.6.2 are similar to those outlined above for the divergence. We will start here with the basic ideas and leave the details for you to go through in Problem 8.113.

We begin by drawing a rectangle in a horizontal plane with the point (x,y,z) at one corner. (Once again, our point should technically lie *inside* the loop, but once you follow our derivation it isn't hard to clean up that detail.) The area of the rectangle is $\Delta A = \Delta x \Delta y$.

Our goal is to find the line integral of a vector field $\vec{v}(x,y)$ around this rectangle in a counterclockwise direction. First we consider the line integrals on the left and right sides of the rectangle. Since a line integral depends on the component of a vector parallel to the line, only f_y will contribute on those sides. If f_y were the same on both sides the two sides would cancel each other out since we are measuring the $-f_y$-component on the left and the $+f_y$-component on the right. So the curl depends on how f_y changes as you go from x to $x + \Delta x$. You can once again spot a derivative on the horizon, but what matters this time is how f_y is changing with respect to x.

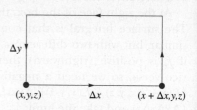

To finish the calculation you write an expression for the sum of the line integrals on the right and left, divide by the area of the rectangle, and take the limit as $\Delta x \to 0$, which gives you a partial derivative. Then you do the same for the top and bottom of the rectangle and add the two to get the z-component of the curl. The calculations of the x- and y-components are almost identical, finally leading to Equation 8.6.2.

8.7.2 Problems: Divergence and Curl II—The Math Behind the Pictures

8.112 In our derivation of Equation 8.6.1 in this section we cheated a little bit since the point is supposed to be *inside* the region, not in one corner of it. Redo the derivation with a rectangular solid centered on the point.

8.113 Finish the calculation of curl presented in this section.

8.114 The electric field $\vec{E}$ around a positive point charge q has magnitude kq/r^2, where r is distance from the point and k is a constant. The field points directly away from the charge. Consider a surface made up of a hemisphere of radius R_1 centered on the point charge, a hemisphere of radius $R_2 > R_1$ centered on the point charge, and a ring connecting the two.

(a) Find the electric flux through the inner hemisphere. (In electromagnetism "flux" means the surface integral of the electric or magnetic field.)

(b) Find the flux of $\vec{E}$ through the outer hemisphere.

(c) Find the flux of $\vec{E}$ through the flat ring connecting the two.

(d) Find the total flux through the surface.

(e) What does your answer tell you about the divergence of $\vec{E}$?

(f) In Cartesian coordinates, $\vec{E} = kq/\left(x^2 + y^2 + z^2\right)^{3/2} \left(x\hat{i} + y\hat{j} + z\hat{k}\right)$. Take the divergence of this field and check that it verifies your prediction.

The prediction and calculation you just did are valid for all points except the location of the point charge. You will consider what's happening at that point in Problem 8.115.

8.115 The electric field $\vec{E}$ around a positive point charge q has magnitude kq/r^2, where r is distance from the point and k is a constant. The field points directly away from the charge.

(a) Calculate the flux of $\vec{E}$ through a sphere of radius R centered on the point charge. (In electromagnetism "flux" means the surface integral of the electric or magnetic field.)

(b) Divide your answer to Part (a) by the volume of the sphere.

(c) Take the limit of your answer to Part (b) as $R \to 0$ to find the divergence of $\vec{E}$ at the location of the point charge. Explain what this answer is telling you about the field.

8.116 A wire carrying a current I in the positive z-direction creates a magnetic field whose magnitude is $|\vec{B}| = \mu_0 I/(2\pi\rho)$ where μ_0 is a constant and ρ is the distance from the wire. The direction of the field points around the wire, counterclockwise as viewed from above. In Section 8.6 Problem 8.102 you computed that the curl of this magnetic field is zero—possibly a surprising result, given that the field looks like it is all curl! In this problem you will consider why that result makes sense.

In the drawings below the wire is not shown, coming out of the page at $(x, y) = (0, 0)$. We arbitrarily choose to consider the magnetic field at the point $(1, 0)$.

(a) As we saw in this section, the curl is defined as the line integral of the field around a loop. The picture below shows a simple circular loop around the point $(1, 0)$. Explain why calculating $\oint \vec{B} \cdot d\vec{s}$ around this loop would be difficult—in other words, why this is the wrong loop to choose.

We choose therefore a more complicated loop that is more tailored to this particular situation.

(b) Two pieces of this loop are lines that radiate out from the origin. Explain how we can tell with no calculation that $\int \vec{B} \cdot d\vec{s}$ along these pieces is zero.

The remaining two pieces are arcs cut from circles around the origin. The $\vec{B}$ field therefore points directly along the outer arc (at radius ρ_2 from the origin), and backward along the inner arc (radius ρ_1).

(c) The outer arc is longer than the inner arc. By what factor?

(d) The magnetic field is stronger at the inner arc than at the outer arc. By what factor?

(e) Using all your results above, explain why $\vec{\nabla} \times \vec{B} = \vec{0}$.

8.8 Vectors in Curvilinear Coordinates

This section begins with some basic considerations to keep in mind when working with vectors in non-Cartesian coordinate systems. We then move on to vector calculus in these systems.

8.8.1 Discovery Exercise: Vectors in Curvilinear Coordinates

We begin our exercise at the point whose Cartesian coordinates are (6, 3).

1. Give the polar coordinates ρ and ϕ for this point.
2. The vector $\hat{\rho}$ is a unit vector in the $+\rho$-direction: that is, a vector with magnitude 1 that points in the direction from the origin to the point. Give the Cartesian coordinates for $\hat{\rho}$ at this point.

See Check Yourself #55 in Appendix L

3. The vector $\hat{\phi}$ is a unit vector in the $+\phi$- direction: that is, a vector of magnitude 1 that points tangent to a circle around the origin. At this particular point, $\hat{\phi} = \left(1/\sqrt{45}\right)\left(-3\hat{i} + 6\hat{j}\right)$.
   (a) In order to qualify as $\hat{\phi}$ our vector must be a unit vector, and it must also be perpendicular to $\hat{\rho}$. (In other words, $|\hat{\phi}| = 1$ and $\hat{\rho} \cdot \hat{\phi} = 0$.) Show that our $\hat{\phi}$ meets both criteria.
   (b) The vector $\hat{\phi} = \left(1/\sqrt{45}\right)\left(3\hat{i} - 6\hat{j}\right)$ also meets both criteria. How do we know that ours is correct and this one is wrong?

We move now to a more general point, the Cartesian (x, y).

4. Give the polar coordinates ρ and ϕ for this point as functions of x and y.
5. Give the Cartesian coordinates for $\hat{\rho}$ at this point.
6. Give the Cartesian coordinates for $\hat{\phi}$ at this point. *Hint*: Look at $\hat{\phi}$ at the point (6, 3) in Part 3 and use that as a guide.

Finally, consider the vector field $\vec{f}(\rho, \phi) = 2\hat{\rho} - \hat{\phi}$. The following questions can be answered graphically—that is, without any difficult computation.

7. At the point $(1, 0)$, $\hat{\rho}$ points to the right and $\hat{\phi}$ points up. Write $\vec{f}(\rho, \phi)$ at this point in Cartesian coordinates and draw it at this point.
8. At the point $(0, 1)$, $\hat{\rho}$ points up and $\hat{\phi}$ points to the left. Write $\vec{f}(\rho, \phi)$ at this point in Cartesian coordinates and draw it at this point.
9. Write $\vec{f}$ in Cartesian coordinates and draw it at three other points.

8.8.2 Explanation: Vector Derivatives in Curvilinear Coordinates

The Cartesian $\hat{i}$ and $\hat{j}$ are unit vectors that point in the $+x$- and $+y$-directions, respectively. These vectors are perpendicular to each other. More importantly, they form a "basis" which means any two-dimensional vector can be represented as a linear combination of them, $\vec{v} = a\hat{i} + b\hat{j}$. All that translates perfectly to polar coordinates: $\hat{\rho}$ and $\hat{\phi}$ are unit vectors[7] that point in the $+\rho$- and $+\phi$-directions, they are perpendicular, and any two-dimensional vector can be represented as $\vec{v} = a\hat{\rho} + b\hat{\phi}$. So far, so good.

But $\hat{\rho}$ points directly away from the origin, and $\hat{\phi}$ points counterclockwise along a circle around the origin. This makes them fundamentally different from their Cartesian counterparts, because *their directions depend on where you are.*

How different from $\hat{i}$ and $\hat{j}$, which are dependable and constant! The vector $\vec{f}(x, y) = 3\hat{i} + 4\hat{j}$ points, conveniently enough, from the origin to the point $x = 3$, $y = 4$. More generally, it defines a vector field that has the same magnitude and direction at all points in space.

---

[7] Some texts use $\hat{e}$ with subscripts to indicate unit basis vectors, such as $\hat{e}_\rho$ where we use $\hat{\rho}$.

By contrast, $\vec{f}(\rho, \phi) = 3\hat{\rho} + 4\hat{\phi}$ does not point from the origin to $\rho = 3$, $\phi = 4$. It isn't a constant either; it defines a vector field whose value at each point depends on the directions of $\hat{\rho}$ and $\hat{\phi}$ at that particular point. This can make interpreting non-Cartesian fields a confusing business, but it can also make life much simpler in some cases.

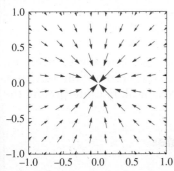

As an example, consider the field $\vec{f}(\rho, \phi) = -(1/\rho^2)\hat{\rho}$ ($\rho \neq 0$). One approach is to consider specific points. At $(1, 0)$ this vector points to the left with a magnitude of 1. At $(1, 1)$ the vector points left-and-down with magnitude $1/2$. At $(0, 2)$ the vector points straight down with magnitude $1/4$, and so on.

But we can see the whole field more generally by noting that $-\hat{\rho}$ gives the direction (always points straight toward the origin), and $1/\rho^2$ gives the magnitude (decreases as you move away from the origin). With that insight you can do the complete sketch quickly.

You may have figured out that this is not an arbitrary example; it can represent the gravitational field exerted by a point mass, or the electric field around an electron.

Of course we could have approached the entire problem in Cartesian coordinates; $\vec{\rho} = \rho\hat{\rho} = x\hat{i} + y\hat{j}$, so we would end up here.

$$\vec{f}(x, y) = -\frac{1}{(x^2 + y^2)^{(3/2)}} \left(x\hat{i} + y\hat{j} \right)$$

That's true, and it can even be useful, but the polar representation is easier for most purposes. It is because $\hat{\rho}$ points in different directions at different points that it lends itself so naturally to this physical situation.

EXAMPLE   A Vector Field in Polar Coordinates

Question: Draw the vector field $\vec{f}(\rho, \phi) = \rho \cos \phi \, \hat{\rho} + \sin^2 \phi \, \hat{\phi}$.

Answer:
One approach is to consider specific points. At $(1, 0)$ this vector points to the right with magnitude 1. At $(0, 1)$ this vector points to the left with magnitude 1. Going through this process slowly and manually can be a valuable way to test and build your facility with polar vectors, but make sure to choose points that make this particular function easy to compute!

But after you've gone through that, look at the big picture. You should recognize the $\hat{\rho}$-component as $\rho \cos \phi = x$. So the radial component is -1 (pointing toward the origin) where $x = -1$, it's zero where $x = 0$, it's 1 (pointing away from the origin) where $x = 1$ and so on. The

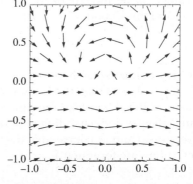

FIGURE 8.15

$\hat{\phi}$-component is zero on the x-axis and is positive (counterclockwise) everywhere else, reaching a maximum on the y-axis. The resulting field is shown in Figure 8.15.

A field like this is easiest to understand in polar coordinates, but if you want a computer to plot it then it's sometimes easier to put it in Cartesian coordinates. See Appendix D for conversions between unit vectors in different coordinate systems.

Vector Calculus in Curvilinear Coordinates

The same tools we use to analyze fields in Cartesian coordinates—gradient, divergence, curl, and Laplacian—apply in other coordinate systems as well. All you need are the formulas, which we list in Appendix F. Applying these formulas is mechanical, but interpreting the results takes practice.

EXAMPLE **A Gradient in Polar Coordinates**

Question: Find the gradient of the function $f(\rho, \phi) = k\phi$ and explain why it came out the way it did.

Answer:
This isn't a trick question; you turn to Appendix F, look up the formula for the gradient in polar coordinates (or look up the cylindrical and ignore the z), and plug in.
$\vec{\nabla} f = (\partial f/\partial \rho)\hat{\rho} + (1/\rho)(\partial f/\partial \phi)\hat{\phi} = (k/\rho)\hat{\phi}$.

But what is all that telling us? Start by thinking about what f looks like. It is zero all along the positive x-axis. It increases as you rotate around: $f = k\pi/4$ along the line $y = x$ in the first quadrant, then $f = k\pi/2$ along the positive y-axis, and so on.

Clearly, if you choose any point and move in the $+\rho$-direction, f does not change; the direction of steepest increase is around the circle in the $+\phi$-direction. That's what the direction of the gradient is telling us.

But you can also easily convince yourself that the rate of change is not always the same. Consider that if you start at the point $(1, 0)$ and go around a circle, f moves from 0 to $2k\pi$. If you start at the point $(3, 0)$, f moves from 0 to $2k\pi$ as you traverse a much bigger circle. In fact, ϕ always goes from 0 to $2k\pi$ as you cover a distance of $2\pi\rho$, which makes the rate of change k/ρ, exactly as the formula predicts. (We'll leave it to you to think about why it makes sense that this formula is undefined at $\rho = 0$.)

The rest of this explanation, and many of the problems, are devoted to deriving these formulas. If you skip all that, the formulas will work anyway, and you'll be no worse off than many students. But if you take the time to work through the derivations, you will be rewarded with a deeper understanding of both curvilinear coordinate systems and the vector calculus operators. (And of course, that deeper understanding may last long after the details of the derivations are forgotten.)

Where Do Those Gradient Formulas Come From?

The gradient $\vec{\nabla} f$ asks "how fast is the scalar function f changing?" If you move in the $+x$-direction, the answer to this question is $\partial f/\partial x$. That's why $\partial f/\partial x$ is the coefficient of $\hat{x}$ in the Cartesian formula for the gradient.

In polar coordinates the gradient looks like this:

$$\vec{\nabla} f(\rho, \phi) = \langle something \rangle \hat{\rho} + \langle something \rangle \hat{\phi}$$

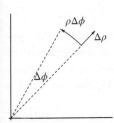

The first $\langle something \rangle$ is the answer to the question: "If I walk a distance Δs in the $\hat{\rho}$-direction (radially outward from the origin), what is the rate of change df/ds?" That's an easy question, because $\Delta \rho$ (the change in the ρ-coordinate) is the same as Δs (the distance moved). So $df/ds = \lim_{\Delta \rho \to 0} \Delta f / \Delta \rho$, which is the just $\partial f / \partial \rho$. That's the first coefficient.

The second $\langle something \rangle$ is the answer to the question: "If I walk a distance Δs in the $\hat{\phi}$-direction (around the origin), what is the rate of change df/ds?" This is a bit trickier, because $\Delta \phi$ is *not* a distance, but an angular change. The distance moved is $\Delta s = \rho \, \Delta \phi$. So the rate of change is the amount the function changes (Δf) divided by the distance ($\rho \, \Delta \phi$), where ρ is constant along the path. So $df/ds = (1/\rho) \lim_{\Delta \phi \to 0} \Delta f / \Delta \phi$, which is $(1/\rho)(\partial f / \partial \phi)$.

That's the second coefficient.

You see that the key to finding the gradient in any coordinate system is asking for each variable, "What distance do I travel as this variable changes?" We've just walked through ρ and ϕ, which covers polar coordinates; adding z to complete the cylindrical set is trivial. You'll tackle the spherical coordinates in Problem 8.142.

Where Do Those Divergence Formulas Come From?

In Section 8.7 we derived the Cartesian formula for the divergence. But we started with another formula that defines the divergence in a way that is independent of what coordinate system you're using.

$$\vec{\nabla} \cdot \vec{f} = \lim_{\Delta V \to 0} \left(\frac{\oiint \vec{f}(x, y, z) \cdot d\vec{A}}{\Delta V} \right) \quad (8.8.1)$$

To turn this definition into a formula, we previously considered a vector function $\vec{f}$ in a rectangular box $\Delta V = \Delta x \Delta y \Delta z$. In cylindrical coordinates, we place a sample point (ρ, ϕ, z) at the corner of a volume $\Delta V = \rho \, \Delta \rho \, \Delta \phi \, \Delta z$, and then integrate through one surface at a time.

We begin with the "outside" surface (farthest from the z-axis). Assuming this surface is small enough to treat $\vec{f}$ as a constant, we compute the surface integral by multiplying the perpendicular component of $\vec{f}$ (the component that points *through* the surface) by the area of the surface: $f_\rho(\rho + \Delta \rho, \phi, z)$ times $(\rho + \Delta \rho) \Delta \phi \, \Delta z$.

The calculation on the inside surface is different in two ways. First, the value of ρ is the ρ of our point, not $\rho + \Delta \rho$. Second, and less obviously, there is a sign change; if f_ρ is positive through this surface then it will flow into our region, creating a negative flux. If the integral through both surfaces were equal they would make no net contribution to the divergence, so the divergence has to depend in some way on how f_ρ is changing as you move in the ρ-direction. That suggests that we're going to end up with a derivative, but the formula is a bit more complicated than it was in the Cartesian case.

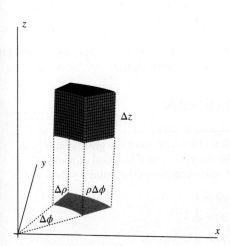

FIGURE 8.16 Differential cylindrical volume.

After we combine these two integrals, Equation 8.8.1 instructs us to divide by ΔV and take the limit:

$$\frac{1}{\rho} \lim_{\Delta \rho \to 0} \frac{(\rho + \Delta \rho) f_\rho (\rho + \Delta \rho, \phi, z) - \rho f_\rho (\rho, \phi, z)}{\Delta \rho}$$

The "aha!" moment comes when you recognize that as the definition of a partial derivative, not for the function f_ρ, but for the function ρf_ρ. And so we arrive at the first part of the divergence in cylindrical coordinates, $(1/\rho)(\partial/\partial \rho)(\rho f_\rho)$. You will look at the remaining four sides, and so finish the cylindrical divergence, in Problem 8.143; the spherical is Problem 8.144.

This is not an easy derivation, but it carries enough insights to be worth your time. Refreshing yourself on the definition of the divergence and where it comes from is a valuable part of understanding that operator. The volume and the surface areas associated with the cylindrical box are vital to working in that coordinate system.

Where Do Those Curl Formulas Come From?

If you followed the divergence work above, nothing here will come as a great surprise. The curl is defined by a line integral:

$$\vec{\nabla} \times \vec{f} = \lim_{\Delta A \to 0} \left(\frac{\oint \vec{f}(x, y, z) \cdot d\vec{s}}{\Delta A} \right)$$

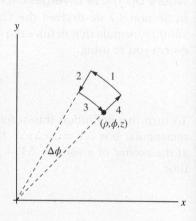

To find the z-coordinate of the curl in cylindrical coordinates, we integrate counterclockwise around a loop whose edges are defined by constant ρ and constant ϕ. (The z-axis in the drawing comes straight out of the page.)

For side 1, f_ϕ points along the curve; the other two components are perpendicular to the curve, and do not contribute. The contribution from side 3 is negative, since we are integrating in the direction of $-\phi$. So the two sides combine to $f_\phi(\rho + \Delta \rho, \phi, z)(\rho + \Delta \rho)\Delta \phi - f_\phi(\rho, \phi, z)\rho \Delta \phi$. Dividing this by $\Delta S = \rho \Delta \rho \Delta \phi$ and taking the limit turns it into $(1/\rho)(\partial/\partial \rho)(\rho f_\phi)$. Sides 2 and 4 (with a bit less work) become $-(1/\rho)(\partial f_\rho / \partial \phi)$, and putting them together produces the z-component in Appendix F.

And Finally, How About That Laplacian?

This one's easy: the Laplacian is the divergence of the gradient, so you can apply the formula you already have. Not everything has to be a conceptual breakthrough, right?

8.8.3 Problems: Vectors in Curvilinear Coordinates

8.117 For the point $x = -\sqrt{3}/2$, $y = 1/2$:
   (a) Draw the point and the two unit vectors $\hat{\rho}$ and $\hat{\phi}$.
   (b) Calculate $\hat{\rho}$ in Cartesian coordinates.
   (c) Calculate $\hat{\rho}$ in Cartesian coordinates at the point $x = -\sqrt{3}$, $y = 1$. *Hint*: the answer is not $-\sqrt{3}\,\hat{i} + \hat{j}$.

In Problems 8.118–8.122, draw the given vector field. Then find its divergence as a function of ρ and ϕ. Finally, find its divergence at the point $\rho = 1$, $\phi = 0$.

8.118 $\vec{f}(\rho, \phi) = \hat{\rho}$

8.119 $\vec{f}(\rho, \phi) = \hat{\phi}$

8.120 $\vec{f}(\rho, \phi) = \hat{\rho} - \hat{\phi}$

8.121 $\vec{f}(\rho, \phi) = \rho\hat{\rho} + (\cos\phi)\hat{\phi}$

8.122 $\vec{f}(\rho, \phi) = (\cos\phi)\hat{\rho} + \rho\hat{\phi}$

In Problems 8.123–8.129, find the gradient and Laplacian of the given scalar field.

8.123 $f(\rho, \phi) = 2\rho + 3\rho\phi$

8.124 $f(\rho, \phi) = \rho \sin\phi$

8.125 $f(\rho, \phi, z) = \rho^2 \cos^2\phi + \rho^2 \sin^2\phi + z^2$

8.126 $f(\rho, \phi, z) = \phi e^{\rho + 2z}$

8.127 $f(r, \theta, \phi) = r(\theta + \phi)$

8.128 $f(r, \theta, \phi) = r \sin\theta \sin\phi$

8.129 $f(r, \theta, \phi) = \ln(\sin\phi/\cos\theta) + 2$

In Problems 8.130–8.138, find the divergence and curl of the given vector field.

8.130 $\vec{f}(\rho, \phi, z) = \hat{\rho}$

8.131 $\vec{f}(\rho, \phi, z) = \hat{\rho} + \rho\hat{\phi} + \hat{k}$

8.132 $\vec{f}(\rho, \phi, z) = (1/\rho)\hat{\rho} + \cos\phi\hat{\phi} - z\hat{k}$

8.133 $\vec{f}(\rho, \phi, z) = z^2\hat{\rho} + \rho^2 \sin^2\phi\hat{\phi} + \rho^2\hat{k}$

8.134 $\vec{f}(r, \theta, \phi) = \hat{\theta}$

8.135 $\vec{f}(r, \theta, \phi) = \hat{\phi}$

8.136 $\vec{f}(r, \theta, \phi) = \hat{r} + \hat{\theta} + \hat{\phi}$

8.137 $\vec{f}(r, \theta, \phi) = r\cos\theta\,\hat{r} + r\cos\theta\,\hat{\theta} + r\cos\theta\,\hat{\phi}$

8.138 $\vec{f}(r, \theta, \phi) = \phi\hat{r} + r\hat{\theta} + \theta\hat{\phi}$

8.139 Make up a field, expressed in polar coordinates, with non-constant ρ and ϕ coordinates, but zero divergence everywhere.

8.140 A positive charge Q creates a potential field $V = kQ/r$ in spherical coordinates.
   (a) Calculate the electric field $\vec{E} = -\vec{\nabla}V$.
   (b) Calculate the divergence and curl of the electric field and explain why these results make sense for this physical situation.

8.141 A wire carrying a current I in the positive z-direction creates a magnetic field $\vec{B} = \mu_0 I/(2\pi\rho)\hat{\phi}$ in cylindrical coordinates. Calculate the divergence and curl of the magnetic field and explain what they tell us about this magnetic field. (The answer to this particular problem is explored in more depth in Section 8.7 Problem 8.116.)

8.142 The gradient in spherical coordinates looks like:
$$\vec{\nabla}f(r, \theta, \phi) = \langle \text{something} \rangle \hat{r} + \langle \text{something} \rangle \hat{\theta} + \langle \text{something} \rangle \hat{\phi}$$
Your job is to find those coefficients. You can of course check your answers against the spherical gradient formula in Appendix F.

   (a) Draw the x, y, and z coordinate axes. Draw a point. Draw a small arrow from that point in the $\hat{r}$-direction.
   (b) If you take a small step in the direction of that arrow, what is its distance ds in terms of r, θ, ϕ, and dr? (The answer is very simple, and should come straight from your drawing.)
   (c) The r-component of the gradient is the change in f (which is df) divided by the distance traveled (which you just wrote). Write this component. A partial derivative will be involved.
   (d) Returning to your drawing, draw another small arrow from the same point in the $\hat{\theta}$-direction.
   (e) If you take a small step in the direction of that arrow, what is its distance ds in terms of r, θ, ϕ, and $d\theta$? (The answer is a bit more complicated than the first one.)
   (f) The θ-component of the gradient is the change in f (which is df) divided by the distance traveled (which you just wrote). Write this component. A partial derivative will be involved.
   (g) Find the ϕ component of the gradient. The process is the same, but the answer will be a bit more complicated than the other two.

8.143 Figure 8.16 shows a cylindrical box with volume $\Delta V = \rho \Delta\rho \Delta\phi \Delta z$. In the Explanation (Section 8.8.2) we found the integral per unit volume through the inner and outer surfaces, ending up with $(1/\rho)(\partial/\partial\rho)(\rho f_\rho)$. In this problem you will work through the other four sides.

   (a) Beginning at the top surface (at $z + \Delta z$), ask yourself three questions. First: which component of $\vec{f}$ points through this surface? Second: when this component is positive, is the resulting flux positive or negative? Third: what is the area of this surface? Based on these three answers, write "*surface integral = component × area*" for this surface.
   (b) Repeat Part (a) at the bottom surface.
   (c) Add your two answers, and divide the result by ΔV. When you cancel and simplify, you should be able to rewrite your answer as a partial derivative, as we did in the explanation.
   (d) Repeat the entire process for the left and right sides to create the final partial derivative. You'll know if you got

8.144 The figure below shows a small box in which each side is defined by holding one of the spherical coordinates constant. The volume of this box is $\Delta V = r^2 \sin\theta \Delta r \Delta\theta \Delta\phi$.

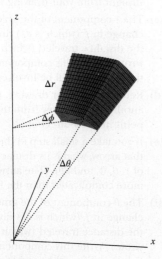

You will use this drawing, along with Equation 8.8.1, to find the formula for the divergence in spherical coordinates.

(a) Beginning at the right-hand surface (at ϕ), ask yourself three questions. First: which component points through this surface, f_r, f_θ, or f_ϕ? Second: when this component is positive, is the resulting flux positive or negative? Third: what is the area of this surface? (That last question will take some thought about the lengths of the sides.) Based on these three answers, write "*surface integral=component×area*" for this surface.

(b) Repeat Part (a) at the left-hand surface (at $\phi + \Delta\phi$).

(c) Add your two answers, and divide the result by ΔV. When you cancel and simplify, you should be able to rewrite your answer as a partial derivative.

(d) Repeat for the inside and outside surfaces (at θ and $\theta + \Delta\theta$).

(e) Repeat for the top and bottom surfaces (at $r + \Delta r$ and r). You'll know if you got it right by checking against the formula in Appendix F for the divergence in spherical coordinates.

8.145 Show that the divergence of the gradient in cylindrical coordinates produces the formula in Appendix F for the Laplacian in cylindrical coordinates.

8.9 The Divergence Theorem

The gradient theorem is a powerful three-dimensional version of the fundamental theorem of calculus, calculating the line integral along an entire curve based on the behavior at the endpoints. In this section we will see another higher dimensional variation of the fundamental theorem. The "divergence theorem" calculates a triple integral throughout a three-dimensional region based on the surface that bounds it.

8.9.1 Discovery Exercise: The Divergence Theorem

The picture shows a three-dimensional region bounded by a closed surface. A vector field $\vec{v}$ is defined everywhere within this region, but we have shown three vectors in particular. Vector (1) is entirely within the region; vector (2) is at the edge of the region, and tangent to the bounding surface; vector (3) is at the edge of the region and points directly out from the bounding surface.

Your job is to add up the divergence of the vector field throughout the region: that is, you want to find $\iiint \left(\vec{\nabla}\cdot\vec{v}\right) dV$. Of course you can't do that without knowing more, but we want to focus on how *these three vectors* contribute to the total.

1. Remember that your goal is to add up the divergence *within the region only*—you don't actually care what happens outside. Given that...

   (a) Will vector (1) contribute a positive number to the total, a negative number, or zero?

(b) Will vector (2) contribute a positive number to the total, a negative number, or zero?
(c) Will vector (3) contribute a positive number to the total, a negative number, or zero?
2. Now, in general, what kinds of vectors will contribute to a positive total divergence within the region?

8.9.2 Explanation: The Divergence Theorem

Vector field $\vec{f}$ is defined everywhere in region V. You can evaluate the surface integral of $\vec{f}$ through the outer surface of V. You can also find the divergence of $\vec{f}$, and then use a triple integral to add that up throughout the region. The "divergence theorem" promises that these two calculations will give the same answer.

That opening paragraph begs two obvious questions. First, those two different calculations seem to have nothing to do with each other; why would they always give the same answer? Second, those two calculations don't look particularly useful; who cares if they give the same answer?

The short answer to both questions is this: the integral of a field through its bounding surface, and the total divergence of a field, are two different ways of asking "How much is this field moving out of this region?" In this section we will expand on that answer, clarifying what the divergence theorem says, why it works, and how it is used.

The Diabolical Dome Disaster

A hemispherical dome of radius R is filled with the noxious chemical Nogoodozine. The evil Simon Bar Sinister makes the dome briefly permeable, allowing some of the gas to escape. The density of Nogoodozine in the dome is a constant, D_0, and the wind velocity is $\left(5kx + py^2\right)\hat{\imath} + ky\hat{\jmath}$, measured relative to the origin at the center of the dome. How much Nogoodozine flows out?

One approach to this question is to focus on where the gas flows out: the dome. The surface integral of the momentum density of the gas through the dome gives its flow rate. This calculation involves parametrizing the surface, calculating the differential area vector $d\vec{A}$, taking the dot product of the flow rate with $d\vec{A}$, and taking a surface integral of that dot product over the surface. You'll carry out that process in Problém 8.157, but there is an easier way.

Section 8.7 defined divergence as the surface integral out of a region per unit volume. Turning that around, the surface integral is divergence times volume—and that suggests our strategy. We're going to find the divergence of the momentum density. When we integrate that divergence throughout the enclosed region, the result will be the flow rate that we want.

The divergence $\vec{\nabla} \cdot \vec{V} = (\partial V_x/\partial x) + (\partial V_y/\partial y) = 5kD_0 + kD_0 = 6kD_0$. Since this is a constant, we can find our integral by multiplying it by the volume $(2/3)\pi R^3$ to find that the rate at which the chemical is leaving (mass per unit time) is $4\pi kD_0 R^3$.

The mathematical equivalence of these two approaches is expressed by the "divergence theorem."

The Divergence Theorem

Suppose the three-dimensional region V is bounded by the closed surface S. For a vector field $\vec{f}$ defined within that region:

$$\iiint_V \left(\vec{\nabla} \cdot \vec{f}\right) dV = \oiint_S \vec{f} \cdot d\vec{A} \qquad (8.9.1)$$

In words: if you add up the divergence everywhere within the region, you get the integral through the bounding surface.

This theorem goes by many names including Gauss's theorem, Green's theorem[8], Ostrogadsky's theorem, and the fundamental theorem for divergences. The theorem tells us that both the total divergence within the region, and the integral through the outer surface, measure how much the vector field is flowing *out* of the region, so we will get the same answer whichever way we calculate it. In the example above the divergence theorem was useful because the surface integral was hard to calculate directly, but the divergence was simple. In the boxed example below the divergence is hard to integrate but the surface integral is easy to find. The divergence theorem says that when you want to find the flow of a vector out of a region, you are free to choose whichever of these two approaches is easier.

EXAMPLE The Divergence Theorem

Question: Find the flow rate of the field $\vec{f} = \rho/\sqrt{a^2 + \rho^2}\,\hat{\rho}$ out of the region bounded by the cylinder $\rho = R$ and the planes $z = 0$ and $z = H$.

Answer:
For the left side of the divergence theorem, we calculate the divergence of the vector field using the formula for divergence in cylindrical coordinates: $\vec{\nabla} \cdot \vec{f} = (1/\rho)(\partial/\partial\rho)\left(\rho f_\rho\right)$. After multiplying by ρ, taking the derivative with the quotient rule, dividing by ρ, simplifying the result, and taking the ϕ and z integrals we eventually end up with the expression

$$2\pi H \int_0^R \frac{\rho\left(2a^2 + \rho^2\right)}{\left(a^2 + \rho^2\right)^{3/2}} \, d\rho$$

Handing this mess to a computer we get the answer $2\pi H R^2/\sqrt{a^2 + R^2}$.

If we don't relish long algebra calculations, though, we could just consider the integral through the outer surface. Since $\vec{f}$ is all in the ρ-direction there is no flux through the top and bottom of the cylinder. Along the sides $\vec{f}$ points directly outward with magnitude $R/\sqrt{a^2 + R^2}$, so we can just multiply that constant by the surface area of the cylinder, $2\pi R H$, to get $2\pi H R^2/\sqrt{a^2 + R^2}$.

It might seem odd that an integral over the surface can be equivalent to an integral throughout the entire volume. If we changed the vector field somewhere inside the region but left it the same on the boundary wouldn't that change the left side of Equation 8.9.1 but not the right side?

To answer that question, consider the effect that one arbitrarily chosen vector has on the left side of the equation, the total divergence. The vector tends to cause positive divergence in the region that it points away from, and negative divergence in the region that it points toward, as shown in Figure 8.17.

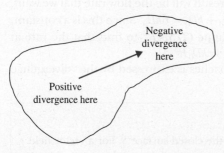

FIGURE 8.17 What effect does one vector have on the divergence?.

---

[8]The name "Green's theorem" is more commonly used for a different but related theorem that we will discuss in Section 8.10. Isn't this fun?

So when we add up the entire divergence, the net effect of any vector is zero, unless that vector is on the surface, and points inward or outward. A vector on the surface pointing outward creates positive divergence inside the region (where we count it) and negative divergence outside (where we don't count it). That's how adding up all the divergence inside the region (left side of Equation 8.9.1) leads us to the flow through the surface (right side).

Below we present three different uses for the divergence theorem: we will derive the equation of continuity, we will discuss field lines, and we will convert Gauss's law from integral to differential form. Each of these is important in its own right, but they also serve as examples of the power and scope of this theorem.

The Equation of Continuity

The following formula describes the balance of any quantity in a well-defined region, such as the mass of a particular chemical in a vat.

$$\text{accumulation} = \text{generation} - \text{consumption} + \text{input} - \text{output} \tag{8.9.2}$$

For a fluid the first two terms on the right represent "sources" and "sinks," meaning places where fluid can be created or destroyed. If the fluid is conserved then these terms equal zero, and the only change in the fluid's mass comes from the flow in and out of the region. Using $\vec{V}$ for the momentum density of the fluid, "input − output" becomes the surface integral of fluid through the boundary of the region, but with a negative sign (do you see why?), and Equation 8.9.2 becomes $\oiint \vec{V} \cdot d\vec{A} = -dm/dt$.

The divergence theorem allows us to rewrite that equation. To avoid confusion with the momentum density $\vec{V}$ we'll use τ for volume and write: $\iiint_\tau \left(\vec{\nabla} \cdot \vec{V} \right) d\tau = -dm/dt$. In the limit where the region becomes infinitesimally small, we can consider the divergence to be constant throughout, and our integral becomes a product: $\left(\vec{\nabla} \cdot \vec{V} \right) \tau = -dm/dt$. Dividing both sides by the volume turns the mass into a density, and the result is called the "equation of continuity." Because we have switched from talking about mass in a region to talking about density, which is a field, the derivative becomes partial.

The Equation of Continuity

If $\vec{V}$ is the momentum density of a fluid in a region where no fluid is created or destroyed and ρ is the density of the fluid, then

$$\vec{\nabla} \cdot \vec{V} = -\frac{\partial \rho}{\partial t} \tag{8.9.3}$$

The equation of continuity is central to fluid mechanics and it also comes up in fields ranging from electricity and magnetism to relativity.

Field Lines

Section 8.2 presented two different tools for visualizing a vector field: vector plots, and field lines (aka streamlines). The latter method is easier to draw and interpret, but cannot be used for all vector fields. We are now at last ready to say exactly when you can use it, and why.

Remember that field lines follow the direction of a vector field, and their density is proportional to the magnitude of the field. For example, looking at Figure 8.18 you can see that the vector field points always to the right, and also points upward when $y > 0$ and downward when $y < 0$. But you can also tell that the magnitude of the field decreases as you move to the right, because the lines become less dense.

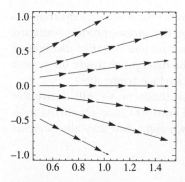

FIGURE 8.18 Field lines.

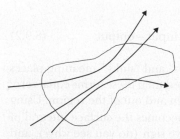

FIGURE 8.19 Field lines go in, and then they go out!.

Now suppose you wanted to draw field lines for $(1/x)\hat{i}$. Because the field always points in the $+x$-direction, the field lines would all be horizontal, so their density would not decrease as you move to the right. That doesn't mean there is anything wrong with $(1/x)\hat{i}$ as a vector field, but it does mean we cannot represent it with field lines; we have to use a vector plot. (Vector plots can be drawn for *any* vector field.)

So what kind of field can have field lines? If you draw a closed surface in a region with no sinks or sources then each field line will enter the region somewhere and leave it somewhere else (Figure 8.19). Thus the surface integral out of any closed region will be zero. That in turn implies—this is where the divergence theorem comes in—that the divergence must be zero everywhere. Fields with zero divergence are called "incompressible vector fields," which makes perfect sense if you look at the equation of continuity. (They are also sometimes called "solenoidal vector fields," which makes less sense.)

We conclude that any vector field that is incompressible in some region can be represented with field lines in that region. This mathematical fact has a physical corollary: the velocity field for an incompressible fluid has a divergence of zero.

The Divergence Theorem and Gauss's Law

Maxwell's equations represent the great triumph of 19th century physics, encapsulating everything then known about electricity, magnetism and light in four equations. We have already seen one of Maxwell's equations, Gauss's law: if you draw any closed surface, the surface integral of the electric field through that surface is proportional to the enclosed charge.

$$\oiint \vec{E} \cdot d\vec{A} = \frac{Q}{\varepsilon_0} \quad \text{Gauss's law in integral form} \tag{8.9.4}$$

In general, any rule that can be expressed as an integral can also be expressed as a derivative. For instance, in one dimension, the relationship between position and velocity can be written in two ways:

$$x = \int v \, dt \quad \text{or, equivalently} \quad v = \frac{dx}{dt}$$

It is the fundamental theorem of calculus—the all-important, remarkable connection between derivatives and integrals—that makes those two very different looking statements equivalent. The divergence theorem provides a three-dimensional version of the fundamental theorem that allows us to rewrite the left side of Gauss's law as follows:

$$\iiint \left(\vec{\nabla} \cdot \vec{E} \right) dV = \frac{Q}{\varepsilon_0}$$

How do you find the total charge inside a region? If you know the charge density $\rho(x, y, z)$, you integrate it through the region:

$$\iiint \left(\vec{\nabla} \cdot \vec{E} \right) dV = \frac{\iiint \rho \, dV}{\varepsilon_0}$$

Some care is required in this next step. Just because two quantities have the same integral over some region does not mean that the two quantities are equal. But if they have the same integral over *any* region—anywhere, any size—then the two quantities must be equal everywhere. This leads us to Gauss's law in differential form:

$$\vec{\nabla} \cdot \vec{E} = \frac{\rho}{\varepsilon_0} \quad \textit{Gauss's law in differential form} \tag{8.9.5}$$

Equations 8.9.4 and 8.9.5 are *not* two of Maxwell's four equations; they are one of the four equations, written in two equivalent ways. Each is more useful than the other in some situations.

8.9.3 Problems: The Divergence Theorem

8.146 A cylinder is defined by $\rho = 3$, $0 \le z \le 4$. If we add the two surfaces $\rho \le 3$ at $z = 0$ and $\rho \le 3$ at $z = 4$ then the result is a closed surface S shaped like a can, enclosing a region V. Your job is to confirm the divergence theorem for the vector field $\vec{f} = 5\rho\hat{\rho} + (z+2)\hat{k}$ in this region.

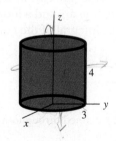

(a) Calculate the divergence $\vec{\nabla} \cdot \vec{f}$.
(b) Calculate the integral $\iiint_V \left(\vec{\nabla} \cdot \vec{f} \right) dV$.
(c) Now, ignoring the work you have already done, calculate $\iint \vec{f} \cdot d\vec{A}$...
  i. ...through the cylindrical surface
  ii. ...through the disk at the top
  iii. ...through the disk at the bottom
(d) Add together these three contributions to find the integral of $\vec{f}$ through the surface S.
(e) If your answers to Parts (b) and (d) match, you have verified the divergence theorem in this case. If they don't match, go back and find the problem!

In Problems 8.147–8.152, confirm the divergence theorem for the given vector field $\vec{f}$ in the region V bounded by the given surface S.

8.147 Vector $\vec{f}$ is $3\hat{k}$ and surface S is a sphere of radius R centered on the origin.

8.148 Vector $\vec{f}$ is $r\hat{r}$ in spherical coordinates, and surface S is a sphere of radius R centered on the origin.

8.149 Vector $\vec{f}$ is $r\hat{r}$ in spherical coordinates, and surface S is a cube with corners at $(0, 0, 0)$ and (L, L, L).

8.150 Vector $\vec{f}$ is $z\hat{k}$ and region V is bounded above by $z = \sqrt{1 - x^2 - y^2}$ and below by $z = 1 - x^2 - y^2$.

8.151 $\vec{V} = x^2 e^{\cos z} \hat{k}$ in the cube with corners at $(0, 0, 0)$ and (π, π, π).

8.152 Vector $\vec{f}$ is $x^2 \hat{i} + y\hat{j} + 2z\hat{k}$ and region V is bounded by the plane $2x + y + 3z = 6$ and the three coordinate planes.

In Problems 8.153–8.155, the vector $\vec{V}$ represents the momentum density of a fluid. Calculate the flow rate of the fluid out of the closed region indicated. In each case you should start by thinking about which side of the divergence theorem will be easier to use.

8.153 $\vec{V} = x\hat{i} + y\hat{j}$ in the sphere $x^2 + y^2 + z^2 \le R^2$

8.154 $\vec{V} = y\hat{i} + x\hat{j}$ in the region between $z = x^2 + y^2$ and $z = 1 - x^2 - y^2$

8.155 $\vec{V} = x^3\hat{i} + y^3\hat{j} + z^3\hat{k}$ in the sphere $x^2 + y^2 + z^2 \le R^2$

8.156 A jet of water is spraying directly down onto a spot on a table, from where the water spreads out radially. Given that water is incompressible, find the velocity $v(\rho)$ of the water as it spreads.

8.157 In the Explanation (Section 8.9.2) we used the divergence theorem to solve for the rate at which the chemical Nogoodozine would leak out of a dome.

(a) Solve for that rate by taking a surface integral of the momentum density through the dome and confirm that you get the same answer. The information you need is in the Explanation (Section 8.9.2).

(b) The divergence theorem only works for closed surfaces, which means you have to consider the ground beneath the dome to be part of the surface. Why didn't you need to include that in your calculation of flow rate in Part (a)?

8.10 Stokes' Theorem

Stokes' theorem does for the curl what the gradient and divergence theorems do for those operators. Specifically, it finds a surface integral based on the curve that bounds the surface. We have seen that the divergence theorem provides a bridge between the differential and integral forms of Gauss's law; Stokes' theorem does the same for Ampere's law, another of Maxwell's equations.

8.10.1 Explanation: Stokes' Theorem

The divergence theorem says that, because divergence is surface integral per unit volume, you can calculate the flow rate out of a region either by integrating the momentum density through the surface, or by integrating the divergence throughout the volume. Stokes' theorem makes a similar statement, starting from the fact that the curl measures circulation per unit area. That means you can either measure circulation around a loop by directly taking a line integral around the loop, or by taking a surface integral of the curl on the surface enclosed by the loop.

Stokes' Theorem

Suppose surface S is bounded by curve C. For a vector field $\vec{f}$ defined throughout that surface:

$$\iint_S \left(\vec{\nabla} \times \vec{f} \right) \cdot d\vec{A} = \oint_C \vec{f} \cdot d\vec{s} \tag{8.10.1}$$

In words: if you add up the curl everywhere within the surface, you get the line integral around the bounding curve.

This theorem goes by many names including Kelvin-Stokes' theorem, the curl theorem, and the fundamental theorem for curls. To make matters more complicated, mathematicians apply the name "Stokes' theorem" to a more general theorem of which both this and the divergence theorem are special cases.

As similar as Stokes' theorem is to the divergence theorem, there is an important difference in the way you apply them. In the divergence theorem, the region bounded by a closed surface is unambiguous; for instance, there is exactly one region bounded by a given sphere. When you apply Stokes' theorem, however, one curve can be the boundary of many different surfaces. Look at just a few of the surfaces that are bounded by a horizontal circle. Stoke's theorem leads to the non-obvious conclusion that the surface integral of the curl will be the same on all these surfaces, and on any other surface bounded by the same horizontal circle.

EXAMPLE | Stokes' Theorem

Question: Confirm Stokes' theorem for the field $\vec{f} = -y\hat{i} + x\hat{j} + (z^2 + 10)\hat{k}$ on the surface $z = \sqrt{9 - x^2 - y^2}$.

Answer:
The left side of Stokes' theorem tells us to integrate the curl $\vec{\nabla} \times \vec{f}$ on the surface. The curl is $2\hat{k}$. The hemisphere can be parameterized $\vec{r} = 3\sin\theta\cos\phi\,\hat{i} + 3\sin\theta\sin\phi\,\hat{j} + 3\cos\theta\,\hat{k}$ for $0 \leq \theta \leq \pi/2$, $0 \leq \phi \leq 2\pi$. We find $d\vec{A}$ as $(\partial\vec{r}/\partial\theta) \times (\partial\vec{r}/\partial\phi)$ and end up with:

$$\left(\vec{\nabla} \times \vec{f}\right) \cdot d\vec{A} = 2\hat{k} \cdot \left(9\sin^2\theta\cos\phi\,\hat{i} + 9\sin^2\theta\sin\phi\,\hat{j} + 9\sin\theta\cos\theta\,\hat{k}\right)\,d\phi\,d\theta$$
$$= 18\sin\theta\cos\theta\,d\phi\,d\theta$$

$$\int_0^{\pi/2}\int_0^{2\pi} 18\sin\theta\cos\theta\,d\phi\,d\theta = 18\pi$$

For the right side of Stokes' theorem, we want a line integral around the curve $x = 3\cos\phi$, $y = 3\sin\phi$, $z = 0$ as ϕ goes from 0 to 2π.

$$\vec{f} \cdot d\vec{s} = \left(-3\sin\phi\hat{i} + 3\cos\phi\hat{j} + 10\hat{k}\right) \cdot \left(-3\sin\phi\,d\phi\,\hat{i} + 3\cos\phi\,d\phi\,\hat{j}\right) = 9$$

And $\int_0^{2\pi} 9\,d\phi = 18\pi$, validating the theorem.

This problem asked us to calculate the line integral and surface integral shown above. But if you wanted to know the circulation of this field around this loop, neither of these calculations would be the easiest way. Remember that Stokes' theorem promises us the same answer if we integrate the curl on *any surface bounded by this curve*. The hemisphere works, but a much simpler surface bounded by the same curve is the disk $x^2 + y^2 \leq 9$, $z = 0$. To integrate $\vec{\nabla} \times \vec{f} = 2\hat{k}$ through that surface $2 \times 9\pi$ gives us 18π with much less work.

Why Stokes' Theorem Works

Section 8.9 explained how the divergence theorem relates the integral through a region to an integral on its surface. Vectors entirely inside the region contribute positively to the divergence behind them and negatively to the divergence ahead of them, so only the ones on the boundary contribute to the net divergence.

A similar argument hold for Stokes' theorem and curl. We begin by considering the left side of Equation 8.10.1: the double integral, over the surface S, of the curl of vector function $\vec{f}$. Remember that curl is a measure of circulation. So in effect we are breaking S into tiny subregions, finding the line integral of $\vec{f}$ around each subregion, and then adding up all the results. This process is illustrated visually in Figure 8.20. The arrows in the drawing do *not* represent vector field $\vec{f}$; they represent the line integrals around which we are evaluating $\vec{f}$. In the limit as each small loop approaches a point, the sum of all those line integrals becomes $\iint_S \left(\vec{\nabla} \times \vec{f}\right) \cdot d\vec{A}$.

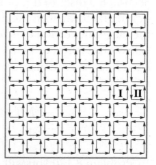

FIGURE 8.20 Integrating the curl over a surface adds up the circulation around many tiny loops.

The vector field itself could be anything. But imagine, for a moment, that $\vec{f}$ happens to point directly upward when you are between the two loops labeled **I** and **II**. As we integrate around loop **I** we will integrate upward through this point, so we will get a positive result. Around loop **II** we will integrate downward through this point, so we will get a negative result. Hence, such a vector will not contribute anything after we sum (integrate) over the whole region.

Does that mean the total curl must be zero? Not exactly. Imagine now that $\vec{f}$ points upward when you are on the *right* side of loop **II**. This vector will contribute positively to the line integral around loop **II**, but this contribution will not be canceled because there *is* no loop on the other side.

By this argument, you can see that the only vectors that contribute to $\iint_S (\vec{\nabla} \times \vec{f}) \cdot d\vec{A}$ are vectors at the very edge of the surface, pointing parallel to that edge. When we add up all those vectors we get a line integral around that edge. Hence, the left side of Equation 8.10.1 has become the right side.

Sign Issues

The surface integral over a region that isn't closed has an ambiguous sign; we can calculate the flow through the surface in either direction. Similarly, a line integral along a curve can be evaluated in either direction along the curve. So how should we evaluate the integrals in Stokes' theorem? The answer is we can do them either way, but once we've chosen the direction of one of the integrals the direction of the other one is determined by the right-hand rule.

Consider the simplest case: a flat, horizontal disk bounded by a circle. If we evaluate the line integral around the circle clockwise, then the right-hand rule tells us to evaluate the surface integral downward. If we choose to evaluate the line integral counterclockwise then we must do the surface integral upward.

If the surface is more complicated than a flat disk, you can still figure out the direction of integration from the right-hand rule. For example, in the example in the gray box above we calculated the flow going out of the hemisphere. Now, picture breaking the hemisphere up into many tiny loops, as we did to a square region in Figure 8.20. Pointing your thumb outward from each little loop, the curl of your fingers dictates that we go around the bounding circle counterclockwise. We did that by having ϕ increase as we integrated.

Ampere's Law

Section 8.9 discussed how Gauss's law for electric fields can be written in either differential or integral form, with the divergence theorem converting between the two. "Ampere's law" is the magnetic version of Gauss's law; it relates the magnetic field to the enclosed current. Like Gauss's law, Ampere's law can be expressed in integral or differential form. In the following equations $\vec{B}$ is the magnetic field, I is the enclosed current, $\vec{J}$ is the current density, and μ_0 is a positive constant.

$$\oint \vec{B} \cdot d\vec{s} = \mu_0 I_{enclosed} \quad \text{Ampere's law in integral form}$$
$$\vec{\nabla} \times \vec{B} = \mu_0 \vec{J} \quad \text{Ampere's law in differential form}$$

In Problem 8.170 you will use Stokes' theorem to show the equivalence of the differential and integral forms of Ampere's law, just as we did in Section 8.9 for Gauss's law.

Green's Theorem

The phrase "positively oriented" in the following definition indicates a counterclockwise traversal of a curve.

Green's Theorem

Let C be a positively oriented, piecewise-smooth, simple closed curve in the plane and let D be the region bounded by C. If the scalar fields $P(x, y)$ and $Q(x, y)$ have continuous partial derivatives on an open region that contains D, then

$$\oint_C (P\, dx + Q\, dy) = \iint_D \left(\frac{\partial Q}{\partial x} - \frac{\partial P}{\partial y} \right) dA$$

This is, on the face of it, one of the more inscrutable theorems in this book. The right side is the double integral over some bounded region in the plane of a scalar function of x and y—a peculiar looking function perhaps, but at least you can see what to do with it. But what is the left side? What does $\int P(x, y)\, dx$ mean?

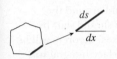

FIGURE 8.21
$P(x, y)\, ds$ and $P(x, y)\, dx$.

As always, such an integral should properly be thought of as a limit. Imagine breaking a curve into a large (but finite) number of small line segments. At each segment you evaluate the value of the function $P(x, y)$. Now, for a "regular" scalar line integral you would multiply $P(x, y)$ times the arc length ds of the small line segment. (That in turn is given by the Pythagorean theorem $ds^2 = dx^2 + dy^2$.) In this case you multiply $P(x, y)$ by the segment's horizontal length dx. (See Figure 8.21.) Add up those products all around the curve. In the limit of this process as the number of pieces approaches infinity, you have $\int P\, dx$.

So now you have some idea of what it means, but it's still not clear what it's doing. It is perhaps more useful to begin by considering a vector $\vec{f} = P(x, y)\hat{i} + Q(x, y)\hat{j}$. Then the left side of Green's theorem becomes something familiar: it is $\oint_C \vec{f} \cdot d\vec{s}$, where $d\vec{s} = dx\,\hat{i} + dy\,\hat{j}$ as always. If you take the curl of that same vector $\vec{f}$ you get the right side of Green's theorem. Viewed this way Green's theorem is just Stokes' theorem, written for the special case of a surface that lies entirely in the xy-plane.

We are not emphasizing Green's theorem; in practice it can be treated as a special case of Stokes' theorem. We have that luxury because we are focusing on intuitive demonstrations of all these theorems, rather than formal proofs. A more rigorous approach works the opposite way: first you prove Green's theorem, and then you use Green's theorem to prove both the divergence theorem and Stokes' theorem.

A Brief Word about Dimensions

Fluid flow and electromagnetic fields, the two most important uses of vector calculus, both tend to take place in three dimensions. But there are times when it is useful to apply these rules to two-dimensional vector fields.

You can take the divergence of a vector field in any number of dimensions, so the divergence theorem can always be applied, but be careful. For a two-dimensional region bounded by a curve, the divergence theorem equates the double integral of the divergence throughout the region to the integral *through* the curve, not around the curve as a line integral does.

The curl technically applies only in three dimensions, so Stokes' theorem cannot be applied in any dimension. It always relates the integral of curl on a 2D surface to a line integral on the boundary of that surface. If the 2D surface lies on a plane the problem will look two dimensional, but the curl will point perpendicular to the plane, so Stokes' theorem always takes place in three dimensions.

8.10.2 Problems: Stokes' Theorem

8.158 The function $\vec{r}(\phi, z) = 3\cos\phi\hat{i} + 3\sin\phi\hat{j} + z\hat{k}$ for $0 \le \phi \le \pi$, $0 \le z \le 4$ defines a half-cylinder. If we add the rectangle with corners $(-3, 0, 0)$ and $(3, 0, 4)$ and the half-disk $0 \le y \le \sqrt{9-x^2}$ at $z = 0$ then our surface S is a can with a bottom but no top. The bounding curve is made up of two curves, both at $z = 4$: the semicircle $y = \sqrt{9-x^2}$ and the line $y = 0$ for $-3 \le x \le 3$. Your job is to confirm Stokes' theorem for the vector field $\vec{f} = -yz\hat{i} + xz\hat{j}$ on this surface.

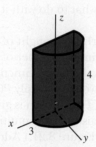

(a) Calculate the curl $\vec{\nabla} \times \vec{f}$.
(b) Calculate $\iint_S \left(\vec{\nabla} \times \vec{f}\right) \cdot d\vec{A}$ pointing *into* the can...
  i. ...on the cylinder
  ii. ...on the back
  iii. ...on the bottom
(c) Add together these three contributions to find the total integral of the curl along surface S.
(d) Now, ignoring the work you have already done, calculate the line integral $\oint \vec{f} \cdot d\vec{s}$ counterclockwise...
  i. ...along the line segment on the top
  ii. ...along the semicircle on the top
(e) Add together these two contributions to find the total line integral along the bounding curve.
(f) If your answers to Parts (c) and (e) match, pat yourself on the back for verifying Stokes' theorem in this case. If they don't match, go back and find the problem!

In Problems 8.159–8.162, confirm Stokes' theorem for the given vector field $\vec{f}$. If a surface S is specified take the surface integral on that surface and the line integral on its bounding curve. If a bounding curve C is specified, you may choose any surface bounded by C that makes the surface integral as easy as possible.

8.159 Vector $\vec{f}$ is $-y\hat{i} + x\hat{j}$ and S is the disk $x^2 + y^2 \le 1$, $z = 0$.

8.160 Vector $\vec{f}$ is $-y\hat{i} + x\hat{j}$ and S is the square with corners at $(0, 0, 0)$ and $(1, 1, 0)$.

8.161 Vector $\vec{f}$ is $xy\hat{i} + xz\hat{j} + yz\hat{k}$ and C is the triangle with vertices at $(1, 0, 0)$, $(0, 1, 0)$, and $(0, 0, 2)$. *Hint*: The easiest surface to use is not the triangle itself.

8.162 Vector $\vec{f}$ is $y\hat{i} + z^2\hat{j} + x^2\hat{k}$ and S is the $z \ge -4$ part of the sphere $x^2 + y^2 + z^2 = 25$.

In Problems 8.163–8.167, the vector $\vec{V}$ represents the momentum density of a fluid. Calculate the absolute value of the circulation of the fluid around the indicated curve. In each case you should start by thinking about which side of Stokes' theorem will be easier to use. If you evaluate a surface integral you should also choose the easiest possible surface.

8.163 $\vec{V} = \ln(\rho)\hat{\phi}$ around the circle $x^2 + y^2 = R^2$ in the xy-plane.

8.164 $\vec{V} = x^2\left(x^2 + (z-1)^2\right)\hat{k}$ around a circle of radius 1 in the xz-plane centered on the point $(0, 0, 1)$.

8.165 $\vec{V} = x\hat{i} + y\hat{j}$ around a circle of radius 1 centered on the origin in the $x = z$ plane.

8.166 $\vec{V} = z\hat{i} - x\hat{k}$ around a circle of radius 1 centered on the origin in the $x = z$ plane.

8.167 $\vec{V} = yz\hat{i} - x\hat{j}$ around a curve made up of the following segments: the quarter-circle $(z-1)^2 + (x-1)^2 = 1$ on the xz-plane from $(1, 0, 0)$ to $(0, 0, 1)$, the quarter-circle $(z-1)^2 + (y-1)^2 = 1$ on the yz-plane from $(0, 0, 1)$ to $(0, 1, 0)$, and the line $x + y = 1$ on the xy-plane from $(0, 1, 0)$ to $(1, 0, 0)$.

8.168 In this problem you will confirm Stokes' theorem for the vector field $\vec{f} = (x + y^2 + z^3)\hat{i}$ on the circle $y^2 + z^2 = 9$ on the yz-plane.

(a) Calculate $\oint \vec{f} \cdot d\vec{s}$ along this circle.

(b) Calculate $\iint_S \left(\vec{\nabla} \times \vec{f}\right) \cdot d\vec{A}$ on the $x \ge 0$ part of the paraboloid $x = 9 - y^2 - z^2$ and confirm that it equals what you got in Part (a).

(c) Choose an easier surface that is also bounded by the same circle and confirm that Stokes' theorem works for that surface as well.

8.169 Why does Stokes' theorem work? To answer this question, consider a vector field defined throughout a bounded two-dimensional surface. Your object is to add up the *curl* of the vector field throughout the surface. The drawing shows three sample vectors within a bounded surface S.

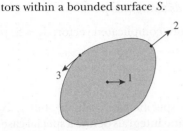

(a) We begin by considering the effects of vector (1), a horizontal vector entirely within the surface. This vector will tend to contribute to the curl at a point immediately below itself. Will this contributed curl point into the page, or out of the page? (If vector 1 represents a fluid flow, and we put a paddlewheel immediately below it, will it tend to turn that paddlewheel clockwise or counterclockwise?)

(b) Vector (1) also tends to contribute to the curl at a point immediately *above* itself. Will this contributed curl point into the page or out of the page?

(c) Taking into account both of your answers, what will the net contribution of vector (1) be to our total curl?

(d) Vector (2) points directly out of the surface. What direction will its total contribution to the curl be, and why?

(e) Vector (3) points tangent to the bounding curve. What direction will its total contribution go the curl be, and why?

(f) Any vector on the boundary of the surface can be broken into a component that points directly out of the surface, and a component tangent to the bounding curve. Which component is relevant to the total curl within the surface?

(g) Explain how this reasoning leads to Stokes' theorem.

8.170 In this problem you will find the differential form of Ampere's law, starting with the integral form:

$$\oint \vec{B} \cdot d\vec{s} = \mu_0 I_{enclosed}$$

(a) Use Stokes' theorem to rewrite the left side of the equation.

(b) The "current density" J is defined as the current per unit area: $J = \lim_{A \to 0}(I/A)$. So to get from current density to current you multiply by area ($I = JA$) if J is constant, or integrate if J varies. Use this fact to rewrite the right side of the equation.

(c) Finish the derivation of the differential form of Ampere's law.

8.171 It turns out that Ampere's law, as we have presented it here, is a special case. Maxwell figured out that a magnetic field could also be created by a changing electric field, and wrote the more general form:

$$\vec{\nabla} \times \vec{B} = \mu_0 \vec{J} + \mu_0 \varepsilon_0 \frac{\partial \vec{E}}{\partial t}$$

Use this equation, along with Gauss's law in differential form, to derive the continuity equation for charge (which mathematically expresses the fact that charge is conserved):

$$\vec{\nabla} \cdot \vec{J} = -\frac{\partial \rho}{\partial t}$$

8.11 Conservative Vector Fields

In Section 8.4 we met the gradient theorem. We saw that for a certain class of vector fields—those that can be written as the gradient of some scalar potential field—line integrals are path independent, and can be evaluated without actually integrating anything. Such vector fields are called "conservative" and this section takes up the question of how to recognize and work with them.

8.11.1 Discovery Exercise: Conservative Vector Fields

We begin by considering a very simple vector: $\vec{v}_1 = x\hat{i} + x\hat{j}$. You are going to find the line integral of this vector from $(1, 0)$ to $(0, 1)$ along three different paths.

1. Path 1 goes left to the origin and then straight up. Evaluate $\int_C \vec{v}_1 \cdot d\vec{s}$ along this path.

   See Check Yourself #56 in Appendix L

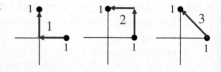

2. Path 2 goes up to $(1,1)$ and then left. Evaluate $\int_C \vec{v}_1 \cdot d\vec{s}$ along this path.
3. Path 3 is a single straight line. Evaluate $\int_C \vec{v}_1 \cdot d\vec{s}$ along this path.

We now repeat the same exercise with a slightly more complicated vector: $\vec{v}_2 = 2xy\hat{i} + x^2\hat{j}$.

4. Evaluate $\int_C \vec{v}_2 \cdot d\vec{s}$ along path 1.
5. Evaluate $\int_C \vec{v}_2 \cdot d\vec{s}$ along path 2.
6. Evaluate $\int_C \vec{v}_2 \cdot d\vec{s}$ along path 3.

You should have found that $\int_C \vec{v}_1 \cdot d\vec{s}$ depends on the specific path you take, but $\int_C \vec{v}_2 \cdot d\vec{s}$ is "path-independent." Vectors with path-independent line integrals define a special case called "conservative" vectors, and you can evaluate their line integrals with the gradient theorem.

7. For the function $f(x,y) = x^2 y$ confirm that $\vec{\nabla} f = \vec{v}_2$.
8. Show that $f(0,1) - f(1,0)$ gives the same answer that you got with all three line integrals.

8.11.2 Explanation: Conservative Vector Fields

In introductory mechanics you may have learned that "conservative" forces (such as gravity) conserve mechanical energy, while non-conservative forces (such as friction) do not. In Section 8.5 we defined a "conservative" vector field as the gradient of a scalar field. In this section we will elaborate on the definition of "conservative" as a mathematical property of a vector field and show how it relates to the physical definition of a conservative force.

Four Definitions of a Conservative Field

We begin by giving four different definitions, the first two of which should look familiar. (We number these definitions, not because they come in any order, but so we can refer back to them later by number.)

Conservative Fields

1. A conservative vector field is the gradient of some scalar field: $\vec{v} = \vec{\nabla} F$. To put the same thing a different way, a conservative vector field has a potential function. (In math texts F would generally be the potential of $\vec{v}$, but here we follow most physics texts and say F is *minus* the potential of $\vec{v}$.)
2. The line integrals of a conservative vector field are "path independent." That is, $\int_C \vec{v} \cdot d\vec{s}$ depends only on the starting and ending points, not on the specific curve drawn between them.
3. The line integral of a conservative vector field around a closed path (one where the starting and ending points are the same) is always zero: $\oint \vec{v} \cdot d\vec{s} = 0$.
4. A conservative vector field is "irrotational," meaning it has no curl: $\vec{\nabla} \times \vec{v} = \vec{0}$. (This is not a perfect definition, as discussed below.)

We are *not* saying "If you want to show that a particular field is conservative, you have to show that it meets all of these criteria." Rather, we are saying that any one of these statements is a sufficient definition—with an exception for Definition 4—because each one implies the others.

For instance, we saw in Section 8.5 how to evaluate the line integral of a gradient based on only the endpoints. The gradient theorem says that for any vector field that can be written as the gradient of some scalar field, the line integrals are path independent. That is, it says that Definition 1 above implies Definition 2.

Our job in this Explanation is to show how all four of these seemingly unrelated definitions connect with each other and with conservation of mechanical energy.

A Tale of Two Forces

You fire a bullet that follows a parabolic arc until it reaches a cliff 100 m high, and you want to know how fast the bullet will hit the cliff. One approach is to write $F = ma$ at every point along the bullet's arc, but you can find the answer more easily if you know that kinetic energy is $(1/2)mv^2$ and gravitational potential energy is mgh. Just set the total energy at the top equal to the total energy the bullet started with and solve for your answer without worrying about the trajectory at all.

If you factor in air resistance, the whole nature of the calculation changes. Now if you want to know how fast the bullet will hit the cliff, you have to take into account the specific path it will travel to get there. We are not saying that potential energy is more complicated in this case; we are saying that the whole concept of a "potential energy for air resistance" makes no sense.

So when we say that the gravitational force is conservative and air resistance is not, we are making three separate but related statements.

- The gravitational force assigns to each object a potential energy based on its position. Air resistance does not.
- An object moving only under the influence of gravity conserves mechanical energy (kinetic plus potential). An object subject to air resistance does not.
- The work done by gravity in moving an object from one point to another—that is, the kinetic energy imparted to the object by gravity—is independent of the path taken between these two points. But the work done by air resistance varies from one path to another.

To relate these points to our definitions of a "conservative field" we use the "work-energy theorem," which tells us that the total work done on an object is the change in kinetic energy of the object.

$$\int_C \vec{F}_{total} \cdot d\vec{s} = \Delta \text{KE} \qquad (8.11.1)$$

Equation 8.11.1 holds for all forces. In the case of non-conservative forces, the result of the calculation depends on the path taken. In the case of conservative forces the calculation depends only on the starting and ending locations. By Definition 2, this means $\vec{F}$ is a conservative vector field if and only if it represents a conservative force. That in turn implies our second bullet point above, namely that for a conservative force we can uniquely define a scalar field PE by the equation $\vec{F} = -\vec{\nabla} PE$.

EXAMPLE: Work, Potential Energy, and Conservative Forces

Question: A mass m is subject to Earth gravity $\vec{F} = -mg\hat{j}$, using y for vertical position and x for horizontal. Find the potential energy at the point (w, h).

Answer:
The question is ambiguous until we choose a point to represent $PE = 0$; we choose (purely for convenience) the origin. The work done in moving our object from $(0, 0)$ to (w, h) is its gain in kinetic energy, and therefore its loss in potential energy. We will calculate the work three times, using the different paths in the picture below.

Since the force is vertical, $\vec{F} \cdot d\vec{s}$ is zero on the horizontal paths.

- Path 1: $W = 0 - mgh = -mgh$
- Path 2: $W = -mgh + 0 = -mgh$
- Path 3: The line is $y = (h/w)x$, so $dy = (h/w)dx$ and the work is $\int_0^w (-mgh/w)\, dx = -mgh$

No matter what path we use to calculate it, we get the same answer. The work done by gravity as the mass m moves from the origin to (w, h) is $-mgh$, so the object's potential energy at (w, h) is mgh.

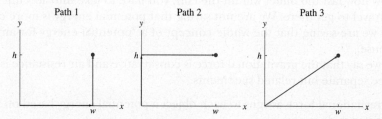

Path 1   Path 2   Path 3

Question: A mass is subject to a different force, $\vec{F} = -kx^2 \hat{j}$. Before you read any further, we encourage you to try calculating PE at the point (w, h) for this new force the same way we did above for gravity.

Answer:
Once again the horizontal segments of the path contribute nothing to the work since the force is entirely vertical.

- Path 1: The force is $-kw^2 \hat{j}$, so the work is $-kw^2 h$
- Path 2: $x = 0$ so $F = 0$ and the work is zero
- Path 3: the work is $W = \int_0^w (-kx^2 h/w)\, dx = -kw^2 h/3$

What happened? The second force is not conservative: the amount of energy it imparts to an object depends on the path along which the object moves. It is therefore impossible to associate a potential energy with that force.

In introductory physics classes the rule for figuring out if a force is conservative is very simple: if the teacher gave you a formula for the potential energy associated with that force, then it must be a conservative force. But to mathematically determine if a particular force is conservative, you need vector calculus.

Those Other Definitions
Above we found the work done by $\vec{F} = -kx^2 \hat{j}$ along three different paths between the same points, and we got different answers. We thereby proved by Definition 2 that this force is not conservative.

We also found the work done by $\vec{F} = -mg\hat{j}$ along three different paths between the same points, and we got the same answer. So it's starting to look conservative, but to really prove it we would need to calculate the line integrals between every possible pair of points in the universe along every possible path. You might be forgiven if you checked a few and assumed that the rest would work, but you might also get the wrong answer.

So here is an alternate definition: a vector field is conservative if its line integral around any *closed loop* (i.e., a path that ends where it started) is zero. The notation $\oint$ is used for such an integral, so $\oint \vec{f} \cdot d\vec{s} = 0$ is Definition 3 of a conservative field. You'll show in Problem 8.184 that this definition is equivalent to the previous one.

That may not sound like a big improvement—you still cannot test every possible closed loop—but Stokes' theorem turns it into a practical test. The circulation of a vector around a closed loop equals the integral of the curl on a surface bounded by that loop. If the circulation is zero around *every* closed loop, then the curl must be zero everywhere. The converse is also true; if the curl of a field is zero everywhere then the circulation around any closed loop must be zero and the field is conservative. Starting from Definition 3, we've thus derived Definition 4. This one gives us a practical test, however. A vector field is conservative if and only if its curl is zero everywhere.

Except it doesn't quite work. In Problem 8.188 you will explore a field that violates Definition 4 and seems to also violate Definition 1. It is the gradient of a scalar function, its curl is zero, and it is manifestly *not* conservative. The problem is not a contrived mathematical curiosity, but a common physical situation, the magnetic field around a straight wire. So exactly how equivalent are these definitions?

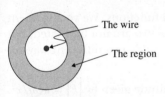

FIGURE 8.22 A region around a current-carrying wire.

Before we can answer that question, we have to raise one technical issue: "conservative" is a property of a field *in a given region of space*. To say that a field is conservative (or non-conservative) at a particular point is meaningless, since both line integrals and gradients require some elbow room. In the case of the magnetic field around a wire, we cannot consider "the whole universe" as our region because $\vec{B}$ blows up at $\rho = 0$, which is to say, on the wire itself. So we ask the question: is the magnetic field conservative on a disk-shaped region around the wire (Figure 8.22)?

In Problem 8.188 you'll find a potential function for $\vec{B}$ that seems to work everywhere in this region, but look at it carefully: even though the magnetic field is defined throughout the region, its potential function is not. You cannot write $\vec{B} = \vec{\nabla} V$ everywhere in the region, so Definition 1, along with Definitions 2 and 3 (as you'll also show), concludes that the magnetic field is not conservative.

Definition 4 is more problematic, as the curl is definitely zero everywhere within this region. This definition only applies when the region in question is "simply connected," however. In 2D that roughly means that the region doesn't have any holes in it. Because this region clearly has a hole in it, Definition 4 does not give us a guarantee.

Why does that hole matter so much? Remember that Definition 4 is directly related to Definition 3 by Stokes' theorem: the line integral around a closed path equals the total curl integrated along any surface bounded by that path. To apply Stokes' theorem in this case we draw a closed loop inside our region (a circle around the wire) and then we draw a surface that is bounded by that loop. But here we run into trouble. Whether we draw a simple disk or an elaborate parachute, the wire—the place where $\vec{B}$ is undefined—is going to cut right through our surface. That means we can't apply Stokes' theorem in our region, which is why Definition 4 gets it wrong.

That leads us to a more precise definition of "simply connected": any closed loop in a simply connected region bounds at least one surface entirely within the region, which means you can apply Stokes' theorem in that region.[9] See Problem 8.175. In 3D this definition is a bit more complicated than "doesn't have any holes." For example, the region in between two concentric spherical shells is simply connected, while a doughnut is not. Make sure you understand why in both cases.

Bottom line? If you want to determine if a field is conservative within a given region, here is our recommendation.

> **Determining if a Field is Conservative Within a Given Region**
> Begin by taking the curl.
>
> 1. If the curl is not zero everywhere in the region, then the field is not conservative—guaranteed.
> 2. If the curl is zero everywhere in the region, and the region is simply connected, then the field is conservative—guaranteed.
> 3. If the curl is zero everywhere in the region, and the region is *not* simply connected, then you're not sure at this stage. In that case take one closed line integral around each of the holes in the region. If they all come out zero the field is conservative. If any of them comes out non-zero the field is not conservative by Definition 3.

Forces, Fields, Potential Energy, and Potential

We've spent a great deal of this chapter talking about potential functions. We've also spent a fair amount of this section talking about potential energy. The two are not unrelated, but they're not identical either. To understand the difference, consider two different ways of expressing Coulomb's law of electrostatics.

- A charge Q exerts a force on another charge q, with magnitude given by $|\vec{F}| = kQq/r^2$.
- A charge Q creates an electric field around itself with magnitude $|\vec{E}| = kQ/r^2$. Another charge q in that electric field experiences a force $\vec{F} = q\vec{E}$.

The second approach separates the process into two steps: first calculate the total electric field created by a constellation of charges, and then calculate the effect of that field on a given charge. The relationship between the electric force and the electric field is mirrored in the relationship between potential energy and electric potential.

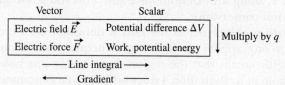

A charge Q creates an electric field (vector, $kQ/r^2\ \hat{r}$) and a potential field (scalar, kQ/r) around itself; these are only functions of position, not of what you put at that position. If you place a charge q into this region it will experience a force (vector, $kQq/r^2\ \hat{r}$) and have a certain potential energy (scalar, kQq/r); these will depend on both the field and the charge q of the particle.

Everything we've said here about electricity works exactly the same with gravity if we replace k with G, write mass instead of charge, and put a minus sign in front of each of the formulas. (The minus sign is because two positive charges repel but two positive masses, which are the only kind there are, attract.)

---

[9]The rigorous definition of "simply connected" also says it can't consist of two disconnected regions, such as two separated disks. This part of the definition is irrelevant for conservative fields since you can still apply Stokes' theorem in such regions.

Stepping Back

If you were to arbitrarily choose three functions of x, y, and z, and call them the three components of a vector field $\vec{v}$, the odds are good that you would *not* end up with a conservative field. For instance, suppose $v_x = x^2 y^2$. If we want $\vec{v} = \vec{\nabla} F$ for some scalar field, that scalar field must have the form $F = (1/3)x^3 y^2$ plus some function of y. That, in turn, means that $v_y = (2/3)x^3 y$ plus some function of y. But you don't get to pick v_y, so you just have to hope it works out! The point is that v_x and v_y (and v_z in three dimensions) have to carefully match each other to give a conservative field.

But if conservative fields are a very restrictive special case, they are also a very important one. They include important real-world examples such as gravitational and electrostatic fields. And once you know that a field is conservative, you can use its potential function and the gradient theorem to simplify calculations tremendously. If the field represents a force, you also know that it will conserve mechanical energy.

8.11.3 Problems: Conservative Vector Fields

8.172 The function $\vec{v} = (1/2)\cos(x/2+y)\hat{i} + \cos(x/2+y)\hat{j}$ is conservative.

(a) *Without* using the gradient theorem, evaluate the line integral of this function from $(0,0)$ to $(\pi/2, \pi/2)$...

  i. ...along a path that goes up to $(0, \pi/2)$ and then over to $(\pi/2, \pi/2)$.

  ii. ...along a path that goes over to $(\pi/2, 0)$ and then up to $(\pi/2, \pi/2)$.

  iii. ...along a single diagonal line.

(b) Confirm that $\vec{v}$ is the gradient of the function $F(x,y) = \sin(x/2+y)$.

(c) Evaluate the line integral from Part (a) using the gradient theorem.

8.173 An infinite wire carrying charge density λ Coulombs/m creates an electric field with magnitude $|\vec{E}| = \lambda/(2\pi\rho\varepsilon_0)$ pointing directly away from the wire, where ρ is the distance to the wire. (You can verify this electric field using Gauss's law, or just take our word for it.)

(a) Show that $\vec{\nabla} \times \vec{E} = \vec{0}$, meaning the field is "irrotational."

(b) Usually you can assume an irrotational field is conservative. Why can you *not* assume that in this case? That means that to show that this field is conservative you will need to find a potential function for it.

(c) Show that you *cannot* take $V = 0$ at infinity. (Try it.)

(d) Write a function $V(\rho)$. Make sure to indicate your choice of a $V = 0$ location.

(e) Verify that $\vec{\nabla} V = -\vec{E}$, thus proving that this field is conservative.

8.174 The xy-plane has constant charge density σ, creating an electric field $\vec{E} = \sigma/(2\varepsilon_0)\hat{k}$ in the region $z > 0$. (You can verify this electric field using Gauss's law, or just take our word for it.)

(a) Show that this is a conservative field in that region.

(b) Show that, although you can define a potential function for this electric field (you just proved that), you *cannot* take $V = 0$ at infinity.

(c) Write a function V. You will have to define the variable(s) it depends on, and where to choose $V = 0$.

8.175 Which of the following regions is simply connected?

(a) All of 3D space

(b) The xy-plane except for the origin

(c) All of 3D space except for the origin

(d) The set of all points with $z > 0$ and distance from the origin between 1 and 2 (half a spherical shell)

(e) The set of all points with distance from the z-axis between 1 and 2 (a cylindrical shell)

8.176 Explain why a spherical shell is simply connected and a doughnut is not.

In Problems 8.177–8.183 step through the process outlined on Page 442 to determine if the given field is conservative in the specified region. You should always start at Step 1 and take the steps in order; however, you will not always need to take all three steps.

8.177 $\vec{v} = 4\hat{i} - 7\hat{j} + 2\hat{k}$ over all space

8.178 $\vec{v} = 3x\hat{i} + 4y\hat{j}$ over the entire xy-plane

8.179 $\vec{v} = 3y\hat{i} + 4x\hat{j}$ over the entire xy-plane

8.180 $\vec{v} = (2x/y)\hat{i} - (x^2/y^2)\hat{j}$ for $y > 1$

8.181 $\vec{v} = \hat{\phi}$ over the entire xy-plane

8.182 $\vec{v} = \phi\hat{\rho} + \hat{\phi}$ over a unit disk in the xy-plane centered on the origin. *Hint*: Where is this field undefined and why?

8.183 $\vec{v} = [z\cos(xz) - y/x]\hat{i} - (\ln x)\hat{j} + x\cos(xz)\hat{k}$ for $x > 0$

8.184 Page 438 contains a number of different definitions of conservative. Show that Definitions 2 and 3 are equivalent. The picture below may help. Consider the fact that the line integral from B to A along path 2 is *minus* the line integral from A to B along that path.

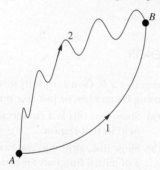

8.185 If a vector field has the property that $\vec{\nabla} \times \vec{v} = \vec{0}$, the field is said to be "irrotational." Page 438 presents this as a definition of a conservative vector field, which is mostly—but not entirely—accurate. Show that if Definition 1 is true for some vector field, then Definition 4 must also be true.

8.186 Prove that any constant vector field is conservative.

8.187 A charge Q creates an electric field around it with magnitude $|\vec{E}| = kQ/r^2$ where k is a positive constant and r is the distance to the charge. The direction of this field is radially outward from the charge. *Hint*: each part of this problem should involve less algebra than the part before.

(a) Write a vector in Cartesian coordinates that represents the electric field from a point charge Q at the origin. Show that this vector field is conservative.

(b) Write a vector in spherical coordinates that represents the electric field from a point charge Q at the origin. Show that this vector field is conservative.

(c) Does this imply that the total electric field from any collection of point charges is conservative? Why or why not?

8.188 A straight wire carrying a constant current I creates a magnetic field around itself given by $\vec{B} = (\mu_0 I/(2\pi\rho))\hat{\phi}$ where μ_0 is a positive constant.

(a) Show that $\vec{B}$ is *not* a conservative field by showing that $\oint \vec{B} \cdot d\vec{s}$ around the loop $\rho = 1$ is not zero.

(b) Show that $\vec{B}$ is irrotational. Why, in this case, does this not imply a conservative field?

(c) Show that $\vec{B} = \vec{\nabla}S$ where $S = \mu_0 I\phi/(2\pi)$. Why, in this case, does this not imply a conservative field?

8.189 In this problem you will replicate the work from Problem 8.188 in Cartesian coordinates, but you do *not* need to have done Problem 8.188 to do this one. A straight wire carrying a constant current I creates a magnetic field around itself given by:

$$\vec{B} = \frac{\mu_0 I}{2\pi}\left(\frac{-y}{x^2 + y^2}\hat{i} + \frac{x}{x^2 + y^2}\hat{j}\right)$$

where μ_0 is a positive constant.

(a) Show that $\vec{B}$ is *not* a conservative field by showing that $\oint \vec{B} \cdot d\vec{s}$ around the loop $x = \cos t$, $y = \sin t$ is not zero.

(b) Show that $\vec{B}$ is irrotational. Why, in this case, does this not imply a conservative field?

(c) Show that $\vec{B} = \vec{\nabla}S$ where $S = \mu_0 I/(2\pi)\tan^{-1}(y/x)$. Why, in this case, does this not imply a conservative field?

8.12 Additional Problems (see felderbooks.com)

CHAPTER 9

Fourier Series and Transforms

> *Before you read this chapter, you should be able to...*
> - identify the amplitude, frequency, period, and phase of the function $A\cos(px + \phi)$.
> - represent oscillatory functions with complex exponentials.
> - represent series, finite or infinite, with $\sum$ notation.
>
> (You don't need to know about Taylor series to read this chapter, but we compare and contrast them to Fourier series in several places. If you're not familiar with Taylor series you'll have to ignore those bits.)
>
> *After you read this chapter, you should be able to...*
> - use a Fourier series to represent a periodic function as a sum of sines and cosines, or equivalently as a sum of complex exponentials.
> - determine if a set of functions is "orthogonal" and if so find the formula for the coefficients of a series built from this set.
> - create a Fourier series for a function defined on a limited domain by using an "even extension" to create a "Fourier cosine series," an "odd extension" to create a "Fourier sine series," or an extension that is neither even nor odd to create a series with sines and cosines.
> - use a Fourier transform to represent a non-periodic function as an integral of complex exponentials with different amplitudes.
> - determine the basic properties of a function from a plot of its Fourier transform, or of a Fourier transform from the plot of the function it represents.
> - use a discrete Fourier transform to model a function defined by a set of data points.
> - create a multivariate Fourier series for a function of more than one variable.

Complicated behavior can often be modeled as the sum of different sine waves of different frequencies. Think of a car going up and down hills. A rock stuck on one of the tires goes up and down quickly due to the rotation of the tire and also goes up and down more slowly due to the hills.

A "Fourier series" mathematically represents a function such as the height of the rock over time as a superposition of individual sine waves. Such a device can reveal the mathematical order underlying seemingly chaotic functions, especially in the case of periodic behavior. Less obviously, Fourier series are also a powerful tool for solving differential equations by adding together different sinusoidal solutions.

9.1 Motivating Exercise: Discovering Extrasolar Planets

We typically imagine the Earth orbiting a stationary sun, but that is only an approximation. While the sun pulls the Earth around, the Earth pulls the sun around too, so they orbit together around their mutual center of mass. The period of this rotation is of course 365 days, so its angular frequency is $\omega = 2\pi/365$ day^{-1}.

Astronomers therefore look for small, regular oscillations in other stars as a way of determining if they have their own planets. Using this technique, Mayor and Queloz in 1995 announced what is now considered the first definitive detection of a planet orbiting a star other than our sun.[1]

In a simpler world we would see a sine wave as the star moves back and forth, and the period of that sine wave would be the orbital period of the planet. But real data is messier because stars can have more than one planet, and because effects such as atmospheric distortion create "noise" in the data.

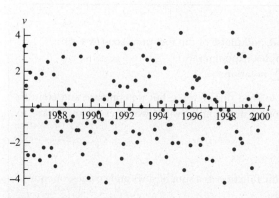

FIGURE 9.1 Velocity of a star vs. time.

Suppose you had been measuring a star carefully for 14 years and collected the data shown in Figure 9.1. Does this star have planets going around it? How many and with what periods? Or did your five-year-old just scribble random dots on your page of data? It's virtually impossible to tell. You need to know if there are regular sinusoidal variations hidden within the random ups-and-downs of those data points.

"Fourier analysis" is designed to answer that question. It allows you to take a function of time and figure out how much it varies with different frequencies. Figure 9.2 shows the "Fourier transform" $\hat{v}(\omega)$ of the data points $v(t)$ shown in Figure 9.1.

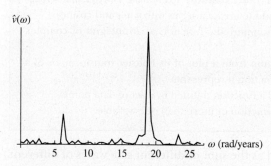

FIGURE 9.2 The "Fourier transform" shows how much variation $v(t)$ has at different frequencies.

It is vital to understand that Figure 9.2 does not represent any new data; it is calculated by analyzing the data in Figure 9.1. You will learn in Section 9.7 (see felderbooks.com) how to perform this analysis using a "discrete Fourier transform." Right now we are handing you Figure 9.2, but you can still interpret it.

1. The Fourier transform shows a sharp spike at approximately $\omega = 19$. This indicates that your star's velocity has a high-amplitude oscillation with angular frequency 19. The only likely explanation is a planet orbiting the star. What is the period of that planet's rotation? (That is, how long is a "year" on that planet?)

   See Check Yourself #57 in Appendix L

2. The Fourier transform also shows a spike at roughly $\omega = 6$, indicating a second planet. How long is a year on that planet?
3. Which planet do you think has the largest mass (and hence the biggest effect on the star's velocity)? Explain how you got your answer from Figure 9.2.

---

[1]M. Mayor, D. Queloz (1995). "A Jupiter-mass companion to a solar-type star". Nature 378 (6555): 355–359. Other claims of "exoplanets" had been claimed earlier, but were considered more controversial. In 1992 Wolszczan and Frail announced the detection of two planets orbiting a pulsar, but Mayor and Queloz's discovery was the first clear planet orbiting a regular star.

9.2 Introduction to Fourier Series

A piano player hits five keys at once, sending a complicated cascade of pressure waves to your ear. A planet spins around its own axis while simultaneously rotating around the sun. Electromagnetic waves carry the broadcasts of dozens of radio stations, all converging on your car antenna at the same time.

What all these scenarios have in common is the superposition of sine waves of different frequencies. When you add $\sin(2x)$ and $\sin(3x)$ together, the sum is not a sine wave; throw in a few more frequencies and the resulting function starts to look almost random. It is often important to *start* with such a sum and work backwards to find the individual frequencies from which it is built. The result—a given function rewritten as a combination of sine waves—is called a "Fourier series."

9.2.1 Discovery Exercise: Introduction to Fourier Series

When you learned about Taylor series you learned that most normal functions can be written as sums of powers of x: $f(x) = c_0 + c_1 x + c_2 x^2 + \ldots$. In this chapter you're going to learn to write functions as sums of sines and cosines, which turns out to be useful in different circumstances. To get you started, let's see what some simple sums of cosines look like. In the first couple of parts you'll generate your own plots, and in the next couple you'll answer questions about the plots below.

1. Sketch $\cos(x)$ and $\cos(10x)$ on the same plot. (Choose a range of x-values big enough to see the behavior of each function clearly.)
2. Sketch $\cos(x) + .1\cos(10x)$. You can see how to do this by taking your $\cos(x)$ sketch and adding small wiggles to it for the $\cos(10x)$ term.
3. The left plot in Figure 9.3 is the sum of two different cosine functions. Estimate the amplitude and period of both cosines. To write this in the form $A\cos(p_1 x) + B\cos(p_2 x)$ remember that the period of $\cos(px)$ is $2\pi/p$.

   See Check Yourself #58 in Appendix L

4. Similarly, try to identify the two cosine waves that make up the right-hand plot in Figure 9.3.

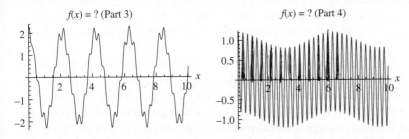

FIGURE 9.3 Each plot is a sum of two cosines. Use these in answering Parts 3 and 4.

The examples we've used so far looked fairly simple because one wave had a much higher amplitude than the other. When you add waves with similar amplitudes, the results tend to look messier.

5. Make a plot of $\cos(5x) + \cos(6x) + \cos(7x)$ and copy it onto your paper. It's a good way to start to see that even simple combinations of cosines can lead to pretty complicated behavior.

9.2.2 Explanation: Introduction to Fourier Series

When you hear a sound, you are detecting oscillations in the air pressure by your eardrum. The frequency of such an oscillation determines the pitch of the sound. For instance, a 262 Hz sine wave means one that oscillates between high and low pressure 262 times each second; that corresponds roughly to the note that piano players call "middle C." Mathematically this would be represented by $P(t) = A\sin(2\pi(262\ t)) \approx A\sin(1646\ t)$, where the amplitude A determines the volume of the sound.

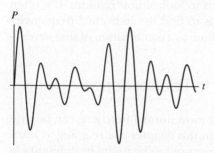

FIGURE 9.4 $P(t) = \sin(2\pi(262\ t)) + \sin(2\pi(330\ t)) + \sin(2\pi(392\ t))$.

Now imagine three notes reaching your ear at once: $\sin(2\pi(262\ t))$, $\sin(2\pi(330\ t))$, and $\sin(2\pi(392\ t))$. Each note is a simple sine wave—the notes C, E, and G, respectively—but Figure 9.4 shows what they look like when you put them together. Virtually nobody could tell by eye what sine waves we used to make this graph, but with musical training you can learn to identify by ear the intervals between the notes. A "Fourier series" does mathematically what your auditory cortex accomplishes biologically; it decomposes a single function into the frequencies and amplitudes of the sine waves that make it up.

In some ways a Fourier series is like a Taylor series: you are given an arbitrary function and your job is to rewrite it by finding the coefficients in an infinite series of other functions. But instead of those other functions being x, x^2, and so on, they are sines and cosines of different frequencies.

$$f(x) = a_0 + a_1 \cos(x) + b_1 \sin(x) + a_2 \cos(2x) + b_2 \sin(2x) + a_3 \cos(3x) + b_3 \sin(3x) + \ldots \tag{9.2.1}$$

In a Taylor series the first term is c_0, a constant, which you can think of as the coefficient of x^0; all other terms are higher powers of x. In a Fourier series the first term is a_0, a constant, which you can think of as the coefficient of $\cos(0x)$; all other terms are sines and cosines.

We shall return to the comparison with Taylor series at the end of this section, but we want to emphasize two important differences up front.

- Fourier series are used to model *periodic* phenomena. You can see from looking at Equation 9.2.1 that $f(x) = f(x + 2\pi)$ for any x-value. In Section 9.4 we will modify Equation 9.2.1 to accommodate periods other than 2π but the resulting series will still be periodic.
- When we find a Taylor series we often make use of the first few terms and throw out the rest. We're finding a simple function, linear or possibly quadratic, that approximates a more complicated function near a given point. With a Fourier series we are less interested in the earliest terms than in the spikes: what frequencies dominate this function? In our musical example, a real piano chord contains a host of higher frequencies ("harmonics" or "overtones"); Fourier analysis will find all of them for you, but it will also tell you that the peak amplitudes are at 262, 330, and 392 Hz.

EXAMPLE   The Use of a Fourier Series

Question: If $P(t)$ represents the power consumption from your home, a graph of $P(t)$ would be an almost unreadable jumble. But if you represented $P(t)$ as a Fourier series, which terms would have high amplitudes? (Assume t is measured in days.)

Answer:
One of the terms in your Fourier series represents oscillation with a period of 1 day. The coefficient of that term—that is, the amplitude associated with that frequency—would be very high. Of course the use of heat, electricity, and hot water is not perfectly periodic (it is not identical from one day to the next) but there are strong predictable daily trends (more usage at 8:00 AM than at 3:00 AM). The high amplitude on this term tells us that $P(t)$ has a predictable 1-day cycle.

There would also be a spike representing a period of one year: you use more heat in the winter than in the spring. So you might reach a reasonable approximation by writing:

$$P(t) = A\sin(2\pi t) + B\sin\left(\frac{2\pi}{365}t\right) \tag{9.2.2}$$

Equation 9.2.2 could never give you a perfectly accurate representation of your power usage. There are undoubtedly many shorter and longer cycles involved, as well as a fair bit of randomness. But if those terms with the right coefficients give you a fairly good approximation, that might enable the power company to create the models they need for prediction and planning purposes.

Give yourself extra credit if you noticed a problem with this example. Your power usage involves highs and lows that never repeat, but a Fourier series is always perfectly periodic. If you built your series from data collected over a ten-year period then the resulting Fourier series would repeat that data for decade after decade. You would therefore see a ten-year period (that does not reflect your actual power usage) as well as the annual and daily cycles (that do). We will return to the issue of Fourier series based on data from a finite domain in Section 9.4.

A Fourier series with sines only (such as the one we wrote in that example), or cosines only, is fairly easy to interpret. But many functions require sine and cosine terms together. (You'll see why in "Even and Odd Functions" below.) This makes interpreting the coefficients a bit trickier.

For instance, suppose you evaluate the Fourier series for a function $f(x)$ and find the following four coefficients. (You also find many other coefficients, but let's just talk about these four.)

$$a_8 = 6 \quad b_8 = 0 \quad a_9 = 3 \quad b_9 = -4$$

The first two numbers tell us about the part of $f(x)$ that oscillates with angular frequency $p = 8$. That part looks like $6\cos(8x)$. Its amplitude of 6 represents the importance of this particular frequency in the overall makeup of $f(x)$.

The other two numbers tell us that the $p = 9$ part of $f(x)$ looks like $3\cos(9x) - 4\sin(9x)$. You will show in Problem 9.24 that the nth frequency terms, $a_n\cos(nx) + b_n\sin(nx)$, represent oscillation with amplitude $\sqrt{a_n^2 + b_n^2}$. So this particular frequency has an amplitude of 5: slightly smaller than the first one.

This result means that $3\cos(nx) + 4\sin(nx)$, $4\cos(nx) + 3\sin(nx)$, and $5\cos(nx)$ all represent oscillations with the same frequency *and* amplitude. The only difference is in the phase. You can compute that phase difference if you need it, but usually it's the amplitudes that matter.

The a_0 term is unique. The sine and cosine terms average out to zero, so when you write a function as a Fourier series the average value of the series is a_0. Turning that around, when you write a Fourier series for a function $f(x)$, a_0 is the average value of $f(x)$. If f is symmetric above and below the y-axis then $a_0 = 0$.

Period and Wavelength, Frequency and Angular Frequency

The graph $y = \sin(3x)$ has a period of $2\pi/3$. Mathematically, this tells us that $y(x) = y(x + 2\pi/3)$ for all x-values; graphically, it tells us that the entire shape repeats itself every $2\pi/3$ x-units. But the physical meaning of the period depends on the meaning of the variables.

Imagine an outlet that delivers voltage $V = 120 \sin(60t)$. Because the independent variable is time, the period of this wave refers to the time elapsed between one peak and the next. Now imagine a standing wave of air pressure $P = 100 \sin(6x)$. This independent variable is position, so the period is the *distance* from one peak to the next, also known as the wavelength. For a spatial wave the terms "period" and "wavelength" are interchangeable.

In Section 9.8 (see felderbooks.com) we will discuss waves that vary in both space and time. In such cases we use the word "wavelength" for the spatial distance between peaks, and generally reserve "period" for the time between peaks.

Returning to our example of the graph $y = \sin(3x)$, there is an unfortunate ambiguity in the word "frequency." We can say that this graph covers 3 radians per x-unit, or we can say that it covers $3/(2\pi)$ full cycles every x-unit. In proper mathematical language the former is called the "angular frequency" and the latter is simply the "frequency." When we spoke above about the note C being represented by $A\sin(2\pi(262\ t)) = A\sin(1646\ t)$, the number 1646 was the angular frequency and 262 the frequency. (The unit "Hertz" means cycles per second, so we can say in this case the angular frequency is 1646 rad/s and the frequency is 262 Hz.)

Following a convention common among physicists and engineers, we will use the shorter term "frequency" when we actually mean "angular frequency" (radians). On the rare occasion that we actually mean the other one (cycles), we'll let you know.

Finally, a few words about letters. When a sine wave is a function of time, its frequency[2] is usually written with the Greek letter ω; when it's a function of space it is typically called p or k. We mostly use p. The "cycles" frequency is generally labeled with either f or the Greek letter ν. We use f. Finally, we use T for period (in time) and λ for wavelength.

$$\text{Time: } \omega = \frac{2\pi}{T} \qquad \text{Space: } p = \frac{2\pi}{\lambda} \qquad f = \frac{\omega}{2\pi} \text{ or } \frac{p}{2\pi}$$

How to Find a Fourier Series

Equation 9.2.1 represents any Fourier series with a period of 2π. Below we give the formulas for the coefficients of such a series. In Section 9.3 we will show where those formulas come from, and in Section 9.4 we'll generalize to functions with periods other than 2π.

The Formula for a Fourier Series

The Fourier series for a function $f(x)$ with period 2π is:

$$f(x) = a_0 + \sum_{n=1}^{\infty} a_n \cos(nx) + \sum_{n=1}^{\infty} b_n \sin(nx)$$

The coefficients are given by the equations:

$$a_0 = \frac{1}{2\pi} \int_{-\pi}^{\pi} f(x)\, dx \qquad a_n = \frac{1}{\pi} \int_{-\pi}^{\pi} f(x) \cos(nx)\, dx \qquad b_n = \frac{1}{\pi} \int_{-\pi}^{\pi} f(x) \sin(nx)\, dx \qquad (9.2.3)$$

The integrals are written here from $-\pi$ to π, but because the function is periodic you can take these integrals over any interval of length 2π and get the same result.

---

[2] ... by which we mean angular frequency, but this is the last time we'll say so.

9.2 | Introduction to Fourier Series

As we always say, an example is clearer than an explanation.

EXAMPLE **Creating a Fourier Series**

Question: The function $f(x)$ is defined to be $x^2 + 3$ in the region $-\pi < x < \pi$ and to repeat periodically everywhere outside that, as shown on the right. Find the Fourier series for $f(x)$.

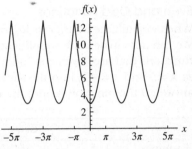

Answer:
Equation 9.2.3 gives us $a_0 = (1/(2\pi)) \int_{-\pi}^{\pi} (x^2 + 3) dx$, $a_n = (1/\pi) \int_{-\pi}^{\pi} (x^2 + 3) \cos(nx) dx$ and $b_n = (1/\pi) \int_{-\pi}^{\pi} (x^2 + 3) \sin(nx) dx$. The first integral is easy, and a_0 comes out $\pi^2/3 + 3$. You can evaluate the other two integrals by parts—remember to treat n as a constant, and integrate with respect to x—or you can use a computer. The result is:

$$a_n = \left(\frac{1}{\pi}\right) \frac{2\left((3+\pi^2)n^2 - 2\right)\sin(\pi n) + 4\pi n \cos(\pi n)}{n^3}$$

Not particularly appealing, is it? But remember that n is our series index, so it's always an integer. That means $\sin(\pi n) = 0$ and $\cos(\pi n) = (-1)^n$. (If that's a new fact to you, take a moment to think about the sine and cosine of π, 2π, 3π and so on.) Meanwhile, integrating $\int_{-\pi}^{\pi} (x^2 + 3) \sin(nx) dx$ gives us $b_n = 0$. We conclude that:

$$f(x) = \frac{\pi^2}{3} + 3 + \sum_{n=1}^{\infty} \frac{(-1)^n 4}{n^2} \cos(nx)$$

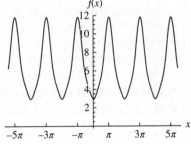

When we asked a computer to graph the 100th partial sum of this series, here's what we got. The series does a good job of replicating $x^2 + 3$ between $-\pi$ and π, and repeats forever from there.

What Functions can be Expressed as Fourier Series?

The set of functions $\sin(nx)$ and $\cos(nx)$ for integer n-values is said to be a "complete" set of functions on the interval $[-\pi, \pi]$. Roughly speaking, this means that you can build almost any function with period 2π as a series from $\sin(nx)$ and $\cos(nx)$ functions. In some senses sines and cosines prove much more flexible than polynomials: a Taylor series can only represent a function on an interval where that function is differentiable, but a Fourier series can represent non-differentiable and even discontinuous functions!

More specifically, you can build a Fourier series for any function $f(x)$ that meets the following three requirements (the "Dirichlet conditions"):

- $f(x)$ must be periodic. (The formulas above assume a period of 2π; we will generalize to arbitrary periods in Section 9.4.)
- $f(x)$ must not have an infinite number of maxima, minima, or discontinuities within one period.
- Finally, $\int |f(x)| dx$ over one full period must be finite.

Virtually every periodic function that would arise in a physical problem satisfies these conditions. Under those circumstances "Dirichlet's theorem" tells us that wherever $f(x)$ is continuous, the Fourier series will converge to $f(x)$. Perhaps more remarkably, if $f(x)$ has a jump discontinuity at some value of x, the Fourier series will converge to the average of the values of f on either side of the jump.

Even and Odd Functions

When we built the Fourier series for $x^2 + 3$ above, we found that all the b_n terms went to zero. We can anticipate such simplifications, and thus save ourselves an integral, by recognizing functions that are "even" or "odd." An easy test is to plug a number and its negative, such as $x = 5$ and $x = -5$, into the same function. If the function is even, you will get the same answer; if the function is odd, the two answers will be opposite.

	Even	Odd	Neither
Example	x^2	x^3	2^x
Demonstration	$5^2 = 25, (-5)^2 = 25$	$5^3 = 125, (-5)^3 = -125$	$2^5 = 32, 2^{-5} = 1/32$
Definition	$f(-x) = f(x)$	$f(-x) = -f(x)$	
Other examples	$x^4, |x|, \cos x, k$	$x, x^5, \sin x$	$x^2 + x$

When you graph an even function, the left side is a mirror image of the right. For an odd function the left is an *upside-down* mirror of the right. By thinking about the "plug in both 5 and -5" test and/or the graphical symmetry, you can convince yourself of the following rules.

- The sum of even functions is an even function; the sum of odd functions is an odd function. When you add an even function to an odd function, the result generally has neither symmetry.
- The product or quotient of two even functions is an even function; the product or quotient of two odd functions is an even function. When you multiply or divide an even and an odd function, the result is an odd function.
- If $f(x)$ is an odd function, then $\int_{-a}^{a} f(x)dx = 0$.
- If $f(x)$ is an even function, then $\int_{-a}^{a} f(x)dx = 2\int_{0}^{a} f(x)dx$.

The above rules apply to any even and odd functions, but they have specific implications for Fourier series.

- If $f(x)$ is even then its Fourier series will have no sine terms. ($b_n = 0$ for all n, as we found for $x^2 + 3$ above.) To find its cosine terms you can integrate over a full period as we did above, or you can save a little time by integrating over *half* the period and doubling the result. (Why? Because if $f(x)$ is even, then $f(x) \cos x$ is also even.)
- If $f(x)$ is odd then its Fourier series will have no cosine or constant terms. ($a_n = 0$ for all n, including a_0.) To find its sine terms you can integrate over a full period or integrate over half the period and then double the result.
- If $f(x)$ is neither even nor odd, then its Fourier series will generally have sines and cosines.

You will prove the most important of these conclusions in Problems 9.28 and 9.29.

Stepping Back

As promised, we conclude our introduction by returning to the analogy with Taylor series. When we write a Maclaurin series, we write a given function as an infinite series of x^n terms:

$$f(x) = c_0 + c_1 x + c_2 x^2 + c_3 x^3 + \ldots$$

Starting with the function $f(x)$, you figure out the coefficients. But you can't list them all, so you have to specify the coefficients as a *function of n*. For instance:

$$e^{-x} = \sum_{n=0}^{\infty} \frac{(-1)^n}{n!} x^n$$

Starting with the function e^{-x}, you can determine that the coefficient of x^n in the Maclaurin series is $(-1)^n/(n!)$. Given those coefficients, you can go the other way to figure out the function.

What we have just established, although you may never have thought about it this way, is a correspondence between the functions $f(x) = e^{-x}$ and $g(x) = (-1)^x/(x!)$. They carry the same information in different forms, and you can convert from either one to the other. But you can only make use of the information if you understand the subtle relationship between the two representations. The fact that $g(3) = -1/6$ does not tell us anything directly about $f(3)$; instead, it tells us how much "cubic function" is in the makeup of $f(x)$.

Similarly, we built a Fourier series above to conclude that:

$$x^2 + 3 = \frac{\pi^2}{3} + 3 + \sum_{n=1}^{\infty} \frac{(-1)^n 4}{n^2} \cos(nx) \qquad -\pi \leq x \leq \pi$$

Again, this establishes an important but subtle relationship between the functions $f(x) = x^2 + 3$ and $g(x) = (-1)^x 4/x^2$. Again, we can convert from one to the other. And again, it's important to understand what the relationship conveys: the fact that $g(3) = -4/9$ tells us how much "cosine with frequency 3" is in $x^2 + 3$.

For this reason we often speak about translating a function from "regular space" to "Fourier space" and back, a topic to which we will return in Section 9.6. In quantum mechanics, the wavefunction in Fourier space represents the probability distribution of momentum just as the wavefunction in regular space represents the probability distribution of position.

9.2.3 Problems: Introduction to Fourier Series

> Appendix G lists all the formulas for finding Fourier series coefficients. It also contains a brief table of some of the integrals that you will need most often in this chapter. Any problem that cannot be solved with u substitution, integration by parts, or with that table will be marked with a computer icon: .
>
> After you integrate remember that n (the summation index) is always an integer, and for any integer n:
>
> $$\sin(\pi n) = 0 \quad \text{and} \quad \cos(\pi n) = (-1)^n$$

9.1 The drawing below is the sum of two sine waves. Estimate their periods and amplitudes and use those estimates to write the two sine functions.

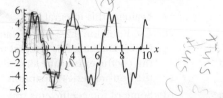

9.2 Sketch a graph showing the sum of two sine waves: one with a small amplitude and low frequency, the other with a much higher amplitude and frequency.

9.3 The function $D(x)$ is odd, and the function $E(x)$ is even. Furthermore, $D(5) = 2$ and $E(5) = 6$. Finally, $Q(x) = D(x)/E(x)$ and $S(x) = D(x) + E(x)$.

(a) $D(-5) =$

(b) $E(-5) =$

(c) $Q(5) =$

(d) $Q(-5) =$

(e) What can you conclude, in general, about the quotient of an odd and an even function?

(f) $S(5) =$

(g) $S(-5) =$

(h) What can you conclude, in general, about the sum of an odd and an even function?

9.4 Identify each function as even, odd, or neither.

(a) $(\sin x) + 3$

(b) $\sin(x + 3)$

(c) $\sin x \cos x$

(d) $(\sin x)/x$

(e) $x^2/(\sin x)$

(f) $\tan x$

(g) e^x

9.5 (a) Draw a simple even function $E(x)$ on the domain $-3 \leq x \leq 3$. (Do not use a constant function. It would work of course, but what fun is that?)

(b) Draw a simple odd function $D(x)$ on the same domain. (Do not use $D(x) = 0$.)

(c) Draw the function $E(x) + D(x)$. Is it even, odd, or neither?

9.6 The function $f(x)$ is shown below.

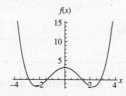

(a) Does this function appear to be even, odd, or neither?

(b) Sketch the derivative of this function. Does $f'(x)$ appear to be even, odd, or neither?

(c) Based on your answers, write a general statement about even and odd functions.

(d) Now prove it, using the definitions of even and odd functions.

In Problems 9.7–9.10 we give you a scenario that can be represented as a function. If this function is written as a Fourier series, there will be one spike—one high coefficient—at the period we indicate. In each problem ...

- Explain why you would see a spike at the period we indicate.
- Suggest one more period that would also have a high coefficient, and explain why.

For instance, the Explanation (Section 9.2.2) gave the example of power usage by a home, and explained why you would expect to see spike periods at 1 day and 1 year.

9.7 The automobile traffic through an intersection has a high term representing a period of 45 seconds.

9.8 The foot traffic through a high school hallway has a high term representing a period of 45 minutes.

9.9 The illumination along an interstate at night has a high term representing a period of 30 yards.

9.10 On a shoreline with a "semidiurnal tide" the sea level has a peak Fourier term with a period of 12 hours.

9.11 Make up a scenario like the ones in Problems 9.7–9.10 and give at least two periods that would appear prominently in its Fourier series.

9.12 **Walk-Through: Building a Fourier Series.** "Square waves" are frequently used in electronics and signal processing. An example is shown below.

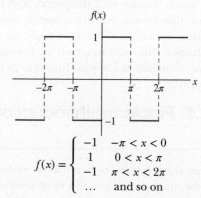

$$f(x) = \begin{cases} -1 & -\pi < x < 0 \\ 1 & 0 < x < \pi \\ -1 & \pi < x < 2\pi \\ \ldots & \text{and so on} \end{cases}$$

(a) Use Equation 9.2.3 to write an integral for a_0, the constant term in the Fourier series expansion. You should be able to evaluate this integral by just looking at the graph.

(b) Returning to Equation 9.2.3, the coefficients of the cosine terms lead us to a piecewise integral:

$$a_n = \frac{1}{\pi}\left(\int_{-\pi}^{0} -\cos(nx)\,dx + \int_{0}^{\pi} \cos(nx)\,dx\right)$$

Evaluate this integral.

(c) Use Equation 9.2.3 to write an integral for b_n. Evaluate this integral to find the coefficients of the sine terms in the Fourier series expansion.

(d) Some of the coefficients you calculated in Parts (a)–(c) should have come out equal to 0. Explain how you could have predicted some of those without doing any calculations, just by looking at the plot of the square wave.

(e) Write the Fourier expansion for this square wave.

9.13 💻 *[This problem depends on Problem 9.12.]* For the series you found in Problem 9.12, have a computer draw the 1st, 5th, 20th, and 100th partial sums. Sketch the results and verify that they are approaching the original square wave. What happens at the x-values for which the original function is discontinuous?

In Problems 9.14–9.20 assume $f(x)$ equals the given function from $x = -\pi$ to π, and is extended periodically beyond that. First, state whether $f(x)$ is even, odd, or neither. Based on that, which (if any) of the coefficients in the Fourier series do you know must be zero? Then calculate the remaining coefficients, simplifying your solution by replacing all occurrences of $\sin(\pi n)$ with 0 and $\cos(\pi n)$ with $(-1)^n$. Write the resulting Fourier series in the simplest possible form.

For problems not marked with a computer icon you should be able to do the integral by hand or with the table in Appendix G. In all cases we urge you to check your final answers by using computer graphs.

9.14 $f(x) = k$ (a constant)

9.15 $f(x) = \begin{cases} 0 & -\pi \leq x < 0 \\ 1 & 0 \leq x < \pi \end{cases}$

9.16 $f(x) = |x|$

9.17 $f(x) = 3x^2 - x$

9.18 $f(x) = x^3$

9.19 $f(x) = e^x$

9.20 The figure below shows a "triangle wave."

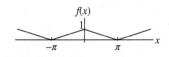

9.21 Find the Fourier series for $f(x) = \sin^2 x$.

9.22 💻 In the example on Page 451 we found the Fourier series for the function $f(x)$ that equals $x^2 + 3$ from $x = -\pi$ to π and is extended periodically from there.

(a) Find the Fourier series for the function $g(x)$ that equals $x^2 + 3$ from 0 to 2π and is extended periodically from there.

(b) Use a computer to draw the 100th partial sum of your series. It should *not* look like the drawing we found for $f(x)$. Why not?

9.23 The function below is the first four terms of a Fourier sine series with half-integer terms.

$$f(x) = b_1 \sin\left(\frac{1}{2}x\right) + b_2 \sin(x) + b_3 \sin\left(\frac{3}{2}x\right) + b_4 \sin(2x)$$

(a) What are the periods of the first, second, third, and fourth terms of this series?

(b) What is the period of the entire series?

9.24 A Fourier series represents each frequency with a sine and a cosine term, but what you typically want to know about each frequency is the amplitude of the wave at that frequency. To see how to get from one to the other, you're going to rewrite the terms $a_n \cos(nx) + b_n \sin(nx)$ as one term of the form $A \cos(nx + \phi)$, where A is the amplitude and ϕ is the phase.

(a) Use a trig identity to rewrite $A\cos(nx + \phi)$ as a sum
<something> $\cos(nx)$+<something>$\sin(nx)$.
Setting this equal to $a_n \cos(nx) + b_n \sin(nx)$, find a_n and b_n in terms of A and ϕ.

(b) Solve for A and ϕ in terms of a_n and b_n.

(c) If the Fourier series of a function $f(x)$ has $a_4 = 10$ and $b_4 = 5$, what are the amplitude and phase of the oscillation with frequency 4?

9.25 💻 Plot each function $g(x)$ given below on the same plot with the function $f(x) = 13\cos(2x)$. For each one describe how the plot of $g(x)$ is similar to or different from the plot of $f(x)$.

(a) $g(x) = 5\cos(2x) + 12\sin(2x)$

(b) $g(x) = 12\cos(2x) + 5\sin(2x)$

(c) $g(x) = 10\cos(2x) + 24\sin(2x)$

(d) $g(x) = 5\cos(4x) + 12\sin(4x)$

9.26 In this problem you will analyze the Fourier series for the function $f(x)$, where $f(x) = 1$ for $-\pi \leq x \leq \pi/2$ and $f(x) = -3$ for $\pi/2 < x \leq \pi$.

(a) Find the Fourier series for $f(x)$.

(b) What is the amplitude of the oscillation with frequency 1? You will need to consider both the sine and cosine terms, but your answer should be a single number.

(c) What is the amplitude of the oscillation with frequency n?

(d) How many frequencies do you need to include if you want a partial sum with all oscillations whose amplitude is at least 1/10 of the amplitude of the $n = 1$ oscillation?

9.27 Dirichlet's theorem guarantees that when a periodic function $f(x)$ has a jump discontinuity, the Fourier series for f evaluated at that point will converge to the average of the values on either side of the jump. This fact can be used to calculate certain infinite series.

(a) Find the Fourier series for $f(x) = \begin{cases} 1 & -\pi/4 < x < \pi/4 \\ 0 & -\pi < x < -\pi/4, \pi/4 < x < \pi \end{cases}$.

Write all the terms up to and including the $\cos(8x)$ term.

(b) Evaluate the Fourier series at $x = \pi/4$. Write your answer in the form $a + bS$, where a and b are numbers and S is the series $1 - (1/3) + (1/5) - (1/7) + \ldots$.

(c) It's not obvious how to calculate the sum S, but we can use this Fourier series to find the answer. Set the quantity $a + bS$ that you wrote down equal to the average of $f(x)$ on the right and left sides of $x = \pi/4$, and solve to find S. You can check your answer by asking a computer to calculate the series S directly.

9.28 In this problem you will show that the Fourier series for even and odd functions have only cosine and sine terms respectively by considering properties of sines and cosines. In Problem 9.29 you'll prove the same thing in a different way.

(a) Is the series $a_0 + a_1 \cos(x) + a_2 \cos(2x) + a_3 \cos(3x) + \ldots$ odd, even, or neither, or does the answer depend on the coefficients you choose?

(b) Is the series $b_1 \sin(x) + b_2 \sin(2x) + b_3 \sin(3x) + \ldots$ odd, even, or neither, or does the answer depend on the coefficients you choose?

(c) How would your answer to Part (a) change if you added the term $b_7 \sin(7x)$ to it?

(d) Using your answers to Part (a)–(c), argue why the Fourier series for an even function can only include cosines and/or a constant, and the Fourier series for an odd function can only include sines.

9.29 In this problem you will show that the Fourier series for even and odd functions have only cosine and sine terms respectively by directly applying Equation 9.2.3. In Problem 9.28 you proved the same thing in a different way.

(a) If you multiply an odd function times an even function, is the result odd, or is it even, or does it depend on the functions?

(b) If $E(x)$ is an even function, and you create a Fourier series for $E(x)$, prove that all the b_n coefficients will come out zero.

(c) Prove in a similar way that for an odd function $O(x)$ all of the a_n coefficients come out to zero.

9.30 The diagram below shows an LC circuit.

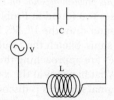

The element labeled V supplies a voltage $V(t)$ to the circuit. For a sinusoidal voltage V, it can be shown that:

If $V(t) = V_0 \sin(\omega t)$ then
$$I(t) = \frac{V_0}{\frac{1}{\omega C} - \omega L} \cos(\omega t)$$

where L and C represent the inductance and capacitance of the circuit, respectively.

(a) Find the current if the voltage is $3\sin(5t)$. Your answer will contain the constants L and C.

(b) Find the current if the voltage is $7\sin(2t)$.

(c) If you add voltages, the resulting currents add. Find the current if the voltage is $3\sin(5t) + 7\sin(2t)$.

(d) Now, suppose the voltage grows and drops as follows. (This is sometimes called a "sawtooth" function.)

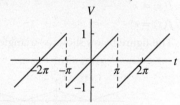

i. Represent the incoming voltage as a Fourier series.

ii. Calculate the resulting current. Your answer will also be in the form of a Fourier series.

9.31 (a) The function $f(x)$ is defined as $\sin(1/x)$ from $x = -\pi$ to $x = \pi$, and then repeats periodically forever. Explain why you cannot build a Fourier series for this function. (*Hint*: it is not because the function is undefined at $x = 0$.)

(b) Write a different periodic function that cannot be modeled by a Fourier series.

9.3 Deriving the Formula for a Fourier Series

You can of course use Equation 9.2.3 without knowing where it came from. But the derivation is more important than you might at first suppose. Once you know the method with sines and cosines, you can use the same trick to build series from many other types of functions.

9.3.1 Explanation: Deriving the Formula for a Fourier Series

The property that allows us to find the coefficients of a Fourier series is that the functions $\sin(nx)$ and $\cos(nx)$ for all integer n-values form an "orthogonal set," as defined below.

> **Definition: Orthogonal (Real) Functions**
>
> Two real functions $f(x)$ and $g(x)$ are "orthogonal" on the domain $a \le x \le b$ if their product integrates to zero on that domain.
>
> $$\int_a^b f(x)g(x)\,dx = 0$$
>
> A set of functions is "orthogonal" if every function in the set is orthogonal to every *other* function in the set.

The set of functions $\sin(nx)$, $\cos(nx)$ is orthogonal from $-\pi$ to π, which is equivalent to the following three statements:

- $\int_{-\pi}^{\pi} \sin(nx)\cos(mx)\,dx = 0$ for any n and m.
- $\int_{-\pi}^{\pi} \sin(nx)\sin(mx)\,dx = 0$ if n does not equal m. (No non-zero function is orthogonal to itself.)
- $\int_{-\pi}^{\pi} \cos(nx)\cos(mx)\,dx = 0$—once again, if $n \ne m$.

Unlike the three integrals above, the two below are not required to make our set of functions orthogonal. But these *are* essential for our derivation of the Fourier coefficients.

- $\int_{-\pi}^{\pi} \sin(nx)\sin(mx)\,dx = \pi$ if n *does* equal m.
- $\int_{-\pi}^{\pi} \cos(nx)\cos(mx)\,dx = \pi$—once again, if $n = m$.

Those integrals are not obvious; you will derive these results in Problem 9.32. But taking these integrals as givens for the moment, we are now ready to return to our original goal of finding the coefficients in a Fourier expansion:

$$f(x) = a_0 + \sum_{n=1}^{\infty} a_n \cos(nx) + \sum_{n=1}^{\infty} b_n \sin(nx)$$

We have to find infinitely many coefficients, but just to demonstrate the method let's find b_4, the coefficient of $\sin(4x)$ in the series. The trick is to multiply both sides by $\sin(4x)$ and then integrate as x goes from $-\pi$ to π.

$$\int_{-\pi}^{\pi} f(x)\sin(4x)\,dx = a_0 \int_{-\pi}^{\pi} \sin(4x)\,dx + \sum_{n=1}^{\infty} a_n \int_{-\pi}^{\pi} \cos(nx)\sin(4x)\,dx$$

$$+ \sum_{n=1}^{\infty} b_n \int_{-\pi}^{\pi} \sin(nx)\sin(4x)\,dx \qquad (9.3.1)$$

The first integral on the right, $\int_{-\pi}^{\pi} \sin(4x)dx$, you can easily confirm for yourself goes to zero. If you're not very comfortable with series notation, take a moment to write out those series explicitly to see what all the other integrals look like. You will see terms such as $\int_{-\pi}^{\pi} \cos(x)\sin(4x)dx$ and $\int_{-\pi}^{\pi} \sin(3x)\sin(4x)dx$; there are infinitely many such terms, but orthogonality promises that *all but one of these integrals evaluates as zero*. The only non-zero integral on the right side of this equation is $\int_{-\pi}^{\pi} \sin^2(4x)dx$ which is π, as we said above.

So Equation 9.3.1 reduces to:

$$\int_{-\pi}^{\pi} f(x)\sin(4x)\,dx = \pi b_4$$

This reasoning leads us to Equation 9.2.3, at least for that one coefficient; you'll use the same trick to find an arbitrary coefficient a_n in Problem 9.35 (and you could just as easily find b_n).

Note that the use of this trick did not depend on any details of the sine and cosine functions, but on the orthogonality of the set. We can therefore use the same trick to represent a function as a series of almost any orthogonal set of functions. You will find the coefficients for a series of "Legendre polynomials" in Problem 9.36, and we will build series from complex exponentials in Section 9.5. In Chapter 12 we will introduce "Sturm-Liouville theory" which proves that many other important sets of functions are orthogonal, and can therefore be used to build series.

9.3.2 Problems: Deriving the Formula for a Fourier Series

9.32 In this problem you will demonstrate that all functions of the form $\sin(nx)$ and $\cos(mx)$ form an orthogonal set from $-\pi$ to π. In Section 9.9 Problem 9.138 (see felderbooks.com) you will generalize this to arbitrary periods.

(a) By considering properties of even and odd functions, argue that $\int_{-\pi}^{\pi} \sin(nx)\cos(mx)dx = 0$ for any integers n and m.

For the remaining parts recall Euler's formula, which says $e^{ix} = \cos x + i \sin x$, and likewise $e^{-ix} = \cos x - i \sin x$.

(b) Using both forms of Euler's formula given above, derive expressions for $\sin x$ and $\cos x$ in terms of e^{ix} and e^{-ix}.

(c) For the next part it will be helpful to know that $\int_{-\pi}^{\pi} e^{ikx}dx = 0$ for any non-zero integer k. Explain why this result holds.

(d) Evaluate $\int_{-\pi}^{\pi} \sin(nx)\sin(mx)dx$.
  i. First replace the sines with exponentials, using your results from Part (b).
  ii. Next evaluate the integral. You can assume that m and n are positive integers, but you will find one answer if $m \neq n$ and a different answer if $m = n$.

(e) Repeat Part (d) for $\int_{-\pi}^{\pi} \cos(nx)\cos(mx)dx$, once again showing that this equals zero for any unequal integers n and m and finding its non-zero value if $m = n$.

9.33 In Problem 9.32 you derived the orthogonality relations for sines and cosines by using complex exponentials. For the case $m = n$ you can evaluate the integrals fairly simply without complex numbers.

(a) Use the identity $\sin^2 x = (1/2)(1 - \cos(2x))$ to evaluate $\int_{-\pi}^{\pi} \sin^2(nx)dx$.

(b) Using your result from Part (a), evaluate $\int_{-\pi}^{\pi} \cos^2(nx)dx$. *Hint*: this should be very quick and easy.

(c) Using the values of those integrals, find the average values of $\sin^2(nx)$ and $\cos^2(nx)$ on the interval $[0, \pi]$.

9.34 (a) Draw graphs of $\sin x$ and $\sin(2x)$ on the same plot, from $x = -\pi$ to $x = \pi$.

(b) Mark the regions of your plot where the two curves have the same sign, and where they have opposite signs.

(c) Use these graphs to explain why $\int_{-\pi}^{\pi} \sin(x)\sin(2x)dx = 0$.

9.35 Use the orthogonality of the trig functions to find a_n, the coefficient of $\cos(nx)$ in a Fourier series.

9.36 We were able to derive the coefficients of a Fourier series because the sine and cosine functions are orthogonal. The derivation is important because the same technique works on other sets of orthogonal functions.

In this problem you will create a series from "Legendre polynomials." There are many different Legendre polynomials and they are called $P_1(x)$, $P_2(x)$, $P_3(x)$ and so on. The following integral gives their orthogonality relationship.

$$\int_{-1}^{1} P_n(x) P_m(x)\, dx = \begin{cases} 0 & m \neq n \\ \frac{2}{2n+1} & m = n \end{cases}$$

The function $f(x)$ is defined from $x = -1$ to 1. Find the coefficient C_k in the series $f(x) = C_1 P_1(x) + C_2 P_2(x) + C_3 P_3(x) + \ldots$

9.4 Different Periods and Finite Domains

In Section 9.2 we showed you how to build a Fourier series for any function $f(x)$ with period 2π. In this section we will extend that method to functions with other periods, and to functions with finite domains.

9.4.1 Discovery Exercise: Different Periods and Finite Domains

1. Consider the function $f(x) = \sin(x) + \sin(3x) + \sin(10x)$.
   (a) Find $f(\pi/6)$, $f(13\pi/6)$, and $f(25\pi/6)$.
   (b) Find another x-value for which $f(x) = f(\pi/6)$.
   (c) What is the period of this function?
2. The function $g(x) = \cos(px)$ has the property that $g(x) = g(x + 5)$ for all x-values.
   (a) The function $g(x)$ might have a period of 5. What value of p would lead to this period?
   (b) The function $g(x)$ might also have a period of $5/2$. What value of p would lead to this period?

   See Check Yourself #59 in Appendix L

   (c) What are all the possible periods of $g(x)$, and their corresponding p-values?
   (d) If you sum a set of functions with all the periods in Part (c), what is the period of the resulting function?

9.4.2 Explanation: Different Periods and Finite Domains

We're going to build this Explanation around the example of a vibrating guitar string, so we need to start with some background. When a guitar string is vibrating in the simplest way possible—one pure sine wave—it has two different frequencies.

- The "spatial frequency" p might be measured in radians per centimeter. It is a property of shape; you can see it in a photograph of the string, or in Figure 9.5. The period $2\pi/p$ is the wavelength, the distance between peaks.
- The "temporal frequency" ω might be measured in radians per second. The period $2\pi/\omega$ is a time, not a distance. The temporal frequency determines the pitch (the note you hear).

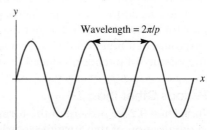

FIGURE 9.5 The temporal period and frequency are not shown because they relate to how quickly the string vibrates. The string will return to precisely this shape after a time $2\pi/\omega$.

460 Chapter 9 Fourier Series and Transforms

These two frequencies are related by $\omega = vp$, but the constant v is based on the specific properties of the string. If the six strings of a guitar all vibrate with the same spatial frequency, they will produce six different notes.

Most of our math will be done on the spatial frequency p. We will be taking Fourier series with respect to x, not with respect to t. But the goal of our calculations will be to determine what notes we expect to hear.

Plucking a D String

If you pluck the D string on a guitar, it produces the note D, but it also produces many other notes more quietly. The particular combination of notes gives the string its distinct sound. Our goal in this section is to mathematically determine what notes will be produced, and at what volumes.

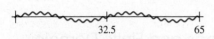

FIGURE 9.6 If a guitar string were pulled into this shape and released it would produce exactly two pitches.

So consider a guitar string 65 cm long, fixed at both ends. If you could somehow pull it into a perfect sine wave and let go, the string would produce only one note. The pitch of that note would depend on the wavelength of the sine wave (as discussed above), and the volume would depend on the amplitude (how high up you pulled the string). If the initial shape of the string were the sum of two sine waves, as shown in Figure 9.6, the string would produce two notes.

But in real life, when you pluck the string you pull it into a triangle shape. What notes will that produce?

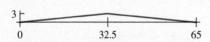

FIGURE 9.7 A real plucked guitar string starts in a triangle shape, and produces many different pitches.

By this point in the chapter you can probably guess that to approach that question, we have to rewrite the shape in Figure 9.7 as a Fourier series. Every wave in that series has a frequency p that determines the pitch of its note and an amplitude A that determines its volume. So we are not building a Fourier series of musical notes, but we are building a Fourier series of guitar-string-shapes that correspond to musical notes.

But there are two problems with this plan. First, the shape of the guitar string doesn't repeat periodically. We're going to solve that problem by making a "periodic extension" of the string's shape and then finding the Fourier series of that periodic function. But that leads us to the second problem, which is that the period of the resulting function won't be 2π.

Below we address these problems in reverse order. First we discuss Fourier series for periodic functions with periods other than 2π. Then we talk about different ways to do a periodic extension of a function defined on a finite domain. At the end we will come back and find the mixture of pitches produced by the plucked guitar string.

Periods Other than 2π

In Section 9.2 we presented the formula for a Fourier series with a period of 2π. Below is a generalization of that formula to arbitrary periods. To make the formulas a bit simpler we wrote them for period $2L$ rather than L, so if you have a function with period 6, be sure to put $L = 3$ into these formulas.

9.4 | Different Periods and Finite Domains

> **The Formula for a Fourier Series with Period $2L$**
>
> The Fourier series for a function $f(x)$ with period $2L$ is:
>
> $$f(x) = a_0 + \sum_{n=1}^{\infty} a_n \cos\left(\frac{n\pi}{L}x\right) + \sum_{n=1}^{\infty} b_n \sin\left(\frac{n\pi}{L}x\right) \quad (9.4.1)$$
>
> The coefficients are given by the equations:
>
> $$a_0 = \frac{1}{2L}\int_{-L}^{L} f(x)\,dx \quad a_n = \frac{1}{L}\int_{-L}^{L} f(x)\cos\left(\frac{n\pi}{L}x\right)dx \quad b_n = \frac{1}{L}\int_{-L}^{L} f(x)\sin\left(\frac{n\pi}{L}x\right)dx \quad (9.4.2)$$
>
> Once again, because the function is periodic, you can take these integrals over any interval of length $2L$ and get the same result.

Equations 9.4.1–9.4.2 give you all you need to calculate the Fourier series for a periodic function with any period. In what follows, we explain why the formulas look the way they do. In particular, where did the $n\pi/L$ inside the sines and cosines come from?

Consider, for example, the Fourier series for a function with period 6. Recall that L is defined as half the period, so we plug $L = 3$ into the formula above. For simplicity we'll assume $f(x)$ is odd so its Fourier series only contains sines, but nothing about the following explanation would change if there were cosines as well.

$$f(x) = \sum_{n=1}^{\infty} b_n \sin\left(\frac{n\pi}{3}x\right) = b_1 \sin\left(\frac{\pi}{3}x\right) + b_2 \sin\left(\frac{2\pi}{3}x\right) + b_3 \sin(\pi x) + b_4 \sin\left(\frac{4\pi}{3}x\right) + \ldots$$

Figure 9.8 shows the first several waves of this series. The first term has a period of 6, the second has a period of 3, then 2, then 3/2. Each period divides evenly into 6—each function repeats itself perfectly every six units—so a sum of *all* these functions has a period of 6. In general, then, if you wish to express a function $f(x)$ with period $2L$ as a sum of sine waves, that sum can only include sines and cosines that repeat every $2L$, which means all the terms have frequency $n\pi/L$ for some integer n.

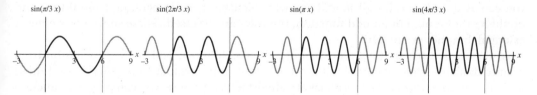

FIGURE 9.8 The sine waves that repeat every 6 are the ones with period 6, 6/2, 6/3, 6/4, and so on, so those are the periods included in a Fourier series for a function with period 6.

You can think of n as a counter, telling us how many "humps" (local minima and maxima) the wave fits between $x = 0$ and $x = 3$ (half a period). The $n = 1$ term has one hump, the $n = 2$ has two, and so on. All of these functions return to zero at $x = 3$, so they could all fit into a guitar string that was tacked down at $x = 0$ and $x = 3$.

Now that we have the equation for Fourier series of functions with periods other than 2π, the remaining piece we need is Fourier series for functions that are only defined on a finite domain. Then we'll be ready to pluck our D string.

A Function on a Finite Domain

It is often useful to create a Fourier series for a function that doesn't repeat at all, but is only defined on a finite domain. For instance, suppose we want to create a Fourier series that looks exactly like the function $y = x + 3$ between $x = 0$ and $x = 5$. The Fourier series will repeat itself forever, but we only care what it does in $[0, 5]$. Given that, we have a choice between three possible approaches, illustrated in Figure 9.9.

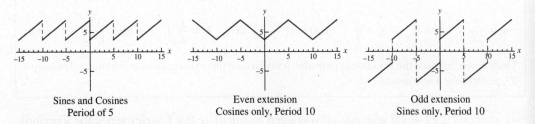

FIGURE 9.9 Three possible periodic extensions of the function $y(x) = x + 3$ with domain $[0, 5]$.

- We can create a periodic extension of $y(x)$ and create a Fourier series for it (involving both sines and cosines) with a period of 5.
- We can create an "even extension" of the function—that is, mirror its behavior on the left—and then create a periodic extension of *that* function. Then we can create a "Fourier cosine series" with a period of 10.
- We can create an "odd extension" of the function—that is, mirror its behavior upside-down on the left—and then create a periodic extension of that. Then we can create a "Fourier sine series" with a period of 10.

Of course these three functions are not the same, but they *are* the same as each other—and as the original function—between $x = 0$ and $x = 5$, which is the only place we care about. All three are useful in different circumstances. We'll show below, for example, why one of these works best for our guitar string problem.

Section 9.2 discussed the Fourier series of odd and even functions. For an even function you can assume $b_n = 0$ for all n, and you can calculate a_n by integrating from 0 to L and doubling the result. For an odd function, the roles are reversed. These shortcuts are never essential, but they are helpful.

In the particular case we are discussing here—building an odd or even extension from a finite domain—someone has built those shortcuts into formulas, and you can find those formulas in Appendix G. The formulas are almost too easy to use; you can plug in a function and integrate without being aware that you are creating an odd or even extension, and then you may be surprised by what you get. But if you understand where they come from, those formulas are the fastest way to find a Fourier series for a function on a finite domain.

Solving the Plucked Guitar String Problem

Through Fourier analysis you can determine what spatial frequencies make up any given shape of a guitar string. Then you have to convert those spatial frequencies into *temporal* frequencies which, in turn, correspond to specific notes. Below we give the relationship between these two frequencies for one particular guitar string; a different string would have a different relationship.

9.4 | Different Periods and Finite Domains   463

> **The pitch produced by a guitar D string**
>
> The following two rules apply very specifically to a 65 cm long guitar D string.
>
> - A wave with spatial frequency $p = \pi/65$ radians/cm produces a note with time frequency $f = 294$ vibrations per second.
> - More generally, a D string in the shape $A \sin(n\pi x/65)$ produces a note with frequency $f = (294 \times n)$ Hz.
>
> (The volume of the note is determined by the amplitude A.)
>
> These two rules are not derivable from Fourier series or anything else in this chapter. They are based on material science and the tension and density of a particular string. We will not be deriving or explaining these rules, but we will be *using* them.

The box above tells us what we will hear from a D string bent into a perfect sine wave. Our goal is to decompose the original shape (Figure 9.7) into a series of sine waves to determine what notes are produced at what volumes.

Since the height is only defined in $0 \le x \le 65$ we need a periodic extension of the string. We could define an odd extension, an even extension, or neither. Mathematically, any of these three approaches would give us the right function between 0 and 65. But physically, a series with cosines is useless to us. Why? Because *a single cosine function cannot represent the state of our string!* (No cosine function goes to zero at $x = 0$.)

You can think of this in the following way: we can write down the note produced if the string starts in any particular sine wave. Thus if we write the initial condition as a sum of sine waves, we know the combination of notes produced by that initial condition. We could also write the initial condition as a sum of cosines. That series would be mathematically valid and it would converge to the right function for the initial condition, but since those don't represent possible states of the string, we wouldn't know what to do with them if we had them.

Because we want only sines in our Fourier series, we begin with an odd extension of the original function.

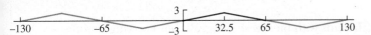

FIGURE 9.10 Odd extension of the shape of a plucked guitar string.

The Fourier coefficients are given by Equation 9.4.2. Because we are considering an odd function we know $a_n = 0$. The period is 130, so we use $L = 65$:

$$b_n = \frac{1}{65} \int_{-65}^{65} f(x) \sin\left(\frac{n\pi}{65} x\right) dx$$

Looking at Figure 9.10 we can see that this integral has to be done piecewise in three segments, $-65 < x < -32.5$, $-32.5 < x < 32.5$, and $32.5 < x < 65$. We can save some effort, however, by noticing that $f(x)$ is odd and $\sin(n\pi x/65)$ is odd, so the entire integrand is even. That means the integral from -65 to 0 will give the same result as the integral from 0 to 65, so we can go from 0 to 65 and double the result.

$$b_n = \frac{2}{65} \int_0^{65} f(x) \sin\left(\frac{n\pi}{65} x\right) dx$$

This integral still has to be broken into two pieces. Looking at Figure 9.10, $f(x) = (3/32.5)x$ from $x = 0$ to 32.5, and $f(x) = 6 - (3/32.5)x$ from $x = 32.5$ to 65.

$$b_n = \frac{2}{65}\left[\int_0^{32.5} \frac{3}{32.5}x\sin\left(\frac{n\pi}{65}x\right)dx + \int_{32.5}^{65}\left(6 - \frac{3}{32.5}x\right)\sin\left(\frac{n\pi}{65}x\right)dx\right]$$

We can evaluate this using the table in Appendix G. The answer is $b_n = (24/n^2\pi^2)\sin(n\pi/2)$. This goes to zero for all even n, and decreases rapidly with increasing n. We can get more intuition for that by writing out the first few non-zero terms.

$$f(x) = 2.4\sin((\pi/65)x) - .27\sin((3\pi/65)x) + .097\sin((5\pi/65)x) + \ldots$$

We can see that the sound is dominated by the $n = 1$ note with frequency 294 Hz, which is the note called D4 (or D over middle C). The next note has frequency 812 Hz, which is slightly higher than G5, but with an amplitude almost 10 times smaller than the dominant note. The remaining notes contribute very little to the overall sound. We can thus see why, when we pluck a guitar string, we think of it as making just one note. Nonetheless, the combination of higher notes (called "overtones") make D4 on a guitar sound different from D4 on a piano or a flute.

EXAMPLE **Fourier Series on a Finite Domain**

Question: Find a Fourier cosine series for the function $y = x + 3$ between $x = 0$ and $x = 5$.

Solution:
The even extension, drawn in Figure 9.9, has the formula:

$$f(x) = \begin{cases} x + 3 & 0 \leq x \leq 5 \\ -x + 3 & -5 \leq x < 0 \end{cases}$$

We could find all the coefficients with piecewise integrals, but knowing that the function is even offers us several shortcuts. First, we know that all the b_n coefficients are zero. Second, we know that the a_n integrals from -5 to 0 will contribute the same as the ones from 0 to 5, so we can just calculate them for positive values of x and then double the answer. (Alternatively we could have just gone straight to the formula for a Fourier cosine series in Appendix G.)

$$a_0 = \frac{1}{5}\int_0^5 (x+3)dx = \frac{11}{2}$$

$$a_n = \frac{2}{5}\int_0^5 (x+3)\cos\left(\frac{n\pi}{5}x\right)dx$$

A bit of integration by parts and/or a computer gives us $a_n = 10[(-1)^n - 1]/(\pi^2 n^2)$. If you think about $(-1)^n - 1$ you realize that it is 0 for even n-values and -2 for odd, giving the final answer:

$$f(x) = \frac{11}{2} - \sum_{n=1}^{\infty} \frac{20}{\pi^2 n^2}\cos\left(\frac{n\pi}{5}x\right) \quad \text{odd } n \text{ only}$$

9.4.3 Problems: Different Periods and Finite Domains

> Appendix G lists all the formulas for finding Fourier series coefficients. It also contains a brief table of some of the integrals that you will need most often in this chapter. Any problem that cannot be solved with u substitution, integration by parts, or with that table will be marked with a computer icon: 🖥.
> After you integrate remember that n (the summation index) is always an integer, and for any integer n:
>
> $$\sin(\pi n) = 0 \quad \text{and} \quad \cos(\pi n) = (-1)^n$$

9.37 Find the period of each function.
   (a) $\sin x$
   (b) $\sin(2x)$
   (c) $\sin(x/5)$
   (d) $\sin(3x) + \sin(6x) + \sin(21x)$
   (e) $\tan x$

9.38 (a) Find the period of the function $\sin(2x) + \sin(10x)$.
   (b) Find the period of the function $\sin(2x) + \sin(5x)$.
   (c) Choose a constant a such that the function $\sin(2x) + \sin(ax)$ is not periodic at all.

9.39 The function $g(x) = \cos(px)$ has the property that $g(x) = g(x + 6)$ for all x-values.
   (a) The function $g(x)$ might have a period of 6. What value of p would lead to this period?
   (b) The function $g(x)$ might also have a period of 3. What value of p would lead to this period?
   (c) What are all the possible periods of $g(x)$, and their corresponding p-values?
   (d) If you sum a set of functions with all the periods in Part (c), what is the period of the resulting function?

9.40 Figure 9.11 shows the $n = 1$, $n = 3$, and $n = 6$ terms of a Fourier sine series. Draw the $n = 2$, $n = 4$, and $n = 5$ terms. You don't need to put numbers on the vertical axis since there's no way for you to know the amplitudes without knowing what function is being modeled. Do include numbers on the horizontal axis, though, because you should be able to show each plot with the correct periods. No math is required here, just a quick sketch!

9.41 The function $f(x)$ is defined as x^2 on the domain $0 \le x \le 3$.
   (a) Draw an even extension of $f(x)$. If your even extension is repeated forever, what is the period of the resulting function?
   (b) Draw an odd extension of $f(x)$. If your odd extension is repeated forever, what is the period of the resulting function?
   (c) Draw a simple periodic extension of $f(x)$, neither even nor odd. What is the period of the resulting function?
   (d) Repeat Parts (a)–(c) for the function $g(x) = e^x$ on the same domain.

9.42 The function $f(x)$ equals x from $x = -4$ to $x = 4$ and is extended periodically from there, with a period of 8.
   (a) How can you know without doing any calculations that the Fourier series for $f(x)$ contains only sine terms?
   (b) Write a sine function with period 8.
   (c) Write a sine function with period 4. This function will also repeat when x increases by 8.
   (d) What is the next largest period that a sine function can have and still repeat when x increases by 8? Write a sine function with that period.
   (e) Write a Fourier series for $f(x)$ without the coefficients filled in: $f(x) = \sum b_n \sin(\text{something } x)$ where b_n is not-yet-determined, but where you do fill in what that "something" is.

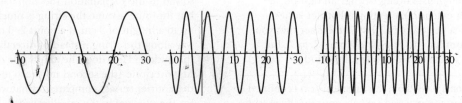

FIGURE 9.11 The $n = 1$, 3, and 6 terms of a Fourier sine series.

(f) Find the coefficients b_n using Equation 9.4.2 and write the Fourier series for $f(x)$. Simplify your answer as much as possible.

9.43 [This problem depends on Problem 9.42.] For the series you found in Problem 9.42, have a computer draw the 1st, 5th, 20th, and 100th partial sums on the domain $[-12, 12]$. Sketch the results and explain why they make sense.

9.44 Walk-Through: Building a Fourier Series (Finite Domain). In this problem you'll find a Fourier series of the function $f(x) = e^x$ on the domain $0 \leq x \leq 5$.

(a) We define a new function $g(x)$ that is an odd extension of $f(x)$. In other words, in the domain $0 \leq x \leq 5$, $g(x) = e^x$. In the domain $-5 < x \leq 0$, $g(x) = -e^{-x}$. Beyond that $g(x)$ repeats with period 10. Sketch a graph of $g(x)$ from $x = -30$ to $x = 30$.

(b) If you create a Fourier series to represent $g(x)$, you know without doing any work that either the a_n or the b_n coefficients will be zero. Which one, and how do you know?

(c) Use Equation 9.4.2 to write down integrals for the non-zero coefficients. Note that you will have to split your integrals into two parts: one going from -5 to 0, and the other from 0 to 5.

(d) Explain how you can know without doing any calculations that both parts of your integral will give the same result. This means you can simply evaluate one of them and multiply the result by 2.

(e) Evaluate your integrals and write the resulting Fourier series. Simplify as much as possible.

9.45 [This problem depends on Problem 9.44.] For the series you found in Problem 9.44, have a computer draw the 1st, 5th, 20th, and 100th partial sums on the domain $[-10, 10]$. Sketch the results and explain why they make sense.

9.46 [This problem depends on Problem 9.44.] Repeat Problem 9.44 with an *even* extension of the function.

9.47 [This problem depends on Problem 9.44.] Repeat Problem 9.44 doing neither an odd nor an even extension, creating a Fourier series with a period of 5. *Hint*: Remember that you do not have to integrate from $-L$ to L, but can integrate along any full period.

For Problems 9.48–9.50, the function $f(x)$ is defined on the given interval and repeats periodically outside of it. Find its Fourier series. You should be able to evaluate the integrals by hand or with the table in Appendix G.

9.48 $f(x) = \begin{cases} -1 & 0 < x < 3 \\ 1 & 3 < x < 6 \end{cases}$

9.49 $f(x) = \begin{cases} 1 & 0 < x < 3 \\ 2 & 3 < x < 6 \end{cases}$

9.50 $f(x) = \begin{cases} -1 & -2 < x < -1 \\ 1 & -1 < x < 1 \\ -1 & 1 < x < 2 \end{cases}$

9.51 $f(x) = x$ on $-2 \leq x \leq 2$

9.52 Appendix G gives the formula for a "Fourier Sine Series." Derive this formula based on Equations 9.4.1–9.4.2. *Hint*: Start by creating an odd extension.

For Problems 9.53–9.57, find a Fourier sine series *and* a Fourier cosine series for the given function on the given domain. For problems not marked with a computer icon you should be able to evaluate the integrals by hand or with the table in Appendix G.

9.53 $f(x) = 3$ on $0 \leq x \leq 1$

9.54 $f(x) = 1 - x$ on $0 \leq x \leq 1$

9.55 $f(x) = x^2$ on $0 \leq x \leq 2$

9.56 $f(x) = x^4$ on $0 \leq x \leq 2$

9.57 $f(x) = \sin x$ on $0 \leq x \leq \pi$

9.58 A square wave is represented by the following function for all $t \geq 0$.

$$f(t) = \begin{cases} 10 & 0 \leq t < 1 \\ -10 & 1 \leq t < 2 \\ 10 & 2 \leq t < 3 \\ \dots & \text{and so on} \end{cases}$$

Write a Fourier series that represents this function. Since the function is only defined for positive t-values, it doesn't matter what your function does for $t < 0$.

9.59 Rework the guitar string problem that we solved in the Explanation (Section 9.4.2), but this time assume the string is plucked up to a height of 3 cm at a point 2/3 of the way along the string instead of directly in the middle. How does the amplitude of the first overtone (the second non-zero term in the Fourier series) compare to what we got for the plucked-in-the-middle string?

9.5 Fourier Series with Complex Exponentials

In this section we rewrite our Fourier series using complex exponentials instead of real-valued sines and cosines. The two representations are entirely equivalent, but in many cases the exponential form is easier to work with.

9.5.1 Discovery Exercise: Fourier Series with Complex Exponentials

"Square waves" are frequently used in electronics and signal processing. An example is shown below, along with its Fourier series.

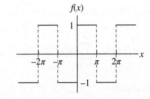

$$f(x) = \frac{4}{\pi}\sin(x) + \frac{4}{3\pi}\sin(3x) + \frac{4}{5\pi}\sin(5x) + \frac{4}{7\pi}\sin(7x) + \ldots$$

Anything that can be expressed with sines and cosines can also be expressed with complex exponential functions. In this exercise you will find the complex exponential series that represents this square wave.

1. Starting with Euler's equations $e^{ix} = \cos x + i\sin x$ and $e^{-ix} = \cos x - i\sin x$, derive a formula for $\sin x$ in terms of e^{ix} and e^{-ix}.
2. Using that formula, replace all the sines in the above expression with complex exponentials.
3. Regroup your answer to find the coefficients in the following series for our square wave.

$$f(x) = c_0 + c_1 e^{ix} + c_{-1} e^{-ix} + c_2 e^{2ix} + c_{-2} e^{-2ix} + c_3 e^{3ix} + c_{-3} e^{-3ix} + \ldots$$

   (a) What is c_0?
   (b) What is c_1?
   (c) What is c_{-3}?

   See Check Yourself #60 in Appendix L

   (d) What is c_n for any even integer n?
   (e) What is c_n for any odd integer n?

9.5.2 Explanation: Fourier Series with Complex Exponentials

Fourier series work because we can represent an oscillation of frequency p by the function $\sin(px)$ or $\cos(px)$. But we can also represent an oscillation of frequency p by the function e^{ipx} or e^{-ipx}. (These two alternate representations are related by Euler's formula $e^{ix} = \cos x + i\sin x$.) So given a complicated periodic function we can decompose it into a series of trig functions, but we can also decompose it into a series of complex exponentials. Sometimes we do this because the function we are modeling is itself complex—for instance, in quantum mechanics it is common to take the Fourier series of a complex wavefunction. But even when the original function is real, exponential functions are easier to work with than sines and cosines.

Interpreting a function expressed as a sum of complex exponentials is much the same as interpreting a sine-and-cosine Fourier series. The function e^{3ix}, just like $\sin(3x)$, oscillates with a wavelength of $2\pi/3$. For example, the heat output across the front of an office building might vary in complicated ways, but the dominant variation would be determined by the length w between the windows, so the coefficients of $e^{2\pi ix/w}$ and $e^{-2\pi ix/w}$ would probably be the biggest Fourier component.

When we gave you the formula for sines and cosines we started with a period of 2π and then presented the more general case. With complex exponentials we will jump straight to the formula for an arbitrary period. Note, however, that once again the formula becomes somewhat simpler if $L = \pi$.

The Formula for a Complex Fourier Series with Period 2L

The Fourier series for a function $f(x)$ with period $2L$, written in terms of complex exponentials, is:

$$f(x) = \sum_{n=-\infty}^{\infty} c_n e^{i(n\pi/L)x} \qquad (9.5.1)$$

The coefficients are given by the equation:

$$c_n = \frac{1}{2L} \int_{-L}^{L} f(x) e^{-i(n\pi/L)x} \, dx \qquad (9.5.2)$$

Just as with the sine-and-cosine version, you can evaluate this integral over any interval of length $2L$.

Notes on this formula.

- We have presented this formula in terms of a spatial variable x but remember that all the same math applies to temporal oscillations. For instance, e^{3it} represents an oscillation repeating every $2\pi/3$ seconds.
- The series goes from $n = -\infty$ to ∞. To write such a series term-by-term you can start at zero and move both up and down:

$$c_0 + c_1 e^{i\left(\frac{\pi}{L}\right)x} + c_{-1} e^{-i\left(\frac{\pi}{L}\right)x} + c_2 e^{i\left(\frac{2\pi}{L}\right)x} + c_{-2} e^{-i\left(\frac{2\pi}{L}\right)x} + c_3 e^{i\left(\frac{3\pi}{L}\right)x} + c_{-3} e^{-i\left(\frac{3\pi}{L}\right)x} + \ldots$$

- In a trig-based Fourier series, the constant term is divided by $2L$ while all the other terms are divided by L. In this version, the constant term c_0 uses the same formula as all the other terms.
- The original function $f(x)$ is not just the real part of the series, the imaginary part of the series, or the modulus of the series; it is the entire series. If $f(x)$ is a real function then the series will add up to a real value. (This means that c_{-n} will be the complex conjugate of c_n.)

EXAMPLE **Complex Fourier Series**

In Section 9.2 we found the Fourier series for the function $x^2 + 3$ on the domain $[-\pi, \pi]$. Here we will find the complex Fourier series for the same function on the same domain.

Equation 9.5.2 gives us:

$$c_n = \frac{1}{2\pi} \int_{-\pi}^{\pi} (x^2 + 3) e^{-inx}\, dx$$

Integration by parts (twice), or a computer, turns this into $c_n = 2(-1)^n/n^2$.

But that formula clearly does not work for $n = 0$. In fact if you went through the integration process you began by finding an antiderivative with an n in the denominator, so that was already invalid for $n = 0$. For this problem and many like it you have to treat that coefficient—the constant term in the series—as a special case.

$$c_0 = \frac{1}{2\pi} \int_{-\pi}^{\pi} (x^2 + 3)\, dx = \frac{\pi^2}{3} + 3$$

With that we have all the terms that make up our series.

$$x^2 + 3 = \frac{\pi^2}{3} + 3 + \sum_{n=-\infty}^{\infty} \frac{2(-1)^n}{n^2} e^{inx} \quad (n \neq 0)$$

Interpreting c_n

Finding a Fourier series with complex exponentials is a bit easier than it is with sines and cosines because you only have one set of coefficients c_n to find, and because integrals with exponentials are typically easier than ones with sines and cosines. It's a bit less obvious how to interpret the results, though. What does it mean to have an amplitude at $p = 3$ and at $p = -3$, and what does it tell you if one of those amplitudes is $3 + 2i$?

You should think of $c_n e^{i(n\pi/L)x} + c_{-n} e^{-i(n\pi/L)x}$ as one term representing an oscillation of frequency $n\pi/L$, just as you would with $a_n \cos((n\pi/L)x) + b_n \sin((n\pi/L)x)$. When $f(x)$ is a real function, the interpretation of its Fourier coefficients is aided by the following useful facts.

Fun Facts About Complex Fourier Series of Real Functions

Suppose c_n are the Fourier coefficients of a periodic function $f(x)$.

- If $f(x)$ is real, then c_{-n} is the complex conjugate of c_n.
- If $f(x)$ is real and even, then c_n and c_{-n} are real and equal.
- If $f(x)$ is real and odd, then c_n and c_{-n} are imaginary, and $c_{-n} = -c_n$.

You'll prove all of these in Problem 9.72. One important consequence of them is that for a real function $f(x)$ you know c_{-n} once you know c_n, so all of the information is in the coefficients for $n \geq 0$. In particular, you'll show in Problem 9.73 that for a real function the amplitude of oscillations at frequency $(\pi/L)n$ is simply $2|c_n|$.

We have seen that when we build a Fourier series for a function defined on a finite domain, we have three ways of building a periodic extension of the function. You can build an "odd" or "even" copy of the function and then build that out periodically, which gives you the advantage of working with only sines or only cosines. But with complex exponentials this advantage disappears, so if you are going to use a complex exponential series you usually just extend the function itself without doubling its period.

Parseval's Theorem

When we express a function in the traditional $f(x)$ format, we say "for *this* value of x the value of the function is *that*" for every possible x-value. When we express the same function with a Fourier series, we convey the same information in a different way: "for *this* frequency the two coefficients of the function are *those*" for every possible frequency. "Parseval's theorem" provides an important connection between these two different representations.

> **Parseval's Theorem**
>
> Suppose c_n are the Fourier coefficients of a periodic function $f(x)$. The average value of $|f(x)|^2$ over one full period equals $\sum_{-\infty}^{\infty} |c_n|^2$.

One way to interpret Parseval's theorem is in terms of the "intensity" of a light wave, which is a measure of how much energy it carries. For the simplest possible light wave—one frequency only—the intensity is proportional to the square of the amplitude. Parseval's theorem gives us mathematical assurance that for a more complicated light wave made up of many different frequencies, the total intensity is the sum of the intensities of the individual frequencies.

When applying Parseval's theorem it is useful to remember that the average value of any function $g(x)$ on the interval $a \leq x \leq b$ is:

$$\frac{1}{b-a} \int_a^b g(x)\, dx \qquad (9.5.3)$$

Deriving the Formula

The derivation of Equation 9.5.2 is similar to the one we did in Section 9.3 for sines and cosines. It begins by generalizing the idea of "orthogonality" to complex functions.

> **Definition: Orthogonal Complex Functions**
>
> The definition of orthogonality for complex functions is the same as the definition we presented on Page 457 for real functions except that you take the complex conjugate of one of the functions. Two functions $f(x)$ and $g(x)$ are orthogonal on the domain $a \leq x \leq b$ if the following equation holds.
>
> $$\int_a^b f^*(x) g(x)\, dx = 0$$
>
> A real function is its own complex conjugate, so our earlier definition can be seen as a special case of this one.

You'll show in Problem 9.74 that the functions $e^{in\pi x/L}$ for all integers n are orthogonal on the interval $-L < x < L$. Specifically:

$$\int_{-L}^{L} \left(e^{i\frac{n\pi}{L}x}\right) \left(e^{-i\frac{m\pi}{L}x}\right) dx = \begin{cases} 0 & n \neq m \\ 2L & n = m \end{cases}$$

Our goal is to find the coefficients in the series below.

$$f(x) = \sum_{n=-\infty}^{\infty} c_n e^{i\left(\frac{n\pi}{L}\right)x}$$

9.5 | Fourier Series with Complex Exponentials

The orthogonality of the set allows us to use the same trick we used for sines and cosines. To find the mth term we multiply both sides of the equation by $e^{-i(m\pi/L)x}$ and then integrate from $-L$ to L.

$$\int_{-L}^{L} f(x) e^{-i\left(\frac{m\pi}{L}\right)x} dx = \sum_{n=-\infty}^{\infty} c_n \int_{-L}^{L} e^{i\left(\frac{n\pi}{L}\right)x} e^{-i\left(\frac{m\pi}{L}\right)x} dx$$

Remember how this trick works? The series on the right has an infinite number of terms, but orthogonality guarantees us that all but one of them—the one where $m = n$—are zero. When m does equal n, the integral is $2L$, as we saw above. So the entire infinite series just becomes:

$$\int_{-L}^{L} f(x) e^{-i\left(\frac{m\pi}{L}\right)x} dx = c_m(2L)$$

which leads us directly to Equation 9.5.2.

9.5.3 Problems: Fourier Series with Complex Exponentials

> Appendix G lists all the formulas for finding Fourier series coefficients. It also contains a brief table of some of the integrals that you will need most often in this chapter. Any problem that cannot be solved with u substitution, integration by parts, or with that table will be marked with a computer icon: 💻.
>
> After you integrate remember that n (the summation index) is always an integer and that for any integer n, positive or negative,
>
> $$e^{in\pi} = (-1)^n$$

9.60 This problem is not directly about Fourier series *per se*, but it gives you a few formulas that will be useful in simplifying complex Fourier series.

(a) Show that for any integer n you can simplify $e^{in\pi}$ to $(-1)^n$. (You can do this by drawing values on the complex plane, or by using Euler's formula.)

(b) If n is an integer then what is $e^{in\pi} - e^{-in\pi}$?

(c) If n is an integer then what is $e^{2in\pi}$?

In Problems 9.61–9.68 find the Fourier series, in complex exponential form, for the given functions. If the function is defined on a finite domain make a periodic extension of it (not odd or even). Simplify your answers as much as possible. For problems not marked with a computer icon you should be able to evaluate the integrals by hand, or with the table in Appendix G.

9.61 The square wave whose function is given below

$$f(x) = \begin{cases} -1 & -\pi < x < 0 \\ 1 & 0 < x < \pi \\ -1 & \pi < x < 2\pi \\ \ldots & \text{and so on} \end{cases}$$

9.62 The "sawtooth function" shown below

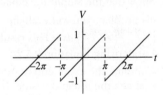

9.63 The "triangle wave" shown below

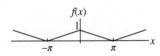

9.64 $y = \sin(3x)$

9.65 $f(x) = x$ on the domain $[0, 5]$

9.66 $f(x) = e^x$ on the domain $[0, 2]$

9.67 $f(x) = x + ix^2$ on $[-1, 1]$

9.68 $f(x) = xe^{ix}$ on $[0, \pi]$

9.69 In the Explanation (Section 9.5.2) we found the complex Fourier series for $x^2 + 3$ on $[-\pi, \pi]$. In Section 9.2 we found the real-valued Fourier series for the same function on the same domain. Show that the two expressions are equivalent.

9.70 Show that if c_{-n} is the complex conjugate of c_n for all n, then the resulting Fourier series represents a real-valued function.

9.71 (a) Show that if c_{-n} and c_n are real and equal to each other for all n, then the resulting Fourier series represents an even real-valued function.

(b) What relationship between c_{-n} and c_n would lead $f(x)$ to be real and odd?

9.72 Prove the "fun facts about complex Fourier series of real functions" on Page 469, using Equation 9.5.2. *Hint*: expand out the complex exponential with Euler's formula. You'll find that c_n comes out imaginary and that $c_{-n} = -c_n$.

9.73 If $f(x)$ is real then its Fourier coefficients obey $c_{-n} = c_n^*$, so the oscillation with frequency $n\pi/L$ can be written as $c_n e^{i(n\pi/L)x} + c_n^* e^{-i(n\pi/L)x}$. Let $c_n = A + iB$, where A and B are real.

(a) Use Euler's formula to rewrite the oscillation without i in it. Your answer should depend on A, B, $n\pi/L$, and x.

(b) Using the fact that $a\cos(px) + b\sin(px)$ is an oscillation with amplitude $\sqrt{a^2 + b^2}$, write the amplitude of this oscillation as a function of A and B.

(c) Show that the amplitude equals $2|c_n|$.

9.74 In this problem you will evaluate $\int_{-L}^{L} \left(e^{i\frac{n\pi}{L}x}\right)\left(e^{-i\frac{m\pi}{L}x}\right) dx$. (This result was used in deriving Equation 9.5.2.)

(a) Combine the exponentials into one and evaluate the integral. Using the properties of complex exponentials and the fact that n and m are integers, show that your answer reduces to 0.

(b) Explain why the argument you made in Part (a) doesn't apply when $n = m$. (You'll redo the calculation for this case in Part (c). Here you just need to explain why the calculation you did in Part (a) fails in this case.)

(c) Simplify the integrand in the case $n = m$ and evaluate the integral.

9.75 You can use Parseval's theorem to sum series that would be hard to calculate otherwise.

(a) Find the coefficients of the Fourier series (in complex exponential form) for $f(x) = x$ from $x = -\pi$ to π, repeated periodically thereafter.

(b) Using your answer to Part (a) write $\sum_{-\infty}^{\infty} |c_n|^2$ in the form *<a constant>* $\sum_{1}^{\infty}(1/n^2)$. *Hint*: Combine $|c_n|^2$ and $|c_{-n}|^2$ into one term.

(c) Find the average value of $f(x)^2$ between $x = -\pi$ and $x = \pi$. (Use Equation 9.5.3.)

(d) Use Parseval's theorem and your answers to Parts (b)–(c) to find $\sum_{1}^{\infty}(1/n^2)$.

9.76 *[This problem depends on Problem 9.75.]* Use the same method you used in Problem 9.75, starting with the Fourier series for $f(x) = -1$ from $x = -1$ to 0, $f(x) = 1$ from $x = 0$ to 1 (repeated periodically afterwards). Your final answer should tell you the sum $\sum_{0}^{\infty}(1/n^2)$ for odd n only.

9.77 *[This problem depends on Problem 9.75.]* Use the same method you used in Problem 9.75, starting with the Fourier series for $f(x) = x^2$ from $x = -\pi$ to π (repeated periodically afterwards). Your final answer should tell you the sum $\sum_{0}^{\infty}(1/n^4)$.

9.78 In quantum mechanics, the state of a particle is described by a complex "wavefunction" $\psi(x)$. If $\psi(x) = Ce^{ikx}$ then the particle has momentum k. If $\psi(x) = Ae^{ikx} + Be^{ipx}$ then the particle *might* have momentum k or it *might* have momentum p; the probability of each possibility is proportional to the amplitude A or B squared. For example, if $\psi(x) = 2e^{7ix} + 3e^{-5ix}$ then the particle either has momentum 7 or -5, and -5 is 9/4 as likely as 7. Since the two probabilities must add to 1, they are 4/13 and 9/13 respectively.

(a) If $\psi(x) = \cos(2x)$ what are its possible momenta and what are the probabilities for each one? (You can rewrite this function as a sum of complex exponentials without using Equation 9.5.2.)

(b) If $\psi(x) = -1$ for $-\pi < x < 0$ and 1 for $0 < x < \pi$ (and is periodic for other values), how many times more likely is it to have momentum 3 than momentum 7?

9.6 Fourier Transforms

Fourier series are used to model periodic functions. To model functions with longer and longer periods requires series with shorter and shorter frequency intervals. In the limit of this process, you can model a non-periodic function by using an integral instead of a series.

9.6.1 Discovery Exercise: Fourier Transforms

What happens to the frequencies of a Fourier series as the period of the function gets longer? Consider a specific example. If $f(x) = e^{-x^2}$ from $-\pi$ to π, repeated periodically thereafter, the frequencies in its Fourier series are 1, 2, 3, ….

1. If we instead define $f(x) = e^{-x^2}$ from -2π to 2π, repeated periodically thereafter, what are the frequencies in its Fourier series?
2. We can get a series that models e^{-x^2} well over a wide range of values if we define $f(x) = e^{-x^2}$ from -500 to 500, and of course make it periodic outside of that. Now what are the frequencies in the Fourier series?

   See Check Yourself #61 in Appendix L

3. For each of the three functions you considered above, how many Fourier modes were there with frequencies between 0 and 10 (inclusive)? (Don't forget to include negative values of n.)
4. In the limit where you consider the entire function e^{-x^2} from $-\infty$ to ∞, what frequencies do you think you would need?

Now let's consider the amplitudes of the Fourier modes.

5. Calculate the coefficients c_0–c_3 of the complex exponential Fourier series for each of the three functions you considered above, with periods 2π, 4π, and 1000. (If you don't have access to a computer while you're doing this exercise, skip this part and the next one and at the end make a prediction as to what you would have seen.)
6. In the limit where you consider the entire function e^{-x^2} from $-\infty$ to ∞, what happens to the amplitudes c_n?
7. Given your answer to Part 3, explain why your answer to Part 6 makes sense.

9.6.2 Explanation: Fourier Transforms

A Fourier series represents a periodic function $f(x)$ as a sum of complex exponentials. (You can equivalently use sines and cosines, but in this section we will only talk about the complex exponential form.) If the period of $f(x)$ is 2π, the frequencies are all the integers: $\omega = 1$ means a period of 2π, $\omega = 2$ a period of π, $\omega = 3$ a period of $2\pi/3$, etc. All these periods fit into 2π, so the resulting series has that period.

If you want to represent a function with a longer period, you need to include more frequencies. For instance, a period of 6π requires frequencies of $1/3$, $2/3$, 1, and so on; three times as many frequencies for three times as much function.

$$\ldots + c_0 \quad + \quad c_1 e^{ix} \quad + \quad c_2 e^{2ix} + \ldots \qquad \text{period} = 2\pi$$

$$\ldots + c_0 + c_1 e^{ix/3} + c_2 e^{2ix/3} + c_3 e^{ix} + c_4 e^{4ix/3} + c_5 e^{5ix/3} + c_6 e^{2ix} + \ldots \qquad \text{period} = 6\pi$$

Suppose you wanted to write a series to represent a complete non-periodic function like $f(x) = e^{-x^2}$. What frequencies would you include? All of them! As the period of f increases the spacing between frequencies in the Fourier series decreases. In the limit where f never repeats, you use frequencies at every real number. The Fourier series becomes a "Fourier integral."

> **Fourier Transforms**
>
> A non-periodic function $f(x)$ can be represented by a Fourier integral:
>
> $$f(x) = \int_{-\infty}^{\infty} \hat{f}(p) e^{ipx} \, dp \qquad (9.6.1)$$
>
> The function $\hat{f}(p)$ is the "Fourier transform" of $f(x)$, and can be calculated from:
>
> $$\hat{f}(p) = \frac{1}{2\pi} \int_{-\infty}^{\infty} f(x) e^{-ipx} \, dx \qquad (9.6.2)$$
>
> The original function $f(x)$ is called the "inverse Fourier transform" of $\hat{f}(p)$.

The Fourier transform $\hat{f}(p)$ plays essentially the same role in a Fourier integral that the coefficients c_n play in a Fourier series. If $\hat{f}(7.2) = 3$ that doesn't mean anything in particular about $f(7.2)$. Instead it means that if you express $f(x)$ as a sum of oscillations, the one with frequency 7.2 will have an amplitude of 3. (Well, it means roughly that. We'll get more precise about it a little later, but if you understand this point qualitatively you've got a good start on understanding Fourier transforms.)

A quick note about terminology may be helpful here. For a periodic function we find the *Fourier coefficients* and use them to write the original function as a *Fourier series*. For a non-periodic function we find the *Fourier transform* and use it to write the original function as a *Fourier integral*. But in an annoying accident of history, people tend to refer to the first process as "finding the Fourier series" (not the coefficients) and the second as "finding the Fourier transform" (not the integral). Neither phrase is a problem by itself, but it would be nice if they were parallel to each other. Sorry about that.

EXAMPLE **Finding A Fourier Transform**

Question: Find the Fourier transform of $f(x) = e^{-|x|}$.

Answer:
Because of the absolute value we have to break the integral in Equation 9.6.2 into two pieces

$$\hat{f}(p) = \frac{1}{2\pi} \left[\int_{-\infty}^{0} e^{x} e^{-ipx} \, dx + \int_{0}^{\infty} e^{-x} e^{-ipx} \, dx \right]$$

We can combine the exponentials within each integral and we're just left with the integral of an exponential. After a bit of algebra we get the following.

$$\hat{f}(p) = \frac{1}{\pi (1 + p^2)}$$

The facts about Fourier series of real functions on Page 469 also apply to Fourier transforms. If $f(x)$ is real that means $\hat{f}(-p)$ is the complex conjugate of $\hat{f}(p)$; if $f(x)$ is real and even that means $\hat{f}(p)$ is real and even; if $f(x)$ is real and odd that means $\hat{f}(p)$ is imaginary and odd.

In some physical contexts the Fourier transform of a function is called its "spectrum." For example, if $A(t)$ represents the amplitude of a light wave hitting a detector (such as your eye)

9.6 | Fourier Transforms

then $\hat{A}(\omega)$ tells you how much of that wave is made up of each possible frequency. Since each frequency of light represents a color, the spectrum tells you how much of each color is in a given light wave. Although this is the most familiar use of "spectrum," scientists sometimes extend the term to Fourier transforms of other types of waves.

Parseval's theorem can also be extended to Fourier transforms: $\int_{-\infty}^{\infty} |f(x)|^2 dx = \int_{-\infty}^{\infty} |\hat{f}(p)|^2 dp$. One important application comes from quantum mechanics, where the wavefunction ψ represents position probabilities and its Fourier transform represents momentum probabilities. A proper wave function should be "normalized"—that is, the sum of all probabilities should be one—and Parseval's theorem assures us that normalizing one representation also normalizes the other.

Plots of Fourier Transforms

To build an intuition for Fourier transforms, it's best to consider some graphical examples. Unfortunately the Fourier transform of a function $f(x)$ is generally complex, which makes graphing difficult. The exception, as we noted above, is that a real and even $f(x)$ has a real $\hat{f}(p)$. We will therefore begin by looking at some real, even functions and their Fourier transforms.

In the examples below we leave out numbers on the vertical axes. (Because $f(x)$ and $\hat{f}(p)$ have different units, the relative y-values wouldn't mean anything.) You should, however, get to the point where you can tell by looking at either plot what the shape of the other one should be, and roughly where it should lie on the horizontal axis. You should also be aware that Fourier transforms are "linear" which means that doubling $f(x)$ causes $\hat{f}(p)$ to double.

FIRST EXAMPLE **A Fourier Transform with Spikes at One Frequency**

Question: Figure 9.12 shows a Fourier transform $\hat{f}(p)$. Sketch the function $f(x)$.

Answer:
The Fourier transform is telling us that $f(x)$ has a sinusoidal oscillation with frequency 2, and nothing else. We know $f(x)$ is even. (How do we know that?) So the plot is:

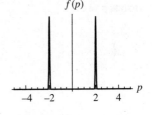

FIGURE 9.12

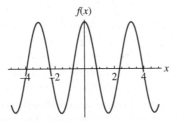

SECOND EXAMPLE **The Fourier Transform of a Superposition of Cosines**

Question: Figure 9.13 shows a function $f(x)$. Sketch its Fourier transform $\hat{f}(p)$.

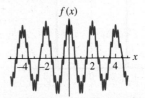

FIGURE 9.13

Answer:
This function sums two waves: one with a large amplitude and a period of 2, and one with a smaller amplitude and a period a tenth of the first. (You can count the number of small peaks in each period of the large wave.) So the Fourier transform has a sharp peak at $p = \pi$ and a smaller peak at $p = 10\pi$.

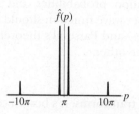

THIRD EXAMPLE — The Fourier Transform of a Constant

Question: Figure 9.14 shows a function $f(x)$. Sketch its Fourier transform $\hat{f}(p)$.

Answer:
Recall that a Fourier series has a constant term c_0 as well as all the oscillatory terms. So this Fourier transform contains that term, the coefficient of e^{0i}, and no other.

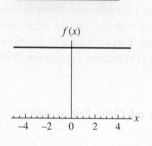

FIGURE 9.14

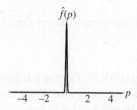

The examples above represent the core skill that we want to develop in this section, which is graphically going from a function to its Fourier transform and vice versa. Please make sure you follow all of them before proceeding.

But now we would like to make a subtler point, which is that the formulas for finding $f(x)$ from $\hat{f}(p)$ (Equation 9.6.1) and finding $\hat{f}(p)$ from $f(x)$ (Equation 9.6.2) are nearly identical. The numerical factors are different and the exponent sign is reversed, but qualitatively both equations describe the same kind of transformation. The symmetry of this transformation gives you another level of insight.

For instance, Figure 9.15 shows a function $f(x)$ that spikes sharply at $x = \pm 3$. What is its Fourier transform? If you try to answer this question directly, by figuring out what oscillatory periods make up such a spike, you probably won't get far. So instead let's ask the opposite question: "$f(x)$ is the Fourier transform of what function?" That's like the first example above, and the answer is a sine wave with frequency 3. We conclude, because of the symmetry of these transformations, that the Fourier transform of Figure 9.15 would look like $\cos(3x)$.

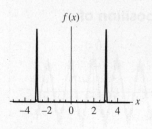

FIGURE 9.15

If $\hat{f}(p)$ spikes at $p = 3$ that means $f(x)$ oscillates with frequency 3, like $\cos(3x)$.

If $f(x)$ spikes at $x = 3$ that means $\hat{f}(p)$ oscillates with frequency 3, like $\cos(3p)$.

Test yourself: If we move the spikes in Figure 9.15 farther away from $x = 0$, what will happen to the Fourier transform? What if we move the spike to $x = 0$? Think about both of these questions backward—pretend the spikes represent the Fourier transform and you want to find the function instead of the other way around—and formulate your answers before reading further.

OK, here are the answers. If you move the spike to higher $|x|$-values then the Fourier transform (which was a frequency-3 oscillation) increases in frequency. If the spike peaks around zero then the Fourier transform is a constant. (This is the reverse of the third example above.) We are presenting these puzzles to develop your *general* understanding of Fourier transforms, but these two *specific* results—"moving a function away from $x = 0$ increases the frequency of its transform" and "the transform of a spike at zero is a constant"—are also worth holding onto as you move forward.

How "sharp" must these peaks be? Strictly speaking, the integral in Equation 9.6.2 doesn't exist for either an infinitely thin spike or a perfect sine or cosine wave, so all the examples above are limiting cases. Figure 9.16 shows what happens to $\hat{f}(p)$ when the peaks that make up $f(x)$ get wider. The transform still looks like a cosine wave near $p = 0$, but the wider the peaks in $f(x)$ the more rapidly $\hat{f}(p)$ falls off with increasing $|p|$. In the limit where the peaks in $f(x)$ are so wide that f is approaching a constant, $\hat{f}(p)$ becomes a thin spike at $p = 0$. In the opposite limit where the peaks are getting infinitely thin, the transform is approaching a perfect sine wave.

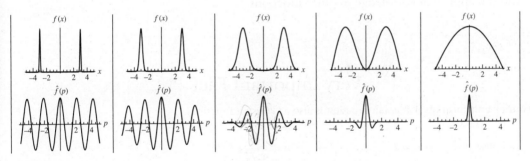

FIGURE 9.16 Each column shows $f(x)$ on top and its Fourier transform $\hat{f}(p)$ below it. When $f(x)$ is sharply peaked at ± 3, $\hat{f}(p)$ is roughly a sine or cosine wave with period $2\pi/3$. When the peaks in $f(x)$ get wider, $\hat{f}(p)$ falls off as it moves away from $p = 0$.

What Functions Have Fourier Transforms?

In Section 9.2 we said that if a function $f(x)$ satisfies the "Dirichlet conditions" it is guaranteed to have a Fourier series. We also noted that almost every periodic function that comes up in physical problems will meet these criteria; being periodic is the significant restriction on which functions have Fourier series.

For Fourier transforms, that condition is relaxed. The "Fourier integral theorem" guarantees that a function $f(x)$ that meets the Dirichlet condition on every finite interval will have a Fourier transform, *provided it also meets the condition that* $\int_{-\infty}^{\infty} |f(x)| dx$ is finite. That new requirement turns out to be pretty restrictive. It implies, among other things, that $\lim_{x \to \pm\infty} f(x)$ must equal zero. You can't find a Fourier transform for x^2 or $\ln x$ or most other functions that you might write down.

You might protest that we have been flagrantly violating that restriction. Our examples above included the Fourier transform of a superposition of cosines and the Fourier transform of a constant, neither of which drops to zero. Both of those cases should properly be viewed as limits. For instance, if you have a function that is constant for a very long time, and then drops to zero at the distant edges, its Fourier transform will approach a spike at $x = 0$. As those edges get farther away, the spike gets thinner, approaching the drawing we showed above.[3]

One of the most important functions that *does* have a Fourier transform is a "Gaussian," such as $f(x) = e^{-x^2}$. We discuss Gaussian functions below.

The Fourier Transform of a Gaussian

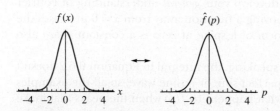

FIGURE 9.17 The Fourier transform of a Gaussian is a Gaussian.

A Gaussian is any function of the form $f(x) = Ae^{-[(x-x_0)/w]^2}$. The constant A controls the height of the peak, x_0 determines where the peak is centered, and w determines how wide it is.

Gaussian functions come up in many applications ranging from quantum mechanics to statistics. The Fourier transform of a Gaussian (Figure 9.17) is important by itself, but it also provides a useful example because you can transform these functions analytically, as you'll show in Problem 9.98. You cannot do that with most functions, but once you know qualitatively how the Fourier transform of a peak behaves, you can apply that knowledge to other functions.

🎺 Very Important Fact 🎺

The Fourier transform of a Gaussian function is also a Gaussian function.

Simplest case: $\quad f(x) = e^{-x^2} \quad \rightarrow \quad \hat{f}(p) = \dfrac{1}{2\sqrt{\pi}} e^{-p^2/4}$

General case: $\quad f(x) = Ae^{-[(x-x_0)/w]^2} \quad \rightarrow \quad \hat{f}(p) = \dfrac{Aw}{2\sqrt{\pi}} e^{ipx_0} e^{-w^2 p^2/4}$

(This formula is reproduced for easy reference in Appendix G.)

So if you start with a Gaussian $f(x)$ you end up with a Gaussian $\hat{f}(p)$. If you change $f(x)$ that causes changes to $\hat{f}(p)$. For instance, if you double A that doubles the height of $f(x)$. This in turn doubles $\hat{f}(p)$ because, as we already said, the Fourier transform is linear. We'll discuss x_0 below, but for now take it to be zero.

---

[3] In the limit where the edges get infinitely far away the spike gets infinitely tall and thin. That limit—an infinitely tall, thin spike—is called a "Dirac delta function" (although it's not technically a function). We discuss Dirac delta functions further in Problem 9.93 of this section, and much more carefully in Chapter 10.

Here we want to focus on w. In the limit as $w \to 0$ the Gaussian becomes an infinitely thin spike at $x = 0$, and its transform approaches a constant. (To keep that constant from approaching zero you would have to let $A \to \infty$ at the same time.) In the opposite limit $w \to \infty$ the Gaussian spreads out infinitely wide and becomes a constant function, and its transform becomes a thin spike at $p = 0$. Finite Gaussian functions fall between these two extremes, with wider-peaked functions having narrower-peaked transforms.

EXAMPLE **The Fourier Transforms of Some Random Bumps**

Question: Sketch the Fourier transforms of $f(x) = 1/(1 + e^{x^2})$ and $g(x) = 1/(1 + e^{(x/4)^2})$.

Answer:
We cannot find the Fourier transforms of these functions analytically. (Try it!) However, a quick sketch shows that $f(x)$ and $g(x)$ are both bumps at $x = 0$, with $g(x)$ four times wider than $f(x)$. So their Fourier transforms will look like bumps at $p = 0$, but with $\hat{f}(p)$ four times wider than $\hat{g}(p)$.

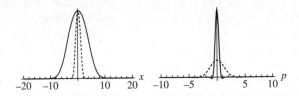

The plot on the left shows the functions $f(x)$ (dashed) and $g(x)$ (solid). The plot on the right shows their Fourier transforms.

The (solid) $g(x)$ is the *wider* of the two functions, so it has the *narrower* of the two transforms. We'll leave it up to you for the moment to think about why the $\hat{f}(p)$ plot is shorter than the $\hat{g}(p)$ plot. See Problem 9.95.

The fact that wide peaks have narrow Fourier transforms and vice versa is critical in quantum mechanics. The "wavefunction" $\psi(x)$ describes the position of the function, and when it is distributed over a wide range of x-values that means the particle's position is very uncertain. The Fourier transform $\hat{\psi}(p)$ similarly describes the particle's momentum. So if the position wavefunction is sharply peaked (small uncertainty in position) the uncertainty in momentum will be large, and vice versa. This result is called the "Heisenberg Uncertainty Principle" and it is used to calculate quantities ranging from the properties of atoms to the sizes of neutron stars.

When Fourier transforms go complex

We have been focusing on real, even functions $f(x)$, which meant real Fourier transforms $\hat{f}(p)$. In all other cases the Fourier transform of a real function is complex. It's also common (for instance in quantum mechanics) to take the Fourier transform of a function that is complex to begin with. The calculations (Equations 9.6.1 and 9.6.2) are no different, but interpretation becomes more difficult.

We begin with a purely mathematical observation: every time you move a function to the left by 1 unit, its Fourier transform multiplies by e^{ip}. (You will prove this in Problem 9.94.) As an example, we have seen that the Fourier transform of the Gaussian function e^{-x^2} is another Gaussian, so we now know that the Fourier transform for $e^{-(x+x_0)^2}$ is a Gaussian multiplied by e^{ix_0p}. OK, but what does *that* look like?

Remember that e^{ix_0p} is an oscillation around the complex plane, completing a full rotation every time p advances by $2\pi/x_0$. When $p = 0$ the multiplier e^{ix_0p} is just 1, which is to say, irrelevant: if our function was real, it's still real. When $p = \pi/(2x_0)$ the multiplier is i, so the function is pure imaginary. At $p = \pi/x_0$ the function is on the negative real axis, and so on.

In summary: if we move $f(x)$ to the left, this does not change the modulus of $\hat{f}(p)$ at all. But it does change its complex phase. If you start with a real even function such as a Gaussian, and move it to the left by x_0 units, the transform looks the same as it did before except that as you move along the p-axis it rotates around the complex plane with a frequency of x_0. We can't easily plot that complex Fourier transform, but we can plot the real and imaginary parts of it. As $\hat{f}(p)$ rotates about the complex plane, the real and imaginary parts both oscillate.

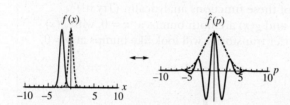

FIGURE 9.18 For the Gaussian on the left centered on zero (dashed curve) the transform is real, so the plot on the right shows $\hat{f}(p)$. For the Gaussian on the left centered on $x = -2$ (solid curve) the transform is complex, so the plot on the right shows the real part of $\hat{f}(p)$. The imaginary part also oscillates with an amplitude that decreases with p, but 90° out of phase with the real part.

Moving the plot of a function $f(x)$ away from $x = 0$ causes its transform to oscillate. The farther away from zero you go, the higher the frequency of oscillation (Figure 9.18). When the argument is time instead of space, this result is called the "time-delay theorem."

A More Detailed Interpretation of $\hat{f}(p)$

Near the beginning of this section we said that $\hat{f}(7)$ in a Fourier transform means roughly the same thing as c_7 in a Fourier series, but not exactly the same. In fact there are two differences between them, one trivial and one meaningful. If you're encountering Fourier transforms for the first time, you may want to just skim this discussion. For those who want to go deeper, however, these differences can be useful to think about.

The trivial difference is a point of notation. In a Fourier transform $\hat{f}(7)$ is the coefficient of e^{7ix}, the oscillation with frequency 7. In a Fourier series c_7 is the coefficient of whatever the 7th term happens to be, which could have any frequency. If the period of the original function is 2π then the 7th term happens to be the one with frequency 7, but in other cases it isn't.

Here is the more important distinction. Suppose we have a Fourier series for a function with period 2π: $f(x) = c_0 + c_1 e^{ix} + c_{-1} e^{-ix} + c_2 e^{2ix} + \dots$. We can meaningfully ask the question "How much does the e^{7ix} term contribute to this sum?" The answer is based on c_7 and its value relative to other coefficients. If $c_7 = 20$ and all the other coefficients are less than 1, the answer is "quite a lot."

Now consider a Fourier integral $f(x) = \int \hat{f}(p) e^{ipx} dp$. How much does the e^{7ix} term contribute to this sum? Nothing. If $\hat{f}(7) = 20$ and the function is less than 1 everywhere else, the answer is still "nothing." We asked the wrong question. "How much does the range of frequencies from $p = 6.9$ to $p = 7.1$ contribute to this integral?" That's the *right* question, and we answer it by integrating.

This issue points to a fundamental difference between a sum and an integral, or between a discrete function and a continuous one. We can summarize by saying that c_7 is the amplitude of an oscillation with frequency 7, while $\hat{f}(7)$ is the amplitude *per unit frequency* of oscillations with frequency close to 7. (Notice that the units work out correctly because of this; the coefficient of e^{7ix} in a Fourier integral is not $\hat{f}(7)$, but $\hat{f}(7)dp$.)

The Obligatory Annoying Section About Conventions

With Fourier transforms, as with so many other mathematical topics, different people use different conventions for no good reason. In the discussion below remember that $\hat{f}(p)$ is the Fourier transform of $f(x)$ and $f(x)$ is the "inverse Fourier transform" of $\hat{f}(p)$. (At least everyone agrees on those terms.)

We defined the Fourier transform with a factor of $1/(2\pi)$ in front of the integral, and the inverse Fourier transform with no factor in front. Some authors reverse that convention and put the $1/(2\pi)$ in the inverse Fourier transform. Still others put $1/\sqrt{2\pi}$ in *each* formula. The only mathematical requirement is that the product of the two factors must equal $1/(2\pi)$. At least that's true when the formulas are expressed in terms of angular frequency. Some authors express everything in terms of frequency instead. What you gain for that is that there's no factor in front of either integral. What you pay is that you have an extra factor of 2π in the exponents because you replace p with $2\pi f$. Finally, sometimes the signs of the exponents are reversed: positive in the Fourier transform formula and negative in the inverse Fourier transform.

We chose our definition because it is one that is widely used, and it shows most clearly the similarities between Fourier transforms and Fourier series. You need to be careful, however, because you can never assume two sources use the same definitions. In particular, if you use a computer program that happens to use a different definition to find Fourier transforms while working problems in this book, your results will not match ours. In that case you will need to look up the conventions used by that program and convert your answers accordingly.

Stepping Back

In this section we've introduced the Fourier transform, which serves essentially the same role for non-periodic functions that Fourier coefficients do for periodic ones. If $\hat{f}(7)$ is large then you know that $f(x)$ varies in some way on the length scale $2\pi/7$. If you know the entire function $\hat{f}(p)$ then you can reconstruct $f(x)$ and vice versa. They contain the same information in two different forms.

If $\hat{f}(p)$ is sharply peaked at one or more frequencies then $f(x)$ varies more or less sinusoidally at those frequencies. A sharp peak at $p = 0$ corresponds to a roughly constant $f(x)$. If $f(x)$ has a bump in its plot, then $\hat{f}(p)$ will also have a bump. The narrower the bump in $f(x)$, the wider the bump in $\hat{f}(p)$.

By now you may be impatiently waiting for us to get past all these plots and intuition-stuff and get into the real business of calculating Fourier transforms for a bunch of functions. We're not going to do that. There are functions you can analytically Fourier transform, and you'll work some examples in the problems. The number of useful ones is not that large, though, and you can look them up or calculate them on a computer when you need them. The hard skill, which you need to focus on, is interpreting the Fourier transform once you have it. Hopefully this section has given you a good start on that.

9.6.3 Problems: Fourier Transforms

For Problems 9.79–9.83 sketch the Fourier transform $\hat{f}(p)$ corresponding to the given function $f(x)$. Label the x-values on the horizontal axis. (You do not need to label the vertical axis.)

9.79
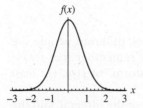

9.80
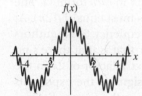

9.81
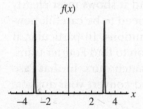

9.82
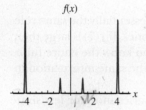

9.83

For Problems 9.84–9.88 calculate the Fourier transform of $f(x)$. Set up the integral by hand, but you may use a computer to evaluate it for you. Then sketch $f(x)$ and $\hat{f}(p)$.

9.84 $f(x) = e^{-3x^2}$

9.85 $f(x) = 5e^{-3x^2}$

9.86 $f(x) = e^{-2|x|}$

9.87 $f(x) = 1/(1 + x^2)$

9.88 $f(x) = (\cos x)e^{-x^2}$

9.89 Let $f(t)$ equal 0 for $t < 0$ and e^{-t/t_0} for $t \geq 0$, where t_0 is a constant. Find the Fourier transform of $f(t)$.

9.90 Let $f(t)$ equal $\sin t$ for $0 \leq t \leq \pi$ and 0 for all other values of t.
  (a) Find the Fourier transform of $f(t)$.
  (b) You should have found an answer that is invalid at $\omega = \pm 1$. Find $\hat{f}(1)$ and $\hat{f}(-1)$, either by evaluating those integrals separately or by taking a limit of your previous answer.

9.91 Answer each of the following questions using Equation 9.6.2.
  (a) Prove that if $f(x)$ is real, $\hat{f}(-p) = \hat{f}^*(p)$.
  (b) Prove that if $f(x)$ is real and even, then $\hat{f}(p)$ is real.
  (c) What can you conclude about $\hat{f}(p)$ if $f(x)$ is real and odd?

9.92 The Frequency Scaling Theorem In this problem you will find and use the "frequency scaling theorem" (also sometimes called the "time/frequency scaling theorem" or "similarity theorem").
  (a) Let $\hat{f}(\omega)$ be the Fourier transform of $f(t)$. Express the Fourier transform of $f(ct)$ in terms of $\hat{f}$, assuming c is a constant. That relationship is the theorem.
  (b) The Fourier transform of $4\sqrt{\pi}xe^{-x^2}$ is $-ipe^{-p^2/4}$. Use the frequency scaling theorem to find the Fourier transform of $4\sqrt{\pi}xe^{-4x^2}$.
  (c) Draw a Gaussian curve that represents the Fourier transform $\hat{g}(p)$ for a particular function $g(x)$. Then on the same graph draw the Fourier transform of $g(x/2)$. (This problem can be done without any calculations; the focus is on the relative shapes of the two curves, not on their numerical values.)

9.93 We said in the Explanation (Section 9.6.2) that in the limit where $f(x)$ approaches a pure sine wave, $\hat{f}(p)$ approaches an infinitely thin spike. No mathematical function can meet that description, so the "Dirac delta function" was invented. The Dirac delta function, written $\delta(x)$, is defined by three characteristics: $\delta(x) = 0$ for any $x \neq 0$, $\delta(0)$ is infinitely large, and $\int_{-\infty}^{\infty} \delta(x) = 1$. The other important property of the Dirac delta function that you need to know is that $\int_{-\infty}^{\infty} f(x)\delta(x)dx = f(0)$. We discuss the Dirac delta function further in Chapter 10.

(a) Sketch a plot of the Dirac delta function. Your drawing will of course be only approximate.

(b) If $\hat{f}(p) = \delta(p)$, calculate $f(x)$.

(c) If $\hat{f}(p) = \delta(p-1)$, calculate $f(x)$.

(d) Explain why both of your results make sense. In other words, if you had been given the functions $f(x)$ that you found, how could you have known that their Fourier transforms would be $\delta(p)$ and $\delta(p-1)$ respectively?

9.94 In this problem you'll show how shifting a function $f(x)$ to the left affects its Fourier transform.

(a) Sketch the functions $f(x) = e^{-x^2}$ and $g(x) = e^{-(x+1)^2}$. How do the two plots differ?

(b) Write an integral for the Fourier transform $\hat{g}(p)$ of $g(x) = e^{-(x+1)^2}$.

(c) Use a u-substitution to rewrite your integral from Part (b) in terms of the Fourier transform $\hat{f}(p)$ of $f(x)$.

(d) Redo Parts (b)–(c) with a generic function $f(x)$ and a shifted version $g(x) = f(x + x_0)$. What effect does the x_0 have on the Fourier transform?

9.95 Look back at the plots in the example "The Fourier Transform of Some Random Bumps" on Page 479. Use Parseval's theorem to explain why the plot of $\hat{g}(p)$ had to be shorter than the plot of $\hat{f}(p)$.

9.96 In this problem you'll see how a Fourier series becomes a Fourier transform in the correct limit.

(a) Let $f(x) = e^{-x^2}$ from $x = -1$ to 1, repeated periodically thereafter. Find the complex exponential Fourier series for $f(x)$ and plot the coefficients as a function of frequency p (not of n), with p going from -10 to 10.

(b) Repeat Part (a) for $f(x) = e^{-x^2}$ from $x = -2$ to 2, repeated periodically thereafter.

(c) Repeat Part (a) for $f(x) = e^{-x^2}$ from $x = -10$ to 10, repeated periodically thereafter.

(d) What is happening to the frequencies of the coefficients as you increase the period? What is happening to the shape of the plot? What is happening to the height of the plot?

(e) Plot the Fourier transform of $f(x) = e^{-x^2}$ as p goes from -10 to 10. How does this plot compare to the previous ones in what frequencies are included, the shape of the plot, and the height of the plot?

9.97 Create an animation showing the plot of the Fourier transform of $f(x) = Ae^{-(x/w)^2}$ as you vary A and w over some ranges of values. Describe the effect each of them has on the plot of $\hat{f}(p)$.

9.98 Exploration: The Fourier Transform of a Gaussian.
Let $f(x) = e^{-x^2}$.

$$\hat{f}(p) = \frac{1}{2\pi} \int_{-\infty}^{\infty} e^{-x^2} e^{-ipx} dx \quad (9.6.3)$$

(a) Differentiate both sides of Equation 9.6.3 with respect to p.

(b) Rewrite your new integrand in the form of (a function of p and x) multiplied by $(d/dx)e^{-x^2}$.

(c) Integrate the resulting integral by parts. Rewrite your result as a multiple of $\hat{f}$ using Equation 9.6.3. (If it's not a multiple of $\hat{f}$ something has gone wrong.)

(d) You should now have an equation for $d\hat{f}/dp$ in terms of $\hat{f}$. Use separation of variables to solve this equation for $\hat{f}(p)$. Your answer will have an arbitrary constant in it.

(e) Evaluate $\hat{f}(0)$ using the formula $\int_{-\infty}^{\infty} e^{-x^2} dx = \sqrt{\pi}$ and use it to find the arbitrary constant, thus giving you $\hat{f}(p)$.

(f) Now let $f(x) = Ae^{-[(x-x_0)/w]^2}$. Use a u-substitution and the result you just found for e^{-x^2} to find $\hat{f}(p)$.

9.7 Discrete Fourier Transforms (see felderbooks.com)

9.8 Multivariate Fourier Series (see felderbooks.com)

9.9 Additional Problems (see felderbooks.com)

CHAPTER 10

Methods of Solving Ordinary Differential Equations

> *Before you read this chapter, you should be able to...*
> - model real-world situations with ordinary differential equations and interpret solutions of these equations to predict physical behavior (see Chapter 1).
> - given a differential equation and a proposed solution, determine if that proposed solution works (see Chapter 1).
> - solve differential equations by "separation of variables" and by "guess and check" (see Chapter 1).
> - use initial conditions to determine the arbitrary constants in a general solution (see Chapter 1).
> - evaluate integrals with u-substitution and integration by parts.
> - evaluate and interpret partial derivatives, and set up equations with differentials (see Chapter 4).
> - use determinants to see if a set of vectors is linearly independent (for Section 10.6 only, see Chapter 6).
>
> *After you read this chapter, you should be able to...*
> - use operator notation and identify linear operators.
> - draw "phase portraits" for second-order ODEs or coupled first-order ODEs.
> - solve linear first-order differential equations.
> - identify and solve "exact" differential equations.
> - use the Wronskian to determine if differential equation solutions are linearly independent, and therefore compose a general solution.
> - use variable substitution to simplify differential equations, including (but not limited to) the Bernoulli equation and the "other kind of homogeneous" equation.
> - find differential equation solutions using the techniques "variation of parameters" and "reduction of order."
> - graph functions based on the Heaviside and Dirac delta functions.
> - solve differential equations by using Laplace transforms.
> - solve differential equations by using Green's functions.

This is intended as a follow-up to Chapter 1, "Introduction to Ordinary Differential Equations." Work through that chapter, or some other source that covers the same material, before beginning this one. We list above some of the skills that we are assuming you already have.

This chapter describes some techniques of solving ordinary differential equations that can't be solved by the methods of Chapter 1. Along the way we introduce two important functions (the Heaviside function and the Dirac delta function) and an operator (the Laplace transform). These topics are presented in the context of ordinary differential equations, but their importance extends beyond that.

The following integrals come up in a number of problems throughout this chapter. We've placed them here for easy reference and will refer back to this spot when they are needed.

$$\int e^{kx} \sin(bx)\,dx = \frac{e^{kx}}{k^2 + b^2}[k\sin(bx) - b\cos(bx)] + C \quad (k^2 + b^2 \neq 0)$$

$$\int e^{kx} \cos(bx)\,dx = \frac{e^{kx}}{k^2 + b^2}[k\cos(bx) + b\sin(bx)] + C \quad (k^2 + b^2 \neq 0)$$

10.1 Motivating Exercise: A Damped, Driven Oscillator

An object with mass m is attached to an ideal spring that exerts a force $F_{spring} = -kx$, but is also slowed down by a drag force $F_{drag} = -bv$.

1. Write the equation of motion for this object. That is, write the differential equation for $x(t)$ based on the given forces and Newton's law $F_{total} = ma$.
2. Now write the equation of motion assuming $m = 1$ kg, $b = 6$ N·s/m, and $k = 9$ N/m.

   See Check Yourself #62 in Appendix L

3. Show that $x = Ae^{-3t} + Bte^{-3t}$ is a valid solution to your differential equation for any values of the constants A and B. You will see in Section 10.2 how we found that solution; right now we are just asking you to verify that it works.
4. What additional information would you need to determine the constants A and B? What kind of behavior does this solution describe, and why does this make sense for our scenario?

Now, in addition to the forces of the spring and drag, there is an external "driving force" F_e acting on the spring. In the following parts you will be writing differential equations that you do not yet know how to solve.

5. Write the equation for $x(t)$ for the driving force $F = t^5 e^{-3t}$. (Assume everything is in SI units. For example the 3 in the exponent is really 3 s^{-1}.) In Section 10.9 (see felderbooks.com) we'll show you how to solve equations like this.
6. Assume the driving force is a constant 3 N from $t = 0$ to $t = 10$ s and zero before and after that. Write the differential equation for $x(t)$. In Section 10.10 we'll show you a more convenient way to write such equations, and in Section 10.11 we'll show you an efficient way to solve them.

In Section 10.13 Problem 10.323 (see felderbooks.com) you'll return to this example and solve it for the two driving functions given above.

10.2 Guess and Check

Chapter 1 presented the method of guess and check and the principle of linear superposition. In this section we review this critical information and fill in some missing details. We also introduce the idea of an "operator" and some attendant notation.

10.2.1 Discovery Exercise: Guess and Check

A linear second-order differential equation starts with the following expression:

$$a_2(x)\frac{d^2y}{dx^2} + a_1(x)\frac{dy}{dx} + a_0(x)y \quad (10.2.1)$$

Suppose the three different functions $c_1(x)$, $c_2(x)$, and $p(x)$ have the following properties: if you plug $y = c_1(x)$ into Equation 10.2.1 you get 0. If you plug in $y = c_2(x)$ in, you again get 0. And if you plug in $y = p(x)$, you get out the function $A(x)$.

1. What do you get if you plug $y = 3c_1(x)$ into Equation 10.2.1?
2. What do you get if you plug $y = 5c_2(x)$ into Equation 10.2.1?
3. What do you get if you plug $y = 10p(x)$ into Equation 10.2.1?
4. What do you get if you plug $y = 13c_1(x) - 12c_2(x) + p(x)$ into Equation 10.2.1?

See Check Yourself #63 in Appendix L

5. Would your answers change if Equation 10.2.1 began with $a_3(x)y'''(x)$? (Assume that the functions $c_1(x)$, $c_2(x)$, and $p(x)$ still gave the same answers as before.)
6. Would your answers change if Equation 10.2.1 contained $a_0(x)y^2$ instead of $a_0(x)y$? (same comment)

10.2.2 Explanation: Guess and Check

This chapter is not an introduction to differential equations. This is intended as a follow-up to Chapter 1, "Introduction to Ordinary Differential Equations." Work through that chapter, or some other source that covers the same material, before beginning this one.

This section in particular follows in the footsteps of Section 1.6, "Guess and Check, and Linear Superposition." We will recap some of the definitions and techniques discussed in that section and fill in some missing details.

We begin by introducing some terminology that makes it easier to talk about differential equations.

Differential Operators and Linear Differential Equations

You know that a "function" is a systematic process that turns one number into another number. In an analogous way, an "operator" is a systematic process that turns one function into another function.

For example, let D stand for the operator "derivative with respect to x" or d/dx. Then we can write $D(x^2) = 2x$ and $D(\sin x) = \cos x$ and so on.

When we write D^2 that means applying the same operator twice. This confuses people enough that it's worth repeating. *Squaring an operator does not mean applying the operator and then squaring the result; it means applying the operator twice.* $D^2 y = d^2y/dx^2$, $D^2 y \neq (dy/dx)^2$.

A "linear operator" L has the property that for any functions f_1 and f_2, and any constants a and b,

$$L(af_1 + bf_2) = aL(f_1) + bL(f_2)$$

EXAMPLE **Linear Operator**

Consider the operator:

$$P = \frac{d^2}{dx^2} + e^x \frac{d}{dx} + 3$$

(The operator 3 means "multiply the function by 3.")

Question: Find Py where $y = \ln x$.

Answer:

$$Py = \frac{d^2y}{dx^2} + e^x \frac{dy}{dx} + 3y = -\frac{1}{x^2} + \frac{e^x}{x} + 3\ln x$$

Question: Show that P is a linear operator.

Answer:

$$P(af_1 + bf_2) = \frac{d^2}{dx^2}(af_1 + bf_2) + e^x \frac{d}{dx}(af_1 + bf_2) + 3(af_1 + bf_2)$$

$$= a\frac{d^2 f_1}{dx^2} + b\frac{d^2 f_2}{dx^2} + e^x a \frac{df_1}{dx} + e^x b \frac{df_2}{dx} + 3af_1 + 3bf_2$$

$$= a\left(\frac{d^2 f_1}{dx^2} + e^x \frac{df_1}{dx} + 3f_1\right) + b\left(\frac{d^2 f_2}{dx^2} + e^x \frac{df_2}{dx} + 3f_2\right)$$

$$= aP(f_1) + bP(f_2)$$

In Chapter 1 we defined a "linear" differential equation. That definition is better expressed in terms of operators.

Linear Differential Equations

Our original definition: In a "linear homogeneous" differential equation for a function $y(x)$ every non-zero term is a function of x or a constant, multiplied by y or one of its derivatives.

New, but completely equivalent, definition: A "linear" differential equation is of the form:

$$Ly = f(x) \tag{10.2.2}$$

where L is a linear differential operator.

If $f(x) = 0$ then the equation is "homogeneous." If $f(x) \neq 0$ then replacing $f(x)$ with zero creates the "complementary homogeneous equation."

One of the useful properties of linear differential equations—not guaranteed for non-linear equations!—is that every nth-order linear equation has a general solution with n arbitrary constants. An even more important property of linear differential equations is that they obey the principle of linear superposition, described below.

Suppose the function y_p is a solution to Equation 10.2.2: in other words, suppose $Ly_p = f(x)$. And suppose the linearly independent functions y_{c1}, y_{c2} ... y_{cn} are all solutions to the complementary homogeneous equation: in other words, $Ly_{ci} = 0$ for all i. The fact that L is a linear operator means that:

$$L(y_p + A_1 y_{c1} + A_2 y_{c2} + \ldots + A_n y_{cn}) = f(x) \tag{10.2.3}$$

Make sure you follow how Equation 10.2.3 follows from the definition of a linear operator. The function y_p is called the "particular solution" because it is just one solution (typically with no arbitrary constants) to the equation we want to solve. The function $A_1 y_{c1} + A_2 y_{c2} + \ldots + A_n y_{cn}$ is called the "complementary solution" because it solves the complementary homogeneous equation. The sum of the particular and complementary solutions is itself a solution to $Ly = f(x)$.

If L is an nth-order differential operator, and if all the y_{ci} functions are linearly independent functions, then we don't just have "a solution": we have the *general* solution, because we have the requisite number of independent arbitrary constants. Section 10.6 (see felderbooks.com) discusses the Wronskian, a tool for determining if a given set of solutions are linearly independent.

The fact that you can make a solution to any linear equation by combining solutions in the form $y_p + A_1 y_{c1} + A_2 y_{c2} + \ldots + A_n y_{cn}$ is called the "principle of linear superposition."

Chapter 10 Methods of Solving Ordinary Differential Equations

Guess and Check

Linearity tells us how to put functions together to find general solutions, but it does not help us find those functions in the first place. Much of this chapter will be dedicated to the process of finding those solutions. (Almost every differential equation we solve in this chapter will be linear.) One of the most valuable techniques is "guess and check": you try a likely looking function with one or more constants left undetermined, and solve for the values of those constants.[1]

But how do you pick the right guess? You can get pretty far with the advice we gave in Chapter 1: "try stuff, but be smart."

EXAMPLE **Guess and Check, and Linear Superposition**

Problem:
Find the general solution to the following differential equation.

$$\frac{d^2y}{dx^2} + 3\frac{dy}{dx} - 4y = 50\cos(2x) \quad \text{or} \quad Ly = 50\cos(2x) \quad \text{where} \quad L = \frac{d^2}{dx^2} + 3\frac{d}{dx} - 4 \quad (10.2.4)$$

Solution:
We begin by solving the complementary homogeneous equation:

$$\frac{d^2y}{dx^2} + 3\frac{dy}{dx} - 4y = 0 \quad \text{or} \quad Ly = 0 \quad (10.2.5)$$

What kind of guess would make sense here? If we tried a sine, then y and y'' would both be sines (and might happily cancel) but y' would be a cosine (and nothing would cancel it). You can similarly convince yourself *without going through the algebra* for $y = x^2$ and $y = \ln x$ and most other basic functions that the terms couldn't possibly cancel. But an exponential function is a reasonable guess, since y and y' and y'' would all look similar. Based on this, we try $y_c = e^{kx}$ in Equation 10.2.5.

$$k^2 e^{kx} + 3k e^{kx} - 4 e^{kx} = 0$$

Factoring out the common e^{kx} (which is never zero) we solve to get $k = -4$ or $k = 1$, meaning that y_c can be e^{-4x} or e^x. We have solved Equation 10.2.5.

We now need to find any particular solution to the original Equation 10.2.4. The right side looks like $\cos(2x)$, but as we said before, if $y = \cos(2x)$ then y' will give us a sine term that no other term will cancel. So instead we try a sine *and* a cosine.

$$y_p = A\sin(2x) + B\cos(2x) \quad (10.2.6)$$

This A and B are not arbitrary constants; we're hoping to find just the right values that will make a solution. Substitute Equation 10.2.6 into Equation 10.2.4.

$$-4A\sin(2x) - 4B\cos(2x) + 6A\cos(2x) - 6B\sin(2x) - 4A\sin(2x) - 4B\cos(2x) = 50\cos(2x)$$

---

[1] The method that we call "Guess and Check" is, under certain circumstances, called "The Method of Undetermined Coefficients." The function that you plug in, which we call a "guess," is sometimes called an "ansatz." An unhealthy obsession with impressive-sounding words is sometimes called "pedantry."

For this equation to be true for all values of x, the coefficients of $\sin(2x)$ on the left side must add up to zero, and the coefficients of $\cos(2x)$ to 50.

$$-4A - 6B - 4A = 0 \quad \text{and} \quad -4B + 6A - 4B = 50$$

Solving, $A = 3$ and $B = -4$ are the values that make Equation 10.2.6 a solution.

We have found one "particular" solution $y_p(x)$ to Equation 10.2.4, and two "complementary" solutions $y_c(x)$ to Equation 10.2.5. Linearity tells us how to put it all together. Because $Ly_p = 50\cos(2x)$, we know that $L(2y_p)$ would give us $100\cos(2x)$, which is not a solution: you cannot multiply your particular solution by an arbitrary constant. But because $Ly_c = 0$ for both y_c functions, we know that any linear combination of these functions will still give us zero, and *adding* such a function to y_p will still give us $50\cos(2x)$. So the general solution, the solution with a full suite of arbitrary constants, is:

$$y = 3\sin(2x) - 4\cos(2x) + C_1 e^{-4x} + C_2 e^x$$

We leave it to you to verify that this solves Equation 10.2.4 for any values of the constants C_1 and C_2.

We stressed in that example, not only how to plug in and work with a guess, but how to begin with a reasonable guess. Below we present some guidelines for specific situations. While the list is by no means exhaustive, it conveys the kinds of things you will look for.

Guessing hints for homogeneous equations

For a homogeneous equation your guess is determined by the functions of x that are multiplied by y, y', y'' and so on. Remember that your goal is to find n linearly independent solutions, where n is the order of the equation, so you can find a general solution with n arbitrary constants. Of course these hints also apply to finding the *complementary* solution to an inhomogeneous equation.

- If all the coefficients are constants, guess e^{kx}.

    Example: $y'' + 3y' - 4y = 0$
    Guess: $y = e^{kx} \to (k^2 + 3k - 4)e^{kx} = 0 \to k = -4, k = 1$
    Solution: $y = C_1 e^{-4x} + C_2 e^x$

    Whenever all the coefficients are constants this guess will lead to a polynomial to solve for k. That polynomial will generally have n independent solutions; in "Some Things That Can Go Wrong" below we discuss cases where it doesn't.

- If the coefficients are descending powers of x (a "Cauchy-Euler equation"), try $y = x^k$.

    Example: $x^5 y'' + 9x^4 y' + 12x^3 y = 0$
    Guess: $y = x^k \to (k(k-1) + 9k + 12)x^{k+3} = 0 \to k = -2, k = -6$
    Solution: $y = C_1 x^{-2} + C_2 x^{-6}$

    Step yourself through the algebra on that one. (We skipped several steps.) Pay attention to how the descending powers of x in the differential equation (x^5, x^4, x^3) led to the common power (x^{k+3}) that then dropped out.

- You should recognize the simple harmonic oscillator equation $y'' + p^2 y = 0$ whenever you see it, and without a moment's hesitation write down the solution $y(x) = C_1 \sin(px) + C_2 \cos(px)$.

$$\text{Example: } \frac{d^2 y}{dx^2} + \frac{3n^2 + 5m^4}{36} y = 0$$

$$\text{Solution: } y = C_1 \sin\left(\frac{\sqrt{3n^2 + 5m^4}}{6} x\right) + C_2 \cos\left(\frac{\sqrt{3n^2 + 5m^4}}{6} x\right)$$

Guessing hints for inhomogeneous equations

For an inhomogeneous equation your guess is usually dictated by the inhomogeneous term ($f(x)$ in Equation 10.2.2). Remember that your goal is just to find one particular solution—any function, the simpler the better!—that works.

- If the right side and the coefficient of y are both constants you can immediately find a constant solution.

$$\text{Example: } (\sin x) \frac{d^2 y}{dx^2} - (\ln x) \frac{dy}{dx} + 3y = 12$$

$$\text{Solution: } y = 4$$

That solution is easy to verify: if y is a constant then all its derivatives are zero, so it works. That gives you one simple solution, and that's all you need for this phase of the process.

- For an exponential function, guess an exponential function with the same power.

$$\text{Example: } \frac{d^2 y}{dx^2} + 3 \frac{dy}{dx} + 2y = e^{3x}$$

$$\text{Guess: } y = Ae^{3x} \quad \rightarrow \quad (9A + 9A + 2A)e^{3x} = e^{3x} \quad \rightarrow \quad A = \frac{1}{20}$$

$$\text{Solution: } y = \frac{1}{20} e^{3x}$$

- For a polynomial, guess a polynomial of the same order.

$$\text{Example: } 3 \frac{d^2 y}{dx^2} + 2 \frac{dy}{dx} + 7y = 14 x^3$$

$$\text{Guess: } y = A_3 x^3 + A_2 x^2 + A_1 x + A_0$$

$$\rightarrow \quad 7 A_3 x^3 + (7 A_2 + 6 A_3) x^2 + (7 A_1 + 4 A_2 + 18 A_3) x + (7 A_0 + 2 A_1 + 6 A_2) = 14 x^3$$

$$\rightarrow \quad 7 A_3 = 14, \; 7 A_2 + 6 A_3 = 0, \; 7 A_1 + 4 A_2 + 18 A_3 = 0, \; 7 A_0 + 2 A_1 + 6 A_2 = 0$$

$$\text{Solution: } y = 2 x^3 - \frac{12}{7} x^2 - \frac{204}{49} x + \frac{912}{343}$$

Your first guess might have been a solution of the form $y = Ax^3$. Do you see why that will generally *not* be successful without the lower order terms?

Also pay particular attention to the last step of that "guess" where we turned one equation into four. If two different third-order polynomials equal each other *for all values of x* then they must have the same coefficient of x^3, and the same coefficient of x^2, and so on—four equations in all.

- When you have sines or cosines, guess both sines *and* cosines.

  Example: $\dfrac{d^2y}{dx^2} - 4\dfrac{dy}{dx} + 5y = 130\cos(2x)$
  Guess: $y = A\sin(2x) + B\cos(2x)$
  $\rightarrow \quad (-4A + 8B + 5A)\sin(2x) + (-4B - 8A + 5B)\cos(2x) = 130\cos(2x)$
  $\rightarrow \quad A + 8B = 0,\ -8A + B = 130$
  Solution: $y = -16\sin(2x) + 2\cos(2x)$

Make sure you see why a cosine alone will generally not work, but sines and cosines together will. Also, as in our previous example, pay attention to the step where one equation becomes two: the coefficient of $\sin(2x)$ on the left equals the coefficient of $\sin(2x)$ on the right, and the same for $\cos(2x)$.

We cannot possibly list every case, so it's better to see the list above as a *way of thinking* rather than a checklist. For instance, suppose the inhomogeneous term were $4e^{2x}\cos(3x)$. What kind of functions would be worth trying? (See Problem 10.22.) Also be aware that the guessing is harder (but sometimes possible) when the coefficients on the left are not constant.

As a final (and very important) note: if the inhomogeneous term is a sum of terms, linearity saves the day once again. For instance, to solve $Ly = 10 + 3\sin x$ you can solve $Ly = 10$ and $Ly = 3\sin x$ separately and then add the two solutions.

Some things that can go wrong, and their solutions

Guessing and checking doesn't always work. (If it did, this would be a very short chapter.) But in some cases where it seems to fail, there are workarounds.

If this happens: Your guess leads to a polynomial in k that has no real roots.
 ...*try this:* Learn to love complex numbers!

  Example: $y'' + 6y' + 13y = 0 \rightarrow y = e^{kx} \rightarrow k^2 + 6k + 13 = 0$
  Solution: $y = C_1 e^{(-3+2i)x} + C_2 e^{(-3-2i)x}$

If you know how to read complex functions, this solution tells you clearly that the system oscillates with a period of π and a decaying amplitude. If you need a real solution, you can convert to $y = e^{-3x}\left[C_1\cos(2x) + C_2\sin(2x)\right]$

This process is covered in more detail in Chapter 3.

If this happens: Your equation has constant coefficients, so you guess an exponential function. When you solve the quadratic function for k you get only one root (a "double root"). So now you have one solution but you need two.
 ...*try this:* Multiply your solution by x to get a second solution.

  Example: $y'' + 6y' + 9y = 0 \rightarrow y = e^{kx} \rightarrow k = -3$
  Solution: $y = C_1 e^{-3x} + C_2 x e^{-3x}$

In Problem 10.23 you will show that this always works in this situation. In Section 10.9 (see felderbooks.com) you will learn the method of "Reduction of Order" which can be used to derive this result.

If this happens: Your equation has descending powers of x (a Cauchy-Euler equation) so you guess x^k. Just as we discussed with exponential functions above, you end up with a "double root" and therefore only one solution.

…try this: Multiply your solution by $\ln x$ to get a second solution.

Example: $x^2 y'' + 5xy' + 4y = 0 \;\rightarrow\; y = x^k \;\rightarrow\; k = -2$
Solution: $y = C_1/x^2 + C_2 \ln x / x^2$

In Problem 10.24 you will show that this always works in this situation. In Section 10.9 (see felderbooks.com) you will learn the method of "Reduction of Order" which can be used to derive this result.

If this happens: You're solving for a particular (inhomogeneous) solution, and you stumble upon a solution to the homogeneous equation.

…try this: Multiply your guess by x. (Yes, we gave you the same advice above, but for a totally different problem.)

Example: $y'' + 3y' + 2y = e^{-2x} \;\rightarrow\; y_p = Ae^{-2x} \;\rightarrow\; 0 = e^{-2x}$
Solution: Guess $y_p = Axe^{-2x}$ and then solve for A

If multiplying by x gives you *another* solution to the homogeneous equation, try multiplying by x^2 instead.

10.2.3 Problems: Guess and Check

10.1 Walk-Through: Guess and Check. In this problem you will solve the equation:

$$\frac{d^2 y}{dx^2} + 3\frac{dy}{dx} + 2y = 6x^2 \qquad (10.2.7)$$

(a) Write the complementary homogeneous equation.

(b) Solve the equation you wrote in Part (a) by guessing a solution of the form $y_c = e^{kx}$ and solving for k. Your final solution should have two arbitrary constants.

(c) To find a particular solution to Equation 10.2.7, begin by plugging in a solution of the form $y_p = Ax^2 + Bx + C$.

(d) Part (c) should have left you with an equation that set two quadratic functions equal to each other. Set the coefficient of x^2 on the left side of this equation equal to the coefficient of x^2 on the right side. Do the same for the coefficient of x and the constant term, and you have three equations to solve for three unknowns.

(e) Solve your equations to find a particular solution.

(f) Write the general solution to Equation 10.2.7. Your solution should have two arbitrary constants.

(g) Verify that your solution works.

10.2 If P is the operator $d^2/dx^2 - 9$, then $Py = d^2 y/dx^2 - 9y$.

(a) Find and simplify $P(\sin x)$.
(b) Find and simplify $P^2 y$.
(c) Write the operator P^2.
(d) Show that P is a linear operator.
(e) Write the differential equation $Py = 0$ and verify that Ae^{3x} is a valid solution.
(f) Find another valid solution to $Py = 0$.

10.3 Let N be the operator defined by $Ny = y(dy/dx)$.

(a) Find Ny where $y = 1/x$.
(b) Show that N is not a linear operator.
(c) Show that $y = \sqrt{x}$ is a valid solution to $Ny = 1/2$.
(d) Show that $y = 3\sqrt{x}$ and $y = \sqrt{x} + 3$ are *not* valid solutions to $Ny = 1/2$.
(e) Find another valid solution to $Ny = 1/2$.

10.4 Equation 10.2.2 represents a linear differential equation, based on a linear operator L.

(a) Suppose y_1 and y_2 are both solutions of this equation. What is $L(y_1 + y_2)$?

(b) In general, $y_1 + y_2$ is *not* a solution to Equation 10.2.2. Under what circumstances *is* it a solution?

10.5 Consider the equation:

$$Ly = 10 \quad \text{where} \quad L = x^3 \frac{d^2}{dx^2} + x^2 \frac{d}{dx} - x \quad (10.2.8)$$

(a) Show that $y_c = x$ and $y_c = 1/x$ are both solutions to the complementary homogeneous equation.

(b) Show that $y = (\ln x)/x$ is not a solution to either the original equation or the complementary homogeneous equation.

(c) Find a solution to Equation 10.2.8. *Hint*: The function that did not work in Part (b), and the fact that L is a linear operator, will enable you to just write down this solution.

(d) Write the general solution to Equation 10.2.8.

10.6 Prove that if L is a linear operator then L^2 is also.

In Problems 10.7–10.22 find the general solution $y(x)$ to the given linear differential equation. Assume any letter other than x or y is a constant.

10.7 $2y'' + 7y' - 4y = 20$

10.8 $2y'' + 7y' - 4y = e^x$

10.9 $2y'' + 7y' - 4y = 3e^{-4x}$

10.10 $y'' = -k^2 y + a \sin(px) \quad (k \neq p)$

10.11 $y'' = -9y + 26 e^{2x}$

10.12 $y' = 2y + 3x + 5$

10.13 $y' + ky = px^2$

10.14 $y'' - 4y' + 4y + 169 \cos(3x) = 0$

10.15 $y'' + 2y' + 5y = 25x$

10.16 $y'' + ay' + by + c = 0$

10.17 $y'' + ay' + by + c + dx = 0$

10.18 $x^2 y'' - 6xy' - 18y + 6 = 0$

10.19 $3x^3 y'' + x^2 y' - 8xy = 24x$

10.20 $ax^3 y' + bx^2 y = c$ *Hint*: for the particular solution start with the guess $y_P = Ax^k$ and solve for A and k.

10.21 $y'' = -9y' - 18y + 27 e^{-x} - 34 \cos(4x)$

10.22 $y'' - 2y' + y = 8 e^{3x} \cos(2x)$

10.23 The Explanation (Section 10.2.2) discusses the unusual case where you plug in the guess e^{kx} and the equation for k has only one solution, a "double root."

(a) We begin with the specific case of $y'' - 10 y' + 25 y = 0$. Guess a function of the form e^{kx} and show that you find one value of k, and therefore one solution with one arbitrary constant.

(b) Multiply your answer from Part (a) by x and show that the resulting function is also a valid solution of the differential equation. Based on this, write the complete solution with two arbitrary constants.

(c) The more general case is $y'' + 2Ry' + R^2 y = 0$ where R is any constant. Repeat Parts (a) and (b) for this differential equation to show that the trick always works.

10.24 The Explanation (Section 10.2.2) discusses the unusual case where you plug in the guess x^k and the equation for k has only one solution, a "double root."

(a) We begin with the specific case of $x^2 y'' + 3xy' + y = 0$. Guess a function of the form x^k and show that you find one value of k, and therefore one solution with one arbitrary constant.

(b) Multiply your answer from Part (a) by $\ln x$ and show that the resulting function is also a valid solution of the differential equation. Based on this, write the complete solution with two arbitrary constants.

(c) The more general case is $x^2 y'' + (2R + 1)xy' + R^2 y = 0$ where R is any constant. Repeat Parts (a) and (b) for this differential equation to show that the trick always works.

10.25 The "Fourier transform" of the function $f(x)$ is defined by the formula:

$$\hat{f}(p) = \frac{1}{2\pi} \int_{-\infty}^{\infty} f(x) e^{-ipx} \, dx$$

Is the operator that turns $f(x)$ into $\hat{f}(p)$ a linear operator or not? Prove your answer.

In Problems 10.26–10.29 a block of mass m is attached to a spring with spring constant k. In addition to the spring force the block experiences a damping force $F_D = -bv$ where b is a constant and v is the block's velocity. For each set of constants given, write the differential equation for the block's motion and find the general solution to it. If initial conditions are given, plug those in and solve for the values of the arbitrary constants. Finally, give a brief qualitative description of the resulting motion.

10.26 $m = 2$ kg, $b = 6$ N·s/m, $k = 4$ N/m.

10.27 $m = 2$ kg, $b = 4$ N·s/m, $k = 4$ N/m,
$x(0) = 0$, $v(0) = -2$ m/s.

10.28 $m = 1$ kg, $b = 2$ N·s/m, $k = 1$ N/m.

10.29 Leave m, b, and k as letters. In your qualitative description you will need to consider different cases, depending on those constants.

10.30 A 2 kg charged metal block is attached to a spring with spring constant 2 N/m. In addition to the spring force the block experiences a damping force $F_D = -bv$ and an electrical force $F_E = F_0 \cos(\omega t)$, where $b = 8$ N·s/m, $F_0 = 4$ N, and $\omega = 10$ s^{-1}.

(a) Write a differential equation for the position $x(t)$ of the block.

(b) Find the general solution to the equation you wrote.

(c) Describe the behavior of the block at late times.

10.31 The current in the circuit shown below obeys the differential equation $LI''(t) + RI'(t) + (1/C)I(t) = V'(t)$. The voltage is $V(t) = V_0 \sin(\omega t)$ where $V_0 = 120$ V and $\omega = 360$ s^{-1}. Initially $I = I' = 0$.

$R = 1000$ Ω
$C = 10^{-6}$ F
$L = 10$ H

(a) Find $I(t)$.

(b) Describe the behavior of the circuit at late times.

10.32 A sound wave in a long thin rod consists of small longitudinal displacements $u(x)$. A standing wave oscillating with angular frequency ω will obey the differential equation $E(x)u''(x) + E'(x)u'(x) + \rho\omega^2 u(x) = 0$, where $E(x)$ is the Young's modulus of the material and ρ is the density.

(a) Solve for the shape of a standing wave in an aluminum rod with $E = 10^{10}$ N/m^2, $\rho = 2700$ kg/m^3, and $\omega = 1000$ s^{-1}.

(b) Solve for the shape of a standing wave assuming a varying Young's modulus $E(x) = kx^2$, where k is constant. Leave ρ and ω as letters. Use the equation $E = kx^2$ to find the units of k, and then check to make sure your answer has correct units.

10.33 **Make Your Own.** Write a second-order inhomogeneous differential equation that isn't in this section (including the problems) and solve it using the techniques from this section.

10.3 Phase Portraits (see felderbooks.com)

10.4 Linear First-Order Differential Equations (see felderbooks.com)

10.5 Exact Differential Equations (see felderbooks.com)

10.6 Linearly Independent Solutions and the Wronskian (see felderbooks.com)

10.7 Variable Substitution

You are familiar with integration by u-substitution: you are looking for an antiderivative in terms of x and you define a new variable, u, for which the antiderivative is easier to find. The same technique is used more generally to solve differential equations. As with integration, figuring out the right substitution is a combination of guesswork, luck, and (mostly) practice.

This is a big and important topic, so we have broken it into two sections. This section discusses variable substitution in general. Section 10.8 focuses on three special cases: Bernoulli's equation, a particular form of equation that is unfortunately called "homogeneous," and equations where the dependent variable appears *only* in its derivatives.

10.7.1 Discovery Exercise: Variable Substitution

Consider the differential equation $dx/dt = (x + t)^2 - 1$. It's not linear, it's not separable, it's not exact, and it doesn't lend itself to any obvious guess, so none of the techniques we have discussed so far will help solve it. The trick to solving it is to rewrite it using different variables. The process is a generalization of what you do when you evaluate an integral using u-substitution. We're going to use the substitution that most readily suggests itself from this equation: $u = x + t$.

1. Recall that u and x are both functions of t. Take the derivative of both sides of $u = x + t$ to get an equation relating du/dt and dx/dt.

   See Check Yourself #65 in Appendix L

2. Use $u = x + t$ and the equation you found for du/dt to rewrite the original differential equation in terms of u and t, with no x in it.

   See Check Yourself #66 in Appendix L

3. Solve the resulting equation for $u(t)$ by separation of variables.
4. Write the general solution for $x(t)$. Your solution should include an arbitrary constant, and should not include the letter u.
5. Verify that your solution solves the original differential equation.

10.7.2 Explanation: Variable Substitution

A differential equation expresses a relationship between a dependent variable and an independent variable. (Throughout this explanation we will use x for the former and t for the latter, although they could of course be anything.) When you do a variable substitution you rewrite the differential equation in terms of a new variable (which we will call u). The hope, of course, is that you get something easier to solve than what you started with. In the end you convert back to the original variables, so u appears in the process but disappears before the end.

As you follow each example below, pay particular attention to how we convert dx/dt and d^2x/dt^2 to the new variable system.

- If you define u as a function of x, then u replaces x, giving you a differential equation for $u(t)$. The same thing happens if you define u as a function of both x and t. In these cases your new equation will have du/dt and d^2u/dt^2, so you must convert your derivatives to those.
- If you define u as a function of t only, then u replaces t, giving you an equation for $x(u)$. In this case you must convert your derivatives to dx/du and d^2x/du^2.

Those two cases involve very different processes, so we shall take them one at a time.

Defining a New Dependent Variable

Consider the differential equation

$$\frac{dx}{dt} = e^{x-2t} + 2 \qquad (10.7.1)$$

You should be able to quickly convince yourself that none of the techniques we've discussed so far will work on this equation. It's not separable, it's not linear, it's not exact, and there's no obvious guess to try. Since the problem is the $x - 2t$ term in the exponential, we'll see if the substitution $u = x - 2t$ simplifies things. (It will.)

When you evaluate an integral with u-substitution you have to find a relationship between du and dx, and use that relationship to replace dx with du in your integral. Similarly, we have to find the relationship to replace dx/dt with du/dt in our equation. To do that we differentiate both sides of $u = x - 2t$ with respect to t. That gives us

$$\frac{du}{dt} = \frac{dx}{dt} - 2$$

We could solve this for dx/dt and plug that into Equation 10.7.1, but it's a bit easier to plug Equation 10.7.1 into this relationship.

$$\frac{du}{dt} = \frac{dx}{dt} - 2 = e^{x-2t} = e^u$$

Now we have a separable equation, which gives us $\int e^{-u} du = \int dt$, which becomes $-e^{-u} = t + C$, and thus $u = -\ln(C - t)$. (We replaced C with $-C$ but since C just means "any number" we kept the same letter.) Finally, we plug in $u = x - 2t$ to get the solution $x(t)$.

$$x = 2t - \ln(C - t)$$

You can easily check that this solves Equation 10.7.1. As always, it's worth taking a moment to consider what this solution is telling us. It says that at some finite time $t = C$ (that depends on the initial condition) x will approach infinity. That's not surprising for a function whose rate of growth goes like the exponent of itself.

More important for our current purpose, though, is how we found that solution. We will show many examples of variable substitution throughout this section, but they will all follow the same pattern. Starting with a differential equation for $x(t)$...

1. Define a new variable u as a function of x and/or t. Use the rules of derivatives to rewrite dx/dt (and any higher order derivatives in your equation) in terms of u.
2. Rewrite the differential equation for that new variable. If you chose the right substitution, the payoff will be an easier differential equation than the one you started with. Which brings us to...
3. Solve it!
4. Finally, convert back to the original variables.

You have used this process every time you evaluated an indefinite integral with u-substitution. First, define $u(x)$. Then rewrite the integrand in terms of u, but be careful to also define du in terms of dx. Integrate the resulting function of u, and finally convert back to x. (In fact u-substitution is a special case of this method since taking an integral is a special case of solving a differential equation; evaluating $\int x \sin(x^2) dx$ means solving the equation $df/dx = x \sin(x^2)$.)

EXAMPLE **Defining a New Dependent Variable**

Question: Solve the differential equation:

$$\frac{dx}{dt} = \frac{1}{t + 3x - 4} \tag{10.7.2}$$

Solution:
Looking at the question, it seems at least reasonable to try this substitution.

$$u = t + 3x - 4 \tag{10.7.3}$$

So the right side of Equation 10.7.2 becomes $1/u$. To convert the left side, we need to take the derivative of both sides of Equation 10.7.3 with respect to t:

$$\frac{du}{dt} = 1 + 3\frac{dx}{dt}$$

With that substitution in place, Equation 10.7.2 becomes:

$$\frac{1}{3}\left(\frac{du}{dt} - 1\right) = \frac{1}{u}$$

We can put this in a separable form and solve it.

$$\frac{du}{dt} = \frac{3+u}{u}$$

$$\int \frac{u}{3+u}\, du = \int dt$$

We can rewrite the left side as $1 - 3/(3+u)$ and integrate to get

$$u - 3\ln|3+u| = t + C$$

Finally, we replace u with $t + 3x - 4$ to find the relationship between t and x that we were looking for originally.

$$t + 3x - 4 - 3\ln|3 + t + 3x - 4| = t + C$$
$$x - \ln|t + 3x - 1| = C$$

That is the answer, in the simplest form we're going to find it. We can't solve it for x. But we can check it by taking the derivative of both sides with respect to t:

$$\frac{dx}{dt} - \frac{1}{t + 3x - 1}\left(1 + 3\frac{dx}{dt}\right) = 0$$

It takes a bit of algebra, but you can solve this for dx/dt and it comes back around to Equation 10.7.2, proving that the solution works.

How did we know to define $u = t + 3x - 4$ in the example above? This is not an exact science, but it looked like a reasonable place to start. As with the method of guess and check, our best advice is "try stuff, but be smart."

A Brief but Important Digression about the Chain Rule

In the examples above we defined a new variable $u(x, t)$. When we define u as a function of t only, the process of converting the derivatives changes considerably. Before we demonstrate that process, we need to pause for a moment on the chain rule.

We assume that you know how to use the chain rule to find the derivative of a composite function. For instance, if $x = \sin(t^2)$ then $dx/dt = \cos(t^2) 2t$.

To understand the chain rule more generally, consider the breakdown of functions in that example. We can write the problem like this:

$$x(u) = \sin(u)$$

$$u(t) = t^2$$

The chain rule then tells us that:

$$\frac{dx}{dt} = \left(\frac{dx}{du}\right)\left(\frac{du}{dt}\right) \quad \text{the chain rule} \tag{10.7.4}$$

Equation 10.7.4 *looks* good—the du factors in the numerator and denominator cancel, leaving dx/dt—but many students have trouble connecting it to the chain rule. Our focus here is not on "how to take the derivative of $\sin(t^2)$" but on how Equation 10.7.4 expresses the steps in that process, as surely as $f(t)g'(t) + f'(t)g(t)$ expresses the steps in the product rule.

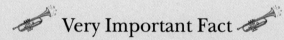

Very Important Fact

When you write the chain rule for single-variable functions, you write a series of fractions that multiply so their numerators and denominators cancel. This is not a new chain rule; it is just a (possibly) new way of writing the chain rule you learned in your first calculus class.

Take a moment to calculate dx/du and du/dt and convince yourself that Equation 10.7.4 does represent the steps you take when you use the chain rule to find the derivative of $x = \sin(t^2)$. If you're not completely convinced ("I'll take their word for it" doesn't count), keep playing with it until you see it.

So what about second derivatives? Students often write $d^2x/dt^2 = (d^2x/du^2)(d^2u/dt^2)$ which has the advantage of being a simple, reasonable extension of Equation 10.7.4 and the disadvantage of being wrong. Instead, you have to remember two key facts.

- The "second derivative" is shorthand for "the derivative of the derivative," so we can express it in terms of first derivatives:

$$\frac{d^2x}{dt^2} \equiv \frac{d\left(\frac{dx}{dt}\right)}{dt}$$

- Equation 10.7.4 applies to *any* function. If you replace x with q or p or r it still works. If you replace x with dx/dt that works too—and it gives us a correct formula for the second derivative.

We have now given you enough information to write the *correct* chain rule for the second derivative. Try it on your own, and then check to make sure you got the same thing we got:

$$\frac{d\left(\frac{dx}{dt}\right)}{dt} = \left(\frac{d\left(\frac{dx}{dt}\right)}{du}\right)\left(\frac{du}{dt}\right) \quad \text{using the chain rule to find a second derivative}$$

10.7 | Variable Substitution

If you got all that, you got the hard part. Now it's time to put it into action.

Defining a New Independent Variable

If you have a differential equation for $x(t)$ and you define a new variable $u(t)$—that is, a new variable that depends only on the independent variable—then u replaces t in your equation, leaving you with a differential equation for $x(u)$.

Consider the equation

$$2t\frac{d^2x}{dt^2} + \left(1 + \sqrt{t}\right)\frac{dx}{dt} - 3x = 0 \qquad (10.7.5)$$

None of the techniques we've described can solve this equation in this form. The problem comes from the extra t and $\sqrt{t}$ floating around, so we'll try the substitution $u = \sqrt{t}$. Rewriting t and $\sqrt{t}$ in terms of u in the differential equation is easy; the hard part is the derivatives. Keep your eye on the following table of what conversions you need.

What you start with	What you want	What you have to work with
$\dfrac{dx}{dt}$ and $\dfrac{d^2x}{dt^2}$	$\dfrac{dx}{du}$ and $\dfrac{d^2x}{du^2}$	$\dfrac{du}{dt}$

The tool for making these conversions is the chain rule, which is why we discussed it above. Based on that, take a moment to write the relationship between dx/dt and du/dt, using the function $u(t) = \sqrt{t}$.

Done? Hopefully you wrote something like this. The first step is the chain rule—you know we got it right because it looks like canceling fractions! In the second step we plug in our known du/dt.

$$\frac{dx}{dt} = \left(\frac{dx}{du}\right)\left(\frac{du}{dt}\right)$$

$$\frac{dx}{dt} = \frac{1}{2\sqrt{t}}\left(\frac{dx}{du}\right) \qquad (10.7.6)$$

That gives us the relationship between the two first derivatives. For the second derivatives we take the derivative of both sides of Equation 10.7.6 with respect to t.

$$\frac{d^2x}{dt^2} = -\frac{1}{4t^{3/2}}\frac{dx}{du} + \frac{1}{2\sqrt{t}}\frac{d\left(\frac{dx}{du}\right)}{dt} \qquad (10.7.7)$$

And what do we do with that awkward last term? This is where the chain rule comes in again. Pay careful attention to how the equation below is exactly the same as Equation 10.7.4 (the generic chain rule), but with different functions.

$$\frac{d\left(\frac{dx}{du}\right)}{dt} = \left(\frac{d\left(\frac{dx}{du}\right)}{du}\right)\left(\frac{du}{dt}\right)$$

$$\frac{d\left(\frac{dx}{du}\right)}{dt} = \frac{1}{2\sqrt{t}}\left(\frac{d^2x}{du^2}\right)$$

Plugging that into Equation 10.7.7 gives us the relationship we need:

$$\frac{d^2x}{dt^2} = -\frac{1}{4t^{3/2}}\left(\frac{dx}{du}\right) + \frac{1}{4t}\left(\frac{d^2x}{du^2}\right)$$

Now that we have the relationships between the derivatives we started with and the derivatives we want, we are ready to plug everything back into Equation 10.7.5.

$$\frac{1}{2}\frac{d^2x}{du^2} - \frac{1}{2\sqrt{t}}\left(\frac{dx}{du}\right) + (1+\sqrt{t})\frac{1}{2\sqrt{t}}\left(\frac{dx}{du}\right) - 3x = 0$$

This simplifies a great deal, suggesting we chose the right substitution.

$$\frac{d^2x}{du^2} + \frac{dx}{du} - 6x = 0$$

The guess $x = e^{ku}$ quickly leads to the solution $x(u) = Ae^{-3u} + Be^{2u}$. Finally, plug back in $u = \sqrt{t}$ to get the general solution.

$$x(t) = Ae^{-3\sqrt{t}} + Be^{2\sqrt{t}}$$

You can (and should) plug this into Equation 10.7.5 and check that it works.

The intimidating part of this process is finding the relationships between the $x(t)$ derivatives and the $x(u)$ derivatives. The good news is that those steps will be essentially identical in every such problem.

1. Evaluate the first derivative using the chain rule.
2. Take the derivative of your result to get the second derivative. This will introduce a mixed partial.
3. Use the chain rule again to relate that mixed partial to the second derivative you want.

Note in the example below how we take all the same steps, in the same order, as in the previous example.

EXAMPLE   **Defining a New Independent Variable**

Question: Use the substitution $u = \sin t$ to solve the following differential equation.

$$\frac{d^2x}{dt^2} + \tan t\frac{dx}{dt} + (\cos^2 t)x = 0 \qquad (10.7.8)$$

Solution:
To perform the substitution we have to rewrite dx/dt and d^2x/dt^2 in terms of u. We can find dx/dt directly from the chain rule.

$$\frac{dx}{dt} = \frac{dx}{du}\frac{du}{dt} = \cos t\frac{dx}{du} \qquad (10.7.9)$$

To find d^2x/dt^2 we begin by taking the derivative of both sides of Equation 10.7.9 with respect to t.

$$\frac{d^2x}{dt^2} = -\sin t \frac{dx}{du} + \cos t \left(\frac{d\left(\frac{dx}{du}\right)}{dt}\right)$$

That mixed partial needs another chain rule.

$$\frac{d\left(\frac{dx}{du}\right)}{dt} = \frac{d\left(\frac{dx}{du}\right)}{du}\left(\frac{du}{dt}\right) = \cos t \left(\frac{d^2x}{du^2}\right)$$

Plugging that back in,

$$\frac{d^2x}{dt^2} = -\sin t \frac{dx}{du} + \cos^2 t \frac{d^2x}{du^2}$$

Putting all these into Equation 10.7.8 gives:

$$\cos^2 t \frac{d^2x}{du^2} - \sin t \frac{dx}{du} + \sin t \frac{dx}{du} + (\cos^2 t)x = 0$$

Cancelling the two middle terms and dividing by $\cos^2 t$ gives us the simple harmonic oscillator equation $x''(u) + x(u) = 0$, with solution $x(u) = A\cos u + B\sin u$. Finally, putting this back in terms of t:

$$x(t) = A\cos(\sin t) + B\sin(\sin t)$$

We encourage you to verify that this solves Equation 10.7.8.

Hopefully you were able to follow each step of that example, and could do the same thing yourself as long as we told you what substitution to do. Unlike our earlier examples, though, this is probably not a substitution that would have occurred to you. It certainly seems reasonable to try a trig function since there are a couple of them sitting in the equation, but it might have taken some trial and error to settle on $u = \sin t$.

Sometimes the best strategy, just as with the guess and check approach, is to leave a k in your substitution and see what the math does. For example if you have an equation with a bunch of exponentials try $u = e^{kt}$ and see if there's any value of k for which the equation simplifies a lot. (See Problem 10.141.) There's no guaranteed formula for finding the right substitution, but as with u substitution for integrals, the more you practice the better your guesses get.

Stepping Back

It's not entirely unfair to think of variable substitution as two completely different techniques. (In the following description we are using x to mean "the dependent variable of your differential equation," t to mean "the independent variable of your differential equation," and u to mean "the substitution variable." Of course different letters can be used!)

- If u is a function of x, or a function of x and t, it replaces x. Your differential equation for $x(t)$ becomes a differential equation for $u(t)$. You have to find the relationships to replace dx/dt and d^2x/dt^2 with du/dt and d^2u/dt^2.
- If u is a function of t only, it replaces t. Your differential equation for $x(t)$ becomes a differential equation for $x(u)$. You have to find the relationships to replace dx/dt and d^2x/dt^2 with dx/du and d^2x/du^2.

The overall outline is the same in both cases: choose a substitution, rewrite the differential equation, solve, and then substitute back. But the "rewrite the differential equation" step—especially replacing the derivatives—is where beginners get lost. For that step all $u(t)$ problems follow our solutions to Equations 10.7.1 and 10.7.2, and all $x(u)$ problems follow our solutions to Equations 10.7.5 and 10.7.8. Of course, you still have to solve what you get—so the more equations you can solve *without* variable substitution, the more use you can make of this technique.

10.7.3 Problems: Variable Substitution

10.120 Walk-Through: Substituting for the Dependent Variable. In this problem you will solve the differential equation $dy/dx = \sec(3y + 9x) - 3$ by using the substitution $u = 3y + 9x$.
(a) Take the derivative of both sides of $u = 3y + 9x$ to find the relationship between du/dx and dy/dx.
(b) Use the relationship you found to replace dy/dx in the differential equation. Also replace $3y + 9x$ with u. Your answer should be a differential equation for $u(x)$ that doesn't have y anywhere in it.
(c) Solve the resulting equation for $u(x)$ using separation of variables.
(d) Plug $u = 3y + 9x$ into your solution and solve for y to find $y(x)$.
(e) Plug your solution $y(x)$ into the original differential equation and verify that it works. (You may want to use a computer or a table of derivatives.)

10.121 Walk-Through: Substituting for the Independent Variable. In this problem you will solve the differential equation $d^2x/dt^2 + [-1/(t+1) + 3(t+1)] dx/dt + 2(t+1)^2 x = 0$ by using the substitution $u = (t+1)^2$.
(a) Find du/dt.
(b) Use the chain rule and the expression you found for du/dt to write an equation for dx/dt in terms of dx/du.
(c) To find d^2x/dt^2 take the derivative with respect to t of both sides of the equation you wrote down in Part (b). Your answer will involve the mixed partial $d\left(\frac{dx}{du}\right)/dt$; you will need to use the chain rule again to relate that to d^2x/du^2.

(d) Use your results to rewrite the original differential equation in terms of u instead of t.
(e) Solve the equation to find the general solution $x(u)$.
(f) Plug $u = (t+1)^2$ into your solution to find the general solution $x(t)$.
(g) Plug your solution into the differential equation and verify that it works.

10.122 Given a differential equation for $x(t)$, one of the simplest substitutions you can make is $u = kt$ where k is a constant. Even in this case, however, you have to use the chain rule several times to find the correct derivatives. Find dx/dt and d^2x/dt^2 in terms of dx/du and d^2x/du^2 for this substitution.

In Problems 10.123–10.130, solve the given differential equation using the indicated substitution. Your answer should be the solution $y(x)$, with the appropriate number of arbitrary constants, and you should check it by plugging it back into the original differential equation.

Remember that if you define a substitution variable $u(y)$ or $u(x, y)$ then u will replace y in your differential equation, following the process we stepped you through in Problem 10.120. If you define a $u(x)$ then u will replace x, following the very different process in Problem 10.121. (And of course we're not really talking about "the letters x and y" but about "the independent and dependent variables" here.)

10.123 $dy/dx = (1 - x - y)/(x + y)$, $u = x + y$

10.124 $dy/dx = e^{ax+y} - a$, $u = ax + y$

10.125 $x(dy/dx) + 4y = x^4 y^2$, $u = 1/y$

10.126 $dy/dx = e^{y/x} + y/x$, $u = y/x$

10.127 $2x\, d^2x/dt^2 + 2(dx/dt)^2 + 49x^2 = 0$, $u = x^2$

10.128 $d^2x/dt^2 + (3 + 3e^{-3t})dx/dt - 18e^{-6t}x = 0$, $u = e^{-3t}$

10.129 $d^2f/dx^2 + (-\cot x + \tan x)df/dx + (\tan^2 x)f = 0$, $u = \cos x$

10.130 $y''(x) - (2/x)y'(x) - 9x^4 y(x) = 0$, $u = x^3$

In Problems 10.131–10.139, solve the given differential equation by substitution. (If you try something that doesn't work, try something else! You may need to try a generic form like $u = t^k$ or $u = e^{kt}$.)

10.131 $dy/dx = (3x - 3y + 1)^2$ (Your substitution will lead you to a separable equation that can be integrated using partial fractions.)

10.132 $t(d^2x/dt^2) - dx/dt + 4t^3 x = 0$

10.133 $dy/dx = e^{6y-2x} + (1/3)$

10.134 $dy/dx = e^{6y-2x}$ *Hint*: this problem ends up involving a somewhat tricky integral, but it can be done with the right u-substitution.

10.135 $d^2x/dt^2 + (3e^{2t} - 2)(dx/dt) + 2e^{4t}x = 0$

10.136 $\dfrac{dx}{dt} = \dfrac{(x-2t)^2 + 2(x-2t) + 1}{x - 2t}$

10.137 $d^2x/dt^2 + [8\sin(2t) - 2\cot(2t)]\, dx/dt + 12\sin^2(2t)x = 0$

10.138 $2y\left(\dfrac{dy}{dt}\right) - 1 = \dfrac{1}{y^2 - t - 3} + y^2 - t - 3$

10.139 $dy/dx = \sec(x^2 - y) + 2x$

10.140 You are working with the linear homogeneous second-order equation $x''(t) + f(t)x'(t) + g(t)x(t) = 0$. You have decided to do a substitution for the independent variable, so you have defined a new function $u(t)$. Your goal is to create a new differential equation relating the following four derivatives: dx/du, d^2x/du^2, du/dt, and d^2u/dt^2. (We will call these four the "allowable derivatives"; all other derivatives must be eliminated from your equations.)

(a) Write the equation for dx/dt in terms of the allowable derivatives.

(b) Write the equation for d^2x/dt^2 in terms of the allowable derivatives.

(c) Rewrite the differential equation in terms of the allowable derivatives.

(d) For some numerical techniques it is desirable to work with an equation that has no first derivative. What relationship must hold involving u, f, and/or g so that the first derivative term $x'(u)$ vanishes in this equation?

10.141 In this problem you will solve the differential equation $x''(t) - 3x'(t) + 36e^{6t}x(t) = 0$.

(a) Substitute $u = e^{kt}$ into this equation, where k is an as-yet-unknown constant. Write the differential equation for $x(u)$, with no t in it.

(b) What must k equal so that the differential equation for $x(u)$ has no first-order derivative term in it?

(c) Write the differential equation for $x(u)$ using the value of k you found in Part (b) and solve it for $x(u)$.

(d) Rewrite your solution in terms of t to find the solution $x(t)$ to the original equation.

(e) Plug in your solution $x(t)$ and verify that it works.

10.142 The "Bessel functions" $J_0(y)$ and $Y_0(y)$ are the solutions $f(y)$ to the differential equation

$$yf''(y) + f'(y) + yf(y) = 0 \qquad (10.7.10)$$

Bessel explored them in detail, but they were originally discovered by Daniel Bernoulli while studying standing waves in a hanging chain. If the chain is suspended from one end and the other end is unattached, the standing waves obey the equation

$$yf''(y) + f'(y) + \dfrac{\omega^2}{g}f(y) = 0 \qquad (10.7.11)$$

Here f is the horizontal displacement of the chain, g is the acceleration due to gravity, and ω is a constant that gives the frequency of oscillation of the standing wave. To solve this, you need to turn Equation 10.7.11 into Equation 10.7.10.

(a) The substitution that suggests itself here is a power law. (Introducing exponentials or trig functions certainly won't make Equation 10.7.11 look like Equation 10.7.10.) Substitute $u = cy^q$ into Equation 10.7.11 and simplify as much as possible. Write everything in terms of u, not y.

(b) What must c and q be in order for the resulting equation to look like Equation 10.7.10? *Hint*: Of course you're allowed to divide both sides of the equation by something to help get it into that form.

(c) Write down the solution to the equation you wrote and put it in terms of y, not

u. Your final answer will involve Bessel functions. It should depend on y, ω, g, and two arbitrary constants.

10.143 The "Legendre polynomials" $P_n(x)$ and $Q_n(x)$ are the solutions $f(x)$ to the differential equation

$$\left(1 - x^2\right) f''(x) - 2x f'(x) + n(n+1) f(x) = 0 \quad (10.7.12)$$

where n is an integer. These functions arise in many problems in spherical coordinates. For example, suppose the potential V on a spherical shell is a known function of the polar angle θ. (In other words, we're assuming it doesn't depend on the azimuthal angle ϕ.) In Chapter 11 we show that the potential inside the shell equals r^n (for some integer n) times a function $f(\theta)$, which is given by the equation:

$$f''(\theta) + \cot(\theta) f'(\theta) + n(n+1) f(\theta) = 0 \quad (10.7.13)$$

Use the substitution $x = \cos\theta$ to transform Equation 10.7.13 into Equation 10.7.12, and then write the general solution $f(\theta)$. Your final answer will include Legendre polynomials, and it should only depend on θ and n.

Problems 10.144–10.147 are about the theory of inflation, which describes the universe a fraction of a second after the big bang. According to this theory the universe at that time was dominated by an exotic form of energy known as the "inflaton field." For these problems all you need to know about the inflaton field is that it's generally designated by the letter ϕ, and that under certain simplifying assumptions it obeys the differential equation

$$\frac{d^2\phi}{dt^2} + \frac{3}{a}\frac{da}{dt}\frac{d\phi}{dt} + \frac{dV}{d\phi} = 0 \quad (10.7.14)$$

The function $V(\phi)$, known as the "potential function" for ϕ, is different in different models of the theory. The function $a(t)$, known as the "scale factor," describes the expansion of the universe and has to be calculated from other equations. Depending on the functions V and a, Equation 10.7.14 can be solved either analytically or numerically, but either way it is easier to solve if you use a variable substitution to eliminate the $d\phi/dt$ term in the middle.

10.144 In one simple model $a(t) = \alpha t^{2/3}$ and $V(\phi) = (1/2) m^2 \phi^2$, where α and m are constants.

(a) Write Equation 10.7.14 using these functions a and V.

(b) Plug in the substitution $u = t^k \phi$ and simplify the resulting equation for $u(t)$.

(c) Find the value of k that eliminates the du/dt term from the differential equation.

(d) Using this value of k, simplify the differential equation for $u(t)$ and find its general solution.

(e) Using your solution for $u(t)$ and the substitution you made, find the general solution for $\phi(t)$ in this model.

(f) Describe how $\phi(t)$ evolves over time.

10.145 In one simple model $a(t) = \alpha t^{1/2}$ and $V(\phi) = (1/4) \lambda \phi^4$, where α and λ are constants. In this problem you will handle this model the same way you attacked Problem 10.144; in Problem 10.146 you will see a better way.

(a) Write Equation 10.7.14 using these functions a and V.

(b) Plug in a substitution $u = t^k \phi$ and simplify the resulting equation for $u(t)$.

(c) Find the value of k that eliminates the du/dt term from the differential equation.

(d) Using this value of k, simplify the differential equation for $u(t)$. This will not be an equation you can solve by hand, but in this form it would be easier for a computer to solve numerically than the original equation.

10.146 In one simple model $a(t) = \alpha t^{1/2}$ and $V(\phi) = (1/4) \lambda \phi^4$, where α and λ are constants. You approached this problem with a dependent variable substitution in Problem 10.145; here you will substitute twice, once for the dependent and once for the independent variable, and make the equation even simpler. It still won't be *solvable* as such, but we don't really expect that in the real world. (You don't need to have done Problem 10.145 to do this one.)

(a) Write Equation 10.7.14 using these functions a and V.

(b) Plug in a substitution $u = t^{1/2} \phi$ and simplify the resulting equation for $u(t)$.

(c) Plug the substitution $\tau = t^k$ into your equation for $u(t)$ to get an equation for $u(\tau)$.

(d) Choose the value of k that simplifies your equation for $u(\tau)$ as much as possible. (You can't use $k = 0$. Do you see why?)

(e) The equation you ended up with is still not simple to solve, but you should be able to look at it and describe how it will behave. Based on your differential equation, sketch a plot of $u(\tau)$. Assume $\lambda > 0$.

(f) Based on the sketch of $u(\tau)$ you just made, sketch a plot of $\phi(t)$.

10.147 In many cases the function $a(t)$ is not known before solving the equation, but a substitution can still simplify the equation. Plug the substitution $u = a^s \phi$ into Equation 10.7.14 and find the value of the constant s that eliminates the first derivative term in the resulting equation for $u(t)$. *Hint*: you will need to use the product rule and the chain rule to find $d\phi/dt$ in terms of du/dt, a, and da/dt.

10.148 **Make your own: dependent variable substitution.** Write a differential equation that can be solved with the substitution $u = t^2 - x$, and solve it. (One way to do this is to start with an equation that has that term in it, make the substitution, and then if it doesn't simplify enough modify the equation you started with as needed.)

10.149 **Make your own: independent variable substitution.** Write a differential equation that can be solved with the substitution $u = t^3$, and solve it. (One way to do this is to start with a differential equation for $x(u)$ that you know you can solve, and then use the substitution $t = u^{1/3}$ to see what $x(t)$ equation needs this substitution.)

10.8 Three Special Cases of Variable Substitution

Our previous examples of variable substitution required educated guesswork in choosing the right function $u(x, t)$. In this explanation we discuss three important cases for which someone has already done that job for you. Rather than inventing a $u(x, t)$ function, your job in such cases is to recognize a given equation as fitting one of these important forms; from there you know what to do. The first case, the Bernoulli equation, requires the techniques of Section 10.4 (see felderbooks.com), but you can still solve the other two types if you haven't gone through that section.

10.8.1 Explanation: Three Special Cases of Variable Substitution

The Bernoulli Equation
Figure 10.5 shows a chain piled up on a table with one end falling off. The linear density of the chain is a constant λ (measured in mass per distance). We are using x to represent the length of chain currently hanging from the table and m for the mass of that segment. Our goal is to find a formula for the velocity v of the chain as a function of x.

The force pulling the chain downward is mg, the force of gravity on the hanging segment.[5] This force is equal to dp/dt, the rate of change of the momentum $p = mv$. Applying the product rule,

$$mg = m\frac{dv}{dt} + v\frac{dm}{dt}$$

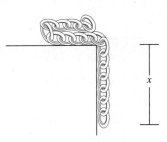

FIGURE 10.5 Falling chain.

If you were wondering why we set the force equal to dp/dt instead of the more familiar ma, the answer is in that equation. The first term on the right is the ma that you're used to setting force equal to, but the second term accounts for the fact that the mass of the falling part of the chain is changing.

---

[5]Technically the force of gravity is exerted on the entire chain, but the normal force of the table pushes back on the piled-up parts, so effectively gravity acts only on the dangling part with mass m.

Because our goal is to write an equation for $v(x)$ we want to get rid of those time derivatives. The first one can be replaced by using the chain rule.[6]

$$\frac{dv}{dt} = \left(\frac{dv}{dx}\right)\left(\frac{dx}{dt}\right) = v\frac{dv}{dx}$$

For the second one we note that $m = \lambda x$ and λ is a constant.

$$\frac{dm}{dt} = \lambda \frac{dx}{dt} = \lambda v$$

Since λ is constant we could just leave it in the equation, but then the other two terms would include the function $m(t)$. If we replace λ with m/x we can cancel m from all of the terms. Our equation now looks like this.

$$g = v\frac{dv}{dx} + \frac{v^2}{x}$$

That's the equation we're looking for. It is an example of a "Bernoulli equation" for $v(x)$. More generally, a Bernoulli equation for a function $y(x)$ is one that can be written in the following form.

> **The Bernoulli Equation**
>
> $$\frac{dy}{dx} + p(x)y = q(x)y^n \tag{10.8.1}$$

For $n = 0$ or $n = 1$ this is a linear equation, and we can use the techniques of Section 10.4 (see felderbooks.com). For any other value of n we can turn this into a linear equation in two steps.

1. Divide both sides of the differential equation by y^n.
2. Use the substitution $u = y^{1-n}$.

In Problem 10.171 you will use this trick to solve for the motion of our falling chain. In Problem 10.172 you will show more generally that this process turns any Bernoulli equation into a linear equation. (Note in particular that if $n = 0$ then both steps do nothing at all!)

This technique only works if the equation is written in the form of Equation 10.8.1. If there is a factor in front of dy/dx, start by dividing it out.

> **EXAMPLE** **Bernoulli Equation**
>
> Question: Solve the equation $x(dy/dx) + 2y = x^7 y^3$.
>
> Solution:
> We divide by x to write this in the standard form for a Bernoulli equation.
>
> $$\frac{dy}{dx} + \frac{2}{x}y = x^6 y^3 \tag{10.8.2}$$
>
> Following the prescription given above, we begin by dividing both sides by y^3.
>
> $$\frac{1}{y^3}\frac{dy}{dx} + \frac{2}{x}\left(\frac{1}{y^2}\right) = x^6$$

---
[6]The chain rule! Get it? We're using the chain rule! Oh, never mind.

Next we apply the substitution $u = y^{-2}$. This means that $du/dx = -2y^{-3}(dy/dx)$, so we will replace $y^{-3}(dy/dx)$ in the equation with $(-1/2)(du/dx)$.

$$-\frac{1}{2}\left(\frac{du}{dx}\right) + \frac{2}{x}u = x^6$$

Now we are working with a first-order linear equation, and we can approach it with the technique discussed in Section 10.4 (see felderbooks.com). First multiply both sides by $-2x^{-4}$.

$$x^{-4}\frac{du}{dx} - 4x^{-5}u = -2x^2$$

The left side is now the derivative of $x^{-4}u$ (confirm that!) so integrating both sides gives us $x^{-4}u = -(2/3)x^3 + C$, or $u = (-2/3)x^7 + Cx^4$. Finally we go back; replace u with y^{-2} and solve the resulting equation for y.

$$y = \frac{1}{\sqrt{-(2/3)x^7 + Cx^4}}$$

We leave it to you to confirm that this is a valid solution to Equation 10.8.2.

The Other Kind of Homogeneous Differential Equations

A moth is on one side of a street when a streetlight on the opposite side turns on. The moth immediately starts flying toward the light at speed v. The moth's trajectory is a straight line across the street.

Suppose, however, that there is a wind blowing at speed s parallel to the street. At each moment the moth points itself straight toward the light, but its actual velocity is the vector sum of its motion toward the light and the wind velocity carrying it down the street. In Problem 10.169 you'll show that in this case its trajectory obeys the following equation.

$$\frac{dy}{dx} = \frac{vy - s\sqrt{x^2 + y^2}}{vx}$$

This is an example of a "homogeneous differential equation."

Chapter 1 introduced the term "homogeneous differential equation." We introduced the term again on Page 487 of this chapter, and it meant the same thing. Sadly this is another use of the phrase "homogeneous differential equation" that is unrelated to the first one. Outside this section, we will use "homogeneous" to mean that a linear differential equation for $y(x)$ has no terms without y in them. In this section, however, it means the following.

Homogeneous Differential Equation

A homogeneous differential equation is a first-order equation that can be written in the following form.

$$\frac{dy}{dx} = f(y/x) \qquad (10.8.3)$$

(Be careful not to confuse this with the more common use of "homogeneous differential equation" described on Page 487.)

You will show in Problem 10.173 that the substitution $u = y/x$ turns any this-kind-of-homogeneous differential equation into a separable equation. In Problem 10.169 you'll show that the moth trajectory equation above can be written in this form and you'll use this substitution to solve that equation.

EXAMPLE: A Homogeneous Differential Equation

Question: Solve

$$\frac{dy}{dx} = \frac{y^2 + xy}{x^2 - xy} \qquad (10.8.4)$$

Solution:
We can rewrite this in the form $dy/dx = f(y/x)$ by dividing the top and bottom of the fraction by x^2.

$$\frac{dy}{dx} = \frac{(y/x)^2 + (y/x)}{1 - (y/x)}$$

Since this equation is homogeneous we make the substitution $u = y/x$. Writing this as $y = xu$ we can rewrite dy/dx using the product rule

$$\frac{dy}{dx} = u + x\frac{du}{dx}$$

Plugging this in, and replacing y/x with u, we get:

$$x\frac{du}{dx} = \frac{u^2 + u}{1 - u} - u = \frac{2u^2}{1 - u}$$

Separating variables:

$$\int \frac{1}{x} dx = \int \frac{1-u}{2u^2} du = \int \left(\frac{1}{2u^2} - \frac{1}{2u}\right) du$$

$$\ln|x| = -\frac{1}{2u} - \frac{1}{2}\ln|u| + C$$

Finally, replace u with y/x. Remembering that $\ln(ab) = \ln a + \ln b$, we can do a little algebra to get:

$$\ln|y| + \ln|x| + \frac{x}{y} = C$$

(We replaced $2C$ with C since they both mean "any constant.") In some problems of this type we will be able to solve the final answer for y. In this case we cannot, but we can still check it by taking the derivative of both sides with respect to x.

$$\frac{1}{y}\frac{dy}{dx} + \frac{1}{x} + \frac{1}{y} - \frac{x}{y^2}\frac{dy}{dx} = 0$$

A little algebra turns this into Equation 10.8.4.

10.8 | Three Special Cases of Variable Substitution

In Section 10.5 (see felderbooks.com) we showed you how to solve equations of the form $P(x,y)\,dx + Q(x,y)\,dy = 0$ if the equation happened to be exact. Another approach to such equations is to write them in the form $dy/dx = -P/Q$ and see if the resulting equation is homogeneous. This method, like several others in this chapter, comes with the following warning: even if an equation is homogeneous and you can therefore make it separable, that doesn't guarantee that you'll be able to integrate what you get.

Substituting for the Derivative

Ben is skydiving and he feels two forces as he falls: the constant force of gravity $F_g = -mg$, and air resistance which can be modeled as $F_a = -bv$. (The faster he falls, the harder the wind pushes back.) His equation of motion is therefore $F_{total} = ma = -mg - bv$, which can be rearranged into:

$$\frac{d^2x}{dt^2} + \frac{b}{m}\frac{dx}{dt} = -g \qquad (10.8.5)$$

This is a second-order linear differential equation, and we could approach it by solving the complementary homogeneous equation first and so on as described in Section 10.2. But it's easier to rewrite the problem as a *first*-order equation for the velocity.

$$\frac{dv}{dt} + \frac{b}{m}v = -g \qquad (10.8.6)$$

With a bit of algebra we can separate and solve that.

$$\frac{dv}{dt} = -\frac{b}{m}\left(v + \frac{mg}{b}\right)$$

$$\int \frac{dv}{v + \frac{mg}{b}} = \int -\frac{b}{m}\,dt$$

$$\ln\left|v + \frac{mg}{b}\right| = -\frac{b}{m}t + C$$

$$v = Ce^{-\frac{b}{m}t} - \frac{mg}{b}$$

To find the position function we were originally looking for, we just integrate the velocity. This adds another arbitrary constant, independent from the first one:

$$x = \int v(t)\,dt = C_1 e^{-\frac{b}{m}t} - \frac{mg}{b}t + C_2$$

You can verify that this is a valid solution to Equation 10.8.5. You can also make sense of these results intuitively. The $v(t)$ function shows that whether Ben starts going fast or slow, his velocity will decay quickly toward $-mg/b$. If you plug that into Equation 10.8.6 you will see that it corresponds to $dv/dt = 0$, the "terminal velocity" where the upward force of the wind perfectly balances the downward force of gravity.

Why that particular substitution? Equation 10.8.5 represents a recognizable special case because it has derivatives of x, but it does not have any term based on x itself. Whenever you see such an equation, you can make your new dependent variable $u = x'(t)$. This gives you a differential equation for $u(t)$ that is one degree lower than your original equation for $x(t)$.

Much less obviously this same substitution turns out to be useful for "autonomous equations," meaning ones in which the *independent* variable doesn't appear. The substitution $v = dx/dt$ will convert a second-order autonomous equation for $x(t)$ into a first-order equation for $v(x)$. See Problem 10.176.

10.8.2 Problems: Three Special Cases of Variable Substitution

10.150 Walk-Through: Bernoulli equation. In this problem you will solve the equation $(dy/dx)/x + 2y/x = 1/y$.
(a) Rewrite this equation in the form of Equation 10.8.1. What is the value of n?
(b) Divide both sides of the resulting equation by y^n.
(c) You are going to solve this equation with the substitution $u = y^{1-n}$. Based on this substitution, what is du/dx?
(d) Solve your answer from Part (c) for dy/dx as a function of u and du/dx.
(e) Rewrite your equation from Part (b) as a differential equation for $u(x)$. The letter y should appear nowhere in your equation.
(f) The resulting differential equation is first-order linear. Multiply both sides by the integrating factor $2e^{4x}$.
(g) Show that the left side of the resulting equation is the derivative with respect to x of ue^{4x}. Using that fact, integrate both sides with respect to x.
(h) Solve the resulting equation for $u(x)$.
(i) Write the solution $y(x)$ to the original differential equation.
(j) Confirm that your solution works.
(k) The power of this technique is that it turned a non-linear equation into a linear equation. At what step in the process did the equation become linear?

In Problems 10.151–10.153, solve the given Bernoulli equation. You may find it helpful to first work through Problem 10.150 as a model.

10.151 $x(dy/dx) + 3y = 6x^4y^4$
10.152 $y(dy/dx) + 2y^2 = e^{2x}$
10.153 $dy/dx + y/x = \sqrt{y}$

10.154 Walk-Through: Homogeneous Equations. In this problem you'll solve the differential equation $dy/dx = (3x^2 + 3y^2)/2xy$.
(a) A homogeneous differential equation is an equation of the form $dy/dx = f(y/x)$. This equation doesn't appear to be in that form. Divide the top and bottom of the fraction by whatever you need to in order to get it in that form.
   Having shown that this equation is homogeneous, you can now solve it with the substitution $u = y/x$, which can be written more usefully in this case as $y = ux$.
(b) Use the product rule to find dy/dx in terms of u, x, and du/dx.
(c) Using your substitutions for y and dy/dx, rewrite the differential equation as an equation for $u(x)$.
(d) Solve that equation using separation of variables to find $u(x)$.
(e) Plug in $y = ux$ to find $y(x)$.
(f) Plug your solution $y(x)$ into the original differential equation to verify that it works.

In Problems 10.155–10.159, solve the given homogeneous equation. You may find it helpful to first work through Problem 10.154 as a model. You may end up with an equation relating x and y rather than a function $y(x)$, but you should still be able to check that it satisfies the differential equation.

10.155 $dy/dx = y^2/x^2 + y/x$
10.156 $dy/dx = (2y^3 + 3xy^2 + x^2y + x^3)/(2xy^2 + 3x^2y + x^3)$
10.157 $\left(x^2 e^{(y/x)^2} + y^2\right) dx - xy\, dy = 0$
10.158 $\left(x^2 + 3y^2\right) dx - 3xy\, dy = 0$
10.159 $dy/dx = (x/y) \sec\left(y^2/x^2\right) + y/x$

In Problems 10.160–10.167, solve the given differential equation by substitution. Start by figuring out if the equation fits into one of the three categories we described in the Explanation (Section 10.8.1): Bernoulli, homogeneous, or $y(x)$ only appears inside derivatives. If it does, you immediately know the right substitution to use. If it doesn't, then you'll have to look at the equation and try to find the right substitution. (If you try something that doesn't work, try something else!)

10.160 $d^2x/dt^2 + (1/t)(dx/dt) = 0$
10.161 $dy/dx + 4x^2y = x^2y^3$
10.162 $dy/dx = \sec(y/x) + y/x$
10.163 $dy/dt = \sec(-t - y) - 1$
10.164 $df/dx = (x^3 - 2f^3)/3xf^2$
10.165 $d^2x/dt^2 + (2/t)dx/dt + (1/t^4)x = 0$
10.166 $dy/dx + y/x = x^5y^5$
10.167 $d^4x/dt^4 + d^2x/dt^2 = 0$

10.168 The equation $dy/dx = y/x + y^3/x^3$ is both a homogeneous equation and a

Bernoulli equation. Solve it both ways and verify that you get the same result.

10.169 In the Explanation (Section 10.8.1) we described a moth flying at speed v across the street toward a streetlight, while a wind blows it along the street at speed s. We put the streetlight at the origin, the moth's original position at $(L, 0)$, and the wind direction as the positive y-direction. To set up the equation for the moth's trajectory $y(x)$, start with the simplest case where $s = 0$, so there is no wind.

(a) The moth's horizontal speed is the x-component of its velocity. Write an equation for dx/dt in terms of its speed v and its position (x, y).

(b) The moth's vertical speed is the y-component of its velocity. Write an equation for dy/dt in terms of its speed v and its position (x, y).

(c) The chain rule says that $dy/dx = (dy/dt)/(dx/dt)$. Based on everything you have written above, write a differential equation for the moth's trajectory $y(x)$. Your answer should only depend on x and y.

(d) Solve that differential equation by separation of variables and show that the resulting motion is a straight line toward the origin.

Now you'll repeat the process with a non-zero wind speed. This has no effect on the moth's x-velocity and simply adds s to its y-velocity.

(e) Redo the calculations until you have a new equation for dy/dx in terms of x, y, v, and s.

(f) Rewrite that equation in the form of Equation 10.8.3.

(g) Solve to find $y(x)$, using the initial condition that the moth starts at position $(L, 0)$. You may find it helpful to know that $\int (1 + x^2)^{-1/2} dx = \ln|x + \sqrt{1 + x^2}| + C$.

(h) Check that your answer to Part (g) includes your answer to Part (d) as a special case.

10.170 [This problem depends on Problem 10.169.] Draw graphs of the moth's trajectory for $v \gg s$, v just a little bigger than s, and $v \ll s$. Explain why the graphs make sense physically in each of those cases.

10.171 In the Explanation (Section 10.8.1) we set up the equation of motion for a chain falling off a table.

(a) Find the general solution $v(x)$ to this equation.

(b) Find the solution $v(x)$ with initial condition $v(0) = 0$.

(c) The acceleration is dv/dt. Find it using the chain rule, so the right-hand side of your answer will have both x and v in it.

(d) Since you already know $v(x)$, plug this in to get an expression for a that doesn't have v in it.

(e) If you chopped off a segment of the chain, it would fall with acceleration g (since we ignored friction and air resistance). From your solution, is this chain falling at that rate, falling faster, or falling slower? Explain physically why.

10.172 Starting with the general form for any Bernoulli equation, Equation 10.8.1, divide both sides by y^n and then substitution $u = y^{1-n}$. Show that the resulting equation for $u(x)$ can be put into the form of Equation 10.4.3 (see felderbooks.com), the generic first-order linear equation.

10.173 In this problem you will prove that the substitution $u = y/x$ always makes a homogeneous differential equation $dy/dx = f(y/x)$ separable.

(a) Writing the substitution as $y = xu$, find dy/dx in terms of u and du/dx.

(b) Plug this in to rewrite $dy/dx = f(y/x)$ in terms of u and x, with no y.

(c) Separate the resulting equation so there is only u on one side and x on the other.

(d) Integrate both sides. You should be able to evaluate the x integral. You will not be able to evaluate the u integral because it will contain the unknown function $f(u)$. Your final answer will be an equation relating a known function of x to an integral with respect to u.

10.174 Show that the differential equation

$$x^a y^b \, dx + x^c y^d \, dy = 0$$

is "homogeneous" by the definition given in this section, which means it can be written in the form $dy/dx = f(y/x)$, if and only if $a + b = c + d$.

10.175 Make Your Own

(a) Write and solve a Bernoulli equation that isn't in this section (including the problems) and solve it by dividing by y^n and using the substitution $u = y^{1-n}$.

(b) Write and solve a homogeneous equation that isn't in this section (including the problems) and solve it by using the substitution $u = y/x$.

10.176 Exploration: An Equation With No x

A differential equation in which the independent variable doesn't appear explicitly is called "autonomous." For example $dy/dx = y^2$ is autonomous but $dy/dx = y^2 + x$ is not. It turns out that a second-order autonomous differential equation for $y(x)$ can be reduced to a first-order equation with the substitution $u = dy/dx$. This is the same substitution we used for equations in which y doesn't appear explicitly, but it works for a very different reason in this case. In that case this substitution turned a second-order equation for $y(x)$ into a first-order equation for $u(x)$. In this case it's going to turn a second-order equation for $y(x)$ into a first-order equation for $u(y)$.

To begin with, consider the generic second-order autonomous equation $d^2y/dx^2 + f(y)dy/dx + g(y) = 0$.

(a) Starting with $u = dy/dx$, use the chain rule to calculate d^2y/dx^2 in terms of u and du/dy.

(b) Rewrite the generic equation above as a first-order differential equation for $u(y)$.

Now you're going to consider the equation for an object falling straight toward Earth, $d^2r/dt^2 = -GM_E/r^2$. Here of course the dependent variable is r and the independent variable is t. Since the new variable is going to be dr/dt, which is the object's velocity, call it v.

(c) Using the substitution $v = dr/dt$, rewrite the equation above as a first-order equation for $v(r)$.

(d) Solve using separation of variables to find an algebra equation relating v and r. Don't bother solving for v.

(e) Multiply both sides of your equation by m, the mass of the object, and rewrite it in the form $(1/2)mv^2 +$ <some function of r> $= C$, where C is the arbitrary constant from your integration. (Since it's an arbitrary constant you can just absorb the factor of m into it.)

(f) What does this equation represent physically? What does the <some function of r> term represent?

At this point you might want to finish the problem by finding $r(t)$. You can plug in $v = dr/dt$ to get a first-order equation for $r(t)$, but in this particular case it has no simple solution. (For some autonomous equations it would.) Nonetheless, knowing the velocity at every position is a useful result by itself.

10.9 Reduction of Order and Variation of Parameters (see felderbooks.com)

10.10 Heaviside, Dirac, and Laplace

We're not going to solve any differential equations in this section, but we are going to introduce three tools. The "Heaviside function" and the "Dirac delta function" give us notation for working with many important discontinuous functions, and the "Laplace transform" allows us to solve differential equations involving such functions. In Section 10.11 we will use these tools to write and solve differential equations.

10.10.1 Discovery Exercise: Heaviside and Dirac

We begin with the following function definition.

$$H(x) = \begin{cases} 0 & x < 0 \\ 1 & x \geq 0 \end{cases}$$

In Parts 1–6 graph each given function. You should be able to draw all 6 graphs in under five minutes with no electronic aid.

1. $H(x)$
2. $H(x-3)$
3. $H(x) - H(x-3)$

   See Check Yourself #68 in Appendix L

4. $x^2 H(x)$
5. $x^2 H(x-3)$
6. $(x-3)^2 H(x-3)$

We now define another function:

$$D(x) = \begin{cases} 0 & x < 0 \\ 1/k & 0 \le x \le k \\ 0 & x > k \end{cases}$$

7. For $k = 4$ draw a graph of $D(x)$ and calculate $\int_{-\infty}^{\infty} D(x)\, dx$.
8. For $k = 1$ draw a graph of $D(x)$ and calculate $\int_{-\infty}^{\infty} D(x)\, dx$.
9. For $k = 1/3$ draw a graph of $D(x)$ and calculate $\int_{-\infty}^{\infty} D(x)\, dx$.

10.10.2 Explanation: Heaviside, Dirac, and Laplace

A mass on a damped spring, driven by an external force F_e, obeys the equation

$$m\frac{d^2x}{dt^2} + b\frac{dx}{dt} + kx = F_e(t)$$

If F_e is an exponential, a trig function, or a power law, this is easy to solve by guess and check. Suppose, however, that F_e is zero until t reaches some constant value t_1 and grows linearly from that point on. Or suppose that F_e represents a blow from a hammer that imparts an impulse Δp to the mass in essentially zero time. How would you even write a function for such a force, never mind solve the differential equation?

To take another example, consider an RLC circuit connected to a power source that supplies a voltage $v(t)$. Assuming the capacitor is initially uncharged, the current i in the circuit will follow the following equation.[8]

$$iR + \frac{1}{C}\int_0^t i\, d\tau + L\frac{di}{dt} = v(t) \tag{10.10.1}$$

No matter what $v(t)$ is, that integral is going to make this equation hard to solve. Furthermore the voltage source, like the external force on our spring above, may be discontinuous: for instance, the power source supplies a constant voltage for t_1 seconds and is then turned off.

Section 10.11 will present the method of Laplace transforms, which will allow us to solve all the differential equations described above. In this section we introduce the three mathematical tools we will need for this method. First we discuss Heaviside functions and Dirac delta functions, which give us a concise notation for some discontinuous functions. Then we introduce Laplace transforms, which turn these differential equations into algebra equations.

---

[8] Usually we use capital letters for current and voltage, but here we need I and V for their Laplace transforms. Also, we are using τ as a dummy variable as we integrate up to the time t.

The Heaviside Function

Figure 10.6 shows a much simpler circuit than the one we described above: just a battery supplying a constant voltage, a resistor, and a switch. When the switch is open (as drawn), no current flows. When you close the switch, a current of $i = v/R$ immediately flows throughout the circuit.[9] If the switch is open until $t = 0$ and closed from then on, the current passing through the circuit can be modeled as $i = (v/R)H(t)$, where H is the "Heaviside function."

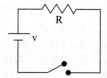

FIGURE 10.6 Open circuit.

Definition: The Heaviside Function

$$H(x) = \begin{cases} 0 & x < 0 \\ 1 & x \geq 0 \end{cases}$$

$H(0)$ is sometimes defined as 0, sometimes as $1/2$, and sometimes is left undefined. We have chosen to define it as 1, but it doesn't really matter.

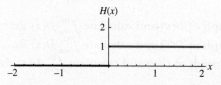

The Heaviside function really is as simple as it looks, but it gives us a notation that applies to a wide variety of discontinuous functions. As with any important function, you want to work with this one enough to be able to "see" graphs in your head with little or no work. The Discovery Exercise (Section 10.10.1) gave you some practice with this skill; the examples below will give you more if you sketch each graph quickly on your own before looking at our answer.

EXAMPLES **The Heaviside Function**

The graphs described below are shown in Figure 10.7 (a–c).

1. **Question:** Graph the function $H(x - 2)$.

   Answer:
   Replacing x with $x - 2$ moves any function two units to the right. So this gives you a Heaviside function that "breaks" at $x = 2$ instead of $x = 0$.

2. **Question:** Graph the function $H(x - 2) - H(x - 3)$.

   Answer:
   You can approach this analytically, but we encourage you to try it visually. Begin with a picture of $H(x - 2)$ and then subtract 1 at all points to the right of $x = 3$.

3. **Question:** Graph the function $x^2[H(x + 1) - H(x - 2)]$.

   Answer:
   Between $x = -1$ and $x = 2$ this Heaviside function is 1, so the product is x^2. Outside that range the function is zero. So the Heaviside function can be used to "turn off" parts of another function.

---

[9] Actually the current cannot *instantaneously* begin to flow everywhere in the circuit, but for most practical purposes we treat it as though it did.

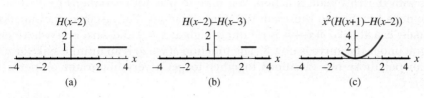

FIGURE 10.7

The Dirac Delta Function

A baseball undergoes a drastic change in momentum (both magnitude and direction) between the moment before it hits the bat and the moment after. This entire change takes place during the few milliseconds when the ball is physically in contact with the bat. The shorter this time interval, the greater the force required to cause the change ($\Delta \vec{p} = \int \vec{F}\, dt$).

Of course the contact period is not literally instantaneous—that would require an infinite force. But the details of what goes on during that brief period are generally not of interest, and would be very difficult to measure or calculate in any case. So we treat the contact period as *effectively* zero, and the force as *effectively* infinite, by taking a limit.

Definition: The Dirac Delta Function

Informally, we imagine the Dirac delta function as:

$$\delta(x) = \begin{cases} 0 & x \neq 0 \\ \infty & x = 0 \end{cases} \quad \text{with the } \infty \text{ well chosen so that} \quad \int_{-\infty}^{\infty} \delta(x)\, dx = 1$$

More properly, we define the Dirac delta function as the limit of a sequence of functions:

$$\delta(x) = \lim_{k \to 0} D(x) \quad \text{where} \quad D(x) = \begin{cases} 0 & x < 0 \\ 1/k & 0 \leq x \leq k \\ 0 & x > k \end{cases}$$

Since $\delta(x)$ is defined by the fact that its integral equals 1 (a unitless number), $\delta(x)$ has units of $1/x$.

So $\delta(x)$ is designed to pack an area of exactly 1 into an infinitesimal width. Strictly speaking $\delta(x)$ is not a function at all, but it is often useful to treat it as one. Mathematicians call it a "generalized function."

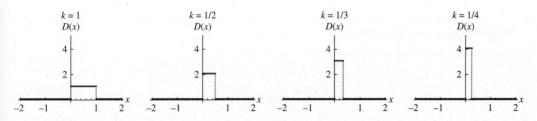

You can see that $\int_{-\infty}^{\infty} \delta(x)\,dx$ and $\int_{-10}^{10} \delta(x)\,dx$ and $\int_{-2}^{0.3} \delta(x)\,dx$ all give the result 1. The integral over any region across $x = 0$ is exactly 1. On the other hand, $\int_{2}^{15} \delta(x)\,dx$ or any such region that does not cross $x = 0$ is zero.

Just as with the Heaviside function, you have to play with variations to see how we use the Dirac delta function to represent different situations. For instance, the function $3\delta(x)$ looks visually exactly like $\delta(x)$—it is infinitely high at $x = 0$ and zero everywhere else—but the integral under the curve is now 3. The function $\delta(x - a)$ is an infinite spike at $x = a$.

You can put those two variations together to get a very important property: for any function $f(x)$,

$$\int_{-\infty}^{\infty} f(x)\delta(x - a)\, dx = f(a) \tag{10.10.2}$$

So just as the Heaviside function can be used to "turn off" a function everywhere except along some interval, the Dirac delta can turn off a function everywhere except at one point. As we will see, Equation 10.10.2 is essential when working with differential equations that involve Dirac delta functions.

A Few Differential Equations

We began this section with damped oscillators and RLC circuits. We can now write the equations for those situations using Heaviside and Dirac delta functions. F_1, V_0, A and t_1 in the following equations represent positive constants.

Case 1: The external force on an oscillator is turned on at $t = t_1$ and grows linearly from then on.

$$m\frac{d^2x}{dt^2} + b\frac{dx}{dt} + kx = F_1\left[(t - t_1)H(t - t_1)\right] \tag{10.10.3}$$

Case 2: The external force on an oscillator is a nearly instantaneous blow.

$$m\frac{d^2x}{dt^2} + b\frac{dx}{dt} + kx = A\delta(t - t_1) \tag{10.10.4}$$

Case 3: The voltage in an RLC circuit is turned on for a time t_1 and then turned off again. The capacitor is initially uncharged.

$$iR + \frac{1}{C}\int_0^t i\, d\tau + L\frac{di}{dt} = V_0\left[H(t) - H(t - t_1)\right] \tag{10.10.5}$$

Below we introduce a tool called "Laplace transforms." In Section 10.11 we will use Laplace transforms to solve Equation 10.10.5, and you will solve the other two in the problems in this section.

Laplace Transforms

Our final topic in this section is an operator, a formula for transforming one function into a different-but-related function. (If you are familiar with Fourier transforms, Laplace transforms are in many ways analogous.)

Definition: Laplace Transform

The "Laplace transform" of a function $f(t)$ is given by:

$$\mathcal{L}\left[f(t)\right] = \int_0^{\infty} f(t)e^{-st}\, dt \tag{10.10.6}$$

We often use a lowercase letter for a function and uppercase for its Laplace transform, so $\mathcal{L}\left[f(t)\right]$ is designated as $F(s)$.

Some points should be made right away.

- The integral in Equation 10.10.6 is evaluated with respect to t, treating s as a constant. Then you evaluate the antiderivative as $t \to \infty$ and at $t = 0$ and subtract. So in the end t is gone but s remains.
- Therefore, the Laplace transform is an operator that starts with a given function $f(t)$ and ends up with a different function $F(s)$.
- The transform is unique: a given $F(s)$ function generally corresponds to only one $f(t)$ function.[10] This is important because it allows for an inverse transform, which we write as $f(t) = \mathcal{L}^{-1}[F(s)]$.
- Because it is based on an integral from 0 to ∞, the Laplace transform depends only on the behavior of a function when $t \geq 0$.
- The variable s can in general take on any *complex* value, but for many input functions the Laplace transform is restricted to a limited domain. For example, the Laplace transform of e^{at} is only defined for $Re(s) > a$.
- Laplace transforms are generally used on functions of time.

One approach to differential equations is to take the Laplace transform of both sides of the original equation, solve the resulting equation, and then transform back to find the function you want. We will demonstrate this approach in Section 10.11, but first we need more background.

Finding a Laplace Transform

There are many ways to find the Laplace transform of a given function. One approach is by directly applying Equation 10.10.6.

EXAMPLE **Laplace Transform by Hand**

Question: Find the Laplace transform of $f(t) = 3t$.

Answer:

$$\mathcal{L}\left[f(t)\right] = \int_0^\infty 3t e^{-st} dt = \left[-\frac{3t}{s} e^{-st} - \frac{3}{s^2} e^{-st}\right]_0^\infty = \frac{3}{s^2}$$

This of course does not mean that $3t$ and $3/s^2$ are the same function, but they are related to each other: the Laplace transform of $3t$ is $3/s^2$.

$$\mathcal{L}[3t] = 3/s^2 \quad \text{or} \quad F(s) = 3/s^2$$

The "inverse Laplace transform" goes the other way.

$$\mathcal{L}^{-1}[3/s^2] = 3t$$

If you want to follow the math in that example, take it slowly. We evaluated the antiderivative using integration by parts; you can replicate that on your own, or confirm our work by taking the derivative of the answer. (In either case, remember that t is the variable and s is a constant!) Then we applied the limits of integration, but even that is not entirely straightforward; an integral to ∞ is called an "improper" integral and must be evaluated with a limit,

---

[10] If $\mathcal{L}\left[f_1(t)\right] = \mathcal{L}\left[f_2(t)\right]$ then f_1 and f_2 can differ at individual points, but cannot differ over any interval of finite length for $t > 0$.

and $\lim_{t \to \infty}(-3t/s)e^{-st}$ requires l'Hôpital's rule. And all that was just for $f(t) = 3t$! You can see why people don't want to go through this very often for complicated functions.

On the other hand, for functions based on the Heaviside or Dirac delta functions, the "by hand" approach is often the way to go.

> **EXAMPLE** **Laplace Transform of a Delta Function**
>
> Question: Find the Laplace transform of $\delta(t - 3)$.
>
> Answer: $\int_0^\infty \delta(t - 3)e^{-st}\,dt$ is the same as $\int_{-\infty}^\infty \delta(t - 3)e^{-st}\,dt$ (do you see why?), which Equation 10.10.2 tells us is e^{-3s}.

Another common approach to finding a Laplace transform is looking it up in a table. There are mathematical techniques—most notably "partial fractions" and "convolution"—that enable you to find the transforms of functions that you might not find directly in the table. We provide a table, a brief discussion of these techniques, and a few practice problems in Appendix H.

But if you're going to look it up in a table, why not just look it up on a computer? Ask any computer algebra system for the Laplace transform of $3t$ and it gives you $3/s^2$ immediately. It is worth your time to take a few Laplace transforms "by hand" to make sure you understand the mechanism, but after that your job is to know how and when to use them.

Those same three approaches—by hand, by table lookup, or by computer—also apply when you're running the transform in the opposite direction. However, a "by hand" inverse Laplace transform requires a contour integral on the complex plane. We'll teach you how to do that in Chapter 13, but in the meantime we'll leave that job to our computers.

Some Properties of Laplace Transforms

The following properties of Laplace transforms can all be derived from Equation 10.10.6, and are vital to their use in solving differential equations.

> **Key Properties of Laplace Transforms**
>
> F and G designate the Laplace transforms of f and g respectively, so $\mathcal{L}[f(t)] = F(s)$.
>
> 1. *The Laplace transform is a linear operator.*
>
> $$\mathcal{L}\left[af(t) + bg(t)\right] = aF(s) + bG(s)$$
>
> This follows from the linearity of the integral operator itself, but its importance cannot be overstated. When we take the Laplace transform of both sides of a differential equation, we will take Laplace transforms one term at a time—an easy step to miss, but it is only allowed because the transform is linear. (If you squared both sides, you couldn't square one term at a time!)
>
> 2. *The Laplace transform turns derivatives into multiplication.*
>
> $$\mathcal{L}[f'(t)] = sF(s) - f(0) \quad (10.10.7)$$
> $$\mathcal{L}[f''(t)] = s^2F(s) - sf(0) - f'(0) \quad (10.10.8)$$

Those constants are $f(0)$ and $f'(0)$ with lowercase "f"s; they are the initial conditions for the original function, not for its transform. You will prove these two rules in Problem 10.228, and extend them to higher order derivatives in Problem 10.229. Because of these rules, the Laplace transform turns a differential equation into an algebra equation.

3. *The Laplace transform turns integrals into division.*

$$\mathcal{L}\left[\int_0^t f(\tau)\,d\tau\right] = \frac{F(s)}{s} \tag{10.10.9}$$

We would be remiss if we did not mention one other important property: a function $f(t)$ will have a Laplace transform if it meets the following conditions.

1. The function must be "piecewise continuous." This means that on any finite interval, you must be able to describe f as a finite number of continuous functions. So a function that is 0 in $[0, 1)$, then 1 in $[1, 2)$, then 0 in $[2, 3)$, then 1 in $[3, 4)$, and so on forever is piecewise continuous. A function that is 0 for all rational numbers and 1 for irrational numbers is not.
2. There must be some constants M and k such that $|f(t)| \leq Me^{kt}$ for all t-values.

Informally, we can say that a function can only have a Laplace transform if it grows *no faster than an exponential function*. This means you cannot take the Laplace transform of e^{t^2}, since it obviously causes the improper integral in Equation 10.10.6 to diverge.[11] But compared to a Taylor series (which requires the function to be continuous and differentiable) or a Fourier transform (which requires the function to approach zero as $x \to \pm\infty$), Laplace transforms are remarkably unrestricted.

Appendix H summarizes much of the information given here on Laplace transforms. It also contains a brief table of Laplace transforms, and discusses the use of "partial fractions" and "convolutions" in the use of such a table.

10.10.3 Problems: Heaviside, Dirac, Laplace

Throughout these problems, $H(x)$ refers to the Heaviside function and $\delta(x)$ to the Dirac delta function.

In Problems 10.199–10.208 graph the given function. Of course you can't draw an infinitely thin spike for a delta function, but you can draw a tall thin spike at the correct x-value. No computers should be needed for this exercise, although you may want to use them to check your work.

10.199 $3H(x-1)$

10.200 $H(x+3) - H(x-2)$

10.201 $H(x) - H(x-1) + H(x-5) - H(x-7)$

10.202 $(\sin x)H(x)$

10.203 $(\sin x)H(x - \pi/2)$

---

[11] We asked Mathematica for $\mathcal{L}\left[e^{t^2}\right]$ and were surprised to get an answer rather than an error. It turns out there is a generalization of the definition that allows for Laplace transforms of such "superexponential" functions. The idea was first proposed in Vignaux, JC. 1939. Sugli integrali di Laplace asintotici. *Atti. Acad. Naz. Lincei, Rend. Cl. Sci. Fis. Mat.*, 29: 396-402 and worked out in greater detail in Deakin, MAB. 1994. Laplace transforms for superexponential functions. *J. Aust. Math. Soc. Ser. B*, 36: 17–25.

520 Chapter 10 Methods of Solving Ordinary Differential Equations

10.204 $\sin(x - \pi/2)H(x - \pi/2)$
10.205 $e^x[H(x) - H(x-1)] + (10-x)[H(x-3) - H(x-5)]$
10.206 $\delta(x+2)$
10.207 $\delta(x+2) + \delta(x-2)$
10.208 $2\delta(x+1)$

In Problems 10.209–10.212 write a function that matches the given graph.

10.209
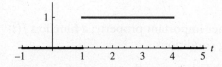

10.210
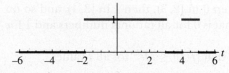

10.211
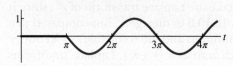

10.212
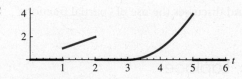

10.213 What are the units of the constant A in Equation 10.10.4?

10.214 In 2D or 3D a Dirac delta function at a point is defined so that the double or triple integral over any region including that point equals 1. Equivalently, $\delta(\vec{x}) = \delta(x)\delta(y)\delta(z)$. Assume a point mass m is located at the origin and no other mass is in the area; write the density function for that region of space. Remember that the integral of density over a region gives you the mass in that region.

In Problems 10.215–10.226 find the Laplace transform of the given function by hand, using Equation 10.10.6. (You may of course want to use a computer to check yourself.)

10.215 k (a constant function)
10.216 e^{kt}
10.217 t
10.218 t^2
10.219 $H(t-2) - H(t-5)$
10.220 $[H(t) - H(t-10)]e^{kt}$
10.221 $[H(t) - H(t-1)]t$
10.222 $\delta(t-10)$
10.223 $\delta(t-3) + \delta(t-4)$
10.224 $(\ln t)\delta(t-4)$
10.225 Problem 10.209.
10.226 Problem 10.211. *Hint*: you may find the integrals on Page 485 useful.

10.227 Prove that the Laplace transform is a linear operator.

10.228 In this problem you will derive the equations governing Laplace transforms of derivatives.
  (a) By definition, $\mathcal{L}[f'(t)] = \int_0^\infty f'(t)e^{-st}dt$. Use integration by parts to turn this formula into Equation 10.10.7. *Hint*: assume $f(t)$ grows slower than an exponential in the limit $t \to \infty$.
  (b) Now, let's rewrite Equation 10.10.7 like this: $\mathcal{L}[f'(t)] = s\mathcal{L}[f(t)] - f(0)$. This rule applies to any function that *has* a Laplace transform: you can replace f with g or h and it is just as true. Instead of these, replace f with g' and show that this leads to Equation 10.10.8.

10.229 In this problem you will generalize Equations 10.10.7 and 10.10.8 to higher order derivatives. (It may be useful to have done Problem 10.228 first, but it's not necessary.)
  (a) Beginning with Equation 10.10.8, replace f with g' and derive a formula for $\mathcal{L}[g'''(t)]$.
  (b) Extrapolating, write down a likely looking rule for $\mathcal{L}[f^{(n)}(t)]$.
  (c) Prove your rule by induction.

10.230 Evaluate the following integrals. This entire problem should take a couple of minutes at most.
(a) $\int_{-3}^{-2} \delta(x)\, dx$
(b) $\int_{-3}^{2} \delta(x)\, dx$
(c) $\int_{-\infty}^{\infty} 4\delta(x)\, dx$
(d) $\int_{-\infty}^{\infty} 4e^x \delta(x)\, dx$

10.231 Evaluate the following integrals. This entire problem should take a couple of minutes at most.
(a) $\int_{0}^{3} \delta(x-2)\, dx$
(b) $\int_{4}^{5} \delta(x-2)\, dx$
(c) $\int_{-\infty}^{\infty} 4\delta(x-2)\, dx$
(d) $\int_{-\infty}^{\infty} 4e^x \delta(x-2)\, dx$

10.232 Write a function $v(t)$ for the voltage in a circuit in each of the following cases. In each case assume the voltage is zero until the first time mentioned. Your functions should be based on the Heaviside function rather than explicitly using piecewise notation.
(a) A constant voltage V_0 is supplied from $t=0$ onward.
(b) A constant voltage V_0 is supplied from $t=-1$ to $t=1$.
(c) The voltage is turned on and grows linearly from 0 at $t=0$ to V_0 at $t=1$, and stays at V_0 from then on.

10.233 A golf club hits a ball of mass m, accelerating it from rest to velocity v. In the limit where you treat the collision as instantaneous, write the force $F(t)$ exerted by the club on the ball.

10.234 A block of mass m, initially at rest, is acted on by a force $F(t) = k\delta(t) + F_1[H(t-t_1) - H(t-t_2)]$, where $t_2 > t_1 > 0$.
(a) What are the units on the constants k and F_1? *Hint*: They are not the same.
(b) Find the block's velocity at $t = t_2$. *Hint*: For any two times t_a and t_b, $v(t_b) - v(t_a) = \int_{t_a}^{t_b} a(t)\, dt$.

10.235 Let $f(t)$ equal 0 for $t < 0$ and $t > 2$. From $t=0$ to $t=1$ it rises linearly from 0 to $f(1) = 1$, and from $t=1$ to $t=2$ it decreases linearly back to 0.
(a) Sketch $f(t)$.
(b) Write $f(t)$ using Heaviside functions.
(c) Find $F(s)$.

10.236 Let $f(t)$ equal $\sin t$ from $t=0$ to $t=\pi$ and 0 everywhere else.
(a) Sketch $f(t)$.
(b) Write $f(t)$ using Heaviside functions.
(c) Find $F(s)$. *Hint*: you may find the integrals on Page 485 useful.

10.237 The integral $\int_{-\infty}^{x} \delta(t)\, dt$ defines a function of x, which we will call for this problem $f(x)$.
(a) Evaluate $f(x)$ at three negative x-values and three positive x-values. (Don't try to find $f(0)$; it is not well defined.)
(b) Graph $f(x)$.
(c) The function $f(x)$ has a name. What is it?

10.238 The Fourier transform of a function $f(x)$ is defined as $\hat{f}(p) = 1/(2\pi) \int_{-\infty}^{\infty} f(x) e^{-ipx}\, dx$. Find the Fourier transform of $f(x) = \delta(x)$. If you've studied Fourier transforms, explain why your answer makes sense.

10.239 Assume that t is measured in seconds and $f(t)$ is measured in "things." Based on Equation 10.10.6, what must be the units of s? (Consider that an exponent must be unitless.) What are the units of $F(s)$?

10.240 (a) Find the Laplace transform of $\cos t$ by hand, using Equation 10.10.6. *Hint*: you may find the integrals on Page 485 useful.
(b) Find the Laplace transform of $\sin t$ by applying Equation 10.10.7 or 10.10.9 to your answer from Part (a).

10.241 (a) Find the Laplace transform of $f(t) = e^t$ by hand, using Equation 10.10.6.
(b) Since e^t is its own derivative, it should work as simultaneously f and f' in Equation 10.10.7. Show that it does.
(c) Find the Laplace transform of $f(t) = e^{-t}$ by hand, using Equation 10.10.6.
(d) Find the Laplace transform of $\sinh t$ by using your previous results and linearity. (If you don't know what $\sinh t$ is, look it up. We will explore it more fully in Chapter 12.)

10.242 Here is a true fact: $\int_0^t \delta(t-2)\, dt = H(t-2)$. (This should make sense if you think about it.) Take the Laplace transform of both sides using Equation 10.10.9 and show that both sides give the same result.

10.11 Using Laplace Transforms to Solve Differential Equations

The properties of Laplace transforms, especially with regard to derivatives, make many differential equations easier to solve.

10.11.1 Explanation: Using Laplace Transforms to Solve Differential Equations

Solving differential equations with Laplace transforms is a three-step process.

1. Take the Laplace transform of both sides of the equation. The properties of Laplace transforms turn the differential equation into an algebra equation.
2. Solve the algebra equation.
3. Take the inverse Laplace transform of the resulting function.

One peculiar feature of this process is the way it handles initial conditions. You do *not* find a general solution with arbitrary constants and then plug in the initial conditions to find a specific solution as you do with every other method we have presented. Instead, the initial conditions are built into your equation when you apply the properties of Laplace transforms (Equation 10.10.7 et al). If you want the general solution you can get it by using generic initial conditions y_0 and $\dot{y}_0$.

In Section 10.10 we presented the following scenario: an RLC circuit is connected to a power source that supplies a constant voltage V_0 for t_1 seconds and is then turned off. To make the problem more specific, let's choose $C = 1\,\text{F}$, $L = 2\,\text{H}$, $R = 3\,\Omega$, $V_0 = 9\,\text{V}$, $t_1 = 3\,\text{s}$, and as an initial condition we'll say there is no current flowing before the power source is turned on. (You'll solve this problem for arbitrary constants R, C, L, V_0 and t_1 in Problem 10.270. The process is the same, but the answer is unwieldy.) Equation 10.10.5 then becomes:

$$3i + \int_0^t i\,d\tau + 2\frac{di}{dt} = 9[H(t) - H(t-3)] \quad i(0) = 0 \quad (10.11.1)$$

There is an integral in the middle of our differential equation (technically making it an "integro-differential equation"), and the function on the right is discontinuous. Both of these features would prove challenging for most of the techniques we have seen, but they pose no difficulty for Laplace transforms.

1. *Take the Laplace transform of both sides of the equation.*

$$\mathcal{L}\left[3i + \int_0^t i\,d\tau + 2\frac{di}{dt}\right] = \mathcal{L}[9[H(t) - H(t-3)]]$$

The Laplace transform on the right is easy to evaluate by hand. It may help if you draw the function; it is 9 on the domain $0 \le t \le 3$ and zero everywhere else.

$$\int_0^\infty V(t)e^{-st}\,dt = 9\int_0^3 e^{-st}\,dt = -\frac{9}{s}e^{-st}\Big|_0^3 = \frac{9}{s}(1 - e^{-3s})$$

The Laplace transform on the left requires the properties of Laplace transforms. In the first line below, we use the property of linearity to separate the terms and pull out constants. In the second line we use the properties of derivatives and integrals to turn the differential equation into an algebra equation. Note our use of $I(s)$ for $\mathcal{L}[i(t)]$.

10.11 | Using Laplace Transforms to Solve Differential Equations

$$3\mathcal{L}[i] + \mathcal{L}\left[\int_0^t i\, d\tau\right] + 2\mathcal{L}\left[\frac{di}{dt}\right] = \frac{9}{s}\left(1 - e^{-3s}\right)$$

$$3I + \frac{I}{s} + 2[sI - i(0)] = \frac{9}{s}\left(1 - e^{-3s}\right)$$

This is the stage where we apply our initial condition and replace $i(0)$ with 0. The result is an algebra equation relating I and s.

$$3I + \frac{I}{s} + 2sI = \frac{9}{s}\left(1 - e^{-3s}\right)$$

2. *Solve the algebra equation.*

$$I = \frac{9\left(1 - e^{-3s}\right)}{3s + 1 + 2s^2}$$

We're done, right? We wanted to find I and we have!

Actually, a lot of electrical engineers would agree with that statement. They often work with the Laplace transforms directly, without bothering to convert back. But for our purposes let's finish the process and find the actual current i instead of its transform I.

3. *Take the inverse Laplace transform.* We plugged that function into a computer, simplified its answer a bit, and got:

$$i(t) = 9\left[e^{-\frac{t}{2}} - e^{-t} + \left(e^{3-t} - e^{\frac{3-t}{2}}\right)H(t - 3)\right] \tag{10.11.2}$$

We *always* urge you to stop and look at your final solution and see what it's telling you (and if it makes sense), but as the methods become more complicated and more computer-dependent this step becomes even more important. Here are a few things we can see from Equation 10.11.2.

- The Heaviside function in the middle means that the current follows one curve while the power source is on (before $t = 3$) and a different curve after the source is switched off. Given the scenario, it would be quite surprising if it were otherwise!
- Because the term in parentheses (multiplied by the Heaviside function) goes to zero at $t = 3$, the current is continuous even when the voltage is not. (The inductor resists sudden changes in current.)
- Over time, the current decays to zero. (The resistor gradually drains off all the energy.)

You can of course check the solution by plugging it back into Equation 10.11.1, but a bit of care is required with the integral. You'll go through that in Problem 10.269.

EXAMPLE   **Solving a Differential Equation Using Laplace Transforms**

Question: Solve this differential equation.[12]

$$2\frac{d^2q}{dt^2} + 3\frac{dq}{dt} + q = \sin^2 t \text{ with initial conditions } q(0) = 4,\, q'(0) = 5$$

---

[12] Eagle-eyed observers may recognize that this models the same circuit as Equation 10.11.1 with a different driving function. We rewrote it as a second-order differential equation for q instead of a first-order integro-differential equation for i, not for any compelling physical reason, but to demonstrate a range of Laplace transform problems.

Solution:
We will follow the same three-step recipe as above.

1. *Take the Laplace transform of both sides of the equation.*

$$\mathcal{L}\left[2\frac{d^2q}{dt^2} + 3\frac{dq}{dt} + q\right] = \mathcal{L}\left[\sin^2 t\right]$$

On the left we will use the properties of Laplace transforms to write everything in terms of Q, the as-yet-unknown Laplace transform of our solution. On the right the computer tells us that $\mathcal{L}\left[\sin^2 t\right] = 2/[s(s^2+4)]$.

$$2\mathcal{L}\left[\frac{d^2q}{dt^2}\right] + 3\mathcal{L}\left[\frac{dq}{dt}\right] + \mathcal{L}[q] = \frac{2}{s(s^2+4)}$$

$$2\left[s^2Q - sq(0) - q'(0)\right] + 3\left[sQ - q(0)\right] + Q = \frac{2}{s(s^2+4)}$$

$$2\left[s^2Q - 4s - 5\right] + 3\left[sQ - 4\right] + Q = \frac{2}{s(s^2+4)}$$

Notice that the properties of Laplace transforms brought in the initial conditions.

2. *Solve the algebra equation.*

$$Q(2s^2 + 3s + 1) - 8s - 10 - 12 = \frac{2}{s(s^2+4)}$$

$$Q(s+1)(2s+1) = \frac{2}{s(s^2+4)} + 8s + 22$$

$$Q = \frac{8s^4 + 22s^3 + 32s^2 + 88s + 2}{s(s+1)(2s+1)(s^2+4)}$$

3. *Take the inverse Laplace transform.* Once again we hit the computer.

$$q(t) = \frac{1}{2} + \frac{1}{170}\left[7\cos(2t) - 6\sin(2t)\right] - \frac{68}{5}e^{-t} + \frac{290}{17}e^{-t/2}$$

If you plug that solution into the original differential equation you end up with $[1 - \cos(2t)]/2$, which is the same as $\sin^2 t$.

Remember that a Laplace transform assumes a domain from 0 to ∞. If you need to model a current that turns on at $t = -2$ consider shifting your time variable. A tricky case arises when the right side of a differential equation contains $\delta(t)$ because this causes a discontinuous change at $t = 0$. The initial conditions at that point are ambiguous—do they occur before or after the instantaneous delta function?—and the Laplace transform method can give wrong answers. One way to solve this is by putting your delta function at a positive time and then taking the limit as that time approaches zero. See Problem 10.265.

Interpreting Laplace Transforms
A Fourier transform can give you insight into the behavior of an otherwise random-looking function. For this reason Chapter 9 stresses *interpreting* Fourier transforms more than *computing* them. The goal is to look at the Fourier transform of a function and say, for instance, "this function has oscillations with periods of 3 and 7."

10.11 | Using Laplace Transforms to Solve Differential Equations

A Laplace transform, by contrast, is more a trick for calculation than a tool for insight. This section is therefore about solving equations rather than interpreting transforms. But consider the following situation, which is unfortunately common. You start with a differential equation. You take the Laplace transform and solve the algebra. Then you ask a computer for the inverse Laplace transform...and there isn't one. "This is the Laplace transform of the solution" is as far as you're going to get. At that point, anything you can learn about the original function is going to come directly from its Laplace transform.

So we're going to give you a quick rule for interpreting Laplace transforms.

> **Interpreting Laplace Transforms**
>
> If the term $s - s_0$ appears in the denominator of a fully simplified Laplace transform $F(s)$, then the function $f(t)$ will have a term that goes like $e^{s_0 t}$.

That rule isn't obvious from anything we've said here, although you can get some idea of why it's true by verifying that the Laplace transform of $e^{s_0 t}$ is $1/(s - s_0)$. (Section 10.10 Problem 10.216.) You will see more generally why that rule works when you learn how to find an inverse Laplace transform in Chapter 13. Here we want to give a few examples to clarify the deliberately vague phrase "goes like."

EXAMPLE   **Interpreting Laplace Transforms**

The Laplace transform $\dfrac{1}{s-3}$ has one vertical asymptote at $s = 3$. It represents the function e^{3t}.

The Laplace transform $\dfrac{s}{s-3}$ is different, but it has the same vertical asymptote. It represents the function $\delta(t) + 3e^{3t}$. We see that changing the numerator does (of course) change the result, but the function still grows like e^{3t}.

The Laplace transform $\dfrac{1}{(s+2)(s-3)}$ has vertical asymptotes at $s = -2$ and $s = 3$. It represents the function $(1/5)\left(e^{3t} - e^{-2t}\right)$.

The Laplace transform $\dfrac{1}{s^2+9}$ may appear to have no such numbers in the denominator, but it factors as $\dfrac{1}{(s+3i)(s-3i)}$. An exponent of $\pm 3i$ indicates sinusoidal oscillation with frequency 3, and sure enough this corresponds to the function $(1/3)\sin(3t)$.

The Laplace transform $\dfrac{1}{s^2+2s+9}$ factors into $\dfrac{1}{(s+1+i)(s+1-i)}$. We therefore expect its inverse transform to involve $e^{(-1\pm i)t}$, but what does that look like? If you rewrite it as $e^{-t}e^{\pm it}$ you can see an oscillation with frequency 1 being multiplied by a decaying exponential. Sure enough, the inverse Laplace transform is $e^{-t}\sin t$.

As you can see, vertical asymptotes—real numbers that are out of domain in the Laplace transform—indicate exponential growth and decay. Imaginary numbers give frequencies of oscillation, just as in a Fourier series. This rule of thumb doesn't tell you whether $f(t)$ is a sine or a cosine or exactly what its amplitude is, but it does tell you the general behavior.

The examples above are all simple cases designed to illustrate how to interpret a Laplace transform, but in those cases it's easy to find the inverse Laplace transform—by hand, with a table, or with a computer—so there's no real need for this interpretation. The example below illustrates the more typical case of a function that cannot be easily inverse Laplace transformed.

> **EXAMPLE** **Interpreting One More Laplace Transform**
>
> The Laplace transform $\dfrac{\sqrt{1+s}}{s^2+2s+9}$ factors into $\dfrac{\sqrt{1+s}}{(s+1+i)(s+1-i)}$, so the denominator has the roots $s = -1 \pm i$. This function has no simple inverse Laplace transform that you can look up in a table, but just by looking at the denominator we can tell that it behaves *qualitatively* like the last example above, oscillating with frequency 1 and an amplitude that decays like e^{-t}. See Problem 10.267.

Coupled Differential Equations

"Coupled" differential equations involve relationships between more than one dependent variable. We introduced coupled equations in Chapter 1 (see felderbooks.com) and we approached them visually in Section 10.3 of this chapter (see felderbooks.com). Many approaches to coupled equations involve decoupling them, by substitution (Chapter 1) or by matrices (Chapter 6).

The example below is taken directly from Chapter 6. There we show how these equations can be used to model two masses on three springs, and how to solve them with linear algebra; here we solve them with Laplace transforms. Quite frankly, the linear algebra approach is better. It involves less algebra, can be done without a computer or lookup table, and gives much more physical insight into the behavior of the system. But Laplace transforms are a valuable technique too, and they do have the advantage that once you have learned to use them on individual equations, you don't have to learn anything new to apply them to coupled equations.

> **EXAMPLE** **Using Laplace Transforms to Solve Coupled Differential Equations**
>
> Problem:
> Find the functions $x_1(t)$ and $x_2(t)$ that solve the following equations subject to the initial conditions $x_1(0) = 2$, $x_2(0) = 1/3$, and $\dot{x}_1(0) = \dot{x}_2(0) = 0$.
>
> $$\frac{d^2 x_1}{dt^2} = -4x_1 + 3x_2$$
> $$\frac{d^2 x_2}{dt^2} = \frac{2}{3}x_1 - 3x_2$$
>
> Solution:
> We begin by taking the Laplace transforms of both equations, using X_1 and X_2 for $\mathcal{L}[x_1]$ and $\mathcal{L}[x_2]$ as usual.
>
> $$s^2 X_1 - s x_1(0) - \dot{x}_1(0) = -4X_1 + 3X_2$$
> $$s^2 X_2 - s x_2(0) - \dot{x}_2(0) = \frac{2}{3}X_1 - 3X_2$$
>
> Plug in the given initial conditions and rearrange into the standard form for simultaneous linear equations.
>
> $$(s^2 + 4)X_1 - 3X_2 = 2s$$
> $$-\frac{2}{3}X_1 + (s^2 + 3)X_2 = \frac{1}{3}s$$

The next step is to solve two simultaneous linear equations for X_1 and X_2. This involves a fair bit of algebra that we're going to skip. (You can use substitution, elimination, or matrices.) After simplification, you end up here.

$$X_1 = \frac{s(2s^2 + 7)}{(s^2 + 2)(s^2 + 5)}$$

$$X_2 = \frac{s(s^2 + 8)}{3(s^2 + 2)(s^2 + 5)}$$

Before reaching for the computer it's worth noting that both denominators telegraph oscillatory solutions with frequencies $\sqrt{2}$ and $\sqrt{5}$, based on the rule we gave above for interpreting Laplace transforms. When we do hit the computer, we find the same answer we found in Chapter 6:

$$x_1(t) = \cos\left(\sqrt{2}\, t\right) + \cos\left(\sqrt{5}\, t\right)$$

$$x_2(t) = \frac{2}{3}\cos\left(\sqrt{2}\, t\right) - \frac{1}{3}\cos\left(\sqrt{5}\, t\right)$$

You can of course confirm—*without* Laplace transforms or matrices—that these functions satisfy the given differential equations and initial conditions.

Stepping Back

After "Guess and Check," this section may be the most important in the chapter. Laplace transforms are a powerful approach to solving differential equations, and their influence is also felt far beyond those applications. (Laplace himself used them in probability theory.) As one electrical engineer put it:[13] "EEs regularly toss around rational polynomials as descriptions of system behavior, and use them to tweak the systems by adding feedback loops, filters, whatever. All these polynomials are Laplace transforms, where the transform is so taken for granted that we don't even mention it."

The method we've presented here is most useful for differential equations with constant coefficients, for which you can apply linearity and the properties of Laplace transforms very directly. In some special cases there are tricks you can use to solve other ODEs. (See Problem 10.277.) The real power of Laplace transforms will become evident, however, when we use them to solve *partial* differential equations in Chapter 11.

10.11.2 Problems: Using Laplace Transforms to Solve Differential Equations

10.243 Walk-Through: Differential Equation by Laplace Transform. In this problem you will solve the following differential equation.

$$3\frac{d^2x}{dt^2} + 10\frac{dx}{dt} - 8x = \delta(t - 5)$$

with initial conditions $x(0) = 2$, $x'(0) = 1$.

(a) Take the Laplace transform of both sides.
  i. On the right side of the equation, find the Laplace transform. You will start with the definition (Equation 10.10.6) and then apply a key property of the Dirac delta function (Equation 10.10.2).

---
[13] Henry Rich, in an email to the authors

ii. On the left side of the equation, apply the properties of Laplace transforms: first linearity, and then the derivative properties (Equations 10.10.7 and 10.10.8). Use X to represent $\mathcal{L}[x]$, and use the values given for $x(0)$ and $x'(0)$.

(b) Solve your resulting algebra equation to find the function $X(s)$.

(c) 🖥 Use a computer to find the inverse Laplace transform of $X(s)$. Write the function $x(t)$ that solves the original equation.

🖥 In Problems 10.244–10.255 solve the given differential equation subject to the given initial conditions. Use the three-step Laplace transform method modeled in Problem 10.243 even if other methods would be easier (which they will be in some cases). Computers should be used only to find Laplace and inverse Laplace transforms as necessary.

10.244 $x''(t) + 9x(t) = 0$; $x(0) = 2$, $x'(0) = 5$

10.245 $x''(t) - 9x(t) = 0$; $x(0) = 2$, $x'(0) = 5$

10.246 $x''(t) - 6x'(t) + 5x(t) = e^{5t}$; $x(0) = x'(0) = 0$

10.247 $y'(t) + \int_0^t y(\tau)d\tau = e^{-t}$; $y(0) = 0$

10.248 $2y''(t) - 7y'(t) - 4y(t) = e^t \sin t$; $y(0) = 2$, $y'(0) = 3$

10.249 $y''(t) + \int_0^t y(\tau)d\tau = 1$; $y(0) = y'(0) = 0$

10.250 $3f''(t) - f'(t) - 4f(t) = H(t-3) - H(t-4)$; $f(0) = 1$, $f'(0) = -1$

10.251 $g''(t) - 25g(t) = \delta(t-1) + 3\delta(t-2)$; $g(0) = 7$, $g'(0) = 5$

10.252 $h''(t) - 6h'(t) + 8h(t) = [H(t-1) - H(t-2)]e^{-t}$; $h(0) = 0$, $h'(0) = 0$

10.253 $h''(t) - 4h(t) = [H(t) - H(t-\pi)]\sin t$; $h(0) = 0$, $h'(0) = 0$

10.254 Equation 10.10.3 with $m = 1$, $b = 4$, $k = 3$, $F_1 = 2$, $t_1 = 5$, $x(0) = 6$, $\dot{x}(0) = 7$

10.255 Equation 10.10.4 with $m = 2$, $b = 7$, $k = 3$, $A = 4$, $t_1 = 3$, $x(0) = 6$, $\dot{x}(0) = 1$

🖥 In Problems 10.256–10.259 use the three-step Laplace transform method to solve the given coupled differential equations and initial conditions, using a computer only to find inverse Laplace transforms as necessary.

10.256 $dx/dt = 3x + 5y$, $dy/dt = x - y$; $x(0) = 4$, $y(0) = 2$

10.257 $dx/dt = 2x + y$, $dy/dt = 3x + 4y$; $x(0) = 1$, $y(0) = -1$

10.258 $d^2x/dt^2 = 5x - 3y$, $d^2y/dt^2 = 4x - 2y$; $x(0) = 6$, $y(0) = 2$, $x'(0) = y'(0) = 0$

10.259 $d^2x/dt^2 = -10x + 3y$, $d^2y/dt^2 = 3x - 2y$; $x(0) = 10$, $y(0) = 40$, $x'(0) = 30$, $y'(0) = -20$

In Problems 10.260–10.264 you will be given a Laplace transform $F(s)$. Based on the Explanation (Section 10.11.1) write a real-valued function $f(t)$ that matches the qualitative behavior of its inverse Laplace transform. (To find the exact function requires a complex integral, but your function should have the exponential functions and cosines in all the right places.) Computers should *not* be used for these problems.

10.260 $3/(s+4)$

10.261 $2s/(s-7)$

10.262 $4e^{-s}/(s^2 + 3s - 10)$

10.263 $1/(s^2 + 25)$

10.264 $1/(8s^2 + 4s + 1)$

10.265 In this problem you're going to use Laplace transforms to solve the equation $x''(t) + 2x(t) = \delta(t)$. The trick is taking the Laplace transform of that delta function.

(a) Find the Laplace transform of $\delta(t-1)$. (This should be quick and easy.)

(b) Show that the Laplace transform of $\delta(t+1)$ is 0.

(c) Explain why the Laplace transform of $\delta(t)$ is ambiguous given the definition of the Laplace transform and the properties of Dirac delta functions.

Since the value of $\mathcal{L}[\delta(t)]$ is ambiguous, you can approach this problem as a limit.

(d) 🖥 Use Laplace transforms to solve the equation $x''(t) + 2x(t) = \delta(t - t_1)$ with initial conditions $x(0) = x'(0) = 0$, where t_1 is a positive constant. Use a computer only for the inverse Laplace transform.

(e) Take the limit of your solution as $t_1 \to 0^+$ to find the solution to the original equation.

(f) Sketch your solution. (This should not require a computer!)

(g) Suppose the differential equation in this problem represents a mass on a spring. First describe the forces acting on this mass based on the differential equation.

Then describe the resulting motion based on your solution, and explain why it makes sense for this scenario.

10.266 In the method of Laplace transforms, unlike other methods, the initial conditions are plugged in as part of the process. But you can still find the general solution by putting in arbitrary values for the initial conditions.

(a) Solve the equation $x''(t) + \omega^2 x(t) = k[H(t - t_1) - H(t - t_2)]$ (where t_1 and t_2 are positive constants) with initial conditions $x(0) = x_0$, $x'(0) = v_0$.

(b) What are the arbitrary constants in your solution? Does it have the correct number to be the general solution to this equation?

10.267 In the example on Page 526 we claimed that the inverse Laplace transform of $\sqrt{1 + s}/(s^2 + 2s + 2)$ is a mess, but that it has the same qualitative behavior as the inverse Laplace transform of $1/(s^2 + 2s + 2)$. Check this by asking a computer for the inverse Laplace transform of $\sqrt{1 + s}/(s^2 + 2s + 2)$. Then have the computer plot the inverse Laplace transforms of these two functions on the same plot. In what ways are they similar, and in what ways are they different? (Note: your inverse transform may appear to be complex, but it is real in the domain $t \geq 0$ which is all that matters.)

10.268 In this problem you'll solve the differential equation $\ddot{y} + \dot{y} + y = (\sin t)/t$. Feel free to start by trying to solve it without Laplace transforms if you really want to appreciate why Laplace transforms are sometimes the best method to use.

(a) Solve this equation with initial conditions $y(0) = \dot{y}(0) = 1$ to find the Laplace transform $Y(s)$. You can use a computer to find $\mathcal{L}\left[(\sin t)/t\right]$ but you should be able to do the rest by hand.

(b) Looking at $Y(s)$, explain how you can know that $y(t)$ should oscillate with a decaying amplitude. What is the period of that oscillation?

(c) Have a computer take the inverse Laplace transform of your solution $Y(s)$ and plot the resulting function $y(t)$. Check that it oscillates with a decaying amplitude and check if its period matches your prediction.

10.269 In the Explanation (Section 10.11.1) we solved Equation 10.11.1 and ended up with Equation 10.11.2. In this problem you will confirm that our answer works.

(a) Confirm that the solution works for $0 < t < 3$.

(b) Now show that the solution works for $t > 3$, but be careful with the integral: $\int_0^3 i(t)\,dt$ represents the charge buildup on the capacitor during the first three seconds and $\int_3^t i(t)\,dt$ represents the charge buildup after that time.

10.270 Equation 10.10.5 represents an RLC circuit with a voltage source that is turned on after t_1 seconds.

(a) Solve the equation, leaving R, L, C, V_0 and t_1 as unknown constants, subject to the initial condition $i(0) = 0$. Your answer should be the Laplace transform $I(s)$. It can be inverse Laplace transformed to give $i(t)$, but the result is too messy to bother with here, so don't.

(b) Here's a different approach to the same problem: start by taking the time derivative of both sides. This turns a first-order integro-differential equation into a second-order differential equation. Solve that. Here are two hints.

- The derivative of a Heaviside function is a Dirac delta function.
- A second-order equation needs two initial conditions, and we only gave you one. You can find the other one by plugging $t = 0$ into both sides of the original integro-differential equation. (Since you are looking for the initial condition, assume $H(t)$ is still zero.)

(c) Show that for $C = 1$, $L = 2$, $R = 3$, $V_0 = 9$, $t_1 = 3$ your answer reduces to the one we found in the Explanation (Section 10.11.1).

10.271 An RLC circuit has $L = 2$, $R = 7$, $C = 1/6$. It starts with zero current and no initial charge on the capacitor. A voltage of $9\sin(t)$ is turned on from $t = 0$ to $t = 2\pi$ and then turned off. Use Equation 10.10.1 to find the Laplace transform of the current. *Hint*: you may find the integrals on Page 485 useful.

10.272 A 1 kg block is attached to a spring with spring constant k. It experiences a damping force $F = -bv$. Take $k = 8$ N/m and $b = 6$ N·s/m.

(a) Write a differential equation for the position $x(t)$ of the block.

Now suppose an 8 N external force is applied to the block for five seconds. Before and after those five seconds the only forces on the block are the spring force and the damping force. (During those five seconds there are three forces on the block: spring, damping, and external.)

(b) Write a differential equation for the block that includes all three forces acting on it.

(c) Assuming the block is at rest at equilibrium prior to the start of the external force, solve the equation you wrote in Part (b). Your answer will be in the form $X(s)$, the Laplace transform of the position.

(d) Without calculating the inverse Laplace transform, does this describe exponential growth, exponential decay, or oscillations? How can you tell?

(e) Evaluate the inverse Laplace transform to find the function $x(t)$ that describes the motion of the block.

(f) Plot $x(t)$. Describe the behavior of the system during the first five seconds and after the first five seconds. Explain why each of these behaviors make sense in the context of the problem.

10.273 The image below shows two objects with mass m suspended between three springs with spring constant k. Call the left object's displacement from equilibrium x_1 and the right object's displacement from equilibrium x_2. The rightmost spring is compressed by an amount x_2, so it exerts a force $-kx_2$ on the right mass.

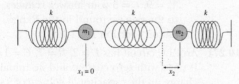

(a) What force does the leftmost spring exert on the left mass? Your answer should depend on k and x_1.

(b) In terms of x_1 and x_2, by what amount is the middle spring stretched?

(c) What is the force of the middle spring on the left mass? On the right mass? Make sure your answers both have the correct signs.

(d) Write the total force on each mass and use Newton's second law to write two coupled differential equations for $x_1(t)$ and $x_2(t)$.

(e) Taking $k = 10$ N/m, $m = 3$ kg, $x_2(0) = 2$ m, and $x_1(0) = x_1'(0) = x_2'(0) = 0$, use Laplace transforms to solve those coupled equations. Your answer will be in the form of two Laplace transforms X_1 and X_2.

(f) Take the inverse Laplace transforms of X_1 and X_2 to find the displacements x_1 and x_2.

10.274 Two coupled pendulums are shown below.

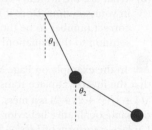

The equations describing this system depend on the angles θ_1 and θ_2, the gravitational acceleration g, and the length L of the strings.

$$2\ddot{\theta}_1 + \ddot{\theta}_2 \cos(\theta_1 - \theta_2) + \dot{\theta}_2^2 \sin(\theta_1 - \theta_2)$$
$$+ 2\frac{g}{L}\sin\theta_1 = 0$$

$$\ddot{\theta}_2 + \ddot{\theta}_1 \cos(\theta_1 - \theta_2) - \dot{\theta}_1^2 \sin(\theta_1 - \theta_2)$$
$$+ \frac{g}{L}\sin\theta_2 = 0$$

These equations are non-linear and cannot be solved using the methods in this section. For small oscillations, however, you can assume θ_1, θ_2, and all of their derivatives remain small, and thus approximate this with a set of linear equations.

(a) Replace all the trig functions with the linear terms of their Maclaurin series expansions. Then eliminate any remaining non-linear terms. The result should be two coupled, linear differential equations.

(b) Now let $\theta_1(0) = \alpha$, $\theta_2(0) = \sqrt{2}\,\alpha$, $\dot{\theta}_1(0) = \dot{\theta}_2(0) = 0$. Take the Laplace transform of your ODEs and use them to get simultaneous algebra equations for $\mathcal{L}[\theta_1(t)]$ and $\mathcal{L}[\theta_2(t)]$.

(c) Solve those algebra equations and then take the inverse Laplace transforms of your answers to find $\theta_1(t)$

and $\theta_2(t)$. (You can do the algebra by hand but it's tedious, so we recommend letting the computer do it for you.)

10.275 The Transfer Function An RLC circuit, a mass on a damped simple harmonic oscillator, and a variety of other systems obey the equation $x''(t) + a_1 x'(t) + a_0 x(t) = f(t)$. The constants a_1 and a_0 define the properties of the system (e.g. resistors, capacitors, and inductors, or damping force and spring constant). The driving term $f(t)$ represents the external input to the system (e.g. voltage source or external force). You can solve this equation for a particular a_0, a_1, and $f(t)$ with some set of initial conditions. For the homogeneous initial conditions $x(0) = x'(0) = 0$, however, you can also derive a general relation between the input $f(t)$ and the output $x(t)$.

(a) Solve this differential equation using Laplace transforms. Your answer will give $X(s)$ as a function of a_0, a_1, and $F(s)$, the Laplace transform of the driving term.

(b) The transfer function $G(s)$ is defined as $X(s)/F(s)$, or (roughly speaking) *output/input*. Calculate $G(s)$ as a function of a_0 and a_1. (The transfer function is more often designated $H(s)$ but we're already using that for the Heaviside function. There just aren't enough letters.)

The transfer function only depends on the properties of the system, not on the driving term $f(t)$. The rest of this problem will focus on a 1 kg block attached to a spring with spring constant 3 N/m, with a damping term $F_d = -bv$, $b = 2$ N·s/m.

(c) Write the differential equation for this system and find its transfer function. We have not told you whether there is an external force or not, because $G(s)$ doesn't depend on it.

(d) For each of the following driving forces f_e, calculate $X(s)$ from the relation $X(s) = G(s)F(s)$. Then use a computer to calculate the inverse Laplace transform $x(t)$.
  i. $f_e = \delta(t-1)$
  ii. $f_e = e^{-2t}$
  iii. $f_e = \sin t$

10.276 *[This problem depends on Problem 10.275.]* The transfer function is only defined for the homogeneous initial conditions $x(0) = x'(0) = 0$. To see why, start with the initial conditions $x(0) = x_0$, $x'(0) = v_0$ and show that you can't find a transfer function that is independent of the driving function.

10.277 Exploration: Laguerre Polynomials The equation $ty''(t) + (1-t)y'(t) + ny(t) = 0$ is called "Laguerre's equation" and its solutions for integer values of n are called "Laguerre polynomials." You can solve this equation using Laplace transforms, but first you need to figure out what to do with the Laplace transform of $ty'(t)$ and $ty''(t)$.

(a) The definition of a Laplace transform is $F(s) = \int_0^\infty f(t)e^{-st}dt$. Differentiate both sides of this equation with respect to s. Use your result to write a formula relating $\mathcal{L}[tf(t)]$ to $F(s)$.

(b) You know that $\mathcal{L}[y'(t)] = sY(s) - y(0)$. Use the formula you just derived to write a formula for $\mathcal{L}[ty'(t)]$. *Hint*: Remember that $y(0)$ is a constant.

(c) Derive a similar formula for $\mathcal{L}[ty''(t)]$.

(d) Take the Laplace transform of both sides of Laguerre's equation. The result should be a first-order ODE for $Y(s)$.

(e) Solve that ODE for $Y(s)$ using separation of variables. The integration step can be done with partial fractions, or you can have a computer evaluate the integral for you.

(f) The first two Laguerre polynomials are $L_0(t) = 1$ and $L_1(t) = 1 - t$. Take the Laplace transform of these two functions and show that they match the solution you found for $n = 0$ and $n = 1$.

10.12 Green's Functions

With the method of "Green's functions" you replace a difficult or unknown function with a succession of Dirac delta functions. If you can solve your equation for each individual delta function, you can add up all the solutions to solve the equation you started with.

10.12.1 Discovery Exercise: Green's Functions

The following differential equation suggests Laplace transforms—and you can solve it that way and get the right answer—but here you're going to use a different approach.

$$\frac{dx}{dt} + x = \delta(t-3) \quad \text{with} \quad x(0) = 0$$

1. For all $t \neq 3$ this equation is just $\dot{x} + x = 0$. Write the general solution to this equation.
2. For $t < 3$ you have the initial condition. Plug it in to find the arbitrary constant and write the solution.
3. At $t = 3$ the function undergoes a discontinuous but finite jump upward. After $t = 3$ your solution from Part 1 applies again, although with a different value of the arbitrary constant. Based on all that information, sketch the solution for $t \geq 0$. Your sketch does not have to be accurate about the size of the $t = 3$ jump, but should be qualitatively correct in all other respects.

You can check yourself by solving the same equation with a Laplace transform!

10.12.2 Explanation: Green's Functions

Suppose you wanted to solve the equation $y''(x) + 6y(x) = 3\sin x + 4e^x + 12$. One of the easiest approaches is to solve three different problems.

$$y_1''(x) + 6y_1(x) = 3\sin x$$
$$y_2''(x) + 6y_2(x) = 4e^x$$
$$y_3''(x) + 6y_3(x) = 12$$

Because the differential equation is linear, the sum $(y_1 + y_2 + y_3)(x)$ will solve your original problem.

Solving smaller problems and then summing the solutions is a common approach to linear equations. To take a more sophisticated example, suppose you wanted to solve $y''(x) + 6y(x) = f(x)$ for some terribly complicated $f(x)$. You might write $f(x)$ as a sum of $\sin(nx)$ terms (a Fourier sine series). The solution to $y''(x) + 6y(x) = \sin(nx)$ is $y = \sin(nx)/(6-n^2)$ for any value of n, so you can sum all those individual functions to find the solution—in the form of an infinite series—of the equation you started with.

With "Green's functions" you write the inhomogeneous term, not as a series of sines, but as a series—actually an integral—of Dirac delta functions. But the idea is very similar to the Fourier approach. If you can solve the differential equation for an arbitrary delta function, you can integrate over all your solutions to solve the equation you started with.

We begin, then, with a discussion of differential equations with Dirac delta functions.

A Differential Equation with a Dirac Delta Function

Consider the following equation on the domain $0 \leq x \leq \pi/2$.

$$\frac{d^2y}{dx^2} + y = 2\delta\left(x - \frac{\pi}{6}\right) \quad \text{with} \quad y(0) = y\left(\frac{\pi}{2}\right) = 0 \quad (10.12.1)$$

The differential equation is an ideal candidate for a Laplace transform, but the boundary conditions are not. A Laplace transform would need $y(0)$ and $y'(0)$, but we have been given $y(0)$ and $y(\pi/2)$. (This is one of the reasons we tend to use Laplace transforms on the time variable rather than on spatial variables.)

10.12 | Green's Functions

So we're going to use a different approach. We begin with the observation that for all x in our domain *except* $x = \pi/6$, Equation 10.12.1 is just the simple harmonic oscillator equation:

$$\frac{d^2y}{dx^2} + y = 0 \quad \left(x \neq \frac{\pi}{6}\right) \qquad (10.12.2)$$

So the delta function divides our domain into three regions.

$0 \leq x < \pi/6$	In this region we are solving Equation 10.12.2 so the solution is $y = A \sin x + B \cos x$. Plugging in the boundary condition $y(0) = 0$ we find that $B = 0$.	$y = A \sin x$
$\pi/6 < x \leq \pi/2$	In this region we are again solving Equation 10.12.2 so the solution is $y = C \sin x + D \cos x$. (The ODE is the same so the general solution is the same. The boundaries are different so the arbitrary constants are different.) Plugging in the boundary condition $y(\pi/2) = 0$ we find that $C = 0$.	$y = D \cos x$
$x = \pi/6$	Right at that spot the driving function is momentarily infinite.	What happens there?

We'll tell you what happens at that particular x-value: $y(x)$ is continuous, but dy/dx discontinuously jumps up by 2. We'll justify those two claims below, but first let's see how they allow us to fill in our solution.

$$y \text{ is continuous}: \quad A \sin \frac{\pi}{6} = D \cos \frac{\pi}{6}$$

$$\frac{dy}{dx} \text{ goes up by 2}: \quad A \cos \frac{\pi}{6} + 2 = -D \sin \frac{\pi}{6}$$

Solving simultaneously we find $A = -\sqrt{3}$ and $D = -1$, which gives us the complete solution.

$$y = \begin{cases} -\sqrt{3} \sin x & x < \pi/6 \\ -\cos x & x > \pi/6 \end{cases}$$

What Happened at $x = \pi/6$?

Now let's fill in the crucial piece left out of our discussion above, the effect of the delta function on the solution. We made two claims above. Below we state them in a more general form.

Effect of a Dirac Delta Function on a Second-Order Differential Equation

Consider a second-order linear differential equation in the following form, where k and p are constants.

$$\frac{d^2y}{dx^2} + a_1(x)\frac{dy}{dx} + a_0(x)y = k\delta(x - p) \qquad (10.12.3)$$

- The delta function will not cause a discontinuity in $y(x)$ at $x = p$.
- The delta function will cause dy/dx to jump discontinuously by k at $x = p$.

As with so many formulas in this chapter, be careful of the form of the differential equation: a coefficient in front of y'' will change the result.

The result above was stated in terms of second-order equations, but it generalizes both up and down. In a first-order equation a delta function causes a discontinuity in the function itself (as in the Discovery Exercise, Section 10.12.1). In a third-order equation the delta function does not cause a discontinuity in the function or its first derivative, but the *second* derivative jumps by k. In general, a delta function in an nth-order equation causes a discontinuous jump in the $(n-1)^{th}$ derivative.

That makes sense if you think about position functions. A first-order differential equation tells you the velocity of an object as a function of its position and time: $v = f(x, t)$. If you introduce a delta function then your velocity spikes momentarily to infinity, causing a discontinuous leap in position. A second-order equation tells you the acceleration: $a = f(v, x, t)$. In this case a delta function causes your acceleration to momentarily spike, so your velocity changes abruptly. (Think of a baseball hitting a bat.)

That last paragraph is a hand-wave designed to convince you that for a second-order differential equation, a delta function causes a discontinuity in the derivative but not in the function itself. You will take up that question more rigorously in Problem 10.291. But it leaves open the question, how much does the first derivative jump?

Remember that a delta function is defined by its integral. So we're going to return to Equation 10.12.1 and integrate both sides *around* the delta function. Defining ϵ as a small Δx we will integrate from $(\pi/6 - \epsilon)$ to $(\pi/6 + \epsilon)$. Then we will let $\epsilon \to 0$ to see what happens in the infinitesimal space around the delta function.

$$\int_{\pi/6-\epsilon}^{\pi/6+\epsilon} \frac{d^2y}{dx^2} \, dx + \int_{\pi/6-\epsilon}^{\pi/6+\epsilon} y \, dx = 2 \int_{\pi/6-\epsilon}^{\pi/6+\epsilon} \delta\left(x - \frac{\pi}{6}\right) dx$$

Let's look at that equation term by term, in the limit as $\epsilon \to 0$.

- The integral of y'' is $y'(\pi/6 + \epsilon) - y'(\pi/6 - \epsilon)$: the change in the derivative around the delta function. In the limit as $\epsilon \to 0$ this becomes the instantaneous change in y' at the exact point of the delta function.
- The second term is the area under the y curve. As long as y itself remains finite this must approach zero as $\epsilon \to 0$.
- The integral over the delta function is 1 for any non-zero ϵ. (Think of how we originally defined the delta function!)

We are left with $\Delta\left(y'\right) = 2$, as we previously claimed. In Problem 10.288 you will repeat this exercise on Equation 10.12.3 to demonstrate our more general claim.

Every Function is an Integral of Delta Functions

We began this section with an analogy to Fourier series. Let's return briefly to that analogy.

When you write a Fourier sine series, you represent an arbitrary function (subject to certain conditions) as a series of sine functions.

$$f(x) = \sum_{n=1}^{\infty} b_n \sin(nx) = b_1 \sin(x) + b_2 \sin(2x) + b_3 \sin(3x) + \ldots$$

In the example above the frequencies are the positive integers $1, 2, 3 \ldots$ You can cover a wider period if you use half-integer frequencies $0.5, 1, 1.5, 2 \ldots$ But to represent an entire non-periodic function you have to use all real numbers as frequencies. Your coefficients now become a Fourier sine transform and your sum becomes an integral.

$$f(x) = \int_0^\infty \hat{f}_s(p) \sin(px) \, dp \qquad (10.12.4)$$

Think of $\hat{f}_s(p)$ as the coefficient of the $\sin(px)$ term, and the integral as adding these terms up across all p-values.

In this section our goal is to represent an arbitrary function $f(x)$ as an integral, not of sines or cosines, but of delta functions. How do we do that? The key is Equation 10.10.2. Below we have copied that equation, but with different letters. Where before we used the letter x we now use p, and where we used a we now use x.

$$f(x) = \int_{-\infty}^\infty f(p) \delta(p - x) \, dp \qquad (10.12.5)$$

Please confirm for yourself that this is just Equation 10.10.2 relabeled; that's how you will know it's true. But then look at it by analogy to Equation 10.12.4. At any given point $x = p$ we write a delta function;[14] its coefficient is the value of $f(x)$ at that particular point. When we integrate to add up these functions across all p-values, we end up with the original function $f(x)$.

Finally, remember why we're doing all that! Imagine a differential equation with an inhomogeneous term $f(x)$ that is difficult to work with. Equation 10.12.5 tells you how to replace $f(x)$ with an integral over delta functions. You solve the differential equation for the delta functions at each point p, and integrate the solutions over all p-values to solve the original equation.

Using Green's Functions to Solve a Differential Equation

Now we're ready to put everything in this section together. Consider the following equation on the domain $0 \leq x \leq \pi/2$, with the inhomogeneous term $f(x)$ left general.

$$\frac{d^2y}{dx^2} + y = f(x) \quad \text{with} \quad y(0) = y\left(\frac{\pi}{2}\right) = 0 \qquad (10.12.6)$$

We now have a strategy for approaching this problem.

1. **Rewrite the inhomogeneous term as an integral over delta functions.** Equation 10.12.5 tells us how to do this: the coefficient of the delta function at every point is the value of the function at that point. Note that we need only consider points between 0 and $\pi/2$ because the problem is restricted to that domain.

$$f(x) = \int_0^{\pi/2} f(p) \delta(p - x) \, dp$$

2. **Write the differential equation with the inhomogeneous term replaced by one arbitrary delta function. The solutions to this equation are called "Green's Functions."** Above we solved Equation 10.12.6 with $f(x)$ replaced by a delta function at $x = \pi/6$. Now we

---

[14] If $\delta(p - x)$ in Equation 10.12.5 is bothering you, feel free to replace it with $\delta(x - p)$. These two are the same because $\delta(x)$ is an even function.

have to repeat that exercise at every x-value in our domain; in other words, we have to solve the following equation for an arbitrary p.

$$\frac{d^2 G}{dx^2} + G = \delta(x-p) \quad \text{with} \quad G(0) = G\left(\frac{\pi}{2}\right) = 0 \qquad (10.12.7)$$

In Problem 10.285 you will solve that equation and end up here.

$$G(x,p) = \begin{cases} -(\cos p)\sin x & x < p \\ -(\sin p)\cos x & x > p \end{cases}$$

For any given p-value (that is, any given spot where you draw the delta function) there is a particular $G(x)$ function that solves this equation. These functions are collectively referred to as the "Green's functions" for this equation.

3. **Integrate over all the Green's functions to solve the original problem.** We have seen that the original $f(x)$ is an integral across all the delta functions, and that each Green's function is the solution for one particular delta function. Because the differential equation is linear, summing (or in this case integrating) all those individual solutions will solve the original equation.

$$y(x) = \int_{-\infty}^{\infty} f(p) G(x,p)\, dp \qquad (10.12.8)$$

In Problem 10.290 you will show that this integral does give you a solution to the differential equation you started with. Our current problem involves only x-values between 0 and $\pi/2$ so those form the limits of integration.

$$y(x) = \int_0^{\pi/2} f(p) G(x,p)\, dp = \int_0^{\pi/2} f(p) \left(\begin{cases} -(\cos p)\sin x & x<p \\ -(\sin p)\cos x & x>p \end{cases} \right) dp$$

To integrate a piecewise function you break it up. Because p is the variable over which we are integrating, we look at the region $0 < p < x$ (the bottom part of the piecewise function above) and then at the region $x < p < \pi/2$ (top part).

$$y(x) = -\cos x \int_0^x f(p) \sin p\, dp - \sin x \int_x^{\pi/2} f(p) \cos p\, dp \qquad (10.12.9)$$

That is the solution to Equation 10.12.6. Of course the problem had an unspecified inhomogeneous term $f(x)$ so that function appears as part of the solution. In Problems 10.286 and 10.287 you will find solutions for a few specific $f(x)$ functions.

> **EXAMPLE** **Using Green's Functions to Solve a Differential Equation**
>
> **Problem:**
> Solve the following equation on the domain $t \geq 0$. Your answer will be a formula involving the unknown function $f(t)$.
>
> $$2\frac{d^2 x}{dt^2} - 7\frac{dx}{dt} - 4x = f(t) \quad \text{with} \quad x(0) = x'(0) = 0 \qquad (10.12.10)$$

Solution:
We will employ the three-step Green's function method discussed above.

1. **Rewrite the inhomogeneous term as an integral over delta functions.** This is just a matter of writing down Equation 10.12.5 on the proper domain.

$$f(t) = \int_0^\infty f(p)\delta(p-t)\,dp$$

2. **Solve the differential equation with the inhomogeneous term replaced by one arbitrary delta function.** The "Green's functions" for this problem are the solutions to the following differential equation.

$$2\frac{d^2 G}{dt^2} - 7\frac{dG}{dt} - 4G = \delta(t-p) \quad \text{with} \quad x(0) = x'(0) = 0 \quad (10.12.11)$$

At all points except $t = p$ we have an equation we can easily solve with guess and check.

$$2\frac{d^2 G}{dt^2} - 7\frac{dG}{dt} - 4G = 0 \quad \rightarrow \quad G = Ae^{4t} + Be^{-t/2}$$

On the left side ($t < p$) the initial conditions $G(0) = G'(0) = 0$ lead to $A = B = 0$, or $G = 0$. So the function stays flat until the delta function "kicks" it. To see what happens there, divide both sides of Equation 10.12.11 by two: $G'' - (7/2)G' - 2G = (1/2)\delta(t-p)$. This is now in the form of Equation 10.12.3 with $k = 1/2$, so the first derivative jumps by $1/2$ at $t = p$. This gives us the initial conditions for the second half of the journey: $G(p) = 0$ and $G'(p) = 1/2$.

$$Ae^{4p} + Be^{-p/2} = 0$$
$$4Ae^{4p} - (1/2)Be^{-p/2} = 1/2$$

Solving, $A = (1/9)e^{-4p}$ and $B = -(1/9)e^{p/2}$, so the solution to Equation 10.12.11 is this.

$$G(t, p) = \begin{cases} 0 & t < p \\ (1/9)\left[e^{4(t-p)} - e^{-(1/2)(t-p)}\right] & t > p \end{cases}$$

3. **Integrate over all the Green's functions to solve the original problem.** Because G vanishes for all $p > t$ we need only integrate it for $p < t$.

$$x(t) = \int_0^t f(p)G(x, p)\,dp = \frac{1}{9}\int_0^t f(p)\left[e^{4(t-p)} - e^{-(1/2)(t-p)}\right]dp \quad (10.12.12)$$

For any given $f(t)$, Equation 10.12.10 represents a specific differential equation and Equation 10.12.12 is the solution.

Stepping Back

If L is a linear operator, Green's functions are one approach to solving the inhomogeneous equation $L(y) = f(x)$. The Green's function $G(x, p)$ is the solution the equation would have if the inhomogeneous term were a delta function $\delta(x-p)$. Physically the Green's function represents the response the system would have to a brief stimulus at one moment of time or point in space.

10.12.3 Problems: Green's Functions

10.278 One crucial step in solving an equation with Green's functions is solving an equation where $\delta(x)$ is the inhomogeneous term. In this problem you'll solve $y''(x) + 4y(x) = 2\delta(x - \pi/8)$ subject to the boundary conditions $y(0) = y(\pi/4) = 0$.

(a) Find the general solution to the differential equation in the region $0 \leq x < \pi/8$. Your solution $y_L(x)$ will of course have two arbitrary constants. Call them A and B.

(b) Use the boundary condition $y(0) = 0$ to eliminate one of the two arbitrary constants.

(c) Find the general solution $y_R(x)$ in the region $\pi/8 < x \leq \pi/4$. Call the arbitrary constants C and D and use the condition $y(\pi/4) = 0$ to eliminate one of them.

(d) The solution $y(x)$ must be continuous, so $y_L(\pi/8) = y_R(\pi/8)$. Use this condition to eliminate one more arbitrary constant.

(e) The effect of the term $2\delta(x - \pi/8)$ on the right is to make $y'(x)$ jump by 2 at $x = \pi/8$. Set $y'_R(\pi/8) = y'_L(\pi/8) + 2$ and solve for the final arbitrary constant.

(f) Write the solution $y(x)$.

10.279 Walk-Through: Green's Functions. In this problem you will solve the equation $y''(x) + 2y'(x) - 15y(x) = f(x)$ on the domain $0 \leq x < \infty$ subject to the conditions $y(0) = 0$ and $y(x)$ is finite in the limit $x \to \infty$.

(a) Rewrite the inhomogeneous term as an integral over delta functions.

(b) Write the differential equation for the Green's function $G(x, p)$. This will be the original differential equation with the inhomogeneous term replaced by an arbitrary delta function at $x = p$.

(c) For all $x \neq p$ the delta function is 0. Solve the differential equation you wrote in Part (b) in this case. Your solution will have two arbitrary constants for $x < p$ and two *different* arbitrary constants for $x > p$, so you will need four letters.

(d) Use the boundary condition $G(0, p) = 0$ to write a relationship between the first two arbitrary constants.

(e) Use the condition that $G(x, p)$ must be finite as $x \to \infty$ to set one of the remaining arbitrary constants to zero.

(f) The Green's function must be continuous at $x = p$. Use this fact to write another relationship between your arbitrary constants.

(g) At $x = p$ the derivative dG/dx jumps discontinuously by 1. Use this fact to write another relationship between your arbitrary constants.

(h) Solve for all three arbitrary constants to find your Green's function $G(x, p)$.

(i) Write the solution $y(x)$ as an integral involving the functions $G(x, p)$ and $f(x)$. (Remember that in order to integrate a piecewise function you break it up!)

(j) Now consider the particular function $f(x) = e^{-x}$. Evaluate the integral from Part (i) to find the solution $y(x)$. Plug it into the differential equation to verify that it works.

In Problems 10.280–10.284 find the Green's function for the given equation subject to the given boundary conditions. You may find it helpful to first work through Problem 10.279 as a model. If the inhomogeneous term is unspecified leave your final answer as an integral with the unknown function in it.

10.280 $y'(x) - (1/x)y(x) = f(x)$ on the domain $x \geq 1$ with $y(1) = 0$

10.281 $y''(x) = x$ on the domain $0 \leq x \leq 10$ with $y(0) = y(10) = 0$

10.282 $y''(x) + k^2 y(x) = \cos^2(kx)$ on the domain $0 \leq x \leq \pi/(2k)$ with $y(0) = y(\pi/2k) = 0$

10.283 $y''(x) + 2y'(x) - 15y(x) = f(x)$ on the domain $-\infty < x < \infty$ with $y(x)$ finite in the limits $x \to \pm\infty$

10.284 $3x''(t) - 8x'(t) + 4x(t) = 4e^{2t}$ on the domain $t \geq 0$ with $x(0) = x'(0) = 0$.

10.285 Solve Equation 10.12.7 from the Explanation (Section 10.12.2). You may find it helpful to go through Problem 10.278 as a guide.

10.286 For the very simple case $f(x) = 1$, use Equation 10.12.9 to find the solution to Equation 10.12.6. Simplify your solution as much as possible, and then verify that it does indeed solve the differential equation.

10.287 For the case $f(x) = \sec x$, use Equation 10.12.9 to find the solution to Equation 10.12.6. Simplify your solution as much as possible, and then verify that it does indeed solve the differential equation.

10.288 Show that for a solution $y(x)$ to Equation 10.12.3, the first derivative dy/dx will experience a jump of k at $x = p$. *Hint*: We showed this in the Explanation (Section 10.12.2) for a specific example. Your more general proof should follow the same form. (You may assume the function is continuous at $x = p$ because otherwise it couldn't obey the differential equation there. You'll rigorously prove that for a particular equation in Problem 10.291.)

10.289 In the Explanation (Section 10.12.2) we solved the equation $2x''(t) - 7x'(t) - 4x = f(t)$ with $x(0) = x'(0) = 0$. Consider this equation in the very simple case of $f(t) = 1$.

(a) Write the differential equation in this case.

(b) Solve it by plugging $f(t) = 1$ into the solution we found. Simplify your answer as much as possible.

(c) Solve the same equation using a Laplace transform. (A computer will be required only for finding the inverse Laplace transform.) Confirm that you get the same answer.

10.290 In the last step of this method you integrate across all the Green's functions to find a solution to the original differential equation. In the Explanation (Section 10.12.2) we justified that step based on the linearity of the differential equation, but you can also plug it in and see that it works.

(a) Begin by plugging Equation 10.12.8 (our hopeful solution) into Equation 10.12.6 (our original problem).

(b) Use Equations 10.12.5 and 10.12.7 to show that the resulting equation is true.

10.291 In the Explanation (Section 10.12.2) we claimed that when you introduce a delta function into a second-order differential equation, the solution remains continuous (although its derivative does not). Here you will show that for the equation $y'' + y = 2\delta(x - \pi/6)$ with boundary conditions $y(0) = y(\pi/2) = 0$. You'll do it using the definition

$$\delta(x - p) = \lim_{\epsilon \to 0} \frac{1}{2\epsilon}[H(x - (p - \epsilon)) - H(x - (p + \epsilon))]$$

(a) First check that definition. Make a sketch of the right-hand side (for some non-zero ϵ). You should be able to see that as $\epsilon \to 0$ it's becoming infinite at $x = p$ and zero everywhere else. Evaluate $\int_{-\infty}^{\infty}$ of the function on the right and show that it equals 1 for any value of ϵ.

(b) To solve the differential equation with this function on the right-hand side, you need to write three solutions. In the region $0 < x < \pi/6 - \epsilon$ the differential equation is $y'' + y = 0$. Find the solution to that equation with the boundary condition $y_1(0) = 0$. Call your one remaining arbitrary constant A.

(c) Similarly, solve for $y_3(x)$ in the region $\pi/6 + \epsilon < x < \pi/2$ subject to $y(\pi/2) = 0$. Call your one remaining arbitrary constant D.

(d) In the region $\pi/6 - \epsilon < x < \pi/6 + \epsilon$ the differential equation is $y'' + y = 1/(2\epsilon)$. Find the solution $y_2(x)$ in this region, with arbitrary constants B and C.

(e) For any non-zero ϵ the driving term is finite, so both $y(x)$ and $y'(x)$ must be continuous and differentiable. Set $y_1 = y_2$ and $y_1' = y_2'$ at the boundary $\pi/6 - \epsilon$, and likewise for y_2 and y_3 at their boundary. The result should be four equations for the four unknown arbitrary constants.

(f) Solve for A, and D, eliminating B and C. You can do this by hand, but you are welcome to use a computer to do the tedious algebra for you if you'd prefer. Either way, simplify as much as possible by using the trig addition and subtraction identities.

(g) Find Δy, the change in y from $\pi/6 - \epsilon$ to $\pi/6 + \epsilon$, and find $\Delta y'$ across the same interval.

(h) Take the limits of Δy and $\Delta y'$ as $\epsilon \to 0$. You should find that in this limit the function is continuous across the jump but its derivative is not.

10.292 In this problem you'll solve the differential equation $y'' - 2y/x^2 = 1$ with boundary conditions $y(0) = 0$, $y(1) = 1$.

(a) Use the Green's function method to solve the differential equation with the homogeneous boundary conditions $y(0) = y(1) = 0$.

(b) Solve the homogeneous differential equation $y'' - 2y/x^2 = 0$ with boundary conditions $y(0) = 0$, $y(1) = 1$.

(c) Add your two solutions to find the complete solution $y(x)$. Plug this solution in and verify that it solves the inhomogeneous differential equation with the inhomogeneous boundary conditions. (You will need to take a limit to verify one of the boundary conditions.)

10.293 Consider an object acted on by two forces, a drag force $F_d = -bv$ and an external force $F_{ext}(t)$. At $t = 0$ the object is at rest at the origin.

(a) Write a differential equation for the position $x(t)$ of the object.

(b) Find the Green's function for the ODE you wrote.

(c) Assume the external force is zero until $t = t_1$ and grows linearly as $F_{ext} = k(t - t_1)$ until it abruptly stops at $t = t_2$. Use the Green's function you found to find $x(t)$ for this force.

10.13 Additional Problems (see felderbooks.com)

CHAPTER 11

Partial Differential Equations

Before you read this chapter, you should be able to...

- solve ordinary differential equations with "initial" or "boundary" conditions (see Chapters 1 and 10).
- evaluate and interpret partial derivatives (see Chapter 4).
- find and use Fourier series—sine series, cosine series, both-sines-and-cosines series, and complex exponential forms, for one or more independent variables (see Chapter 9). Fourier transforms are required for Section 11.10 only.
- find and use Laplace transforms (for Section 11.11 only, see Chapter 10).

After you read this chapter, you should be able to...

- model real-world situations with partial differential equations and interpret solutions of these equations to predict physical behavior.
- use the technique "separation of variables" to find the normal modes for a partial differential equation. We focus particularly on trig functions, Bessel functions, Legendre polynomials, and spherical harmonics. You will learn about each of these in turn, but you will also learn to work more generally with any unfamiliar functions you may encounter.
- use those normal modes to create a general solution for the equation in the form of a series, and match that general solution to boundary and initial conditions.
- solve problems that have multiple inhomogeneous conditions by breaking them into subproblems with only one inhomogeneous condition each.
- solve partial differential equations by using the technique of "eigenfunction expansions."
- solve partial differential equations by using Fourier or Laplace transforms.

The differential equations we have used so far have been "ordinary differential equations" (ODEs) meaning they model systems with one independent variable. In this chapter we will use "partial differential equations" (PDEs) to model systems involving more than one independent variable.

We will begin by discussing the *idea* of a partial differential equation. How do you set up a partial differential equation to model a physical situation? Once you have solved such an equation, how can you interpret the results to understand and predict the behavior of a system? Just as with ordinary differential equations, you may find this part as or more challenging than the mechanics of finding a solution, but you will also find that understanding the equations is more important than solving them. The good news is, the work you have already done in understanding ordinary differential equations and partial derivatives will provide a strong foundation in understanding these new types of equations.

In the first three sections we will discuss a few key concepts including arbitrary functions, boundary and initial conditions, and normal modes. These discussions should be seen as extensions of work you have already done in earlier chapters. For instance, you know that the general solution to an ordinary differential equation (ODE) involves arbitrary constants.

A partial differential equation (PDE) may have an *infinite number* of arbitrary constants, or (equivalently) an *arbitrary function,* in the general solution. These constants or functions are determined from the initial and/or boundary conditions.

In the middle of the chapter—the largest part—we will deal with the engineer's and physicist's favorite tool for solving partial differential equations: "separation of variables." This technique replaces one *partial* differential equation with two or more *ordinary* differential equations that can be solved by hand or with a computer. The solutions to those equations will lead us to work with a wide variety of functions, some familiar and some new.

We will then discuss techniques that can be used when separation of variables fails. Some partial differential equations that cannot be solved by separation of variables can be solved with "eigenfunction expansion," which involves expanding your function into a series before you solve. If one of your variables has an infinite domain you can still use this trick, but it's called the "method of transforms" because you take a Fourier or Laplace transform instead of using a series expansion. Appendix I guides you through the process of looking at a new PDE and deciding which technique to use.

As you explore these techniques you will encounter many of the most important equations in engineering and physics: equations governing light and sound, heat, electric potential, and more. After this chapter you will be prepared to understand these equations, to solve them, and to interpret the solutions.

11.1 Motivating Exercise: The Heat Equation

A coin at temperature u_c is placed in a room at a constant temperature u_r.[1] "Newton's Law of Heating and Cooling" states that the rate of change of the coin's temperature is proportional to the *difference* between the temperatures of the room and the coin. (In other words, the coin will cool down faster in a freezer than in a refrigerator.) We can express this law as a differential equation: $du_c/dt = k(u_r - u_c)$ where k is a positive constant.

1. What does this differential equation predict will happen if a cold coin is placed in a hot room? Explain how you could get this answer from this differential equation, even if you didn't already know the answer on physical grounds.
2. Verify that $u_c(t) = u_r + Ce^{-kt}$ is the solution to this differential equation.

Now, suppose we replace the coin with a long insulated metal bar. We assumed above that the coin had a uniform temperature u_c, changing with time but not with position. A long bar, on the other hand, can have different temperatures at the left end, the right end, and every point in between. That means that temperature is now a function of time and position along the bar: $u(x, t)$.

To write an equation for $u(x, t)$, consider how a small piece of the bar (call it P) at position x will behave in some simple cases.

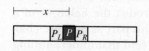

We assume that piece P is so small that its temperature is roughly uniform. However, the pieces to the left and right of it (P_L and P_R) have their own temperatures. Piece P interacts with these adjacent pieces in the same way the coin interacted with the room: the rate of heat transfer between P and the pieces on each side depends on the temperature difference between them. We also assume that heat transfer between the different parts of the bar is fast enough that we can ignore any heat transfer between the bar and the surrounding air.

---

[1] Throughout this chapter we will use the letter u for temperature. Both u and T are commonly used in thermodynamics, but we need T for something different.

3. First, suppose we start with the temperature of the bar uniform. We use $u(x, 0)$ to indicate the *initial temperature* of the bar—that is, the temperature at time $t = 0$—so we can express the condition "the initial temperature is a constant" by writing $u(x, 0) = u_0$.

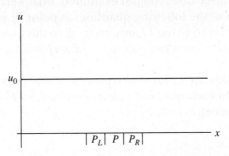

   (a) Will P give heat to P_L, absorb heat from P_L, or neither?
   (b) Will P give heat to P_R, absorb heat from P_R, or neither?
   (c) Will the temperature of P go up, go down, or stay constant?

4. Now consider a linearly increasing initial temperature: $u(x, 0) = mx + b$, with $m > 0$.

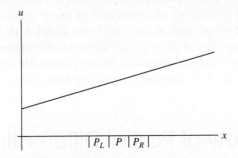

   (a) Will P give heat to P_L, absorb heat from P_L, or neither?
   (b) Will P give heat to P_R, absorb heat from P_R, or neither?
   (c) Will the rate of heat transfer between P and P_L be faster, slower, or the same as the rate of heat transfer between P and P_R?
   (d) Will the temperature of P go up, go down, or stay constant?

5. Now consider a parabolic initial temperature: $u(x, 0) = ax^2 + bx + c$. Assume $a > 0$ so the parabola is concave up, and assume that P is on the increasing side of the parabola, as shown in Figure 11.1.

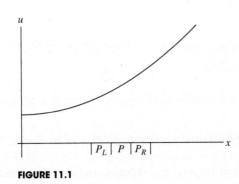

FIGURE 11.1

   (a) Will P give heat to P_L, absorb heat from P_L, or neither?
   (b) Will P give heat to P_R, absorb heat from P_R, or neither?

(c) Will the rate of heat transfer between P and P_L be faster, slower, or the same as the rate of heat transfer between P and P_R?
(d) Will the temperature of P go up, go down, or stay constant?

6. For each of the three cases you just examined, what were the signs (negative, positive, or zero) of each of the following quantities at point P: u, $\partial u/\partial x$, $\partial^2 u/\partial x^2$, and $\partial u/\partial t$? Just to be clear, you're giving 12 answers in all to this question. You'll get the signs of u, $\partial u/\partial x$, and $\partial^2 u/\partial x^2$ from our pictures, and $\partial u/\partial t$ from what was happening at point P in each case.

7. Which of the following differential equations would be consistent with the answers you gave to Part 6? In each one, k is a (real) constant, so k^2 is a positive number and $-k^2$ is a negative number.
   (a) $\partial u/\partial t = k^2 u$
   (b) $\partial u/\partial t = -k^2 u$
   (c) $\partial u/\partial t = k^2(\partial u/\partial x)$
   (d) $\partial u/\partial t = -k^2(\partial u/\partial x)$
   (e) $\partial u/\partial t = k^2(\partial^2 u/\partial x^2)$
   (f) $\partial u/\partial t = -k^2(\partial^2 u/\partial x^2)$

The "heat equation" you just found is an example of a "partial differential equation," which involves partial derivatives of a function of more than one variable. In this case, it involves derivatives of $u(x, t)$ with respect to both x and t. In general, partial differential equations are harder to solve than ordinary differential equations, but there are systematic approaches that enable you to solve many linear partial differential equations such as the heat equation analytically. For non-linear partial differential equations, the best approach is often numerical.

11.2 Overview of Partial Differential Equations

In Chapter 1 we discussed what differential equations are, how they represent physical situations, and what it means to solve them. We showed how the "general" solution has arbitrary constants which are filled in based on initial conditions to find a "particular" solution: a function.

When multiple independent variables are involved the derivatives become *partial* derivatives and the differential equations become *partial differential equations*. In this section we give an overview of these equations, showing how they are like ordinary differential equations and how they are different. The rest of the chapter will focus on techniques for solving these equations.

11.2.1 Discovery Exercise: Overview of Partial Differential Equations

We begin with an ordinary differential equation: that is, a differential equation with only one independent variable.

1. Consider the differential equation $dy/dx = y$. In words, "the function $y(x)$ is its own derivative."
   (a) Verify that $y = 2e^x$ is a valid solution to this differential equation.
   (b) Write another solution to this equation.
   (c) Write the *general* solution to this equation. It should have one arbitrary constant in it.
   (d) Find the only *specific* solution that meets the condition $y(0) = 7$.

11.2 | Overview of Partial Differential Equations

Things become more complicated when differential equations involve functions of more than one variable, and therefore partial derivatives. For the following questions, suppose that z is a function of two independent variables x and y.

2. Consider the differential equation $\frac{\partial}{\partial x} z(x, y) = z(x, y)$. In words, "if you start with the function $z(x, y)$ and take its partial derivative with respect to x, you get the same function you started with." Note that $\partial z/\partial y$ is not specified by this differential equation, and may therefore be anything at all.
   (a) Which of the following functions are valid solutions to this differential equation? Check all that apply.
      i. $z = 5$
      ii. $z = e^x$
      iii. $z = e^y$
      iv. $z = e^x e^y$
      v. $z = y e^x$
      vi. $z = x e^y$
      vii. $z = e^x \sin y$
   (b) Write a *general* solution to the differential equation $\partial z/\partial x = z$. Your solution will have an *arbitrary function* in it.
   (c) Find the only *specific* solution that meets the condition $z(0, y) = \sin(y)$.

   See Check Yourself #69 in Appendix L

3. Consider the differential equation $\frac{\partial}{\partial x} z(x, y) = \frac{\partial}{\partial y} z(x, y)$.
   (a) Express this differential equation in words.
   (b) Which of the following functions are valid solutions to this differential equation? Check all that apply.
      i. $z = 5$
      ii. $z = e^x$
      iii. $z = e^y$
      iv. $z = e^{x+y}$
      v. $z = \sin(x + y)$
      vi. $z = \sin(x - y)$
      vii. $z = \ln(x + y)$
   (c) Parts i, iv, v, and vii above are all specific examples of the general form $z = f(x + y)$. By plugging $z = f(x + y)$ into the original differential equation $\partial z/\partial x = \partial z/\partial y$, show that any function of this form provides a valid solution.
   (d) Find the only *specific* solution that meets the condition $z(0, y) = \cos y$. *Hint*: It's not in the list above.

11.2.2 Explanation: Overview of Partial Differential Equations

Chapter 1 stressed the importance of differential equations in modeling physical situations. Chapter 4 stressed the importance of *multivariate* functions: functions that depend on two or more variables.

Put the two together and you have differential equations with multiple independent variables. Because these equations are built from partial derivatives, they are called "partial differential equations." Investigate almost any field in physics and you will find a partial differential equation at the core: Maxwell's equations in electrodynamics, the diffusion equation in mass transfer operations, the Navier–Stokes equation in fluid dynamics, Schrödinger's equation in quantum mechanics, and the wave equation in optics, to name a few.

The acronym PDE is often used for "partial differential equation," as opposed to a single-variable "ordinary differential equation" or ODE.

Multivariate functions

Consider a guitar string, pinned to the x-axis at $x = 0$ and $x = 4\pi$, but free to move up and down between the two ends.

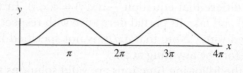

We can describe the motion by writing the height y as a function of the horizontal position x and the time t. We can look at such a $y(x, t)$ function in three different ways:

- At any given moment t there is a particular $y(x)$ function that describes the entire string. The string's motion is an evolution over time from one $y(x)$ to the next.
- Any given point x on the string oscillates according to a particular $y(t)$ function. The next point over (at $x + \Delta x$) oscillates according to a slightly different $y(t)$, and all the different $y(t)$ functions together describe the motion of the entire string.
- Finally, we can treat t as a spatial variable and plot y on the xt-plane.

The function $y(x, t)$ has two derivatives at any given point: $\partial y/\partial x$ gives the slope of the string at a given point, and $\partial y/\partial t$ gives the velocity of the string at that point. (A dot is often used for a time derivative, so $\dot{y}$ means $\partial y/\partial t$.) There are therefore four second derivatives. $\partial^2 y/\partial x^2$ is concavity, and $\partial^2 y/\partial t^2$ is acceleration. The "mixed partials" $\partial^2 y/\partial x \partial t$ and $\partial^2 y/\partial t \partial x$ generally come out the same.

All this is a quick reminder of how to think about multivariate functions and derivatives. It does not, however, address the question of what function a guitar string would actually follow. The function $y(x, t) = (1 - \cos x) \cos t$ looks fine, doesn't it? But in fact, no free guitar string would actually do that. In order to show that, and to find what it *would* do, we have to start with the equation that governs its motion.

Understanding partial differential equations

A guitar string will generally obey a partial differential equation called "the wave equation":

$$\frac{\partial^2 y}{\partial x^2} = \frac{1}{v^2} \frac{\partial^2 y}{\partial t^2} \qquad \text{The wave equation} \qquad (11.2.1)$$

where v is a constant. You will explore where this equation comes from in Problems 11.32 and 11.45, but here we want to focus on what it tells us. Let's begin by considering the function we proposed earlier.

EXAMPLE Checking a Possible Solution to the Wave Equation

Question: Does $y(x, t) = (1 - \cos x) \cos t$ satisfy the wave equation?

Answer:
We can answer this by taking the partial derivatives.

$$y(x, t) = (1 - \cos x) \cos t \quad \rightarrow \quad \frac{\partial^2 y}{\partial x^2} = \cos x \cos t \quad \text{and} \quad \frac{\partial^2 y}{\partial t^2} = -(1 - \cos x) \cos t$$

We see that $\partial^2 y/\partial x^2$ is not the same as $(1/v^2)(\partial^2 y/\partial t^2)$ (no matter what the constant v happens to be), so this function does not satisfy the wave equation. Left to its own devices, a guitar string will not follow that function.

We see with PDEs—just as we saw with ODEs in previous chapters—that it may be difficult to *find* a solution, but it is easy to *verify* that a solution does (or in this case does not) work.

We also saw with ODEs that we can often predict the behavior of a system directly from the differential equation, without ever finding a solution. This is also an important skill to develop with PDEs. Let's see what we can learn by looking at the wave equation.

The second derivative with respect to position, $\partial^2 y/\partial x^2$, gives the concavity: it is related to the shape of the string at one frozen moment in time. The second derivative of the height with respect to time, $\partial^2 y/\partial t^2$, is vertical acceleration: it describes how one point on the string is moving up and down, independent of the rest of the string. So Equation 11.2.1 tells us that wherever the string is concave up, it will accelerate upward; wherever the string is concave down, it will accelerate downward.

As an example, suppose our guitar string starts at rest in the shape $y(x, 0) = 1 - \cos x$. (Don't ask how it got there.) This function has points of inflection at $\pi/2$, $3\pi/2$, $5\pi/2$, and $7\pi/2$. So the two peaks, which are concave down, will accelerate downward; the middle, which is concave up, will accelerate upward. On the left and right are other concave up regions that will move upward, but remember that the guitar string is pinned down at $x = 0$ and $x = 4\pi$, so the very ends cannot move regardless of the shape. Based on all these considerations, we predict something like Figure 11.2.

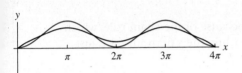

FIGURE 11.2 The blue plot shows the string at time $t = 0$ and the black plot shows the string a short time later.

The drawing shows the curve moving up where it was concave up, and down where it was concave down, while remaining fixed at the ends. We can state with confidence that the string will move from the blue curve to something kind of like the black curve, more or less. For exact solutions we have to actually find the function that matches the wave equation and all the conditions of this scenario. Within the next few sections you'll know how to do all that. But if you followed how we made that drawing, then you'll also know how to see if your answers make sense.

Boundary and Initial Conditions

In order to predict the behavior of a string, you need more than just the differential equation that governs its motion.

- You need the "initial conditions": that is, you need both the position and the velocity of every point on the string when it starts (at $t = 0$). In the example above we gave you the initial position as $y = 1 - \cos x$, and told you that the string started at rest.
- You also need the "boundary conditions." In the example above the guitar string was fixed at $y = 0$ for all time at the left and right sides. A different boundary condition (such as a moving end) would lead to different behavior over time, even if the initial conditions were unchanged.

When you solve a linear ODE you need one condition—one fact—for each arbitrary constant. For instance a second-order linear ODE has two arbitrary constants in the general solution, so you need two extra facts to find a specific solution. These might be the values of the function at two different points, or the value and derivative of the function at one point.

A PDE, on the other hand, requires an infinite number of facts. In our example the initial state occurs at an infinite number of x-positions and the boundary conditions occur at an infinite number of times. To match such conditions the solution must have an infinite number of arbitrary variables: in other words, an entire *arbitrary function*.

> **EXAMPLE** **A Solution with an Arbitrary Function**
>
> For the partial differential equation $\partial u/\partial x + x(\partial u/\partial y) = 0$, the general solution is $u = f\left(y - x^2/2\right)$.
>
> **Question:** Give three *specific* examples of solutions to this equation.
>
> **Answer:**
> $\sqrt{y - x^2/2}$, $\sin\left(y - x^2/2\right)$, and $5/\left[2 + \ln(y - x^2/2)\right]$ are all solutions.
>
> **Question:** Verify that *one* of the functions you just wrote solves this differential equation.
>
> **Answer:**
> $$u(x, y) = \sin\left(y - \frac{x^2}{2}\right) \to \frac{\partial u}{\partial x} = \cos\left(y - \frac{x^2}{2}\right)(-x), \quad \frac{\partial u}{\partial y} = \cos\left(y - \frac{x^2}{2}\right)$$
> Since $\partial u/\partial x$ is $\partial u/\partial y$ multiplied by $-x$, the combination $\partial u/\partial x + x(\partial u/\partial y)$ is equal to zero.
>
> **Question:** Prove that all functions of the form $u = f\left(y - x^2/2\right)$ solve this differential equation.
>
> **Answer:**
> The specific solution we tested above was valid because of the chain rule, which required the x derivative to be multiplied by $-x$ while the y derivative was just multiplied by 1. This generalizes to any function f.
> $$u(x, y) = f\left(y - \frac{x^2}{2}\right) \to \frac{\partial u}{\partial x} = f'\left(y - \frac{x^2}{2}\right)(-x), \quad \frac{\partial u}{\partial y} = f'\left(y - \frac{x^2}{2}\right)$$
> $$\frac{\partial u}{\partial x} + x\frac{\partial u}{\partial y} = 0$$
>
> Several students have objected to our use of the notation f' in this example: it obviously means a derivative, but with respect to what? (This is the kind of question that only excellent students ask.) In this case the derivative is being taken with respect to $\left(y - x^2/2\right)$ but more generally it means "the derivative of the f function." For instance in the previous example f was a sine so f' was a cosine.

The question of exactly what information you need in order to specify the arbitrary function and find a specific solution to a PDE turns out to be surprisingly complicated. We will return to that question when we discuss Sturm-Liouville theory in Chapter 12. Here we will offer just one guideline, which is to look at the order of the equation. For instance, the heat equation $\partial u/\partial t = \alpha(\partial^2 u/\partial x^2)$ is second order in space, and therefore requires two boundary conditions (such as the function $u(t)$ at each end). It is first order in time, and therefore requires only one initial condition (usually the function $u(x)$ at $t = 0$). The wave equation $\partial^2 y/\partial x^2 = (1/v^2)(\partial^2 y/\partial t^2)$, on the other hand, requires two boundary conditions *and* two initial conditions (usually position and velocity).

Remember that a linear ODE is referred to as "homogeneous" if every term in the equation includes the dependent variable or one of its derivatives. If a linear equation is homogeneous then a linear combination of solutions is itself a solution: a very helpful property, when you have it! The same rule applies to PDEs: Laplace's equation $\partial^2 V/\partial x^2 + \partial^2 V/\partial y^2 + \partial^2 V/\partial z^2 = 0$ is linear and homogeneous, so any linear combination of solutions is itself a solution. Poisson's equation $\partial^2 V/\partial x^2 + \partial^2 V/\partial y^2 + \partial^2 V/\partial z^2 = f(x, y, z)$ is inhomogeneous as long as $f \neq 0$.

However, *unlike* with ODEs, we will now be making the same distinction with our boundary and initial conditions.

> **Definition: Homogeneous**
> A linear differential equation, boundary condition, or initial condition is referred to as "homogeneous" if it has the following property:
> If the functions f and g are both valid solutions to the equation or condition, then the function $Af + Bg$ is also a valid solution for any constants A and B.

This definition should be clearer with an example:

EXAMPLE **Homogeneous and Inhomogeneous Boundary Conditions**

If the end of our waving string is fixed at $y(0, t) = 0$, we have a *homogeneous boundary condition*. In other words, if $f(0, t) = 0$ and $g(0, t) = 0$, then $(f + g)(0, t) = 0$ too.

However, $y(0, t) = 1$ represents an *inhomogeneous* condition. If $f(0, t) = 1$ and $g(0, t) = 1$, then $(f + g)(0, t) = 2$, so the function $f + g$ does not meet the boundary condition.

As a final note on conditions, you may be surprised that we distinguish between "initial" (time) and "boundary" (space) conditions. You can graph a $y(x, t)$ function on the xt-plane, so doesn't time act mathematically like just another spatial dimension? Sometimes it does, but initial conditions often look different from boundary conditions, and must be treated differently. In our waving string example the differential equation is second order in both space and time, so we need two spatial conditions and two temporal—but look at the ones we got! The boundary conditions specify y on both the left and right ends of the string. The initial conditions, on the other hand, say nothing about the "end time": instead, they specify both y and $\dot{y}$ at the beginning. This common (though not universal) pattern—boundary conditions on both ends, initial conditions on only one end—leads to significant differences in the ways initial and boundary conditions affect our solutions.

A Few Important PDEs

As you read through this chapter you will probably notice that we keep coming back to the same partial differential equations. Our purpose is *not* to make you memorize a laundry list of PDEs, or to treat each one as a different case: on the contrary, we want to give you a set of tools that you can use on just about *any* linear PDE you come across.

Nonetheless, the examples are important. Each of the equations listed below comes up in many different contexts, and their solutions describe many of the most important quantities in engineering and physics.

$$\frac{\partial^2 y}{\partial x^2} = \frac{1}{v^2} \frac{\partial^2 y}{\partial t^2} \quad \text{the wave equation} \tag{11.2.2}$$

We discussed the one-dimensional wave equation above. In two dimensions we could write $\partial^2 \Psi / \partial x^2 + \partial^2 \Psi / \partial y^2 = (1/v^2)(\partial^2 \Psi / \partial t^2)$ and in three dimensions we would add a third spatial derivative term. We can write this equation very generally using the "Laplacian" operator as $\nabla^2 \Psi = (1/v^2)(\partial^2 \Psi / \partial t^2)$, which is represented by different differential equations as we change dimensions and coordinate systems. You may recall the Laplacian from vector calculus but we will supply the necessary equations as we go through this chapter.

The wave equation describes the propagation of light waves through space, of sound waves through a medium, and many other physical phenomena. You will explore where it comes from in Problems 11.32 and 11.45.

$$\frac{\partial u}{\partial t} = \alpha \frac{\partial^2 u}{\partial x^2} \quad (\alpha > 0) \quad \text{the heat equation} \quad (11.2.3)$$

Like Equation 11.2.2, Equation 11.2.3 (which you derived in the motivating exercise) is a one-dimensional version of a more general equation where the second derivative is replaced by a Laplacian. The equation is used to model both conduction of heat and diffusion of chemical species. (In the latter case it's called the "diffusion equation.")

$$\nabla^2 V = f(x, y, z) \quad \text{Poisson's equation} \quad (11.2.4)$$

Poisson's equation is used to model spatial variation of electric potential, gravitational potential, and temperature. In general, f is a function of position—for instance, it may represent electrical charge distribution. Poisson's equation therefore represents many *different* PDEs with different solutions, depending on the function $f(x, y, z)$. In the special case $f = 0$ it reduces to Laplace's equation:

$$\nabla^2 V = 0 \quad \text{Laplace's equation} \quad (11.2.5)$$

We have chosen to express Equations 11.2.4 and 11.2.5 in their general multidimensional form: the one-dimensional versions are not very interesting, and in fact are not partial differential equations at all. These equations relate different spatial derivatives, but no time derivative. Problems involving these equations therefore have boundary conditions but no initial conditions.

$$-\frac{\hbar^2}{2m} \nabla^2 \Psi + V(\vec{x}) \Psi = i\hbar \frac{d\Psi}{dt} \quad \text{Schrödinger's equation} \quad (11.2.6)$$

Schrödinger's equation serves a role in quantum mechanics analogous to $\vec{F} = m\vec{a}$ in Newtonian mechanics: it is the starting point for solving almost any problem. Like Poisson's equation, Schrödinger's equation actually represents a wide variety of differential equations with different solutions, depending in this case on the potential function $V(\vec{x})$.

11.2.3 Problems: Overview of Partial Differential Equations

For Problems 11.1–11.6 indicate which of the listed functions are solutions to the given PDE. *Choose all of the valid solutions; there may be more than one.*

11.1 $\partial z/\partial y + \partial^2 z/\partial x^2 = 0$
 (a) $z(x, y) = x^2 - y$
 (b) $z(x, y) = x^2 - 2y$
 (c) $z(x, y) = e^x e^{-y}$
 (d) $z(x, y) = e^{-x} e^y$

11.2 $\partial^2 u/\partial y^2 + k(\partial u/\partial y) + \alpha(\partial^2 u/\partial x^2) = 0$
 (a) $u(x, y) = x^2 - 2\alpha y/k$
 (b) $u(x, y) = x^2 - 2\alpha y/k + C$
 (c) $u(x, y) = \sin\left(kx\sqrt{2/\alpha}\right) e^{ky}$
 (d) $u(x, y) = e^{kx\sqrt{2/\alpha}} e^{ky}$

11.3 $\partial u/\partial t = \alpha \left(\partial^4 u/\partial x^4\right)$
 (a) $u(x, t) = \alpha t^2 - x^5/120$
 (b) $u(x, t) = e^{\alpha t} + e^x$
 (c) $u(x, t) = e^{\alpha t} e^x$
 (d) $u(x, t) = e^{\alpha t} \sin x$

11.4 $t^2 \left(\partial^2 f/\partial t^2\right) + t \left(\partial f/\partial t\right) - 2x^2 \left(\partial^2 f/\partial x^2\right) = 0$
 (a) $f(x, t) = x^2 t^2$
 (b) $f(x, t) = x^2 t^2 + C$
 (c) $f(x, t) = e^t e^x$
 (d) $f(x, t) = e^t + e^x$

11.5 $\partial^2 f/\partial t^2 - v^2 (\partial^2 f/\partial x^2) = 0$
  (a) $f(x,t) = (x+vt)^5$
  (b) $f(x,t) = (x+vt)^5 + C$
  (c) $f(x,t) = (x+vt)^5 + kt^2/2$
  (d) $f(x,t) = e^{vt}e^x - kx^2/(2v^2)$
  (e) $f(x,t) = \ln(vt)\ln(x) - kx^2/(2v^2)$

11.6 $\partial^2 f/\partial t^2 - v^2 (\partial^2 f/\partial x^2) = k$
  (a) $f(x,t) = (x+vt)^5$
  (b) $f(x,t) = (x+vt)^5 + kt^2/2$
  (c) $f(x,t) = (x+vt)^5 + kt^2/2 + C$
  (d) $f(x,t) = e^{vt}e^x - kx^2/(2v^2)$
  (e) $f(x,t) = \ln(vt)\ln(x) - kx^2/(2v^2)$

For Problems 11.7–11.8 indicate which of the listed functions are solutions to the given PDE *and* the given boundary conditions. You will need to know that $\sinh x = (e^x - e^{-x})/2$. List all of the valid solutions; there may be more than one.

11.7 $\partial^2 z/\partial y^2 + \partial^2 z/\partial x^2 = 0$, $z(x,0) = z(x,\pi) = z(0,y) = 0$, $z(\pi,y) = \sin y$
  (a) $z(x,y) = \sin x \sin y$
  (b) $z(x,y) = \sinh x \sin y$
  (c) $z(x,y) = \sinh x \sin y / \sinh \pi$
  (d) $z(x,y) = \sinh x \sin y / \sinh \pi + C$

11.8 $\partial^2 y/\partial t^2 - v^2 (\partial^2 y/\partial x^2) = 0$, $y(0,t) = y(L,t) = 0$
  (a) $y(x,t) = \sin x \sin(vt)$
  (b) $y(x,t) = \sinh(\pi x) \sin(\pi vt)$
  (c) $y(x,t) = \sin(\pi x/L) \sin(\pi vt/L)$
  (d) $y(x,t) = \sin(\pi x/L) \sin(\pi vt/L) + C$

11.9 For the differential equation $x(\partial z/\partial x) + y(\partial z/\partial y) = 0$:
  (a) Show that $z = \ln x - \ln y$ is a valid solution.
  (b) Show that $z = \sin(x/y)$ is a valid solution.
  (c) Show that, for any function f, the function $z = f(x/y)$ is a valid solution.
  (d) Is $z = f(x/y) + C$ a solution? Why or why not?
  (e) Is $z = f(y/x)$ a solution? Why or why not?

11.10 For the differential equation $2y(\partial^2 z/\partial x^2) + \partial^2 z/(\partial x \partial y) = 0$:
  (a) Show that $z = f(x - y^2)$ is a valid solution to for any function f.
  (b) Is $z = f(x^2 - y)$ a solution? Why or why not?
  (c) Is $z = Af(y^2 - x)$ a solution? Why or why not?

11.11 For the differential equation $\frac{\partial}{\partial x}(xz) = \partial z/\partial y + z$:
  (a) Show that $z = \ln x + y$ is a valid solution.
  (b) Show that $z = f(xe^y)$ is a solution for any function f.
  (c) Is $z = f(x^2 e^{2y}) + C$ a solution? Why or why not?
  (d) Is $z = Cf(e^x y)$ a solution? Why or why not?

The sketches below show possible initial values of a function, each with a corresponding set of boundary conditions. For Problems 11.12–11.18, copy each of these initial condition sketches onto your paper and then, on the same sketch, show what the function will look like a short time later. Clearly label which sketch is the initial condition and which is the function at a later time. When an initial velocity is needed, assume that it is zero everywhere.

See Figure 11.2 for an example of what an answer should look like.

(a)
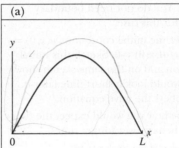
The function is fixed at zero at the two ends.

(b)
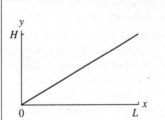
The function is fixed at zero on the left and at H on the right.

(c)
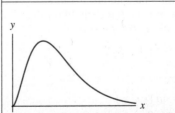
The function is fixed at zero on the left. The domain extends to infinity on the right.

11.12 $\partial^2 y/\partial x^2 = (1/v^2)(\partial^2 y/\partial t^2)$
11.13 $\partial y/\partial t = -\alpha^2 (\partial^2 y/\partial x^2)$
11.14 $\partial^2 y/\partial t^2 = -\alpha^2 x^2 (\partial^2 y/\partial x^2)$

11.15 $\partial y/\partial t = a^2 y \left(\partial^2 y/\partial x^2\right)$

11.16 Equation 11.2.3 (the heat equation)

11.17 $\partial y/\partial t = -k^2 \left(\partial y/\partial x\right)$

11.18 $\partial^2 y/\partial t^2 = k^2 \left(\partial y/\partial x\right)$

11.19 Figure (a) above shows the initial *velocity* of a string attached at $y = 0$ on both ends. (For this problem pretend the vertical axis is labeled $\partial y/\partial t$ instead of y.) The initial position is $y = 0$ everywhere. For this problem you do not need to copy the sketch. Instead draw a sketch that shows the position of the string a short time after the initial moment, assuming the string obeys the wave equation, Equation 11.2.2.

11.20 An infinite string obeys the wave equation everywhere. You do not need boundary conditions for this case.
   (a) Consider the initial conditions $u(x, 0) = \sin x$, $\partial u/\partial t(x, 0) = 0$. Sketch this initial condition and on the same sketch show how it would look a short time later if $u(x, t)$ obeys the wave equation.
   (b) Describe how you would expect the function to behave over longer times.
   (c) Repeat Parts (a) and (b) for the initial condition $u(x, 0) = 1 + \sin x$, $\partial u/\partial t(x, 0) = 0$. What are the similarities and the differences between the long term behavior in these two cases?

11.21 [*This problem depends on Problem 11.20.*] The "Klein-Gordon" equation $(1/v^2)\left(\partial^2 u/\partial t^2\right) - \partial^2 u/\partial x^2 + \omega^2 u = 0$ arises frequently in field theory. It's similar to the wave equation, but with an added term. Repeat Problem 11.20 for the Klein-Gordon equation. In what ways is the behavior described by these two equations similar and in what ways is it different?

11.22 State whether the given condition is homogeneous or inhomogeneous. Assume that the function $f(x, t)$ is defined on the domain $0 \leq x \leq x_f$, $0 \leq t < \infty$.
   (a) $f(0, t) = f(x_f, t) = 0$
   (b) $f(0, t) = 3$
   (c) $f(x, 0) = 0$
   (d) $f(x, 0) = \sin\left(\pi x/x_f\right)$
   (e) $\lim_{t \to \infty} f(x, t) = 0$
   (f) $\lim_{t \to \infty} f(x, t) = 1$
   (g) $\lim_{t \to \infty} f(x, t) = \infty$

11.23 The function $z(x, y)$ is defined on $x \in [3, 10]$, $y \in [2, 5]$. The boundary conditions are: $z(3, y) = 0$, $z(x, 2) = 1$, $z(10, y) = 2$, and $z(x, 5) = x^2$.
Which of these four boundary conditions are homogeneous, and which are not?

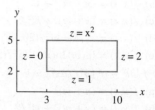

11.24 The function f is defined for all real values of x, and is *periodic:* that is, it is subject to the condition $f(x, t) = f(x + 2\pi, t)$ for all x. Is this a homogeneous condition, or not? Explain.

11.25 The function g is subject to the condition that $g(0)$ must be finite. (This condition turns out to be both common and important.) Is this a homogeneous condition, or not? Explain.

In Problems 11.26–11.28 you will be given a partial differential equation.
   (a) Suppose f is defined on the domain $-\infty < x < \infty$. Give one example of a sufficient set of initial conditions for this equation. At least one of your initial conditions must be non-zero somewhere.
   (b) Describe in words how $f(x, t)$ will behave for a short time after $t = 0$ for that set of initial conditions.

11.26 $\partial f/\partial t = k^2 \left(\partial f/\partial x\right)$

11.27 $\partial f/\partial t = -k^2 \left(\partial f/\partial x\right)$

11.28 $\partial^2 f/\partial t^2 = k^2 \left(\partial f/\partial x\right)$

Problems 11.29–11.35 depend on the Motivating Exercise (Section 11.1).

11.29 In the Motivating Exercise you used physical arguments to write the equation for the evolution of the temperature distribution in a thin bar. In that problem we ignored the surrounding air, assuming that heat transfer within the bar takes place much faster than transfer with the environment. Now write a different PDE for the temperature $u(x, t)$ in a thin bar with an external heat source that provides heat proportional to distance from the left end of the bar.

11.30 The "specific heat" of a material measures how much the temperature changes in response to heat flowing in or out. If the same amount of heat is supplied to two bricks of the same size, one of which has twice as high a specific heat as the other, the one with the higher specific heat will increase its temperature by half as much as the other one.

(Be careful about that difference; *higher* specific heat means *smaller* change in temperature.)

The motivating exercise tacitly assumed that the specific heat of the bar was constant. Now consider instead a bar whose specific heat is proportional to distance from the left edge of the bar: $h = kx$. Derive the PDE for the temperature $u(x, t)$ across the bar.

11.31 The Motivating Exercise was based on a practically one-dimensional object. Now consider heat flowing through a three-dimensional object. The outer surface of that object is held at a fixed temperature distribution. (One simple example would be a cube, where the top is held at temperature $u = 100°$ and the other five sides are all held at $u = 0°$. But this problem refers to the *general* three-dimensional heat equation with fixed boundary conditions, not to any specific example.)

(a) Write the heat equation: the partial differential equation that governs the temperature $u(x, y, z, t)$ inside the object.

(b) Under these circumstances, the heat equation will approach a "steady state" that, if reached, will never change. Write a partial differential equation for the steady-state solution $u_0(x, y, z)$.

11.32 **The wave equation**: In the Motivating Exercise you used physical arguments to write down the heat equation. In this problem you'll use similar arguments to explain the form of the wave equation. Consider a string with a tension T. As you did in the motivating exercise, you should focus on a small piece of string (P) at some position x and how it interacts with the pieces to the left and right of it (P_L and P_R). *You should justify your answers to all the parts of this problem based on the physical description, not based on the wave equation, since that equation is what we're trying to derive.*

(a) If the string is initially flat, $y(x, 0) =$ <constant>, will P_L exert an upward, downward, or zero force on P? What about the force of P_R on P? Will the net force on P be upward, downward, or zero?

(b) Repeat Part (a) if the initial shape of the string is linear, $y(x, 0) = mx + b$. (Assume $m > 0$.)

(c) Repeat Part (a) if the initial shape of the string is parabolic, $y(x, 0) = ax^2 + bx + c$ as shown in Figure 11.1 (the increasing part of a concave up curve).

(d) Does the net force on P depend on y, $\partial y/\partial x$, or $\partial^2 y/\partial x^2$?

(e) Does the force on P determine y, $\partial y/\partial t$, or $\partial^2 y/\partial t^2$?

(f) Explain in words why Equation 11.2.2 is the correct description for the motion of a string.

11.33 *[This problem depends on Problem 11.32.]* Write the PDE for a string with a drag force, e.g. a string vibrating underwater. The drag force on each piece of the string is proportional to the velocity of the string at that point, but opposite in direction. *Hint*: Go through the derivation in Problem 11.32 and see at what point the drag force would be added to what you did there.

11.34 *[This problem depends on Problem 11.32.]* Write the PDE for a string whose density is some function $\rho(x)$. (*Hint*: Think about what this implies for the mass of each small segment, and what that means for the acceleration of that segment.)

11.35 The chemical gas Wonderflonium has accumulated in a pipe. When the density of Wonderflonium is even throughout the pipe it stays constant, but if there's more Wonderflonium in one part of the pipe than another, it will tend to flow from the region of high Wonderflonium density to the region of low Wonderflonium density.

(a) Write a PDE that could describe the concentration of Wonderflonium in the pipe.

(b) Give the sign and units of any constants in your equation.

For Problems 11.36–11.44 use a computer to numerically solve the given PDEs and graph the results. You can make plots of $f(x)$ at different times, make a 3D plot of $f(x, t)$, or do an animation of $f(x)$ evolving over time. For each one describe the late term behavior (steady state, oscillating, growing without bound), based on your computer results. Then explain, based on the given equations, why you would expect that behavior even if you didn't have a computer.

11.36 $\partial^2 f/\partial t^2 = \partial^2 f/\partial x^2$, $0 < x < 1$, $0 < t < 10$, $f(0, t) = 0$, $f(1, t) = e - 1$, $f(x, 0) = e^x - 1$, $\dot{f}(x, 0) = 0$

11.37 $\partial^2 f/\partial t^2 = \partial^2 f/\partial x^2$, $0 < x < 1$, $0 < t < 10$, $f(0, t) = 0$, $f(1, t) = 0$, $f(x, 0) = 0$, $\dot{f}(x, 0) = \sin(\pi x)$

11.38 $\partial f/\partial t = \partial^2 f/\partial x^2$, $0 < x < 1$, $0 < t < 10$, $f(0,t) = 0$, $f(1,t) = e-1$, $f(x,0) = e^x - 1$

11.39 If you are unable to numerically solve this equation explain why.
$\partial^2 f/\partial t^2 = -\partial^2 f/\partial x^2$, $0 < x < 1$, $0 < t < 0.5$, $f(0,t) = 0$, $f(1,t) = e-1$, $f(x,0) = e^x - 1$, $\dot{f}(x,0) = 0$

11.40 $\partial f/\partial t = \partial f/\partial x$, $0 < x < 1$, $0 < t < 1$, $f(1,t) = 0$, $f(x,0) = e^{-20(x-.5)^2}$

11.41 $\partial f/\partial t = \partial f/\partial x$, $0 < x < 1$, $0 < t < 3$, $f(1,t) = \sin(10t)$, $f(x,0) = 0$

11.42 $\partial f/\partial t = -\partial f/\partial x$, $0 < x < 1$, $0 < t < 1$, $f(0,t) = 0$, $f(x,0) = e^{-20(x-.5)^2}$

11.43 Try solving this out to several different final times and see if you get consistent behavior. If you are unable to numerically solve this equation explain why.
$\partial f/\partial t = -\partial f/\partial x$, $0 < x < 1$, $f(1,t) = 0$, $f(x,0) = e^{-20(x-.5)^2}$

11.44 $\partial f/\partial t = x(\partial f/\partial x)$, $0 < x < 1$, $0 < t < 1$, $f(1,t) = 0$, $f(x,0) = e^{-20(x-.5)^2}$

11.45 Exploration: The Wave Equation, Quantitatively

In Problem 11.32 you gave qualitative arguments for the form of the wave equation. In this problem you will derive it more rigorously. Consider a string of uniform tension T and density (mass per unit length) λ. At some moment t the string has shape $y(x)$. Focus on a small piece of string at position x with length dx. As before we will call this small piece of string P.

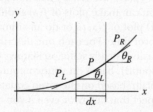

(a) The string to the right of P exerts a force T_R on P. Find the y-component of this force. Your answer will depend on the angle θ_R in the figure above.

(b) The angle θ_R is related to the slope of $y(x)$ at the right edge of the string. Rewrite your answer to Part (a) in terms of this slope: $(\partial y/\partial x)_R$.

(c) The wave equation we discuss in this chapter is only valid for small displacements of a string. Assuming the slope is small, find the linear terms of the Maclaurin series for your answer to Part (b). Use this approximate expression for the rest of the problem. If you need a derivative that you don't know you can look it up in a book or online. Or, if you prefer, you can simply have a computer generate the terms you need of the Maclaurin series.

(d) Write the y-component of the force exerted on P by the string to the left of it. You should once again assume that the slope is small and your answer should again be a linear function of the slope: $(\partial y/\partial x)_L$.

(e) Write the y-component of the net force on P.

(f) What is the mass of P in terms of quantities given in this problem?

(g) Use your answers to Parts (e) and (f) to write an equation for the vertical acceleration of P.

(h) Take the limit of your answer as $dx \to 0$ and show that this reduces to Equation 11.2.2. Express v in that equation as a function of T and λ.

11.46 Exploration: Large Waves [This problem depends on Problem 11.45.] In this problem you will redo Problem 11.45 *without* the assumption of small vibrations.

(a) In Part (b) of Problem 11.45 you derived the expression for the y-component of the force on P from the segment of string to the right of it. Simplify this expression as much as possible without assuming that the slope is small. Eliminate all trig and inverse trig functions from the expression.

(b) Repeat steps d-g in Problem 11.45 without assuming the slopes are small. The resulting expression for acceleration should have dx in the denominator and a complicated function of $(\partial y/\partial x)_R$ and $(\partial y/\partial x)_L$ in the numerator. This expression should not include any trig functions.

(c) Even though the slope may be anything, dx is small, so $(\partial y/\partial x)_R \approx (\partial y/\partial x)_L$.

Replace every occurrence of $(\partial y/\partial x)_L$ in your equation with $(\partial y/\partial x)_R - \epsilon$. Since ϵ must approach zero in the limit $dx \to 0$, you can take a Maclaurin series in ϵ of your acceleration equation and only keep the linear term. The terms that don't have ϵ should cancel, leaving $\epsilon/(dx)$ times an expression that doesn't contain any infinitesimal terms.

(d) Recalling that $\epsilon = (\partial y/\partial x)_R - (\partial y/\partial x)_L$, what is $\lim_{dx \to 0} \epsilon/(dx)$?

(e) Write the differential equation for large amplitude vibrations of a string. Since we are taking the limit $dx \to 0$, you can drop the subscripts on the slopes now and just write them as $\partial y/\partial x$. The result should be a non-linear differential equation involving $\partial^2 y/\partial t^2$, $\partial^2 y/\partial x^2$, and $\partial y/\partial x$.

(f) Show that your equation reduces to the wave equation when $\partial y/\partial x = 0$. How small does the slope have to be for the right hand side of this non-linear equation to be within 1% of the right hand side of the wave equation?

(g) The equation you just derived is more general than the wave equation because you dropped the assumption of small slopes. There are still some important approximations being used in this derivation, however. List at least two assumptions/approximations you made in deriving this equation. (*Note*: Saying that dx is small is *not* an assumption. It's part of the definition of dx.)

11.3 Normal Modes

We have seen that a vibrating string can be represented by a function $y(x,t)$ that obeys a partial differential equation called the "wave equation." The resulting motion depends on the string's initial position and velocity, and can sometimes be so complicated that it appears almost random. This section presents two different ways to see order behind the chaos. In the first approach the behavior is seen to be the sum of two different functions: one that holds its shape but moves to the left, and one that holds its shape but moves to the right.

The second approach—which will dominate, not only this section, but most of the rest of the chapter—builds up the behavior from special solutions called "normal modes." Finding simple normal modes that match your boundary conditions, and then summing those normal modes to describe more complicated behavior, is the key to understanding a wide variety of systems.

11.3.1 Discovery Exercise: Normal Modes

A guitar string extends from $x = 0$ to $x = \pi$. It is fixed at both ends, but free to vibrate in between.

The position of the string $y(x,t)$ is subject to the equation:

$$\frac{\partial^2 y}{\partial x^2} = \frac{1}{9}\frac{\partial^2 y}{\partial t^2} \qquad (11.3.1)$$

The phrase "fixed at both ends" in the problem statement gives us the boundary conditions:

$$y(0,t) = y(\pi,t) = 0 \qquad (11.3.2)$$

1. One solution to this problem is $y = 5\sin(2x)\cos(6t)$. (Later in the chapter you will find such solutions for yourself; right now we are focusing on understanding the equation and its solutions.)
   (a) Confirm that this solution solves the differential equation by plugging $y(x, t)$ into both sides of Equation 11.3.1 and showing that you get the same answer.
   (b) Confirm that this solution meets the boundary conditions, Equation 11.3.2.
   (c) Draw graphs of the shape of the string between $x = 0$ and $x = \pi$ at times $t = 0$, $t = \pi/12$, $t = \pi/6$, $t = \pi/4$, and $t = \pi/3$. Then describe the resulting motion in words.
   (d) Which of the three numbers in this solution—the 5, the 2, and the 6—is an arbitrary constant? That is, if you change that number to any other constant, the function will still solve the differential equation and meet the boundary conditions.
2. Another solution to this equation is $y = -(1/2)\sin(10x)\cos(kt)$, if you choose the correct value of k.
   (a) Plug this function into both sides of Equation 11.3.1 and solve for k.
   (b) Does the resulting function also meet the boundary conditions?
   (c) What is the period of this function in space (i.e. the distance between adjacent peaks, or the wavelength)?
   (d) What is the period of this function in time (i.e. how long do you have to wait before the string returns to its initial position)?

   See Check Yourself #70 in Appendix L

3. For the value of k that you calculated in 2, is the function $y = 5\sin(2x)\cos(6t) - (1/2)\sin(10x)\cos(kt)$ a valid solution? (Make sure to check whether it meets both the differential equation and the boundary conditions!)
4. Next consider the solution $y = A\sin(px)\cos(kt)$.
   (a) For what values of k will this function solve Equation 11.3.1? Your answer will depend on p.
   (b) For what values of p will this solution match the boundary conditions?
5. Write the solution to this differential equation in the most general form you can.

11.3.2 Explanation: Normal Modes

A vibrating string such as a guitar string obeys the *one-dimensional wave equation*:

$$\frac{\partial^2 y}{\partial x^2} = \frac{1}{v^2}\frac{\partial^2 y}{\partial t^2} \qquad (11.3.3)$$

The dependent variable y represents the displacement of the string from its relaxed height. (Note that y can be positive, negative, or zero.) The constant v is related to the tension and linear density of the string.

When you are solving an Ordinary Differential Equation (ODE) you often try to find the "general solution," a closed-form function that represents all possible solutions, with a few arbitrary constants to be filled in based on initial conditions. The same can sometimes be done for a Partial Differential Equation (PDE), and below we present the general solution to the wave equation, valid for all possible initial and boundary conditions. However, it is not

11.3 | Normal Modes

always practical to solve problems using this general solution. Instead we will find a special set of particular solutions known as "normal modes" and build up solutions for different initial and boundary conditions using these normal modes.

In this section we'll simply present the normal modes for the wave equation and use combinations of them to build other solutions. In later sections we'll show you how to find the normal modes for other PDEs.

The General Solution, aka d'Alembert's Solution

The functions $\sin\left[(x+vt)^2\right]$, $3/(x+vt)$, and $6\ln(x+vt)$ are all valid solutions to the wave equation. You may not have any idea how we just came up with them, but you can easily verify that they work. More generally, *any* function of the form $f(x+vt)$ will solve the wave equation: you can confirm this in general just as you can for the specific cases.

$$y = f(x+vt) \quad \rightarrow \quad \frac{\partial^2 y}{\partial x^2} = f''(x+vt), \quad \frac{\partial^2 y}{\partial t^2} = v^2 f''(x+vt) \quad \rightarrow \quad \frac{\partial^2 y}{\partial x^2} = \frac{1}{v^2}\frac{\partial^2 y}{\partial t^2}$$

For similar reasons, any function $g(x-vt)$ will also be a solution. Any sum of such functions will be a solution as well, so we can write the general solution of this equation as:

$$y(x,t) = f(x+vt) + g(x-vt)$$

(We don't need arbitrary constants in front because f and g can include any constants.) This form of the general solution to the wave equation is known as "d'Alembert's Solution." The constant v in this solution is not arbitrary; it was part of the original differential equation. However, f and g are *arbitrary functions:* replace them with any functions at all and you have a solution. For instance, the function $y = 1/(x+vt)^2 - 3\ln(x-vt)$ solves the equation (as you will demonstrate in Problem 11.47).

What does all that tell us about vibrating strings? $g(x-vt)$ represents a function that does not change its shape over time: it only moves to the right with speed v. For instance, the function $y(x,t) = e^{-(x-vt)^2}$ describes the behavior you might see if you grab one end of a rope and give it a quick jerk.

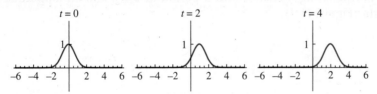

Similarly, $f(x+vt)$ represents an arbitrary curve moving steadily to the left. But the general solution $f(x+vt) + g(x-vt)$ does *not* describe a static curve that moves: it can change its shape over time in ways that might surprise you. For instance, we show here the function $\cos\left[(x+t)^2\right] + e^{-(x-t-1)^2}$.

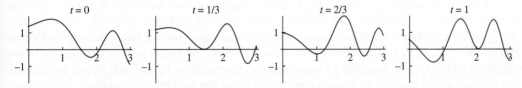

The correct combination of functions f and g describe any possible solution to the wave equation. But while f retains its shape while moving to the left, and g retains its shape while moving to the right, a combination of the two may evolve in complicated ways.

Simple Solutions for Simple Cases

Now that we have the general solution, it's just a matter of matching it to the boundary and initial conditions, and we can solve any wave equation problem, right? In principle, that's correct. In practice, it can be hard. For instance, below we consider a string that is fixed at both ends (simple boundary conditions), and begins at rest in the shape of a sine wave (simple initial conditions). Before we present the solution, you may want to try to find a function of the form $f(x+vt) + g(x-vt)$ that fits these conditions. If you don't get very far, you may be interested to hear about a very different approach—more practical and, surprisingly, no less general in the end.

Consider a string subject to the following conditions.

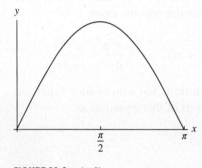

FIGURE 11.3 $y(x, 0)$.

1. The string is fixed at both ends. If the length of the string is π this imposes the conditions $y(0, t) = 0$ and $y(\pi, t) = 0$.
2. The initial shape of the string is half a sine wave $y(x, 0) = \sin x$ as shown in Figure 11.3, and the initial velocity is zero.

You can determine the motion of such a string experimentally by plucking a tight rubber band. (This is easy to do, but you have to watch carefully.) Even without such an experiment you can imagine the behavior based on your physical intuition. The sine wave will decrease in amplitude until it momentarily flattens out, and then begin to open up again on the negative side until it reaches an upside-down half-wave $y = -\sin x$. Then it will start moving up again, and so on.

What function would describe that kind of motion? You may be able to guess the answer yourself, at least to within a constant or two. We'll fill in those constants and tell you that the solution is:

$$y = \sin(x)\cos(vt) \tag{11.3.4}$$

Every point on the string oscillates with a period of $2\pi/v$ and an amplitude given by its original height (Figure 11.4).

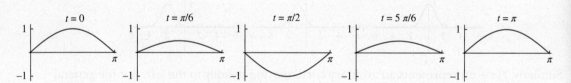

FIGURE 11.4 The function $y = \sin(x)\cos(vt)$ with $v = 2$.

Is it really that simple? The "rubber band" argument may not be sufficiently rigorous for you; in fact this solution may remind you of a similar function that *didn't* work in the previous section. But the initial condition is different now, and our new solution works perfectly. We leave it to you to confirm that the function $y = \sin(x)\cos(vt)$ satisfies the wave equation (11.3.3), our boundary conditions $y(0, t) = y(\pi, t) = 0$, and our initial conditions $y(x, 0) = \sin x$ and $\dot{y}(x, 0) = 0$. Those confirmations are the acid test of a solution, however we arrived at it.

And what about our general solution? Can this function be rewritten in the form $y(x, t) = f(x+vt) + g(x-vt)$? It must be possible, because *all* solutions to the wave equation must have this form. In this particular case it can be done using trig identities, but it isn't necessary. It was easier to start with a physically motivated guess than to find the right functions f and g.

Making an educated guess works great if the initial condition happens to be $y = \sin x$, but can we do it for other cases? You probably won't be surprised to hear that guessing a solution is just as easy if $y(x, 0) = \sin(2x)$, or $\sin(10x)$, or any other sine wave that fits comfortably into the total length: the string just oscillates, retaining its original shape but changing amplitude. You'll show in the problems that if the initial conditions are zero velocity and $y(x, 0) = \sin(nx)$ (where n is an integer), then the solution is $y(x, t) = \sin(nx)\cos(nvt)$.

Solutions of this form are called the "normal modes" of the string.

> **Definition: Normal Mode**
> A "normal mode" is a solution that evolves in time by changing its amplitude while leaving its basic shape unchanged. For instance, if all points x on a string oscillate with the same frequency and phase as each other, the result is a standing wave that grows and shrinks.

To see more clearly what this definition means, consider the example $y(x, t) = \sin(x)\cos(vt)$ discussed above. At $x = \pi/6$ this becomes $y(\pi/6, t) = .5\cos(vt)$, which is an oscillation with frequency $v/(2\pi)$ and amplitude .5. (Remember that *frequency* just means one over the period.) At the point $x = \pi/3$ the string oscillates with frequency $v/(2\pi)$ and amplitude $\sqrt{3}/2$. Since every point on the string oscillates at the same frequency, this solution is a normal mode of the system. This may remind you of the "normal modes" of coupled oscillators in Chapter 6. In that case the system consisted of a finite number of discrete oscillators instead of an infinite number of oscillating points, but the definition of normal mode is the same in both cases.

If the string starts out in the curve $y(x, 0) = \sin(nx)$, we know exactly how it will evolve over time. Surprisingly, that insight turns out to be the key to the motion of our string under *any* initial conditions.

More Complicated Solutions for More Complicated Cases

What if the string doesn't happen to start in a sine wave? The bad news is that, in general, the string will *not* keep its overall shape while stretching and compressing vertically. The good news is that we can apply our finding from one very special case—the normal modes—to find the solution for almost any initial condition. The key, as it often is, is the ability to write a general solution as a *sum* of specific solutions.

> **EXAMPLE   A Sum of Two Normal Modes**
>
> **Problem:**
> Consider a string subject to the same boundary condition we used above, $y(0, t) = y(\pi, t) = 0$, but starting in the initial form $y(x, 0) = 7\sin x + 2\sin(8x)$. How will such a string evolve over time?
>
> **Solution:**
> The wave equation is linear and homogeneous, which means that any linear combination of solutions is also a solution. Our boundary conditions are also homogeneous, which means that any sum of solutions will match the

boundary conditions as well. Since we know that $y(x, t) = \sin(x)\cos(vt)$ and $y(x, t) = \sin(8x)\cos(8vt)$ are both solutions, it must also be true that:

$$y(x, t) = 7\sin(x)\cos(vt) + 2\sin(8x)\cos(8vt)$$

is a solution. Since it matches the boundary and initial conditions, it is *the* solution for this case.[2]

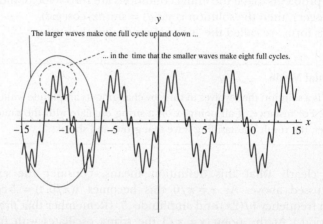

Once again, this should look familiar if you've studied coupled oscillators. There too you can solve for any initial condition by writing it as a sum of normal modes.

You may object that our example was too easy: the initial condition wasn't exactly a normal mode, but it was the next best thing. How do we find the solution for an initial condition that doesn't just "happen" to be the sum of a few sine waves?

The answer is that practically *any* function on a finite domain "happens" to be the sum of sine waves—or at least, we can write it as the sum of sine waves if we want to. That's what Fourier series are all about! And that insight leads us to a general approach. First you decompose your initial condition into a sum of sine waves (or, more generally, into a sum of normal modes). Then you write your solution as a sum of individual solutions for each of these normal modes.

EXAMPLE **A Plucked Guitar String**

Problem:
Solve the wave equation with boundary conditions $y(0, t) = 0$ and $y(\pi, t) = 0$, and initial conditions $\dot{y}(x, 0) = 0$, $y(x, 0) = \begin{cases} x/2 & 0 < x < \pi/2 \\ (\pi - x)/2, & \pi/2 < x < \pi \end{cases}$

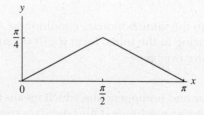

---

[2] As we said earlier, we are not going to formally discuss exactly what boundary and initial conditions are sufficient to conclude that a solution is unique. In this case, however, we can make the argument on purely physical grounds: we know everything there is to know about this particular string, and it can only do one thing!

Solution:
We begin by writing the initial condition as a sum of normal modes. In other words, we write the Fourier sine series for the function. The answer (after some calculations) is:

$$y(x,0) = \sum_{n=1}^{\infty} (-1)^{(n-1)/2} \frac{2}{n^2 \pi} \sin(nx), \; n \text{ odd} = \frac{2}{\pi} \sin(x) - \frac{1}{9\pi} \sin(3x) + \frac{2}{25\pi} \sin(5x) + \ldots$$

That infinite sum may look intimidating, but if you look term by term, you can see that for each value of n this is just a constant times a sine function. We can do the same thing we did above with the sum of just two sine functions; we multiply each sine function of x by the corresponding cosine function of t. So the solution is:

$$y(x,t) = \sum_{n=1}^{\infty} (-1)^{(n-1)/2} \frac{2}{n^2 \pi} \sin(nx) \cos(nvt), \; n \text{ odd}$$
$$= \frac{2}{\pi} \sin(x) \cos(vt) - \frac{1}{9\pi} \sin(3x) \cos(3vt) + \ldots \quad (11.3.5)$$

If you had trouble with the step where we took the Fourier sine series of $y(x,0)$ you should review Fourier series now; we're going to use them a lot in this chapter. All the relevant formulas are in Appendix G.

Even if you didn't have trouble with the calculation, you may still find an infinite series to be an unsatisfying answer. However, that $1/n^2$ will make the series converge pretty quickly, so for most purposes the first few terms will give you a pretty good approximate answer. As we go through the chapter, you'll see that most analytical solutions for PDEs come in the form of series; expressions in closed form are the exceptions.

Now you know how to solve for the motion of a vibrating string, at least if:

- The ends of the string are at $x = 0$ and $x = \pi$.
- The ends of the string are held fixed at $y = 0$.
- The initial velocity of the string is zero everywhere.

In the problems you will use the same technique with these three conditions remaining constant—only the initial position, and therefore the Fourier series, will vary. Then you will do other problems that change all three of these conditions, and you will find that although the forms of the solutions vary, the basic idea carries through. The example below, for instance, is the same in length and initial velocity, but different in boundary conditions.

EXAMPLE **Air in a Flute**

The air inside a flute obeys the wave equation $\partial^2 s/\partial x^2 = (1/c_s^2)(\partial^2 s/\partial t^2)$, where $s(x,t)$ is displacement of the air and the constant c_s is the speed of sound. For reasons we're not going to get into here, s is not constrained to go to zero at the edges, but $\partial s/\partial x$ is. Hence, we are solving the same differential equation with different boundary conditions.

For consistency, let's consider a flute that extends from $x = 0$ to $x = \pi$, and let's take as a simple initial condition $s(x,0) = \cos x$, $\partial s/\partial t(x,0) = 0$. (Why can't we start with the same initial condition we used for our string above? Because it doesn't meet our new boundary conditions!)

The solution to our new system is $s(x, t) = \cos(x)\cos(c_s t)$. You can confirm this solution by verifying the following requirements:

1. It satisfies the wave equation $\partial^2 s/\partial x^2 = (1/c_s^2)(\partial^2 s/\partial t^2)$.
2. It satisfies the boundary condition $\partial s/\partial x = 0$ at $x = 0$ and $x = \pi$.
3. It satisfies the initial condition $s(x, 0) = \cos x$. (Did you expect our solution to have $\sin(c_s t)$ instead of the cosine? This would meet the first two requirements, but not the initial condition. Try it!)

This solution is a normal mode since every point in the flute vibrates with the same frequency. More generally, $s(x, t) = \cos(nx)\cos(nc_s t)$ is the normal mode of this system for the initial condition $s(x, 0) = \cos(nx)$, $\partial s/\partial t(x, 0) = 0$.

Therefore, if the system happened to start in the state $s(x, 0) = 10\cos(3x) + 2\cos(8x)$, $\partial s/\partial t(x, 0) = 0$, its motion would be described by the function $s(x, t) = 10\cos(3x)\cos(3c_s t) + 2\cos(8x)\cos(8c_s t)$. More generally, we can decompose any initial condition into a Fourier cosine series, and simply write the solution from there. (For a function defined on a finite interval you create an "odd extension" of that function to write a Fourier sine series or an "even extension" to write a Fourier cosine series: see Chapter 9.)

Normal modes provide a very general approach that can be used to solve many problems in partial differential equations, provided you can take two key steps.

1. Figure out what the normal modes are. In this case we figured them out using physical intuition—or, as it may seem to you, improbably lucky guesswork.
2. Rewrite any initial condition as a sum of normal modes. In this case we used Fourier series.

In the next section we will see a more general approach to the first step, *finding* the normal modes. In the sections that follow, we will see that the second step is often possible even when the normal modes do not involve sines and cosines.

11.3.3 Problems: Normal Modes

11.47 Demonstrate that the function $y = 1/(x + vt)^2 - 3\ln(x - vt)$ is a valid solution to the wave equation (11.3.3).

11.48 Demonstrate that any function of the form $f(x - vt) + g(x + vt)$ is a valid solution to the wave equation (11.3.3).

11.49 Consider the function $y = (x - vt)^2$.
   (a) For the case $v = 1$, draw graphs of this function at times $t = 0$, $t = 1$, and $t = 2$.
   (b) In general, how does this function vary over time?
   (c) Repeat parts (a) and (b) for $v = 2$.
   (d) In general, how does the constant v affect the behavior of this function?

11.50 Consider the function $g(x + 3t)$. At time $t = 0$, the function looks like this, stretching along the *x*-axis toward infinity in both directions.

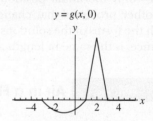

   (a) Draw graphs of this function at times $t = 1$, $t = 2$, and $t = 3$.
   (b) In general, how does this function evolve over time?

11.51 A string that obeys the wave equation (11.3.3) is tacked down at the ends, so $y(0, t) = y(\pi, t) = 0$. The string starts at rest, so $\dot{y}(x, 0) = 0$. If the initial position of the string happens to be $y(x, 0) = \sin(2x)$, then the string will follow the function $y = \sin(2x)\cos(2vt)$.

(a) Guess at a solution to the same differential equation with the same boundary conditions, but with initial position $y(x, 0) = \sin(5x)$.

(b) Confirm that your guess to Part (a) satisfies the differential Equation 11.3.3, the boundary conditions $y(0, t) = y(\pi, t) = 0$, and the initial conditions $y(x, 0) = \sin(5x)$ and $\dot{y}(x, 0) = 0$. (If it doesn't, keep guessing until you find one that does.)

(c) Now guess at the solution $y(x, t)$ for a string that is identical to the strings above, except that its initial position is $y(x, 0) = 3\sin(2x)$. Once again, confirm that your solution solves the wave equation and all required conditions.

(d) Solve the wave equation subject to all the same conditions above, except that this time, $y(x, 0) = 6\sin(x) - 5\sin(7x)$. Once again, confirm that your solution solves the wave equation and all required conditions.

(e) Solve the wave equation for the same conditions one last time, but this time with the initial position $y(x, 0) = \sum_{n=1}^{\infty}(1/n^2)\sin(nx)$. Leave your answer in the form of a series. (This part should not take much more work than the other ones.) You do not need to verify this answer.

11.52 A string of length π is fixed at both ends, so $y(0, t) = y(\pi, t) = 0$. You pull the string up at two points and then let go, so the initial conditions are:

$$y(x, 0) = \begin{cases} x & x < \pi/3 \\ \pi/3 & \pi/3 \leq x \leq 2\pi/3 \\ \pi - x & 2\pi/3 < x \end{cases}$$

and $\dfrac{\partial y}{\partial t}(x, 0) = 0$

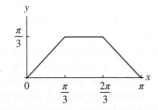

(a) Rewrite the initial condition as a Fourier sine series.

(b) Write the solution $y(x, t)$. Your answer will be expressed as a series and will include a constant v.

(c) Take $v = 2$ and have a computer calculate the 20th partial sum of the solution you found Part (b). Plot the solution at a series of times and describe its evolution.

11.53 A great deal of what we want you to learn in this section can be expressed in one concise mathematical statement: "Any function of the form $y(x, t) = A\sin(x)\cos(vt) + B\sin(2x)\cos(2vt) + C\sin(3x)\cos(3vt) + \ldots$ is a solution of the wave equation (11.3.3) and the boundary conditions $y(0, t) = y(\pi, t) = 0$."

(a) Prove that this result is true.

(b) Explain why this result can be used to solve the wave equation for almost any initial conditions, whether they are sinusoidal or not.

In Problems 11.54–11.59 a string of length π, initially at rest, has boundary conditions $y(0, t) = y(\pi, t) = 0$. For the initial shape given in the problem:

(a) Find the solution $y(x, t)$ to the wave equation (11.3.3), taking $v = 2$. Your answer will be in the form of a series.

(b) If your answer is in the form of an infinite series make plots of the 1st partial sum, the 3rd partial sum, and the 20th partial sum of the series solution at several times.

11.54 $y(x, 0) = \sin(10x)$

11.55 $y(x, 0) = \sin(2x) + (1/10)\sin(10x)$

11.56 $y(x, 0) = (1/10)\sin(2x) + \sin(10x)$

11.57 $y(x, 0) = \begin{cases} 1 & \pi/3 < x < 2\pi/3 \\ 0 & \text{elsewhere} \end{cases}$

11.58 $y(x, 0) = \pi^2/4 - (x - \pi/2)^2$

11.59 Make up an initial position $y(x, 0)$. You may use any function that obeys the boundary conditions *except* the trivial case $y(x, 0) = 0$, or any function we have already used in the Explanation (Section 11.3.2) or the problems above.

11.60 Section 11.2 looked at a string pinned to the x-axis at $x = 0$ and $x = 4\pi$, with initial position $y(x, 0) = 1 - \cos x$. We found that this is not a normal mode; it will not just oscillate. Now you find out what such a string actually will do. (*Hint*: when finding the Fourier series you will be faced with a difficult integral. The easiest approach is rewriting the trig functions as complex exponentials.)

11.61 In the Explanation (Section 11.3.2), we make a big deal of the fact that functions of the form $y(x, t) = \sin(nx)\cos(nvt)$ are "normal modes" of the wave equation (11.3.3). This does *not* simply mean that these functions are valid solutions of the differential equation; it means something much stronger than that. Explain in your own words what a "normal mode" means, and why it is important.

11.62 A string of length 1 is fixed at both ends and obeys the wave equation (11.3.3) with $v = 2$. For each of the initial conditions given below assume the initial velocity of the string is zero.

(a) Have a computer numerically solve the wave equation for this string with initial condition $y(x, 0) = \sin(2\pi x)$ and animate the resulting motion of the string. Solve to a late enough time to see the string oscillate at least twice, using trial and error if necessary. Describe the resulting motion.

(b) Have a computer numerically solve for and animate the motion of the string for initial condition $y(x, 0) = \sin(20\pi x)$, using the same final time you used in Part (a). How is this motion different from what you found in Part (a)?

(c) Consider the initial condition $y(x, 0) = \sin(2\pi x) + (.1) \sin(20\pi x)$. What would you expect the motion of the string to look like in this case? Solve the wave equation with this initial condition numerically and animate the results. Did the results match your prediction?

(d) Finally, make an animation of the solution to the wave equation for the case $y(x, 0) = .2 \sin(2\pi x) \left[7 + 6x - 100x^2 + 100x^3 + \cos(36x) - e^{-36x^2} \right]$. (We chose this simply because it looks like a crazy, random mess.) Describe the resulting motion.

(e) How is the evolution of a string that starts in a normal mode different from the evolution of a string that starts in a different shape?

11.63 In the Explanation (Section 11.3.2), we discussed a string of length π fixed at both ends with initial shape $y(x, 0) = \sin x$ and no initial velocity. Based on physical arguments we guessed that the initial shape of the string would oscillate sinusoidally in time. We then jumped to the exact correct function with very little justification. (Did you notice?) In this problem, your job is to fill in the missing steps. Start with a "guess" that represents the initial function oscillating: $y(x, t) = \sin(x) (A \sin(\alpha t) + B \cos(\beta t))$. Plug this guess into the wave equation (11.3.3) along with the initial conditions $y(x, 0) = \sin x$ and $\dot{y}(x, 0) = 0$, and solve for A, B, α, and β.

In Problems 11.64–11.68 you will solve for the displacement of air inside a flute of length π. The displacement $s(x, t)$ obeys the wave equation $\partial^2 s / \partial x^2 = (1/c_s^2)(\partial^2 s / \partial t^2)$, just like a vibrating string, but the boundary conditions for the flute are $\partial s / \partial x(0, t) = \partial s / \partial x(\pi, t) = 0$. This leads to a different set of normal modes, which we found in the example on Page 561. The initial condition for $s(x, 0)$ is given below, and you should assume in each case that $\partial s / \partial t(x, 0) = 0$. For each of these initial conditions find the solution $s(x, t)$ to the wave equation, taking $c_s = 3$.

11.64 $s(x, 0) = \cos(5x)$

11.65 $s(x, 0) = \cos(x) + (1/10) \cos(10x)$

11.66 $s(x, 0) = (1/10) \cos(x) + \cos(10x)$

11.67 $s(x, 0) = (x^2 - \pi^2)^2$. Your answer will be an infinite series. Make plots of the 20th partial sum of the series solution at three or more different times. Describe how the function $s(x)$ is evolving over time.

11.68 Make up an initial position $s(x, 0)$. You may use any function that obeys the boundary conditions $\partial s / \partial x(0, t) = \partial s / \partial x(\pi, t) = 0$ *except* the trivial case $s(x, 0) = 0$, or any function we have already used in the Explanation (Section 11.3.2) or the problems above.

11.69 In the Explanation (Section 11.3.2), we considered a string of length π with fixed ends and zero initial velocity. We wrote an expression for the normal modes of this system and showed how to solve for the motion if the string started in one of the normal modes. More importantly, we showed how to write any other given initial condition as a sum of normal modes using Fourier series and thus solve the wave equation. In this problem you will perform a similar analysis for a string of length L.

(a) If a string starts in the initial position $y(x, 0) = \sin(nx)$ where n is any integer, it is guaranteed to meet the boundary conditions $y(0, 0) = y(\pi, 0) = 0$. What must be true of the constant k if the initial position meets the boundary conditions $y(0, 0) = y(L, 0) = 0$? (Your answer will once again end in the phrase "where n is any integer.")

(b) For a string of length π with fixed ends and zero initial velocity, the normal modes can be written as $y(x, t) = \sin(nx)\cos(nvt)$, $n = 0, 1, 2, \ldots$. Based on the initial position you wrote in Part (a), write a similar expression for the normal modes $y(x, t)$ for a string of length L. Confirm that your solution solves the wave equation (11.3.3).

(c) Find the solution $y(x, t)$ if the string starts at rest in position $y(x, 0) = \sin(3\pi x/L)$. Make sure your solution satisfies the wave equation and matches the initial and boundary conditions.

(d) Find the solution $y(x, t)$ if the string starts at rest in position $y(x, 0) = 2\sin(3\pi x/L) + 5\sin(8\pi x/L)$.

(e) Now find the solution if the string starts at rest in position
$$y(x, 0) = \begin{cases} 1 & L/3 < x < 2L/3 \\ 0 & \text{elsewhere} \end{cases}.$$
(*Hint*: You will need to start by expanding this initial function in a Fourier sine series.) Your answer will be in the form of an infinite series.

11.70 In the Explanation (Section 11.3.2), we considered a string with zero initial velocity and non-zero initial displacement. We wrote an expression for the normal modes of this system and showed how to solve for the motion if the string started in one of the normal modes. More importantly, we showed how to write a more complicated initial position as a sum of normal modes using Fourier series and thus solve the wave equation.

In this problem you will perform a similar analysis for a string with the same boundary conditions, $y(0, t) = y(\pi, t) = 0$, but with different initial conditions: your string has zero initial *displacement* and non-zero initial *velocity*.

(a) Regardless of its initial velocity, this string has zero initial acceleration. How do we know that?

(b) Suppose that the string has initial velocity $\partial y/\partial t(x, 0) = \sin x$. Sketch the shape of the string a short time later after the initial time.

(c) Describe in words how you would expect the string to evolve over time. (To answer this, you will need to take into account not only the initial velocity, but also the wave equation (11.3.3) that dictates its acceleration over time.)

(d) Express your guess as a mathematical function $y(x, t)$ and verify that it solves the wave equation and matches the initial and boundary conditions. (If at first your guess doesn't succeed: try, try again.)

(e) Next consider the initial velocity $\partial y/\partial t(x, 0) = \sin(3x)$. Find the solution $y(x, t)$ for this case and make sure your solution satisfies the wave equation and the initial and boundary conditions.

(f) In the Explanation (Section 11.3.2), we found that the normal modes for a string of length π with fixed ends and zero initial velocity can all be written as $y(x, t) = \sin(nx)\cos(nvt)$, $n = 0, 1, 2, \ldots$. Write a similar expression for the normal modes of a string of length π with fixed ends and zero initial *displacement.*

(g) Find the solution $y(x, t)$ if the string starts at $y(x, 0) = 0$ with initial velocity $\partial y/\partial t(x, 0) = 2\sin(3x) + 4\sin(5x)$.

(h) Find the solution if the string starts at $y(x, 0) = 0$ with initial velocity
$$\partial y/\partial t(x, 0) = \begin{cases} 1 & \pi/3 < x < 2\pi/3 \\ 0 & \text{elsewhere} \end{cases}.$$
This might occur if the middle of the string were suddenly struck with a hammer. (*Hint*: You will need to start by expanding this initial function in a Fourier sine series.) Your answer will be in the form of an infinite series.

11.71 In the Explanation (Section 11.3.2), we found the normal modes for a vibrating string that obeys the wave equation (11.3.3) with fixed boundaries and zero initial velocity. If the string is infinitely long we no longer have those boundary conditions. What are all the possible normal modes for a system obeying the wave equation (11.3.3) with zero initial velocity on the real line: $-\infty < x < \infty$? To answer this you should look at the Explanation and see what restriction the boundary conditions imposed on our normal modes, and then remove that restriction.

11.72 A function $y(x, t)$ may be said to be a "normal mode" if the initial shape $y(x, 0)$ evolves in time by changing amplitude—that is, by stretching or compressing vertically—but does not change in any other way. We have seen that the normal modes for the wave equation are sines and cosines, but different equations may have very different normal modes. Express with no words (just a simple equation) the statement "$y(x, t)$ is a normal mode" as defined above.

11.73 A string of length 1 obeys the wave equation with $v = 2$. The string is initially at rest at $y = 0$. The right side of the string is fixed, but the left side is given a quick jerk: $y(0, t) = e^{-100(t-1)^2}$. Solve the wave equation numerically with this boundary condition and animate the results out to $t = 5$. Describe the motion of the string.

11.74 A string of length π obeys the wave equation with $v = 2$. The right side of the string is fixed.

Suppose the string starts at rest at $y = 0$ and you excite it by vibrating the left end of it: $y(0, t) = \sin(11t)$. Notice that this is *not* one of the normal mode frequencies.

(a) Solve the wave equation numerically with this boundary condition out to $t = 10$. Describe the resulting motion.

Next suppose the left end vibrates according to: $y(0, t) = \sin(10t)$.

(b) Is this oscillation occurring at one of the normal mode frequencies given by Equation 11.3.4?

(c) Solve the wave equation numerically with this boundary condition out to $t = 10$. How is the resulting motion different from what you found in Part (a)?

11.75 Two different general solutions. The general solution for a vibrating string of length π with fixed ends and zero initial velocity is

$$y(x, t) = \sum_{n=0}^{\infty} b_n \sin(nx) \cos(nvt)$$

We know, however, that any solution to the wave equation can be written in the form of d'Alembert's solution

$$y(x, t) = f(x + vt) + g(x - vt)$$

so it must be possible to rewrite the normal mode solution in this form.

(a) Use trig identities to rewrite $\sin(nx) \cos(nvt)$ in terms of $(x + vt)$ and $(x - vt)$.

(b) Find the functions f and g such that $f(x + vt) + g(x - vt) = \sum_{n=0}^{\infty} b_n \sin(nx) \cos(nvt)$.

11.76 Exploration: Wind Instruments
When air blows across an opening in a pipe it excites the air inside the pipe at many different frequencies, but the only ones that get amplified by the pipe are the normal modes of the pipe. So the normal modes of a wind instrument determine the notes you hear. Generally the longer wavelength modes are louder, so the dominant tone you hear from the pipe is that of the normal mode with the longest wavelength (lowest frequency). This is called the "fundamental note" or "fundamental frequency" of a wind instrument.

Sound waves in a pipe obey the wave equation $\partial^2 s / \partial x^2 = (1/c_s^2)(\partial^2 s / \partial t^2)$, where $c_s = 345 m/s$ is the speed of sound. When the end of the pipe is open s is not constrained to go to zero at the edges of the pipe, but $\partial s / \partial x$ is. Real wind instruments change the effective length of the pipe by opening and closing valves or holes in different places, but for this problem we will consider the simplest case of a cylindrical instrument such as a flute or clarinet with all the holes closed except at the ends.

(a) A flute is open at both ends. What are all of the possible normal modes for a flute of length L? (This is similar to the example on Page 561, but this time the flute is of length L instead of length π. How does this change the normal modes?) Be sure to include both the space and time parts of the normal mode. For simplicity you can assume that $(\partial s / \partial t)(x, 0) = 0$.

(b) A typical flute might be 66 cm. What is the fundamental frequency of such a flute? Remember that frequency is defined as $1/\text{period}$, so you will need to start by figuring out the period of the normal mode.

(c) What is the frequency of the next normal mode in the series? Look up what notes these frequencies correspond to. (For example, a frequency of 2500 Hz is roughly a note of E, two octaves above middle C.)

(d) What are all of the possible normal modes for a pipe of length L that is closed at one end ($s(0, t) = 0$) and open at the other ($\partial s / \partial x(L, t) = 0$)? A clarinet is a typical example.

(e) If a clarinet and a flute were the same length, which one would play a higher fundamental note and why?

(f) One common type of clarinet is 60 cm long. What is the fundamental frequency of such a clarinet? What is the frequency of the next normal mode in the series? Look up what notes these frequencies correspond to.

11.4 Separation of Variables—The Basic Method

This section is the heart of the chapter. By reducing a *partial* differential equation to two or more *ordinary* differential equations, the technique we introduce here allows you to find the normal modes of the system and thereby a solution.

The three sections that follow this are all elaborations of the basic method presented here. If you carefully follow the algebra in this section and see how the steps fit together to form a big picture, you will be well prepared for much of the rest of this chapter. At the end, in an unusually long "Stepping Back," we discuss the variations we will present so you can see how the chapter fits together.

11.4.1 Discovery Exercise: Separation of Variables—The Basic Method

Consider a bar going from $x = 0$ to $x = L$. The "heat equation" (which you derived in Section 11.1) governs the evolution of the temperature distribution:

$$\frac{\partial u}{\partial t} = \alpha \frac{\partial^2 u}{\partial x^2} \qquad (11.4.1)$$

where $u(x, t)$ is the temperature and α is a positive constant called "thermal diffusivity" that reflects how efficiently the bar conducts heat. Both ends are immersed in ice water, so $u(0, t) = u(L, t) = 0$, but the initial temperature may be non-zero at other interior points. You are going to solve for the temperature $u(x, t)$. Your strategy will be to guess a solution of the form $u(x, t) = X(x)T(t)$ where X is a function of the variable x (but not of t), and T is a function of the variable t (but not of x).

1. Plug the function $u(x, t) = X(x)T(t)$ into Equation 11.4.1. Note that if $u(x, t) = X(x)T(t)$ then $\partial u/\partial x = X'(x)T(t)$.
2. "Separate the variables" in your answer to Part 1. In other words, algebraically rearrange the equation so that the left side contains all the functions that depend on t, and the right side contains all the functions that depend on x. In this problem—in fact in many problems—you can separate the variables by dividing both sides of the equation by $X(x)T(t)$. You should also divide both sides of the equation by α, which will bring α to the side with t; this step is not necessary but it will make the equations a bit simpler later on.
3. The next step relies on this key result: if the left side of the equation depends on t (but not on x), and the right side depends on x (but not on t), then both sides of the equation *must equal a constant*. Explain why this result must be true.
4. Based on the result from Part 3, you can now turn your *partial* differential equation from Part 1 into two *ordinary* differential equations: "the left side of the equation equals a constant" and "the right side of the equation equals *the same* constant." Write both equations. Call the constant P.

See Check Yourself #71 in Appendix L

The next step, finding real solutions to these two ordinary differential equations, depends on the sign of the constant P. We shall therefore handle the three cases separately. We'll start by solving the equation for $X(x)$. Remember that the constant α is necessarily positive. Avoid complex answers.

5. Assuming $P > 0$, we can replace P with k^2 where k is any real number. Solve the equation for $X(x)$ for this case. Your solution will have two arbitrary constants in it: call them A and B.
6. Assuming $P = 0$, solve the equation for $X(x)$, once again using A and B for the arbitrary constants.
7. Assuming $P < 0$, we can replace P with $-k^2$ where k is any real number. Solve the equation for $X(x)$ for this case, once again using A and B for the arbitrary constants.
8. For two of the three solutions you just found the only way to match the boundary conditions is with the "trivial" solution $X(x) = 0$. Only one of the three allows for non-trivial solutions that match the boundary conditions. Based on that fact, what must the sign of P be?

   See Check Yourself #72 in Appendix L

9. The boundary condition $u(0, t) = 0$ implies that $X(0) = 0$. Plug this into your solution for $X(x)$ to find the value of one of the two arbitrary constants.
10. Find the values of k that match the second boundary condition $u(L, t) = 0$. *Hint*: there are infinitely many such values. We will return to these values when we match initial conditions. For the rest of this exercise we will continue to just write k.

Having found $X(x)$, we now turn our attention to the ODE you wrote for $T(t)$ way back in Part 4.

11. Replace P with $-k^2$ in the $T(t)$ differential equation, as you did with the $X(x)$ one. (Why? Because both differential equations were set equal to the *same constant P*.)
12. Having done this replacement, solve the equation for $T(t)$. Your solution will introduce a new arbitrary constant: call it C.
13. Write the solution $u(x, t) = X(x)T(t)$ based on your answers. This solution should depend on k.
14. Explain why, when you combine your $X(x)$ and $T(t)$ functions into one $u(x, t)$ function, you can combine the arbitrary constants from the two functions into one arbitrary constant.

The solution you just found is a normal mode of this system. In the Explanation that follows we will write the general solution to such an equation as a linear superposition of all the normal modes, and use initial conditions to solve for the arbitrary constants.

11.4.2 Explanation: Separation of Variables—The Basic Method

In the last section we found that, if a vibrating string with fixed ends happens to start at rest in a sinusoidal shape, it will evolve very simply over time, changing amplitude only. We called such a solution a "normal mode." If the string doesn't happen to start in such a fortuitous position, we can model its motion by writing the position as a sum of these normal modes (a Fourier series).

That's a great result, but it all started with a lucky guess. How could we have solved that problem if we hadn't thought of using sine waves? More importantly, how do you solve other problems?

It turns out that you can solve a wide variety of partial differential equations by writing the general solution as a sum of normal modes. "Separation of variables" is the most important technique for finding normal modes and solving partial differential equations. (You may recall a technique called "separation of variables" for solving ordinary differential equations. Both techniques involve some variables that have to be separated from each other, but beyond that, they have nothing to do with each other. Sorry about that.)

Before we dive into the details, let's start with an analogy to a familiar problem. To solve the *ordinary* differential equation $y'' - 6y' + 5y = 0$, we might take the following steps:

- *Guess a solution with an unknown constant.* In Chapter 1 we saw that the correct guess for such an equation would be $y = e^{bx}$.
- *Plug your guess into the original equation, to solve for the unknown constant.* Plugging $y = e^{bx}$ into the original differential equation, we can solve to find two solutions $y = e^x$ and $y = e^{5x}$. (Try it.)
- *Sum the solutions.* If a differential equation is linear and homogeneous, then any linear combination of solutions is also a solution. So we write $y = Ae^x + Be^{5x}$. Since we are solving a second-order linear equation and we have a solution with two arbitrary constants, this is the general solution.
- *Plug in initial conditions.* "Initial conditions" in this context means two additional pieces of information beside the original differential equation. For instance, if we know that $f(2) = 3$ then we can write $3 = Ae^2 + Be^{10}$. If we know one other piece of information—such as another point, or the derivative at that point—we can solve for the arbitrary constants, finding the specific solution we want.

Plugging in "guesses" in this way reduces ordinary differential equations to algebra equations, which are generally easier to solve.

Separation of variables allows you to solve partial differential equations in much the same way. We will show below that by "guessing" a solution that changes in amplitude only, you turn your partial differential equation into several ordinary differential equations. Solving these equations gives you your normal modes with an infinite number of arbitrary constants to match based on "initial" (time) and "boundary" (space) conditions. With the variables separated, you can approach these two kinds of conditions separately.

The Problem
We're going to demonstrate this new technique with a familiar problem. When we arrive at the solution, we will recognize it from the previous sections. But this time we will derive the solution in a way that can be applied to different problems.

Here, then, is our familiar problem: a string of length L obeys the wave equation:

$$\frac{\partial^2 y}{\partial x^2} = \frac{1}{v^2} \frac{\partial^2 y}{\partial t^2} \tag{11.4.2}$$

The string is fixed at both ends, which gives us our boundary conditions:

$$y(0, t) = y(L, t) = 0 \tag{11.4.3}$$

To fully solve for the motion of the string, we need to know its initial position, and its initial velocity. For the moment we will keep both of those generic:

$$y(x, 0) = f(x) \quad \text{and} \quad \frac{dy}{dt}(x, 0) = g(x) \tag{11.4.4}$$

Solve for the motion of the string.

Step 1: The Guess
You could imagine y as any function of x and t, such as x^t or $3\ln(xt)/(x+t)$ or even more hideous-looking combinations, but almost none of them would work as solutions to the wave equation. Our hopeful guess is a solution of the form:

$$y(x, t) = X(x)T(t)$$

where X is a function of the variable x (but not of t), and T is a function of the variable t (but not of x).

It's worth taking a moment to consider what sort of functions we are looking at. $X(x)$ represents the shape of the string at a given moment t. Since we have placed no restrictions on the form of that function, it could turn out to be simple or complicated.

But how does that function evolve over time? $T(t)$ is simply a number, positive or negative, at any given moment. When you multiply $X(x)$ by a number, you stretch it vertically. You may also turn it upside down. But *beyond those simple transformations, the function $X(x)$ will not alter in any way.*

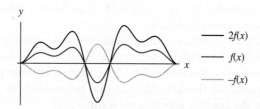

Wherever $X(x) = 0$, it will stay zero forever: that is to say, the places where the string crosses the x-axis will never change. (The only exception is that, whenever $T(t)$ happens to be zero, the entire string is uniformly flat.) The x-values of the critical points, where the string reaches a local maximum or minimum, will likewise never change.

A function of that form—changing in amplitude, but not in shape—is called a "normal mode." In guessing a solution of the form $y(x, t) = X(x)T(t)$, we are asserting "There exists a set of normal modes for this differential equation." We will then plug in this function to find the normal modes. If the initial state does not happen to correspond perfectly to a normal mode (which it usually doesn't) we will build it up as a *sum* of normal modes; since we understand how each of them evolves in time, we can compute how the entire assemblage evolves.[3]

Step 2: Plug In the Guess

What do we get if we plug $y(x, t) = X(x)T(t)$ into Equation 11.4.2? When we take the derivative with respect to x, the function $T(t)$ (which by definition has no x-dependence) acts as a constant. Similarly, $X(x)$ is a constant when we take derivatives with respect to t. So the equation becomes:

$$X''(x)T(t) = \frac{1}{v^2}X(x)T''(t)$$

Now comes the step that gives the technique of "separation of variables" its name. We rearrange the equation so that all the t-dependency is on the right, and all the x-dependency is on the left. We accomplish this (in this case and in fact in many cases) by dividing both sides by $X(x)T(t)$.

$$\frac{X''(x)}{X(x)} = \frac{1}{v^2}\frac{T''(t)}{T(t)} \qquad (11.4.5)$$

If x changes, does the right side of this equation change? The answer must be "no" since $T(t)$ has no x-dependency. That means—since the two sides of the equation must stay equal for *all values* of x and t—that changing x cannot change the left side of the equation either.

Similarly, changing t has no effect on the left side of the equation, and must therefore have no effect on the right side. If both sides of the equation are equal, and neither one

---
[3] When can you build up your initial conditions as a sum of normal modes? The answer, to make a long story short, is "almost always." We'll make that story long again when we discuss Sturm-Liouville theory in Chapter 12.

depends on x or t, then they must be...*(drum roll please)*...a constant! Calling this constant P, we write:

$$\frac{X''(x)}{X(x)} = P \text{ and } \frac{1}{v^2}\frac{T''(t)}{T(t)} = P \quad (11.4.6)$$

Step 3: Solve the Spatial Equation and Match the Boundary Conditions

Instead of one partial differential equation, we now have two *ordinary* differential equations—and pretty easy ones at that. We start with the spatial equation:

$$X''(x) = PX(x)$$

The solution to this equation depends on the sign of P. In the following analysis we define a new real number k. Because k is real we use k^2 to mean "any positive P" and $-k^2$ to mean "any negative P."

$P > 0$	$P = k^2$	$X''(x) = k^2 X(x)$	$X(x) = Ae^{kx} + Be^{-kx}$
$P = 0$		$X''(x) = 0$	$X(x) = mx + b$
$P < 0$	$P = -k^2$	$X''(x) = -k^2 X(x)$	$X(x) = A\sin(kx) + B\cos(kx)$

The exponential[4] and linear solutions can only satisfy our boundary condition ($y = 0$ at both ends) with the "trivial" solution $X(x) = 0$. Unless our initial conditions place the string in an unmoving horizontal line, this solution is inadequate.

On the other hand, sines and cosines are flexible enough to meet all our demands. We can tailor them to meet our boundary conditions, and then we can sum the result to meet whatever initial condition the string throws at us. We therefore declare that P is negative and replace it with $-k^2$.

Our first boundary condition is $y(0, t) = 0$. Plugging $x = 0$ into $A\sin(kx) + B\cos(kx)$ and setting it equal to zero gives $B = 0$, so we have a sine without a cosine.

The next condition is $y(L, t) = 0$.

$$A\sin(kL) = 0$$

One way to match this condition would be to set $A = 0$, which would bring us back to the trivial $y(x, t) = 0$. The alternative is $\sin(kL) = 0$, which can only be satisfied if $kL = n\pi$ where n is an integer. The case $n = 0$ gives us the trivial solution again, and negative values of n give us the same solutions as positive values (just with different values of the arbitrary constant A), so the complete set of non-trivial functions $X(x)$ is:

$$X(x) = A\sin(kx) \quad (k = \frac{n\pi}{L} \text{ for all positive integers } n) \quad (11.4.7)$$

We need to stress that we are *not* saying that our string must, at any given time, take the shape $x = A\sin(n\pi x/L)$. We are saying, instead, that Equation 11.4.7 defines the normal modes of the string. If the string is described by that equation then it will evolve simply in time—we'll figure out exactly how in a moment. If the string is not in such a shape then we will write it as a sum of such functions and then evolve each one independently.

---

[4]The exponential solution can also be written with hyperbolic trig functions as $A\sinh(kx) + B\cosh(kx)$. This is often more convenient for matching boundary conditions, but it does not change anything fundamental such as the inability to meet these particular conditions.

Step 4: Solve the Time Equation
Equation 11.4.6 introduced one new constant P, not two. We found that P must be negative and wrote $P = -k^2$ in the $X(x)$ equation; the P in the $T(t)$ equation is the same variable.

$$\frac{1}{v^2} \frac{T''(t)}{T(t)} = -k^2$$

We can quickly rewrite this as $T''(t) = -k^2 v^2 T(t)$, which has a similar form—and therefore a similar solution—to our spatial equation.

$$T(t) = C\sin(vkt) + D\cos(vkt)$$

We found that to match the boundary conditions for $X(x)$ we needed $k = n\pi/L$, and (again) this is the same k. The solution we are looking for is a product of the $X(x)$ and $T(t)$ functions.

$$X(x)T(t) = \sin\left(\frac{n\pi}{L}x\right)\left[C\sin\left(\frac{nv\pi}{L}t\right) + D\cos\left(\frac{nv\pi}{L}t\right)\right] \quad \text{where} \quad n = 1, 2, 3 \ldots \quad (11.4.8)$$

We have absorbed the arbitrary constant A into the arbitrary constants C and D. Why can we do that? Remember that A simply means "you can put any constant here" and C also means "you can put any constant here," so AC simply stands for "any constant." We can call this new constant C with no loss of generality.

For any positive integer n and any values of C and D the function 11.4.8 is a valid solution to the differential equation 11.4.2 and matches the boundary conditions 11.4.3.

Step 5: Sum the Solutions
Equation 11.4.8 is a family of different solutions, each representing a normal mode of the string. For instance one normal mode looks like this.

$$y(x, t) = \sin\left(\frac{3\pi}{L}x\right)\cos\left(\frac{3v\pi}{L}t\right) \quad (11.4.9)$$

Equation 11.4.9 has two different frequencies[5] and it's important not to get them confused. $3\pi/L$ is a quantity in space, not time—it means that the wavelength of the string is $2L/3$. The wavelengths of our normal modes were determined by the boundary conditions $y(0, t) = y(L, t) = 0$. $3v\pi/L$ is a quantity in time, not space—it means that the string will return to its starting position every $2L/(3v)$ seconds. The fact that this frequency is v times the spatial one is a result of the PDE we are solving.

Equation 11.4.9 tells us that if the string starts at rest in a perfect sine wave with spatial frequency $3\pi/L$ then it is in a normal mode and will retain its basic shape while its amplitude oscillates with temporal frequency $3v\pi/L$.

That was a normal mode with $n = 3$. Every positive integer n corresponds to a different wavelength. Every such wavelength fits our boundary conditions, and every such mode will oscillate with a different period in time. The amplitude of each normal mode can be anything, represented by the arbitrary constants C and D.

Because we are solving a linear homogeneous PDE with homogeneous boundary conditions, any linear combination of these solutions is itself a solution. So $\sin(\pi x/L)\left[3\sin(v\pi t/L) + 4\cos(v\pi t/L)\right]$ is a solution, and $\sin(2\pi x/L)\left[5\sin(2v\pi t/L) - 8\cos(2v\pi t/L)\right]$

---

[5] Our word "frequency," whether in space or time, is shorthand for the quantity that should more properly be termed "angular frequency."

is another solution, and if you sum those two functions you get yet another solution. To find the *general* solution we sum all possible solutions.

$$y(x,t) = \sum_{n=1}^{\infty} \sin\left(\frac{n\pi}{L}x\right)\left[C_n \sin\left(\frac{nv\pi}{L}t\right) + D_n \cos\left(\frac{nv\pi}{L}t\right)\right] \quad (11.4.10)$$

We've added a subscript to the constants C and D because they can take on different values for each value of n.

Step 6: Match the Initial Conditions

Finally—*after* we write one solution that is a sum of all the solutions we have found so far—we impose the initial conditions to find the values of the arbitrary constants. Note that in the heat equation that you solved in the Discovery Exercise (Section 11.4.1) the time dependence was first order, so there was only one initial condition and one arbitrary constant per solution. The wave equation is second order in time, and so requires two initial conditions (such as initial position and velocity) and two arbitrary constants per solution (C and D). But since we have an infinite number of solutions, "two constants per solution" is actually an *infinite number* of arbitrary constants: C_1, D_1, C_2, D_2, and so on. We have to find them all to match our initial conditions $y(x,0) = f(x)$ and $dy/dt(x,0) = g(x)$.

From Equation 11.4.10 we can write:

$$\frac{\partial y}{\partial t} = \sum_{n=1}^{\infty} \sin\left(\frac{n\pi}{L}x\right)\left[C_n\left(\frac{nv\pi}{L}\right)\cos\left(\frac{nv\pi}{L}t\right) - D_n\left(\frac{nv\pi}{L}\right)\sin\left(\frac{nv\pi}{L}t\right)\right] \quad (11.4.11)$$

Plugging $t = 0$ into Equations 11.4.10 and 11.4.11,

$$y(x,0) = \sum_{n=1}^{\infty} D_n \sin\left(\frac{n\pi}{L}x\right) = f(x) \quad (11.4.12)$$

$$\frac{\partial y}{\partial t}(x,0) = \sum_{n=1}^{\infty} C_n\left(\frac{nv\pi}{L}\right)\sin\left(\frac{n\pi}{L}x\right) = g(x) \quad (11.4.13)$$

It is now time to relate these equations to the results we saw in the previous section. We said that if the string is in a *normal mode*, its motion will be very simple. If it is not in a normal mode, we can understand its motion by building the initial conditions as a sum of normal modes.

Let's start with a very simple example: the initial position $f(x) = 3\sin(5\pi x/L)$, and the initial velocity $g(x) = 0$. Can you see what this does to our coefficients in Equations 11.4.12 and 11.4.13? It means that $D_5 = 3$, that $D_n = 0$ for all n other than 5, and that $C_n = 0$ for *all* n. In other words, $y(x,t) = 3\cos(5v\pi t/L)\sin(5\pi x/L)$.

On the other hand, what if $f(x)$ is not quite so convenient? We are left with using Equation 11.4.12 to solve for all the D_n coefficients to match the initial position. But that is exactly what we do when we create a Fourier sine series: we find the coefficients to build an arbitrary function as a series of sine waves. You may want to quickly review that process in Chapter 9. Here we are going to jump straight to the answer: $D_n = (2/L)\int_0^L f(x)\sin(n\pi x/L)\,dx$.

Similar arguments apply to the initial velocity and C_n, with the important caveat that the Fourier coefficients in Equation 11.4.13 are $C_n(nv\pi/L)$. So we can write $C_n(nv\pi/L) = (2/L)\int_0^L g(x)\sin(n\pi x/L)\,dx$.

We're done! The complete solution to Equation 11.4.2 subject to the boundary conditions 11.4.3 and the initial conditions 11.4.4 is:

$$y(x, t) = \sum_{n=1}^{\infty} \sin\left(\frac{n\pi}{L}x\right) \left[C_n \sin\left(\frac{nv\pi}{L}t\right) + D_n \cos\left(\frac{nv\pi}{L}t\right)\right] \quad (11.4.14)$$

$$C_n = \frac{2}{nv\pi} \int_0^L g(x) \sin\left(\frac{n\pi}{L}x\right) dx, \quad D_n = \frac{2}{L} \int_0^L f(x) \sin\left(\frac{n\pi}{L}x\right) dx$$

In the problems you'll evaluate this solution analytically and numerically for various initial conditions $f(x)$ and $g(x)$. For some simple initial conditions you may be able to evaluate this sum explicitly, but for others you can use partial sums to get numerical answers to whatever accuracy you need.

EXAMPLE Separation of Variables

Solve the partial differential equation

$$4\frac{\partial z}{\partial t} - 9\frac{\partial^2 z}{\partial x^2} - 5z = 0$$

on the domain $0 \leq x \leq 6$, $t \geq 0$ subject to the boundary conditions $z(0, t) = z(6, t) = 0$ and the initial condition $z(x, 0) = \sin^2(\pi x/6)$.

1. Assume a solution of the form $z(x, t) = X(x)T(t)$.
2. Plug this solution into the original differential equation and obtain

$$4X(x)T'(t) - 9X''(x)T(t) - 5X(x)T(t) = 0$$

Rearrange to separate the variables:

$$\frac{4T'(t)}{T(t)} - 5 = \frac{9X''(x)}{X(x)}$$

Both sides must now equal a constant:

$$\frac{4T'(t)}{T(t)} - 5 = P \quad \text{and} \quad \frac{9X''(x)}{X(x)} = P$$

3. If $P > 0$ the solution is exponential and if $P = 0$ the solution is linear. Neither of these can match the boundary conditions except in the trivial case $X(x) = 0$. We therefore conclude that $P < 0$, replace it with $-k^2$, and obtain the solution $X(x) = A\sin(kx/3) + B\cos(kx/3)$. The boundary condition $z(0, t) = 0$ implies $B = 0$. The boundary condition $z(6, t) = 0$ implies $k = \pi n/2$. So $X(x) = A\sin(\pi n x/6)$.
4. The time equation can be rewritten as $T'(t) = (P+5)T/4$ and the solution is $T(t) = Ce^{(P+5)t/4}$. We can combine this with our spatial solution and replace P with $-(\pi n/2)^2$ to get

$$X(x)T(t) = D\sin\left(\frac{\pi n}{6}x\right) e^{(-\pi^2 n^2 + 20)t/16}$$

5. Because both our equation and our boundary conditions are homogeneous, any linear combination of solutions is a solution:

$$z(x,t) = \sum_{n=1}^{\infty} D_n \sin\left(\frac{\pi n}{6}x\right) e^{(-\pi^2 n^2 + 20)t/16}$$

6. Finally, our initial condition gives

$$\sin^2\left(\frac{\pi x}{6}\right) = \sum_{n=1}^{\infty} D_n \sin\left(\frac{\pi n}{6}x\right)$$

This is a Fourier sine series. The formula for the coefficients is in Appendix G. The resulting integral looks messy but it can be solved by using Euler's formula and some trig identities or by just plugging it into a computer program. The result is

$$D_n = \frac{2}{6}\int_0^6 \sin^2\left(\frac{\pi x}{6}\right)\sin\left(\frac{n\pi x}{6}\right) dx = \begin{cases} 8/(4n\pi - n^3\pi) & n \text{ odd} \\ 0 & n \text{ even} \end{cases}$$

So the solution is

$$z(x,t) = \sum_{n=1}^{\infty} \frac{8}{4n\pi - n^3\pi} \sin\left(\frac{\pi n}{6}x\right) e^{(-\pi^2 n^2 + 20)t/16}, \quad n \text{ odd}$$

Stepping Back Part I: Solving Problems that are Just Like this One

It may seem like the process above involved a number of tricks that might be hard to apply to other problems. But the method of separation of variables is actually very general, and the steps you take are pretty consistent.

1. Begin by *assuming* a solution which is a simple product of single-variable functions: in other words, a normal mode.
2. Substitute this assumed solution into the original PDE. Then try to "separate the variables," algebraically rearranging so that each side of the equation depends on only one variable. (This often involves dividing both sides of the equation by $X(x)T(t)$.) Set both sides of the equation equal to a constant, thus turning one partial differential equation into two ordinary differential equations. Depending on the boundary conditions, you may be able to quickly identify the sign of the separation constant.
3. Solve the spatial equation and substitute in your boundary conditions to restrict your solutions. This step may eliminate some solutions (such as the cosines in the example above) and/or restrict the possible values of the separation constant.
4. Solve the time equation and multiply it by your spatial solution. Since both solutions will have arbitrary constants in front, you can often absorb the arbitrary constants from the spatial solution into those from the time solution. The resulting functions are the normal modes of your system.
5. Since your equation is linear and homogeneous, the linear combination of all these normal modes provides the most general solution that satisfies the differential equation and the boundary conditions. This solution will be an infinite series.
6. Finally, match your solution to your initial conditions to find the coefficients in that series. For instance, if your normal modes were sines and cosines, then you are building the initial conditions as Fourier series.

In the examples above the boundary conditions were homogeneous while the initial conditions were not. If two solutions y_1 and y_2 satisfy the boundary conditions $y(0, t) = y(L, t) = 0$ then the sum $y_1 + y_2$ also satisfies these boundary conditions. That's why we can apply these boundary conditions to each normal mode individually, confident that the sum of all normal modes will meet the same conditions.

By contrast if two solutions y_1 and y_2 individually satisfy the initial condition $y(x, 0) = f(x)$, their sum $(y_1 + y_2)(x, 0)$ will add up to $2f(x)$. So we have to apply the condition $y(x, 0) = f(x)$ to the entire series *after* summing, rather than to the individual parts.

It's fairly common for the boundary conditions to be homogeneous and the initial conditions inhomogeneous. In such cases, you will follow the pattern we gave here: first apply the boundary conditions to $X(x)$, then sum all solutions into a series, and then apply the initial conditions to $\sum X(x)T(t)$.

Stepping Back Part II: Solving Problems that are a Little Bit Different

Many differential equations *cannot* be solved by the exact method described above. We list here the primary variations you may encounter, many of which are discussed later in this chapter.

- **You can't separate the variables.** If you cannot algebraically separate your variables—for instance, if the differential equation is inhomogeneous—this particular technique will obviously not work. You may still be able to solve the differential equation using other techniques, some of which are discussed later in this chapter and some of which are discussed in textbooks on partial differential equations.
- **Your boundary conditions are inhomogeneous.** Just as with ODEs, you can add a "particular" solution to a "complementary" solution. We discuss this situation in Section 11.8.
- **Your initial conditions are homogeneous.** You'll show in Problem 11.97 why separation of variables doesn't generally work for homogeneous initial conditions. This problem forces you to find an alternative method, some of which are discussed in this chapter.
- **You have more than two independent variables.** Do separation of variables multiple times until you have one ordinary differential equation for each independent variable. We'll solve such an example in Section 11.5.
- **You have more than one spatial variable and no time variable.** If you have inhomogeneous boundary conditions along one boundary, and homogeneous boundary conditions along the others, you can treat the inhomogeneous boundary the way we treated the initial conditions above. (Apply the homogeneous boundary conditions, then sum, and then apply the inhomogeneous boundary condition.) You already know everything you need for such a problem, so you'll try your hand at it in Problem 11.96. If you have more than one inhomogeneous boundary condition, you need to apply the method described in Section 11.8.
- **Your normal modes are not sines and cosines.** Separation of variables gives you an ordinary differential equation that you can solve to find the normal modes of the partial differential equation you started with. Besides trig functions these normal modes may take the form of Bessel functions, Legendre polynomials, spherical harmonics, and many other categories of functions. If you can find the normal modes, and if you can build your initial conditions as a sum of normal modes, the overall process remains exactly the same. See Sections 11.6 and 11.7.
- **Your differential equation is non-linear.** You're pretty much out of luck. Separation of variables will not work for a non-linear equation because it relies on writing the solution as a linear combination of normal modes. In fact, there are very few non-linear partial differential equations that can be analytically solved by any method. When scientists and engineers need to solve a non-linear partial differential equation, which they do

11.4.3 Problems: Separation of Variables—The Basic Method

In the Explanation (Section 11.4.2) we found the general solution to the wave equation 11.4.2 with boundary conditions $y(0, t) = y(L, t) = 0$. In Problems 11.77–11.81 plug the given initial conditions into the general solution 11.4.14 and solve for the coefficients C_n and D_n to get the complete solution $y(x, t)$. In some cases your answers will be in the form of infinite series.

11.77 $y(x, 0) = F \sin(2\pi x/L), \partial y/\partial t(x, 0) = 0$

11.78 $y(x, 0) = 0, \partial y/\partial t(x, 0) = c \sin(\pi x/L)$

11.79 $y(x, 0) = H \sin(2\pi x/L), \partial y/\partial t(x, 0) = c \sin(\pi x/L)$

11.80 $y(x, 0) = \begin{cases} x & 0 < x < L/2 \\ L - x & L/2 < x < L \end{cases}, \frac{\partial y}{\partial t}(x, 0) = 0$

11.81 $y(x, 0) = \begin{cases} x & 0 < x < L/2 \\ L - x & L/2 < x < L \end{cases},$

$\frac{\partial y}{\partial t}(x, 0) = \begin{cases} 0 & 0 < x < L/3 \\ c & L/3 < x < 2L/3 \\ 0 & 2L/3 < x < L \end{cases}$

11.82 **Walk-Through: Separation of Variables.** In this problem you will use separation of variables to solve the equation $\partial^2 y/\partial t^2 - \partial^2 y/\partial x^2 + y = 0$ subject to the boundary conditions $y(0, t) = y(1, t) = 0$.

(a) Begin by guessing a separable solution $y = X(x)T(t)$. Plug this guess into the differential equation. Then divide both sides by $X(x)T(t)$ and separate variables so that all the t dependence is on the left and all the x dependence is on the right. Put the constant term on the t-dependent side. (You could put it on the other side, but the math would get a bit messier later on.)

(b) Explain in your own words why both sides of the separated equation you just wrote must equal a constant.

(c) Set the x-equation equal to a constant named P and find the general solution to the resulting ODE three times: for $P > 0$, $P = 0$, and $P < 0$. Explain why you cannot match the boundary conditions with the solutions for $P > 0$ or $P = 0$ unless you set $X(x) = 0$. Since you now know P must be negative, you can call it $-k^2$. Write your solution for $X(x)$ in terms of k.

(d) Apply the boundary condition $y(0, t) = 0$ to show that one of the arbitrary constants in your solution from Part (c) must be 0.

(e) Apply the boundary condition $y(1, t) = 0$ to find all the possible values for k. There will be an infinite number of them, but you should be able to write them in terms of a new constant n, which can be any positive integer.

(f) Solve the ODE for $T(t)$, expressing your answer in terms of n.

(g) Multiply $X(x)$ times $T(t)$ to find the normal modes of this system. You should be able to combine your three arbitrary constants into two. Write the general solution $y(x, t)$ as a sum over these normal modes. Your arbitrary constants should include a subscript n to indicate that they can take different values for each value of n.

11.83 *[This problem depends on Problem 11.82.]* In this problem you will plug the initial conditions

$y(x, 0) = \begin{cases} 1 & 1/3 < x < 2/3 \\ 0 & \text{elsewhere} \end{cases}, \frac{\partial y}{\partial t}(x, 0) = 0$

into the solution you found to Problem 11.82.

(a) The condition $\partial y/\partial t(x, 0) = 0$ should allow you to set one of your arbitrary constants to zero. The remaining condition should give you the equation $y(x, 0) = \sum_{n=1}^{\infty} D_n \sin(n\pi x)$, which is a Fourier sine series for the function $y(x, 0)$. Find the coefficients D_n. The appropriate formula is in Appendix G. (*Of course you may not have called this constant D in your solution, but you should get an equation of this form.*)

(b) Plug the coefficients that you found into your general solution to write the complete solution $y(x, t)$ for this problem. The result should be an infinite series.

(c) Have a computer calculate the 100th partial sum of your solution and plot it at a variety of times. Describe how the function $y(x)$ is evolving over time.

Solve Problems 11.84–11.90 using separation of variables. For each problem use the boundary conditions $y(0, t) = y(L, t) = 0$. When the initial conditions are given as arbitrary functions, write the solution as a series and write expressions for the coefficients in the series, as we did for the wave equation. When specific initial conditions are given, solve for the coefficients. The solution may still be in the form of a series. It may help to first work through Problem 11.82 as a model.

11.84 $\partial y/\partial t = c^2(\partial^2 y/\partial x^2)$, $y(x,0) = f(x)$

11.85 $\partial y/\partial t = c^2(\partial^2 y/\partial x^2)$, $y(x,0) = \sin(\pi x/L)$

11.86 $\partial y/\partial t + y = c^2(\partial^2 y/\partial x^2)$, $y(x,0) = f(x)$

11.87 $\partial^2 y/\partial t^2 + y = c^2(\partial^2 y/\partial x^2)$,
$y(x,0) = f(x)$, $\partial y/\partial t(x,0) = 0$

11.88 $\dfrac{\partial^2 y}{\partial t^2} + y = c^2 \dfrac{\partial^2 y}{\partial x^2}$,
$y(x,0) = \begin{cases} x & 0 \le x \le L/2 \\ L - x & L/2 < x \le L \end{cases}$,
$\dfrac{\partial y}{\partial t}(x,0) = \begin{cases} -x & 0 \le x \le L/2 \\ x - L & L/2 < x \le L \end{cases}$

11.89 $\partial^4 y/\partial t^4 = -c^2(\partial^2 y/\partial x^2)$, $y(x,0) = \dot{y}(x,0) = \ddot{y}(x,0) = 0$, $\partial^3 y/\partial t^3(x,0) = \sin(3\pi x/L)$ *Hint*: the algebra in this problem will be a little easier if you use hyperbolic trig functions. If you aren't familiar with them you can still do the problem without them.

11.90 $\partial y/\partial t = c^2 t(\partial^2 y/\partial x^2)$, $y(x,0) = y_0 \sin(\pi x/L)$

Problems 11.91–11.94 refer to a rod with temperature held fixed at the ends: $u(0,t) = u(L,t) = 0$. For each set of initial conditions write the complete solution $u(x,t)$. You will need to begin by using separation of variables to solve the heat equation (11.2.3), but if you do more than one of these you should just find the general solution once.

11.91 $u(x,0) = u_0 \sin(2\pi x/L)$

11.92 $u(x,0) = \begin{cases} cx & 0 < x < L/2 \\ c(L-x) & L/2 < x < L \end{cases}$

11.93 $u(x,0) = \begin{cases} 0 & 0 < x < L/3 \\ u_0 & L/3 < x < 2L/3 \\ 0 & 2L/3 < x < L \end{cases}$

11.94 $u(x,0) = 0$. Show how you can find the solution to this the same way you would for Problems 11.91–11.93, and also explain how you could have predicted this solution without doing any calculations.

11.95 As we solved the "string" problem in the Explanation (Section 11.4.2) we determined that a positive separation constant P leads to the solution $X(x) = Ae^{kx} + Be^{-kx}$. We then discarded this solution based on the argument that it cannot meet the boundary conditions $X(0) = X(L) = 0$. Show that there is no possible way for that solution to meet those boundary conditions unless $A = B = 0$. Then explain why we discard *that* solution too.

11.96 Given enough time, any isolated region of space will tend to approach a steady state where the temperature at each point is unchanging. In this case the time derivative in the heat equation becomes zero and the temperature obeys Laplace's equation 11.2.5. In two dimensions this can be written as $\partial^2 u/\partial x^2 + \partial^2 u/\partial y^2 = 0$. In this problem you will use separation of variables to solve for the steady-state temperature $u(x,y)$ on a rectangular slab subject to the boundary conditions $u(0,y) = u(x_f,y) = u(x,0) = 0$, $u(x,y_f) = u_0$.

(a) Separate variables to get ordinary differential equations for the functions $X(x)$ and $Y(y)$.

(b) Solve the equation for $X(x)$ subject to the boundary conditions $X(0) = X(x_f) = 0$.

(c) Solve the equation for $Y(y)$ subject to the homogeneous boundary condition $Y(0) = 0$. Your solution should have one undetermined constant in it corresponding to the one boundary condition you have not yet imposed.

(d) Write the solution $u(x,y)$ as a sum of normal modes.

(e) Use the final boundary condition $u(x,y_f) = u_0$ to solve for the remaining coefficients and find the general solution.

(f) Check your solution by verifying that it solves Laplace's equation and meets each of the homogeneous boundary conditions given above.

(g) Use a computer to plot the 40th partial sum of your solution. (You will have to choose some values for the constants in the problem.) Looking at your plot, describe how the temperature depends on y for a fixed value of x (other than $x = 0$ or $x = x_f$). You should see that it goes from $u = 0$ at $y = 0$ to $u = u_0$ at $y = y_f$. Does it increase linearly? If not describe what it does.

11.97 We said in the Explanation (Section 11.4.2) that separation of variables generally doesn't work for an initial value problem when the initial conditions are homogeneous. To illustrate why, consider once again the wave equation 11.4.2 for a string stretched from $x = 0$ to $x = L$. Assume the initial position and velocity of the string are both zero.

(a) If the boundary conditions are homogeneous, $y(0, t) = y(L, t) = 0$, what is the solution? (You can figure this one out without doing any math: just think about it.)

Note that this case is easy to solve but not very interesting or useful. For the rest of the problem we will therefore assume that the boundary conditions are not entirely homogeneous, and for simplicity we'll take the boundary conditions $y(0, t) = 0$, $y(L, t) = H$.

(b) Explain why you need to apply the initial conditions to each $T(t)$ function separately, and then apply the boundary conditions to the entire sum.

(c) Separate variables and write the resulting ODEs for $X(x)$ and $T(t)$.

(d) Solve the equation for $T(t)$ with the initial conditions given above. (Choose the sign of the separation constant to lead to sinusoidal solutions, not exponential or linear.) Using your solution, explain why separation of variables cannot be used to find any non-trivial solutions to this problem.

11.98 Solve Problem 11.97 using the heat equation (11.2.3) instead of the wave equation (11.4.2).

11.99 In quantum mechanics a particle is described by a "wavefunction" $\Psi(x, t)$ which obeys the Schrödinger equation:

$$-\frac{\hbar^2}{2m}\frac{\partial^2 \Psi}{\partial x^2} + V(x)\Psi = i\hbar\frac{\partial \Psi}{\partial t}$$

(We are considering only one spatial dimension for simplicity.) This is really a whole family of PDEs, one for each possible potential function $V(x)$. Nonetheless we can make good progress without knowing anything about the potential function.

(a) Plug in a trial solution of the form $\Psi(x, t) = \psi(x)T(t)$ and separate variables. Call the separation constant E. (It turns out that each normal mode—called an "energy eigenstate" in quantum mechanics—represents the state of the particle with energy E.)

(b) Solve the ODE for $T(t)$. Describe in words the time dependence of an energy eigenstate.

Your work above applies to any one-dimensional Schrödinger equation problem; further work depends on the particular potential function. For the rest of the problem you'll consider a "particle in a box" that experiences no force inside the box but cannot move out of the box. In one dimension that means $V(x) = 0$ on $0 \leq x \leq L$ with the boundary conditions $\psi(0) = \psi(L) = 0$,

(c) Solve your separated equation using this $V(x)$ and boundary conditions to find all the allowable values of E—in other words find the energy levels of this system.

(d) Write the general solution for $\psi(x)$ as a sum over all individual solutions. Your answer will not explicitly involve E.

(e) Write the general solution for $\Psi(x, t)$. Use the values of E you found so that your answer has n in it but not E.

(f) Find $\Psi(x, t)$ for the initial condition $\Psi(x, 0) = \psi_0$ (a constant) for $L/4 \leq x \leq 3L/4$ and 0 elsewhere.

11.100 Exploration: Laplace's Equation on a Disk
The steady-state temperature in an isolated region of space obeys Laplace's equation (11.2.5). In this problem you are going to solve for the steady-state temperature on a disk of radius a, where the temperature on the boundary of the disk is given by $u(a, \phi) = u_0 \sin(k\phi)$. Laplace's equation will be easiest to solve in polar coordinates, so you are looking for a function $u(\rho, \phi)$. (Feel free to try this problem in Cartesian coordinates. You'll feel great about it until you try to apply the boundary conditions.) In polar coordinates Laplace's equation can be written as:

$$\frac{\partial^2 u}{\partial \rho^2} + \frac{1}{\rho^2}\frac{\partial^2 u}{\partial \phi^2} + \frac{1}{\rho}\frac{\partial u}{\partial \rho} = 0$$

(a) This problem has an "implicit," or unstated, boundary condition that $u(\phi + 2\pi) = u(\phi)$. Explain how we know this boundary condition must be followed even though it was never stated in the problem. Because of that implicit boundary condition the constant k in the boundary conditions cannot be just any real number. What values of k are allowed, and why?

(b) Plug in the initial guess $R(\rho)\Phi(\phi)$ and separate variables in Laplace's equation.

(c) Setting both sides of the separated equation equal to a constant P, find the general real solution to the

ODE for $\Phi(\phi)$ for the three cases $P > 0$, $P = 0$, and $P < 0$.

(d) Use the implicit boundary condition from Part (a) to determine which of the three solutions you found is the correct one. Now that you know what sign the separation constant must have, rename it either p^2 or $-p^2$. Use the period of 2π to constrain the possible values of p.

(e) Write the ODE for $R(\rho)$. Solve it by plugging in a guess of the form $R(\rho) = \rho^c$ and solve for the two possible values of c. Since the ODE is linear you can then combine these two solutions with arbitrary constants in front. Your answer should have p in it.

(f) There is another implicit condition: $R(0)$ has to be finite. Use that condition to show that one of the two arbitrary constants in your solution for $R(\rho)$ must be zero. (You can assume that $p > 0$.)

(g) Multiply your solutions for $\Phi(\phi)$ and $R(\rho)$, combining arbitrary constants as much as possible, and write the general solution $u(\rho, \phi)$ as an infinite series.

(h) Plug in the boundary condition $u(a, \phi) = u_0 \sin(k\phi)$ and use it to find the values of the arbitrary constants. (You could use the formula for the coefficients of a Fourier sine series but you can do it more simply by inspection.)

(i) Write the solution $u(\rho, \phi)$.

11.5 Separation of Variables—More than Two Variables

We have seen that with one independent variable you have an ordinary differential equation; with two independent variables you have a partial differential equation. What about three independent variables? (Or four or five or eleven?) Does each of those need its own chapter?

Fortunately, the process scales up. With each new variable you have to separate one more time, resulting in many ordinary differential equations. In the end your solution is expressed as a series over more than one variable.

11.5.1 Discovery Exercise: Separation of Variables—More than Two Variables

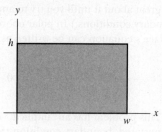

A thin rubber sheet is stretched on a rectangular frame. The sheet is glued to the frame; however, it is free to vibrate inside the frame. (This system functions as a rectangular drum; the more common circular drum will be taken up in Section 11.6.) For convenience we place our axes in the plane of the frame, with the lower-left-hand corner of the rectangle at the origin.

The motion of this sheet can be described by a function $z(x, y, t)$ which will obey the wave equation in two dimensions:

$$\frac{\partial^2 z}{\partial x^2} + \frac{\partial^2 z}{\partial y^2} = \frac{1}{v^2}\frac{\partial^2 z}{\partial t^2} \quad \text{the wave equation in two dimensions} \quad (11.5.1)$$

The boundary condition is that $z = 0$ on all four edges of the rectangle. The initial conditions are $z(x, y, 0) = z_0(x, y)$, $\dot{z}(x, y, 0) = v_0(x, y)$.

1. Begin with a "guess" of the form $z(x, y, t) = X(x)Y(y)T(t)$. Substitute this expression into Equation 11.5.1.
2. Separate the variables so that the left side of the equation depends on x (not on y or t), and the right side depends on y and t (not on x). (This will require two steps.)
3. Explain why both sides of this equation must now equal a constant.

4. The function $X(x)$ must satisfy the conditions $X(0) = 0$ and $X(w) = 0$. Based on this restriction, is the separation constant positive or negative? Explain your reasoning. If positive, call it k^2; if negative, $-k^2$.
5. Solve the resulting differential equation for $X(x)$.
6. Plug in both boundary conditions on $X(x)$. The result should allow you to solve for one of the arbitrary constants and express the real number k in terms of an integer-valued n.

   See Check Yourself #73 in Appendix L

7. You have another equation involving $Y(y)$ and $T(t)$ as well as k. Separate the variables in this equation so that the $Y(y)$ terms are on one side, and the $T(t)$ and k terms on the other. Both sides of this equation must equal a constant, but that constant is *not* the same as our previous constant.
8. The function $Y(y)$ must satisfy the conditions $Y(0) = 0$ and $Y(h) = 0$. Based on this restriction, is the *new* separation constant positive or negative? If positive, call it p^2; if negative, $-p^2$.
9. Solve the resulting differential equation for $Y(y)$.
10. Plug in both boundary conditions on $Y(y)$. The result should allow you to solve for one of the arbitrary constants and express the real number p in terms of an integer-valued m.

    Important: p is not necessarily the same as k, and m is not necessarily the same as n. These new constants are unrelated to the old ones.

11. Solve the differential equation for $T(t)$. (The result will be expressed in terms of both k and p, or in terms of both n and m.)
12. Write the solution $X(x)Y(y)T(t)$. Combine arbitrary constants as much as possible. If your answer has any real-valued k or p constants, replace them with the integer-valued n and m constants by using the formulas you found from boundary conditions.
13. The general solution is a sum over *all possible* n- and m-values. For instance, there is one solution where $n = 3$ and $m = 5$, and another solution where $n = 12$ and $m = 5$, and so on. Write the double sum that represents the general solution to this equation.

    See Check Yourself #74 in Appendix L

14. Write equations relating the initial conditions $z_0(x, y)$ and $v_0(x, y)$ to your arbitrary constants. These equations will involve sums. Solving them requires finding the coefficients of a double Fourier series, but for now it's enough to just write the equations.

11.5.2 Explanation: Separation of Variables—More than Two Variables

When you have more than two independent variables you have to separate variables more than once to isolate them. Each time you separate variables you introduce a new arbitrary constant, resulting in multiple series.

The Problem

The steady-state electric potential in a region with no charged particles follows Laplace's equation $\nabla^2 V = 0$. In three-dimensional Cartesian coordinates, this equation can be written:

$$\frac{\partial^2 V}{\partial x^2} + \frac{\partial^2 V}{\partial y^2} + \frac{\partial^2 V}{\partial z^2} = 0 \tag{11.5.2}$$

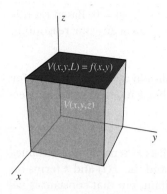

Consider a cubic box of side length L. Five sides of this box are grounded, which means they are held at potential $V = 0$. The sixth side has some arbitrary potential function that is also held fixed over time. There are no charged particles inside the box. What is the potential $V(x, y, z)$ at all points inside the box?

For convenience we choose our axes so that one corner of the box is at the origin and the non-grounded side is the top ($z = L$). We can therefore describe the potential on that side by some function $V(x, y, L) = f(x, y)$.

Note that this problem has five homogeneous boundary conditions and one inhomogeneous boundary condition. As we discussed in the last section, we will apply the homogeneous conditions before we sum all the solutions, and the inhomogeneous one afterwards. In other words, we will treat the inhomogeneous boundary condition much as we treated the *initial* conditions for the vibrating string.

Separating the Variables, and Solving for the First One

Our guess is a fully separated function, involving three functions of one variable each.

$$V(x, y, z) = X(x)Y(y)Z(z)$$

We plug that guess into Laplace's equation (11.5.2) and divide both sides by $X(x)Y(y)Z(z)$ to separate variables.

$$X''(x)Y(y)Z(z) + X(x)Y''(y)Z(z) + X(x)Y(y)Z''(z) = 0 \quad \rightarrow \quad \frac{X''(x)}{X(x)} + \frac{Y''(y)}{Y(y)} + \frac{Z''(z)}{Z(z)} = 0$$

We now bring one of these terms to the other side. We've chosen the y-term below, but we could have chosen x just as easily. It would be a bit tougher if we chose z (the variable with an inhomogeneous boundary condition); we'll explain why in a moment.

$$\frac{Y''(y)}{Y(y)} = -\frac{X''(x)}{X(x)} - \frac{Z''(z)}{Z(z)} \qquad (11.5.3)$$

You may now object that the variables are not entirely separated. (How could they be, with three variables and only two sides of the equation?) But the essential relationship still holds: the left side of the equation depends only on y, and the right side of the equation depends only on x and z, so the only function both sides can equal is a constant.

When you set the left side of Equation 11.5.3 equal to a constant you get the same problem we solved for $X(x)$ in Section 11.4, so let's just briefly review where it goes.

- We are solving $Y''(y)/Y(y) =$ <*a constant*>, with the boundary conditions $Y(0) = Y(L) = 0$.
- A positive constant would lead to an exponential solution and a zero constant to a linear solution, neither of which could meet those boundary conditions.
- We therefore call the constant $-k^2$ and after a bit of algebra arrive at the solution.

$$Y(y) = A\sin(ky) \text{ where } k = \frac{n\pi}{L} \text{ where } n \text{ can be any positive integer}$$

Now you can see why we start by isolating a variable with homogeneous boundary conditions. The inhomogeneous condition for $Z(z)$ cannot be applied until after we build our series, so it would not have determined the sign of our separation constant.

Separating Again, and Solving for the Second Variable

The left side of Equation 11.5.3 equals $-k^2$, so the right side must equal the same constant. We rearrange terms in that equation to separate variables a second time.

$$-\frac{X''(x)}{X(x)} - \frac{Z''(z)}{Z(z)} = -k^2 \quad \rightarrow \quad \frac{X''(x)}{X(x)} = k^2 - \frac{Z''(z)}{Z(z)}$$

Both sides of this equation must equal a constant, and this constant must be negative. (Can you explain why?) However, this new constant does *not* have to equal our existing $-k^2$; the two are entirely independent. We will call the new constant $-p^2$.

$$\frac{X''(x)}{X(x)} = -p^2 \quad \text{and} \quad k^2 - \frac{Z''(z)}{Z(z)} = -p^2$$

The $X(x)$ differential equation and boundary conditions are the same as the $Y(y)$ problem that we solved above. The solution is therefore also the same, with one twist: p cannot also equal $n\pi/L$ because that would make the two constants the same. Instead we introduce a new variable m that must be a positive integer, but not necessarily the same integer as n.

$$X(x) = B\sin(px) \text{ where } p = \frac{m\pi}{L} \text{ where } m \text{ can be any positive integer}$$

The Variable with an Inhomogeneous Boundary Condition

We have one ordinary differential equation left.

$$k^2 - \frac{Z''(z)}{Z(z)} = -p^2$$

Rewriting this as $Z''(z) = (k^2 + p^2) Z(z)$, we see that the positive constant requires an exponential solution.

$$Z(z) = Ce^{\sqrt{k^2+p^2}\,z} + De^{-\sqrt{k^2+p^2}\,z}$$

Our fifth *homogeneous* boundary condition $Z(0) = 0$ leads to $C + D = 0$, so $D = -C$. We can therefore write:

$$Z(z) = C\left(e^{\sqrt{k^2+p^2}\,z} - e^{-\sqrt{k^2+p^2}\,z}\right) = C\left(e^{(\pi/L)\sqrt{n^2+m^2}\,z} - e^{-(\pi/L)\sqrt{n^2+m^2}\,z}\right)$$

(You can write $\sinh\left(\sqrt{k^2+p^2}\,z\right)$ instead of $e^{\sqrt{k^2+p^2}\,z} - e^{-\sqrt{k^2+p^2}\,z}$. It amounts to the same thing with a minor change in the arbitrary constant.)

When we solved for $X(x)$ and $Y(y)$, we applied their relevant boundary conditions as soon as we had solved the equations. Remember, however, that the boundary condition at $z = L$ is inhomogeneous; we therefore cannot apply it until we combine the three solutions and write a series.

The Solution so far, and a Series

When we multiply all three solutions, the three arbitrary constants combine into one.

$$X(x)Y(y)Z(z) = A\sin\left(\frac{m\pi}{L}x\right)\sin\left(\frac{n\pi}{L}y\right)\left(e^{(\pi/L)\sqrt{n^2+m^2}\,z} - e^{-(\pi/L)\sqrt{n^2+m^2}\,z}\right)$$

Remember that this is a solution for any positive integer n and any positive integer m. For instance, there is one solution with $n = 5$ and $m = 3$ and this solution can have any coefficient A. There is another solution with $n = 23$ and $m = 2$, which could have a different coefficient A. In order to sum all possible solutions, we therefore need a *double* sum.

$$V(x, y, z) = \sum_{n=1}^{\infty} \sum_{m=1}^{\infty} A_{mn} \sin\left(\frac{m\pi}{L}x\right) \sin\left(\frac{n\pi}{L}y\right) \left(e^{(\pi/L)\sqrt{n^2+m^2}z} - e^{-(\pi/L)\sqrt{n^2+m^2}z}\right) \quad (11.5.4)$$

We have introduced the subscript A_{mn} to indicate that for each choice of m and n there is a different free choice of A.

The Last Boundary Condition

Applying our last boundary condition $V(x, y, L) = f(x, y)$ to the entire series, we write:

$$\sum_{n=1}^{\infty} \sum_{m=1}^{\infty} A_{mn} \sin\left(\frac{m\pi}{L}x\right) \sin\left(\frac{n\pi}{L}y\right) \left(e^{\pi\sqrt{n^2+m^2}} - e^{-\pi\sqrt{n^2+m^2}}\right) = f(x, y)$$

At this point in our previous section, building up the (one-variable) function on the right as a series of sines involved writing a Fourier Series. In this case we are building a two-variable function as a double Fourier Series. Once again the formula is in Appendix G.

$$A_{mn}\left(e^{\pi\sqrt{n^2+m^2}} - e^{-\pi\sqrt{n^2+m^2}}\right) = \frac{4}{L^2} \int_0^L \int_0^L f(x, y) \sin\left(\frac{m\pi}{L}x\right) \sin\left(\frac{n\pi}{L}y\right) dx\,dy \quad (11.5.5)$$

As before, if $f(x, y)$ is particularly simple you can sometimes find the coefficients A_{mn} explicitly. In other cases you can use numerical approaches. You'll work examples of each type for different boundary conditions in the problems.

EXAMPLE ### The Two-Dimensional Heat Equation

Problem:
Solve the two-dimensional heat equation:

$$\frac{\partial u}{\partial t} = \alpha \left(\frac{\partial^2 u}{\partial x^2} + \frac{\partial^2 u}{\partial y^2}\right)$$

on the domain $0 \le x \le 1$, $0 \le y \le 1$, $t \ge 0$ subject to the boundary conditions

$$u(0, y, t) = u(1, y, t) = u(x, 0, t) = u(x, 1, t) = 0$$

and the initial condition

$$u(x, y, 0) = u_0(x, y).$$

Solution:
We begin by assuming a solution of the form:

$$u(x, y, t) = X(x)Y(y)T(t)$$

Plugging this into the differential equation yields:

$$X(x)Y(y)T'(t) = \alpha\left[X''(x)Y(y)T(t) + X(x)Y''(y)T(t)\right]$$

Dividing through by $X(x)Y(y)T(t)$, we have:

$$\frac{T'(t)}{T(t)} = \alpha\left[\frac{X''(x)}{X(x)} + \frac{Y''(y)}{Y(y)}\right]$$

It might seem natural at this point to work with $T(t)$ first: it is easier by virtue of being first order, and it is already separated. But you always have to start with the variables that have homogeneous boundary conditions. As usual, we will handle the boundary conditions *before* building a series, and tackle the initial condition last. We must separate one of the spatial variables: we choose X quite arbitrarily.

$$\frac{T'(t)}{T(t)} - \alpha\frac{Y''(y)}{Y(y)} = \alpha\frac{X''(x)}{X(x)}$$

The separation constant must be negative. (Do you see why?) Calling it $-k^2$, we solve the equation and boundary conditions for X to find: $X(x) = A\sin\left(kx/\sqrt{\alpha}\right)$ where $k = n\pi\sqrt{\alpha}$ for $n = 1, 2, 3\ldots$

Meanwhile we separate the other equation.

$$\frac{T'(t)}{T(t)} - \alpha\frac{Y''(y)}{Y(y)} = -k^2 \quad\rightarrow\quad \frac{T'(t)}{T(t)} + k^2 = \alpha\frac{Y''(y)}{Y(y)}$$

Once again the separation constant must be negative: we will call it $-p^2$. The equation for Y looks just like the previous equation for X, yielding $Y(y) = B\sin\left(py/\sqrt{\alpha}\right)$ where $p = m\pi\sqrt{\alpha}$ for $m = 1, 2, 3\ldots$

Finally, we have $T'(t) = -\left(k^2 + p^2\right)T(t)$, so

$$T(t) = Ce^{-(k^2+p^2)t}$$

Writing a complete solution, collecting arbitrary constants, and summing, we have:

$$u(x, y, t) = \sum_{n=1}^{\infty}\sum_{m=1}^{\infty} A_{mn}\sin(n\pi x)\sin(m\pi y)e^{-\pi^2\alpha(n^2+m^2)t}$$

Finally we are ready to plug in our inhomogeneous initial condition, which tells us that

$$\sum_{n=1}^{\infty}\sum_{m=1}^{\infty} A_{mn}\sin(n\pi x)\sin(m\pi y) = u_0(x, y)$$

and therefore, once again using a formula from Appendix G:

$$A_{mn} = 4\int_0^1\int_0^1 u_0(x, y)\sin(n\pi x)\sin(m\pi y)\,dx\,dy \tag{11.5.6}$$

For example, suppose the initial temperature is 5 in the region $1/4 < x < 3/4, 1/4 < y < 3/4$ and 0 everywhere outside it. From Equation 11.5.6:

$$A_{mn} = 4\int_0^1\int_0^1 u_0(x, y)\sin(n\pi x)\sin(m\pi y)\,dx\,dy = 20\int_{1/4}^{3/4}\int_{1/4}^{3/4}\sin(n\pi x)\sin(m\pi y)\,dx\,dy$$

This integral is simple to evaluate, and after a bit of algebra gives

$$A_{mn} = \frac{80}{mn\pi^2} \sin\left(\frac{m\pi}{4}\right) \sin\left(\frac{m\pi}{2}\right) \sin\left(\frac{n\pi}{4}\right) \sin\left(\frac{n\pi}{2}\right)$$

Writing all the terms with $n, m < 4$ gives

$$u(x, y, t) = \frac{80}{\pi^2}\left(\frac{1}{2}\sin(\pi x)\sin(\pi y)e^{-2\pi^2\alpha t} - \frac{1}{6}\sin(\pi x)\sin(3\pi y)e^{-10\pi^2\alpha t}\right.$$
$$\left. -\frac{1}{6}\sin(3\pi x)\sin(\pi y)e^{-10\pi^2\alpha t} + \frac{1}{18}\sin(3\pi x)\sin(3\pi y)e^{-18\pi^2\alpha t} + \ldots\right)$$

If you plot the first few terms you can see that they represent a bump in the middle of the domain that flattens out over time due to the decaying exponentials. With computers, however, you can go much farther. The plots below show this series (with $\alpha = 1$) with all the terms up to $n, m = 100$ (10,000 terms in all, although many of them equal zero).

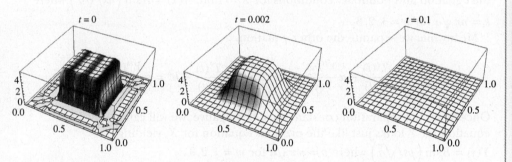

At $t = 0$ you can see that the double Fourier series we constructed accurately models the initial conditions. A short time later the temperature is starting to even out more, and at much later times it relaxes towards zero everywhere. (Do you see why we call $t = 0.1$ "much later"? See Problem 11.105.)

11.5.3 Problems: Separation of Variables—More than Two Variables

Problems 11.101–11.105 follow up on the example of Laplace's equation in a cubic region from the Explanation (Section 11.5.2).

11.101 The solution is written in terms of sines of x and y and exponentials of z. Does this mean that for any given boundary condition the solution $V(x)$ at a fixed y and z will look sinusoidal? If so, explain how you know. If not, explain what you can conclude about what $V(x)$ at fixed y and z will look like from this solution?

11.102 Solve Laplace's equation $\partial^2 V/\partial x^2 + \partial^2 V/\partial y^2 + \partial^2 V/\partial z^2 = 0$ in the cubic region $x, y, z \in [0, L]$ with $V(x, y, L) = V_0$ and $V = 0$ on the other five sides.

11.103 (a) Solve Laplace's equation in the cubic region $0 \leq x, y, z \leq L$ with $V(x, y, L) = \sin(\pi x/L)\sin(2\pi y/L) + \sin(2\pi x/L)\sin(\pi y/L)$ and $V = 0$ on the other five sides.

(b) Sketch $V(x, L/2, L)$ and $V(x, L/4, L)$, and $V(x, 0, L)$ as functions of x. How does changing the y-value affect the plot?

(c) At $z = L$ sketch how V depends on x for many values of y. You can do this by making an animation or a series of still images, but either way you should have enough to see if it follows the behavior you predicted in Part (b). How would your sketches

11.104 Solve Laplace's equation for $V(x, y, z)$ in the region $0 \leq x, y, z \leq L$ with $V(x, y, 0) = f(x, y)$ and $V = 0$ on the other five sides. (This is the same problem that was solved in the Explanation (Section 11.5.2) except that the side with non-zero potential is at $z = 0$ instead of $z = L$.)

11.105 The example on Page 584 ended with a series, which we plotted at several times. From those plots it's clear that for $\alpha = 1$ the initial temperature profile hadn't changed much by $t = 0.002$ but had mostly relaxed towards zero by $t = 0.1$. How could you have predicted this by looking at the series solution?

11.106 **Walk-Through: Separation of Variables—More Than Two Variables.** In this problem you will use separation of variables to solve the equation $\partial^2 u/\partial t^2 - \partial^2 u/\partial x^2 - \partial^2 u/\partial y^2 + u = 0$ subject to the boundary conditions $u(0, y, t) = u(1, y, t) = u(x, 0, t) = u(x, 1, t) = 0$.

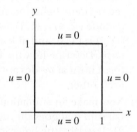

(a) Begin by guessing a separable solution $u = X(x)Y(y)T(t)$. Plug this guess into the differential equation. Then divide both sides by $X(x)Y(y)T(t)$ and separate variables so that all the y and t dependence is on the left and the x dependence is on the right. Put the constant term on the left side. (You could put it on the right side, but the math would get a bit messier later on.)

(b) Set both sides of that equation equal to a constant. Given the boundary conditions above for u, what are the boundary conditions for $X(x)$? Use these boundary conditions to explain why the separation constant must be negative. You can therefore call it $-k^2$.

(c) Separate variables in the remaining ODE, putting the t-dependent terms and the constant terms on the left, and the y-dependent term on the right. Set both sides equal to a constant and explain why this separation constant also must be negative. Call it $-p^2$.

(d) Solve the equation for $X(x)$. Use the boundary condition at $x = 0$ to show that one of the arbitrary constants must be 0 and apply the boundary condition at $x = 1$ to find all the possible values for k. There will be an infinite number of them, but you should be able to write them in terms of a new constant m, which can be any positive integer.

(e) Solve the equation for $Y(y)$ and apply the boundary conditions to eliminate one arbitrary constant and to write the constant p in terms of a new integer n.

(f) Solve the equation for $T(t)$. Your answer should depend on m and n and should have two arbitrary constants.

(g) Multiply $X(x)$, $Y(y)$, and $T(t)$ to find the normal modes of this system. You should be able to combine your four arbitrary constants into two. Write the general solution $u(x, y, t)$ as a sum over these normal modes. Your arbitrary constants should include a subscript mn to indicate that they can take different values for each combination of m and n.

11.107 *[This problem depends on Problem 11.106.]* In this problem you will plug the initial conditions
$$u(x, y, 0) = \begin{cases} 1 & 1/3 < x < 2/3,\ 1/3 < y < 2/3 \\ 0 & \text{elsewhere} \end{cases},$$
$\dfrac{\partial u}{\partial t}(x, y, 0) = 0$ into the solution you found to Problem 11.106.

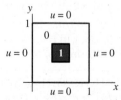

(a) Use the condition $\partial u/\partial t(x, y, 0) = 0$ to show that one of your arbitrary constants must equal zero.

(b) The remaining condition should give you an equation that looks like $u(x, y, 0) = \sum_{m=1}^{\infty} \sum_{n=1}^{\infty} F_{mn} \sin(m\pi x) \sin(n\pi y)$ (with a different letter if you didn't use F_{mn} for the same arbitrary constant we did). This is a double Fourier sine series for the function $u(x, y, 0)$. Find the coefficients F_{mn}. The appropriate formula is in Appendix G.

(c) 🖥 Have a computer calculate a partial sum of your solution including all terms up to $m, n = 20$ (400 terms in all) and plot it at a variety of times. Describe how the function u throughout the region is evolving over time.

Solve Problems 11.108–11.112 using separation of variables. For each problem use the boundary conditions $u(0, y, t) = u(L, y, t) = u(x, 0, t) = u(x, H, t) = 0$. When the initial conditions are given as arbitrary functions, write the solution as a series and write expressions for the coefficients in the series. When specific initial conditions are given, solve for the coefficients. It may help to first work through Problem 11.106 as a model.

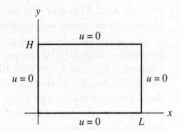

11.108 $\partial u/\partial t = a^2(\partial^2 u/\partial x^2) + b^2(\partial^2 u/\partial y^2)$, $u(x, y, 0) = f(x, y)$

11.109 $\partial u/\partial t = a^2(\partial^2 u/\partial x^2) + b^2(\partial^2 u/\partial y^2)$, $u(x, y, 0) = \sin(2\pi x/L)\sin(3\pi y/H)$

11.110 $\partial u/\partial t + u = a^2(\partial^2 u/\partial x^2) + b^2(\partial^2 u/\partial y^2)$, $u(x, y, 0) = f(x, y)$

11.111 $\partial^2 u/\partial t^2 + u = \partial^2 u/\partial x^2 + \partial^2 u/\partial y^2$, $u(x, y, 0) = 0, \partial u/\partial t(x, y, 0) = g(x, y)$

11.112 $\partial^2 u/\partial t^2 + \partial u/\partial t - \partial^2 u/\partial x^2 - \partial^2 u/\partial y^2 + u = 0$, $u(x, y, 0) = 0, \partial u/\partial t(x, y, 0) = \sin(\pi x/L)\sin(2\pi y/H)$

For Problems 11.113–11.114 solve the 2D wave equation $\partial^2 z/\partial x^2 + \partial^2 z/\partial y^2 = (1/v^2)(\partial^2 z/\partial t^2)$ on the rectangle $x, y \in [0, L]$ with $z = 0$ on all four sides and initial conditions given below. If you have not done the Discovery Exercise (Section 11.5.1) you will need to begin by using separation of variables to solve the wave equation.

11.113 $z(x, y, 0) = \sin(\pi x/L)\sin(\pi y/L), \dot{z}(x, y, 0) = 0$

11.114 $z(x, y, 0) = 0$,
$\dot{z}(x, y, 0) = \begin{cases} c & L/3 < x < 2L/3, L/3 < y < 2L \\ 0 & \text{otherwise} \end{cases}$

11.115 🖥 *[This problem depends on Problem 11.114.]* The equation you solved in Problem 11.114 might represent an oscillating square plate that was given a sudden blow in a region in the middle. This might represent a square drumhead hit by a square drumstick.[6] The solution you got, however, was a fairly complicated looking double sum.

(a) Have a computer plot the initial function $\dot{z}(x, y, 0)$ for the partial sum that goes up through $m = n = 5$, then again up through $m = n = 11$, and finally up through $m = n = 21$. As you add terms you should see the partial sums converging towards the shape of the initial conditions that you were solving for.

(b) Take several of the non-zero terms in the series (the individual terms, not the partial sums) and for each one use a computer to make an animation of the shape of the drumhead (z, not $\dot{z}$) evolving over time. You should see periodic behavior. You should include $(m, n) = (1, 1)$, $(m, n) = (1, 3)$, $(m, n) = (3, 3)$, and at least one other term. Describe how the behaviors of these normal modes are different from each other.

(c) Now make an animation of the partial sum that goes through $m = n = 21$. Describe the behavior. How is it similar to or different from the behavior of the individual terms you saw in the previous part?

(d) How would your answers to Parts (b) and (c) have looked different if we had used a different set of initial conditions?

Solve Problems 11.116–11.118 using separation of variables. Write the solution as a series and write expressions for the coefficients in the series, as we did for Laplace's equation in the Explanation (Section 11.5.2), Equations 11.5.4 and 11.5.5.

11.116 Solve the wave equation $\partial^2 u/\partial x^2 + \partial^2 u/\partial y^2 + \partial^2 u/\partial z^2 = (1/v^2)(\partial^2 u/\partial t^2)$ in a 3D cube of side L with $u = 0$ on all six sides and initial conditions $u(x, y, z, 0) = f(x, y, z), \dot{u}(x, y, z, 0) = 0$.

---

[6]If "square drumhead" sounds a bit artificial, don't worry. In Section 11.6 we'll solve the wave equation on a more conventional circular drumhead.

11.117 In the wave equation the parameter v is the sound speed, meaning the speed at which waves propagate in that medium. "Anisotropic" crystals have a different sound speed in different directions. Solve the anisotropic wave equation $\partial^2 u/\partial t^2 = v_x^2(\partial^2 u/\partial x^2) + v_y^2(\partial^2 u/\partial y^2)$ in a square box of side length L, with $u = 0$ on all four sides and initial conditions $u(x, y, 0) = 0$, $\dot{u}(x, y, 0) = g(x, y)$.

11.118 Solve the heat equation $\partial u/\partial t = \alpha \left(\partial^2 u/\partial x^2 + \partial^2 u/\partial y^2 + \partial^2 u/\partial z^2 \right)$ in a 3D cube of side L with $u = 0$ on all six sides and initial condition $u(x, y, z, 0) = f(x, y, z)$.

11.6 Separation of Variables—Polar Coordinates and Bessel Functions

We have seen how separation of variables finds the "normal modes" of an equation. If the initial conditions happen to match the normal modes, the solution will evolve simply in time; if the initial conditions are more complicated, we build them as sums of normal modes.

In the examples we have seen so far, the normal modes have been sines and cosines. In general, the normal modes may be Bessel functions, Legendre polynomials, spherical harmonics, associated Laguerre polynomials, Hankel Functions, and many others. This section will provide a brief introduction to Bessel functions, and show how they arise in solutions to PDEs in polar and cylindrical coordinates. The next section will go through a similar process, showing how PDEs in spherical coordinates lead to Legendre polynomials. In both sections our main point will be that the normal modes may change, but the process of separating variables remains consistent.

11.6.1 Explanation: Bessel Functions—The Unjustified Essentials

The solutions to many important PDEs in polar and cylindrical coordinates involve "Bessel functions"—a class of functions that may be totally unfamiliar to you. Our presentation of Bessel functions comes in two parts. In this chapter we present all the properties you need to use Bessel functions in solving PDEs; in Chapter 12 we will show where those properties come from.

As an analogy, consider how little information about sines and cosines you have actually needed to make it through this chapter so far.

- The ordinary differential equation $d^2y/dx^2 + k^2 y = 0$ has two real, linearly independent solutions, which are called $\sin(kx)$ and $\cos(kx)$. Because the equation is linear and homogeneous, its general solution is $A\sin(kx) + B\cos(kx)$.
- The function $\sin(kx)$ has zeros at $x = n\pi/k$ where n is any integer. The function $\cos(kx)$ has zeros at $(\pi/2 + n\pi)/k$. (These facts are important for matching boundary conditions.)
- Sines and cosines form a "complete basis," which is a fancy way of saying that you can build up almost any function $f(x)$ as a linear combination of sines and cosines (a Fourier series). All you need is the formula for the coefficients. (This information is important for matching initial conditions.)

As an introduction to trigonometry, those three bullet points are hopelessly inadequate. They don't say a word about how the sine and cosine functions are defined (SOHCAHTOA or the unit circle, for instance). They don't discuss why these functions are "orthogonal," and how you can use that fact to *derive* the coefficients of a Fourier series. The list above, without any further explanation, feels like a grab-bag of random facts without any real math. Nonetheless, it gives us enough to solve the PDEs that happen to have trigonometric normal modes.

Here's our point: you can look up a few key facts about a function you've never even heard of and then use that function to solve a PDE. Later you can do more research to better understand the solutions you have found.

A Differential Equation and its Solutions

Bessel's Equation and Its Solutions

The ordinary differential equation

$$x^2 y'' + xy' + (k^2 x^2 - p^2)y = 0 \qquad (11.6.1)$$

has two real, linearly independent solutions, which are called "Bessel functions" and designated $J_p(kx)$ and $Y_p(kx)$. Because the equation is linear and homogeneous, its general solution is $AJ_p(kx) + BY_p(kx)$.

It's important to note that A and B are arbitrary constants (determined by initial conditions), while p and k are specific numbers that appear in the differential equation. p determines what functions solve the equation, and k stretches or compresses those functions. For instance:

- The ordinary differential equation $x^2 y'' + xy' + (x^2 - 9)y = 0$ has two real, linearly independent solutions called $J_3(x)$ and $Y_3(x)$. Because the equation is linear and homogeneous, its general solution is $AJ_3(x) + BY_3(x)$.
- The equation $x^2 y'' + xy' + (x^2 - 1/4)y = 0$ has general solution $AJ_{1/2}(x) + BY_{1/2}(x)$. The $J_{1/2}(x)$ that solves this equation is a completely different function from the $J_3(x)$ that solved the previous.
- The equation $x^2 y'' + xy' + (25x^2 - 9)y = 0$ has solutions $J_3(5x)$ and $Y_3(5x)$. Of course $J_3(5x)$ is the same function as $J_3(x)$ but compressed horizontally.

We are not concerned here with negative values of p, although we may encounter fractional values.

A Few Key Properties

The functions $J_p(x)$ are called "Bessel functions of the first kind," or sometimes just "Bessel functions." Here is a graph and a few of their key properties.

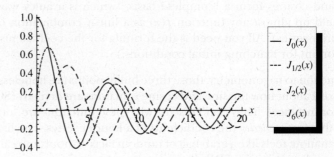

11.6 | Separation of Variables—Polar Coordinates and Bessel Functions

- $J_0(0) = 1$. For all non-zero p-values, $J_p(0) = 0$.
- There's no easy way to figure out the actual values of $J_p(x)$ by hand: you look them up in a table or use a mathematical software program, most of which have built-in routines for calculating Bessel functions. (The same statement can be made for sines and cosines. Do you know a better way to find sin 2?)
- All the $J_p(x)$ functions have an infinite number of zeros. We shall have more to say about these below.

The functions $Y_p(x)$ are called "Bessel functions of the second kind." Here is a graph and a few of their key properties.

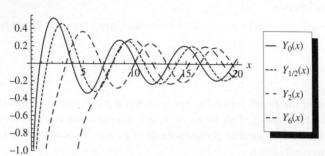

- For all p-values, $\lim_{x \to 0^+} Y_p(x) = -\infty$. Therefore $Y_p(x)$ solutions are discarded whenever boundary conditions require a finite answer at $x = 0$.
- All the $Y_p(x)$ functions have an infinite number of zeros.
- Some textbooks refer to these as "Neumann functions," and still others as "Weber functions." Some use the notation $N_p(x)$ instead of $Y_p(x)$. (This may not technically qualify as a "key property" but we had to mention it somewhere.)

More about Those Zeros (Brought to You by the Letter "alpha")

The zeros of the Bessel functions are difficult to calculate, but they are listed in Bessel function tables and can be generated by mathematical software.

Because there is no simple formula for the zeros of a Bessel function, and because the zeros play an important role in many formulas, they are given their own symbols. By convention, $\alpha_{p,n}$ represents the nth positive zero of the Bessel function J_p. For instance, the first four positive zeros of the function J_3 are roughly 6.4, 9.8, 13.0, and 16.2, so we write $\alpha_{3,1} = 6.4$, $\alpha_{3,2} = 9.8$, and so on.

Fourier-Bessel Series Expansions

We start with a quick reminder of a series expansion you should already be familiar with. Given a function $f(x)$ defined on a finite interval $[0, a]$ with $f(0) = f(a) = 0$, you can write the function as a Fourier sine series.

$$f(x) = b_1 \sin\left(\frac{\pi}{a}x\right) + b_2 \sin\left(\frac{2\pi}{a}x\right) + b_3 \sin\left(\frac{3\pi}{a}x\right) + \ldots$$

The terms in that series are all the function $\sin(kx)$ where the restriction $k = n\pi/a$ comes from the condition $f(a) = 0$ (since $\sin(n\pi) = 0$ for all integer n). The only other thing you need to know is the formula for the coefficients, $b_n = (2/a) \int_0^a f(x) \sin(n\pi x/a) dx$.

Moving to the less familiar, you can instead choose to represent the same function $f(x)$ as a series of (for instance) $J_7(kx)$ terms.

$$f(x) = A_1 J_7\left(\frac{\alpha_{7,1}}{a}x\right) + A_2 J_7\left(\frac{\alpha_{7,2}}{a}x\right) + A_3 J_7\left(\frac{\alpha_{7,3}}{a}x\right) + \ldots$$

Chapter 11 Partial Differential Equations

The terms in that series are all the function $J_7(kx)$ where the restriction $k = \alpha_{7,n}/a$ comes from the condition $f(a) = 0$ (since $J_7(\alpha_{7,n}) = 0$ for all n by definition). The only other thing you need is the formula for the coefficients.

Of course, there's nothing special about the number 7. You could build the same function as a series of $J_{5/3}(kx)$ functions if you wanted to. The allowable values of k would be different and the coefficients would be different; it would be a completely different series that adds up to the same function.

Fourier-Bessel Series

On the interval $0 \leq x \leq a$, we can expand a function into a series of Bessel functions by writing

$$f(x) = \sum_{n=1}^{\infty} A_n J_p\left(\frac{\alpha_{p,n}}{a} x\right)$$

- J_p is one of the Bessel functions. For instance, if $p = 3$, you are expanding $f(x)$ in terms of the Bessel function J_3. This creates a "Fourier-Bessel series expansion of order 3." We could expand the same function $f(x)$ into terms of J_5, which would give us a different series with different coefficients.
- $\alpha_{p,n}$ is defined in the section on "zeros" above.[7]
- The formula for the coefficients A_n is given in Appendix J (in the section on Bessel functions).

Do you see the importance of all this? Separation of variables previously led us to $d^2y/dx^2 + k^2y = 0$, which gave us the normal modes $\sin(kx)$ and $\cos(kx)$. We used the known zeros of those functions to match boundary conditions. And because we know the formula for the coefficients of a Fourier series, we were able to build arbitrary initial conditions as *series* of sines and cosines.

Now suppose that separation of variables on some new PDE gives us $J_5(kx)$ for our normal modes. We will need to match boundary conditions, which requires knowing the zeros: we now know them, or at least we have names for them ($\alpha_{5,1}$, $\alpha_{5,2}$ and so on) and can look them up. Then we will need to write the initial condition as a sum of normal modes: we now know the formula for the coefficients of a Fourier-Bessel expansion of order 5. Later in this section you'll see examples where we solve PDEs in this way.

Appendix J lists just the key facts given above—the ODE, the function that solves the ODE, the zeros of that function, and the coefficients of its series expansion—for Bessel functions and other functions that are normal modes of common PDEs. With those facts in hand, you can solve PDEs with a wide variety of normal modes. But, as we warned in the beginning, nothing in that appendix justifies those mathematical facts in any way. For the mathematical background and derivations, see Chapter 12.

Not-Quite-Bessel Functions

Many equations look a lot like Equation 11.6.1 but don't match it perfectly. There are two approaches you might try in such a case. The first is to let $u = x^2$ or $u = \sqrt{x}$ or $u = \cos x$ or some such and see if you end up with a perfect match. (Such substitutions are discussed in Chapter 10.) The second is to peak into Appendix J and see if you have a known variation of Bessel functions.

---

[7] Some textbooks define $\lambda_{p,n} = \alpha_{p,n}/a$ to make these equations look simpler.

11.6 | Separation of Variables—Polar Coordinates and Bessel Functions

As an example, consider the following.

$$x^2 y'' + xy' - (k^2 x^2 + p^2)y = 0 \tag{11.6.2}$$

This is just like Equation 11.6.1 except for the sign of the k^2 term. You can turn this into Equation 11.6.1 with a change of variables, but you have to introduce complex numbers to do so. The end result, however, is a pair of functions that give you real answers for any real input x. The two linearly independent solutions to this equation are called "modified Bessel function," generally written $I_p(kx)$ and $K_p(kx)$. The important facts you need to know about these functions when you're solving PDEs are that $K_p(x)$ diverges at $x = 0$ (just as $Y_p(x)$ does), and that none of the modified Bessel functions have any zeroes at any point except $x = 0$. Those facts are all listed in Appendix J.

11.6.2 Discovery Exercise: Separation of Variables—Polar Coordinates and Bessel Functions

A vertical cylinder of radius a and length L has no charge inside. The cylinder wall and bottom are grounded: that is, held at zero potential. The potential at the top is given by the function $f(\rho)$. Find the potential inside the cylinder.

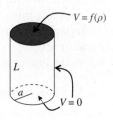

The potential in such a situation will follow Laplace's equation $\nabla^2 V = 0$. The system geometry suggests that we write this equation in cylindrical coordinates. The lack of any ϕ-dependence in the boundary conditions lends the problem azimuthal symmetry: that is, $\partial V/\partial \phi = 0$. We can therefore write Laplace's equation as:

$$\frac{\partial^2 V}{\partial \rho^2} + \frac{1}{\rho}\frac{\partial V}{\partial \rho} + \frac{\partial^2 V}{\partial z^2} = 0 \tag{11.6.3}$$

with the boundary conditions:

$$V(\rho, 0) = 0, \quad V(a, z) = 0, \quad V(\rho, L) = f(\rho)$$

1. To solve this, assume a solution of the form $V(\rho, z) = R(\rho)Z(z)$. Plug this guess into Equation 11.6.3.
2. Separate the variables in the resulting equation so that all the ρ-dependence is on the left and the z-dependence on the right.

See Check Yourself #75 in Appendix L

3. Let the separation constant equal zero.
   (a) Verify that the function $R(\rho) = A\ln(\rho) + B$ is the general solution for the resulting equation for $R(\rho)$. (You can verify this by simply plugging our solution into the ODE, of course. Alternatively, you can find the solution for yourself by letting $S(\rho) = R'(\rho)$ and solving the resulting first-order separable ODE for $S(\rho)$, and then integrating to find $R(\rho)$.)
   (b) We now note that within the cylinder, the potential V must always be finite. (This type of restriction is sometimes called an "implicit" boundary condition, and will be discussed further in the Explanation, Section 11.6.3.) What does this restriction say about our constants A and B?
   (c) Match your solution to the boundary condition $R(a) = 0$, and show that this reduces your solution to the trivial case $R(\rho) = 0$ everywhere. We conclude that, for any boundary condition at the top other than $f(\rho) = 0$, the separation constant is not zero.

4. Returning to your equation in Part 2, let the separation constant equal k^2.
   (a) Solve for $R(\rho)$ by matching with either Equation 11.6.1 or 11.6.2 for the proper choice of p.
   (b) Use the implicit boundary condition that $R(0)$ is finite to eliminate one of the two solutions.
   (c) Explain why the remaining solution cannot match the boundary condition $R(a) = 0$. You can do this by looking up the properties of the function in your solution or you can plot the remaining solution, choosing arbitrary positive values for k and the arbitrary constant, and explain from this plot why the solution cannot match $R(a) = 0$.
5. Returning to your equation in Part 2, let the separation constant equal $-k^2$.
   (a) Solve for $R(\rho)$ by matching with either Equation 11.6.1 or 11.6.2 for the proper choice of p.
   (b) Use the implicit boundary condition that $R(0)$ is finite to eliminate one of the two solutions.
   (c) Use the boundary condition $R(a) = 0$ to restrict the possible values of k.
6. Based on your results from Parts 3–5 you should have concluded that the separation constant must be negative to match the boundary conditions on $R(\rho)$. Calling the separation constant $-k^2$, solve the equation for $Z(z)$, using the boundary condition $Z(0) = 0$.
7. Write $V(\rho, z)$ as an infinite series.

See Check Yourself #76 in Appendix L

The process you have just gone through is the same as in the previous section, but this ODE led to Bessel functions instead of sines and cosines. Bessel functions, like sines and cosines, form a "complete basis"—you can use them in series to build almost any function by the right choice of coefficients. In the problems you will choose appropriate coefficients to match a variety of boundary conditions for the upper surface in this exercise. You will also see how Bessel functions come up in a variety of different physical situations.

11.6.3 Explanation: Separation of Variables—Polar Coordinates and Bessel Functions

We're going to solve the wave equation by using separation of variables again. Like our example in Section 11.5, we will have three independent variables instead of two: this will require separating variables twice, resulting in a double series. Nothing new so far.

However, because we begin this time in polar coordinates, the ordinary differential equation that results from separating variables will *not* be $d^2y/dx^2 + k^2y = 0$. A different equation will lead us to a different solution: Bessel functions, rather than sines and cosines, will be our normal modes.

Despite this important difference, it's important to note how much this problem has in common with the previous problems we have worked through. Matching boundary conditions, summing up the remaining solutions, and then matching initial conditions all follow the same basic pattern. And once again we can sum our normal modes to represent any initial state of the system. Once you are comfortable with this general outline, you can solve a surprisingly wide variety of partial differential equations, even though the details—especially the forms of the normal modes—may vary from one problem to the next.

11.6 | Separation of Variables—Polar Coordinates and Bessel Functions

The Problem
Consider a simple drum: a thin rubber sheet stretched along a circular frame of radius a. If the sheet is struck with some object—let's call it a "drumstick"—it will vibrate according to the wave equation.

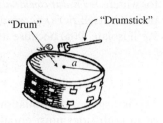

The wave equation can be written as $\nabla^2 z = (1/v^2)(\partial^2 z/\partial t^2)$, where z is vertical displacement. In rectangular coordinates this equation becomes $\partial^2 z/\partial x^2 + \partial^2 z/\partial y^2 = (1/v^2)(\partial^2 z/\partial t^2)$, but using x and y to specify boundary conditions on a circle is an ugly business. Instead we will use the Laplacian in polar coordinates (Appendix F), which makes our wave equation:

$$\frac{\partial^2 z}{\partial t^2} = v^2 \left(\frac{\partial^2 z}{\partial \rho^2} + \frac{1}{\rho}\frac{\partial z}{\partial \rho} + \frac{1}{\rho^2}\frac{\partial^2 z}{\partial \phi^2} \right) \tag{11.6.4}$$

You can probably figure out for yourself that the first boundary condition, the one that led us to polar coordinates, is $z(a, \phi, t) = 0$. This is an "explicit" boundary condition, meaning it comes from a restriction that is explicitly stated in the problem. (The drumhead is attached to the frame.) There are also two "implicit" boundary conditions, restrictions that are imposed by the geometry of the situation.

- In many cases, certainly including this one, z has to be finite throughout the region. This includes the center of the circle, so z cannot blow up at $\rho = 0$. (This is going to be more important than it may sound.)
- The point $\rho = 2, \phi = \pi/3$ is the *same physical place* as $\rho = 2, \phi = 7\pi/3$, so it must give rise to the same z-value. More generally, the function must be periodic such that $z(\rho, \phi, t)$ always equals $z(\rho, \phi + 2\pi, t)$.

These conditions apply because we are using polar coordinates to solve a problem on a disk around the origin, not because of the details of this particular scenario. If you find that your initial and boundary conditions aren't sufficient to determine your arbitrary constants, you should always check whether you missed some implicit boundary conditions.

The problem also needs to specify as initial conditions both $z(\rho, \phi, 0)$ and $\partial z/\partial t(\rho, \phi, 0)$. After we solve the problem with its boundary conditions, we'll look at a couple of possible initial conditions.

Separation of Variables
We begin by guessing a solution of the form

$$z(\rho, \phi, t) = R(\rho)\Phi(\phi)T(t)$$

Substituting this expression directly into the polar wave equation (11.6.4) yields:

$$R(\rho)\Phi(\phi)T''(t) = v^2 \left(R''(\rho)\Phi(\phi)T(t) + \frac{1}{\rho}R'(\rho)\Phi(\phi)T(t) + \frac{1}{\rho^2}R(\rho)\Phi''(\phi)T(t) \right)$$

As always, we need to start by isolating a variable with homogeneous boundary conditions. ρ has the obviously homogeneous condition $R(a) = 0$. But the boundary condition for ϕ, periodicity, is also homogeneous. (You may think of "homogeneous" as a fancy way of saying "something is zero," but that isn't quite true: a homogeneous equation or condition is one for which *any linear combination of solutions is itself a solution*. A sum of functions with the same period is itself periodic, so $z(\rho, \phi, t) = z(\rho, \phi + 2\pi, t)$ is a homogeneous boundary condition.) We'll isolate $\Phi(\phi)$ because its differential equation is simpler.

$$\frac{\rho^2}{v^2} \frac{T''(t)}{T(t)} - \rho^2 \frac{R''(\rho)}{R(\rho)} - \rho \frac{R'(\rho)}{R(\rho)} = \frac{\Phi''(\phi)}{\Phi(\phi)}$$

The left side of the equation depends only on ρ and t, and the right side depends only on ϕ, so both sides must equal a constant. A positive constant would lead to an exponential $\Phi(\phi)$ function, which couldn't be periodic, so we call the constant $-p^2$ which must be zero or negative. (Our choice of the letter p is motivated by a sneaky foreknowledge of where this particular constant is going to show up in Bessel's equation, but of course any other letter would serve just as well.)

$$\frac{\rho^2}{v^2} \frac{T''(t)}{T(t)} - \rho^2 \frac{R''(\rho)}{R(\rho)} - \rho \frac{R'(\rho)}{R(\rho)} = -p^2$$

$$\frac{\Phi''(\phi)}{\Phi(\phi)} = -p^2 \tag{11.6.5}$$

The ϕ equation is now fully separated, and we will return to it later, but first we want to finish separating variables. This turns the remaining equation into:

$$\frac{1}{v^2} \frac{T''(t)}{T(t)} = \frac{R''(\rho)}{R(\rho)} + \frac{1}{\rho} \frac{R'(\rho)}{R(\rho)} - \frac{p^2}{\rho^2}$$

Both sides must equal a constant, and this new constant is not related to p. We logically expect $T(t)$ to oscillate, rather than growing exponentially or linearly—this is a vibrating drumhead, after all!—so we will choose a negative separation constant again and call it $-k^2$. (If we missed this physical argument we would find that a positive constant would be unable to match the boundary conditions.)

$$\frac{1}{v^2} \frac{T''(t)}{T(t)} = -k^2 \tag{11.6.6}$$

$$\frac{R''(\rho)}{R(\rho)} + \frac{1}{\rho} \frac{R'(\rho)}{R(\rho)} - \frac{p^2}{\rho^2} = -k^2 \tag{11.6.7}$$

One partial differential equation has been replaced by three ordinary differential equations: two familiar, and one possibly less so.

The Variables with Homogeneous Boundary Conditions

The solution to Equation 11.6.5 is:

$$\Phi(\phi) = A \sin(p\phi) + B \cos(p\phi) \tag{11.6.8}$$

The condition that $\Phi(\phi)$ must repeat itself every 2π requires that p be an integer. Negative values of p would be redundant, so we need only consider $p = 0, 1, 2, \ldots$.

Moving on to $R(\rho)$, Equation 11.6.7 can be written as:

$$\rho^2 R''(\rho) + \rho R'(\rho) + \left(k^2 \rho^2 - p^2\right) R(\rho) = 0 \tag{11.6.9}$$

You can plug that equation into a computer or look it up in Appendix J. (The latter is a wonderful resource if we do say so ourselves, and we hope you will familiarize yourself with it.) Instead we will note that it matches Equation 11.6.1 in Section 11.6.1 and write down the answer we gave there.

$$R(\rho) = A J_p(k\rho) + B Y_p(k\rho) \tag{11.6.10}$$

Next we apply the boundary conditions on $R(\rho)$. We begin by recalling that all functions $Y_p(x)$ blow up as $x \to 0$. Because our function must be finite at the center of the drum we discard such solutions, leaving only the J_p functions.

Our other boundary condition on ρ was that $R(a) = 0$. Using $\alpha_{p,n}$ to represent the nth zero of J_p we see that this condition is satisfied if $ka = \alpha_{p,n}$.

$$R(\rho) = AJ_p\left(\frac{\alpha_{p,n}}{a}\rho\right)$$

Remember that $J_1(x)$ is one specific function—you could look it up, or graph it with a computer. $J_2(x)$ is a different function, and so on. Each of these functions has an infinite number of zeros. When we write $\alpha_{3,7}$ we mean "the seventh zero of the function $J_3(x)$." The index n, which identifies one of these zeros by count, can be any positive integer.

The Last Equation, and a Series

The equation $(1/v^2)(T''(t)/T(t)) = -k^2$ becomes $T''(t) = -k^2 v^2 T(t)$ which we can solve by inspection:

$$T(t) = C\sin(kvt) + D\cos(kvt)$$

When we introduced p and k they could each have been, so far as we knew, any real number. The boundary conditions have left us with possibilities that are still infinite, but *discrete*: p must be an integer, and ka must be a zero of J_p. So we can write the solution as a series over all possible n- and p-values. As we do so we add the subscript pn to our arbitrary constants, indicating that they may be different for each term. We also absorb the constant A into our other arbitrary constants.

$$z = \sum_{p=0}^{\infty}\sum_{n=1}^{\infty} J_p\left(\frac{\alpha_{p,n}}{a}\rho\right)\left[C_{pn}\sin\left(\frac{\alpha_{p,n}}{a}vt\right) + D_{pn}\cos\left(\frac{\alpha_{p,n}}{a}vt\right)\right]\left[E_{pn}\sin(p\phi) + F_{pn}\cos(p\phi)\right]$$
(11.6.11)

If you aren't paying extremely close attention, those subscripts are going to throw you. $\alpha_{p,n}$ is the nth zero of the function J_p: you can look up $\alpha_{3,5}$ in a table of Bessel function zeros. (It's roughly 19.4.) On the other hand, C_{pn} is one of our arbitrary constants, which we will choose to meet the initial conditions of our specific problem. C_{35} is the constant that we will be multiplying specifically by $\sin\left(\alpha_{3,5}vt/a\right)$.

Can you see why p starts at 0 and n starts at 1? Plug $p = 0$ into Equation 11.6.8 and you get a valid solution: $\Phi(\phi)$ equals a constant. But the first zero of a Bessel function is conventionally labeled $n = 1$, so only positive values of n make sense.

Initial Conditions

We haven't said much about specific initial conditions up to this point, because our goal is to show how you can match *any* reasonable initial conditions in such a situation.

The initial conditions would be some given functions $z_0(\rho, \phi)$ and $\dot{z}_0(\rho, \phi)$. From the solution above:

$$z_0 = \sum_{p=0}^{\infty}\sum_{n=1}^{\infty} D_{pn} J_p\left(\frac{\alpha_{p,n}}{a}\rho\right)\left[E_{pn}\sin(p\phi) + F_{pn}\cos(p\phi)\right]$$

$$\dot{z}_0 = \sum_{p=0}^{\infty}\sum_{n=1}^{\infty} \left(\frac{\alpha_{p,n}}{a}v\right) C_{pn} J_p\left(\frac{\alpha_{p,n}}{a}\rho\right)\left[E_{pn}\sin(p\phi) + F_{pn}\cos(p\phi)\right]$$

We can consider the possible initial conditions in four categories.

1. If our initial conditions specify that z_0 and $\dot{z}_0$ are both zero, then we are left with the "trivial" solution $z = 0$ and the drumhead never moves.

2. If $z_0 = 0$ and $\dot{z}_0 \neq 0$, then $D_{pn} = 0$. Since D_{pn} was the coefficient of $\cos\left(\alpha_{p,n}vt/a\right)$ in our solution, we are left with only the sinusoidal time oscillation, $C_{pn}\sin\left(\alpha_{p,n}vt/a\right)$.
3. If $z_0 \neq 0$ and $\dot{z}_0 = 0$, then $C_{pn} = 0$. This has the opposite effect on our solution, leaving only the cosine term for the time oscillation.
4. The most complicated case, $z_0 \neq 0$ and $\dot{z}_0 \neq 0$, requires a special trick. We find one solution where $z_0 = 0$ and $\dot{z}_0 \neq 0$ (case two above), and another solution where $z_0 \neq 0$ and $\dot{z}_0 = 0$ (case three above). When we *add* the two solutions, the result still solves our differential equation (because it was homogeneous), and also has the correct z_0 and the correct $\dot{z}_0$.

In *all* cases we rely on the "completeness" of our normal modes—which is a fancy way of saying, we rely on the ability of trig functions and Bessel functions to combine in series to create any desired initial function that we wish to match.

Below we solve two different problems of the second type, where $z_0 = 0$ and $\dot{z}_0 \neq 0$. In the problems you'll work examples of the third type. You'll also try your hand at a few problems of the fourth type, but we will take those up more generally in Section 11.8.

First Sample Initial Condition: A Symmetric Blow

Solve for the motion of a circular drumhead that is struck in the center with a circular mallet, so that

$$z_0 = 0, \quad \dot{z}_0 = \begin{cases} s & \rho < \rho_0 \\ 0 & \rho_0 < \rho < a \end{cases}$$

As noted above, we can immediately say that because z_0 is zero, $D_{pn} = 0$.

Next we note that both the problem and the initial conditions have "azimuthal symmetry": the solution will have no ϕ dependence. How can a function that is multiplied by $E_{pn}\sin(p\phi) + F_{pn}\cos(p\phi)$ possibly wind up independent of ϕ? The answer is, if $E_{pn} = 0$ for all values of p and n, and $F_{pn} = 0$ for all values except $p = 0$. Then $\sum_{p=0}^{\infty} E_{pn}\sin(p\phi) + F_{pn}\cos(p\phi)$ simply becomes the constant F_{0n}.

We can now absorb that constant term into the constant C_{0n}. Since the only value of p left in our sum is $p = 0$ we replace p with zero everywhere in the equation and drop it from our subscripts:

$$\dot{z}_0 = \sum_{n=1}^{\infty} \left(\frac{\alpha_{0,n}}{a}v\right) C_n J_0\left(\frac{\alpha_{0,n}}{a}\rho\right)$$

As always, we now build up our initial condition from the series by choosing the appropriate coefficients. If our normal modes were sines and cosines, we would be building a Fourier series. In this case we are building a Fourier-Bessel series, but the principle is the same, and the formula in Appendix J allows us to simply plug in and find the answer:

$$\left(\frac{\alpha_{0,n}}{a}v\right) C_n = \frac{2}{a^2 J_1^2(\alpha_{0,n})} \int_0^a \dot{z}_0(\rho) J_0\left(\frac{\alpha_{0,n}}{a}\rho\right) \rho\, d\rho = \frac{2s}{a^2 J_1^2(\alpha_{0,n})} \int_0^{\rho_0} J_0\left(\frac{\alpha_{0,n}}{a}\rho\right) \rho\, d\rho$$

$$C_n = \frac{2s}{av\alpha_{0,n} J_1^2(\alpha_{0,n})} \int_0^{\rho_0} J_0\left(\frac{\alpha_{0,n}}{a}\rho\right) \rho\, d\rho$$

This integral can be looked up in a table or plugged into a computer, with the result

$$C_n = \frac{2s}{av\alpha_{0,n} J_1^2(\alpha_{0,n})} \frac{\rho_0 a J_1\left(\frac{\alpha_{0,n}}{a}\rho_0\right)}{\alpha_{0,n}} = \frac{2s\rho_0}{v\alpha_{0,n}^2} \frac{J_1\left(\frac{\alpha_{0,n}}{a}\rho_0\right)}{J_1^2(\alpha_{0,n})}$$

11.6 | Separation of Variables—Polar Coordinates and Bessel Functions

Plugging this into the series expansion for z gives the final answer we've been looking for:

$$z = \frac{2s\rho_0}{v} \sum_{n=1}^{\infty} \frac{J_1\left(\frac{\alpha_{0,n}}{a}\rho_0\right)}{\alpha_{0,n}^2 J_1^2(\alpha_{0,n})} J_0\left(\frac{\alpha_{0,n}}{a}\rho\right) \sin\left(\frac{\alpha_{0,n}}{a}vt\right) \quad (11.6.12)$$

That hideous-looking answer actually tells us a great deal about the behavior of the drumhead. The constants outside the sum set the overall scale of the vibrations. The big fraction inside the sum is a number for each value of n that tells us the relative importance of each normal mode. For $\rho_0 = a/3$ the first few coefficients are roughly 0.23, 0.16, and 0.069, and they only get smaller from there. So only the first two modes contribute significantly. The first and most important term represents the shape on the left of Figure 11.5 oscillating up and down. The second term is a slightly more complicated shape. Higher order terms will have more wiggles (i.e. more critical points between $\rho = 0$ and $\rho = a$).

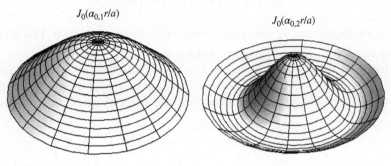

Figure 11.5

EXAMPLE **Partial Differential Equation with Bessel Function Normal Modes**

Problem:
Solve the equation:

$$4x\frac{\partial^2 z}{\partial x^2} + 4\frac{\partial z}{\partial x} + \frac{\partial^2 z}{\partial y^2} - \frac{9}{x}z = 0$$

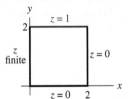

on the domain $0 \leq x \leq 2$, $0 \leq y \leq 2$ subject
to the boundary conditions $z(2, y) = z(x, 0) = 0$ and
$z(x, 2) = 1$ and the requirement that $z(0, y)$ is finite.

Solution:
Writing $z(x, y) = X(x)Y(y)$ leads us to:

$$4xX''(x)Y(y) + 4X'(x)Y(y) + X(x)Y''(y) - \frac{9}{x}X(x)Y(y) = 0$$

Divide both sides by $X(x)Y(y)$ and separate the variables:

$$4x\frac{X''(x)}{X(x)} + 4\frac{X'(x)}{X(x)} - \frac{9}{x} = -\frac{Y''(y)}{Y(y)}$$

In the problems you will show that we cannot match the boundary conditions with a positive or zero separation constant, so we call the constant $-k^2$ and we get:

$$4xX''(x) + 4X'(x) + \left(k^2 - \frac{9}{x}\right)X(x) = 0 \tag{11.6.13}$$

$$Y''(y) = k^2 Y(y) \tag{11.6.14}$$

We begin with $X(x)$ because its boundary conditions are homogenous.
Equation 11.6.13 is unfamiliar enough that you may just want to pop it into a computer. (If you want to solve it by hand, look for the variable substitution that turns it into Equation 11.6.1.)

$$X(x) = A J_3\left(k\sqrt{x}\right) + B Y_3\left(k\sqrt{x}\right)$$

As before, we discard the Y_p solutions because they blow up at $x = 0$. The second boundary condition, $X(2) = 0$, means that $k = \alpha_{3,n}/\sqrt{2}$.
Turning to the other equation we get $Y(y) = Ce^{ky} + De^{-ky}$. The boundary condition $Y(0) = 0$ gives $C = -D$ so $Y(y) = C\left(e^{ky} - e^{-ky}\right)$ (or equivalently $Y(y) = E \sinh(ky)$). Putting the solutions together, and absorbing one arbitrary constant into another, we get

$$z(x, y) = \sum_{n=1}^{\infty} A_n \left(e^{\alpha_{3,n}y/\sqrt{2}} - e^{-\alpha_{3,n}y/\sqrt{2}}\right) J_3\left(\frac{\alpha_{3,n}}{\sqrt{2}}\sqrt{x}\right)$$

Finally, the inhomogeneous boundary condition $z(x, 2) = 1$ gives

$$1 = \sum_{n=1}^{\infty} A_n \left(e^{\alpha_{3,n}\sqrt{2}} - e^{-\alpha_{3,n}\sqrt{2}}\right) J_3\left(\frac{\alpha_{3,n}}{\sqrt{2}}\sqrt{x}\right)$$

This is a Fourier-Bessel series, but the argument of the Bessel function is $\sqrt{x}$ instead of x. So we define $u = \sqrt{x}$ and write this.

$$1 = \sum_{n=1}^{\infty} A_n \left(e^{\alpha_{3,n}\sqrt{2}} - e^{-\alpha_{3,n}\sqrt{2}}\right) J_3\left(\frac{\alpha_{3,n}}{\sqrt{2}}u\right)$$

Then we look up the equation for Fourier-Bessel coefficients in Appendix J.

$$A_n \left(e^{\alpha_{3,n}\sqrt{2}} - e^{-\alpha_{3,n}\sqrt{2}}\right) = \frac{1}{J_4^2(\alpha_{3,n})} \int_0^{\sqrt{2}} J_3\left(\frac{\alpha_{3,n}}{\sqrt{2}}u\right) u\, du$$

The analytic solution to that integral is no more enlightening to look at than the integral itself, so we simply leave it as is:

$$z(x, y) = \sum_{n=1}^{\infty} \frac{1}{J_4^2(\alpha_{3,n})} \left(\int_0^{\sqrt{2}} J_3\left(\frac{\alpha_{3,n}}{\sqrt{2}}u\right) u\, du\right) \frac{e^{\alpha_{3,n}y/\sqrt{2}} - e^{-\alpha_{3,n}y/\sqrt{2}}}{e^{\alpha_{3,n}\sqrt{2}} - e^{-\alpha_{3,n}\sqrt{2}}} J_3\left(\frac{\alpha_{3,n}}{\sqrt{2}}\sqrt{x}\right)$$

You can plug that whole mess into a computer, as is, and plot the 20th partial sum of the series. It matches the homogeneous boundary conditions perfectly, matches the inhomogeneous one well, and smoothly interpolates between them everywhere else.

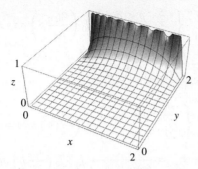

Second Sample Initial Condition: An Asymmetric Blow

You might want to skip this part, work some problems with PDEs that lead to Bessel functions, and then come back here. This will show you how to handle multivariate series with different normal modes: in this example, trig functions of one variable and Bessel functions of another.

Solve for the motion of a circular drumhead that is struck by a mallet at an angle, so it only hits on one side.

$$z_0 = 0, \quad \dot{z}_0 = \begin{cases} s & \rho < \rho_0, 0 < \phi < \pi \\ 0 & \text{otherwise} \end{cases}$$

Once again the condition $z_0 = 0$ leads to $D_{pn} = 0$. This time, however, there is ϕ dependence, so we can't throw out as much of our equation.

$$\dot{z}_0(\rho, \phi) = \sum_{p=0}^{\infty} \sum_{n=1}^{\infty} \left(\frac{\alpha_{p,n}}{a} v\right) J_p\left(\frac{\alpha_{p,n}}{a}\rho\right) \left[E_{pn} \sin(p\phi) + F_{pn} \cos(p\phi)\right]$$

(We've absorbed C_{pn} into the other constants of integration.) In Chapter 9 we found double series expansions based entirely on sines and cosines. The principle is the same here, but we are now expanding over both trig and Bessel functions. First note that you can rewrite this as:

$$\dot{z}_0(\rho, \phi) = \sum_{n=1}^{\infty} \left(\frac{\alpha_{0,n}}{a} v\right) J_0\left(\frac{\alpha_{0,n}}{a}\rho\right) F_{0n} + \sum_{p=1}^{\infty} \left\{ \sum_{n=1}^{\infty} \left(\frac{\alpha_{p,n}}{a} v\right) J_p\left(\frac{\alpha_{p,n}}{a}\rho\right) F_{pn} \right\} \cos(p\phi)$$
$$+ \sum_{p=1}^{\infty} \left\{ \sum_{n=1}^{\infty} \left(\frac{\alpha_{p,n}}{a} v\right) J_p\left(\frac{\alpha_{p,n}}{a}\rho\right) E_{pn} \right\} \sin(p\phi)$$

This is now in the form of a Fourier series, where:

$$\sum_{n=1}^{\infty} \left(\frac{\alpha_{0,n}}{a} v\right) J_0\left(\frac{\alpha_{0,n}}{a}\rho\right) F_{0n} = a_0 = \frac{1}{2\pi} \int_0^{2\pi} \dot{z}_0(\rho, \phi) d\phi = \begin{cases} \frac{s}{2} & \rho < \rho_0 \\ 0 & \rho > \rho_0 \end{cases}$$

$$\sum_{n=1}^{\infty} \left(\frac{\alpha_{p,n}}{a} v\right) J_p\left(\frac{\alpha_{p,n}}{a}\rho\right) F_{pn} = a_p = \frac{1}{\pi} \int_0^{2\pi} \dot{z}_0(\rho, \phi) \cos(p\phi) d\phi = 0$$

$$\sum_{n=1}^{\infty} \left(\frac{\alpha_{p,n}}{a} v\right) J_p\left(\frac{\alpha_{p,n}}{a}\rho\right) E_{pn} = b_p = \frac{1}{\pi} \int_0^{2\pi} \dot{z}_0(\rho, \phi) \sin(p\phi) d\phi = \begin{cases} \frac{2s}{p} & \rho < \rho_0, p \text{ odd} \\ 0 & \text{otherwise} \end{cases}$$

Chapter 11 Partial Differential Equations

Now we can turn this around, and view the left side of each of these equations as a Fourier-Bessel series expansion of the function on the right side. We see that $F_{pn} = 0$ for all $p > 0$ and $E_{pn} = 0$ for even values of p. To find the other coefficients we return to Appendix J for the coefficients of a Fourier-Bessel series expansion.

$$\left(\frac{\alpha_{0,n}}{a}v\right) F_{0n} = \frac{2}{a^2 J_1^2(\alpha_{0,n})} \int_0^a \left\{ \begin{array}{ll} \frac{s}{2} & \rho < \rho_0 \\ 0 & \rho > \rho_0 \end{array} \right\} J_0\left(\frac{\alpha_{0,n}}{a}\rho\right) \rho \, d\rho = \frac{\rho_0 s}{a\alpha_{0,n} J_1^2(\alpha_{0,n})} J_1\left(\frac{\alpha_{0,n}}{a}\rho_0\right)$$

$$\rightarrow F_{0n} = \frac{\rho_0 s}{v\alpha_{0,n}^2 J_1^2(\alpha_{0,n})} J_1\left(\frac{\alpha_{0,n}}{a}\rho_0\right)$$

$$\left(\frac{\alpha_{p,n}}{a}v\right) E_{pn} = \frac{2}{a^2 J_{p+1}^2(\alpha_{p,n})} \int_0^a \left\{ \begin{array}{ll} \frac{2s}{p} & \rho < \rho_0 \\ 0 & \rho > \rho_0 \end{array} \right\} J_p\left(\frac{\alpha_{p,n}}{a}\rho\right) \rho \, d\rho = \frac{4s}{pa^2 J_1^2(\alpha_{p,n})} \int_0^{\rho_0} J_p\left(\frac{\alpha_{p,n}}{a}\rho\right) \rho \, d\rho$$

$$\rightarrow E_{pn} = \frac{4s}{pav\alpha_{p,n} J_1^2(\alpha_{p,n})} \int_0^{\rho_0} J_p\left(\frac{\alpha_{p,n}}{a}\rho\right) \rho \, d\rho \quad \text{odd } p \text{ only}$$

The integral in the last equation can be evaluated for a general p, but the result is an ugly expression involving hypergeometric functions, so it's better to leave it as is and evaluate it for specific values of p when you are calculating partial sums. Putting all of this together, the solution is:

$$z = \sum_{n=1}^{\infty} \left\{ \frac{\rho_0 s}{v\alpha_{0,n}^2 J_1^2(\alpha_{0,n})} J_1\left(\frac{\alpha_{0,n}}{a}\rho\right) J_0\left(\frac{\alpha_{0,n}}{a}\rho\right) \sin\left(\frac{\alpha_{0,n}}{a}vt\right) \right. \tag{11.6.15}$$

$$\left. + \sum_{p=1}^{\infty} \left[\frac{4s}{pav\alpha_{p,n} J_1^2(\alpha_{p,n})} \left(\int_0^{\rho_0} J_p\left(\frac{\alpha_{p,n}}{a}\rho\right) \rho \, d\rho \right) J_p\left(\frac{\alpha_{p,n}}{a}\rho\right) \sin\left(\frac{\alpha_{p,n}}{a}vt\right) \sin(p\phi) \right] \quad \text{odd } p \text{ only} \right\}$$

Even if you're still reading at this point, you're probably skimming past the equations thinking "I'll take their word for it." But we'd like to draw your attention back to that last equation, because it's not as bad as it looks at first blush. The first term in curly braces is just the solution we found in the previous example. The dominant mode is shown on the left in Figure 11.5, the next mode is on the right in that figure, and higher order terms with more wiggles have much smaller amplitude. The new feature in this solution is the sum over p. Each of these terms is again a Bessel function that oscillates in time, but now multiplied by $\sin(p\phi)$, so it oscillates as you move around the disk. One such mode is shown in Figure 11.6.

$J_4(\alpha_{4,2}r/a) \sin(4\phi)$

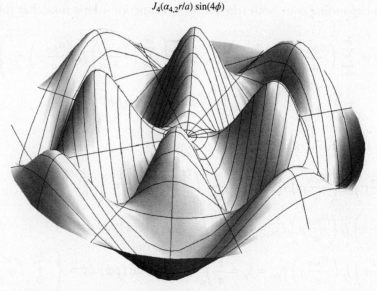

Figure 11.6

11.6 | Separation of Variables—Polar Coordinates and Bessel Functions

The higher the values of n and p the more wiggles a mode will have, but the smaller its amplitude will be. It's precisely because it can support so many different oscillatory modes, each with its own frequency, that a drum can make such a rich sound. At the same time, the frequency of the dominant mode determines the pitch you associate with a particular drum.

Stepping Back

When you read through a long, ugly derivation—and in this book, they don't get much longer and uglier than the "circular drum" above—it's easy to get lost in the details and miss the big picture. Here is everything we went through.

1. We were given the differential equation (the wave equation), the boundary conditions (the tacked-down edge of the drum), and a couple of sample initial conditions (first the symmetric blow, and then the asymmetric). In addition to the explicit boundary conditions, we recognized that the geometry of the problem led to some *implicit* boundary conditions. Without them, we would have found in step 3 that we didn't have enough conditions to determine our arbitrary constants.
2. Recognizing the circular nature of the boundary conditions, we wrote the wave equation in polar coordinates.
3. We separated variables, writing the solution as $z(\rho, \phi, t) = R(\rho)\Phi(\phi)T(t)$, but otherwise proceeding exactly as before: plug in, separate (twice, because there are three variables), solve the ODEs, apply homogeneous boundary conditions, build a series. Not all of this is easy, but it should mostly be familiar by now; the only new part was that one of the ODEs led us to Bessel functions.
4. To match initial conditions, we used a Fourier-Bessel series expansion. This step takes advantage of the fact that Bessel functions, like sines and cosines, form a "complete basis": you can sum them to build up almost any function.
5. Finally—possibly the most intimidating step, but also possibly the most important!—we *interpreted* our answer. We looked at the coefficients to see how quickly they were dropping, so we could focus on only the first few terms of the series. We used computers to draw the normal modes: in the problems you will go farther with computers, and actually view the motion. (Remember that it is almost never possible to explicitly sum an infinite series, but—assuming it converges—you can always approximate it with partial sums.)

It's important to get used to working with a wide variety of normal modes; trig and Bessel functions are just the beginning, but the process always remains the same. It's also important to get used to big, ugly equations that you can't solve or visualize without the aid of a computer. Mathematical software is as indispensable to the modern scientist or engineer as slide rules were to a previous generation.

11.6.4 Problems: Separation of Variables—Polar Coordinates and Bessel Functions

The differential equations in Problems 11.119–11.121 can be converted to the form of Equation 11.6.1 by an appropriate change of variables. For each one do the variable substitution to get the right equation, solve, and then substitute back to get the answer in terms of x. In the first problem the correct variable substitution is given for you; in the others you may have to do some trial and error.

11.119 $x^2(d^2y/dx^2) + x(dy/dx) + (4x^4 - 16)y = 0$. Use the substitution $u = x^2$.

11.120 $x^2(d^2y/dx^2) + x(dy/dx) + (x^6 - 9)y = 0$.

11.121 $d^2y/dx^2 + e^{2x}y = 0$. (*Hint*: to properly match Bessel's Equation, you need to end up with u^2 in front of the y term.)

11.122 Plot the first five Bessel functions $J_0(x) - J_4(x)$ from $x = 0$ to $x = 10$.

11.123 The Bessel functions are defined as

$$J_p(x) = \sum_{m=0}^{\infty} \frac{(-1)^m}{m!\,\Gamma(p+m+1)} \left(\frac{x}{2}\right)^{2m+p}$$

Plot $J_1(x)$ and the first 10 partial sums of this series expansion from $x = 0$ to $x = 10$ and show that the partial sums are converging to the Bessel function. (You may need to look up the gamma function $\Gamma(x)$ in the online help for your mathematical software to see how to enter it.)

11.124 Repeat Problem 11.123 for $J_{1/2}(x)$.

For Problems 11.125–11.127 expand the given function in a Fourier-Bessel series expansion of the given order. You should use a computer to evaluate the integrals, either analytically or numerically. Then plot the first ten partial sums of the expansion on the same plot as the original function so you can see the partial sums converging to the function. Be sure to clearly mark the plot of the function in some way (a different color, a thicker line) so it's clear on the plot which one it is.

11.125 $f(x) = 1, 0 < x < 1, p = 1$

11.126 $f(x) = x, 0 < x < \pi, p = 0$

11.127 $f(x) = \sin x, 0 < x < \pi, p = 1/2$

Problems 11.128–11.131 refer to a vibrating drumhead (a circle of radius a) with a fixed edge: $z(a, \phi, t) = 0$. In the Explanation (Section 11.6.3) we showed that the solution to the wave equation for this system is Equation 11.6.11. The arbitrary constants in that general solution are determined by the initial conditions. For each set of initial conditions write the complete solution $z(\rho, \phi, t)$. In some cases your answers will contain integrals that cannot be evaluated analytically.

11.128 $z(\rho, \phi, 0) = a - \rho$, $\dot{z}(\rho, \phi, 0) = 0$. This represents pulling the drumhead up by a string attached to its center and then letting go.

11.129 $z(\rho, \phi, 0) = 0$, $\dot{z}(\rho, \phi, 0) = c \sin(3\pi\rho/a)$

11.130 $z(\rho, \phi, 0) = 0$,
$$\dot{z}(\rho, \phi, 0) = \begin{cases} c & \rho < a/2 \\ 0 & a/2 < \rho < a \end{cases}$$

11.131 $z(\rho, \phi, 0) = (a - \rho)\sin(\phi)$, $\dot{z}(\rho, \phi, 0) = 0$

11.132 In the Explanation (Section 11.6.3), we found the Solution 11.6.12 for an oscillating drumhead that was given a sudden blow in a region in the middle. This might represent a drumhead hit by a drumstick. The solution we got, however, was a fairly complicated looking sum. In this problem you will make plots of this solution; you will need to choose some values for all of the constants in the problem.

(a) Have a computer plot the initial function $\dot{z}(r, 0)$ for the first 20 partial sums of this series. As you add terms you should see the partial sums converging towards the shape of the initial conditions that you were solving for.

(b) Take the first three terms in the series (the individual terms, not the partial sums) and for each one use a computer to make an animation of the shape of the drumhead (z, not $\dot{z}$) evolving over time. You should see periodic behavior. Describe how the behavior of these three normal modes is different from each other.

(c) Now make an animation (or just a sequence of plots at different times) of the 20th partial sum of the solutions. Describe the behavior. How is it similar to or different from the behavior of the individual terms you saw in the previous part? If you've done Problem 11.115 from Section 11.5 compare these results to the animation you got for the "square drumhead." Can you see any qualitative differences?

11.133 Equation 11.6.15 gives the solution for the vibrations of a drumhead struck with an asymmetric blow. Make an animation of this solution showing that motion using the partial sum that goes up to $n = 10, p = 11$.

11.134 In the Explanation (Section 11.6.3), we found the Solution 11.6.12 for an oscillating drumhead that was given a sudden blow in a region in the middle and we discussed how the different modes oscillating at different frequencies correspond to pitches produced by the drum. To find the frequency of a mode you have to look at the time dependence and recall that frequency is one over period.

(a) For a rubber drum with a sound speed of $v = 100$m/s and a radius of $a = 0.3$m find the frequency of the dominant mode. Look up what note this corresponds to. Then find the notes corresponding to the next two modes. (The modes other than the dominant one are called "overtones.")

(b) Find the dominant pitch for a rubber drum with sound speed $v = 100$m/s and a radius of 1.0m.

(c) By stretching the drum more tightly you can increase the sound speed. Find the dominant pitch for a more tightly stretched rubber drum with $v = 200$m/s and a radius of $a = 0.3$m.

(d) Explain in your own words why these different drums sound so different. How would you design a drum if you wanted it to make a low, booming sound?

11.135 Walk-Through: Differential Equation with Bessel Normal Modes. In this problem you will solve the partial differential equation $\partial y/\partial t - x^{3/2}(\partial^2 y/\partial x^2) - x^{1/2}(\partial y/\partial x) = 0$ subject to the boundary condition $y(1, t) = 0$ and the requirement that $y(0, t)$ be finite.

(a) Begin by guessing a separable solution $y = X(x)T(t)$. Plug this guess into the differential equation. Then divide both sides by $X(x)T(t)$ and separate variables.

(b) Find the general solution to the resulting ODE for $X(x)$ three times: with a positive separation constant k^2, a negative separation constant $-k^2$, and a zero separation constant. For each case you can solve the ODE by hand using the variable substitution $u = x^{1/4}$ or you can use a computer. Show which one of your solutions can match the boundary conditions without requiring $X(x) = 0$.

(c) Apply the condition that $y(0, t)$ is finite to show that one of the arbitrary constants in your solution from Part (b) must be 0.

(d) Apply the boundary condition $y(1, t) = 0$ to find all the possible values for k. There will be an infinite number of them, but you should be able to write them in terms of a new constant n that can be any positive integer. Writing k in terms of n will involve $\alpha_{p,n}$, the zeros of the Bessel functions.

(e) Solve the ODE for $T(t)$, expressing your answer in terms of n.

(f) Multiply $X(x)$ times $T(t)$ to find the normal modes of this system. You should be able to combine your two arbitrary constants into one. Write the general solution $y(x, t)$ as a sum over these normal modes. Your arbitrary constant should include a subscript n to indicate that they can take different values for each value of n.

(g) Use the initial condition $y(x, 0) = \sin(\pi x)$ to find the arbitrary constants in your solution, using the equations for a Fourier-Bessel series in Appendix J. Your answer will be in the form of an integral that you will not be able to evaluate. *Hint*: You may have to define a new variable to get the resulting equation to look like the usual form of a Fourier-Bessel series.

11.136 *[This problem depends on Problem 11.135.]*

(a) Have a computer calculate the 10th partial sum of your solution to Problem 11.135 Part (g). Describe how the function $y(x)$ is evolving over time.

(b) You should have found that the function at $x = 0$ starts at zero, then rises slightly, and then asymptotically approaches zero again. Explain why it does that. *Hint*: think about why all the normal modes cancel out at that point initially, and what is happening to each of them over time.

In Problems 11.137–11.140 you will be given a PDE and a set of boundary and initial conditions.

(a) Solve the PDE with the given boundary conditions using separation of variables. You may solve the ODEs you get by hand or with a computer. The solution to the PDE should be an infinite series with undetermined coefficients.

(b) Plug in the given initial condition. The result should be a Fourier-Bessel series. Write an equation for the coefficients in your series using the equations in Appendix J. This equation will involve an integral that you may not be able to evaluate.

(c) Have a computer evaluate the integrals in Part (b) either analytically or numerically to calculate the 20th partial sum of your series solution and either plot the result at several times or make a 3D plot of $y(x, t)$. Describe how the function is evolving over time.

It may help to first work through Problem 11.135 as a model.

11.137 $\dfrac{\partial y}{\partial t} = \dfrac{\partial^2 y}{\partial x^2} + \dfrac{1}{x}\dfrac{\partial y}{\partial x} - \dfrac{y}{x^2}$, $y(3, t) = 0$,

$y(x, 0) = \begin{cases} 1 & 1 \leq x \leq 2 \\ 0 & \text{elsewhere} \end{cases}$. Assume

y is finite for $0 \leq x \leq 3$.

11.138 $\dfrac{\partial^2 y}{\partial t^2} = \dfrac{\partial^2 y}{\partial x^2} + \dfrac{1}{x}\dfrac{\partial y}{\partial x} - \dfrac{y}{x^2}$, $y(3, t) = 0$,

$y(x, 0) = 0$, $\dfrac{\partial y}{\partial t}(x, 0) = \begin{cases} 1 & 1 \le x \le 2 \\ 0 & \text{elsewhere} \end{cases}$.

Assume y is finite for $0 \le x \le 3$.

11.139 $\partial z/\partial t - \partial^2 z/\partial x^2 - (1/x)(\partial z/\partial x) - (1/x^2)(\partial^2 z/\partial y^2) = 0$, $z(1, y, t) = z(x, 0, t) = z(x, 1, t) = 0$, $z(x, y, 0) = J_{2\pi}(\alpha_{2\pi,3} x) \sin(2\pi y)$.
Assume z is finite throughout $0 \le x \le 1$.

11.140 $\sin^2 x \cos^2 x (\partial z/\partial t) - \sin^2 x (\partial^2 z/\partial x^2) - \tan x (\partial z/\partial x) + 4(\cos^2 x) z = 0$, $z(\pi/2, t) = 0$, $z(x, 0) = \cos x$. Assume z is finite throughout $0 \le x \le \pi/2$. After you separate variables you will use the substitution $u = \sin x$ to turn an unfamiliar equation into Bessel's equation. (If your separation constant is the wrong sign you will get modified Bessel functions, which cannot meet the boundary conditions.)

11.141 Daniel Bernoulli first discovered Bessel functions in 1732 while working on solutions to the "hanging chain problem." A chain is suspended at $y = L$ and hangs freely with the bottom just reaching the point $y = 0$. The sideways motions of the chain $u(y, t)$ are described by the equation $\partial^2 u/\partial t^2 = g\left(y(\partial^2 u/\partial y^2) + \partial u/\partial y\right)$.

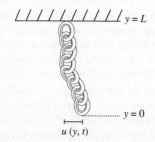

(a) Separate variables and solve the equation for $T(t)$. Choose the sign of the separation constant that gives you oscillatory solutions.

(b) The equation for $Y(y)$ can be turned into Bessel's equation with a substitution. You can start with $u = cy^q$ and find what values of c and q work, but we'll save you some algebra and tell you the correct substitution is $u = cy^{1/2}$. Plug that in and find the value of c needed to turn your $Y(y)$ equation into Bessel's equation.

(c) Solve the equation for $Y(u)$ and plug the substitution you found back in to get a solution for $Y(y)$ subject to the boundary condition $u(L, t) = 0$ and the condition that Y remain finite in the range $0 \le y \le L$. You should find that your solutions are Bessel functions and that you can restrict the possible values of the separation constant.

(d) Write the solution $u(y, t)$ as an infinite series and use the initial conditions $u(y, 0) = f(y)$, $\dot{u}(y, 0) = h(y)$ to find the coefficients in this series.

11.142 *[This problem depends on Problem 11.141.]* In this problem you'll use the solution you derived in Problem 11.141 to model the motion of a hanging chain. For this problem you can take $g = 9.8$ m/s^2, $L = 1$m.

(a) Calculate the first five coefficients of the series you derived for $u(y, t)$ in Problem 11.141 using the initial conditions $u(y, 0) = d - (d/L)y$, $\dot{u}(y, 0) = 0$ where $d = .5$m You can do this analytically by hand, use a computer to find it analytically, or use a computer to do it numerically. However you do it, though, you should get numbers for the five coefficients.

(b) Using the fifth partial sum to approximate $u(y, t)$ make an animation showing $u(y)$ at different times or a 3D plot showing $u(y, t)$ at times ranging from $t = 0$ to $t = 5$. Does the behavior look reasonable for a hanging chain?

11.143 Solve the heat equation (11.2.3) on a circular disk of radius a with the temperature of the edge held at zero and initial condition $u(\rho, \phi, 0) = T_0 J_2(\alpha_{2,3}\rho/a)\sin(2\phi)$. (You will need to use the formula for the Laplacian in polar coordinates. Because α_{mn} is going to show up in the solution you should use D instead of α in the heat equation.) How long will it take for the point $\rho = a/2$, $\phi = \pi/2$ to drop to half its original temperature?

11.144 In the Explanation (Section 11.6.3), we solved for a vibrating drumhead with generic initial conditions, and then plugged in specific initial conditions with and without azimuthal symmetry. If you know from the beginning of your problem that you have azimuthal symmetry you can eliminate the ϕ dependence from the differential equation before solving. In that case the drumhead will obey the equation

$\partial^2 z/\partial t^2 = v^2 \left(\partial^2 z/\partial \rho^2 + (1/\rho)(\partial z/\partial \rho)\right)$. Solve this equation to find $z(\rho, t)$ with a fixed edge $z(a, t) = 0$ and arbitrary initial conditions $z(\rho, 0) = f(\rho)$, $\dot{z}(\rho, 0) = g(\rho)$.

11.145 *[This problem depends on Problem 11.144.]* Solve for the displacement $z(\rho, t)$ of a drum with initial displacement $z(r, 0) = a - \rho$ and initial velocity $\dot{z}(\rho, 0) = c$.

11.7 Separation of Variables—Spherical Coordinates and Legendre Polynomials

Just as we saw that some differential equations in polar coordinates led to normal modes in the form of Bessel functions, we will see here that some differential equations in spherical coordinates lead to normal modes in the form of Legendre polynomials. Once again we will emphasize that you do not have to learn a new process or become an expert on a new function; the process is the same, and you can look up the information you need about the functions as they come up.

We will also discuss another function—spherical harmonics—that can be used as a shorthand for functions of θ and ϕ.

11.7.1 Explanation: Separation of Variables—Spherical Coordinates and Legendre Polynomials

We can summarize almost everything we have done so far as a four-step process.

1. Separate variables to turn one *partial* differential equation into several *ordinary* differential equations.
2. Solve the ordinary differential equations. The product of these solutions is one solution to the original equation: a special solution called a *normal mode*.
3. Match any homogeneous boundary conditions.
4. Sum all the solutions and match the inhomogeneous boundary or initial conditions. This requires that the normal modes form a "complete basis" so you can sum them to meet any given conditions.

Step 1—and, to a large extent, step 4—are much the same from one problem to the next. The two middle steps, on the other hand, depend on the differential equations you end up with. Mathematicians have been studying and cataloguing solutions to ordinary differential equations for centuries. We have seen solutions in the forms of trig functions and Bessel functions. In this section we will see another important form, Legendre polynomials. There are many more.

So how do you solve problems when each new differential equation might require a function you've never seen? One approach is to look each differential equation up in a table such as Appendix J, being ready with a variable substitution or two if the equations don't quite match. Equivalently, you can type your differential equation into the computer and see what it comes up with. We used the "table" approach in Section 11.6 and we will use a computer here. Our point is not to suggest that polar coordinates require a by-hand approach and spherical coordinates are somehow more suitable to a computer; we could just as easily have done it the other way. Our real point is that you need to be ready to use either approach as the situation demands. And in either case, the real skill demanded of you is working with an unfamiliar function once you get it.

Mathematically, both approaches may leave you unsatisfied. Both tell you that the general solution to $x^2 y'' + xy' + (49x^2 - 36)y = 0$ is $AJ_6(7x) + BY_6(7x)$, but neither one tells you where this solution comes from. We will address that important question in Chapter 12 for Bessel functions, Legendre polynomials, and some other important examples. Our focus in this chapter is *using* those ODE solutions to solve PDEs.

The Problem
A spherical shell of radius a surrounds a region of space with no charge; therefore, inside the sphere, the potential obeys Laplace's equation 11.2.5. On the surface of the shell, the given boundary condition is a potential function $V(a, \theta) = V_0(\theta)$. Find the potential inside the sphere.

Laplace's equation in spherical coordinates is derived in Chapter 8:

$$\frac{\partial^2 V}{\partial r^2} + \frac{2}{r}\frac{\partial V}{\partial r} + \frac{1}{r^2}\left(\frac{\partial^2 V}{\partial \theta^2} + \frac{\cos(\theta)}{\sin(\theta)}\frac{\partial V}{\partial \theta} + \frac{1}{\sin^2(\theta)}\frac{\partial^2 V}{\partial \phi^2}\right) = 0$$

Many problems, including this one, have "azimuthal symmetry": the answer will not depend on the angle ϕ. If V has no ϕ-dependency then $\partial V/\partial \phi = 0$, reducing the equation to:

$$\frac{\partial^2 V}{\partial r^2} + \frac{2}{r}\frac{\partial V}{\partial r} + \frac{1}{r^2}\left(\frac{\partial^2 V}{\partial \theta^2} + \frac{\cos(\theta)}{\sin(\theta)}\frac{\partial V}{\partial \theta}\right) = 0$$

where $0 \leq r \leq a$ and $0 \leq \theta \leq \pi$. (We will solve this problem without azimuthal symmetry in the next section.)

As we have done before we notice a key "implicit" boundary condition, which is that the potential function must not blow up anywhere inside the sphere.

Separating the Variables
We begin by assuming a function of the form $V(r, \theta) = R(r)\Theta(\theta)$. We've left the next steps to you (Problem 11.157) but choosing k as the separation constant you should end up with:

$$r^2 R''(r) + 2r R'(r) - k R(r) = 0 \qquad (11.7.1)$$

$$\Theta''(\theta) + \frac{\cos(\theta)}{\sin(\theta)}\Theta'(\theta) + k\Theta(\theta) = 0 \qquad (11.7.2)$$

Solving for $\Theta(\theta)$
Equation 11.7.2 does not readily evoke any of our standard differential equations. We could find a convenient variable substitution to make the equation look like one of the standard forms in Appendix J as we did in the previous section. Instead, just to highlight another important approach, we're going to pop the equation into Mathematica. (You could just as easily use another program like Matlab or Maple.)

```
In[1]:= DSolve[f''[θ] + Cos[θ]/Sin[θ] f'[θ] + k f[θ] == 0, f[θ], θ]

Out[1]:= {{f[θ] → C[1] LegendreP[1/2(-1 + √(1 + 4k)), Cos[θ]] +
          c[2] LegendreQ[1/2(-1 + √(1 + 4k)), Cos[θ]]}}
```

Oh no, it's a whole new kind of function that we haven't encountered yet in this chapter! Don't panic: the main point of this section, and one of the main points of this whole chapter, is that you can attack this kind of problem the same way no matter what function you find for the normal modes.

11.7 | Separation of Variables—Spherical Coordinates and Legendre Polynomials

So you look up "LegendreP" and "LegendreQ" in the Mathematica online help, and you find that these are "Legendre polynomials of the first and second kind," respectively. If Mathematica writes "LegendreP(3,x)," the online help tells you, standard notation would be $P_3(x)$. So the above solution can be written as:

$$\Theta(\theta) = AP_l(\cos\theta) + BQ_l(\cos\theta) \text{ where } l = \frac{1}{2}\left(-1 + \sqrt{1+4k}\right)$$

What else do you need? You need to know a few properties of Legendre polynomials in order to match the boundary conditions (implicit and explicit), and later you'll need to use a Legendre polynomial expansion to match initial conditions, just as we did earlier with trig functions and Bessel functions. So take a moment to look up Legrendre polynomials in Appendix J: all the information we need for this problem is right there.

In this case, the only boundary condition on θ is that the function should be bounded everywhere. Since the argument of P_l and Q_l in this solution is $\cos\theta$, the Legendre polynomials need to be bounded in the domain $[-1, 1]$. That's not true for any of the Q_l functions, and it's only true for P_l when l is a non-negative integer, so we can write the complete set of solutions as

$$\Theta(\theta) = AP_l(\cos\theta), \quad l = 0, 1, 2, 3, \ldots$$

From the definition of l above we can solve for k to find $k = l(l+1)$. In other words, the only values of k for which this equation has a bounded solution are 0, 2, 6, 12, 20, etc.

Solving for $R(r)$

Remember that when you separate variables each side equals the same constant, so we can substitute $k = l(l+1)$ into Equation 11.7.1 to give

$$r^2 R''(r) + 2rR'(r) - l(l+1)R(r) = 0$$

You'll solve this equation (sometimes called a "Cauchy–Euler Equation") in the problems, both by hand and with a computer. The result is

$$R(r) = Br^l + Cr^{-l-1}$$

Since r^{-l-1} blows up at $r = 0$ for all non-negative integers l, we discard this solution based on our implicit boundary condition. Combining our two arbitrary constants, we find that:

$$V(r,\theta) = R(r)\Theta(\theta) = Ar^l P_l(\cos\theta)(l = 0, 1, 2 \ldots)$$

As always, since our original equation was linear and homogeneous, we write a general solution as a sum of all the particular solutions, each with its own arbitrary constant:

$$V(r,\theta) = \sum_{l=0}^{\infty} A_l r^l P_l(\cos\theta) \tag{11.7.3}$$

We must now meet the boundary condition $V(a,\theta) = V_0(\theta)$.

$$V_0(\theta) = \sum_{l=0}^{\infty} A_l a^l P_l(\cos\theta)$$

We are guaranteed that we can meet this condition, because Legendre polynomials—like the trig functions and Bessel functions that we have seen before—constitute a *complete* set of functions on the interval $-1 \leq x \leq 1$. Appendix J gives the coefficients necessary to build an arbitrary function as a Legendre series expansion:

$$A_l a^l = \frac{2l+1}{2} \int_0^\pi V_0(\theta) P_l(\cos\theta) \sin\theta \, d\theta$$

So the complete solution is

$$V(r,\theta) = \sum_{l=0}^\infty A_l r^l P_l(\cos\theta) \tag{11.7.4}$$

where

$$A_l = \frac{2l+1}{2a^l} \int_0^\pi V_0(\theta) P_l(\cos\theta) \sin\theta \, d\theta \tag{11.7.5}$$

11.7.2 Explanation: Spherical Harmonics

In Section 11.7.1 we found the potential distribution inside a spherical shell held at a known potential $V_0(\theta)$. Because the problem had no ϕ-dependence we had to separate variables only once, solve two ordinary differential equations for r and θ, and write the solution as a single series. If we had to work with r, θ, and ϕ as three independent variables we would in general have to separate variables twice, solve three ordinary differential equations, and write the solution as a double series, as we did in Section 11.5.

However, there is a shortcut in spherical coordinates: separate variables only once, giving you one ordinary differential equation for the independent variable r and one partial differential equation for the two variables θ and ϕ. The solution to this partial differential equation will often be a special function called a "spherical harmonic."

The Problem, and the Solution

A spherical shell of radius a contains no charge within the shell; therefore, within the shell, the potential obeys Laplace's equation (11.2.5). On the surface of the shell the potential $V(a, \theta, \phi)$ is given by $V_0(\theta, \phi)$. Find the potential inside the sphere.

Note as always the "implicit" boundary conditions on this problem. The potential must be finite in the domain $0 \leq r \leq a, 0 \leq \theta \leq \pi$. The solution must also be 2π-periodic in ϕ.

Laplace's equation in spherical coordinates is derived in Chapter 8:

$$\frac{\partial^2 V}{\partial r^2} + \frac{2}{r}\frac{\partial V}{\partial r} + \frac{1}{r^2}\left(\frac{\partial^2 V}{\partial \theta^2} + \frac{\cos\theta}{\sin\theta}\frac{\partial V}{\partial \theta} + \frac{1}{\sin^2\theta}\frac{\partial^2 V}{\partial \phi^2}\right) = 0 \tag{11.7.6}$$

Anticipating that we're only going to separate variables once, we look for a solution of the form:

$$V(r,\theta,\phi) = R(r)\Omega(\theta,\phi)$$

We plug this into the original differential equation, multiply through by $r^2/[R(r)\Omega(\theta,\phi)]$, separate out the r-dependent terms, and use the separation constant k to arrive at $r^2 R''(r) + 2rR'(r) - kR(r) = 0$ and

$$\frac{\partial^2 \Omega}{\partial \theta^2} + \frac{\cos\theta}{\sin\theta}\frac{\partial \Omega}{\partial \theta} + \frac{1}{\sin^2\theta}\frac{\partial^2 \Omega}{\partial \phi^2} + k\Omega = 0 \tag{11.7.7}$$

In Problem 11.158 you'll solve Equation 11.7.7 using separation of variables. The solutions will of course be products of functions of ϕ with functions of θ. This PDE comes up so often

11.7 | Separation of Variables—Spherical Coordinates and Legendre Polynomials

in spherical coordinates, however, that its solution has a name. Looking up Equation 11.7.7 in Appendix J we find that the solutions are the "spherical harmonics" $Y_l^m(\theta, \phi)$, where l is a non-negative integer such that $k = l(l + 1)$ and m is an integer satisfying $|m| \leq l$. (The requirement that l and m be integers comes from the implicit boundary conditions on θ and ϕ.)

The equation for $R(r)$ is a Cauchy–Euler equation, just like in the last section. If you plug in $k = l(l + 1)$ and use the requirement that it be bounded at $r = 0$ the solution is $R(r) = Cr^l$. The solution to Equation 11.7.6 is thus

$$V(r, \theta, \phi) = \sum_{l=0}^{\infty} \sum_{m=-l}^{l} C_{lm} r^l Y_l^m(\theta, \phi)$$

Finally, we match the inhomogeneous boundary condition $V(a, \theta, \phi) = V_0(\theta, \phi)$.

$$V_0(\theta, \phi) = \sum_{l=0}^{\infty} \sum_{m=-l}^{l} C_{lm} a^l Y_l^m(\theta, \phi)$$

From the formula in Appendix J for the coefficients of a spherical harmonic expansion we can write

$$C_{lm} = \frac{1}{a^l} \int_0^{2\pi} \int_0^{\pi} V_0(\theta, \phi) \left[Y_l^m(\theta, \phi)\right]^* \sin \theta \, d\theta \, d\phi$$

where $\left[Y_l^m(\theta, \phi)\right]^*$ designates the complex conjugate of the spherical harmonic function $Y_l^m(\theta, \phi)$. This completes our solution for the potential in a hollow sphere.

Spherical Harmonics

Spherical harmonics are a computational convenience, not a new method of solving differential equations. We could have solved the problem above by separating all three variables: we would have found a complex exponential function in ϕ and an associated Legendre polynomial in θ and then multiplied the two. ("Associated Legendre polynomials" are yet another special function, different from "Legendre polynomials." As you might guess, they are in Appendix J.) All we did here was skip a step by *defining* a spherical harmonic as precisely that product: a complex exponential in ϕ, multiplied by an associated Legendre polynomial in θ.

Complex exponentials form a complete basis (Fourier series), and so do associated Legendre polynomials. By multiplying them we create a complete basis for functions of θ and ϕ. Essentially any function that depends only on direction, and not on distance, can be expanded as a sum of spherical harmonics. That makes them useful for solving partial differential equations in spherical coordinates, but also for a wide variety of other problems ranging from analyzing radiation coming to us from space to characterizing lesions in multiple sclerosis patients.[8]

11.7.3 Stepping Back: An Overview of Separation of Variables

The last four sections of this chapter have all been on solving partial differential equations by separation of variables. The topic deserves that much space: partial differential equations come up in almost every aspect of physics and engineering, and separation of variables is the most common way of handling them.

---

[8] Goldberg-Zimring, Daniel, et. al., "Application of spherical harmonics derived space rotation invariant indices to the analysis of multiple sclerosis lesions' geometry by MRI," Magnetic Resonance Imaging, Volume 22, Issue 6, July 2004, Pages 815-825.

On the other hand there's a real danger that, after going through four sections on one technique, you will feel like there is a mountain of trivia to master. We started this section by listing the four steps of every separation of variables problem. Let's return to that outline, but fill in all the "gotchas" and forks in the road. You may be surprised at how few there are.

1. **Separate variables to turn one *partial* differential equation into several *ordinary* differential equations.**

    In this step you "guess" a solution that is separated into functions of one variable each. (Recall that such a solution is a *normal mode*.) Then you do a bit of algebra to isolate the dependency: for instance, one side depends only on θ, the other does not depend at all on θ. Finally you set both sides of the equation equal to a constant—the same constant for both sides!—often using k^2 if you can determine up front that the constant is positive, or $-k^2$ if negative.

    If separating the variables turns out to be impossible (as it will for almost any inhomogeneous equation for instance), you can't use this technique. The last few sections of this chapter will present alternative techniques you can use in such cases.

    If there are three variables you go through this process twice, introducing two constants…and so on for higher numbers of variables. (Spherical harmonics represent an exception to this rule in one specific but important special case.)

2. **Solve the ordinary differential equations. The product of these solutions is one solution to the original equation.**

    Sometimes the "solving" step can be done by inspection. When the solution is less obvious you may use a variable substitution and table lookup, or you may use a computer.

    The number of possible functions you might get is the most daunting part of the process. Trig functions (regular and hyperbolic), exponential functions (real and complex), Bessel functions, Legendre polynomials…no matter how many we show you in this chapter, you may encounter a new one next week. But as long as you can look up the function's general behavior, its zeros, and how to use it in series to build up other functions, you can work with it to find a solution. That being said, we have not chosen arbitrarily which functions to showcase: trig functions, Bessel functions, Legendre polynomials, and spherical harmonics are the most important examples.

3. **Match any homogeneous boundary conditions.**

    Matching the homogeneous boundary conditions sometimes tells you the value of one or more arbitrary constants, and sometimes limits the possible values of the constants of separation. If you have homogeneous *initial* conditions, this method generally won't work.

4. **Sum up all solutions into a series, and then match the inhomogeneous boundary and initial conditions.**

    First you write a series solution by summing up all the solutions you have previously found. (The original differential equation must be linear and homogeneous, so a linear combination of solutions—a series—is itself a solution.) Then you set that series equal to your inhomogeneous boundary or initial condition. (Your normal modes must form a "complete basis," so you can add them up to meet any arbitrary conditions.)

    If you had three independent variables—and therefore two separation constants—the result is a double series. This trend continues upward as the number of variables climbs.

    If you have more than one inhomogeneous condition, you create subproblems with one inhomogeneous condition each: this will be the subject of the next section.

11.7 | Separation of Variables—Spherical Coordinates and Legendre Polynomials

The rest of the chapter discusses additional techniques that can be used instead of, or sometimes alongside, separation of variables for different types of problems. Appendix I gives a flow chart for deciding which techniques to use for which problems. If you go through that process and decide that separation of variables is the right technique to use, you might find it helpful to refer to the list above as a reminder of the key steps in the process.

11.7.4 Problems: Separation of Variables—Spherical Coordinates and Legendre Polynomials

In the Explanation (Section 11.7.1) we found that the potential $V(r, \theta)$ inside a hollow sphere is given by Equation 11.7.3, with the coefficients A_l given by Equation 11.7.5. In Problems 11.146–11.148 you will be given a particular potential $V_0(\theta)$ on the surface of the sphere to plug into our general solution.

11.146 $V_0(\theta) = c$ on the upper half of the sphere and 0 on the lower half.
  (a) Write the expression for A_l. Have a computer evaluate the integral and plug it in to get an expression for $V(r, \theta)$ as an infinite sum.
  (b) Evaluate your solution at $r = a$ and plot the 20th partial sum of the resulting function $V(a, \theta)$ on the same plot as the given boundary condition $V_0(\theta)$. If they don't match well go back and figure out where you made a mistake.

11.147 $V_0(\theta) = \sin\theta$
  (a) Write the expression for A_l. Leave the integral unevaluated.
  (b) Evaluate your solution at $r = a$ by numerically integrating for the necessary coefficients. Plot the 20th partial sum of the resulting function $V(a, \theta)$ on the same plot as the given boundary condition $V_0(\theta)$. If they don't match well go back and figure out where you made a mistake.

11.148 $V_0(\theta) = P_3(\cos\theta)$. Write the expression for A_l. Use the orthogonality of the Legendre polynomials and the fact that $\int_{-1}^{1} P_l^2(u)\,du = 2/(2l+1)$ to evaluate the integral and write a closed-form solution for $V(r, \theta)$.

11.149 Walk-Through: Differential Equation with Legendre Normal Modes. In this problem you will solve the partial differential equation $\partial y/\partial t - (1 - x^2)(\partial^2 y/\partial x^2) + 2x(\partial y/\partial x) = 0$ on the domain $-1 \leq x \leq 1$. Surprisingly, the only boundary condition you need for this problem is that y is finite on the interval $[-1, 1]$.

  (a) Begin by guessing a separable solution $y = X(x)T(t)$. Plug this guess into the differential equation. Then divide both sides by $X(x)T(t)$ and separate variables.
  (b) Find the general solution to the resulting ODE for $X(x)$. Using the requirement that y is finite on the domain $[-1, 1]$ show that one of your arbitrary constants must be zero and give the possible values for the separation constant. There will be an infinite number of them, but you should be able to write them in terms of a new constant l that can be any non-negative integer.
  (c) Solve the ODE for $T(t)$, expressing your answer in terms of l.
  (d) Multiply $X(x)$ times $T(t)$ to find the normal modes of this system. You should be able to combine your two arbitrary constants into one. Write the general solution $y(x, t)$ as a sum over these normal modes. Your arbitrary constants should include a subscript l to indicate that they can take different values for each value of l.

For the rest of the problem you will plug the initial condition $y(x, 0) = x$ into the solution you found.

  (e) Plugging $t = 0$ into your solution to Problem 11.149 should give you a Fourier-Legendre series for the function $y(x, 0)$. Setting this equal to x, use the equation for A_l from Appendix J to find the coefficients in the form of an integral.
  (f) Here are two facts about Legendre polynomials: $P_1(x) = x$, and the Legendre polynomials are *orthogonal*, meaning that $\int_{-1}^{1} P_l(x)P_m(x)\,dx = 0$ if $l \neq m$. Using those facts, you can analytically integrate your answer from Part 11.149(e). (For most initial conditions you wouldn't be able to evaluate this integral explicitly.) Use your result for this integral to write the solution $y(x, t)$ in closed form.

(g) Demonstrate that your solution satisfies the original differential equation and the initial condition.

In Problems 11.150–11.153 you will be given a PDE, a domain, and a set of initial conditions. You should assume in each case that the function is finite in the given domain.

(a) Solve the PDE using separation of variables. You may solve the ODEs you get by hand or with a computer. The solution to the PDE should be an infinite series with undetermined coefficients.

(b) Plug in the given initial condition. The result should be a Fourier-Legendre series. Use this series to write an equation for the coefficients in your solution. In most cases, unlike the problem above, you will not be able to evaluate this integral analytically: your solution will be in the form $\sum_{l=1}^{\infty} A_l$ <something> where A_l is defined as an integral. See Equations 11.7.4–11.7.5 for an example.

(c) 🖥 Have a computer evaluate the integral in Part (b) either analytically or numerically to calculate the 20th partial sum of your series solution and plot the result at several times. Describe how the function is evolving over time.

11.150 $\partial y/\partial t - (1 - x^2)(\partial^2 y/\partial x^2) + 2x(\partial y/\partial x) = 0$, $-1 \le x \le 1, y(x, 0) = 1$, and $y(x)$ is finite everywhere in that domain

11.151 $\partial y/\partial t = (9 - x^2)(\partial^2 y/\partial x^2) - 2x(\partial y/\partial x)$, $-3 \le x \le 3, y(x, 0) = \sin(\pi x/3)$

11.152 $\partial y/\partial t = \partial^2 y/\partial \theta^2 + \cot \theta (\partial y/\partial \theta)$, $0 \le \theta \le \pi, y(\theta, 0) = \sin^2 \theta$

11.153 $\partial^2 y/\partial t^2 = (9 - x^2)(\partial^2 y/\partial x^2) - 2x(\partial y/\partial x), -3 \le x \le 3, y(x, 0) = 4x, (\partial y/\partial t)(x, 0) = 0$. Using the fact that $P_1(x) = x$ you should be able to get a closed-form solution (no sum) and explicitly check that it solves the original PDE. You don't need to do the computer part for this problem.

In Problems 11.154–11.156 solve the given PDE using the domain, boundary conditions, and initial conditions given in the problem. You should assume in each case that the function is finite in the given domain and that it is periodic in ϕ. Your solutions will be in the form of series involving spherical harmonics. You should write integral expressions for the coefficients. (*Hint*: You may need a variable substitution to get the spherical harmonic equation. If you can't find one that works try separating out all the variables and you should get solutions involving complex exponentials and associated Legendre polynomials that you can recombine into spherical harmonics.)

11.154 $\partial y/\partial t = \partial^2 y/\partial \theta^2 + \cot \theta (\partial y/\partial \theta) + \csc^2 \theta (\partial^2 y/\partial \phi^2), 0 \le \theta \le \pi, 0 \le \phi \le 2\pi$, $y(\theta, \phi, 0) = \theta(\pi - \theta) \cos(\phi)$

11.155 $\dfrac{\partial y}{\partial t} = (1 - x^2) \dfrac{\partial^2 y}{\partial x^2} - 2x \dfrac{\partial y}{\partial x} + \dfrac{1}{1 - x^2} \dfrac{\partial^2 y}{\partial \phi^2}$, $-1 \le x \le 1, 0 \le \phi \le 2\pi$, $y(x, \phi, 0) = (1 - x^2) \cos \phi$

11.156 $\dfrac{\partial z}{\partial t} = (1 - x^2) \dfrac{\partial^2 z}{\partial x^2} - 2x \dfrac{\partial z}{\partial x} + \dfrac{1}{1 - x^2} \left(2 \dfrac{\partial z}{\partial \phi} + 4\phi \dfrac{\partial^2 z}{\partial \phi^2} \right), -1 \le x \le 1$, $0 \le \phi \le 4\pi^2, y(x, \phi, 0) = (1 - x^2) \cos\left(\sqrt{\phi}\right)$

11.157 In the Explanation (Section 11.7.1) we encountered the differential equation $r^2 R''(r) + 2rR'(r) - l(l + 1)R(r) = 0$, sometimes called a "Cauchy–Euler Equation."

(a) To solve this ordinary differential equation assume a solution of the form $R(r) = r^p$ where the constant p may depend on l but not on r. Plug this solution into the differential equation, solve for p, and write the general solution $R(r)$. Make sure your answer has two arbitrary constants!

(b) 🖥 Find the same solution by plugging the differential equation into a computer. (You may need to tell the computer program that l is a positive integer to get it to simplify the answer. You may also in that process discover that some differential equations are easier to solve by hand than with a computer.)

11.158 In the Explanation (Section 11.7.2) we derived a PDE for $\Omega(\theta, \phi)$, which we said led to spherical harmonics. Use separation of variables to find the normal modes of this PDE, using Appendix J for any necessary ODE solutions. Your solution should include using the implicit boundary conditions to limit l and m. Use complex exponentials instead of trig functions. Your final answer should be a formula for $Y_l^m(\theta, \phi)$ in terms of other functions.

11.159 In the Explanation (Section 11.7.2) we derived the formula for the gravitational potential in the interior of a thin spherical shell of radius a with potential $V_0(\theta, \phi)$ on the surface of the shell.

(a) Use the formula we derived to find the gravitational potential inside a spherical shell that is held at a constant potential V_0. Simplify your answer as much as possible. *Hint*: the integral $\int_0^{2\pi} \int_0^{\pi} [Y_l^m(\theta,\phi)]^* \sin\theta\, d\theta\, d\phi$ equals $2\sqrt{\pi}$ for $l = m = 0$ and 0 for all other l and m.

(b) The gravitational field is given by $\vec{g} = -\vec{\nabla} V$. Use your result from Part (a) to find the gravitational field in the interior of the sphere.

11.160 Solve the equation

$$\frac{\partial^2 y}{\partial t^2} = (4-x^2)\frac{\partial^2 y}{\partial x^2} - 2x\frac{\partial y}{\partial x} - \frac{4}{4-x^2}y$$

on the domain $-2 \le x \le 2$ with initial conditions $y(x,0) = 4-x^2$, $\partial y/\partial t(x,0) = 0$.

11.161 The equation $\nabla^2 u = -\lambda^2 u$ is known as the "Helmholtz equation." For this problem you will solve it in a sphere of radius a with boundary condition $u(a,\theta,\phi) = 0$. You should also use the implicit boundary condition that the function u is finite everywhere inside the sphere. Note that you will need to use the formula for the Laplacian in spherical coordinates.

(a) Separate variables, putting $R(r)$ on one side and $\Omega(\theta,\phi)$ on the other. Solve the equation for $\Omega(\theta,\phi)$ *without* separating a second time. You should find that this equation only has non-trivial solutions for some values of the separation constant.

(b) Using the values of the separation constant you found in the last part, solve the equation for $R(r)$, applying the boundary condition. You should find that only some values of λ allow you to satisfy the boundary condition. Clearly indicate the allowed values of λ.

(c) Combine your answers to write the general solution to the Helmholtz equation in spherical coordinates. You should find that the boundary conditions given in the problem were not sufficient to specify the solution. Instead the general solution will be a double sum over two indices with arbitrary coefficients in front of each term in the sum.

11.162 A quantum mechanical particle moving freely inside a spherical container of radius a can be described by a wavefunction Ψ that obeys Schrödinger's equation (11.2.6) with $V(\vec{x}) = 0$ and boundary condition $\Psi(a,\theta,\phi,t) = 0$. In this problem you will find the possible energies for such a particle.

(a) Write Schrödinger's equation with $V(\vec{x}) = 0$ in spherical coordinates. Plug in a trial solution $\Psi(r,\theta,\phi,t) = T(t)\psi(r,\theta,\phi)$ and separate variables to get an equation for $T(t)$. Verify that $T(t) = Ae^{-iEt/\hbar}$ is the solution to your equation for $T(t)$, where E is the separation constant. That constant represents the energy of the atom.

(b) Plug $\psi(r,\theta,\phi) = R(r)\Omega(\theta,\phi)$ into the remaining equation, separate, and solve for $\Omega(\theta,\phi)$. Using the implicit boundary conditions that Ψ must be finite everywhere inside the sphere and periodic in ϕ, show that the separation constant must be $l(l+1)$ where l is an integer. (*Hint*: If you get a different set of allowed values for the separation constant you should think about how you can simplify the equation and which side the constant E should go on to get the answer to come out this way.)

(c) Solve the remaining equation for $R(r)$. If you write the separation constant in the form $l(l+1)$ you should be able to recognize the equation as being similar to one of the ones in Appendix J. You can get it in the right form with a variable substitution or solve it on a computer. You should find that the implicit boundary condition that Ψ is finite allows you to set one arbitrary constant to zero. The explicit boundary condition $\Psi(a,\theta,\phi,t) = 0$ should allow you to specify the allowed values of E. These are the possible energies for a quantum particle confined to a sphere.

When a large star collapses at the end of its life it becomes a dense sphere of particles known as a neutron star. Knowing the possible energies of a particle in a sphere allows astronomers to predict the behavior of these objects.

11.163 Exploration: Potential Inside a Hemisphere In this problem you will find the electric potential inside a hollow hemisphere of radius a with a constant potential $V = c$ on the curved upper surface of the hemisphere and $V = 0$ on the flat lower surface.

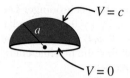

Because there is no charge inside the hemisphere, the potential will obey Laplace's equation. Because the problem has "azimuthal symmetry" (no ϕ-dependence) Laplace's equation can be written as

$$\frac{\partial^2 V}{\partial r^2} + \frac{2}{r}\frac{\partial V}{\partial r} + \frac{1}{r^2}\left(\frac{\partial^2 V}{\partial \theta^2} + \frac{\cos\theta}{\sin\theta}\frac{\partial V}{\partial \theta}\right) = 0$$

Since the PDE is the same as the one we solved in the Explanation (Section 11.7.1), separation of variables will go the same way, leading to Equation 11.7.3. But from that point the problem will be different because of the different domain and boundary conditions.

(a) Write the domains for r and θ and boundary conditions for this problem, including all implicit boundary conditions.

(b) We now define a new variable $u = \cos\theta$. Rewrite your answers to Part (a) (both domain and boundary conditions) in terms of r and u instead of r and θ.

(c) We found in the Explanation that the general solution to our differential equation is $V(r,\theta) = \sum_{l=0}^{\infty} A_l r^l P_l(\cos\theta)$ or, with our new variables, $V(r,u) = \sum_{l=0}^{\infty} A_l r^l P_l(u)$. Plug into this equation the condition "$V = c$ everywhere on the curved upper surface of the hemisphere."

Our next step is to find the constants A_l by treating your answer to Part (c) as a Fourier-Legendre expansion in u. But u goes from 1 to 0, and Fourier-Legendre expansions are for functions with domain $[-1, 1]$. To get around this problem, we take the boundary condition "$V = c$ as u goes from 0 to 1" and create an odd extension of it, called $W(u)$. This means that the Fourier-Legendre series for $W(u)$ will converge to c on the interval $[0, 1]$.

(d) Write the function $W(u)$.

(e) Now that you can write a Fourier-Legendre expansion of W, use it to find the constants A_l. Your answer will be in the form of an integral.

(f) For even values of l, the Legendre polynomials are even functions. Explain why we can use this fact to discard all even powers of l.

(g) For odd values of l, the Legendre polynomials are odd functions. Use this fact to rewrite your answer to Part (e) in a way that uses c instead of W for odd values of l. (You can leave your result in integral form, or evaluate the resulting integral pretty easily with a table or computer lookup.)

(h) Plug these constants into Equation 11.7.3 to get the solution $V(r, \theta)$.

(i) Show that this solution matches the boundary condition that $V = 0$ on the flat lower surface of the hemisphere.

(j) 🖥 Have a computer plot the fifth partial sum as a function of r and θ. You will need to choose values for a and c. You should be able to see that it roughly matches the boundary conditions.

11.8 Inhomogeneous Boundary Conditions

Recall that we solved inhomogeneous ODEs by finding a "particular" solution (that could not meet the initial conditions) and a "complementary" solution (to a different differential equation). When we added these two solutions, we found the general function that solved the original differential equation and could meet the initial conditions.

In this section, we will show how you can use the same approach to solve PDEs with inhomogeneous boundary conditions. You can also use this method on inhomogeneous PDEs; we will guide you through that process in Problem 11.176.

11.8.1 Discovery Exercise: Inhomogeneous Boundary Conditions

1. In each of the problems we have worked so far, there has been only one inhomogeneous boundary *or* initial condition. Explain why the technique of Separation of Variables, as we have described it, relies on this limitation.

   Now consider a problem with three inhomogeneous boundary conditions. The cylinder to the right has no charge inside. The potential therefore obeys Laplace's equation, which in cylindrical coordinates is:

   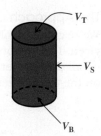

   $$\frac{\partial^2 V}{\partial \rho^2} + \frac{1}{\rho}\frac{\partial V}{\partial \rho} + \frac{1}{\rho^2}\frac{\partial^2 V}{\partial \phi^2} + \frac{\partial^2 V}{\partial z^2} = 0$$

   The potential on top is given by V_T, which could be a constant or a function of ρ or even a function of ρ and ϕ. (For our present purpose it doesn't really matter.) The potential on the bottom is given by V_B. The potential on the side is V_S which could be a function of both z and ϕ.

2. The approach to such a problem is to begin by solving three *different* problems. Each problem is the same differential equation as the original, but each has only one inhomogeneous boundary condition. In the first such subproblem, we take $V = V_T$ on top, but $V = 0$ on the side and bottom. What are the other two subproblems?
3. If you solved all three subproblems and added the solutions, would the resulting function solve the original differential equation? Would it solve all the original boundary conditions?

11.8.2 Explanation: Inhomogeneous Boundary Conditions

Every example we have worked so far we has either been an initial value problem with homogeneous boundary conditions or a boundary value problem with one inhomogeneous boundary condition. We applied the homogeneous conditions before we summed; after we summed we could find the coefficients of our series solution to match the inhomogeneous condition at the one remaining boundary.

If we have multiple inhomogeneous conditions, we bring back an old friend from Chapter 1. To solve inhomogeneous ordinary differential equations we found two different solutions—a "particular" solution and a "complementary" solution—that summed to the general solution we were looking for. The same technique applies to the world of partial differential equations. In Problem 11.176 you'll apply this to solve an inhomogeneous PDE just as you did with ODEs. In this section, however, we'll show you how to use the same method to solve *linear, homogeneous differential equations* with *multiple inhomogeneous boundary conditions*.

The two examples below are different in several ways. In both cases, however, our approach will involve finding two different functions that we can add to solve the original problem we were given.

The First Problem: Multiple Inhomogeneous Boundary Conditions

If a two-dimensional surface is allowed to come to a steady-state temperature distribution, that distribution will obey Laplace's equation:

$$\frac{\partial^2 u}{\partial x^2} + \frac{\partial^2 u}{\partial y^2} = 0$$

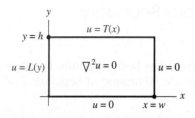

Consider a rectangle with length w and height h. For convenience we place the origin of our coordinate system at the lower-left-hand corner of the rectangle, so the upper-right-hand corner is at (w, h). The boundary conditions are $u(x, 0) = u(w, y) = 0$, $u(0, y) = L(y)$, $u(x, h) = T(x)$.

The Approach

We're going to solve two different problems, neither of which is exactly the problem we were given. The first problem is to find a solution such that $u_L(x, y)$ is $L(y)$ on the left side and zero on all three other sides. The second problem is to find a function such that $u_T(x, y)$ is $T(x)$ on the top and zero on all three other sides.

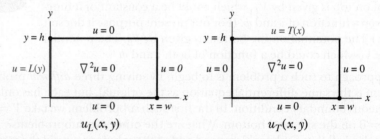

Each subproblem has only one inhomogeneous condition, so we can take our familiar approach: separate variables and apply the homogeneous conditions, then create a series, and finally match the inhomogeneous condition. Of course, neither u_L nor u_T is a solution to our original problem! However, their sum $(u_L + u_T)(x, y)$ will satisfy Laplace's equation and fit the boundary conditions on all four sides.

Solving the Individual Problems

We begin by finding the first function, $u_L(x, y)$. Separation of variables leads us quickly to:

$$\frac{X''(x)}{X(x)} = -\frac{Y''(y)}{Y(y)}$$

Since we wish $u_L(x, y)$ to go to zero at $y = 0$ and $y = h$, the separation constant must be positive, so we shall call it k^2. Solving the second equation and applying the boundary conditions:

$$Y(y) = A \sin(ky)$$

where $k = n\pi/h$. The first equation comes out quite differently. The positive separation constant leads to an exponential solution:

$$X(x) = Ce^{kx} + De^{-kx}$$

The requirement $X(w) = 0$ means $Ce^{kw} + De^{-kw} = 0$, which leads to the unpleasant-looking:

$$X(x) = C\left(e^{kx} - e^{k(2w-x)}\right)$$

Having plugged in our three homogeneous conditions, we now combine the two functions, absorb C into A, and write a sum:

$$u_L(x, y) = \sum_{n=1}^{\infty} A_n \sin(ky) \left(e^{kx} - e^{k(2w-x)}\right)$$

where $k = n\pi/h$.

11.8 | Inhomogeneous Boundary Conditions

Our last and inhomogeneous condition, $u_L(0, y) = L(y)$, becomes:

$$\sum_{n=1}^{\infty} A_n \sin\left(\frac{n\pi}{h}y\right)\left(1 - e^{2kw}\right) = L(y)$$

This is a Fourier sine series. You can find the formula for the coefficients in Appendix G.

$$A_n \left(1 - e^{2kw}\right) = \frac{2}{h} \int_0^h L(y) \sin\left(\frac{n\pi}{h}y\right) dy$$

For the second function $u_T(x, y)$ we start the process over, separating the variables and using a negative separation constant $-p^2$ this time. You will go through this process in Problem 11.164 and arrive at:

$$u_T(x, y) = \sum_{m=1}^{\infty} B_m \sin(px)\left(e^{py} - e^{-py}\right) \quad (11.8.1)$$

where $p = m\pi/w$. Since $u_T(x, y) = T(x)$, we have:

$$\sum_{m=1}^{\infty} B_m \sin(\frac{m\pi}{w}x)\left(e^{py} - e^{-py}\right) = T(x)$$

So the coefficients B_m are determined from

$$B_m \left(e^{ph} - e^{-ph}\right) = \frac{2}{w} \int_0^w T(x) \sin\left(\frac{m\pi}{w}x\right) dx$$

Solving the Actual Problem

None of the "separation of variables" math above is new, and we skipped a lot of hopefully familiar steps along the way. Only the final step is new: having found the two functions $u_L(x, y)$ and $u_T(x, y)$, we add them, and declare the sum $(u_L + u_T)(x, y)$ to be the solution to our original problem:

$$u(x, y) = \sum_{n=1}^{\infty} A_n \sin(ky)\left(e^{kx} - e^{k(2w-x)}\right) + \sum_{m=1}^{\infty} B_m \sin(px)\left(e^{py} - e^{-py}\right)$$

where $k = n\pi/h$, $p = m\pi/w$, $A_n \left(1 - e^{2kw}\right) = (2/h) \int_0^h L(y) \sin\left(n\pi y/h\right) dy$, and $B_m \left(e^{ph} - e^{-ph}\right) = (2/w) \int_0^w T(x) \sin\left(m\pi x/w\right) dx$.

That is the step you need to think about! *Assuming* that u_L and u_T solve the individual problems they were designed to solve…

- Can you convince yourself that $u_L + u_T$ is still a solution to the differential equation $\partial^2 u/\partial x^2 + \partial^2 u/\partial y^2 = 0$? (This would not work for *all* differential equations: why must it for this one?)
- Can you convince yourself that $u_L + u_T$ meets all the proper boundary conditions? (These are different from the boundary conditions that either function meets alone!)

As a final thought, before we move on to the next example: suppose all four boundary conditions had been inhomogeneous instead of only two. Can you outline the approach we would use to find a general solution? What individual problems would we have to solve first?

The Second Problem: Initial-Value Problem with Inhomogeneous Boundary

A rod stretches from $x = 0$ to $x = L$ with each end held at a fixed non-zero temperature: $u(0, t) = T_L$ and $u(L, t) = T_R$. The initial temperature distribution of the rod is represented by the function $u(x, 0) = T_0(x)$. Thereafter the temperature in the rod obeys the heat equation:

$$\frac{\partial u}{\partial t} = \alpha \frac{\partial^2 u}{\partial x^2} \; (\alpha > 0)$$

Find the temperature distribution in the rod as a function of time.

The Approach

Once again we are going to break the initial problem into two subproblems that will sum to our solution. In this case we will find a "particular" solution and a "complementary" solution, just as we did for ordinary differential equations.

The "particular" solution $u_P(x, t)$ will solve our differential equation with the original boundary conditions. It will be a very simple, very specific solution with no arbitrary constants.

The "complementary" solution will solve our differential equation with homogeneous boundary conditions $u_C(0, t) = u_C(L, t) = 0$. It will have the arbitrary constants of a general solution: therefore, after we add it to our particular solution, we will be able to match initial conditions.

The Particular Solution

Our goal for a particular solution is to find *any* function $u_P(x, t)$ that solves our original differential equation and matches our boundary conditions. To find a simple solution, we make a simple assumption: for our particular solution, $\partial u_P / \partial t$ will equal zero.

Where does that leave us? $\partial^2 u_P / \partial x^2 = 0$ is easy to solve by inspection; the solution is any line $u_P = mx + b$. Our boundary conditions $u(0, t) = T_L$ and $u(L, t) = T_R$ allow us to solve quickly for the slope and y-intercept, bringing us to:

$$u_P = \frac{T_R - T_L}{L} x + T_L$$

Before we move on, stop to consider what that function represents. It certainly solves $\partial u / \partial t = \alpha(\partial^2 u / \partial x^2)$, since it makes both sides of the differential equation zero. It also matches the boundary conditions properly. On the other hand, we can't possibly make this function solve the *initial* conditions: there are no arbitrary constants to play with, and in fact no time dependence at all. That's where the complementary solution comes in.

The Complementary Solution

For the complementary solution we're going to solve the original differential equation, but with boundary conditions $u_C(0, t) = u_C(L, t) = 0$. With homogeneous boundary conditions and an inhomogeneous initial condition we have the perfect candidate for separation of variables. So we set $u_C(x, t) = X(x)T(t)$, plug in, simplify, and end up with:

$$\frac{T'(t)}{T(t)} = \alpha \frac{X''(x)}{X(x)}$$

The process at this point is familiar. A positive separation constant would make $X(x)$ exponential, and a zero separation constant would make $X(x)$ linear. Neither solution could reach zero at both ends with non-zero values in the middle, so we choose a negative separation constant $-k^2$. In Problem 11.170 you'll solve for $X(x)$ and apply the boundary conditions $X(0) = X(L) = 0$, then solve for $T(t)$, and end up here.

$$u_C(x, t) = Ae^{-k^2 t} \sin\left(\frac{k}{\sqrt{\alpha}} x\right) \text{ where } k = (n\pi/L)\sqrt{\alpha}, \ n = 1, 2, 3, \ldots \quad (11.8.2)$$

Since u_C solves a homogeneous differential equation with homogeneous boundary conditions, any sum of solutions is itself a solution, so the general solution is:

$$u_C(x, t) = \sum_{n=1}^{\infty} A_n e^{-\alpha(n^2\pi^2/L^2)t} \sin\left(\frac{n\pi}{L} x\right)$$

The Actual Solution

The functions u_P and u_C both solve the differential equation $\partial u/\partial t = \alpha(\partial^2 u/\partial x^2)$. They do *not* meet the same boundary conditions: u_P meets the original boundary conditions, and u_C goes to zero on both ends. When we add them, we have a solution to both the original differential equation and the original boundary conditions:

$$u(x, t) = \frac{T_R - T_L}{L} x + T_L + \sum_{n=1}^{\infty} A_n e^{-\alpha(n^2\pi^2/L^2)t} \sin\left(\frac{n\pi}{L} x\right)$$

As a final step, of course, we must make this *entire function* match the initial condition $u(x, 0) = T_0(x)$. Plugging $t = 0$ into our function gives us:

$$u(x, 0) = \frac{T_R - T_L}{L} x + T_L + \sum_{n=1}^{\infty} A_n \sin\left(\frac{n\pi}{L} x\right) = T_0(x)$$

As usual we appeal to the power of a Fourier sine series to represent any function on a finite domain. In this case we have to choose our coefficients A_n so that:

$$\sum_{n=1}^{\infty} A_n \sin\left(\frac{n\pi}{L} x\right) = T_0(x) - \frac{T_R - T_L}{L} x - T_L$$

The solution is straightforward to write down in general, even though it may be intimidating to calculate for a given function T_0:

$$A_n = \frac{2}{L} \int_0^L \left(T_0(x) - \frac{T_R - T_L}{L} x - T_L\right) \sin\left(\frac{n\pi}{L} x\right) dx$$

A Physical Look at the "Particular" Solution

We presented the process above as nothing more than a mathematical trick. We made a simple mathematical assumption—in this case, $\partial u/\partial t = 0$—so we could find a function that would match both our differential equation and our boundary conditions. If we had been solving a different PDE we might have used a different assumption for the same purpose.

But $\partial u/\partial t = 0$ is not just any assumption: it says "I want to find the solution to this equation that will *never change*," the "steady-state" solution. If the heat in the rod ever happens to assume the distribution $u = [(T_R - T_L)/L]x + T_L$, it will stay that way forever.

And what of the complementary solution? Because we were forced by the second-order equation for $X(x)$ to choose a negative separation constant, the first-order equation for $T(t)$ led to a decaying exponential function $e^{-k^2 t}$. No matter what the initial conditions, the complementary solution will gradually die down toward zero; the total solution $u_P + u_C$ will approach the solution represented by u_P. In the language of physics, u_P represents a *stable* solution: the system will always trend toward that state.

In some situations you will find a steady-state solution that the system does not approach over time, analogous to an unstable equilibrium point in mechanics. In many cases such as this one, though, the steady-state solution represents the late-time behavior of the system and the complementary solution represents the transient response to the initial conditions.

The fact that this physical system approaches the steady-state solution is not surprising if you think about it. If the ends of the rod are held at constant temperature for a long time, the rod will move toward the simplest possible temperature distribution, which is a linear change from the left temperature to the right. We mention this to remind you of where we began the very first chapter of the book. *Solving* equations is a valuable skill, but computer solutions are becoming faster and more accurate all the time. *Understanding the solutions,* on the other hand, will require human intervention for the foreseeable future. To put it bluntly, interpreting solutions is the part that someone might pay you to do.

11.8.3 Problems: Inhomogeneous Boundary Conditions

11.164 Derive Equation 11.8.1 in the Explanation (Section 11.8.2) for $u_T(x, y)$. The process will be very similar to the one shown for deriving u_L.

11.165 Walk-Through: Inhomogeneous Boundary Conditions. In this problem you will solve the equation $2\partial^2 f/\partial x^2 - \partial^2 f/\partial y^2 = 0$ on a square extending from the origin to the point (π, π) with boundary conditions $f(x, 0) = f(0, y) = 0$, $f(x, \pi) = 4\sin(5x)$, and $f(\pi, y) = 3\sin(7y)$.

(a) Use separation of variables to find a function $f_1(x, y)$ that satisfies this PDE with boundary conditions $f_1(x, 0) = f_1(x, \pi) = f_1(0, y) = 0$, and $f_1(\pi, y) = 3\sin(7y)$.

(b) Use separation of variables to find a function $f_2(x, y)$ that satisfies this PDE with boundary conditions $f_1(x, 0) = f_2(0, y) = f_2(\pi, y) = 0$ and $f_2(x, \pi) = 4\sin(5x)$.

(c) Demonstrate that the function $f(x, y) = f_1(x, y) + f_2(x, y)$ satisfies the original PDE and the original boundary conditions.

For Problems 11.166–11.169 solve the given PDEs subject to the given boundary conditions.

11.166 $\partial^2 V/\partial x^2 + \partial^2 V/\partial y^2 + \partial^2 V/\partial z^2 = 0$,
$V(0, y, z) = V(x, 0, z) = V(x, y, 0) = V(x, y, L) = 0$, $V(L, y, z) = V_0$, $V(x, L, z) = 2V_0$

11.167 $\partial^2 u/\partial x^2 + \partial^2 u/\partial y^2 - \gamma^2 u = 0$, $u(0, y) = u(x, L) = 0$, $u(L, y) = u_0$, $u(x, 0) = u_0 \sin(2\pi x/L)$

11.168 $y^2 \left(\partial^2 u/\partial x^2 - \partial^2 u/\partial y^2\right) - y(\partial u/\partial y) + u = 0$, $u(0, y) = u(x, 0) = 0$, $u(L, y) = \sin(\pi y/H)$, $u(x, H) = u_0$. Your answer should be an infinite sum. The formula for the coefficients will include an integral that you will not be able to evaluate analytically, so you should simply leave it as an integral.

11.169 $\partial^2 u/\partial x^2 + \partial^2 u/\partial y^2 - \partial u/\partial y = 0$ in the rectangular region $0 \le x \le 1, 0 \le y \le 3$ subject to the boundary conditions $u(0, y) = u(1, y) = 0$, $u(x, 0) = 3\sin(2\pi x)$, $u(x, 3) = \sin(\pi x)$. (*Warning*: The answer will be somewhat messy. You should, however, be able to get a solution with no sums or integrals in it.)

11.170 Derive Equation 11.8.2 in the Explanation (Section 11.8.2) for $u_C(x, t)$.

11.171 Walk-Through: Particular and Complementary Solutions. In this problem you'll solve the equation $\partial u/\partial t = \partial^2 u/\partial x^2 - u$ with the boundary conditions $u(0, t) = 0$, $u(1, t) = 1$ and the initial condition $u(x, 0) = x$.

(a) Find the steady-state solution $u_{ss}(x)$ by solving $\partial^2 u_{ss}/\partial x^2 - u_{ss} = 0$ subject to the boundary conditions above. (You can solve this ODE by inspection, and plug in the boundary conditions to find the arbitrary constants.) The solution $u_{ss}(x)$ should solve our original PDE and boundary conditions, but since it has no arbitrary constants it cannot be made to satisfy the initial condition.

(b) Find a complementary solution to the equation $\partial u_C/\partial t = \partial^2 u_C/\partial x^2 - u_C$ subject to the homogeneous boundary conditions $u(0, t) = u(1, t) = 0$. (You can solve this PDE by separation of variables.) This complementary solution should have an arbitrary constant, but it will not satisfy the original boundary conditions.

(c) Add these two solutions to get the general solution to the original PDE with the original boundary conditions.

(d) Apply the initial conditions to this general solution and solve for the arbitrary constants to find the complete solution $u(x, t)$ to this problem.

(e) What function is your solution approaching as $t \to \infty$?

For Problems 11.172–11.174 solve the given PDEs subject to the given boundary and initial conditions. For each one you will need to find a steady-state solution and a complementary solution, add the two, and then apply the initial conditions. For each problem, will the general solution approach the steady-state solution or not? Explain. It may help to first work through Problem 11.171 as a model.

11.172 $\partial^2 u/\partial t^2 = v^2(\partial^2 u/\partial x^2)$, $u(0, t) = 0$, $u(1, t) = 2$, $u(x, 0) = \partial u/\partial t(x, 0) = 0$

11.173 $\partial u/\partial t = \partial^2 u/\partial \theta^2 + u$, $u(0, t) = 0$, $u(\pi/2, t) = \alpha$, $u(\theta, 0) = \alpha\,[\sin(\theta) + \sin(4\theta)]$.

11.174 $\partial u/\partial t = \partial^2 u/\partial x^2 + (1/x)(\partial u/\partial x) - (1/x^2)u$, $u(0, t) = 0$, $u(1, t) = 1$, $u(x, 0) = 0$

11.175 The left end of a rod at $x = 0$ is immersed in ice water that holds the end at $0°$ C and the right edge at $x = 1$ is immersed in boiling water that keeps it at $100°$ C. (You may assume throughout the problem that all temperatures are measured in degrees Celsius and all distances are measured in meters.)

(a) What is the steady-state temperature $u_{SS}(x)$ of the rod at late times?

(b) Solve the heat equation (11.2.3) with these boundary conditions and with initial condition $u(x, 0) = 100x(2 - x)$ to find the temperature $u(x, t)$. You can use your steady-state solution as a "particular" solution.

(c) The value of α for a copper rod is about $\alpha = 10^{-4}$ m^2/s. If a 1 meter copper rod started with the initial temperature given in this problem how long it would take for the point $x = 1/2$ to get within 1% of its steady-state temperature? Although your solution for $u(x, t)$ is an infinite series, you should answer this question by neglecting all of the terms except u_{SS} and the first non-zero term of u_C.

11.176 Exploration: An Inhomogeneous PDE
We have used the technique of finding a particular and a complementary solution in two different contexts: inhomogeneous ODEs, and homogeneous PDEs with multiple inhomogeneous conditions. We can use the same technique for inhomogeneous PDEs. Consider as an example the equation $\partial u/\partial t - \partial^2 u/\partial x^2 = \kappa$ subject to the boundary conditions $u(0, t) = 0$, $u(1, t) = 1$ and the initial condition $u(x, 0) = (1 + \kappa/2)\,x - (\kappa/2)x^2 + \sin(3\pi x)$.

(a) Find a steady-state solution u_{SS} that solves $-\partial^2 u_{SS}/\partial x^2 = \kappa$ subject to the boundary conditions given above. Because this will be a *particular* solution you do not need any arbitrary constants.

(b) Find the general complementary solution that solves $\partial u_C/\partial t - \partial^2 u_C/\partial x^2 = 0$ subject to the boundary conditions $u(0, t) = u(1, t) = 0$. You don't need to apply the initial conditions yet.

(c) Add the two to find the general solution to the original PDE subject to the original boundary conditions. Apply the initial condition and solve for the arbitrary constants to get the complete solution.

(d) Verify by direct substitution that this solution satisfies the PDE, the boundary conditions, and the initial condition.

11.9 The Method of Eigenfunction Expansion

The rest of the chapter will be devoted to three different techniques: "eigenfunction expansion," "the method of Fourier transforms," and "the method of Laplace transforms." All three can solve some PDEs that cannot be solved by separation of variables.

All three techniques start by writing the given differential equation in a different form. The specific transformation is chosen to turn a derivative into a multiplication, thereby turning a PDE into an ODE.

As you read through this section, watch how a derivative turns into a multiplication, and how the resulting differential equation can be solved to find the function we were looking for—in the form of a Fourier series. After you follow the details, step back and consider what it was that made a Fourier series helpful for this particular differential equation, and you will be well set for the other transformations we will discuss.

11.9.1 Discovery Exercise: The Method of Eigenfunction Expansion

The temperature in a bar obeys the heat equation, with its ends fixed at zero temperature:

$$\frac{\partial u}{\partial t} = \alpha \frac{\partial^2 u}{\partial x^2} \tag{11.9.1}$$

$$u(0, t) = u(L, t) = 0$$

We previously found the temperature $u(x, t)$ of such a bar by using separation of variables; you are now going to solve the same problem (and hopefully get the same answer!) using a different technique, "eigenfunction expansion." Later we will see that eigenfunction expansion can be used in some situations where separation of variables cannot—most notably in solving some inhomogeneous equations.

To begin with, replace the unknown function $u(x, t)$ with its unknown Fourier sine expansion in x.

$$u(x, t) = \sum_{n=1}^{\infty} b_n(t) \sin\left(\frac{n\pi}{L} x\right) \tag{11.9.2}$$

1. Does this particular choice (a Fourier series with no cosines) guarantee that you meet your boundary conditions? If so, explain why. If not, explain what further steps will be taken later to meet them.
2. What is the second derivative with respect to x of $b_n(t) \sin(n\pi x/L)$?
3. What is the first derivative with respect to t of $b_n(t) \sin(n\pi x/L)$?
4. Replacing $u(x, t)$ with its Fourier sine series as shown in Equation 11.9.2, rewrite Equation 11.9.1.

See Check Yourself #77 in Appendix L

Now we use one of the key mathematical facts that makes this technique work: if two Fourier sine series with the same frequencies are equal to each other, then the coefficients must equal each other. For instance, the coefficient of $\sin(3x)$ in the first series must equal the coefficient of $\sin(3x)$ in the second series, and so on.

5. Set the nth coefficient on the left side of your answer to Part 4 equal to the nth coefficient on the right. The result should be an *ordinary* differential equation for the function $b_n(t)$.
6. Solve your equation to find the function $b_n(t)$.
7. Write the function $u(x, t)$ as a Fourier sine series, with the coefficients properly filled in.
8. How will your temperature function behave after a long time? Answer this question based on your answer to Part 7; then explain why this answer makes sense in light of the physical situation.

11.9.2 Explanation: The Method of Eigenfunction Expansions

Separation of variables is a powerful approach to solving partial differential equations, but it is not a universal one. Its most obvious limitation is that you may find it algebraically impossible to separate the variables: for example, separation of variables can never solve an *inhomogeneous* partial differential equation (although it may still prove useful in finding the "complementary" solution). Another limitation that you may recall from Section 11.4 is that you generally cannot apply separation of variables to a problem with homogeneous initial conditions.

The "method of eigenfunction expansions" may succeed in some cases where separation of variables fails.

The Problem

Let us return to the first problem we solved with separation of variables: a string fixed at the points $(0,0)$ and $(L,0)$ but free to vibrate between those points. The initial position $y(x, 0) = f(x)$ and initial velocity $(dy/dt)(x, 0) = g(x)$ are specified as before.

In this case, however, there is a "driving function"—which is to say, the string is subjected to an external force that pushes it up or down. This could be a very simple force such as gravity (pulling down equally at all times and places), or it could be something much more complicated such as shifting air pressure around the string. Our differential equation is now inhomogeneous:

$$\frac{\partial^2 y}{\partial x^2} - \frac{1}{v^2}\frac{\partial^2 y}{\partial t^2} = q(x, t) \tag{11.9.3}$$

where $q(x, t)$ is proportional to the external force on the string. In Problem 11.177 you will show that separation of variables doesn't work on this problem. Below we demonstrate a method that does.

Overview of the Method

Here's the plan. We're going to replace all the functions in this problem—$y(x, t)$, $q(x, t)$ and even the initial functions $f(x)$ and $g(x)$—with their respective Fourier sine series. (A physicist would say we are translating the problem into "Fourier space.") The resulting differential equation will give us the coefficients of the Fourier expansion of $y(x, t)$.

That may sound pointlessly roundabout, so it's worth discussing a few obvious questions before we jump into the math.

- **Can you really do that?** What makes this technique valid is that Fourier series are unique: when you find a Fourier sine series for a given function, you have found the *only* Fourier sine series for that function with those frequencies. Put another way, if two different Fourier sine series with the same frequencies equal each other for all values of x,

$$a_1 \sin(x) + a_2 \sin(2x) + a_3 \sin(3x) + \ldots = b_1 \sin(x) + b_2 \sin(2x) + b_3 \sin(3x) + \ldots$$

then a_1 must equal b_1, and so on. So after we turn both sides of our equation into Fourier series, we will confidently assert that the corresponding coefficients must be equal, and solve for them.
- **How does this make the problem easier?** Suppose some function $f(x)$ is expressed as a Fourier sine series:

$$f(x) = b_1 \sin(x) + b_2 \sin(2x) + b_3 \sin(3x) + \ldots$$

What happens, term by term, when you take the second derivative? $b_1 \sin(x)$ is just multiplied by -1. $b_2 \sin(2x)$ is multiplied by -4, the next term by -9, and so on. In general, for any function $b_n \sin(nx)$, *taking a second derivative is the same as multiplying by* $-n^2$.

So, what happens to a differential equation when you replace a derivative with a multiplication? If you started with an ordinary differential equation, you are left with an algebra equation. In our example, starting with a two-variable partial differential equation, we will be left with an ordinary differential equation. If we had started with a three-variable PDE, we would be down to two…and so on.

This will all (hopefully) become clear as you look through the example, but now you know what you're looking for.

- **Do you always use a Fourier sine series?** No. Below we discuss why that type of series is the best choice for this particular problem.

A note about notation: There is no standard notation for representing different Fourier series in the same problem, but we can't just use b_3 to mean "the coefficient of $\sin(3x)$" when we have four different series with different coefficients. So we're going to use b_{y3} for "the coefficient of $\sin(3x)$ in the Fourier series for $y(x, t)$," b_{q3} for the third coefficient in the expansion of $q(x, t)$, and similarly for $f(x)$ and $g(x)$.

The Expansion

The first step is to replace $y(x, t)$ with its Fourier expansion in x. Recalling that $y(x, t)$ is defined on the interval $0 \leq x \leq L$, we have three options: a sine-and-cosine series with period L, a sine-only expansion with period $2L$ based on an odd extension on the negative side, or a cosine-only expansion with period $2L$ based on an even extension on the negative side. Chapter 9 discussed these alternatives and why the boundary conditions $y(0, t) = y(L, t) = 0$ lend themselves to a sine expansion. We therefore choose to create an odd extension of our function, and write:

$$y(x, t) = \sum_{n=1}^{\infty} b_{yn}(t) \sin\left(\frac{n\pi}{L}x\right) \qquad (11.9.4)$$

with the frequency $n\pi/L$ chosen to provide the necessary period $2L$.

This differs from Chapter 9 because we are expanding a multivariate function (x and t) in one variable only (x). You can think about it this way: when $t = 2$, Equation 11.9.4 represents a specific function $y(x)$ being expanded into a Fourier series with certain (constant) coefficients. When $t = 3$ a different $y(x)$ is being expanded, with different coefficients…and so on, for all relevant t-values. So the coefficients b_{yn} are constants with respect to x, but vary with respect to time.

We will similarly expand all the other functions in the problem:

$$q(x,t) = \sum_{n=1}^{\infty} b_{qn}(t) \sin\left(\frac{n\pi}{L}x\right), \quad f(x) = \sum_{n=1}^{\infty} b_{fn} \sin\left(\frac{n\pi}{L}x\right), \quad g(x) = \sum_{n=1}^{\infty} b_{gn} \sin\left(\frac{n\pi}{L}x\right)$$

Since the initial functions $f(x)$ and $g(x)$ have no time dependence, b_{fn} and b_{gn} are constants.

Plugging In and Solving

We now plug Equation 11.9.4 into Equation 11.9.3 and get:

$$\frac{\partial^2}{\partial x^2}\left[\sum_{n=1}^{\infty} b_{yn}(t) \sin\left(\frac{n\pi}{L}x\right)\right] - \frac{1}{v^2}\frac{\partial^2}{\partial t^2}\left[\sum_{n=1}^{\infty} b_{yn}(t) \sin\left(\frac{n\pi}{L}x\right)\right] = \sum_{n=1}^{\infty} b_{qn}(t) \sin\left(\frac{n\pi}{L}x\right)$$

11.9 | The Method of Eigenfunction Expansion

The next step—rearranging the terms—may look suspicious if you know the properties of infinite series, and we will discuss later why it is valid in this case. But if you put aside your doubts, the algebra is straightforward.

$$\sum_{n=1}^{\infty} \left[\frac{\partial^2}{\partial x^2} \left(b_{yn}(t) \sin\left(\frac{n\pi}{L}x\right) \right) - \frac{1}{v^2} \frac{\partial^2}{\partial t^2} \left(b_{yn}(t) \sin\left(\frac{n\pi}{L}x\right) \right) \right] = \sum_{n=1}^{\infty} b_{qn}(t) \sin\left(\frac{n\pi}{L}x\right)$$

Now we take those derivatives. Pay particular attention, because this is the step where the equation becomes easier to work with.

$$\sum_{n=1}^{\infty} \left[-\frac{n^2\pi^2}{L^2} b_{yn}(t) \sin\left(\frac{n\pi}{L}x\right) - \frac{1}{v^2} \frac{d^2 b_{yn}(t)}{dt^2} \sin\left(\frac{n\pi}{L}x\right) \right] = \sum_{n=1}^{\infty} b_{qn}(t) \sin\left(\frac{n\pi}{L}x\right)$$

As we discussed earlier, the $\partial^2/\partial x^2$ operator has been replaced by a multiplied constant. At the same time, the partial derivative with respect to time has been replaced by an *ordinary* derivative, since b_{yn} has no x-dependence.

We can now rewrite our original differential equation 11.9.3 with both sides Fourier expanded:

$$\sum_{n=1}^{\infty} \left[-\frac{n^2\pi^2}{L^2} b_{yn}(t) - \frac{1}{v^2} \frac{d^2 b_{yn}(t)}{dt^2} \right] \sin\left(\frac{n\pi}{L}x\right) = \sum_{n=1}^{\infty} b_{qn}(t) \sin\left(\frac{n\pi}{L}x\right)$$

This is where the uniqueness of Fourier series comes into play: if the Fourier sine series on the left equals the Fourier sine series on the right, then their corresponding coefficients must be the same.

$$-\frac{n^2\pi^2}{L^2} b_{yn}(t) - \frac{1}{v^2} \frac{d^2 b_{yn}(t)}{dt^2} = b_{qn}(t) \tag{11.9.5}$$

It doesn't look pretty, but consider what we are now being asked to do. For any given $q(x,t)$ function, we find its Fourier sine series: that is, we find the coefficients $b_{qn}(t)$. That leaves us with an *ordinary* second-order differential equation for the coefficients $b_{yn}(t)$. We may be able to solve that equation by hand, or we may hand it over to a computer.[9] Either way we will have the original function $y(x,t)$ we were looking for—but we will have it in the form of a Fourier series.

In Problem 11.178 you will address the simplest possible case, that of $q(x,t) = 0$, to show that the solution matches the result we found using separation of variables. Below we show the result of a constant force such as gravity. In Problem 11.198 you will tackle the problem more generally using the technique of variation of parameters.

Of course the solution will contain two arbitrary constants, which bring us to our initial conditions. Our first condition is $y(x, 0) = f(x)$. Taking the Fourier expansion of both sides of *that* equation—and remembering once again the uniqueness of Fourier series—we see that $b_{yn}(0) = b_{fn}$. Our other condition, $dy/dt(x, 0) = g(x)$, tells us that $\dot{b}_{yn}(0) = b_{gn}$. So we can use our initial conditions on y to find the initial conditions for b: or to put it another way, we translate our initial conditions into Fourier space.

Once we have solved our ordinary differential equation with the proper initial conditions, we have the functions $b_{yn}(t)$. These are the coefficients of the Fourier series for y, so we now have our final solution in the form of a Fourier sine series.

---

[9]Fortunately, computers—which are still poor at solving partial differential equations—do a great job of finding Fourier coefficients and solving ordinary differential equations.

A Sample Driving Function

As a sample force let's consider the downward pull of gravity, so $q(x, t)$ is a constant Q. The Fourier sine series for a constant on the domain $0 \leq x \leq L$ is:

$$Q = \sum_{\text{odd } n} \frac{4Q}{\pi n} \sin\left(\frac{n\pi}{L}x\right)$$

For even values of n, b_{qn} is zero. A common mistake is to ignore those values entirely, but they are not irrelevant; instead, they turn Equation 11.9.5 into

$$\frac{d^2 b_{yn}(t)}{dt^2} = -\frac{v^2 n^2 \pi^2}{L^2} b_{yn}(t) \quad \text{even } n \tag{11.9.6}$$

We can solve this by inspection:

$$b_{yn} = A_n \sin\left(\frac{v n \pi}{L}t\right) + B_n \cos\left(\frac{v n \pi}{L}t\right) \quad \text{even } n \tag{11.9.7}$$

For odd values of n Equation 11.9.5 becomes:

$$\frac{d^2 b_{yn}}{dt^2} + C b_{yn} = D \text{ where } C = \frac{v^2 n^2 \pi^2}{L^2} \text{ and } D = -\frac{4Qv^2}{\pi n} \quad \text{odd } n$$

This is an inhomogeneous second-order ODE and we've already solved its complementary equation: the question was Equation 11.9.6 and our answer was Equation 11.9.7. That leaves us only to find a particular solution, and $b_{yn,particular} = D/C$ readily presents itself. So we can write:

$$b_{yn} = A_n \sin\left(\frac{v n \pi}{L}t\right) + B_n \cos\left(\frac{v n \pi}{L}t\right) - \left[\frac{4QL^2}{\pi^3 n^3}, n \text{ odd}\right] \tag{11.9.8}$$

Next we solve for A_n and B_n using the initial conditions $b_{yn}(0) = b_{fn}$ and $db_{yn}/dt(0) = b_{gn}$. You'll do this for different sets of initial conditions in the problems. The solution to our original PDE is

$$y(x, t) = \sum_{n=1}^{\infty} b_{yn}(t) \sin\left(\frac{n\pi}{L}x\right)$$

with b_{yn} defined by Equation 11.9.8.

EXAMPLE Eigenfunction Expansion

Solve the differential equation

$$-9\frac{\partial^2 y}{\partial x^2} + 4\frac{\partial y}{\partial t} + 5y = x$$

on the domain $0 \leq x \leq 1$ for all $t \geq 0$ subject to the boundary conditions $\partial y/\partial x(0, t) = \partial y/\partial x(1, t) = 0$ and the initial condition $y(x, 0) = f(x) = 2x^3 - 3x^2$.

Before we start solving this problem, let's think about what kind of solution we expect. The PDE is simplest to interpret if we write it as $4(\partial y/\partial t) = 9(\partial^2 y/\partial x^2) - 5y + x$, which we can read as "the vertical velocity depends on…." The first term says that y tends to increase when the concavity is positive and decrease when it is negative. The second two terms taken together say that y tends to increase if $-5y + x > 0$: that is, it will tend to move up if it is below the line $y = x/5$ and move down if it is above it. Both of these effects are always at work. For instance, if the curve is concave up and below the line $y = x/5$ both effects will push y upward; if it is concave down below the line the two effects will push in opposite directions.

The initial condition is shown below:

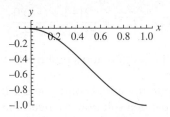

Since the initial function is negative over the whole domain, the $-5y + x$ terms will push it upward. Meanwhile, the concavity will push it down on the left and up on the right. You can easily confirm that the concavity will "win" on the left, so y should initially decrease at small values of x and increase at larger values. However, as the values on the left become more negative, they will eventually be pushed back up again.

To actually solve the equation, we note that the boundary conditions can be most easily met with a Fourier cosine series. So we replace y with its Fourier cosine expansion:

$$y(x,t) = \frac{a_{y0}(t)}{2} + \sum_{n=1}^{\infty} a_{yn}(t) \cos(n\pi x)$$

On the right side of our differential equation is the function x. Remember that we only care about this function from $x = 0$ to $x = 1$, but we need to create an even extension to find a cosine expansion. While you can do this entirely by hand (using integration by parts) or entirely with a computer, you may find that the easiest path is a hybrid: you first determine by hand that $a_n = 2 \int_0^1 x \cos(n\pi x) dx$ and then hand that integral to a computer, or use one of the integrals in Appendix G. (For the case $n = 0$ in particular, it's easiest to do the integral yourself.) You find that:

$$x = \frac{1}{2} + \sum_{n=1}^{\infty} \frac{2(-1+(-1)^n)}{\pi^2 n^2} \cos(n\pi x)$$

Plugging our expansions for y and x into the original problem yields:

$$2a'_{y0}(t) + \frac{5a_{y0}(t)}{2} + \sum_{n=1}^{\infty} \left[9n^2\pi^2 a_{yn}(t) \cos(n\pi x) + 4a'_{yn}(t) \cos(n\pi x) + 5a_{yn}(t) \cos(n\pi x) \right]$$
$$= \frac{1}{2} + \sum_{n=1}^{\infty} \frac{2[-1+(-1)^n]}{\pi^2 n^2} \cos(n\pi x)$$

Setting the Fourier coefficients on the left equal to the coefficients on the right and rearranging leads to the differential equations:

$$a'_{y0}(t) = -\frac{5}{4} a_{y0}(t) + \frac{1}{4}$$
$$a'_{yn}(t) = \left(\frac{-9n^2\pi^2 - 5}{4} \right) a_{yn}(t) + \frac{[-1+(-1)^n]}{2\pi^2 n^2}, \quad n > 0$$

Each of these is just the separable first-order differential equation $df/dx = Af + B$ with uglier-looking constants. The solution is $f = Ce^{Ax} - B/A$, or, in our case,

$$a_{y0}(t) = C_0 e^{-(5/4)t} + \frac{1}{5}$$

$$a_{yn}(t) = C_n e^{\frac{9n^2\pi^2 - 5}{4}t} + \frac{2[-1 + (-1)^n]}{\pi^2 n^2 (9n^2\pi^2 + 5)}, \quad n > 0$$

Note that the numerator of the last fraction simplifies to 0 for even n and -4 for odd n.

We now turn our attention to the initial condition $y(x, 0) = f(x) = 2x^3 - 3x^2$. To match this to our Fourier-expanded solution, we need the Fourier expansion of this function. Remember that we need a Fourier cosine expansion that is valid on the domain $0 \le x \le 1$. We leave it to you to confirm the result (which we obtained once again by figuring out by hand what integral to take and then handing that integral over to a computer):

$$a_{f0} = -1, \, a_{fn} = \frac{48}{\pi^4 n^4} \quad \text{(odd } n \text{ only)}$$

Setting $a_{y0}(0) = a_{f0}$ leads to $C_0 = -6/5$ and doing the same for higher values of n gives:

$$C_n = \frac{48}{\pi^4 n^4} + \frac{4}{\pi^2 n^2 (9n^2\pi^2 + 5)} \quad \text{(odd } n \text{ only)}$$

We have now solved the entire problem—initial conditions and all—in Fourier space. That is, we have found the a_n-values which are the coefficients of the Fourier series for $y(x, t)$. So the solution to our original problem is:

$$y(x, t) = \frac{1}{10} - \frac{3}{5} e^{-(5/4)t} + \sum_{\text{odd } n} \left[C_n e^{-(9n^2\pi^2 + 5)t/4} - \frac{4}{\pi^2 n^2 (9n^2\pi^2 + 5)} \right] \cos(n\pi x)$$

with the C_n coefficients defined as above.

It's an intimidating-looking formula to make sense of, but we can use a computer to evaluate partial sums and see how it behaves.

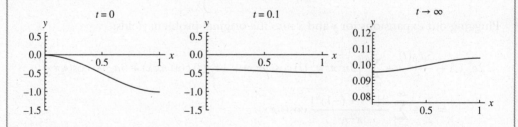

The figures above show the solution (using the 10th partial sum) at three times. As predicted the function initially moves down on the left and up on the right, but eventually moves up everywhere until the terms all cancel out to give $\partial y / \partial t = 0$.

Stepping Back

We can break the process above into the following steps.

1. Replace all the functions—for instance $y(x, t)$, $q(x, t)$, $f(x)$, and $g(x)$ in our first example above—with their respective Fourier series.
2. Plug the Fourier series back into the original equation and simplify.
3. You now have an equation of the form <*this Fourier series*>=<*that Fourier series*>. Setting the corresponding coefficients equal to each other, you get an equation to solve for the coefficients—one that should be easier than the original equation, because some of the derivatives have been replaced with multiplications.
4. Solve the resulting equation to find the Fourier coefficients of the function you are looking for—with arbitrary constants, of course. You find these arbitrary constants based on the Fourier coefficients of the initial conditions. You now have the function you were originally looking for, expressed in the form of a Fourier series.
5. In some cases, you may be able to explicitly sum the resulting Fourier series to find your original function in closed form. More often, you will use computers to plot or analyze the partial sums and/or the dominant terms of the Fourier series.

As always, it is important to understand some of the subtleties.

- **Why did we expand in x instead of in t?** There are two reasons, based on two very general limitations on this technique.

  First, *you can only do an eigenfunction expansion for a variable with homogeneous boundary conditions*. To understand why, think about how we dealt with the boundary conditions and the initial conditions in the problems above. For the initial condition—the condition on t, which we did *not* expand in—we expanded the condition itself into a Fourier series in x and matched that to our equation, coefficient by coefficient. But for the boundary condition—the condition on x, which we *did* expand in—we chose our sine or cosine series so that each individual term would match the boundary condition, knowing that this would make the entire series match the homogeneous condition.

  Second, as you may remember from Chapter 9, *you can only take a Fourier series for a function that is periodic, or defined on a finite domain*. In many problems, including the ones we worked above, the domain of x is finite but the domain of t is unlimited. For non-periodic functions you need to do a *transform* instead of a series expansion. Solving partial differential equations by transforms is the subject of Sections 11.10–11.11.

- **The "suspicious step"—moving a derivative in and out of a series.** In general, $\frac{d}{dx}\left(\sum f_n(x)\right)$ is *not* the same as $\sum (df_n/dx)$. For instance, you will show in Problem 11.199 that if you take the Fourier series for x, and take the derivative term by term, you do *not* end up with the Fourier series for 1. However, moving a derivative inside a Fourier series—which is vital to this method—*is* valid when the periodic extension of the function is continuous.[10] Since our function was equal on both ends its extension is everywhere continuous, so the step is valid.

- **Which Fourier series?** The choice of a Fourier sine or cosine series is dictated by the boundary conditions. In the "wave equation" problem above the terms in the expansion had to be sines in order to match the boundary conditions $y(0, t) = y(L, t) = 0$. In the example that started on Page 628 we needed cosines to match the condition $\partial y/\partial x(0, t) = \partial y/\partial x(1, t) = 0$.

- **Why a Fourier series at all?** We used a Fourier series because our differential equations were based on second derivatives with respect to x.

---

[10] See A. E. Taylor's "Differentiation of Fourier Series and Integrals" in *The American Mathematical Monthly*, Vol. 51, No. 1.

A function $f(x)$ is an "eigenfunction" of the operator L if $L[f(x)] = kf(x)$ for some constant k. That is the property we used in these problems: $b_n \sin(nx)$ and $b_n \cos(nx)$ are eigenfunctions of the second derivative operator. When $y(x, t)$ is expressed as a series of sinusoidal terms, the second derivative acting on each term is replaced with a simple multiplication by $-n^2$.

Different equations will require different expansions. For example, in Problem 11.205 you will show that the eigenfunctions of the Laplacian in polar coordinates involve Bessel functions and then use an expansion in Bessel functions to solve an inhomogeneous wave equation on a disk.

11.9.3 Problems: The Method of Eigenfunction Expansion

11.177 Try to solve the equation $\partial^2 y/\partial x^2 - (1/v^2)(\partial^2 y/\partial t^2) = xt$ using separation of variables. Explain why it doesn't work. (Your answer should not depend on the initial or boundary conditions.)

In the Explanation (Section 11.9.2) we found the general solution to the inhomogeneous wave equation 11.9.3 with boundary conditions $y(0, t) = y(L, t) = 0$. For Problems 11.178–11.183 you should use the given inhomogeneous term $q(x, t)$ and initial conditions $y(x, 0) = f(x)$, $\partial y/\partial t(x, 0) = g(x)$ to solve Equation 11.9.5 for $b_{yn}(t)$ and plug it into Equation 11.9.4 to get the complete solution $y(x, t)$. Your answers will be in the form of infinite series.

11.178 $q(x, t) = 0$. For this problem you should use generic initial conditions $f(x)$ and $g(x)$ and show that the solution you get from the method of eigenfunction expansion is the same one we got in Section 11.4 from separation of variables.

11.179 $q(x, t) = \kappa \sin(\omega t)$, $f(x) = g(x) = 0$

11.180 $q(x, t) = \kappa \sin(p\pi x/L)$, $f(x) = g(x) = 0$ (p an integer)

11.181 $q(x, t) = \kappa \sin(p\pi x/L) e^{-\omega t}$, $f(x) = g(x) = 0$ (p an integer)

11.182 $q(x, t) = \begin{cases} 1 & L/3 < x < 2L/3 \\ 0 & \text{elsewhere} \end{cases}$,

$f(x) = \sin\left(\frac{2\pi}{L}x\right)$, $g(x) = 0$

11.183 $q(x, t) = \kappa e^{-\omega t}$, $f(x) = \begin{cases} x & 0 \le x \le L/2 \\ L-x & L/2 < x \le L \end{cases}$, $g(x) = 0$

11.184 Walk-Through: Eigenfunction Expansion. In this problem you will solve the partial differential equation $\partial u/\partial t - \partial^2 u/\partial x^2 = xt$ with boundary conditions $u(0, t) = u(\pi, t) = 0$ using the method of eigenfunction expansions.

(a) In the first step, you replace the function $u(x, t)$ with the series $\sum_{n=1}^{\infty} b_{un}(t) \sin(nx)$. Explain why it's necessary to use a Fourier series with sines only (no cosines).

(b) Write the right side of the PDE as a Fourier sine series in x and find the coefficients $b_{qn}(t)$.

(c) Plug the Fourier sine expansions into both sides of the PDE. The x-derivatives should turn into multiplications. The result should look like:

$$\sum_{n=1}^{\infty} \left(\text{an expression involving } b_{un}(t) \text{ and } \frac{\partial b_{un}}{\partial t}\right) \sin(nx) = \sum_{n=1}^{\infty} (\text{a function of } n \text{ and } t) \sin(nx)$$

(d) If two Fourier series are equal to each other then each coefficient of one must equal the corresponding coefficient of the other. This means you can set the expressions in parentheses on left and right in Part (c) equal. The result should be an ODE for $b_{un}(t)$.

(e) Find the general solution to the ODE you wrote for $b_{un}(t)$ in Part (d) and use this to write the solution $u(x, t)$ as an infinite series. The answer should involve an arbitrary coefficient A_n inside the sum.

11.185 *[This problem depends on Problem 11.184.]* In this problem you will plug the initial condition

$u(x, 0) = \begin{cases} 1 & \pi/3 < x < 2\pi/3 \\ 0 & \text{elsewhere} \end{cases}$ into the

solution you found to Problem 11.184.

(a) Expand the given initial condition into a Fourier sine series.

(b) Plug $t = 0$ into your general solution to Problem 11.184 and set it equal to the Fourier-expanded initial condition you wrote in Part (a). Setting the coefficients on the left equal to the corresponding coefficients on the right, solve to find the coefficients A_n.

(c) Plugging the solution you just found for A_n into the solution you found in Problem 11.184 gives you a series for $u(x, t)$. Plot the 20th partial sum of this series solution at several times and describe its behavior.

Problems 11.186–11.192 are initial value problems that can be solved by the method of eigenfunction expansion. For each problem use the boundary conditions $y(0, t) = y(L, t) = 0$, and assume that time goes from 0 to ∞. When the initial conditions are given as arbitrary functions, write the solution as a series and write expressions for the coefficients in the series. When specific initial conditions are given, solve for the coefficients. The solution may still be in the form of a series. It may help to first work through Problems 11.184–11.185 as a model.

11.186 $\partial y/\partial t - (\partial^2 y/\partial x^2) + y = \kappa$, $y(x, 0) = 0$

11.187 $\partial^2 y/\partial t^2 - (\partial^2 y/\partial x^2) + y = \sin(\pi x/L)$, $y(x, 0) = \partial y/\partial t(x, 0) = 0$

11.188 $\partial^2 y/\partial t^2 - (\partial^2 y/\partial x^2) + y = \sin(\pi x/L)\cos(\omega t)$, $y(x, 0) = 0$, $\partial y/\partial t(x, 0) = g(x)$

11.189 $\partial y/\partial t - (\partial^2 y/\partial x^2) = e^{-t}$, $y(x, 0) = f(x)$

11.190 $\partial y/\partial t - (\partial^2 y/\partial x^2) = e^{-t}$, $y(x, 0) = \sin(3\pi x/L)$

11.191 $\dfrac{\partial^3 y}{\partial t \, \partial x^2} + y = \kappa$, $y(x, 0) = 0$.

11.192 $\dfrac{5}{4}\dfrac{\partial^2 y}{\partial t^2} - \dfrac{\partial^2 y}{\partial x^2} - 3\dfrac{\partial^3 y}{\partial t \, \partial x^2} = (\sin x)e^{-t}$, $y(x, 0) = 0$, $\dfrac{\partial y}{\partial t}(x, 0) = 0$. Take $L = \pi$. You should be able to express your answer in closed form (with no series).

Problems 11.193–11.197 are boundary problems that can be solved by the method of eigenfunction expansion. In all cases, x goes from 0 to L and y goes from 0 to H.

11.193 $\partial^2 u/\partial x^2 + \partial^2 u/\partial y^2 = y$, $u(0, y) = u(L, y) = u(x, H) = u(x, 0) = 0$

11.194 $\partial^2 u/\partial x^2 + \partial^2 u/\partial y^2 + \partial u/\partial x = y$, $u(0, y) = u(L, y) = u(x, H) = u(x, 0) = 0$

11.195 $\partial^2 u/\partial x^2 + \partial^2 u/\partial y^2 - u = \kappa$, $u(0, y) = u(L, y) = u(x, H) = u(x, 0) = 0$

11.196 $\partial^2 u/\partial x^2 + \partial^2 u/\partial y^2 = \sin(3\pi x/L)$, $u(x, 0) = u(0, y) = u(L, y) = 0$, $u(x, H) = \sin(2\pi x/L)$

11.197 $\partial^2 u/\partial x^2 + x(\partial^2 u/\partial y^2) + \partial u/\partial x = 0$, $u(L, y) = u(x, H) = u(x, 0) = 0$, $u(0, y) = \kappa$. Use eigenfunction expansion to reduce this to an ODE and use a computer to solve it with the appropriate boundary conditions. The solution $u(x, y)$ will be an infinite sum whose terms are hideous messes. Verify that it's correct by showing that each term individually obeys the original PDE, and by plotting a large enough partial sum of the series as a function of x and y to show that it matches the boundary conditions. (Making the plot will require choosing specific values for L, H, and κ.)

11.198 In the Explanation (Section 11.9.2) we showed that the problem of a bounded driven string could be solved by solving the ordinary differential equation 11.9.5. For particular driving functions $q(x, t)$ you might use a variety of techniques to solve this equation, but this ODE can be solved for a generic q with the technique *variation of parameters*, which we discussed in Chapter 10.

(a) Begin by solving the complementary homogeneous equation $-(n^2\pi^2/L^2)b_{yn}(t) - (1/v^2)(d^2 b_{yn}(t)/dt^2) = 0$ by inspection. You should end up with two linearly independent solutions $y_1(t)$ and $y_2(t)$. The general solution to the complementary equation is therefore $Ay_1(t) + By_2(t)$.

Our goal is now to find a solution—any particular solution!—to the original equation. We can then add this particular solution to $Ay_1(t) + By_2(t)$ to find the general solution.

(b) Use variation of parameters to find a particular solution to this equation. (You will need to begin by putting Equation 11.9.5 into the correct form for this technique.) Your solution will involve integrals based on the unknown function $b_{qn}(t)$, the Fourier coefficients of $q(x, t)$.

(c) As an example, consider the driving function $q(x, t) = t$. (The force is uniform across the string, but increases over time.) For that given driving force, take a Fourier sine series to find $b_{qn}(t)$.

(d) Plug that $b_{qn}(t)$ into your formulas and integrate to find $u(t)$ and $v(t)$. (You do *not* need an arbitrary constant when you integrate; remember, all we need is one working solution!) Put them together with your complementary solution to find the general solution to this problem.

(e) Demonstrate that your solution correctly solves the differential equation $-(n^2\pi^2/L^2)b_{yn}(t) - (1/v^2)(d^2 b_{yn}(t)/dt^2) = 4t/(n\pi)$.

11.199 Derivatives of Fourier sine series. The method of eigenfunction expansion relies on taking the derivative of an infinite series term by term. For example, it assumes that $\frac{d}{dx}\left(\sum b_n \sin x\right) = \sum \frac{d}{dx}(b_n \sin x)$ (so the derivative of a Fourier sine series is a Fourier cosine series). This step can safely be taken for a function that is continuous on $(-\infty, \infty)$.[11] Otherwise, it can get you into trouble!

Consider, as an example, the function $y = x$ on $0 \le x \le 1$. To find a Fourier sine series for this function, we create an odd extension on the interval $-1 \le x \le 0$ and then extend out periodically. Define $f(x)$ as this odd, extended version of the function $y = x$.

(a) Make a sketch of $f(x)$ from $x = -3$ to $x = 3$. Is it everywhere continuous?

(b) Find the Fourier series of the function $f(x)$ that you drew in Part (a). This is equivalent to taking the Fourier sine series of the original function, $y = x$ on $0 \le x \le 1$.

(c) What is $f'(x)$?

(d) Is $f'(x)$ odd, even, or neither? What does that tell you about its Fourier series?

(e) Find the Fourier series of $f'(x)$. (You can ignore the discontinuities in $f'(x)$: "holes" do not change a Fourier series as long as there is a finite number of them per period. Ignoring them, finding this Fourier series should be trivial and require no calculations.)

(f) Take the term-by-term derivative of the Fourier series for $f(x)$—the series you found in Part (b). Do you get the Fourier series for $f'(x)$, the series you find in Part (e)? Is the term-by-term derivative of the Fourier series for $f(x)$ a convergent series?

(g) Your work above should convince you that the derivative of a Fourier series is not always the Fourier series of the derivative. It is therefore important to know if we are dealing with a continuous function! Fortunately, it isn't hard to tell. In general, for any function $f(x)$ defined on a finite interval $0 \le f(x) \le L$, if you make an odd extension of the function and extend it periodically over the real line, the resulting function will be continuous if and only if $f(x)$ is continuous on the interval $0 < x < L$ and $f(0) = f(L) = 0$. Explain why these conditions are necessary and sufficient for the extension to be continuous.

11.200 *[This problem depends on Problem 11.199.]* **Derivatives of other Fourier series:** Suppose the function $f(x)$ is defined on the domain $0 < x < L$ and is continuous within that domain. Recall that we said you can take the derivative of a Fourier series term by term if the function is continuous on the entire real line.

(a) What are the conditions on $f(x)$ between 0 and L under which you can differentiate its Fourier cosine series term by term to find the Fourier cosine series of $\partial f/\partial t$? (In the notation of this section, what are the conditions under which $a_{(\partial f/\partial t)n} = \partial a_{fn}/\partial t$?) Recall that a Fourier cosine series involves an even extension of the function on the interval $-L < x < 0$. Include a brief explanation of why your answer is correct.

(b) What are the conditions on $f(x)$ between 0 and L under which you can differentiate the regular Fourier series (the one with sines and cosines) term by term to find the Fourier series of $\partial f/\partial t$? Include a brief explanation of why your answer is correct.

11.201 The air inside a flute obeys the wave equation with boundary conditions $\partial s/\partial x(0, t) = \partial s/\partial x(L, t) = 0$. The wave equation in this case is typically inhomogeneous because of someone blowing across an opening, creating a driving force that varies with x (position in the flute). Over a small period of time, it is reasonable to treat this driving function as a constant with respect to time. In this problem you will solve the equation $\partial^2 s/\partial x^2 - (1/c_s^2)(\partial^2 s/\partial t^2) = q(x)$ with the initial conditions $s(x, 0) = \partial s/\partial t(x, 0) = 0$.

(a) Explain why, for this problem, a cosine expansion will be easier to work with than a sine expansion.

---

[11] Strictly speaking your function must also be "piecewise smooth," meaning the derivative exists and is continuous at all but a finite number of points per period. Most functions you will encounter pass this test with no problem, but being continuous is a more serious issue, as the example in this problem illustrates.

(b) Expanding $s(x, t)$ and $q(x)$ into Fourier cosine series, write a differential equation for $a_{sn}(t)$.

(c) Find the general solution to this differential equation. You should find that the solution for general n doesn't work for $n = 0$ and you'll have to treat that case separately. Remember that $q(x)$ has no time dependence, so each a_{qn} is a constant.

(d) Now plug in the initial conditions to find the arbitrary constants and write the series solution $s(x, t)$.

(e) ![computer] As a specific example, consider $q(x) = ke^{-(x-L/2)^2/x_0^2}$. This driving term represents a constant force that is strongest at the middle of the tube and rapidly drops off as you move towards the ends. (This is not realistic in several ways, the most obvious of which is that the hole in a concert flute is not at the middle, but it nonetheless gives a qualitative idea of some of the behavior of flutes and other open tube instruments.) Use a computer to find the Fourier cosine expansion of this function and plot the 20th partial sum of $s(x, t)$ as a function of x at several times t (choosing values for the constants). Describe how the function behaves over time. Based on the results you find, explain why this equation could not be a good model of the air in a flute for more than a short time.

11.202 A thin pipe of length L being uniformly heated along its length obeys the inhomogeneous heat equation $\partial u/\partial t = \alpha(\partial^2 u/\partial x^2) + Q$ where Q is a constant. The ends of the pipe are held at zero degrees and the pipe is initially at zero degrees everywhere. (Assume temperatures are in celsius.)

(a) Solve the PDE with these boundary and initial conditions.

(b) Take the limit as $t \to \infty$ of your answer to get the steady-state solution. If you neglect all terms whose amplitude is less than 1% of the amplitude of the first term, how many terms are left in your series? Sketch the shape of the steady-state solution using only those non-negligible terms.

11.203 Exploration: Poisson's Equation—Part I
The electric potential in a region with charges obeys Poisson's equation, which in Cartesian coordinates can be written as $\partial^2 V/\partial x^2 + \partial^2 V/\partial y^2 + \partial^2 V/\partial z^2 = -(1/\varepsilon_0)\rho(x, y, z)$. In this problem you will solve Poisson's equation in a cube with boundary conditions $V(0, y, z) = V(L, y, z) = V(x, 0, z) = V(x, L, z) = V(x, y, 0) = V(x, y, L) = 0$. The charge distribution is given by $\rho(x, y, z) = \sin(\pi x/L)\sin(2\pi y/L)\sin(3\pi z/L)$.

(a) Write Poisson's equation with this charge distribution. Next, expand V in a Fourier sine series in x. When you plug this into Poisson's equation you should get a PDE for the Fourier coefficients $b_{vn}(y, z)$. Explain why the solution to this PDE will be $b_{vn} = 0$ for all but one value of n. Write the PDE for $b_{vn}(y, z)$ for that one value.

(b) Do a Fourier sine expansion of your b-variable in y. The notation becomes a bit strained at this point, but you can call the coefficients of this new expansion b_{bvn}. The result of this expansion should be to turn the equation from Part (a) into an ODE for $b_{bvn}(z)$. Once again you should find that the solution is $b_{bvn} = 0$ for all but one value of n. Write the ODE for that one value.

(c) Finally, do a Fourier sine expansion in z and solve the problem to find $V(x, y, z)$. Your answer should be in closed form, not a series. Plug this answer back in to Poisson's equation and show that it is a solution to the PDE and to the boundary conditions.

11.204 Exploration: Poisson's Equation—Part II
[This problem depends on Problem 11.203.]
Solve Poisson's equation for the charge distribution $\rho(x, y, z) = \sin(\pi x/L)\sin(2\pi y/L) z$. The process will be the same as in the last problem, except that for the last sine series you will have to expand both the right and left-hand sides of the equation, and your final answer will be in the form of a series.

11.205 Exploration: A driven drum
In Section 11.6 we solved the wave equation on a circular drum of radius a in polar coordinates, and we found that the normal modes were Bessel functions. If the drum is being excited by an external source (imagine such a thing!) then it obeys an inhomogeneous wave equation

$$\frac{\partial^2 z}{\partial t^2} - v^2\left(\frac{\partial^2 z}{\partial \rho^2} + \frac{1}{\rho}\frac{\partial z}{\partial \rho}\right)$$
$$= \begin{cases} \kappa\cos(\omega t) & 0 \leq \rho \leq a/2 \\ 0 & \rho > a/2 \end{cases}$$

Since everything in the problem depends only on ρ we eliminated the ϕ-dependence from the Laplacian. The drum is clamped down so $z = 0$ at the outer edge, and the initial conditions are $z(\rho, 0) = \partial z/\partial t(\rho, 0) = 0$.

(a) To use eigenfunction expansion, we need the right eigenfunctions: a set of normal modes $R_n(\rho)$ with the property that $d^2 R_n/d\rho^2 + (1/\rho)(dR_n/d\rho) = qR_n(\rho)$ for some proportionality constant q. For positive q-values, this leads to modified (or "hyperbolic") Bessel functions. Explain why these functions cannot be valid solutions for our drum.

(b) We therefore assume a negative proportionality constant: replace q with $-s^2$ and solve the resulting ODE for the functions $R_n(\rho)$. Use the implicit boundary condition to eliminate one arbitrary constant, and the explicit boundary condition to constrain the values of s. You do *not* need to apply the initial conditions at this stage. The Bessel functions you are left with as solutions for $R_n(\rho)$ are the eigenfunctions you will use for the method of eigenfunction expansion.

(c) Now return to the inhomogeneous wave equation. Expand both sides in a Fourier-Bessel expansion using the eigenfunctions you found in Part (b). The result should be an ODE for the coefficients $A_{zn}(t)$.

(d) The resulting differential equation looks much less intimidating if you rewrite the right side in terms of a new constant:

$$\gamma_n = \frac{2}{J_1^2(\alpha_{0,n})} \int_0^{1/2} J_0(\alpha_{0,n} u) u \, du$$

There is no simple analytical answer for this integral, but you can find a numerical value for γ_n for any particular n. Rewrite your answer to Part (c) using γ_n.

(e) Solve this ODE with the initial conditions $A_{zn}(0) = \partial A_{zn}/\partial t(0) = 0$.

(f) Plug your answer for $A_{zn}(t)$ into your Fourier-Bessel expansion for $z(\rho, t)$ to get the solution.

(g) Use the 20th partial sum of your series solution to make a 3D plot of the shape of the drumhead at various times. Describe how it evolves in time.

11.10 The Method of Fourier Transforms

The previous section used a Fourier series to turn a derivative into a multiplication, which turned a PDE into an ODE. For a non-periodic function on an infinite domain, you can accomplish the same thing using a Fourier transform instead of a Fourier series.

11.10.1 Discovery Exercise: The Method of Fourier Transforms

We have seen that the temperature in a bar obeys the heat equation:

$$\frac{\partial u}{\partial t} = \alpha \frac{\partial^2 u}{\partial x^2} \qquad (11.10.1)$$

Now consider the temperature $u(x, t)$ of an *infinitely long* bar.

1. When you used the method of eigenfunction expansions to solve this problem for a finite bar (Exercise 11.9.1), you began by expanding the unknown solution $u(x, t)$ in a Fourier series. Explain why you cannot do the same thing in this case.

You can take an approach that is similar to the method of eigenfunction expansion, but in this case you will use a Fourier *transform* instead of a Fourier series. You begin by taking a Fourier transform of both sides of Equation 11.10.1. Using $\mathcal{F}$ to designate a Fourier transform with respect to x, this gives:

$$\mathcal{F}\left[\frac{\partial u}{\partial t}\right] = \mathcal{F}\left[\alpha \frac{\partial^2 u}{\partial x^2}\right] \qquad (11.10.2)$$

2. The u you are looking for is a function of x and t. When you solve Equation 11.10.2 you will find a new function $\mathcal{F}[u]$. What will that be a function of? (It's not a function of u. That's the original function you're taking the Fourier transform of.)
3. One property of Fourier transforms is "linearity" which tells us that, in general, $\mathcal{F}[af + bg] = a\mathcal{F}(f) + b\mathcal{F}(g)$. Another property of Fourier transforms is that $\mathcal{F}\left[\partial^2 f/\partial x^2\right] = -p^2 \mathcal{F}[f]$. Apply these properties (in order) to the right side of Equation 11.10.2.
4. Another property of Fourier transforms is that, if the Fourier transform is with respect to x and the derivative is with respect to t, you can move the derivative in and out of the transform: $\mathcal{F}\left[\partial f/\partial t\right] = \frac{\partial}{\partial t}\mathcal{F}[f]$. Apply this property to the left side of Equation 11.10.2.

See Check Yourself #78 in Appendix L

5. Solve this first-order differential equation to find $\mathcal{F}[u]$ as a function of p and t. Your solution will involve an arbitrary function $g(p)$.
6. Use the formula for an inverse Fourier transform to write the general solution $u(x, t)$ as an integral. (Do not evaluate the integral.) See Appendix G for the Fourier transform and inverse transform formulas. Your answer should depend on x and t. (Even though p appears in the answer, it only appears inside a definite integral, so the answer is *not* a function of p.)
7. What additional information would you need to solve for the arbitrary function $g(p)$ and thus get a particular solution to this PDE?

11.10.2 Explanation: The Method of Fourier Transforms

Recall from Chapter 9 that a Fourier *series* always represents a periodic function. If a function is defined on a finite domain, you can make a Fourier series for it by periodically extending it over the whole real line. But for a non-periodic function on an infinite domain, a Fourier *transform* is needed instead.

In the last section we expanded partial differential equations in Fourier series; in this section we use the "method of transforms." Watch how the example below parallels the method of eigenfunction expansion, but solves a problem that is defined on an infinite domain.

Notation and Properties of Fourier Transforms

Fourier transforms are discussed in Section 9.6, and the formulas are collected in Appendix G. But we need to raise a few issues that were not mentioned in that chapter. First, we need a bit of new notation. We will use $\hat{f}(p)$ for the Fourier transform of $f(x)$ just as we did in Chapter 9, but we also need a way of representing the Fourier transform of a larger expression. We will use the symbol $\mathcal{F}$.

More substantially, in this section we will be taking Fourier transforms of multivariate functions. These are not really multivariate Fourier transforms; we are taking the Fourier transform with respect to x, treating t as a constant.

$$\mathcal{F}\left[f(x, t)\right] = \hat{f}(p, t)$$

Most importantly, our work here requires a few properties of Fourier transforms. Given that a Fourier transform represents a function as an integral over terms of the form $\hat{f}(p, t)e^{ipx}$, the following two properties are not too surprising:

$$\mathcal{F}\left[\frac{\partial^{(n)} f}{\partial t^{(n)}}\right] = \frac{\partial^{(n)}}{\partial t^{(n)}} \mathcal{F}[f] \qquad (11.10.3)$$

$$\mathcal{F}\left[\frac{\partial^{(n)} f}{\partial x^{(n)}}\right] = (ip)^n \mathcal{F}[f] \qquad (11.10.4)$$

Chapter 11 Partial Differential Equations

The second formula turns a derivative with respect to x into a multiplication, and therefore turns a partial differential equation into an ordinary differential equation, just as Fourier *series* did in the previous section.

Finally, we will need the "linearity" property of Fourier transforms:

$$\mathcal{F}[af + bg] = a F(f) + b F(g)$$

The Problem

In order to highlight the similarity between this method and the previous one, we're going to solve essentially the same problem: a wave on a one-dimensional string driven by an arbitrary force function.

$$\frac{\partial^2 y}{\partial x^2} - \frac{1}{v^2}\frac{\partial^2 y}{\partial t^2} = q(x, t) \tag{11.10.5}$$

$$y(x, 0) = f(x), \quad \frac{\partial y}{\partial t}(x, 0) = g(x)$$

However, our new string is infinitely long. This change pushes us from a Fourier *series* to a Fourier *transform*.

We are going to Fourier transform all three of these equations before we're through. Of course, we can only use this method in this way if it is possible to Fourier transform all the relevant functions! At the end of this section we will talk about some of the limitations that restriction imposes.

The Solution

We begin by taking the Fourier transform of both sides of Equation 11.10.5:

$$\mathcal{F}\left[\frac{\partial^2 y}{\partial x^2} - \frac{1}{v^2}\frac{\partial^2 y}{\partial t^2}\right] = \mathcal{F}[q]$$

Applying the linearity property first, we write:

$$\mathcal{F}\left[\frac{\partial^2 y}{\partial x^2}\right] - \frac{1}{v^2}\mathcal{F}\left[\frac{\partial^2 y}{\partial t^2}\right] = \mathcal{F}[q] \tag{11.10.6}$$

Now our derivative properties come into play. Equations 11.10.3 and 11.10.4 turn this differential equation into $-p^2 \hat{y} - (1/v^2)(\partial^2 \hat{y}/\partial t^2) = \hat{q}$. This is the key step: a second derivative with respect to x has become a multiplication by p^2. This equation can be written more simply as:

$$\frac{\partial^2 \hat{y}}{\partial t^2} + v^2 p^2 \hat{y} = -v^2 \hat{q} \tag{11.10.7}$$

It may look like we have just traded our old $y(x, t)$ PDE for a $\hat{y}(p, t)$ PDE. But our new equation has derivatives only with respect to t. The variable p in this equation acts like n in the eigenfunction expansion: for any given value of p, we have an ODE in t that we can solve by hand or by computer. The result will be the function $\hat{y}(p, t)$, the Fourier transform of the function we are looking for.

Just as you can use a series solution by evaluating as many partial sums as needed, you can often use a Fourier transform solution by finding numerical approximations to the inverse

11.10 | The Method of Fourier Transforms

Fourier transform. In some cases you will be able to evaluate the integral explicitly to find a closed-form solution for $y(x, t)$.

A Sample Driving Function

As a sample force, consider the effect of hanging a small weight from our infinitely long string:

$$\frac{\partial^2 y}{\partial x^2} - \frac{1}{v^2}\frac{\partial^2 y}{\partial t^2} = \begin{cases} Q & -L < x < L \\ 0 & \text{elsewhere} \end{cases} \tag{11.10.8}$$

The driving force in this case is independent of time. Notice that we are neglecting any forces other than this small weight, so the rest of the string will not fall until it is pulled down by the weighted part. Before starting the problem take a moment to think about what you would expect the solution to look like in the simplest case, where the string starts off perfectly still and horizontal.

Throughout this calculation there will be a number of steps such as taking a Fourier transform or solving an ODE that you could solve either by hand or on a computer. We'll just show the results as needed and in Problem 11.206 you'll fill in the missing calculations.

We take the Fourier transform of the right hand side of Equation 11.10.8 and plug it into Equation 11.10.7.

$$\frac{\partial^2 \hat{y}}{\partial t^2} + v^2 p^2 \hat{y} = -\frac{Qv^2}{\pi p}\sin(Lp) \tag{11.10.9}$$

Since the inhomogeneous part of this equation has no t dependence, it's really just the equation $\hat{y}''(t) + a\hat{y}(t) = b$ where a and b act as constants. (They depend on p but not on t.) You can solve it with guess and check and end up here.

$$\hat{y}(p, t) = A(p)\sin(pvt) + B(p)\cos(pvt) - \frac{Q}{\pi p^3}\sin(Lp) \tag{11.10.10}$$

The arbitrary "constants" A and B are constants with respect to t, but they are functions of p and we have labeled them as such. They will be determined by the initial conditions, just as A_n and B_n were in the series expansions of the previous section. In Problem 11.206 you will solve this for the simplest case $y(x, 0) = \partial y/\partial t(x, 0) = 0$ and show that $B(p) = (Q/\pi p^3)\sin(Lp)$ and $A(p) = 0$. So $\hat{y}(p, t) = (Q/\pi p^3)\sin(Lp)\big(\cos(pvt) - 1\big)$. This can be simplified with a trig identity to become $\hat{y}(p, t) = -(2Q/\pi p^3)\sin(Lp)\sin^2\big(pvt/2\big)$. The solution $y(x, t)$ is the inverse Fourier transform of $\hat{y}(p, t)$.

$$y(x, t) = -\frac{2Q}{\pi}\int_{-\infty}^{\infty}\frac{\sin(Lp)}{p^3}\sin^2\left(\frac{pvt}{2}\right)e^{ipx}\,dp \tag{11.10.11}$$

This inverse Fourier transform can be calculated analytically, but the result is messy because you get different functions in different domains. With the aid of a computer, however, we can get a clear—and physically unsurprising—picture of the result. At early times the weight pulls the region around it down into a parabola. At later times the weighted part makes a triangle, with the straight lines at the top and edges smoothly connected by small parabolas. In the regions $x > d + vt$ and $x < -d - vt$, the function $y(x, t)$ is zero because the effect of the weight hasn't yet reached the string.

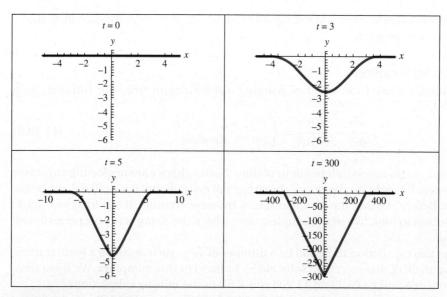

FIGURE 11.7 The solution for an infinite string continually pulled down by a weight near the origin. These figures represent the solution given above with $Q = L = v = 1$.

> **EXAMPLE** **Using a Fourier Transform to Solve a PDE**

Solve the differential equation:

$$-9\frac{\partial^2 y}{\partial x^2} + 4\frac{\partial y}{\partial t} + 5y = \begin{cases} 1 & -1 \leq x \leq 1 \\ 0 & |x| > 1 \end{cases}$$

on the domain $-\infty \leq x \leq \infty$ for all $t \geq 0$ with initial condition $y(x, 0) = f(x) = e^{-x^2}$.

Before solving this let's consider what kind of behavior we expect. If we rewrite this as

$$4\frac{\partial y}{\partial t} = 9\frac{\partial^2 y}{\partial x^2} - 5y + \begin{cases} 1 & -1 \leq x \leq 1 \\ 0 & |x| > 1 \end{cases}$$

then we can see the function will tend to decrease when it is concave down and/or positive. In addition, it will always have a tendency to increase in the range $-1 \leq x \leq 1$.

The initial position is shown below.

The initial function $y(x, 0) = f(x) = e^{-x^2}$

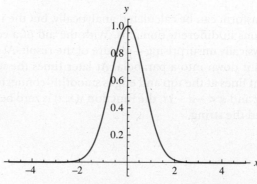

11.10 | The Method of Fourier Transforms

Near $x = 0$, the negative concavity and positive values will push downward more strongly than the driving term pushes upward, so the peak in the middle will decrease until these forces cancel. At larger values of x the upward push from the positive concavity is larger than the downward push from the positive values of y (you can check this), so the function will initially increase there. At very large values of x there is no driving force, and the concavity and y-values are near zero, so there will be little initial movement.

Now let's find the actual solution. We begin by taking the Fourier transform of both sides.

$$\mathcal{F}\left[-9\frac{\partial^2 y}{\partial x^2} + 4\frac{\partial y}{\partial t} + 5y\right] = \mathcal{F}\left[\begin{cases} 1 & -1 \leq x \leq 1 \\ 0 & |x| > 1 \end{cases}\right]$$

Apply the linearity of Fourier transforms on the left.

$$-9\mathcal{F}\left[\frac{\partial^2 y}{\partial x^2}\right] + 4\mathcal{F}\left[\frac{\partial y}{\partial t}\right] + 5\mathcal{F}[y] = \mathcal{F}\left[\begin{cases} 1 & -1 \leq x \leq 1 \\ 0 & |x| > 1 \end{cases}\right]$$

Next come our derivative properties.

$$9p^2\hat{y} + 4\frac{\partial \hat{y}}{\partial t} + 5\hat{y} = \mathcal{F}\left[\begin{cases} 1 & -1 \leq x \leq 1 \\ 0 & |x| > 1 \end{cases}\right]$$

which we can rearrange as:

$$4\frac{\partial \hat{y}}{\partial t} + (5 + 9p^2)\hat{y} = \mathcal{F}\left[\begin{cases} 1 & -1 \leq x \leq 1 \\ 0 & |x| > 1 \end{cases}\right]$$

The right side of this equation is an easy enough Fourier transform to evaluate, and we're skipping the integration steps here.

$$4\frac{\partial \hat{y}}{\partial t} + (5 + 9p^2)\hat{y} = \frac{\sin p}{\pi p}$$

Although the constants are ugly, this is just a separable first-order ODE. Once again we can solve it by hand or by software.

$$\hat{y} = C(p)e^{-(9p^2+5)t/4} + \frac{\sin p}{\pi p(9p^2 + 5)} \tag{11.10.12}$$

Next we apply the initial condition to find $C(p)$. You can find the Fourier transform of the initial condition in Appendix G.

$$\hat{f}(p) = \frac{1}{2\sqrt{\pi}} e^{-p^2/4}$$

Setting $\hat{y}(p, 0) = \hat{f}(p)$ gives:

$$C(p) = \frac{1}{2\sqrt{\pi}} e^{-p^2/4} - \frac{\sin p}{\pi p(9p^2 + 5)} \tag{11.10.13}$$

Finally, we can get the solution $y(x, t)$ by plugging Equation 11.10.13 into 11.10.12 and taking the inverse Fourier transform.

$$y(x, t) = \int_{-\infty}^{\infty} \left[\left(\frac{1}{2\sqrt{\pi}} e^{-p^2/4} - \frac{\sin p}{\pi p(9p^2 + 5)} \right) e^{-(9p^2+5)t/4} + \frac{\sin p}{\pi p(9p^2 + 5)} \right] e^{ipx} dp$$

There's no simple way to evaluate this integral in the general case, but we can understand its behavior by looking at the time dependence. At $t = 0$ the last two terms cancel and we are left with the inverse Fourier transform of $e^{-p^2/4}/(2\sqrt{\pi})$, which reproduces the initial condition $y(x, 0) = e^{-x^2}$. (If the solution didn't reproduce the initial conditions when $t = 0$ we would know we had made a mistake.) At late times the term $e^{-\frac{9p^2+5}{4}t}$ goes to zero and we are left with the inverse Fourier transform of the last term, which can be analytically evaluated on a computer to give

$$\lim_{t \to \infty} y(x, t) = \begin{cases} \frac{1}{10} \left(e^{\sqrt{5}/3} - e^{-\sqrt{5}/3} \right) e^{(\sqrt{5}/3)x} & x < -1 \\ \frac{1}{10} \left[2 - e^{-\sqrt{5}/3} \left(e^{(\sqrt{5}/3)x} + e^{-(\sqrt{5}/3)x} \right) \right] & -1 \leq x \leq 1 \\ \frac{1}{10} \left(e^{\sqrt{5}/3} - e^{-\sqrt{5}/3} \right) e^{-(\sqrt{5}/3)x} & x > 1 \end{cases} \quad (11.10.14)$$

As complicated as that looks, it is just some numbers multiplied by some exponential functions.

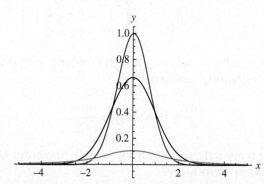

The solution 11.10.14 at $t = 0$ (blue), $t = .1$ (black), and in the limit $t \to \infty$ (gray).

This picture generally confirms the predictions we made earlier. The peak at $x = 0$ shrinks and the tail at large $|x|$ initially grows. We were not able to predict ahead of time that in some places the function would grow for a while and then come back down some. (Look for example at $x = 2$.) Moreover, we now have an exact function with numerical values for the late time limit of the function.

Stepping Back
The method of transforms boils down to a five-step process.
1. Take the Fourier transform of both sides of the differential equation.
2. Use the rules of Fourier transforms to simplify the resulting equation, which should turn one derivative operation into a multiplication. If you started with a two-variable partial differential equation, you are now effectively left with a one-variable, or ordinary, differential equation.

3. Solve the resulting differential equation.
4. Plug in the initial conditions. (You will need to Fourier transform these as well.)
5. If possible, take an *inverse* Fourier transform to find the function you were originally looking for. If it is not possible to do this analytically, you can still approximate the relevant integral numerically.

If a function is defined on the entire number line, and has no special symmetry, its Fourier transform will be expressed in terms of sines and cosines—or, equivalently, in complex exponentials as we did above. If a function is defined on *half* the number line, from 0 to 8, then you can create an "odd extension" of that function and use a Fourier sine transform, or an "even extension" with a Fourier cosine transform. You'll work through an example of this technique in Problem 11.222.

The method of transforms can be used in situations where separation of variables cannot, such as inhomogeneous equations. It can also be used in situations where series expansions cannot: namely, infinite non-periodic functions. There are, however, two basic requirements that must be met in order to use a Fourier transform with respect to a variable x.

First, x must not appear in any of the coefficients of your equation. We've seen the simple formulas for $\mathcal{F}\left[\partial f/\partial x\right]$ and $\mathcal{F}\left[\partial f/\partial t\right]$. The corresponding formulas for terms like $\mathcal{F}\left[xf\right]$ or $\mathcal{F}\left[x(\partial f/\partial x)\right]$ are not simple, and they are not useful for solving PDEs.

Second, you must be able to take the Fourier transform. In Chapter 9 we discuss the conditions required for $f(x)$ to have a Fourier transform. The most important is that $\int_{-\infty}^{\infty} |f(x)|\, dx$ must be defined; that restriction, in turn, means that $f(x)$ must approach zero as x approaches $\pm\infty$. This means that some problems cannot be approached with a Fourier transform. It also means that we almost always do our Fourier transforms in x rather than t, since we are rarely guaranteed that a function will approach zero as $t \to \infty$. To turn time derivatives into multiplications you can often use a "Laplace transform," which is the subject of the next section.

Finally, we should note that a Fourier transform is particularly useful for simplifying equations like the wave equation because the only spatial derivative is second order. When you take the second derivative of e^{ipx} you get $-p^2 e^{ipx}$ which gives you a simple, *real* ODE for $\hat{f}$. For a first-order spatial derivative, the Fourier transform would bring down an imaginary coefficient. You can still use this technique for such equations, but you have to work harder to physically interpret the results: see Problem 11.219.

11.10.3 Problems: The Method of Fourier Transforms

For some of the Fourier transforms in this section, you should be able to evaluate them by hand. (See Appendix G for the formula.) For some you will need a computer. (Such problems are marked with a computer icon.) And for some, the following formulas will be useful. (If you're not familiar with the Dirac delta function, see Appendix K.)

$$\mathcal{F}[1] = \delta(p) \text{ (the Dirac delta function)}$$
$$\mathcal{F}\left[e^{-(x/k)^2}\right] = \frac{k}{2\sqrt{\pi}} e^{-(kp/2)^2} \tag{11.10.15}$$

It will also help to keep in mind that a "constant" depends on what variable you're working with. If you are taking a Fourier transform with respect to x, then $t^2 \sin t$ acts as a constant. If you are taking a derivative with respect to t, then $\delta(x)$ acts as a constant.

Unless otherwise specified, your final answer will be the Fourier transform of the PDE solution. Remember that if your answer has a delta function in it you can simplify it by replacing anything of the form $f(p)\delta(p)$ with $f(0)\delta(p)$.

11.206 In this problem you'll fill in some of the calculations from the Explanation (Section 11.10.2).

(a) To derive Equation 11.10.9 we needed the Fourier transform of the right hand side of Equation 11.10.8. Evaluate this Fourier transform directly using the formula for a Fourier transform in Appendix G.

(b) Verify that Equation 11.10.10 is a solution to Equation 11.10.9.

(c) The initial conditions $y(x, 0) = \partial y/\partial t(x, 0) = 0$ can trivially be Fourier transformed into $\hat{y}(p, 0) = \partial \hat{y}/\partial t(p, 0) = 0$. Plug those conditions into Equation 11.10.10 and derive the formulas for $A(p)$ and $B(p)$ given in the Explanation.

11.207 [*This problem depends on Problem 11.206.*] Use a computer to evaluate the inverse Fourier transform 11.10.11. (You can do it by hand if you prefer, but it's a bit of a mess.) Assume $t > 2L/v$ and simplify the expression for $y(x, t)$ in each of the following regions: $0 < x < L$, $L < x < vt - L$, $vt - L < x < vt + L$, $x > vt + L$. In each case you should find a polynomial of degree 2 or less in x. As a check on your answers, reproduce the last frame of Figure 11.7.

11.208 Walk-Through: The Method of Fourier Transforms. In this problem you will solve the following partial differential equation on the domain $-\infty < x < \infty$, $0 \leq t < \infty$.

$$\frac{\partial u}{\partial t} - \frac{\partial^2 u}{\partial x^2} + u = e^{-x^2}$$

$$u(x, 0) = \begin{cases} 1 + x & -1 \leq x \leq 0 \\ 1 - x & 0 < x \leq 1 \\ 0 & \text{elsewhere} \end{cases}$$

(a) Take the Fourier transform of both sides of this PDE. On the left side you will use Equations 11.10.3–11.10.4 to get an expression that depends on $\hat{u}(p, t)$ and $\partial \hat{u}/\partial t$. On the right you should get a function of p and/or t. Equations 11.10.15 may be helpful.

(b) Take the Fourier transform of the initial condition to find the initial condition for $\hat{u}(p, t)$.

(c) Solve the differential equation you wrote in Part (a) with the initial condition you found in Part (b) to get the solution $\hat{u}(p, t)$. (*Warning: the answer will be long and messy.*)

(d) Write the solution $u(x, t)$ as an integral over p.

Solve Problems 11.209–11.213 on the domain $-\infty < x < \infty$ using the method of transforms. When the initial conditions are given as arbitrary functions the Fourier transforms of those functions will appear as part of your solution. It may help to first work through Problem 11.208 as a model.

11.209 $\partial y/\partial t - c^2(\partial^2 y/\partial x^2) = e^{-t}$, $y(x, 0) = f(x)$

11.210 $\partial y/\partial t - c^2(\partial^2 y/\partial x^2) = e^{-t}$, $y(x, 0) = e^{-x^2}$

11.211 $\partial y/\partial t - c^2(\partial^2 y/\partial x^2) + y = \kappa$, $y(x, 0) = 0$. (You should be able to inverse Fourier transform your solution and give your answer as a function $y(x, t)$.)

11.212 $\dfrac{\partial y}{\partial t} - c^2 \dfrac{\partial^2 y}{\partial x^2} + y$
$= \begin{cases} Q & -1 < x < 1 \\ 0 & \text{elsewhere} \end{cases}$, $y(x, 0) = 0$

11.213 $\dfrac{\partial^2 y}{\partial t^2} - c^2 \dfrac{\partial^2 y}{\partial x^2} + y = e^{-x^2} \cos(\omega t)$, $y(x, 0) = 0$,

$\dfrac{\partial y}{\partial t}(x, 0) = \begin{cases} x & 0 \leq x \leq 1/2 \\ 1 - x & 1/2 < x \leq 1 \\ 0 & \text{elsewhere} \end{cases}$

For Problems 11.214–11.217 solve the equation $\partial^2 y/\partial x^2 - (1/v^2)(\partial^2 y/\partial t^2) = q(x, t)$ with initial conditions $y(x, 0) = f(x)$, $\dot{y}(x, 0) = g(x)$. Equations 11.10.15 may be needed.

11.214 $q(x, t) = 0$, $f(x) = \begin{cases} F & -1 < x < 1 \\ 0 & \text{otherwise} \end{cases}$, $g(x) = 0$

11.215 $q(x, t) = \begin{cases} -Q & -1 \leq x < 0 \\ Q & 0 \leq x \leq 1 \\ 0 & \text{elsewhere} \end{cases}$,

$f(x) = g(x) = 0$

11.216 $q(x, t) = \kappa e^{-(x/d)^2} \sin(\omega t)$, $f(x) = 0$, $g(x) = 0$

11.217 $q(x, t) = -Q$, $f(x) = 0$, $g(x) = Ge^{-(x/d)^2}$. The letters Q and G stand for constants. In general any answer with $f(p)\delta(p)$ in it can be simplified by replacing $f(p)$ with the value $f(0)$, since $\delta(p) = 0$ for all $p \neq 0$. In this case, however, you should have terms with $\delta(p)/p^2$, which is undefined at $p = 0$. Expand $\cos(vpt)$ in your answer in a Maclaurin series in p and simplify the result. You should get something where the coefficient in front of $\delta(p)$ is non-singular, and you can replace that coefficient with its value at $p = 0$.

11.218 (a) Solve the equation $\partial^2 y/\partial x^2 - (1/v^2)(\partial^2 y/\partial t^2) = 0$ with initial conditions $y(x, 0) = Fe^{-(x/d)^2}$, $dy/dt(x, 0) = 0$. Your answer should be an equation for $\hat{y}(p, t)$ with no arbitrary constants.

(b) You're going to take the inverse Fourier transform of your solution, but to do so it helps to start with the following trick. Rewrite the second half of Equation 11.10.15 in the form $e^{-(x/h)^2} = $<an integral>.

(c) Now take the inverse Fourier transform of your answer from Part (a) to find $y(x, t)$. Start by writing the integral for the Fourier transform. Then do a variable substitution to make it look like the integral you wrote in Part (b). Use that to evaluate the integral and then reverse your substitution to get a final answer in terms of x and t.

(d) Verify that your solution $y(x, t)$ satisfies the differential equation and initial conditions.

11.219 The complex exponential function e^{ipx} is an eigenfunction of $\partial y/\partial x$, but with an imaginary eigenvalue. This makes it harder to interpret the results of the method of transforms when single derivatives are involved. To illustrate this, use the method of transforms to solve the equation

$$\frac{\partial y}{\partial t} - \frac{\partial^2 y}{\partial x^2} - \frac{\partial y}{\partial x} = \begin{cases} 1 & -1 < x < 1 \\ 0 & \text{elsewhere} \end{cases}$$

with initial condition $y(x, 0) = 0$

Your final result should be the Fourier transform $\hat{y}(p, t)$, which will be a complex function that cannot easily be inverse Fourier transformed and admits no obvious physical interpretation.

11.220 An infinite rod being continually heated by a localized source at the origin obeys the differential equation $\partial u/\partial t - \alpha(\partial^2 u/\partial x^2) = ce^{-(x/d)^2}$ with initial condition $u(x, 0) = 0$.

(a) Solve for the temperature $u(x, t)$. Your answer will be in the form of a Fourier transform $\hat{u}(p, t)$.

(b) 🖥️ Take the inverse Fourier transform of your answer to get the function $u(x, t)$. Plot the temperature distribution at several different times and describe how it is evolving over time.

11.221 The electric potential in a region is given by Poisson's equation $\partial^2 V/\partial x^2 + \partial^2 V/\partial y^2 + \partial^2 V/\partial z^2 = (1/\varepsilon_0)\rho(x, y, z)$. An infinitely long bar, $0 \leq x \leq 1$, $0 \leq y \leq 1$, $-\infty < z < \infty$ has charge density $\rho(x, y, z) = \sin(\pi x)\sin(\pi y)e^{-z^2}$. Assume the edges of the bar are grounded so $V(0, y, z) = V(1, y, z) = V(x, 0, z) = V(x, 1, z) = 0$ and assume the potential goes to zero in the limits $z \to \infty$ and $z \to -\infty$.

(a) Write Poisson's equation for this charge distribution and take the Fourier transform of both sides to turn it into a PDE for $\hat{V}(x, y, p)$.

(b) Plug in a guess of the form $\hat{V} = C(p)\sin(\pi x)\sin(\pi y)$ and solve for $C(p)$.

11.222 Exploration: Using Fourier Sine Transforms
For a variable with an infinite domain $-\infty < x < \infty$ you can take a Fourier transform as described in this section. For a variable with a semi infinite domain $0 < x < \infty$, however, it is often more useful to take a Fourier sine transform. The formulas for a Fourier sine transform and its inverse are in Appendix G. As long as the boundary condition on f is that $f(0, t) = 0$ the rules for Fourier sine transforms of second derivatives are the same as the ones for regular Fourier transforms.

$$F_s\left[\frac{\partial^2 f}{\partial x^2}\right] = -p^2 F_s[f], \quad F_s\left[\frac{\partial^2 f}{\partial t^2}\right] = \frac{\partial^2 F_s[f]}{\partial t^2}$$

In this problem you will use a Fourier sine transform to solve the wave equation $\partial^2 y/\partial x^2 - (1/v^2)(\partial^2 y/\partial t^2) = 0$ on the interval $0 < x < \infty$ with initial conditions

$$y(x, 0) = \begin{cases} x & 0 < x < d \\ 2d - x & d < x < 2d \\ 0 & x > 2d \end{cases}, \quad \dot{y}(x, 0) = 0$$

and boundary condition $y(0, t) = 0$.

(a) Take the Fourier sine transform of the wave equation to rewrite it as an ordinary differential equation for $\hat{y}_s(p, t)$.

(b) Find the Fourier sine transform of the initial conditions. These will be the initial conditions for the ODE you derived in Part (a).

(c) Solve the ODE using these initial conditions and use the result to write the solution $\hat{y}_s(t)$.

11.11 The Method of Laplace Transforms

Laplace transforms are an essential tool in many branches of engineering. When we use them to solve PDEs they act much like Fourier transforms, turning a derivative into a multiplication, but they generally operate on the time variable rather than a spatial variable.

This section completes our chapter on PDEs. There are many techniques we did not discuss, but we believe we have presented you with the most important approaches. Going through this section as you have gone through the ones before, pay attention to the details so you can apply the technique yourself. But after the details, the "Stepping Back" will help you figure out which technique to apply to which problem. This important question is addressed more broadly in Appendix I, so you can look at a new equation and sort through the various methods that you have mastered.

11.11.1 Explanation: The Method of Laplace Transforms

We have seen that when a function is defined on a finite domain, it is sometimes useful to replace that function with a series expansion. We have also seen that when a function is defined on an *infinite* domain, we can use a transform—an integral—instead of a series.

Just as with series, the right transform may involve sines, cosines, and/or complex exponentials (Fourier), but other transforms may be called for in other circumstances. In this section we look at one of the most important examples, the Laplace transform. We also briefly discuss the general topic of finding the right transform for any given problem.

Laplace Transforms

Chapter 10 introduces Laplace transforms and their use in solving ODEs; Appendix H contains a table of Laplace transforms and some table-looking-up techniques. We're not going to review all that information here, so you may want to refresh yourself.

Solving a PDE by Laplace transform is very similar to solving a PDE by Fourier transform, except that we tend to use Laplace transforms on the time variable instead of spatial variables. Based on our work in the previous sections, you can probably see how we are going to put the following rules to use.

$$\mathcal{L}\left[\frac{\partial^{(n)} f(x,t)}{\partial x^{(n)}}\right] = \frac{\partial^{(n)}}{\partial x^{(n)}}(F(x,s)) \tag{11.11.1}$$

$$\mathcal{L}\left[\frac{\partial^{(n)} f(x,t)}{\partial t^{(n)}}\right] = s^n F(x,s) - s^{n-1} f(x,0) - s^{n-2}\frac{\partial f}{\partial t}(x,0) - s^{n-3}\frac{\partial^2 f}{\partial t^2}(x,0) - \ldots - \frac{\partial^{(n-1)} f}{\partial t^{(n-1)}}(x,0) \tag{11.11.2}$$

The first rule is not surprising: since we are using $\mathcal{L}[\,]$ to indicate a Laplace transform in the time variable, a derivative with respect to x can move in and out of the transform. (When we took Fourier transforms with respect to x, derivatives with respect to t followed a similar rule.)

The second rule is analogous to an eigenfunction relationship. It shows that inside a Laplace transform, taking the nth time derivative is equivalent to multiplying by s^n. So the Laplace transform serves the same purpose as the Fourier transform in the previous section, and the Fourier series in the section before that: by turning a derivative into a multiplication, it turns a PDE into an ODE. In this case, however, there are correction terms on the boundary that bring initial conditions into the calculations. (Similar corrective terms come into play with Fourier transforms on semi-infinite intervals.)

As with the Fourier transform, the final property we need is linearity: $\mathcal{L}[af + bg] = a\mathcal{L}[f] + b\mathcal{L}[g]$ where a and b are constants, f and g functions of time.

Temperature on a Semi-Infinite Bar

A bar extends from $x = 0$ to $x = \infty$. The left side of the bar is kept at temperature $u(0, t) = u_L$, and the rest of the bar starts at $u(x, 0) = 0$. Find the temperature distribution in the bar as a function of time, assuming it obeys the heat equation:

$$\frac{\partial u}{\partial t} = \alpha \frac{\partial^2 u}{\partial x^2}$$

As often occurs in problems with infinite domains, there is an "implicit" boundary condition that is never stated in the problem. The entire bar starts at 0°, and the heat applied to the left side must propagate at a finite speed through the bar; therefore, it is safe to assume that $\lim_{x \to \infty} u(x, t) = 0$.

We begin by taking the Laplace transform of both sides of the equation, using the linearity property to leave the constant α outside.

$$\mathcal{L}\left[\frac{\partial u}{\partial t}\right] = \alpha \mathcal{L}\left[\frac{\partial^2 u}{\partial x^2}\right]$$

Next we apply our derivative rules to both sides.

$$sU(x, s) - u(x, 0) = \alpha \frac{\partial^2}{\partial x^2} U(x, s)$$

(Note that the second term is the original function u, not the Laplace transform U.) Plugging in our initial condition $u(x, 0) = 0$, we can rewrite the problem as:

$$\frac{\partial^2 U}{\partial x^2} = \frac{s}{\alpha} U$$

Just as we saw with Fourier transforms in the previous section, a partial differential equation has turned into an effectively *ordinary* differential equation, since the derivative with respect to time has been replaced with a multiplication. We can solve this by inspection.

$$U(x, s) = A(s) e^{\sqrt{s/\alpha}\, x} + B(s) e^{-\sqrt{s/\alpha}\, x}$$

A and B must be constants with respect to x, but may be functions of s.

The "implicit" boundary condition says that in the limit as $x \to \infty$ the function $u(x, t)$—and therefore the transformed function $U(x, s)$—must approach zero. This condition kills the growing exponential term. Before we can apply the explicit boundary condition we must Laplace transform it as well. The Laplace transform of any constant function k is k/s, so $u(0, t) = u_L$ becomes $U(0, s) = u_L/s$, which gives $B(s) = u_L/s$. That allows us to write the answer $U(x, s)$.

$$U(x, s) = \frac{u_L}{s} e^{-\sqrt{s/\alpha}\, x} \qquad (11.11.3)$$

Equation 11.11.3 is the Laplace transform of the function $u(x, t)$ that we're looking for. What can you do with *that*?

In some cases you stop there. Engineers who are used to working with Laplace transforms can read a lot into a solution in that form, and we provide some tips for interpreting Laplace transforms in Chapter 10. In some cases you can apply an inverse Laplace transform to find the actual function you're looking for. This process requires integration on the complex plane, so we defer it to Chapter 13. Our approach in this section will be electronic: we asked our computer for the inverse Laplace transform of Equation 11.11.3, and thus got the solution to this problem.

$$u(x, t) = u_L \operatorname{erfc}\left(\frac{x}{2\sqrt{\alpha t}}\right)$$

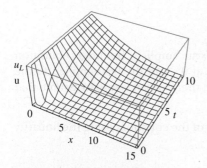

FIGURE 11.8 Temperature on a semi-infinite rod with zero initial temperature and a fixed boundary temperature. The effect of the boundary spreads from left to right, pulling the temperature at each point asymptotically up towards u_L.

As always we urge you not to panic at the sight of an unfamiliar function! Even without knowing anything about the "complementary error function" $\operatorname{erfc}(x)$ (which you can look up in Appendix K, or in Chapter 12 for more details), you can describe the temperature in the bar by looking at a computer-generated graph of the solution (Figure 11.8).

- At $x = 0$ the plot shows a temperature of u_L for all t-values. This was our boundary condition: the temperature at the left side of the bar is held constant.
- At $x = 15$ the temperature is uniformly zero. It will rise eventually, but in the domain of our picture ($0 \le t \le 10$) the heat from the left side has not yet had time to propagate that far.
- The in-between values are the most interesting. For instance, at $x = 5$, we see that the temperature stays at zero for a few seconds, until the heat from the left side reaches it. The temperature then starts to rise dramatically. Eventually it will approach u_L, as we can see happening already at lower x-values.

Before we leave this problem we should point out a peculiarity that you may not have noticed: the initial and boundary conditions contradict each other. The problem stipulated that $u(0, t) = u_L$ at all times, and also that $u(x, 0) = 0$ at all x-values, but $u(0, 0)$ cannot possibly be two different values! Maybe the entire bar was at $u = 0$ when the left side was suddenly brought into contact with an object of temperature u_L; it is only an approximation to say that the temperature at the boundary will rise instantaneously to match. As you have seen, we can solve the problem analytically despite this contradiction. But mathematical software packages sometimes get confused by such conflicts.

EXAMPLE Laplace Transform

Problem:
Solve

$$\frac{\partial v}{\partial t} + \frac{\partial v}{\partial x} = 1 \qquad (11.11.4)$$

on the domain $0 \le x < \infty$, $0 \le t < \infty$ with initial condition $v(x, 0) = 0$ and boundary condition $v(0, t) = 0$.

Solution:
We begin by taking the Laplace transform of both sides, applying the linearity and derivative properties on the left. The Laplace transform of 1 is $1/s$, so

$$sV(x, s) + \frac{\partial V(x, s)}{\partial x} = \frac{1}{s}$$

You can solve this by finding a complementary and a particular solution, with the result

$$V(x, s) = \frac{1}{s^2} + C(s)e^{-sx}$$

The boundary condition $v(0, t) = 0$ becomes $V(0, s) = 0$, which means

$$V(x, s) = \frac{1 - e^{-sx}}{s^2} \qquad (11.11.5)$$

You can easily find the inverse Laplace transform of Equation 11.11.5 on a computer:

$$v(x, t) = \begin{cases} t & t < x \\ x & t \geq x \end{cases} \qquad (11.11.6)$$

It's an odd-looking solution, isn't it? You'll investigate it in Problem 11.230.

Choosing the Right Transform, Part 1: Remember the Eigenfunction!

We have seen that for a variable that is defined on a finite domain (or that is periodic) we can expand into a series; on an infinite domain we use a "transform," expanding into an integral. But we have also seen that there are different *kinds* of series and transforms, and you have to start by picking the right one.

The most important rule, whether you are doing series or transforms, is to remember the eigenfunction. The Laplace transform is based on a simple exponential, which is an eigenfunction of derivatives of any order. For instance, if you take the Laplace transform of the operator $\alpha(\partial^3 v/\partial t^3) + \beta(\partial^2 v/\partial t^2) + \gamma(\partial v/\partial t) + \delta v$, you get $(-\alpha s^3 + \beta s^2 - \gamma s + \delta)V$ plus some boundary terms that don't depend on t. Can you see the point of that simple exercise? If your problem includes derivatives of any order, multiplied by only constants, then a Laplace transform will turn the entire differential operator into a multiplication.

The Fourier transform is also based on an exponential function, albeit a complex one, and is therefore an eigenfunction of the same operators. So Fourier and Laplace transforms both work on the same differential operators: you choose one or the other based on the boundary conditions, as we will discuss below.

However, not all equations involve just derivatives multiplied by constants, so Fourier and Laplace transforms are not always useful. For instance, consider an equation based on the Laplacian in polar coordinates. We have seen that in the homogeneous case, such an equation leads to Bessel's equation, whose normal modes—found by separating variables and solving an ordinary differential equation—are Bessel functions. In the *inhomogeneous* case, where separation of variables doesn't work, we can expand such a problem in a "Hankel transform": an integral over Bessel functions. The Hankel transform has the property that $H_0\left(f'' + f'/x\right) = s^2 H_0(f)$, which allows it to simplify differential equations based on that operator.

Hankel transforms and many others are discussed in texts on partial differential equations.[12] We have chosen to focus on the two most important transforms, Fourier and Laplace. But if you understand the underlying principles that make these transforms useful, it isn't difficult to make a pretty good guess, given an unfamiliar problem, about what transform will be most likely to help.

---

[12] One that we got a lot out of is Asmar, Nakhle, *Partial Differential Equations with Fourier Series and Boundary Value Problems* 2nd edition, Prentice Hall, 2004.

Choosing the Right Transform, Part 2: Fourier or Laplace?

As discussed above, Fourier and Laplace transforms both apply to the same differential operators. Nonetheless, one of these transforms will often succeed where the other fails, so it's generally important to choose the right one. We'll start with one big rule of thumb: not guaranteed, but easy to apply and generally useful.

We usually take Fourier transforms in space, and Laplace transforms in time.

Now we'll offer some specific limitations of both transforms. Along the way, we'll show how these limitations lead to that rule of thumb. That way, you'll know why the rule works—and when you need to make an exception.

1. **Laplace transforms can be defined on semi-infinite intervals (such as $0 \leq x < \infty$) only; Fourier transforms can be defined on semi-infinite or infinite ($-\infty < x < \infty$) intervals.** The domain of time is often semi-infinite, whereas spatial variables are defined on all kinds of domains.
2. **A Fourier transform requires the integral $\int_{-\infty}^{\infty} f(x)\,dx$ to exist, which in turn requires $f(x)$ to approach zero at infinity.** It is common (although certainly not universal) for a function to approach zero as $x \to \infty$, but it is rare to get a guarantee up front about what will happen as $t \to \infty$. The Laplace transform, by contrast, exists as long as $f(t)$ grows exponentially or slower at late times. This is not very restrictive: it's rare to find a function that grows faster than e^t does!
3. **When you use a Laplace transform for an nth-order derivative, you need to know the function's value and its first $(n-1)$ derivatives at $t = 0$.** That's exactly what you tend to get from initial conditions. For instance, if you are solving the wave equation on a string, your initial conditions are generally the position and velocity at $t = 0$, which is just what you need for a Laplace transform. Boundary conditions don't go that way: you may get y on both ends, or $\partial y/\partial x$ on both ends, but you are less likely to get both variables on the same end.

There are a lot of rules to guide you in choosing the right method for a given differential equation. Our goal has been to present these rules in a common-sense way: you will work with these rules better, and remember them longer, if you understand where they come from. In the summary flow chart in Appendix I we forget all the "why" questions and just list the rules.

11.11.2 Problems: The Method of Laplace Transforms

11.223 Walk-Through: The Method of Laplace Transforms. In this problem you will solve the partial differential equation $\partial^2 u/\partial t^2 - \partial^2 u/\partial x^2 = \cos(t)\sin(\pi x)$ in the domain $0 \leq x \leq 1$, $0 \leq t < \infty$ with boundary conditions $u(0, t) = 0$, $u(1, t) = t$ and initial condition $u(x, 0) = -\sin(\pi x)$, $\partial u/\partial t(x, 0) = 0$ using the method of Laplace transforms.

(a) Take the Laplace transform of both sides of this PDE. On the left side you will use Equations 11.11.1–11.11.2 to get an expression that depends on $U(x, s)$ and its spatial derivatives. On the right you should get a function of x and s.

(b) Take the Laplace transform of the boundary conditions to find the boundary conditions for $U(x, s)$.

(c) Solve the differential equation you wrote in Part (a) with the boundary conditions you found in Part (b) to get the solution $U(x, s)$.

Solve Problems 11.224–11.229 on the domain $0 \leq x \leq 1$ using the method of transforms. Your answer in most cases will be the Laplace transform $Y(x, s)$ of the solution $y(x, t)$. It may help to first work through Problem 11.223 as a model.

11.224 $\partial y/\partial t - c^2(\partial^2 y/\partial x^2) = e^{-t}$, $y(x, 0) = y(0, t) = y(1, t) = 0$

11.225 $\partial^2 y/\partial t^2 - c^2(\partial^2 y/\partial x^2) = \sin t$, $y(x, 0) = \partial y/\partial t(x, 0) = y(0, t) = y(1, t) = 0$

11.226 $\partial^2 y/\partial t^2 + (\partial^2 y/\partial x^2) = e^{-t}$, $y(x, 0) = \sin(\pi x)$, $\partial y/\partial t(x, 0) = 0$, $y(0, t) = y(1, t) = 0$

11.227 $\partial y/\partial t - (\partial^2 y/\partial x^2) = 0$, $y(x, 0) = 0$,
$y(0, t) = t$, $y(1, t) = 0$

11.228 $\partial^2 y/\partial t^2 - \partial y/\partial t - \partial^2 y/\partial x^2 = \sin(t)\sin(\pi x)$,
$y(0, t) = te^{-t}$, $y(1, t) = 0$, $y(x, 0) = \sin(\pi x)$, $\partial y/\partial t(x, 0) = 0$

11.229 $\partial y/\partial t - c^2(\partial^2 y/\partial x^2) = xt$, $y(x, 0) = 0$,
$y(0, t) = y(1, t) = 0$

11.230 In the Explanation (Section 11.11.1) we solved a differential equation and ended up with Equation 11.11.6. In this problem you'll consider what that solution looks like.

(a) Draw sketches of $v(x)$ on the domain $0 \leq x \leq 10$ at $t = 5$ and $t = 6$. (This should be pretty quick and easy.)

(b) What is $\partial v/\partial x$ at the point $x = 4$ in both sketches? (You can see the answer by looking.)

(c) How does v at the point $x = 4$ move or change between the two sketches? Based on that, what is $\partial v/\partial t$ at that point?

(d) What is $\partial v/\partial x$ at the point $x = 7$ in both sketches?

(e) How does v at the point $x = 7$ move or change between the two sketches? Based on that, what is $\partial v/\partial t$ at that point?

(f) Explain why this function solves Equation 11.11.4.

(g) Describe the life story of the point $x = x_0$ (where $x_0 > 0$) as t goes from 0 to ∞.

11.231 A string that starts at $x = 0$ and extends infinitely far to the right obeys the wave equation 11.2.2. The string starts at rest with no vertical displacement but the end of it is being shaken so $y(0, t) = y_0 \sin(\omega t)$.

(a) Using words and a few sketches of how the string will look at different times, predict how the string will behave—not by solving any equations (yet), but by physically thinking about the situation.

(b) Convert the wave equation and the initial conditions into an ODE for the Laplace transform $Y(x, s)$.

(c) Find the general solution to this ODE.

(d) It takes time for waves to physically propagate along a string. Since the string starts with zero displacement and velocity and is only being excited at $x = 0$, it should obey the implicit boundary condition $\lim_{x \to \infty} y(x, t) = 0$. Use this boundary condition to solve for one of the arbitrary constants in your solution.

(e) Use the explicit boundary condition at $x = 0$ to solve for the other arbitrary constant and thus find the Laplace transform $Y(x, s)$.

(f) 🖥 Take the inverse Laplace transform to find the solution $y(x, t)$. Your answer will involve the "Heaviside step function" $H(x)$ (see Appendix K).

(g) Explain in words what the string is doing. Does the solution you found behave like the sketches you made in Part (a)? If not, explain what is different.

11.12 Additional Problems (see felderbooks.com)

CHAPTER 12

Special Functions and ODE Series Solutions

> *Before you read this chapter, you should be able to...*
> - model real-world situations with ordinary differential equations and interpret solutions of these equations to predict physical behavior (see Chapter 1).
> - given a differential equation and a proposed solution, determine if that proposed solution works (see Chapter 1).
> - solve differential equations by "separation of variables" and by "guess and check" (see Chapter 1).
> - use initial conditions to determine the arbitrary constants in a general solution (see Chapter 1).
> - use and interpret series notation.
> - find the Taylor series for a given function around a given point (see Chapter 2).
> - find the Fourier series for a given function (see Chapter 9).
>
> *After you read this chapter, you should be able to...*
> - recognize and use the hyperbolic trig functions, the error function, and the gamma function.
> - solve differential equations by the method of power series and the method of Frobenius.
> - derive and use the equations for Legendre polynomials and Bessel functions.
> - use Sturm-Liouville theory to recognize ODEs that have complete, orthogonal sets of solutions and to classify the types of boundary conditions needed for these ODEs.

If you have gone through Chapter 11 then you are familiar with the process of solving a partial differential equation (PDE) by separation of variables. This process leads to an ordinary differential equation (ODE) whose solution may be sines and cosines, but it may also be less familiar functions: Bessel functions, Legendre polynomials, and others. Solving a PDE often requires you to recognize an ODE and its solutions, work with the properties of those solutions, and build a series from those solutions.

In this chapter you will see a lot of the mathematical background that we glossed over in Chapter 11. You will see how ordinary differential equations can lead to those Bessel functions, Legendre polynomials, and others. Along the way you will learn an important technique for solving ordinary differential equations by assuming a solution in the form of a power series (or a "generalized" power series). And you will encounter a number of important functions with which you may have been previously unfamiliar.

12.1 Motivating Exercise: The Circular Drum

In this exercise you will solve a partial differential equation. This process may be new to you or it may be review, depending on whether you have gone through Chapter 11. One purpose of this exercise is to teach and/or remind of you this process. But focus your attention on

12.1 | Motivating Exercise: The Circular Drum

the steps we skip—the times we just look up an answer and accept it instead of solving for it. Many of those are the same steps we skipped in Chapter 11, and the rest of this chapter is meant to fill in those gaps.

A circular drumhead of radius a is allowed to vibrate. If the initial state of the drum has "azimuthal symmetry" (no ϕ dependence) then the drum will continue to have azimuthal symmetry over time. Under that circumstance the height z is a function of the polar distance ρ and the time t, and is governed by the following PDE (where v is a constant).

$$\frac{\partial^2 z}{\partial t^2} = v^2 \left(\frac{\partial^2 z}{\partial \rho^2} + \frac{1}{\rho} \frac{\partial z}{\partial \rho} \right) \quad \text{the wave equation in polar coordinates with azimuthal symmetry} \quad (12.1.1)$$

The technique called "Separation of Variables" tells us to write the solution as the product of two different functions.

$$z(\rho, t) = R(\rho) T(t)$$

The next few steps—covered in Chapter 11 but not relevant for our purpose here—take us to the two equations below. Note the introduction of a new constant k.

$$T''(t) = -k^2 v^2 T(t) \quad (12.1.2)$$

$$\rho^2 R''(\rho) + \rho R'(\rho) + k^2 \rho^2 R(\rho) = 0 \quad (12.1.3)$$

1. Find the general solution to Equation 12.1.2. You should be able to do this by inspection.
2. Write the general solution to Equation 12.1.3 by matching it to one of the ODEs in Appendix J. *Hint*: some of the ODEs in the appendix are actually classes of ODEs identified by different values of a parameter p. Equation 12.1.3 is the $p = 0$ version of one of the listed ODEs.

See Check Yourself #79 in Appendix L

3. An "implicit boundary condition" that $R(\rho)$ must obey is that it must be finite at $\rho = 0$. Look up in Appendix J the basic properties of the functions you found in Part 2, and use this implicit boundary condition to set one of your two arbitrary constants to zero.
4. The other boundary condition is $R(a) = 0$ because the outer edge of the drum can't move up and down. Use that boundary condition to limit the possible values of k, once again using information from Appendix J. There are infinitely many possible values for k so your answer will contain a new parameter n, where n can be any positive integer.

See Check Yourself #80 in Appendix L

This should all make sense if you're willing to simply accept the information in Appendix J, but where did that information come from? Sections 12.4 and 12.6 will show you two (closely related) techniques for solving ODEs such as Equation 12.1.3, and Sections 12.5 and 12.7 will apply those techniques to two common ODEs to derive the properties of two of the functions in Appendix J.

There is another gap, maybe harder to see. At the $\rho = 0$ boundary we did not specify a value of R; we merely asserted that it must be finite. At the $\rho = a$ boundary we needed a specific value. In Section 12.8 we will see a rule for determining what kind of condition is needed at a boundary.

You're not done yet, though. You still need to combine these functions into a solution $z(\rho, t)$ and match initial conditions.

5. Multiply your solutions $R(\rho)$ and $T(t)$ to find a solution $z(\rho, t)$. You should be able to absorb the remaining arbitrary constants in $R(\rho)$ into the arbitrary constants in $T(t)$ to give you a solution $z(\rho, t)$ with two arbitrary constants. Don't write k in your answer; use the value of k that you found above.

   See Check Yourself #81 in Appendix L

6. The solution you just found is called a "normal mode" of the PDE. There are infinitely many such normal modes, one for each value of n. Write the general solution $z(\rho, t)$ as a sum of all of these normal modes. Since the arbitrary constants can take different values for each n, write a subscript n on them.

7. For simplicity we'll use as one initial condition $\dot{z}(\rho, 0) = 0$. Plug this into your solution and show that one of the two arbitrary constants must be zero. This should leave you with an infinite series with just one undetermined set of constants B_n. (Of course you may have used a different letter, but we'll refer to it as B_n from here on.)

8. Our other initial condition will be an unspecified function $z(\rho, 0) = f(\rho)$. Plugging $t = 0$ into your solution for z write $f(\rho)$ as an infinite series.

   See Check Yourself #82 in Appendix L

9. Use Appendix J to find how to determine the coefficients of a series of the type you just wrote. Express B_n as an integral involving the unknown function $f(\rho)$.

Once again it all worked because we just invoked Appendix J. Can any initial condition $f(\rho)$ can be written in a series of this form? It turns out the functions that you found in the appendix have a property called "completeness," which essentially means yes, any $f(\rho)$ can be written as a sum of these functions. Section 12.8 will explain under what circumstances the solutions of a given ODE are complete in this sense, and will also explain how to derive the formula you used for finding the coefficients B_n of such a series.

At the end of Section 12.8 we will return to our circular drum and see how the material in the chapter has filled in all the gaps we left in this section.

12.2 Some Handy Summation Tricks

We assume you are already familiar with the notation $\sum_{n=1}^{3} n^2$ to mean $1 + 4 + 9$. However, there are some tricks that can help you write sums, and occasionally products, in concise and useful ways. You don't want a little confusion about series notation to prevent you from following the actual material in this chapter.

12.2.1 Explanation: Some Handy Summation Tricks

This section introduces a few useful tricks you should know when writing series.

1. *Alternating series*
   To write a series whose terms alternate between negative and positive, insert $(-1)^n$. For example:

   $$\sum_{n=1}^{\infty} (-1)^n n^2 = -1 + 4 - 9 + 16 - \ldots$$

   If we wanted the first term to be positive we would replace $(-1)^n$ with $(-1)^{n+1}$.

   When you are generating a Fourier series you may end up with the term $\cos(\pi n)$. This is another alternator because $\cos(0) = \cos(2\pi) = \cos(4\pi) = 1$, while $\cos(\pi) = \cos(3\pi) = \cos(5\pi) = -1$. So for integer n you can always replace $\cos(\pi n)$ with $(-1)^n$ (and you should).

2. *Series that only include odd or even terms*
   Sometimes when we write a series we explicitly say that it only includes odd or even values of the index. For example:[1]

   $$\sum_{\text{odd } n} x^n \quad \text{means} \quad \sum_{n=1}^{\infty} x^n \text{ (odd } n \text{ only)} \quad \text{means} \quad x + x^3 + x^5 + \ldots$$

   Sometimes it is useful, however, to write such series without the extra note. Replace n with $2n+1$ and let n go over all integers from 0 to ∞. The $2n+1$ will then pick out only the odd integer values. For example, the series above can be written as:

   $$\sum_{n=0}^{\infty} x^{2n+1} = x + x^3 + x^5 + \ldots$$

   Similarly, replacing n with $2n$ picks out the even terms. In unusual cases where we want every third term we could replace n with $3n$ (or $3n+1$ or $3n+2$).

3. *Product notation*
   Just as a capital Sigma, Σ, is used for a sum, a capital Pi, $\prod$, is used for a product. For example:

   $$\prod_{n=1}^{3}(x+n)^2 = (x+1)^2(x+2)^2(x+3)^2$$

4. *Recurrence relations and explicit definitions*
   The individual terms in a sequence, or the coefficients in a power series, can be described in two different but equivalent ways. As an example, consider the Maclaurin series for e^x.

   $$e^x = 1 + x + \frac{1}{2!}x^2 + \frac{1}{3!}x^3 + \ldots$$

   - A "recursive definition" or "recurrence relation" gives a rule for finding each term from the term or terms before it. In this case, $c_n = c_{n-1}/n$ tells us that we can get to c_3 by dividing c_2 by 3, and so on. Along with a recurrence relation we also need to specify starting conditions: in this case, $c_0 = 1$.
   - An "explicit definition" gives the rule for finding each term without finding the terms before it. In this case, $c_n = 1/(n!)$ tells us that the first coefficient is 1, but it also tells us quickly that the 30th term is $1/(30!)$. An explicit definition is what you need if you want to write a series in series notation.

5. *Reindexing series to start at a different value of the index*
   For our purposes in this chapter, this is the most important and possibly least familiar of these tricks. Can you see why the following statement is true?

   $$\sum_{n=1}^{6} n^2 \text{ is the same series as } \sum_{n=-1}^{4}(n+2)^2$$

   The first series is $1 + 4 + 9 + 16 + 25 + 36$, and the second series is term by term the same. We can say that we got the second series by replacing n with $n+2$ everywhere in the first series, so n^2 became $(n+2)^2$, the starting condition $n=1$ became $(n+2)=1$, and similarly for the ending condition $n=6$.

---

[1] The first form should only be used if the starting value of n is unambiguous from context.

Rewriting a series in this way does not seem to have an official name that we could find, so we call it "reindexing" the series. This trick is useful when we want to combine sums. For example, suppose we wanted to add these two infinite series.[2]

$$f(x) = \sum_{n=0}^{\infty} a_n x^n + \sum_{n=0}^{\infty} b_n x^{n+2} \tag{12.2.1}$$

Rewriting this as $\sum_{n=0}^{\infty} \left(a_n x^n + b_n x^{n+2} \right)$ is perfectly valid but not very helpful. It's more useful to reindex the second sum so that both series are based on the same power of x.

$$f(x) = \sum_{n=0}^{\infty} a_n x^n + \sum_{n=2}^{\infty} b_{n-2} x^n \tag{12.2.2}$$

The most important thing here is to convince yourself that we haven't changed anything. The second series in Equation 12.2.1 is $b_0 x^2 + b_1 x^3 + \ldots$ and the second series in Equation 12.2.2 is *also* $b_0 x^2 + b_1 x^3 + \ldots$ Whenever you are unsure of a reindexing step, check the first few terms to make sure they haven't changed.

Now we can put together Equation 12.2.2 in a more useful form.

- The $n = 0$ term (corresponding to x^0) and the $n = 1$ term (corresponding to x^1) exist only in the first series, so we will write them separately.
- All subsequent terms exist in both series, so we will combine them into one series.

$$f(x) = a_0 + a_1 x + \sum_{n=2}^{\infty} (a_n + b_{n-2}) x^n \tag{12.2.3}$$

Equation 12.2.2 is preferable to Equation 12.2.1 because it expresses both series in terms of the same power of x, which then allows us to combine them into one series. It would be just as good if both series were based on x^{n-2} or x^{n+26}, as long as they are both the same. We will use this trick many times in this chapter.

12.2.2 Problems: Some Handy Summation Tricks

For Problems 12.1–12.7 write the given sum or product using summation (or product) notation. Do not actually evaluate the sum (or product).

12.1 $1 + 2 + 3 + \ldots + 57$

12.2 $1 + 3 + 5 + 7 + \ldots + 57$

12.3 $100!$

12.4 $\sin(x) - \sin(2x) + \sin(3x) - \ldots - \sin(100x)$

12.5 $\sin(x) - \sin(3x) + \sin(5x) - \sin(7x) + \ldots + \sin(57x)$

12.6 $1 + 4 + 7 + 10 + \ldots + 100$

12.7 $1 + 2^2 + 3 + 4^2 + 5 + 6^2 + 7 + \ldots + 56^2 + 57$

12.8 Write $e^{1+4+9+\ldots+100^2}$ two ways.
  - (a) Using summation notation
  - (b) Using product notation

12.9 Consider the infinite series $14 + 17 + 20 + 23 + \ldots$
  - (a) Write a recurrence relation for finding the term a_n from a_{n-1}.
  - (b) Assume the first term (14) is a_1. Write an explicit formula for a_n and use it to calculate a_{100}.
  - (c) Write the series in series notation.

---

[2] Assume both series are absolutely convergent so we can rearrange terms.

12.10 Here is the Maclaurin expansion of the sine.

$$\sin x = x - \frac{1}{3!}x^3 + \frac{1}{5!}x^5 - \frac{1}{7!}x^7 + \ldots$$

We use c_n to represent the coefficient of the x^n term so $c_0 = 0$, $c_1 = 1$, $c_2 = 0$, $c_3 = -1/(3!)$ and so on.

(a) Write a recurrence relation for finding the coefficient c_n from c_{n-2} for all $n \geq 2$.
(b) What initial values do you need to specify, in addition to that recurrence relation, to specify the entire series?
(c) Write an explicit formula for c_n. *Hint*: you will give one formula for odd n and a different formula for even n.
(d) Write the series in series notation. *Hint*: you can indicate a sum that includes only odd values of n by writing $\sum_{\text{odd } n}$.

12.11 The series $\sum_{n=13}^{18} \ln n$ is the same as the series $\sum_{n=11}^{16} \ln(n+2)$.

(a) Show the equivalence of these two series by writing out the terms of each.
(b) If the series $\sum_{n=16}^{21} f(n)$ is also identical to both of these series, find the function $f(n)$.
(c) If the series $\sum_{n=a}^{b} \ln(n+10)$ is also identical to both of these series, find a and b.

12.12 The series $\sum_{n=0}^{\infty} c_n x^n$ is the same as the series $\sum_{n=-2}^{\infty} c_{n+2} x^{n+2}$.

(a) Show the equivalence of these two series by writing the fifth partial sum of each series.
(b) If the series $\sum_{n=3}^{\infty} f(n)$ is also identical to both of these series, find the function $f(n)$.
(c) If the series $\sum_{n=a}^{\infty} c_{n-7} x^{n-7}$ is also identical to both of these series, find a.

12.13 If $y = \sum_{n=0}^{\infty} c_n x^n$, find the coefficient of x^n in the Maclaurin expansion of y'. *Hint*: assume the series are convergent, which allows you to differentiate a Taylor series term by term.

For Problems 12.14–12.18, reindex one or more of the given sums so that they all appear with the same power of x, then combine them into one sum. You may have one or more terms added outside the sum.

12.14 $\sum_{n=0}^{\infty} x^n + \sum_{n=0}^{\infty} 2x^{n+1}$

12.15 $\sum_{n=0}^{\infty} x^n + \sum_{n=0}^{\infty} 2x^{n+1} + \sum_{n=0}^{\infty} 3x^{n+2}$

12.16 $\sum_{n=0}^{\infty} n(n-1)x^{n-2} + \sum_{n=0}^{\infty} nx^{n-1} + \sum_{n=0}^{\infty} x^n$

12.17 $\sum_{n=0}^{\infty} n(n-1)x^{n-2} + \sum_{n=0}^{\infty} (1-x)nx^{n-1} + \sum_{n=0}^{\infty} x^n$ *Hint*: break the middle sum into two sums before reindexing.

12.18 $\sum_{n=0}^{\infty} c_n n(n-1)x^{n-2} + \sum_{n=0}^{\infty} c_n(1-x)nx^{n-1} + \sum_{n=0}^{\infty} c_n x^n$ *Hint*: break the middle sum into two sums before reindexing.

12.19 Later in the chapter, while solving differential equations, we'll encounter equations something like the following.

$$\sum_{n=0}^{\infty} c_n(n+3)(1-x^2)x^{n-2} + \sum_{n=0}^{\infty} c_n(n-7)x^{n-1}$$

$$+ \sum_{n=0}^{\infty} c_n x^n = \frac{8}{x} \quad (12.2.4)$$

(a) Break the first sum into two parts, each going from $n=0$ to infinity, but with different powers of x.
(b) With the first sum broken up in this way you should now have a total of four sums. Reindex all of them so that they have x^n in them.
(c) Rewrite Equation 12.2.4 in the form *<some terms>* $+ \sum$ *<something>* $x^n = 8/x$.
The only way this equation can hold for every value of x is if the coefficient of each power of x on the left equals the corresponding power on the right.
(d) Write an equation asserting that the coefficient of x^{-2} on the left equals zero, and solve to find c_0.
(e) Write an equation asserting that the coefficient of x^{-1} on the left equals 8, and solve to find c_1.
(f) Write an equation asserting that the coefficient of x^n in your equation for $n \geq 0$ equals zero. Solve this to give you a recurrence relation, an equation for c_{n+2} in terms of c_n and c_{n+1}.
(g) By plugging $n=0$ into your recurrence relation, find c_2.
(h) Use your recurrence relation to find c_3 and c_4.

12.3 A Few Special Functions

In Chapter 10 we introduced two functions that you may have been unfamiliar with, the Heaviside function and the Dirac delta function (with a note that the latter is not technically a function at all). In this section we will introduce a few more.

12.3.1 Explanation: A Few Special Functions

If you try to name all the common functions you are used to working with you might come up with constants, powers, exponentials, logarithms, and trig functions. If you expand that list to include anything that comes from combining those with addition, subtraction, multiplication, division, and composition (plugging one function into another one), you get the set of "elementary functions."[3] For example, $e^{\tan x^2}$ is an elementary function.

It may seem that *any* function can made from those building blocks, but in fact many functions that come up frequently in physics and engineering aren't in the short list above. The term "special function" has no formal definition, but it is loosely used to mean a function that is useful enough for people to have given it a name and studied its properties. Below we introduce a few important ones: the hyperbolic trig functions (which are elementary) and the error and gamma functions (which are not). Later in this chapter we will meet a number of others in both categories. All of these are listed in Appendices J and K.

The Hyperbolic Trig Functions

The "hyperbolic sine" (written sinh which is pronounced "sinsh") and "hyperbolic cosine" (written cosh which is pronounced just like it looks) are elementary functions, because they are defined in terms of exponentials.

Definition: Hyperbolic Trig Functions

$$\sinh x = \frac{e^x - e^{-x}}{2} \qquad \cosh x = \frac{e^x + e^{-x}}{2} \qquad (12.3.1)$$

Those are relatively simple function definitions, but it may not be obvious what they have to do with sines and cosines (or with hyperbolas for that matter). Below we compare the "normal" trig functions to their hyperbolic cousins. You will prove a number of these properties in the problems.

Trig Functions	Hyperbolic Trig Functions
The general solution to the differential equation $x''(t) = -k^2 x$ is $x = A\sin(kt) + B\cos(kt)$.	The general solution to the differential equation $x''(t) = k^2 x$ is $x = A\sinh(kt) + B\cosh(kt)$.
$\sin 0 = 0$ and $\cos 0 = 1$	$\sinh 0 = 0$ and $\cosh 0 = 1$
$d/dx(\sin x) = \cos x$ and $d/dx(\cos x) = -\sin x$	$d/dx(\sinh x) = \cosh x$ and $d/dx(\cosh x) = \sinh x$ (no negative sign!)
$\cos^2 x + \sin^2 x = 1$	$\cosh^2 x - \sinh^2 x = 1$
The equations $x = \cos t$, $y = \sin t$ describe a circle.	The equations $x = \cosh t$, $y = \sinh t$ describe a hyperbola.

---

[3] Strictly speaking we could have left off trig functions since you can get them by plugging complex arguments into exponentials. Less obviously you can construct the inverse trig functions from logarithms with complex arguments, so they are also elementary.

12.3 | A Few Special Functions

The functions tan x, cot x, sec x and csc x have hyperbolic equivalents based on those two: for instance, the hyperbolic tangent is tanh x = sinh $x/$ cosh x.

For our purposes, the most important use of these functions is to solve differential equations. The hyperbolic trig functions often match initial conditions more easily than exponential functions.

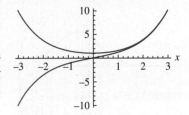

FIGURE 12.1 cosh x (upper plot) and sinh x (lower plot).

EXAMPLE **Hyperbolic Trig functions**

Problem:
Solve the equation $x''(t) = k^2 x(t)$ subject to the initial conditions $x(0) = 3$ and $x'(0) = 5$.

One approach: The general solution to this differential equation can be written $x(t) = Ae^{kt} + Be^{-kt}$. The initial conditions in this case become $A + B = 3$ and $Ak - Bk = 5$. To find the constants you solve these two equations simultaneously. For instance, you might rewrite the first equation as $B = 3 - A$ and substitute into the second equation to get $Ak - (3 - A)k = 5$, which solves to $A = (3k + 5)/(2k)$. A similar method leads to $B = (3k - 5)/(2k)$.

Better approach: An alternative way of writing the general solution is $x(t) = C \sinh(kt) + D \cosh(kt)$. The initial conditions are now $D = 3$ and $Ck = 5$.

As a final note, replacing x with ix turns the definitions of sinh x and cosh x into formulas for sin x and cos x. Bear in mind, however, that all these functions are real-valued for any real input.

The Error Function
The error function[4] is *not* an elementary function. It is not defined by combining or composing the basic functions, but by integrating one.

Definition: The Error Function

$$\operatorname{erf} x = \frac{2}{\sqrt{\pi}} \int_0^x e^{-t^2} dt \qquad (12.3.2)$$

Take a moment to convince yourself that Equation 12.3.2 does in fact define a function of x, not a function of t. If you choose $x = 1$ you can evaluate that integral (using for instance a Riemann sum with a lot of slices) and you will get a number. If you choose $x = 2$ you will get a different number. The letter t is a "dummy variable" that disappears before the calculation is done. For any given x-value the error function gives you a number, which represents (apart from a few constants thrown in for convenience) the area under the "Gaussian" curve e^{-t^2} between 0 and x.

---

[4] The name "error function" has the unfortunate side effect that when you ask a computer to solve a differential equation and it responds with an error function, it looks like the computer is reporting a mistake.

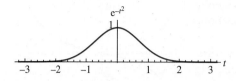

FIGURE 12.2 A Gaussian Function.

As an example, take a moment (without a computer) to find $\int_0^1 e^{-t^2}\, dt$. Go on, try it... we'll wait.

Got stuck, didn't you? You can see the area we want on the drawing. You can see that the integral is finite and positive. But you can't put together sines, logs, and exponents into a function whose derivative comes out e^{-t^2}. That's why we define a new function, called the error function, whose only definition is that it *is* that antiderivative.

Here are a few things to know about that integral and that function.

- The total area under e^{-t^2} between 0 and ∞ is $\sqrt{\pi}/2$, as you will prove in Problem 12.49. The error function is therefore defined with $2/\sqrt{\pi}$ in front, so $\lim_{x\to\infty} \text{erf}\, x = 1$.
- Because e^{-t^2} approaches zero very quickly, the vast majority of the area under the curve occurs very near the *y*-axis. For instance, erf 1 is over 0.8 and erf 2 is over 0.995. That means that less than 0.5% of the total area under this curve is found to the right of $x = 2$.
- And by the way, there's a name for that. Just as the error function erf x measures the area under the curve to the left of x, the "complementary error function" erfc x measures the area under the curve to the *right* of x. You can see that erfc $x = 1-\text{erf}\, x$.
- You will show in Problem 12.26 that erf is an odd function, meaning $\text{erf}(-x) = -\text{erf}\, x$.

The error function has many uses but the one that gives it its name comes from measurement theory, where it tells you the probability of getting a particular "error" when you make a measurement. This is because so many probability functions can be described by "Gaussian" functions, which means variations of e^{-x^2}.

For example, the height of adult U.S. men in 2003–2006 falls along a Gaussian curve with its center (the average height) at 176.3 cm.[5] If you pick an adult man at random in the U.S. and measure his height, he will almost certainly not be exactly 176.3 cm tall. In fact, he's probably not within 5 cm of that height. On the other hand, he probably is within 10 cm of the average, and almost certainly within 30. The function $P(x)$ below gives the probability that a randomly selected American man falls within x cm of 176.3.

$$P(x) = \frac{1}{8\sqrt{\pi}} \int_0^x e^{-(t/16)^2}\, dt \tag{12.3.3}$$

For example, $P(3)$ represents the probability that a randomly chosen man will have a height between 173.3 cm and 179.3 cm. In Problem 12.30 you'll rewrite this probability function in terms of erf x.

The Factorial and Gamma Functions

We presume you are familiar with the factorial function.

$$9! = 1 \times 2 \times 3 \times 4 \times 5 \times 6 \times 7 \times 8 \times 9 = 362{,}880$$

The value of 0! is defined to be 1. The factorial function is not defined for negative numbers or non-integers. For many purposes, including some Taylor series and probability calculations, a factorial function that is restricted to non-negative integers is enough.

---

[5] "Anthropometric Reference Data for Children and Adults: United States, 2003–2006," Margaret McDowell et. al., National Health Statistics Reports, Number 10, October 22, 2008. All height and weight statistics cited in this chapter come from that report.

In other circumstances, however, it is useful to generalize the factorial to a broader domain. The way to do this is to find a different function that matches the factorial at integer values but smoothly interpolates between them. You will show in Problem 12.38 that $x! = \int_0^\infty t^x e^{-t} dt$ for all positive integers x. That motivates the following definition.

Definition: The Gamma Function

$$\Gamma(x) = \int_0^\infty t^{x-1} e^{-t} dt \quad x > 0 \qquad (12.3.4)$$

This definition has $x - 1$ instead of x, so the relation between the gamma function and the factorial function is off by one. That is, $\Gamma(7)$ is not the same as $7!$ but instead is $6!$.

$$\Gamma(x) = (x - 1)!$$

We can think of factorials in terms of the recurrence relation they follow as they move up. For instance, we said above that $9! = 362,880$. If we now asked you for $10!$ you could immediately answer $3,628,800$. The equation $n! = n \times (n-1)!$ expresses this relationship in general form. The recurrence relation for the gamma function looks the same except that it is offset by one.

$$\Gamma(x) = (x - 1) \times \Gamma(x - 1) \qquad (12.3.5)$$

In Problem 12.39 you will show that this recurrence relation follows from Equation 12.3.4. (This amounts to another proof that the gamma function tracks the factorial function at the integers.) In other problems you will use this recurrence relation to investigate the behavior of the gamma function, such as how it acts on negative numbers.

We shall see the gamma function appearing in other special functions in this chapter such as Bessel functions.

As a final note, for very large x the factorial function can be approximated by "Stirling's formula" $x^x e^{-x} \sqrt{2\pi x}$. Because this approximates $x!$ it also approximates $\Gamma(x + 1)$. This approximation can be useful when factorials or gamma functions appear in functions that need to be differentiated or integrated.

Stepping Back

You might expect us to say at this point "hyperbolic trig functions, the error function, and the gamma function are all very important, and you should be intimately familiar with them." These functions are important—and you may encounter any or all of them in various applications—but there are a lot of other important functions out there. We can't possibly list them all, and it wouldn't necessarily do you a lot of good if we did.

So we'd rather say this. You may frequently have the experience of typing an equation into a computer and getting back a function you've never heard of. Don't panic! Look up the function you found—in the online help for your math software, or on the Web, or in the appendices to this book—and find the properties you need to know. Does this function have zeros? What does its graph look like? How does it behave as $x \to \infty$? Whatever question you started with, you can probably answer it based on the properties of your function.

12.3.2 Problems: A Few Special Functions

12.20 Show that the functions $\sinh x$ and $\cosh x$ are each other's derivatives.

12.21 For each question below, show how your answer comes from Equations 12.3.1.

(a) Is $\cosh x$ an odd function, an even function, or neither?

(b) Is $\sinh x$ an odd function, an even function, or neither?

(c) Calculate $\lim\limits_{x \to \infty} \cosh x$ and $\lim\limits_{x \to -\infty} \cosh x$.

(d) Calculate $\lim\limits_{x \to \infty} \sinh x$ and $\lim\limits_{x \to -\infty} \sinh x$.

(e) Calculate $\cosh 0$ and $\sinh 0$.

(f) Based on all your answers, draw sketches of $y = \cosh x$ and $y = \sinh x$.

(g) Calculate $\lim\limits_{x \to \infty} \tanh x$ and $\lim\limits_{x \to -\infty} \tanh x$.

12.22 In this problem you will investigate the relationship between the hyperbolic trig functions and hyperbolas. It will help to know the standard forms for a hyperbola:

$$\left(\frac{x-h}{a}\right)^2 - \left(\frac{y-k}{b}\right)^2 = 1 \quad \text{horizontal hyperbola centered at } (h, k)$$

$$\left(\frac{y-k}{a}\right)^2 - \left(\frac{x-h}{b}\right)^2 = 1 \quad \text{vertical hyperbola centered at } (h, k)$$

(a) Based on Equations 12.3.1 calculate $\cosh^2 t - \sinh^2 t$. Simplify your answer as much as possible.

(b) Based on your answer, what can you say about the curve represented by the parametric equations $x = \cosh t$, $y = \sinh t$?

(c) 🖥 Use a computer to graph $x = \cosh t$, $y = \sinh t$. How does the resulting curve match your prediction, and how does it surprise you?

You should be able to answer the remaining questions *without* a computer based on what you have found above.

(d) Draw the curve represented by $x = \sinh t$, $y = \cosh t$.

(e) Draw the curve represented by $x = \cosh(3t)$, $y = \sinh(3t)$.

(f) Draw the curve represented by $x = 2\cosh t + 4$, $y = 5\sinh t - 1$.

12.23 Consider an object whose position x follows the differential equation $x''(t) = \omega^2 x(t)$ for some real constant ω.

(a) Show that $x = Ae^{\omega t} + Be^{-\omega t}$ solves this differential equation for any values of the constants A and B.

(b) Solve for A and B to match the initial conditions $x(0) = x_0$ and $x'(0) = v_0$.

(c) Now show that $x = C\sinh(\omega t) + D\cosh(\omega t)$ solves the same differential equation for any values of the constants C and D.

(d) Solve for C and D to match the initial conditions $x(0) = 0$ and $x'(0) = v_0$.

(e) Solve for C and D to match the initial conditions $x(0) = x_0$ and $x'(0) = 0$.

(f) Solve for C and D to match the initial conditions $x(0) = x_0$ and $x'(0) = v_0$.

12.24 Find the Maclaurin series expansions of $\sinh x$ and $\cosh x$.

12.25 Use the error function to express "the area under the graph of e^{-x^2} between $x = 5$ and $x = 7$." Your final answer should be expressed in terms of the error function, not in terms of approximate decimals.

12.26 In this problem you will demonstrate that $\text{erf } x$ is an odd function. For the purposes of this problem you can use the approximation $\text{erf } 1 = 0.8$.

(a) Make a quick sketch of the Gaussian function and shade in the area represented by erf 1.

(b) What is $\text{erf}(-1)$? Explain how you can reach your conclusion from your drawing.

(c) Verify your conclusion by plugging $-x$ into the definition of $\text{erf } x$ and using a u substitution to rewrite that expression for $\text{erf}(-x)$ in terms of $\text{erf } x$.

(d) Is erfc x an even function, an odd function, or neither? Explain how you can reach this conclusion from the work done above.

12.27 For each question below, show how your answer comes from Equation 12.3.2 and Figure 12.2. No computer should be required until Part (e).

(a) Calculate erf 0.

(b) For $x > 0$ when is erf x increasing and when is it decreasing?

(c) For $x > 0$ when is erf x concave up and when is it concave down?

(d) In Problem 12.26 you showed that erf x is an odd function, and in Problem 12.49

you will show that $\lim_{x \to \infty} \text{erf } x = 1$. Based on those facts and your conclusions above, draw a sketch of $y = \text{erf } x$. You don't need to have done those problems to answer this one.

(e) Use a computer to graph $y = \text{erf } x$ as x goes from -6 to 6. If the result does not match your drawing, figure out what went wrong.

12.28 Find $\frac{d}{dx}(\text{erf } x)$.

12.29 Find the Maclaurin series expansion of erf x up to third order.

12.30 Equation 12.3.3 represents the probability of a randomly chosen American man being within x cm of the average height of 176.3.

(a) With a computer you can determine that $P(5) \approx 0.34$. Explain what this result means in terms that a typical 7th grade student could understand.

(b) Use a u-substitution to express $P(x)$ in terms of an error function.

12.31 Standard Deviation When you measure a quantity Q that varies among many different individuals, such as height, weight, IQ, hair length, etc., you find an average value $<Q>$ and some range of variation around that value. Many such properties are "normally distributed," which means that the probability of an individual being within x of the average value equals

$$P(x) = \frac{\sqrt{2}}{s\sqrt{\pi}} \int_0^x e^{-(1/2)(Q/s)^2} dQ$$

for some constant s, which is known as the "standard deviation" of Q.

(a) Rewrite $P(x)$ in terms of an error function.

(b) What is $\lim_{x \to \infty} P(x)$? Explain why this result makes sense in terms of what $P(x)$ means.

(c) Roughly what percentage of individuals will be within s of the mean value?

(d) Using data from the Explanation (Section 12.3.1), what is the standard deviation of adult American men's heights? *Hint*: it's not 16 cm.

12.32 [This problem depends on Problem 12.31.]

(a) The average height of U.S. adult women is 162 cm with a standard deviation of 11 cm. What fraction of U.S. adult women are between 151 cm and 173 cm?

(b) What fraction of U.S. adult women are taller than 184 cm? Write your answer in terms of an error function. *Hint*: start by finding the probability that a woman is outside the range 140 cm–184 cm and then halve that result.

(c) Express the answer to Part (b) as a percentage by asking a computer for the value of the erf function you found.

(d) Your goal now is to ask the question "What fraction of U.S. adult *men* are taller than ___ cm?" and have the answer come out the same as it did in Part (b). What height should you use?

12.33 Calculate $\Gamma(3)$ by starting with Equation 12.3.4 and using integration by parts.

12.34 Simplify $\Gamma\left(3\frac{1}{2}\right)/\Gamma\left(2\frac{1}{2}\right)$. *Hint*: you should not have to find $\Gamma\left(3\frac{1}{2}\right)$ to find this ratio.

12.35 Simplify $\Gamma(13.3)/\Gamma(11.3)$. *Hint*: you should not have to find $\Gamma(13.3)$ to find this ratio.

12.36 Simplify $\Gamma(x+17)/\Gamma(x+13)$.

12.37 Write $(x+4)(x+5)(x+6)\ldots(x+20)$ as the ratio of two gamma functions.

12.38 In this problem you're going to show that $n! = \int_0^\infty t^n e^{-t} dt$.

(a) Show that this formula works for $n = 0$.

(b) Using integration by parts, show that $\int_0^\infty t^{n+1} e^{-t} dt = (n+1) \int_0^\infty t^n e^{-t} dt$.

(c) Using your results from Parts (a) and (b), prove that $n! = \int_0^\infty t^n e^{-t} dt$ for all positive integers n.

12.39 Using integration by parts and the definition of the gamma function, prove that $\Gamma(x+1) = x\Gamma(x)$.

12.40 In this problem you will examine the behavior of the gamma function for negative x-values based on the recurrence relation (Equation 12.3.5). No computer should be required until Part (e).

(a) We begin with the fact that $\Gamma(1/2) = \sqrt{\pi}$. (You will prove that in Problem 12.42; for this problem, take it as a given.) Given that fact and the recurrence relation, calculate $\Gamma(-1/2)$, $\Gamma(-3/2)$, $\Gamma(-5/2)$, and $\Gamma(-7/2)$.

(b) It is also true (and much more obvious) that $\Gamma(1) = 1$. Given that and the recurrence relation, what can you conclude about $\Gamma(0)$? From that, what can you conclude about $\Gamma(-1)$, $\Gamma(-2)$, and other negative integers?

(c) $\Gamma(x)$ is positive for any $x > 0$. Given that, for what negative x-values is $\Gamma(x)$ positive, and for what values is it negative?

(d) Based on all your conclusions above, draw a graph of the gamma function for $x < 0$.

(e) Use a computer to graph $y = \Gamma(x)$ as x goes from -6 to 0. If the result does not match your drawing, figure out what went wrong.

12.41 The integral that defines the gamma function is improper in two ways: the upper limit is infinity, and the integrand blows up as $x \to 0$ for $p < 1$. In this problem you will prove that this integral converges for all $p > 0$.

(a) First, prove that $\int_1^\infty x^{p-1} e^{-x} dx$ converges for all p. One way to do this is to use the integral test (Chapter 2) which promises that for any continuous, non-increasing function $f(x)$, $\int_1^\infty f(x)\, dx$ converges if and only if $\sum_{n=1}^\infty f(n)$ converges.

(b) Second, explain how you know that $\int_0^1 x^{p-1} e^{-x} dx$ must be finite for $p \geq 1$. (This should be trivial. If you're stuck start by graphing the function.)

(c) Finally, prove that $\int_0^1 x^{p-1} e^{-x} dx$ converges for $0 < p < 1$.
  i. Prove that $\int_0^1 x^{p-1} dx$ converges for $0 < p < 1$.
  ii. Using that result, argue that $\int_0^1 x^{p-1} e^{-x} dx$ must also converge.

12.42 Evaluate $\Gamma(1/2)$. *Hint*: use a u-substitution to turn it into an integral of a Gaussian, and remember that erf $\infty = 1$.

12.43 In thermodynamics the "entropy" of a system is $S = k_B \ln \Omega$, where $k_B = 1.38 \times 10^{-23}$ m^{2}kg s^{-2} and Ω, known as the "multiplicity," is the number of states the system can be in. A crystalline solid with N atoms has multiplicity $\Omega = \Gamma(q+N)/[\Gamma(q+1)\Gamma(N)]$ where q is an integer proportional to the total energy of the system.[6]

(a) Assuming q and N are much larger than 1 (as they are for any macroscopic solid), use Stirling's approximation on Ω and use the result to calculate the entropy S.

(b) The temperature is related to the entropy by $T = 1/(dS/dU)$, where U is the energy of the system. As a first step towards that calculate dS/dq and simplify your result as much as possible. *Hint*: it will be easiest to take the derivative if you simplify as much as possible before differentiating, including using the laws of logs to separate out logs of products and ratios.

(c) For a macroscopic crystal N is of the order of 10^{23} or more, so simplify your answer by replacing $N - 1$ with N everywhere.

(d) As large as N is, it's still true that at room temperature $q \gg N$. That means you can treat N as a regular number and do a Maclaurin series in $1/q$.
  i. Define $x = 1/q$ and rewrite your expression for dS/dq as a function of N and x.
  ii. Take the Maclaurin series of that function with respect to x, keeping only linear terms. (This will involve creating Maclaurin series for two separate functions. You will discover that one of them doesn't *have* linear terms, which shows that for sufficiently small x this function is irrelevant.)
  iii. Finally, substitute $x = 1/q$ into the result to get it back in terms of N and q again.

(e) We said that the energy U is proportional to q. Let $U = \epsilon q$ and use your result for dS/dq to find dS/dU using the chain rule.

(f) Find the temperature, $T = 1/(dS/dU)$, as a function of N and U. If you have studied thermodynamics you may recognize your result as an example of the "equipartition theorem," which relates temperature and energy.

12.44 The "beta function" is defined for positive x and y as $B(x, y) = \int_0^1 t^{x-1}(1-t)^{y-1} dt$. It can be shown that $B(x, y) = \Gamma(x)\Gamma(y)/\Gamma(x+y)$. (It's not at all obvious.) In this problem you'll consider the function $B(x, 3)$.

(a) Write a formula for $B(4, 3)$ using three factorial functions. Then write out the individual factorials and cancel the terms on the top and bottom to show that $B(4, 3) = (1 \times 2)/(4 \times 5 \times 6)$.

(b) Go through the same process to find a formula for $B(5, 3)$.

(c) Find a formula for $B(x, 3)$ where x is any integer greater than 3. Your final answer should not involve any special functions: just a finite number of terms being multiplied and divided.

(d) Based on your answer, what is $\lim_{x \to \infty} B(x, 3)$?

(e) Based on your answers above, what is $\lim_{x \to \infty} B(x+1, 3)/B(x, 3)$?

12.45 Problem 12.44 introduced the "beta function" $B(x, y)$. (You don't need to have done that problem to do this one.) A

---

[6] Strictly speaking N in this formula is the number of ways the atoms can oscillate, which is typically three times the number of atoms in a 3D crystal.

true graph of the beta function is a surface, but we can get a sense for its behavior from a sequence of curves.

(a) 🖥 Draw graphs of $y = B(x, 1)$, $y = B(x, 3)$, and $y = B(x, 10)$. Your graphs should be confined to the region $x > 0$.

(b) What is $\lim_{x \to \infty} B(x, y)$ in each case?

(c) The three functions are similar in many ways. How are they different?

12.46 You are using a differential equation to model an important real-valued quantity starting at time $t = 0$ and a computer has just told you that your quantity follows an "Airy function" (sometimes called an "Airy function of the first kind") $Ai(t)$. You may never have heard of an Airy function. Look it up and answer the following questions about your quantity.

(a) What limit will it approach as $t \to \infty$?

(b) Will it oscillate, grow monotonically, decay monotonically, or behave in some other way?

(c) Will it ever equal zero? If so, will it do it a finite number of times or an infinite number?

(d) Does this seem like a plausible model for the height of a bouncing rubber ball? The mass of a melting ball of ice? The speed of an orbiting planet? In each case say why or why not.

12.47 You are using a differential equation to model an important real-valued quantity starting at time $t = 0$ and a computer has just told you that your quantity follows an "Airy function" (sometimes called an "Airy function of the first kind") $Ai(-t)$. (If you did Problem 12.46 note how this one is different: t is positive but the argument of Ai is negative.) You may never have heard of an Airy function. Look it up and answer the following questions about your quantity.

(a) What limit will it approach as $t \to \infty$?

(b) Will it oscillate, grow monotonically, decay monotonically, or behave in some other way?

(c) Will it ever equal zero? If so, will it do it a finite number of times or an infinite number?

(d) Does this seem like a plausible model for the height of a bouncing rubber ball? The mass of a melting ball of ice? The speed of an orbiting planet? In each case say why or why not.

12.48 Exploration: Fresnel Integrals. Two special functions that we did not discuss are the "Fresnel integrals." (These are two different functions, often denoted $C(x)$ and $S(x)$ in the literature.) Add them to this section. You should write an explanation, including the definition of the functions and what they are used for. You should also write a problem or two designed to help a student graph these functions by hand (and solve your own problems).

12.49 Exploration: The Error Function at Infinity

$$\lim_{x \to \infty} \operatorname{erf} x = \frac{2}{\sqrt{\pi}} \int_0^\infty e^{-t^2} dt$$

It's not obvious that that integral will even give us a finite answer, much less what that answer is. And we can't find it by evaluating the antiderivative, because the error function is the only antiderivative we're going to get. But Gauss discovered a clever trick that can be used to find this integral.

We start by defining a constant that's equal to the integral we are looking for: $S = \int_0^\infty e^{-t^2} dt$. (You'll multiply it by $2/\sqrt{\pi}$ later.) Therefore $S^2 = \left(\int_0^\infty e^{-t^2} dt \right) \left(\int_0^\infty e^{-t^2} dt \right)$, the product of two integrals.

(a) The next step is to replace every t in the first integral with x, and every t in the second integral with y. Explain briefly how we know that this does not change the answer.

(b) Combine the two integrals in your expression for S^2 into one double integral. (The resulting integral is "separable" which would turn it back into two separate integrals. We don't want to do that, but it gives you a quick check on your answer.)

(c) Use the laws of exponents to combine the two exponential functions into one.

(d) Looking at your limits of integration, what 2D region are you integrating over?

(e) Rewrite your double integral in polar coordinates. This involves rewriting the integrand in terms of ρ and ϕ and also replacing $dx\, dy$ with $\rho\, d\rho\, d\phi$. Use your answer to Part (d) to figure out the limits of integration in polar coordinates.

(f) You started with an integral that was impossible to evaluate. In polar coordinates you should now have one that is easy to evaluate. Do so and find the value of S^2. Then take the square root to get S, the integral you were originally looking for.

(g) So what is $\lim_{x \to \infty} \operatorname{erf} x$?

12.4 Solving Differential Equations with Power Series

In this section we will use Maclaurin series to solve differential equations. If the solution to a differential equation is $y = \sin x$, this method will instead tell us that the solution is $y = x - x^3/(3!) + x^5/(5!) - \ldots$. This technique succeeds where many traditional techniques, designed to find solutions in closed form, cannot. But it also serves as an introduction to two broader topics. One topic is "using series to solve differential equations"; once we have done this with Maclaurin series, we can generalize to other kinds of series. The other topic is special functions defined by differential equations and series. The special functions called "Hermite polynomials" and "Legendre polynomials" can be derived by the technique presented here, and other types of special functions can be derived from the generalizations of the technique we present later in the chapter.

12.4.1 Discovery Exercise: Solving Differential Equations with Power Series

In this exercise you are going to find a particular solution to the following differential equation. (You may already know how to solve this equation, but our purpose here is to demonstrate a technique.)

$$\frac{dy}{dx} = y + \sin x \qquad (12.4.1)$$

Our goal is to find a function $y(x)$ that satisfies Equation 12.4.1. Now, suppose we found such a function and then expanded it out into a Maclaurin series. Then it would look like this.

$$y = c_0 + c_1 x + c_2 x^2 + c_3 x^3 + c_4 x^4 + c_5 x^5 + \ldots \qquad (12.4.2)$$

So instead of saying "we need to find the mystery function" we can reframe the question as "we need to find the mystery coefficients" (which, in turn, define the function itself).

1. Based on Equation 12.4.2 write a formula for dy/dx.

    See Check Yourself #83 in Appendix L

    Just to save you a bit of time, we will tell you—although we hope you could rapidly figure this out yourself, or possibly even have it memorized, that:

    $$\sin x = x - \frac{x^3}{3!} + \frac{x^5}{5!} - \ldots$$

2. Plug Equation 12.4.2, and your answer to Part 1, and the Maclaurin expansion of $\sin x$, into Equation 12.4.1.
3. Combine like powers of x on the right side of your answer to Part 2.
4. Your equation in Part 3 has a constant term on the left and a constant term on the right. Write an equation asserting that these two must be equal.
5. Your equation in Part 3 has a coefficient of x on the left and a coefficient of x on the right. Write an equation asserting that these two must be equal.

    See Check Yourself #84 in Appendix L

6. In like manner, write equations setting the coefficients of x^2, x^3, and x^4 equal to each other. You now have five equations relating the coefficients.

7. Choose $c_0 = 0$. Based on the equations you have written, find all the other coefficients up to c_5.
8. Based on your coefficients, write down a solution to Equation 12.4.1. It will not be an exact answer, but it will be the power series of the solution up to the fifth term. *Hint*: several of the terms will be zero, and some won't.

12.4.2 Explanation: Solving Differential Equations with Power Series

You know how to find the Maclaurin series for a given function, and you may have some common Maclaurin series memorized. For instance, you may know off the top of your head that:

$$e^x = \sum_{n=0}^{\infty} \frac{x^n}{n!} = 1 + x + \frac{x^2}{2!} + \frac{x^3}{3!} + \frac{x^4}{4!} + \ldots \quad (12.4.3)$$

Such a series gives us one way of identifying a function. So instead of saying "I'm thinking of the function that raises e to a given power," you can instead say "I'm thinking of the function whose nth Maclaurin coefficient is $1/(n!)$." Equation 12.4.3 tells us that these are two equivalent statements.

So, one approach we can use to solve differential equations is to look for the solution in power series form. Instead of saying "I'm looking for this function in closed form" we say "I'm looking for the coefficients of the power series for this function."

Population Growth

We begin with one of the simplest and most important differential equations, the equation for unconstrained population growth: "The more rabbits (or trees or fish or amoebas) you have in this generation, the more you add in the next."

$$\frac{dP}{dt} = kP \quad (12.4.4)$$

We hope you know how to solve that equation by separating variables or even by just looking at it. But we're going to use a guess-and-check approach where our guess is the generic Maclaurin series.

$$P(t) = c_0 + c_1 t + c_2 t^2 + c_3 t^3 + c_4 t^4 + \ldots \quad (12.4.5)$$

As always, we will plug the guess into the differential equation and solve for the constants; if we find a working solution, that retroactively proves we made a good guess. This differs from our "guess and check" examples in earlier chapters only because we now have an infinite number of constants to solve for.

From Equation 12.4.5 we conclude that

$$dP/dt = c_1 + 2c_2 t + 3c_3 t^2 + 4c_4 t^3 + \ldots \quad (12.4.6)$$

and with that we are ready to plug into Equation 12.4.4.

$$c_1 + 2c_2 t + 3c_3 t^2 + 4c_4 t^3 + \ldots = kc_0 + kc_1 t + kc_2 t^2 + kc_3 t^3 + kc_4 t^4 + \ldots \quad (12.4.7)$$

Now comes the key step. If these two polynomials equal each other for all values of t, their coefficients must all be the same. (To put it another way, one function cannot have two different Maclaurin series.) So Equation 12.4.7 requires that the constant term on the left equal the constant term on the right, and the coefficient of t on the left equal the coefficient of t on the right, and so on. This gives us the following equations.

$$\begin{aligned}
\text{constant terms:} &\quad c_1 = kc_0 \\
\text{coefficients of } t: &\quad 2c_2 = kc_1 \quad \rightarrow \quad c_2 = (k/2)c_1 \\
\text{coefficients of } t^2: &\quad 3c_3 = kc_2 \quad \rightarrow \quad c_3 = (k/3)c_2 \\
\text{coefficients of } t^3: &\quad 4c_4 = kc_3 \quad \rightarrow \quad c_4 = (k/4)c_3
\end{aligned} \qquad (12.4.8)$$

... and so on. There are infinitely many such equations, but we can summarize them all in two rules.

1. The constant term, c_0, can be anything.
2. For any $n > 0$, the coefficients must follow the relationship $c_n = (k/n)c_{n-1}$.

The second rule is a recurrence relation: it defines each coefficient in terms of its predecessor. As an example, if we happen to start with $c_0 = 1$, Equations 12.4.8 supply all the other coefficients one at a time.

$$c_0 = 1 \quad \rightarrow \quad c_1 = k \times 1 = k \quad \rightarrow \quad c_2 = (k/2)k = k^2/2 \quad \rightarrow \quad c_3 = (k/3)(k^2/2) = k^3/(2 \times 3) \ldots$$

Don't forget what all those coefficients mean! They go into Equation 12.4.5 to create one valid solution to the differential equation.

$$P(t) = 1 + kt + \frac{k^2}{2}t^2 + \frac{k^3}{2 \times 3}t^3 + \ldots$$

Our recurrence relation tells us that the next coefficient will be $k^4/(4!)$ and so on. If you know your Maclaurin series, you may recognize this function as e^{kt}. But that last step is not essential, and in fact is generally only possible with toy examples like this one. More often you will leave your final answer in the form of a Maclaurin series with a recurrence relation. You can then approximate the function with arbitrary accuracy by evaluating partial sums.

As a final note before we move on, c_0 is the arbitrary constant in this solution. In the example above we chose $c_0 = 1$, but any other number would lead to a valid solution. In applications the arbitrary constant comes from initial conditions, and in fact you can see by plugging $t = 0$ into both sides of Equation 12.4.5 that $c_0 = P(0)$.

The Same Problem in Series Notation

In our example above we wrote out all our series term by term. We will now see how the same problem looks in series notation. After this we will generally stick to series notation in the chapter, but if you get confused it is always a good idea to write out the terms.

The first step, where we replaced P with its Maclaurin series, now looks like this. If you haven't worked with series notation in a while take a minute to work out the first few terms of this sum and convince yourself that the following series reproduces Equation 12.4.5.

$$P(t) = \sum_{n=0}^{\infty} c_n t^n$$

Now that we have P, we need $P'(t)$. We can take the derivative *inside* the sum, which amounts to taking the derivative of each individual term in the series.

$$\frac{dP}{dt} = \sum_{n=0}^{\infty} n c_n t^{n-1}$$

This time if you plug in $n = 0$ you get 0. (This corresponds to the fact that the first term of our Maclaurin series, c_0, drops out when you take the derivative.) Plug in $n = 1$ and then $n = 2$ and you will see Equation 12.4.6 begin to emerge.

12.4 | Solving Differential Equations with Power Series

The next step, just as before, is to plug the series representations of P and P' into the equation we want to solve (Equation 12.4.4). We bring both series to one side of the equation because in a moment we're going to combine them.

$$\sum_{n=0}^{\infty} n c_n t^{n-1} - \sum_{n=0}^{\infty} k c_n t^n = 0$$

We want the powers of t to match, so we reindex one of the two series. (If this trick is unfamiliar to you, please see Section 12.2; reindexing will be a vital skill in this chapter.) We arbitrarily chose the first one.

$$\sum_{n=-1}^{\infty} (n+1) c_{n+1} t^n - \sum_{n=0}^{\infty} k c_n t^n = 0$$

The first series has a t^{-1} term that the second series does not, but that term turns out to be zero so we drop it. All $n \geq 0$ terms appear in both series, so we write them under one sum.

$$\sum_{n=0}^{\infty} \left[(n+1) c_{n+1} - k c_n \right] t^n = 0$$

For this Maclaurin series to equal zero, every individual term—that is, every coefficient—must equal zero. So we set the coefficient of t^n equal to zero and get $(n+1)c_{n+1} = kc_n$, which is the recurrence relation we found before. This relation holds for all $n \geq 0$. Plug in $n = 0$ and it gives us c_1 in terms of c_0. Plug in $n = 1$ and it gives us c_2 in terms of c_1... and so on. But nothing we have found constrains c_0 in any way; it is the arbitrary constant in this problem.

We have now solved this problem twice with the same technique but different notation. As you go through the example below you may want to write out each step term by term: that will help you understand both series notation and the power series technique. (Or your professor may require you to do so by assigning Problem 12.53.)

EXAMPLE Solving a Differential Equation with a Power Series

Problem:
Find the particular solution to $\dfrac{d^2 y}{dx^2} - x^2 \dfrac{dy}{dx} + y = \dfrac{1}{x+2}$ with $y(0) = 0$ and $y'(0) = 1$.

Solution:
We begin with our guess, which is the generic power series, and its derivatives. This step is the same in all power series problems.

$$y = \sum_{n=0}^{\infty} c_n x^n \quad \to \quad y' = \sum_{n=0}^{\infty} n c_n x^{n-1} \quad \to \quad y'' = \sum_{n=0}^{\infty} n(n-1) c_n x^{n-2}$$

To represent the entire equation in power series form we need the Maclaurin series for $1/(x+2)$. If you don't know how to find that on your own, please review Taylor series; it's an important skill. (You may find the list of common Maclaurin series in Appendix B a useful starting point for finding some series.)

$$\frac{1}{x+2} = \frac{1}{2} - \frac{x}{4} + \frac{x^2}{8} - \frac{x^3}{16} + \ldots = \sum_{n=0}^{\infty} \frac{(-1)^n}{2^{1+n}} x^n$$

With that in place we can rewrite our entire differential equation in terms of Maclaurin series.

$$\sum_{n=0}^{\infty} n(n-1)c_n x^{n-2} - \sum_{n=0}^{\infty} nc_n x^{n+1} + \sum_{n=0}^{\infty} c_n x^n = \sum_{n=0}^{\infty} \frac{(-1)^n}{2^{1+n}} x^n$$

We want to compare coefficients of x^n so we reindex those derivatives, replacing n with $n+1$ in the first derivative series and $n-1$ in the second. (Plug in the first few n-values to convince yourself that these series are the same, term by term, as the ones we had before.)

$$\sum_{n=-2}^{\infty} (n+2)(n+1)c_{n+2} x^n - \sum_{n=1}^{\infty} (n-1)c_{n-1} x^n + \sum_{n=0}^{\infty} c_n x^n - \sum_{n=0}^{\infty} \frac{(-1)^n}{2^{1+n}} x^n = 0$$

The $n=-2$ and $n=-1$ terms appear only in the first series, and both equal zero. The $n=0$ term appears in three out of four of the series, so we write it separately. All $n \geq 1$ terms appear in all four series, so we group them under one sum.

$$\left(2c_2 + c_0 - \frac{1}{2}\right) x^0 + \sum_{n=1}^{\infty} \left[(n+2)(n+1)c_{n+2} - (n-1)c_{n-1} + c_n - \frac{(-1)^n}{2^{1+n}}\right] x^n = 0$$

Of course x^0 is just 1, but we wrote it explicitly to make it clear that these represent all the coefficients of x^0 in the sums that included that term. Because this entire sum must equal zero, we can write two relationships. First, the total coefficient of x^0 must be zero. Second, inside the sum, the total coefficient of x^n for all $n \geq 1$ must be zero.

$$2c_2 + c_0 - \frac{1}{2} = 0 \quad \rightarrow \quad c_2 = \frac{1}{4} - \frac{c_0}{2} \quad \quad (12.4.9)$$

$$(n+2)(n+1)c_{n+2} - (n-1)c_{n-1} + c_n - \frac{(-1)^n}{2^{1+n}} = 0 \quad \rightarrow$$

$$c_{n+2} = \frac{(n-1)c_{n-1} - c_n + (-1)^n/2^{1+n}}{(n+2)(n+1)} \quad (n \geq 1) \quad \quad (12.4.10)$$

Those two equations correspond to the general solution that we normally look for. The first two constants are determined by initial conditions: c_0 is $y(0)$ and c_1 is $y'(0)$. Given those two values, Equation 12.4.9 gives us c_2 and Equation 12.4.10 gives us c_3 and beyond.

In this particular problem, $c_0 = 0$ and $c_1 = 1$. Equation 12.4.9 tells us that $c_2 = 1/4 - 0/2 = 1/4$. Plugging $n=1$ into Equation 12.4.10 tells us how to find c_3 from c_0 and c_1.

$$c_3 = \frac{0c_0 - c_1 + (-1)^1/2^2}{(3)(2)} = \frac{-1 - 1/4}{6} = -\frac{5}{24}$$

We can then plug $n=2$ into the recurrence relation to find c_4 from c_2 and c_3, and so on up. Here is the solution we find, up to the fifth power.

$$y = x + \frac{1}{4}x^2 - \frac{5}{24}x^3 + \frac{7}{96}x^4 + \frac{31}{960}x^5 + \cdots$$

Plugging that solution into the left-hand-side of the differential equation gives us $(1/2) - (x/4) + (x^2/8) + \ldots$. It would take infinitely many terms to reproduce $1/(x+2)$ exactly, but recreating the first few terms is a good sign.

When Can You Use This Method?

To start with, the method of power series can only be used on a linear differential equation. For a second-order equation, this means the following form.

$$\frac{d^2y}{dx^2} + a_1(x)\frac{dy}{dx} + a_0(x)y = f(x) \tag{12.4.11}$$

But there is a further restriction: the functions $a_1(x)$, $a_0(x)$ and $f(x)$ must be "analytic," which means they can be represented in power series form. If that is the case then the solution $y(x)$ will also be analytic, and this method will find the coefficients of its power series.

As an example, consider a differential equation that includes $\sqrt{x}$. That function is not differentiable at $x = 0$ so it cannot be represented by a Maclaurin series. You could still use this method, but you would have to use a Taylor series centered on some other value such as $x = 1$.

Note also that some Taylor series converge only within a given interval. The function $1/(1-x)$ can be represented by $1 + x + x^2 + x^3 + \ldots$ but only within the interval $-1 < x < 1$. So a differential equation containing this function could be solved by the method of power series, but the solution would only be valid on that domain.

Section 12.6 will present a generalization of the power series method that avoids some of these limitations.

Stepping Back

It takes time to get used to working with series, especially expressed in $\sum$ notation, but beyond that the process we've described here is mechanical.

1. Begin with $y = \sum c_n x^n$ and find y' and y'' and so on as needed.
2. Convert any functions of x into power series.
3. Set like powers of x equal to each other and the result will be a recurrence relation.

Of course the answer must involve arbitrary constants! For a first-order linear differential equation you should find that c_0 is unconstrained, and the recurrence relation will define each coefficient in terms of the previous one. For a second-order equation both c_0 and c_1 should be free variables, and the recurrence relation will define each coefficient in terms of its *two* predecessors. As always, the arbitrary constants will be determined by initial or boundary conditions, not by the differential equation itself.

In some cases you will find that all the coefficients past a certain point are zero. Then the solution you were looking for is a finite polynomial, and you have it all. (You will encounter two particularly important examples of this, "Legendre polynomials" and "Hermite polynomials," in Section 12.5.) It is also possible that you will find an infinite Taylor series that you can convert back to a function in closed form, as in our e^{kx} example above.

But in most cases you will have nothing but a recurrence relation—and that's OK. The great virtue of power series is that you can use them to make arbitrarily accurate approximations. The 5th partial sum or the 10th partial sum may give you all the accuracy you need. And if they don't, you can get the 100th partial sum just as easily from a computer.

12.4.3 Problems: Solving Differential Equations with Power Series

12.50 Walk-Through: Power Series. In this problem you will find a solution to the equation $y'' - 4y = e^{3x}$ in the form of a power series.

(a) Starting with $y = \sum_{n=0}^{\infty} c_n x^n$, find $y''(x)$.

(b) Find the Maclaurin series for the function e^{3x}.

(c) Using both of your answers, rewrite the original differential equation with three power series.

(d) Reindex so all of the sums are expressed in terms of x^n.

(e) To check your reindexing, write down the first two non-zero terms of each series that you reindexed, both before and after the reindexing. If they are not a perfect match, see what went wrong!

(f) The $n = -2$ and $n = -1$ terms appear only in the first series. Explain why we can ignore them in this case.

(g) All terms with $n \geq 0$ appear in all three series. Set the coefficient of x^n (for all $n \geq 0$) equal to zero and solve for c_{n+2} to get the general recurrence relation for c_{n+2} in terms of c_n.

(h) Plug $n = 0$ and $n = 1$ into your recurrence relation. The result should be an equation for c_2 in terms of c_0 and an equation for c_3 in terms of c_1.

(i) Explain why c_0 and c_1 are arbitrary constants.

(j) For the particular case of $c_0 = 0$ and $c_1 = 1$ find the fourth partial sum of the solution $y = \sum_{n=0}^{\infty} c_n x^n$.

12.51 *[This problem depends on Problem 12.50.]* Solve the differential equation in Problem 12.50 without using power series. You can find the general solution to the complementary homogeneous equation, and a particular solution to the original equation, by guess and check. Then apply the initial conditions $y(0) = 0$ and $y'(0) = 1$. Finally find the Maclaurin series for your solution and confirm that it matches your solution to Problem 12.50.

12.52 In the Explanation (Section 12.4.2) we found a recurrence relation for the coefficents of the solution to $dP/dt = kP$, and used that recurrence relation to find a specific solution with $c_0 = 1$. (In each part below leave your answer in the form of a series up to the fourth power.)

(a) Using the same recurrence relation, find the solution that begins with $c_0 = 0$.

(b) Using the same recurrence relation, find the solution that begins with $c_0 = 3$.

(c) Using the same recurrence relation, find the general solution by leaving c_0 as a constant.

12.53 On Page 669 we solved the differential equation $y'' - x^2 y' + y = 1/(x + 2)$ by using power series. We represented all our steps in series notation.

(a) Solve that problem writing out the first few terms of each series term by term—starting with the initial guess—so the symbol $\sum$ never appears in your solution. Your final answer will be a formula for c_2 as a function of c_0, a formula for c_3 as a function of c_1, and a formula for c_4 as a function of c_1 and c_2.

(b) Now assuming $y(0) = 0$ and $y'(0) = 1$ find the power series up to the fourth power.

12.54 If $y = \sum_{n=0}^{\infty} c_n x^n$, find the coefficient of x^n in the Maclaurin expansion of y'''.

In Problems 12.55–12.61, assume a solution to the given differential equation in the form $y = \sum c_n x^n$.

- Find the recurrence relation that gives each coefficient in the power series in terms of one or more previous coefficients.
- Find a particular solution, up to the fourth power, that matches the given condition(s).

It may help to first work through Problem 12.50 as a model.

12.55 $dy/dx = y + x^2$, $y(0) = -2$

12.56 $2y'(x) + 3y(x) = \sqrt{x+1}$, $y(0) = 1$

12.57 $dy/dx + xy = 1$, $y(0) = 2$

12.58 $y'' = p^2 y$, $y(0) = 2$, $y'(0) = 0$

12.59 $2\dfrac{d^2 x}{dt^2} + \dfrac{dx}{dt} - x = \dfrac{1}{1-t}$, $x(0) = 1$, $x'(0) = 0$

12.60 $\ddot{x} - 2x = \cos t$, $\dot{x}(0) = x(0) = 0$

12.61 $d^3 y/dx^3 + x^2 (dy/dx) + y = 0$, $y(0) = y'(0) = y''(0) = 24$

12.62 The function $y = \ln x$ cannot be expanded out into a Maclaurin series (why?), but it can be expanded into a Taylor series around $x = 1$. Beginning with $y = \sum c_n(x-1)^n$ find the particular solution, up to the fourth power, of the equation $4y'' + y = \ln x$ with conditions $y(1) = 8$, $y'(1) = 1$.

12.63 [This problem depends on Problem 12.62.]

(a) Have a computer get the exact solution to $4y'' + y = \ln x$ with initial conditions $y(1) = 0$, $y'(1) = 1$. Comment on why a power series approach might be better suited to this problem, even though there is an exact solution.

(b) Plot the fourth-order Taylor series solution that you found in Problem 12.62 and the exact solution together on the same plot, starting at $t = 0$. Experiment with the final value of t until you can see at what value the two solutions start to diverge significantly from each other. Estimate this value.

12.5 Legendre Polynomials

One of the most important uses of the technique of power series is to solve the Legendre differential equation. The solutions, called Legendre polynomials, are useful in a variety of applications, most notably solving differential equations in spherical coordinates.

12.5.1 Discovery Exercise: Legendre Polynomials

Consider the following differential equation where μ is a constant.

$$(1 - x^2)y'' - 2xy' + \mu y = 0 \quad -1 < x < 1 \quad (12.5.1)$$

1. Assuming a power series solution of the form $y(x) = \sum c_n x^n$, find the recurrence relation for c_{n+2} in terms of c_n, n, and the constant μ.

   See Check Yourself #85 in Appendix L

Now consider the particular solution for the constants $\mu = 20$, $c_0 = -3$, and $c_1 = 0$.

2. Show that all the odd coefficients will be zero.
3. Find c_2, c_4, and so on until you reach a point where all subsequent even coefficients will be zero.
4. Write down the resulting polynomial and verify that it solves Equation 12.5.1 for $\mu = 20$.

12.5.2 Explanation: Legendre Polynomials

In Chapter 11, in the middle of solving an important electrostatics problem in spherical coordinates, we encountered the following differential equation.

$$\frac{d^2\Theta}{d\theta^2} + \frac{\cos\theta}{\sin\theta}\frac{d\Theta}{d\theta} + k\Theta = 0 \quad (12.5.2)$$

In Problem 12.81 you will apply the substitution $u = \cos\theta$ to this equation. The result is a particularly important equation, in part because partial differential equations in spherical coordinates come up fairly often. So this equation and its solutions have been named and studied.

> **The Legendre Differential Equation and its Solutions**
>
> The Legendre Differential Equation: $(1-x^2)y'' - 2xy' + l(l+1)y = 0 \quad -1 < x < 1 \quad (12.5.3)$
>
> The General Solution: $y = AP_l(x) + BQ_l(x) \quad (12.5.4)$

Equation 12.5.4 is what we expect to get from a second-order homogeneous differential equation: two independent solutions ($P_l(x)$ and $Q_l(x)$) multiplied by arbitrary constants and added. Much of this section will be dedicated to finding and exploring those two solutions.

The expression $l(l+1)$ represents "any constant" and we could just give it its own letter, as we did in Equation 12.5.1. But the most important uses of this equation occur when that constant can be written as $l(l+1)$ for some non-negative integer l. So Equation 12.5.3 represents an infinite number of related equations such as the following.

The Legendre Differential Equation for $l = 3$: $(1-x^2)y'' - 2xy' + 12y = 0 \quad -1 < x < 1$

The General Solution: $y = AP_3(x) + BQ_3(x)$

Solving the Legendre Equation

In the Discovery Exercise (Section 12.5.1) you used the method of power series on Equation 12.5.1. You found that the solution can be written as $y = \sum c_n x^n$ with the following recurrence relation.

$$c_{n+2} = \frac{n^2 + n - \mu}{(n+1)(n+2)} c_n \quad (12.5.5)$$

You can think of this section as a study of how much we can learn about a function from its recurrence relation. We begin with the following observations.

- The constant c_0 is arbitrary. (It may be determined by initial conditions, but it is not determined by the differential equation.) All the other even-numbered coefficients are based on that constant.
- In particular, if $c_0 = 0$ then all the even-numbered coefficients are zero.
- The constant c_1 is also arbitrary, and the odd-numbered coefficients are based on it. If $c_1 = 0$ then all odd-numbered coefficients are zero.

So we have two series, one based on c_0 and one based on c_1, each of which solves the differential equation independently. The total solution is the sum of these two solutions, as we expect from a second-order equation.

If $\mu = l(l+1)$ for some non-negative integer l then the numerator of our recurrence relation becomes factorable.

$$c_{n+2} = \frac{n^2 + n - \mu}{(n+1)(n+2)} c_n = \frac{n^2 + n - l(l+1)}{(n+1)(n+2)} c_n = \frac{(n+l+1)(n-l)}{(n+1)(n+2)} c_n \quad (12.5.6)$$

Starting from c_0 and c_1 you use Equation 12.5.6 to generate coefficients. But when you reach $n = l$ one of your two series (the odd or even-numbered coefficients) comes to an end, as we see in the following example.

EXAMPLE The Legendre Equation with $l = 5$

Problem:
Here is the Legendre equation for $l = 5$.

$$(1 - x^2)y'' - 2xy' + 30y = 0 \qquad (12.5.7)$$

Find the solution with $c_0 = 1$ and $c_1 = 1$.

Solution:
The solution begins with the method of power series, but we've already gone through that and arrived at Equation 12.5.6. Using this recurrence relation to generate odd coefficients gives us the following.

$$c_1 = 1 \rightarrow c_3 = \frac{7 \times -4}{2 \times 3}(1) = -\frac{14}{3} \rightarrow c_5 = \frac{9 \times -2}{4 \times 5}\left(-\frac{14}{3}\right) = \frac{21}{5} \rightarrow c_7 = \frac{11 \times 0}{6 \times 7}\left(\frac{21}{5}\right) = 0$$

We don't have to go any further. $c_7 = 0$ will make $c_9 = 0$, and that in turn will make $c_{11} = 0$. So for $l = 5$ the odd-numbered coefficients stop after x^5.

$$y_{\text{odd}} = x - \frac{14}{3}x^3 + \frac{21}{5}x^5 \qquad (12.5.8)$$

This is, all by itself, a valid solution to Equation 12.5.7. But it does not meet the condition $c_0 = 1$ that was stipulated in the problem. That condition leads to the following.

$$c_0 = 1 \rightarrow c_2 = \frac{6 \times -5}{1 \times 2}(1) = -15 \rightarrow c_4 = \frac{8 \times -3}{3 \times 4}(-15) = 30 \rightarrow c_6 = \frac{10 \times -1}{5 \times 6}(30) = -10$$

This series does not terminate, but we have found the first few terms.

$$y_{\text{even}} = 1 - 15x^2 + 30x^4 - 10x^6 \ldots \qquad (12.5.9)$$

That infinite series is also a solution to Equation 12.5.7 (when the series converges at all, as we discuss below). The solution that meets both given conditions is the sum of these two solutions.

You can confirm for yourself that Equation 12.5.8 is a valid solution to Equation 12.5.7. But it's more important to see what happened, and how it generalizes. The recurrence relation Equation 12.5.6 has a factor of $(n - l)$ in the numerator, or in this case $(n - 5)$. When we were counting off odd-numbered coefficients, that factor killed the c_7 coefficient and all the other coefficients based on it, leaving us with a fifth-order polynomial.

On the other hand, when we were counting up even-numbered coefficients, we never reached $n = 5$ exactly. Consequently, none of those coefficients came out zero, so the polynomial kept going forever.

If you followed all that, you are ready for a few more general observations about the solutions to the Legendre equation.

- If l is an odd integer then the solution based on c_1 will be a polynomial of order l, but the solution based on c_0 will not terminate. Therefore, if l is an odd integer and $c_0 = 0$ the entire solution will be a polynomial of order l.

- Similarly, if l is an even integer then the solution based on c_0 will be a polynomial of order l. This will be the complete solution if $c_1 = 0$.
- Finally, if l is a non-integer then both solutions will be infinite series.

Legendre Polynomials of the First Kind

We have seen that the Legendre equation always has two independent solutions. One solution is an lth-order polynomial. (If l is odd then this solution is based on odd powers of x; if l is even then it is based on even powers.) The other solution is a non-terminating polynomial, an infinite series. The general solution is the sum of these two solutions.

The finite polynomial is referred to as a "Legendre polynomial of the first kind," designated as $P_l(x)$. So what is $P_5(x)$? If you have followed everything we have said so far, you may be ready to answer "$P_5(x)$ is the finite solution to the Legendre equation with $l = 5$, so it is Equation 12.5.8." But that equation is the particular finite solution generated with $c_1 = 1$. If we had chosen $c_1 = 2$ that would have doubled the second coefficient, which in turn would have doubled the third coefficient, and so on. Because the Legendre equation is linear and homogeneous, a doubled solution would still be a solution.

So we can define $P_5(x)$ based on any c_1 we like. (Remember that the general solution involves $AP_l(x)$ where A is an arbitrary constant, so there is no loss of generality in choosing a particular c_0 or c_1 to be part of $P_l(x)$ itself.) A common convention is to choose the starting coefficients such that $P_l(1) = 1$. With this convention, the first five Legendre polynomials are:

$$P_0(x) = 1$$
$$P_1(x) = x$$
$$P_2(x) = \frac{1}{2}\left(3x^2 - 1\right)$$
$$P_3(x) = \frac{1}{2}\left(5x^3 - 3x\right) \quad (12.5.10)$$
$$P_4(x) = \frac{1}{8}\left(35x^4 - 30x^2 + 3\right)$$
$$P_5(x) = \frac{1}{8}\left(63x^5 - 70x^3 + 15x\right)$$

If you should need $P_6(x)$ you can find it by solving Legendre's equation with $l = 6$ just as we did above for $l = 5$. You can also look it up in a table or ask a computer for it. But even before you look you know that it will be a sixth-order polynomial involving only even powers of x.

People often use the phrase "Legendre polynomials" as shorthand for "Legendre polynomials of the first kind." This reflects the fact that the second kind, discussed below, is less important in applications.

Legendre Polynomials of the Second Kind

The second solution to Legendre's equation, the infinite series, is referred to as a "Legendre polynomial of the second kind" and designated as $Q_l(x)$. For instance, Equation 12.5.9 gives the first few terms of $Q_5(x)$ with $c_0 = 1$.

As with any infinite series, we must ask first where this series converges at all. We apply the ratio test (discussed in Chapter 2), noting that our recurrence relation gives us the ratio we need.

$$\lim_{n \to \infty} \left| \frac{(n+l+1)(n-l)}{(n+1)(n+2)} x^2 \right| = x^2$$

Based on this calculation the ratio test tells us that $Q_l(x)$ converges for all $-1 < x < 1$, and that it diverges when $x < -1$ or $x > 1$. The ratio test tells us nothing about what happens for $x = \pm 1$, but the series is generally not convergent there. (You will show this for $l = 1$ in Problem 12.82.)

12.5 | Legendre Polynomials

This result is not surprising if you remember where we started. We said in Section 12.4 that the method of power series supplies a solution to Equation 12.4.11 if $a_1(x)$, $a_0(x)$ and $f(x)$ are all analytic functions. The Legendre equation in this form looks like the following.

$$\frac{d^2y}{dx^2} - \frac{2x}{1-x^2}\frac{dy}{dx} + \frac{l(l+1)}{1-x^2}y = 0$$

The coefficients of y' and y are only analytic for $-1 < x < 1$, so we are only guaranteed to find power series solutions in that domain.

The divergence of the $Q_l(x)$ functions is one of the most important facts about these functions. In many applications we are looking for a function that is finite at $x = \pm 1$, and on this basis we restrict our solution to the Legendre functions of the first kind.

Non-Recursive Definitions

Equation 12.5.6 gives us a "recursive definition" of a series; it defines each coefficient in terms of the coefficients before it. Such a definition can tell us a great deal about a function (and sometimes it's the best we can find), but it's a slow way to figure out c_{326}. In many cases, fortunately including Legendre polynomials, we can find an "explicit definition" for each coefficient based only on the initials ones.

From Equation 12.5.6 we get:

$$c_2 = \frac{(l+1)(-l)}{(1)(2)}c_0 \qquad (12.5.11)$$

$$c_4 = \frac{(l+3)(2-l)}{(3)(4)}c_2 \qquad (12.5.12)$$

$$c_6 = \frac{(l+5)(4-l)}{(5)(6)}c_4 \qquad (12.5.13)$$

... and so on. Plug Equation 12.5.11 into Equation 12.5.12.

$$c_4 = \frac{(l+1)(-l)(l+3)(2-l)}{(1)(2)(3)(4)}c_0 = \frac{l(l+1)(l-2)(l+3)}{4!}c_0$$

Now plug that into Equation 12.5.13.

$$c_6 = -\frac{l(l+1)(l-2)(l+3)(l-4)(l+5)}{6!}c_0$$

And so we arrive at a general form for the even-powered solution to the Legendre equation.

$$y_{even} = c_0 \left[1 - \frac{l(l+1)}{2!}x^2 + \frac{l(l+1)(l-2)(l+3)}{4!}x^4 - \frac{l(l+1)(l-2)(l+3)(l-4)(l+5)}{6!}x^6 + \ldots \right]$$
(12.5.14)

Is that a formula for $P_l(x)$ or $Q_l(x)$? The answer depends on l. For an even l that formula gives us a finite polynomial (can you see why?) which is $P_l(x)$. For an odd l it gives us the infinite series $Q_l(x)$.

In Problem 12.80 you will find the comparable formula for the odd-powered series.

A different non-recursive formula for the Legendre polynomials is called "Rodrigues' formula."

$$P_l(x) = \frac{1}{2^l l!} \frac{d^l}{dx^l}\left[(x^2-1)^l\right] \quad \text{Rodrigues' formula} \qquad (12.5.15)$$

In Problem 12.85 you will prove Rodrigues' formula by connecting it to Equation 12.5.14. In subsequent problems you will use Rodrigues' formula to derive important properties of the Legendre polynomials.

Fourier-Legendre Series

In the Motivating Exercise (Section 12.1) you reviewed the process of solving a PDE. In the last step you have to rewrite the initial state of your system—which could in principle be almost any function—as a series based on the normal modes you have found. In many cases involving spherical coordinates, those normal modes are Legendre polynomials. So it is important to be able to write any given function as a "Fourier-Legendre series."

$$f(x) = \sum_{n=0}^{\infty} A_n P_n(x) = A_0 P_0(x) + A_1 P_1(x) + A_2 P_2(x) + \ldots$$

Appendix J gives the formula for finding the coefficients A_n, and Section 12.8 explains how this formula was derived. Here we simply illustrate the process.

EXAMPLE Legendre Series Expansion

Question: Represent the function $\sqrt[3]{x}$ as a series of Legendre polynomials.

Answer:
The formula for the coefficients of a Fourier-Legendre series is given in Appendix J. We find our first coefficient by plugging $l = 0$ into that formula.

$$A_0 = \frac{2(0)+1}{2} \int_{-1}^{1} \sqrt[3]{x} P_0(x) dx$$

Now we use a table of Legendre polynomials (such as Equation 12.5.10) or a computer to find that $P_0(x) = 1$. Evaluating, $A_0 = (1/2) \int_{-1}^{1} \sqrt[3]{x}\, dx = 0$. Plugging in $l = 1$ gives:

$$A_1 = \frac{2(1)+1}{2} \int_{-1}^{1} \sqrt[3]{x} P_1(x) dx = \frac{3}{2} \int_{-1}^{1} \sqrt[3]{x}(x) dx = \frac{9}{7}$$

We can similarly find as many of the other coefficients as we want. The integrands will only involve powers of x so they are easy, if tedious, to integrate. For this function all the even coefficients are zero. (Do you see why?) The next non-zero coefficient is $A_3 = -6/13$, so we can write:

$$\sqrt[3]{x} = \frac{9}{7} P_1(x) - \frac{6}{13} P_3(x) + \ldots$$

Using only those two terms we get a passable approximation of $\sqrt[3]{x}$ between $x = -1$ and $x = 1$. Adding one more term makes it significantly more accurate, and by the time we get through P_{19} the approximation almost perfectly tracks the function within this domain. For $|x| > 1$ the Fourier-Legendre series looks nothing like $\sqrt[3]{x}$.

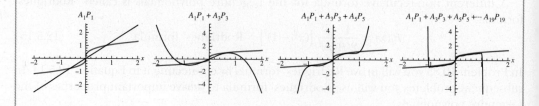

12.5 | Legendre Polynomials

It is common for normal modes to be in the form $P_l(\cos\theta)$, and Appendix J also lists the formulas for expanding a function $f(\theta)$ as a sum of such terms.

Stepping Back

A great deal more can be said—and has been said—about the Legendre polynomials. There are recurrence relations that give $P_l(x)$ in terms of $P_{l-1}(x)$ and $P_{l-2}(x)$, formulas for $P_l'(x)$, and so on. And there are surprising applications of Legendre polynomials; for instance, they can be used in finding polynomial approximations to curves that work better than Taylor series under some circumstances.

But we have focused on using Legendre polynomials to solve partial differential equations.

- You need to know that the Legendre polynomials solve Legendre's differential equation. We stated that fact in Chapter 11 and now we have proven it.
- You need to know that Legendre polynomials of the first kind are finite series when l is an integer, while Legendre polynomials of the second kind are always infinite series and generally divergent at $x = \pm 1$, and are therefore discarded in many problems.
- Finally, you need to know how to write the initial conditions as a sum of Legendre polynomials. We have shown you how to use the formula; you will derive that formula in Section 12.8.

Section 12.7 will go through a similar treatment of Bessel functions.

12.5.3 Problems: Legendre Polynomials

12.64 Hermite's equation. $y''(x) - 2xy'(x) + 2ky(x) = 0$ (where k is a constant) arises in calculations involving probability, the quantum harmonic oscillator, and others. The solutions are "Hermite polynomials," an important class of functions in their own right. We present them here because the derivation of the Hermite polynomials is very similar to the derivation we have demonstrated for the Legendre polynomials.

(a) Assuming a power series solution of the form $y(x) = \sum c_n x^n$, find the recurrence relation for c_{n+2} in terms of c_n, n, and the constant k.

(b) Consider the particular solution for the constants $k = 5$, $c_0 = 0$, and $c_1 = 15$.

   i. Show that all the even coefficients will be zero.
   ii. Find c_3 and c_5.
   iii. Show that all odd coefficients for $n > 5$ will be zero.
   iv. The particular solution you have found is called a "Hermite polynomial of order 5," sometimes written $H_5(x)$. Write this function and show that it satisfies Hermite's equation for $k = 5$.

(c) The general solution to Hermite's equation involves a sum of two series, an even function based on c_0 and an odd function based on c_1. For what values of k will the even function be a finite polynomial? For what values of k will the odd function be a finite polynomial?

(d) Choose values of k, and c_1 such that the solution to Hermite's equation will be a fourth-order polynomial. Find that polynomial using $c_0 = 12$.

12.65 *[This problem depends on Problem 12.64.]* The general solution to Hermite's equation involves a sum of two series, an even function based on c_0 and an odd function based on c_1. Find a non-recursive definition for the even function, similar to Equation 12.5.14 for the even-powered solution to the Legendre equation.

12.66 Are Legendre polynomials of the first kind even functions, odd functions, or neither? (*Hint*: the answer begins with "It depends...")

12.67 We said that the most important uses of Legendre's differential equation are for non-negative integers l. Later we explained that for non-integer values of l the solutions would diverge at one or both of $x = \pm 1$, so they are generally discarded. You may have been wondering about negative integer values, though. Looking at Equation 12.5.3, explain why negative integers for l would simply lead to the same Legendre polynomials $P_l(x)$ we already found for positive integers. *Hint*: try plugging in a few negative integer values and see what the equation looks like.

12.68 Walk-Through: Legendre Series. In this problem you will represent the function $x^2 + 3$ as a series of Legendre polynomials, $\sum A_l P_l(x)$.

(a) To find the first coefficient, plug $l = 0$ into the equation in Appendix J for the coefficients of a Legendre series. This equation includes an integral that involves $P_0(x)$. You can find $P_0(x)$ and the other Legendre polynomials you need in Equation 12.5.10. Evaluate the resulting integral and find A_0.

(b) Similarly find coefficients A_1–A_5.

(c) You should find that all but two of the coefficients came out zero. (If you keep going you will find nothing but zeros from here on up.) Write the function $x^2 + 3$ as a sum of two Legendre polynomials.

(d) Confirm that your answer gives $x^2 + 3$ exactly.

In Problems 12.69–12.76 find the Legendre series expansion for the given function through the $P_3(x)$ term. Your answer should be in the form $A_0 P_0(x) + A_1 P_1(x) + A_2 P_2(x) + A_3 P_3(x)$, with specific values for the A_l. Unless the problem is marked with a computer icon (![computer]) all of the necessary integrals can be done with u-substitution and/or integration by parts.

12.69 x^3

12.70 $\sqrt[5]{x}$

12.71 $x^{-1/3}$

12.72 $f(x) = \begin{cases} -1 & x < 0 \\ 1 & x \geq 0 \end{cases}$

12.73 $f(x) = \begin{cases} 1+x & x < 0 \\ 1-x & x \geq 0 \end{cases}$

12.74 e^x

12.75 $\cos x$

12.76 ![computer] e^{-x^2}

12.77 ![computer] Have a computer calculate the Legendre series for $f(x) = x^{-1/3}$ up to the eleventh order. Plot that partial sum and the function $f(x)$ on the same plot from $x = -2$ to $x = 2$. Describe how the Legendre series approximates the function near $x = 0$, in the region $0 < |x| < 1$, and in the region $|x| > 1$.

12.78 In this problem you're going to compare the Legendre series and the Maclaurin series for the function $f(x) = \cos(2x)$.

(a) Calculate the second-order Maclaurin series for $\cos(2x)$.

(b) Calculate the second-order Legendre series for $\cos(2x)$. Expand the Legendre polynomials in your answer and gather like powers of x so it's in the form of a quadratic function of x, just like the Maclaurin series.

(c) ![computer] On one plot, show $\cos(2x)$ and the two quadratic approximations you just calculated from $x = -0.1$ to $x = 0.1$. Which quadratic better approximates the function in this interval?

(d) ![computer] On one plot, show $\cos(2x)$ and the two quadratic approximations you just calculated from $x = -1$ to $x = 1$. Which quadratic better approximates the function in this interval?

(e) Taylor series and Legendre series are two different ways of approximating a function with a polynomial. Based on your results, what are Taylor series more useful for and what are Legendre series more useful for?

12.79 Starting from Equation 12.5.6 explain why Legendre's differential equation will have one finite polynomial solution and one infinite series solution for integer l, and two infinite series solutions for non-integer l.

12.80 In the Explanation (Section 12.5.2) we derive Equation 12.5.14, an explicit formula for the Legendre polynomials based on even powers of x. Use a similar process, starting from Equation 12.5.6, to find a formula for the polynomials based on odd powers of x.

12.81 Equation 12.5.2 comes up in the solution of important partial differential equations in spherical coordinates. Apply the substitution $u = \cos \theta$ to rewrite this differential equation. Remember that this is a substitution for the independent variable, so you will have to write $d\Theta/d\theta$ and $d^2\Theta/d\theta^2$ in terms of $d\Theta/du$ and $d^2\Theta/du^2$.

(a) Show that the final result can be written in the form of Equation 12.5.3.

(b) Write the solution to Equation 12.5.2. (You have already done all the work.)

12.82 In the Explanation (Section 12.5.2) we use the ratio test to show that the Legendre functions of the second kind converge for $-1 < x < 1$ and diverge when $x < -1$ or $x > 1$. That leaves open the question of convergence at $x = \pm 1$.

In this problem you will examine the convergence of $Q_1(x)$ at $x = 1$. ($P_l(x)$ converges for all finite x since it's a finite polynomial.)

(a) Write the first five terms of the series $Q_1(1)$. (One way to do this is to plug $l = 1$ and $x = 1$ into Equation 12.5.14, although you will need to add the x^8 term to the ones we gave.) Simplify each term in fractional form as much as possible, but do not add the terms.

(b) Write a simple closed form expression for the nth term of this series.

(c) Prove that the resulting series converges or diverges. (The one thing you already know, before you start, is that the ratio test will be no help here. That leaves all the other tests.)

12.83 In Chapter 11 we solved for the electric potential inside a hollow spherical shell of radius R with potential $V_0(\theta)$ on the shell. (We assumed the potential was independent of the azimuthal angle ϕ.) We found that inside the shell the potential could be written

$$V(r, \theta) = \sum_{n=0}^{\infty} A_n r^n P_n(\cos \theta)$$

If $V_0 = \cos^2 \theta$, find the coefficients A_n.

(a) Use Appendix J to help you write a general formula (in terms of an integral) for the coefficient A_n.

(b) Use a u substitution to simplify your answer as much as possible. Your answer will still be an integral representation of A_n.

(c) Find the first four coefficients A_0–A_3.

12.84 Use Rodrigues' formula (Equation 12.5.15) to find $P_3(x)$.

12.85 In this problem you will derive Equation 12.5.15, Rodrigues' Formula, by showing that it can be written in the form of Equation 12.5.14 for even l. (The calculation for odd l is very similar.) The starting point is the binomial expansion, which says

$$(x^2 - 1)^l = \sum_{i=0}^{l} (-1)^i \frac{l!}{i!(l-i)!} x^{2l-2i}$$

(a) Explain why you can ignore all the terms in this sum with $i > l/2$. *Hint:* look at what you're going to do to $(x^2 - 1)^l$ in Equation 12.5.15.

(b) It will be convenient to examine the terms in the sum in a different order. Define $j = l/2 - i$ and rewrite the sum as a sum over j instead of i.

(c) Write out the terms for $j = 0, 1,$ and 2, simplifying as much as possible. Your answer will still contain factorials.

(d) Take the lth derivative of the terms you just wrote, once again expressing your answer in terms of factorials.

(e) The first of your terms should be a constant. Factor that constant out of all three terms and simplify as much as possible. You should get Equation 12.5.14.

Of course you just proved this for the first three terms in the sum. It is possible to go through a similar process to find the jth term and verify that it matches the jth term in $P_n(x)$, but the algebra is a bit messy.

12.86 Showing that a set of functions is "orthogonal" is a key step in finding the formula for the coefficients of a series expansion, as we will see in Section 12.8. In this problem you will use Rodrigues' formula (Equation 12.5.15) to prove the orthogonality of the Legendre polynomials: that is, you will show that $\int_{-1}^{1} P_l(x) P_m(x) dx = 0$ for all $l \neq m$.

(a) First, prove that $\int_{-1}^{1} f(x) P_l(x) dx = (-1)^l / (2^l l!) \int_{-1}^{1} f^{(l)}(x)(x^2 - 1)^l dx$, where $f^{(l)}(x)$ means the lth derivative of $f(x)$.

   i. Substitute for $P_l(x)$ using Rodrigues' formula and integrate the left side by parts.

   ii. The term outside the integral should involve the $(l-1)^{\text{th}}$ derivative of $(x^2 - 1)^l$. Explain why this term must equal zero for $x = \pm 1$. *Hint:* to take the $(l-1)^{\text{th}}$ derivative would involve repeated use of the chain rule and the product rule, but every term would have something in common.[7]

   iii. Show that after l integrations by parts you're left with the formula above.

(b) Given the result you just proved, it follows that $\int_{-1}^{1} x^m P_l(x) dx = 0$ for $m < l$. Use this fact to argue that the Legendre polynomials must be orthogonal.

12.87 Calculating the "norm" of a set of functions is a key step in finding the formula for the coefficients of a series expansion, as we will see in Section 12.8. In this

---

[7]We know $(l-1)^{\text{th}}$ looks awkward, but $(l-1)^{\text{st}}$ looked even worse. You see the things you have to worry about when you write a math textbook?

problem you will use Rodrigues' formula (Equation 12.5.15) to derive the norm of the Legendre polynomials: $\int_{-1}^{1} [P_l(x)]^2 \, dx$.

(a) Substitute for $P_l(x)$ using Rodrigues' formula. Pull the constant out of the integral.
(b) Integrate by parts and argue that the term outside the integral must equal zero. *Hint*: you may find it helpful to look at Problem 12.86.
(c) Find the integral you're left with after l integrations by parts.
(d) That integral involves the $(2l)^{\text{th}}$ derivative of $(x^2 - 1)^l$. Evaluate that derivative. *Hint*: if you expand $(x^2 - 1)^l$ all but one of the terms will vanish when you take the derivative.
(e) 💻 Evaluate the integral and compute the norm of the Legendre polynomials.
(f) 💻 The norm of the Legendre polynomials is most simply expressed as $\int_{-1}^{1} [P_l(x)]^2 \, dx = 2/(2l + 1)$. Your computer may have given you an answer that looks different from this, but you can check it by calculating some of the terms. Calculate $\int_{-1}^{1} [P_l(x)]^2 \, dx$ for $l = 1$, 2, and 3 using the answer you got and, if it looks different, the formula we just gave you. Make sure they match.

12.6 The Method of Frobenius

The "method of Frobenius" is a generalization of the idea of solving a differential equation with a power series. Just as the power series method gives us the Hermite and Legendre polynomials, the method of Frobenius will allow us to find an important group of functions called Bessel functions.

12.6.1 Discovery Exercise: The Method of Frobenius

A "generalized power series" is a function in the following form, where r is any constant.

$$y = x^r(c_0 + c_1 x + c_2 x^2 + c_3 x^3 + \ldots) = c_0 x^r + c_1 x^{r+1} + c_2 x^{r+2} + c_3 x^{r+3} + \ldots$$

1. Based on this form, find y' and y''.
2. Plug the given y, y', and y'' into the differential equation $x^2 y'' + 8xy' + 12y = 0$.
3. Set the coefficients of x^r on the left and right sides of your equation equal to each other, and solve the resulting equation for r.

See Check Yourself #86 in Appendix L

12.6.2 Explanation: The Method of Frobenius

We begin with a definition.

Definition: Generalized Power Series

A "generalized power series" is a function in the following form, where r and all the c_n coefficients are constants.

$$y = x^r \sum_{n=0}^{\infty} c_n x^n = \sum_{n=0}^{\infty} c_n x^{r+n} = c_0 x^r + c_1 x^{r+1} + c_2 x^{r+2} + c_3 x^{r+3} + \ldots \quad (c_0 \neq 0) \quad (12.6.1)$$

The restriction $c_0 \neq 0$ is not a limitation: it just means r is the exponent of the lowest-order term with a non-zero coefficient.

If $r = 0$, the "generalized" power series is just a "plain old" power series. If r is a positive integer this is still a power series, in which the first few coefficients happen to be zero. But if r is negative, fractional, or complex then this can express functions that cannot be represented by a Maclaurin series. For instance, $e^x/\sqrt{x}$ has no Maclaurin series because it is undefined at $x = 0$, but it can be represented as $x^{-1/2} \left[1 + x + x^2/(2!) + \ldots\right]$.

The "method of Frobenius" solves a differential equation by assuming a solution in the form of Equation 12.6.1. From there, the process goes like this.

1. Begin by making both sides of the equation look more or less like Equation 12.6.1.
2. Choose the *lowest possible power* of x and set its coefficient on the left equal to its coefficient on the right. The resulting equation (the "indicial equation") can be solved for r. For an nth-order equation you will usually find n-values of r.
3. For each value of r, find the recurrence relations that define the coefficients of its series. You should in general find that only c_0 is undetermined, so each value of r corresponds to one series with one arbitrary constant.

To illustrate this method, we will solve the following equation.

$$2\frac{d^2y}{dx^2} + \frac{1}{x}\left(\frac{dy}{dx}\right) - \left(2 + \frac{3}{x}\right)y = 0 \qquad (12.6.2)$$

We will work almost entirely in series notation, but as before we encourage you to work in parallel without it: for each step we take, write out the first few terms of each series.

Finding r by Writing the "Indicial Equation"

We begin with Equation 12.6.1 (our guess) and its derivatives.

$$y = \sum_{n=0}^{\infty} c_n x^{r+n} \quad \to \quad y' = \sum_{n=0}^{\infty} c_n (r+n) x^{r+n-1} \quad \to \quad y'' = \sum_{n=0}^{\infty} c_n (r+n-1)(r+n) x^{r+n-2}$$

When we plug these into the differential equation the term $(2 + 3/x)y$ becomes two different series, one for $2y$ and one for $(3/x)y$, so we have a total of four series.

$$\sum_{n=0}^{\infty} 2c_n(r+n-1)(r+n)x^{r+n-2} + \sum_{n=0}^{\infty} c_n(r+n)x^{r+n-2} - \sum_{n=0}^{\infty} 2c_n x^{r+n} - \sum_{n=0}^{\infty} 3c_n x^{r+n-1} = 0$$

As in Section 12.4 we reindex those series so they all reflect the same power of x. We could, for instance, rewrite them all with the exponent $r + n$. Instead we will rewrite them in terms of x^{r+n-2}, just because two of them already look like that. As long as they are all the same we will be able to compare.

$$\sum_{n=0}^{\infty} 2c_n(r+n-1)(r+n)x^{r+n-2} + \sum_{n=0}^{\infty} c_n(r+n)x^{r+n-2} - \sum_{n=2}^{\infty} 2c_{n-2} x^{r+n-2} - \sum_{n=1}^{\infty} 3c_{n-1} x^{r+n-2} = 0$$

As always, check the first few terms of each series to convince yourself that we haven't changed anything!

The $n = 0$ and $n = 1$ terms appear in only some of the series, so we write them separately. All $n \geq 2$ terms appear in all four series so we combine them into one sum. After a bit of simplification we arrive here.

$$c_0 r(2r-1)x^{r-2} + [c_1(2r+1)(r+1) - 3c_0]x^{r-1}$$
$$+ \sum_{n=2}^{\infty} \left[c_n(2r+2n-1)(r+n) - 2c_{n-2} - 3c_{n-1}\right] x^{r+n-2} = 0 \qquad (12.6.3)$$

We always begin with the lowest power of x that is represented in the equation. Setting the coefficient of x^{r-2} equal to zero gives us $c_0 r(2r - 1) = 0$. Since $c_0 \neq 0$ by the definition of our series, we can divide by c_0 and get an equation for r. This is called the "indicial equation" for this differential equation.

$$r(2r - 1) = 0 \quad \text{the indicial equation for Equation 12.6.2} \quad (12.6.4)$$

This can trivially be solved to give $r = 0$ and $r = 1/2$. We therefore expect that the two solutions to Equation 12.6.2 will be in the form of Equation 12.6.1, one with $r = 0$ and the other with $r = 1/2$. The general solution will be any linear combination of these two solutions.

Finding the Recurrence Relations

The x^{r-2} terms in Equation 12.6.3 led us to the indicial equation. The x^{r-1} term gives us the relationship between c_1 and c_0.

$$c_1 = \frac{3c_0}{(2r + 1)(r + 1)} \quad (12.6.5)$$

The general recurrence relation comes from setting the coefficient of x^{r+n-2} inside the sum equal to zero.

$$c_n = \frac{2c_{n-2} + 3c_{n-1}}{(2r + 2n - 1)(r + n)} \quad (n \geq 2) \quad (12.6.6)$$

Given any particular value of c_0 (based on initial conditions) we use Equation 12.6.5 to find c_1 and Equation 12.6.6 to find all the rest.

It Doesn't Look Like It, But We're Pretty Much Done

Equation 12.6.2 is a second-order linear homogeneous differential equation. Our goal is therefore to find two linearly independent solutions; any linear combination of those solutions will be a solution, and we will therefore have our general solution.

We began by guessing a solution of the form Equation 12.6.1. Plugging this into the differential equation, we first found and solved Equation 12.6.4, the indicial equation for this equation. We found two solutions, $r = 0$ and $r = 1/2$.

For $r = 0$ our solution looks like this.

$$y_0 = \sum_{n=0}^{\infty} c_n x^n \quad \text{where} \quad c_1 = 3c_0 \quad \text{and} \quad c_n = \frac{2c_{n-2} + 3c_{n-1}}{n(2n - 1)} \quad (n \geq 2)$$

For the $r = 1/2$ solution we will use d rather than c to make it clear that the coefficients of the series below are not the same as the coefficients of the series above.

$$y_{1/2} = \sqrt{x} \sum_{n=0}^{\infty} d_n x^n \quad \text{where} \quad d_1 = d_0 \quad \text{and} \quad d_n = \frac{2d_{n-2} + 3d_{n-1}}{n(2n + 1)} \quad (n \geq 2)$$

The general solution is a sum of these two series, with arbitrary constants supplied by c_0 (the constant term in the $r = 0$ series) and d_0 (the constant term in the $r = 1/2$ series). You'll find solutions for specific initial and boundary conditions in Problem 12.96.

The most important use of the Frobenius method is to solve "Bessel's equation" $x^2 y'' + xy' + (x^2 - p^2) y = 0$. We shall devote much more space to this equation and its important solutions in Section 12.7, but here we're just going to use it as an example of the method of Frobenius. Bessel's equation is actually many different equations because of the constant p; in the example below $p = 1/3$.

12.6 | The Method of Frobenius

EXAMPLE **The Method of Frobenius**

Problem:
Solve the differential equation $x^2 y'' + xy' + (x^2 - 1/9) y = 0$.

Solution:
The method of Frobenius always begins with the same guess.

$$y = \sum_{n=0}^{\infty} c_n x^{r+n} \quad \to \quad y' = \sum_{n=0}^{\infty} c_n(r+n) x^{r+n-1} \quad \to \quad y'' = \sum_{n=0}^{\infty} c_n(r+n-1)(r+n) x^{r+n-2}$$

Plug this into the differential equation. The first series comes from multiplying the y'' expansion above by x^2 (so the exponent goes up by two). Similarly, the second series is xy', followed by $x^2 y$ and finally $-y/9$.

$$\sum_{n=0}^{\infty} c_n(r+n-1)(r+n) x^{r+n} + \sum_{n=0}^{\infty} c_n(r+n) x^{r+n} + \sum_{n=0}^{\infty} c_n x^{r+n+2} - \sum_{n=0}^{\infty} \frac{1}{9} c_n x^{r+n} = 0$$

Reindex the series so that they all have the same power of x. Since three of them already look like x^{r+n} we'll change the other series to match.

$$\sum_{n=0}^{\infty} c_n(r+n-1)(r+n) x^{r+n} + \sum_{n=0}^{\infty} c_n(r+n) x^{r+n} + \sum_{n=2}^{\infty} c_{n-2} x^{r+n} - \sum_{n=0}^{\infty} \frac{1}{9} c_n x^{r+n} = 0$$

The x^r and x^{r+1} terms appear in some series; subsequent terms appear in all series.

$$\left[c_0(r-1)r + c_0 r - \frac{1}{9} c_0 \right] x^r + \left[c_1 r(r+1) + c_1(r+1) - \frac{1}{9} c_1 \right] x^{r+1}$$
$$+ \sum_{n=2}^{\infty} \left[c_n(r+n-1)(r+n) x^{r+n} + c_n(r+n) x^{r+n} + c_{n-2} x^{r+n} - \frac{1}{9} c_n x^{r+n} \right] = 0$$

The x^r terms give us the indicial equation.

$$c_0(r-1)r + c_0 r - \frac{1}{9} c_0 = 0 \quad \to \quad r^2 - \frac{1}{9} = 0 \quad \to \quad r = \pm \frac{1}{3} \quad (12.6.7)$$

The x^{r+1} terms give us an equation for c_1.

$$c_1 r(r+1) + c_1(r+1) - \frac{1}{9} c_1 = 0 \quad \to \quad \left(r^2 + 2r - \frac{8}{9} \right) c_1 = 0 \quad \to \quad c_1 = 0 \quad (12.6.8)$$

Note a difference in our treatment of Equations 12.6.7 and 12.6.8. In the former case—the indicial equation—we assume by hypothesis that $c_0 \neq 0$, and therefore we divide it out and solve for r. In the latter case we can make no such assumption. In fact, since we already know that $r = \pm 1/3$ the quadratic in r cannot be zero, so we conclude that $c_1 = 0$.

The general recurrence relation comes from setting the coefficients of x^{r+n} equal in the sum.

$$c_n(r+n-1)(r+n) + c_n(r+n) + c_{n-2} - \frac{1}{9} c_n = 0 \quad \to \quad c_n = \frac{c_{n-2}}{1/9 - (r+n)^2} \quad \text{for } n \geq 2$$

Because we know that $c_1 = 0$ this relation promises us that $c_n = 0$ for all odd n. So we have two different series, one with $r = 1/3$ and one with $r = -1/3$. Each has its own arbitrary c_0 (based on the two initial conditions) and a recurrence relation for all subsequent even-numbered n-values. In Section 12.7 we will explore the properties of these solutions.

When can you use the method of Frobenius?

The method of Frobenius can be used on equations in the following form.

$$\frac{d^2y}{dx^2} + \frac{b_1(x)}{x}\left(\frac{dy}{dx}\right) + \frac{b_0(x)}{x^2}y = 0 \tag{12.6.9}$$

The functions $b_1(x)$ and $b_0(x)$ must be "analytic functions," meaning that they can be represented by Maclaurin series.[8] But in some cases $b_1(x)/x$ and $b_2(x)/x^2$ may *not* be analytic functions, so there are cases where you can use the method of Frobenius but you cannot use the method of power series.

On the other hand, Equation 12.6.9 is necessarily homogeneous, so there are also cases where you can use the method of power series but not the method of Frobenius.

Even if your problem fits Equation 12.6.9 perfectly, you can run into trouble. One potential difficulty is an indicial equation with a "double root." (For instance, if $r^2 - 6r + 9 = 0$ then $r = 3$.) From this you get one generalized power series, but a second-order linear differential equation requires *two* linearly independent solutions. You can find the second one by using "reduction of order," a technique discussed in Chapter 10. When you apply this technique to a Frobenius problem you end up with another infinite series added to a logarithmic function.

A less obvious difficulty is an indicial equation with two roots that differ by an integer (such as $r = 1/3$ and $r = 7/3$). Sometimes that works out fine, and you get two linearly independent solutions (see Problem 12.97). But sometimes the recurrence relation for the series with lower r blows up, so that solution has to be discarded. If your first solution can be written in closed form, this problem has the same resolution as the double root problem we discussed above: reduction of order can find you a second solution involving a log (see Problem 12.98). In Section 12.7 we'll see a completely different solution designed specifically for the important case of Bessel functions.

12.6.3 Problems: The Method of Frobenius

12.88 Walk-Through: The Method of Frobenius. In this problem you will solve the equation $2xy'' + y' + 2(1 + x)y = 0$.

(a) Beginning with a guess of $y = \sum_{n=0}^{\infty} c_n x^{r+n}$ calculate y' and y''.

(b) Plug your series for y, y', and y'' into the differential equation. Note that you will end up with four different series: one for $2xy''$, one for y', one for $2y$, and one for $2xy$.

(c) You should find in this case that two of your series are expressed in terms of x^{r+n-1}. Reindex the other two so that all four series contain the same power of x.

(d) Combine all four series into one, pulling out any terms with powers of x that don't appear in all four series.

---

[8]You can also use the method of Frobenius on an equation of the form $y'' + y'b_1/(x-c) + yb_2/(x-c)^2$ if b_1 and b_0 can be represented in Taylor series about $x = c$. This doesn't seem to come up very often.

(e) Write an equation setting the coefficient of the lowest power of x equal to zero. (This will come from a term outside the sum.)

(f) Because we assume for a generalized power series that $c_0 \neq 0$, you can divide both sides of your equation from Part (e) by c_0. The result is the "indicial equation." Solve it for r.

(g) For this particular problem there should be one other power of x outside the sum. Set the coefficient of that power equal to zero to get an equation for c_1 in terms of c_0.

(h) Setting the coefficient inside the sum equal to zero, write a recurrence relation for c_n in terms of c_{n-1} and c_{n-2} valid for all $n \geq 2$.

(i) Now plug in the values of r you found. For each one:
   i. Write the formula for c_1 in terms of c_0.
   ii. Write the recurrence relation for c_n for $n \geq 2$.
   iii. Write the third partial sum of this series. Include the powers of x.

12.89 *[This problem depends on Problem 12.88.]* In Problem 12.88 you found two series solutions to the differential equation $2xy'' + y' + 2(1 + x)y = 0$, each one with an arbitrary constant.

(a) If you are told that $y'(0)$ is finite, you can eliminate one of your series solutions. Which one, and why?

(b) If you are also told that $y(0) = 3$ then you can solve for the arbitrary constant in the remaining solution. Write the third partial sum of the resulting series.

In Problems 12.90–12.93 use the method of Frobenius to solve the given differential equation. Your final answer should be in the form of two third-order series in the form $y(x) = c_0 [x^r + ax^{r+1} + bx^{r+2}]$. You'll fill in the constants r, a, and b for each series, but c_0 will be left as an arbitrary constant. (Call it d_0 in the second solution.) It may help to first work through Problem 12.88 as a model.

12.90 $2x^2y'' + xy' + (x - 3)y = 0$

12.91 $4xy'' + (x - 2)y' = 0$

12.92 $x^2y'' + xy' + (x^2 - 1/25)y = 0$ (Bessel's equation with $p = 1/5$)

12.93 $x^2y'' + xy' + (4x^2 - 1/25)y = 0$ (Slight variation on Bessel's equation with $p = 1/5$)

12.94 Solve the equation $2x^2y'' + xy' + (\sin x)y = 0$ using the method of Frobenius, calculating each series through the x^{r+3} term. Start by expanding $\sin x$ in a Maclaurin series through the x^5 term. From there it might be easier to just write out the lowest several terms of each series rather than using series notation.

12.95 In this problem you'll solve the equation $3x^2y'' + 7xy' - 4y = 0$.

(a) Go through the method of Frobenius far enough to get the indicial equation and find the two values of r.

(b) Set the coefficient of the next power of x equal to zero. Given that you know the only two possible values for r, what does this equation let you conclude about c_1?

(c) Following similar logic, what can you conclude about all of the coefficients c_n, for $n > 2$?

(d) Write the general solution to this equation in closed form.

12.96 In the Explanation (Section 12.6.2) we found the general solution to Equation 12.6.2 in the form of a sum of two series, with arbitrary constants c_0 and d_0. In this problem you're going to find specific solutions for some initial and boundary conditions. First consider some initial conditions.

(a) Set $y(0) = 1$. What does that tell you about the constants c_0 and d_0?

(b) Calculate $y'(0)$. What must be true about the coefficients in order for $y'(0)$ to be finite?

(c) Write the fourth partial sum of the specific solution for $y(0) = 1$ with $y'(0)$ finite.

You may be surprised that "$y'(0)$ is finite" is a sufficient initial condition. If we know that $y(0) = 1$ and $y'(0)$ is finite then we don't get to pick $y'(0)$; it must be the same as $y(0)$. We will see more generally when such a condition is sufficient when we take up Sturm-Liouville theory later in the chapter.

Next you'll treat this as a boundary value problem.

(d) Set $y(0) = 0$. What does that tell you about the coefficients in the problem?

(e) Calculate the fourth partial sum of the solution with $y(0) = 0$. Your answer will still have one arbitrary constant in it.

(f) Using that fourth partial sum, set $y(1) = 1$ and calculate the remaining arbitrary constant. (This will only be an approximate value since you only used the fourth partial sum, but it should be a very good approximation.)

12.97 In this problem and Problem 12.98 you will encounter indicial roots that differ by an integer. The two problems will resolve themselves very differently, however. In this problem you will solve the equation $y'' + y' - (2/x^2)y = 0$.

(a) Proceed in the usual way with the method of Frobenius until you have written and solved the indicial equation and written the recurrence relation for $n \geq 1$.

(b) You should have found that the indicial equation has two roots that differ by an integer. When this happens, the *higher* of the two roots never causes a problem. Show in this particular case that the recurrence relation for the higher r-value never blows up.

(c) For the *lower* of the two r-values, the recurrence relation blows up at what n-value?

(d) Start generating terms for the series with the lower r-value. You should find that your series terminates before it runs into trouble.

(e) Is that really valid? Can we ignore the fact that this series seems destined to blow up? As with any differential equation, the solution will bear itself out. Write your terminating solution—arbitrary c_0 constant and all—and verify that it solves the differential equation.

12.98 The equation $(x^2 + x)y'' - xy' + y = 0$ leads to an indicial equation whose roots differ by an integer. You'll see how that can lead to a problem, and what you can do about it.

(a) Find the two allowed values of r and the recurrence relation for each.

(b) You should have found that the indicial equation has two roots that differ by an integer. When this happens, the higher of the two roots never causes a problem, but the lower can. Show in this particular case that the recurrence relation for the higher r-value never blows up but that the lower r-value must be discarded.

(c) Show that the valid solution terminates after a finite number of terms. Write that solution in closed form and verify that it works. We'll call it y_1.

(d) Since the other series had to be discarded, you need to find the second solution using reduction of order. Plug the guess $y_2 = u(x)y_1(x)$ into the differential equation and simplify as much as possible. Solve the resulting equation to find $u(x)$. *Hint*: solving the equation will involve a substitution to turn a second-order equation into a first-order equation, then separation of variables, and finally partial fractions.

(e) Write the general solution to the differential equation and verify that it works.

12.99 Solve the equation $y'' + (1/x)y' + (1/x^2)y = 0$ using the method of Frobenius. You should find that both series terminate so you can write your solution in closed form. *Hint*: you will not get a normal recursion relation, but simply an equation relating c_n and r. As always, assume $c_0 \neq 0$ and solve for the two allowed values of r. Then figure out what the equation tells you about all the other coefficients c_n. At the end, plug your closed form answer back in and check that it satisfies the differential equation.

12.100 The "Hypergeometric equation" $x(1-x)y'' + [C - (A + B + 1)x]y' - ABy = 0$ (where A, B, and C are constants) has been much studied because its solutions, "hypergeometric functions," include many other classes of special functions.

(a) Write and solve the indicial equation. You should find $r = 0$ and one other solution.

(b) Find the equation for c_{n+1} as a function of c_n.

(c) Simplify the recurrence relation for the case of $r = 0$.

12.7 Bessel Functions

Preliminary to writing this section we searched Amazon.com for books with "Bessel Functions" in the title. We found 220 books with names like *Bessel Functions for Engineers* and *Introduction to Bessel Functions*. The classic in the field, G. N. Watson's 800-page *Treatise On the Theory of Bessel Functions*, was first published in 1922 and most recently reprinted in 2012. Expanding our search to look for "Bessel functions" in the keywords instead of the title increased the result to over 11,000 books.

12.7 | Bessel Functions

Our point is that Bessel functions are a big deal. They appear in electromagnetism, thermodynamics, quantum mechanics, and many other fields, and there is a lot to learn about them. But we can cover the basics in a relatively short space, and from there you can look up more details on computers or reference tables as you need them.

We recommend first reading the section "Bessel Functions—The Unjustified Essentials" in Chapter 11. That section is our best introduction to the subject, covering the information you need to use these functions in PDEs and other contexts. Then come back to this section to see where Bessel functions come from and why they have the properties they do.

12.7.1 Discovery Exercise: Bessel Functions

This exercise requires a computer. The problems in this exercise refer to "Bessel functions of the first kind" $J_p(x)$ and "Bessel functions of the second kind" $Y_p(x)$. Refer to your software's documentation to find the proper syntax for entering these functions. For all the following questions, use the domain $0 \leq x \leq 50$.

1. Graph the function $J_1(x)$. Answer in words: how is $J_1(x)$ like a sine function? How does it differ from a sine function?
2. How many zeros does $J_1(x)$ have in this domain?
3. The first three positive zeros of this function are called $\alpha_{1,1}$, $\alpha_{1,2}$, and $\alpha_{1,3}$. Find their values. Your answers should be accurate to the second decimal place.
4. Find the absolute maximum of $J_1(x)$.
5. Graph the functions $J_{1.2}(x)$, $J_2(x)$, and $J_{15}(x)$. Answer in words: how are these functions alike? How do they differ?
6. Graph the function $Y_1(x)$. Answer in words: how is this function like $J_1(x)$? How is it different?

12.7.2 Explanation: Bessel Functions

In the Motivating Exercise (Section 12.1) you examined the following equation for the normal modes of a simple drum.

$$\rho^2 R''(\rho) + \rho R'(\rho) + (k^2 \rho^2 - p^2) R(\rho) = 0 \tag{12.7.1}$$

In that exercise you looked up this equation in Appendix J and found that the solutions are called "Bessel functions." In this section we will solve this equation with the method of Frobenius and thus derive the properties of Bessel functions that you found in that appendix. The first step is the substitution $x = k\rho$. You'll show in Problem 12.109 that this turns Equation 12.7.1 into Bessel's equation (defined below).

Bessel's Differential Equation and its Solutions

The following is called Bessel's differential equation.

$$x^2 y'' + xy' + (x^2 - p^2)y = 0 \tag{12.7.2}$$

The general solution to this equation is $y = AJ_p(x) + BY_p(x)$, where J_p and Y_p are called Bessel functions of the first and second kind, respectively, both of order p.

Notice that each p represents a different differential equation with different solutions. For example, the general solution to $x^2 y'' + xy' + (x^2 - 9/4)y = 0$ is $y = AJ_{3/2}(x) + BY_{3/2}(x)$.

Finding the Recurrence Relation

We outline the key steps here; you will fill in the details in Problem 12.101.

We begin by plugging Equation 12.6.1 (the generalized power series) into Equation 12.7.2 (Bessel's equation). After reindexing so that all four series are expressed in terms of x^{r+n} we use the lowest value of n to find that the indicial equation is $r^2 - p^2 = 0$, leading as usual to two values of r.

$$r = \pm p$$

The next n tells us that $c_1 = 0$. Finally, using all four series, we find the recurrence relation.

$$c_n = \frac{c_{n-2}}{p^2 - (r+n)^2}$$

Because every coefficient is a multiple of the coefficient two below it, $c_1 = 0$ means that all odd coefficients are zero. That means the entire series is made up from even values of n, so we will write $n = 2m$ where m runs over all non-negative integers.

$$c_{2m} = \frac{c_{2m-2}}{p^2 - (r+2m)^2} \qquad (12.7.3)$$

The First Solution

We have seen that Bessel's equation has two solutions, one with $r = p$ and one with $r = -p$, and we have found a recurrence relation that applies to both solutions. Now we focus on the first of the two solutions. Letting $r = p$ in Equation 12.7.3:

$$c_{2m} = \frac{c_{2m-2}}{p^2 - (p+2m)^2} = \frac{c_{2m-2}}{p^2 - (p^2 + 4mp + 4m^2)} = -\frac{c_{2m-2}}{4m(p+m)} \qquad (12.7.4)$$

It's not obvious where that leads, so let's start generating terms.

$$m = 1 \rightarrow c_2 = -\frac{1}{4(p+1)} c_0 = -\frac{1}{1!\, 2^2 (p+1)} c_0$$

$$m = 2 \rightarrow c_4 = -\frac{1}{4(2)(p+2)} c_2 = \frac{1}{2!\, 2^4 (p+1)(p+2)} c_0$$

$$m = 3 \rightarrow c_6 = -\frac{1}{4(3)(p+3)} c_2 = -\frac{1}{3!\, 2^6 (p+1)(p+2)(p+3)} c_0$$

You can see the pattern here in three parts.

- The negative sign in Equation 12.7.4 makes the coefficients alternate sign, which we represent by a factor of $(-1)^m$.
- It may seem a bit perverse to write the number 4 as $1! \times 2^2$ or the number 32 as $2! \times 2^4$ but when you see the third term the pattern starts to emerge. The 4 in the denominator of Equation 12.7.4 adds two more powers of 2 at each step, and the m in the denominator keeps the factorial going.
- Finally, $(p+1)(p+2)\ldots(p+m)$ can be written as $(p+m)!/p!$ (think about it). But p, unlike m, is not constrained to be an integer. We therefore use the gamma function instead of the factorial, remembering that $x! = \Gamma(x+1)$.

Putting it all together gives us the following formula for the coefficients.

$$c_{2m} = \frac{(-1)^m \Gamma(p+1)}{m!\, 2^{2m} \Gamma(p+m+1)} c_0$$

Applying those coefficients to the generalized power series gives us one solution to Bessel's equation.

$$y = x^p \sum_{m=0}^{\infty} \frac{c_0(-1)^m \Gamma(p+1)}{m! \, 2^{2m} \Gamma(p+m+1)} x^{2m} = c_0 x^p \Gamma(p+1) \sum_{m=0}^{\infty} \frac{(-1)^m}{m! \, \Gamma(p+m+1)} \left(\frac{x}{2}\right)^{2m}$$

Because m counts by ones and x is raised to the $2m$ we are still generating only even-numbered terms, consistent with our early conclusion that all odd-numbered terms are zero.

But just as we did with Legendre polynomials, we now get to choose a convenient c_0 to define the Bessel functions. It seems clear that a choice with $\Gamma(p+1)$ in the denominator will make the formula simpler. It is also conventional to put a 2^p in the denominator; this turns x^p into $(x/2)^p$ so it combines with the other power of x. So we set $c_0 = \dfrac{1}{2^p \Gamma(p+1)}$ and arrive at the all-important formula we have been looking for.

Definition: Bessel Functions of the First Kind

$$J_p(x) = \sum_{m=0}^{\infty} \frac{(-1)^m}{m! \, \Gamma(p+m+1)} \left(\frac{x}{2}\right)^{2m+p} \tag{12.7.5}$$

Bessel Functions of the First Kind

Figure 12.3 shows a graph of $J_{4.5}(x)$. It looks something like $\sin x$ but with two important differences. First: the distance between consecutive zeros in a sine wave is always π, but the distance between zeroes of a Bessel function varies. For large x the spacing between zeroes asymptotically approaches π. Second: the amplitude of the Bessel function decreases approximately like $1/\sqrt{x}$.

We chose the number 4.5 arbitrarily for that graph because it doesn't matter. $J_\pi(x)$ and $J_{17}(x)$ are different from $J_{4.5}(x)$ and from each other, but they all have the same basic form: a generally sinusoidal shape with the distance between zeros approaching π and the amplitude decreasing. They all start at the origin except $J_0(x)$ which starts at $(0,1)$ (and therefore looks more like a cosine than a sine).

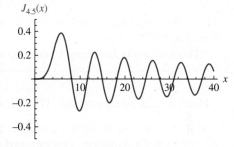

FIGURE 12.3

By convention, $\alpha_{p,n}$ represents the nth positive zero of the Bessel function J_p. You can look these up in tables or in most computer programs. For instance, $\alpha_{4.5,3} \approx 15.04$ indicates that the third positive zero of $J_{4.5}(x)$ occurs at roughly 15.04, as you can see on the graph above. (What is $\alpha_{4.5101} - \alpha_{4.5100}$? We have given you enough information in the previous few paragraphs to answer this question fairly accurately without a computer.)

The Second Solution

Remember that our indicial equation gave us $r = p$ and $r = -p$. To find the first solution we plugged $r = p$ into Equation 12.7.3 and started generating terms. In Problem 12.103 you will go through the same process for $r = -p$ and end up here. (You can also arrive at this formula by replacing p with $-p$ in Equation 12.7.5.)

$$J_{-p}(x) = \sum_{m=0}^{\infty} \frac{(-1)^m}{m! \, \Gamma(-p+m+1)} \left(\frac{x}{2}\right)^{2m-p} \tag{12.7.6}$$

If you were paying close attention in Section 12.6 you may spot a potential snag. If the two values of r differ by an integer we saw that sometimes you get two valid solutions and sometimes you don't. For Bessel's equation the two values of r are p and $-p$, which differ by an integer if p is an integer or a half-integer. As it turns out, half-integer values of p do not lead to trouble here but integer values do. In Problem 12.104 you will apply the method of Frobenius to Bessel's equation for $p = 2$ and show that you don't get two valid, independent solutions. In Problem 12.105 you will start from the formulas we've derived for J_p and J_{-p} and show that the following holds for any integer p:

$$J_{-p}(x) = (-1)^p J_p(x) \quad \text{integer } p \text{ only}$$

Do you see why that matters? For non-integer values of p we have the two solutions we need, and we can write the general solution to Bessel's equation as $AJ_p(x) + BJ_{-p}(x)$. But for integer p we have two linearly dependent functions, so $AJ_p(x) + BJ_{-p}(x)$ is really just one solution $CJ_p(x)$. We do not have a general solution in that case.

If you understand the problem as we have described it, you can appreciate the following very clever solution.

Definition: Bessel Functions of the Second Kind

$$Y_p(x) = \frac{\cos(\pi p) J_p(x) - J_{-p}(x)}{\sin(\pi p)} \qquad (12.7.7)$$

This is the definition for non-integer values of p. For integer n, $Y_n(x)$ is defined as the limit of Equation 12.7.7 as $p \to n$.

(This function is sometimes labeled $N_p(x)$ but we will stick with Y.)

That's a strange concoction. What does it give you?

- If p is not an integer then $J_p(x)$ and $J_{-p}(x)$ are two linearly independent solutions of Bessel's equation. $\cos(\pi p)$ and $\sin(\pi p)$ are just numbers, and $\sin(\pi p)$ is not zero. Equation 12.7.7 represents a linear combination of two solutions, and therefore it is itself a solution.
- If p is an integer, $\cos(\pi p) = (-1)^p$ and $\sin(\pi p) = 0$, so Equation 12.7.7 evaluates as the indeterminate form 0/0. But in the limit as p *approaches* an integer, Equation 12.7.7 again represents a solution to Bessel's equation. We know this because, as we have already stated, this function does supply a solution for any non-integer p.
- It turns out that $Y_p(x)$ is linearly independent of $J_p(x)$ whether p is an integer or not. We are not going to prove this in general, but you'll demonstrate it for several p-values in Problem 12.106.

The general solution to Bessel's equation for any value of p can thus be written as $y = AJ_p(x) + BY_p(x)$.

Bessel Functions of the Second Kind

Figure 12.4 shows a graph of $Y_{4.5}(x)$. Like a Bessel function of the first kind, this function oscillates with gradually decreasing amplitude and has an infinite number of irregularly spaced zeros. Once again, if you graph $Y_1(x)$ or $Y_{36}(x)$ you will see a different curve with the same general shape.

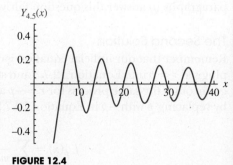

FIGURE 12.4

But there is one big difference: for all p-values, $\lim_{x \to 0^+} Y_p(x) = -\infty$. For this reason, whenever you solve Bessel's equation on a region that requires a finite answer at $x = 0$, you immediately discard the $Y_p(x)$ function.

Fourier-Bessel Series

By analogy to Legendre series you might expect us to expand a function into a series of Bessel functions like $A_0 J_0(x) + A_1 J_1(x) + \ldots$ Well, it doesn't work that way. Instead, you start with a function $f(x)$ that you want to expand *and* a particular Bessel function $J_p(x)$ that you want to expand it into. A Fourier-Bessel series gives you a series representation of $f(x)$ in terms of the Bessel function you specified. For a function $f(x)$ with domain $0 \leq x \leq a$, a Fourier-Bessel series expansion in terms of J_p looks like this.

$$f(x) = \sum_{n=1}^{\infty} A_n J_p\left(\frac{\alpha_{p,n}}{a} x\right)$$

An analogy to Fourier sine series may be helpful here. If a function is defined from 0 to 1 then you can build a series from sines of different frequencies: $f(x) = b_1 \sin(\pi x) + b_2 \sin(2\pi x) + \ldots$. If the function is defined from 0 to a then the series becomes $f(x) = b_1 \sin(\pi x/L) + b_2 \sin(2\pi x/L) + \ldots$. In both cases the function is the same in each term (sine), but the argument changes so that L lands on the first zero of the first term, the second zero of the second term, and so on. Similarly, the $p = 6$ Fourier-Bessel series $f(x) = A_1 J_6(\alpha_{6,1} x/a) + A_1 J_6(\alpha_{6,2} x/a) + \ldots$ has terms where the coefficient changes so that $x = a$ always falls on one of the zeroes of the J_6.

The limitation of that analogy is that a sine is just a sine. In this case, you had to pick one specific Bessel function (such as J_6) to expand your function into. If you expanded the same function into J_7 functions then you would get a different series with different coefficients.

The formula for finding the coefficients A_n are given in Appendix J. As we did with Fourier-Legendre series we're going to defer the derivation to Section 12.8 and focus here on using the formula.

EXAMPLE **Fourier-Bessel Series Expansion**

Question: Expand the function $f(x) = \sin x$ on the domain $0 \leq x \leq 2$ into a series based on the function $J_8(x)$.

Solution:
Using the equation in Appendix J the first coefficient is:

$$A_1 = \frac{2}{4 J_9^2(\alpha_{8,1})} \int_0^2 \sin(x) J_8\left(\frac{\alpha_{8,1}}{2} x\right) x \, dx$$

At this point we reach for a computer, both to find that $\alpha_{8,1} \approx 12.2$ and to numerically evaluate the integral to get $A_1 = 3.94$. The computer then tells us that $\alpha_{8,2} \approx 16.0$ and then another integral gives $A_2 = 0.706$, and so on. Putting these together, we get the $p = 8$-order Fourier-Bessel series for $\sin x$ on the interval $[0, 2]$.

$$\sin x = 3.94 J_8\left(\frac{12.2}{2} x\right) + 0.706 J_8\left(\frac{16.0}{2} x\right) + 1.68 J_8\left(\frac{19.6}{2} x\right) + \ldots$$

As we add more terms the series gradually converges to sin x within the domain $[0, 2]$. Outside that domain it doesn't match at all.

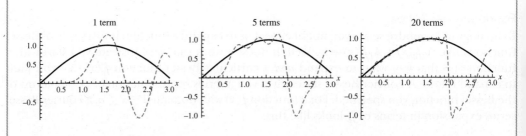

Stepping Back

Bessel's equation arises in many problems, including the solutions of PDEs in cylindrical coordinates. The general solution is $AJ_p(x) + BY_p(x)$. In the most common cases, $x = 0$ is part of the domain of the problem and the solution is constrained to be finite at that point, so we set $B = 0$.

In many applications, however, the ODE that arises is not precisely Equation 12.7.2, but some close cousin of it. These other equations can often be turned in Bessel's equation by a simple change of variables, but when we convert back to the original variables we end up with Bessel functions with different arguments and/or different coefficients in front. Many of these functions arise in applications enough to have their own names. In the Problems you will use variable substitution and Bessel functions to derive the "modified Bessel functions," the "Kelvin functions," the "spherical Bessel functions" and the "Riccati–Bessel functions." You can find the most commonly used of these functions listed in terms of the differential equations they solve in Appendix J.

12.7.3 Problems: Bessel Functions

12.101 The Explanation (Section 12.7.2) outlined the steps to finding the recurrence relation for Bessel's equation $x^2 y'' + xy' + (x^2 - p^2)y = 0$. In this problem you will go through the entire process.

(a) Explain how you can tell that Bessel's equation is not appropriate for the method of power series, but can be solved by the method of Frobenius.

(b) Substitute $y = \sum_{n=0}^{\infty} c_n x^{r+n}$ into the equation. Your resulting equation should contain four different series, treating $x^2 y$ and $-p^2 y$ as two different terms in the original equation.

(c) Three of your series contain x^{r+n} but one contains x^{r+n+2}. Reindex that series so that all four series contain x^{r+n}. Pull out terms that only appear in some of the series and combine the four sums into one.

(d) Set the coefficient of x^r equal to zero to find the indicial equation and solve for both values of r.

(e) Set the coefficient of x^{r+1} equal to zero and find c_1. (The case $p = \pm 1/2$ is special, and you'll address it in Problem 12.102. For now ignore it.)

(f) Set the coefficient of x^{r+n} equal to zero and solve to find the recurrence relation for c_n in terms of c_{n-2}.

(g) Explain how you know that all the terms x^{r+n} for odd n must be zero. (Again, assume $p \neq \pm 1/2$.)

12.102 *[This problem depends on Problem 12.101.]* In Problem 12.101 you used the method of Frobenius to write $J_p(x)$ as a generalized power series. You showed that each coefficient c_n depends on c_{n-2} and that $c_1 = 0$, so all the odd coefficients are zero. The argument that gave you $c_1 = 0$ didn't work, however, for the case $p = \pm 1/2$. In this problem you'll consider the case $p = 1/2$. (The argument for $p = -1/2$ is exactly the same.)

(a) For $p = 1/2$ what are the possible values of r?

(b) Write the equation for c_1. For which value of r can you conclude that $c_1 = 0$ and for which value of r can you *not* conclude this?

(c) Write the recurrence relation for c_n in terms of c_{n-2}, valid for $n \geq 2$.

(d) For the value of r for which you *cannot* conclude that $c_1 = 0$, write the first three terms of the series solution that begins with c_1. Your answer will have c_1 in it as an undetermined constant.

(e) For the value of r for which you *can* conclude that $c_1 = 0$, write the first three terms of the series solution. Your answer will have c_0 in it as an undetermined constant.

(f) Looking at your answers to Parts (d)–(e), why are you allowed to set $c_1 = 0$ for $p = 1/2$ without losing any generality?

12.103 In the Explanation (Section 12.7.2) we found that Bessel's equation can be solved with generalized power series using $r = \pm p$. We then used $r = p$ to derive the formula for Bessel functions of the first kind $J_p(x)$. In this process you will derive the formula for $J_{-p}(x)$.

(a) Plug $r = -p$ into the recurrence relation Equation 12.7.3. Simplify your answer as much as possible.

(b) Use your recurrence relation to find c_2, c_4, and c_6 in terms of c_0.

(c) Write the general formula for c_{2m} in terms of c_0. Depending on how you answered the previous part this may be very easy or it may require a lot of thinking about where the numbers came from.

(d) Your answer to Part (c) should include the product $(p-1)(p-2)(p-3)\ldots(p-m)$. (If it doesn't, figure out how to make it include that product!) If you factor -1 out of each term it reverses all the subtractions and leaves a new factor of $(-1)^m$ outside. You cannot rewrite the resulting product with factorials in this case (do you see why?) so rewrite it using gamma functions.

(e) Write the resulting series solution to Bessel's equation. Show that your answer can be written as Equation 12.7.6 with the proper choice of c_0.

12.104 The equation $x^2 y'' + xy' + (x^2 - 4)y = 0$ is Bessel's equation with $p = 2$. Starting at the beginning with the method of Frobenius show that you get two r-values that differ by an integer, that the higher r-values leads to a valid solution, and that the lower r-value does not.

12.105 In this problem you will prove that $J_{-p}(x) = (-1)^p J_p(x)$ for positive integers p.[9] Your starting point will be Equation 12.7.6.

(a) Because $\Gamma(x)$ is infinite when x is zero or a negative integer, the sum begins with one or more terms that equal zero. Rewrite the lower limit of the sum so instead of starting at $m = 0$ it starts at the first non-zero term.

(b) Reindex the sum so it begins at $m = 0$ again, but this time with all the terms non-zero.

(c) Pull $(-1)^p$ out of the sum. What other manipulations do you have to do to this equation to show that it equals $(-1)^p J_p(x)$? *Hint*: remember how gamma functions are related to factorials.

(d) You have now shown that $J_{-p}(x)$ and $J_p(x)$ are linearly dependent, but that's only true for integer p-values. Name at least one step in your solution that was invalid for non-integer values of p, and explain why.

12.106 The Explanation (Section 12.7.2) discussed the fact that $J_p(x)$ and $J_{-p}(x)$ are linearly dependent for integer values of p (a fact that you proved in Problem 12.105). The function $Y_p(x)$ is therefore constructed to be the second solution to Bessel's equation. The Explanation showed that $Y_p(x)$ solves the equation; in this problem you will show for a few specific cases that $J_p(x)$ and $Y_p(x)$ are linearly independent.

(a) Choose two positive numbers, one integer and the other non-integer. Plot $J_p(x)$ and $Y_p(x)$ for $p = 0$ and for p equal to each of the two numbers you chose, six plots in all. Choose domains for the plots that allow you to see the behavior for both small and large positive values of x. (You do not need to include negative values of x in your plot.)

---

[9] Clearly if it's true for positive integers p it's also true for negative integers, but the proof is a bit easier if you assume p is positive.

(b) Based on your plots, describe how each of these six functions behaves in the limit $x \to 0^+$.

(c) Based on your answer to Part (b) argue that J_p and Y_p cannot be linearly dependent for the three values of p you considered.

12.107 Hankel Functions. The solution to Bessel's equation is sometimes written in the form $AH_p^{(1)}(x) + BH_p^{(2)}(x)$, where $H_p^{(1)}$ and $H_p^{(2)}$ are "Hankel functions" of the first and second kind, respectively, defined as $H_p^{(1)} = J_p(x) + iY_p(x)$ and $H_p^{(2)} = J_p(x) - iY_p(x)$.

(a) Show that this solution is equivalent to the general solution $CJ_p(x) + DY_p(x)$ by finding what values of the arbitrary constants C and D make these two solutions equal.

(b) Express the Hankel functions in terms of $J_p(x)$ and $J_{-p}(x)$.

12.108 In this problem you'll examine the behavior of J_1, J_5, and $J_{1.5}$.

(a) Plot all three functions from $x = 0$ to $x = 100$. Describe the plots. What is happening to the amplitude? To the frequency?

(b) For each function, make a list of the first 100 zeroes. Then, for each function, make a list of the differences between successive zeroes. (For example, if you were doing this for $\sin x$ the second list would be $(\pi, \pi, \pi, \ldots)$.) Describe what is happening to the distance between zeroes in each case.

(c) Repeat Parts (a)–(b) for $Y_{.1}$, $Y_{.5}$, and $Y_{1.5}$.

12.109 Equation 12.7.1 comes up in the solution of important partial differential equations in polar coordinates. Apply the substitution $x = k\rho$ to rewrite this differential equation. Remember that this is a substitution for the independent variable, so you will have to write $dR/d\rho$ and $d^2R/d\rho^2$ in terms of dR/dx and d^2R/dx^2. Show that the final result can be written in the form of Equation 12.7.2.

12.110 Modified (or Hyperbolic) Bessel Functions. In this problem you will solve the equation $x^2y'' + xy' - (x^2 + p^2)y = 0$. Rather than solving from scratch with the method of Frobenius, you can use a variable substitution to turn it into Bessel's equation and use the solution we have already worked out.

(a) The substitution will be of the form $u = ax$ where a is a constant. (Part of your job is to determine the right constant to use.) Based on that substitution find the formulas you need to replace x, dy/dx, and d^2y/dx^2 with u, dy/du and d^2y/du^2.

(b) Substitute to rewrite the differential equation in terms of $y(u)$, with the variable x completely gone.

(c) Choose the constant a to make your equation look like Bessel's equation. Based on your equation write the solutions $y(x)$ to the original differential equation, the "modified Bessel functions."

12.111 Kelvin Functions. Solve the equation $y'' + (1/x)y' - iy = 0$ by making the substitution $u = ax$ and choosing the appropriate value of a to turn this into Bessel's equation. If you are stuck you may want to first work through Problem 12.110; the solution is different but the process is the same.

12.112 Spherical Bessel Functions. In this problem you will solve the equation $x^2y'' + 2xy' + [x^2 - n(n+1)]y = 0$. Rather than solving from scratch with the method of Frobenius, you can use a variable substitution to turn it into Bessel's equation and use the solution we have already worked out.

(a) The substitution will be of the form $v = x^s y$ where s is a constant. (Part of your job is to determine the right constant to use.) Based on that substitution find the formulas you need to replace y, dy/dx, and d^2y/dx^2 with the appropriate functions of v.

(b) Plug all that into the equation. The result should be a differential equation for $v(x)$, with the variable y completely gone. Simplify your answer as much as possible.

(c) Multiply both sides of your equation by the same thing so that your equation, like Bessel's equation, will have x^2 as the coefficient of the second derivative.

(d) Choose the right value of the constant s so that your equation, like Bessel's equation, will have x as the coefficient of the first derivative.

(e) You should recognize your equation now as Bessel's equation, with a different constant playing the role of p. Based on your equation and the substitution you used, write the solution $y(x)$ to the original differential equation. These functions are called "spherical Bessel functions."

12.113 Riccati–Bessel Functions. Solve the equation $x^2 y'' + [x^2 - n(n+1)]y = 0$ by making the substitution $v = x^s y$ and choosing the appropriate value of s to turn this equation into Bessel's equation. If you are stuck you may want to first work through Problem 12.112; the solution is different but the process is the same.

12.114 The last few problems featured modified Bessel functions, Kelvin functions, Spherical Bessel functions, and Riccati–Bessel functions. All these functions are special cases of the solution to the following differential equation.

$$y'' + \frac{1-2a}{x} y' + \left[(bcx^{c-1})^2 + \frac{a^2 - p^2 c^2}{x^2} \right] y = 0$$

(a) Begin by making the substitution $y(x) = x^a u(x)$. You are replacing the dependent variable y with a new dependent variable u, so you will end up with a second-order differential equation for $u(x)$. Simplify as much as possible.

(b) Now apply the substitution $z = bx^c$. This time you are replacing the independent variable x with a new independent variable z, so you will end up with a second-order differential equation for $u(z)$. Simplify as much as possible.

(c) Write the solution to the original equation.

12.8 Sturm-Liouville Theory and Series Expansions

In Section 12.1 we solved a PDE by separating variables. The last step required us to rewrite an initial condition as a series of $J_0(k\rho)$ functions (the zero-order Bessel functions). When is it possible to do that, and how do you find the coefficients of such a series?

Sturm-Liouville theory promises us that the $J_0(kx)$ functions are a "complete set," meaning that we can use a series of them to represent almost any function. The theory also tells us that the $J_0(kx)$ functions are an "orthogonal set," which points toward the method of finding the coefficients.

If that were the only result of Sturm-Liouville theory, it would give us one of the most important properties of the J_0 function. Then we would need a different theory for the J_1 function, and another for Legendre polynomials, and so on for every possible normal mode. But the results of Sturm-Liouville theory apply to a very wide class of functions, including all the normal modes of common ODEs, so this theory provides one of the cornerstones of the whole PDE solving process.

12.8.1 Discovery Exercise: Sturm-Liouville Theory and Series Expansions

Consider a particle moving freely inside a sphere of radius a. According to quantum mechanics the particle is described by a "wavefunction" $\psi(x, y, z)$ that obeys a partial differential equation called "Schrödinger's equation." You don't need to know any quantum mechanics to do this exercise; you can just take our word that when you separate variables in this PDE you get the following ODE, where r is distance from the center of the sphere.[10]

$$r^2 R''(r) + 2r R'(r) + \frac{2mE}{\hbar^2} r^2 R(r) = 0 \quad (12.8.1)$$

Both $\hbar$ (a physical constant of the universe) and m (the mass of the particle) are given constants in this problem. E (the energy of the particle) is *not* given—it comes from the solution.

---

[10] Actually the ODE is more complicated because it involves terms related to the particle's angular momentum. Here we're considering particles with zero angular momentum.

In classical mechanics the energy could be any non-negative real number; for any particular energy, you could write an equation for the motion. In quantum mechanics, as you will see, only certain energy levels are possible; for any allowable energy level, you can write an equation for the wavefunction.

1. Look up the equation for "spherical Bessel functions" in Appendix J. Do a variable substitution of the form $u = kr$ and find the value of k that turns Equation 12.8.1 into that equation. (The spherical Bessel equation has a parameter p in it. Equation 12.8.1 should match it for one particular value of p.)
2. Write the general solution to Equation 12.8.1 in terms of spherical Bessel functions. Write your answer in terms of r, not in terms of u.

   See Check Yourself #87 in Appendix L

3. One boundary condition says that $R(r)$ must be finite at $r = 0$. Look up the properties of spherical Bessel functions in Appendix J and use that boundary condition to set one of the arbitrary constants in your general solution equal to zero.
4. The second boundary condition says that $R(a) = 0$. Using this boundary condition, what are the possible values of E?
5. This problem—which is not just Equation 12.8.1, but also includes the two boundary conditions—cannot be solved for all values of E. It can only be solved for the values of E that you listed in Part 4. Each allowable E is called an "eigenvalue" of this problem. Is there a finite number of eigenvalues, or an infinite number? Is there a lowest eigenvalue? Is there a highest eigenvalue?
6. Every eigenvalue E is associated with a particular solution to this problem, which is called its "eigenfunction." Write the first two eigenfunctions. If you are doing this exercise as homework (or in a classroom with computers), choose any positive values for m and a and graph those eigenfunctions.

12.8.2 Explanation: Sturm-Liouville Theory and Series Expansions

Let's return to the very end of Section 12.1—or, for that matter, the very end of almost any problem involving separation of PDE variables. We have a function that was given in the problem, possibly an initial position or velocity. We need to express that function as a series based on different forms of the normal modes (such as sines and cosines, or Bessel functions, or Legendre polynomials). Can we do that? If so, how do we find the coefficients?

It turns out that the answer is almost always "yes we can," and there is a universal way of finding the formula for the coefficients. This is one of the remarkable results of "Sturm-Liouville theory." In this section we will present some of the most important results of Sturm-Liouville theory. In Section 12.9 (see felderbooks.com) we will prove one of those results.

The Sturm-Liouville Problem and Its Solutions

A "Sturm-Liouville problem" is basically just a homogeneous ODE with homogeneous boundary conditions. But the results of Sturm-Liouville theory are expressed most easily when the problem is represented in the form given below.

> **The Sturm-Liouville Problem**
>
> A Sturm-Liouville problem is to find the solutions to the equation:
>
> $$\frac{d}{dx}\left(p(x)\frac{dy}{dx}\right) + q(x)y(x) + \lambda w(x)y(x) = 0 \qquad (12.8.2)$$

on the interval $a \leq x \leq b$ subject to the boundary conditions:

$$c_1 y(a) + c_2 y'(a) = 0$$
$$c_3 y(b) + c_4 y'(b) = 0 \qquad (12.8.3)$$

A particular Sturm-Liouville problem is defined by specific functions $p(x)$, $q(x)$ and $w(x)$ and specific constants c_1–c_4. The problem definition does *not* specify the constant λ. Rather, your goal is to find the constants λ for which solutions exist. These λ-values are called the "eigenvalues" of the differential equation. Each such λ is associated with a particular solution, its "eigenfunction" $y_\lambda(x)$.

The functions p, p', q, and w should all be real and continuous, and both p and w should be positive on the interval (a, b).

Equation 12.8.2 may look unfamiliar, but it is just an unusual way of expressing any second-order linear homogeneous ODE (as you will show in Problem 12.125). Equations 12.8.3 may also look unfamiliar, but again, they are a general way of writing the sorts of boundary conditions we are used to. Our purpose in this section is not to introduce a new type of equation, but to discuss a general class that includes many of the differential equations we have seen.

But unlike many differential equations, Equation 12.8.2 has an unspecified constant. Your job is not only to solve the differential equation, but to find the λ-values for which solutions exist. This type of problem is very common: when you separate variables in a partial differential equation, you get ordinary differential equations with unspecified constants. So Sturm-Liouville theory is most useful for describing the normal modes of a PDE.

Even in that case, Sturm-Liouville theory doesn't always apply. Both Equation 12.8.2 and Equations 12.8.3 are necessarily homogeneous, so they do not apply to many important ODEs and conditions. In addition, Equations 12.8.3 represent one condition on each end of the boundary. This is the sort of conditions we usually get for spatial variables (as in the example below, where the boundary conditions are $y(0) = y(L) = 0$). With time variables we often get separate values for $y(0)$ and $y'(0)$, which cannot be expressed in this form.

But despite these limitations, Sturm-Liouville theory applies to a broad class of important problems. In the example below we solve an old favorite example in order to acquaint you with this new formalism.

EXAMPLE **The Simplest Sturm-Liouville Problem**

Let $p(x) = w(x) = 1$ and $q(x) = 0$. Let $a = 0$ and $b = L$. Let $c_1 = c_3 = 1$ and $c_2 = c_4 = 0$. Then our Sturm-Liouville problem becomes this.

$$\text{Solve } \frac{d^2 y}{dx^2} + \lambda y(x) = 0 \text{ on } 0 \leq x \leq L \text{ with } y(0) = y(L) = 0 \qquad (12.8.4)$$

You know how to do this, but let's walk quickly through the steps with Sturm-Liouville terminology.

1. **Find the solution or solutions to the differential equation.** The real-valued solution in this case breaks into three categories based on λ.

$$y = \begin{cases} A e^{\sqrt{-\lambda}\, x} + B e^{-\sqrt{-\lambda}\, x} & : \lambda < 0 \\ A \cos\left(\sqrt{\lambda}\, x\right) + B \sin\left(\sqrt{\lambda}\, x\right) & : \lambda > 0 \\ A x + B & : \lambda = 0 \end{cases}$$

2. **Apply the boundary conditions. This will in general restrict the possible values of λ, as well as eliminating some solutions.** In this case the $\lambda < 0$ solutions cannot possibly meet the boundary conditions (Problem 12.128), and the $\lambda = 0$ solution can only meet them with the trivial solution $y(x) = 0$. So we focus on positive λ-values and their sine-and-cosine solutions. The condition $y(0) = 0$ means that $A = 0$. The condition $y(L) = 0$ means that $\sqrt{\lambda}\, L = n\pi$ for any positive integer n.

What have we found?

- If we choose $n = 1$ then $\lambda = \pi^2/L^2$. This is the first eigenvalue of the problem. It gives us the differential equation $y'' + \left(\pi^2/L^2\right) y = 0$ whose solution, the eigenfunction associated with that eigenvalue, is $y_{\pi^2/L^2} = B \sin(\pi x/L)$.
- If we choose $n = 2$ then $\lambda = 4\pi^2/L^2$. This is the second eigenvalue of the problem. It gives us the differential equation $y'' + \left(4\pi^2/L^2\right) y = 0$ whose solution, the eigenfunction associated with that eigenvalue, is $y_{4\pi^2/L^2} = B \sin(2\pi x/L)$.
- More generally, every eigenvalue of this problem is of the form $\lambda = n^2 \pi^2 / L^2$ for a positive integer n. Each such value is associated with an eigenfunction $y_\lambda = B \sin(n\pi x/L)$.

Usually when you solve a second-order ODE with two boundary conditions you expect to get a unique answer, but with a Sturm-Liouville problem you never will. That's because the ODE and the boundary conditions are linear and homogeneous, so you can stick an arbitrary constant in front of any solution. Between the two boundary conditions you can solve for one arbitrary constant (setting $A = 0$ in the example above) and restrict the possible values of λ. For $\lambda = 9\pi^2/L^2$ the problem above has infinitely many solutions, and for $\lambda = 3\pi^2/L^2$ it has none.

If Equation 12.8.4 had come from separating variables in a PDE, every eigenfunction $y_n = B \sin(n\pi x/L)$ would represent one normal mode of the system. (The system in this case might be a vibrating string tacked down at the ends.) We would then need to build the initial position of the string as a series of those functions.

You don't need Sturm-Liouville theory to tell you we can sum these eigenfunctions to match almost any given initial condition; you just need to have studied Fourier series. You can build any function on a finite domain from sines (by building an odd extension), subject to the not-very-restrictive Dirichlet conditions. All you need is the formula for the coefficients, and you can derive that formula based on the orthogonality of the trig functions. (If you don't know all that, you may want to review Chapter 9.)

But Sturm-Liouville theory promises us similar results over a very wide class of solutions to Equation 12.8.2. It tells us that...

- A list of the eigenvalues from lowest to highest will start somewhere (the lowest eigenvalue), but it will generally go on infinitely up from there (no highest eigenvalue). In our example above we saw that the eigenvalues were π^2/L^2, $4\pi^2/L^2$, $9\pi^2/L^2$ and so on.
- The nth eigenfunction will have $n - 1$ zeros on the interval (a, b). In our example above the nth eigenfunction $B \sin(n\pi x/L)$ crosses the x-axis exactly $n - 1$ times between 0 and L (not including the endpoints).
- If $p(a) = 0$ then "y must remain finite at $x = a$" is a sufficient condition at that boundary. (This also means you *cannot* specify a particular numerical value at that boundary; you solve the problem and find out what value y must have at the boundary.) If $p(a) \neq 0$ then a specific numerical boundary condition must be specified at that boundary, and the constant(s) can be chosen to meet that condition.
- And of course the same is true at the other boundary $x = b$.

But the most important result of Sturm-Liouville theory is worth putting in its own box.

12.8 | Sturm-Liouville Theory and Series Expansions

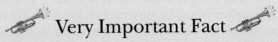

Very Important Fact

If $p(x)$, $p'(x)$, $q(x)$, and $w(x)$ are all real, continuous functions and both $p(x)$ and $w(x)$ are positive on the interval $a < x < b$, then the solutions of Equations 12.8.2–12.8.3 are complete and orthogonal with weight $w(x)$ on that interval.

The property of "completeness" promises us that we can build any initial condition from our normal modes, and "orthogonality" tells us how to do it. At this point we should define those terms a bit more carefully.

Definitions: Completeness and Orthogonality

A set of functions $y_n(x)$ is "complete" on an interval $a \leq x \leq b$ if any well-behaved function $f(x)$ on that interval can be expressed as a sum of the functions y_n:

$$f(x) = \sum_n c_n y_n(x)$$

A set of functions $y_n(x)$ is "orthogonal" on an interval $a \leq x \leq b$ with respect to the "weight function" $w(x)$ if

$$\int_a^b w(x) y_m^*(x) y_n(x) \, dx = 0 \quad \text{for all } m \neq n$$

We have several comments on these definitions:

- We're being deliberately vague about the phrase "well-behaved." There are different definitions of completeness used in different contexts. For our purposes we are talking about functions that are square-integrable on the domain $[a, b]$. In practice this restriction is almost never violated in physical problems, so when we say a set of functions is complete on a given interval that means you can take whatever function $f(x)$ appears in your problem and write it as a sum (possibly containing infinitely many terms) of the functions in that set.
- We took the complex conjugate of one of the functions $y_m(x)$ in our definition of orthogonality. For real functions such as sines this doesn't matter, but you have to remember to include it when you expand in complex functions like e^{inx}.
- Finally, we added a "weight function" $w(x)$. Sines and cosines don't need a weight function (or equivalently their weight function is $w(x) = 1$). But Fourier's brilliant trick for finding the coefficients of a Fourier series works for any orthogonal set, with or without a weight function, as we will see below.

Let's look more closely at the issue of orthogonality and how it leads us to the coefficients of a series.

Orthogonality and the Kronecker Delta

The following fact is not hard to prove. (The easiest way is perhaps by rewriting sines as complex exponentials.)

$$\int_0^L \sin\left(\frac{m\pi x}{L}\right) \sin\left(\frac{n\pi x}{L}\right) dx = 0 \text{ for any unequal positive integers } m \text{ and } n$$

The sentence "the functions $\sin(n\pi x/L)$ are orthogonal on the interval from 0 to L" is just a way of asserting that that integral comes out zero when $m \neq n$. But orthogonality isn't quite enough to find the coefficients of a Fourier series. We also need to know the "norm" of these sine functions, meaning the value of the integral when m does equal n. In this case it equals $L/2$. We can express both the $m \neq n$ and $m = n$ results in a piecewise function, but it is more common to use the "Kronecker delta."

> **Definition: Kronecker Delta**
> $$\delta_{mn} = \begin{cases} 0 & : \quad m \neq n \\ 1 & : \quad m = n \end{cases}$$

Using this symbol we can express the integral of the sine products concisely.

$$\int_0^L \sin\left(\frac{m\pi}{L}x\right) \sin\left(\frac{n\pi}{L}x\right) dx = \frac{L}{2}\delta_{mn} \qquad (12.8.5)$$

Now suppose we want to represent an arbitrary function on the interval $[0, L]$ as a series of sine functions.

$$y_0(x) = \sum_{n=1}^{\infty} b_n \sin\left(\frac{n\pi}{L}x\right) \qquad (12.8.6)$$

We can find the coefficients using Fourier's trick, which relies only on the orthogonality relationship (12.8.5). Begin by multiplying both sides of Equation 12.8.6 by $\sin(m\pi x/L)$ and integrating.

$$\int_0^L y_0(x) \sin\left(\frac{m\pi}{L}x\right) dx = \sum_{n=1}^{\infty} b_n \int_0^L \sin\left(\frac{m\pi}{L}x\right) \sin\left(\frac{n\pi}{L}x\right) dx$$

Because of orthogonality, all the terms in the sum vanish except for the term $n = m$, which gives $(L/2)b_n$. Multiplying both sides by $(2/L)$, $b_n = (2/L)\int_0^L y_0(x) \sin(n\pi x/L) dx$.

Our point here is not to re-teach you Fourier series. Our point is that we were able to find the coefficients of a Fourier series because we knew that the sine and cosine functions are orthogonal. Sturm-Liouville theory guarantees us that a much broader class of functions is orthogonal, and in doing so it gives us the way to derive their series expansion formulas.

> **EXAMPLE** **Finding the Coefficients of a Series Expansion**
>
> **Problem:**
> The "Chebyshev polynomials" of the first kind, $T_n(x)$, are complete on the interval $[-1, 1]$ with orthogonality relation
>
> $$\int_{-1}^{1} \frac{1}{\sqrt{1-x^2}} T_n(x) T_m(x) dx = \begin{cases} 0 & n \neq m \\ \pi & n = m = 0 \\ \pi/2 & n = m \neq 0 \end{cases}$$
>
> Find the fourth partial sum of the series expansion of e^x in Chebyshev polynomials.

Solution:
We begin by writing the expansion.

$$e^x = \sum_{n=0}^{\infty} c_n T_n(x)$$

To find the coefficients we multiply both sides of the equation by $T_m(x)$ and by the weight function $1/\sqrt{1-x^2}$ and then integrate.

$$\int_{-1}^{1} \frac{1}{\sqrt{1-x^2}} T_m(x) e^x \, dx = \sum_{n=0}^{\infty} c_n \int_{-1}^{1} \frac{1}{\sqrt{1-x^2}} T_n(x) T_m(x) \, dx = c_m \begin{cases} \pi & m = 0 \\ \pi/2 & m \neq 0 \end{cases}$$

From this equation we can get the numerical values of the coefficients on a computer.

$$c_0 = \frac{1}{\pi} \int_{-1}^{1} \frac{1}{\sqrt{1-x^2}} T_0(x) e^x \, dx \approx 1.266$$

$$c_1 = \frac{2}{\pi} \int_{-1}^{1} \frac{1}{\sqrt{1-x^2}} T_1(x) e^x \, dx \approx 1.130$$

$$c_2 = \frac{2}{\pi} \int_{-1}^{1} \frac{1}{\sqrt{1-x^2}} T_2(x) e^x \, dx \approx 0.271$$

$$c_3 = \frac{2}{\pi} \int_{-1}^{1} \frac{1}{\sqrt{1-x^2}} T_2(x) e^x \, dx \approx 0.044$$

The fourth partial sum is thus:

$$e^x \approx 1.266 T_0(x) + 1.130 T_1(x) + 0.271 T_2(x) + 0.044 T_3(x)$$

It's clear from these numbers that the coefficients are rapidly shrinking. This is further confirmed by plotting each of the first four partial sums. Each plot below shows e^x (solid) and a partial sum of its Chebyshev series (dashed). By the fourth partial sum the plots are indistinguishable within the interval $[-1, 1]$.

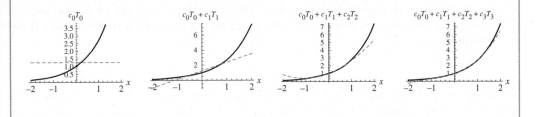

The most important lesson from the example above is that you didn't need to have heard of a Chebyshev polynomial. You needed to know a few key properties that you could easily look up, and ask a computer to calculate some numerical integrals.

The properties you need to know about any given set of functions, if you want to build a series from it, are its orthogonality relationship (which Sturm-Liouville theory often guarantees must exist) and its norm. Finding the norm is done on a case by case basis. It's pretty easy for sines and cosines. In Section 12.5 Problem 12.87 you derived the norm for the Legendre polynomials.

The Circular Drum Problem Revisited

In the Motivating Exercise (Section 12.1) you solved for the normal modes of a circular drum of radius a, starting from the equation $\rho^2 R''(\rho) + \rho R'(\rho) + k^2 \rho^2 R(\rho) = 0$. The solution involved a number of "magic" steps where you just had to take our word for certain results. We can now revisit this problem and explain how different calculations from this chapter are used in solving the drum problem and many others like it.

1. Recognize this as a Sturm-Liouville equation, $(\rho R')' + k^2 \rho R = 0$.
   Explanation: We noted that most second-order ODEs you encounter in physical problems can be written in Sturm-Liouville form. In this case $p = \rho$, $q = 0$, $w = \rho$, and the eigenvalue λ is called k^2. (Sometimes you can figure out how to write an ODE in this form just by playing with it, but Problem 12.125 shows you a systematic way to do it.) The equation is still easiest to solve in the original form, but writing it in Sturm-Liouville form lets us use the results from this section. Note that Sturm-Liouville theory applies only when $p(\rho)$ and $w(\rho)$ are positive on the interval $0 < \rho < a$, which they clearly are.

2. Write the general solution $R(\rho) = CJ_0(k\rho) + DY_0(k\rho)$.
   Explanation: The equation for $R(\rho)$ is a slight variant of Bessel's equation with $p = 0$. In Section 12.1 we had you look up the solution in Appendix J. That's a great habit to cultivate, but having gone through Section 12.7 you now know where those solutions came from.

3. Write the boundary conditions: $R(a) = 0$ and $R(0)$ must be finite.
   Explanation: These boundary conditions make sense physically, but how do we know that mathematically the solution to this equation needs a specific boundary condition at $\rho = a$ and just a requirement of finiteness at $\rho = 0$? The answer comes from $p(\rho)$. Since $p(0) = 0$, it's sufficient to specify that $R(0)$ is finite. Since $p(a) \neq 0$, you have to specify a value.

4. Use the boundary condition at $\rho = 0$ to set $D = 0$.
   Explanation: You saw in Section 12.7 that $Y_p(x)$ blows up at $x = 0$, so it is often thrown out in solutions to physical problems.

5. Use the boundary condition at $\rho = a$ to set $k = \alpha_{0,n}/a$.
   Explanation: Sturm-Liouville theory guaranteed that there would be an infinite sequence of eigenvalues k^2, with a minimum value but no maximum. Moreover, it said that the nth eigenfunction of the ODE would have $n - 1$ zeroes on the interval $(0, a)$. These values of k and the corresponding solutions $AJ_p(k\rho)$ match both these requirements.

Assuming the drum starts in some shape $f(\rho)$ and is released from rest, its motion is described by the equation $z(\rho, t) = \sum_{n=1}^{\infty} B_n J_0(\alpha_{0,n}\rho/a) \cos(v\alpha_{0,n}t/a)$. The derivation of those cosine functions is not part of this chapter (see Chapter 11 for that part of the story). But once you have the solution in this form you can write the initial condition as $f(\rho) = \sum_{n=1}^{\infty} A_n J_0(\alpha_{0,n}\rho/a)$, at which point Sturm-Liouville theory comes into play again.

6. Write the coefficients $B_n = \dfrac{2}{a^2 J_1^2(\alpha_{0,n})} \int_0^a f(\rho) J_0\left(\dfrac{\alpha_{0,n}}{a}\rho\right) \rho \, d\rho$.

   Explanation: Sturm-Liouville theory says that the eigenfunctions $J_0(\alpha_{0,n}\rho/a)$ must form a complete set, so any initial condition $f(\rho)$ can be written in this form. The orthogonality of the eigenfunctions, also promised by Sturm-Liouville theory, enables us to calculate the formula for the coefficients in Appendix J. You'll do that calculation in Problem 12.117 (taking $a = 1$ for simplicity).

12.8.3 Problems: Sturm-Liouville Theory and Series Expansions

12.115 Consider Equation 12.8.2 with $p(x) = 1 - x^2$, $q(x) = 0$, and $w(x) = 1$.

(a) The results of Sturm-Liouville theory apply only when $p(x)$ and $w(x)$ are positive everywhere except the endpoints. What domain does that restrict our conclusions to?

(b) Equation 12.8.2 includes the derivative of $p(x)y'(x)$. Expand out this derivative and write the resulting differential equation.

(c) Write the general solution to this differential equation. (Appendix J may be helpful here, or you may recognize the differential equation from earlier work in the chapter, hint hint.)

(d) How can you know from the ODE (not the solutions) that it is sufficient to say $y(x)$ must be finite at $x = -1$ and $x = 1$ rather than needing to specify values at the boundaries?

(e) Assuming $y(x)$ is finite at both endpoints of the interval, identify the eigenvalues and eigenfunctions. Note the promise of Sturm-Liouville theory that there should be a lowest eigenvalue but not a highest.

(f) Sturm-Liouville theory claims that the nth eigenfunction should have $n - 1$ zeros in the region of interest. Find the 2 zeroes of the 3rd eigenfunction.

12.116 *[This problem depends on Problem 12.115.]* Sturm-Liouville theory promises us that the solutions you found in Problem 12.115 are orthogonal, but it does not tell us what the norm is. So we will tell you.

$$\int_{-1}^{1} P_m(x)P_n(x)\,dx = \frac{2\delta_{mn}}{2m+1}$$

(a) What is the weight function in this orthogonality relationship?

(b) Your goal is to find the coefficients of the series $f(x) = \sum_{n=0}^{\infty} A_n P_n(x)$. Start by multiplying both sides of this equation by $P_m(x)$ times the weight function.

(c) Integrate both sides of the equation from $x = -1$ to $x = 1$. You should be able to use the orthogonality relation to eliminate all but one of the terms in the infinite sum.

(d) Solve for the coefficient A_m. Your answer should include an integral that involves the unknown function $f(x)$.

12.117 Walk-Through: Series Expansions. In this problem you'll derive the coefficients for a zero-order Fourier-Bessel series. In other words, you'll figure out how to write an arbitrary function $f(x)$ on the interval $[0, 1]$ as a sum of the functions $J_0(\alpha_{n,0}x)$, where J_0 is a "zero-order Bessel function" and $\alpha_{0,n}$ is one of the zeroes of J_0 (meaning $J_0(\alpha_{n,0}) = 0$). The best part is, you could do all this even if you'd never heard of a Bessel function. Your starting point is the orthogonality relation $\int_0^1 xJ_0(\alpha_{0,m}x)J_0(\alpha_{0,n}x)\,dx = (1/2)J_1^2(\alpha_{0,m})\delta_{mn}$. (We're not going to prove that here, but remember that Sturm-Liouville theory promises us that such a relationship must exist.)

(a) What is the weight function in this orthogonality relationship?

(b) Your goal is to find the coefficients of the series $f(x) = \sum_{n=1}^{\infty} c_n J_0(\alpha_{0,n}x)$. Start by multiplying both sides of this equation by $J_0(\alpha_{0,m}x)$ times the weight function.

(c) Integrate both sides of the equation from $x = 0$ to $x = 1$. You should be able to use the orthogonality relation to eliminate all but one of the terms in the infinite sum.

(d) Solve for the coefficient c_m. Your answer should include an integral that involves the unknown function $f(x)$.

12.118 *[This problem depends on Problem 12.117.]* Find the first three coefficients of the zero-order Fourier-Bessel series for $f(x) = x$ on the interval $[0, 1]$. Use a computer to evaluate the necessary integrals numerically.

For Problems 12.119–12.121 you will be given an orthogonality relation for a set of functions on a given interval (specified by the limits of integration). Use that orthogonality relation to derive a formula for the coefficients of a series expansion in which an arbitrary function $f(x)$ on that interval is written as a sum of the orthogonal functions. It may help to first work through Problem 12.117 as a model. *Don't worry if you've never heard of the function in the problem. You don't need to have!*

12.119 First-order Bessel functions: $\int_0^1 xJ_1(\alpha_{1,m}x)J_1(\alpha_{1,n}x)\,dx = (1/2)J_2^2(\alpha_{1,m})\delta_{mn}$, where $\alpha_{1,m}$ and $\alpha_{1,n}$ are two zeroes of $J_1(x)$.

12.120 Zero-order spherical Bessel functions: $\int_0^1 x^2 j_0(\alpha_{1/2,m} x) j_0(\alpha_{1/2,n} x) dx = (1/2) j_1^2(\alpha_{1/2,m}) \delta_{mn}$, where $\alpha_{1/2,m}$ and $\alpha_{1/2,n}$ are zeroes of $J_{1/2}(x)$.

12.121 Hermite polynomials:
$\int_{-\infty}^{\infty} H_m(x) H_n(x) e^{-x^2} dx = \sqrt{\pi}\, 2^m(m!) \delta_{mn}$

12.122 In this problem you'll derive the coefficients for a Fourier series of complex exponentials.

(a) Prove that the functions $y_n(x) = e^{inx}$ are orthogonal on the interval $[-\pi, \pi]$ by evaluating $\int_{-\pi}^{\pi} y_m^*(x) y_n(x) dx$ for $m \neq n$.

(b) Evaluate $\int_{-\pi}^{\pi} y_n^*(x) y_n(x) dx$. (This is the same integral as before, but this time with $m = n$.)

(c) Write one concise equation with a Kronecker delta to summarize your results so far in this problem.

(d) We want to write a function $f(x)$ as a series $f(x) = \sum_{n=-\infty}^{\infty} c_n e^{inx}$. Use the orthogonality relation you have just found to derive a formula for the coefficients c_n.

12.123 *[This problem depends on Problem 12.122.]* Express the function $f(x) = |x|$ on the interval $[-\pi, \pi]$ as a sum of complex exponentials. Simplify your answer as much as possible, remembering that for an integer n, $e^{in\pi} = e^{-in\pi} = (-1)^n$.

12.124 *[This problem depends on Problem 12.122.]* In the Explanation (Section 12.8.2) we said that the functions $\sin(n\pi x/L)$ for positive integers n were orthogonal and complete on the interval $0 \le x \le L$. Why did we only need positive integers for that case, but we need all integers for the complex exponentials in Problem 12.122? *Hint*: a good answer should involve the phrase "linearly independent."

12.125 A generic second-order, linear, homogeneous ODE can be written in the form

$$a(x)y''(x) + b(x)y'(x) + c(x)y(x) = \lambda y(x)\ (a(x) \neq 0)$$

(a) Start with Equation 12.8.2 and use the substitutions $p = e^{\int (b/a) dx}$, $q = pc/a$, $w = -p/a$ to get it into this form.

(b) A Sturm-Liouville problem also requires p and w to be positive. Suppose $p < 0$ and $w > 0$. Define a new function $\tilde{p} = -p$ and a new constant $\tilde{\lambda} = -\lambda$ and rewrite the equation in Sturm-Liouville form in terms of these new letters. It's easy to similarly handle $p > 0$, $w < 0$ and $p < 0$, $w < 0$. (Don't bother.)

If p or w switches sign partway through the domain, however, then the problem is not a Sturm-Liouville problem.

12.126 Hermite's differential equation is $f''(x) - 2xf'(x) + \lambda f(x) = 0$. Write this in Sturm-Liouville form, identifying the functions p, q, and w. *Hint*: if you're stuck, you may find Problem 12.125 helpful.

12.127 Laguerre's differential equation is $xf''(x) + (1-x)f'(x) + \lambda f(x) = 0$. Write this in Sturm-Liouville form, identifying the functions p, q, and w. *Hint*: if you're stuck, you may find Problem 12.125 helpful.

12.128 (a) Show that the function $y = Ae^{kx} + Be^{-kx}$ cannot meet the conditions $y(0) = y(L) = 0$ for any real constants A, B, k, and L unless $L = 0$ or $A = B = 0$.

(b) Explain the importance of this fact in Sturm-Liouville problems.

Problems 12.129–12.132 are about the circular drum problem discussed in the Explanation (Section 12.8.2).

12.129 Explain why it was essential to apply the boundary conditions before the initial conditions in this problem.

12.130 Show that $k = \alpha_{0,n}/a$ is guaranteed to meet the boundary condition $z(a, t) = 0$.

12.131 Find the solution for the drumhead to match the initial condition $z(\rho, 0) = a^2 - \rho^2$.

12.132 *If you have worked through Chapter 11, this problem will be a good review. If you have not studied PDEs, either in Chapter 11 or elsewhere, this problem will probably be too much.* Our treatment of this problem assumed "azimuthal symmetry" (no ϕ dependence) from the beginning. Without that assumption the PDE is:

$$\frac{\partial^2 z}{\partial t^2} = v^2 \left(\frac{\partial^2 z}{\partial \rho^2} + \frac{1}{\rho} \frac{\partial z}{\partial \rho} + \frac{1}{\rho^2} \frac{\partial^2 z}{\partial \phi^2} \right)$$

the wave equation in polar coordinates

In addition to the two boundary conditions we gave, the third boundary condition is that the function must be periodic in ϕ. That is, $z(\rho, \phi, t) = z(\rho, \phi + 2\pi, t)$. Separation of variables tells us to write $z(\rho, \phi, t) = \Phi(\phi) R(\rho) T(t)$ where these functions obey three ODEs with two shared constants p and k.

$$\Phi''(\phi) = -p^2 \Phi(\phi)$$
$$\rho^2 R''(\rho) + \rho R'(\rho) + (k^2 \rho^2 - p^2) R(\rho) = 0$$
$$T''(t) = -k^2 v^2 T(t)$$

(a) Begin by solving for $\Phi(\phi)$. The condition of periodicity will restrict p to a discrete-but-still-infinite set of allowed values.

(b) Use a variable substitution to put the $R(\rho)$ equation into a familiar form and then solve it. The solution will be more general than the one we found with azimuthal symmetry because it will have a p in it. Make sure to apply its two boundary conditions and note the resulting restriction on k.

(c) How could you have known from looking at the ODE (not at the solution or the physical setup) that "R is finite" is a sufficient boundary condition at $\rho = 0$, while you need a specific value at $\rho = a$?

(d) Solve for $T(t)$.

(e) Multiply your three separate functions to create one solution $z(\rho, \phi, t)$. You will at this point have five arbitrary constants, but you can absorb one into the other four.

(f) Write the general solution as a series over the indices n and p.

(g) Show that if you apply the assumptions of azimuthal symmetry and $\dot{z}(\rho, \phi, 0) = 0$ to your final solution, it becomes the solution we found.

12.9 Proof of the Orthgonality of Sturm-Liouville Eigenfunctions (see felderbooks.com)

12.10 Special Application: The Quantum Harmonic Oscillator and Ladder Operators (see felderbooks.com)

12.11 Additional Problems (see felderbooks.com)

CHAPTER 13

Calculus with Complex Numbers

Before you read this chapter, you should be able to...

- work algebraically with expressions involving $i \equiv \sqrt{-1}$.
- use the "complex conjugate" and other tricks to rewrite a complex number in the form $z = x + iy$.
- find the modulus $|z|$ and phase ϕ of a complex number, and use them to write it in the form $z = |z|e^{i\phi}$.
- locate numbers on the complex plane, and multiply complex numbers by multiplying their moduli and adding their phases.
- given a function $\mathbf{X}e^{i\omega t}$ identify the amplitude, frequency, and phase of its oscillation.
- solve real-valued problems (such as differential equations) by representing the solutions as the real or imaginary parts of complex functions.
- evaluate and interpret partial derivatives (see Chapter 4).

This chapter is *not* an introduction to complex numbers. Chapter 3 presented all of the skills above (except the last one). It is particularly important that you are comfortable visualizing numbers on the complex plane and working with oscillatory functions in complex form. If those skills feel rusty, please review them before diving into this chapter.

After you read this chapter, you should be able to...

- graph a complex function in several different ways, including a "mapping" from one region of the complex plane to another.
- identify branch cuts and singularities (especially poles) of a complex function.
- find the derivative of a complex function, or prove that none exists.
- determine if a given function is analytic.
- use analytic functions to solve Laplace's equation with boundary conditions, both with and without conformal mapping.
- evaluate contour integrals, especially around closed loops.
- use contour integrals to evaluate real integrals, and to find inverse Laplace transforms.
- find the Taylor series or Laurent series for a complex function around a given point, and find the radius of convergence of such a series.
- find the residue of a complex function at a given pole, with or without a Laurent series.

In Chapter 3 we worked with many complex-valued functions, especially those of the form $\mathbf{X}e^{i\omega t}$. But although the *results* of those functions were complex, the *arguments* (t in this example) were generally real. This chapter will focus on complex functions of complex variables, the study of which is called "complex analysis." This will allow us to move from complex algebra to complex calculus, extending the ideas of differentiation and integration to complex functions. As we found with complex algebra, doing calculus on complex functions is often the easiest way to answer difficult questions about real functions.

13.1 Motivating Exercise: Laplace's Equation

In a region with no heat sources or sinks the steady-state temperature T obeys Laplace's equation: $\partial^2 T/\partial x^2 + \partial^2 T/\partial y^2 = 0$. For some boundary conditions this is relatively easy to solve by inspection. For example, consider the semi-infinite slab shown in Figure 13.1. The left edge at $x = 0$ is held at $T = 1$ (in some units). The right edge at $x = \pi/2$ is held at $T = 0$. The bottom at $y = 0$ is insulated, which means $\partial T/\partial y = 0$ along that edge.[1]

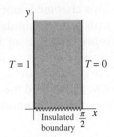

FIGURE 13.1

1. The symmetry of these boundary conditions means that the solution $T(x, y)$ must be independent of y. Using that fact, rewrite Laplace's equation as an ODE for $T(x)$ and solve it with the boundary conditions $T(0) = 1$ and $T(\pi/2) = 0$.

   See Check Yourself #90 in Appendix L

Next consider Laplace's equation in the entire first quadrant, with boundary conditions $T = 0$ on the positive x-axis and $T = 1$ on the positive y-axis. For these boundary conditions, it's more convenient to write Laplace's equation in polar coordinates.

$$\frac{\partial^2 T}{\partial \rho^2} + \frac{1}{\rho}\frac{\partial T}{\partial \rho} + \frac{1}{\rho^2}\frac{\partial^2 T}{\partial \phi^2} = 0$$

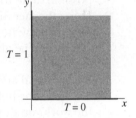

FIGURE 13.2

2. Once again, from symmetry, we can conjecture that T only depends on one of the two polar coordinates. Which one, and why?

3. Using that fact, rewrite Laplace's equation as an ODE and solve it with the correct boundary conditions.

   See Check Yourself #91 in Appendix L

Those problems may have required dusting off a few skills with partial derivatives and ODEs, but hopefully nothing about them seemed new or unsolvable. Now try this one on for size. Consider the same semi-infinite slab we started with, but this time assume the bottom and right edges are at $T = 0$ and the left edge is at $T = 1$. (Figure 13.3.) You're welcome to try to solve this, but unless you've studied complex analysis you're going to find it a lot harder than the others.

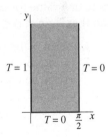

FIGURE 13.3

4. Even though you can't solve Laplace's equation with these boundary conditions, you can still predict a fair amount about the behavior from what you know. What is the temperature range inside the slab? Where is it hottest, and where coldest? What other qualitative predictions can you make?

We'll go ahead and solve the equation for you: $T(x, y) = (2/\pi) \tan^{-1}\left[\cot(x)(e^y - e^{-y})/(e^y + e^{-y})\right]$. Sorry to give that away when you were just on the verge of guessing it! While this problem looks superficially similar to the slab problem we started with, we can use complex analysis to show that it is actually mathematically equivalent to the problem in Figure 13.2. Once you have the solution to the first quadrant problem, 5–10 minutes of easy algebra can convert it to this hideous-looking solution to the second slab problem. You won't get to that point until Section 13.9, though. First, we have to start with the idea of a complex function...

---

[1] For the problems in this exercise to have unique solutions we have to assume the implicit boundary condition that T remains finite as $y \to \infty$.

Chapter 13 Calculus with Complex Numbers

13.2 Functions of Complex Numbers

This chapter is not an introduction to complex algebra. We assume you are comfortable with basic manipulation of complex numbers, including Cartesian ($x + iy$) and polar ($|z|e^{i\phi}$) representations of numbers in the complex plane. You should also be comfortable with some common complex functions, including complex exponentials $e^{it} = \cos t + i \sin t$ and the complex conjugate $z^* = x - iy$. This chapter will explore complex functions more generally, and extend the ideas of calculus to them. We begin with a general discussion of complex functions of complex variables.

13.2.1 Discovery Exercise: Functions of Complex Numbers

Recall that any complex number can be written in the "polar" form $z = |z|e^{i\phi}$, where the modulus $|z|$ is the distance from the origin to the point in the complex plane, and the phase ϕ is the polar angle. For example, the number $2i$ can be written as $2e^{i\pi/2}$.

1. Find a number that can correctly fill in this blank: $e^{\underline{\ ?\ }} = i$. In other words, find one possible value for $\ln i$.
2. Using the fact that the phases ϕ and $\phi + 2\pi$ represent the same direction in the complex plane, write all of the possible values of $\ln i$. There are infinitely many of them, but you should be able to write them all in a simple form using n to stand for an arbitrary integer.
3. Find the real and imaginary parts of $\ln z$, where $z = |z|e^{i\phi}$ for real $|z|$ and ϕ. The rules of logs will be helpful here, but to find *all* the answers you will also need to think about the work you did above.
   See Check Yourself #92 in Appendix L

You've just shown that $\ln z$ is a "multiple-valued function," meaning that for any complex number z the function $\ln z$ yields more than one answer (infinitely many in this case). To define a single-valued version of the log function, you need to restrict the values of the phase ϕ to one value for each direction in the complex plane. The most common choice for how to do that is $-\pi < \phi \le \pi$. The resulting values are called the "principal values" of the logarithm.

4. What is the principal value of $\ln(-i)$?
5. If you move around the complex plane taking the principal value of $\ln z$ at each point, it will be continuous everywhere except when you cross one ray (half a line). What is that ray?
6. You could define a single-valued logarithm function by restricting the phase to $0 < \phi \le 2\pi$. If you do that, where will the discontinuity in the function occur?

13.2.2 Explanation: Functions of Complex Numbers

Many complex functions are obvious extrapolations from real functions. The real function $f(x) = x^2$ multiplies any real number by itself to give a new real number, so $f(5) = 25$. The complex function $f(z) = z^2$ multiplies any complex number by itself to give a new complex number, so $f(3 + 2i) = 9 + 12i - 4 = 5 + 12i$.

In other cases the extension is not so clear. Before we try to calculate $\sin i$ we need some idea of what the question means. It clearly does *not* mean, for instance, "draw a right triangle with one angle $i°$ and divide the opposite side by the hypotenuse." We must define the function $\sin z$ in such a way that $\sin x$ for a real x gives the same answer it always did, and its properties extend in a logical way across the complex plane. As we go through this section you'll see how we do that for a variety of basic functions.

Graphing Complex Functions

One of the most powerful ways to understand any function is by graphing, and here we encounter a problem. The graph of $f(x) = x^2$ uses one axis for the argument x and one for the function f, giving us a 2D plot. A plot of $f(z) = z^2$ would require four dimensions, which is a little impractical.

There are a number of workarounds for this problem. If the argument is complex but the output is real, you can graph the function in three dimensions: the horizontal plane represents the argument, and the vertical axis[2] the function. For instance, Figure 13.4 shows the graph of $f(z) = |z|$, the modulus function.

If the argument and the function are both complex, you can still visualize some aspects of the behavior by plotting the modulus $|f|$ as a function of z on a 3D plot. You can also graph $Re(f)$ and $Im(f)$ on two separate graphs. Both approaches (Figure 13.5) have advantages and disadvantages, and we will use them as appropriate in the chapter. In Section 13.8 we will introduce another method called "mapping" in which we plot z on one graph and $f(z)$ on another, showing what complex outputs correspond to what complex inputs.

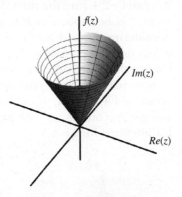

FIGURE 13.4 A plot of $|z|$ vs. z.

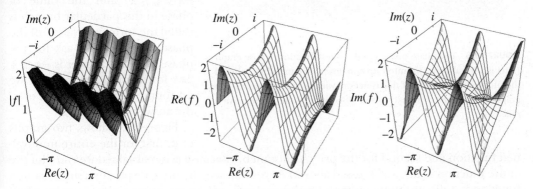

FIGURE 13.5 Plots of the modulus, real part, and imaginary part of $f(z) = \sin z$. The function oscillates in the real direction and grows exponentially in both the positive and negative imaginary directions.

Over the years you have studied elements of real graphs such as holes, asymptotes, and jump discontinuities. These are not just visual effects; they are visual ways of describing and understanding the behavior of a function. Complex graphs have their own peculiar phenomena, and as you learn about them you will learn more about complex functions. Below we introduce two of the most important such properties, "branch cuts" and "poles."

Multiple-Valued Functions and Branches

By definition a real function can only give one answer for a given input value. Consider, for instance, $f(x) = \sqrt{x}$. If we define this function as "the number that you square to get x" then $\sqrt{25}$ has two answers, 5 and -5. That would violate the definition of a real function. We therefore define $\sqrt{x}$ as "the *non-negative* number that you square to get x" so $\sqrt{25} = 5$. If we want both answers we can write $\pm\sqrt{25}$ but the square root itself is still single-valued.

---

[2]Calling it the z-axis seemed hopelessly confusing in this context.

The complex function $f(z) = \sqrt{z}$ gives two answers for any non-zero complex number z. For example, the two square roots of $3 - 4i$ are $2 - i$ and $-2 + i$. (You can easily check this.) No simple property such as "non-negative" distinguishes these two numbers. Which one should we choose?

The solution is to extend the definition of "function" to allow multiple answers; we say that $2 - i$ and $-2 + i$ are the two values of $\sqrt{3 - 4i}$. Every non-zero complex number has 2 square roots, 3 cube roots, 8 eighth roots, and so on. So in most areas of math where we are only considering functions of real variables, the phrase "single-valued function" is redundant: a function is *by definition* single-valued. In complex analysis, however, some functions are single-valued and others are multivalued.

One particularly important multivalued function is the function ϕ, the phase of z. The phase of $1 + i$ is $\pi/4$, or equivalently $\pi/4 + 2\pi$, or more generally $\pi/4 + 2\pi k$ where k is any integer. As you can see, the phase of z has infinitely many values for any non-zero z. (Phase is undefined for $z = 0$.)

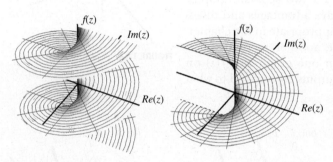

FIGURE 13.6 The phase of z is a multivalued function, whose graph (left) is a spiral going infinitely up and down. The principal value of the phase is one branch of that function, with the range $(-\pi, \pi]$. The graph of that branch (right) shows a discontinuity on the negative real axis.

Nonetheless, there are times when you ask a question and you just want one answer. For any multivalued function we can define a "branch" with one value for each z. For the phase, the most commonly used branch is $-\pi < \phi \leq \pi$, and the value of phase in this particular branch is called the "principal value" of the phase of z. So if we ask for the phase of $-i$ the answer is $-\pi/2 + 2\pi k$ for any integer k, but if we ask for the "principal" phase of $-i$ the answer is only $-\pi/2$.

Figure 13.6 shows two graphs of ϕ, first for the entire multivalued function and second for the principal branch. (Because ϕ is real valued we can plot the entire function in 3D, as discussed above.) As you can see, the price we pay for a single-valued function is a discontinuity on the negative real axis, where the phase jumps from π on the line to $-\pi$ below it. A line where a branch of a function is discontinuous is called a "branch cut." While $(-\pi, \pi]$ is the range used to define the principal value of the phase, there might be circumstances where it was convenient to use another branch, say $[0, 2\pi)$. The branch cut for that branch would be on the positive real axis.

> **EXAMPLE** **A Multivalued Function and its Branch Cut**
>
> Question: Find all three complex answers to $\sqrt[3]{-8}$.
>
> Answer:
> You can solve this question algebraically by setting $(x + iy)^3 = -8$, expanding the left side, and setting the real part on the left equal to -8 and the imaginary part equal to 0. But it's much easier on the complex plane.

When you multiply two complex numbers, their moduli multiply and their phases add. So when you cube a complex number, its modulus cubes and its phase triples. That tells us how to find a cube root. The number -8 has a modulus of 8 and a phase of π, so its cube root must have a modulus of 2 and a phase of $\pi/3$. That gives us the number $2e^{(\pi/3)i}$, aka $1 + \sqrt{3}i$.

But that's only one answer; we're supposed to get three! Well, we said above that the number -8 has a phase of π, but remember that the phase itself is multivalued. We can also describe -8 as having a modulus of 8 and a phase of 3π. Now its cube root has a modulus of 2 and a phase of π, giving us a new answer, $\sqrt[3]{-8} = -2$. (That answer at least comes as no surprise.) We can *also* describe -8 as having a phase of 5π, so $\sqrt[3]{-8} = 2e^{(5\pi/3)i} = 1 - \sqrt{3}i$. There are our three answers. (We'll leave it to you to figure out what happens when you consider the phases 7π, 9π, and so on.)

Question: If we create a branch so that $\sqrt[3]{z}$ is a single valued function, where is the branch cut? (Assume here that the phase function is defined as going from $-\pi$ to π.)

Answer:
If you start with a number z whose phase is slightly less than π, its cube root will have a phase slightly less than $\pi/3$. Now move z down just a bit in the complex plane and—because of the way we have defined phase—suddenly its phase undergoes a discontinuous jump to a number slightly above $-\pi$, so its cube root becomes something completely different.

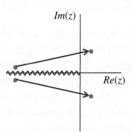

The wavy line shows the branch cut for $\sqrt[3]{z}$, which is the same as the branch cut for the phase. The arrows show that when you take the cube root of a number just above the branch cut its phase goes from π to $\pi/3$, while the cube root of a number just below the branch cut takes its phase from $-\pi$ to $-\pi/3$.

The moral of the story is that functions that depend on phase inherit the branch cut from the phase function. For fractional powers such as $x^{1/3}$, and (as we'll see below) for the logarithm, the principal value has a branch cut on the negative real axis.

Poles and Singularities

Journey back to high school for a moment and consider the following real-valued function of a real variable.

$$f(x) = \frac{(x-7)(x-1)}{(x-7)(x+4)}$$

At $x = -4$ and $x = 7$ this function is undefined, but *around* those values the behavior is very different. $x = 7$ is a "removable discontinuity" or "hole" meaning that the undefined value has no effect on the nearby behavior of the function. $x = -4$ is a "vertical asymptote" meaning that the function blows up around this value, or $\lim_{x \to -4} |f(x)| = \infty$. (The absolute value is required because the function might approach ∞, $-\infty$, or both.)

Now consider the following complex-valued function of a complex variable.

$$f(z) = \frac{(z-4)(z-i)}{(z-4)(z-7+2i)}$$

This function is undefined at $z = 4$ and $z = 7 - 2i$, but once again we see very different behavior around these points. The function is completely unaffected by the $z - 4$ terms in the numerator and denominator except at $z = 4$ itself. But around $z = 7 - 2i$ the function blows up: $\lim_{z \to 7-2i} |f(z)| = \infty$ where the modulus of this function serves roughly the same purpose that the absolute value served for the previous one.

Both $z = 4$ and $z = 7 - 2i$ are called "singularities" of $f(z)$. The former is a "removable singularity" or "hole." The latter is a "pole." Later in the chapter we'll define the term "pole" more precisely and introduce a third type, an "essential" singularity, but the most important thing to know about poles is that the modulus of the function approaches infinity as you approach them.

A Brief Roundup of Basic Functions

Three important functions of a complex number z are $|z|$ (its modulus), ϕ_z (its phase), and z^* (its complex conjugate). Beyond those, the basic functions form the same list as the functions of real numbers. We list them below to note some of their properties on the complex plane.

- $f(z) = z^n$ for a positive integer n means the same thing it does for real numbers: you multiply z by itself n times. This function is single-valued and continuous everywhere.
- $f(z) = z^{1/n}$ for a positive integer $n > 1$ is a root. As we discuss above, this function gives n answers for any non-zero z. Its principal branch has a branch cut on the negative real axis. The same is true for $z^{m/n}$, where m/n is a fraction reduced to its lowest terms.
- $f(z) = z^{-n}$ for a positive integer n is single-valued, but has a pole at $z = 0$.
- $f(z) = e^z$ is periodic with period $2\pi i$, meaning $f(z + 2\pi i) = f(z)$ for all z. (Take a moment to picture what this means on the complex plane.) It's exponential as you move along the real axis and oscillatory in the imaginary direction. If we break z into its real and imaginary parts we can write $e^{x+iy} = e^x e^{iy}$ which we recognize as the standard form $|z|e^{i\phi}$. Thus we see that the real part of z provides the modulus of e^z, the imaginary part the phase. The function is single-valued and everywhere continuous.
- $f(z) = \ln z$ is best understood by writing z in terms of its modulus and phase. $\ln\left(|z|e^{i\phi}\right) = \ln |z| + \ln e^{i\phi} = \ln |z| + i\phi$ which we recognize as the standard form $x + iy$. So the modulus of z determines the real part of its log, and the phase becomes the imaginary part. (Compare this last sentence to the corresponding statement we made about e^z.) Like fractional powers, this is either a multivalued function or inherits the branch cut of the phase. See below for a note about notation.
- The functions $\sin z$ and $\cos z$ are single-valued and continuous. They are both periodic with period 2π, just like their real counterparts. You will explore the trig functions more thoroughly in the problems.

The identities and properties of the real functions generally hold for their complex extensions. For example, $e^{a+b} = e^a e^b$ and $\sin^2 z + \cos^2 z = 1$. Even Euler's formula $e^{iz} = \cos z + i \sin z$ holds for complex z, although the cosine and sine no longer represent the real and imaginary parts of the number.

For a real-valued function $f(x)$, the function $f(x) + 3$ represents a vertical shift that adds three to every value of f. The function $f(x + 3)$ is a horizontal shift that moves the graph to

13.2.3 Problems: Functions of Complex Numbers

13.1 Let $f(z) = z^2$.

(a) We can write the input z in real and imaginary parts as $z = x + iy$, and the output f in real and imaginary parts as $f(z) = u + iv$. Then the complex function $f(z)$ can be expressed as two real functions $u(x, y)$ and $v(x, y)$. Write these two functions for $f(z) = z^2$.

(b) Alternatively we can write the input z in polar form as $|z|e^{i\phi_z}$ and the output f also in polar form as $|f|e^{i\phi_f}$. Now you can express $f(z)$ as two real functions $|f|(|z|, \phi_z)$ and $\phi_f(|z|, \phi_z)$. Write these two functions for $f(z) = z^2$.

(c) Describe what squaring a number does to its location on the complex plane. Your answers above may help here.

13.2 Let $f(z)$ be the principal branch of $\ln z$.

(a) The easiest way to understand this function is to write the input in polar form as $z = |z|e^{i\phi}$, but the output in Cartesian form as $f(z) = u + iv$. In this way the complex function $f(z) = \ln z$ can be described by two real functions $u(|z|, \phi)$ and $v(|z|, \phi)$. Write these two functions.

(b) If you start at $z = 0$ and move upward along the imaginary axis, what is $\lim_{z \to i\infty} u$? What is $\lim_{z \to i\infty} v$?

(c) What are $\lim_{z \to 0} u$ and $\lim_{z \to 0} v$ if you approach the origin along the negative imaginary axis?

(d) Alternatively we can describe z in real and imaginary parts as $x + iy$. Then $f(z)$ can be described by the two real functions $u(x, y)$ and $v(x, y)$. Write these two functions.

13.3 🖥 Consider the principal branch of $\ln z$. For each of the plots below take as your domain the square going from $-2 - 2i$ to $2 + 2i$.

(a) Create a three-dimensional graph with z on the horizontal plane and the modulus $|f(z)|$ on the vertical axis. What does your plot show $|f|$ doing on the branch cut? What does it show $|f|$ doing as $z \to 0$? Where is the minimum value of $|f|$ on your plot?

(b) Plot $Re(f)$ and $Im(f)$ in two side-by-side plots. What happens to each one on the branch cut? In the limit $z \to 0$?

(c) The formula $z = e^{i\phi}$ for all $-\pi < \phi \leq \pi$ describes every point on the unit circle. Sketch the set of complex points $f(z)$ for all z along this circle. On the same plot make a sketch of $f(z)$ for all points on the larger circle $z = 2e^{i\phi}$. What shapes are your plots? (This part does not require a computer, although you may use one if you wish.)

In Problems 13.4–13.9 you will consider the functions $\cos z$ and $\sin z$ in different ways. Most of them involve Equation 13.2.1, which gives the real and imaginary parts of the complex cosine function.

$$\cos(x + iy) = \frac{1}{2}\cos x \left(e^y + e^{-y}\right) + \frac{1}{2}i\sin x \left(e^{-y} - e^y\right)$$

(13.2.1)

13.4 You know Euler's formula $e^{ix} = \cos x + i \sin x$.

(a) Knowing that the cosine is an even function and the sine an odd function, find the real and imaginary parts of e^{-ix} where x is a real number.

(b) Use the formulas for e^{ix} and e^{-ix} to find a formula for $\cos x$ in terms of exponential functions.

(c) Your formula from Part (b) is valid if x is a complex number, so replace x with $x + iy$ in your formula and derive Equation 13.2.1.

13.5 Find the real and imaginary parts of the function $\sin(x + iy)$.

13.6 🖥 Use Equation 13.2.1 to derive the formula $|\cos z| = (1/2)\sqrt{e^{2y} + e^{-2y} + 2\left(\cos^2 x - \sin^2 x\right)}$. Then create a three-dimensional plot of $|\cos(x + iy)|$ as a function of x and y. List three facts that you can see about the function cos z by looking at your graph.

13.7 Consider Equation 13.2.1 for the special case $y = 0$.
   (a) In this case, what are the real and imaginary parts of cos z?
   (b) If you were to follow along the complex plane from $x = 0$ to $x \to \infty$, always holding y at zero, what behavior would you see in the cosine function?
   (c) How do your answers change if y is held at a non-zero constant y_0?

13.8 Consider Equation 13.2.1 for the special case $x = 0$.
   (a) In this case, what are the real and imaginary parts of cos z?
   (b) If you were to follow along the complex plane from $y = 0$ to $y \to \infty$, always holding x at zero, what behavior would you see in the cosine function?
   (c) How do your answers change if x is held at a non-zero constant x_0?

13.9 For complex numbers, just as for real numbers, $\sin z = \cos(z - \pi/2)$. Once you know the behavior of the cosine function on the complex plane, how does that behavior translate to the behavior of the sine function? *Hint*: remember that $\pi/2$ is a real number!

13.10 🖥 For each of the functions below, make 3D plots of $|f|$, $Re(f)$, and $Im(f)$. In each case the horizontal plane will represent z and the vertical axis will be the real quantity you are plotting. Choose your domains so that you can clearly see how the function behaves. Describe some property of each function that is most easily seen in one of the plots, and what about that plot reveals that property.
   (a) $f(z) = z^3$
   (b) $f(z) = z^*$
   (c) $f(z) = \tan z$
   (d) $f(z) = \ln(\cos z)$

13.11 (a) If the complex number z has modulus $|z|$ and phase ϕ_z, what are the modulus and phase of $z^{1/2}$?
   (b) What are the two values of $(4i)^{1/2}$?
   (c) We can define the principal value of $z^{1/2}$ by using the principal value of ϕ in your formula from Part (a). Where is the branch cut for the principal value of $z^{1/2}$?

13.12 Which of the following functions have branch cuts?
   (a) z^3
   (b) $z^{1/5}$
   (c) $z^{2/5}$
   (d) e^z
   (e) $\ln |z|$

13.13 Identify all the zeros (places where $f(z) = 0$) and poles of the function
$$f(z) = \frac{(z+1)(z-2i)}{(z-3)^3(z+5-2i)}.$$

13.14 The real function $\dfrac{1}{x^2 + 9}$ has no vertical asymptotes, but the complex function $\dfrac{1}{z^2 + 9}$ has two poles. Where are they?

13.15 The function $f(z) = \dfrac{z + 1 + i}{(z + 4i)(z + 1 + i)}$ has a removable singularity and a pole.
   (a) Identify the removable singularity. How does $f(z)$ behave for z-values close to that number?
   (b) Identify the pole. How does $f(z)$ behave for z-values close to that number?

13.16 The Explanation (Section 13.2.2) shows that the principal branch of $\sqrt[3]{z}$ has a branch cut on the negative real axis. Using this fact you should be able to answer the following questions without too much work.
   (a) Where is the branch cut for the function $\sqrt[3]{z - 2 - 3i}$?
   (b) Where is the branch cut for the function $\sqrt[3]{z - 2}$?
   (c) Where is the branch cut for the function $\sqrt[3]{z - 3i}$?
   (d) Where is the branch cut for the function $\sqrt[3]{z - 2 - 3i}$?

13.3 Derivatives, Analytic Functions, and Laplace's Equation

We begin complex calculus by introducing "analytic functions." Roughly speaking an analytic function is one that is differentiable, but that turns out to be a more restrictive condition for complex functions than it is for real ones.

13.3 | Derivatives, Analytic Functions, and Laplace's Equation

13.3.1 Discovery Exercise: Derivatives and Analytic Functions

A derivative is the limit of a ratio. For real numbers, we can write this.

$$\frac{df}{dx} = \lim_{\Delta x \to 0} \left(\frac{\Delta f}{\Delta x} \right)$$

So df/dx does not just ask how much f changed; it divides that by how much x changed, which measures how far you traveled along the x-axis.

The derivative of a complex function can be written the same way.

$$\frac{df}{dz} = \lim_{\Delta z \to 0} \left(\frac{\Delta f}{\Delta z} \right)$$

So df/dz divides the change in f by how much z changed. But Δz is *not* a measure of how far you traveled; it is a complex number that measures how much z changed as you traveled in whatever distance you did your traveling. To consider what all that means, we present you with the following fact.

The derivative of $f(z) = z^2$ at the point $z = 5$ is 10.

This tells us that if we start at $z = 5$ and move off by a reasonably small Δz the function should change by roughly 10 Δz *no matter what direction we move in*. Let's see how that pans out.

1. What is $f(5)$?
2. Suppose you move away from $z = 5$ by one unit in the positive real direction. What is the new z, and what is Δz? What is the new f, and what is Δf? Is their ratio approximately 10?
3. Suppose you move away from $z = 5$ by one unit in the positive imaginary direction. What is the new z, and what is Δz? What is the new f, and what is Δf? Is their ratio approximately 10?

   See Check Yourself #93 in Appendix L

4. Suppose you move away from $z = 5$ by one unit in the negative real direction. What is the new z, and what is Δz? What is the new f, and what is Δf? Is their ratio approximately 10?
5. Suppose $\Delta z = 1 + i$. What is the new z? What is the new f, and what is Δf? Is it still approximately $10 \Delta z$?
6. In every case above you should have found that Δf was approximately 10 times Δz, but never exactly. If $f'(5) = 10$ exactly (which it is), why didn't your answers come out exactly that way?

13.3.2 Explanation: Derivatives and Analytic Functions

The derivative of a complex function is defined the same way it is for real functions.

$$\frac{df}{dz} \equiv \lim_{\Delta z \to 0} \frac{f(z + \Delta z) - f(z)}{\Delta z} \quad \text{Definition of the derivative}$$

For a real function $f(x)$ we have two choices for how to take the limit $\Delta x \to 0$: from the right or from the left. For example, applying the definition of the derivative to the function $f(x) = |x|$ at $x = 0$ gives $f'(x) = 1$ if we take the limit from the right and $f'(x) = -1$ if we take it from the left. We therefore say the derivative of $|x|$ is undefined at $x = 0$.

For a complex function z can approach z_0 in an infinite number of ways, coming in along any line or curve on the complex plane. The derivative $f'(z_0)$ is only defined if *all* of those derivatives are equal (Figure 13.7).

The Discovery Exercise (Section 13.3.1) focuses on this issue of definition. In that example we claim that a particular function has a derivative of 10 at a particular point. This means that if you step away from that point by some dz the value of the function will change by $10\,dz$ no matter what direction you move in. If you move straight down by a small amount ds, so $dz = -i\,ds$, then the function will change by $-10i\,ds$ and so on. If those directions don't all yield the same ratio df/dz then the derivative is undefined.

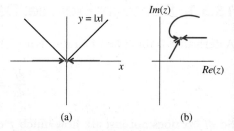

FIGURE 13.7 For a real function (a) the derivative is only defined if the slope is the same from the left and right. For a complex function (b) we can approach a given point z_0 in infinitely many ways, and the derivative is only defined if all of them give the same answer. (The complex plot above, unlike the real one, doesn't show any particular function $f(z)$.)

EXAMPLE — The Derivative of the Complex Conjugate Function

Question: Prove that $f(z) = z^*$ is not differentiable at $z = i$.

Answer:

At $z = i$, $f(z) = -i$. To use the definition of the derivative we write the increment Δz in real and imaginary parts: $\Delta z = \Delta x + i\Delta y$. First we'll take the derivative in the real direction, $\Delta z = \Delta x$, and then in the imaginary direction, $\Delta z = i\Delta y$.

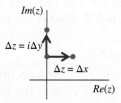

Real direction: From the definition of the complex conjugate, $f(i + \Delta x) = -i + \Delta x$, so the definition of the derivative gives

$$\lim_{\Delta x \to 0} \frac{f(i + \Delta x) - f(i)}{\Delta x} = \lim_{\Delta x \to 0} \frac{-i + \Delta x - (-i)}{\Delta x} = 1$$

Imaginary direction: This time $\Delta z = i\Delta y$ and $f(i + i\Delta y) = -i - i\Delta y$, so the definition of the derivative gives

$$\lim_{\Delta y \to 0} \frac{f(i + i\Delta y) - f(i)}{i\Delta y} = \lim_{\Delta y \to 0} \frac{-i - i\Delta y - (-i)}{i\Delta y} = -1$$

We can make sense of these numbers by remembering that the complex conjugate always mirrors over the real axis. If we start at $z = i$ and move to the right, z^* moves to the right with us; that's why $\Delta f/\Delta z = 1$ in the real direction. If we start at $z = i$ and move up, z^* moves down; that's why $\Delta f/\Delta z = -1$ in the imaginary direction.

13.3 | Derivatives, Analytic Functions, and Laplace's Equation

Since the limit gives a different answer in these two directions, z^* is not differentiable at $z = i$. (As we will discuss below, z^* is not differentiable anywhere.)

The example above shows that z^* has no derivative at $z = i$ because the derivatives in the real and complex directions are not the same. If those two derivatives *had* come out the same, that still would not prove that the derivative was defined—it might come out different along a diagonal line, or along a parabola, or along any other path that leads to $z = i$. A function is only differentiable if all possible paths give the same answer. This is, in many cases, easier to prove than it sounds.

EXAMPLE **The Derivative of the Complex Function $f(z) = z^2$**

Question: Find the derivative of $f(z) = z^2$.

Answer:
From the definition of the derivative:

$$f'(z) = \lim_{\Delta z \to 0} \frac{(z + \Delta z)^2 - z^2}{\Delta z} = \lim_{\Delta z \to 0} \frac{z^2 + 2z\Delta z + (\Delta z)^2 - z^2}{\Delta z}$$

$$= \lim_{\Delta z \to 0} \frac{2z\Delta z + (\Delta z)^2}{\Delta z} = \lim_{\Delta z \to 0} (2z + \Delta z) = 2z$$

The above bit of algebra shows that the derivative of z^2 is $2z$, but it also proves that the complex function z^2 is differentiable at all.

The algebra in this case looks identical, step by step, to the same problem with real numbers. We did not take the limit until the last step, by which point it is clear that as Δz approaches zero—no matter how it approaches it!—the derivative is $2z$. This enabled us to confidently claim in the Discovery Exercise that $f'(5) = 10$. In Problem 13.18 you'll try to use the same formula on z^* and discover that it doesn't work.

Analytic Functions

The field of complex analysis is mostly the study of "analytic functions," which is roughly synonymous with "differentiable functions." More precisely, we say a function is "analytic" at $z = z_0$ if there is some disk centered on z_0 in which the function is differentiable. The mathematical term for such a disk is a "neighborhood" of z_0.

Definition: Analytic Function

A function $f(z)$ is analytic at $z = z_0$ if f is differentiable at every point in some neighborhood of z_0.

If f is differentiable at every point in the complex plane then it is analytic everywhere. Such a function is called "entire."

Some sources define "analytic" to mean that a function has a Taylor series that converges to the value of the function, and use "holomorphic" for a function that is differentiable. Fortunately there is a theorem proving that the two are equivalent, so this definition of "analytic" will serve our purposes.

What Functions are Analytic?

You'll prove in Problem 13.17 that z^n for any positive integer n is entire, and in Problem 13.19 that any linear combination of analytic functions is analytic. Those two innocuous results take us farther than you might think. Together they prove that any polynomial is analytic. And what about more complicated functions like e^z and $\sin z$? Such functions can be defined by their Maclaurin series. For example, we know that for real x, $e^x = 1 + x + x^2/2 + x^3/6 + \ldots$, so we define $e^z = 1 + z + z^2/2 + z^3/6 + \ldots$ for complex z. Wherever the series converges, the resulting function is analytic. The series for the exponential and trig functions converge everywhere, so these functions are entire.

The product of two analytic functions is analytic, and the ratio of two analytic functions is analytic except where the denominator is zero. This implies, among other things, that any rational expression is analytic except at the poles.

Fractional powers and the logarithm, as you may recall, are multivalued functions. We make them single-valued by isolating the principal branch, which has a branch cut on the negative real axis. These functions are analytic except on their branch cuts and at the origin.

The composition of two analytic functions is also analytic, which gives us functions like $e^{\sqrt{z}}$ and $\sin(z^2 + e^z)$. Finally, the derivative of an analytic function is analytic. We summarize these results below.

 Very Important Fact: Common Analytic Functions

The functions z^n, e^z, $\sin z$, $\cos z$, and $\ln z$, and combinations of these functions, are analytic except at discontinuities (branch cuts and poles).

We would love to summarize this even more concisely by saying "practically all functions are analytic outside of obvious discontinuities," but that isn't true. For example the complex conjugate, a simple and very important function, is continuous everywhere and differentiable nowhere. Fortunately there is a rigorous and easy test for determining if any function is analytic.

A complex function starts with a complex number $z = x + iy$ and produces a new complex number $f(z) = u + iv$. So every complex function is defined by two real functions $u(x, y)$ and $v(x, y)$. For instance, $z^2 = (x + iy)^2 = x^2 + 2ixy - y^2$, so we can describe it as $u(x, y) = x^2 - y^2$ and $v(x, y) = 2xy$. The "Cauchy-Riemann equations" determine if $f(z)$ is analytic based on the properties of its two constituent functions.

The Cauchy-Riemann Equations

Let x and y be the real and imaginary parts of the complex number z, and let $u(x, y)$ and $v(x, y)$ be the real and imaginary parts of a function $f(z)$. If f is analytic, then u and v satisfy the Cauchy-Riemann equations:

$$\frac{\partial u}{\partial x} = \frac{\partial v}{\partial y} \quad \text{and} \quad \frac{\partial u}{\partial y} = -\frac{\partial v}{\partial x} \qquad (13.3.1)$$

This rule is reversible: if $u(x, y)$ and $v(x, y)$ have continuous partial derivatives and satisfy the Cauchy-Riemann equations in some region, then $f(z)$ is analytic in that region.

You may want to confirm for yourself, as an example, that $f(z) = z^2$ satisfies the Cauchy-Riemann equations.

These equations show that being analytic is actually a very restrictive condition. If you choose a given function $u(x, y)$, Equations 13.3.1 tell you what $\partial v/\partial x$ and $\partial v/\partial y$ must be, and therefore determine $v(x, y)$ to within an additive constant. If your v function does not happen to be a perfect match for your u function, the resulting $f(z)$ will not be analytic.

Below we will see another property of analytic functions, easily derivable from the Cauchy-Riemann conditions: they give us solutions to an important differential equation called Laplace's equation.

13.3.3 Explanation: Solving Laplace's Equation with Analytic Functions

The circle $x^2 + y^2 = R^2$ is held at a constant temperature $T = T_0$, and the circle $x^2 + y^2 = 4R^2$ is held at $T = 2T_0$. Given enough time the temperature between these two circles will settle down to a steady-state temperature distribution that obeys Laplace's equation $\nabla^2 T = 0$. Find this distribution.

The above problem is an example of a "Dirichlet problem." You are given the value of a function on the boundary of a region and ask to find a solution to a PDE for that function in the interior of that region, matching the given boundary conditions. The Dirichlet problem for Laplace's equation arises, not only in thermodynamics, but in electrostatics (because electric potential V obeys the same mathematical law as temperature T) and other fields. In these cases you are looking for a function that has two properties.

- The function $T(x, y)$ obeys Laplace's equation inside a given region. In two dimensional Cartesian coordinates Laplace's equation looks like this.

$$\frac{\partial^2 T}{\partial x^2} + \frac{\partial^2 T}{\partial y^2} = 0 \quad \text{Laplace's equation in two dimensions} \quad (13.3.2)$$

Any function that satisfies Laplace's equation is called "harmonic."
- The function meets certain specified values along the boundary of that region. In our example above we need $T(x, y) = T_0$ on the inner circle and $T(x, y) = 2T_0$ on the outer.

Such a problem has nothing to do with complex numbers. But one of the most powerful approaches to such problems begins with a remarkable property of analytic functions.

🎺 Very Important Fact: Analytic Functions and Laplace's Equation 🎺

Suppose a complex function $f(x + iy)$ is analytic. If this function is broken down into its real and imaginary parts

$$f(x + iy) = u(x, y) + iv(x, y)$$

then $u(x, y)$ and $v(x, y)$ are both harmonic functions. That is, they are both solutions to Laplace's equation.

It's a relatively short hop from the Cauchy-Riemann conditions to that fact, as you will demonstrate in Problem 13.31. But here let's focus on what that fact tells us. We said in Section 13.3.2 that the function z^2 is analytic. We also said that its real part is $u(x, y) = x^2 - y^2$ and its imaginary part is $v(x, y) = 2xy$. We therefore have a guarantee that the function $x^2 - y^2$ satisfies Laplace's equation (13.3.2). We also have a guarantee that the function $2xy$ satisfies Laplace's equation. We can use those facts if we happen to be looking for harmonic functions in the right regions.

EXAMPLE   Laplace's equation on a very contrived region

Problem:
Find a function $V(x, y)$ that satisfies Laplace's equation with boundary conditions $V(x, y) = 1$ along the curve $x^2 - y^2 = 1$ and $V(x, y) = 4$ along the curve $x^2 - y^2 = 4$.

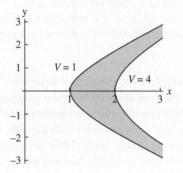

Solution:
The function $V(x, y) = x^2 - y^2$ works perfectly. It clearly meets those two boundary conditions, and we know it is harmonic because it is the real part of the analytic function z^2.

Most problems don't work out quite that easily, of course. But because any composition of analytic functions is itself an analytic function, you can often stretch your solution to meet less contrived boundary conditions. This brings us to the problem we started with, the temperature between two concentric circles.

EXAMPLE   Steady-State Temperature Between Two Concentric Circles

Problem:
Find the steady-state temperature in the region between the circles $x^2 + y^2 = R^2$ and $x^2 + y^2 = 4R^2$ if the temperature on the inner circle is held constant at $T = T_0$ and the temperature on the outer circle is held constant at $T = 2T_0$. (Assume the region is flat and insulated on top and bottom so we can ignore the third dimension.)

13.3 | Derivatives, Analytic Functions, and Laplace's Equation

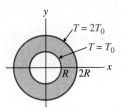

Solution:
The temperature distribution must obey two rules: it must satisfy Laplace's equation inside our region, and it must meet the given conditions on the boundary of our region.

We said above that $\ln z$ is an analytic function (except along its branch cut) and we said on Page 714 that $\ln z = \ln |z| + i\phi$. That means its real part $u = \ln\left(\sqrt{x^2 + y^2}\right)$ and its imaginary part $v = \tan^{-1}(y/x)$ must both be harmonic functions. The first of these only depends on distance from the origin, so it is constant along the inner circle and along the outer circle.

That's a good start, but to match the boundary conditions we need some arbitrary constants. Recall that the composition of analytic functions is still analytic, so if we multiply by a constant, then take the log, and then multiply by another constant, it's still analytic: $f(z) = A \ln(Bz)$. If we take A and B to be real then the real part of f is $A \ln(B|z|) = A \ln\left(B\sqrt{x^2 + y^2}\right)$. Now we have a harmonic function with two arbitrary constants that is constant along circles centered on the origin, and we can use the boundary conditions to find those constants: $A \ln(BR) = T_0$, $A \ln(2BR) = 2T_0$. A little algebra gives $A = T_0/\ln 2$, $B = 2/R$, so

$$T(x, y) = \frac{T_0}{\ln 2} \ln\left(\frac{2}{R}\sqrt{x^2 + y^2}\right)$$

(The argument of the log is unitless and the coefficient in front has units of temperature, so the units work.)

You may well have two objections to this process. First: how do we know that the function we found is the right one? The answer lies in a uniqueness theorem. Given a closed bounded region with a function V specified everywhere on the boundary, there is only one function V that satisfies Laplace's equation inside the region and meets the boundary condition around the region. (You will prove this uniqueness theorem in Section 13.11 Problem 13.158, see felderbooks.com.) So once we find a working solution by any means we know it is the right solution.

Second: how did we happen to think of using a log, and how are you supposed to think of the right functions? You have to be familiar with your basic analytic functions (there aren't many of them) and their properties. There are three shapes in particular to keep your eye on.

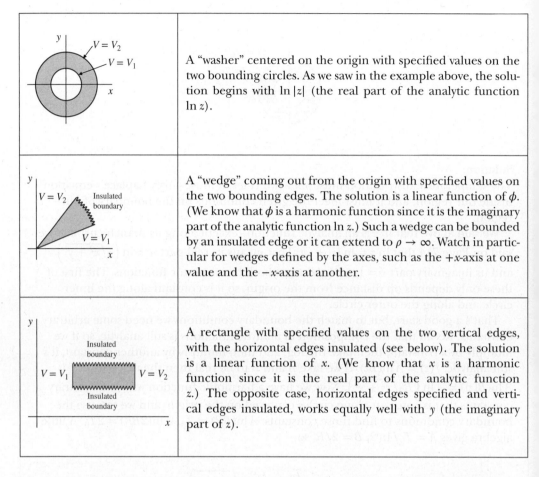

	A "washer" centered on the origin with specified values on the two bounding circles. As we saw in the example above, the solution begins with $\ln|z|$ (the real part of the analytic function $\ln z$).
	A "wedge" coming out from the origin with specified values on the two bounding edges. The solution is a linear function of ϕ. (We know that ϕ is a harmonic function since it is the imaginary part of the analytic function $\ln z$.) Such a wedge can be bounded by an insulated edge or it can extend to $\rho \to \infty$. Watch in particular for wedges defined by the axes, such as the $+x$-axis at one value and the $-x$-axis at another.
	A rectangle with specified values on the two vertical edges, with the horizontal edges insulated (see below). The solution is a linear function of x. (We know that x is a harmonic function since it is the real part of the analytic function z.) The opposite case, horizontal edges specified and vertical edges insulated, works equally well with y (the imaginary part of z).

A common boundary condition is that one or more edges are "insulated." The solution function must have a derivative of zero perpendicular to such a boundary. In a wedge for instance your solution will be a function of ϕ, and any function of ϕ has a zero derivative perpendicular to a circle around the origin.

The above list is not exhaustive, of course. In our first example above we needed to match a boundary curve shaped like a hyperbola, which lines up with the real or imaginary parts of z^2. But the three shapes in the above list are the most useful. In Section 13.9 we'll show you a technique for matching a somewhat broader set of boundary conditions than you can find by just guessing at analytic functions. But that technique is based on mapping more complicated regions to familiar ones, so it is still useful to have a bank of solved regions to draw on.

Stepping Back

Analytic functions have a number of not-obviously-related properties. If a function $f(z)$ is analytic…

- The derivative $f'(z)$ is defined, which is to say, it is the same in all directions.
- The Taylor series for $f(z)$ converges to $f(z)$.
- The real part $u(x, y)$ and the imaginary part $v(x, y)$ obey the Cauchy-Riemann conditions.
- The real part $u(x, y)$ and the imaginary part $v(x, y)$ are both, separately, solutions to Laplace's equation.

If any of the first three criteria are met then f is analytic and you're guaranteed that all four are true.

The idea of an analytic function is useful in many contexts, but in this section we have seen one of the most important: you can easily determine that many basic functions are analytic, and based on that find solutions to Laplace's equation.

13.3.4 Problems: Derivatives, Analytic Functions, and Laplace's Equation

13.17 The Explanation (Section 13.3.2) used the definition of the derivative to prove that the derivative of z^2 is $2z$—which, in turn, proves that the function z^2 is entire. Repeat this procedure to find $f'(z_0)$ where $f(z) = z^n$ for any positive integer n. (One approach is to use the binomial expansion $(a+b)^n = a^n + na^{n-1}b + \ldots$)

13.18 The Explanation (Section 13.3.2) used the definition of the derivative to prove that the derivative of z^2 is $2z$. Attempt to use the same procedure to find the derivative of z^*. What goes wrong?

13.19 Use the definition of the derivative to show that the derivative of $af(z) + bg(z)$ (where a and b are constants) is $af'(z) + bg'(z)$, and thus that any linear combination of analytic functions is analytic.

13.20 In this problem you'll analyze the function $f(z) = |z|$.

(a) Draw a complex plane and draw on it the level curves $f = 1$, $f = 2$, and $f = 3$.

(b) Mark the point $z = 1$, $f = 1$ on your plot. Draw a small arrow starting at this point and moving in the positive real direction. What is the derivative of f in that direction? No calculation should be required; just think about how much f changes per unit change in z as you move in this direction.

(c) Now add a small arrow starting at $z = 1$ and moving in the positive imaginary direction. How fast does f change per unit change in z as you move in this direction? Once again, the answer should be clear from looking at your contour plot.

(d) Is f analytic? Explain in terms of your answers above.

In Problems 13.21–13.24 you will determine if the given function is analytic.

- Calculate $f'(z)$ in the real and imaginary directions. (You may find the example on Page 718 helpful for this step.) If you get different answers, f is not analytic.
- If you get the same answer in both directions, try to calculate $f'(z)$ directly from the definition of the derivative.
- After you've finished, verify your answer by checking if the function satisfies the Cauchy-Riemann equations.

13.21 $f(z) = z$

13.22 $f(z) = Re(z)$

13.23 $f(z) = az$, where a is an arbitrary constant

13.24 $f(z) = 1/z$ (at points other than $z = 0$)

13.25 Use the Cauchy-Riemann equations to show that the function e^z is analytic for any complex z.

13.26 **Walk-Through: Laplace's Equation.** The electric potential is held fixed at $V = 0$ on the positive x-axis and $V = V_0$ on the line $y = x$, $x > 0$. (The discontinuity at the origin won't affect the problem.) Find the electric potential in the region between these two boundaries, assuming it obeys Laplace's equation in that region.

(a) Begin by writing the function that is the imaginary part of $\ln z$.

(b) How do you know (without doing any calculations) that your answer to Part (a) satisfies Laplace's equation?

(c) How do you know that your answer to Part (a) is a good candidate function for this problem, but is not quite right?

(d) Write the function that is the imaginary part of $A \ln z + Bi$ where A and B are real constants.

(e) How do you know (without doing any work) that your answer to Part (d) still satisfies Laplace's equation?

(f) Find the constants A and B to make your answer to Part (d) match the boundary conditions of this problem.

(g) The problem as stated takes place in an infinitely long region. Now suppose the region is terminated by the curve $\rho = R$, and that this curve is insulated—meaning that $\partial V/\partial \rho = 0$ along this border. Does the same solution still work? If it does not, adjust it properly for this changed scenario. If it does, explain why.

In Problems 13.27–13.29 find the solution $V(x, y)$ to Laplace's equation in the given region subject to the given boundary conditions.

13.27 $V(x, y) = 0$ on the circle $x^2 + y^2 = R^2$ and $V(x, y) = V_0$ on the circle $x^2 + y^2 = 9R^2$. Solve for V in between them.

13.28 $V = V_U$ along the y-axis for $y > 0$ and $V = V_L$ along the y-axis for $y < 0$. Solve for V in the region $x \geq 0$.

13.29 A rectangle. The bottom at $y = 3$ is at V_1 and the top at $y = 5$ is at V_2. The left and right sides are insulated, so $\partial V/\partial x = 0$ on the edges (and in fact everywhere). You can answer this even though we didn't tell you where the right and left edges are.

13.30 The circle $x^2 + y^2 = R^2$ is held at constant potential $V(x, y) = V_0$. The potential in the interior of the circle obeys Laplace's equation.

(a) We have used logs to solve similar problems. Explain why an answer that involves $\ln \sqrt{x^2 + y^2}$ cannot possibly work for this problem.

(b) Just by examining the problem find another, simpler solution that will work.

13.31 Show that if a function $f(z)$ is analytic, where $z = x + iy$ and $f = u + iv$, then u and v must obey Laplace's equation (Equation 13.3.2). *Hint*: any analytic function must obey the Cauchy-Riemann equations.

13.32 In the Explanation (Section 13.3.3) we solved $\nabla^2 T = 0$ in between the circles $x^2 + y^2 = R^2$ with $T = T_0$ and $x^2 + y^2 = 4R^2$ with $T = 2T_0$. It's also possible to solve that problem directly without complex functions.

(a) This problem is easiest in polar coordinates. Based on symmetry, what should be true about the derivatives of T with respect to ϕ?

(b) In polar coordinates $\nabla^2 T = \partial^2 T/\partial \rho^2 + (1/\rho)\partial T/\partial \rho + (1/\rho^2)\partial^2 T/\partial \phi^2$. Using your answer to Part (a), write Laplace's equation in polar coordinates as an ODE for $T(\rho)$.

(c) Find the general solution to that ODE.

(d) Plug in the boundary conditions to find the specific solution that we found in the explanation.

13.33 The function $f(z) = az + b$ with a and b real is analytic. (It would still be analytic if a and b weren't real, but choosing them to be real will be more useful for our purposes.)

(a) Since $f(z)$ is analytic, $u(x, y) = Re(f)$ is harmonic. On what curves is u constant?

(b) If we used u as the solution to an electric potential problem, what would the boundary conditions for that problem be?

(c) Specify some particular boundary conditions for the electric potential of the type you described in Part (b). Define curves and give the value of the potential on those curves. You may use numbers or letters. Then use $u(x, y)$ to solve Laplace's equation with the boundary conditions you specified.

13.34 On Page 722 we used the real part of the analytic function $f(z) = \ln z$ to solve Laplace's equation in a region between two concentric circles. Using the same technique, write a problem about steady-state temperature that can be solved using the imaginary part of z^2. Solve the problem you wrote.

13.35 If the number z has a modulus of $|z|$ and a phase of ϕ then the function $\sqrt{z}$ has a modulus of $\sqrt{|z|}$ and a phase of $\phi/2$.

(a) Find the real and imaginary parts of $\sqrt{z}$.

(b) What shape is described by the level curves of the real part of this function? (This tells us what kind of region might lead us to use this function in a Laplace's equation problem.)

13.4 Contour Integration

Integrating real functions is hard. Most functions can't be integrated and those that can require techniques ranging from simple u-substitutions to combinations of partial fractions and trig substitutions. It turns out that in some very important cases, definite integrals along paths in the complex plane (or "contours") can be easy to evaluate.

Our organization of this topic is one that we generally avoid, but it seemed cleanest in this case: we present the pure math here, and a few practical applications in Section 13.5.

13.4.1 Explanation: Contour Integration

The integral of a complex function is in some respects like a line integral. As with a line integral, you begin with a function to integrate and a path to integrate along. As with a line integral, the path matters: as you'll show in Problem 13.50, the integral of $f(z) = Im(z)$ from $z = 0$ to $z = 1 + i$ can take different values depending on the path you take between them. A specific path through the complex plane is called a "contour,"[3] and the integral along such a path is a "contour integral."

But there are important differences too. When you evaluate the line integral of a scalar function you multiply the value of the function at each point by ds, the arclength. When you evaluate the line integral of a vector function you *dot* the function with $\vec{ds}$, or multiply the along-the-path component of the function by ds. In either case the end result is a scalar value. By contrast, a contour integral multiplies the value of the function by dz, just as we saw with derivatives, and the final result is a complex number.

To take a concrete example, the line integral of $f(x, y) = 1$ along any curve adds up all the differential arclengths, so it gives you the total arclength of the curve. The integral of $f(z) = 1$ along any contour adds up all the differential changes in z, so it gives you $z_{end} - z_{start}$.

The integral of $f(z)$ can be defined by a Riemann sum. We break the contour into N segments at points $z_0, z_1, \ldots, z_N$, calculate $\sum_{k=0}^{N-1} f(z_k)(z_{k+1} - z_k)$, and then take the limit as the points get arbitrarily close together.

Now, suppose $f(z)$ is entire. This implies that its antiderivative $F(z)$ is also entire.[4] When we evaluate $\int f(z)dz$, each step along the contour multiplies $f(z) = dF/dz$ times dz, giving us the small change dF. Adding up those changes along the full contour gives us $F(z_{end}) - F(z_{start})$. So the fundamental theorem applies to analytic complex functions much as it does to real functions.

You might be surprised to learn that we're not going to use that result very much. "Suppose $f(z)$ is entire"—no branch cuts, no poles, everywhere differentiable—turns out to be a relatively uninteresting special case. There is a more general formula for integrating a complex function along a parameterized contour, but we're not going to talk about that either. Instead, we're going to focus on two special cases.

1. **Integrals along a closed contour.** A contour is "closed" if it ends where it began. Most of our applications will apply to integrals around closed contours, because—as we will see below—they are especially easy to evaluate. We define the positive direction around a closed curve to be the one where the interior of the curve is always on your left as you move around. (For simple closed curves this is equivalent to saying you go around counterclockwise.)
2. **Integrals along the real axis.** When you evaluate a contour integral of a function $f(z)$ along the real axis from $z = a$ to $z = b$ and $f(z)$ happens to be real along that part of the axis, the contour integral reduces to the real integral we are used to taking. In the next section we'll use that fact to solve real integrals by evaluating complex integrals.

---

[3]More precisely, a contour is a directed curve that can be defined by a parametrization $z(t)$ that is piecewise continuous and differentiable and that is one-to-one with the possible exception of ending at the same point it started. For most purposes you can just think of it as a curve with a specified direction that you move along.
[4]That leap isn't obvious; it's another wonderful property of analytic functions.

Analytic Functions around Closed Loops

We're going to present (without proof) some theorems that, in the right circumstances, make closed contour integrals of analytic functions easy to calculate. Then we'll show you how to apply these formulas to a variety of real problems.

We begin with the easiest case of all.

If $f(z)$ is analytic everywhere on and inside closed contour C, then $\oint_C f(z)\,dz = 0$.

Want to know the integral of e^{z^2} around a unit circle centered on $z = 2$? It's zero! This fact follows trivially from the fundamental theorem that we presented above: $F(z_{end}) - F(z_{start}) = 0$ for a closed loop because z_{end} is z_{start}.

But that result requires $f(z)$ to be analytic, not only along the contour, but everywhere inside it. To see why, consider the function $f(z) = 1/z$. This function is analytic everywhere except the origin, but its antiderivative $\ln z$ has a branch cut. Any closed contour that surrounds the origin must pass through that branch cut. Take a moment to re-read our demonstration of the fundamental theorem earlier and convince yourself that it is not valid across such a discontinuity.

Practically everything we want to know about closed contour integrals can be seen as extensions of this one case, an integral of $1/z$ along a closed contour that surrounds the origin. In Problem 13.53 you will calculate $\oint (1/z)\,dz$ for such a loop, but here we simply present the result, generalized to a slightly broader class of functions.

Cauchy's Integral Formula

Let C be a closed contour that encloses the point $z = z_0$. Let $g(z)$ be a function that is analytic everywhere on and inside that contour. And let $f(z) = g(z)/(z - z_0)$ where z_0 is a point inside C. Because the quotient of analytic functions is analytic where defined, $f(z)$ is analytic everywhere inside C except at the pole $z = z_0$. "Cauchy's integral formula" tells us the integral of $f(z)$ around C.

$$\oint_C f(z)\,dz = \oint_C \frac{g(z)}{z - z_0}\,dz = 2\pi i\, g(z_0) \qquad (13.4.1)$$

Be careful when applying Equation 13.4.1: the bottom must be $z - z_0$ for some constant z_0. If the bottom were $2z - 7 + i$, for instance, you would have to divide the top and bottom by 2 before applying this formula.

It's easy to see that the integral of $1/z$ around any loop that encloses the origin is a special case of Cauchy's integral formula, where $g(z) = 1$ and $z_0 = 0$. What isn't quite as obvious is that any such integral can be viewed as a variation of that case, so when you prove the integral for $1/z$ (Problem 13.53) you have done most of the work to prove Cauchy's formula.

Slightly more complicated formulas apply when there are higher powers in the denominator. Given the same conditions stated above,

$$\oint_C \frac{g(z)}{(z - z_0)^2}\,dz = 2\pi i\, g'(z_0) \text{ and } \oint_C \frac{g(z)}{(z - z_0)^3}\,dz = \frac{2\pi i}{2!} g''(z_0) \text{ and } \oint_C \frac{g(z)}{(z - z_0)^4}\,dz = \frac{2\pi i}{3!} g'''(z_0)$$

and so on. These are examples of the "generalized Cauchy's integral formula":

$$\oint_C \frac{g(z)}{(z - z_0)^n}\,dz = \frac{2\pi i}{(n-1)!}\left(\frac{d^{n-1}g}{dz^{n-1}}(z_0)\right)$$

Read the last part of this formula as "the $(n-1)^{\text{th}}$ derivative of $g(z)$ evaluated at $z = z_0$."

Before showing you examples of how to use these formulas, we should note that they are often presented as formulas for the value of some analytic function g at an arbitrary point inside the contour. For instance, Equation 13.4.1 is often written $g(z_0) = 1/(2\pi i) \oint g(z)/(z - z_0)\, dz$. While this is clearly the same formula, it is written to say that the values of $g(z)$ inside a region can be determined from its values on the boundary. That's a fascinating fact about analytic functions, but we will be using Cauchy's formulas to calculate contour integrals, not to find values of $g(z)$ at interior points.

EXAMPLE Contour Integrals

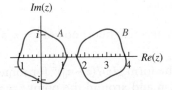

Two contours in the complex plane

Question: Evaluate $\oint e^{z^2}\, dz$ around each of the two contours shown above.

Answer:
The function e^{z^2} is analytic everywhere (can you see how we know that?) so the integral around either contour is zero.

Question: Evaluate $\oint (\cos z)/(z - \pi)\, dz$ around each of the two contours shown above.

Answer:
The function $\cos z$ is analytic everywhere, so we can use Cauchy's integral formula. Since contour A doesn't enclose the point $z = \pi$ the integral around it is zero. Since contour B does enclose $z = \pi$, the integral around it is $2\pi i \cos \pi = -2\pi i$.

Question: Evaluate $\oint 1/[(2z - 1)(z - 2)]\, dz$ around each of the two contours shown above.

Answer:
The integrand has two poles, at $z = 1/2$ and $z = 2$. The integral around contour B encloses the second pole. For that contour $g(z) = 1/(2z - 1)$, which is analytic everywhere on and inside B. Since $g(2) = 1/3$, the integral around B is $(2/3)\pi i$. The integral around contour A encloses the pole at $z = 1/2$, but we can't use Cauchy's integral formula until we write the integrand in the form $g(z)/(z - z_0)$, which means we have to factor a 2 out of $2z - 1$:

$$f(z) = \frac{1}{2(z - 2)} \frac{1}{z - 1/2} \quad \rightarrow \quad g(z) = \frac{1}{2(z - 2)} \quad \rightarrow \quad g(1/2) = -1/3$$

So the integral around A is $-(2/3)\pi i$

Question: Evaluate $\oint e^{2z}/(z - 3 - (i\pi/4))^3\, dz$ around each of the two contours shown above.

Answer:
Since e^{2z} is analytic everywhere we can use the generalized Cauchy's integral formula with $n = 3$. Once again, the integral around contour A is zero since it doesn't

surround the point $z = 3 + i\pi/4$. Looking at the plot, the point $3 + i\pi/4 \approx 3 + 0.8i$ is inside contour B. To find the integral around B we take the second derivative of e^{2z}, which is $4e^{2z}$, and evaluate it at $z = 3 + i\pi/4$ to get $4e^6 e^{i\pi/2} = 4ie^6$. Plugging this into the generalized Cauchy's integral formula we get $(2\pi i)/2!(4ie^6) = -4\pi e^6$.

Some Terminology: Poles and Residues

We discussed the idea of a "pole" in Section 13.2. Here we define that word more carefully, along with some related terminology.

- If a function $f(z)$ is analytic in a neighborhood of $z = z_0$ and $\lim_{z \to z_0} |f| = \infty$, then the point $z = z_0$ is a pole of $f(z)$.
- If $f(z)$ can be written in the form $g(z)/(z - z_0)^n$ for some positive integer n, where $g(z)$ is analytic and non-zero on and around the point $z = z_0$, then $z = z_0$ is an "nth-order pole." A first-order pole is also called a "simple pole."
- If $f(z)$ has a simple pole at $z = z_0$ the "residue" of $f(z)$ at that pole is simply $g(z_0)$, or equivalently $\lim_{z \to z_0}(z - z_0)f(z)$.
- If $f(z)$ has an nth-order pole at $z = z_0$ the "residue" of $f(z)$ at that pole is $g^{(n-1)}(z_0)/(n-1)!$, where $g^{(n-1)}(z_0)$ means the $(n-1)^{\text{st}}$ derivative of g evaluated at $z = z_0$.

With that terminology we can state Cauchy's integral formula more concisely than we did above. The formulation below is also more general, because it can handle a contour that encloses multiple poles.

The Residue Theorem

If $f(z)$ is analytic in some domain except at a finite number of points, then a closed contour integral of f in that domain equals $2\pi i$ times the sum of the residues of f inside that contour.

(We're assuming the contour doesn't touch any of the poles. Integrals through poles are discussed in Section 13.6 (see felderbooks.com).)

See if you can convince yourself by means of a few drawings that the one-pole result implies the many-pole result.

We've defined the terms "pole" and "residue" here using functions of the form $g(z)/(z - z_0)^n$ where g is analytic and non-zero. When we discuss Laurent series in Section 13.7 we'll give more general definitions and techniques for finding poles and residues for functions that aren't of this form, but these definitions are useful in most practical cases.

Some Special Contour Integrals That Equal Zero

We said we would focus on two important types of contour integrals: closed integrals and integrals on the real axis. We've discussed closed integrals here, and in Section 13.5 we'll apply these ideas to calculate real integrals, but we need to mention a few other integrals that will be useful in Section 13.5.

Two Very Important Facts

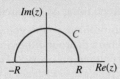

Let C be the open semicircular contour shown above.

Fact 1: If $p_1(z)$ and $p_2(z)$ are polynomials and p_2 is at least two orders higher than p_1, then:

$$\lim_{R \to \infty} \int_C \frac{p_1(z)}{p_2(z)} dz = 0 \qquad (13.4.2)$$

This result also holds if C is any arc or a complete circle about the origin (still with $R \to \infty$), but the semicircular case is the one we'll make use of.

Fact 2: If $p_1(z)$ and $p_2(z)$ are polynomials and p_2 is at least *one* order higher than p_1, then:

$$\lim_{R \to \infty} \int_C \frac{p_1(z)}{p_2(z)} e^{iz} dz = 0 \qquad (13.4.3)$$

This result only holds for the semicircle in the upper half-plane (or any portion of it) because e^{iz} decays in the positive imaginary direction, whereas it oscillates in the real direction and grows in the negative imaginary direction. More generally, e^{iz}, e^{-iz}, e^{z}, or e^{-z} each decays in one direction. For example, if we use e^z the semicircle has to go in the left half-plane.

We can justify Equation 13.4.2 in a rough way by noting that the modulus of the integrand falls like $1/R^2$ (or faster) and the length of the contour is πR, so as $R \to \infty$ the integral must approach zero. In Problem 13.55 you'll fill in some of the steps in that argument. In Problem 13.56 you'll extend the argument to Equation 13.4.3.[5]

13.4.2 Problems: Contour Integration

13.36 Walk-Through: Contour Integral. In this problem you will evaluate $\oint_C f(z)\, dz$ where $f(z) = (2z+i)/(z^2+1)$ and C is the circle of radius 1 centered on the point $z = i$.

(a) Draw the curve C.

(b) Explain how you can know that $f(z)$ is analytic except at its poles, which are the points where its denominator is zero.

(c) What are the two poles of $f(z)$?

(d) Which of these poles is enclosed by the contour C? Your answer could be neither, both, or specifying one of them.

(e) To evaluate the integral you first need to calculate the residue at each pole enclosed by the contour. To find the residue of $f(z)$ at a simple pole $z = z_0$ you rewrite $f(z)$ in the form $g(z)/(z - z_0)$, and $g(z_0)$ is the residue. Find the residue for each pole enclosed by C. (Each residue will require a different function $g(z)$.)

(f) The contour integral of $f(z)$ around C is $2\pi i$ times the sum of the residues of the poles enclosed by C. What is $\oint f(z)\, dz$?

13.37 In this problem you will evaluate $\oint_C f(z)\, dz$ where $f(z) = e^z/(z^3 + 2z^2 + z)$ and C is the circle of radius 2 centered on the point $z = -2 + i$.

(a) Draw the curve C.

(b) If f can be written in the form $g(z)/(z - z_0)^n$ where g is analytic and n is a positive integer, then f has an nth-order pole at $z = z_0$. What are the poles of f, and what order is each of them? *Hint:* Start by factoring the denominator of f as much as possible.

---

[5] For a more complete derivation of Equations 13.4.2–13.4.3 see any standard book on complex analysis, e.g. "Visual Complex Analysis," Tristan Needham, Clarendon Press, 1997.

(c) Which poles are enclosed by the contour C?

(d) Find the residue for each pole enclosed by C.

(e) What is $\oint_C f(z)\, dz$?

(f) Curve C_1 is the circle with the same center as C, but with radius 1. Evaluate $\oint_{C_1} f(z)\, dz$.

(g) Curve C_3 is the circle with the same center as C, but with radius 3. Evaluate $\oint_{C_3} f(z)\, dz$.

For Problems 13.38–13.49 evaluate $\oint_C f(z)\, dz$.

13.38 $f(z) = 1/(z - i - 3)$, C is the circle of radius 10 centered on the origin.

13.39 $f(z) = 1/(2z - 3 - i)$, C is the square with corners at $z = -1 - i$ and $z = 1 + i$.

13.40 $f(z) = 1/[z(2z - i)(z + 2)]$, C is the square with corners at $z = -1 - i$ and $z = 1 + i$.

13.41 $f(z) = (z + 1)/(z^2 - 1)$, C is the circle of radius 3 centered on the origin.

13.42 $f(z) = (\cos z)/z^2$, C is the square with corners at $z = -1 - i$ and $z = 1 + i$.

13.43 $f(z) = e^{z^3}$, C is the ellipse $x^2 + 2y^2 = 1$ where $x = Re(z)$, $y = Im(z)$.

13.44 $f(z) = e^z \sum_{n=-3}^{3} \dfrac{1}{z - in}$, C is the circle of radius 1 centered on the point $z = (3/2)i$.

13.45 $f(z) = (\ln z)/(z^2 - 3zi)$, C is the rectangle with corners at $z = -1 + i$ and $z = 1 + 10i$.

13.46 $f(z) = (\cos z)/(2z^2 - z)$, C is the circle of radius 10 centered on the origin.

13.47 $f(z) = 1/(2z^2 + z(1 - 2i) - i)$, C is the circle of radius 10 centered on the origin.

13.48 $f(z)$ is a branch of $\ln z$ defined with the branch cut on the *positive* real axis. C is the circle of radius 1 centered on the point $z = -3$.

13.49 $f(z) = (\sin z)/(z^2 + 2iz - 1)^3$, C is the circle of radius 2 centered on $z = -i$.

13.50 In this problem you will show that $\int Im(z)\, dz$ from $z = 0$ to $z = 1 + i$ is path-dependent.

(a) First, evaluate the integral along the path $0 \to 1 \to 1 + i$. Draw this path in the complex plane, integrate $Im(z)$ along each of the two line segments that make it up, and add the results.

(b) Similarly, draw the path $0 \to i \to 1 + i$ and evaluate $\int Im(z)\, dz$ along it. You should get a different answer than you got in Part (a).

13.51 Argue, using the definition of a contour integral, that the contour integral of a constant function around any closed loop must be zero. (Don't say "because it's analytic"—yes, we told you this is true of any analytic function, but don't assume that going in. Instead, base your answer on the fact that the contour integral is defined using Δz, not arclength.)

13.52 In this problem you're going to evaluate $\oint \tan z\, dz$ around the circle of radius $\pi/2$ centered on $z = \pi/2$. The function has a pole at $z = \pi/2$, but is it a simple pole and what is its residue? Try answering those questions yourself and then look at the trick we describe below.

(a) One trick for finding simple poles and their residues is to evaluate $\lim_{z \to z_0}(z - z_0)f(z)$. If $f(z_0)$ is undefined, but this limit is finite and non-zero, then f has a simple pole at $z = z_0$ and the limit equals its residue. Use the trick introduced in this problem to show that $f(z) = \tan z$ has a simple pole at $z = \pi/2$ and find its residue.

(b) Evaluate $\oint \tan z\, dz$ around the circle described above.

13.53 In this problem you will prove Cauchy's integral formula for the simple case of a contour around the unit circle and the function $f(z) = 1/z$. (It's not hard to generalize this proof to circles of other radii or centers other than the origin.)

(a) The unit circle can be defined as all the points $z = e^{i\phi}$ for $0 \leq \phi < 2\pi$. Write a Riemann sum for $\oint f(z)\, dz$ using steps of size $\Delta \phi$. Your answer should be a sum involving nothing but numbers and the variables ϕ and $\Delta \phi$. Simplify your answer as much as possible.

(b) You should have found that the "summand" (the quantity you are summing) didn't depend on ϕ. Given that, you can evaluate the sum by simply multiplying the summand by the number of steps, $2\pi/\Delta\phi$. Evaluate $\oint f(z)\, dz$ as the limit as $\Delta\phi \to 0$ of the resulting quantity.

13.54 In this problem you will evaluate $\int (1/z)\, dz$ along a semicircle of radius R around the origin.

(a) Argue from symmetry that $\int(1/z)\, dz$ around a semicircle in the upper half-plane equals $\int(1/z)\, dz$ around a semicircle in the lower half-plane. *Hint*: consider the modulus and phase of z and dz at different points around the circle.

(b) What is $\oint(1/z)\, dz$ around the entire circle of radius R around the origin?

(c) Using your answers to Parts (a)–(b), what is $\int (1/z)dz$ around a semicircle of radius R centered on the origin in the upper half-plane?

13.55 In this problem you will derive Equation 13.4.2 for the special case where $p_1(z) = z^m$ and $p_2(z) = z^n$. Recall that C is a semicircle of radius R around the origin.

(a) Along the contour C, what is $|p_1/p_2|$? Your answer should be in terms of R.

(b) What is L, the length of the contour?

(c) For any function $f(z)$, if $|f(z)| \leq M$ everywhere along contour C then $|\int_C f(z)\, dz| \leq ML$, where L is the length of C. (You should be able to convince yourself of this by thinking of the integral as a Riemann sum.) Use this result and what you've derived in this problem to prove that $\lim_{R\to\infty} \int_C (p_1/p_2)\, dz = 0$.

13.56 Exploration: Deriving Equation 13.4.3.
Despite the name of this problem, you will only derive Equation 13.4.3 for the special case $p_1/p_2 = 1/z$, so you'll evaluate $I = \lim_{R\to\infty} \int (e^{iz}/z)dz$. It's not hard to generalize this argument to any rational expression p_1/p_2 where the order of the denominator is at least one higher than the order of the numerator.

(a) For points on the contour C, $z = Re^{i\phi}$ where ϕ, the phase of z, goes from 0 to π. Rewrite I as an integral over ϕ. (This will include finding dz in terms of $d\phi$.)

(b) Using Euler's formula to rewrite $e^{i\phi}$ in terms of sines and cosines, simplify your answer to Part (a) as much as possible.

(c) To find an upper bound on $|I|$, take the modulus of the integrand. We'll call the resulting integral J (just because it comes after I).

You've now identified a real integral J whose value is an upper bound on $|I|$, so the goal from here is to show that $\lim_{R\to\infty} J = 0$. Unfortunately, J still isn't an integral you can evaluate, so now you're going to find an upper bound for J, and then prove that *that* integral approaches zero. To make life easier you'll rewrite J as 2 times the integral from 0 to $\pi/2$. (You should be able to convince yourself that this is valid if you have the correct form for J.)

(d) Your formula for J should involve $\sin\phi$. Draw a plot of $y = \sin\phi$ from $\phi = 0$ to $\phi = \pi/2$ and on the same plot draw a line $y(\phi)$ from $(0,0)$ to $(\pi/2, 1)$. You can see that the line is always below the sine curve. Find the equation of that line. Argue that replacing $\sin\phi$ with that linear function of ϕ in your integral increases the value of the integral. Let's call this new integral K.

(e) Evaluate the integral K.

(f) Show that $\lim_{R\to\infty} K = 0$. Since $K > J > |I|$, this proves that $\lim_{R\to\infty} I = 0$. (If the modulus of a complex number approaches 0 then the number itself must approach 0.)

13.5 Some Uses of Contour Integration

Section 13.4 discussed how to take integrals around loops in the complex plane. In this section we will present two practical uses for that skill.

13.5.1 Explanation: Some Uses of Contour Integration

As we stressed in Chapter 3, complex numbers are a shortcut for getting from a real question to a real answer. In this section we will see two such uses for contour integrals. First we will see some integrals of real functions, on the real number line, that can most easily be evaluated on the complex plane. Then we will look at the Laplace transform—a real transformation that turns one real function into another real function, but that can only be inverted by a contour integral.

Preface: Improper Integrals and the Cauchy Principal Value
This "preface" has nothing to do with imaginary numbers.

If your limits of integration involve $\pm\infty$, or if your region of integration includes a vertical asymptote, then you have an "improper integral." Such an integral is defined as a limit, and is said to "converge" (the limit exists) or "diverge" (the limit does not exist). The following examples both converge.

$$\int_1^\infty \frac{1}{x^2}\,dx = \lim_{a\to\infty}\int_1^a \frac{1}{x^2}\,dx = \lim_{a\to\infty}\left.-\frac{1}{x}\right|_1^a = \lim_{a\to\infty}\left(-\frac{1}{a}+1\right) = 1$$

$$\int_0^1 \ln x\,dx = \lim_{a\to 0^+}\int_a^1 \ln x\,dx = \lim_{a\to 0^+}\left. x\ln x - x\right|_a^1 = \lim_{a\to 0^+}(-1 - a\ln a + a) = -1$$

That last step requires a bit of algebra followed by l'Hôpital's rule, but the important part here is the first step in which the improper integral is rewritten as a limit.

The integral in Figure 13.8 is less clear. You have to break up the limit at $x = 0$ (the vertical asymptote) and evaluate the two sides separately, each with a limit. But do you evaluate the limit before you add the two pieces, or after?

$$\int_{-1}^1 \frac{1}{x^3}\,dx = \lim_{a\to 0^-}\int_{-1}^a \frac{1}{x^3}\,dx + \lim_{b\to 0^+}\int_b^1 \frac{1}{x^3}\,dx \quad \begin{array}{l}\text{the standard definition}\\ \text{(limits before addition)}\end{array} \quad (13.5.1)$$

$$\int_{-1}^1 \frac{1}{x^3}\,dx = \lim_{c\to 0^+}\left(\int_{-1}^{-c} \frac{1}{x^3}\,dx + \int_c^1 \frac{1}{x^3}\,dx\right) \quad \begin{array}{l}\text{the Cauchy principal value}\\ \text{(addition before limits)}\end{array} \quad (13.5.2)$$

In Problem 13.57 you will show that this particular integral diverges under the first definition but converges under the second. In effect the Cauchy principal value allows the infinite negative value on the left to cancel the infinite positive value on the right. In Problem 13.58 you will show that $\int_{-\infty}^\infty x\,dx$ leads to the same two answers.

We don't want to overstate this case. Sometimes both the standard and Cauchy definitions lead to convergence, and sometimes both lead to divergence. But we are going to demonstrate techniques for using contour integrals to evaluate real integrals, and what we're going to find—especially once we start integrating around poles in Section 13.6 (see felderbooks.com)—will be the Cauchy principal value. If you have learned the standard definition, or if you ask your favorite software for these integrals, you may find that our answers don't match. Now you know why.

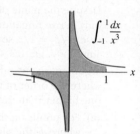

FIGURE 13.8 The integral $\int_{-1}^1 dx/x^3$ is divergent by the standard definition, but its Cauchy principal value is not.

Using Contour Integrals to Evaluate Real Integrals

The following problem has nothing to do with imaginary numbers.

$$\int_{-\infty}^\infty \frac{1}{x^2+1}\,dx$$

You may know how to evaluate that using trig substitution. You may also know off the top of your head that the antiderivative is $\tan^{-1}x$, so by plugging in the limits of integration you'll get the answer $\pi/2 - (-\pi/2) = \pi$. But set all that aside for the moment, because we're going to use contour integrals to get the same answer you would have gotten with those methods. This same technique can then be applied to many integrals that are difficult or impossible to evaluate in other ways.

We begin by replacing our real problem with a complex one. Remember that the contour integral of a real function, evaluated along the real axis, is the same as the real integral of that function. In other words the following contour integral problem is the same as the real integral problem we started with.

$$\int_C \frac{1}{z^2+1}\,dz \quad \text{where } C \text{ is the real axis traversed from } -\infty \text{ to } \infty$$

13.5 | Some Uses of Contour Integration

As we have seen, the easiest contour integrals to evaluate are around closed loops. So we draw a closed loop, a semicircular contour of radius R centered on the origin (Figure 13.9). Now, that loop is the sum of two parts: a curve (that we don't care about) and a line segment (that approaches our line C as $R \to \infty$). If we can determine the integral around the entire closed loop *and* the integral around the curved part, we can subtract them to get the integral we're looking for.

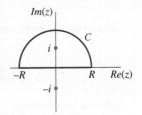

FIGURE 13.9 A semicircular contour. The gray circles at $\pm i$ are poles of $f(z)$.

The integral around the curved part sounds like the hardest step, but remember that R is approaching ∞. In that limit the curved part of the integral is 0 by Equation 13.4.2.

So now let's tackle the closed loop. The function f is analytic everywhere except at the two poles $z = i$ (enclosed) and $z = -i$ (not enclosed), so $\oint f(z)\,dz$ is $2\pi i$ times the residue at $z = i$. Rewriting f as $1/[(z - i)(z + i)]$ we see that the residue at $z = i$ is $1/(2i)$. So the integral around the closed loop, and thus the integral along the real axis, and thus the real integral we started with, is $2\pi i/(2i) = \pi$, just as we expected it to be.

EXAMPLE A Real Integral from $-\infty$ to ∞

Question: Evaluate $\int_{-\infty}^{\infty} (x^2 + 1)/(x^4 + 4)\,dx$.

Answer:
As we did above, we consider this real integral to be one leg of a semicircular contour integral $\oint f(z)\,dz$, where $f(z) = (z^2 + 1)/(z^4 + 4)$. Once again, in the limit where the radius of the semicircle approaches ∞, the modulus of the integrand along the curve part decreases like $1/R^2$ and the length grows like R, so the curved part of the contour integral vanishes by Equation 13.4.2 and the real integral equals the contour integral.

To find the poles of $f(z)$ we need to find the roots of $z^4 + 4 = 0$, or equivalently find the fourth roots of -4. They all have modulus $\sqrt{2}$. Their phases must be 1/4 the phase of -4, which can be π, 3π, 5π, or 7π, so the roots have phases $\pi/4$, $3\pi/4$, $5\pi/4$, and $7\pi/4$. (We could keep going to include, say, $9\pi/4$, but it would be the same angle as $\pi/4$. A fourth root has only four solutions.) The complex numbers with these phases and modulus $\sqrt{2}$ are $1 + i$, $-1 + i$, $-1 - i$, and $1 - i$. (These points are marked in the figure above.) That tells us how to factor our integrand.

$$f(z) = \frac{z^2 + 1}{(z - (1 + i))(z - (-1 + i))(z - (-1 - i))(z - (1 - i))}$$

The contour encloses the two simple poles at $z = \pm 1 + i$. The residue at $1 + i$ is

$$\frac{(1+i)^2 + 1}{((1+i) - (-1+i))((1+i) - (-1-i))((1+i) - (1-i))}$$

After a bit of algebra this can be simplified to $1/16 - (3/16)i$. The residue at $-1 + i$ can similarly be calculated to be $-1/16 - (3/16)i$, so the sum of the residues is $-(3/8)i$. The final answer is $2\pi i$ times the sum of the residues, or $(3/4)\pi$.

This technique of using an infinitely large semicircular contour to evaluate a real integral works when the integral goes from $-\infty$ to ∞ and the function $f(z)$ falls off quickly enough at large z that the curved part of the integral goes to zero.

As a slightly more complicated example, consider the integral $\int_{-\infty}^{\infty} (\cos x)/(x^2 + 1) \, dx$. We could try the same thing we did above, extending this into an infinitely large semicircular contour in the complex plane. The problem is that $\cos z$ blows up as z moves infinitely up the imaginary axis. To see why this is true, and how to get around the problem, we can use Euler's formula to rewrite $\cos z$ as $(1/2)(e^{iz} + e^{-iz})$. (If that formula isn't familiar, take a minute to expand those exponentials using Euler's formula and prove that this equals $\cos z$.) As $z \to i\infty$ the first exponential decays but the second one approaches ∞. That fact shows why we have a problem, but it also suggests a solution. We can rewrite our original integral as a sum of two integrals.

$$\int_{-\infty}^{\infty} \frac{\cos x}{x^2 + 1} \, dx = \frac{1}{2} \left[\int_{-\infty}^{\infty} \frac{e^{ix}}{x^2 + 1} \, dx + \int_{-\infty}^{\infty} \frac{e^{-ix}}{x^2 + 1} \, dx \right]$$

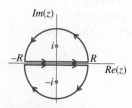

FIGURE 13.10 One contour is in the upper half-plane and the other in the lower half-plane. The gray circles show the poles at $\pm i$.

We can evaluate the first integral by closing the contour in the upper half of the complex plane, and evaluate the second one by closing it off in the lower half. Each of these contour integrals goes to zero along the curved part by Equation 13.4.3,[6] so the integral along the real axis equals the entire contour integral. If we rewrite the first integrand as $e^{iz}/[(z-i)(z+i)]$ we immediately see that its residue at $z = i$ is $e^{-1}/(2i)$, so the contour integral equals π/e.

The second integral proceeds similarly, but there's a subtlety we have to be careful about. Rewriting the integrand as $e^{-iz}/[(z-i)(z+i)]$ we see that the residue at $z = -i$ is $-e^{-1}/(2i)$. However, from Figure 13.10 we can see that this contour goes around *clockwise*. That means the contour integral equals $-2\pi i$ times the residue, so once again it gives us π/e. Putting these together the final answer is $(1/2)(\pi/e + \pi/e) = \pi/e$.

Inverse Laplace Transforms

If you have gone through Chapter 10 or 11 then you have glimpsed the importance of Laplace transforms. You can transform a differential equation for $f(t)$ into an algebra equation for its Laplace transform $F(s)$ and solve that easily. But then you need an *inverse* Laplace transform to recover the function $f(t)$ that you were originally looking for. Textbooks generally punt this step—in some cases to table lookups, and in our chapters to the computer. That's because an inverse Laplace transform requires a contour integral.

---

[6] well, technically by the paragraph below Equation 13.4.3

13.5 | Some Uses of Contour Integration

So now at last we are ready to see how that works.

The Inverse Laplace Transform

If $F(s)$ is the Laplace transform of $f(t)$ then

$$f(t) = \frac{1}{2\pi i} \int_{a-i\infty}^{a+i\infty} F(s) e^{st} \, ds \quad \text{"the Bromwich integral"} \quad (13.5.3)$$

The integral is evaluated along a vertical straight line in the complex plane. You can choose any value for the real number a so long as it pushes this vertical line to the right of all the poles of $F(s)$.

Problems involving Laplace transforms typically take $t = 0$ as the initial time, so only positive values of t matter. In Problem 13.82 you'll show that this formula always gives $f(t) = 0$ for $t < 0$.

Below we give an example of how to use this formula. In Section 13.5.2 we will show where the formula comes from.

EXAMPLE   **An Inverse Laplace Transform**

Question: Find the function $f(t)$ whose Laplace transform is $F(s) = 1/(s-2)$

Answer:
Clearly $F(s)$ has a single pole at $s = 2$, so for our contour we can choose any vertical line to the right of that. We'll see shortly that it doesn't matter which one. We need to evaluate

$$\frac{1}{2\pi i} \int \frac{e^{st}}{s-2} \, ds \quad (13.5.4)$$

To do so we'll use the same method we used for the real integrals above: closing the contour with an infinitely large semicircle. As always, the trick is making sure that the curved part of the integral equals zero. We are only interested in positive values of t, so e^{st} approaches 0 as we move to the left, and we therefore close the contour on that side and Equation 13.4.3 guarantees that the curved integral is 0.

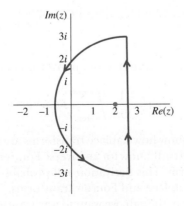

This closed contour encloses the pole, whose residue is e^{2t}, so the integral equals $2\pi i e^{2t}$. Plugging this into Equation 13.5.4 we get $f(t) = e^{2t}$. You can confirm this easily enough by taking its Laplace transform: $F(s) = \int_0^\infty e^{2t} e^{-st} dt$. Did you get what you expected?

Stepping Back

Mathematics is an infinitely creative field, with new techniques always waiting to be discovered. That's an exciting thought for a mathematician, but it can be intimidating when you have a test next week. "I followed your examples, but I never would have thought of those steps on my own. How am I going to do the next problem?"

It's not as scary as it looks. You might feel that our first example above would be a lot harder if the limits of integration were 0 to 10 instead of $-\infty$ to ∞, and you're right. In fact the whole trick wouldn't work in that case, so we aren't going to ask that question.

On the topic of "using contour integrals to evaluate real integrals," all the problems you're likely to encounter fall into three categories.

- For $\int_{-\infty}^{\infty} f(z)\, dz$ where $|f(z)|$ drops sufficiently quickly as $iz \to \pm\infty$ you draw a semicircle—upward if $|f(z)|$ drops in that direction, downward if it drops there. Our first example above fell into this category.
- Some functions can be broken into two functions, one that drops as you move up and one that drops as you move down. You break such a function up and draw two semicircles. Our cosine example above fell into this category.
- The third category, which we have not demonstrated, is rational functions of $\sin x$ or $\cos x$ integrated over a full period. You begin by rewriting your sines and cosines as complex exponentials, and the substitution $u = e^{ix}$ maps a segment on the real number line to a closed loop on the complex plane. We will step you through such a situation in Problem 13.71. It doesn't look like the examples we've worked here, but it is always the same u, always the same du...once you know how to do that problem, you have the final piece of the puzzle.

Inverse Laplace transforms fall into only one category. Equation 13.5.3 sets up an integral along a vertical line, with a chosen to place this line to the right of all poles of $F(s)$. You then close the contour to the left and get $f(t)$ from the residues of $e^{st} F(s)$ for all positive values of t.

13.5.2 Explanation: Deriving the Formula for an Inverse Laplace Transform

In Section 13.5.1 we presented the formula for an inverse Laplace transform:

$$f(t) = \frac{1}{2\pi i} \int_{a-i\infty}^{a+i\infty} F(s) e^{st} ds$$

To derive this formula we will show how Laplace transforms are related to Fourier transforms and then use the formula we already know for an inverse Fourier transform to get the formula for an inverse Laplace transform. This derivation will probably not make much sense if you are not already familiar with Laplace and Fourier transforms.

Before starting the derivation, though, we want to warn you about one complication we're going to run into. Some functions have Laplace transforms but don't have Fourier transforms. First we'll go through the derivation for a function $f(t)$ that can be both Fourier and Laplace transformed. Then we'll show how to modify the derivation when $f(t)$ can't be Fourier transformed.

Functions that can be Fourier transformed

Consider a function $f(t)$ that is zero for $t < 0$. The Fourier transform $\hat{f}(\omega)$ and the Laplace transform $F(s)$ of $f(t)$ are given by:

$$\hat{f}(\omega) = \frac{1}{2\pi} \int_0^\infty f(t) e^{-i\omega t}\, dt, \qquad F(s) = \int_0^\infty f(t) e^{-st}\, dt$$

The two formulas look similar, but with three differences.

1. There's a factor of 2π in one and not the other, which is purely a matter of convention. Many authors leave out the 2π in the definition of $\hat{f}$, which makes it look even more like $F(s)$.
2. The lower limit of integration is always 0 for a Laplace transform, but it's usually $-\infty$ for a Fourier transform. This reflects the fact that we only take Laplace transforms of functions where $f(t) = 0$ for $t < 0$. We will only consider such functions here, so we can take the Fourier integral from 0 to ∞ as well.
3. The important difference is that the Fourier integral has $e^{-i\omega t}$ while the Laplace integral has e^{-st}.

If we make the variable substitution $s = i\omega$ into the Fourier integral formula, we see that $F = 2\pi \hat{f}$. *A Laplace transform is just a Fourier transform expressed in terms of a new variable $s = i\omega$.*

To get the inverse Laplace transform we can plug this substitution into the formula for the inverse Fourier transform.

$$f(t) = \int_{-\infty}^\infty \hat{f}(\omega) e^{i\omega t}\, d\omega = \int_{-i\infty}^{i\infty} \frac{F}{2\pi} e^{st} \frac{ds}{i} = \frac{1}{2\pi i} \int_{-i\infty}^{i\infty} F(s) e^{st}\, ds$$

Our original integral took place along the entire real axis. The substitution $s = i\omega$ rotates it, so our new integral is a contour integral straight up the imaginary axis.

Functions that cannot be Fourier transformed

We began our derivation with the formula for a Fourier transform, but we can only find a Fourier transform if $\int_0^\infty f(t)\,dt$ exists, which requires $\lim_{t\to\infty} f(t) = 0$. Laplace transforms are much less restrictive. We can take the Laplace transform of any sufficiently smooth function that grows exponentially, or slower than exponentially—which covers most of them. One way to write that criterion is to say that if a function $f(t)$ can be Laplace transformed then we can define a function $g(t) = e^{-at} f(t)$ such that $\lim_{t\to\infty} g(t) = 0$, as long as we make a large enough.

The useful thing about writing the function in this way is that the new function $g(t)$ can be Fourier transformed as well as Laplace transformed, and we can go through the same argument we made before to conclude that:

$$g(t) = \frac{1}{2\pi i} \int_{-i\infty}^{i\infty} G(s) e^{st}\, ds$$

We need one more fact about Laplace transforms to complete the derivation. For any function $f(t)$ with Laplace transform $F(s)$, the Laplace transform of $e^{-at} f(t)$ is $F(s+a)$. In other words, multiplying $f(t)$ by an exponential simply shifts $F(s)$ to the left. (You'll prove this in

Problem 13.83.) In our case this means $G(s) = F(s+a)$. Plugging in that relationship and the definition $g(t) = e^{-at}f(t)$, we get:

$$f(t) = \frac{1}{2\pi i} \int_{-i\infty}^{i\infty} F(s+a)e^{(s+a)t}\, ds$$

(We were able to move e^{at} inside the integral since it doesn't depend on s.) The last step is the substitution $u = s + a$. The s integral was along the imaginary axis. Since $u = s + a$ where a is a real number, the integral over u goes along a vertical axis shifted to the right by an amount a. As a shorthand, we write that as

$$f(t) = \frac{1}{2\pi i} \int_{a-i\infty}^{a+i\infty} F(u)e^{ut}\, du$$

Since u is just a variable of integration we are free to rename it s, which gets us back to Equation 13.5.3.

13.5.3 Problems: Some Uses of Contour Integration

Problems 13.57–13.61 do not involve complex numbers.

13.57 In this problem you will evaluate $\int_{-1}^{1}(1/x^3)dx$ using two different definitions. Both involve breaking the integral up at the vertical asymptote $(x = 0)$.

(a) Equation 13.5.1 represents the standard definition. The integral converges only if the two separate integrals converge. Show that they do not.

(b) Equation 13.5.2 represents the Cauchy principal value. Show that in this case the limit exists, and find the answer.

13.58 In this problem you will evaluate $\int_{-\infty}^{\infty} x\, dx$ using two different definitions.

(a) In the standard definition you break up the integral at any finite value. (We have chosen zero.)

$$\int_{-\infty}^{\infty} x\, dx = \int_{-\infty}^{0} x\, dx + \int_{0}^{\infty} x\, dx$$

Rewrite each integral as a limit and then evaluate them separately. The integral converges only if the two separate integrals converge.

(b) The Cauchy principal value rewrites the integral as one limit.

$$\int_{-\infty}^{\infty} x\, dx = \lim_{a\to\infty} \int_{-a}^{a} x\, dx$$

Evaluate the integral this way.

(c) Draw a graph of the area you are finding, and explain why both results came out the way they did.

13.59 Evaluate $\int_{-1}^{1}\left(1/\sqrt[3]{x}\right) dx$ using both the standard definition and the Cauchy principal value. You should show exactly what integrals you are evaluating at what limits in both cases.

13.60 Evaluate $\int_{-1}^{1}\left(1/x^2\right) dx$ using both the standard definition and the Cauchy principal value. You should show exactly what integrals you are evaluating at what limits in both cases.

13.61 Find an improper integral where the standard definition and the Cauchy principal value differ, and evaluate both of them. (If you've already found one in one of the previous problems, you may use that one. If not, figure one out.) Then use a computer to evaluate the integral. Which answer does the computer give?

13.62 Walk-through: Using a Contour Integral to Evaluate a Real Integral. In this problem you will evaluate the real integral $\int_{-\infty}^{\infty} x^2/[(x^2+1)(x^2+9)]dx$.

(a) We define a complex function $f(z) = z^2/[(z^2+1)(z^2+9)]$. Find all the poles of f. *Hint*: Start by factoring the denominator as far as possible.

(b) Draw a semicircular contour of radius R in the upper half-plane that encloses all of the poles of f with positive imaginary parts.

(c) Explain how we know, for this particular function $f(z)$, that in the limit $R \to \infty$ the contour integral around the curved part of your semicircle approaches 0. (That in turn implies that the closed contour integral equals the contour integral along the real axis, which is what we wanted to find.)

(d) Find the residues of all of the poles of f enclosed by your contour, and use those residues to calculate the closed contour integral. This gives you your answer for the real integral as well.

For Problems 13.63–13.69 use a closed contour integral to find the value of the real integral. Some problems may require using two contours, as illustrated in the Explanation (Section 13.5.1).

13.63 $\int_{-\infty}^{\infty} 1/(x^2 + 9) \, dx$

13.64 $\int_{-\infty}^{\infty} 1/(x^2 + 1)^2 \, dx$

13.65 $\int_{-\infty}^{\infty} (2x - 5)/[(x^2 + 1)(x^2 + 4)] \, dx$

13.66 $\int_{-\infty}^{\infty} 1/(x^2 - 2x + 2) \, dx$

13.67 $\int_{-\infty}^{\infty} \cos(5x)/(x^2 + 1) \, dx$

13.68 $\int_{-\infty}^{\infty} (\sin^2 x)/(x^4 + 4) \, dx$

13.69 $\int_0^{\infty} \cos(2x)/(x^2 + 4) \, dx$ Hint: You'll need to start by relating this to an integral from $-\infty$ to ∞.

13.70 In the Explanation (Section 13.5.1) we showed how to evaluate integrals like $\int_{-\infty}^{\infty} (x^2 + 1)/(x^4 + 4) \, dx$ using semicircular contour integrals.

(a) Explain why that technique would not work for $\int_{-\infty}^{\infty} (x^3 + 1)/(x^4 + 4) \, dx$.

(b) Explain why that technique would not work for $\int_0^{10} (x^2 + 1)/(x^4 + 4) \, dx$.

(c) Explain how you could easily adapt the technique to find $\int_0^{\infty} (x^2 + 1)/(x^4 + 4) \, dx$.

13.71 Walk-Through: Turning a Real Integral of Trig Functions into a Complex Contour Integral. In this problem you will evaluate $\int_0^{2\pi} dx/(5 + 4 \sin x)$ by using a complex u-substitution to turn it into a contour integral.

(a) Use the identity $\sin x = (e^{ix} - e^{-ix})/(2i)$ to rewrite this integral in terms of complex exponentials, simplifying as much as possible.

(b) Do a u-substitution: $u = e^{ix}$. As always, remember to substitute for dx as well as x. Once again, simplify the integral as much as possible, but don't worry yet about the limits of integration.

(c) Your integrand should have a quadratic function of u in the denominator. If you haven't done so already, factor it. (You may be able to do so by inspection, but if you're having trouble you can use the quadratic formula to find the roots.)

(d) Ok, now worry about the limits of integration. Since u is a complex number, the integral now traces out a contour in the complex plane. Draw the contour followed by u as x goes from 0 to 2π.

(e) Evaluate this contour integral to get a final answer for $\int_0^{2\pi} dx/(5 + 4 \sin x)$.

For problems 13.72–13.75 use the method we outlined in Problem 13.71 to calculate $\int_0^{2\pi} f(x) \, dx$ by turning it into a contour integral. (You can check your answers by evaluating the real integrals on a computer, or in some cases by hand, but you should do each of them using a contour integral.)

13.72 $f(x) = (\sin x)/(5 + 3 \cos x)$

13.73 $f(x) = 2/(25 + 7 \sin x)$

13.74 $f(x) = 3 \sin x/(17 + 8 \sin x)$

13.75 $f(x) = 1/(2 + \sin x)$

13.76 Walk-Through: A Bromwich Integral. In this problem you will calculate the inverse Laplace transform of $F(s) = 1/(s - 1)^2$.

(a) Use Equation 13.5.3 to write an integral for the inverse Laplace transform $f(t)$. Leave the number a unspecified.

(b) Mark on the complex plane all the poles of the integrand, and draw a vertical contour anywhere to the right of those poles.

(c) Add to your drawing a semicircular contour that encloses the vertical one you drew. Explain how you can know that the integral around the curved part of this contour approaches 0 as the radius of the semicircle approaches ∞, provided we assume $t > 0$. (Getting this right will depend on how you closed the contour.)

(d) Evaluate the closed contour integral to find $f(t)$ for $t > 0$.

For Problems 13.77–13.81 use a Bromwich integral to calculate the inverse Laplace transform of $F(s)$. You only need to calculate $f(t)$ for $t > 0$. (You may want to first work through Problem 13.76 as a model.)

13.77 $F(s) = 1/(s - 1)$

13.78 $F(s) = 1/[(s + 1)(s - 2)]$

13.79 $F(s) = 1/s$

13.80 $F(s) = 1/(s + 5)^2$

13.81 $F(s) = (s + 1)(s + 2)/(s + 3)^3$

13.82 In the example on Page 737 we found the inverse Laplace transform $f(t)$ for a given function $F(s)$, focusing only on $t > 0$. For $t < 0$ we start from the same formula, evaluating $\int_{a-i\infty}^{a+i\infty} F(s)e^{st}\,ds$.

(a) Explain why, for negative values of t, you need to close the contour on the right in order to evaluate this integral.

(b) Calculate $\int_{a-i\infty}^{a+i\infty} F(s)e^{st}\,ds$ using that closed contour.

(c) Explain why you will get the same result for any function $F(s)$.

13.83 Starting from the equation $F(s) = \int_0^\infty e^{-st}f(t)\,dt$, prove that the Laplace transform of $e^{-at}f(t)$ is $F(s+a)$.

13.6 Integrating Along Branch Cuts and Through Poles (see felderbooks.com)

13.7 Complex Power Series

This section will briefly discuss Taylor series of analytic complex functions, and how they can give us some insight into the convergence intervals for Taylor series of real functions. We will then generalize the idea of a Taylor series to a series that includes negative powers of z, called a "Laurent series." These series can be used to approximate functions around singularities, where normal Taylor series cannot. More importantly, we will see that Laurent series provide a shortcut for finding poles and calculating residues.

13.7.1 Explanation: Complex Power Series

Taylor Series

Suppose you find the Taylor series for a real function $f(x)$ about $x = x_0$. Before you use that series you have to test it for convergence. The ratio test will give you one of three possible results: the series converges only at x_0, it converges everywhere, or it converges in some interval centered on x_0. In the latter case you then apply other tests (there are quite a few to choose from) to determine if the series converges at the endpoints of that interval. And even if the series converges everywhere, it may not converge to the function you started with—a detail we only briefly mentioned in Chapter 2, but it can happen.

Now suppose you find the Taylor series for a complex function $f(z)$ about $z = z_0$, and suppose $f(z)$ is analytic at z_0. Your job here is much easier. If $f(z)$ has one or more non-removable singularities then the series converges everywhere in a disk centered on z_0, and the radius of that disk is the distance from z_0 to the nearest such singularity. It diverges everywhere outside that disk. (As with real functions, you need other tests for points on the edge of the disk.) If there are no such singularities, then the series converges everywhere. And by the way it's guaranteed to converge to $f(z)$, not to some other function![7]

Determining convergence is so much easier on the complex plane that in many cases the best way to analyze the convergence of a real series is to generalize it to a complex series.

---

[7]We said you have to find the distance to a "non-removable" singularity to cover a small exception. For instance, the Maclaurin series for $(z-3i)/(z-3i)$ converges everywhere, not just until you get to $3i$. Of course you could also violate this rule by defining a function that is z^2 near the origin and $Re(z)$ far away, and then your series would not converge to the function you started with. Don't do that.

EXAMPLE Radius of Convergence

Question: Where does the Maclaurin series for the real function $f(x) = 1/(x^4 + 2)$ converge?

Answer:
The real function $f(x)$ is the complex function $f(z) = 1/(z^4 + 2)$ evaluated on the real line. This $f(z)$ is analytic everywhere except the poles at $z = (-2)^{1/4}$. The fourth roots of -2 have phases $\pi/4$, $3\pi/4$, $5\pi/4$ and $7\pi/4$, but more importantly they all have magnitude $2^{1/4}$, so that is the distance from the origin to the nearest singularity. We therefore draw a disk around the origin with radius $2^{1/4}$. The series converges everywhere inside this disk and diverges everywhere outside it.

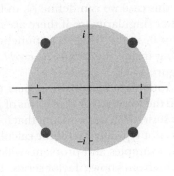

We now return to our original real-valued function. It doesn't matter that none of the poles lies on the real axis. The complex function converges everywhere in the disk drawn above, so the real function converges everywhere on the real axis within that disk: in other words, between $-\sqrt[4]{2}$ and $\sqrt[4]{2}$.

For both the real and complex series we say that the "radius of convergence" is $2^{1/4}$. On the complex plane this is the radius of a disk; on the real number line it is the distance from 0 to either edge of the interval of convergence.

We get no guarantee of what happens to our complex series on the border of its disk of convergence, or to our real series at the two endpoints of its interval of convergence. If we wanted to know we would need to find the Maclaurin series for $f(x)$ and use tests of series convergence on it.

Laurent Series

We said above that all analytic functions have well-behaved Taylor series, but as we've seen in this chapter the most interesting behavior of a function often occurs around the points where it is *not* analytic. It is possible to do a series expansion in powers of z about such a point, but the series must include negative powers of z.[8] Such a series is called a "Laurent Series."

---

[8] Technically a "power series" only involves positive integer powers, so this section is not accurately named. But who wants to read a section called "Power Series and Other Series that are a Lot Like Them but Don't Meet the Technical Definition"?

> **Definition: Laurent Series**
>
> Let $f(z)$ be analytic in the region between two circles of radius R_1 and R_2 centered on the point $z = z_0$. There is a series of the form
>
> $$\sum_{n=-\infty}^{\infty} c_n(z-z_0)^n \qquad (13.7.1)$$
>
> that converges to $f(z)$ at every point z in that ring-shaped region. That series is a Laurent series for $f(z)$.

If f is analytic everywhere inside the outer radius R_2 then all of the c_n for negative n will be zero and the Laurent series will simply be the Taylor series for f. Things get more interesting, though, when $z = z_0$ is an isolated singularity of $f(z)$. In this case we can define R_2 to be the distance to the nearest other singularity (∞ if there are no other singularities) and let $R_1 = 0$, so the "ring" is a "punctured disk" that includes everything inside the radius R_2 *except* the point $z = z_0$, as shown in Figure 13.12.

FIGURE 13.12 The Laurent series for a function $f(z)$ around an isolated singularity converges to f everywhere inside a disk around $z = z_0$ except for the point $z = z_0$ itself.

You might reasonably expect that we are now going to give you the formula for c_n in terms of $f(z)$, just as we can do with Taylor series. You will work out that formula in Problem 13.104, but it is not typically useful for calculating Laurent series. Instead, our examples and problems will focus on calculating Laurent series from known Taylor series. That is easy in many cases, and it gives you a lot of information. In fact Laurent series give us a way to generalize some of the definitions in Section 13.4, as well as a surprisingly quick way to compute residues. (For some types of functions partial fractions are useful for finding Laurent series. See Problems 13.98–13.101.)

> **Definitions: Pole, Residue, Essential Singularity**
>
> Suppose the Laurent series $f(z) = \sum_{n=-\infty}^{\infty} c_n(z-z_0)^n$ converges in a punctured disk around the point $z = z_0$.
>
> - If $c_n \neq 0$ for some negative value of n but it does equal zero for all values of n below that, then the function $f(z)$ has a "pole" of order $|n|$ at the point $z = z_0$.
> - If $c_n \neq 0$ for an infinite number of negative values of n, then the function $f(z)$ has an "essential singularity" at the point $z = z_0$.
> - If either of the above is true then the "residue" of $f(z)$ at $z = z_0$ is the coefficient c_{-1}.

You can relate this definition of "pole" to our earlier definition by considering a function like this.

$$\frac{1}{(z-10)^3} + \frac{7}{(z-10)^2} + \frac{4}{z-10} + 12 + 6(z-10) + 2(z-10)^2$$

Our new definition tells us that this function has a third-order pole at $z = 10$. We could combine all these terms into one fraction with a polynomial (and therefore analytic) numerator and a denominator of $(z-10)^3$ and then our previous definition would tell us the same thing.

Our new definition also introduces a new term, "essential singularity," for a point where the negative powers just keep on going. You can think of the order of a pole as telling you how quickly the function blows up as you approach the singularity. An nth-order pole acts like $(z - z_0)^{-n}$ near the singularity because the term with that power of z grows infinitely larger than all the other terms. An essential singularity blows up faster than any inverse power of z.

Perhaps most surprising is our new definition of "residue," which is just the coefficient of the -1-order term for a pole or an essential singularity. In fact this is the general definition, and the derivative-and-factorial formulas in Section 13.4 were a special case. In Problem 13.102 you'll try both definitions on a specific function, and in Problem 13.103 you will show more generally that they are equivalent.

EXAMPLE   Laurent Series

Question: Find the Laurent series for $f(z) = (1 - e^{z^2})/z^5$ about the origin and use it to characterize the singularity at the origin and find its residue.

Answer:
We can find the series by expanding e^{z^2} in a Maclaurin series. Since $e^x = 1 + x + x^2/(2!) + \ldots$, we know $e^{z^2} = 1 + z^2 + z^4/(2!) + \ldots$.

$$f(z) = \frac{1 - \left(1 + z^2 + z^4/(2!) + z^6/(3!) + \ldots\right)}{z^5} = -z^{-3} - \frac{1}{2}z^{-1} - \frac{1}{6}z - \ldots$$

From this we can immediately see that the origin is a third-order pole with residue $-(1/2)$. It follows that a contour integral of $f(z)$ around the origin would give $-\pi i$.

In Section 13.2 we listed three types of singularity. If a function $f(z)$ has any kind of singularity at $z = z_0$ then you can safely assert that $f(z_0)$ is undefined, but the three types behave differently *near* z_0. As you approach a removable singularity the function approaches some finite limit. As you approach a pole the function blows up, which is to say, $\lim_{z \to z_0} |f(z)| = \infty$.

For an essential singularity $\lim_{z \to z_0} |f(z)|$ is undefined, and here's why: in any neighborhood around z_0, no matter how small, $f(z)$ must take on every complex value, with at most one exception. See Problem 13.105. This is honestly not the most useful fact you will ever encounter, but it's so remarkable that we didn't want you to go to your grave never having heard it.

13.7.2 Problems: Complex Power Series

13.93 In this problem you will calculate $\oint e^{1/z} dz$ around a unit circle around the origin.

(a) Explain how you can know that $f(z) = e^{1/z}$ has a singularity at the origin and is analytic everywhere else.

(b) Find the Laurent series for $f(z)$ about the origin by using the Maclaurin series for e^x.

(c) Identify the singularity at the origin as an essential singularity or a pole, and if it's a pole specify its order.

(d) What is the residue of f at the origin?

(e) What's the value of $\oint e^{1/z} dz$?

13.94 Calculate $\oint (z^5 + 1) e^{1/z} dz$ about the unit circle around the origin. *Hint*: Use a Laurent series to find the residue.

13.95 Use a complex extension of each real function to find the radius of convergence of its (obviously real) Maclaurin series.

(a) $f(x) = 1/(x^2 + 1)$
(b) $f(x) = e^x/[(x-1)^2 + 1]$
(c) $f(x) = 1/(e^x + 1)$

13.96

(a) Find the radius of convergence of the Maclaurin series for $f(z) = 1/(e^z + 1)$. (If you did Problem 13.95 you have already solved this part.)

(b) Now consider the real function $f(x)$, whose Maclaurin series should have the same radius of convergence as that of $f(z)$. Plot $f(x)$ and the 100th partial sum on the same plot. Include vertical lines at the boundaries of the interval of convergence you found. How does the partial sum behave inside this interval? Outside it?

13.97 The function $(\sin z)/(z - \pi/4)$ has a pole at $z = \pi/4$.

(a) Find the first three terms of the Laurent series for $(\sin z)/(z - \pi/4)$ about the point $z = \pi/4$.

(b) Find the residue of $(\sin z)/(z - \pi/4)$ at $z = \pi/4$. *Hint*: You should be able to just write it down.

(c) Find the contour integral of $(\sin z)/(z - \pi/4)$ around the unit circle.

13.98 In this problem you will find the Laurent series for $f(z) = 1/[z(z-2)]$ about the point $z = 0$.

(a) You may remember the method of "partial fractions" for rewriting fractions of this type. Show that you can write f in the form $a/z + b/(z-2)$ by finding the correct constants a and b. (You should be able to easily check that your answer is correct by adding the fractions you get.)

(b) Use this expansion to write the first four non-zero terms of the Laurent series in the region around $z = 0$. *Hint*: use the Maclaurin series for $1/(z-2)$.

(c) What is the radius of convergence of this Laurent series?

(d) Use the Laurent series you found to find the residue of $f(z)$ at $z = 0$.

(e) Specify a closed contour that encloses the pole at $z = 0$ and lies entirely in the region of convergence of your Laurent series, and evaluate the integral of $f(z)$ around that contour.

13.99 Use partial fractions to find the first three non-zero terms of the Laurent series for $f(z) = 1/[z(z^2 - 1)]$ around $z = 0$. What is the radius of convergence of that series? If you're stuck you may find it helpful to look at Problem 13.98.

13.100 Use partial fractions to find the first three non-zero terms of the Laurent series for $f(z) = e^z/[(z-1)(z-3)]$ around $z = 1$. What is the radius of convergence of that series? If you're stuck you may find it helpful to look at Problem 13.98.

13.101 A function with a singularity at $z = z_0$ has a Laurent series about that point valid out to a radius equal to the distance to the nearest other singularity. It also has a Laurent series about $z = z_0$ that is valid beyond that other singularity (out to the distance to the next one). In this problem you'll derive both Laurent series for $f(z) = 1/[z(z-1)]$ valid about $z = 0$.

(a) Use partial fractions to rewrite f in the form $a/z + b/(z-1)$. (You need to find the constants a and b.)

(b) Use the Maclaurin series for $1/(z-1)$ to write the first four non-zero terms of the Laurent series for f valid in the region $|z| < 1$.

(c) Explain how you can know that this series doesn't converge to f when $|z| > 1$.

(d) Rewrite $1/(z-1)$ as $1/[z(1 - 1/z)]$ and expand $1/(1 - 1/z)$ in powers of $1/z$ to find a Laurent series for f. This one is valid in the region $|z| > 1$. Write your final answer in summation notation.

(e) Explain how you can know that this series *does* converge for $|z| > 1$ and doesn't for $|z| < 1$.

13.102 Section 13.4 and this section gave two very different ways of calculating residues. In this problem you will try them both on $f(z) = 1/[z(z-1)]$.

(a) First, without calculating the Laurent series, find the residue of $f(z)$ at $z = 0$.

(b) Write $1/(z-1)$ as a Maclaurin series. You can calculate this directly or use the binomial expansion.

(c) Multiply that Maclaurin series by $1/z$ to find the Laurent series for $f(z)$ about $z = 0$. Confirm that the series gives you the same value for the residue that you found in Part (a).

13.103 Suppose $f(z) = g(z)/(z-z_0)^n$, where $g(z)$ is analytic and non-zero at $z = z_0$. Find the Laurent series for f by expanding g in a Taylor series about $z = z_0$, and use your result to find the residue of f at $z = z_0$.

13.104 In this problem you will derive the formula for the coefficients c_n of a Laurent series, Equation 13.7.1. For simplicity we will consider a Laurent series in a punctured disk about the point $z = z_0$, and we will assume that the contours we use in the problem enclose $z = z_0$ (which may or may not be singular) but do not touch or enclose any other singularities. Recall that the residue of $f(z)$ at $z = z_0$ is c_{-1}.

(a) What is $\oint f(z)\, dz$? Your answer should contain c_{-1}. Rearrange it to get a formula for c_{-1} in terms of this contour integral.

(b) What is $\oint f(z)(z - z_0)\, dz$? (*Hint*: Think about what the Laurent series for $f(z)(z - z_0)$ is.) Your answer will contain a different coefficient c this time. Once again, solve your equation for that coefficient.

(c) Evaluate $\oint f(z)(z - z_0)^m$ and solve the resulting equation for the c coefficient that appears in it.

(d) Use your result from Part (c) to write a general formula for c_n.

13.105 "Picard's theorem" says that in any neighborhood of an essential singularity a function $f(z)$ must take on every possible complex value, with at most one exception.

(a) Find the Laurent series for $f(z) = e^{1/z^2}$ about the origin and use it to prove that f has an essential singularity at the origin.

(b) Plot $f(\rho e^{i\phi})$ for fixed ρ and $0 < \phi < 2\pi$. Do this for several decreasing values of ρ so you can see how f behaves around the origin as you get closer to it. Describe your results.

(c) Find the one value f never takes as you get close to the origin.

13.8 Mapping Curves and Regions

In Section 13.2 we discussed several ways to graph complex functions. This section introduces another one, called "mapping." Here we present mapping as a tool for visualizing the behavior of complex functions. In Section 13.9 we will show how it can be used to solve Laplace's equation in regions where it is difficult to solve by other methods.

13.8.1 Discovery Exercise: Mapping Curves and Regions

A complex function can be described as a "mapping" from the complex plane to itself. That means that for every complex number that we put into the function we get a complex number out. To see what this means, consider the function $f(z) = z^2$.

1. If $z = 1 + i$ what is $f(z)$?
   See Check Yourself #94 in Appendix L

2. Draw a complex plane representing z. Mark on that plane the grid of points 0, 1, 2, i, $1+i$, $2+i$, $2i$, $1+2i$, and $2+2i$. Label the points "A," "B," and so on. Then draw another complex plane and mark the points $f(z)$ for each of the nine z-values above, using the same labels.

As you can see, this technique represents one complex function with two different complex planes: one for the input z, and one for the output $f(z)$. Now draw two *more* complex planes, once again representing the input and output of $f(z) = z^2$. You will use these two for Parts 3–6.

3. On the z graph mark the line segment going from $z = 0$ to $z = 2$ (on the positive real axis). On the $f(z)$ graph mark the set of points f corresponding to all the z-values on that line segment.
   See Check Yourself #95 in Appendix L

4. Add a line segment from $z = 0$ to $z = 2i$ to the z graph and the corresponding set of points f on the f graph.
5. Add the curve $|z| = 2$ in the first quadrant to the z graph and the corresponding curve to the f graph.
6. The three curves you just drew on your z graph define a quarter of a disk. What shape does $f(z) = z^2$ map this quarter-disk to?

13.8.2 Explanation: Mapping Curves and Regions

We noted in Section 13.2 that with a two dimensional input and a two dimensional output, a complete graph of a complex function would require four dimensions. We also discussed various workarounds for this problem. One graphing technique we did not discuss is "mapping," which means you draw the input z on one plane and the output $f(z)$ on a different one.

In the Discovery Exercise (Section 13.8.1) you created a mapping to show how one region of the complex plane is mapped by the function z^2 to another region. Below we show a few examples of mappings of the function e^z. No single mapping can convey the entire function, but a few well-chosen mappings can lend significant visual intuition to your understanding of a function. As we will see in Section 13.9, mapping also provides a powerful technique for solving certain problems involving Laplace's equation, such as the problem in the Motivating Exercise (Section 13.1).

EXAMPLE **Mapping**

Problem:
Show how the function e^z maps points in the first quadrant.

Solution:
We've selected a grid of nine points, which will be enough to show the basic behavior of the function. We can calculate e^z for each of these points either by hand or on a computer, and plot the results.

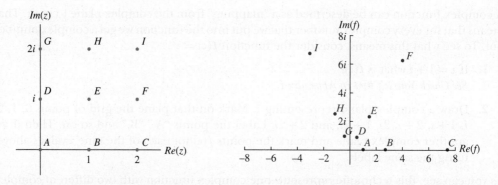

Of course, more important than drawing the picture is interpreting what it tells us. As z moves to the right ($A \to B \to C$, $D \to E \to F$, or $G \to H \to I$), e^z moves directly outward from the origin. In other words, increasing the real part of z increases the modulus of e^z. Similarly, increasing the imaginary part of z (moving up through the z points) increases the phase of e^z without affecting the modulus.

If we write $z = x + iy$ we get $e^z = e^x e^{iy}$, so e^z has modulus e^x and phase y, confirming algebraically what we just concluded graphically.

Curves in the Complex Plane

The example above showed the mapping of a grid of points. In more complicated cases it is often useful to map curves or regions in the complex plane rather than just points. So how do you express a *curve* on the complex plane? We'll address that question here, and then come back to the mappings of those curves.

You can probably figure out a few complex curves if you think about it. If we write $\text{Im}(z) = -3$ that means "show me every number whose imaginary part is -3": a horizontal line. Similarly, $|z| = 4$ holds the modulus constant, tracing out a circle.

A more general approach is to specify a point z with a free variable involved. For instance, consider $z = 3a + 4ai$ where a can be any real number. That specifies an infinite number of points, but it doesn't specify *all* points, since the imaginary part has to be exactly $4/3$ the real part. So it draws the line that, on the real xy-plane, would be represented as $y = (4/3)x$. The example below shows a few more curves on the complex plane.

EXAMPLE   **Complex Curves**

Question: Draw each of the following curves: a) e^{ia} where $0 \leq a \leq 2\pi$, b) $ae^{(\pi/6)i}$ where $a \geq 0$, c) ae^{ia} where $0 \leq a \leq 2\pi$.

Answer:
The first is the unit circle. The second is a half-line that starts at the origin and goes up at an angle of 30° from the real axis. The last one consists of points whose moduli equal their phases. It starts at the origin and as it moves around counterclockwise it also moves out, ending with a phase of 2π (the positive real axis) and a modulus of 2π (hence the point $z = 2\pi$).

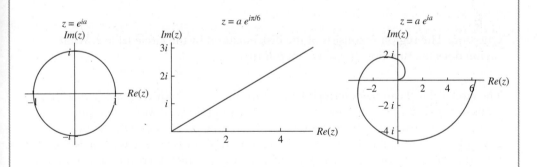

Mapping Curves and Regions

If a function $f(z)$ acts on a specific point, the result is a different point. (Or more than one point, but let's assume single values for the moment.) We say the function "maps" or "transforms" the point z to the point $f(z)$. If a function acts on a curve, the result is a different curve. Both a point and a curve are mathematically expressed as a formula $z = $<*something*>, the only difference being that the formula for a curve has a free variable in it, so the process of mapping a curve is mathematically identical to the process of mapping a point.

But most often we are interesting in mapping *regions*. A region can be defined by a set of bounding curves, so the easiest approach is usually to map each of these curves. Sometimes the new curves will bound more than one region, in which case we can choose any point in the original region and see which new region it maps to, as illustrated in the second example below.

EXAMPLE **Mapping a Region**

Question: The region R consists of the infinite vertical band between $x = 1$ and $x = 2$. What region does the function $f(z) = z^2$ map R to?

Answer:
First we find the formulas for the two lines that bound this region. Each line has a constant real part and an imaginary part that can be any real number, so they are $z = 1 + iy$ and $z = 2 + iy$, where $-\infty < y < \infty$. To map these curves we plug their formulas into f. For the first curve $f(z) = (1 + iy)^2 = (1 - y^2) + 2iy$. If we write $f = u + iv$, then $u = 1 - y^2$ and $v = 2y$. We can eliminate y from these equations to get a formula for the curve in the uv-plane: $u = 1 - (v/2)^2$. This describes a left-facing parabola. A similar calculation gives $u = 4 - (y/4)^2$ for the second curve.

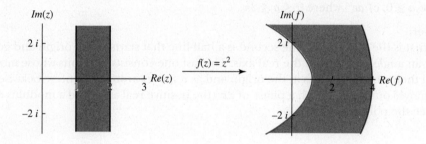

Question: The region R consists of the disk enclosed by the circle $|z| = 2$. What region does the function $f(z) = 1/z$ map R to?

Answer:
The circle $|z| = 2$ can be described by the formula $z = 2e^{i\phi}$ with $-\pi < \phi \leq \pi$, which maps to $z = (1/2)e^{-i\phi}$. This new curve is a circle of radius $1/2$. We're not quite done, though. In the example above there was only one region bounded by the two parabolas, so that had to be the mapped region. In this case the new circle is the boundary of its interior but also of its exterior. To see which is correct pick a point z inside region R. For example, if $z = 1$ then $f(z) = 1$, which is *outside* the new circle, so the mapped region is the exterior of $|z| = 1/2$.

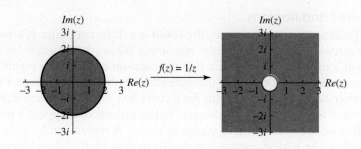

Möbius Transformations

We conclude this section with one particularly important class of transformations. Transformations of the form

$$f(z) = \frac{a + bz}{c + dz} \quad (13.8.1)$$

are known as Möbius transformations.[9] Consider, for example, the Möbius transformation $f(z) = 1/z$. You'll show in Problem 13.119 that this transformation reverses the phase of its argument ($\phi \to -\phi$) and inverts the modulus ($|z| \to 1/|z|$). If you use this to map the unit circle you just get back the unit circle. That doesn't mean each point on the unit circle is mapped to itself; points on the unit circle are mapped to their complex conjugates. For points not on the unit circle z is not mapped to z^*, but points in the upper half-plane do move to the lower half-plane and vice versa.

To check your understanding of complex curves and mapping, we encourage you to try and answer each of the following questions about the transformation $1/z$ before reading our answers.

- What does the circle of radius R centered on the origin map to? *Answer*: It maps to the circle of radius $1/R$ centered on the origin.
- What happens to a circle of radius $1/4$ that is not centered on the origin? *Answer*: It's a bit of a trick question because the answer depends on where the circle is. If the circle extends from $|z| = 1/4$ to $|z| = 3/4$ then the mapped circle will go from $|z| = 4/3$ to $|z| = 4$, so its radius and distance from the origin will increase. If the circle goes from $|z| = 100$ to $|z| = 100.5$ then it will map to a tiny circle very near the origin. More generally, regions close to the origin get moved far away and become large, while regions far from the origin get moved closer and shrink.
- What happens in the limit where one edge of the circle approaches the origin? *Answer*: The points infinitesimally close to the origin get mapped infinitely far away. If the circle is very close to touching the origin it will get mapped to an enormous circle. If one edge of the original circle touches the origin, the circle will get mapped to a line. See Figure 13.13.

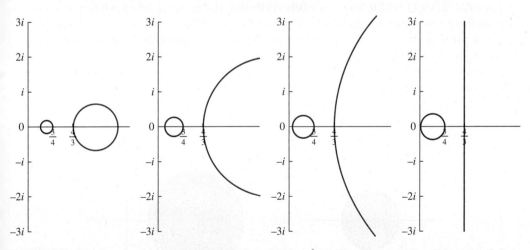

FIGURE 13.13 The function $1/z$ maps the black circle to the blue curve. The closer one edge of the original circle is to the pole at $z = 0$ the larger the mapped circle is. In the limit where the original circle touches the pole, it maps to a line.

---

[9] Yes, it's the same guy who thought of the Möbius strip.

752 Chapter 13 Calculus with Complex Numbers

We've discussed all this for the simple example $f(z) = 1/z$, but the same principles apply to any transformation described by Equation 13.8.1.

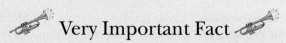

Very Important Fact

Every Möbius transformation (Equation 13.8.1) has a singularity at $z = -c/d$. Any line or circle that passes through that singularity is mapped to a line; any line or circle that doesn't is mapped to a circle.

EXAMPLE A Möbius Transformation

Question: What does the function $f(z) = (-6 - z)/(4 + 3z)$ map the unit disk about the origin to?

Answer:
The unit circle can be described as $z = e^{i\phi}$ for $0 \le \phi \le 2\pi$. Plugging this into the function, the mapped curve is the set of points $f = (-6 - e^{i\phi})/(4 + 3e^{i\phi})$.

Instead of doing a lot of algebra to get this into a recognizable form, we can take a shortcut. We know this is a Möbius transformation, and the unit circle does not touch the pole at $z = -4/3$, so it must map to a circle. Moreover, since all the coefficients are real we know it will map real numbers to real numbers, so the points $z = 1$ and $z = -1$ will remain on the real axis. So we calculate $f(1) = -1$ and $f(-1) = -5$. The resulting circle has radius 2 and is centered on $f = -3$.

As a test we can plug in $z = i$, which after a bit of calculation gives $f = -(27/25) + (14/25)i$. You can easily verify that the distance from $f(i)$ to $f = -3$ is 2, so this point is also on the circle of radius 2 centered on -3. In Problem 13.120 you'll prove more generally that all the points on the original circle map to the circle we've just described.

Finally, by noting that $z = 0$ maps to $z = -3/2$, we see that the region interior to the original circle is mapped to the region interior to the new one.

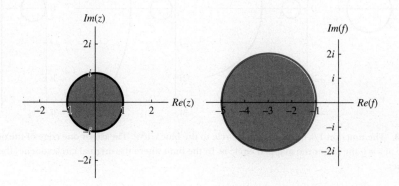

13.8.3 Problems: Mapping Curves and Regions

13.106 The equation $Re(z) = 5$ indicates all the complex numbers whose real part is 5.

(a) Draw this shape on the complex plane.

(b) Write an equation in the form $z =$ <something> that describes this shape. Your formula will have a free variable b that can be any real number.

(c) What shape do you get from that same formula if you restrict b to positive numbers?

13.107 Draw the curve represented by $z = 2e^{i\phi}$ where…

(a) $0 \leq \phi \leq 2\pi$

(b) $0 \leq \phi \leq 3\pi$

(c) $\pi/2 \leq \phi \leq \pi$

13.108 Draw the curve represented by $z = a + a^2 i$ where a can be any real number.

13.109 Consider the multivalued function $f(z) = z^{1/3}$. Throughout this problem you will only use values of z that have $|z| = 1$.

(a) Draw two complex planes, one for z and one for $f(z)$. On the z-plane draw a curve with $-\pi \leq \phi < \pi$. (Remember you're taking $|z| = 1$.) Draw the corresponding curve on the f-plane.

(b) Using a different color pen, draw another circle for $\pi \leq \phi < 3\pi$. (This will be more or less on top of the first circle you drew.) Using that same color, draw the corresponding curve on the f-plane.

(c) Using a third color (if you have it) add curves to the z and f-planes for $3\pi \leq \phi < 5\pi$.

(d) You don't need to draw a fourth curve for $5\pi \leq \phi < 7\pi$, but describe what that curve would look like on the f-plane if you did.

(e) The first color you used represents the principal value of $z^{1/3}$. Describe what would happen on the f-plane if you rotated z around the unit circle three times, only using the principal value of f.

13.110 Walk-Through: Mapping Curves and Regions. In this problem you will consider the unit square with corners at $z = 0$ and $z = 1 + i$ and find the region that $f(z) = e^z$ maps this square to. *Hint:* for much of this problem it will be helpful to write $e^z = e^{x+iy} = e^x e^{iy}$.

(a) Draw the z-plane with the square on it and draw another set of axes for the f-plane. At this point the second graph is just a blank set of axes.

(b) Explain how we know that the bottom line segment on the z-plane, which lies along the real axis, will map to a segment along the real axis of $f(z)$. (Note that the segment on the imaginary axis of z will *not* map to the imaginary axis of f.)

(c) What points do $z = 0$ and $z = 1$ map to on the f-plane?

(d) What curve does the line segment from $z = 0$ to $z = 1$ map to on the f-plane? Answer in words and also add this curve to your second graph.

(e) What curve does the line segment from $z = 0$ to $z = i$ map to on the f-plane? Again, answer in words and add the curve to your second graph. (*Hint*: You can write $f(z)$ in terms of the real and imaginary parts of z as $f(z) = e^x e^{iy}$. Think about what happens to the point f as you hold x constant at 0 and vary y from 0 to 1.)

(f) In words and on your graph, what curve does the line segment from $z = i$ to $z = 1 + i$ map to?

(g) In words and on your graph, what curve does the line segment from $z = 1$ to $z = 1 + i$ map to?

Your f-plane graph should now show four curves that enclose the region that $f(z)$ maps this unit square to. The inside of the original square maps either to the inside of this new region, or to the outside of this new region.

(h) Choose a point z inside the original square (not on its boundary!) and find what point $f(z)$ it maps to.

(i) What does your result tell us about the mapping of this region?

In Problems 13.111–13.116 find what $f(z)$ maps the given curve or region to. If $f(z)$ is multivalued only use the principal value.

13.111 $f(z) = z^2$. The region is the half-disk $|z| \leq 1$, $Re(z) > 0$.

13.112 $f(z) = z^*$. The region is above the parabola $Im(z) = Re(z)^2$ and below the line $Im(z) = 2$.

13.113 $f(z) = z^{1/4}$. The region is the unit circle centered on the origin.

13.114 $f(z) = \ln z$. The region is the unit circle centered on the origin.

13.115 $f(z) = 1/z$. The region is in the first quadrant between the circles of radius 1 and 2 centered on the origin.

13.116 $f(z) = e^z$. The curve is the line segment from $z = -1 - i$ to $z = 1 + i$.

13.117 $f(z) = (1 + 3z)/(5 + z)$. The curve is a unit circle centered on the point $z = 2 + 2i$.

13.118 🖥 $f(z) = e^z$. The curve is a quarter-circle of radius 2 in the first quadrant, centered on the origin.

13.119 The Explanation (Section 13.8.2) made the claim that the transformation $f(z) = 1/z$ reverses the phase and inverts the modulus of its argument. Prove it. (This requires only a small bit of algebra.)

13.120 In the example on Page 752 we applied the mapping $f(z) = (-6 - z)/(4 + 3z)$ to the unit circle. We never proved that it would map to a circle: we just accepted that fact, and went on to find that the resulting circle must have radius 2 and center -3. This is generally the easiest approach for Möbius transformations, but in this problem you will prove that the result in this case actually is a circle.

(a) If our final conclusion was correct, then the transformation $g(z) = (f(z) + 3)/2$ maps the unit circle to itself. (This doesn't mean it's the identity transformation; it could map each point on the unit circle to a different point on the unit circle.) Explain how we know that.

(b) You have just shown that the statement "$f(z)$ maps the unit circle to a circle" is equivalent to the statement "$g(z)$ maps the unit circle to itself." To prove the former you will prove the latter, by showing that for any $z = e^{i\phi}$ the resulting $g(z)$ has modulus $|g(z)| = 1$. (We recommend you start by simplifying $g(z)$ as much as possible.)

13.9 Conformal Mapping and Laplace's Equation

In Section 13.8 we introduced "mapping" to visualize complex functions. Here we use a particular type of mapping called "conformal mapping" to solve Laplace's equation with more complicated boundaries than we could handle in Section 13.3.

13.9.1 Explanation: Conformal Mapping and Laplace's Equation

In Section 13.3 we used analytic functions to solve Laplace's equation subject to certain boundary conditions. Every time you write an analytic function you know that its real and imaginary parts are both harmonic. So you use and combine your known list of analytic functions to find a harmonic function that lines up along your boundary curves.

That's a powerful technique, but as we saw, its primary use is limited to regions that happen to correspond to the basic analytic functions. The function z^2 is great for hyperbolic boundaries, you can handle a perfect circular boundary with $\ln z$, and so on. If your boundary doesn't happen to line up with one of these functions, you're stuck.

"Conformal mapping" extends that technique to more regions. It is based on the idea of a mapping, as explained in Section 13.8: a given function maps a given region on the z-plane to a different region on the $f(z)$ plane. If the mapping is conformal, which we define below, then you can relate the solution of Laplace's equation in the mapped region to the solution in the original region.

> **Definition: Conformal Mapping**
>
> A conformal mapping is one that preserves the angles between intersecting lines or curves.
>
> The mapping performed by a function $f(z)$ is conformal at every point where $f(z)$ is analytic and $f'(z) \neq 0$.

13.9 | Conformal Mapping and Laplace's Equation

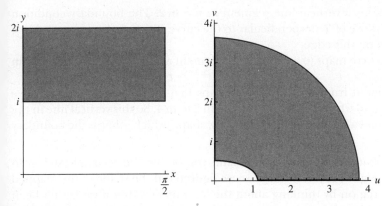

FIGURE 13.14 A rectangle mapped by the function sin z demonstrates how angles are preserved in a conformal mapping.

To illustrate what this definition means, Figure 13.14 shows how $f(z) = \sin z$ maps a rectangular region. Two of the straight edges remain straight, one becomes an arc, and one becomes a more complicated looking curve, but the intersections of these boundaries are all still at right angles.

We introduced mapping as a technique for visualizing the behavior of a complex function. Here we will be using it for quite a different purpose. Remember that the technique we introduced earlier for solving Laplace's equation required that you guess an analytic function whose real or imaginary part would work for the region of interest. With conformal mapping, if the region doesn't suggest an obvious solution, you can find an analytic function that maps it to a more convenient shape and solve the problem there. We demonstrate this process below.

The Problem

The two dimensional region shown in Figure 13.15 has no heat sources or sinks. The boundary on the x-axis is held at temperature $T = 0$ and the boundary on the y-axis at $T = 100$. The two curved boundaries are insulated, which imposes the boundary condition that the derivative of T normal to the boundary is zero on those curves. Find the steady-state temperature in this region.

With a bit of cleverness you might be able to solve this problem right now, but it will demonstrate the mapping technique very clearly. Then we will work an example where a conformal mapping is more necessary.

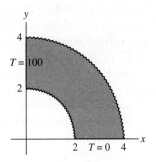

FIGURE 13.15 Find the temperature in this region.

The Mapping

The technique begins by selecting an analytic function that will map Figure 13.15 to a simpler region. We have seen before that the log function works well with circles, so let's see what it does to our region.

Recall that $\ln z = \ln |z| + i\phi$. In other words the real part of f comes from the modulus of z and the imaginary part is the phase of z. Using that fact we will go through each curve on the boundary of the original figure. We will end up defining a new shape, and we will carry our boundary conditions with us from the old one.

- The line segment from $z = 2$ to $z = 4$ consists of positive real numbers, so it maps to a line segment on the real axis going from $f = \ln 2$ to $f = \ln 4$. The boundary condition for this segment is $T = 0$.
- The inner curve consists of numbers with $|z| = 2$ and ϕ going from 0 to $\pi/2$, so it maps to a set of numbers f whose real part is $\ln 2$ and whose imaginary part goes from 0 to

$\pi/2$. In other words it's a vertical line segment at $u = \ln 2$. The boundary condition here is that the derivative of T perpendicular to the curve is zero, so on our new shape $\partial T/\partial u$ must be zero on this edge.
- Similarly, the outer curve maps to a vertical line segment at $u = \ln 4$, where once again $\partial T/\partial u = 0$.
- Finally the line segment from $z = 2i$ to $z = 4i$ maps to a set of points f that all have imaginary part $\pi/2$, and whose real parts go from $\ln 2$ to $\ln 4$. So this vertical line in the original shape becomes a horizontal line in the new shape, which inherits the boundary condition $T = 100$.

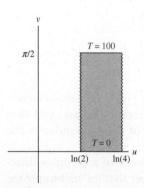

FIGURE 13.16 The mapped region.

Put all of those together and you define the rectangle shown in Figure 13.16. In this case we figured out the shape of the mapped region by thinking about the log function, but if we hadn't been able to figure it out we could have written each curve as a function $z(a)$ and plugged that into $\ln z$ to find the curve $f(a)$.

Before we proceed, make sure you understand what we just determined. Figure 13.15 shows the region we are interested in: it is a collection of points (x, y), each one at a particular temperature T. If we think of those points as $x + iy$ on the complex plane, then the function $\ln z$ maps each of those points to a different point $u + iv$ in the region shown in Figure 13.16. We define the temperature at each point in the new region to be the same as the temperature at the corresponding point in the old region. (That's why we inherit the same boundary conditions.)

Finding the temperature distribution in a rectangle is easier than the problem we were originally presented with. So we are going to solve that problem, and then, because of the mapping, we can use that solution to solve our original problem.

Solving Laplace's Equation in the Mapped Region

We need a function $T(u, v)$ that satisfies $T(u, 0) = 0$ and $T(u, \pi/2) = 100$. (Technically it only has to satisfy those conditions on $\ln 2 \leq u \leq \ln 4$.) It must satisfy $\partial T/\partial u = 0$ (at least along the vertical edges). And finally, it must be harmonic: that is, it must satisfy Laplace's equation $\partial^2 T/\partial u^2 + \partial^2 T/\partial v^2 = 0$.

Well, we can do all that with a linear function in v! The slope is $100/(\pi/2)$ and just like that we have our solution: $T(u, v) = 200v/\pi$. You can easily check that this satisfies Laplace's equation (as any linear function does) and meets all four boundary conditions.

We're not quite done yet—after all, that wasn't actually the problem we set out to solve. But the final step, converting this answer to the original region, is going to be easier than you might think.

Mapping Back

We now have a function $T(u, v) = (200/\pi)v$ that works inside the rectangle. Remember what it stands for: the temperature at each point in this region is the same as the temperature $T(x, y)$ in the corresponding point in the original washer-shaped region.

But we also have the function that maps from our original (x, y) points to the new (u, v) points. This function is $f(z) = \ln z$, or in other words $u = \ln(\sqrt{x^2 + y^2})$ and $v = \tan^{-1}(y/x)$. Putting all that together, we can just write down the function we were looking for.

$$T(x, y) = T(u(x, y), v(x, y)) = \frac{200}{\pi} \tan^{-1}\left(\frac{y}{x}\right)$$

That last step is almost too easy. We were looking for a function that was harmonic (satisfied Laplace's equation) inside the original region, and that satisfied all the conditions on the original boundaries. Are we guaranteed that we have found one?

- Does the new function meet the right boundary conditions? At every point on the top of our rectangle, $T(u, v) = 100$. And we know that $\ln z$ maps those points to the y-axis boundary of our original shape. So $T(u(x, y), v(x, y))$ along that boundary must also give 100. In other words, every point in Figure 13.15 has the same temperature as the corresponding point in Figure 13.16. So if we had all the right boundary temperatures in the rectangle, we still have them in the washer.
- Is the new function harmonic? Inside the rectangle we found an analytic function. The mapping from the first region to the second was based on an analytic function, i.e. it was a conformal mapping. And the composition of two analytic functions is another analytic function. So our new function is still analytic, which means it must be made of harmonic functions.
- Finally, and most subtly, what about the insulated boundaries? The derivative of T perpendicular to the vertical edges of the rectangle was zero; does that mean the same is true for the curved edges of the washer? It does, because conformal mappings preserve all angles. If ∇T was perpendicular to the boundaries of the rectangle, it is also perpendicular to the corresponding boundaries of the washer.

Of course you can (and should) verify that our solution satisfies Laplace's equation and meets the boundary conditions of this particular problem. But the three arguments above make the more general case that *whenever* you use a conformal mapping, the solution in the simple shape maps to a valid solution on the original shape. And remember that all of this rests on a uniqueness theorem; if you have found a harmonic function that satisfies the boundary conditions, you have found the one and only solution.

You may have noticed that the solution to this example, $T(x, y) \propto \tan^{-1}(y/x)$, is simply proportional to the phase of z. We could have solved the problem simply by recognizing that the phase of z is the imaginary part of $\ln z$ and thus harmonic, and that with appropriate constants it can satisfy all our boundary conditions. Even though this particular problem could have been solved without using a conformal mapping, we chose to start with it to clearly show how the method works. As promised, however, we can now give you an example of a problem that you would not want to solve without a conformal mapping.

EXAMPLE   Potential on a Circle

Problem:
The unit circle has electric potential $V = 1$ along the top and $V = -1$ along the bottom. It encloses no charges, so the potential inside obeys Laplace's equation. Find the potential function inside the circle using the mapping $f(z) = (1 - z)/(1 + z)$. (After all we've said about logs and circles, you might be surprised that we're not using the log here. We invite you to try that; it doesn't make the problem any easier.)

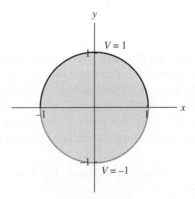

Solution:
We begin by defining our boundary mathematically. The unit circle on the complex plane consists of all the points $z = e^{i\phi}$. On the bottom half of the circle $-\pi < \phi < 0$, and on the top half $0 < \phi < \pi$; we have to treat these two curves separately because they have different boundary conditions.

Now we apply our mapping function to our boundary curves to find the new region. Plugging $z = e^{i\phi}$ into our mapping function gives $f(z) = (1 - e^{i\phi})/(1 + e^{i\phi})$. After a bit of algebra (multiplying the top and bottom by $1 + e^{-i\phi}$ is one possible approach) and a few uses of Euler's formula we end up here.

$$f(z) = \frac{-\sin\phi}{1 + \cos\phi} i$$

We notice immediately that there is no real part, so this lies entirely along the imaginary axis. As ϕ goes from 0 to π, f goes from 0 to $-\infty$, so the top half of the circle (where $V = 1$) maps to the negative imaginary axis. Similarly, the bottom half of the circle at $-\pi < \phi < 0$ maps to the positive imaginary axis, which thus inherits the boundary condition $V = -1$.

That circle was the boundary of our original region. The imaginary axis is the boundary of our new region, so our finite disk has been mapped to an infinite half-plane. But which half? One way to figure that out is to pick a sample point in the disk and try it. The easiest point is $z = 0$ which leads to $f(z) = 1$, so our points fall on the right side of the plane. (If that seems shady, plug in $z = \rho e^{i\phi}$. You will find that the real part of $f(z)$ is positive for $\rho < 1$ and negative for $\rho > 1$, showing again that our disk maps to the right side of the plane.)

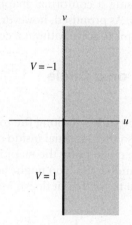

Now we have to solve Laplace's equation in that region. The key insight you need here is that this region is a big wedge, of the sort that we discussed in Section 13.3. We know that the phase ϕ is the imaginary part of $\ln z$ and therefore harmonic. This region can be described fully by $-\pi/2 \leq \phi \leq \pi/2$, with the boundary conditions $\phi(-\pi/2) = 1$ and $\phi(\pi/2) = -1$. So the function $V = -(2/\pi)\phi$ is harmonic and satisfies the boundary conditions, giving us the solution in the mapped region.

$$V(u, v) = -\frac{2}{\pi} \tan^{-1}\left(\frac{v}{u}\right)$$

To map this back to the original region we use our $f(z)$ function, but we have to break it down into a real part $u(x, y)$ and an imaginary part $v(x, y)$. So we write $f(z) = (1 - x - iy)/(1 + x + iy)$, multiply the top and bottom by $1 + x - iy$ and go through a bit more algebra, and end up here.

$$u(x, y) = \frac{1 - x^2 - y^2}{(1 + x)^2 + y^2} \quad \text{and} \quad v(x, y) = \frac{-2y}{(1 + x)^2 + y^2}$$

We use these mapping functions to turn our solution on the half-plane into the solution we wanted for the disk.

$$V(x, y) = -\frac{2}{\pi} \tan^{-1}\left(\frac{2y}{x^2 + y^2 - 1}\right)$$

As always, it's worth checking that this solution does satisfy Laplace's equation and the boundary conditions. (If you plug in $x^2 + y^2 = 1$ this will blow up, but if you take a limit as it approaches 1 you should get $V = 1$ on the upper half-circle and $V = -1$ on the lower one. More simply, you can just sample a bunch of points very close to the edge of the disk on the top and bottom.)

Finding the Right Transformation

The one remaining piece of the puzzle is choosing your mapping. You start with a problem: "solve Laplace's equation in this strange ugly region with these boundary conditions." You want to find a conformal mapping—an analytic function—that turns your region into a rectangle, or a pair of concentric circles around the origin, or a wedge, or something else that you know how to handle. Given an infinite number of analytic functions you can try, how do you find the right one?

Unfortunately there is no systematic method. Like the simpler Laplace technique in Section 13.3, conformal mapping is an answer searching for a question. What you often do—and what you will do in Problem 13.123 and several others—is work backward.

- You start with a simple region that you know how to solve.
- You apply a mapping to turn it into a complicated region.
- You find the *inverse* of that mapping so you have an analytic function that turns the complicated region into the simple one.
- And presto! You now have one complicated region for which you know how to solve Laplace's equation, because you know how to map it onto a simpler region.

That may sound like a trick that is useful only for textbook writers, but with time and practice you build up a library of regions that you know how to work with. Then a new region may remind you, even if only approximately, of something that you have solved before. Problems that follow this pattern are called "inverse problems," and many of the most useful conformal mapping solutions have been found in this way.

There is, however, a class of problems for which you can systematically find the right mapping. If your region involves lines and circles, you can use a Möbius transformation to map it to a different set of lines and circles. There are many configurations of lines and circles we know how to solve: concentric circles, parallel lines, wedges. There are many others we can't easily solve: non-concentric circles, boundaries with both circles and lines. The example we worked above fell into this category. The original problem involved a circle, but it was hard to solve because the top and bottom had different boundary conditions. We knew we could find a Möbius transformation that would turn that circle into a line, and a line with two different boundaries is a problem we can solve (because it's a special case of a wedge). You'll see other examples in the problems where a Möbius transformation will make the problem solvable, and you just have to do some algebra to find the right Möbius transformation to use.

Having come this far, you are now in a position to solve the problem we posed in the Motivating Exercise (Section 13.1). See Problem 13.131.

Stepping Back
Assuming you understand every step of the process we've outlined in this function, it's still easy to lose track of the details. It may be helpful to keep Figure 13.17 around for quick reference. What you want to know is $V(x, y)$: the solution to Laplace's equation in some region in the xy-plane. You find a function $f(z)$ that maps that region to a simpler region where you are able to solve Laplace's equation with the mapped boundary conditions. That gives you $V(u, v)$, and you can then plug in the functions $u(x, y)$ and $v(x, y)$ to get the solution $V(x, y)$ that you wanted.

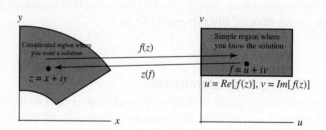

FIGURE 13.17 Mapping between a complicated region z where you want a solution to Laplace's equation and a simple region f where you can actually solve it.

For finding the regions, finding the new boundaries, and translating the solution back in the end, it's the forward mapping $f(z)$ that you need. Generally the only time you need to backwards mapping $z(f)$ is when you are doing an inverse problem: start with a known simple solution and apply a mapping $z(f)$ to find a complicated problem that you can solve with it.

13.9.2 Problems: Conformal Mapping and Laplace's Equation

13.121 Pretend you wanted to know the steady-state temperature in the 2D region shown below, which we'll call A.

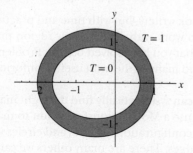

The inner boundary, at $T = 0$, is given by the curve $(x^2 + y^2)^2 + 2x^3 + 2xy^2 + x^2 + y^2 = 1$. The outer curve, at $T = 1$, is $(x^2 + y^2)^2 +$ $2x^3 + 2xy^2 + x^2 + y^2 = 4$. If you can solve this problem by inspection you have our permission to skip the rest of the problems in this section.

(a) You're going to solve this problem with the mapping $f(z) = z^2 + z$. Plug in $z = x + iy$ and calculate $|f|^2$. Simplify your answer as much as possible. Show that $|f|^2$ equals 1 on the inner curve and 4 on the outer one.

(b) Draw the region in the uv-plane that A is mapped to by the function $f(z)$. We'll call that new region B.

(c) What are the boundary conditions for $T(u, v)$ in region B?

(d) Solve Laplace's equation with the correct boundary conditions to find $T(u, v)$ in region B.
(e) Write $f(z) = u + iv$ as two real functions $u(x, y)$ and $v(x, y)$.
(f) Using the function $u(x, y)$ and $v(x, y)$ and the solution $T(u, v)$, find the solution $T(x, y)$ for region A.
(g) Verify that your solution satisfies the boundary conditions of the original problem. (Your solution to the last part should be long and complicated, and this part should nonetheless be very easy to do without a computer.)
(h) The point $(4/5, 0)$ is part of the original region A.
  i. What temperature is it? Answer based on the temperature function you found for region A.
  ii. What point does it correspond to in the mapped region B?
  iii. What temperature is *that* point in? Answer based on the temperature function you found for region B.
  iv. If the two temperatures did not come out the same, worry!

13.122 [This problem depends on Problem 13.121.]
(a) Plot the region described in Problem 13.121. On the same plot show contour lines of your solution $T(x, y)$ and verify that the boundaries of the region lie on contours.
(b) Verify that your solution to Problem 13.121 satisfies Laplace's equation. (This would be tedious without a computer.)

13.123 Walk-through: Inverse Problems. This is an inverse conformal mapping problem of the sort that we discussed in the Explanation (Section 13.9.1). You will start on the f-plane where the real axis is u and the imaginary axis is v, and solve a straightforward Laplace problem there. Then we will give you the function $z(f)$ to map this simple region to a complicated region on the z-plane where the real axis is x and the imaginary axis is y. You will end up with a difficult problem on the z-plane that you can solve by mapping it to the f-plane.
  Your starting point will be a problem you already know how to solve: the potential between two concentric circles.
(a) Consider two concentric circles centered on the origin, one of radius $1/2$ and the other of radius 1. If the potential on the inner circle is -1 and the potential on the outer circle is 0, find the potential in the region between the two circles (assuming this region is free of charge). Call this plane the f plane where $f = u + iv$, so the axes are u and v and your answer should be a function $V(u, v)$.

The next step is a mapping that turns this easy problem into a hard one. For this problem you'll use $z(f) = \sqrt{f + 1}$. This turns both circles into more complicated closed loop shapes in the z-plane.

(b) Find the equations of these complicated loop shapes. We encourage you to plot them, but it's not easy without a computer so we are not requiring it until Problem 13.124.
(c) Invert the mapping function we gave you to find the function that maps the complicated shapes back to the concentric circles.
(d) Write the function $f(z)$ in the form $u(x, y) + iv(x, y)$.
(e) Now you are ready to make up a problem. Your problem will look something like this: "Find the potential function in the complicated region described by _____ with boundary conditions _____ by using the mapping _____." If you make up exactly the right problem, then you have already done almost all the work to solve it.
(f) Solve the problem you just wrote.

13.124 [This problem depends on Problem 13.123.] In Problem 13.123 you solved Laplace's equation in the region in between the two curves generated by mapping concentric circles with the function $z(f) = \sqrt{f + 1}$. In this problem you'll see what that region and your solution look like.
(a) Plot the region on which you solved the problem. (This is *not* the region between the two concentric circles we gave you; it is a more complicated region that you solved by mapping it to those concentric circles.)
(b) Using the solution $V(x, y)$ you found, make a contour plot showing the equipotential lines (lines of constant V) in the region between these two

curves. Each of the boundary curves should be an equipotential.

Problems 13.125–13.127 are inverse problems of the kind that we discussed in the Explanation (Section 13.9.1). You might want to work through Problem 13.123 as a model before tackling these. In each problem you will be given a simple region R in the $f = u + iv$ plane, a set of boundary conditions for a function $T(u, v)$, and a transformation that maps R to a more complicated region in the $z = x + iy$ plane. For each problem...

(a) Solve Laplace's equation in the region R to find $T(u, v)$ with the given boundary conditions.

(b) Use $z(f)$ to find the region in the z-plane that corresponds to region R in the f-plane. (In some cases you can easily sketch it; in other cases you will only have a formula that is not easy to plot without a computer.)

(c) Invert $z(f)$ to find the function $f(z)$. Use that function and your solution from Part (a) to find the solution $T(x, y)$ to Laplace's equation in the region you found in Part (b).

(d) 🖥 Plot the boundaries of the xy region and plot the isotherms of $T(x, y)$ in that region.

13.125 R is the ring between a circle of radius 1 and a circle of radius 2, both centered on the origin. $T = 1$ on the inner circle and $T = 2$ on the outer one. $z(f) = 1/(3 + f)$. *Hint*: Because this is a Möbius transformation the circles will get mapped to circles. The new circles both have centers on the real line, so you can easily figure out what circles they are.

13.126 R is an upper-right quarter-disk of radius 3 centered on $f = 1$. $T = 0$ on the lower edge, $T = 1$ on the left edge, and the curved edge is insulating. (The derivative normal to the curved boundary equals 0.) $z(f) = \sqrt{f}$. *Hint*: To find $T(u, v)$ in region R first find it in a quarter-circle centered on the origin and then shift your answer by 1 in the u-direction.

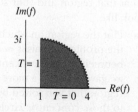

13.127 Use the same region and boundary conditions as Problem 13.126, but with the transformation $z(f) = f^2$. You do not need to have done that problem to do this one. (*Warning*: The answer will be messy.)

13.128 A circle of radius R_1 is inside a larger circle of radius R_2, and the two circles are tangent to each other (meaning they touch at one point). The inner circle is at temperature T_1 and the outer one at temperature T_2. Find the steady-state temperature in the region between them. *Hint*: Remember that a Möbius transformation whose pole touches a circle turns it into a line, so you can place your uv-axes in such a way that you can do a simple Möbius transformation to turn both of these circles into lines.

13.129 The Möbius transformation $f(z) = 1/(z - 2)$ maps the circle $|z| = 1$ to a circle that we will call C and the circle $|z| = 2$ to a line that we will call L.

(a) How do we know that $|z| = 1$ maps to a circle, and $|z| = 2$ to a line? *Hint*: It isn't obvious looking at these curves; it is a property of Möbius transformations.

(b) Plugging $z = e^{i\phi}$ into the transformation will give you C but it's easier to start knowing you're going to get a circle. Find the two points at the edges of the diameter of $|z| = 1$ along the real axis, see where they map to, and that will give you two points that define a diameter of C. Based on this, draw circle C.

(c) It's a bit easier to find line L; just find two points! Based on this, draw L on the same graph where you drew C.

(d) If C is held at potential V_C and L is held at potential V_L find the electric potential in the region exterior to C and to the left of L.

13.130 🖥 *[This problem depends on Problem 13.129.]* Use your solution to Problem 13.129 to make a plot showing the curves of constant potential ("equipotentials") in the region bounded by the circle and the line.

13.131 The region R is the semi infinite slab $0 \leq x \leq \pi/2, 0 \leq y < \infty$. The temperature is held at 0 on the bottom and right edges, and at 1 on the left edge. Use the transformation $f(z) = \sin z$ to map R to a different region, solve Laplace's equation with appropriate boundary conditions on the new shape, and transform your solution back to find $T(x, y)$ in R. (The hardest calculation in this problem should be writing the real and imaginary parts of $f(z) = \sin z$ in the form $u(x, y)$ and $v(x, y)$. One way to do it is to write $\sin z = (1/2i)(e^{iz} - e^{-iz})$, plug in $z = x + iy$, and then use Euler's formula.)

13.132 [*This problem depends on Problem 13.131.*] Use your solution to Problem 13.131 to make a plot showing the curves of constant temperature ("isotherms") in R. Make sure your plot confirms that the temperature is constant at one value along the bottom and right edges, and constant at a different value along the left edge.

13.133 Exploration: Fringing at the Edge of a Parallel Plate Capacitor. A parallel plate capacitor consists of two parallel slabs, each held at a fixed potential. If we assume the plates are infinitely long, which is usually a good approximation near the center of the plates, the potential in between them is linear. Near one edge of the plates, it is useful to approximate the plates as "half-infinite," extending infinitely far in one direction but not the other. In this problem you will find a solution for the half-plates by starting with the full-plates solution. As usual we will use $z = x + iy$ for the complex plane with the problem you want to solve (the half-plates) and $f = u + iv$ for the complex plane with the problem you can solve (the full plates).

(a) If the horizontal line $v = \pi$ is at $V = 1$ and the horizontal line $v = -\pi$ is at $V = -1$, what is the potential $V(u, v)$ in the region between these two lines?

(b) Show that the transformation $z(f) = f + e^f$ maps the line $\mathrm{Im}(f) = \pi$ to a half-line (or "ray" as mathematicians would say).
   i. Start with the line $f = a + i\pi$, where a is real. Show that $z(f)$ has a constant imaginary part.
   ii. Write an expression for $\mathrm{Re}(z)$ as a function of a and find the maximum value it takes. Show that it has no minimum, meaning it extends infinitely far to the left.

(c) Repeat the process for the line $f = a - i\pi$.

If we could invert the transformation $z(f)$ we could use that to find the solution $V(x, y)$ for the two half plates. However this function cannot be inverted. (Try!) So instead of finding $V(x, y)$ what you can do is find the equipotentials, the curves in the xy-plane along which V is constant. From your solution $V(u, v)$ you know these correspond to curves of constant v.

(d) For $v = 0$ use the function $z(f) = f + e^f$ to express x and y as functions of u. (Recall that x and y are the real and imaginary parts of z.) The functions $x(u)$ and $y(u)$ give a parametric representation of an equipotential curve. Sketch this equipotential curve in the xy-plane, labeled with the appropriate V-value.

(e) For $v = 3\pi/4$ write the parametric functions $x(u)$ and $y(u)$. To sketch the resulting curve by hand, consider what x and y are doing for very large negative values of u and for very large positive values of u. Add this equipotential to your sketch, labeled with the appropriate V-value.

(f) Find $x(u)$ and $y(u)$ for $v = 0$, $v = \pm\pi/4$, $v = \pm\pi/2$, and $v = \pm 3\pi/4$. You've already done two of these; add the other five curves to your sketch, each labeled with the appropriate V-value.

(g) In words, describe how the equipotentials behave far inside the capacitor (between the two rays), near the edge of the capacitor, and far away from the capacitor.

(h) Have a computer plot a large number of equipotentials using your parametric representation. Check if the behavior matches what you determined from your rough sketches.

13.10 Special Application: Fluid Flow (see felderbooks.com)

13.11 Additional Problems (see felderbooks.com)

APPENDIX A

Different Types of Differential Equations

Linear vs. Non-linear

A *linear* differential equation is a linear combination of the dependent variable and its derivatives. For instance, a linear third-order ODE can be written as

$$a_3(x)\frac{d^3y}{dx^3} + a_2(x)\frac{d^2y}{dx^2} + a_1(x)\frac{dy}{dx} + a_0(x)y = f(x)$$

where the $a_n(x)$ functions and $f(x)$ can be any functions of x but cannot contain y. This equation would be *non-linear* if it contained y^2, or $\sqrt{\frac{dy}{dx}}$, or $y\frac{dy}{dx}$, for instance.

Why do you care?

- Non-linear differential equations are very difficult to solve in general. Most solutions involve approximating them with linear equations, or using computers to find approximate numerical solutions.

Ordinary vs. Partial

An *ordinary* differential equation (ODE) has one dependent variable, one independent variable, and constants.

A *partial* differential equation (PDE) has one dependent variable, two or more independent variables, and constants.

Why do you care?

- The techniques for solving the two are different.
- We often solve PDEs by turning them into ODEs.
- The general solution to an ODE has a finite number of arbitrary *constants* (more on this below). The general solution to a PDE has arbitrary *functions* (or, equivalently, an infinite number of arbitrary constants—such as the coefficients of a Fourier series).

Order of the Equation

The *order* of a differential equation is the *highest-order derivative* in the equation. For instance:

$$\frac{d^3y}{dx^3} + e^x \frac{dy}{dx} - 3y = 4x^5$$

is a third-order differential equation, because it has a third derivative, and it has no fourth derivatives, fifth derivatives, *etc.*

Why do you care?

- Every nth-order linear ODE has a general solution with n arbitrary constants.
- Because an nth-order ODE requires n arbitrary constants, it generally requires n conditions to get a specific solution. For example, for the equation above you could specify $y(0)$, $y'(0)$, and $y''(0)$, or you could specify $y(0)$, $y'(0)$, and $y(5)$.
- For PDEs, a similar rule applies as a general guideline. For instance, the heat equation $\partial u/\partial t = \alpha(\partial^2 u/\partial x^2)$ is first order in t and second order in x, so it will generally require one initial condition and two boundary conditions.

Homogeneous vs. Inhomogeneous

A linear differential equation is *homogeneous* if it has no term that does not contain the dependent variable. For example,

$$x\frac{d^2y}{dx^2} + (\cos x)y = f(x)$$

is homogeneous if $f(x) = 0$, but inhomogeneous for any other value of $f(x)$ (including a constant).

Why do you care?

- If a linear differential equation is *homogeneous* then any linear combinations of solutions is itself a solution. For example, if $\sin(3x)$ and $\ln(x+5)$ are both solutions, then $16\sin(3x) - \pi \ln(x+5)$ is also a solution.
- The fact that linear combinations of solutions are also solutions gives you a strategy for finding the general solution to an nth-order ODE. If you have found n linearly independent solutions $y_1(x), y_2(x), \ldots, y_n(x)$ then the general solution is $y(x) = C_1 y_1(x) + C_2 y_2(x) + \ldots + C_n y_n(x)$ where the C_i are all arbitrary constants.
- If a differential equation is linear but *inhomogeneous*, it can sometimes be solved by finding the general solution to the "complementary" equation that you get by replacing $f(x)$ with zero, and then adding any particular solution of the original inhomogeneous equation.

Note that this is one of two unrelated meanings of "homogeneous differential equation." The other, discussed in Section 10.8, refers to an ODE that can be written in the form $y'(x) = f(y/x)$ for any function f.

Appendix A Different Types of Differential Equations

Hyperbolic, Elliptic, Parabolic

In algebra we learn that the second-order equation:

$$ax^2 + by^2 + cx + dy + e = 0$$

describes a *parabola* if a or b is zero, an *ellipse* if a and b have the same sign, and a *hyperbola* if a and b have opposite signs. (If a and b are both zero then it isn't a second-order equation at all: it's a line.)

By analogy, the second-order, linear partial differential equation:

$$A\frac{\partial^2 z}{\partial x^2} + B\frac{\partial^2 z}{\partial y^2} + C\frac{\partial z}{\partial x} + D\frac{\partial z}{\partial y} + E = 0$$

is described as *parabolic* if A or B is zero, *elliptic* if they have the same sign, and *hyperbolic* if they have different signs. (If A and B are both zero then it isn't a second-order differential equation at all.)

Why do you care?

- An elliptic equation typically involves only spatial derivatives, and its boundary conditions are usually specified on all borders of the domain of interest. A parabolic or hyperbolic equation typically involves both time and space. The spatial boundary conditions are usually given on all spatial borders, but the initial conditions are typically given at only one value of time. These are not guaranteed mathematical truths, but they are good rules of thumb.
- A parabolic equation is second order in one variable and first order in the other. When separated, it is likely to show exponential growth or decay in the first-order variable, but something more complicated (trig functions, Bessel functions, *etc*) in the second-order variable.

Note: If a second-order algebra equation also has an *xy* term, you can do a linear transformation to a new set of coordinates where that term vanishes and then use the rule above to determine which shape it describes. Likewise, if a second-order PDE has $\frac{\partial^2 z}{\partial x \, \partial y}$ you can do a linear transformation to a new set of independent variables for which that term will vanish. We do not discuss this technique in this book: see a textbook on PDEs for more details.

APPENDIX B

Taylor Series

Some Common Maclaurin Series

$$e^x = \sum_{n=0}^{\infty} \frac{x^n}{n!} = 1 + x + \frac{x^2}{2!} + \frac{x^3}{3!} + \frac{x^4}{4!} + \ldots$$

$$\sin x = \sum_{n=0}^{\infty} (-1)^n \frac{x^{2n+1}}{(2n+1)!} = x - \frac{x^3}{3!} + \frac{x^5}{5!} - \frac{x^7}{7!} + \ldots$$

$$\cos x = \sum_{n=0}^{\infty} (-1)^n \frac{x^{2n}}{(2n)!} = 1 - \frac{x^2}{2!} + \frac{x^4}{4!} - \frac{x^6}{6!} + \ldots$$

$$(1+x)^p \quad \text{for } |x| < 1 = 1 + px + \frac{p(p-1)x^2}{2!} + \frac{p(p-1)(p-2)x^3}{3!} + \ldots$$

The last expansion above is valid for any value of p, but for positive integer values of p we can write it in summation notation as $\sum_{n=0}^{p} p! x^n / [(p-n)! n!]$.

Memorization tip: If you remember that sine and cosine both alternate, but you can't remember which starts with 1 and which starts with x, just remember that $\sin 0 = 0$ and $\cos 0 = 1$.

The General Form for a Taylor Series

Maclaurin series (Taylor series around $x = 0$):

$$f(x) = \sum_{n=0}^{\infty} \frac{f^{(n)}(0)}{n!} x^n = f(0) + f'(0)x + \frac{f''(0)}{2!} x^2 + \frac{f'''(0)}{3!} x^3 + \ldots$$

Taylor series about $x = a$:

$$f(x) = \sum_{n=0}^{\infty} \frac{f^{(n)}(a)}{n!} (x-a)^n = f(a) + f'(a)(x-a) + \frac{f''(a)}{2!} (x-a)^2 + \frac{f'''(a)}{3!} (x-a)^3 + \ldots$$

Estimating the Possible Error in a Truncated Taylor Series

Suppose you use an nth-order Taylor series about a point $x = a$ to approximate a function f at some point $x \neq a$. The "remainder" or "error" R_n is the difference between the approximation and the actual value $f(x)$. There are several techniques you can use to find bounds on that error. We present three of them below.

1. If the series has alternating terms with monotonically decreasing absolute value then the remainder is guaranteed to be less than or equal in magnitude to the next term in the series.

$$R_n \leq \left| c_{n+1} x^{n+1} \right| \quad \text{alternating series only}$$

Note that your ability to use this rule depends, not only on the Taylor series you are using, but on the value you are plugging in. For instance the series $1 + x + x^2 + x^3 + \ldots$ alternates for $x < 0$ but not for $x > 0$.

2. A more general formula—in that it works whether the series alternates or not—is the "Lagrange remainder."

$$R_n = \frac{(x-a)^{n+1}}{(n+1)!} f^{(n+1)}(z) \quad \text{The Lagrange remainder}$$

Here $f^{(n+1)}$ is the $(n+1)^{st}$ derivative of f and z is a number between x and a. Note that if $z = x$ then the Lagrange remainder is the next term in the Taylor series—the same upper bound we saw for alternating series above.

In practice you can't generally use this to find the error because you don't know z, but you can use it to place an upper bound on the error by finding the value of z in that domain that maximizes this formula.

3. If a Maclaurin series converges on the interval $|x| < 1$ and has monotonically decreasing absolute value then $R_n \leq \left| c_{n+1} x^{n+1} \right| / (1 - |x|)$.

APPENDIX C

Summary of Tests for Series Convergence

Infinite series are discussed in Chapter 2.

The nth partial sum of an infinite series is the sum of the first n terms.

$$S_n = \sum_{i=1}^{n} a_i = a_1 + a_2 + a_3 \ldots a_n$$

The sum of an infinite series is defined as the limit of the partial sums.

$$\sum_{n=1}^{\infty} a_n = \lim_{n \to \infty} S_n$$

In some rare cases such as "telescoping series" you can find a closed-form expression for the nth partial sum and then use this definition directly to determine the convergence or divergence of a series. For other cases we have the tests.

Name	Converge or Diverge?	Notes
Ratio Test	If $\lim_{n \to \infty} \left\lvert \frac{a_{n+1}}{a_n} \right\rvert$ exists and is: • <1 the series absolutely converges • >1 the series diverges • = 1 this test is inconclusive	This is the first step in evaluating almost any power series. It may also be the last step. It may, however, show that the series converges within an interval and diverges outside that interval—in which case other tests are used for the endpoints of the interval.
nth term test	• If $\lim_{n \to \infty} a_n \neq 0$ then the series diverges. • If $\lim_{n \to \infty} a_n = 0$ this test is inconclusive.	1. Because it is quick and easy, this is generally the first test to try for a strictly numerical series. 2. This test can *only prove divergence*, never convergence!
Geometric Series $a_n = a_1 r^{n-1}$	• If $\lvert r \rvert < 1$ it converges to $\frac{a_1}{1-r}$ • If $\lvert r \rvert \geq 1$ it diverges	
p-series $a_n = \frac{1}{n^p}$	• If $p > 1$ the series converges • If $p \leq 1$ the series diverges	Don't get these confused with geometric series! $1/2^n$ is geometric; $1/n^2$ is a p-series.

Appendix C Summary of Tests for Series Convergence

Name	Converge or Diverge?	Notes
Integral Test	If the improper integral $\int_1^\infty f(x)\,dx$ exists—that is to say, if the integral converges to some finite value—then the series $\sum_{n=1}^{\infty} f(n)$ converges. If the integral diverges, the series diverges.	Works for any continuous, positive-term, decreasing function that you can integrate. The primary limitation is that some functions—most notably those involving factorials—cannot be integrated.
Basic Comparison Test	• If $a_n \le b_n$ for all n, and $\sum b_n$ converges then $\sum a_n$ converges. • If $a_n \ge b_n$ for all n, and $\sum b_n$ diverges then $\sum a_n$ diverges.	Usually used to compare a given series to a geometric or p-series, since we know what they do. The comparison test applies only to positive-term series. Here are a few useful comparisons: $\|\sin n\| \le 1$, $\|\cos n\| \le 1$, $1 < \ln n < n$ (after a certain point)
Limit Comparison Test	If $\lim\limits_{n\to\infty} \dfrac{a_n}{b_n}$ exists and is not zero then $\sum a_n$ and $\sum b_n$ either *both converge* or *both diverge*.	Like the comparison test, this requires finding some other series to compare to. You find some simple series that you suspect grows like your series for large n; then you use this test to make that intuition formal.
Absolute Convergence	If $\sum \|a_n\|$ converges, then $\sum a_n$ converges. In this case it is said to "absolutely converge."	Sometimes the best approach to an alternating series is to show that it *would converge* even if it didn't alternate! In that case it certainly converges when it does alternate.
Alternating Series Test	If the alternating series $\sum_{n=1}^{\infty}(-1)^{n-1} b_n = b_1 - b_2 + b_3 - b_4 + \ldots$ ($b_n \ge 0$) satisfies… 1. $b_{n+1} \le b_n$ for all n, and 2. $\lim\limits_{n\to\infty} b_n = 0$ then the series is convergent.	Note that the b_n terms here are *not* the individual terms of the original series; they are the terms with the alternator removed.

The discussion above applies to real series. Section 13.7 discusses convergence of complex power series. It includes a method for using complex series to find the interval of convergence of real power series, as an alternative to using the ratio test.

APPENDIX D

Curvilinear Coordinates

The following table offers conversions between the three coordinate systems. All of these conversions can be quickly derived from basic right triangle geometry. Note that we have taken the "shortcut" of writing $\cos\theta = z/r$ when converting to spherical—in other words, you calculate r before θ.

Everything in this appendix applies to polar coordinates if you simply use the formulas for cylindrical coordinates with $z = 0$.

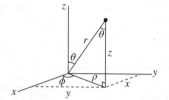

From \ To	Cartesian	Cylindrical	Spherical
Cartesian		$\rho^2 = x^2 + y^2$ $\tan\phi = y/x$ $z = z$	$r^2 = x^2 + y^2 + z^2$ $\cos\theta = z/r$ $\tan\phi = y/x$
Cylindrical	$x = \rho\cos\phi$ $y = \rho\sin\phi$ $z = z$		$r^2 = \rho^2 + z^2$ $\cos\theta = z/r$ $\phi = \phi$
Spherical	$x = r\sin\theta\cos\phi$ $y = r\sin\theta\sin\phi$ $z = r\cos\theta$	$\rho = r\sin\theta$ $\phi = \phi$ $z = r\cos\theta$	

Caution: the equation $\phi = \tan^{-1}(y/x)$ will always give you an answer in the range $-\pi/2 \leq \phi \leq \pi$. For a point in the second or third quadrant, add π to get the correct ϕ.

Differentials and Jacobians

A differential move in the x-direction has length dx, and the same rule applies for all of the length coordinates: a differential move in the ρ-direction has length $d\rho$ and so on.

It's different for the angles. A differential move in the ϕ-direction has length $\rho\,d\phi$ in cylindrical coordinates, or equivalently $r\sin\theta\,d\phi$ in spherical coordinates.

A differential move in the θ-direction has length $r\,d\theta$.

You can summarize all of this with vector equations.

$$d\vec{s}_{Cartesian} = dx\,\hat{i} + dy\,\hat{j} + dz\,\hat{k}$$
$$d\vec{s}_{cylindrical} = d\rho\,\hat{\rho} + \rho\,d\phi\,\hat{\phi} + dz\,\hat{k}$$
$$d\vec{s}_{spherical} = dr\,\hat{r} + r\,d\theta\,\hat{\theta} + r\sin\theta\,\hat{\phi}$$

Appendix D Curvilinear Coordinates

From these differential lengths you can immediately get differential volumes.

$$dV_{Cartesian} = dx\,dy\,dz$$
$$dV_{cylindrical} = \rho\,d\rho\,d\phi\,dz$$
$$dV_{spherical} = r^2 \sin\theta\,dr\,d\theta\,d\phi$$

The Jacobians for these coordinate systems are just these formulas without the differentials, e.g. $r^2 \sin\theta$ for spherical coordinates.

In polar coordinates there's no dz so $dA_{polar} = \rho\,d\rho\,d\phi$.

Unit Vectors

The symbols $\hat{i}$, $\hat{j}$, and $\hat{k}$ represent unit vectors (vectors of magnitude 1) in the x-, y-, and z-directions, respectively. The Cartesian unit vectors are invariant: that is, $\hat{i}$ is the same vector no matter where you are. By contrast, curvilinear unit vectors are position dependent. For instance, $\hat{r}$ means a unit vector in the r-direction, which means it points directly away from the origin—so $\hat{r}$ at one point is not the same vector as $\hat{r}$ at another point.

The following table gives formulas for the curvilinear unit vectors as a function of position, expressed in Cartesian and the native coordinate systems. Once again polar coordinate are simply cylindrical coordinates without the $\hat{k}$.

Cylindrical	Spherical
$\hat{\rho} = \dfrac{x\hat{i} + y\hat{j}}{\sqrt{x^2 + y^2}} = \cos\phi\,\hat{i} + \sin\phi\,\hat{j}$	$\hat{r} = \dfrac{x\hat{i} + y\hat{j} + z\hat{k}}{\sqrt{x^2 + y^2 + z^2}} = \sin\theta\cos\phi\,\hat{i} + \sin\theta\sin\phi\,\hat{j} + \cos\theta\,\hat{k}$
$\hat{\phi} = \dfrac{-y\hat{i} + x\hat{j}}{\sqrt{x^2 + y^2}} = -\sin\phi\,\hat{i} + \cos\phi\,\hat{j}$	$\hat{\phi} = \dfrac{-y\hat{i} + x\hat{j}}{\sqrt{x^2 + y^2}} = -\sin\phi\,\hat{i} + \cos\phi\,\hat{j}$
$\hat{k} = \hat{k}$	$\hat{\theta} = \dfrac{xz\hat{i} + yz\hat{j} - (x^2 + y^2)\hat{k}}{\sqrt{x^2 + y^2}\sqrt{x^2 + y^2 + z^2}} = \cos\theta\cos\phi\,\hat{i} + \cos\theta\sin\phi\,\hat{j} - \sin\theta\,\hat{k}$

APPENDIX E

Matrices

The 2 × 2 Rotation Matrix

In two dimensions, the matrix that rotates vectors counterclockwise by an angle $\Delta\phi$ is:

$$\begin{pmatrix} \cos(\Delta\phi) & -\sin(\Delta\phi) \\ \sin(\Delta\phi) & \cos(\Delta\phi) \end{pmatrix}$$

For a clockwise rotation you replace $\Delta\phi$ with $-\Delta\phi$, which just switches the signs on the $\sin(\Delta\phi)$ and $-\sin(\Delta\phi)$ terms. Remember that rotating the axes clockwise (a passive transformation) is accomplished with the same matrix as rotating the vectors counterclockwise (an active transformation).

Linear Algebra Glossary

In all the definitions below **R** and **S** refer to the following two matrices, while **M** refers to any generic matrix.

$$\mathbf{R} = \begin{pmatrix} 3 & 2 & 4 \\ 1 & 5 & 8 \end{pmatrix} \quad \mathbf{S} = \begin{pmatrix} 6 & 2 \\ 3 & 5 \end{pmatrix}$$

The dimensions of a matrix are the number of rows, and the number of columns, in that order. (**R** has dimensions 2 × 3.) Each element is referred to by its row followed by its column, written as a subscript. ($R_{23} = 8$.)

Linear Algebra Terms That Can Apply to Either Square or Not-Square Matrices

Adjoint The adjoint $\mathbf{M}^\dagger$ of a matrix **M** is obtained by taking the "transpose" and then taking the complex conjugate of each element. ($\mathbf{M}^\dagger_{ij} = \mathbf{M}^*_{ji}$.) For a real-valued matrix such as **R** or **S**, adjoint and transpose mean the same thing. *Section 7.3.*

Column A column is a vertical list of numbers. Matrix **R** has three columns, **S** two. *Section 6.2.*

Column Matrix A column matrix is a matrix with just one column. *Section 6.2.*

Rank The rank of a matrix is the number of independent rows. It can be shown that this is equal to the number of independent columns. *Section 7.4* (see felderbooks.com).

Row A row is a horizontal list of numbers. Matrices **R** and **S** have two rows each. *Section 6.2.*

Row Matrix A row matrix is a matrix with just one row. *Section 6.2.*

Transpose The "transpose" $\mathbf{M}^T$ of a matrix $\mathbf{M}$ is obtained by turning all its rows into columns. ($\mathbf{M}^T_{ij} = \mathbf{M}_{ji}$.) Section 7.1.

$$\mathbf{R}^T = \begin{pmatrix} 3 & 1 \\ 2 & 5 \\ 4 & 8 \end{pmatrix} \qquad \mathbf{S}^T = \begin{pmatrix} 6 & 3 \\ 2 & 5 \end{pmatrix}$$

Linear Algebra Terms That Apply Only to Square Matrices

Block Diagonal A block diagonal matrix can be written as a set of square matrices going from the upper left to the lower right, with all zeroes outside of them. Section 7.1.

$$\begin{pmatrix} 2 & 3 & 0 & 0 & 0 \\ 1 & 4 & 0 & 0 & 0 \\ 0 & 0 & 1 & 3 & 1 \\ 0 & 0 & 2 & 4 & 4 \\ 0 & 0 & 3 & 1 & 5 \end{pmatrix} \quad \text{A block diagonal matrix}$$

Determinant The determinant is the product of the "eigenvalues" of a matrix. The determinant is zero if the rows of the matrix are linearly dependent. (If the rows are then the columns will be also and vice-versa.) Section 6.7.

Diagonal Matrix In a diagonal matrix every element is zero except the ones on the "main diagonal." Section 6.2.

Eigenvectors and Eigenvalues A column matrix $\mathbf{v}$ is an eigenvector of a square matrix $\mathbf{M}$ if $\mathbf{Mv} = \lambda \mathbf{v}$ for some non-zero constant λ. The constant λ is the corresponding eigenvalue. Section 6.8.

Hermitian A complex square matrix $\mathbf{M}$ is Hermitian (sometimes called self-adjoint) if it equals its "adjoint." ($\mathbf{M} = \mathbf{M}^\dagger$.) For a real matrix, "symmetric" and Hermitian mean the same thing. A Hermitian matrix has orthogonal "eigenvectors." (Two vectors $\vec{v}$ and $\vec{u}$ are orthogonal if $\sum_i v_i u_i^* = 0$. For spatial vectors, orthogonal and perpendicular mean the same thing. Don't confuse this with "orthogonal matrices.") Section 7.3.

Identity Matrix The identity matrix $\mathbf{I}$ has 1s along the "main diagonal" and 0s everywhere else. For any matrix $\mathbf{M}$ with the same dimensions as a given identity matrix, $\mathbf{MI} = \mathbf{IM} = \mathbf{M}$. Section 6.6.

Inverse Matrix The inverse matrix $\mathbf{M}^{-1}$ of a square matrix $\mathbf{M}$ is defined by the property $\mathbf{MM}^{-1} = \mathbf{M}^{-1}\mathbf{M} = \mathbf{I}$ (where $\mathbf{I}$ is the "identity matrix"). Section 6.6.

$$\text{The inverse of } \begin{pmatrix} a & c \\ b & d \end{pmatrix} \text{ is } \frac{1}{ad - bc} \begin{pmatrix} d & -c \\ -b & a \end{pmatrix}.$$

Main Diagonal *(also principal diagonal, primary diagonal, or leading diagonal)* The main diagonal is the list of items going from the upper left to the lower right (M_{ii}). Section 6.2.

Orthogonal A real square matrix is orthogonal if it preserves magnitudes of vectors. In other words if $\mathbf{M}$ is orthogonal and $\mathbf{M}\vec{v} = \vec{u}$ then $|\vec{u}| = |\vec{v}|$. A matrix is orthogonal if and only if its "transpose" is its "inverse." ($\mathbf{M}^T = \mathbf{M}^{-1}$). An equivalent condition is that a matrix is orthogonal if and only if each of its columns is a vector of magnitude 1 and these vectors are all perpendicular to each other. Section 7.1.

Square Matrix A square matrix has the same number of rows as columns. Matrix **S** is a square matrix, **R** is not. *Section 6.2.*

Symmetric A square matrix **M** is symmetric if the matrix equals its "transpose." ($\mathbf{M} = \mathbf{M}^T$.) A real symmetric matrix has perpendicular "eigenvectors." *Section 7.3.*

Trace The trace is the sum of the terms on the "main diagonal" $\left(\sum_i M_{ii}\right)$.

Unitary A complex square matrix is unitary if it preserves magnitudes of complex vectors. (The magnitude of a complex vector is $\sqrt{\sum_i |v_i|^2}$.) A matrix is unitary if and only if $\mathbf{M}^\dagger = \mathbf{M}^{-1}$. For a real matrix unitary and "orthogonal" mean the same thing. *Section 7.3.*

APPENDIX F

Vector Calculus

Vector Calculus Operators

Operator	Notation	What kind of field goes in?	What kind of field comes out?
gradient	$\vec{\nabla} f$	scalar	vector
divergence	$\vec{\nabla} \cdot \vec{f}$	vector	scalar
curl	$\vec{\nabla} \times \vec{f}$	vector	vector
Laplacian	$\vec{\nabla}^2 f$	scalar	scalar

Cartesian Coordinates

Differential Length: $d\vec{s} = dx\,\hat{i} + dy\,\hat{j} + dz\,\hat{k}$

Gradient of f: $\dfrac{\partial f}{\partial x}\hat{i} + \dfrac{\partial f}{\partial y}\hat{j} + \dfrac{\partial f}{\partial z}\hat{k}$

Divergence of $\vec{f}$: $\dfrac{\partial f_x}{\partial x} + \dfrac{\partial f_y}{\partial y} + \dfrac{\partial f_z}{\partial z}$

Curl of $\vec{f}$: $\left(\dfrac{\partial f_z}{\partial y} - \dfrac{\partial f_y}{\partial z}\right)\hat{i} + \left(\dfrac{\partial f_x}{\partial z} - \dfrac{\partial f_z}{\partial x}\right)\hat{j} + \left(\dfrac{\partial f_y}{\partial x} - \dfrac{\partial f_x}{\partial y}\right)\hat{k}$

Laplacian of f: $\dfrac{\partial^2 f}{\partial x^2} + \dfrac{\partial^2 f}{\partial y^2} + \dfrac{\partial^2 f}{\partial z^2}$

Polar and Cylindrical Coordinates

Differential Length: $d\vec{s} = d\rho\,\hat{\rho} + \rho\,d\phi\,\hat{\phi} + dz\,\hat{k}$

Gradient of f: $\dfrac{\partial f}{\partial \rho}\hat{\rho} + \dfrac{1}{\rho}\dfrac{\partial f}{\partial \phi}\hat{\phi} + \dfrac{\partial f}{\partial z}\hat{k}$

Divergence of $\vec{f}$: $\dfrac{1}{\rho}\dfrac{\partial}{\partial \rho}(\rho f_\rho) + \dfrac{1}{\rho}\dfrac{\partial f_\phi}{\partial \phi} + \dfrac{\partial f_z}{\partial z}$

Curl of $\vec{f}$: $\left(\dfrac{1}{\rho}\dfrac{\partial f_z}{\partial \phi} - \dfrac{\partial f_\phi}{\partial z}\right)\hat{\rho} + \left(\dfrac{\partial f_\rho}{\partial z} - \dfrac{\partial f_z}{\partial \rho}\right)\hat{\phi} + \dfrac{1}{\rho}\left[\dfrac{\partial}{\partial \rho}(\rho f_\phi) - \dfrac{\partial f_\rho}{\partial \phi}\right]\hat{k}$

Laplacian of f: $\dfrac{\partial^2 f}{\partial \rho^2} + \dfrac{1}{\rho}\dfrac{\partial f}{\partial \rho} + \dfrac{1}{\rho^2}\dfrac{\partial^2 f}{\partial \phi^2} + \dfrac{\partial^2 f}{\partial z^2}$

Spherical Coordinates

Differential Length: $d\vec{s} = dr\,\hat{r} + r\,d\theta\,\hat{\theta} + r\sin\theta\,d\phi\,\hat{\phi}$

Gradient of f: $\dfrac{\partial f}{\partial r}\hat{r} + \dfrac{1}{r}\dfrac{\partial f}{\partial \theta}\hat{\theta} + \dfrac{1}{r\sin\theta}\dfrac{\partial f}{\partial \phi}\hat{\phi}$

Divergence of $\vec{f}$: $\dfrac{1}{r^2}\dfrac{\partial}{\partial r}(r^2 f_r) + \dfrac{1}{r\sin\theta}\dfrac{\partial}{\partial \theta}(f_\theta \sin\theta) + \dfrac{1}{r\sin\theta}\dfrac{\partial f_\phi}{\partial \phi}$

Curl of $\vec{f}$: $\dfrac{1}{r\sin\theta}\left[\dfrac{\partial}{\partial \theta}(f_\phi \sin\theta) - \dfrac{\partial f_\theta}{\partial \phi}\right]\hat{r} + \dfrac{1}{r}\left[\dfrac{1}{\sin\theta}\dfrac{\partial f_r}{\partial \phi} - \dfrac{\partial}{\partial r}(rf_\phi)\right]\hat{\theta}$
$+ \dfrac{1}{r}\left[\dfrac{\partial}{\partial r}(rf_\theta) - \dfrac{\partial f_r}{\partial \theta}\right]\hat{\phi}$

Laplacian of f: $\dfrac{\partial^2 f}{\partial r^2} + \dfrac{2}{r}\dfrac{\partial f}{\partial r} + \dfrac{1}{r^2 \sin^2\theta}\dfrac{\partial^2 f}{\partial \phi^2} + \dfrac{\cos\theta}{r^2\sin\theta}\dfrac{\partial f}{\partial \theta} + \dfrac{1}{r^2}\dfrac{\partial^2 f}{\partial \theta^2}$

Fundamental Theorems

Name	Given	Formula	In Words
Gradient Theorem	A vector function $\vec{f}$, a scalar function F such that $\vec{\nabla} F = \vec{f}$, and a curve that runs from P_1 to P_2	$\int_C \vec{f}\cdot d\vec{s} = F(P_2) - F(P_1)$	You can evaluate the line integral of a conservative vector field along any curve from the values of its potential function at the endpoints.
Divergence Theorem	A vector function $\vec{f}$ and a three dimensional region V bounded by the closed surface S.	$\iiint_V (\vec{\nabla}\cdot\vec{f})\,dV = \oiint_S \vec{f}\cdot d\vec{A}$	If you add up the divergence everywhere within the region, you get the integral through the bounding surface.
Stokes' Theorem	A vector function $\vec{f}$ and a surface S bounded by the curve C.	$\iint_S (\vec{\nabla}\times\vec{f})\cdot d\vec{A} = \oint_C \vec{f}\cdot d\vec{s}$	If you add up the curl everywhere within the surface, you get the line integral around the bounding curve.

Stokes' theorem is intrinsically three dimensional, but the divergence theorem can apply in any number of dimensions. In 2D $\iint_S (\vec{\nabla}\cdot\vec{f})\,dA = \oint_C \vec{f}\cdot d\vec{s}$ where S is a surface and C is its boundary.

APPENDIX G

Fourier Series and Transforms

Some Integrals that are Useful in Finding Fourier Series

You can figure out the following using integration by parts, but they come up often enough that it is helpful to have a simple list.

$$\int x \sin(ax)\,dx = -\frac{x}{a}\cos(ax) + \frac{1}{a^2}\sin(ax) + C \quad (a \neq 0)$$

$$\int x \cos(ax)\,dx = \frac{x}{a}\sin(ax) + \frac{1}{a^2}\cos(ax) + C \quad (a \neq 0)$$

$$\int x e^{iax}\,dx = \left(-\frac{ix}{a} + \frac{1}{a^2}\right)e^{iax} + C \quad (a \neq 0)$$

$$\int e^{kx}\sin(bx)\,dx = \frac{e^{kx}}{k^2 + b^2}[k\sin(bx) - b\cos(bx)] + C \quad (k^2 + b^2 \neq 0)$$

$$\int e^{kx}\cos(bx)\,dx = \frac{e^{kx}}{k^2 + b^2}[k\cos(bx) + b\sin(bx)] + C \quad (k^2 + b^2 \neq 0)$$

After you integrate remember that n (the summation index) is always an integer, and for any integer n:

$$\sin(\pi n) = 0, \qquad \cos(\pi n) = (-1)^n, \qquad e^{i\pi n} = (-1)^n$$

Fourier Series Formulas

The sine-cosine Fourier series for a function $f(x)$ with period $2L$ is:

$$f(x) = a_0 + \sum_{n=1}^{\infty} a_n \cos\left(\frac{n\pi}{L}x\right) + \sum_{n=1}^{\infty} b_n \sin\left(\frac{n\pi}{L}x\right) \quad \text{where}$$

$$a_0 = \frac{1}{2L}\int_{-L}^{L} f(x)\,dx \qquad a_n = \frac{1}{L}\int_{-L}^{L} f(x)\cos\left(\frac{n\pi}{L}x\right)dx \qquad b_n = \frac{1}{L}\int_{-L}^{L} f(x)\sin\left(\frac{n\pi}{L}x\right)dx$$

The complex exponential Fourier series for a function $f(x)$ with period $2L$ is:

$$f(x) = \sum_{n=-\infty}^{\infty} c_n e^{i\left(\frac{n\pi}{L}\right)x} \quad \text{where} \quad c_n = \frac{1}{2L}\int_{-L}^{L} f(x) e^{-i\left(\frac{n\pi}{L}\right)x}\,dx$$

You can evaluate any of these integrals over any full period. $\left(\int_{-L}^{L} \text{ is the same as } \int_{0}^{2L}.\right)$

Fourier Sine and Cosine Series

If a function $f(x)$ is only defined from $x = 0$ to $x = L$ then it can be represented by a Fourier sine series:

$$f(x) = \sum_{n=1}^{\infty} b_n \sin\left(\frac{n\pi}{L}x\right) \quad \text{where} \quad b_n = \frac{2}{L} \int_0^L f(x) \sin\left(\frac{n\pi}{L}x\right) dx$$

or equivalently by a Fourier cosine series:

$$f(x) = a_0 + \sum_{n=1}^{\infty} a_n \cos\left(\frac{n\pi}{L}x\right)$$

$$\text{where} \quad a_0 = \frac{1}{L} \int_0^L f(x) \, dx \quad \text{and} \quad a_n = \frac{2}{L} \int_0^L f(x) \cos\left(\frac{n\pi}{L}x\right) dx$$

Either one of these series will in general converge to $f(x)$ at all points on the interval $0 < x < L$ where f is continuous. On the interval $-L < x < 0$ the sine series will create an "odd extension" of the function and the cosine series an "even extension."

Multivariate Fourier Series

The sine-cosine Fourier series for a function $f(x, y)$ with periods $2L$ and $2W$ in the x-and y-directions respectively is:

$$\sum_{m=0}^{\infty}\sum_{n=0}^{\infty} A_{mn} \cos\left(\frac{m\pi}{L}x\right)\cos\left(\frac{n\pi}{W}y\right) + B_{mn}\cos\left(\frac{m\pi}{L}x\right)\sin\left(\frac{n\pi}{W}y\right) +$$
$$C_{mn}\sin\left(\frac{m\pi}{L}x\right)\cos\left(\frac{n\pi}{W}y\right) + D_{mn}\sin\left(\frac{m\pi}{L}x\right)\sin\left(\frac{n\pi}{W}y\right)$$

where

$$A_{mn} = \frac{1}{QLW}\int_{-W}^{W}\int_{-L}^{L} f(x,y) \cos\left(\frac{m\pi}{L}x\right)\cos\left(\frac{n\pi}{W}y\right) dx\, dy,$$
$$B_{mn} = \frac{1}{QLW}\int_{-W}^{W}\int_{-L}^{L} f(x,y) \cos\left(\frac{m\pi}{L}x\right)\sin\left(\frac{n\pi}{W}y\right) dx\, dy$$

and similarly for C and D. The Q in the denominator is 4 if m and n are both zero, 2 if one of them is zero, and 1 if neither one is zero. (This sounds strange but it's a consistent extrapolation of a_0 in a single-variable series.)

If the function is even in x then all terms with $\sin x$ will go to zero, if a function is odd in x then all terms with $\cos x$ will go to zero, and the same is true of course for y.

The complex exponential form is:

$$f(x,y) = \sum_{m=-\infty}^{\infty}\sum_{n=-\infty}^{\infty} c_{mn} e^{i\left(\frac{m\pi}{L}\right)x} e^{i\left(\frac{n\pi}{W}\right)y}$$

$$\text{where} \quad c_{mn} = \frac{1}{4LW}\int_{-W}^{W}\int_{-L}^{L} f(x,y) e^{-i\left(\frac{m\pi}{L}\right)x} e^{-i\left(\frac{n\pi}{W}\right)y} dx\, dy$$

If a function is only defined on the domain $0 \le x \le L$, $0 \le y \le W$ you can make an odd or even extensions of each variable. The most useful of these is an odd extension in both variables, so we present the formulas for that. It's not hard to generalize to the formulas for other extensions.

$$f(x,y) = \sum_{m=0}^{\infty} \sum_{n=0}^{\infty} D_{mn} \sin\left(\frac{m\pi}{L}x\right) \sin\left(\frac{n\pi}{W}y\right)$$

$$\text{where } D_{mn} = \frac{4}{LW} \int_0^W \int_0^L f(x,y) \sin\left(\frac{m\pi}{L}x\right) \sin\left(\frac{n\pi}{W}y\right) dx\, dy$$

Fourier Transforms

Warning: Different sources use slightly different equations for Fourier transforms. Your formulas will work as long as you stick consistently to the conventions we present here, or to any other self-consistent set. But if you Fourier transform with our equation, and then inverse Fourier transform with someone else's, you may get an incorrect answer.

A non-periodic function $f(x)$ can be represented by a Fourier integral:

$$f(x) = \int_{-\infty}^{\infty} \hat{f}(p) e^{ipx}\, dp$$

The function $\hat{f}(p)$ is the "Fourier transform" of $f(x)$, and can be calculated from:

$$\hat{f}(p) = \frac{1}{2\pi} \int_{-\infty}^{\infty} f(x) e^{-ipx}\, dx$$

The corresponding formulas for a multivariate function $f(x,y)$ are

$$f(x,y) = \int_{-\infty}^{\infty} \int_{-\infty}^{\infty} \hat{f}(p,k) e^{ipx} e^{iky}\, dp\, dk \quad \text{where} \quad \hat{f}(p,k) = \frac{1}{4\pi^2} \int_{-\infty}^{\infty} \int_{-\infty}^{\infty} f(x,y) e^{-ipx} e^{-iky}\, dx\, dy$$

For a function defined on the half-infinite interval $[0, \infty)$ you can define a Fourier sine or cosine integral by making an odd or even extension to negative values.

$$f(x) = \int_0^{\infty} \hat{f}_s(p) \sin(px)\, dp \quad \text{where} \quad \hat{f}_s(p) = \frac{2}{\pi} \int_0^{\infty} f(x) \sin(px)\, dx$$

or

$$f(x) = \int_0^{\infty} \hat{f}_c(p) \cos(px)\, dp \quad \text{where} \quad \hat{f}_c(p) = \frac{2}{\pi} \int_0^{\infty} f(x) \cos(px)\, dx$$

A particularly useful Fourier transform that comes up frequently is a Gaussian.

Simplest case: $f(x) = e^{-x^2} \quad \to \quad \hat{f}(p) = \dfrac{1}{2\sqrt{\pi}} e^{-p^2/4}$

General case: $f(x) = Ae^{-[(x-x_0)/w]^2} \quad \to \quad \hat{f}(p) = \dfrac{Aw}{2\sqrt{\pi}} e^{ipx_0} e^{-w^2 p^2/4}$

APPENDIX H

Laplace Transforms

The "Laplace transform" of a function $f(t)$ is given by:

$$\mathcal{L}\left[f(t)\right] = F(s) = \int_0^\infty f(t)e^{-st}\,dt$$

Chapter 10 introduces Laplace transforms and uses them to solve ordinary differential equations and coupled differential equations. Chapter 11 uses Laplace transforms to solve partial differential equations. To find an inverse Laplace transform by hand requires a contour integral on the complex plane, so this is discussed in Chapter 13.

The Laplace transform is a linear operator.

$$\mathcal{L}\left[af(t) + bg(t)\right] = aF(s) + bG(s)$$

The Laplace transform turns derivatives and integrals into multiplication and division.

$$\mathcal{L}[f'(t)] = sF(s) - f(0)$$

$$\mathcal{L}[f''(t)] = s^2 F(s) - sf(0) - f'(0)$$

$$\mathcal{L}\left[\frac{\partial^{(n)} f}{\partial t^{(n)}}\right] = s^n F(s) - s^{n-1} f(0) - s^{n-2}\frac{\partial f}{\partial t}(0) - s^{n-3}\frac{\partial^2 f}{\partial t^2}(0) - \ldots - \frac{\partial^{(n-1)} f}{\partial t^{(n-1)}}(0)$$

$$\mathcal{L}\left[\int_0^t f(\tau)\,d\tau\right] = \frac{F(s)}{s}$$

Some Useful Laplace Transforms

$f(t)$	$\mathcal{L}[f(t)]$, or $F(s)$	Domain	Other Notes
a (a constant)	a/s	$Re(s) > 0$	
$\delta(t - t_0)$ (Dirac delta)	$e^{-t_0 s}$	All s	$t_0 > 0$
$H(t - t_0)$ (Heaviside)	$e^{-t_0 s}/s$	$Re(s) > 0$	$t_0 \geq 0$
t	$1/s^2$	$Re(s) > 0$	
t^p	$p!/s^{p+1}$	$Re(s) > 0$	$p = 0, 1, 2, \ldots$
t^p	$\Gamma(p+1)/s^{p+1}$	$Re(s) > 0$	$p > -1$
e^{-at}	$1/(s+a)$	$Re(s) > Re(-a)$	
$\sin(\omega t)$	$\omega/(s^2 + \omega^2)$	$Re(s) > |Im(\omega)|$	
$\cos(\omega t)$	$s/(s^2 + \omega^2)$	$Re(s) > |Im(\omega)|$	
$t^n g(t)$	$(-1)^n (d^n G/ds^n)$		
$e^{-at} g(t)$	$G(s + a)$		

You can find much more extensive tables online, or ask your favorite mathematical software package for Laplace and inverse Laplace transforms. When using a table, don't forget the property of linearity! For instance, you can tell immediately from the table above that:

$$\mathcal{L}\left[3\sin(5t) + 7e^{11t}\right] = 15/(s^2 + 25) + 7/(s - 11)$$

Other cases may take more work. For instance, suppose you want $\mathcal{L}^{-1}\left[1/(2s+5)^7\right]$. Your first step might be to pull out the 2 on the bottom.

$$\mathcal{L}^{-1}\left[\frac{1}{2^7} \frac{1}{(s + 5/2)^7}\right]$$

That looks like the formula for $t^n e^{-at}$ with $n = 6$ and $a = 5/2$. It's missing the 6! in the numerator, but we can put that in too.

$$\mathcal{L}^{-1}\left[\frac{1}{(2s+5)^7}\right] = \frac{1}{(2^7)(6!)} \mathcal{L}^{-1}\left[\frac{6!}{(s+5/2)^7}\right] = \frac{t^6 e^{-5t/2}}{(2^7)(6!)}$$

Try a few! It's easy to check your answers (to all the problems in this appendix) if you have a computer that can take Laplace transforms.

1. $\mathcal{L}[5t^2 + 6t + 7]$
2. $\mathcal{L}[4H(t - 1) - 4H(t - 2)]$
3. $\mathcal{L}[\sinh t]$ Hint: $\sinh t = \left(e^t - e^{-t}\right)/2$.
4. $\mathcal{L}^{-1}[1/s^3]$
5. $\mathcal{L}^{-1}[1/(3s + 4)^2]$
6. $\mathcal{L}^{-1}[1/(cs + d)^5]$
7. $\mathcal{L}^{-1}[1/(2s^2 + 18)]$

Partial Fractions

You may have learned the technique of partial fractions as a trick for finding integrals, but it is also very useful for looking up inverse Laplace transforms.

EXAMPLE   **Partial Fractions and Inverse Laplace Transforms**

Problem:
Use partial fractions and the table above to find $\mathcal{L}^{-1}\left[11/(4s^2 + 5s - 6)\right]$.

Solution:
Noting that the bottom factors into $(4s - 3)(s + 2)$, we can rewrite the fraction. (This is not a property of Laplace transforms, it's just algebra.)

$$\frac{11}{(4s-3)(s+2)} = \frac{A}{4s-3} + \frac{B}{s+2}$$

Now we add the fractions on the right.

$$\frac{A}{4s-3} + \frac{B}{s+2} = \frac{A(s+2) + B(4s-3)}{(4s-3)(s+2)} = \frac{As + 2A + 4Bs - 3B}{(4s-3)(s+2)} = \frac{(A+4B)s + (2A-3B)}{(4s-3)(s+2)}$$

We want this to be the same as the fraction we started with. The denominators are the same so the numerators must match term by term.

$$A + 4B = 0 \quad \text{and} \quad 2A - 3B = 11$$

You can solve these with substitution, elimination or matrices to find $A = 4$ and $B = -1$. So we can rewrite our original problem in a different form. *Then* we can use linearity to look up each of these in the table individually and combine the answers.

$$\mathcal{L}^{-1}\left[\frac{11}{4s^2+5s-6}\right] = \mathcal{L}^{-1}\left[\frac{4}{4s-3} - \frac{1}{s+2}\right] = \mathcal{L}^{-1}\left[\frac{1}{s-3/4} - \frac{1}{s+2}\right] = e^{3t/4} - e^{-2t}$$

Try a few!

8. $\mathcal{L}^{-1}\left[(s+22)/(s^2+9s+14)\right]$
9. $\mathcal{L}^{-1}\left[(9s+6)/(s^2-2s-24)\right]$
10. $\mathcal{L}^{-1}\left[13/(2s^2+11s-6)\right]$
11. $\mathcal{L}^{-1}\left[\dfrac{3s^2-13s+34}{(s+1)(s-4)^2}\right]$ *Hint*: You need to break this up into three fractions: one with denominator $s+1$, one with $s-4$, and one with $(s-4)^2$.

Convolution

The property of linearity applies to the sum of two functions. "Convolution" gives you a way of finding the inverse Laplace transform of a *product* of two functions.

Convolution

Given two functions $f(t)$ and $g(t)$, the "convolution" of these two functions is defined as:

$$(f * g)(t) = \int_0^t f(\tau)g(t-\tau)d\tau$$

The Laplace transform of a convolution is the product of the Laplace transforms of the individual functions.

$$\mathcal{L}\left[(f * g)(t)\right] = \mathcal{L}[f(t)]\mathcal{L}[g(t)]$$

Convolution is commutative, meaning $f * g = g * f$. (Problem 18.) For our purposes this means that you can apply the $t - \tau$ to whichever function is convenient.

In the example below we use convolution to evaluate the same inverse Laplace transform that we found above with partial fractions.

EXAMPLE **Convolution and Inverse Laplace Transforms**

Problem:
Use convolution and the table above to find $\mathcal{L}^{-1}\left[11/(4s^2 + 5s - 6)\right]$.

Solution:
Once again we begin by noting that we can factor the denominator. We can also factor out the 11 (linearity again) and rewrite the problem like this.

$$11\,\mathcal{L}^{-1}\left[\left(\frac{1}{4s-3}\right)\left(\frac{1}{s+2}\right)\right]$$

We can find the Laplace transforms of $1/(4s-3)$ and $1/(s+2)$ using the table above.

$$\mathcal{L}^{-1}\left[\frac{1}{4s-3}\right] = \frac{1}{4}\mathcal{L}^{-1}\left[\frac{1}{s-3/4}\right] = \frac{1}{4}e^{3t/4} \quad \text{and} \quad \mathcal{L}^{-1}\left[\frac{1}{s+2}\right] = e^{-2t}$$

The inverse Laplace transform of the product of these two functions is the convolution of those two functions.

$$11\int_0^t \frac{1}{4}e^{3\tau/4}e^{-2(t-\tau)}d\tau = \frac{11}{4}e^{-2t}\int_0^t e^{11\tau/4}d\tau = e^{-2t}\left(e^{11t/4} - 1\right) = e^{3t/4} - e^{-2t}$$

Hey, that looks familiar!

Try a few!

12. $\mathcal{L}^{-1}\left[(s+22)/(s^2+9s+14)\right]$ (Yes, this is Problem 8 again. If you did it before with partial fractions, make sure you get the same answer with convolution.)

13. $\mathcal{L}^{-1}\left[\dfrac{1}{s(s^2+9)}\right]$

14. $\mathcal{L}^{-1}\left[\dfrac{1}{s^2(s^2+9)}\right]$

15. $\mathcal{L}^{-1}\left[\left(\dfrac{a}{s^2+a^2}\right)\left(\dfrac{s}{s^2+a^2}\right)\right]$ *Hint*: You will need a few trig identities.

16. Use convolution to show that if $\mathcal{L}[f(t)] = F(s)$ then $\mathcal{L}^{-1}\left[F(s)(1/s)\right] = \int_0^t f(\tau)d\tau$. (This is the property that makes Laplace transforms useful for integro-differential equations.)

17. The table above gives the Laplace transforms of a and $\delta(t-t_0)$. Use those facts and convolution to show that $\mathcal{L}^{-1}\left[e^{-t_0 s}/s\right] = H(t-t_0)$.

18. Prove that convolution is commutive. *Hint*: substitute $u = t - \tau$.

APPENDIX I

Summary: Which PDE Technique Do I Use?

Solving differential equations is in some ways like integration. You learn a battery of techniques, each with a specific set of steps and conditions. Faced with a new problem, you have to start by figuring out *which technique to use*. There are no surefire rules for this task—in fact, for some problems, several different techniques may apply. But below we have collected the major guidelines for the techniques presented in Chapter 11.

If you go through this process and decide to use separation of variables, you may find it helpful to review the summary of that process on Page 611.

1. **Is your equation non-linear?**
   You cannot use the techniques in Chapter 11. You may be able to find a linear approximation to your equation (Chapter 2). Otherwise you need to solve the equation numerically.

The rest of this appendix will focus on linear equations only.

---

2. **First consider separation of variables.**
   This is the simplest technique, and therefore the first to try. For a wide variety of differential equations this technique will not only find a solution, but will also identify the normal modes of your system, which is a useful goal in itself.
   (a) Start with a set of simple questions.
      i. **Is your differential equation homogeneous?** *and*
      ii. **Does it have only two independent variables, and can you algebraically separate them?** *and*
      iii. **Does it have inhomogeneous initial conditions and homogeneous boundary conditions,** *or* **no initial conditions and exactly one inhomogeneous boundary condition?**

      If you can answer "yes" to *all* of these questions, you may be a winner! Try separation of variables in its simplest form (Section 11.4). However, be ready for any of a wide variety of normal modes (Sections 11.6 and 11.7).
   (b) **Does your problem meet all the conditions from Part (a) except that you have more than two independent variables?** As long as you can separate them, the technique should still work. However, you will have to solve more ordinary differential equations, and you will end up with more series. (Section 11.5).
   (c) **Does your problem meet all the conditions from Part (a) except that you have initial conditions and inhomogeneous boundary conditions?** You may be able to find a particular solution that meets the boundary conditions and a complementary

solution that solves the PDE and the now-modified initial conditions with homogeneous boundary conditions (Section 11.8). If the boundary conditions are difficult to match (e.g. time-dependent boundary conditions) you can use an extra trick that we illustrate in Section 11.12 Problem 11.260 (see felderbooks.com).

(d) **Does your problem meet all the conditions from Part (a) except that you have too many inhomogeneous boundary conditions?** You may be able to create "subproblems" with fewer inhomogeneous conditions, and then sum the solutions to solve the problem you were given (Section 11.8).

(e) **Is your differential equation inhomogeneous?** The easiest approach to an inhomogeneous equation is to find *one particular solution* to the equation, often by making a simplifying assumption. ("Hey, a line will work" or "What if I look for a steady-state solution by setting $\partial f/\partial t$ equal to zero?") Such a solution has no arbitrary constants and therefore cannot match your initial conditions, but you can then add it to a solution to the "complementary" homogeneous equation, which you may be able to find with separation of variables (Section 11.8). If that approach does not work you discard separation of variables altogether and move on to the techniques discussed below.

As powerful as separation of variables is, it is not universal. For example, it cannot be used on any equation for which you cannot separate the variables (which includes, but is not limited to, inhomogeneous equations). When separation of variables is unusable for any reason, the following techniques may serve.

These techniques apply to individual variables, and can be done in combination. For example, if you have a problem where $0 < x < L$, $-\infty < y < \infty$, and $0 < t < \infty$, you might be able to simplify the problem by applying an eigenfunction expansion to x, a Fourier transform to y, or a Laplace transform to t. It may even prove useful to do all three!

3. **If you have a spatial variable with homogeneous boundary conditions and a finite domain, consider eigenfunction expansion (Section 11.9).**
   To apply this method, you have to be able to find the eigenfunctions of your differential operator. The most common example is the operator $\partial^2/\partial x^2$ for which the eigenfunctions are sines and cosines, so you expand into a Fourier series.

4. **Consider using a Fourier transform (Section 11.10) if you have a spatial variable with an infinite or semi-infinite domain that appears only in a second derivative and (optionally) in an inhomogeneous term. In other words, consider this method if you can rewrite your PDE as**

$$\frac{\partial^2 f}{\partial x^2} + (\text{term that may include } x \text{ but doesn't depend on } f)$$
$$+ (\text{other terms that don't involve } x) = 0$$

We specified that this was a *spatial* variable because time usually doesn't have boundary conditions at infinity, which we need as described below.

(a) **Does your variable have an infinite domain with** $\lim_{x \to \infty} f(x, \ldots) = \lim_{x \to -\infty} f(x, \ldots) = 0$?
Use a regular Fourier transform with complex exponentials.

(b) **Does your variable have a semi-infinite domain (e.g. $0 < x < \infty$), with $\lim_{x \to \infty} f(x, \ldots) = 0$, and an explicit boundary condition that specifies f at $x = 0$?**
Use a Fourier sine transform.

(c) **Does your variable have a semi-infinite domain (e.g. $0 < x < \infty$), with $\lim_{x \to \infty} f(x, \ldots) = 0$, and an explicit boundary condition that specifies $\partial f/\partial x$ at $x = 0$?**
Use a Fourier cosine transform.

5. **If time appears in your equation only in the derivatives, not in the coefficients, consider a Laplace transform (Section 11.11).**
   This method only works if t is defined on a semi-infinite domain $0 < t < \infty$ and the conditions on this variable are all defined at $t = 0$. Those criteria rarely apply to spatial variables.
6. **If you have a variable with an infinite or semi-infinite domain that does not meet the criteria for Parts 4 or 5, consider a different transform.**
   The transform you choose is based on your differential operator and boundary conditions. The transform should be an expansion of your solution in eigenfunctions of the differential operator in your equation. Other transforms are not covered in this book, but by stressing the commonalities of all transform methods, we hope we have prepared you to select and use different transforms for different problems.

This summary does not include every known analytical technique for solving partial differential equations, but it does cover the techniques that are most commonly used in simple situations. It is not uncommon to encounter PDEs that cannot be solved by any of the methods described above, or PDEs that cannot be analytically solved at all! This brings us to the last step of our summary:

7. **If you need to solve a partial differential equation that cannot be solved by any of these techniques, consider solving it numerically.**

APPENDIX J

Some Common Differential Equations and Their Solutions

The table below lists solutions to some common differential equations. The sections below the table give the minimum amount of information you need to use these functions to solve ordinary differential equations, or to find the normal modes for partial differential equations. More information about many of these functions can be found in Chapter 12. Some special functions that are not primarily defined as solutions to particular ODEs can be found in Appendix K.

You may need to do a bit of algebra to make your problem match an equation in this table. Watch in particular for variable substitutions (Section 10.7).

Example: $(k^2 - x^2)y'' - 2xy + l(l+1)y = 0$

Substitute: $u = x/k$

After some algebra the equation becomes: $(1 - u^2)y'' - 2uy' + l(l+1)y = 0$

Look that up below to get the solution: $y = AP_n(x/k) + BQ_n(x/k)$

Equation	Solution	Name
$y'' + k^2 y = 0$	$A\sin(kx) + B\cos(kx)$ or, equivalently $A\cos(kx + \phi)$	Trigonometric functions
$y'' - k^2 y = 0$	$Ae^{kx} + Be^{-kx}$ or, equivalently $A\sinh(kx) + B\cosh(kx)$	Exponential functions or, equivalently Hyperbolic trig functions
$(1 - x^2)y'' - 2xy' + l(l+1)y = 0$	$AP_l(x) + BQ_l(x)$	Legendre polynomials
$(1 - x^2)y'' - 2xy' + \left(l(l+1) - \dfrac{m^2}{1 - x^2}\right)y = 0$	$AP_l^m(x)$	Associated Legendre functions
$x^2 y'' + xy' + (k^2 x^2 - p^2)y = 0$	$AJ_p(kx) + BY_p(kx)$	Bessel functions
$x^2 y'' + xy' - (k^2 x^2 + p^2)y = 0$	$AI_p(kx) + BK_p(kx)$	Modified Bessel functions
$x^2 y'' + 2xy' + (k^2 x^2 - p(p+1))y = 0$	$Aj_p(kx) + By_p(kx)$	Spherical Bessel functions
$y'' - k^3 xy = 0$	$A\,Ai(kx) + B\,Bi(kx)$	Airy functions
$xy'' + (1 - x)y' + ny = 0$	$AL_n(x)$	Laguerre polynomials
$xy'' + (\alpha + 1 - x)y' + ny = 0$	$AL_n^\alpha(x)$	Associated Laguerre polynomials
$y'' - 2xy' + 2ny = 0$	$AH_n(x)$	Hermite polynomials
$\dfrac{\partial^2 y}{\partial \theta^2} + \dfrac{\cos\theta}{\sin\theta}\dfrac{\partial y}{\partial \theta} + \dfrac{1}{\sin^2\theta}\dfrac{\partial^2 y}{\partial \phi^2} + l(l+1)y = 0$	$Y_l^m(\theta, \phi)$	Spherical harmonics

Trigonometric Functions

We assume you know the basic properties of trig functions, but we include them to emphasize the parallels with many less familiar functions. To solve a PDE whose normal modes are sines and cosines, you generally only need the facts in this appendix—and the same is true for the other functions listed here.

- $\sin 0 = 0$; $\cos 0 = 1$.
- $\sin x$ has an infinite number of zeros. They occur at $n\pi$ where n is any integer. $\cos x$ has an infinite number of zeros that occur at $\pi/2 + n\pi$.
- Both $\sin x$ and $\cos x$ are periodic, with period 2π.
- $\sin x$ has odd symmetry, $\cos x$ even.
- $\frac{d}{dx} \sin x = \cos x$; $\frac{d}{dx} \cos x = -\sin x$. (Like many of the other facts listed here, this only holds if x is measured in radians.)
- If a function $f(x)$ on the domain $[-L, L]$ is expanded into a "Fourier series" of the form:

$$f(x) = a_0 + \sum_{n=1}^{\infty} \left[a_n \cos\left(\frac{n\pi}{L}x\right) + b_n \sin\left(\frac{n\pi}{L}x\right) \right]$$

then the coefficients are given by:

$$a_0 = \frac{1}{2L} \int_{-L}^{L} f(x)\, dx \quad a_n = \frac{1}{L} \int_{-L}^{L} f(x) \cos\left(\frac{n\pi}{L}x\right) dx \quad b_n = \frac{1}{L} \int_{-L}^{L} f(x) \sin\left(\frac{n\pi}{L}x\right) dx$$

- Sines and cosines are orthogonal. For integer n and m:

$$\int_{-L}^{L} \sin\left(\frac{m\pi}{L}x\right) \sin\left(\frac{n\pi}{L}x\right) dx = \int_{-L}^{L} \cos\left(\frac{m\pi}{L}x\right) \cos\left(\frac{n\pi}{L}x\right) dx = L\delta_{mn}$$

$$\int_{-L}^{L} \sin\left(\frac{m\pi}{L}x\right) \cos\left(\frac{n\pi}{L}x\right) dx = 0$$

More information on Fourier series can be found in Chapter 9 and Appendix G.

Exponential Functions

- The function e^x has no zeros, and is not symmetric.
- $\lim_{x \to \infty} e^x = \infty$ and $\lim_{x \to -\infty} e^x = 0$. The reverse is of course true for e^{-x}.
- The function $A\sinh(kx) + B\cosh(kx)$ (discussed below) is fully equivalent to $Ae^{kx} + Be^{-kx}$, but with different values of the arbitrary constants. The hyperbolic trig form is easier for matching given initial conditions $y(0)$ and $y'(0)$.

Hyperbolic Trig Functions

The hyperbolic trig functions are "elementary functions" because they are defined in terms of exponentials.

$$\sinh x = \frac{e^x - e^{-x}}{2} \quad \cosh x = \frac{e^x + e^{-x}}{2}$$

The other four hyperbolic trig functions are defined in parallel to the regular trig functions: for instance, $\tanh x = \sinh x / \cosh x$.

You can easily work out all the properties below from the definitions above. It is more important to note how these properties do, and do not, parallel the properties of the sine and cosine. And it is most important to note how these properties fit into solving differential equations.

- $\sinh 0 = 0$; $\cosh 0 = 1$.
- The origin is the only zero of the hyperbolic sine. The hyperbolic cosine has no zeros.
- $\lim_{x \to \infty} \cosh x = \lim_{x \to -\infty} \cosh x = \lim_{x \to \infty} \sinh x = \infty$; $\lim_{x \to -\infty} \sinh x = -\infty$.
- $\sinh x$ has odd symmetry, $\cosh x$ even.
- $\frac{d}{dx} \sinh x = \cosh x$; $\frac{d}{dx} \cosh x = \sinh x$. (No negative sign!)

The hyperbolic trig functions are discussed in more detail in Section 12.3.

Legendre Polynomials and Associated Legendre Functions

- The functions $P_l(x)$, "Legendre polynomials of the first kind," are lth-order polynomials. The first few are:

$$P_0(x) = 1, \quad P_1(x) = x, \quad P_2(x) = \frac{1}{2}(3x^2 - 1), \quad P_3(x) = \frac{1}{2}(5x^3 - 3x)$$

- The functions $Q_l(x)$, "Legendre polynomials of the second kind," are non-terminating polynomials. They converge on the interval $-1 < x < 1$ and diverge elsewhere. For applications that require convergence on a larger interval, including at $x = \pm 1$, these functions are discarded.
- If a function $f(x)$ in the domain $[-1, 1]$ is expanded into a "Fourier-Legendre series expansion" of the form:

$$f(x) = \sum_{l=0}^{\infty} A_l P_l(x)$$

then the coefficients are given by:

$$A_l = \frac{2l+1}{2} \int_{-1}^{1} f(x) P_l(x) dx$$

- The Legendre functions are orthogonal on the domain $[-1, 1]$.

$$\int_{-1}^{1} P_l(x) P_m(x) dx = \frac{2}{2l+1} \delta_{lm}$$

- It is often convenient to expand functions in Legendre polynomials of cosines:

$$f(\theta) = \sum_{l=0}^{\infty} A_l P_l(\cos \theta)$$

In this case the coefficients are given by

$$A_l = \frac{2l+1}{2} \int_0^{\pi} f(\theta) P_l(\cos \theta) \sin \theta \, d\theta$$

- The functions $P_l^m(x)$ are called "Associated Legendre Functions." The Legendre polynomials are special cases of the associated Legendre functions: $P_l(x) = P_l^0(x)$.
- If a function $f(x)$ in the domain $[-1, 1]$ is expanded into an "Fourier-associated Legendre series expansion" of order m, of the form:

$$f(x) = \sum_{l=m}^{\infty} A_l P_l^m(x)$$

then the coefficients are given by:

$$A_l = \frac{2l+1}{2}\frac{(l-m)!}{(l+m)!}\int_{-1}^{1} f(x)P_l^m(x)\,dx \qquad l = (m, m+1, m+2\ldots)$$

- The associated Legendre functions for any given m are orthogonal.

$$\int_{-1}^{1} P_a^m(x)P_b^m(x)\,dx = \frac{2}{2a+1}\frac{(a+m)!}{(a-m)!}\delta_{ab}$$

- The functions $P_l^m(x)$ are sometimes called "associated Legendre polynomials" even though they are not generally polynomials. Such is the perversity of the human spirit.

Legendre polynomials are used in solving PDEs in Section 11.7, and are derived in Section 12.5.

Bessel Functions

The Bessel functions are two different families of functions. $J_p(x)$ stands for the "Bessel functions of the first kind of order p" and $Y_p(x)$ are "Bessel functions of the second kind of order p."

- $J_0(0) = 1$. For $p > 0$, $J_p(0) = 0$.
- All the $J_p(x)$ functions have an infinite number of zeros. These are difficult to calculate, but they are listed in many mathematical tables and computer systems. By convention $\alpha_{p,n}$ represents the nth zero of the Bessel function J_p; for instance, $\alpha_{0,4}$ is the fourth positive zero of the function J_0.[1]
- For $p \geq 0$, $\lim_{x \to 0} Y_p(x)$ diverges. For many problems, this makes the $J_p(x)$ functions the only solutions.
- All the $Y_p(x)$ functions have an infinite number of zeros.
- If a function $f(x)$ in the domain $[0, a]$ is expanded into a "Fourier-Bessel series expansion" of the form:

$$f(x) = \sum_{n=1}^{\infty} A_n J_p\left(\frac{\alpha_{p,n}}{a}x\right)$$

then the coefficients are given by:

$$A_n = \frac{2}{a^2 J_{p+1}^2(\alpha_{p,n})}\int_0^a x f(x) J_p\left(\frac{\alpha_{p,n}}{a}x\right) dx$$

Note that this involves a choice of p. For instance, you may take a given function and expand it into third-order Bessel functions $J_3(\alpha_{3,n}x/a)$, or ninth-order Bessel functions $J_9(\alpha_{9,n}x/a)$, or any other order. All these will produce different series with different coefficients.

- The Bessel functions of a given order are orthogonal to each other with weight x. For integer m and n:

$$\int_0^a x J_p\left(\frac{\alpha_{p,m}}{a}x\right)J_p\left(\frac{\alpha_{p,n}}{a}x\right) dx = \frac{a^2}{2}J_{p+1}^2(\alpha_{p,m})\delta_{mn}$$

---

[1] When a Bessel function is being examined on a finite interval $[0, a]$, some textbooks use $\lambda_{p,n}$ to mean $\frac{\alpha_{p,n}}{a}$.

- The solutions to Bessel's equation can also be written in terms of "Hankel functions," also known as "Bessel functions of the Third Kind."

$$H_p^{(1)}(x) = J_p(x) + iY_p(x), \qquad H_p^{(2)}(x) = J_p(x) - iY_p(x)$$

Bessel functions are introduced and used to solve PDEs in Section 11.6, and derived in Section 12.7.

Modified (or Hyperbolic) Bessel functions

The modified Bessel functions $I_p(x)$ and $K_p(x)$ bear a similar relationship to the Bessel functions as the hyperbolic trig functions bear to the regular trig functions: they solve the same equation but with a different sign for one of the constants.

- $I_0(0) = 1$. For $p > 0$, $I_p(0) = 0$.
- For all values of p, $\lim_{x \to 0} K_p(x)$ diverges.
- For $p \geq 0$ neither $I_p(x)$ nor $K_p(x)$ has any zeroes for positive values of x.
- The modified Bessel functions are not orthogonal or complete, so you cannot make a "Fourier-modifed Bessel series expansion" out of them.

Spherical Bessel functions

- $j_0(0) = 1$. For $p > 0$, $j_p(0) = 0$.
- For all values of p, $\lim_{x \to 0} y_p(x)$ diverges.
- The spherical Bessel functions are related to the ordinary Bessel functions by

$$j_p(x) = \sqrt{\frac{\pi}{2x}} J_{p+1/2}(x), \qquad y_p(x) = \sqrt{\frac{\pi}{2x}} Y_{p+1/2}(x)$$

Therefore the zeros of $j_p(x)$ are the same as the zeros of $J_{p+1/2}(x)$ and likewise for $y_p(x)$ and $Y_{p+1/2}(x)$.

- If a function $f(x)$ in the domain $[0, a]$ is expanded into a "Fourier-spherical Bessel series expansion" of the form:

$$f(x) = \sum_{n=1}^{\infty} A_n j_p\left(\frac{\alpha_{p+1/2,n}}{a} x\right)$$

then the coefficients are given by:

$$A_n = \frac{2}{a^3 j_{p+1}^2(\alpha_{p+1/2,n})} \int_0^a x^2 f(x) j_p\left(\frac{\alpha_{p+1/2,n}}{a} x\right) dx$$

As with regular Bessel functions, you can choose to expand a function $f(x)$ in whatever order (value of p) spherical Bessel function you want.

- The spherical Bessel functions of a given order are orthogonal to each other with weight x^2. For integer m and n:

$$\int_0^a x^2 j_p\left(\frac{\alpha_{p+1/2,m}}{a} x\right) j_p\left(\frac{\alpha_{p+1/2,n}}{a} x\right) dx = \frac{a^3}{2} j_{p+1}^2(\alpha_{p+1/2,m}) \delta_{mn}$$

Airy functions

- The values at $x = 0$ can be expressed in terms of gamma functions.

$$Ai(0) = \frac{1}{3^{2/3} \Gamma(2/3)}, \qquad Bi(0) = \frac{1}{3^{1/6} \Gamma(2/3)}, \qquad Ai'(0) = -\frac{1}{3^{1/3} \Gamma(1/3)}, \qquad Bi'(0) = \frac{3^{1/6}}{\Gamma(1/3)}$$

Appendix J Some Common Differential Equations and Their Solutions

(The gamma function $\Gamma(x)$ function is described in Section 12.3, and in Appendix K.)
- $\lim\limits_{x \to \infty} Ai(x) = \lim\limits_{x \to -\infty} Ai(x) = \lim\limits_{x \to -\infty} Bi(x) = 0$. $\lim\limits_{x \to \infty} Bi(x)$ diverges.
- Neither Airy function has any zeroes for positive x, and both have infinitely many zeroes for negative x.
- For positive x the Airy functions can be written in terms of modified Bessel functions.

$$Ai(x) = \frac{1}{\pi}\sqrt{\frac{x}{3}} K_{1/3}\left(\frac{2}{3}x^{3/2}\right), \quad Bi(x) = \sqrt{\frac{x}{3}}\left[I_{1/3}\left(\frac{2}{3}x^{3/2}\right) + I_{-1/3}\left(\frac{2}{3}x^{3/2}\right)\right]$$

- For negative x the Airy functions can be written in terms of regular Bessel functions.

$$Ai(-x) = \sqrt{\frac{x}{9}}\left[J_{1/3}\left(\frac{2}{3}x^{3/2}\right) + J_{-1/3}\left(\frac{2}{3}x^{3/2}\right)\right],$$

$$Bi(-x) = \sqrt{\frac{x}{9}}\left[J_{1/3}\left(\frac{2}{3}x^{3/2}\right) - J_{-1/3}\left(\frac{2}{3}x^{3/2}\right)\right]$$

You can therefore find the zeroes of the Airy functions in terms of the zeroes of $J_{1/3}(x)$.

Laguerre Polynomials and Associated Laguerre Polynomials
- All the formulas given here for the associated Laguerre polynomials $L_n^\alpha(x)$ apply to the Laguerre polynomials $L_n(x)$ if you set $\alpha = 0$.
- The boundary conditions that typically go with the Laguerre equations are a bit complicated, but in practice they restrict n and α to be integers and they eliminate the second solution to the equations.
- The Laguerre polynomials can be defined by the following formula.

$$L_n(x) = \frac{1}{n!} e^x \frac{d^n}{dx^n}(x^n e^{-x}) = \sum_{m=0}^{n}(-1)^m \frac{n!}{(n-m)!(m!)^2} x^m$$

With this definition $L_n(0) = 1$ for all n. Sometimes the $1/n!$ is omitted, in which case $L_n(0) = n!$.

Warning: The formulas given below apply to the definition for the Laguerre polynomials given above. Some sources omit the $1/n!$ coefficient; the resulting functions solve the same differential equation, but the formulas must be modified accordingly.

- $L_n(x)$ is an nth-order polynomial. The first few are:

$$L_0(x) = 1, \quad L_1(x) = 1 - x, \quad L_2(x) = 1 - 2x + \frac{1}{2}x^2, \quad L_3(x) = 1 - 3x + \frac{3}{2}x^2 - \frac{1}{6}x^3$$

- The associated Laguerre polynomials can be defined by $L_n^\alpha(x) = (-1)^\alpha (d^\alpha/dx^\alpha)L_{n+\alpha}(x)$.
- If a function $f(x)$ in the domain $[0, \infty)$ is expanded into a "Fourier-associated Laguerre series expansion" of the form:

$$f(x) = \sum_{n=0}^{\infty} A_n L_n^\alpha(x)$$

then the coefficients are given by:

$$A_n = \frac{n!}{(n+\alpha)!} \int_0^\infty f(x) x^\alpha e^{-x} L_n^\alpha(x) dx$$

- The associated Laguerre polynomials for any given α are orthogonal with weight $x^\alpha e^{-x}$.

$$\int_0^\infty x^\alpha e^{-x} L_m^\alpha(x) L_n^\alpha(x) dx = \frac{(n+\alpha)!}{n!} \delta_{mn}$$

Hermite Polynomials

> **Warning:** The differential equation and properties given here for Hermite polynomials are standard in physics. In probability theory different conventions are typically used.

- The boundary conditions that typically go with the Hermite equations are a bit complicated, but in practice they generally restrict n to be an integer and they eliminate the second solution to the equations.
- The Hermite polynomials can be defined by the following formula.

$$H_n(x) = (-1)^n e^{x^2} \frac{d^n}{dx^n} \left(e^{-x^2} \right)$$

- $H_n(x)$ is an nth-order polynomial. The first few are:

$$H_0(x) = 1, \quad H_1(x) = 2x, \quad H_2(x) = 4x^2 - 2, \quad H_3(x) = 8x^3 - 12x$$

- If a function $f(x)$ in the domain $(-\infty, \infty)$ is expanded into a "Fourier-Hermite series expansion" of the form:

$$f(x) = \sum_{n=0}^\infty A_n H_n(x)$$

then the coefficients are given by:

$$A_n = \frac{1}{\sqrt{\pi}\, 2^n n!} \int_{-\infty}^\infty f(x) e^{-x^2} H_n(x) dx$$

- The Hermite polynomials are orthogonal with weight e^{-x^2}.

$$\int_{-\infty}^\infty e^{-x^2} H_m(x) H_n(x) dx = \sqrt{\pi}\, 2^n\, n!\, \delta_{mn}$$

Spherical Harmonics

Spherical harmonics are unique within this appendix. All the other solutions are functions of one variable which we have chosen here to call x. Spherical harmonics are a multivariate function of the spherical coordinates θ and ϕ. (See Appendix D.) As such they provide a general representation for any function that depends on direction but not on distance.

Appendix J Some Common Differential Equations and Their Solutions

Specifically, spherical harmonics are products of complex exponentials in ϕ and associated Legendre polynomials in θ.

$$Y_l^m(\theta, \phi) = \sqrt{\frac{2l+1}{4\pi} \frac{(l-m)!}{(l+m)!}} P_l^m(\cos\theta) e^{im\phi} \quad \text{where } l \text{ and } m \text{ are integers and } -l \leq m \leq l$$

> **Warning:** Different textbooks use different conventions for the constant in front of $P_l^m(\cos\theta)e^{im\phi}$. The formula given above is common in physics, but be sure to look at the convention in any other source you are using. The formula below for the coefficients of an expansion in spherical harmonics is only valid for this convention.

Any piecewise smooth function of θ and ϕ on the domain $0 \leq \theta \leq \pi$ that is 2π periodic in ϕ can be expanded in spherical harmonics

$$f(\theta, \phi) = \sum_{l=0}^{\infty} \sum_{m=-l}^{l} A_{lm} Y_l^m(\theta, \phi)$$

with coefficients

$$A_{lm} = \int_0^{2\pi} \int_0^{\pi} f(\theta, \phi) \left[Y_l^m\right]^*(\theta, \phi) \sin\theta \, d\theta \, d\phi$$

where $\left[Y_l^m\right]^*$ is the complex conjugate of Y_l^m.

Spherical harmonics are discussed and used to solve partial differential equations in Section 11.7.2.

APPENDIX K

Special Functions

The functions in this appendix are listed in alphabetic order. Functions that are primarily defined as solutions to particular ODEs (including Bessel functions, Legendre polynomials, hyperbolic trig functions, and others) are listed in Appendix J.

Beta Function

$$B(x, y) = \int_0^1 t^{x-1}(1 - t)^{y-1}\, dt \quad x > 0, y > 0$$

$$B(x, y) = \frac{\Gamma(x)\Gamma(y)}{\Gamma(x + y)}$$

The beta function is "symmetric" meaning that $B(x, y) = B(y, x)$. It is used in probability and statistics. It is not discussed in this book.

Dirac Delta Function

Informally, we imagine the Dirac delta function as:

$$\delta(x) = \begin{cases} 0 & x \neq 0 \\ \infty & x = 0 \end{cases} \quad \text{with the } \infty \text{ well chosen so that} \quad \int_{-\infty}^{\infty} \delta(x)\, dx = 1$$

More properly, we define the Dirac delta function as the limit of a sequence of functions.

$$\delta(x) = \lim_{k \to 0} D(x) \quad \text{where} \quad D(x) = \begin{cases} 0 & x < 0 \\ 1/k & 0 \leq x \leq k \\ 0 & x > k \end{cases}$$

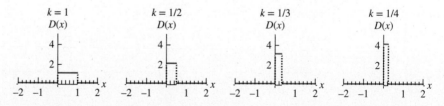

- Strictly speaking $\delta(x)$ is not a function at all, but it is often useful to treat it as one. Mathematicians call it a "generalized function."
- For any function $f(x)$ that is continuous around 0, $f(x)\delta(x)$ is the same thing as $f(0)\delta(x)$. For instance, $(x^2 + 2\cos x)\delta(x) = 2\delta(x)$.
- $\int_{-\infty}^{\infty} f(x)\delta(x - a)\, dx = f(a)$.

That last bullet point follows logically from the one before it, but it is the most important; that integral is the key to using the Dirac delta in both Laplace transforms and Green's functions.

The Dirac delta function is discussed primarily in Chapter 10. Section 10.10 defines and introduces this function; Section 10.11 uses $\delta(x)$ in differential equations to model fast changes, and then solves those equations with Laplace transforms; Section 10.12 uses "Greeen's functions" to introduce $\delta(x)$ into other differential equations as a mechanism for solving them.

Error Function

$$\operatorname{erf} x = \frac{2}{\sqrt{\pi}} \int_0^x e^{-t^2} dt$$

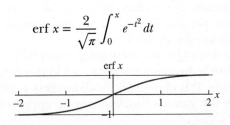

- $\lim_{x \to \infty} \operatorname{erf} x = 1$.
- $\operatorname{erf} x$ is an odd function.
- The "complementary area function" $\operatorname{erfc} x = 1 - \operatorname{erf} x$.

The error function and its use in statistics are discussed in Section 12.3.

Factorial

$$n! = \prod_{i=1}^n i = 1 \times 2 \times 3 \times \ldots \times n$$

- The factorial function is only defined for non-negative integer n-values. A generalization of the factorial function onto the real domain is the "gamma function" $\Gamma(x)$.
- Factorials obey the recurrence relation $n! = (n-1)! \times n$.
- $0!$ is defined as 1, based on the recurrence relation above.
- For very large x the factorial function can be approximated by "Stirling's formula" $x! \approx x^x e^{-x} \sqrt{2\pi x}$.

Factorials are used extensively in Taylor series (Chapter 2). They are discussed as part of the build-up to the gamma function in Section 12.3.

Fresnel Integrals

$$S(x) = \int_0^x \sin\left(\frac{\pi}{2} t^2\right) dt \qquad C(x) = \int_0^x \cos\left(\frac{\pi}{2} t^2\right) dt$$

- Both Fresnel integrals are odd functions.
- $\lim_{x \to \infty} S(x) = \lim_{x \to \infty} C(x) = 1/2$.
- Some books use the definitions $S(x) = \int_0^x \sin(t^2) dt$ and $C(x) = \int_0^x \cos(t^2) dt$.

The Fresnel integrals are not discussed in this book.

Appendix K Special Functions

Gamma Function

$$\Gamma(x) = \int_0^\infty t^{x-1} e^{-t} dt \quad x > 0$$

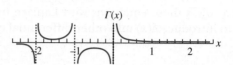

- The gamma function generalizes the factorial function. The two are related by $\Gamma(x) = (x-1)!$ for positive integer x, but the gamma function is also defined for negative and non-integer x-values.
- The gamma function obeys the recurrence relation $\Gamma(x) = (x-1) \times \Gamma(x-1)$.
- The integral above is divergent for $x < 0$, but $\Gamma(x)$ is defined for negative values by its recurrence relation. It is undefined for all integers $x \leq 0$.
- For very large x the gamma function can be approximated by "Stirling's formula" $\Gamma(x+1) \approx x^x e^{-x} \sqrt{2\pi x}$.
- $\Gamma(1/2) = \sqrt{\pi}$.

The gamma function is discussed in Section 12.3.

Heaviside Function

$$H(x) = \begin{cases} 0 & x < 0 \\ 1 & x \geq 0 \end{cases}$$

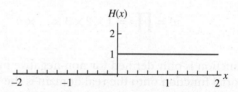

The Heaviside function really is as simple as it looks, but it gives us a notation that applies to a wide variety of discontinuous functions. Section 10.10 defines and introduces this function; Section 10.11 uses $H(x)$ in differential equations to model discontinuous changes, and then solves those equations with Laplace transforms.

APPENDIX L

Answers to "Check Yourself" in Exercises

1. The force is towards the equilibrium point: left if the object is on the right side and right if it's on the left. The resulting motion will be an oscillation.
2. $d^2x/dt^2 = -(k/m)x$. In words, we might read this equation as follows: "The position function $x(t)$ has the property that when you take its second derivative, you get the same function you started with, multiplied by $-k/m$." As a quick check, the left side of this equation has units m/s^2 and the right side has $N/m/kg\ m$, or N/kg. Since a Newton is a kg m/s², this becomes m/s², so the units match on the two sides.
3. One possible answer is $y(x) = 3x^2 + 2$.
4. $dy/dx = y$
5. $y = \pm e^{x^3/3+C}$.
6. $u_1(x) = e^{-x^2}$, $u_2(x) = e^{2x^2}$
7. $dR/dt = 5R - 10F$
8. $f(x_0 + \Delta x) \approx f(x_0) + f'(x_0)\Delta x$. This formula should not come as any great surprise. Since $f'(x_0)$ represents the slope (or "rise over run"), we multiply it by Δx (the "run") to estimate how much the function has gone *up* from its original value of $f(x_0)$.
9. 605.
10. If you plug $a_1 = 5$, $r = 3$, and $n = 5$ into your formula, you reproduce our original series. Your formula should therefore yield the answer 605. If it doesn't, figure out what went wrong!
11. 1/2
12. $x = Ae^{-10t} + Be^{-4t}$
13. $x = Ae^{\left(-2+\sqrt{-36}\right)t} + Be^{\left(-2-\sqrt{-36}\right)t}$.
14. 34
15. $B = i$
16. $\dfrac{1}{1+2i}e^{(1+2i)x} + C$
17. $y = 1/2$
18. A line has zero concavity. Because $\partial^2 y/\partial x^2 = 0$ the equation predicts $\partial^2 y/\partial t^2 = 0$: no acceleration. Because it started at rest and is not accelerating, the string will never move.
19. positive
20. $y(50) = 130$
21. $-\infty$
22. down, 2
23. $\dfrac{\partial f}{\partial x}u_x + \dfrac{\partial f}{\partial y}u_y = 1\left(\dfrac{1}{\sqrt{1+c^2}}\right) + 2\left(\dfrac{c}{\sqrt{1+c^2}}\right) = \dfrac{1+2c}{\sqrt{1+c^2}}$

24. $a = \dfrac{\partial f}{\partial x}(x_0, y_0)$ (We'll leave b and c for you to determine.)
25. $V = -\dfrac{2+\sqrt{2}}{4}\dfrac{GM}{d}$. (Even if your answer is correct it may take some simplification to make it look like ours. You can easily check your answer by punching it into a calculator. Ours comes out to $-0.85 GM/d$.)
26. 56 miles
27. $M = \int_0^W (kx)(Hdx) = (1/2)kW^2H$. Since kx has units of mass per distance squared, k must have units of mass per distance cubed, so this answer does have units of mass.
28. $M = (1/4)qW^2H^2$
29. $m = 2kR^5/15$. Since σ is mass over distance squared k must have units of mass over distance to the fifth, so the units are correct.
30. $\rho = 5$, $\phi = \tan^{-1}(4/3)$, $z = 10$
31. $r = \sqrt{125}$, $\theta = \cos^{-1}(10/\sqrt{125})$, $\phi = \tan^{-1}(4/3)$
32. $(2\hat{\imath} + 3\hat{\jmath}) \cdot (4\hat{\imath} + 24\hat{\jmath}) = 80$
33. $19{,}544/15 \approx 1303$
34. You should have graphed the curve $y = x^2$.
35. For $v = 2$ you get the parabola $y = (x-4)^2$ on the plane $z = 4$.
36. The key is the surface area of the top. The water accumulated is the water that has passed through the top of the bucket.
37. (rate of rainfall) × (top area of bucket). This might be written $W = rA$.
38. 180 and 122,000
39. Vector $\vec{v}_2 = \vec{A} - 2\vec{B}$. Count yourself correct if you had something reasonably close to that.
40. $820{,}000{,}008$ and $6{,}800$
41. $w = 3$. We'll leave it to you to find the others. (Of course the real way to check yourself is to see if $\mathbf{AB} = \mathbf{I}$!)
42. $x = 49/8$, $y = -7/4$
43. Matrix $\mathbf{B}$ stretches any shape by a factor of 3 in the x-direction and a factor of 5 in the y-direction.
44. $x_f = \rho_0 \cos(\phi_0 + \Delta\phi)$, $y_f = \rho_0 \sin(\phi_0 + \Delta\phi)$
45. $\mathbf{C} = \begin{pmatrix} 1/\sqrt{2} & 1/\sqrt{2} \\ -1/\sqrt{2} & 1/\sqrt{2} \end{pmatrix}$
46. $\mathbf{M} = \mathbf{C}^{-1}\mathbf{M'}\mathbf{C} = \begin{pmatrix} 3/2 & 1/2 \\ 1/2 & 3/2 \end{pmatrix}$
47. $(3, 2)$
48. $\hat{\imath} + 9\hat{\jmath}$
49. the x-component
50. The depth increases as you walk. The farther you go, the faster it increases.
51. The first leaf moves in a line at constant speed. The second leaf begins with a velocity of $\vec{v} = \hat{\imath} + 3\hat{\jmath}$ (speed of $\sqrt{10}$) and gradually turns to the right. It keeps moving in both the x- and y-directions forever, but its direction gets closer and closer to horizontal. The third leaf's long-term behavior (in the next part) is quite different.
52. It will roll to the left, gaining speed as it approaches A and then losing speed as it continues to the left. It will come to a stop when it reaches the same height it started at. Then it will begin rolling back to the right. Over time it will oscillate between these two points (with different x-values but the same $V(x)$-value) forever.
53. positive x, no y-movement
54. neither
55. $\hat{\rho}_x = 6/\sqrt{45}$, $\hat{\rho}_y = 3/\sqrt{45}$

Appendix L Answers to "Check Yourself" in Exercises

56. $-1/2$
57. $2\pi/19$ Earth years
58. By eye, it's roughly $2\cos(3x) + 0.3\cos(20x)$
59. $4\pi/5$
60. $2i/(3\pi)$
61. $\pi/500, 2\pi/500, 3\pi/500 \ldots$
62. $d^2x/dt^2 + 6(dx/dt) + 9x = 0$
63. $A(x)$
64. $A\sin(kx) + B\cos(kx), A\sin(kx + \phi), Ae^{ikx} + B\sin(kx), Ae^{ikx} + Be^{-ikx}$
65. $du/dt = dx/dt + 1$
66. $du/dt = u^2$
67. $xu''(x) - u'(x) = 0$
68.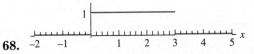
69. $z = e^x \sin y$.
70. $\pi/15$.
71. One of your equations should be rewriteable as $X''(x) = PX(x)$.
72. Your solutions in Parts 5, 6, and 7 should have been exponential, linear, and sinusoidal, respectively. Since an exponential or linear function cannot reach two zeros, you should have concluded that P must be negative. We will now replace P with $-k^2$, which allows for all possible negative values.
73. $X(x) = A\sin(n\pi x/w)$
74. $z(x, y, t) = \sum_{n=1}^{\infty}\sum_{m=1}^{\infty} \sin(kx)\sin(py)\left[C_{nm}\sin\left(v\sqrt{p^2 + k^2}\, t\right) + D_{nm}\cos\left(v\sqrt{p^2 + k^2}\, t\right)\right]$
    where $k = n\pi/w$ and $p = m\pi/h$.
75. $\dfrac{R''(\rho)}{R(\rho)} + \dfrac{1}{\rho}\dfrac{R'(\rho)}{R(\rho)} = \dfrac{-Z''(z)}{Z(z)}$. Both sides of your equation must now equal a constant. Because the boundary conditions on $R(\rho)$ are homogeneous, we consider the ODE for $R(\rho)$ first to determine the sign of the separation constant.
76. $V(\rho, z) = \sum_{n=1}^{\infty} A_n J_0(\alpha_{0,n}\rho/a)\left(e^{(\alpha_{0,n}/a)z} - e^{-(\alpha_{0,n}/a)z}\right)$
77. $\sum_{n=1}^{\infty} b'_n(t)\sin\left(\dfrac{n\pi}{L}x\right) = \sum_{n=1}^{\infty} -\dfrac{\alpha b_n(t)n^2\pi^2}{L^2}\sin\left(\dfrac{n\pi}{L}x\right)$
78. $(\partial \mathcal{F}[u]/\partial t) = -\alpha p^2 \mathcal{F}[u]$
79. $R(\rho) = CJ_0(k\rho) + DY_0(k\rho)$
80. $k = \alpha_{0,n}/a$, where $\alpha_{0,n}$ is the nth zero of J_0.
81. $z = J_0\left(\dfrac{\alpha_{0,n}}{a}\rho\right)\left[A\sin\left(\dfrac{\alpha_{0,n}}{a}vt\right) + B\cos\left(\dfrac{\alpha_{0,n}}{a}vt\right)\right]$
82. $f(\rho) = \sum_{n=1}^{\infty} B_n J_0\left(\dfrac{\alpha_{0,n}}{a}\rho\right)$
83. $dy/dx = c_1 + 2c_2 x + 3c_3 x^2 + 4c_4 x^3 + 5c_5 x^4 + \ldots$
84. $2c_2 = c_1 + 1$
85. $c_{n+2} = \dfrac{n^2 + n - \mu}{(n+1)(n+2)} c_n$
86. $r = -3, r = -4$
87. $R(r) = Aj_0\left(\dfrac{\sqrt{2mE}}{\hbar}r\right) + By_0\left(\dfrac{\sqrt{2mE}}{\hbar}r\right)$

88. $\cos x + x \sin x$
89. $D^2 - x^2 - 1$
90. $T = 1 - (2/\pi)x$
91. $T = (2/\pi)\phi$
92. $\ln z = \ln |z| + i(\phi + 2\pi n)$, where n is any integer.
93. $z = 5 + i$, $\Delta z = i$, $f(z + \Delta z) = (5 + i)^2 = 24 + 10i$, $\Delta f = -1 + 10i$, so $\Delta f/\Delta z = 10 + i$.
94. $2i$
95. The line segment from $z = 0$ to $z = 2$ maps to a line segment from $f = 0$ to $f = 4$, also on the positive real axis.

INDEX

Regular numbers refer to page numbers in this printed book. Numbers with the symbol § and the word "Online" refer to online sections located at www.felderbooks.com.

A

Absolute Convergence, *see* Series
Absolute Maximum and Minimum, 173
Active Transformation, 354
Adding Waves, 130
Adiabatic Process, §4.10(online) 14
Adjoint, 372, 774
Airy Functions, 665, 794, §12.11(online) 13
Aliasing, §9.7(online) 6
Alpha (Bessel Function Zeros), 591
Alternating Series, *see* Series
Alternating Series Test, *see* Series
Ampère's Law, 248, 434
Amplitude, 20, 24
Analytic Function, 671, 719, 720
Angular Frequency, 450
Anisotropic Crystal, 588
Ants, 347
Arbitrary Constants, *see* Differential Equation
Arbitrary Function, 547
Arctic Sea Ice Extent, 38
Argand Plane, 115
Arrhenius Equation, §4.11(online) 18
Assignment Problem, §7.5(online) 24
Associative Property of Matrix Multiplication, 300, 303
Asymptotic Expansions, *see* Series
Atmospheric Density, 202
Atoms, 295, 306
Attractor, §10.3(online) 4
Augmented Matrix, §7.4(online) 2

B

Babylon, 5, 239
Barometric Formula, §1.9(online) 22
Basic Comparison Test, *see* Series
Basic Feasible Solution, §7.5(online) 10
Basic Variable, §7.5(online) 13
Basis Vectors, **286**, 287, 307, 311, 318, 370
Beehive, §6.10(online) 1
Bernoulli Equation, 505
Bessel Functions, 503, 688, 793
  Hyperbolic or Modified, 593, 598, 696, 794
  of the First Kind, 691
  of the Second Kind, 692
  Spherical, 696, 697, 706, 794
  Third Kind (Hankel Functions), 794
  Zeros, 591
Bessel Series, 591, 693, 793
Bessel's Equation, 689
Beta Function, 664, 798
Black Holes, §1.9(online) 21
Blackbody Radiation, 143
Bland's Rule, §7.5(online) 14
Block on a Ramp, 286, 359
Blood Types, 285
Bombelli, Rafael, 108
Books (Matrix Example), 280, 281, 290, 298, 310, 320
Boundary Conditions, 547, 576, 787
  Implicit, 595
  Inhomogeneous, 616
  Time Dependent, §11.12(online) 4

Branch Cut, 711
Brightness, 401
Bromwich Integral, 737
Bullet, 439
Buoyancy, 150

C

Cauchy Principal Value, 733
Cauchy's Integral Formula, 728
  Proof, 732
Cauchy-Euler Equation, 46, 489, 614, §10.9(online) 34
Cauchy-Riemann Equations, 720
Cell Phones, 384
Center of Mass, 180, 192, 201
Chain Rule, 145, 497
Chaos Theory, 92
Chebyshev Polynomials, 702
Chemical Kinetics, §1.9(online) 23, §4.11(online) 18
Chemical Reaction, 324
Chocolate River, 145
Choosing the Right Transform, 649, 787
Circuit, 26, 47, 126, 136, 142, 150, 197, 344, 456, 494, 529, 531, **§3.6(online) 1**, §4.10(online) 16, §6.10(online) 3, §10.4(online) 16, §10.9(online) 40, §10.13(online) 42
Circular Drum, 652, 704
Circulation, 417, 432
Clairaut's Theorem, 141
Cobb-Douglas Production Function, 157, 186

Coleman–Weinberg Equation, 58
Column Matrix, 278, 774
Commutator, §12.10(online) 4, 8
Complementary Error Function, 660, §2.8(online) 2
Completeness, 701
Completing the Square, §10.4(online) 13
Complex Conjugate, 110
Complex Exponential Function, 118, 714
Complex Number, 106
Complex Plane, 114
Compound Interest, 6, 11, 14, 15
Computer Graphics, 298
Conductor, 171
Conditional Convergence, *see* Series
Conformal Mapping, 754
Conservative Field, 402, 403, 437, **438**, 778, §10.5(online) 22
Constraint, *see* Optimization
Continuity Equation, 429
Contour Integral, 726
  Along Branch Cuts, §13.6(online) 1
  for Evaluating Real Integrals, 734
Contour Lines, 400
Contour Plot, 381
Convolution, 518, 785
Coordinate Systems, 221, 229, 772
  Cylindrical, 419, 777
  Polar, 419, 777
  Spherical, 419, 777
cosh (Hyperbolic Cosine), 658, 791
Cosine (Complex), 715
Cosmological Constant, 38
Coulomb's Law, 180, 394
Cramer's Rule, §6.10(online) 5
Critical Point, 173, §10.3(online) 3
Cross Product, 366
Crystal, 51, 396
Curl, 407, **408**, 416, 777
  in Curvilinear Coordinates, 419, 777
Curvilinear Coordinates, 772

Cyclone Zone, §13.10(online) 12
Cylindrical Coordinates, **230**, 772

D
d'Alembert's Solution, 557, 566
Damped Oscillator, 122
Darwin, 353
Day Trading, 11
Degenerate Eigenvalues, 350
Del, 411
Derivative (Complex), 717
Determinant, *see* Matrix
Diagonal Matrix, 278, 330, 775
Diagonalizing a Matrix, 329, 339
Diet Problem, §7.6(online) 29
Differential Equation
  Analytical Solution, §1.8(online) 10, 14
  Arbitrary Constants, 15, 36, §1.7(online) 3
  Autonomous, 509, 512
  Complementary Solution, 43, 487
  Coupled, 267, 336, 494, 526, §1.7(online) 1, **49**, §10.3(online) 1
  Elliptic, 767
  Euler's Method, §1.8(online) 18
  Exact, 494, §10.5(online) 18, 19
  General Solution, **18**, §1.7(online) 3
  Guess and Check, 39, 485
  Guessing Hints, 489
  Homogeneous (all terms involve $y(x)$), **41**, 487, 766, §1.7(online) 3
  Homogeneous (dy/dx equals a function of y/x), 507
  Hyperbolic, 767
  Inhomogeneous, 43, 766
  Linear, **41**, 487, 494, 765, §1.7(online) 3, §10.4(online) 12
  Linear Superposition, 39, **42**, 485, §1.7(online) 3
  Numerical Solution, §1.8(online) 11, 14
  on Computers, 49, §1.8(online) 10
  Order, **18**, 766, §1.7(online) 3
  Ordinary Differential Equations, **1**, 545, 765
  Parabolic, 767

Partial Differential Equation, 541, 544, 765, 787
  Inhomogeneous, 623, 625
  Variable Substitution, 494, 505
Differentials, 149, 772, §10.5(online) 18
Diffraction Grating, 131, 135
Diffusion Equation, 550
Dimensional Analysis, 8
Dipole, 171, 386, 395, 402, §2.9(online) 8
Dirac Delta Function, 482, 512, **515**, 531, 798
Dirac, Paul, §12.10(online) 5
Directional Derivative, 158
Dirichlet Conditions, 451
Dirichlet Problem, 721
  Uniqueness of Solutions, §13.11(online) 17
Discrete Fourier Transform, 483, §9.7(online) 1
  Formula, §9.7(online) 2
Divergence, 407, **408**, 416, 777
  in Curvilinear Coordinates, 419, 777
Divergence Theorem, 426, **427**, 778
Dot Product, 371
Double Integral, 204, §5.11(online) 3, 4
Double Pendulum, 274, 344
Double Root, 491–493, §10.9(online) 34, 38
Double-Angle Formulas, 120
Driven Oscillator, 128
Driving Function, 625
Drum, 580, 595, 635
Drumhead, §9.8(online) 9
dx, 149, 191

E
Earth–Moon System, 388, 397, 400, §10.13(online) 42
Eiffel Tower, §5.12(online) 8
Eigenfunction, 649, 698
Eigenfunction Expansion, 623, 788
Eigenvectors and Eigenvalues, *see* Matrix
Einstein Summation Notation, 364
Elementary Function, **658**
Elementary Operations, §7.4(online) 3
Elephants, 14
Elimination, §7.4(online) 1

Enthalpy, §4.10(online) 17
Entire Function, 719
Entropy, 180, 664
Equation of Continuity, 429
Equation of State,
  §2.9(online) 9
Equatorial Bulge,
  §4.11(online) 22
Equilibrium, 28, 393
Equipartition Theorem, 664,
  §4.10(online) 11
Equipotential, 171, 400
erf (Error Function), 659, 799,
  §2.8(online) 4
erfc (Complementary Error
  Function), 660,
  §2.8(online) 2
Essential Singularity, 744
Euler's Formula, 118
Euler–Mascheroni Constant,
  §2.8(online) 4
Even Extension, 462, 780
Even Function, 452
Expanding Universe, 38, 504,
  §10.3(online) 10,
  §8.12(online) 2
Expansion by Minors, 317
Expectation Value, 92
Explicit Definition of a Series,
  655
Exponential Integral,
  §2.8(online) 4
Extrasolar Planets, 445

F
Factorial, 660, 799
Faraday's Law, 264
Fast Fourier Transform,
  §9.7(online) 6
Feasible Region, §7.5(online) 10
Feldor, §3.6(online) 6
Field
  Electric, 394, 406, 414, 418,
    419, 425, 442–444
  Gravitational, 388, 397
  Magnetic, 415, 419, 425, 444
  Scalar, **379**, 380
  Vector, **379**, 381
Field Lines, 382, 429
Finches, 353
First Law of Thermodynamics,
  §4.10(online) 9
Flatland, 263
Flow Rate, **256**, 407, 408, 436,
  §8.12(online) 2

Fluid Flow,
  §13.10(online) 7
Flux, **256**, 427
Fourier Series, 447, 779,
  §3.6(online) 7
  Derivation of Formula, 457
  Derivatives of, 634
  for Partial Differential
    Equations, 560, 623
  Fourier Cosine Series, 462,
    780
  Fourier Sine Series, 462, 780
  Multivariate, 483, 780,
    §9.8(online) 9, 10
  with Complex Exponentials,
    467, 468, 779
  with Sines and Cosines, 461,
    779
Fourier Sine Transform
  for Partial Differential
    Equations, 645, 788,
    §11.12(online) 3
Fourier Transform, 472, 474, 781
  for Partial Differential
    Equations, 636, 788
Fourier-Bessel Series, 591,
  693, 793
Fourier-Legendre Series,
  678, 792
Frequency, 20, 450
Frequency Scaling Theorem,
  482
Fresnel Integral, 57, 665, 799
Frobenius, Method of, 682
Fubini's Theorem, 206

G
Galilean Relativity, see Relativity
Gamma Function, 660, 800
Gauss's Law, 262, 430
Gaussian, 32, 79, 478, 659,
  §5.12(online) 10
General Relativity, see Relativity
Generalized Cauchy's Integral
  Formula, 728
Generalized Power Series, 682
Generalized Vectors, 290
Geological Maps, 383
Geometric Series, 131
Geometric Transformations, 347
Gibbs Phenomenon,
  §9.9(online) 17
Gin and Vermouth,
  §10.4(online) 12
Global Maximum and
  Minimum, 173
Gluppity-Glup, §10.4(online) 16

Google PageRank,
  §6.10(online) 4
Gradient, **163**, 164, 396, 399,
  399, 402, 438, 777
  in Curvilinear Coordinates,
    419, 777
  for Optimization, 172
Gradient Theorem, 402,
  403, 778
Gram-Schmidt Process, 376
Graveyards, 301, 310
Gravity, §12.11(online) 14
Green's Functions, 531
Green's Theorem, 434, 435
Guitar String, 460, 546, 555, 560,
  563, §9.8(online) 12

H
Half-Life, §1.9(online) 22
Hanging Chain Problem,
  606
Hankel Functions, 696, 794
Hankel Transform, 649
Harmonic Function, 721
Harmonic Series, see Series
Harry Potter, 277
Hawking Radiation,
  §1.9(online) 21
Heat Capacity, §4.10(online) 10
Heat Equation, 550
  Derivation, 542, 552
Heaviside Function, 512,
  514, 800
Height Distribution, 660
Helmholtz Equation, 615
Hermione, 415
Hermite Polynomials, 796
Hermite's Differential Equation,
  679, 706
Hermitian Matrix, see Matrix
Hessian Matrix, 176
Hole (Type of Singularity), 713
Homogeneous Algebra
  Equation, 314
Homogeneous Condition, 549
Hooke's Law, 2, 12
Hotel Bathroom, 280, 285
Hubble's Law, 38,
  §8.12(online) 2
Hydrogen Atom, §12.10(online)
  12, §12.11(online) 13
Hydrogen Molecule, 75
Hyperbolic Bessel Functions,
  696
Hyperbolic Trig Functions, **658**,
  791
Hypergeometric Functions, 688

I

i, 106
Ice Cream, 237
Ideal Gas Equation,
  §4.10(online) 11
Identity Matrix, see Matrix
Illumination, §10.5(online) 25
Im, 106
Imaginary Number, 105, **106**
Impedance, §3.6(online) 2
Implicit Differentiation, 153
Improper Integral, 733
Impulse, 197
Inconsistent Equations, 313
Index Notation for Matrices,
  278, 363, 366
Indicial Equation, 683
Inertia Tensor, §7.6(online) 28
Infinite Series, see Series
Inflationary Cosmology, 504,
  §10.3(online) 10
Initial Conditions, 547, 576
Inner Product, 371
Insulated Boundary, 724
Integral Table, 779
Integral Test, see Series
Integrating Factor,
  §10.5(online) 22
Integro-Differential Equation,
  522, 529
Interstate, 40, 139
Inverse Laplace Transform, 737
  Derivation, 738
Inverse Matrix, see Matrix
Irrotational, 438
Irrotational Flow,
  §13.10(online) 7
Isobaric Process,
  §4.10(online) 15
Isothermal Process,
  §4.10(online) 14

J

Jacobian, 222, 224, 228, 773
Jones Matrix, 375
Joukowski Airfoil,
  §13.10(online) 15

K

Kelvin Functions, 696
Klein-Gordon Equation, 552
Kronecker Delta, 366, 702

L

Ladder Operators, 707,
  §12.10(online) 4
Lagrange Multipliers, 181
Lagrange Point,
  §10.13(online) 42
Lagrange Remainder, 66, 769,
  §2.9(online) 8
Laguerre Differential Equation,
  531, 706, §12.11(online) 13
Laguerre Polynomials, 795
Laminar Flow, §13.10(online) 12
Laplace Transform, 516, 782
  for Partial Differential
    Equations, 646, 789
  for Solving Coupled
    Differential Equations, 526
  for Solving Ordinary
    Differential Equations, 522
  Interpreting, 524
  Properties, 518, 520, 782
  Table, 783
Laplace's Equation, 550, 709,
  721, §9.9(online) 16
  on a Disk, 579
  Uniqueness of Solutions,
    §13.11(online) 17
Laplacian, 407, **412**, 777
  in Curvilinear Coordinates,
    419, 777
Laurent Series, 743, 744
Learning Function,
  §1.9(online) 21
Least Squares Method, 181
Legendre Differential Equation,
  674
Legendre Polynomials, 370, 376,
  377, 458, 504, 607, **673**, 792
  Associated, 792
Legendre Series, 678, 792
Length Contraction, 369
Level Curves, 381, 400
Level Surfaces, 381, 400
Levi-Civita Symbol, 367
Light, 401
Limaçon, 228
Limit Comparison Test, see
  Series
Line Integral, **240**, 402, 417,
  432, 438
  of a Scalar Function, 243
  of a Vector Function, 240
Linear Approximation, 52
Linear Density, 191
Linear Operator, 486
Linear Programming, 184, 377,
  §4.11(online) 20,
  §7.5(online) 8
Linearly Dependent
  Equations, 313, 494,
  §10.6(online) 27, 28
Linearly Dependent Vectors,
  318
Liouville's Theorem,
  §13.11(online) 18
Lissajous Surface, 260
Local Maximum and Minimum,
  173
Logistic Equation, 14, 33,
  §1.9(online) 23
Lorentz Factor, 368
Lotka–Volterra Equations,
  §1.7(online) 8
Lowering Operator,
  §12.10(online) 9

M

Möbius Transformation, 751
Maclaurin Series,
  see Taylor Series
Magnetic Field, 198, 248
Magnetic Induction, 264
Main Diagonal, 278, 775
Mapping Complex Functions,
  747
Matrix
  Adjoint, 774
  Block Diagonal, 352, 775
  Column Matrix, 774
  Determinant, **312**, 316, 351,
    410, 775
  and Inverse Matrix, 319
  Eigenvectors and Eigenvalues,
    325, **327**, 698, 775
    Degenerate, 350
  Hermitian, 372, 775
  Inverse, 305, 775
  Index Notation, 280, 774
  Orthogonal, 353, 372, 775
  Rank, 774, §7.4(online) 4
  Row Matrix, 774
  Square, 278, 776
  Symmetric, 372, 776
  Trace, §6.10(online) 1
  Transpose, 353, 372, 775
  Unitary, 372, 776
Matrix Algebra, 307
Matrix Multiplication,
  280, 294
Matrix of Coefficients, 315
Maxwell Relations,
  §4.10(online) 17
Maxwell's Equations, 430, 434
Meteoroid, 141
Middle C, 448

Milkmaid Problem, 185
Minimum Ratio Rule,
§7.5(online) 14
Mites, 285
Mixed Partials, 140
Mixing Problem, 22,
§1.7(online) 6
Modified Bessel Functions, 696
Modulus, 110, 115
Molecules, 295, 306
Moment Generating Function, 92
Moment of Inertia, 180, 192, 200
Momentum Density, 256, 408
Monge, Gaspard, §7.5(online) 24
Moore's Law, 90
Mordor, 401
Morse Potential, 75
Multiple-Valued Function, 710, 711
Multivariate Function, 546

N

nth Term Test, see Series
Nabla (∇), 411
Natural Basis, 330
Neighborhood, 719
Newton's Law of Heating and Cooling, 34, 38, 542,
§1.7(online) 7,
§1.9(online) 22
Newton's Law of Universal Gravitation, 75, 188, 238, 267, §5.11(online) 1
Nezzer Chocolate Factory,
§4.11(online) 20
Nonbasic Variable,
§7.5(online) 13
Norm, 371, 681, 702
Normal Form, §7.5(online) 12
Normal Mode, **268**, 320, 331, **555**, 559
Normalize, 371
Nuclear Facility, 378
Nutrition, §7.6(online) 29
Nyquist Frequency,
§9.7(online) 6

O

Objective Function, 174, 182
Oceania, §1.7(online) 5
Odd Extension, 462, 780
Odd Function, 452
ODE, see Differential Equation

Operator, 411, **486**,
§12.10(online) 4, 6
Linear, §12.10(online) 6
Optimization
Constraint, 174, 182
Gradient, 172
Lagrange Multipliers, 181
Simplex Method, 377,
§7.5(online) 8
Ordinary Differential Equation, see Differential Equation
Orientation of a Shape, 357
Orthogonal Basis, 371
Orthogonal Functions, **457**, 470, 701,
§12.9(online) 1
Orthogonal Matrix, see Matrix
Orthogonal Trajectories,
§10.13(online) 43
Orthonormal Basis, 371

P

p-Series, see Series
PageRank, §6.10(online) 4
Parabolic Approximation, 59
Parametric Surfaces, 249
Parseval's Theorem, 470
Partial Derivative, 136,
§4.0(online) 1
Definition, 138
Partial Differential Equation, see Differential Equation
Partial Fractions, 518, 784
Partial Sum, 84
Passive Transformation, 354
Path-Independent, 438
PDE, see Differential Equation
Pendulums, Coupled,
§4.7(online) 7
Period, 20, 24, 450
Petri Dish, 402
Phase, 20, 24, 115
Phase Portrait, 494,
§10.3(online) 1
Phase Space,
§10.3(online) 7
Phasor, §3.6(online) 4
Picard's Theorem, 747
Pitch, Dependence on Temperature,
§2.9(online) 9
Planck's Law of Blackbody Radiation, 143
Point Matrix, 298
Poisson's Equation, 550, 635

Polar Coordinates, 221, 248, 589, 772
Pole, 713, 730, 744
Population Growth, 14, 22, 35, 37, §1.7(online) 1, 8,
§1.9(online) 19, 20
Positive-Definiteness, 371, 374
Potential, **387**, 402, 406, 438, 778
Electric, 383, 394, 406, 425, 442
Gravitational, 388, 397
in Multiple Dimensions, 396
in One Dimension, 387
Magnetic, 425
Potential Energy, 439, 442
Power Factor, §3.6(online) 6
Power Series, see Taylor Series
Predator-Prey System,
§1.7(online) 1, 8
Pressure, 401
Pressure Gradient, 170,
§4.11(online) 20
Price Elasticity of Demand, 157
Principal Axes,
§7.6(online) 29
Principal Value, 710, 711
Product Notation (Π), 655
Projectile Motion, 186
Pseudovector, 361, 366

Q

Quantum Harmonic Oscillator,
§12.10(online) 4,
§12.11(online) 13
Quantum Mechanics, 375, 472, 579, 615,
§1.8(online) 17,
§6.10(online) 3

R

Radioactive Decay,
§1.9(online) 22
Radius of Convergence, 743
Raising Operator,
§12.10(online) 9
Rank of a Matrix, see Matrix
Ratio Test, 770
Re, 106
Recurrence Relation, 655, 668
Recursive Definition of a Series, see Series
Recursion, see Recursion
Reduction of Order,
§10.9(online) 33
Reindexing a Series, 655

Relativity
  Galilean, 368
  General, §8.12(online) 2
  Special, 68, 143, 151, 152, 368,
    §4.11(online) 21,
    §4.7(online) 7,
    §10.5(online) 24
Removable Singularity, 713
Repulsor, §10.3(online) 4
Residue, 730, 744
Residue Theorem, 730
Restricted Normal Form,
  §7.5(online) 13
Riccati–Bessel
  Functions, 697
Riemann Sum, 90, 104
Rock in a Stream,
  §13.10(online) 9
Rodrigues' Formula, 677, 681
Romeo and Juliet,
  §1.7(online) 2, 8,
  §10.3(online) 1
Rotation Matrix, 347, 349, 774
Row Echelon Form,
  §7.4(online) 4
Row Matrix, 278, 774
Row Reduction, 377,
  §7.4(online) 1
Rubber Band, 558
Rubber Woman, 361, 365

S
Saddle Point, 176
Sand Moving Problem,
  §7.5(online) 24
Scalar, 360
Schrödinger's Equation, 550,
  579, 615, 697,
  §12.10(online) 5,
  §12.11(online) 13,
  §6.10(online) 3
Schwarz Inequality,
  §7.6(online) 29
Schwarz' Theorem, 141
Second Derivatives
  Test, 176
Self-Adjoint, 775
Separable Integrals, 208
Separation of Variables
  Ordinary Differential
    Equations, 34
  Partial Differential Equations,
    567, 611, 787
    More than Two Variables,
      580, 787
Separatrix, §10.3(online) 3

Sequence, 82, 83, 91
Series, 83, 768, 770
  Absolute Convergence,
    100, 771
  Alternating Series, 100, 654
  Alternating Series Test,
    101, 771
  Asymptotic Expansion,
    §2.8 (online) 1
  Basic Comparison Test, 98,
    771
  Conditional Convergence,
    101
  Convergence, 84, 770
  Explicit Definition, 655
  Geometric, 83, 770
  Harmonic, 87
  Infinite, 84
  Integral Test, 96, 771
  Limit Comparison Test, 99,
    771
  nth-Term Test, 86, 770
  p-Series, 97, 770
  Ratio Test, 93
  Recursive Definition, 655
  Remainder, 768
  Telescoping, 85, 770
Shear Matrix, 311
Similar Matrices, 289, 352
Similarity Theorem, 482
Simple Harmonic Oscillator, 2,
  13, **19**, 21, 23, 24, 485, 490,
  §10.3(online) 6,
  §2.9(online) 9
Simple Pole, 730
Simplex Method,
  see Optimization
Simply Connected, 441
Simultaneous Linear Equations,
  307, 310
Singularity, 713, 744
sinh (Hyperbolic Sine),
  658, 791
Slack Variable,
  §7.5(online) 12
Slope Fields, **25**, 26
Snowball, 12
Soil Transport Problem,
  §7.5(online) 24
Solar Cell, 395
Sound Wave, 494
Spacely Sprockets,
  §10.5(online) 26
Span, 287, 370
Special Functions, 658,
  790, 798

Special Relativity,
  see Relativity
Species Concentration,
  §7.4(online) 6
Specific Heat, 552
Spectrum, 474
Spherical Bessel Functions, 696,
  697, 706
Spherical Coordinates, **232**,
  607, 772
Spherical Harmonics, 610, 796
Spherical Pendulum, 72
Square Matrix, see Matrix
Square Wave, 454, 467
Stable Equilibrium, 28
Standard Deviation, 663
Starbucks, 384
Steady State Solution, 621
Stefan-Boltzmann Law, 151
Stirling's Approximation, 661,
  799, 800
Stokes' Theorem, 432, **432**,
  441, 778
Stream Function,
  §13.10(online) 7
Streamlines, 382
Stress-Energy Tensor, 369
Sturm-Liouville Theory,
  458, 697
Superman, §1.8(online) 17
Surface Integral, 253
  of a Scalar Function, 259
  of a Vector Function, 255
Surplus Variable,
  §7.5(online) 12
Symmetric Matrix, see Matrix

T
Tangent Plane,
  §4.7(online) 1
Taylor Series, **50**, 448, 452, 666,
  742, 768
  $\mathcal{O}$, 79
  Complex, 742
  Formula, 71, 768
  for Solving Differential
    Equations, 666,
    §12.11(online) 14
  Generalized Power Series, 682
  Maclaurin Series (Around
    Zero), **60**, 666, 768
  Multivariate, 172,
    §4.7(online) 1, 4
  Not Around Zero, 70
  Used to Define Complex
    Functions, 112

Telescoping Series, *see* Series
Tensor, 358, 362, **363**
Tetrahydrofrefurol, 11
Thermodynamic Identity,
    §10.5(online) 25,
    §4.10(online) 10
Thermodynamics, 187, 664,
    §4.10(online) 9
Thin Lens Formula,
    151, 157
Three Spring Problem,
    266, 283, 291, 294,
    308, 311, 320, 331, 336,
    §1.7(online) 6
Three Stooges, 18
Three-Petaled Rose, 227
Thurstone, Louis,
    §1.9(online) 21
Time Delay Theorem, 480
Time Dilation, 369
Topo Map, 384
Trace, 369, 776
Traffic Density, 139
Traffic Flow, 324
Trajectory, §10.3(online) 2
Transfer Function, 531
Translation by Matrix
    Multiplication,
    §7.6(online) 27
Transportation Problem,
    §7.5(online) 24
Transpose, 353, 372, 775

Tribbles, §1.9(online) 19
Triple-Angle Formulas, 125

U
Underdamped Harmonic
    Oscillator, 104
Uniqueness Theorem,
    §13.11(online) 17
Unit Vectors
    in Different Coordinate
    Systems, 420, 773
Units, 8
Unstable Equilibrium, 28
Unitary Matrix, *see* Matrix

V
Valentine Basket, 300
Van der Waals Equation of State,
    156, §2.9(online) 9
Variation of Parameters, 633,
    §10.9(online) 33
Vector, 360
Vector Calculus, 378, 777,
    §8.0(online) 1
Vector Space, 369
    Complex, 372
    Definition, 371
    Dimension, 370
Velocity Gradient, 363, 367
Virial Expansion,
    §2.9(online) 9

W
Water Gas Shift Process,
    §7.4(online) 7
Wave Equation, 136,
    546, 549, 569,
    §4.11(online) 18
    Derivation, 553, 554
    General Solution,
    557, 566
Wavelength, 450
Weight Function, 701
Wet 'n Wild,
    §13.10(online) 12
Whitney Umbrella, 252, 260
Wind Chill,
    §4.11(online) 21
Wind Instruments, 561, 564,
    566, 634
Wolfram Alpha,
    §1.8(online) 10
Wonderflonium, 553
Work–Energy Theorem, 439
Wronskian,
    §10.6(online) 27, 28

Y
Young's Modulus, 494

Z
Zeno's Paradox, 92